Kostoris Library Chr

R02842P498

D1757460

THIS BOOK BELONGS TO:
Medical Library
Christie Hospital NHS Trust
Manchester
M20 4BX
Phone: 0161 446 3452

MOLECULAR TARGETING IN ONCOLOGY

CANCER DRUG DISCOVERY AND DEVELOPMENT

BEVERLY A. TEICHER, SERIES EDITOR

MOLECULAR TARGETING IN ONCOLOGY

Edited by

HOWARD L. KAUFMAN

Columbia University
New York, NY

SCOTT WADLER

Cornell University
New York, NY

KAREN ANTMAN

Boston University School of Medicine
Boston, MA

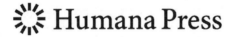 Humana Press

Editors

Howard L. Kaufman
Division of Surgical Oncology
Columbia University
New York, NY

Scott Wadler
Division of Hematology and Medical Oncology
Department of Medicine
Weill Medical College
Cornell University
New York, NY

Karen Antman
Boston University Medical Campus
Boston University School of Medicine
Boston, MA

Series Editor
Beverly A. Teicher, PhD
Genzyme Corporation
Framingham, MA

ISBN: 978-1-58829-577-4   e-ISBN: 978-1-59745-337-0

Library of Congress Control Number: 2007932637

©2008 Humana Press, a part of Springer Science+Business Media, LLC
All rights reserved. This work may not be translated or copied in whole or in part without the written permission of the publisher (Humana Press, 999 Riverview Drive, Suite 208, Totowa, NJ 07512 USA), except for brief excerpts in connection with reviews or scholarly analysis. Use in connection with any form of information storage and retrieval, electronic adaptation, computer software, or by similar or dissimilar methodology now known or hereafter developed is forbidden.
The use in this publication of trade names, trademarks, service marks, and similar terms, even if they are not identified as such, is not to be taken as an expression of opinion as to whether or not they are subject to proprietary rights.
While the advice and information in this book are believed to be true and accurate at the date of going to press, neither the authors nor the editors nor the publisher can accept any legal responsibility for any errors or omissions that may be made. The publisher makes no warranty, express or implied, with respect to the material contained herein.

Printed on acid-free paper

9 8 7 6 5 4 3 2 1

springer.com

We dedicate this book to Scott Wadler, MD—our mentor, colleague and friend.

Dr. Scott Wadler was a gifted clinician, skilled educator and an expert thought leader in Oncology. His foresight and intuition suggested the need for a book on molecular targeting long before this became a buzz word in cancer research. His efforts forged the foundation of this book, and its structural organization is a reflection of his vision for how the field should think about the new era of molecularly targeted agents. His contributions were many, and he will be missed by patients, students and colleagues alike.

INTRODUCTION

In contrast to the premise that drug development in oncology has been empirically based, history shows that anticancer drugs have been targeted from the very beginning. The earliest anticancer drugs were targeted antimetabolites (purine and pyrimidine analogs and antifolates). Next came the alkylating agents. Certainly, 1-phenylalanine mustard (melphalan) was targeted to melanin metabolism but was found to be more effective in myeloma than in melanoma. Cyclophosphamide was designed as a prodrug to be selectively metabolized in tumor cells. Clinical trials, however, demonstrated antitumor activity, but its metabolism was through the hepatic p450 system. The next class introduced, hormone antagonists, proved effective with less toxicity. Thus, the concepts were brilliant in the past, but our knowledge of the science was as yet inadequate. Our understanding of the biology of cancer only now has permitted much more elegant and effective therapeutic interventions such as Herceptin®, Gleevec®, and Avastin®.

In the 1980s and 1990s, starting with the development of rituximab (Rituxan) for B cell lymphomas, trastuzumab (Herceptin) for Her-2-positive breast cancer, and imatinib (Gleevec) for the treatment of chronic myelogenous leukemia (CML) and GI stromal tumors (GIST), the pharmaceutical industry has regrouped to rationally design drugs with novel mechanisms of action. Epidermal growth factor receptor (EGFR) inhibitors, Iressa® and Tarceva®, to inhibit signal transduction pathways, inhibitors of mammalian target of rapamycin (mTOR), inhibitors of histone deacetylases, and drugs that promote apoptosis are in clinical trials or have recently completed clinical trials.

In *Molecular Targeting in Oncology*, we have attempted to present an overview of the development of targeted therapies for the treatment of cancer with an emphasis on clinical application. Five sections cover the most important elements of drug development: General Strategies for Molecular Targeting in Oncology, Molecular Targeting for Specific Disease Sites, Classes of Drugs for Molecular Targeting in Oncology, Specific Drugs for Molecular Targeting in Oncology, and Challenges in Molecular Targeting in Oncology. These sections present different perspectives on how targeted therapeutics are being evaluated. The "Strategies" section focuses on approaches using targeted therapies to inhibit cell growth. The section on "Disease Sites" describes how clinicians are evaluating targeted therapies in specific organ systems. The third section on "Classes" of targeted therapies illustrates how various classes of pharmacologic and immunologic agents are developed for individual molecular targets. The "Drugs" section focuses on selected new drugs that have novel mechanisms of action. The final section deals with "Challenges" for the future of targeted therapeutics and includes chapters on appropriate patient selection, use of combination therapy, how to deal with tumor cell resistance, advances in targeted imaging, measurement of clinical effects, clinical trial design, and preclinical development of targeted agents. Although the structure of this book guarantees some overlap between chapters and sections, readers might start with the chapters that most interest them and use the supporting chapters to gain a better understanding of how targeted drug development is being viewed by basic science investigators, industry representatives, and government scientists.

The structure of *Molecular Targeting in Oncology* is designed to cover the flavor of the rapidly developing area of targeted therapies for the treatment of patients with cancer. Targeted therapies will likely continue as focus for future drug development as the molecular pathways mediating tumor initiation, progression, and metastasis are better defined and the ability to rationally design drugs using high-throughput technology becomes more firmly established. Therapeutic activity for some targeted agents, such as Gleevec, is remarkable, whereas others significantly reduce toxicity in cancer patients while providing comparable clinical response rates compared with conventional cytotoxic drugs. As further knowledge of the biology of the cancer cell expands over the next few years, the number of targeted agents will likely increase. Our challenge will be to determine how best to use these agents to improve the outcome for our patients with cancer.

Howard L. Kaufman, MD

CONTENTS

ix

CONTRIBUTORS

ALEX A. ADJEI, MD, PHD • *Division of Medical Oncology, Mayo Clinic and Foundation, Rochester, MN*

KAREN ANTMAN, MD • *Boston University Medical Campus, Boston University School of Medicine, Boston, MA*

IGOR ASTSATUROV, MD, PHD • *Department of Medical Oncology, Division of Medical Sciences, Fox Chase Cancer Center, Philadelphia, PA*

DEENA M. ATIEH, MD • *Memorial Sloan-Kettering Cancer Center at Basking Ridge, Basking Ridge, NJ*

MICHAEL B. ATKINS, MD • *Biologic Therapeutics and Cutaneous Oncology Programs, Beth Israel Deaconess Medical, Center, Boston, MA*

NILOFER S. AZAD, MD • *Medical Ovarian Cancer Team, Laboratory of Pathology, Center for Cancer Research, National Cancer Institute, Bethesda, MD*

ALFONSO BELLACOSA, MD, PHD • *Popular Science Division, Fox Chase Cancer Center, Philadelphia, PA*

MONICA BERTAGNOLLI, MD • *Division of Surgical Oncology, Department of Surgery, Brigham & Women's Hospital, Dana Farber Cancer Institute, Harvard University, Boston, MA*

LIAT BINYAMIN, PHD • *Department of Medical Oncology, Division of Medical Sciences, Fox Chase Cancer Center, Philadelphia, PA*

HOSSEIN BORGHAEI, DO, MS • *Department of Medical Oncology, Division of Medical Sciences, Fox Chase Cancer Center, Philadelphia, PA*

JAN BORNSCHEIN, MD • *Gastrointestinal Pathobiology Group, Yale University School of Medicine, New Haven, CT*

MICHAEL A. CARDUCCI, MD • *Sidney Kimmel Comprehensive Cancer Center, Johns Hopkins University, Baltimore, MD*

HELEN X. CHEN, MD • *Investigational Drug Branch, Cancer Therapy Evaluation Program, Division of Cancer Treatment and Diagnosis, National Cancer Institute, Bethesda, MD*

JANET E. DANCEY, MD • *Investigational Drug Branch, Cancer Therapy Evaluation Program, Division of Cancer Treatment and Diagnosis, National Cancer Institute, Bethesda, MD*

ANGELA M. DAVIES, MD • *Division of Hematology and Oncology, Department of Internal Medicine, University of California Davis Cancer Center, Sacramento, CA*

M. J. A. DE JONGE, MD, PHD • *Department of Medical Oncology, Erasmus University Medical Centre, Rotterdam, The Netherlands*

GERALD V. DENIS, PHD • *Cancer Research Center, Boston University School of Medicine, Boston, MA*

JAMES H. DOROSHOW, MD • *Division of Cancer Treatment and Diagnosis, National Cancer Institute, Bethesda, MD*

IGNAT DROZDOV, BS • *Gastrointestinal Pathobiology Group, Yale University School of Medicine, New Haven, CT*

SARITA DUBEY, MD • *Department of Medicine, Comprehensive Cancer Center, University of California, San Francisco, CA*

GRACE K. DY, MD • *Division of Medical Oncology, Mayo Clinic and Foundation, Rochester, MN*

JOHN B. EASTON, PHD • *Department of Molecular Pharmacology, St. Jude Children's Research Hospital, Memphis, TN*

DOUGLAS V. FALLER, MD, PHD • *Genetics Program and Cancer Research Center, Department of Genetics & Genomics, Boston University School of Medicine, Boston, MA*

DAVID R. GANDARA, MD • *Division of Hematology and Oncology, Department of Internal Medicine, University of California Davis Cancer Center, Sacramento, CA*

O. GAUTSCHI, MD • *Division of Hematology and Oncology, Department of Internal Medicine, University of California Davis Cancer Center, Sacramento, CA*

ANUPAMA GOEL, MD • *Hematology-Oncology Division, Continuum Cancer Centers of NY, St. Luke's Roosevelt Hospital, New York, NY*

ALAN M. GEWIRTZ, MD • *Department of Hematology/Oncology, Hematologic Mallgnancies Program, School of Medicine, University of Pennsylvania, Philadelphia, PA*

PAUL H. GUMERLOCK, MD • *Division of Hematology and Oncology, Department of Internal Medicine, University of California Davis Cancer Center, Sacramento, CA*

AKSHAY GUPTA, MD • *Department of Medicine, University of Pittsburgh, Pittsburgh, PA*

OLWEN HAHN, MD • *Section of Hematology/Oncology, Departments of Medicine, University of Chicago, Chicago, IL*

ERNEST T. HAWK, MD, MPH • *Office of Centers, Training, & Resources, National Cancer Institute, Bethesda, MD*

ELISABETH I. HEATH, MD • *Barbara Ann Karmanos Cancer Institute, Wayne State University, Detroit, MI*

CHERYL HO, MD • *University of California Davis Cancer Center, Sacramento, CA*

PETER J. HOUGHTON, PHD • *Department of Molecular Pharmacology, St. Jude Children's Research Hospital, Memphis, TN*

HERBERT HURWITZ, MD • *Division of Medical Oncology, Department of Medicine, Duke University, Durham, NC*

PERCY IVY, MD • *National Cancer Institute, Bethesda, MD*

HOWARD L. KAUFMAN, MD • *Division of Surgical Oncology, Columbia University, New York, NY*

MARK KIDD, PHD • *Gastrointestinal Pathobiology Group, Yale University School of Medicine, New Haven, CT*

JOHN M. KIRKWOOD, MD • *Department of Medicine, University of Pittsburgh, Pittsburgh, PA*

J. J. E. M. KITZEN, MD • *Department of Medical Oncology, Erasmus University Medical Centre, Rotterdam, The Netherlands*

CHAD D. KNIGHTS, PHD • *Department of Oncology, Lombardi Comprehensive Cancer Center, Georgetown University Medical Center, Washington, DC*

ELISE C. KOHN, MD • *Medical Ovarian Cancer Team, Laboratory of Pathology, Center for Cancer Research, National Cancer Institute, Bethesda, MD*

HENRY B. KOON, MD • *Biologic Therapeutics and Cutaneous Oncology Programs, Beth Israel Deaconess Medical Center, Boston, MA*

PETER KOZUCH, MD • *Hematology-Oncology Division, Continuum Cancer Centers of NY, St. Luke's Roosevelt Hospital, New York, NY*

PRIMO N. LARA, MD • *University of California Davis Cancer Center, Sacramento, CA*

IGOR LATICH, MD • *Gastrointestinal Pathobiology Group, Yale University School of Medicine, New Haven, CT*

JOSEP M. LLOVET, MD • *Liver Cancer Research Program, The Mount Sinai School of Medicine, New York, NY*

P. C. MACK, MD • *University of California Davis Cancer Center, Sacramento, CA*

JOHN MENDELSOHN, MD • *The University of Texas M. D. Anderson Cancer Center, Houston, TX*

ANNE MENKENS, PHD • *Cancer Imaging Program, National Cancer Institute, Bethesda, MD*

IRVIN M. MODLIN, MD, PHD, DSC • *Gastrointestinal Pathobiology Group, Yale University School of Medicine, New Haven, CT*

NEAL J. MEROPOL, MD • *Gastrointestinal Cancer Program, Gastrointestine Tumor Risk Assessment Program, Fox Chase Cancer Center, Philadelphia, PA*

TAKAFUMI NAKAMURA, PHD • *Molecular Medicine Program, Mayo Clinic College of Medicine, Rochester, MN*

LEN NECKERS, PHD • *Urologic Oncology Branch, National Cancer Institute, Bethesda, MD*

MIGUEL-ANGEL PERALES, MD • *Department of Medicine, Memorial Sloan Kettering Cancer Center, New York, NY*

RICHARD G. PESTELL, MBBS, PHD, FRACP • *Kimmel Cancer Center, Department of Cancer Biology, The Thomas Jefferson University, Philadelphia, PA*

STEPHEN J. RUSSELL, MD, PHD • *Molecular Medicine Program, Mayo Clinic College of Medicine, Rochester, MN*

SHERMINI SAINI, MD • *The Division of Pediatric Hematology-Oncology, Department of Pediatrics, Duke University, Durham, NC*

GISELE SAROSY, MD • *Medical Ovarian Cancer Team, Laboratory of Pathology, Center for Cancer Research, National Cancer Institute, Bethesda, MD*

JOAN H. SCHILLER, MD • *Division of Hematology/Oncology, Department of Internal Medicine, The Harold C. Simmons Comprehensive Cancer Center, UT Southwestern Medical Center, Dallas, TX*

JONATHAN D. SCHWARTZ, MD • *Division of Hematology and Oncology, Department of Medicine, The Mount Sinai School of Medicine, New York, NY*

LALITHA K. SHANKAR, MD, PHD • *Cancer Imaging Program, National Cancer Institute, Bethesda, MD*

JOSEPH A. SPARANO, MD • *Albert Einstein College of Medicine and Breast Evaluation Center, Montefiore Medical Center, Bronx, NY*

WALTER STADLER, MD, FACP • *Sections of Hematology/Oncology and Urology, Departments of Medicine and Surgery University of Chicago, Chicago, IL*

DANIEL C. SULLIVAN, MD • *Cancer Imaging Program, National Cancer Institute, Bethesda, MD*

MARK L. SUNDERMEYER, MD • *Fox Chase Cancer Center, Philadelphia, PA*

SAM THIAGALINGAM, PHD • *Genetics Program and Cancer Research Center, Department of Medicine, Boston University School of Medicine, Boston, MA*

JOSEPH E. TOMASZEWSKI, PHD • *Toxicology & Pharmacology Branch, Developmental Therapeutics Program and Division of Cancer Treatment and Diagnosis, National Cancer Institute, Bethesda, MD*

LINDA T. VAHDAT, MD • *Breast Cancer Research Program, Division of Hematology and Medical Oncology, Weill Medical College of Cornell University, New York, NY*

J. VERWEIJ, MD, PHD • *Department of Medical Oncology, Erasmus University Medical Centre, Rotterdam, The Netherlands*

JAYE L. VINER, MD, MA • *Office of Centers, Training, & Resources, National Cancer Institute, Bethesda, MD*

DAN T. VOGL, MD • *Hematologic Mallgnancies Program, School of Medicine, University of Pennsylvania, Philadelphia, PA*

SCOTT WADLER, MD • *Division of Hematology and Medical Oncology, Department of Medicine, Weill Medical College, Cornell University, New York, NY*

LOUIS M. WEINER, MD • *Department of Medical Oncology, Division of Medical Sciences, Fox Chase Cancer Center, Philadelphia, PA*

JEDD D. WOLCHOK, MD, PHD • *Department of Medicine, Memorial Sloan Kettering Cancer Center, New York, NY*

I GENERAL STRATEGIES FOR MOLECULAR TARGETING IN ONCOLOGY

1 The Cell Cycle

Therapeutic Targeting of Cell Cycle Regulatory Components and Effector Pathways in Cancer

Chad D. Knights, PhD, and Richard G. Pestell, MBBS, MD, PhD, FRACP

SUMMARY

Dysregulation of cell cycle signaling is a pathognomonic feature of tumor initiation and progression. An understanding of the key cell cycle components dysregulated in cancer and the molecular mechanisms responsible has led to the generation of new targeted therapeutics. The development of therapies which selectively inactivate key genetic drivers in specific tumors and the use of molecular abnormalities within the cancer to selectively activate therapies exemplify mechanism-based therapies. The tyrosine kinase inhibitor signal transduction inhibitor-571 (STI-571) is a proto-typic molecular-targeted therapy, which selectively targets aberrant Bcr-Abl kinase activity and produces a highly specific anti-cancer effect in chronic myelogenous leukemia (CML) patients. Novel therapies include inhibitors of tyrosine kinases, cyclin-dependent kinases or histone deacetylases, lytic viruses that kill cells with defective p53 function, and molecular mimics that induce or recapitulate endogenous tumor suppressors. These new approaches are derived from an understanding that dysregulated cell cycle control components drive tumorigenesis. Components of the cell cycle often play distinct roles in the biological processes of normal development, normal cell cycle progression in the adult animal, and during the process of tumori-genesis. The realization that cell cycle components play redundant roles in the cell cycle of embryogenesis, but are required for tumorigenesis, provides an additional, compelling rationale for targeting the aberrant cell cycle in cancer. Ultimately, the continued study of the mechanisms used by cancerous cells to evade cell cycle checkpoint control provides the groundwork for the development of rational cancer therapies aimed at improving both the efficacy of treatment and the quality of patient life.

Key Words: Cyclin-dependent kinase; CDK inhibitors; cell cycle; therapy; acety-lation; p53; cyclin D1; EGFR; HDAC; STI-571; flavopiridol; CDK2 inhibitor.

From: *Cancer Drug Discovery and Development: Molecular Targeting in Oncology*
Edited by: H. L. Kaufman, S. Wadler, and K. Antman © Humana Press, Totowa, NJ

1. INTRODUCTION

A greater understanding of the molecular genetic changes within an individual patient's tumor has led to an alternative therapeutic approach, in which the key genetic drivers of tumorigenesis serve as targets for therapy. Aberrant function and expression of the cell cycle is a uniform feature of human tumors. Targeted therapies directed to abnormal cell cycle control protein function has led to the development of effective new therapies. Herein, we describe the components regulating the cell cycle and highlight recent progress in the field. We discuss preclinical and clinical therapeutic advances, targeting the cell cycle and new types of therapeutics under development.

2. THE CELL CYCLE IN NORMAL CELL DIVISION

Eukaryotic cells, upon stimulation by mitogenic signals, pass through a highly regulated sequence of events referred to as the cell cycle. The cell cycle is marked by four distinct phases: G_1 (Gap1) phase, S (DNA synthesis) phase, G_2 (Gap2) phase, and M (mitosis) phase. During the G_1 phase, the abundance of mitogenic signals determines DNA synthesis, apoptosis, or progression to a quiescent state (G_0 phase). After the commitment to cellular division, cells undergo DNA replication in S phase, passage through G_2 phase, cellular division in M phase, and ultimately return to G_1 phase.

The orderly progression of the cell cycle is controlled primarily by the cyclin-dependent kinases (CDKs). CDKs are serine/threonine-specific protein kinases whose catalytic activity is positively regulated by cyclins and negatively regulated by CDK inhibitors (CDKIs), the expression of which is tightly temporally regulated during cell division. The cyclin-CDK holoenzymes phosphorylate diverse substrates including the retinoblastoma tumor suppressor protein (pRb) and the related p130 and p107 proteins. Many of the cyclin-CDKs regulate cell cycle "checkpoints" that protect a cell from erroneous DNA replication and ensure the accuracy and precision of cell division. Since the discovery of the first CDK by Timothy Hunt in 1983 (CDC2 in yeast), at least 13 human CDKs have been identified which function in cell cycle regulation and other cellular processes (Table 1).

Passage through the G_1 restriction point is regulated by the expression of two G_1 cyclin families, the cyclin D family (D1, D2, and D3), and the cyclin E family (E1 and E2). The D cyclins interact with CDK4 and CDK6, and the E cyclins interact with CDK2, forming heterodimeric holoenzymes, which can phosphorylate pRb rendering it inactive and allowing passage from G_1 into S phase (Fig. 1). The D cyclins, in particular cyclin D1 that is the rate-limiting subunit in the formation of CDK4/6 holoenzymes, are sensitive to mitogenic stimuli and link extracellular proliferation cues to the underlying cell cycle program. Hyper-phosphorylation of pRb by cyclin D-CDK4/6 holoenzymes during mid-G_1 phase results in the release of E2F family members that direct the transcription of the E cyclins and components that are necessary for DNA replication in late G_1 phase. Cyclin E–CDK2 complexes lead to further pRb phosphorylation forming a positive feedback loop that precipitates entry into S phase. Some redundancy in cyclin D/E function may also exist, as transgenic expression of cyclin E in cyclin D1-deficient mice can rescue approximately one-third of the mice from cyclin D1-deficient phenotypes, suggesting complex partial redundancy in cyclin function *(1)*.

Table 1
The Cyclin–Dependent Kinases and their Heterodimeric Regulatory Cyclin Partners.
The Proposed Functions are Shown on the Right.

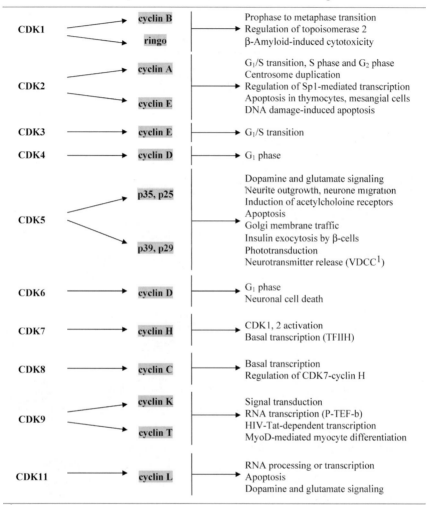

CDK	Cyclin Partner	Proposed Functions
CDK1	cyclin B, ringo	Prophase to metaphase transition Regulation of topoisomerase 2 β-Amyloid-induced cytotoxicity
CDK2	cyclin A, cyclin E	G₁/S transition, S phase and G₂ phase Centrosome duplication Regulation of Sp1-mediated transcription Apoptosis in thymocytes, mesangial cells DNA damage-induced apoptosis
CDK3	cyclin E	G₁/S transition
CDK4	cyclin D	G₁ phase
CDK5	p35, p25, p39, p29	Dopamine and glutamate signaling Neurite outgrowth, neurone migration Induction of acetylcholoine receptors Apoptosis Golgi membrane traffic Insulin exocytosis by β-cells Phototransduction Neurotransmitter release (VDCC[1])
CDK6	cyclin D	G₁ phase Neuronal cell death
CDK7	cyclin H	CDK1, 2 activation Basal transcription (TFIIH)
CDK8	cyclin C	Basal transcription Regulation of CDK7-cyclin H
CDK9	cyclin K, cyclin T	Signal transduction RNA transcription (P-TEF-b) HIV-Tat-dependent transcription MyoD-mediated myocyte differentiation
CDK11	cyclin L	RNA processing or transcription Apoptosis Dopamine and glutamate signaling

[1]VDCC - voltage-dependent Ca2+ channel. Reproduced with permission from (74)

Cyclin D1 expression is critical in the proliferation of numerous cell types including hematopoietic, fibroblast, myocytes, and epithelial cells (2,3). Cellular levels of cyclin D1 can be influenced by a number of mitogenic and oncogenic signals including mutations of Ras, Src, Rac, and ErbB2 (HER-2/neu) (4–7). The phosphatidylinositol-3-kinase (PI3K)/Akt signaling pathway both induces cyclin D1 expression and stabilizes the abundance of cyclin D1 (8,9).

The successful completion of DNA replication during S phase leads to the G₂/M checkpoint, which is controlled by cyclin B and CDK1 (Fig. 1). The cyclin B-CDK1 heterodimer forms during the S to G₂ phase transition but is rendered inactive in early G₂ by phosphorylation. The CDC25 phosphatase dephosphorylates the cyclin B–CDK1

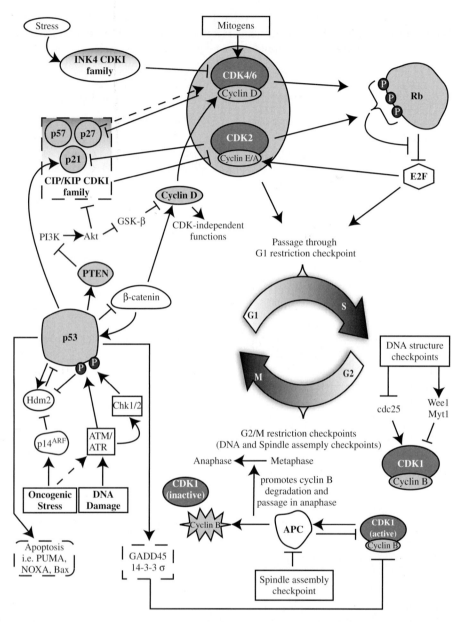

Fig. 1. The checks and balances of the cell cycle. Mitogenic stimuli (e.g., growth factors) and activation of survival pathways (e.g., Akt kinase) enhance the expression of cyclin D family members, which activate cyclin-dependent kinase (CDK) 4/6 holoenzymes, resulting in the phosphorylation and inactivation of retinoblastoma tumor suppressor protein (pRb). The subsequent release of E2F from pRb leads to increased levels of cyclin A/E and CDK2 activity thus perpetuating pRb phosphorylation and progression from G_1 to S phase. The INK4 and CIP/KIP families of CDKIs respond to various stress conditions, including the activation of p53, and work to prevent the activation of cyclin-CDK holoenzymes thereby arresting the cell cycle. Passage through the G_2-DNA structure checkpoint into M phase is accomplished by dephosphorylation of the cyclin B-CDK1 holoenzyme, which results when the activity of the cdc25 phosphatase outpaces that of the Wee/Myt1 kinases. During M phase, the APC is activated by cyclin B-CDK1 and targets cyclin B for degradation allowing passage into anaphase and the subsequent completion of mitosis.

complex resulting in its sustained activation through the completion of metaphase after which cyclin B is ubiquitylated and targeted for degradation by the anaphase promoting complex (APC). In a feedback mechanism, phosphorylation of APC by cyclin B-CDK1 is required for cyclin B proteolysis and transition out of G_2 into interphase (Fig. 1).

In contrast to cyclins, CDKIs were initially described through their ability to negatively regulate CDK function. The CDKIs group into two distinct families—the INK4 family and the CIP/KIP family. The INK4 family (p16^{INK4A}, p15^{INK4B}, p18^{INK4C}, and p19^{INK4D}) binds directly to and inhibits CDK4 and CDK6. The CIP/KIP family (p21^{CIP1}, p27^{KIP1}, and p57^{KIP2}) share structural homology and can bind and form ternary complexes with cyclin–CDK complexes (cyclin B1-CDK1, cyclin A/E-CDK2). p21^{CIP1} and p27^{KIP1} also promote the assembly and activity of cyclin D–CDK4/6 complexes (10,11). The INK4 and CIP/KIP families of CDKIs thus control pRb phosphorylation indirectly through their effect on CDKs, thereby regulating passage through the G_1 restriction point.

The CIP/KIP family of CDKIs functions in a concentration-dependent manner and are subject to proteasome-mediated degradation through ubiquitin-dependent (p21^{CIP1} and p27^{KIP1}) and ubiquitin-independent (p21^{CIP1}) pathways (12–18). Regulation of p21^{CIP1} and p27^{KIP1} activity and accumulation are primarily controlled by post-translational modifications. In response to mitogenic stimuli, p27^{KIP1} is phosphorylated on threonine 187 by cyclin E-CDK2. This phosphorylation creates a docking site for the substrate recognition factor Skp2, which is part of the larger Skp1-cullin-F-box ubiquitin ligase complex that promotes the ubiquitin-mediated proteolysis of p27^{KIP1}. Skp2-independent degradation of p27^{KIP1} involves the Kip1 ubiquitination promoting complex (19). Both p21^{CIP1} and p27^{KIP1} are the target of Akt-induced phosphorylation that results in sequestration of the proteins in the cytoplasm thereby maintaining nuclear CDK2 activity.

3. THE CELL CYCLE IN DEVELOPMENT

The role of cyclins, CDKs, and CDKIs has been analyzed in transgenic mice. Given the importance of each of these components in normal cell cycle progression, the functional redundancy demonstrated in these experiments was surprising. *Cyclin D1$^{-/-}$* mice are viable with fatty liver, defects in mammary epithelial cell differentiation, retinal apoptosis, and poorly migrating macrophages (20–29). *Cyclin D2$^{-/-}$* mice displayed defective ovarian granulosa cell development and hypoplastic testes and reduced proliferation of B cells in granule neurons (22,26). *Cyclin D3$^{-/-}$* mice display a hypoplastic thymus (24), and *cyclin D1, D2, and D3$^{-/-}$* mice show a hematopoietic defect reflecting defective proliferative capacity of hematopoietic stem cells, dying at day 16.5 during embryonic development (2). Deletion of either cyclin E1 or cyclin E2 alone has no effect on mouse development or cellular proliferation in vitro (30,31). Deletion of cyclin E1 and cyclin E2 results in placental and cardiac defects, with death at embryonic day 11.5 due to failure of endoreduplication of placental trophoblasts (30,31). Mouse embryonic fibroblasts (MEFs) proliferate more slowly, with a failure to reenter the cell cycle due to failure of loading mini-chromosome maintenance proteins onto prereplication origins (30,31). CDK2-null mice are viable although sterile (32,33). Mice deleted of CDK4 are viable, with mild defects in hematopoiesis and thymic and splenic hypoplasia *(34)*. $CDK4^{-/-}$ mice are viable though sterile *(35–38)*. Curiously, the animals demonstrated insulin-dependent diabetes due to abnormal development of β-islet cells. Mice deleted of *cyclin A1 (CCNA1)* have a normal phenotype other than a defect in male meiosis. *Cyclin A2* deletion results in embryonic lethality *(39)*. Collectively, these studies demonstrated an important role for specific cell cycle components in distinct compartments, but were more surprising for their relatively benign effect on cell cycle progression.

Analysis of CDKI function in development also suggested a relatively unimportant role for these proteins individually in normal development. Disruption of individual *INK4* genes in the mouse germ line (p16$^{\mathrm{INK4a}}$, p15$^{\mathrm{INK4b}}$, p18$^{\mathrm{INK4c}}$, and p19$^{\mathrm{INK4d}}$) resulted in viable and fertile mice. Mice developed relatively normally, suggesting no one family member was essential for cell cycle control.

Subsequent analysis of transgenic mice with deleted cell cycle components has provided important insight into cell cycle function in response to oncogenic stimuli. Mice lacking p16$^{\mathrm{INK4a}}$ are particularly tumor-prone and develop a wide spectrum of cancers, particularly when exposed to chemical carcinogens or X-rays *(40,41)*. Cyclin D1-deficient mice are resistant to tumorigenesis induced by oncogenic Ras targeted to either the breast or the skin *(42,43)*, but have increased mammary tumors induced by activating β-catenin *(44)*. Mice either completely deficient or heterozygous for *cyclin D1* are resistant to colonic tumorigenesis induced by activation of the *Apc^{Min}* gene mutation *(45)*. Mice haploinsufficient for CDKN1B (p27$^{\mathrm{KIP1}}$) develop pituitary tumorigenesis and enhanced tumorigenic response to 7,12-dimethylbenz[a]anthracene *(46)*. In addition, mammary tumorigenesis induced by ErbB2 was accelerated in the CDKN1B heterozygous background *(47)*. MEFs derived from mice deleted of *cyclin* and *CDK* genes demonstrated resistance to oncogenic transformation and reduced ability to enter the cell cycle from quiescence in a subset of experiments. Additionally, *cyclin D1, cyclin D2* , and *cyclin D3^{-/-}* cells show reduced induction of DNA synthesis *(2,5)*. *Cyclin D1, cyclin D2* , and *cyclin D3^{-/-}* MEFs also show reduced susceptibility to transformation by Ras, Myc, E1A, or dominant negative p53, as do *CDK4^{-/-}* and *cyclin E1 and cyclin E2^{-/-}* MEFs *(31)*. *CDK2^{-/-}* MEFs can be transformed with oncogenic Ras and E1A, but less efficiently than wild-type cells *(32,33)*. Mice deficient in either INK4a or ARF *(48)* are significantly more tumor-prone than wild-type animals, but less tumor-prone than either *p53^{-/-}* or *INK4a/ARF^{-/-}* animals while displaying different tumor spectra suggesting varying roles for the INK4a/ARF proteins. Collectively, these studies are consistent with an important role for the CDKI proteins as tumor suppressors in vivo.

4. THE CELL CYCLE IN CANCER

Tumorigenesis in vivo involves a multi-step process within the primary cell of origin and requires heterotypic signals from the local environment, including angiogenic cues. Inactivation of recessive tumor suppressor genes results from somatic mutations or inherited defects. Tumor suppressor genes include *TP53, RB1, INK4a, ARF, APC, PTCH, PTEN, SMAD4, DPC4, TFC1, NF1, WT1, MSH2, MLH1, ATM, MBS1, CHK2, BRCA1, BRCA2, FA* genes, and *VHL (49)*. However, dysregulation of cell cycle components are a common feature. The steps governing initiation and commitment to tumorigenesis may be distinct. Prototypic tumor suppressors are recessive. Their functions are diverse governing a wide range of normal cellular activities, including cell cycle checkpoint control, mitogenic signaling pathways, protein turnover, DNA damage, hypoxia, and other stress responses (reviewed in ref. *49*). The transition from benign to malignant disease is associated with increases in chromosomal aberrations. Tumor mediators drive bridge fusion breakage cycles that facilitate genomic instability, promoting the molecular genetic aberrations required for full malignant transition *(50)*. The contributions of telomerase activity in the initiation and progression of cancer is complex, and, although a target of cell cycle control, the role of telomerase activity and the therapeutic target role of telomerase as a target in cancer therapy is complex *(51)*.

Deregulation of pRb signaling pathways is a common hallmark found in up to 90% of all human cancers, which leads to unchecked progression into S phase. Deletion of pRb results in tumors of the retina and an increased predisposition to osteosarcomas, while inactivation of CDKIs and/or overexpression of cyclins that predominantly regulate the G_1–S transition are displayed in a broad spectrum of tumors. One paradigm that distills much of the working knowledge of the cell cycle in tumors proposes two parallel pathways of cell cycle surveillance exist. One arm is composed of the CDK/p16^{INK4A}/pRb pathway, and the other is composed of the p53/HDM2/p14^{ARF} pathway. Deregulation of any point within a pathway is sufficient to inactivate the pathway. Thus, overexpression of cyclin D1 or deletion of the p16^{INK4A} would inactivate the pRb pathway, resulting in unchecked cell cycle progression. Inactivation of both pathways is required for tumorigenesis by eliminating the checks and balances used by a cell to maintain fidelity of cell cycle control. Inactivation of the p53 arm is commonly accomplished by human double minute 2 (*HDM2*) gene amplification or p53 gene mutations that occur in approximately 50% of all human cancers.

It has been predicted that nearly all human cancers carry at least one alteration in the p53 surveillance pathway. The tumor suppressor p53 protein is an essential regulator of the G_1 checkpoint and can respond to multiple types of cellular stress including DNA damage, oncogenic signaling [possibly via DNA damage *(52)*], and hypoxia *(53)*. Tight regulation of p53 function is required and provided by HDM2, an E3 ubiquitin ligase. HDM2 directs p53 ubiquitylation and subsequent proteasome-mediated degradation, and is a transcriptional target of p53, forming a negative feedback loop. DNA damage dissociates the p53–HDM2 interaction by inducing a kinase cascade that results in the phosphorylation of the HDM2 binding site on the N-terminus of p53 *(54,55)*, while oncogenic stimuli results in the induction of p14^{ARF}, which sequesters HDM2 into nucleoli *(56,57)*. The p53/HDM2/p14^{ARF} pathway is disabled by mutation or repression of ARF by other proteins (Twist and TBX2). Although oncogenes such as Myc activate ARF gene expression, p19^{ARF} (murine homolog of p14^{ARF}) can also negatively regulate Myc's transcriptional activity through a direct physical interaction independent of Mdm2 and p53 *(58)*.

Phosphorylation regulates the half-life of p53 and facilitates its acetylation and activation of apoptosis. The acetylation of p53, like p53 phosphorylation, is responsible for directing the function of p53 in response to stress *(59,60)*. Acetylation of p53 on lysine 373 and 382 by the histone acetyltransferases (HATs) p300 or CREB-binding protein (CBP) can induce apoptosis, while the acetylation of lysine 320 by p300/CBP-associated factor (P/CAF) has been linked to nonapoptotic stimuli *(61,62)*. Numerous other post-translational modifications have been described to act on p53 including sumoylation and methylation. While highly complex and as yet ill-defined, these various signals are likely to control the diverse functions of the p53 tumor suppressor protein, which include the induction of senescence, the induction of cell

Table 2
Cyclin-Dependent Kinase (CDK)-Independent Functions of Cyclin D1

Signaling targets	Functional significance	Reference
Transcription factors		
ERα	Cyclin D1 (not D2 or D3) recruits SRC1 to ERα potentiating activation of unliganded ERα	178,179
C/EBPβ	Cyclin D1 activates transcriptional activity	180
AR	Cyclin D1 represses ligand-bound AR by interfering with P/CAF association and recruiting HDAC1 and HDAC3 to AR	70,72,181
TR	Cyclin D1 represses both unliganded TR and liganded TR activity	182
PPARγ	Cyclin D1 represses PPARγ induction and transcriptional activity	29,68
Myb	Cyclin D1 inhibits transcriptional activity	183
	Cyclin D1 inhibits Myb p300-dependent actylation	184
DMP1	Cyclin D1 inhibits transcriptional activity	185
BETA2/NeuroD	Cyclin D1 represses transcriptional activity	186
STAT3	Cyclin D1 represses STAT3 activation	187
MyoD	Cyclin D1 represses transcriptional activity	188
Sp1	Cyclin D1 represses transcriptional activity	189
Brg1	Cyclin D1 and E may regulate SWI/SNF complex through Brg1	190
AIB-1	p160 family of co-activators; inhibited	178
GRIP-1	p160 family of co-activators; inhibited	191
Transcriptional co-factors		
NcoA/SRC1a	Cyclin D1 recruits SRC1 to ERα potentiating activation	178
P/CAF	Cyclin D1 represses HAT activity	70
p300/CBP	Cyclin D1 represses HAT activity	UO
	Phosphorylated by cyclin E-CDK2 increasing E2F association	192
	Cyclin D1 prevents p300 dependent induction of PPARγ	UO
HDAC1	Cyclin D1 recruits HDAC1 to AR and PPARγ	68 and UO
HDAC3	Cyclin D1 recruits HDAC1 to TR and PPAR	68,182
TAF250	Cyclin D1 represses SP-1-mediated transcription	193
Additional targets		
BRCA1	Cyclin D1 overcomes BRCA1-mediated inhibition of liganded ERα	194
BARD1	Phosphorylated by cyclin E1/A1-CDK2; reduces BRCA1: BRAD ubiquitin ligase activity	195
NPAT	Phosphorylated by cyclin E-CDK2; induces histone expression	196

UO, Unpublished observation.

cycle arrest, and the induction of apoptosis through both transcriptional-dependent and transcriptional-independent pathways *(63,64)*. p53 primarily induces a G_1 growth arrest through the induction of p21^{CIP1} and a G_2 arrest by promoting the transcription of GADD45 and 14-3-3σ that interfere with cyclin B-CDK1 activation. Furthermore, p53 induction of the PTEN tumor suppressor protein can indirectly regulate cyclin D-CDK4/6 function by inhibiting the PI3K/Akt kinase cascade increasing the activity of the CIP/KIP family of CDKIs and destabilizing cyclin D1.

The cyclin D1 gene is amplified and overexpressed in a broad spectrum of human malignancies ranging from breast carcinomas to soft tissue sarcomas *(65)*. Of interest, cyclin E is also overexpressed in human breast cancer, and a subset of human tumors display a cleaved form of cyclin E that correlates with a poor prognosis *(66)*. CDK-independent functions of cyclins contribute to gene expression, cellular differenti-ation, and growth *(67)*. Cyclin D1 alters the function of more than 30 transcription factors including v-Myb, MyoD, ERα and STAT3 (Table 2) through physical inter-action with co-activators (p300/pCAF) and histone deacetylases (HDACs) *(68)*. Cyclin D1 repression of p300 coactivator function has been linked to the inhibition of p300-autoacetylation by cyclin D1 *(69)*. Alternatively, cyclin D1 can repress the ligand-dependent activity of the androgen receptor (AR) by recruiting HDACs including HDAC3 and by competing with P/CAF for binding to AR *(70–72)*. Cyclin D1 represses the transcriptional activity of peroxisome proliferator-activated receptor γ (PPARγ) through recruitment of HDACs, HP1α and the SUV39 methyltransferase to the PPARγ response element to silence transcription in the context of the local chromatin structure *(68)*. In addition to inactivating the tumor suppressor pRb, cyclin D1 blocks the function of BRCA1 and the estrogen receptor *(73)*. Thus, cyclin D1 regulates the cell cycle through CDK activity, regulates chromatin topography at the sites of transcription, and blocks function of tumor suppressors such as BRCA1 and many transcription factors.

5. PHARMACOLOGIC CYCLIN-DEPENDENT KINASE INHIBITORS

Based on the essential functions cyclins and CDKs perform in cell cycle progression and tumorigenesis, pharmacologic inhibitors of the cell cycle machinery have been investigated *(74)*. Inhibitors of CDKs interfere with cyclin binding, compete with ATP for binding to the kinase-ATP binding site, or stimulate natural CDKIs. Of the approx-imately 50 inhibitors that have been described to date, most are low molecular weight, flat hydrophobic heterocycles that compete for the CDK-ATP binding site. Many of these inhibitors work at nanomolar concentrations and have been co-crystallized with CDK2 or modeled with CDKs *(75)*. While numerous new classes of inhibitors have been characterized (Table 3), flavopiridol and the staurosporine derivative UCN-01 have progressed to clinical trials and demonstrate promise in a wide array of human cancers.

5.1. *Flavopiridol*

Flavopiridol, which is a semisynthetic flavonoid derived from rohitukine, is a broad spectrum CDKI with activity against CDK1, CDK2, CDK4, and CDK6 [IC$_{50}$ (inhibitory concentration 50%) of ~100nM and against CDK7 [IC$_{50}$ of ~300 nM *(76)*]. The antitumor activities of flavopiridol include inhibition of growth and proliferation, induction of apoptosis, and inhibition of angiogenesis. Flavopiridol competitively and

Table 3
Direct Cyclin-Dependent Kinase (CDK) Modulators

Targeted CDKs	Compound	Targeted CDKs	Compound
CDK1/CDK2/CDK5	Roscovitine and CYC202	Cdk4	Pyrrolo-carbazoles
	Olomucine		Indolocarbazoles
	CVT-313		Tryaminopyrimidine (CINK4) *(202)*
	Butyrolactone I		Fascaplysin
	Purvalanol		PD0183812 *(203)*
	BMS-387032		PD0332991 *(204)*
	Aloisines		Cynnamaldehydes
	Indirubins		Dioxobenzothiazoles
	Hymenialdisine		Pyrazol-3-ylurea (compound 15b)
	Pyrazolo-piridines		Bicyclic 2-anilinopyrimidines (<20 nM) *(203,205)*
	Pyrazolo-quinoxalines		2-Anilinopyrimidines (7 nM) *(206)*
	Indenopyrazoles (9 nM) *(197)*		2,4-bis anilinopyrimidines (10 nM) *(207,208)*
	SU9516		
	Nitrosopirimidines	Nonspecific cdk	Flavopiridol
	Paullones		Staurosporine
	Diaminotriazole (2 nM) *(198)*		UCN-01
	Aminoimadazole (28 nM) *(199)*		Oxyndoles
			Quinazolines
	Oxindoles (6 nM) *(200,201)*	Unknown	Toyocamycin
			Myricetin

References are shown in italics and IC_{50} amounts are shown within parentheses. Reproduced with permission in part from ref. *(95)*.

reversibly inhibits the CDK-ATP-binding site and represses the expression of cyclin D1, cyclin D3, and CDK4 *(77)*. Flavopiridol also inhibits P-TEFb (cyclin T1-CDK9), independently of ATP binding *(78)*, which is critical for the function of RNAP II and transcription elongation. Flavopiridol inhibits proliferation of hematopoietic cells *(79,80)* and human umbilical vein endothelial cells *(81)*, prevents the induction of vascular endothelial growth factor (VEGF) by hypoxia in human monocytes *(82)* and induces apoptosis *(77,79)*.

Initial pharmacokinetic studies of flavopiridol in rodents displayed poor oral bioavailability, so subsequent treatments involved intravenous or intraperitoneal drug administration, where the major toxicities seen involved the bone marrow and gastrointestinal tract *(83)*. Mice treated with boluses of flavopiridol for 5 days or continuously with a 72-h infusion of flavopiridol both demonstrated antitumor activities indicating that repetitive high peak plasma concentrations were desirable for the most effective treatment course *(80,84)*. Synergy is seen with a number of cytotoxic chemotherapeutics and typically requires that the chemotherapeutic treatment precede flavopiridol dosing *(85)*.

Four phase I clinical trials have been completed to date using flavopiridol in monotherapy treatment. The first of these studies was completed by the U.S. National Cancer Institute (NCI) and enrolled 76 patients with refractory malignancies and evidence of prior disease progression *(86)*. Flavopiridol was administered as a 72-h continuous infusion every 2 weeks during which a maximum tolerated dose (MTD) of 50 mg/m^2/day over 3 days was identified with a dose-limiting toxicity (DLT) of secretory diarrhea. In the presence of antidiarrheal prophylaxis, the MTD was escalated to 78 mg/m^2/day and was limited by the occurrence of hypotension and proinflammatory syndrome that included local tumor pain. A second phase I trial that employed a similar treatment regimen corroborated the NCI's findings with an MTD of 40 mg/m^2/day and a DLT of secretory diarrhea *(87)*. Both studies reported patient plasma concentrations between 300 and 500 nM, which can inhibit CDK activity in vitro. Minor responses were observed in patients with non-Hodgkin's lymphoma, colon cancer, renal cell carcinoma (RCC), and prostate cancer, although follow-up phase II trials using a similar regimen demonstrated no significant antitumor effect with only modest activity against metastatic RCC *(88–91)*. However, studies of individuals with stage IV non-small-cell lung cancer (NSCLC) and refractory mantle cell lymphoma yielded encouraging results. Patients with refractory mantle cell lymphoma ($n = 30$) had an overall response rate of 11% with 71% of the patients attaining stable disease with a 3.4 month duration of response *(92)*. In those with NSCLC, the median overall survival for the 20 enrolled patients was approximately 7.5 months *(93)*, which is comparable to the median survival following chemotherapy containing platinum analogs in combination with taxanes or gemcitabine *(94)*. This has prompted the initiation of a phase III clinical trial comparing standard combination chemotherapy versus combination chemotherapy with flavopiridol *(95)*.

In a fourth phase I trial ($n = 26$), flavopiridol was administered as a 24-h continuous infusion every 2 weeks to patients with previously treated chronic lymphocytic leukemia (CLL) *(96)*. A MTD of 140 mg/m^2 was achieved with thrombocytopenia and diarrhea being the most common toxicities observed. Despite the ability to achieve flavopiridol concentrations capable of inducing apoptosis in cultured CLL cells *(79,97)*, there were no partial or complete responses noted in this phase I trial. From this study, a keen observation has recently been made concerning flavopiridol bioavailability in that flavopiridol has a much higher binding affinity for human plasma proteins compared to fetal calf serum (FCS), which was predominantly used in all of the preclinical studies. Substitution of human serum for FCS in vitro results in a decrease of free drug from 63–100% to 5–8%. Taking this into consideration, the dose schedule of flavopiridol has been optimized and is currently being reevaluated in phase I trials *(96,98)*. Preclinical trials have demonstrated the efficacy of post-flavopiridol treatment in combination with a number of different chemotherapeutic drugs, including the microtubule stabilizing drug paclitaxel *(99)* and irinotecan, which stabilizes DNA–topoisomerase complexes *(100)*.

5.2. UCN-01

UCN-01 (7-hydroxystaurosporine) is a staurosporine analog that induces G$_1$ cell cycle arrest, and abrogation of the G$_2$/M checkpoint resulting in apoptosis. Abrogation of the G$_2$/M checkpoint by UCN-01 in the presence of DNA damage is accomplished by activation of CDK1 and by increasing cdc25 phosphatase activity. The G$_1$ growth arrest induced by UCN-01 may involve a loss of CDK2 activity due to increased p21^{CIP1}/p27^{KIP1} interaction with CDK2 *(101)*. Furthermore, UCN-01 can alter the PI3K/Akt survival pathway by inhibiting PDK1, an upstream kinase that is required for sustained Akt activation *(102)*.

Based on preclinical findings, the first phase I trial of UCN-01 was conducted by administering a 72-h continuous infusion every 2 weeks *(103)*. Unexpectedly, UCN-01 displayed a long half-life (30 days), which was approximately 100 times longer than preclinical models suggested. Following this observation, protocols were adjusted to supply UCN-01 once every 4 weeks using a 36-h continuous infusion. Dose-limiting toxicities of nausea/vomiting, hyperglycemia, and pulmonary toxicity were observed and led to the phase II recommendation of 42.5 mg/m^2/day given by a 72-hour continuous infusion. During this trial, a patient with metastatic melanoma had a partial response that lasted approximately 8 months while a patient with refractory anaplastic large-cell lymphoma had a complete regression and was disease-free 4 years after treatment.

5.3. Outlook—Pharmacologic Cyclin-Dependent Kinase Inhibitors

In addition to flavopiridol and UCN-01, CYC202 (R-roscovitine), BMS-387032, and E7070 have demonstrated strong therapeutic potential in preclinical studies (Table 3). CYC202 is a purine analog and BMS-387032 is a 2-aminothiazole that both target CDK2 for inhibition by competing for ATP binding. Both agents have demonstrated antiproliferative effects in a number of tumor cell lines associated with a reduction in pRb phosphorylation, most likely as a result of CDK2 inhibition *(104,105)*. Phase I clinical trials have been initiated for BMS-387032 while phase I trials with CYC202 are under way but have yet to yield an objective response. E7070 is a sulfonamide that has antitumor activity in a range of in vivo and in vitro models and has been shown to inhibit CDK2 activity, upregulate p53, and induce apoptosis. Both phase I and II trials have been conducted using E7070, and while the phase I trials did not demonstrate a therapeutic response, phase II trials have provided more promising results *(98)*. A growing number of selective cyclin or CDKIs have been developed, with selectivity to either CDK2 or CDK4 kinase (Table 3) (reviewed in ref. *106*).

6. THERAPEUTIC TARGETING OF HDACS

The control of histone acetylation through HATs and deacetylases (HDACs) is central to the regulation of gene transcription through the alteration of chromatin topography and promoter accessibility. Non-histone proteins are acetylated, including transcription factors, signal mediators, co-activators, and structural proteins (Table 4). The process of acetylation involves the transfer of an acetyl group to the ε amino group of a lysine residue thereby neutralizing lysine's positive charge within the targeted substrate. Either single or multiple acetylations of protein factors can control a variety of functional activities such as DNA–protein interactions, protein–protein interactions, and subcellular localization, thus altering function *(107)*.

Importantly, like phosphorylation, acetylation of transcription factors has been shown to directly regulate contact-independent growth *(108)*. Proteins involved in controlling the cell cycle are acetylated *(109)* or associate with either HATs or HDACs providing new targets for therapeutic intervention in cancer *(59,70,71,110)*. Point

Table 4
Tumor-Associated Proteins Whose Transcriptional Expression is Altered in Response to Histone Deacetylase (HDAC) Inhibitor Treatment of Cells

Regulated protein	Regulated protein function (oncogeneic or antioncogenic/tumor supressing)
Downregulated by HDAC inhibitors	
erbB2 (HER2/neu)	Growth factor receptor (EGFR class)
TGF-β	Regulates cell proliferation and differentiation through TGF-b type II receptor
Thioredoxin	Disulfide reductase, cytokine activity, can inhibit apoptosis
Telomerase	Prevents telomere erosion
RECK	Regulates matrix metalloproteinases
VEGF	Angiogenic factor
β-FGF	Angiogenic factor
Myb/c-MyBL2	Oncogenic transcription factor-regulation of transfromation and differentiation
raf-1	Effector of Ras
cyclin A	Cell cycle regulator
cyclin B	Cell cycle regulator
DAF	Complement inhibitory protein
Abl	Growth factor receptor, component of bcr/abl chimeric kinase
DEK	Putative role in regulating chromatin structure and postsplicing events
Proteasome	Degradation of misfolded or oxidized proteins
Upregulated by HDAC inhibitors	
Fas/Fas ligand	Proapoptotic
Bcl2	Proapoptotic
p53	Proapoptotic
Bak, Bax, Bim	Proapoptotic
c-myc	Inhibitor of differentiation
Caspase 3	Cysteine protease involved in apoptosis, proapoptotic
CPA3	Carboxypeptidase, putative role in regulating differentiation
RECK	Negatively regulates matrix metalloproteinases
p21^{CIP1}	Cell cycle regulation
Gelsolin	Regulation of cell morphology
ERα	Estrogen-activated nuclear receptor regulates transcription of estrogen responsive genes
TSSC3	Regulates Fas-mediated apoptosis
IGFPB-3	Augments IGF actions, promotes apoptosis, and inhibits cell growth
TBP-2	Inhibits thiol-reducing activity of thioredoxin

Bak, Bcl2 antagonist killer; Bax, Bcl2-associated X protein; CPA3, carboxypeptidase A3; DAF, decay-accelerating factor; TBP-2, thioredoxin binding protein; TSSC3, tumor supressing subtransferable candidate. Reprinted with permission from the Annual Review of Pharmacology and Toxicology (by Annual Reviews, http://www.annualreviews.org) *(115)* and references therein.

mutations have been identified in transcription factors at their site of acetylation, arising as somatic mutations, including the ERα and AR in breast and prostate cancer, respectively (73,111–113).

Protein deacetylation is regulated by either trichostatin A (TSA) or nicotinamide-adenine-dinucleotide (NAD)-dependent HDACs (109,114,115). HDACs repress transcription through recruitment and association with large multiple protein corepressor complexes. While not required for activity, HDACs commonly associate in larger multi-protein complexes with either mSin3 proteins or Mi-2-NuRD. To date, 18 mammalian HDACs have been identified and are grouped into three class based on their conserved sequence homolog with yeast HDACs. Class I is comprised of HDAC1, 2, 3, 8, and 11 and display homology to the yeast *Rpd3*; Class II is comprised of HDAC4, 5, 6, 7, 9, and 10 and are homologous to the yeast *Hda1*; and Class III is comprised of SIRT1-SIRT7 and share homology with the *Sir2* family of yeast deacetylases. Class I and II HDACs function in a Zn^{2+}-dependent manner, while class III HDACs are dependent on the availability of NAD (109).

HDAC inhibitors inhibit cancer cell growth (through cell cycle arrest at both G_1 and G_2/M checkpoints), induction of differentiation, and/or induction of apoptosis (Table 5). HDAC inhibitors lead to the hyperacetylation of histones of the chromatin around the $p21^{CIP1}$ promoter inducing p21 gene expression and inhibiting CDK activity required for cell cycle progression. HDAC inhibitors repress the expression of growth-promoting genes such as cyclin D1. HDAC inhibitors can be classified into structural groups including hydroxamic acids [e.g., TSA, suberoylanilide hydroxamic acid (SAHA), pyroxamide, and oxamflatin], short chain fatty acids (e.g., valproic acid and sodium butyrate), benzamides (e.g., MS-275), and cyclic tetrapeptides (e.g., trapoxin, apicidin, and depsipeptide). Of these agents, depsipeptide (FR901228, FK228, NSC 630176) has significant preclinical and clinical potential.

6.1. Depsipeptide—Preclinical Studies

Depsipeptide can induce a $p21^{CIP1}$-dependent G_1 arrest associated with repression of cyclin D1 and a $p21^{CIP1}$-independent G_2/M arrest (116,117). In culture, depsipeptide effectively inhibited the proliferation of human tumor cell lines and had less effect on non-transformed cultured cells (118), inhibiting human B-cell CLL (B-CLL) cells and B-cell prolymphocytic leukemia (B-PLL) cells while sparing peripheral blood mononuclear cells (119,120). In addition, the same B-PLL cells failed to respond to treatment with F-araA, gemcitabine, flavopiridol, or UCN-01. The mean lethal concentration to 50% (LC_{50}) in B-CLL and B-PLL cells after 96 h of in vitro exposure to depsipeptide ranged between 0.2 and 15 nM.

Two phase I dose escalation trials of depsipeptide have been completed (121,122). Depsipeptide was administered as a 4-hour infusion either biweekly (day 1 and 5) every 3 weeks, or 3 times a week every 4 weeks. These studies used a starting dose of 1.0 mg/m², which was defined as 1/3 the toxic dose low in preclinical rat studies. The MTD achieved was 13.3 mg/m² and 17.8 mg/m², respectively. The most common DLTs were thrombocytopenia and progressive fatigue. In a phase II study with peripheral or cutaneous T-cell lymphoma, objective responses were observed in 11 patients including one complete response in a peripheral T-cell lymphoma at a dose level of 12.7 mg/m².

Patients with CLL and acute myeloid leukemia in a phase I trial of depsipeptide (123) using a dose of 13.3 mg/m² administered by a 4-hour infusion three times every

Table 5
Non-Histone Acetyl-Transferase Substrates and Their Accompaning Factor
Acetyltransferases (FATs)

Substrates for FAT	FAT	Possible effects on transcription
General transcriptional factors		
TFIIF	P300/CBP, P/CAF	Unknown
TFIIEB	P300/CBP, P/CAF, TAFII 250	Unknown
TAF(I)68	P/CAF	Up
UBF	CBP	Up
CIITA	P/CAF	Up
Transcriptional effectors		
P53	P300/CBP, P/CAF	Up
GATA-1,-3	P300/CBP, P/CAF	Up
EKLF	P300/CBP	Up
TCF	P300/CBP	Down
C-Myb	P300/CBP, GCN5	Up
HIV-1tat	P300/CBP, P/CAF	Up
E2F1,2	P300/CBP, P/CAF	Up
E2F,4	TRRAP, P/CAF	Up
TR-RXR	P300/CBP	Unknown
MyoD	P300/CBP, P/CAF	Up
TAL1/SCL	P300/CBP, P/CAF	Up
AR	P300/CBP, P/CAF, TIP6 0	Down
SF-1	GCN5	Up
ERα	P300/CBP	Down
Sp1	P300/CBP	Up
E1A	P300/CBP	Up
YY1	P300/CBP, P/CAF	Down
RelA	P300/CBP	Nuclear import
STAT6	P300/CBP	Up
IRF-1,2	P300/CBP, P/CAF	Up
NF-E2	P300/CBP	Up
pRb	P300/CBP	Up
Nuclear receptor coactivators		
P300/CBP	P300/CBP	Down
P/CAF	P/CAF	Unknown
ACTR	P300/CBP	Down
SRC-1	P300/CBP	Unknown
TIF2	P300/CBP	Unknown
Rip140	P300/CBP	Up
PC4	P300	Up
Nonhistone chromatin proteins		
HMG1	P300/CBP	Unknown
HMG2	–	Unknown
HMG14	P300/CBP	Down
HMG17	P/CAF	Unknown
HMGI (Y)	P300/CBP,P/CAF	Up (P/CAF), Down (p300)
Sin1	GCN5	Unknown
Fen-1	P300	Reduce DNA binding and nuclease activity
Others		
α-Tubulin	P300	Unknown
Importin-α7	P300/CBP	Unknown
CDP/cut	P300/CBP,P/CAF	Reduce DNA binding

Reproduced with permission from ref. *(107)*.

15 days (day 1, 8, and 15) over a 4-week cycle received no objective responses based on NCI criteria. Nausea, fatigue, and anorexia prevented more than two cycles of treatment in 85% of the enrolled patients. Of the CLL patients who began therapy with elevated leukocyte counts ($n = 7$), all demonstrated improvement in peripheral leukocyte counts, suggesting further studies are warranted.

6.2. Outlook—HDAC Inhibitors

Two promising HDAC inhibitors are the hydroxamic acid-based hybrid polar SAHA and the benzamide derivative MS-275. SAHA has good oral bioavailability and in phase II trials demonstrated antitumor activities in patients with solid tumors and with Hodgkin's disease *(124)*. SAHA is scheduled for phase III trials in patients with advanced T-cell lymphoma and relapsed diffuse large B-cell lymphoma. MS-275 has antitumor activity against human xenografts *(125)* and is currently being investigated in phase I trials.

7. TARGETING THE EPIDERMAL GROWTH FACTOR RECEPTOR FAMILY

Several different inhibitors directed to growth factor receptors have been developed and are now either in clinical trials or part of clinical management. Small molecular inhibitors of receptors [Imatinib, for BCR-ABL, KIT and platelet-derived growth factor receptor β (PDGFRβ), Gefitinib, and Erlotinib for EGFR] and humanized monoclonal antibodies (Trastuzumab and Bevacizumab for VEGF-A and Cetuzimab for EGFR) are the best examples. Transporter and symporters' such as the sodium iodide symporter *(126)* may also serve as useful targets in the future. The epidermal growth factor receptor (EGFR) family includes tyrosine kinase receptors that have been implicated in the development and progression of cancer. The EGFR family is comprised of four different receptor members: (i) EGFR (ErbB1 or HER1); (ii) ErbB2 (Her2/Neu); (iii) ErbB3 (HER3); and (iv) ErbB4 (HER4). These members share a similar transmembrane ligand-binding protein structure with a conserved cytoplasmic tyrosine kinase and sarcoma (SRC)-homology domains and divergent extracellular ligand-binding domains. The primary activating ligands for EGFR are EGF, TGFα, amphiregulin, heparin-binding EGF, betacellulin, and epiregulin *(127,128)*. Upon ligand binding, the EGFRs can form either homodimers or heterodimers that activate their intrinsic tyrosine kinase activity leading to autophosphorylation or transphosphorylation, or phosphorylation by SRC-related kinase. Phosphorylation of C-terminal tyrosines in the EGFRs orchestrates the activation of a number of downstream effector pathways including the Ras-Raf-MAPK, the PI3K/Akt, and JAK/STAT signaling pathways that promote cell division and survival *(129)*.

The aberrant activation of the EGFR family is accomplished by receptor and ligand overexpression, gene amplification, and activating mutations *(129)*. A wide array of human cancers including NSCLC, head and neck, breast, ovarian, prostate, bladder, colorectal, and malignant gliomas display a high level of cell surface expression of EGFR due to overexpression *(130–132)*. Several synthetic EGFR inhibitors have been developed, including ZD1829 (Gefitinib, Iressa®), a tyrosine kinase inhibitor with good preclinical responses.

7.1. ZD1839

ZD1839 is an EGFR-specific, orally bioavailable, synthetic anilinoquinazoline tyrosine kinase inhibitor that competes with ATP for binding to the intracellular tyrosine kinase catalytic domain. Inhibition of EGFR with ZD1839 induced accumulation of p27^{KIP1} and a G$_1$ cell cycle arrest in addition to changes in activity and phosphorylation of downstream biomarkers such as MAPK, Akt, and c-Fos (133–137). ZD1839 has antiproliferative activity in a wide range of human cancer cell lines and/or xenografts including breast, ovarian, vulvar, prostate, colon, small cell lung, NSCLC, and gastric carcinomas (74). Combinational therapies with ZD1839 and cytotoxic chemotherapy agents or radiation showed an additive or synergistic effect on growth inhibition or apoptosis in breast, prostate, head and neck, and colon cancers (74). Animal studies demonstrated very little toxicity with doses ranging from 12.5 to 200 mg/kg daily. However following treatment withdrawal, the regrowth rates of some tumor models were comparable to that of untreated animals (138,139), suggesting that continuous patient dosing may be required to prevent a relapse of tumor progression.

While ZD1839 was designed to specifically inhibit the function of EGFR, it can also indirectly interfere with other receptor family members including ErbB2 by trapping them in inactive heterodimers. ZD1839 induces unphosphorylated, inactive EGFR/ErbB2 heterodimers, blocking ErbB2-specific signaling pathways (137,140–142). Treatment with ZD1839 and the anti-ErbB2 receptor antibody Trastuzumab increased growth inhibition and apoptosis in ErbB2-positive breast cancer cells.

Five phase I clinical trials have been completed using ZD1839 (143–147). The pharmacokinetic analysis supported once-daily oral dosing with plasma concentrations reaching a steady-state level by day 7 that were above the level required to inhibit at least 90% of EGF-stimulated growth in cultured epidermal carcinoma KB cells (144, 145). Analysis of in vivo biomarkers using paired biopsies of patient skin presystemic and postsystemic exposure to ZD1839 showed decreased levels of activated EGFR, Akt, MAPK, and ERK with upregulation of p27^{KIP1} (143).

ZD1839 was well tolerated in a cohort of nearly 260 patients with an MTD of 700–1000 mg/day and DLTs of diarrhea and nausea. Objective responses were observed in patients with NSCLC, head and neck cancer, colorectal cancer, breast cancer, ovarian cancer, and hormone-refractory prostate cancer, characterized by prolonged disease stabilization, a decline in tumor markers, and/or relief of disease-related symptoms (143–147), but required continuous exposure to ZD1839.

IDEAL-1 and IDEAL-2 were randomized, double blind, parallel-group, multi-institutional phase II trials established to evaluate the efficacy and safety of daily administered doses of 250 mg/day or 500 mg/day of ZD1839 in patients with advanced stage III or IV NSCLC (148–150). IDEAL-1 was an international trial containing 210 patients that had been exposed to one or two prior chemotherapy regimens, with at least one therapy containing platinum. IDEAL-2 was conducted in the United States and enrolled 221 patients that had previously received either two or more chemotherapy treatments that contained platinum and docetaxel either as a single agent or in combination. Results from both trials using either dose schedule of ZD1893 demonstrated objective antitumor activity with clinical benefits as a second-line, third-line, four-line, and even fifth-line therapy. In addition to antitumor activity, ZD1893 provided symptom relief and manageable toxicity in patients with NSCLC, who had undergone multiple rounds of previous therapy.

Two INTACT (Iressa® Non-Small Cell Lung Cancer Trial Assessing Combination Treatment) phase III trials were established to determine the efficacy of ZD1893 as a first-line treatment in combination with either gemcitabine/cisplatin (INTACT1) or paclitaxel/carboplatin (INTACT2) *(151,152)*. Although neither trial resulted in a survival benefit with ZD1893, activating mutations in the ATP-binding site of EGFR were identified in patients' tumors that confer susceptibility to ZD1893 in NSCLC *(153)*, raising the possibility that subset of NSCLC tumors may be selected that will be responsive to ZD1893 in future studies.

7.2. Outlook—EGFR Tyrosine Kinase Inhibitors

In addition to ZD1893, a second small molecule inhibitor of EGFR tyrosine kinase activity is currently entering clinical trials. OSI-774 (Erlotinib, Tarceva™) has demonstrated modest activity in NSCLC patients as a single agent and has currently been entered into larger combinational trials with traditional chemotherapy regimens *(154)*. With further studies pending, both ZD1893 and OSI-774 have shown potential to be effective in first-line and second-line therapies against NSCLC.

8. BCR-ABL TYROSINE KINASE INHIBITORS

The Philadelphia (Ph) chromosome is the product of a reciprocal translocation between chromosomes 9 and 22 [t(9:22)] and is the cytogenetic hallmark of CML *(155)*. This translocation juxtaposes the c-abl oncogene with the *bcr* gene resulting in the expression of a Bcr-Abl fusion protein with constitutive tyrosine kinase activity. Bcr-Abl can stimulate multiple signaling pathways including Ras, PI3K, JAK2, STAT5, MAPK, and nuclear factor-κB. Approximately 95% of CML patients harbor the Ph chromosome, and the natural progression of CML is largely dependent on the oncogenic potential of the resulting Bcr-Abl fusion protein *(156)*. This Bcr-Abl kinase is expressed solely in CML cells providing a select target for therapy directed specifically at tumorigenic cells.

8.1. STI-571

Signal transduction inhibitor-571 (STI-571) (Gleevec® or imatinib mesylate) is an excellent example of a molecular-targeted therapy for the treatment of specific malignancies. STI-571 is a potent inhibitor of c-Abl, c-Kit, and platelet-derived PDGFRβ (IC_{50} values of less than 400 nM), while interfering with other kinases only at concentrations greater than 10,000 nM *(157)*. STI-571 competitively inhibits the phosphorylation of target substrates and demonstrated antiproliferative effects on Bcr-Abl-positive CML cells with little or no effect on normal hematopoietic cells *(158)*. In vivo studies identified the need for continuous infusion of STI-571 for the strongest antitumor activity.

STI-571 phase I dose escalation trials initially recruited patients with CML in chronic phase who were unresponsive to interferon-based therapy but later enrolled CML patients in blast crisis and Ph-positive acute lymphoblastic leukemia patients *(159)*. STI-571, which has a high oral bioavailability, was administered orally once daily with a report of only grade 1 and grade 2 nonhematologic toxicities. STI-571 has a t½ of 13–16 h with peak serum concentration obtained approximately 4 h after treatment. Remarkably, following treatment with ≥300 mg/day, 98% of patients with chronic phase CML and 55% of subjects with CML in blast crisis demonstrated complete responses with 96% and 18% of those responses lasting more than 1 year, respectively. Completion of these phase I studies led to the recommendation of 400 mg/day for patients with CML in chronic phase and 600 mg/day for patients with CML in blast phase. These findings prompted the immediate development of phase II and III trials.

In less than 1 year over 1000 subjects, fitting similar criteria as the phase I subjects, had enrolled in STI-571 phase II trials (*160–162*). As stipulated by the phase I data, subjects with chronic phase CML were given 400 mg/day, and patients with advanced CML or CML in blast crisis were given 600 mg/day. After 18 months of observation, 95% of the patients with CML in chronic phase demonstrated a complete response, with 89% of subjects demonstrating 18-month progression-free survival. While subjects with accelerated CML only had a 34% complete response, this group still demonstrated an 82% overall response with 59% having progression-free survival. Subjects with CML in blast phase had a 52% overall response rate with 8% displaying complete hematologic remission with ≤5% residual blasts.

In phase III studies, 400 mg/day of STI-571 was compared to standard doses of interferon plus cytarabine in a large randomized trial that enrolled approximately 1100 newly diagnosed chronic phase CML patients (*163*). A considerable statistical advantage was demonstrated by STI-571 on all facets of response, including complete hematologic response rate, major cytogenetic response rate, complete cytogenetic response rate, reported toxicity, and freedom from disease progression. In addition, patients failing on the standard chemotherapy regimen were allowed to crossover and displayed not only higher levels of therapeutic efficacy with STI-571 but also a much higher quality of life (*164*).

8.2. Outlook—STI-571

STI-571 has changed the standard of care for CML patients and has curative effects in other forms of cancer, including gastrointestinal stromal tumors that commonly display altered c-kit activity. Resistance to STI-571 occurs by mutations in the Bcr-Abl kinase. Further studies will determine if subsets of patients are prone to relapse and might benefit from alternate treatment courses (*165,166*).

9. FUTURE THERAPEUTIC STRATEGIES

9.1. Sulfonamide Carbonic Anhydrase Inhibitors

Metabolic changes that occur during tumorigenesis are maintained by altered gene expression that advantage cellular growth, with consequent alterations in glucose uptake (*167*) and metabolic signaling that can be used to image and treat tumors. Carbonic anhydrases (CAs) catalyze the hydration of CO_2 to bicarbonate, which is required for gluconeogenesis, amino acid synthesis, lipogenesis, and pyrimidine synthesis. Tumor cells can gain an advantage by overexpression of CAs. CA IX is a hypoxia-inducible factor associated with aggressive tumors (*168*). Cancers predominantly express CA IX and CA XII isoenzymes in contrast to their normal cell counterparts thus providing a therapeutic target unique to cancer cells. CA inhibitors inhibit the growth of human cancer cells in culture as well as inhibit the metastatic potential of a number of renal cancer cell lines *(169)*. Sulfonamide CA inhibitors (SCAIs), either aromatic or heterocyclic sulfonamides, are potent antitumor agents *(170)*. Modifying the "tail" attached to the aromatic or heterocyclic ring results in increased antitumor activity. SCAIs include chloroquinoxaline sulfonamides (E7070), sulfonylureas, and a more general sulfonamides/sulfonyl group *(171)*. E7070 inhibits G_1 cell cycle and is under phase II clinical investigation. Sulfonylureas have antitumor effects against rat tumors and human xenografts in vivo. SCAIs destabilize tubulin polymerization through an interaction with β-tubulin like several other antitumor drugs (vinca alkaloids, vincristine, and vinblastine) and target HDAC1 and CDK2.

9.2. Therapeutic Targeting of the p53 Tumor Suppressor Protein

The inactivation of p53 protein occurs by mutation in approximately 50% of all cancers, while the remaining cancers have been shown, or are predicted to have, defects in upstream signaling or downstream effector pathways involving p53. p53-targeted therapies are being developed based on restructuring mutant p53 proteins or stimulation of wtp53 function.

Gene therapy with p53 gene transfer by intratumoral injection of adenoviral p53 (Ad-p53) was effective in some of patients with NSCLC or head and neck squamous cell carcinoma that received treatment *(172,173)* without significant toxicity. Ad-p53 with radiotherapy (60 Gy) induced complete or partial responses in patients with NSCLC *(174)*, implying p53 may require an activating agent such as DNA damage to elicit a significant response. While limited by the requirement of direct tumor injection, new nanotechnology-based delivery systems may be valuable for systemic delivery.

PRIMA-1 was identified in a large-scale screening of low molecular weight compounds that selectively inhibited the growth of cells expressing mutant $p53^{His245}$. PRIMA-1 can restore wtp53 conformation that is essential for p53 transcriptional activation and suppress the growth of human tumor xenografts by either intratumor or intravenous administration *(175)*. Nutlin-2 is a small molecule agonist of the p53–HDM2 interaction which activates wtp53 *(176)* and may be useful for soft tissue sarcomas and osteosarcomas that frequently display HDM2 gene amplification. Preclinical trials for both of these molecules have been encouraging and warrant further clinical evaluation.

10. CONCLUSIONS

The identification of molecular genetic targets involved in the initiation and progression of human cancer through tumor and serum analyses is being driven by advances in proteomics and bioinformatics *(66)*. Improved delivery systems using nanotechnology-based systems will likely revolutionize the clinical practice of oncology. The ability to now induce gene expression at a single cell level using light-activated gene therapy *(177)* emphasizes a paradigm shift that will likely change the therapeutic index of cancer treatment several orders of magnitude. The shift in thinking from empiricism to mechanism, targeting cell cycle abnormalities in tumors, is a journey, not a destination, and a most exciting journey it is.

ACKNOWLEDGMENTS

This work was supported in part by awards from the Susan Komen Breast Cancer Foundation, Breast Cancer Alliance Inc., R01CA70896, R01CA75503, R01CA86072, R01CA86071 (R.G.P.).

REFERENCES

1. Geng Y, Whoriskey W, Park MY, Bronson RT, Medema RH, Li T et al. Rescue of cyclin D1 deficiency by knockin cyclin E. Cell 1999; 97(6):767–777.
2. Kozar K, Ciemerych MA, Rebel VI, Shigematsu H, Zagozdzon A, Sicinska E et al. Mouse development and cell proliferation in the absence of D-cyclins. Cell 2004; 118(4):477–491.
3. Pestell RG, Albanese C, Reutens AT, Segall JE, Lee RJ, Arnold A. The cyclins and cyclin-dependent kinase inhibitors in hormonal regulation of proliferation and differentiation. Endocr Rev 1999; 20(4):501–534.
4. Lee RJ, Albanese C, Fu M, D'Amico M, Lin B, Watanabe G et al. Cyclin D1 is required for transformation by activated Neu and is induced through an E2F-dependent signaling pathway. Mol Cell Biol 2000; 20(2):672–683.
5. Albanese C, D'Amico M, Reutens AT, Fu M, Watanabe G, Lee RJ et al. Activation of the cyclin D1 gene by the E1A-associated protein p300 through AP-1 inhibits cellular apoptosis. J Biol Chem 1999; 274(48):34186–34195.
6. Joyce D, Bouzahzah B, Fu M, Albanese C, D'Amico M, Steer J et al. Integration of Rac-dependent regulation of cyclin D1 transcription through a nuclear factor-kappaB-dependent pathway. J Biol Chem 1999; 274(36):25245–25249.
7. Lee RJ, Albanese C, Stenger RJ, Watanabe G, Inghirami G, Haines GK, III et al. pp60(v-src) induction of cyclin D1 requires collaborative interactions between the extracellular signal-regulated kinase, p38, and Jun kinase pathways. A role for cAMP response element-binding protein and activating transcription factor-2 in pp60(v-src) signaling in breast cancer cells. J Biol Chem 1999; 274(11):7341–7350.
8. Sears RC, Nevins JR. Signaling networks that link cell proliferation and cell fate. J Biol Chem 2002; 277(14):11617–11620.
9. Albanese C, Wu K, D'Amico M, Jarrett C, Joyce D, Hughes J et al. IKKalpha regulates mitogenic signaling through transcriptional induction of cyclin D1 via Tcf. Mol Biol Cell 2003; 14(2):585–599.
10. Sherr CJ, Roberts JM. CDK inhibitors: positive and negative regulators of G1-phase progression. Genes Dev 1999; 13(12):1501–1512.
11. Olashaw N, Bagui TK, Pledger WJ. Cell cycle control: a complex issue. Cell Cycle 2004; 3(3): 263–264.
12. Carrano AC, Eytan E, Hershko A, Pagano M. SKP2 is required for ubiquitin-mediated degradation of the CDK inhibitor p27. Nat Cell Biol 1999; 1(4):193–199.
13. Sutterluty H, Chatelain E, Marti A, Wirbelauer C, Senften M, Muller U et al. p45SKP2 promotes p27Kip1 degradation and induces S phase in quiescent cells. Nat Cell Biol 1999; 1(4):207–214.
14. Tomoda K, Kubota Y, Kato J. Degradation of the cyclin-dependent-kinase inhibitor p27Kip1 is instigated by Jab1. Nature 1999; 398(6723):160–165.
15. Yeh KH, Kondo T, Zheng J, Tsvetkov LM, Blair J, Zhang H. The F-box protein SKP2 binds to the phosphorylated threonine 380 in cyclin E and regulates ubiquitin-dependent degradation of cyclin E. Biochem Biophys Res Commun 2001; 281(4):884–890.
16. Nakayama K, Nagahama H, Minamishima YA, Matsumoto M, Nakamichi I, Kitagawa K et al. Targeted disruption of Skp2 results in accumulation of cyclin E and p27(Kip1), polyploidy and centrosome overduplication. EMBO J 2000; 19(9):2069–2081.
17. Rodier G, Montagnoli A, Di Marcotullio L, Coulombe P, Draetta GF, Pagano M et al. p27 cytoplasmic localization is regulated by phosphorylation on Ser10 and is not a prerequisite for its proteolysis. EMBO J 2001; 20(23):6672–6682.
18. Ishida N, Hara T, Kamura T, Yoshida M, Nakayama K, Nakayama KI. Phosphorylation of p27Kip1 on serine 10 is required for its binding to CRM1 and nuclear export. J Biol Chem 2002; 277(17):14355–14358.

19. Kamura T, Hara T, Matsumoto M, Ishida N, Okumura F, Hatakeyama S et al. Cytoplasmic ubiquitin ligase KPC regulates proteolysis of p27(Kip1) at G1 phase. Nat Cell Biol 2004; 6(12):1229–1235.

20. Ciemerych MA, Kenney AM, Sicinska E, Kalaszczynska I, Bronson RT, Rowitch DH et al. Development of mice expressing a single D-type cyclin. Genes Dev 2002; 16(24):3277–3289.

21. Fantl V, Stamp G, Andrews A, Rosewell I, Dickson C. Mice lacking cyclin D1 are small and show defects in eye and mammary gland development. Genes Dev 1995; 9(19):2364–2372.

22. Huard JM, Forster CC, Carter ML, Sicinski P, Ross ME. Cerebellar histogenesis is disturbed in mice lacking cyclin D2. Development 1999; 126(9):1927–1935.

23. Lam EW, Glassford J, Banerji L, Thomas NS, Sicinski P, Klaus GG. Cyclin D3 compensates for loss of cyclin D2 in mouse B-lymphocytes activated via the antigen receptor and CD40. J Biol Chem 2000; 275(5):3479–3484.

24. Sicinska E, Aifantis I, Le Cam L, Swat W, Borowski C, Yu Q et al. Requirement for cyclin D3 in lymphocyte development and T cell leukemias. Cancer Cell 2003; 4(6):451–461.

25. Sicinski P, Donaher JL, Parker SB, Li T, Fazeli A, Gardner H et al. Cyclin D1 provides a link between development and oncogenesis in the retina and breast. Cell 1995; 82(4):621–630.

26. Sicinski P, Donaher JL, Geng Y, Parker SB, Gardner H, Park MY et al. Cyclin D2 is an FSH-responsive gene involved in gonadal cell proliferation and oncogenesis. Nature 1996; 384(6608):470–474.

27. Solvason N, Wu WW, Parry D, Mahony D, Lam EW, Glassford J et al. Cyclin D2 is essential for BCR-mediated proliferation and CD5 B cell development. Int Immunol 2000; 12(5):631–638.

28. Neumeister P, Pixley FJ, Xiong Y, Xie H, Wu K, Ashton A et al. Cyclin D1 governs adhesion and motility of macrophages. Mol Biol Cell 2003; 14(5):2005–2015.

29. Wang C, Pattabiraman N, Zhou JN, Fu M, Sakamaki T, Albanese C et al. Cyclin D1 repression of peroxisome proliferator-activated receptor gamma expression and transactivation. Mol Cell Biol 2003; 23(17):6159–6173.

30. Geng Y, Yu Q, Sicinska E, Das M, Schneider JE, Bhattacharya S et al. Cyclin E ablation in the mouse. Cell 2003; 114(4):431–443.

31. Parisi T, Beck AR, Rougier N, McNeil T, Lucian L, Werb Z et al. Cyclins E1 and E2 are required for endoreplication in placental trophoblast giant cells. EMBO J 2003; 22(18):4794–4803.

32. Berthet C, Aleem E, Coppola V, Tessarollo L, Kaldis P. Cdk2 knockout mice are viable. Curr Biol 2003; 13(20):1775–1785.

33. Ortega S, Prieto I, Odajima J, Martin A, Dubus P, Sotillo R et al. Cyclin-dependent kinase 2 is essential for meiosis but not for mitotic cell division in mice. Nat Genet 2003; 35(1):25–31.

34. Malumbres M, Sotillo R, Santamaria D, Galan J, Cerezo A, Ortega S et al. Mammalian cells cycle without the D-type cyclin-dependent kinases Cdk4 and Cdk6. Cell 2004; 118(4):493–504.

35. Moons DS, Jirawatnotai S, Parlow AF, Gibori G, Kineman RD, Kiyokawa H. Pituitary hypoplasia and lactotroph dysfunction in mice deficient for cyclin-dependent kinase-4. Endocrinology 2002; 143(8):3001–3008.

36. Rane SG, Dubus P, Mettus RV, Galbreath EJ, Boden G, Reddy EP et al. Loss of Cdk4 expression causes insulin-deficient diabetes and Cdk4 activation results in beta-islet cell hyperplasia. Nat Genet 1999; 22(1):44–52.

37. Tsutsui T, Hesabi B, Moons DS, Pandolfi PP, Hansel KS, Koff A et al. Targeted disruption of CDK4 delays cell cycle entry with enhanced p27(Kip1) activity. Mol Cell Biol 1999; 19(10):7011–7019.

38. Zou X, Ray D, Aziyu A, Christov K, Boiko AD, Gudkov AV et al. Cdk4 disruption renders primary mouse cells resistant to oncogenic transformation, leading to Arf/p53-independent senescence. Genes Dev 2002; 16(22):2923–2934.

39. Murphy M, Stinnakre MG, Senamaud-Beaufort C, Winston NJ, Sweeney C, Kubelka M et al. Delayed early embryonic lethality following disruption of the murine cyclin A2 gene. Nat Genet 1997; 15(1):83–86.

40. Krimpenfort P, Quon KC, Mooi WJ, Loonstra A, Berns A. Loss of p16Ink4a confers susceptibility to metastatic melanoma in mice. Nature 2001; 413(6851):83–86.

41. Sharpless NE, Bardeesy N, Lee KH, Carrasco D, Castrillon DH, Aguirre AJ et al. Loss of p16Ink4a with retention of p19Arf predisposes mice to tumorigenesis. Nature 2001; 413(6851):86–91.

42. Yu Q, Geng Y, Sicinski P. Specific protection against breast cancers by cyclin D1 ablation. Nature 2001; 411(6841):1017–1021.

43. Robles AI, Rodriguez-Puebla ML, Glick AB, Trempus C, Hansen L, Sicinski P et al. Reduced skin tumor development in cyclin D1-deficient mice highlights the oncogenic ras pathway in vivo. Genes Dev 1998; 12(16):2469–2474.

44. Rowlands TM, Pechenkina IV, Hatsell SJ, Pestell RG, Cowin P. Dissecting the roles of beta-catenin and cyclin D1 during mammary development and neoplasia. Proc Natl Acad Sci USA 2003; 100(20):11400–11405.

45. Hulit J, Wang C, Li Z, Albanese C, Rao M, Di Vizio D et al. Cyclin D1 genetic heterozygosity regulates colonic epithelial cell differentiation and tumor number in ApcMin mice. Mol Cell Biol 2004; 24(17):7598–7611.

46. Fero ML, Randel E, Gurley KE, Roberts JM, Kemp CJ. The murine gene p27Kip1 is haplo-insufficient for tumour suppression. Nature 1998; 396(6707):177–180.

47. Muraoka RS, Lenferink AE, Law B, Hamilton E, Brantley DM, Roebuck LR et al. ErbB2/Neu-induced, cyclin D1-dependent transformation is accelerated in p27-haploinsufficient mammary epithelial cells but impaired in p27-null cells. Mol Cell Biol 2002; 22(7):2204–2219.

48. Sharpless NE, Ramsey MR, Balasubramanian P, Castrillon DH, DePinho RA. The differential impact of p16(INK4a) or p19(ARF) deficiency on cell growth and tumorigenesis. Oncogene 2004; 23(2):379–385.

49. Sherr CJ. Principles of tumor suppression. Cell 2004; 116(2):235–246.

50. Chin K, de Solorzano CO, Knowles D, Jones A, Chou W, Rodriguez EG et al. In situ analyses of genome instability in breast cancer. Nat Genet 2004; 36(9):984–988.

51. Maser RS, DePinho RA. Connecting chromosomes, crisis, and cancer. Science 2002; 297(5581):565–569.

52. Bartkova J, Horejsi Z, Koed K, Kramer A, Tort F, Zieger K et al. DNA damage response as a candidate anti-cancer barrier in early human tumorigenesis. Nature 2005; 434(7035):864–870.

53. Harris SL, Levine AJ. The p53 pathway: positive and negative feedback loops. Oncogene 2005; 24(17):2899–2908.

54. Knights CD, Liu Y, Appella E, Kulesz-Martin M. Defective p53 post-translational modification required for wild type p53 inactivation in malignant epithelial cells with mdm2 gene amplification. J Biol Chem 2003; 278(52):52890–52900.

55. Bode AM, Dong Z. Post-translational modification of p53 in tumorigenesis. Nat Rev Cancer 2004; 4(10):793–805.

56. Weber J, Taylor L, Roussel M, Sherr C, Bar-Sagi D. Nucleolar Arf sequesters Mdm2 and activates p53. Nat Cell Biol 1999; 1:20–26.

57. Tao W, Levine AJ. P19(ARF) stabilizes p53 by blocking nucleo-cytoplasmic shuttling of Mdm2. Proc Natl Acad Sci USA 1999; 96(12):6937–6941.

58. Qi Y, Gregory MA, Li Z, Brousal JP, West K, Hann SR. p19ARF directly and differentially controls the functions of c-Myc independently of p53. Nature 2004; 431(7009):712–717.

59. Avantagiatti ML, Ogryzko V, Gardner K, Giordano A, Levine AS, Kelly K. Recruitment of p300/CBP in p53-dependent signal pathways. Cell 1997; 89:1175–1184.

60. Lill NL, Grossman SR, Ginsberg D, DeCaprio J, Livingston DM. Binding and modulation of p53 by p300/CBP coactivators. Nature 1997; 387(6635):823–827.

61. Di Stefano V, Soddu S, Sacchi A, D'Orazi G. HIPK2 contributes to PCAF-mediated p53 acetylation and selective transactivation of p21Waf1 after nonapoptotic DNA damage. 2005.

62. Knights CD, Catania J, Di Giovanni S, Muratoglu S, Perez R, Swartzbeck A et al. Distinct p53 acetylation cassettes differentially influence gene-expression patterns and cell fate. J cell Biol 2006; 173(4):533–544.

63. Nister M, Tang M, Zhang XQ, Yin C, Beeche M, Hu X et al. p53 must be competent for transcriptional regulation to suppress tumor formation. Oncogene 2005; 24(22):3563–3573.

64. Erster S, Moll UM. Stress-induced p53 runs a transcription-independent death program. Biochem Biophys Res Commun 2005; 331(3):843–850.

65. Deshpande A, Sicinski P, Hinds PW. Cyclins and cdks in development and cancer: a perspective. Oncogene 2005; 24(17):2909–2915.

66. Hedenfalk I, Duggan D, Chen Y, Radmacher M, Bittner M, Simon R et al. Gene-expression profiles in hereditary breast cancer. N Engl J Med 2001; 344(8):539–548.

67. Fu M, Wang C, Li Z, Sakamaki T, Pestell RG. Minireview: cyclin D1: normal and abnormal functions. Endocrinology 2004; 145(12):5439–5447.

68. Fu M, Rao M, Bouras T, Wang C, Wu K, Zhang X et al. Cyclin D1 inhibits peroxisome proliferator-activated receptor gamma-mediated adipogenesis through histone deacetylase recruitment. J Biol Chem 2005; 280(17):16934–16941.

69. Fu M, Wang C, Rao M, Wu X, Bouras T, Zhang X et al. Cyclin D1 represses p300 transactivation through a CDK-independent mechanism. J Biol Chem 2005; 280(33):29728–29742.

70. Reutens AT, Fu M, Wang C, Albanese C, McPhaul MJ, Sun Z et al. Cyclin D1 binds the androgen receptor and regulates hormone-dependent signaling in a p300/CBP-associated factor (P/CAF)-dependent manner. Mol Endocrinol 2001; 15(5):797–811.
71. McMahon C, Suthiphongchai T, DiRenzo J, Ewen ME. P/CAF associates with cyclin D1 and potentiates its activation of the estrogen receptor. Proc Natl Acad Sci USA 1999; 96(10): 5382–5387.
72. Petre-Draviam CE, Williams EB, Burd CJ, Gladden A, Moghadam H, Meller J et al. A central domain of cyclin D1 mediates nuclear receptor corepressor activity. Oncogene 2004; 24(3):431–444.
73. Wang C, Fan S, Li Z, Fu M, Rao M, Ma Y et al. Cyclin D1 antagonizes BRCA1 repression of estrogen receptor {alpha} activity. Cancer Res 2005; 65(15):6557–6567.
74. Liu MC, Marshall JL, Pestell RG. Novel strategies in cancer therapeutics: targeting enzymes involved in cell cycle regulation and cellular proliferation. Curr Cancer Drug Targets 2004; 4(5):403–424.
75. Knockaert M, Greengard P, Meijer L. Pharmacological inhibitors of cyclin-dependent kinases. Trends Pharmacol Sci 2002; 23(9):417–425.
76. Sedlacek HH. Mechanisms of action of flavopiridol. Crit Rev Oncol Hematol 2001; 38(2):139–170.
77. Carlson B, Lahusen T, Singh S, Loaiza-Perez A, Worland PJ, Pestell R et al. Down-regulation of cyclin D1 by transcriptional repression in MCF-7 human breast carcinoma cells induced by flavopiridol. Cancer Res 1999; 59(18):4634–4641.
78. Chao SH, Price DH. Flavopiridol inactivates P-TEFb and blocks most RNA polymerase II transcription in vivo. J Biol Chem 2001; 276(34):31793–31799.
79. Byrd JC, Shinn C, Waselenko JK, Fuchs EJ, Lehman TA, Nguyen PL et al. Flavopiridol induces apoptosis in chronic lymphocytic leukemia cells via activation of caspase-3 without evidence of bcl-2 modulation or dependence on functional p53. Blood 1998; 92(10):3804–3816.
80. Arguello F, Alexander M, Sterry JA, Tudor G, Smith EM, Kalavar NT et al. Flavopiridol induces apoptosis of normal lymphoid cells, causes immunosuppression, and has potent antitumor activity in vivo against human leukemia and lymphoma xenografts. Blood 1998; 91(7):2482–2490.
81. Brusselbach S, Nettelbeck DM, Sedlacek HH, Muller R. Cell cycle-independent induction of apoptosis by the anti-tumor drug Flavopiridol in endothelial cells. Int J Cancer 1998; 77(1): 146–152.
82. Melillo G, Sausville EA, Cloud K, Lahusen T, Varesio L, Senderowicz AM. Flavopiridol, a protein kinase inhibitor, down-regulates hypoxic induction of vascular endothelial growth factor expression in human monocytes. Cancer Res 1999; 59(21):5433–5437.
83. Kelland LR. Flavopiridol, the first cyclin-dependent kinase inhibitor to enter the clinic: current status. Expert Opin Investig Drugs 2000; 9(12):2903–2911.
84. Sedlacek HH, Czech J, Naik R. (L86 8275; NSC 649890), a new kinase inhibitor for tumor therapy. Int J Oncol 1996; 9:1143–1168.
85. Bible KC, Kaufmann SH. Cytotoxic synergy between flavopiridol (NSC 649890, L86–8275) and various antineoplastic agents: the importance of sequence of administration. Cancer Res 1997; 57(16):3375–3380.
86. Senderowicz AM, Headlee D, Stinson SF, Lush RM, Kalil N, Villalba L et al. Phase I trial of continuous infusion flavopiridol, a novel cyclin-dependent kinase inhibitor, in patients with refractory neoplasms. J Clin Oncol 1998; 16(9):2986–2999.
87. Thomas JP, Tutsch KD, Cleary JF, Bailey HH, Arzoomanian R, Alberti D et al. Phase I clinical and pharmacokinetic trial of the cyclin-dependent kinase inhibitor flavopiridol. Cancer Chemother Pharmacol 2002; 50(6):465–472.
88. Schwartz GK, Ilson D, Saltz L, O'Reilly E, Tong W, Maslak P et al. Phase II study of the cyclin-dependent kinase inhibitor flavopiridol administered to patients with advanced gastric carcinoma. J Clin Oncol 2001; 19(7):1985–1992.
89. Stadler WM, Vogelzang NJ, Amato R, Sosman J, Taber D, Liebowitz D et al. Flavopiridol, a novel cyclin-dependent kinase inhibitor, in metastatic renal cancer: a University of Chicago Phase II Consortium study. J Clin Oncol 2000; 18(2):371–375.
90. Burdette-Radoux S, Tozer RG, Lohmann RC, Quirt I, Ernst DS, Walsh W et al. Phase II trial of flavopiridol, a cyclin dependent kinase inhibitor, in untreated metastatic malignant melanoma. Invest New Drugs 2004; 22(3):315–322.
91. Liu G, Gandara DR, Lara PN, Jr., Raghavan D, Doroshow JH, Twardowski P et al. A phase II trial of flavopiridol (NSC #649890) in patients with previously untreated metastatic androgen-independent prostate cancer. Clin Cancer Res 2004; 10(3):924–928.

92. Kouroukis CT, Belch A, Crump M, Eisenhauer E, Gascoyne RD, Meyer R et al. Flavopiridol in untreated or relapsed mantle-cell lymphoma: results of a phase II study of the National Cancer Institute of Canada Clinical Trials Group. J Clin Oncol 2003; 21(9):1740–1745.

93. Shapiro GI, Supko JG, Patterson A, Lynch C, Lucca J, Zacarola PF et al. A phase II trial of the cyclin-dependent kinase inhibitor flavopiridol in patients with previously untreated stage IV non-small cell lung cancer. Clin Cancer Res 2001; 7(6):1590–1599.

94. Schiller JH, Harrington D, Belani CP, Langer C, Sandler A, Krook J et al. Comparison of four chemotherapy regimens for advanced non-small-cell lung cancer. N Engl J Med 2002; 346(2): 92–98.

95. Senderowicz AM. Small-molecule cyclin-dependent kinase modulators. Oncogene 2003; 22(42):6609–6620.

96. Flinn IW, Byrd JC, Bartlett N, Kipps T, Gribben J, Thomas D et al. Flavopiridol administered as a 24-hour continuous infusion in chronic lymphocytic leukemia lacks clinical activity. Leuk Res 2005; 29(11):1253–1257.

97. Konig A, Schwartz GK, Mohammad RM, Al Katib A, Gabrilove JL. The novel cyclin-dependent kinase inhibitor flavopiridol downregulates Bcl-2 and induces growth arrest and apoptosis in chronic B-cell leukemia lines. Blood 1997; 90(11):4307–4312.

98. Benson C, Kaye S, Workman P, Garrett M, Walton M, de Bono J. Clinical anticancer drug development: targeting the cyclin-dependent kinases. Br J Cancer 2005; 92(1):7–12.

99. Schwartz GK, O'Reilly E, Ilson D, Saltz L, Sharma S, Tong W et al. Phase I study of the cyclin-dependent kinase inhibitor flavopiridol in combination with paclitaxel in patients with advanced solid tumors. J Clin Oncol 2002; 20(8):2157–2170.

100. Shah MA, Kortmansky J, Motwani M, Drobnjak M, Gonen M, Yi S et al. A phase i clinical trial of the sequential combination of irinotecan followed by flavopiridol. Clin Cancer Res 2005; 11(10):3836–3845.

101. Mani S, Wang C, Wu K, Francis R, Pestell R. Cyclin-dependent kinase inhibitors: novel anticancer agents. Expert Opin Investig Drugs 2000; 9(8):1849–1870.

102. Sato S, Fujita N, Tsuruo T. Interference with PDK1-Akt survival signaling pathway by UCN-01 (7-hydroxystaurosporine). Oncogene 2002; 21(11):1727–1738.

103. Sausville EA, Arbuck SG, Messmann R, Headlee D, Bauer KS, Lush RM et al. Phase I trial of 72-hour continuous infusion UCN-01 in patients with refractory neoplasms. J Clin Oncol 2001; 19(8):2319–2333.

104. Whittaker SR, Walton MI, Garrett MD, Workman P. The cyclin-dependent kinase inhibitor CYC202 (R-Roscovitine) inhibits retinoblastoma protein phosphorylation, causes loss of cyclin D1, and activates the mitogen-activated protein kinase pathway. Cancer Res 2004; 64(1):262–272.

105. Kim KS, Kimball SD, Misra RN, Rawlins DB, Hunt JT, Xiao HY et al. Discovery of aminothiazole inhibitors of cyclin-dependent kinase 2: synthesis, X-ray crystallographic analysis, and biological activities. J Med Chem 2002; 45(18):3905–3927.

106. Sridhar J, Pattabiraman N, Rosen EM, Pestell RG. CDK inhibitors as anticancer agents: Perspectives for the future. In: Yue EW, Smith PJ, eds. Inhibitors of cyclin-dependent kinases as anti-tumor agents. United States: CRC Press, 2006:389–408.

107. Wang C, Fu M, Mani S, Wadler S, Senderowicz AM, Pestell RG. Histone acetylation and the cell-cycle in cancer. Front Biosci 2001; 6:D610–D629.

108. Fu M, Rao M, Wang C, Sakamaki T, Wang J, Di Vizio D et al. Acetylation of androgen receptor enhances coactivator binding and promotes prostate cancer cell growth. Mol Cell Biol 2003; 23(23):8563–8575.

109. Fu M, Wang C, Wang J, Zafonte BT, Lisanti MP, Pestell RG. Acetylation in hormone signaling and the cell cycle. Cytokine Growth Factor Rev 2002; 13(3):259–276.

110. Imbriano C, Gurtner A, Cocchiarella F, Di Agostino S, Basile V, Gostissa M et al. Direct p53 transcriptional repression: in vivo analysis of CCAAT-containing g2/M promoters. Mol Cell Biol 2005; 25(9):3737–3751.

111. The Androgen Receptor Gene Mutations Database World Wide Web Server. Montreal, Quebec: The Lady Davis Institute for Medical Research, Sir Mortimer B. Davis-Jewish General Hospital. (Accessed June 15, 2005 at http://www.androgendb.mcgill.ca)

112. Cui Y, Zhang M, Pestell R, Curran EM, Welshons WV, Fuqua SA. Phosphorylation of estrogen receptor alpha blocks its acetylation and regulates estrogen sensitivity. Cancer Res 2004; 64(24):9199–9208.

113. Wang C, Fu M, Angeletti RH, Siconolfi-Baez L, Reutens AT, Albanese C et al. Direct acetylation of the estrogen receptor alpha hinge region by p300 regulates transactivation and hormone sensitivity. J Biol Chem 2001; 276(21):18375–18383.

114. Vigushin DM, Coombes RC. Targeted histone deacetylase inhibition for cancer therapy. Curr Cancer Drug Targets 2004; 4(2):205–218.

115. Drummond DC, Noble CO, Kirpotin DB, Guo Z, Scott GK, Benz CC. Clinical development of histone deacetylase inhibitors as anticancer agents. Annu Rev Pharmacol Toxicol 2005; 45: 495–528.

116. Sandor V, Robbins AR, Robey R, Myers T, Sausville E, Bates SE et al. FR901228 causes mitotic arrest but does not alter microtubule polymerization. Anticancer Drugs 2000; 11(6):445–454.

117. Sandor V, Senderowicz A, Mertins S, Sackett D, Sausville E, Blagosklonny MV et al. P21-dependent g(1)arrest with downregulation of cyclin D1 and upregulation of cyclin E by the histone deacetylase inhibitor FR901228. Br J Cancer 2000; 83(6):817–825.

118. Ueda H, Nakajima H, Hori Y, Fujita T, Nishimura M, Goto T et al. FR901228, a novel antitumor bicyclic depsipeptide produced by Chromobacterium violaceum No. 968. I. Taxonomy, fermentation, isolation, physico-chemical and biological properties, and antitumor activity. J Antibiot (Tokyo) 1994; 47(3):301–310.

119. Byrd JC, Shinn C, Ravi R, Willis CR, Waselenko JK, Flinn IW et al. Depsipeptide (FR901228): a novel therapeutic agent with selective, in vitro activity against human B-cell chronic lymphocytic leukemia cells. Blood 1999; 94(4):1401–1408.

120. Ueda H, Manda T, Matsumoto S, Mukumoto S, Nishigaki F, Kawamura I et al. FR901228, a novel antitumor bicyclic depsipeptide produced by Chromobacterium violaceum No. 968. III. Antitumor activities on experimental tumors in mice. J Antibiot (Tokyo) 1994; 47(3):315–323.

121. Marshall JL, Rizvi N, Kauh J, Dahut W, Figuera M, Kang MH et al. A phase I trial of depsipeptide (FR901228) in patients with advanced cancer. J Exp Ther Oncol 2002; 2(6):325–332.

122. Sandor V, Bakke S, Robey RW, Kang MH, Blagosklonny MV, Bender J et al. Phase I trial of the histone deacetylase inhibitor, depsipeptide (FR901228, NSC 630176), in patients with refractory neoplasms. Clin Cancer Res 2002; 8(3):718–728.

123. Byrd JC, Marcucci G, Parthun MR, Xiao JJ, Klisovic RB, Moran M et al. A phase 1 and pharmacodynamic study of depsipeptide (FK228) in chronic lymphocytic leukemia and acute myeloid leukemia. Blood 2005; 105(3):959–967.

124. Marks PA, Richon VM, Rifkind RA. Histone deacetylase inhibitors: inducers of differentiation or apoptosis of transformed cells. J Natl Cancer Inst 2000; 92(15):1210–1216.

125. Saito A, Yamashita T, Mariko Y, Nosaka Y, Tsuchiya K, Ando T et al. A synthetic inhibitor of histone deacetylase, MS-27–275, with marked in vivo antitumor activity against human tumors. Proc Natl Acad Sci USA 1999; 96(8):4592–4597.

126. Tazebay UH, Wapnir IL, Levy O, Dohan O, Zuckier LS, Zhao QH et al. The mammary gland iodide transporter is expressed during lactation and in breast cancer. Nat Med 2000; 6(8): 871–878.

127. Salomon DS, Brandt R, Ciardiello F, Normanno N. Epidermal growth factor-related peptides and their receptors in human malignancies. Crit Rev Oncol Hematol 1995; 19(3):183–232.

128. Yarden Y, Sliwkowski MX. Untangling the ErbB signalling network. Nat Rev Mol Cell Biol 2001; 2(2):127–137.

129. Grandis JR, Sok JC. Signaling through the epidermal growth factor receptor during the development of malignancy. Pharmacol Ther 2004; 102(1):37–46.

130. Porebska I, Harlozinska A, Bojarowski T. Expression of the tyrosine kinase activity growth factor receptors (EGFR, ERB B2, ERB B3) in colorectal adenocarcinomas and adenomas. Tumour Biol 2000; 21(2):105–115.

131. Salomon DS, Normanno N, Ciardiello F, Brandt R, Shoyab M, Todaro GJ. The role of amphiregulin in breast cancer. Breast Cancer Res Treat 1995; 33(2):103–114.

132. Sung T, Miller DC, Hayes RL, Alonso M, Yee H, Newcomb EW. Preferential inactivation of the p53 tumor suppressor pathway and lack of EGFR amplification distinguish de novo high grade pediatric astrocytomas from de novo adult astrocytomas. Brain Pathol 2000; 10(2):249–259.

133. Moyer JD, Barbacci EG, Iwata KK, Arnold L, Boman B, Cunningham A et al. Induction of apoptosis and cell cycle arrest by CP-358,774, an inhibitor of epidermal growth factor receptor tyrosine kinase. Cancer Res 1997; 57(21):4838–4848.

134. Albanell J, Rojo F, Averbuch S, Feyereislova A, Mascaro JM, Herbst R et al. Pharmacodynamic studies of the epidermal growth factor receptor inhibitor ZD1839 in skin from cancer patients: histopathologic and molecular consequences of receptor inhibition. J Clin Oncol 2002; 20(1): 110–124.

135. Di Gennaro E, Barbarino M, Bruzzese F, De Lorenzo S, Caraglia M, Abbruzzese A et al. Critical role of both p27KIP1 and p21CIP1/WAF1 in the antiproliferative effect of ZD1839 ('Iressa'), an epidermal growth factor receptor tyrosine kinase inhibitor, in head and neck squamous carcinoma cells. J Cell Physiol 2003; 195(1):139–150.

136. Barnes CJ, Bagheri-Yarmand R, Mandal M, Yang Z, Clayman GL, Hong WK et al. Suppression of epidermal growth factor receptor, mitogen-activated protein kinase, and Pak1 pathways and invasiveness of human cutaneous squamous cancer cells by the tyrosine kinase inhibitor ZD1839 (Iressa). Mol Cancer Ther 2003; 2(4):345–351.

137. Normanno N, Campiglio M, De LA, Somenzi G, Maiello M, Ciardiello F et al. Cooperative inhibitory effect of ZD1839 (Iressa) in combination with trastuzumab (Herceptin) on human breast cancer cell growth. Ann Oncol 2002; 13(1):65–72.

138. Ciardiello F, Caputo R, Bianco R, Damiano V, Pomatico G, De Placido S et al. Antitumor effect and potentiation of cytotoxic drugs activity in human cancer cells by ZD-1839 (Iressa), an epidermal growth factor receptor-selective tyrosine kinase inhibitor. Clin Cancer Res 2000; 6(5):2053–2063.

139. Wakeling AE, Guy SP, Woodburn JR, Ashton SE, Curry BJ, Barker AJ et al. ZD1839 (Iressa): an orally active inhibitor of epidermal growth factor signaling with potential for cancer therapy. Cancer Res 2002; 62(20):5749–5754.

140. Anido J, Matar P, Albanell J, Guzman M, Rojo F, Arribas J et al. ZD1839, a specific epidermal growth factor receptor (EGFR) tyrosine kinase inhibitor, induces the formation of inactive EGFR/HER2 and EGFR/HER3 heterodimers and prevents heregulin signaling in HER2-overexpressing breast cancer cells. Clin Cancer Res 2003; 9(4):1274–1283.

141. Moasser MM, Basso A, Averbuch SD, Rosen N. The tyrosine kinase inhibitor ZD1839 ("Iressa") inhibits HER2-driven signaling and suppresses the growth of HER2-overexpressing tumor cells. Cancer Res 2001; 61(19):7184–7188.

142. Moulder SL, Yakes FM, Muthuswamy SK, Bianco R, Simpson JF, Arteaga CL. Epidermal growth factor receptor (HER1) tyrosine kinase inhibitor ZD1839 (Iressa) inhibits HER2/neu (erbB2)-overexpressing breast cancer cells in vitro and in vivo. Cancer Res 2001; 61(24):8887–8895.

143. Goss G, Hirte H, Miller WH, Lorimer IAJ, Stewart D, Batist G et al. A phase I study of oral ZD 1839 given daily in patients with solid tumors: IND.122, a study of the Investigational New Drug Program of the National Cancer Institute of Canada Clinical Trials Group. Invest New Drugs 2005; 23(2):147–155.

144. Baselga J, Rischin D, Ranson M, Calvert H, Raymond E, Kieback DG et al. Phase I safety, pharmacokinetic, and pharmacodynamic trial of ZD1839, a selective oral epidermal growth factor receptor tyrosine kinase inhibitor, in patients with five selected solid tumor types. J Clin Oncol 2002; 20(21):4292–4302.

145. Herbst RS, Maddox AM, Rothenberg ML, Small EJ, Rubin EH, Baselga J et al. Selective oral epidermal growth factor receptor tyrosine kinase inhibitor ZD1839 is generally well-tolerated and has activity in non-small-cell lung cancer and other solid tumors: results of a phase I trial. J Clin Oncol 2002; 20(18):3815–3825.

146. Ranson M, Hammond LA, Ferry D, Kris M, Tullo A, Murray PI et al. ZD1839, a selective oral epidermal growth factor receptor-tyrosine kinase inhibitor, is well tolerated and active in patients with solid, malignant tumors: results of a phase I trial. J Clin Oncol 2002; 20(9):2240–2250.

147. Nakagawa K, Tamura T, Negoro S, Kudoh S, Yamamoto N, Yamamoto N et al. Phase I pharmacokinetic trial of the selective oral epidermal growth factor receptor tyrosine kinase inhibitor gefitinib ('Iressa', ZD1839) in Japanese patients with solid malignant tumors. Ann Oncol 2003; 14(6):922–930.

148. Fukuoka M, Yano S, Giaccone G, Tamura T, Nakagawa K, Douillard JY et al. Multi-Institutional Randomized Phase II Trial of gefitinib for previously treated patients with advanced non-small-cell lung cancer. J Clin Oncol 2003; 21(12):2237–2246.

149. Cella D, Herbst RS, Lynch TJ, Prager D, Belani CP, Schiller JH et al. Clinically meaningful improvement in symptoms and quality of life for patients with non-small-cell lung cancer receiving gefitinib in a randomized controlled trial. J Clin Oncol 2005; 23(13):2946–2954.

150. Herbst RS. Dose-comparative monotherapy trials of ZD1839 in previously treated non-small cell lung cancer patients. Semin Oncol 2003; 30(1 Suppl 1):30–38.

151. Herbst RS, Giaccone G, Schiller JH, Natale RB, Miller V, Manegold C et al. Gefitinib in combination with paclitaxel and carboplatin in advanced non-small-cell lung cancer: a phase III trial–INTACT 2. J Clin Oncol 2004; 22(5):785–794.

152. Giaccone G, Herbst RS, Manegold C, Scagliotti G, Rosell R, Miller V et al. Gefitinib in combination with gemcitabine and cisplatin in advanced non-small-cell lung cancer: a phase III trial–INTACT 1. J Clin Oncol 2004; 22(5):777–784.

153. Lynch TJ, Bell DW, Sordella R, Gurubhagavatula S, Okimoto RA, Brannigan BW et al. Activating mutations in the epidermal growth factor receptor underlying responsiveness of non-small-cell lung cancer to gefitinib. N Engl J Med 2004; 350(21):2129–2139.

154. Tibes R, Trent J, Kurzrock R. Tyrosine kinase inhibitors and the dawn of molecular cancer therapeutics. Annu Rev Pharmacol Toxicol 2005; 45:357–384.

155. Rudkin CT, Hungerford DA, Nowell PC. DNA contents of chromosome Ph1 and chromosome 21 in human chronic granulocytic leukemia. Science 1964; 144:1229–1231.

156. Daley GQ, Van Etten RA, Baltimore D. Induction of chronic myelogenous leukemia in mice by the P210bcr/abl gene of the Philadelphia chromosome. Science 1990; 247(4944):824–830.

157. Manley PW, Cowan-Jacob SW, Buchdunger E, Fabbro D, Fendrich G, Furet P et al. Imatinib: a selective tyrosine kinase inhibitor. Eur J Cancer 2002; 38 Suppl 5:S19–S27.

158. Druker BJ, Tamura S, Buchdunger E, Ohno S, Segal GM, Fanning S et al. Effects of a selective inhibitor of the Abl tyrosine kinase on the growth of Bcr-Abl positive cells. Nat Med 1996; 2(5):561–566.

159. Druker BJ, Talpaz M, Resta DJ, Peng B, Buchdunger E, Ford JM et al. Efficacy and safety of a specific inhibitor of the BCR-ABL tyrosine kinase in chronic myeloid leukemia. N Engl J Med 2001; 344(14):1031–1037.

160. Kantarjian H, Sawyers C, Hochhaus A, Guilhot F, Schiffer C, Gambacorti-Passerini C et al. Hematologic and cytogenetic responses to imatinib mesylate in chronic myelogenous leukemia. N Engl J Med 2002; 346(9):645–652.

161. Sawyers CL, Hochhaus A, Feldman E, Goldman JM, Miller CB, Ottmann OG et al. Imatinib induces hematologic and cytogenetic responses in patients with chronic myelogenous leukemia in myeloid blast crisis: results of a phase II study. Blood 2002; 99(10):3530–3539.

162. Talpaz M, Silver RT, Druker BJ, Goldman JM, Gambacorti-Passerini C, Guilhot F et al. Imatinib induces durable hematologic and cytogenetic responses in patients with accelerated phase chronic myeloid leukemia: results of a phase 2 study. Blood 2002; 99(6):1928–1937.

163. O'Brien SG, Guilhot F, Larson RA, Gathmann I, Baccarani M, Cervantes F et al. Imatinib compared with interferon and low-dose cytarabine for newly diagnosed chronic-phase chronic myeloid leukemia. N Engl J Med 2003; 348(11):994–1004.

164. Hahn EA, Glendenning GA, Sorensen MV, Hudgens SA, Druker BJ, Guilhot F et al. Quality of life in patients with newly diagnosed chronic phase chronic myeloid leukemia on imatinib versus interferon alfa plus low-dose cytarabine: results from the IRIS study. J Clin Oncol 2003; 21(11):213–82146.

165. Soverini S, Martinelli G, Rosti G, Bassi S, Amabile M, Poerio A et al. ABL mutations in late chronic phase chronic myeloid leukemia patients with up-front cytogenetic resistance to imatinib are associated with a greater likelihood of progression to blast crisis and shorter survival: a study by the GIMEMA working party on chronic myeloid leukemia. J Clin Oncol 2005; 23(18):4100–4109.

166. Druker BJ. Perspectives on the development of a molecularly targeted agent. Cancer Cell 2002; 1(1):31–36.

167. Moadel RM, Weldon RH, Katz EB, Lu P, Mani J, Stahl M et al. Positherapy: targeted nuclear therapy of breast cancer with 18F-2-deoxy-2-fluoro-D-glucose. Cancer Res 2005; 65(3):698–702.

168. Koukourakis MI, Giatromanolaki A, Sivridis E, Simopoulos K, Pastorek J, Wykoff CC et al. Hypoxia-regulated carbonic anhydrase-9 (CA9) relates to poor vascularization and resistance of squamous cell head and neck cancer to chemoradiotherapy. Clin Cancer Res 2001; 7(11): 3399–3403.

169. Chegwidden WR, Dodgson SJ, Spencer IM. The roles of carbonic anhydrase in metabolism, cell growth and cancer in animals. EXS 2000;(90):343–363.

170. Supuran CT, Scozzafava A. Carbonic anhydrase inhibitors–Part 94. 1,3,4-thiadiazole-2-sulfonamidederivatives as antitumor agents? Eur J Med Chem 2000; 35(9):867–874.

171. Scozzafava A, Owa T, Mastrolorenzo A, Supuran CT. Anticancer and antiviral sulfonamides. Curr Med Chem 2003; 10(11):925–953.
172. Schuler M, Rochlitz C, Horowitz JA, Schlegel J, Perruchoud AP, Kommoss F et al. A phase I study of adenovirus-mediated wild-type p53 gene transfer in patients with advanced non-small cell lung cancer. Hum Gene Ther 1998; 9(14):2075–2082.
173. Clayman GL, El Naggar AK, Lippman SM, Henderson YC, Frederick M, Merritt JA et al. Adenovirus-mediated p53 gene transfer in patients with advanced recurrent head and neck squamous cell carcinoma. J Clin Oncol 1998; 16(6):2221–2232.
174. Schuler M, Herrmann R, De Greve JL, Stewart AK, Gatzemeier U, Stewart DJ et al. Adenovirus-mediated wild-type p53 gene transfer in patients receiving chemotherapy for advanced non-small-cell lung cancer: results of a multicenter phase II study. J Clin Oncol 2001; 19(6):1750–1758.
175. Bykov VJ, Issaeva N, Shilov A, Hultcrantz M, Pugacheva E, Chumakov P et al. Restoration of the tumor suppressor function to mutant p53 by a low-molecular-weight compound. Nat Med 2002; 8(3):282–288.
176. Vassilev LT, Vu BT, Graves B, Carvajal D, Podlaski F, Filipovic Z et al. In vivo activation of the p53 pathway by small-molecule antagonists of MDM2. Science 2004; 303(5659):844–848.
177. Lin W, Albanese C, Pestell RG, Lawrence DS. Spatially discrete, light-driven protein expression. Chem Biol 2002; 9(12):1347–1353.
178. Zwijsen RM, Buckle RS, Hijmans EM, Loomans CJ, Bernards R. Ligand-independent recruitment of steroid receptor coactivators to estrogen receptor by cyclin D1. Genes Dev 1998; 12(22): 3488–3498.
179. Neuman E, Ladha MH, Lin N, Upton TM, Miller SJ, DiRenzo J et al. Cyclin D1 stimulation of estrogen receptor transcriptional activity independent of cdk4. Mol Cell Biol 1997; 17(9): 5338–5347.
180. Lamb J, Ramaswamy S, Ford HL, Contreras B, Martinez RV, Kittrell FS et al. A mechanism of cyclin D1 action encoded in the patterns of gene expression in human cancer. Cell 2003; 114(3):323–334.
181. Knudsen KE, Cavenee WK, Arden KC. D-type cyclins complex with the androgen receptor and inhibit its transcriptional transactivation ability. Cancer Res 1999; 59(10):2297–2301.
182. Lin HM, Zhao L, Cheng SY. Cyclin D1 is a ligand-independent co-repressor for thyroid hormone receptors. J Biol Chem 2002; 277(32):28733–28741.
183. Ganter B, Fu S, Lipsick JS. D-type cyclins repress transcriptional activation by the v-Myb but not the c-Myb DNA-binding domain. EMBO J 1998; 17(1):255–268.
184. Schubert S, Horstmann S, Bartusel T, Klempnauer KH. The cooperation of B-Myb with the coactivator p300 is orchestrated by cyclins A and D1. Oncogene 2004; 23(7):1392–1404.
185. Inoue K, Sherr CJ. Gene expression and cell cycle arrest mediated by transcription factor DMP1 is antagonized by D-type cyclins through a cyclin-dependent-kinase-independent mechanism. Mol Cell Biol 1998; 18(3):1590–1600.
186. Ratineau C, Petry MW, Mutoh H, Leiter AB. Cyclin D1 represses the basic helix-loop-helix transcription factor, BETA2/NeuroD. J Biol Chem 2002; 277(11):8847–8853.
187. Bienvenu F, Gascan H, Coqueret O. Cyclin D1 represses STAT3 activation through a Cdk4-independent mechanism. J Biol Chem 2001; 276(20):16840–16847.
188. Skapek SX, Rhee J, Kim PS, Novitch BG, Lassar AB. Cyclin-mediated inhibition of muscle gene expression via a mechanism that is independent of pRb hyperphosphorylation. Mol Cell Biol 1996; 16(12):7043–7053.
189. Opitz OG, Rustgi AK. Interaction between Sp1 and cell cycle regulatory proteins is important in transactivation of a differentiation-related gene. Cancer Res 2000; 60(11):2825–2830.
190. Shanahan F, Seghezzi W, Parry D, Mahony D, Lees E. Cyclin E associates with BAF155 and BRG1, components of the mammalian SWI-SNF complex, and alters the ability of BRG1 to induce growth arrest. Mol Cell Biol 1999; 19(2):1460–1469.
191. Lazaro JB, Bailey PJ, Lassar AB. Cyclin D-cdk4 activity modulates the subnuclear localization and interaction of MEF2 with SRC-family coactivators during skeletal muscle differentiation. Genes Dev 2002; 16(14):1792–1805.
192. Morris L, Allen KE, La Thangue NB. Regulation of E2F transcription by cyclin E-Cdk2 kinase mediated through p300/CBP co-activators. Nat Cell Biol 2000; 2(4):232–239.
193. Adnane J, Shao Z, Robbins PD. Cyclin D1 associates with the TBP-associated factor TAF(II)250 to regulate Sp1-mediated transcription. Oncogene 1999; 18(1):239–247.

194. Wang H, Shao N, Ding QM, Cui J, Reddy ES, Rao VN. BRCA1 proteins are transported to the nucleus in the absence of serum and splice variants BRCA1a, BRCA1b are tyrosine phosphoproteins that associate with E2F, cyclins and cyclin dependent kinases. Oncogene 1997; 15(2):143–157.

195. Hayami R, Sato K, Wu W, Nishikawa T, Hiroi J, Ohtani-Kaneko R et al. Down-regulation of BRCA1-BARD1 ubiquitin ligase by CDK2. Cancer Res 2005; 65(1):6–10.

196. Zhao J, Kennedy BK, Lawrence BD, Barbie DA, Matera AG, Fletcher JA et al. NPAT links cyclin E-Cdk2 to the regulation of replication-dependent histone gene transcription. Genes Dev 2000; 14(18):2283–2297.

197. Yue EW, DiMeo SV, Higley CA, Markwalder JA, Burton CR, Benfield PA et al. Synthesis and evaluation of indenopyrazoles as cyclin-dependent kinase inhibitors. Part 4: heterocycles at C3. Bioorg Med Chem Lett 2004; 14(2):343–346.

198. Emanuel SL, Rugg C, Lin R, Connolly P, Napier C, Hollister B et al. Evaluation of the CDK inhibitor JNJ-7706621 as a targeted antitumor agent. 95th Annual Meeting Proceedings of American Association Cancer Research 833. 2004.

199. Hamdouchi C, Keyser H, Collins E, Jaramillo C, De Diego JE, Spencer CD et al. The discovery of a new structural class of cyclin-dependent kinase inhibitors, aminoimidazo[1,2-a]pyridines. Mol Cancer Ther 2004; 3(1):1–9.

200. Luk KC, Simcox ME, Schutt A, Rowan K, Thompson T, Chen Y et al. A new series of potent oxindole inhibitors of CDK2. Bioorg Med Chem Lett 2004; 14(4):913–917.

201. Dermatakis A, Luk KC, DePinto W. Synthesis of potent oxindole CDK2 inhibitors. Bioorg Med Chem 2003; 11(8):1873–1881.

202. Soni R, O'Reilly T, Furet P, Muller L, Stephan C, Zumstein-Mecker S et al. Selective in vivo and in vitro effects of a small molecule inhibitor of cyclin-dependent kinase 4. J Natl Cancer Inst 2001; 93(6):436–446.

203. Fry DW, Bedford DC, Harvey PH, Fritsch A, Keller PR, Wu Z et al. Cell cycle and biochemical effects of PD 0183812. A potent inhibitor of the cyclin D-dependent kinases CDK4 and CDK6. J Biol Chem 2001; 276(20):16617–16623.

204. Fry DW, Harvey PJ, Keller PR, Elliott WL, Meade M, Trachet E et al. Specific inhibition of cyclin-dependent kinase 4/6 by PD 0332991 and associated antitumor activity in human tumor xenografts. Mol Cancer Ther 2004; 3(11):1427–1438.

205. Hirai H, Kawanishi N, Iwasawa, Y. Recent advances in the development of selective small molecule inhibitors for cyclin-dependent kinases. Curr TOP Med chem 2005; 5(2):167–179.

206. McInnes C, Wang S, Anderson S, O'Boyle J, Jackson W, Kontopidis G et al. Structural determinants of CDK4 inhibition and design of selective ATP competitive inhibitors. Chem Biol 2004; 11(4): 525–534.

207. Breault GA, Ellston RP, Green S, James SR, Jewsbury PJ, Midgley CJ et al. Cyclin-dependent kinase 4 inhibitors as a treatment for cancer. Part 2: identification and optimisation of substituted 2,4-bis anilino pyrimidines. Bioorg Med Chem Lett 2003; 13(18):2961–2966.

208. Beattie JF, Breault GA, Ellston RP, Green S, Jewsbury PJ, Midgley CJ et al. Cyclin-dependent kinase 4 inhibitors as a treatment for cancer. Part 1: identification and optimisation of substituted 4,6-bis anilino pyrimidines. Bioorg Med Chem Lett 2003; 13(18):2955–2960.

2 mTOR

Properties and Therapeutics

John B. Easton, PhD, and Peter J. Houghton, PhD

SUMMARY

The mammalian target of rapamycin (mTOR) is a serine threonine kinase that regulates cell growth in response to growth factor-mediated activation of receptor tyrosine kinase signaling or to cellular stresses such as the deprivation of nutrients, energy, or oxygen (hypoxia). A number of proteins responsible for both the activation and inhibition of mTOR, as well as some targets of mTOR kinase activity, are modified in cancer. The growth of a number of tumor cell lines is inhibited by treatment with rapamycin, the naturally occurring specific inhibitor of mTOR function. A role for mTOR in angiogenesis has also been proposed. These observations have resulted in the development of additional small molecule inhibitors directed against mTOR and the initiation of a number of clinical trials to evaluate the therapeutic potential of these compounds. Initial studies indicate these compounds may have some benefit for certain subsets of cancer. However, progress in predicting which patients will benefit has been hampered somewhat by trial design as well as an incomplete understanding of mTOR function and regulation. All of the current inhibitors of mTOR are close derivatives of rapamycin and inhibit mTOR using the same mechanism. The potential for development of alternatives to the current generation of mTOR inhibitors depends in part on the results of a number of current clinical trials.

Key Words: mTOR; inhibitors; rapamycin; cancer; clinical trials.

1. INTRODUCTION

The identification of Tor [the yeast ortholog of mammalian target of rapamycin (mTOR)] was the result of a selective screen with the macrocyclic lactone rapamycin in the budding yeast *Saccharomyces cereviseae*. In yeast, there are two forms of Tor, Tor1 and Tor2 *(1)*. The Tor1 protein interacts with a number of other proteins to form Tor complex 1 (TORC1). TORC1 is involved in the regulation of G1 cell cycle progression, protein synthesis, and amino acid transport *(2)*. Tor2 is able to substitute

From: *Cancer Drug Discovery and Development: Molecular Targeting in Oncology*
Edited by: H. L. Kaufman, S. Wadler, and K. Antman © Humana Press, Totowa, NJ

for Tor1 in the TORC1 complex, but it also is able to form a distinct protein complex (TORC2) that functions to regulate the yeast cytoskeletal structure and is not affected by treatment with rapamycin *(3,4)*. In higher organisms, only one TOR protein has been identified. Until recently, the only characterized function of this protein was in the context of a rapamycin-sensitive TORC1 complex with functions that closely mirrored those originally observed for yeast TORC1. Recently, however, a TORC2 complex has been characterized in mammalian cells that is not affected by rapamycin and regulates cytoskeletal structure through the control of f-actin polymerization *(5,6)*.

2. mTOR, A PIKK FAMILY MEMBER

Several laboratories identified mTOR (FRAP, RAFT1, SEP) around the same time *(7–10)*. Cloning, sequence, and functional analysis demonstrated that mTOR was a 289-KDa serine threonine kinase with a number of conserved protein motifs (Fig. 1). The homology in the c-terminus of mTOR with the catalytic kinase domain (CD) of the lipid kinase phosphoinositide-3 kinase (PI3K) led to its characterization as a phosphatidylinositol 3′ kinase related protein kinase (PIKK) family member *(11)*. Besides mTOR, PIKK members include TEL1, ATM, ATR, DNA-dependent protein kinase (DNA-PK), and TRRAP.

There are several conserved motifs in the mTOR sequence, including a number of HEAT repeats (Huntington, EF3, A subunit of PP2A, TOR1), present in the amino terminal half of the protein that are thought to mediate interaction with HEAT-binding proteins *(12)*. Among proteins that are known to interact with HEAT repeats are the importins, which are involved in the transport of proteins into the nucleus. mTOR is found both in the nucleus and in the cytoplasm; however, it is not currently known if the HEAT repeats mediate nuclear localization.

After the HEAT repeats is a region of amino acid sequence unique to mTOR. This sequence is followed by a region conserved among some of the PIKK family members termed the FAT (FRAP, ATM, TRAPP) domain. PIKK family members that contain a FAT domain also contain a c-terminal domain termed FATC. Because of the

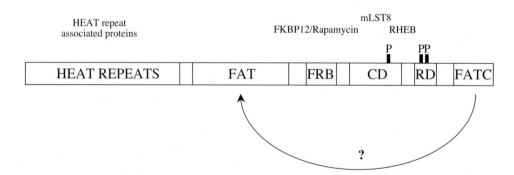

Fig. 1. Functional regions of the phosphotidylinositol 3′ kinase related protein kinase family member mammalian target of rapamycin (mTOR). mTOR contains a number of motifs which are found in other proteins. Shown above the block are the proteins known to bind to these domains. The '?' indicates Potential interdomain interactions between the FAT and FATC domains within mTOR. The currently identified phosphorylation sites are designated with a 'P' (see text for details).

conservation of both domains among the PIKK family members, it has been proposed that they may interact with each other to regulate kinase activity *(13)*.

The FKBP12 rapamycin-binding domain (FRB) is required for the binding of the rapamycin–FKBP12 complex to mTOR *(14,15)*. The ternary complex containing FKBP12 (FK506-binding protein), rapamycin, and mTOR is required for the cell cycle and growth control effects observed in cells treated with rapamycin. Point mutations introduced in the FRB result in rapamycin-resistant forms of mTOR *(14)*.

The catalytic domain (CD) contains autophosphorylation sites as well as the ATP-binding site. The regulatory domain region of mTOR contains sites that are phosphorylated in response to growth factors, although it is not clear exactly how this phosphorylation affects mTOR function *(16–18)*.

3. mTOR ACTIVATION

A number of growth factors have been shown to activate mTOR signaling, including epithelial growth factor (EGF), platelet-derived growth factor (PDGF), insulin-like growth factors 1 (IGF-1) and IGF-2, and insulin *(19)*. These growth factors bind to their cognate growth factor receptors, which results in the activation of the receptor by autophosphorylation *(20)*. A number of downstream signals are then transmitted as a result of proteins binding to the phosphorylated receptor (Fig. 2). One of the central proteins that is recruited to the activated receptors through its regulatory subunit is PI3K *(21)*. As a result of being recruited to the cell membrane, PI3K phosphorylates the membrane lipid phosphatidyl inositol 4-5 bisphosphate to generate phosphatidyl inositol 3, 4, 5, trisphosphate (PIP3) *(22)*. PIP3 in turn recruits the phosphoinositide-dependent kinase 1 (PDK1) to the cell membrane.

Marine thymoma viral oncogene homolog 1 (AKT), a substrate of PDK1, also binds to PIP3 in the cell membrane through the plextrin homology domain present in the amino terminus of the protein. Complete activation of AKT requires phosphorylation on two sites within the protein. The first site, T308, is phosphorylated by PDK1. There is some controversy as to which kinase or kinases phosphorylate the second site at S473. The list of candidates includes PDK1, integrin-linked kinase (ILK), DNA-PK, and the TOR complex 2 (mTORC2) that is defined as the complex containing mTOR, mAVO3 (also known as RICTOR), and mLST8 (GβL) *(23–27)*. It is interesting to note that mTOR as part of mTORC2 combines characteristics of two of the other kinases implicated in S473 phosphorylation. Both mTOR and DNA-PK are PIKK family members, while both mTOR (in the context of the TORC2 complex) and ILK regulate components of the cytoskeleton. Possibly, more than one kinase is capable of phosphorylating S473 of AKT in vivo. The signaling context and perhaps the tissue type may determine the kinase responsible for phosphorylating AKT on S473. Until the details are worked out, the kinase that phosphorylates S473 on AKT is referred to generically as PDK2 *(28)*.

After activation, AKT then phosphorylates and inhibits the activity of the GTPase-activating protein (GAP), TSC2 (tuberin) *(29–31)*. There is some evidence that the inhibition of TSC2 is the result of sequestration by binding to a 14-3-3 family member. The association of 14-3-3 with TSC2 is the result of phosphorylation of TSC2 on S939 by AKT *(32,33)*. This mechanism of inactivation by AKT is consistent with a number of other AKT substrates, such as the apoptotic protein BAD, that are phosphorylated

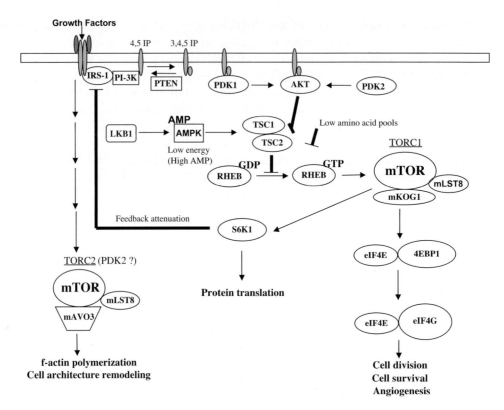

Fig. 2. Conditions regulating mammalian target of rapamycin (mTOR) and the proteins involved in the mTOR pathway. Low amino acid pools, low energy levels, and growth factor deprivation all downregulate mTOR through a series of kinase cascades. The arrows indicate activation of the target. The blocked lines indicate inhibition of the target (see text for details).

on a 14-3-3 binding sites by AKT *(34,35)*. Recent data indicate that TSC2 is also inactivated by phosphorylation at position S664 by extracellular signal-regulated kinase 2 (ERK2) *(36)*. However, unlike the AKT-mediated inhibition of TSC2, the mechanism of TSC2 inhibition by ERK2 appears to be 14-3-3 independent *(36)*.

Inhibition of TSC2 results in the stabilization of the small G protein, Ras Homology Enriched in Brain (RHEB), in its active GTP-bound form *(37)*. GTP-bound RHEB then activates mTOR signaling *(37–41)*. RHEB may accomplish this acitivation by a conformational change resulting from GTP binding to RHEB that is directly bound to mTOR *(42)*. Phosphorylation of mTOR at position 2448 is also AKT dependent but may be mediated through another kinase *(43)*. Furthermore, the effect of this phosphorylation is not understood. Another protein that significantly enhances mTOR activity by binding is mLST8 (GβL), which is part of both the TORC1 and TORC2 complexes *(44,45)*.

Activated mTOR phosphorylates downstream targets that include the eukaryotic translation initiation factor 4E binding protein 1 (4EBP1) and p70 S6 kinase (S6K1) *(46,47)*. These phosphorylation events are mediated by sequences present on both 4EBP1 and S6K1 termed TOS (target of rapamycin signaling) motifs *(48)*. The protein mKOG1 (RAPTOR), a component of the TORC1 complex, facilitates phosphoryation of 4EBP1 by binding to the TOS domain of 4EBP1 and bringing it into proximity with mTOR (*49–52*). Phosphorylation of 4EBP1 by mTOR and other kinases results in the release of eukaryotic translation initiation factor 4E (eIF4E) from 4EBP1 (*53*). After being released from 4EBP1, eIF4E binds to eIF4G and in association with other translation initiation factors translates RNA transcripts containing a 5′ m^{7}GTP cap (*54*). This includes the transcripts for c-MYC, cyclin D1, ornithine decarboxylase 1 (ODC1), and hypoxia inducible factor 1α (HIF1α) (*55–60*). Phosphorylation of S6K1 by mTOR leads to its activation. Activation of S6K1 results in an increase in the levels of protein translation, but also eventually leads to inhibition of receptor signaling, and a decrease in activation of receptor-dependent kinases such as AKT. The mechanism of receptor inhibition by S6K1 is through S6K1-dependent phosphorylation of insulin receptor substrate 1, a protein that mediates PI3K signaling (*61,62*).

4. EXTRACELLULAR FACTORS THAT INHIBIT mTOR FUNCTION

A great deal of information has been derived from studies conducted in yeast examining the function of TOR in its various protein complexes. In yeast, Tor activity is regulated in response to the availability of nutrients such as amino acids and glucose (*63*). As the identification and characterization of TOR complexes in higher organisms has progressed, it has become clear that in addition to its function as a sensor for nutrients and energy, mTOR responds to signals unique to multicellular organisms, such as responses from extracellular growth factors, and tissue hypoxia (*64*).

The deprivation of amino acids rapidly inactivates mTOR. The exact mechanism is not understood, but the evidence indicates that this function is mediated through the G protein RHEB. *Drosophila* starved for amino acids during development normally have reduced cell size in the fat bodies, but in *Drosophila* that overexpress RHEB, cells in the fat body reach normal size despite the starvation conditions (*65*). Increased expression of RHEB in mammalian cells results in a failure of mTOR activity to be inhibited in response to amino acid deprivation (*66*), and in at least one study, RHEB was shown to disassociate from mTOR in response to amino acid deprivation (*66*). Initial studies indicated that the specific target regulating RHEB is the TSC1/TSC2 complex (*39,67*), as the loss of TSC1/TSC2 results in an mTOR-dependent increase in S6K1 activity and blocks the inactivation of S6K1 in response to the removal of amino acids from the growth media (*68*). A more recent publication, however, reported that TSC2 was not directly regulating RHEB in response to amino acid deprivation (*69*).

Another condition that leads to the downregulation of mTOR activity is insufficient energy levels within the cell (*70*). A change in energy levels is sensed by a change in the concentration of intracellular AMP, the low energy product resulting from the cleavage of ATP, the major high energy store in the cell (*70*). Elevated concentrations of AMP lead to the association of AMP with AMP-dependent protein kinase (AMPK). AMPK association with AMP converts the enzyme into a favorable substrate for phosphorylation by the kinase LKB1. Phosphorylation of AMPK by LKB1 on T172 results in the activation of AMPK (*71,72*). Activated AMPK then directly phosphorylates TSC2 and activates the GAP function of TSC2. The activation of TSC2 results in an increase in the population of the inactive GDP-bound form of RHEB and results in downregulation of mTOR activity (*70*).

If cells in culture are exposed to hypoxic conditions, mTOR activity is rapidly inhibited, as measured by both autophosphorylation of mTOR and phosphorylation of the mTOR substrates S6K1 and 4EBP1 (73). The mechanism responsible for this inhibition is unknown but may involve either TSC2 or RHEB, because other upstream components such as AKT and AMPK do not appear to be targeted (73).

The withdrawal of growth factors results in an inhibition of mTOR activity due to removal of the ligand-mediated receptor stimulation and consequent phosphorylation cascade described in Section 3. However, the attenuation of mTOR activity after the withdrawal of growth factors is likely to be facilitated by the activity of phosphatases that actively dephosphorylate pathway proteins, as well as protein turnover of m^7GTP cap-dependent translation products.

Collectively, these observations indicate that RHEB functions as a restriction point that can only be negotiated when the various sensors of the cellular environment indicate conditions are favorable to growth. If any of the sensors detect a deficiency in growth conditions, mTOR activity is inhibited. Increasing the availability of other conditions favorable to growth (e.g., insulin) does not overcome this inhibition.

Most of the studies examining regulators of mTOR activity have focused on the TORC1-dependent functions of mTOR. However, morphological changes are observed in many cell lines in response to the deprivation of growth factors, suggesting that mTOR in the context of the TORC2 complex may also be regulated in response to some or all of these effectors of mTOR activity (5).

5. THE EFFECTS OF INHIBITING mTOR FUNCTION

5.1. Cell Cycle Progression

There are a number of events that can occur as a result of the inhibition of mTOR at both the cellular and tissue levels. Deletion of TOR1 in yeast leads to a G1 arrest phenotype (1). In mammalian cells, a similar effect is observed when mTOR in TORC1 is inhibited by treatment with rapamycin (74). In its activated state, mTOR, as part of TORC1, phosphorylates 4EBP1 allowing for efficient m^7GTP cap-dependent translation. Inhibiting mTOR results in a block of mTOR-dependent phosphorylatation of 4EBP1. Non-phosphorylated 4EBP1 remains bound to eIF4E, leading to a failure of eIF4E to bind to the m^7GTP cap-containing RNA, greatly reducing translation of m^7GTP cap-containing transcripts. Many of these cap-containing transcripts are for proteins required for cell cycle progression including c-MYC, ODC, and cyclin D. Several lines of evidence indicate that this mechanism of translational control contributes to the cell cycle effects of mTOR inhibition. For instance, cell lines that overexpress eIF4E are able to partially overcome the effects of the G1 delay induced by rapamycin (75). In cell lines selected for resistance to rapamycin, 4EBP1 levels are reduced, creating a stoichiometric deficit with eIF4E, thereby relieving inhibition of eIF4E (76). Cells that have reverted back to rapamycin sensitivity have levels of 4EBP1 similar to those observed in wildtype cells (77).

The inhibition of mTOR not only reduces the translation of m^7GTP cap-dependent proteins involved in cell cycle progression such as MYC, ODC, and cyclin D; it also results in increased levels of the CDK2 inhibitor p27[kip1], an inhibitor of G1-S phase cell cycle progression. This inhibitor binds to CDK2 and blocks its kinase activity, which is required for progression into S-phase. Although p27[kip1] levels are upregulated as a result of inhibition of m^7GTP cap-dependent translation, the mechanism responsible for this upregulation has not been determined (76,77). Induction of p27[kip1] appears to be a significant contributor to the block in cell proliferation observed in vivo and in some cell lines *(75–77)*. In support of a role for p27[kip1] in mTOR-mediated cell cycle control is the observation that expression of a constitutively active form of 4EBP1 in the adenocarcinoma cell line (MCF7) results in increased levels of p27[kip1] and inhibition of cell proliferation. The dependence of the increased levels of p27[kip1] on the overexpression 4EBP1 indicates that cap-dependent translation is either directly or indirectly involved in this process *(79)*.

The extent of the G1 arrest varies among different cell lines. This is likely to be a function of both the cell type and the genetic modifications present in the cell line, as modification of downstream targets of mTOR are likely to reduce the effects on cell cycle progression observed as a result of inhibiting mTOR.

5.2. Apoptosis

Although the principal effect of rapamycin in most normal cells and many cell lines is growth arrest, for certain cell types, treatment with rapamycin leads to apoptosis. Among these cell types are certain populations of dendritic cells and renal tubular cells *(80,81)*. Two proteins in the mTOR signaling pathway are associated with apoptosis and, therefore, may contribute to rapamycin-mediated apoptosis in these cells.

The first protein, c-MYC, has been implicated in both anti-apoptotic and pro-apoptotic functions *(82)*. The translation of c-MYC is regulated in part by cap-dependent mechanisms and is therefore dependent on eIF4E and sensitive to inhibitors of mTOR.

The second protein, AKT, is known to phosphorylate the pro-apoptotic protein BAD *(83)*. Phosphorylation of BAD by AKT results in its sequestration by 14-3-3, thereby blocking its apoptotic function *(84)*. However, because AKT is an activating component of mTOR, it also contributes to downstream events such as the regulation of c-MYC. Events such as the amplification of AKT2 that increase mTOR activity also affect c-MYC-mediated apoptosis or survival.

The mTOR component of AKT-dependent apoptosis may be more significant than it was believed previously. In a study of lymphomas, it was determined that inhibition of mTOR reverses the chemoresistance in lymphomas expressing AKT *(85)*. Overexpression of eIF4E mimicked the chemoresistance observed with AKT. Rapamycin could reverse the chemoresistance in cells overexpressing AKT, but not the cells overexpressing eIF4E.

The relationship between TOR signaling, p53, and apoptosis has also been examined in cultured cells. Under serum-free conditions, rapamycin treatment causes apoptosis in tumor cell lines with mutated p53. Overexpression of wildtype p53 or p21[Cip1] protects against these cells from rapamycin-dependent apoptosis *(86)*. The apoptotic effect from rapamycin treatment on the p53 mutant cells is a result of stress-dependent activation of the c-Jun N-terminal kinase (JNK) pathway. Activation of the JNK pathway, in turn, is dependent on the presence of 4EBP1 and its ability to bind eIF4E when mTOR is inhibited by rapamycin *(87)*.

The above observations indicate that the activation of cap-dependent translation has an important role in blocking apoptosis as well as promoting growth and survival of transformed cells.

MEF cells from PTEN$^{+/-}$ mice have increased levels of phosphorylated 4EBP1 and active S6K, which is consistent with increased mTOR signaling *(88)*. In some instances, PTEN$^{-/-}$ cells are extremely sensitive to rapamycin *(88–90)*. It is believed that as a result of the increased PI3K activity in PTEN$^{-/-}$ cells during development, the cells become more dependent on mTOR function. In contrast to wildtype cells, the mRNAs for cyclin D1 and c-MYC become associated with the monosomal fraction in rapamycin-treated cells lacking PTEN.

5.3. Angiogenesis

Treatment of endothelial cells with rapamycin significantly reduces the production of vascular endothelial growth factor (VEGF) *(91)*. Activation of the mTOR pathway is involved in angiogenesis through regulation of the levels of the transcription factor HIF1α *(92–96)*. HIF1α is a primary activator of VEGF *(95)*. The expression level of HIF1α is in part regulated in response to the activation of mTOR. Presumably, this effect is through eIF4E, as HIF1α is a m^{7}GTP cap-containing transcript *(96,97)*.

6. CANCER AND THE mTOR PATHWAY

Pathways upstream of mTOR are activated in many human cancers. The dual function phosphatase that negatively regulates PI3K, PTEN, is mutated, silenced, or deleted in a number of tumor types including glioblastoma, hepatocellular carcinoma, lung carcinoma, melanoma, endometrial carcinomas, and prostate cancer *(98–100)*. The net effect of loss of function is an upregulation or constitutive activation of AKT and consequently mTOR signaling. Activating mutations of AKT2, or gene amplification of AKT2, are frequently observed in some types of cancer.

Mutations in TSC proteins, associated with the tuberous sclerosis syndrome, are associated with well-vascularized hamartomas (benign lesions), but also with an increased risk of renal cell carcinoma. Inactivating mutations in the LKB1 kinase gene are associated with the Peutz–Jeghers cancer prone syndrome *(101)*. The inactivation of LKB1 results in a block in the ability of AMPK to activate TSC2 as would normally occur under conditions of energy deprivation. Because TSC2 inhibits the mTOR-activating protein Rheb, the net effect of LKB1 mutations is a higher level of mTOR activity under low-energy conditions.

Cancer-related changes in pathways downstream of mTOR are also reported *(71,72)*. For instance, S6K1 is overexpressed or constitutively active in tumor cell lines and in early stages of transformation in ovarian surface epithelium associated with BRCA1 mutations *(102)*. S6K1 is also amplified in some breast carcinomas *(103)*. Generally, for tumors that have an amplification of S6K1, there is a corresponding increase in the level of S6K1 protein *(103)*.

The gene coding for eIF4E is altered in a number of tumors. Progressive amplification of the *eIF4E* gene is associated with late stage head and neck carcinoma, ductal cell breast carcinoma, and thyroid carcinoma *(104–106)*. Levels of eI4E are elevated in some colon carcinomas in comparison to normal colon cells *(107,108)*. The levels of eIF4E are also increased in some bladder and breast cancers that have a poor outcome *(109,110)*. In these cancers, a corresponding increase in VEGF was also observed *(110,111)*.

Another recent study examining lymphomagenesis using em-MYC mice found that eIF4E cooperates with c-MYC in B-cell lymphomagenesis accelerating the formation of tumors *(112)*. Of note, the incidence of em–myc lymphomas is increased when these mice are backcrossed to p53 heterozygous mice, and tumors arising in these mice result from the loss of the remaining wildtype p53 allele. In contrast when eIF4E is overexpressed, the wildtype p53 allele is not mutated; thus, a mild increase in the level of eIF4E appears to abrogate the requirement for suppressing p53-mediated apoptosis in this model system. These data point to the possibility that at least under some conditions eIF4E may able to act as an oncogene.

The levels of 4EBP1 or the presence of inactivating mutations have not been rigorously investigated in tumor samples. But based on its inhibitory effect on eIF4E, one might predict that such alterations in 4EBP1 might also be observed in some cancers. It is likely that the ratio 4EBP1 to eIF4E is not only an important determinant in tumor progression, but also may predict the extent to which inhibition of mTOR will prove to be an effective treatment. In support of this cell lines in which 4EBP1 is overexpressed become more sensitive to rapamycin, whereas cells overexpressing eIF4E are less sensitive to rapamycin (75). Currently, the strongest clinical data to support this hypothesis have been observed in colon carcinoma. In these tumors, both EIF4e and 4EBP1 are overexpressed, but the 4EBP1 levels are higher in patients without significant metastatic disease (113).

Although a daunting task, ultimately, the molecular properties of individual patient's tumors are likely to prove important for the successful application of mTOR inhibitors.

7. PROPERTIES OF mTOR INHIBITORS

Currently, rapamycin and its analogs are the only specific small molecule inhibitors of mTOR yet described. Rapamycin forms a ternary complex with FKBP12 and mTOR (Fig. 3). The formation of this ternary complex results in a potent inhibition of mTOR signaling. Structural studies indicate that there are relatively few contacts between the two proteins (114). This has led to speculation that FKBP12 binding to rapamycin may lock rapamycin into a favorable conformation for binding to mTOR. Although other inhibitors of mTOR besides rapamycin have been developed, to date all of these compounds are the result of relatively minor modifications to the structure of rapamycin. The principal advantage of these compounds has been increased solubility and stability. As is illustrated in Fig. 3, FKBP12 binds to one face of the compound while mTOR binds to the adjacent hydrophobic face. Given the fact that the two proteins are bound to adjacent sides of the same ring structure, it is not surprising that it has been difficult to identify substitutions that maintain the functional properties of the parent compound.

To date, all the compounds that have progressed to clinical trials involve substitution at the C40 hydroxyl position of rapamycin. The majority of these compounds substitute esters or ethers. Ariad, however, has developed a C40 derivative, AP23573, which substitutes a phosphonate for the C40 hydroxyl (Fig. 3) (115). Although this compound is not a pro-drug, it inhibits mTOR activity and binds FKBP12 at concentrations similar to that of rapamycin. CCI-779 acts as a pro-drug which is metabolized to rapamcyin in the body (Fig. 3). Rapamycin is already approved as an immunosuppressant for organ transplantation and marketed under rapamune (Wyeth). The additional rapamycin/sirolimus derivatives in clinical development include CCI-779/temsirolimus (Wyeth), RAD-001/everolimus (Novartis), AP23573 (Ariad), AP23481 (Ariad), and ABT-578(Abbot).

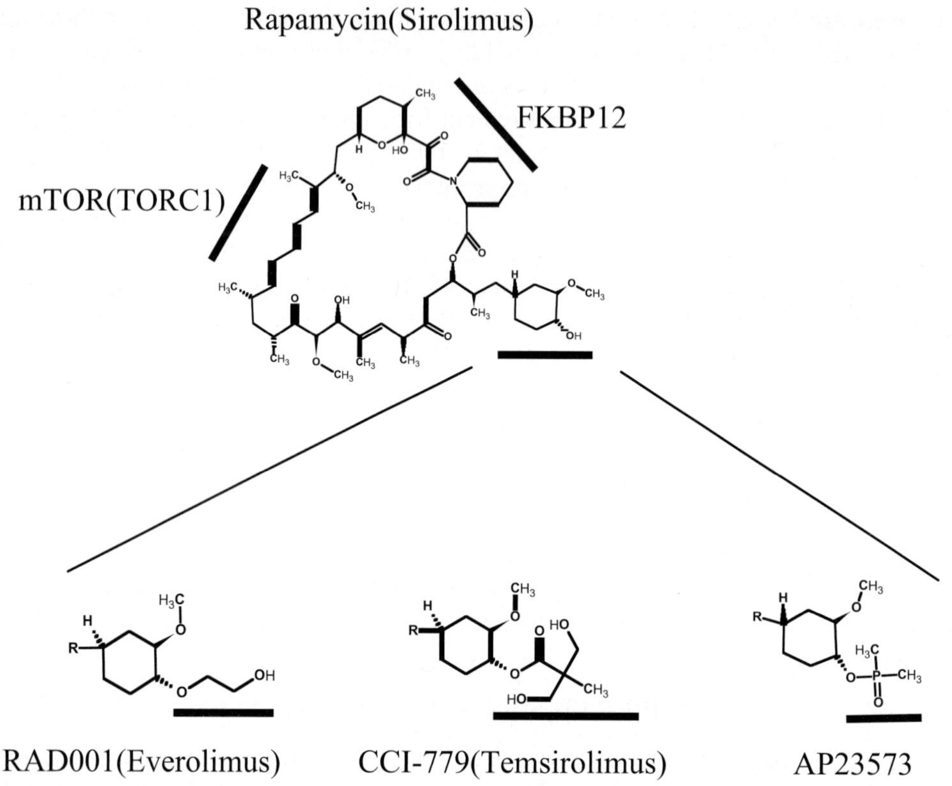

Fig. 3. Chemical structure of rapamycin and related analogs. Shown is the structure of rapamycin in the conformation known to bind FKBP12 and the FKBP12 rapamycin-binding domain of the mammalian target of rapamycin in the TORC1 complex. The bottom portion of the figure shows a portion of rapamycin with the modification specific to the derived compound underlined.

8. mTOR INHIBITORS: PRECLINICAL DATA

Since the original identification of the tumor suppressing properties of rapamycin in the NCI in vivo cancer screen, rapamycin and later the analogs CCI-779 and RAD001 have been tested for their effects on a number of tumor-derived cell lines and mouse xenograft tumor models (*116–118*). Treatment of cells with rapamycin or its analogs inhibits proliferation in a large number of cell lines, and in some instances, treatment leads to apoptosis. These cell lines are derived from a number of tumor types including rhabdomyosarcoma, neuroblastoma, glioblastoma, small cell lung carcinoma, osteosarcoma, pancreatic carcinoma, renal cell carcinoma, Ewing sarcoma, prostate cancer, and breast cancer (*119–134*).

The breadth of tumor types that are affected is impressive but not surprising given the role of mTOR as a nutrient sensor, cell cycle regulator, and growth regulator, all of which are important for tumor progression. The results also provide a cautionary note, because for most of the tumor types only a subset of the cell lines respond to rapamycin, and for some of these tumor types, it is a relatively small subset. This illustrates the critical importance of understanding the molecular characteristics of individual tumors and which, if any, proteins in the mTOR signaling pathways are disregulated or modified.

9. mTOR INHIBITORS: CLINICAL TRIALS

9.1. Results from Completed Trials

Based on the preclinical data, a phase I trial evaluating the safety of CCI-779 was implemented (135). The results indicated that with daily intravenous (iv) treatment, there are significant grade 3 toxicities, including hypocalcaemia, vomiting, thrombocytopenia, and increase in the level of hepatic transaminases. One patient had an objective response (non-small-cell carcinoma), while a number of patients had minor responses or stable disease (cervical carcinoma, uterine carcinoma, renal cell carcinoma, and soft tissue sarcoma). On a weekly treatment schedule, patients experienced no grade 3 toxicities regardless of dosage. Three patients had a partial tumor regression (renal cell, neuroendecrine, and breast carcinomas).

Subsequently, based on the phase I results, a number of phase II trials were initiated to study the effects treating advanced stage refractory renal cell carcinoma, refractory mantle cell lymphoma, and refractory metastatic breast cancer with CCI-779 (136–138). The results of the two completed trials for renal cell carcinoma were positive with an objective response rate of 5–7%, a minor response rate of 26–29%, and stable disease in approximately 40% of the patients. The phase II trial for mantle cell lymphoma consisted of a weekly treatment with 250 mg (fixed dose) of CCI-779 iv, which resulted in an overall response rate of 38% (3% complete response and 35% partial response) with a median time to progression of 6.9 months in responders versus 6.5 months for all patients treated. For the phase II trial in pretreated patients with advanced breast cancer, dosages of 75 mg or 250 mg of CCI-779 were given iv weekly. The overall response rate of 9.2% (all partial responses) was similar for both dosage levels, but toxicity was decreased in the patients treated with the 75-mg dosage.

The initial results of the RAD001 phase I MTD trial using a fixed dosing schedule indicated mild toxicity with tumor responses observed in several patients (139,140). Clinical data are lacking for the more recently developed inhibitors from Ariad Pharmaceuticals.

9.2. Open Clinical Trials

Results from the initial dose finding and safety trials have prompted the development of a number of additional phase I and phase II trials for various types of cancer. These include trials for breast cancer, prostate cancer, pancreatic cancer, malignant gliomas, leukemia, lymphoma, muliple myeloma, melanoma, and renal cell carcinoma (summarized in Table 1).

For a few of these studies such as a prostate cancer study in which CCI-779 is used as a neoadjuvant to shrink tumors prior to prostatectomy, a comprehensive examination of the phosphorylation status and expression levels of the various components of the mTOR signaling pathway will be performed, including examination of the PTEN status in these tumors. This study will also monitor the S6K1 activity in PBMCs, because data indicate that this activity may serve as an indirect biomarker of the drug activity within the tumors (140).

However, many of the other studies will not collect comprehensive data about the mTOR pathway. This is unfortunate, as it is likely that such information would prove valuable in the future for targeting patient populations more likely to respond to CCI-779 treatment. This would however require a change in the design of clinical trials from a tumor classification approach to a biomarker-based approach.

Currently, phase III trials are in progress to test the efficacy of CCI-779 either alone or in combination with interferon-α as a first-line treatment of renal cell carcinoma and the use of the oral form of CCI-779 in combination with the aromatase inhibitor letrozole for the treatment of locally advanced or metastatic breast cancer.

As a result of the development of CCI-779, there are few clinical cancer trials testing rapamycin. Currently, there is an ongoing phase II trial examining the effect of rapamycin treatment on refractory renal cell carcinoma. There is also a phase I trial establishing safety in treatment of pediatric patients with refractory acute leukemia or lymphoma.

For AP23573, there is currently a phase I trial for multiple myelomas and a phase II trial of patients with taxane-resistant androgen-independent prostate cancer. The current clinical trials that are open and recruiting for the various analogs of rapamycin are summarized in Table 1.

Table 1
Summary of Current Clinical Trials using Rapamycin or its Analogs

Drug	Additional therapy	Disease	Phase
CCI-779	Surgery	High risk prostate cancer	II
CCI-779 (oral)	Letrozole	Advanced breast cancer, recurrent breast cancer	III
CCI-779	Single agent	Pancreatic cancer	II
CCI-779	Single agent	Recurrent adult brain tumors	I-II
CCI-779	Interferon-α	Renal cell carcinoma, kidney neoplasms	III
CCI-779	Single agent	Renal cell carcinoma, kidney neoplasms	III
CCI-779	Single agent	Advanced stage small cell lung cancer	II
CCI-779	Single agent	Stage IV melanoma, recurrent Melanoma	II
Rapamycin	Single agent	Recurrent adult brain tumor adult glioblastoma multiforme	I–II
Rapamycin	Single agent	Recurrent childhood lymphoma, recurrent childhood leukemia	I
AP23573	Single agent	Advanced recurrent lymphoma, multiple myeloma	I
AP23573	Single agent	Taxane-resistant, androgen-independent, prostate cancer	II

10. THE FUTURE OF mTOR AS A TARGET FOR CHEMOTHERAPY

10.1. Combination Therapy and Trial Design

The determination of the efficacy of inhibiting mTOR in the treatment of various types of cancer is still being evaluated, and there are many possibilities that can be explored in identifying areas where rapamycin might be an effective treatment for cancer.

Most of the studies examining the analogs of rapamycin have focused on establishing the safety of these compounds as a single agent, usually in patients previously treated with chemotherapy. However, like many past regimens developed to treat cancer, it is quite likely that compounds that target mTOR will prove more effective in combination with other chemotherapeutic agents directed against alternative molecular targets. For example, in multiple myeloma cell lines and cells from patients, treatment with the combination of CC-5013 (lenalidomide) and rapamycin resulted in a synergistic apoptotic effect. This combination of drugs was also able to overcome growth advantages conferred by the addition of growth factors, growth on stromal cells, or drug resistance. Although, CCI-5013 and rapamycin appear to have somewhat similar physiological effects (immune modulator and anti-angiogenic), the mechanism of action of these drugs appears to be through different pathways, perhaps accounting for their synergistic effects *(141)*.

A number of possible approaches with combination therapy targeting different signaling pathways could be imagined. Treatment with rapamycin followed by the timed addition of drugs targeting S-phase such as irinotecan may have an additive effect in tumors. For tumors where treatment with mTOR inhibitors may cause a general slowing of growth without a significant accumulation of G1 phase cells, concomitant therapy with compounds such as interferon-α or other compounds inducing general apoptosis may prove more appropriate. This combination is currently a component of phase III trials to determine efficacy as a treatment for renal cell carcinoma.

Targeting multiple proteins in the same pathway may also prove effective. This is especially true where mutations have generated drug resistance that can be bypassed by treatment with another agent. Chronic myelogenous leukemia is defined by the presence of the tyrosine kinase fusion product BCR-ABL. The function of BCR-ABL is required for proliferation of the leukemic cells. Patients who develop resistance to the BCR-ABL kinase inhibitor, imatinib, usually have developed point mutations in BCR-ABL *(142)*. The effects of BCR-ABL are dependent on upregulation of PI3K activity and its downstream effectors, including mTOR. Therefore, mTOR inhibitors might be predicted to have a significant effect on CML progression. Recent preclinical data have added support to this hypothesis *(143)*.

Another possible approach to targeting the same pathway would be to target both proteins simultaneously in first-line treatment to reduce the likelihood of developing resistant tumor cells. In the case of imatinib, this would be accomplished by slowing proliferation of the initial tumor cell population by co-treatment with rapamycin.

When targeting multiple proteins in the same pathway to bypass drug resistance, it is important to determine if the second drug eliminates the effects of the mutation creating the resistance. A good example of the importance of this is illustrated by pre-clinical studies using the EGF receptor inhibitor, iressa, and rapamycin in renal cell carcinoma cell lines. In this study, it was discovered that these two agents acted synergistically to inhibit cell growth only in cell lines that contained the wildtype E3 ubiquitin ligase complex protein, VHL *(144)*.

Combining rapamycin with other treatment modalities may also prove effective in cancer treatment. Previously, it has been demonstrated that inhibition of the PI3K/Akt signaling pathway sensitizes tumor vasculator to radiation *(145,146)*. More recently, it has been determined that mTOR inhibitors also sensitize tumor vasculature to ionizing radiation *(147)*. Indicating that for some types of cancer combining mTOR inhibitors with radiotherapy may have some efficacy.

Regardless of which therapies are considered, careful pre-clinical studies in tumor models such as the mouse xenograft or transgenic model should be conducted to study various dosing and timing considerations as well as the effects on the various molecular components of mTOR signaling before the use of these new combination therapies are applied in the clinic.

10.2. Developing Novel Inhibitors of mTOR

The success of the ATP mimetic imanitib in targeting the BCR-ABL tyrosine kinase proves that the use of agents targeting the ATP-binding domain of kinases can be a successful strategy in diseases where a clearly defined kinase function is required for tumor viability. An interesting discovery as a result of the development of imanitib is that the ATP mimetic does not have to be absolutely specific. Imanitib also targets activated c-KIT and the PDGF receptor *(148)*.

For mTOR, the situation is more complicated. As described in Section 2., the rapamycin–FKBP12 complex, although a specific inhibitor of mTOR function, does not appear to inhibit kinase activity directly. Instead, rapamycin inhibition of mTOR is mediated by allosterically interfering with the association of mTOR to other proteins, specifically mKOG1 (RAPTOR). Current reports indicate that inhibition by rapamycin only occurs when mTOR is part of the TORC1 complex. So to date, all of the clinical trials using rapamycin-derived inhibitors have been generating data resulting from targeting a portion of mTOR function. ATP mimetics against mTOR would almost certainly inhibit the function of mTOR in the context of both TORC1 and TORC2. In yeast, preventing the formation of both TORC1 and TORC2 is lethal, and in mice, deletion of mTOR results in an embryonic lethal phenotype *(2,149,150)*. Therefore, both the therapeutic potential and the activity of ATP mimetics directed against mTOR are less certain and may result in unacceptable biological effects. Potent inhibitors of the kinase activity may result in widespread apoptosis rather than the cytostatic effect most frequently observed with rapamycins in normal cells. Currently there are no reported ATP mimetic inhibitors of mTOR in development. However, if the rapamycin analogs prove to be clinically successful in defined tumor types, this is likely to change, as there are a limited number of alterations that can be made to rapamycin.

Another potential approach to inhibit mTOR function would be to disrupt the active mTOR complex (TORC1 or TORC2). In theory, this mode of inhibition would be similar in action to that observed for rapamycin. However, both the target and the mechanism could be different. For example, small interfering RNAs against mKOG1 would be predicted to prevent the formation of active TORC1 complexes, whereas siRNAs directed against mLST8 would be predicted to disrupt both the TORC1 and TORC2 complexes. Small molecules binding to mTOR at the mLST8-binding site would also be predicted to inhibit the function of TORC1 and TORC2.

Although not directly inhibiting mTOR per se, employing small molecule inhibitors that bind to mTOR substrates and are directed against a conserved motif within these substrates such as the TOS motif may also prove effective in blocking tumor cell growth.

In the final analysis whether or not these types of compounds will be developed depends in large part on what future discoveries reveal about the role of mTOR in cancer and the success of currently ongoing clinical trials of the rapamycin analogs.

ACKNOWLEDGMENTS

This work was supported by USPHS Awards CA23099, CA77776, CA96966, and CA21765 (Cancer Center Support Grant) and by American, Lebanese, Syrian, Associated Charities (ALSAC).

REFERENCES

1. Cafferkey R, McLaughlin MM, Young PR, Johnson RK, Livi GP. Yeast TOR (DRR) proteins: amino-acid sequence alignment and identification of structural motifs. Gene 1994;141:133–136
2. Helliwell SB, Wagner P, Kunz J, Deuter-Reinhard M, Henriquez R, Hall MN. TOR1 and TOR2 are structurally and functionally similar but not identical phosphatidylinositol kinase homologues in yeast. Mol Biol Cell 1994;5:105–118.
3. Zheng XF, Florentino D, Chen J, Crabtree GR, Schreiber SL. TOR kinase domains are required for two distinct functions, only one of which is inhibited by rapamycin. Cell 1995;82:121–130.
4. Schmidt A, Kunz J, Hall MN. TOR2 is required for organization of the actin cytoskeleton in yeast. Proc Natl Acad Sci USA 1996;93:13780–13785.
5. Jacinto E, Loewith R, Schmidt A, Lin S, Ruegg MA, Hall A, Hall MN. Mammalian TOR complex 2 controls the actin cytoskeleton and is rapamycin insensitive. Nat Cell Biol 2004;6:1122–1128.
6. Sarbassov DD, et al. Rictor, a novel binding partner of mTOR, defines a rapamycin-insensitive and raptor-independent pathway that regulates the cytoskeleton. Curr Biol 2004;14:1296–1302.
7. Brown EJ, Albers MW, Shin TB, Ichikawa K, Keith CT, Lane WS, Schreiber SL. A mammalian protein targeted by G1-arresting rapamycin-receptor complex. Nature 1994;369:756–758.
8. Sabatini DM, Erdjument-Bromage H, Lui M, Tempst P, Snyder SH. Raft1: a mammalian protein that binds to FKBP12 in a rapamycin-dependent fashion and is homologous to yeast TORs. Cell 1994;78:35–43.
9. Sabers CJ, Martin MM, Brunn GJ, Williams JM, Dumont FJ, Wiederrecht G, Abraham RT. Isolation of a protein target of the FKBP12-rapamycin complex in mammalian cells. J Biol Chem 1995;270:815–822.
10. Chen Y, et al A putative sirolimus (rapamycin) effector protein. Biochem Biophys Res Commun 1994;203:1–7.
11. Abraham RT. Cell cycle checkpoint signaling through the ATM and ATR kinases. Genes Dev 2001;15:2177–2196.
12. Andrade MA, Bork P. HEAT repeats in the Huntington's disease protein. Nat Genet 1995; 11:115–116.
13. Bosotti R, Isacchi A, Sonnhammer EL. FAT: a novel domain in PIK-related kinases. Trends Biochem Sci 2000;25:225–227.
14. Chen J, Zheng XF, Brown EJ, Schreiber SL. Identification of an 11-kDa FKBP12-rapamycin-binding domain within the 289-kDa FKBP12-rapamycin-associated protein and characterization of a critical serine residue. Proc Natl Acad Sci USA 1995;92:4947–4951.
15. Choi J, Chen J, Schreiber SL, Clardy J. Structure of the FKBP12-rapamycin complex interacting with the binding domain of human FRAP. Science 1996;273:239–242.
16. Scott PH, Brunn GJ, Kohn AD, Roth RA, Lawrence JC. Evidence of insulin-stimulated phosphorylation and activation of the mammalian target of rapamycin mediated by a protein kinase B signaling pathway. Proc Natl Acad Sci USA 1998;95:7772–7777.

17. Nave B, Ouwens M, Withers DJ, Alessi DR, Shepherd PR. Mammalian target of rapamycin is a direct target for protein kinase B: identification of a convergence point for opposing effects of insulin and amino acid deficiency on protein translation. Biochem J 1999;344:427–431.

18. Sekulic A, Hudson CC, Homme JL, Yin P, Otterness DM, Karnitz LM, Abraham RT. A direct linkage between the phosphoinositide 3-kinase-AKT signaling pathway and the mammalian target of rapamycin (mTOR) in mitogen-stimulated and transformed cells. Cancer Res 2000;60:3504–3513.

19. Chung J, Grammer TC, Lemon KP, Kazlauskas A, Blenis J. PDGF- and insulin-dependent pp70^{S6k} activation mediated by phosphatidylinositol-3-OH kinase. Nature 1994;370:71–75.

20. Ullrich A, Schlessinger J. Signal transduction by receptors with tyrosine kinase activity. Cell 1990;61:203–212.

21. Ueki K, Algenstaedt P, Mauvais-Jarvis F, Kahn CR. Positive and negative regulation of phospho-inositide 3-kinase-dependent signaling pathways by three different gene products of the p85a regulatory subunit. Mol Cell Biol 2000;20:8035–8046.

22. Whitman M, Downes CP, Keeler M, Keller T, Cantley L. Type I phosphatidylinositol kinase makes a novel inositol phospholipid, phosphatidylinositol-3-phosphate. Nature 1988;332:644–646.

23. Chan T, Rittenhouse S, Tsichlis P. AKT/PKB and other D3 phosphoinostide regulated kinases: kinase acitivation by phosphoimositide-dependent phosporylation. Annu Rev Biochem 1998;68:965–1014.

24. Delcommenne M, Tan C, Gray V, Rue L, Woodgett J, Dedhar S. Phosphoinositide-3-OH kinase-dependent regulation of glycogen synthase kinase 3 and protein kinase B/AKT by the integrin-linked kinase. Proc Natl Acad Sci USA 1998;95:11211–11216.

25. Persad S, et al. Regulation of protein kinase B/Akt-serine 473 phosphorylation by integrin-linked kinase: critical roles for kinase activity and amino acids arginine 211 and serine 343. J Biol Chem 2001;276:27462–27469.

26. Feng J, Park J, Cron P, Hess D, Hemmings BA. Identification of a PKB/Akt hydrophobic motif Ser-473 kinase as DNA-dependent protein kinase. J Biol Chem 2004;279:41189–41196.

27. Sarbassov DD, Guertin DA, Ali SM, Sabatini DM. Phosphorylation and regulation of Akt/PKB by the rictor-mTOR complex. Science 2005;307:1098–1101.

28. Jacinto E, Hall MN. Tor signalling in bugs, brain and brawn. Nat Rev Mol Cell Biol 2003;4:117–126.

29. Inoki K, Li Y, Zhu T, Wu J, Guan KL. TSC2 is phosphorylated and inhibited by Akt and suppresses mTOR signalling. Nat Cell Biol 2002;4:648–657.

30. Potter CJ, Pedraza LG, Xu T. Akt regulates growth by directly phosphorylating Tsc2. Nat Cell Biol 2002;4:658–665.

31. Manning BD, Tee AR, Logsdon MN, Blenis J, Cantley LC. Identification of the tuberous sclerosis complex-2 tumor suppressor gene product tuberin as a target of the phoshpoinositide-3-kinase/akt pathway. Mol Cell 2002;10:151–162.

32. Liu MY, Cai S, Espejo A, Bedford MT, Walker CL. 14-3-3 interacts with the tumor suppressor tuberin at Akt phosphorylation site(s). Cancer Res 2002;62:6475–6480.

33. Li Y, Inoki K, Yeung R, Guan KL. Regulation of TSC2 by 14-3-3 binding. J Biol Chem 2002;277:44593–44596.

34. del Peso L, Gonzalez-Garcia M, Page C, Herrera R, Nunez G. Interleukin-3 phosphorylation of BAD through the protein kinase Akt. Science 1997;278:687–689.

35. Datta SR, Dudek H, Tao X, Masters S, Fu H, Gotoh Y, Greenberg ME. Akt phosphorylation of BAD couples survival signals to the cell-intrinsic death machinery. Cell 1997;91:231–241.

36. Ma L, Chen Z, Erdjument-Bromage H, Tempst P, Pandolfi P. Phosphorylation and functional inactivation of TSC2 by Erk: implications for tuberous sclerosis and cancer pathogenesis. Cell 2005;121:179–193.

37. Tee AR, et al. Tuberous sclerosis complex-1 and -2 gene products function together to inhibit mammalian target of rapamycin (mTOR)-mediated downstream signaling. Proc Natl Acad Sci USA 2002 99:13571–13576.

38. Zhang Y, et al. Rheb is a direct target of the tuberous sclerosis tumour suppressor proteins. Nat Cell Biol 2003;5:578–581.

39. Saucedo LJ, et al. Rheb promotes cell growth as a component of the insulin/TOR signalling network. Nat Cell Biol 2003;5:566–571.

40. Stocker H, et al. Rheb is an essential regulator of S6K in controlling cell growth in Drosophila. Nat Cell Biol 2003;5:559–565.

41. Inoki K, Li Y, Xu T, Guan KL. Rheb GTPase is a direct target of TSC2 GAP activity and regulates mTOR signaling. Genes Dev 2003;15:1829–1834.
42. Long X, Lin Y, Ortiz-Vega S, Yonezawa K, Avruch J. Rheb binds and regulates the mTOR kinase. Curr Biol 2005;15:702–713.
43. Holz MK, Blenis J. Identification of S6 kinase 1 as a novel mammalian target of rapamycin (mTOR)-phosphorylating kinase. J Biol Chem 2005;280:26089–26093.
44. Kim DH, et al. GβL, a positive regulator of the rapamycin-sensitive pathway required for the nutrient-sensitive interaction between raptor and mTOR. Mol Cell 2003;4:895–904.
45. Loewith R, Jacinto E, Wullschleger S, Lorberg A, Crespo JL, Bonenfant D, Oppliger W, Jenoe P, Hall MN. Two TOR complexes, only one of which is rapamycin sensitive, have distinct roles in cell growth control. Mol Cell 2002; 10:457–468.
46. Brunn SJ, Hudson CC, Sekulic A, Williams JM, Hosoi H, Houghton PJ, Lawrence JC Jr, Abraham RT. Phosphorylation of the translational repressor PHAS-I by the mammalian target of rapamycin. Science 1997; 277:99–101.
47. Burnett PE, Barrow RK, Cohen NA, Snyder SH, Sabatini DM. RAFT1 phosphorylation of the translational regulators p70 S6 kinase and 4E-BP1. Proc Natl Acad Sci USA 1998;95:1432–1437.
48. Schalm SS, Blenis J. Identification of a conserved motif required for mTOR signaling. Curr Biol 2002;12:632–639.
49. Hara K, Maruki Y, Long X, Yoshino K, Oshiro N, Hidayat S, Tokunaga C, Avruch J, Yonezawa K. Raptor, a binding partner of target of rapamycin (TOR), mediates TOR action. Cell 2002;110:177–189.
50. Kim DH, Sarbassov DD, Ali SM, King JE, Latek RR, Erdjument-Bromage H, Tempst P, Sabatini DM. mTOR interacts with raptor to form a nutrient-sensitive complex that signals to the cell growth machinery. Cell 2002;110:163–175.
51. Schalm SS, Fingar DC, Sabatini DM, Blenis J. TOS motif-mediated raptor binding regulates 4E-BP1 multisite phosphorylation and function. Curr Biol 2003;13:797–806.
52. Nojima H, et al. The mammalian target of rapamycin (mTOR) partner, raptor, binds the mTOR substrates p70 S6 kinase and 4E-BP1 through their TOR signaling (TOS) motif. J Biol Chem 2003;278:15461–15464.
53. Mader S, Lee H, Pause A, Sonenberg N. The translation initiation factor eIF-4E binds to a common motif shared by the translation factor eIF-4G and the translational repressor 4E-binding proteins. Mol Cell Biol 1995;15:4990–4997.
54. Pause A, Belsham GJ, Gingras AC, Donze O, Lin TA, Lawrence JC, Sonenberg N. Insulin-dependent stimulation of protein synthesis by phosphorylation of a regulator of 5′-cap function. Nature 1994;371:762–767.
55. Rosenwald IB, Kaspar R, Rousseau D, Gehrke L, Leboulch P, Chen JJ, Schmidt EV, Sonenberg N, London IM. Eukaryotic translation initiation factor 4E regulates expression of cyclin D1 at transcriptional and post-transcriptional levels. J Biol Chem 1995;270:21176–21180.
56. Hashemolhosseini S, Nagamine Y, Morley SJ, Desrivieres S, Mercep L, Ferrari S. Rapamycin inhibition of the G1 to S transition is mediated by effects on cyclin D1 mRNA and protein stability. J Biol Chem 1998; 273:14424–14429.
57. Shantz LM, Pegg AE. Overproduction of ornithine decarboxylase caused by relief of translational repression is associated with neoplastic transformation. Cancer Res 1994;54:2313–2316.
58. DeBenedetti A, Joshi B, Graff JR, Zimmer SG. CHO cells transformed by the translation factor eIF4E display increased c-myc expression but require overexpression of Max for tumorigenicity. Mol Cell Differ 1994;2:347–371.
59. Zhong H, Chiles K, Feldser D, Laughner E, Hanrahan C, Georgescu MM, Simons JW, Semenza GL. Modulation of hypoxia-inducible factor 1 alpha expression by the epidermal growth factor/phosphatidylinositol3-kinase/PTEN/AKT/FRAP pathway in human prostate cancer cells: implications for tumor angiogenesis and therapeutics. Cancer Res 2000;60:1541–1545.
60. Mayerhofer M, Valent P, Sperr WR, Griffin JD, Sillaber C. BCR/ABL induces expression of vascular endothelial growth factor and its transcriptional activator, hypoxia inducible factor-1alpha, through a pathway involving phosphoinositide 3-kinase and the mammalian target of rapamycin. Blood 2002;100:3767–3775.
61. Wang X, et al. Regulation of elongation factor 2 kinase by p90(RSK1) and p70 S6 kinase. EMBO J 2001;20:4370–4379.

62. Harrington LS, et al The TSC1–2 tumor suppressor controls insulin-P13K signaling via regulation of IRS proteins. J Cell Biol 2004;166:213–223.
63. Thomas G, Sabatini DM Hall MN (editors). Tor (Target of Rapamycin). In: Current Topics in Microbiology and Immunology, Vol 279 (2004). Nutrient Signaling Through TOR Kinases Controls Gene Expression and Cellular Differentiation in Fungi J.R. Rohde, M.E. Cardenas p. 53–72.
64. Huang S, Bjornsti MA, Houghton PJ. Rapamycins: Mechanisms of action and cellular resistance. Cancer Biol Ther 2003;2:222–232.
65. Stocker H, et al. Rheb is an essential regulator of S6K in controlling cell growth in Drosophila. Nat Cell Biol 2003;5:559–565.
66. Long X, Ortiz-Vega S, Lin Y, Avruch J. Rheb binding to mammalian target of rapamycin (mTOR) is regulated by amino acid sufficiency. J Biol Chem 2005;280:23433–23436.
67. Tee AR, Manning BD, Roux PP, Cantley LC, Blenis J. Tuberous sclerosis complex gene products, Tuberin and Hamartin, control mTOR signaling by acting as a GTPase-activating protein complex toward Rheb. Curr Biol 2003;13:1259–1268.
68. Gao X, et al. Tsc tumour suppressor proteins antagonize amino-acid-TOR signalling. Nat Cell Biol 2002;4:699–704.
69. Smith EM, Finn SG, Tee AR, Browne GJ, Proud CG. The tuberous sclerosis protein TSC2 is not required for the regulation of the mammalian target of rapamycin by amino acids and certain cellular stresses. J Biol Chem 2005;280:18717–18727.
70. Inoki K, Zhu T, Guan KL. TSC2 mediates cellular energy response to control cell growth and survival. Cell 2003;115:577–590.
71. Woods A, et al. LKB1 is the upstream kinase in the AMP-activated protein kinase cascade. Curr Biol 2003;13:2004–2008.
72. Shaw RJ, Kosmatka M, Bardeesy N, Hurley RL, Witters LA, DePinho RA, Cantley LC. The tumor suppressor LKB1 kinase directly activates AMP-activated kinase and regulates apoptosis in response to energy stress. Proc Natl Acad Sci USA 2004 Mar 9;101(10):3329–3335.
73. Arsham AM, Howell JJ, Simon MC. A novel hypoxia-inducible factor-independent hypoxic response regulating mammalian target of rapamycin and its targets. J Biol Chem 2003:278:29655–29660.
74. Fingar DC, Richardson CJ, Tee AR, Cheatham L, Tsou C, Blenis J. mTOR controls cell cycle progression through its cell growth effectors S6K1 and 4E-BP1/eukaryotic translation initiation factor 4E. Mol Cell Biol 2004;24:200–216.
75. Dilling, MB, et al. 4E-binding proteins, the suppressors of eukaryotic initiation factor 4E, are down-regulated in cells with acquired or intrinsic resistance to rapamycin. J Biol Chem 2002;277:13907–13917.
76. Nourse J, et al. Interleukin-2-mediated elimination of the p27Kip1 cyclin-dependent kinase inhibitor prevented by rapamycin. Nature 1994;372: 570–573.
77. Barata JT, Cardoso AA, Nadler LM, and Boussiotis VA. Interleukin-7 promotes survival and cell cycle progression of T-cell acute lymphoblastic leukemia cells by down-regulating the cyclin-dependent kinase inhibitor p27(kip1). Blood 2001;98:1524–1531.
78. Law BK, et al. Rapamycin potentiates transforming growth factor beta-induced growth arrest in nontransformed, oncogene-transformed, and human cancer cells. Mol Cell Biol 2002;22:8184–8198.
79. Jiang H, Coleman J, Miskimins R, Miskimins WK. Expression of constitutively active 4EBP-1 enhances p27Kip1 expression and inhibits proliferation of MCF7 breast cancer cells. Cancer Cell Int 2003;3:2.
80. Lieberthal W, et al. Rapamycin impairs recovery from acute renal failure: role of cell-cycle arrest and apoptosis of tubular cells. Am J Physiol Renal Physiol 2001;281:693–706.
81. Woltman AM, et al. Rapamycin specifically interferes with GM-CSF signaling in human dendritic cells, leading to apoptosis via increased p27KIP1 expression. Blood 2003;101:1439–1445.
82. Secombe J, Pierce SB, Eisenman RN. Myc: a weapon of mass destruction. Cell 2004;117:153–156.
83. Datta SR, et al. Akt phosphorylation of BAD couples survival signals to the cell-intrinsic death machinery. Cell 1997;91:231–241.
84. Zha J, Harada H, Yang E, Jockel J, Korsmeyer Sj. Serine phosphorylation of death agonist BAD in response to survival factor results in binding to 14–3-3 not BCL-X(L). Cell 1996;87:619–628.
85. Wendel HG, et al. Survival signalling by Akt and eIF4E in oncogenesis and cancer therapy. Nature 2004;428:332–337.

86. Huang S, Liu LN, Hosoi H, Dilling MB, Shikata T, Houghton PJ. p53/p21(CIP1) cooperate in enforcing rapamycin-induced G(1) arrest and determine the cellular response to rapamycin. Cancer Res 2001;61:3373–3381.

87. Huang S, et al. Sustained activation of the JNK cascade and rapamycin-induced apoptosis are suppressed by p53/p21(Cip1). Mol Cell 2003;11:1491–1501.

88. Podsypanina K, et al. An inhibitor of mTOR reduces neoplasia and normalizes p70/S6 kinase activity in Pten+/- mice. Proc Natl Acad Sci USA 2001;98:10320–10325.

89. Neshat MS, et al. Enhanced sensitivity of PTEN-deficient tumors to inhibition of FRAP/mTOR. Proc Natl Acad Sci USA 2001;98:10314–10319.

90. Shi Y, et al. Enhanced sensitivity of multiple myeloma cells containing PTEN mutations to CCI-779. Cancer Res 2002;62:5027–5034.

91. Guba M, et al. Rapamycin inhibits primary and metastatic tumor growth by antiangiogenesis: involvement of vascular endothelial growth factor. Nat Med 2002;8:128–135.

92. Treins C, Giorgetti-Peraldi S, Murdaca J, Semenza GL, Van Obberghen E. Insulin stimulates hypoxia-inducible factor 1 through a phosphatidylinositol 3-kinase/target of rapamycin-dependent signaling pathway. J Biol Chem 2002;277:27975–27981.

93. Zhong H, et al. Modulation of hypoxia-inducible factor 1alpha expression by the epidermal growth factor/phosphatidylinositol 3-kinase/PTEN/AKT/FRAP pathway in human prostate cancer cells: implications for tumor angiogenesis and therapeutics. Cancer Res 2000;60:1541–1545.

94. Zundel W, et al. Loss of PTEN facilitates HIF-1-mediated gene expression. Genes Dev 2000;14:391–396.

95. Jiang BH, et al. Phosphatidylinositol 3-kinase signaling controls levels of hypoxia-inducible factor 1. Cell Growth Differ 2001;12:363–369.

96. Laughner E, Taghavi P, Chiles K, Mahon PC, Semenza GL. HER2 (neu) signaling increases the rate of hypoxia-inducible factor 1alpha (HIF-1alpha) synthesis: novel mechanism for HIF-1-mediated vascular endothelial growth factor expression. Mol Cell Biol 2001;21:3995–4004.

97. Mayerhofer M, Valent P, Sperr WR, Griffin JD, Sillaber C. BCR/ABL induces expression of vascular endothelial growth factor and its transcriptional activator, hypoxia inducible factor-1alpha, through a pathway involving phosphoinositide 3-kinase and the mammalian target of rapamycin. Blood 2002;100:3767–3775.

98. Li J, et al. PTEN, a putative protein tyrosine phosphatase gene mutated in human, brain, breast, and prostate cancer. Science 1997;275:1943–1947.

99. Steck PA, et al. Identification of a candidate tumour suppressor gene, MMAC1, at chromosome 10q23.3 that is mutated in multiple advanced cancers. Nat Genet 1997;15:356–362.

100. Risinger JI, Hayes AK, Berchuck A, Barrett JC. PTEN/MMAC1 mutations in endometrial cancers. Cancer Res 1997;57:4736–4738.

101. Park WS, Moon YW, Yang YM, Kim YS, Kim YD, Fuller BG, Vortmeyer AO, Fogt F, Lubensky IA, Zhuang Z. Mutations of the STK11 gene in sporadic gastric carcinoma. Int J Oncol 1998;3:601–604.

102. Wong AS, Kim SO, Leung PC, Auersperg N, Pelech SL. Profiling of protein kinases in the neoplastic transformation of human ovarian surface epithelium. Gynecol Oncol 2001;82:305–311.

103. Couch FJ, Wang XY, Wu GJ, Qian J, Jenkins RB, James CD. Localization of PS6K to chromosomal region 17q23 and determination of its amplification in breast cancer. Cancer Res 1999;59:1408–1411.

104. Haydon MS, Googe JD, Sorrells DS, Ghali GE, Li BD. Progression of eIF4e gene amplification and overexpression in benign and malignant tumors of the head and neck. Cancer 2000;88:2803–2810.

105. Sorrells DL, Meschonat C, Black D, Li BD. Pattern of amplification and overexpression of the eukaryotic initiation factor 4E gene in solid tumor. J Surg Res 1999;85;37–42.

106. Wang S, et al. Expression of eukaryotic translation initiation factors 4E and 2alpha correlates with the progression of thyroid carcinoma. Thyroid 2001;1:1101–1107.

107. Berkel HJ, Turbat-Herrera EA, Shi R, de Benedetti A. Expression of the translation initiation factor eIF4E in the polyp-cancer sequence in the colon. Cancer Epidemiol Biomarkers Prev 2001;10:663–666.

108. Rosenwald IB, et al. Upregulation of protein synthesis initiation factor eIF-4E is an early event during colon carcinogenesis. Oncogene 1999;18: 2507–2517.

109. Li BD, et al. Prospective study of eukaryotic initiation factor 4E protein elevation and breast cancer outcome. Ann Surg 2002;235:732–738.
110. Crew JP, et al. Eukaryotic initiation factor-4E in superficial and muscle invasive bladder cancer and its correlation with vascular endothelial growth factor expression and tumour progression. Br J Cancer 2000;82: 161–166.
111. Scott PA, et al. Differential expression of vascular endothelial growth factor mRNA vs protein isoform expression in human breast cancer and relationship to eIF-4E. Br J Cancer 1998;77:2120–2128.
112. Ruggero D, et al. The translation factor eIF-4E promotes tumor formation and cooperates with c-Myc in lymphomagenesis. Nat Med 2004;10:484–486.
113. Martin ME, et al. 4E binding protein 1 expression is inversely correlated to the progression of gastrointestinal cancers. Int J Biochem Cell Biol 2000;32:633–642.
114. Choi J, Chen J, Schreiber SL, Clardy, J. Structure of the FKBP12-rapamycin complex interacting with the binding domain of human FRAP. Science 1996;273:239–242.
115. Metcalf CEA, et al. Structure-based dosing of AP23573, a phosphorous-containing analog of rapamycin for anti-tumor therapy. Proc Am Assoc Cancer Res. Orlando, Florida. 2004;ab:2476.
116. Douros J, Suffness M. New antitumor substances of natural origin. Cancer Treat Rev 1981;8:63–87.
117. Houchens DP, Ovejera AA, Riblet SM, Slagel DE. Human brain tumor xenografts in nude mice as a chemotherapy model. Eur J Cancer Clin Oncol 1983;19:799–805.
118. Eng CP, Sehgal SN, Vezina C. Activity of rapamycin (AY-22,989) against transplanted tumors. J Antibiot (Tokyo) 1984;37:1231–1237.
119. Dilling MB, Dias P, Shapiro DN, Germain GS, Johnson RK, Houghton PJ. Rapamycin selectively inhibits the growth of childhood rhabdomyosarcoma cells through inhibition of signaling via the type I insulin-like growth factor receptor. Cancer Res 1994;54:903–907.
120. Shi Y, Frankel A, Radvanyi LG, Penn LZ, Miller RG, Mills GB. Rapamycin enhances apoptosis and increases sensitivity to cisplatin in vitro. Cancer Res 1995;55:1982–1988.
121. Seufferlein T, Rozengurt E. Rapamycin inhibits constitutive $p70^{s6k}$ phosphorylation, cell proliferation, and colony formation in small cell lung cancer cells. Cancer Res 1996;56:3895–3897.
122. Hosoi H, Dilling MB, Liu LN, Danks MK, Shikata T, Sekulic A, Abraham RT, Lawrence JC Jr, Houghton PJ. Studies on the mechanism of resistance to rapamycin in human cancer cells. Mol Pharmacol 1998;54:815–824.
123. Hosoi H, Dilling MB, Shikata T, Liu LN, Shu L, Ashmun RA, Germain GS, Abraham RT, Houghton PJ. Rapamycin causes poorly reversible inhibition of mTOR and induces p53-independent apoptosis in human rhabdomyosarcoma cells. Cancer Res 1999;59:886–894.
124. Geoerger B, Kerr K, Tang CB, Fung KM, Powell B, Sutton LN, Phillips PC, Janss AJ. Antitumor activity of the rapamycin analog CCI-779 in human primitive neuroectodermal tumor/medulloblastoma models as single agent and in combination chemotherapy. Cancer Res 2001;61:1527–1532.
125. Ogawa T, Tokuda M, Tomizawa K, Matsui H, Itano T, Konishi R, Nagahata S, Hatase O. Osteoblastic differentiation is enhanced by rapamycin in rat osteoblast-like osteosarcoma (ROS 17/2.8) cells. Biochem Biophys Res Commun 1998;249:226–230.
126. Grewe M, Gansauge F, Schmid RM, Adler G, Seufferlein T. Regulation of cell growth and cyclin D1 expression by the constitutively active FRAP-S6K1 pathway in human pancreatic cancer cells. Cancer Res 1999; 59:3581–3587.
127. Shah SA, Potter MW, Ricciardi R, Perugini RA, Callery MP. Frap-S6K1 signaling is required for pancreatic cancer cell proliferation. J Surg Res 2001;97:123–130.
128. Gibbons JJ, Discafani C, Peterson R, Hernandez R, Skotnicki J, Frost P. The effect of CCI-779, a novel macrolide anti-tumor agent, on the growth of human tumor cells in vitro and in nude mouse xenografts in vivo. Proc Am Assoc Cancer Res 1999;40:301.
129. Yu K, Zhang W, Lucas J, Toral-Barza L, Peterson R, Skotnicki J, Frost P, Gibbons J. Deregulated PI3K/AKT/TOR pathway in PTEN-deficient tumor cells correlates with an increased growth inhibition sensitivity to a TOR kinase inhibitor CCI-779. Proc Am Assoc Cancer Res 2001;42:802.
130. Yu K, Toral-Barza L, Discafani C, Zhang WG, Skotnicki J, Frost P, Gibbons JJ. mTOR, a novel target in breast cancer: the effect of CCI-779, an mTOR inhibitor, in preclinical models of breast cancer. Endocr Relat Cancer 2001;8:249–258.

131. Hultsch T, Martin R, Hohman RJ. The effect of the immunophilin ligands rapamycin and FK506 on proliferation of mast cells and other hematopoietic cell lines. Mol Biol Cell 1992;3:981–987.

132. Gottschalk AR, Boise LH, Thompson CB, Quintans J. Identification of immunosuppressant-induced apoptosis in a murine B-cell line and its prevention by bcl-x but not bcl-2. Proc Natl Acad Sci USA 1994;91:7350–7354.

133. Muthukkumar S, Ramesh TM, Bondada S. Rapamycin, A potent immunosuppressive drug, causes programmed cell death in B lymphoma cells. Transplantation 1995;60:264–270.

134. Mateo-Lozano S, Tirado OM, Notario V. Rapamycin induces the fusion-type independent downregulation of the EWS/FLI-1 proteins and inhibits Ewing's sarcoma cell proliferation. Oncogene 2003;22:9282–9287.

135. Dancey JE. Clinical development of mammalian target of rapamycin inhibitors. Hematol Oncol Clin North Am 2002;16:1101–1114.

136. Atkins MB, et al. Randomized phase II study of multiple dose levels of CCI-779, a novel mammalian target of rapamycin kinase inhibitor, in patients with advanced refractory renal cell carcinoma. J Clin Oncol 2004;22:909–918.

137. Witzig TE, et al. Phase II trial of single-agent temsirolimus (CCI-779) for relapsed mantle cell lymphoma. J Clin Oncol 2005;23:5347–5356.

138. Chan S, et al. Phase II study of temsirolimus (CCI-779), a novel inhibitor of mTOR, in heavily pretreated patients with locally advanced or metastatic breast cancer. J Clin Oncol 2005;23:5314–5322.

139. O'Donnell A, et al. A phase I study of the oral mTOR inhibitor RAD001 as monotherapy to identify the optimal biologically effective dose using toxicity, pharmacokinetic (PK) and pharmacodynamic (PD) endpoints in patients with solid tumours. Proc Am Soc Clin Oncol 2003;22:200.

140. Boulay A, et al. Antitumor efficacy of intermittent treatment schedules with the rapamycin derivative RAD001 correlates with prolonged inactivation of ribosomal protein S6 kinase 1 in peripheral blood mononuclear cells. Cancer Res 2004;64:252–261.

141. Raje N, et al. Combination of the mTOR inhibitor rapamycin and CC-5013 has synergistic activity in multiple myeloma. Blood 2004;104:4188–4193.

142. Gorre ME, Mohammed M, Ellwood K, Hsu N, Paquette R, Rao PN, Sawyers CL. Clinical resistance to STI-571 cancer therapy caused by BCR-ABL gene mutation or amplification. Science 2001;293:876–880.

143. Mohi MG, et al. Combination of rapamycin and protein tyrosine kinase (PTK) inhibitors for the treatment of leukemias caused by oncogenic PTKs. Proc Natl Acad Sci USA 2004;101:3130–3135.

144. Gemmill RM, Zhou M, Costa L, Korch C, Bukowski RM, Drabkin HA. Synergistic growth inhibition by Iressa and Rapamycin is modulated by VHL mutations in renal cell carcinoma. Br J Cancer 2005;92:2266–2277.

145. Edwards E, Geng L, Tan J, Onishko H, Donnelly E, Hallahan DE. Phosphatidylinositol 3-kinase/Akt signaling in the response of vascular endothelium to ionizing radiation. Cancer Res 2002;62:4671–4677.

146. Tan J, Hallahan DE. Growth factor-independent activation of protein kinase B contributes to the inherent resistance of vascular endothelium to radiation-induced apoptotic response. Cancer Res 2003;63:7663–7667.

147. Shinohara ET, Cao C, Niermann K, Mu Y, Zeng F, Hallahan DE, Lu B. Enhanced radiation damage of tumor vasculature by mTOR inhibitors. Oncogene 2005;24:1–9.

148. Sawyers CL. Finding the next Gleevec: FLT3 targeted kinase inhibitor therapy for acute myeloid leukemia. Cancer Cell 2002;5:413–415.

149. Hentges KE, et al. FRAP/mTOR is required for proliferation and patterning during embryonic development in the mouse. Proc Natl Acad Sci USA 2001;98:13796–13801.

150. Hentges K, Thompson K, Peterson A. The flat-top gene is required for the expansion and regionalization of the telencephalic primordium. Development 1999;126:1601–1609.

3   Ras/Raf/MEK Inhibitors

Joseph A. Sparano, MD

SUMMARY

Signal transduction is a complex process that involves a network of molecules that facilitate communication within, between, and among cells and their environment. Molecules involved in signal transduction include receptor tyrosine kinases (RTKs), non-receptor tysrosine kinases (NTRKs), serine-threonine kinases (STKs), and G proteins. These molecules influence a variety of cell processes critical for cellular proliferation, apoptosis, motility, and other biological processes. This has led to the rational development of inhibitors targeting specific pathways, including Ras, Raf, and MEK proteins. Tipifarnib, an inhibitor of the Ras (a G protein), has activity in acute leukemia, and may enhance the effects of cytotoxic therapy in patients with locally advanced breast cancer. Sorafenib, an inhibitor of the Raf (an STK), has activity in advanced renal cell carcinoma, although its effects may be mediated in part by its ability to inhibit other pathways. Inhibitors of MEK (an STK) enhance the effects of cytotoxic therapy in preclinical systems. These and other agents in development hold promise for application in a variety of disease types, used alone or in combination with other therapies.

Key Words: Ras; Raf; MEK; inhibitors.

1. INTRODUCTION

Intracellular signaling pathways transduce signals from cell surface to the nucleus that modulate cell proliferation, cell death (apoptosis), and a variety of other biological processes *(1,2)*. Many cancers have acquired genetic abnormalities that result in aberrant expression and activity of many signaling molecules *(3–5)*. Receptor protein tyrosine kinases, non-receptor protein tyrosine kinases, serine/threonine kinases (STKs), and G proteins are among the most important mediators of signal transduction (Table 1). Many of the key proteins involved in signal transduction have kinase activity, and the protein kinase complement of the human genome has recently been catalogued *(6,7)*. Protein kinases mediate most signal transduction in eukaryotic cells by modifying substrate activity, thereby influencing a variety of cellular processes including metabolism, transcription, cell cycle progression, cytoskeletal arrangement, cell motility, apoptosis, and differentiation. They also play a key role in intercellular communication during normal physiologic process such as the development of nervous

From: *Cancer Drug Discovery and Development: Molecular Targeting in Oncology*
Edited by: H. L. Kaufman, S. Wadler, and K. Antman © Humana Press, Totowa, NJ

Table 1
Overview of Signaling Molecules

Class	Ligand-binding domain	Transmembrane domain	Intra-cellular domain	Examples
Receptor tyrosine kinases	+	+	+	EGFR, Her/2, Her3, Her4, VEFGR, PDGFR
Non-receptor tyrosine kinases	–	–	+	Src, abl, JAK
Serine/threonine kinases	–	–	+	Raf, Akt, MEK
G proteins	–	–	+	Ras, Rho, Rab, Sar1/Arf, Ran

EGFR, epidermal growth factor receptor; HER, human epidermal growth factor receptor; PDGFR, platelet-derived growth factor receptor; VEGFR, vascular endothelial growth factor receptor.

and immune systems. Growth promoting signals typically activate signaling through the small G protein Ras, resulting in sequential activation of Raf, mitogen-activated protein kinase (MAPK), and finally extracellular signal-regulated kinase (ERK). This pathway is often referred to as the "Ras-MEK-ERK" pathway *(8)*.

Receptor tyrosine kinases (RTKs) are glycoproteins that have an extracellular ligand-binding domain, a hydrophobic transmembrane domain, and an intracellular catalytic domain. There are currently at least 19 known families, with examples including the human epidermal growth factor receptor family (HER 1–4), vascular endothelial growth factor receptor (VEGFR), and platelet-derived growth factor receptor (PDGFR). Ligand binding to the extracellular domain results in receptor homodimerization and/or heterodimerization, which in turn leads to tyrosine phosphorylation that promotes downstream signaling. Nonreceptor tyrosine kinases (NRTKs) are cytoplasmic proteins that lack transmembrane domains. NRTKs transmit extracellular signals to downstream intermediates by binding to activated receptors. There are currently 10 known families of NRTKs, with examples including Src, abl, and JAK. STKs are almost all intracellular and include key mediators of carcinogenesis such as Raf and AKT/protein kinase B (PKB). Finally, the G protein superfamily is structurally classified into at least five families, including the Ras, Rho, Rab, Sar1/Arf, and Ran families. Ras proteins function as intracellular molecular switches linking RTK and NRTK-mediated activation to downstream cytoplasmic and nuclear events.

A simplified schema of critical pathways involved in signal transduction is shown in Fig. 1. Such pathways that offer the potential for therapeutic intervention include (i) the Ras-Raf-MEK pathway, which promotes cell proliferation induced by estrogen, insulin-like growth factors (IGFs), and epidermal growth factor (EGF) families; (ii) the stress response pathways mediated by the stress-activated protein kinase *c-jun* amino-terminal kinase (JNK) and p38 MAPKs; and (iii) the phosphatidylinositol 3′-kinase (PI3K) and AKT (also known as PKB) pathway, which promotes cell survival in response to growth factors (including the IGFs and members of the EGF family). Many rationally designed anticancer therapies that target specific aberrant elements in

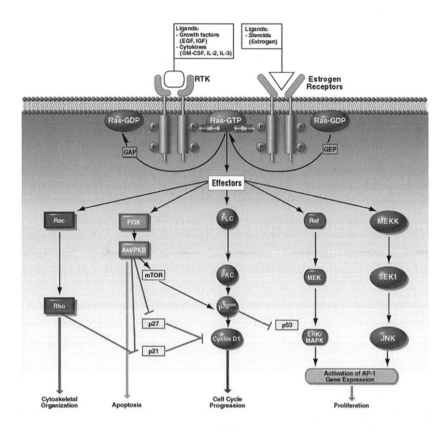

Fig. 1. Simplified schema of signal transduction pathways.

signal transduction pathways are currently in various stages of preclinical or clinical evaluation *(9,10)*. This chapter will focus on the Ras/Raf/MEK pathway as a potential target for therapeutic intervention, and those agents that have undergone phase II and III testing in the clinic.

2. RAS

2.1. Role of Ras and Other G Proteins in Signal Transduction

Ras proteins are guanine nucleotide-binding proteins that belong to the small guanosine triphosphate (GTP)-binding protein (G protein) superfamily *(11)*. This G protein superfamily consists of more than 100 members that are structurally classified into at least five families, including the Ras, Rho, Rab, Sar1/Arf, and Ran families. Most of small G proteins are widely distributed in mammalian cells, although expression levels vary in different cell types. All G proteins have consensus amino acid sequences for specific binding with guanosine diphosphate (GDP) and GTP, for GTPase activity (which hydrolyzes bound GTP to GDP and Pi) and for interacting with downstream effectors. These small G proteins regulate a wide variety of cellular functions. Specifically, the Ras family regulates gene expression in normal cell growth and differentiation, the Rho family regulates cytoskeletal reorganization and gene expression, the Rab and Sar1/Arf families regulate vesicle trafficking, and the Ran family regulates nucleocytoplasmic transport and microtubule organization. Many upstream regulators and downstream effectors of small G proteins have been identified, and their modes of activation and cross-talk have gradually been elucidated. They are not only closely related structurally but also can be physiologically regulated by the same signals *(11,12)*.

One of the five G protein families is Ras, which consists of four 21-kilodalton proteins (H-Ras, N-Ras, and K-Ras4A and K-Ras4B) that are encoded by three *ras* proto-oncogenes: the *H-ras* gene (homologous to the oncogene of the Harvey murine sarcoma virus), the *K-ras* gene (homologous to the oncogene of the Kirsten murine sarcoma virus), and the *N-ras* gene (which does not have a retroviral homologue and was first isolated from a neuroblastoma cell line). Ras proteins contain 188 or 189 amino acids that have 50–55% sequence homology from various species. The first 86 amino acids are identical across the mammalian species, the next 78 amino acids have 79% homology, and the following 25 amino acids are highly variable but are important for posttranslational modification *(11)*. Ras proteins are synthesized in the cytosol as an inactive precursor. Posttranslational modification is crucial for their action. Members of Ras, Rho/Rac/Cdc42, and Rab families have sequences at their carboxyl termini that undergo posttranslational modification by proteolysis and addition of lipids (i.e., prenylation). Prenylation is the most important posttranslational modification that renders the Ras protein hydrophobic and favors localization to the inner surface of the plasma membrane, where it mediates its biological effects. Prenylation is the covalent addition of either a farnesyl (15-carbon) or a geranylgeranyl (20-carbon) group to the cysteine residue located in the so-called CAAX tetrapeptide that is found at the carboxy terminus of small G proteins (Fig. 2). Three classes of isoprenyltransferase enzymes have been identified in mammalian cells, including protein farnesyltransferase (FTase), type I protein geranylgeranyltransferase (GGTase-I), and type II protein geranylgeranyltransferase (GGTase-II). In the CAAX tetrapeptide, C represents a cysteine residue, A represents aliphatic amino acids (usually valine, leucine, or isoleucine), and X represents may represent different amino acids that influence how the protein is modified. For example, FTase catalyzes farnesylation of proteins in which X is methionine, serine, alanine, glutamine, or cysteine (such as Ras, Lamin B, Rho B) and GGTase-I catalyzes geranylgeranylation of proteins in which X is leucine, isoleucine, or phenylalanine (such as Rho, Rap, and Rac). GGTase-II catalyzes the geranylgeranylation of sequences CXC, CCX, or XXCC (e.g., Rab proteins) *(13–15)*.

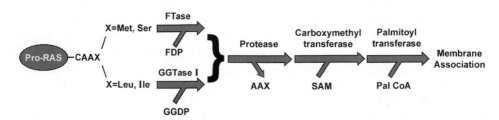

Fig. 2. Prenylation pathways of Ras.

2.2. Regulation of Ras Signaling

The biochemical output of Ras proteins is tightly regulated by their ability to cycle between an active GTP-bound state (Ras-GTP) and an inactive GDP-bound state (Ras-GDP). Stimulation by extracellular growth factors is required in normal cells to maintain wildtype Ras in an activated state; otherwise, it reverts rapidly to the inactive form. In normal cells, the activity of Ras protein is tightly regulated by guanine nucleotide exchange proteins (GEPs) and GTPase-activating proteins (GAPs) in response to a large variety of extracellular signals, such as growth factors, cytokines, and hormones. Wildtype Ras has low intrinsic GTPase activity that can be enhanced by GAPs to convert active Ras to the inactive form. GAPs are therefore negative regulators of Ras signaling, whereas GEPs positively regulate Ras signaling. Losses of GAP function have been implicated in some human disease states (e.g., von Recklinghausen neurofibromatosis) *(11)*.

Ras-GTP activates several downstream effector pathways including the Raf/MEK/ERK, MEKK/JNK, PI3K/Akt, PLC/PKC/cyclin D1, and Rac/Rho cascades. Activation by the Ras pathway is modulated by extracellular signals. Typically, the cell surface receptors for the growth factors proximal to Ras are RTKs that are dimerized and autophosphorylated following the binding of growth factors. One of the phosphorylated tyrosine residues on the cytoplasmic domain of activated receptor binds to growth factor receptor-binding protein (Grb2) that couples with Ras GEPs through its *src*-homology 2 (SH2)-binding and *src*-homology 3 (SH3)-binding domains. The stable complex of phosphorylated tyrosine residue, Grb2, and Ras GEPs couples the Ras protein to the plasma membrane-bound RTKs.

2.3. Mutations in Ras Proteins in Human Cancer

Oncogenic mutations of the three known human *ras* genes are found in 30% of all human cancers *(16,17)*. Mutations are common in certain types of gastrointestinal cancers (e.g., pancreas 90% and colorectal 50%), uncommon in other cancer types (e.g., cervical 6% and breast 2%), and intermediate in other types (e.g., selective hematologic malignancies 20–65%). Most of the mutations in *ras* genes are missense mutations at codons 12, 13, and 61 and in exons 1 and 2. Each of these amino acid residues participates in GTP binding, and amino acid substitutions result in stabilization of the active GTP-bound form of Ras. These mutated Ras proteins have an intrinsic defect in GTP hydrolysis and are markedly resistant to degradation by GAPs, thus leading to constitutive activation of the Ras pathway that confers a proliferative advantage to the cell. In preclinical models, expression of these mutant *ras* genes is oncogenic *(16–18)*.

In addition to activation by point mutations in Ras genes, many cancer-related and non-cancer-related mutations can also induce hyperactive Ras. For example, although the frequency of Ras mutations in breast cancer is very low (<2%) *(19,20)*, aberrant function of the Ras signal transduction pathway occurs in HER-2/*neu* overexpressing *(21)* and estrogen-dependent *(22)* breast cancer models. In addition, Ras protein overexpression (not associated with *ras* mutations) is associated with poor prognosis of breast cancer *(23,24)* and may represent a late event that occurs due to perturbation of other pathways *(25)*. Furthermore, upstream events may lead to activation of the Ras pathway without Ras protein overexpression *(26–28)*.

2.4. Rationale and Mechanisms for Targeting the Ras Pathway and Farnesyl Transferase

As a hyperactive Ras pathway is common in human cancer, there is obvious rationale for targeting this pathway for therapeutic intervention *(12,13,17,29)*. At least three different strategies have been developed targeting an activated Ras pathway, including (i) blocking upstream activation of Ras at the cell surface receptors [such as estrogen receptor (ER), HER2/*neu*, and RTK]; (ii) targeting Ras itself by inhibiting either *ras* gene expression (i.e., antisense molecules) or interrupting protein processing [such as FTase inhibitor (FTI) or a geranylgeranyl transferase inhibitor]; and (iii) inhibiting downstream effector pathways (e.g., Raf kinase and MEK inhibitors) *(30)*. Several of these approaches are currently being evaluated in clinical trials *(12,13,17,29)*.

Most studies to date have focused on targeting Ras by inhibiting farnesylation of Ras protein using a FTI. FTase is a heterodimeric zinc metalloenzyme composed of a 48-kilodalton alpha-subunit and 46-kilodalton beta-subunit. There are many substrates for FTase in the mammalian cells. Most prenylated proteins are members of signal transduction cascades that play important roles in normal cellular functions. A few examples include proteins involved in cytoskeletal organization (Rho), nuclear membrane structure (lamins A and B), chaperone function (heat shock protein 40, also called HDJ-2, for human homolog of the DNA-J heat shock protein found in *Escherichia coli*), mitosis [centromere-associated proteins E and F], proteins involved in visual function (cyclic guanosine monophosphate phosphodiesterase alpha, rhodopsin kinase, and transducin gamma), platelet function (Rap2), and skeletal muscle function (phosphorylase kinase alpha and beta) *(13,29,31–33)*. Therefore, inhibitors of farnesyl transferase not only have the potential for altering other signaling pathways within cancer cells, but also have the potential for producing organ-specific toxicity. Several commonly used drugs, such as statins (an inhibitor of HMG-CoA reductase)*(34–37)* and aminobisphosphonates (such as alendronate, pamidronate, zoledronate, risedronate, and ibandronate) *(38,39)*, can also inhibit isoprenylation of small G proteins, including Ras, without prohibitive toxicity.

2.5. Farnesyl Transferase Inhibitors in the Clinic

FTIs have been classified into three subclasses, including (i) farnesyl pyrophosphate analogs (non-peptidomimetics), which compete with the isoprenoid substrates for FTase; (ii) peptidomimetic inhibitors, which mimic the structure of CAAX portion of Ras and compete with Ras for FTase; and (iii) bisubstrate analogs, which combine the properties of both *(13,17,31)*. These drugs are in various stages of clinical development. Among them, two oral FTIs that have been most extensively studied in clinical trials ranging from phase I to phase III trials included tipifarnib (formerly R115777, Zarnestra™; Johnson & Johnson Pharmaceutical Research and Development, L.L.C., Raritan, NJ) (Fig. 3) and lonafarnib (formerly SCH66336, Sarasar®; Schering-Plough, Inc. Kenilworth, NJ) (Fig. 4). A number of other FTIs that have been evaluated in phase I and II trials are no longer being developed *(40)*.

2.6. Phase I Trials of FTIs

Zujewski et al. reported the first phase I trial of tipifarnib given for 5 consecutive days every 2 weeks *(44)*. The maximum tolerated dose was 500 mg twice daily, with

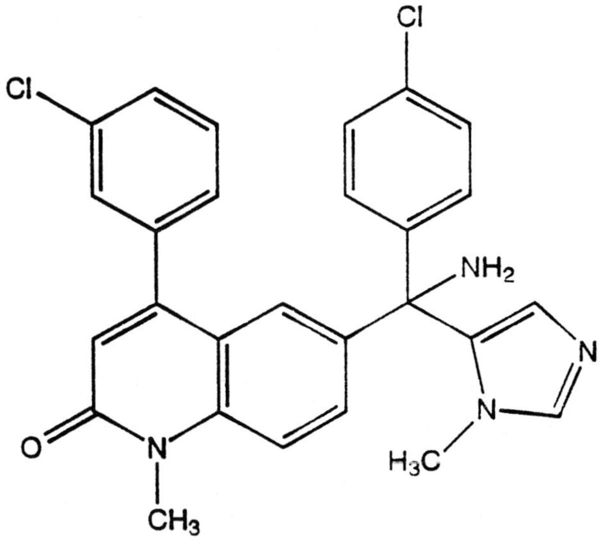

Fig. 3. Chemical structure of tipifarnib. Reproduced with permission from ref. 44.

nausea, vomiting, headache, fatigue, anemia, and hypotension being dose limiting. Peak plasma concentration occurred 0.5–4.0 h after oral administration, elimination was biphasic (sequential half-lives of about 5 and 16 h), and steady state concentrations were achieved in 2–3 days with little drug accumulation. Many phase I trials with single-agent FTIs with different dosing and schedules have been subsequently published and reviewed extensively elsewhere *(64)*. The recommended phase II doses (RPTDs) for tipifarnib is dependent on schedule: (i) 300 mg twice daily either 3 weeks on/1 week off, or continuously *(41)*, or higher doses (600 mg twice daily) given continuously if neutropenia was not considered dose limiting but rather desirable (e.g., leukemia) *(42)*, (ii) 240 mg/m^2 twice daily for 21 consecutive days *(43)*, or (iii) 500 mg twice daily for 5 consecutive days every 14 days *(44)*, or *(4)* 600 mg twice a day for 7 consecutive days every 14 days. The RPTD for lonafarnib is 200 mg twice a day continuously

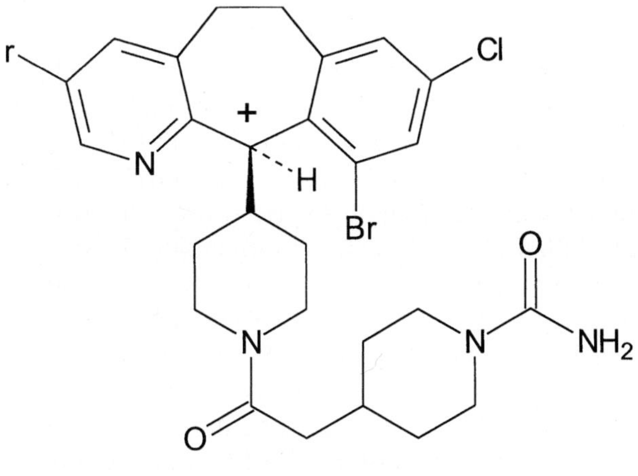

Fig. 4. Chemical structure of lonafarnb. Reproduced with permission from ref.45.

(45), 300 mg once a day continuously *(46)*, 350 mg twice a day for 7 consecutive days every 21 days *(47)*, or 245 mg/m^2 intravenously once weekly *(48)*. Thus, there is a trend of delivering higher tolerated dose of FTIs when the FTIs are given intermittently rather than continuously. In general, dose-limiting toxicities include neutropenia, thrombocytopenia, gastrointestinal toxicity, peripheral neuropathy, and fatigue. For tipifarnib, intermittent dosing is associated with less toxicity than continuous dosing *(49)*. Clinical response has been observed in a variety of tumor types in phase I trials, including acute myelogenous leukemia, myeloproliferative disorders, breast carcinoma, glioma, pancreatic carcinoma, colon, and non-small cell lung cancer *(40)*. The myelosuppression caused by FTIs in some studies has limited the dose escalation of the FTI and/or the cytotoxic agent(s). In addition, tipifarnib has also been shown to inhibit in vitro the metabolism of specific CYP3A4, CYP2D6, and CYP2C8/9/10 isoenzymes, possibly indicating a potential interaction with co-administrated drugs that are primarily metabolized by cytochrome P450 *(50,51)*.

Several phase I studies have evaluated surrogates for FT inhibition. Preclinical data suggested that the increased ratios of unfarnesylated to farnesylated proteins (e.g., heat shock protein HDJ-2 or the intranuclear intermediate filament protein lamin A) are indicative of impaired farnesylation in response to FTI treatment *(52)*. This has been borne out in some clinical trials. For example, Britten et al. *(53)* reported that L-778,123 administered intravenously in patients with advanced cancer produced a dose-dependent increase in the mean percentage of unprenylated HDJ2 in peripheral blood mononuclear cells (PBMCs), increasing from only 5% at baseline to as high as 35% at the highest doses. Likewise, Haas et al. *(54)* reported that tipifarnib resulted in an increase in unprenylated HDJ2 in PBMC to as high as 50%. Kelland et al. *(55)* reported san increase in prelamin A in PBMCs from a patient with metastatic breast cancer who responded to tipifarnib, but no such increase in non-responders.

2.7. Phase II Trials of FTIs

A number of phase II trials have been performed in a variety of diseases. Responses have been noted in patients with breast cancer *(49)*, malignant glioma *(56)*, multiple myeloma *(57,58)*, myelodyspastic syndrome *(59)*, and acute myelogenous leukemia *(60)*; no activity has been observed in carcinomas of the lung (small cell and non-small cell lung) *(61,62)*, bladder *(63)*, and prostate *(54)*.

A promising level of activity has been noted for FTIs in myelodysplastic syndrome (MDS), elderly patients with acute myeloid leukemia, and breast cancer. Kurzrock *(59)* treated 27 assessable patients with MDS with tipifarnib (600 mg orally bid in cycles of 4 weeks of therapy followed by a 2-week rest period); there were three responses including two complete and one partial response. Two of the responders had a diploid karyotype and one had multiple cytogenetic abnormalities including monosomy 5 and 7. The starting dose of 600 mg PO bid resulted in side effects (myelosuppression, fatigue, neurotoxicity, rash, or leg pain) necessitating dose reduction (*n* = 4) or discontinuation of therapy (*n* = 7) in 11 (41%) of 27 patients during the induction period (12 weeks). Lower doses of 300 mg PO bid were well tolerated. All responses occurred in patients who had been reduced to this dose level during the initial two cycles. Additional studies are ongoing evaluating the role of tipifarnib in MDS and elderly patients with acute leukemia *(64)*.

With regard to breast cancer, Johnston reported a phase II trial of tipifarnib in patients with ER-positive metastatic breast cancer who had progressive disease after second-line endocrine therapy or with ER-negative disease *(49)*. Seventy-six patients received either 400 mg (*n* = 7) or 300 mg (*n* = 34) twice daily on a continuous schedule, or 300 mg bid using a 3-week on, 1-week off intermittent schedule (*n* = 35). The clinical benefit rate (partial response or stable for at least 24 weeks) was comparable in the continuous (25%) and intermittent schedules (24%). There was no statistical association between response to tipifarnib and tumor characteristics (such as the status of ER, HER2, and mutation in three *ras* genes). Sites of response occurred in liver, lung, pleura, lymph nodes, breast, and skin nodules. There was significantly less toxicity associated with the intermittent compared with the continuous schedule, including neutropenia, thrombocytopenia, and neurotoxicity. Although there was high interpersonal variability in pharmacokinetics of tipifarnib, no significant differences were observed between the two dosing regimens. Daily area under the curve plasma concentration was found to be a better predictor for severe neutropenia than the administered daily dose. Based upon these results, and preclinical data suggesting that FTIs may enhance the effects of antiestrogenic therapy *(65)*, a multicenter, randomized phase II trial was performed in 121 postmenopausal patients with advanced ER-positive breast cancer that had progressed after tamoxifen; patients were randomized (2:1) to receive the aromatase inhibitor letrozole (2.5 mg daily) in combination with either a placebo (*n* = 40) or in combination with tipifarnib (*n* = 81) *(66)*. The dose and schedule of tipifarnib was 300 mg bid given for 21 of 28 days. Seventy percent of patients relapsed while receiving adjuvant tamoxifen, and 30% patients progressed on tamoxifen therapy as first-line treatment for metastatic disease. The dominant site of metastasis was visceral in 57%, soft tissue in 36%, and bone in 7% of patients. Objective response rate occurred in 38% [95% confidence interval (95% CI); 23%, 55%] for the letrozole arm and 26% (95% CI; 16%, 37%) in the tipifarnib-letrozole arm. The median duration of objective response was similar in the two treatment arms (16.0 vs. 14.8 months), and an equal proportion had stable disease for at least 24 weeks (23% in both groups). The clinical benefit rate was 62% (95% CI; 45%, 77%) in the letrozole alone arm and 49% (95% CI; 37%, 61%) in the tipifarnib-letrozole arm. There was no difference in time to disease progression or overall survival, and 30 patients remained on treatment at the time of the analysis. Although the results of these trials do not demonstrate a sufficient level of single agent activity for tipifarnib, nor benefit when combined with antiestrogen therapy, other evidence suggests that tipifarnib may increase the chance of having a pathological complete response (pCR) after preoperative chemotherapy for operable breast cancer. pCR is a surrogate endpoint that has been associated with improved disease-free survival and overall survival following neoadjuvant breast cancer chemotherapy *(67)*. Sparano et al. *(68)* reported that 7 of 21 patients (33%; 95% CI, 15%, 55%) with locally advanced breast cancer treated with standard doxorubicin (60 mg/m^2) and cyclophosphamide (600 mg/m^2) every 2 weeks plus tipifarnib (200 mg PO bid on days 2–7 of each cycle) plus granulocyte-colony stimulating factor, exceeding the expected 5–10% pCR rate that would have been expected in this combination with chemotherapy alone. Five patients who underwent serial biopsy of the breast tumor before and after tipifarnib demonstrated at least 50% FTase enzyme inhibition in the primary tumor (median 100%, range 55–100%) after 6 days of tipifarnib administration, indicating that

Table 2
Randomized Phase III Trials of Farnesyl Transferase Inhibitors

Reference	Disease	Phase	Agents	No.	Median PFS (months)	Median OS (months)
Van Cutsem, 2002 (69)	Pancreatic carcinoma	III	Gemcitabine + Tipifarnib Gemcitabine + Placebo	341 347	3.7 3.6	6.4 6.1
Rao, 2004 (70)	Colorectal carcinoma	III	Tipifarnib Placebo	235 133	2.7 2.7	5.7 6.1

PFS, progression-free survival; OS, overall survival.

the FTI was effectively inhibiting the target enzyme in vivo. Additional confirmatory studies are ongoing.

2.8. Phase III Trials of FTIs

Two phase III trial studies have been reported for tipifarnib (Table 2). Gastrointestinal cancers were selected in these trials because of the high prevalence of Ras mutations in pancreatic (90%) and colorectal cancers (50%). In one trial, gemcitabine was given with either in combination with tipifarnib (200 mg twice a day continuously) or a placebo in 688 patients with untreated, locally advanced, or metastatic pancreatic carcinoma (69). There was no difference in median progression-free survival (PFS) (3.7 versus 3.6 months; $p = 0.72$) or overall survival (6.4 versus 6.1 months, $p = 0.75$). A second trial compared tipifarnib (300 mg twice a day given for 21 consecutive days every 28 days) with a placebo in 368 patients with metastatic colorectal carcinoma who had progressive disease after two prior chemotherapy regimens (70). There was no significant difference in median PFS (2.7 months in both arms) or overall survival (5.7 months versus 6.1 months, $p = 0.396$). Another study is currently evaluating standard carboplatin/paclitaxel chemotherapy used alone or in combination with lonafarnib in patients with advanced non-small cell lung cancer.

3. RAF

3.1. Raf Signaling and Mutations in Human Cancer

Raf is an STK that is the principal effector of Ras. Raf may also be activated constitutively (by mutation) or by Ras-independent elements (as reviewed in ref. (71)). There are three 68-kilodalton to 74-kilodalton cytosolic Raf proteins (A-Raf, B-Raf, and C-Raf). Raf proteins are encoded by the *raf* proto-oncogenes, *A-raf*, *B-raf*, and *C-raf*, which are located on chromosomes Xp11, 7q32, and 3p25, respectively. They share highly conserved regions including CR1 (adjacent to the amino terminus), CR2, and CR3 (adjacent to the carboxyl terminus). The CR1 region contains the regulatory domain and the CR3 region the kinase domain; there is also an activation loop. GTP-bound Ras interacts directly with Raf through its Ras-binding domain in the amino-terminal regulatory region; there is an adjacent zinc-binding cysteine-rich domain of CR1, facilitating recruitment of Raf to the cell membrane for activation. A-Raf is overexpressed in urogenital tissues (e.g., kidney, ovary, prostate, and epididymis), B-Raf is overexpressed in neural, testicular, splenic, and hematopoietic tissues, and C-Raf is ubiquitously expressed in most tissues. There are constitutively active mutant Raf proteins that occur due to point (missense) mutations, deletions, amplification, and rearrangements of *raf*. The mutations have been identified in malignant melanoma, hematopoietic cancers, and a variety of solid tumors (including thyroid, breast, kidney, liver, larynx, and biliary tract cancers). Mutations of *B-raf* and *ras* only rarely occur in the same tumor. C-raf mutations are less common and may be produced by a variety of genetic alternations.

3.2. Phase I Studies of Raf Kinase Inhibitors

Sorafenib (previously known as BAY43-9006) is a novel bi-aryl urea inhibitor of Raf-1, suppressing both wildtype and V599E mutant BRAF activity in vitro, and several RTKs involved in angiogenesis and tumor progression, including VEGFR-2, VEGFR-3, PDGFR-beta, Flt-3, and c-KIT *(72)*. Sorafenib (Fig. 5) also inhibits the mitogen-activated protein kinase pathway in a variety of cancer cell lines. In addition, it impairs angiogenesis in xenograft models, and there is correlation between impaired tumor growth with inhibition of the ERKs 1/2 phosphorylation *(73)*. The drug has been evaluated using the following daily oral schedules: (i) 7 days every 15 days *(74)*; (ii) 21 days every 28 days *(75)*; (iii) 28 days every 35 days; and (iv) continuous treatment *(76)*. Pharmacokinetic studies indicated dose proportionality up to 600 mg twice daily and high interpatient variability, but not intrapatient variability. Steady-state blood concentrations are achieved by 7 days, with terminal half-life values range from 30 to 45 h *(76)*. The most common dose-related toxicities were diarrhea, vomiting, skin rash, fatigue, hypertension, and palmar-plantar erythrodysesthesia (hand-foot syndrome). Hand-foot syndrome is characterized by desquamation and discomfort of the digits, which is reversible. Uncommon laboratory abnormalities included elevations in serum amylase and lipase, lymphopenia, and anemia. At sorafenib doses exceeding 400 mg twice daily on a continuous schedule, the incidence of diarrhea, hand-foot syndrome, and other toxicities is unacceptably high. The Recommended Phase II Tolerated Dose (RPTD) selected for development was 400 mg bid given continuously. Tumor regression was noted with several schedules, particularly when doses exceeded 200 mg twice daily. In a pooled analysis of phase I trials that included patients treated at the RPTD identified for phase II/III trials (400 mg bid continuously), 15% experienced grade 2/3 hand-foot skin reaction, 24% experienced grade 2/3 diarrhea; sorafenib

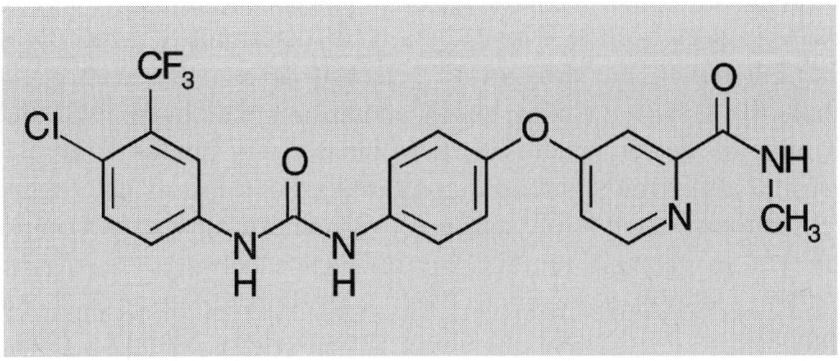

Fig. 5. Chemical structure of sorafenib. Reproduced with permission from ref.71

Table 3
Randomized Phase II and III Trials of Sorafenib in Renal Cell Cancer

Reference	Patient selection	Arms	No.	Median PFS (weeks)	Hazard ratio
Ratain, 2006 (84)	Cytokine-refractory disease; 202 patients, of whom 65 (32%) had stable disease after a 12-week course of sorafenib	Sorafenib Placebo	32 33	23 6	0.29 $p = 0.0001$
Gore, 2006 (85)	Cytokine refractory disease	Sorafenib Placebo	384 385	24 12	0.44 (0.35, 0.55) $p = 0.000001$

PFS, progression-free survival; OS, overall survival.

induced stable disease for 6 months in 12% of patients (6% stabilized for 1 year), and patients who experienced skin toxicity and/or diarrhea had significantly prolongation of time to disease progression compared with patients without such toxicity ($p < 0.05$) *(77)*. Other side effects include fatigue, nausea, and hypertension. In the phase I trials, clinical evidence of tumor regression or prolonged stable disease was observed in renal cell, colorectal, hepatocellular, ovarian, and breast carcinoma *(72)*. Sorafenib has been combined with a variety of cytotoxic and biological agents, including doxorubicin *(78)*, gemcitabine *(79)*, oxaliplatin *(80)*, and interferon *(81)*.

3.3. Randomized Phase II and III Studies of Raf Kinase Inhibitors

Sorafenib was evaluated in a series of large phase II studies that employed a randomized discontinuation design. Briefly, this trial design enriches for potentially responsive tumors by powering the trial to identify a sufficient number of patients with responding or stable disease after a prespecified time point, usually 8–12 weeks *(82)*. Patients who are responding at this point continue the agent, whereas those who have progressive disease discontinue treatment. Individuals with stable disease are randomized to continue active drug or a placebo, followed by reevaluation at 8–12 weeks. This design allows investigators to determine if apparent slow tumor growth is attributable to the drug or to selection of patients with naturally slow-growing tumors. By selecting a more homogeneous population, the randomized portion of the study requires fewer patients than would a study randomizing all patients at entry. The design also avoids potential confounding because of heterogeneous tumor growth. Because the two randomly assigned treatment groups each comprise patients with apparently slow growing tumors, any difference between the groups in disease progression after randomization is more likely a result of the study drug and less likely a result of imbalance with respect to tumor growth rates. Stopping rules during the initial open-label stage and the subsequent randomized trial stage allow one to reduce the overall sample size. This study was successfully applied to the evaluation of a putative antiangiogenic agent carboxyaminoimidazole in renal cell carcinoma *(83)*.

Using this trial design, Ratain et al. *(84)* evaluated 202 patients with advanced renal cell carcinoma in a phase II trial of sorafenib (400 mg po bid) given continuously in patients with cytokine refractory renal cell carcinoma (Table 3). After 12 weeks, 65 patients (32%) who had stable disease (bidimensional tumor measurements remaining within 25% of baseline) were then randomized in a double-blind fashion to continue sorafenib ($n = 32$) or placebo ($n = 33$) for an additional 12 weeks *(84)*. At 24 weeks, 6 patients (18%) taking placebo were progression-free compared with 16 patients (50%) taking sorafenib ($p = 0.0077$). Median progression free survival (PFS) after randomization was greater with sorafenib (23 versus 6 weeks, $p = 0.0001$, hazard ratio 0.29). Sorafenib was restarted in 25 patients who progressed on placebo after a median time from randomization of 7 weeks. Median PFS after restarting sorafenib in these 25 patients was 24 weeks, with 13 of 25 patients continuing therapy. The most common drug-related adverse events were rash (62%), hand–foot skin reaction (61%), and fatigue (56%), but led to discontinuation of therapy in only 2%. Grade 3/4 drug-related adverse events occurred in 47% of patients, and the most common were hypertension (24%), hand–foot skin reaction (13%), and fatigue (5%).

Based upon the promising results from this large phase II trial, in a phase III study, 769 patients with unresectable and/or measurable renal cell carcinoma who have received at least one prior systemic therapy were randomly assigned to treatment with either sorafenib ($n = 384$) or a placebo ($n = 385$) *(85)*. All patients were required to have an Eastern Cooperative Oncology Group (ECOG) performance status of 0 or 1 and were required to be low or intermediate risk by the Memorial Sloan Kettering Criteria *(86)*. The primary and secondary end points were overall survival and progression-free survival, respectively. The median PFS for patients in the sorafenib arm was 24 weeks, compared with 12 weeks in the placebo arm [hazard ratio 0.44 (95% CI, 0.35–0.55); $p = 0.\ 000001$]. An exploratory analysis indicated that the results were consistent across patient subsets evaluated by age (<65 versus > 65 years), ECOG performance status (0 or 1), and risk category (low or intermediate). Of 672 patients evaluable for response, only seven partial responses (2%) were noted in the sorafenib arm compared with none in the placebo arm. Disease stabilization was seen in 261 (78%) versus 186 (55%) patients in the sorafenib and placebo arms, respectively, and disease progression was noted in 29 (9%) versus 102 (30%) patients, respectively. In a planned interim analysis after 220 deaths, the rate of overall survival was longer with sorafenib (hazard ratio 0.72; 95% CI, 0.55–0.95), although this did not meet the pre-specified criterion for statistical significance at this time point; additional analyses are planned as the survival data mature. Based upon the results of this trial, sorafenib was approved for cytokine-refractory metastatic renal cell carcinoma by the U.S. Food and Drug Administration.

4. MEK

4.1. MEK Signaling and Mutations

Constitutive activation of the MEK and ERK have been observed in about one-third of tumor cell lines, which is due to upstream dysregulation of Raf-1, Ras, or other signaling molecules *(87)*. Although MEK has not been identified as an oncogene product, it is a downstream focal point of many signaling pathways activated by known oncogenes. Potent small-molecule inhibitors targeting the components of the ERK pathway have been developed, including PD184352 (or CI-1040), PD0325901, and ARRY-142886 (as reviewed in refs (88–90)). Combination of MEK inhibitors with antitublin agents such as paclitaxel has demonstrated synergistic activity in human lung cancer heterotransplants (91).

4.2. Phase I and II trials of MEK Inhibitors

CI-1040 is the first MEK-targeted agent to enter clinical trials (Fig. 6). It is a highly potent and selective non-competitive inhibitor of both MEK isoforms, MEK1 and MEK2. MEK inhibitors such as CI-1040 bind to hydrophobic binding pocket of MEK1 and MEK2 that is adjacent to but distinct from the magnesium ATP-binding site, which induces a conformational change in unnphosphorylated MEK that locks it into a closed but catalytically inactive form (92). CI-1040 inhibits the clonogenic growth of a panel of tumor cell lines of diverse origin; there is also significant antitumor activity in both mouse and human xenograft models including a variety of tumor types (93).

LoRusso and colleagues (94) performed a phase I trial of CI-1040 using multiple daily-dosing frequencies administered for 21 of 28 days, then continuously. Single dose and steady-state pharmacokinetics were assessed during cycle 1, and phosphorylated extracellular receptor kinase (pERK) levels were assessed in leukocytes and also in tumor tissue from selected patients. Seventy-seven patients with a variety of cancer types received CI-1040 doses ranging from 100 mg once daily to 800 mg thrice daily. Grade 3 asthenia was dose limiting at the highest dose level tested. Ninety-eight percent of all drug-related adverse events were grade 1 or 2 in severity; most common toxicities included diarrhea, asthenia, rash, nausea, and vomiting. Plasma concentrations of CI-1040 and its active metabolite, PD 0184264, increased in a less than dose proportional manner from 100 to 800 mg QD. Administration with a high-fat meal resulted in an increase in drug exposure. The RPTD was 800 mg bid administered with food. Of the 66 patients assessable for response, one partial response was achieved in a patient with pancreatic cancer, and 19 patients (28%) had stable disease lasting a median of 5.5 months (range, 4–17 months). Inhibition of tumor pERK (median, 73%; range, 46–100%) was demonstrated in 10 patients.

Rinehart (95) subsequently performed a multicenter phase II trial of CI-1040 in patients with advanced colorectal, non-small cell lung, breast, or pancreatic cancer. Patients received oral CI-1040 continuously at a dose of 800 mg twice daily. All patients

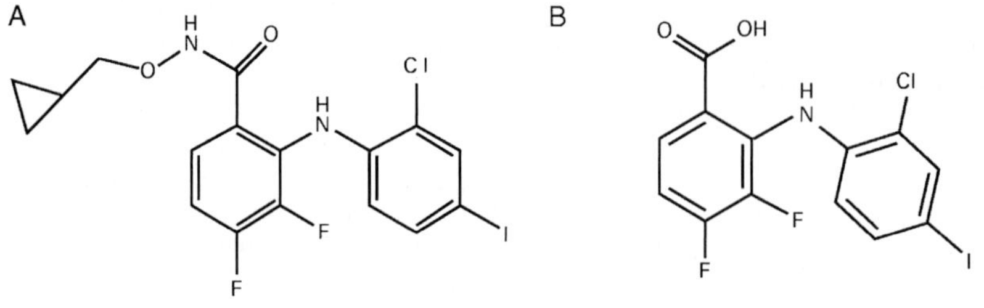

Fig. 6. Chemical structure of CI-1040 (PD 0184352) and (B) PD 0184264, the acid metabolite. Reproduced within permission from ref. 94.

had measurable disease at baseline, an ECOG performance status of 2 or less, and adequate bone marrow, liver, and renal function. Expression of pERK, pAkt, and Ki-67 was assessed in archived tumor specimens by quantitative immunohistochemistry. Sixty-seven patients with breast ($n = 14$), colon ($n = 20$), lung ($n = 18$), and pancreatic ($n = 15$) cancer received a total of 194 courses of treatment (median, 2.0 courses; range, 1–14 courses). No complete or partial responses were observed. Stable disease lasting a median of 4.4 months (range, 4–18 months) was confirmed in eight patients (one breast, two colon, two pancreas, and three lung cancer patients). Treatment was well tolerated, with 81% of patients experiencing toxicities of grade 2 or less severity. The most common toxicities included diarrhea, nausea, asthenia, and rash. A mild association ($p = 0.055$) between baseline pERK expression in archived tumor specimens and stable disease was observed. The authors concluded that CI-1040 demonstrated insufficient antitumor activity to warrant further development in the four tumors tested, but suggested further studies of PD 0325901, a second generation MEK inhibitor with significantly improved pharmacologic and pharmaceutical properties.

5. CONCLUSIONS

The mitogen-activated protein kinase (MAPK) signaling pathway plays a critical role in transmitting signals from cell surface molecules to the nucleus, thereby influencing proliferation, cell survival, and other biological processes. Proteins that are critical components of the pathway include Ras, Raf, and MEK. Inhibitors of specific component of these pathways have been developed and extensively evaluated in the clinic, some producing disappointing results, others demonstrating clinical useful activity that has resulted in approval of these agents and their commercial availability. Inhibitors of Ras and Raf were initially developed to target tumor harboring mutations in *ras* or *raf* genes that were oncogenic, although clinical response has not correlated with mutation status. Interference in these pathways influences downstream pathways that regulate tumor proliferation, apoptosis, stress response, cytoskeletal organization, and membrane trafficking. Challenges that remain in clinical development of these agents include identifying more effective agents, defining optimal dose and schedule, identifying a "molecular signature" of the tumor predictive of response to these agents, identify how these drugs can best be combined with standard therapy, and to define optimal surrogate markers that are indicative of producing the desired biological effect.

REFERENCES

1. Brivanlou AH, Darnell JE, Jr. Signal transduction and the control of gene expression. Science 2002; 295:813–8.
2. Green DR, Evan GI. A matter of life and death. Cancer Cell 2002; 1:19–30.
3. Hanahan D, Weinberg RA. The hallmarks of cancer. Cell 2000; 100:57–70
4. Blume-Jensen P, Hunter T. Oncogenic kinase signalling. Nature 2001 411(6835):355–65.
5. Hahn WC, Weinberg RA. Rules for making human tumor cells. N Engl J Med 2002; 347:1593–603.
6. Manning G, Whyte DB, Martinez R, Hunter T, Sudarsanam S. The protein kinase complement of the human genome. Science 2002; 298:1912–34.
7. Milanesi L, Petrillo M, Sepe L, et al. Systematic analysis of human kinase genes: a large number of genes and alternative splicing events result in functional and structural diversity. BMC Bioinformatics 2005; 6 Suppl 4:S20.
8. Sebolt-Leopold JS, Herrera R. Targeting the mitogen-activated protein kinase cascade to treat cancer. Nat Rev Cancer 2004; 4:937–47.

 9. Gibbs JB. Mechanism-based target identification and drug discovery in cancer. Science 2000; 287:1969–73.
10. Elsayed YA, Sausville EA. Selected novel anticancer treatments targeting cell signaling proteins. Oncologist 2001; 6:517–37.
11. Takai Y, Sasaki T, Matozaki T. Small GTP-binding proteins. Physiol Rev 2001; 81:153–208.
12. Kolch W. Ras/Raf signalling and emerging pharmacotherapeutic targets. Expert Opin Pharmacother 2002; 3:709–18.
13. Sebti SM, Hamilton AD. Inhibition of Ras prenylation: a novel approach to cancer chemotherapy. Pharmacol Ther 1997; 74:103–14.
14. Sebti SM, Hamilton AD. Farnesyltransferase and geranylgeranyltransferase I inhibitors in cancer therapy: important mechanistic and bench to bedside issues. Expert Opin Investig Drugs 2000; 9:2767–82.
15. Thoma NH, Iakovenko A, Owen D, et al. Phosphoisoprenoid binding specificity of geranylgeranyl-transferase type II. Biochemistry 2000; 39:12043–52.
16. Gibbs JB, Oliff A. The potential of farnesyltransferase inhibitors as cancer chemotherapeutics. Annu Rev Pharmacol Toxicol 1997; 37:143–66.
17. Rowinsky EK, Windle JJ, Von Hoff DD. Ras protein farnesyltransferase: a strategic target for anticancer therapeutic development. J Clin Oncol 1999; 17:3631–52.
18. Oldham SM, Cox AD, Reynolds ER, Sizemore NS, Coffey RJ, Jr., Der CJ. Ras, but not Src, transformation of RIE-1 epithelial cells is dependent on activation of the mitogen-activated protein kinase cascade. Oncogene 1998; 16:2565–73.
19. Rochlitz CF, Scott GK, Dodson JM, et al. Incidence of activating ras oncogene mutations associated with primary and metastatic human breast cancer. Cancer Res 1989; 49:357–60.
20. Thor A, Ohuchi N, Hand PH, et al. Ras gene alterations and enhanced levels of Ras p21 expression in a spectrum of benign and malignant human mammary tissues. Lab Invest 1986; 55:603–15.
21. Norgaard P, Law B, Joseph H, et al. Treatment with farnesyl-protein transferase inhibitor induces regression of mammary tumors in transforming growth factor (TGF) alpha and TGF alpha/neu transgenic mice by inhibition of mitogenic activity and induction of apoptosis. Clin Cancer Res 1999; 5:35–42.
22. Clark GJ, Der CJ. Aberrant function of the Ras signal transduction pathway in human breast cancer. Breast Cancer Res Treat 1995; 35:133–44.
23. Malaney S, Daly RJ. The ras signaling pathway in mammary tumorigenesis and metastasis. J Mammary Gland Biol Neoplasia 2001; 6:101–13.
24. Theillet C, Lidereau R, Escot C, et al. Loss of a c-H-ras-1 allele and aggressive human primary breast carcinomas. Cancer Res 1986; 46:4776–81.
25. Smith CA, Pollice AA, Gu LP, et al. Correlations among p53, Her-2/neu, and ras overexpression and aneuploidy by multiparameter flow cytometry in human breast cancer: evidence for a common phenotypic evolutionary pattern in infiltrating ductal carcinomas. Clin Cancer Res 2000; 6:112–26.
26. Kato S, Masuhiro Y, Watanabe M, et al. Molecular mechanism of a cross-talk between oestrogen and growth factor signalling pathways. Genes Cells 2000; 5:593–601.
27. Bunone G, Briand PA, Miksicek RJ, Picard D. Activation of the unliganded estrogen receptor by EGF involves the MAP kinase pathway and direct phosphorylation. EMBO J 1996; 15:2174–83.
28. Park DS, Lee H, Riedel C, et al. Prolactin negatively regulates caveolin-1 gene expression in the mammary gland during lactation, via a Ras-dependent mechanism. J Biol Chem 2001; 276:48389–97.
29. Gibbs JB, Oliff A, Kohl NE. Farnesyltransferase inhibitors: Ras research yields a potential cancer therapeutic. Cell 1994; 77:175–8.
30. Marshall CJ. Cell signalling. Raf gets it together. Nature 1996; 383:127–8.
31. Adjei AA. Blocking oncogenic Ras signaling for cancer therapy. J Natl Cancer Inst 2001; 93:1062–74.
32. Tamanoi F, Gau CL, Jiang C, Edamatsu H, Kato-Stankiewicz J. Protein farnesylation in mammalian cells: effects of farnesyltransferase inhibitors on cancer cells. Cell Mol Life Sci 2001; 58:1636–49.
33. Capeans C, Pineiro A, Pardo M, et al. Role of inhibitors of isoprenylation in proliferation, phenotype and apoptosis of human retinal pigment epithelium. Graefes Arch Clin Exp Ophthalmol 2001; 239:188–98.
34. Miller AC, Samid D. Tumor resistance to oxidative stress: association with ras oncogene expression and reversal by lovastatin, an inhibitor of p21ras isoprenylation. Int J Cancer 1995; 60:249–54.

35. Tatsuta M, Iishi H, Baba M, et al. Suppression by pravastatin, an inhibitor of p21ras isoprenylation, of hepatocarcinogenesis induced by N-nitrosomorpholine in Sprague-Dawley rats. Br J Cancer 1998; 77:581–7.

36. Bouterfa HL, Sattelmeyer V, Czub S, Vordermark D, Roosen K, Tonn JC. Inhibition of Ras farnesylation by lovastatin leads to downregulation of proliferation and migration in primary cultured human glioblastoma cells. Anticancer Res 2000; 20:2761–71.

37. Liao JK. Isoprenoids as mediators of the biological effects of statins. J Clin Invest 2002; 110:285–8.

38. Luckman SP, Hughes DE, Coxon FP, Graham R, Russell G, Rogers MJ. Nitrogen-containing bisphosphonates inhibit the mevalonate pathway and prevent post-translational prenylation of GTP-binding proteins, including Ras. J Bone Miner Res 1998; 13:581–9.

39. Rogers MJ, Gordon S, Benford HL, et al. Cellular and molecular mechanisms of action of bisphosphonates. Cancer 2000; 88:2961–78.

40. Sebti SM, Adjei AA. Farnesyltransferase inhibitors. Semin Oncol 2004; 31:28–39.

41. Crul M, de Klerk GJ, Swart M, et al. Phase I clinical and pharmacologic study of chronic oral administration of the farnesyl protein transferase inhibitor R115777 in advanced cancer. J Clin Oncol 2002; 20:2726–35.

42. Karp JE, Lancet JE, Kaufmann SH, et al. Clinical and biologic activity of the farnesyltransferase inhibitor R115777 in adults with refractory and relapsed acute leukemias: a phase 1 clinical-laboratory correlative trial. Blood 2001; 97:3361–9.

43. Hudes GR, Schol J, Baab J, et al. Phase I clinical and pharmacokinetic trial of the farnesyltransferase inhibitor R115777 on a 21-day dosing schedule. Proc ASCO 1999: 18; 156a(abstr 501).

44. Zujewski J, Horak ID, Bol CJ, et al. Phase I and pharmacokinetic study of farnesyl protein transferase inhibitor R115777 in advanced cancer. J Clin Oncol 2000; 18:927–41.

45. Eskens FA, Awada A, Cutler DL, et al. Phase I and pharmacokinetic study of the oral farnesyl transferase inhibitor SCH 66336 given twice daily to patients with advanced solid tumors. J Clin Oncol 2001; 19:1167–75.

46. Awada A, Eskens FA, Piccart M, et al. Phase I and pharmacological study of the oral farnesyltransferase inhibitor SCH 66336 given once daily to patients with advanced solid tumours. Eur J Cancer 2002; 38:2272–8.

47. Adjei AA, Erlichman C, Davis JN, et al. A Phase I trial of the farnesyl transferase inhibitor SCH66336: evidence for biological and clinical activity. Cancer Res 2000; 60:1871–7.

48. Voi M, Tabernero J, Cooper M, et al. A phase I study of the farnesyltransferase (FT) inhibitor BMS-214662 administered as a weekly 1-hour infusion in patients (Pts) with advanced solid tumors: clinical findings. Proc Am Soc Clin Oncol 2001; 20:(Abstract 312).

49. Johnston S, Hickish, T, Houston, S, et al. Efficacy and tolerability of the two dosing regimens of R115777 (Zarnestra), a farnesyl protein transferase inhibitor, in patients with advanced breast cancer. Proc Am Soc Clin Oncol 2002; 21:(Abstract 138).

50. Johnson & Johnson Pharmaceutical Corporation: Clinical Brochure tipifarnib.

51. Henderson MC, Miranda CL, Stevens JF, Deinzer ML, Buhler DR. In vitro inhibition of human P450 enzymes by prenylated flavonoids from hops, Humulus lupulus. Xenobiotica 2000; 30:235–51.

52. Adjei AA, Davis JN, Erlichman C, Svingen PA, Kaufmann SH. Comparison of potential markers of farnesyltransferase inhibition. Clin Cancer Res 2000; 6:2318–25.

53. Britten CD, Rowinsky EK, Soignet S, et al. A phase I and pharmacological study of the farnesyl protein transferase inhibitor L-778,123 in patients with solid malignancies. Clin Cancer Res 2001; 7:3894–903.

54. Haas N, Peereboom D, Ranganathan S, et al. Phase II trial of R115777, an inhibitor of farnesyltransferase, in patients with hormone refractory prostate cancer. Proc Am Soc Clin Oncol 2002; 21:(Abstract 271).

55. Kelland LR, Smith V, Valenti M, et al. Preclinical antitumor activity and pharmacodynamic studies with the farnesyl protein transferase inhibitor R115777 in human breast cancer. Clin Cancer Res 2001; 7:3544–50.

56. Cloughesy T, Kuhn J, Wen P, et al. Phase II trial of R115777 (Zarnestra) in patients with recurrent glioma not taking enzyme inducing antiepileptic drugs (EIAED): a North American Brain Tumor Consortium (NABTC) report. Proc Am Soc Clin Oncol 2002; 21:(Abstract 317).

57. Alsina M, Overton R, Belle N, Wilson EF, Sullivan D, Djulbegovic B, Fonseca R, Dalton WS, Sebti SM. Farnesyl transferase inhibitor FTI-R115777 is well tolerated, induces stabilization of

disease and inhibits farnesylation and oncogenic/tumor survival pathways in patients with advanced mutiple myeloma (Abstract). Proc Am Assoc Cancer Res 2002; 2002:4960.

58. Alsina M, Fonseca R, Wilson EF, et al. Farnesyltransferase inhibitor tipifarnib is well tolerated, induces stabilization of disease, and inhibits farnesylation and oncogenic/tumor survival pathways in patients with advanced multiple myeloma. Blood 2004; 103:3271–7.

59. Kurzrock R, Kantarjian HM, Cortes JE, et al. Farnesyltransferase inhibitor R115777 in myelodysplastic syndrome: clinical and biologic activities in the phase 1 setting. Blood 2003; 102:4527–34.

60. Zimmerman TM, Harlin H, Odenike OM, et al. Dose-ranging pharmacodynamic study of tipifarnib (R115777) in patients with relapsed and refractory hematologic malignancies. J Clin Oncol 2004; 22:4764–70.

61. Adjei A, Mauer A, Marks R, et al. A phase II study of the farnesyltransferase inhibitor R115777 in patients with advanced non-small cell lung cancer. Proc Am Soc Clin Oncol 2002; 21:(Abstract 1156).

62. Evans T, Fidias P, Skarin A, et al. A phase II study of efficacy and tolerability of the farnesyl-protein transferase inhibitor L-778,123 as first-line therapy in patients with advanced non-small cell lung cancer (NSCLC). Proc Am Soc Clin Oncol 2002; 21:(Abstract 1861).

63. Rosenberg JE, von der Maase H, Seigne JD, et al. A phase II trial of R115777, an oral farnesyl transferase inhibitor, in patients with advanced urothelial tract transitional cell carcinoma. Cancer 2005; 103:2035–41.

64. Mesa RA. Tipifarnib: farnesyl transferase inhibition at a crossroads. Expert Rev Anticancer Ther 2006; 6:313–9.

65. Ellis CA, Vos MD, Wickline M, et al. Tamoxifen and the farnesyl transferase inhibitor FTI-277 synergize to inhibit growth in estrogen receptor-positive breast tumor cell lines. Breast Cancer Res Treat 2003; 78:59–67.

66. Johnston S, Semiglazov V, Manikhas G, Spaeth D, Romieu G, Dodwell D, Wardley A, Neven P, Bessems A, Ma Y-W, Howes AJ. A randomised, blinded, phase II study of tipifarnib (Zarnestra) combined with letrozole in the treatment of advanced breast cancer after antiestrogen therapy. San Antonio Breast Cancer Symposium 2005 (Abstract 5087).

67. Fisher B, Brown A, Mamounas E, et al. Effect of preoperative chemotherapy on local-regional disease in women with operable breast cancer: findings from National Surgical Adjuvant Breast and Bowel Project B-18. J Clin Oncol 1997; 15:2483–93.

68. Sparano JA, Moulder S, Kazi A, et al. Targeted inhibition of farnesyltransferase in locally advanced breast cancer: a phase I and II trial of tipifarnib plus dose-dense doxorubicin and cyclophosphamide. J Clin Oncol. 2006 Jul 1; 24(19):3013–8. Epub 2006 Jun 12.S.

69. van Cutsem E, van de Velde H, Karasek P, Oettle H, Vervenne WL, Szawlowski A, Schoffski P, Post S, Verslype C, Neumann H, Safran H, Humblet Y, Perez Ruixo J, Ma Y, von Hoff D. Phase III trial of gemcitabine plus tipifarnib compared with gemcitabine plus placebo in advanced pancreatic cancer. J Clin Oncol. 2004 Apr 15; 22(8):1430–8.

70. Rao S, Cunningham D, de Gramont A, et al. Phase III double-blind placebo-controlled study of farnesyl transferase inhibitor R115777 in patients with refractory advanced colorectal cancer. J Clin Oncol 2004; 22:3950–7.

71. Beeram M, Patnaik A, Rowinsky EK. Raf: a strategic target for therapeutic development against cancer. J Clin Oncol 2005; 23:6771–90.

72. Strumberg D. Preclinical and clinical development of the oral multikinase inhibitor sorafenib in cancer treatment. Drugs Today (Barc) 2005; 41:773–84.

73. Wilhelm SM, Carter C, Tang L, et al. BAY 43–9006 exhibits broad spectrum oral antitumor activity and targets the RAF/MEK/ERK pathway and receptor tyrosine kinases involved in tumor progression and angiogenesis. Cancer Res 2004; 64:7099–109.

74. Clark JW, Eder JP, Ryan D, Lathia C, Lenz HJ. Safety and pharmacokinetics of the dual action Raf kinase and vascular endothelial growth factor receptor inhibitor, BAY 43–9006, in patients with advanced, refractory solid tumors. Clin Cancer Res 2005; 11:5472–80.

75. Moore M, Hirte HW, Siu L, et al. Phase I study to determine the safety and pharmacokinetics of the novel Raf kinase and VEGFR inhibitor BAY 43–9006, administered for 28 days on/7 days off in patients with advanced, refractory solid tumors. Ann Oncol 2005; 16:1688–94.

76. Strumberg D, Richly H, Hilger RA, et al. Phase I clinical and pharmacokinetic study of the Novel Raf kinase and vascular endothelial growth factor receptor inhibitor BAY 43–9006 in patients with advanced refractory solid tumors. J Clin Oncol 2005; 23:965–72.

77. Strumberg D, Awada A, Hirte H, et al. Pooled safety analysis of BAY 43–9006 (sorafenib) monotherapy in patients with advanced solid tumours: Is rash associated with treatment outcome? Eur J Cancer 2006; 42:548–56.

78. Richly H, Henning BF, Kupsch P, et al. Results of a phase I trial of sorafenib (BAY 43–9006) in combination with doxorubicin in patients with refractory solid tumors. Ann Oncol 2006; 17:866–73.

79. Siu LL, Awada A, Takimoto CH, et al. Phase I trial of sorafenib and gemcitabine in advanced solid tumors with an expanded cohort in advanced pancreatic cancer. Clin Cancer Res 2006; 12:144–51.

80. Kupsch P, Henning BF, Passarge K, et al. Results of a phase I trial of sorafenib (BAY 43–9006) in combination with oxaliplatin in patients with refractory solid tumors, including colorectal cancer. Clin Colorectal Cancer 2005; 5:188–96.

81. Reddy GK, Bukowski RM. Sorafenib: recent update on activity as a single agent and in combination with interferon-alpha2 in patients with advanced-stage renal cell carcinoma. Clin Genitourin Cancer 2006; 4:246–8.

82. Rosner GL, Stadler W, Ratain MJ. Randomized discontinuation design: application to cytostatic antineoplastic agents. J Clin Oncol 2002; 20:4478–84.

83. Stadler WM, Rosner G, Small E, et al. Successful implementation of the randomized discontinuation trial design: an application to the study of the putative antiangiogenic agent carboxyaminoimidazole in renal cell carcinoma–CALGB 69901. J Clin Oncol 2005; 23:3726–32.

84. Ratain MJ, Eisen T, Stadler WM, et al. Phase II placebo-controlled randomized discontinuation trial of sorafenib in patients with metastatic renal cell carcinoma. J Clin Oncol. 2006 Jun 1; 24(16): 2505–12. Epub 2006 Apr 24.

85. Gore ME, Escudier B. Emerging efficacy endpoints for targeted therapies in advanced renal cell carcinoma. Oncology (Williston Park) 2006; 20:19–24.

86. Motzer RJ, Bacik J, Mazumdar M. Prognostic factors for survival of patients with stage IV renal cell carcinoma: memorial sloan-kettering cancer center experience. Clin Cancer Res 2004; 10:6302S–3S.

87. Hoshino R, Chatani Y, Yamori T, et al. Constitutive activation of the 41-/43-kDa mitogen-activated protein kinase signaling pathway in human tumors. Oncogene 1999; 18:813–22.

88. Kohno M, Pouyssegur J. Targeting the ERK signaling pathway in cancer therapy. Ann Med 2006; 38:200–11.

89. Sosman JA, Puzanov I. Molecular targets in melanoma from angiogenesis to apoptosis. Clin Cancer Res 2006; 12:2376s–2383s.

90. Thompson N, Lyons J. Recent progress in targeting the Raf/MEK/ERK pathway with inhibitors in cancer drug discovery. Curr Opin Pharmacol 2005; 5:350–6.

91. McDaid HM, Lopez-Barcons L, Grossman A, et al. Enhancement of the therapeutic efficacy of taxol by the mitogen-activated protein kinase kinase inhibitor CI-1040 in nude mice bearing human heterotransplants. Cancer Res 2005; 65:2854–60.

92. Ohren JF, Chen H, Pavlovsky A, et al. Structures of human MAP kinase kinase 1 (MEK1) and MEK2 describe novel noncompetitive kinase inhibition. Nat Struct Mol Biol 2004; 11:1192–7.

93. Sebolt-Leopold JS, Dudley DT, Herrera R, et al. Blockade of the MAP kinase pathway suppresses growth of colon tumors in vivo. Nat Med 1999; 5:810–6.

94. Lorusso PM, Adjei AA, Varterasian M, et al. Phase I and pharmacodynamic study of the oral MEK inhibitor CI-1040 in patients with advanced malignancies. J Clin Oncol 2005; 23:5281–93.

95. Rinehart J, Adjei AA, Lorusso PM, et al. Multicenter phase II study of the oral MEK inhibitor, CI-1040, in patients with advanced non-small-cell lung, breast, colon, and pancreatic cancer. J Clin Oncol 2004; 22:4456–62.

4

17-AAG

Targeting the Molecular Chaperone Heat Shock Protein 90

Len Neckers, PhD, and Percy Ivy, MD

SUMMARY

Heat shock protein 90 (Hsp90) is a molecular chaperone required for the stability and function of a number of conditionally activated and/or expressed signaling proteins, as well as multiple mutated, chimeric, and/or over-expressed signaling proteins, that promote cancer cell growth and/or survival. Hsp90 inhibitors, by interacting specifically with a single molecular target, cause the inactivation, destabilization, and eventual degradation of Hsp90 client proteins, and they have shown promising anti-tumor activity in preclinical model systems. One Hsp90 inhibitor, 17-allylamino-17-demethoxygeldanamycin, has completed phase I clinical trials, and several phase II trials of this agent are planned or are in progress. Phase I testing of a related Hsp90 inhibitor, 17-dimethylaminoethylamino-17-demethoxygeldanamycin, is currently in progress. Hsp90 inhibitors are unique in that, although they are directed toward a specific molecular target, they simultaneously inhibit multiple signaling pathways that frequently interact to promote cancer cell survival. Furthermore, by inhibiting nodal points in multiple overlapping survival pathways utilized by cancer cells, combination of an Hsp90 inhibitor with standard chemotherapeutic agents may dramatically increase the in vivo efficacy of the standard agent. Hsp90 inhibitors may circumvent the characteristic genetic plasticity that has allowed cancer cells to eventually evade the toxic effects of most molecularly targeted agents. The mechanism-based use of Hsp90 inhibitors, both alone and in combination with other drugs, should be effective toward multiple forms of cancer.

Key Words: Heat shock protein 90; cancer; molecular chaperone; molecularly targeted therapeutics; genetic plasticity; oncogene; geldanamycin; benzoquinone ansamycin.

1. INTRODUCTION

Cancer is a disease of genetic instability. Although only a few specific alterations seem to be required for generation of the malignant phenotype, at least in colon carcinoma, there are approximately 10,000 estimated mutations at the time of diagnosis *(1,2)*. This genetic plasticity of cancer cells allows them to frequently escape the precise

From: *Cancer Drug Discovery and Development: Molecular Targeting in Oncology*
Edited by: H. L. Kaufman, S. Wadler, and K. Antman © Humana Press, Totowa, NJ

molecular targeting of a single signaling node or pathway, making them ultimately non-responsive to molecularly targeted therapeutics. Even Gleevec™ (Novartis Pharma-ceuticals Corp.), a well-recognized clinically active Bcr-Abl tyrosine kinase inhibitor, can eventually lose its effectiveness under intense, drug-dependent selective pressure, due to either mutation of the drug interaction site or expansion of a previously existing resistant clone *(3)*. Most solid tumors at the time of detection are already sufficiently genetically diverse to resist single-agent molecularly targeted therapy *(4)*. Thus, a simultaneous attack on multiple nodes of a cancer cell's web of overlapping signaling pathways should be more likely to affect survival than would inhibition of one or even a few individual signaling nodes. Given the number of key nodal proteins that are heat shock protein 90 (Hsp90) clients (see the website maintained by D. Picard, http://www.picard.ch/downloads/Hsp90interactors.pdf), inhibition of Hsp90 may serve the purpose of collapsing, or significantly weakening, a cancer cell's safety net. Indeed, following a hypothesis first proposed by Hanahan and Weinberg several years ago *(5)*, genetic instability allows a cell to eventually acquire six capabilities that are charac-teristic of most if not all cancers. These are (i) self-sufficiency in growth signaling; (ii) insensitivity to anti-growth signaling; (iii) ability to evade apoptosis; (iv) sustained angiogenesis; (v) tissue invasion and metastasis; and (vi) limitless replicative potential. As is highlighted in Fig. 1, Hsp90 plays a pivotal role in acquisition and maintenance of each of these capabilities. Several excellent reviews provide an in depth description of the many signaling nodes regulated by Hsp90 *(6–12)*.

Cancer cells survive in the face of frequently extreme environmental stress, such as hypoxia and acidosis, as well as in the face of the exogenously applied environmental stresses of chemotherapy or radiation. These stresses tend to generate free radicals that can cause significant physical damage to cellular proteins. Given the combined protective role of molecular chaperones toward damaged proteins and the dependence of multiple signal transduction pathways on Hsp90, it is therefore not surprising that molecular chaperones in general, and Hsp90 in particular, are highly expressed in most

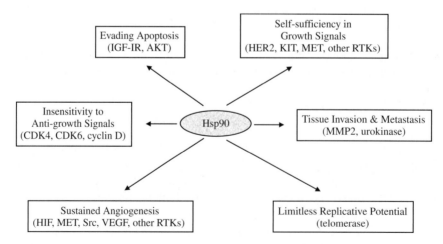

Fig. 1. Heat shock protein 90 (Hsp90) function is implicated in establishment of each of the hallmarks of cancer as first proposed by Hanahan and Weinberg *(5)*. Importantly, Hsp90 function may also permit the genetic instability on which acquisition of the six hallmarks depends.

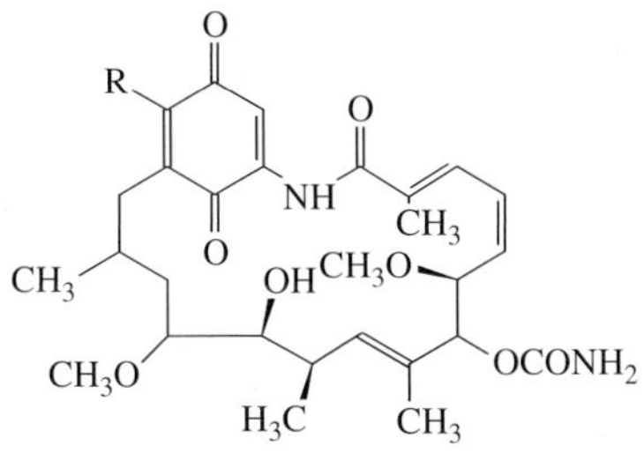

Compound	R Group
17-Allylaminogeldanamycin (17-AAG)	$CH_2=CHCH_2NH-$
17-Aminogeldanamycin (17-AG)	NH_2
17-Dimethylaminoethylamino-17-demethoxygeldamycin (17-DMAG)	$(CH_2)_2NCH_2CH_2NH-$
Geldanamycin	CH_3O-

Fig. 2. The chemical structures of geldanamycin, 17-AAG, its biologically active metabolite 17-AG, and 17-DMAG, highlighting the unique substitutions to the quinone moiety of the pharmacophore that characterize each molecule.

tumor cells. However, Hsp90 may be elevated in tumor cells and may provide a unique molecular target therein for an additional reason. Using *Drosophila* and *Arabidopsis* as model systems, Lindquist and colleagues *(13,14)* have shown that an ancient function of Hsp90 may be to permit accumulation at the protein level of inherent genetic mutations, and thus the chaperone may play a pivotal role in the evolutionary process itself. Extrapolating this hypothesis to genetically unstable cancer cells, it is not a great leap to think that Hsp90 may be critical to their ability to survive in the presence of an aberrantly high mutation rate.

The benzoquinoid ansamycin antibiotics, first isolated from the actinomycete, *Streptomyces hygroscopicus var.* geldanus var. nova *(15)*, include geldanamycin (GA) and its semi-synthetic derivatives, 17-allylamino-17-demethoxygeldanamycin (17-AAG) and the more water-soluble 17-dimethylaminoethylamino-17-demethoxygeldanamycin (17-DMAG) (see Fig. 2). These small molecules inhibit the chaperone function of the heat shock protein Hsp90 *(16)* and are currently being evaluated in phase I and II clinical trials. The parent compound, GA, is broadly cytotoxic in the National Cancer Institute (NCI) 60-cell line screen *(17)*; its poor solubility and unacceptable liver toxicity in dogs precluded testing in humans. Because 17-AAG is less toxic than GA in rats *(18)* and caused growth inhibition in breast *(19)*, melanoma *(20)*, and ovarian mouse xenograft models, the NCI initiated phase I trials in 1999.

2. Hsp90: A CHAPERONE OF ONCOGENES

Several recent, excellently detailed reviews of the mechanics of Hsp90 function are in the scientific literature *(7,9,10,21–23)*. For the purposes of the current update on Hsp90-directed therapeutics, suffice it to say that Hsp90 is a conformationally flexible protein that associates with a distinct set of co-chaperones in dependence on nucleotide (ATP or ADP) occupancy of an amino-terminal-binding pocket in Hsp90. Nucleotide exchange and ATP hydrolysis (by Hsp90 itself, with the assistance of co-chaperones) drive the so-called Hsp90 chaperone machine to bind, chaperone, and release client proteins. The Hsp90 inhibitors currently in clinical trial (17-AAG and 17-DMAG), as well as those under development, all share the property of displacing nucleotide from the amino terminal pocket in Hsp90, and therefore short-circuiting the Hsp90 chaperone machine, much as one would stop the rotation of a bicycle wheel by inserting a stick between the spokes. Cycling of the chaperone machine is critical to its function. The Hsp90 inhibitors, by preventing nucleotide-dependent cycling, interfere with the chaperone activity of Hsp90, resulting in targeting of client proteins to the proteasome, the cell's garbage disposal, where they are degraded *(24)*. Even if the proteasome is inhibited, client proteins are not rescued from Hsp90 inhibition, but instead accumulate in a misfolded, inactive form in detergent-insoluble subcellular complexes *(25)*.

2.1. Hsp90 Inhibitors Target Mutated and Chimeric Proteins Uniquely Expressed in Certain Cancers

Hsp90 characteristically chaperones a number of mutated or chimeric kinases that are key mediators of disease. Thus, anaplastic large cell lymphomas are characterized by expression of the chimeric protein NPM-ALK, which originates from a fusion of the nucleophosmin *(NPM)* and the membrane receptor anaplastic lymphoma kinase (ALK) genes. The chimeric kinase is constitutively active and capable of causing malignant transformation *(26)*. Bonvini and colleagues *(27)* have shown that NPM-ALK kinase is an Hsp90 client protein and that GA and 17-AAG destabilize the kinase and promote its proteasome-mediated degradation in several anaplastic large cell lymphoma cell lines.

FLT3 is a receptor tyrosine kinase that regulates proliferation, differentiation, and survival of hematopoietic cells. FLT3 is frequently expressed in acute myeloid leukemia, and in 20% of patients with this cancer, the tumor cells express a FLT3 protein harboring an internal tandem duplication in the juxtamembrane domain. This mutation is correlated with leukocytosis and a poor prognosis *(28)*. Minami and colleagues have reported that Hsp90 inhibitors cause selective apoptosis of leukemia cells expressing tandemly dublicated FLT3. Furthermore, these investigators reported that mutated FLT3 was an Hsp90 client protein and that brief treatment with multiple Hsp90 inhibitors resulted in the rapid dissociation of Hsp90 from the kinase, accompanied by the rapid loss of kinase activity together with loss of activity of several downstream FLT3 targets including MAP kinase, Akt, and Stat5a *(29)*. Minami et al. propose that Hsp90 inhibitors should be considered as promising compounds for the treatment of acute myeloid leukemia characterized by tandemly duplicated FLT3 expression.

BCR-ABL (p210$^{Bcr-Abl}$) is an Hsp90 client protein that is also effectively inhibited by the novel tyrosine kinase inhibitor imatinib *(25,30,31)*. While imatinib has proven very effective in initial treatment of patients with chronic myelogenous leukemia, a majority of patients who are treated when their disease is in blast crisis stage (e.g., advanced) eventually relapse despite continued therapy *(32)*. Relapse is correlated with loss of BCR-ABL inhibition by imatinib, due to either gene amplification or specific point mutations in the kinase domain that preclude association of imatinib with the kinase *(33)*. Gorre and colleagues have reported the very exciting finding that BCR-ABL protein that was resistant to imatinib remained dependent on Hsp90 chaperoning activity and thus retained sensitivity to Hsp90 inhibitors, including GA and 17-AAG. Both compounds induced the degradation of "wild-type" and mutant BCR-ABL, with a trend indicating more potent activity toward mutated imatinib-resistant forms of the kinase *(34)*. These findings were recently confirmed by other investigators *(35)*, thus providing a rationale for the use of 17-AAG in treatment of imatinib-resistant chronic myelogenous leukemia.

Mutations in the proto-oncogene c-*kit* cause constitutive kinase activity of its product, KIT protein, and are associated with human mastocytosis and gastrointestinal stromal tumors (GISTs). Although currently available tyrosine kinase inhibitors are effective in the treatment of GIST, there has been limited success in the treatment of mastocytosis. Treatment with 17-AAG of the mast cell line human mast cell (HMC)-1.2, harboring the Asp816Val and Val560Gly KIT mutations, and the cell line HMC-1.1, harboring a single Val560Gly mutation, causes both the level and activity of KIT and downstream signaling molecules AKT and STAT3 to be down-regulated following drug exposure *(36)*. These data were validated using Cos-7 cells transfected with wild-type and mutated KIT. 17-AAG promotes cell death of both HMC mast cell lines. In addition, neoplastic mast cells isolated from patients with mastocytosis and incubated with 17-AAG ex vivo are selectively sensitive to Hsp90 inhibition as compared to the mononuclear fraction as a whole. These data provide compelling evidence that 17-AAG may be effective in the treatment of c-*kit*-related diseases including mastocytosis, GIST, mast cell leukemia, sub-types of acute myelogenous leukemia, and testicular cancer.

2.2. Hsp90 Inhibitors Target the Androgen Receptor in Prostate Cancer

Androgen receptor continues to be expressed in the majority of hormone-independent prostate cancers, suggesting that it remains important for tumor growth and survival. Receptor over-expression, mutation, and/or post-translational modification may all be mechanisms by which androgen receptor can remain responsive either to low levels of circulating androgen or to anti-androgens. Vanaja et al. *(37)* have shown that Hsp90 association is essential for the function and stability of the androgen receptor in prostate cancer cells. These investigators reported that androgen receptor levels in LNCaP cells were markedly reduced by the Hsp90 inhibitor GA, as was the ability of the receptor to become transcriptionally active in the presence of synthetic androgen. In addition, Georget et al. *(38)* have shown that GA preferentially destabilized androgen receptor bound to anti-androgen, thus suggesting that the clinical efficacy of anti-androgens may be enhanced by combination with an Hsp90 inhibitor. These investigators also reported that GA prevented the nuclear translocation of ligand-bound androgen receptor and inhibited the transcriptional activity of nuclear-targeted receptors, implicating Hsp90 in multiple facets of androgen receptor activity. Finally, Solit and colleagues *(39)* have reported that 17-AAG caused degradation of both wild-type and mutant androgen receptors and inhibited both androgen-dependent and androgen-independent prostate tumor growth in nude mice. Importantly, these investigators also demonstrated the loss of Her2 and Akt proteins, two Hsp90 clients that are upstream post-translational activators of the androgen receptor, in the tumor xenografts taken from 17-AAG-treated animals.

2.3. Hsp90 Inhibitors Exert Anti-Angiogenic Activity by Promoting Oxygen-Independent and VHL-Independent Inactivation and Degradation of Hypoxia-Inducible Factor-1α Leading to Inhibition of VEGF Expression

Hypoxia-inducible factor-1α (HIF-1α) is a nuclear transcription factor involved in the transactivation of numerous target genes, many of which are implicated in the promotion of angiogenesis and adaptation to hypoxia (for a review, see ref. 40). Although these proteins are normally labile and expressed at low levels in normoxic cells, their stability and activation increase several-fold in hypoxia. The molecular basis for the instability of these proteins in normoxia depends upon VHL, the substrate recognition component of an E3 ubiquitin ligase complex that targets HIF-1α for proteasome-dependent degradation (41). Hypoxia normally impairs VHL function, thus allowing HIF to accumulate. HIF-1α expression has been documented in diverse epithelial cancers and most certainly supports survival in the oxygen-depleted environment inhabited by most solid tumors.

VHL can also be directly inactivated by mutation or hyper-methylation, resulting in constitutive over-expression of HIF in normoxic cells. In hereditary von Hippel–Lindau disease, there is a genetic loss of VHL, and affected individuals are predisposed to an increased risk of developing highly vascular tumors in a number of organs. This is due, in large part, to deregulated HIF expression and the corresponding up-regulation of the HIF target gene vascular endothelial growth factor (VEGF). A common manifestation of VHL disease is the development of clear cell renal cell carcinoma (CC-RCC) (42). VHL inactivation also occurs in nonhereditary, sporadic CC-RCC.

HIF-1α interacts with Hsp90 (43), and both GA and another Hsp90 inhibitor, radicicol, reduce HIF-dependent transcriptional activity (44,45). Hur et al. demonstrated that HIF protein from radicicol-treated cells was unable to bind DNA, suggesting that Hsp90 is necessary for mediating the proper conformation of HIF and/or recruiting additional cofactors. Likewise, Isaacs et al. reported GA-dependent, transcriptional inhibition of VEGF. Additionally, GA down-regulated HIF-1α protein expression by stimulating VHL-independent HIF-1α proteasomal degradation (45,46).

HIF-1α induction and VEGF expression have been associated with migration of glioblastoma cells in vitro and metastasis of glioblastoma in vivo. Zagzag et al. (47), in agreement with the findings described above, have reported that GA blocks HIF-1α induction and VEGF expression in glioblastoma cell lines. Furthermore, these investigators have shown that GA blocks glioblastoma cell migration, using an in vitro assay at non-toxic concentrations. This effect on tumor cell motility was independent of p53 and PTEN status, which makes Hsp90 inhibition an attractive modality in glioblastoma, where mutations in *p53* and *PTEN* genes are common and where tumor invasiveness is a major therapeutic challenge.

Dias et al. (48) have recently reported that VEGF promotes elevated Bcl2 protein levels and inhibits activity of the pro-apoptotic caspase-activating protein Apaf in normal endothelial cells and in leukemia cells bearing receptors for VEGF. Intriguingly, these investigators show that both phenomena require VEGF-stimulated Hsp90 association (e.g., with Bcl2 and Apaf) and that GA reverses both processes. Thus, GA blocked the pro-survival effects of VEGF by both preventing accumulation of anti-apoptotic Bcl2 and blocking the inhibition of pro-apoptotic Apaf.

2.4. Hsp90 Inhibitors Target Met and RET Receptor Tyrosine Kinases

The Met receptor tyrosine kinase is frequently over-expressed in cancer and is involved in angiogenesis as well as in the survival and invasive ability of cancer cells. A recent report by Maulik et al. *(49)* has demonstrated a role for Met in migration and survival of small cell lung cancer. Met is an Hsp90 client protein, and these investigators went on to show that GA-antagonized Met activity reduced the Met protein level and promoted apoptosis in several small cell lung cancer cell lines, even in the presence of excess Met ligand.

Hypoxia potentiates the invasive and metastatic potential of tumor cells. In an important recent study, Pennacchietti and colleagues reported that hypoxia (via two HIF-1α response elements) transcriptionally activated the Met gene and synergized with Met ligand in promoting tumor invasion. Furthermore, they showed that the pro-invasive effects of hypoxia were mimicked by Met over-expression and that inhibition of Met expression prevented hypoxia-induced tumor invasion *(50)*. Coupled with an earlier report describing induction of HIF-1 transcriptional activity by Met ligand *(51)*, these data identify the HIF–VEGF–Met axis as a critical target for intervention using Hsp90 inhibitors, either alone or in conjunction with other inhibitors of angiogenesis. As Bottaro and Liotta *(52)* recently pointed out, the sole use of angiogenesis inhibitors to deprive tumors of oxygen might produce an unexpectedly aggressive phenotype in those cells that survived the treatment. These authors speculated that combination of Met inhibitors with anti-angiogenesis agents should therefore be beneficial. We would suggest that combination of an anti-angiogenesis drug with an Hsp90 inhibitor would not only potentiate the anti-tumor effects obtained by inhibiting angiogenesis but would also break the HIF–Met axis by simultaneously targeting both Hsp90-dependent signaling proteins.

Mutation of a related receptor tyrosine kinase, RET, is associated with human cancer and several human neuroendocrine diseases. Point mutations of RET are responsible for multiple endocrine neoplasia types 2A and 2B (familial medullary thyroid carcinoma). Somatic gene rearrangements juxtaposing the TK domain of RET to heterologous gene partners are found in papillary carcinomas of the thyroid *(53–55)*.

Possible effects of 17-AAG on RET activity and cell growth of the TT MTC cell line have been examined *(56)*. Following treatment with 17-AAG, RET tyrosine kinase activity was inhibited by nearly 80%, as was the rate of cell growth. Thus, 17-AAG should be considered as an attractive pharmacologic agent for use as systemic therapy in patients with recurrent metastatic MTC for which non-surgical therapy has been ineffective.

2.5. Combined Inhibition of Hsp90 and the Proteasome Disrupt the Endoplasmic Reticulum and Demonstrate Enhanced Toxicity Toward Cancer Cells

Proteasome-mediated degradation is the common fate of Hsp90 client proteins in cells treated with Hsp90 inhibitors *(57,58)*. Proteasome inhibition does not protect

Hsp90 clients in the face of chaperone inhibition—instead client proteins become insoluble *(25,59)*. Because the deposition of insoluble proteins can be toxic to cells *(60,61)*, interest has arisen in combining proteasome inhibition with inhibition of Hsp90, the idea being that dual treatment will lead to enhanced accumulation of insoluble proteins and trigger apoptosis. This hypothesis is particularly appealing as a small molecule proteasome inhibitor has demonstrated efficacy in early clinical trials *(62,63)*. Initial experimental support for such an hypothesis was provided by Mitsiades et al. *(64)*, who reported that Hsp90 inhibitors enhanced multiple myeloma cell sensitivity to proteasome inhibition. Importantly, transformed cells are more sensitive to the cytotoxic effects of this drug combination than are non-transformed cells. Thus, 3T3 fibroblasts are fully resistant to combined administration of 17-AAG and Velcade™ at concentrations that prove cytotoxic to 3T3 cells transformed by *HPV16* virus encoding viral proteins E6 and E7 *(65)*. In the same study, Mimnaugh et al. demonstrated that the endoplasmic reticulum is one of the main targets of this drug combination. In the presence of combined doses of both agents that show synergistic cytotoxicity, these investigators noted a nearly complete disruption of the architecture of the endoplasmic reticulum. Because all secreted and transmembrane proteins must pass through this organelle on their route to the extracellular space, it is not surprising that a highly secretory cancer such as multiple myeloma would be particularly sensitive to combined inhibition of Hsp90 and the proteasome. One might speculate that other highly secretory cancers, including hepatocellular carcinoma and pancreatic carcinoma, would also respond favorably to this drug combination.

2.6. Hsp90 Inhibitors Sensitize Cancer Cells to Radiation

Gius and colleagues *(66)* have reported that 17-AAG potentiates both the in vitro and the in vivo radiation response of cervical carcinoma cells. An enhanced radiation response was noted when cells were exposed to radiation within 6–48 h after drug treatment. Importantly, at 17-AAG concentrations that were themselves non-toxic, Hsp90 inhibition enhanced cell kill in response to an otherwise ineffective radiation exposure (2 Gy) by more than one log. Even at moderately effective levels of radiation exposure (4–6 Gy), addition of non-toxic amounts of 17-AAG enhanced cell kill by more than one log. Importantly, the sensitizing effects of 17-AAG observed in the cervical carcinoma cells were not seen in 3T3 cells but were observed in *HPV16-E6-* and *HPV16-E7*-transformed 3T3 cells. The authors demonstrated convincingly that the effect of 17-AAG was multi-factorial, as several pro-survival Hsp90 client proteins were rapidly down-regulated upon drug treatment. In vitro findings were confirmed by a murine xenograft study in which the anti-tumor activity of both single and fractionated radiation exposure was dramatically enhanced by treatment with 17-AAG, either 16 h prior to single radiation exposure or on days 1 and 4 of a 6-day period during which the animals received fractionated radiation exposure. Machida and colleagues *(67)* reported similar findings for lung carcinoma and colon adenocarcinoma cells in vitro. Thus, 17-AAG has been validated as a potential therapeutic agent that can be used at clinically relevant doses to enhance cancer cell sensitivity to radiation. It is reasonable to expect that other Hsp90 inhibitors will have a similar utility.

2.7. *Targeting Hsp90 on the Cancer Cell Surface*

Recently, Becker and colleagues *(68)* reported that Hsp90 expression is dramatically up-regulated in malignant melanoma cells as compared to benign melanocytic lesions, and that Hsp90 is expressed on the surface of seven of eight melanoma metastases. Eustace et al. *(69,70)* have identified cell surface Hsp90 to be crucial for the invasiveness of HT-1080 fibrosarcoma cells in vitro. Taken together, these data implicate Hsp90 as an important determinant of tumor cell invasion and metastasis. Indeed, in the Eustace et al. study, the investigators demonstrated that GA covalently affixed to cell-impermeable beads was able to significantly impair cell invasion across a Matrigel-coated membrane. These findings have been confirmed using a polar (and thus cell impermeable) derivative of 17-DMAG in place of GA beads (Neckers et al., unpublished observations). Coincident with its inhibitory effects on cell invasiveness, cell-impermeable GA also antagonized the maturation, through proteolytic self-processing, of the metalloproteinase 2 (MMP2), a cell surface enzyme whose activity has been previously demonstrated as essential to cell invasion. Furthermore, these investigators demonstrated that Hsp90 could be found in association with MMP2 in the culture medium bathing the HT-1080 cells. It is intriguing to speculate that association with Hsp90 on the cell surface is necessary for the self-proteolysis of MMP2. Thus, a possible chaperone function for cell surface Hsp90 may be directly implicated in tumor cell invasiveness and metastasis. As such, cell surface Hsp90 may represent a novel, perhaps cancer-specific target for cell-impermeant Hsp90 inhibitors.

3. METABOLISM OF 17-AAG AND 17-DMAG IN VIVO

In human or murine hepatic microsome assays, 17-aminogeldanamycin (17-AG), a diol, and an epoxide are the three major metabolites of 17-AAG *(71)*. The 17-AAG diol was the major metabolite in human hepatic microsomes, followed by 17-AG; in contrast, 17-AG was the most abundant metabolite in murine microsomes. Acrolein, a nephrotoxin, is a potential by-product of the 17-AG metabolite. Finally, the epoxide is probably formed by addition of oxygen across the double bond of the allylamino side chain. CYP3A4 enzymatic metabolism is responsible for 17-AG and epoxide formation. Microsomal epoxide hydrolase catalyzes the conversion of the diol to 17-AG, which does not undergo further microsomal metabolism.

17-AAG metabolites are active and may have clinical significance. The biologically active epoxides and acrolein may induce toxic effects in humans *(71)*. Pharmacodynamic studies show that the 17-AG metabolite (see Fig. 2) is as active as 17-AAG in decreasing cellular p185[erbB2] in human breast cancer SKBr3 cells in culture *(72)*. 17-AG caused growth inhibition in six human colon cancer lines and three ovarian cancer cell lines *(73)*.

The quinone-metabolizing enzyme DT-diaphorase may alter 17-AAG's antitumor activity and toxicologic properties *(73)*. 17-AAG growth inhibitory activity was increased 32-fold by transfection of the active DT-diaphorase gene NQO1 into the DT-diaphorase-deficient BE human colon carcinoma cell line, and concomitant depletion of Raf-1 and mutant p53 protein confirmed the Hsp90 inhibition mechanism of action. Increased growth inhibition was not observed with the parent compound, GA. The increased sensitivity to 17-AAG in cell lines transfected with NQO1 was also seen in xenograft models.

In contrast to 17-AAG, 17-DMAG appears to be only minimally metabolized by CYP3A4 *(74)*. Therefore, intestinal CYP3A4 should not impede 17-DMAG's oral activity. 17-AG does not appear to be a metabolite of 17-DMAG based on the lack of conversion at the 17th position of the compound. The marked metabolic differences between 17-AAG and 17-DMAG suggest that they may have distinct toxicity profiles and therapeutic indices.

4. TOXICITY OF 17-AAG AND 17-DMAG IN ANIMALS

Single-dose range finding, multiple-dose range finding, 5-day daily dose, and multiple-dose/dimethyl sulfoxide (DMSO) formulation toxicity studies have been conducted in rats. Additionally, single-dose range-finding/microdispersed formulation and multiple-dose range finding reconstituted lyophilized formulation, and 5-day daily dose with microdispersed and DMSO-formulated 17-AAG have been conducted. In those studies, the following trends were noted.

1. Doses of GA exceeding 5 mg/kg in rats were generally toxic, leading to death.
2. A single dose of microdispersed 17-AAG could be given in doses up to 25 mg/kg in both rats and dogs; the maximum tolerated dose (MTD) when given daily for 5 days was 25 mg/kg/day for rats and 7.5 mg/kg/day for dogs.
3. Lyophylized 17-AAG was tolerated in rats at doses up to 30 mg/kg when given daily or twice daily and at 10 mg/kg/day in dogs.
4. Hepatotoxicity, renal failure, and gastrointestinal (mainly emesis and diarrhea) toxicities were the dose-limiting toxicities (DLTs) in both species. Dogs also experienced gallbladder toxicities.

For 17-DMAG, similar studies were conducted with the following results.

1. When given intravenously (IV) or PO, the maximum daily dose was 12–15 mg/m^2/day in rats and 8 mg/m^2/day in dogs.
2. The main DLTs in both species were renal, gastrointestinal, hepatobiliary, and bone marrow effects.

5. PHARMACOLOGY OF 17-AAG AND 17-DMAG

5.1. Preclinical Pharmacokinetics

Plasma pharmacokinetics (PK) of 17-AAG were measured by high performance liquid chromatography (HPLC) following IV administration of 27, 40, or 60 mg/kg to CD2F1 mice *(75)*. By non-compartmental analysis, area under the curve (AUCs) (402, 625, and 1739 µg/mL min, respectively) increased proportionally for the lower doses, but a greater-than-linear increase was observed at the highest dose. Analysis with the trapezoidal function gave more linear AUCs: 375, 624, and 1373 µg/mL min, respectively for the doses used. Total body clearance varied from 34.5 to 66.3 mg/min/kg. The plasma data were best approximated by a 2-compartment, open, linear model. Terminal half-lives ($t_{1/2}$) were 73, 87, and 361 min following doses of 27, 40, and 60 mg/kg, respectively. In dogs given 1 h IV infusions of 2–10 mg/kg/day × 5 days, the mean $t_{1/2}$ was dose independent and ranged from 46 to 73 min *(76)*.

In a preliminary report of studies performed in normal mice and SCID mice bearing MDA-MB-453 xenografts, both 17-AAG and 17-AG levels were below detection in normal tissues 7 h after a single injection of 40 mg/kg 17-AAG *(77)*. However, 17-AAG and 17-AG levels were 0.5–1 µg/g in tumor tissue for more than 48 h.

The pharmacokinetic-pharmacodynamic relationships for 17-AAG were investigated in nude mice bearing human ovarian cancer xenografts CH1 and A2780 *(78,79)*. Following a single dose of 80 mg/kg IP, the half-lives in plasma, liver, and tumor were 0.88, 0.86, and 7.5 h in the A2780-bearing mice and 0.89, 1.73, and 3.86 h in the CH1-bearing mice, respectively, confirming other reports of differential drug accumulation in tumor *(78)*. There was no tumor response on the single-dose regimen, and western blotting showed a minimal induction of Hsp70 at 16–24 h in A2780, but not in CH1, xenografts. Tumor response was obtained with multiple dosing. The growth delay was greater in A2780 tumors (6.8 days) than in H1 tumors (2 days), and tumor growth resumed 2–4 days after dosing ceased. On day 4, expression of Raf-1, Lck, and Cdk4 was reduced, and expression of Hsp70 was increased in the mouse peripheral blood leukocytes (PBLs) *(79)*. With the exception of Lck, which is not expressed in the A2780 tumors, these changes were mirrored in the tumor tissue. These preliminary reports suggest that the use of PBLs to measure pharmacodynamic endpoints may be possible in clinical trials. The same markers are being used to guide the phase I study of 17-AAG at the investigator's institution.

5.2. Clinical Pharmacology

Plasma PK were described in patients who entered on a phase I trial of 17-AAG given daily for 5 days every 3 weeks *(80–82)*. One patient each was treated at dose levels of 10, 14, 20, and 28 mg/m^2, eight patients at 40 mg/m^2, and seven patients at 56 mg/m^2. A two-compartment, open model best fit the pharmacokinetic data *(81)*. Mean values for terminal $t_{1/2}$, clearance, and steady-state volume of distribution (VD_{ss}) were 2.5 ± 0.5 h, 41.0 ± 13.5 L/h, and 86.6 ± 34.6 L/m^2, respectively. Peak plasma concentrations reached 3170 ± 1310 nM at 56 mg/m^2 *(81)*. Using non-compartmental analysis of data from patients treated with 56 mg/m^2, average values for 17-AAG and 17-AG, respectively, were C_{max} equal to 2080 and 770 nM, AUC equal to 6708 and 5558 nM h, and terminal $t_{1/2}$ equal to 3.8 and 8.6 h *(82)*. Clearances of 17-AAG and 17-AG were 19.9 and 30.8 L/h/m^2, and VD_{ss} were 92 and 203 L/m^2, respectively. Over all dose levels, the total amount of drug recovered in urine was 10.6% for 17-AAG and 7.8% for 17-AG. There were no significant differences between day 1 and day 5 PK values. The MTD was 40 mg/m^2, a dose at which Hsp90 inhibition would be expected. Another phase I trial, which used the same daily × 5 schedule, provided PK data for the 80 mg/m^2 dose *(83)*. The $t_{1/2}$ was 1.5 h, and the peak plasma level was 2700 nM at 30 min. Plasma levels at 1, 6, 24, 72, and 96 h were 1930, 190, 36, 63, and 57 nM, respectively. For the active metabolite, 17-AG, the $t_{1/2}$ was 1.75 h, and the peak plasma level was 607 nM at 1 h. 17-AG plasma levels at 0.5, 6, 24, 72, and 96 h were 262, 138, 46, 101, and 39 nM, respectively. Thus, concentrations exceeded in vitro and xenograft concentrations of 10–500 nM for cell kill.

Preliminary PK and pharmacodynamic data have also been reported from a phase I trial of 17-AAG given weekly for 3 weeks, and every 4 weeks have been reported *(84,85)*. The median clearance of 17-AAG from plasma samples (n = 9) drawn on day 1 was 412 (range 208–4885) mL/min/m^2. C_{max} increased linearly with dose, and the $t_{1/2}$ was 166 ± 115 min *(85)*. The $t_{1/2}$ for 17-AG was 277 min (4.6 h). 17-AAG was a substrate for both the CYP3A4 and CYP3A5 enzyme systems *(85)*.

6. CLINICAL TRIAL DATA

A phase I Institute of Cancer Research (UK) phase I trial of 17-AAG in malignant melanoma used a once-weekly administration schedule. The starting dose was 10 mg/m^2/week administered IV once weekly in a cohort of three patients. Doses were doubled in each succeeding cohort *(86)*. Adverse events included grade 1/2 nausea and grade 1/2 fatigue in 3 and 9 of the first 15 patients, respectively. One patient experienced grade 3 vomiting at the 80 mg/m^2/week dose. Grade 3 nausea and vomiting occurred in two of six patients treated at the 320 mg/m^2/week dose, following which the dose was escalated by 40% to 450 mg/m^2/week *(87)*. The DLT at 450 mg/m^2/week was grade 3/4 elevation of AST/ALT in one of six patients *(88)*. A total of 28 patients have been treated to date on this trial. Among the six patients treated at the 320–450 mg/m^2/week dose range, two patients showed SD for 27 and 91 weeks, respectively.

PD marker analysis of tumor biopsies done before and 24 h after treatment in nine patients showed depletion of c-Raf in four of seven samples (where the marker was expressed) and cdk4 depletion and Hsp70 induction in eight of the nine samples *(88)*.

At the highest dose level, PK analysis indicated a $t_{1/2}$ of 5.8 ± 1.9 h, VD_{ss} of 274 ± 108 L, clearance of 35.5 ± 16.6 L/h, and C_{max} of 16.2 ± 6.3 μM *(88)*, which is above the levels of 375 nM–10 μM reported to inhibit Hsp90 in vitro *(89)*. Although an MTD was not established in this trial, the RP2D is likely to be 450 mg/m^2/week, as there was evidence of tumor target inhibition at that dose level *(88)*. Updated results of this phase I trial have recently been published *(90)*.

The NCI has sponsored 17 phase I studies (seven single agent and ten combinations) to evaluate 17-AAG. An overview of trials conducted under this IND, regardless of status, is presented in Table 1 and 2. Two NCI-sponsored phase I studies have been completed, and the data have been published *(91,92)*. Table 3 covers the four trials currently being planned or conducted with 17-DMAG, again, regardless of status. Of note, dosing was adjusted based on results from early phase I work. In one study, patients with advanced solid tumors were treated with a 60-min IV infusion for 5 consecutive days every 3 weeks. An MTD of 40 mg/m^2/dose was established *(81)*. In a second study, patients with advanced solid tumors who received daily doses for 5 days every 3 weeks reached an MTD of 80 mg/m^2/dose *(83)*. Increasing the dosing interval to days 1, 8, and 15 of a 3-week cycle resulted in an MTD of 308 mg/m^2/dose *(84)*, and this protocol was amended to alter the dosing to days 1, 4, 8, and 11 based on pharmacokinetic endpoints. Additionally, when patients were dosed weekly for 4 weeks, dosing could be escalated to 450 mg/m^2/dose *(86,87)*.

The Hsp90 inhibitors are a class of agents that affect a diverse group of client proteins involved in oncogenesis. Many of these clients are expressed in a disease-specific fashion. The development of these inhibitors as biomodulators is complex and not necessarily governed by standard approaches. The clinical approach taken with the Hsp90 inhibitors was to proceed simultaneously with single-agent phase II studies as well as disease-specific combinations that would be used to evaluate the biomodulatory effects of 17-AAG and 17-DMAG. The ongoing clinical trials outlined in Tables 1–3

Table 1

National Cancer Institute Sponsored 17-Allylanino-17-Demelnoxygdadanamyan (17-AAG) Single Agent and Combination Phase 1 Clinical Trials

Study number	No. of patients/disease type	Agent(s)	Dose/schedule	Toxicities
6323	70/Solid tumor	17-AAG	Dose escalation from 150 mg/m² at level 1 to 480 mg/m² at level 5 IV twice per week for 2 weeks followed by a 1 week rest for patients with solid tumors. In adult leukemia patients, the rest week is ornitted.	
ADVL0316	36/Solid tumor	17-AAG	Dose escalation from 150 mg/m² at level 1 to 480 mg/m² at level 5 IV twice per week for 2 weeks followed by a 1 week rest for patients with solid tumors. In adult leukemia patients, the rest week is omitted.	
T98-0075	38/Solid tumor	17-AAG	Dose escalations of 40 mg/m²/day at level 1 to 301 mg/m²/day at level 7 given on days 1, 4, 15, and 18 of a 4 week cycle.	Grade 2 hepatitis, grade 3 nausea, grade 3 dypsnea
T99-0035	96/Solid tumor and refractory hematological malignancies	17-AAG	150 mg/m² twice a week for 12 weeks and escalated by 40% with each cohort. An MTD is defined independently for each population.	Nausea, vomiting secondary to pancreatitis, and grade 3 fatigue
T99-0038	24/Solid tumor	17-AAG	220 mg/m²/week for 12 weeks, escalating to 700 mg/m²/week	Grade 3 reversible hepatitis
T99-0058	130/Solid tumor	17-AAG	Cohort 1: From 10 mg/m²/dose to 603 mg/m²/dose on days 1, 8, and 15 in a 28-day cycle.Cohort 2: From 10 mg/m²/dose to 603 mg/m²/dose on days 1, 4, 8, and 11 in a 21-day cycle.	Grade 4 elevated SGOT, dypsnea, hypoxia

(Continued)

Table 1
(Continued)

Study number	No. of patients/disease type	Agent(s)	Dose/schedule	Toxicities
5291	66/ Solid tumor	17-AAG, gemcitabine, and cisplatin	Cohort A: 17-AAG 154 mg/m² IV over 1 h on days 1 and 8, every 21 days; gemcitabine 500 mg/m² IV over 30 mins on days 1 and 8, every 21 days; cisplatin 30 mg/m² IV over 2 h on days 1 and 8, every 21 days B, C, D An MTD of 17-AAG 154 mg/m² IV over 1 h on days 1 and 8, every 21 days; gemcitabine 750 mg/m² IV over 30 mins on days 1 and 8, every 21 days; cisplatin 40 mg/m² IV over 2 h on days 1 and 8, every 21 days	
5878	30/ Solid tumor	17-AAG and docetaxel	Schedule 1: Docetaxel 55–75 mg/m² IV over 1hr on day 1, every 21 days; 17-AAG 80–650 mg/m² IV over 1 hr on day 1, every 21 daysSchedule 2: Docetaxel 35 mg/m² IV over 1hr on days 1, 8, and 15 every 28 days; 17-AAG 160–450 mg/m² IV over 1 hr on day 1, 8, and 15 every 28 days	
6494	35/ Solid tumor	17-AAG and paclitaxel	17-AAG 100–225 mg/m² IV over 1 hr on days 1, 4, 8, 11, 15, and 18 every 28 daysPaclitaxel 80 mg/m² IV over 1hr on days 1, 8, and 15 every 28 days	
5932	18/CML	17-AAG and imatinib	Imatinab 600 mg/day PO once a day started 4-5 days prior to the first 17-AAG treatment. 17-AAG 20–60 mg/m² days 1 and 4 of weeks 1 and 2, each 3 weeks	Just opening

Study	Disease	Combination	Regimen
6383	36/ALL	17-AAG and cytarabine	Cytarabine 400 mg/m²/day continuous infusion days 1 through 5; 17-AAG 100 mg/m² to 400 mg/m² over 60 min on days 3 and 6; Repeat 30 ± 5 days after marrow recovery or hospital discharge
6518	30/CLL	17-AAG, fludarabine and rituximab	17-AAG 100–360 mg/m² IV over 60 minutes on days 1, 4, 8, 11, 15, and 18 of a 28-day cycle; fludarabine 25 mg/m² IVPB will be administered days 1–5; rituximab day 1, cycle 1 100 mg IVPB over 4 h; day 3, cycle 1 375 mg/m² using standard escalation; day 5, cycle 1 of therapy 375 mg/m² using standard escalation.
6520	74/Hematologic unspecified	17-AGG and bortezomib	17-AAG 100 mg/m² to 250 mg/m² administered over 1 h immediately prior to PS-341 0.7 to 1.3 mg/m² on days 1, 4, 8, and 11 of each cycle
6121	42/Solid tumor	17-AAG and bortezomib	17-AAG 100–250 mg/m² administered over 1 h immediately prior to PS-341 0.7 – 1.3 mg/m² on days 1, 4, 8, and 11 of each cycle
6972	27/Solid tumor	17-AAG and BAY 43-9006	BAY 43-9006 400 mg BID starting 2 weeks prior to 17-AAG 100–250 mg/m² days 1, 8, and 15 every 28 days
7009	46/Solid tumor	17-AAG and irinotecan	Irinotecan 85–125 mg/m² followed by 17-AAG 220–450 mg/m² once weekly for 2 weeks in a 21-day cycle.

IV, intravenously; ALL, acute lymphoblastic leukemia; CLL, chronic lymphocytic leukemia; CML, chronic myelogenous leukemia.

Table 2
National Cancer Institute Sponsored 17-Allylanino-17-Demelnoxygdadanamycin (17-AAG)
Phase 2 Clinical Trials

6307	40/Ovarian epithelial cancer stage IV	17-AAG	17-AAG 220 mg/m² IV over 1 h on days 1, 4, 8, and 11, every 21 days
6399	26/Clear cell carcinoma of the kidney	17-AAG	17 AAG: 300 mg/m² IV over 1–2 h. on days 1, 8, 15, q28 days.
6454	36/malignant mast cell neoplasm	17-AAG	17-AAG 220 mg/m² IV over 1 h on days 1, 4, 8, and 11, every 3 weeks
6479	58/ Renal cell carcinoma stage IV	17-AAG	17-AAG: 220 mg/m² IV over 60–90 min twice weekly for 2 weeks. Cycle = 21 days.
6480	50/Malignant melanoma stage IV	17-AAG	7-AAG 450 mg/m² IV over 1–2 h q week x 6 weeks, every 8 weeks
6482	72/Medullary thyroid cancer	17-AAG	17-AAG 220 mg/m² IV over 1 h on days 1, 4, 8, and 11, q21 days
6500	25/Malignant melanoma stage IV	17-AAG	17-AAG 450 mg/m² IV over 1 h , once every 7 days, for 12 weeks
6552	41/Breast cancer stage IV	17-AAG	17-AAG 220 mg/m² IV over 1 h on days 1, 4, 8, and 11, q21 days
6651	28/Prostate cancer stage IV	17-AAG	17 AAG: 300 mg/m² IV over 1–2 h. on days 1,8,15, q28 days.
6936	70/Mantle cell lymphoma	17-AAG	17-AAG 220 mg/m² IV over 1 h on days 1, 4, 8, and 11, q21 days

IV, intravenously.

Table 3
National Cancer Institute Sponsored 17-Allylanino-17-Demelnoxygdadanamycin (17-AAG)
Phase 1 Clinical Trials

6542	30/ Solid tumor	17-DMAG	17-DMAG 2.5–40 mg/m² IV weekly x 3
6544	40/Solid tumor	17-DMAG	(Cycle = 4 weeks): 17-DMAG 1–40 mg/m² IV over 1 h on days 1 and 4 of each week
6547	30/Solid tumor N	17-DMAG	(Cycle = 4 weeks): 17-DMAG 1.25–10 mg/m² IV over 1 hour each week
6548	60/Solid tumor N	17-DMAG	17-DMAG 1.5 – 19 mg/m² IV over 1 h, daily x 5, q 21 days

will be used to assess activity of the agents in a disease-specific fashion and to provide a response comparison for the phase I combinations to proceed into disease-specific phase II investigations. As these studies mature and reach completion, the role of Hsp90 inhibitors in the treatment of cancer should be better defined with regard to their activity and molecular targeted effects.

REFERENCES

1. Stoler DL, Chen N, Basik M, et al. The onset and extent of genomic instability in sporadic colorectal tumor progression. Proc Natl Acad Sci USA 1999;96(26):15121–15126.
2. Hahn WC, Weinberg RA. Modelling the molecular circuitry of cancer. Nat Rev Cancer 2002;2(5):331–341.
3. La Rosee P, O'Dwyer ME, Druker BJ. Insights from pre-clinical studies for new combination treatment regimens with the Bcr-Abl kinase inhibitor imatinib mesylate (Gleevec/Glivec) in chronic myelogenous leukemia: a translational perspective. Leukemia 2002;16(7):1213 1219.
4. Kitano H. Cancer robustness: tumour tactics. Nature 2003;426(6963):125.
5. Hanahan D, Weinberg RA. The hallmarks of cancer. Cell 2000;100(1):57–70.
6. Isaacs JS. Heat-shock protein 90 inhibitors in antineoplastic therapy: is it all wrapped up? Expert Opin Investig Drugs 2005;14:569–589.
7. Chiosis G, Vilenchik M, Kim J, Solit D. Hsp90: the vulnerable chaperone. Drug Discov Today 2004;9(20):881–888.
8. Workman P. Combinatorial attack on multistep oncogenesis by inhibiting the Hsp90 molecular chaperone. Cancer Lett 2004;206(2):149–157.
9. Bagatell R, Whitesell L. Altered Hsp90 function in cancer: a unique therapeutic opportunity. Mol Cancer Ther 2004;3(8):1021–1030.
10. Zhang H, Burrows F. Targeting multiple signal transduction pathways through inhibition of Hsp90. J Mol Med 2004;82(8):488–499.
11. Goetz MP, Toft DO, Ames MM, Erlichman C. The Hsp90 chaperone complex as a novel target for cancer therapy. Ann Oncol 2003;14(8):1169–1176.
12. Isaacs JS, Xu W, Neckers L. Heat shock protein 90 as a molecular target for cancer therapeutics. Cancer Cell 2003;3(3):213–217.
13. Rutherford SL, Lindquist S. Hsp90 as a capacitor for morphological evolution. Nature 1998;396(6709):336–342.
14. Queitsch C, Sangster TA, Lindquist S. Hsp90 as a capacitor of phenotypic variation. Nature 2002;417(6889):618–624.
15. DeBoer C, Meulman PA, Wnuk RJ, Peterson DH. Geldanamycin, a new antibiotic. J Antibiot (Tokyo) 1970;23(9):442–447.
16. Schulte TW, Neckers LM. The benzoquinone ansamycin 17-allylamino-17-demethoxygeldanamycin binds to HSP90 and shares important biologic activities with geldanamycin. Cancer Chemother Pharmacol 1998;42(4):273–279.
17. Supko JG, Hickman RL, Grever MR, Malspeis L. Preclinical pharmacologic evaluation of geldanamycin as an antitumor agent. Cancer Chemother Pharmacol 1995;36(4):305–315.
18. Page J, Heath J, Fulton R, et al. Comparison of geldanamycin (NSC-122750) and 17-allylaminogeldanamycin (NSC-330507D) toxicity in rats. Proc Am Assoc Cancer Res 1997;38:abstract 2067.
19. Paine-Murrieta G, Cook P, Taylor CW, Whitesell L. The anti-tumor activity of 17-allylaminogeldanamycin is associated with modulation of target protien levels in vivo. Proc Am Assoc Cancer Res 1999;40:abstract 119.
20. Burger AM, Fiebig HH, Newman DJ, Camalier RF, Sausville EA. Antitumor activity of 17-allylaminogeldanamycin (NSC 330507) in melanoma xenografts is associated with decline in Hsp90 protein expression. 10th NCI-EORTC Symposium on New Drugs in Cancer Therapy 1998: abstract 504.
21. Siligardi G, Hu B, Panaretou B, Piper PW, Pearl LH, Prodromou C. Co-chaperone regulation of conformational switching in the Hsp90 ATPase cycle. J Biol Chem 2004; 279(50):51989–51998.
22. Wegele H, Muller L, Buchner J. Hsp70 and Hsp90–a relay team for protein folding. Rev Physiol Biochem Pharmacol 2004;151:1–44.

23. Prodromou C, Pearl LH. Structure and functional relationships of Hsp90. Curr Cancer Drug Targets 2003;3(5):301–323.

24. Neckers L. Hsp90 inhibitors as novel cancer chemotherapeutic agents. Trends Mol Med 2002;8 (4 Suppl):S55–S61.

25. An WG, Schulte TW, Neckers LM. The heat shock protein 90 antagonist geldanamycin alters chaperone association with p210bcr-abl and v-src proteins before their degradation by the proteasome. Cell Growth Differ 2000;11(7):355–360.

26. Fujimoto J, Shiota M, Iwahara T, et al. Characterization of the transforming activity of p80, a hyperphosphorylated protein in a Ki-1 lymphoma cell line with chromosomal translocation t(2;5). Proc Natl Acad Sci USA 1996;93(9):4181–4186.

27. Bonvini P, Gastaldi T, Falini B, Rosolen A. Nucleophosmin-anaplastic lymphoma kinase (NPM-ALK), a novel Hsp90- client tyrosine kinase: down-regulation of NPM-ALK expression and tyrosine phosphorylation in ALK(+) CD30(+) lymphoma cells by the Hsp90 antagonist 17-allylamino,17-demethoxygeldanamycin. Cancer Res 2002;62(5):1559–1566.

28. Naoe T, Kiyoe H, Yamamoto Y, et al. FLT3 tyrosine kinase as a target molecule for selective antileukemia therapy. Cancer Chemother Pharmacol 2001;48(Suppl 1):S27–S30.

29. Minami Y, Kiyoi H, Yamamoto Y, et al. Selective apoptosis of tandemly duplicated FLT3-transformed leukemia cells by Hsp90 inhibitors. Leukemia 2002;16(8):1535–1540.

30. Shiotsu Y, Neckers LM, Wortman I, et al. Novel oxime derivatives of radicicol induce erythroid differentiation associated with preferential G(1) phase accumulation against chronic myelogenous leukemia cells through destabilization of Bcr-Abl with Hsp90 complex. Blood 2000;96(6): 2284–2291.

31. Druker BJ, Tamura S, Buchdunger E, et al. Effects of a selective inhibitor of the Abl tyrosine kinase on the growth of Bcr-Abl positive cells. Nat Med 1996;2(5):561–566.

32. Sawyers CL, Hochhaus A, Feldman E, et al. Imatinib induces hematologic and cytogenetic responses in patients with chronic myelogenous leukemia in myeloid blast crisis: results of a phase II study. Blood 2002;99(10):3530–3539.

33. Shah NP, Nicoll JM, Nagar B, et al. Multiple BCR-ABL kinase domain mutations confer polyclonal resistance to the tyrosine kinase inhibitor imatinib (STI571) in chronic phase and blast crisis chronic myeloid leukemia. Cancer Cell 2002;2(2):117–125.

34. Gorre ME, Ellwood-Yen K, Chiosis G, Rosen N, Sawyers CL. BCR-ABL point mutants isolated from patients with STI571-resistant chronic myeloid leukemia remain sensitive to inhibitors of the BCR-ABL chaperone heat shock protein 90. Blood 2002;100:3041–3044.

35. Nimmanapalli R, O'Bryan E, Huang M, et al. Molecular characterization and sensitivity of STI-571 (imatinib mesylate, Gleevec)-resistant, Bcr-Abl-positive, human acute leukemia cells to SRC kinase inhibitor PD180970 and 17-allylamino-17-demethoxygeldanamycin. Cancer Res 2002;62(20): 5761–5769.

36. Fumo G, Akin C, Metcalfe DD, Neckers L. 17-Allylamino-17-demethoxygeldanamycin (17-AAG) is effective in down-regulating mutated, constitutively activated KIT protein in human mast cells. Blood 2004;103:1078–1084.

37. Vanaja DK, Mitchell SH, Toft DO, Young CYF. Effect of geldanamycin on androgen receptor function and stability. Cell Stress Chaperones 2002;7:55–64.

38. Georget V, Terouanne B, Nicolas J-C, Sultan C. Mechanism of antiandrogen action: key role of Hsp90 in conformational change and transcriptional activity of the androgen receptor. Biochemistry 2002;41:11824–11831.

39. Solit D, Zheng F, Drobnjak M, et al. 17-allylamino-17-demthoxygeldanamycin induces the degradation of androgen receptor and HER-2/neu and inhibits the growth of prostate cancer xenografts. Clin Cancer Res 2002:986–993.

40. Harris AL. Hypoxia- a key regulatory factor in tumor growth. Nat Rev Cancer 2002;2:38–47.

41. Maxwell PH, Wiesener MS, Chang G-W, et al. The tumor suppressor protein VHL targets hypoxia-inducible factors for oxygen-dependent proteolysis. Nature 1999;399:271–275.

42. Seizinger BR, Rouleau GA, Ozelius LJ, et al. Von Hippel-Lindau disease maps to the region of chromosome 3 associated with renal cell carcinoma. Nature 1988;332(6161):268–269.

43. Gradin K, McGuire J, Wenger RH, et al. Functional interference between hypoxia and dioxin signal transduction pathways: competition for recruitment of the Arnt transcription factor. Mol Cell Biol 1996;16(10):5221–5231.

44. Hur E, Kim HH, Choi SM, et al. Reduction of hypoxia-induced transcription through the repression of hypoxia-inducible factor-1alpha/aryl hydrocarbon receptor nuclear translocator DNA binding by the 90-kDa heat-shock protein inhibitor radicicol. Mol Pharmacol 2002;62(5):975–982.

45. Isaacs JS, Jung YJ, Mimnaugh EG, Martinez A, Cuttitta F, Neckers LM. Hsp90 regulates a von Hippel Lindau-independent hypoxia-inducible factor-1 alpha-degradative pathway. J Biol Chem 2002;277(33):29936–29944.

46. Mabjeesh NJ, Post DE, Willard MT, et al. Geldanamycin induces degradation of hypoxia-inducible factor 1α protein via the proteasome pathway in prostate cancer cells. Cancer Res 2002;62: 2478–2482.

47. Zagzag D, Nomura M, Friedlander DR, et al. Geldanamycin inhibits migration of glioma cells in vitro: a potential role for hypoxia-inducible factor (HIF-1alpha) in glioma cell invasion. J Cell Physiol 2003;196(2):394–402.

48. Dias S, Shmelkov SV, Lam G, Rafii S. VEGF(165) promotes survival of leukemic cells by Hsp90-mediated induction of Bcl-2 expression and apoptosis inhibition. Blood 2002;99(7):2532–2540.

49. Maulik G, Kijima T, Ma PC, et al. Modulation of the c-Met/hepatocyte growth factor pathway in small cell lung cancer. Clin Cancer Res 2002;8(2):620–627.

50. Pennacchietti S, Michieli P, Galluzzo M, Mazzone M, Giordano S, Comoglio PM. Hypoxia promotes invasive growth by transcriptional activation of the met protooncogene. Cancer Cell 2003;3(4): 347–361.

51. Tacchini L, Dansi P, Matteucci E, Desiderio MA. Hepatocyte growth factor signalling stimulates hypoxia inducible factor-1 (HIF-1) activity in HepG2 hepatoma cells. Carcinogenesis 2001;22(9):1363–1371.

52. Bottaro DP, Liotta LA. Out of air is not out of action. Nature 2003;423:593–595.

53. Santoro M, Melillo RM, Carlomagno F, Fusco A, Vecchio G. Molecular mechanisms of RET activation in human cancer. Ann N Y Acad Sci 2002;963:116–121.

54. Jhiang SM. The RET proto-oncogene in human cancers. Oncogene 2000;19(49):5590–5597.

55. Ichihara M, Murakumo Y, Takahashi M. RET and neuroendocrine tumors. Cancer Lett 2004;204(2):197–211.

56. Cohen MS, Hussain HB, Moley JF. Inhibition of medullary thyroid carcinoma cell proliferation and RET phosphorylation by tyrosine kinase inhibitors. Surgery 2002;132(6):960–966.

57. Mimnaugh EG, Chavany C, Neckers L. Polyubiquitination and proteasomal degradation of the p185c-erbB-2 receptor protein-tyrosine kinase induced by geldanamycin. J Biol Chem 1996;271(37): 22796–22801.

58. Schneider C, Sepp-Lorenzino L, Nimmesgern E, et al. Pharmacologic shifting of a balance between protein refolding and degradation mediated by Hsp90. Proc Natl Acad Sci USA 1996;93(25): 14536–14541.

59. Basso AD, Solit DB, Chiosis G, Giri B, Tsichlis P, Rosen N. Akt forms an intracellular complex with heat shock protein 90 (Hsp90) and Cdc37 and is destabilized by inhibitors of Hsp90 function. J Biol Chem 2002;277(42):39858–39866.

60. French BA, van Leeuwen F, Riley NE, et al. Aggresome formation in liver cells in response to different toxic mechanisms: role of the ubiquitin-proteasome pathway and the frameshift mutant of ubiquitin. Exp Mol Pathol 2001;71(3):241–246.

61. Waelter S, Boeddrich A, Lurz R, et al. Accumulation of mutant huntingtin fragments in aggresome-like inclusion bodies as a result of insufficient protein degradation. Mol Biol Cell 2001;12(5): 1393–1407.

62. Aghajanian C, Soignet S, Dizon DS, et al. A phase I trial of the novel proteasome inhibitor PS341 in advanced solid tumor malignancies. Clin Cancer Res 2002;8(8):2505–2511.

63. L'Allemain G. [Update on. the proteasome inhibitor PS341]. Bull Cancer 2002;89(1):29–30.

64. Mitsiades N, Mitsiades CS, Poulaki V, et al. Molecular sequelae of proteasome inhibition in human multiple myeloma cells. Proc Natl Acad Sci USA 2002;99(22):14374–14379.

65. Mimnaugh EG, Xu W, Vos M, et al. Simultaneous inhibition of hsp 90 and the proteasome promotes protein ubiquitination, causes endoplasmic reticulum-derived cytosolic vacuolization, and enhances antitumor activity. Mol Cancer Ther 2004;3(5):551–566.

66. Bisht KS, Bradbury CM, Mattson D, et al. Geldanamycin and 17-allylamino-17-demethoxygeldanamycin potentiate the in vitro and in vivo radiation response of cervical tumor cells via the heat shock protein 90-mediated intracellular signaling and cytotoxicity. Cancer Res 2003;63(24): 8984–8995.

67. Machida H, Matsumoto Y, Shirai M, Kubota N. Geldanamycin, an inhibitor of Hsp90, sensitizes human tumour cells to radiation. Int J Radiat Biol 2003;79(12):973–980.

68. Becker B, Multhoff G, Farkas B, et al. Induction of Hsp90 protein expression in malignant melanomas and melanoma metastases. Exp Dermatol 2004;13(1):27–32.

69. Eustace BK, Sakurai T, Stewart JK, et al. Functional proteomic screens reveal an essential extracellular role for hsp90 alpha in cancer cell invasiveness. Nat Cell Biol 2004;6(6):507–514.

70. Eustace BK, Jay DG. Extracellular roles for the molecular chaperone, hsp90. Cell Cycle 2004;3(9):1098–1100.

71. Egorin MJ, Rosen DM, Wolff JH, Callery PS, Musser SM, Eiseman JL. Metabolism of 17-(allylamino)-17-demethoxygeldanamycin (NSC 330507) by murine and human hepatic preparations. Cancer Res 1998;58(11):2385–2396.

72. Schnur RC, Corman ML, Gallaschun RJ, et al. erbB-2 oncogene inhibition by geldanamycin derivatives: synthesis, mechanism of action, and structure-activity relationships. J Med Chem 1995;38(19):3813–3820.

73. Kelland LR, Sharp SY, Rogers PM, Myers TG, Workman P. DT-Diaphorase expression and tumor cell sensitivity to 17-allylamino, 17-demethoxygeldanamycin, an inhibitor of heat shock protein 90. J Natl Cancer Inst 1999;91(22):1940–1949.

74. Egorin MJ, Lagattuta TF, Hamburger DR, et al. Pharmacokinetics, tissue distribution, and metabolism of 17-(dimethylaminoethylamino)-17-demethoxygeldanamycin (NSC 707545) in CD2F1 mice and Fischer 344 rats. Cancer Chemother Pharmacol 2002;49(1):7–19.

75. Eiseman JL, Sentz DL, Zuhowski EG. Plasma pharmacokinetics and tissue distribution of 17-allylaminogeldanamycin (NSC 330507), a prodrug for geldanamycin, in CD2F1 mice and Fisher 344 rats. Proc Am Assoc Cancer Res 1997;38:abstract 2063.

76. Noker PE, Thompson RB, Smith AC, et al. Toxicity and pharmacokinetics of 17-allylaminogeldanamycin (17-AAG, NSC-330507) in dogs. Proc Am Assoc Cancer Res 1999;40:abstract 804.

77. Eiseman JL, Grimm A, Sentz DL, et al. Pharmacokinetics and tissue distribution of 17-allylamino(17demethoxy)geldanamycin in SCID mice bearing MDA-MB-453 xenografts and alterations in the expression of p185^{erbB2} in xenografts following treatment. AACR-NCI-EORTC International Conference on Molecular Targets and Cancer Therapeutixs 1999;abstract 536.

78. Banerji U, Maloney A, Asad Y, et al. Pharmacokinetic-pharmacodynamic (PK-PD) relationships for the HSP90 molecular chaperone inhibitor 17-allylamino-17demethoxygeldanamycin (17AAG) in human ovarian cancer xenografts. Proc 11th NCI-EORTC Symposium on New Drugs in Cancer Therapy 2001;abstract 395.

79. Banerji U, Walton M, Raynauld F, et al. Validation of pharmacodynamic endpoints for the HSP90 molecular chaperone inhibitor 17-allylamino 17-demethoxygeldanamycin (17AAG) in a human tumor xenograft model. Proc Am Assoc Cancer Res 2001;42:abstract 4473.

80. Agnew EB, Neckers LM, Hehman HE, et al. Human plasma pharmacokinetics of the novel antitumor agent, 17-allylaminogeldanamycin (AAG) using a new HPLC-based analytic assay. Proc Am Assoc Cancer Res 2000;41:abstract 4458.

81. Wilson RH, Takimoto CH, Agnew EB. Phase I pharmacologic study of 17-(Allylamino)-17-demethoxygeldanamycin (AAG) in adult patients with advanced solid tumors. Proc Am Soc Clin Oncol 2001;20:abstract 325.

82. Agnew EB, Wilson RH, Morrison G, et al. Clinical pharmacokinetics of 17-(allylamino)-17-demethoxygeldanamycin and the active metabolite 17-(amino)-17-demethoxygeldanamycin given as a one-hour infusion daily for 5 days. Proc Am Assoc Cancer Res 2002;43:abstract 1349.

83. Munster PN, Tong W, Schwartz L, et al. Phase I trial of 17-(allylamino)-17-demethoxygeldanamycin (17-AAG) in patients with advanced solid malignancies. Proc Am Soc Clin Oncol 2001;20; abstract 326.

84. Erlichman C, Toft D, Reid J. A phase I trial of 17-allyl-amino-geldanamycin in patients with advanced cancer. Proc Am Assoc Cancer Res 2001;42:abstract 4474.

85. Goetz M, Toft D, Reid J. A phase I trial of 17-allyl-amino-geldanamycin (17-AAG) in patients with advanced cancer. Eur J Cancer 2002;38 (Suppl 7):S54–S55, abstract 170.

86. Banerji U, O'Donnell A, Scurr M, et al. Phase I trial of the heat shock protein 90 (HSP90) inhibitor 17-allylamino 17-demethoxygeldanamycin 17aag. Pharmacokinetic (PK) profile and pharmacodynamic (PD) endpoints. Proc Am Soc Clin Oncol 2001;20:abstract 326.

87. Banerji U, O'Donnell A, Scurr M, et al. A pharmacokinetically (Pk) - pharmacodynami-cally (Pd) driven phase I trial of the Hsp90 molecular chaperone inhibitor 17-allyamino 17-demethoxygeldanamycin (17AAG). Proc 93rd Annu Meet Am Assoc Cancer Res 2002;43: abstract 1352.

88. Banerji U, O'Donnell A, Scurr M, et al. A pharmacokinetically (PK) - pharmacodynamically (PD) guided phase I trial of the heat shock protein 90 (HSP90) inhibitor 17-allylamino,17-demethoxygeldanamycin (17AAG). Proc Am Soc Clin Oncol 2003;22:abstract 797.

89. Burger AM, Sausville EA, Carmalier RF, Newman DJ, Fiebig HH. Response of human melanomas to 17-AAG is associated with modulation of the molecular chaperone function of Hsp90. Proc Am Assoc Cancer Res 2000;41:abstract 2844.

90. Banerji U, O'Donnell A, Scurr M, et al. Phase I pharmacokinetic and pharmacodynamic study of 17-allylamino,17-demethoxygeldanamycin in patients with advanced malignancies. J Clin Oncol 2005;23:4152–4161.

91. Goetz M, Toft D, Reid J, et al. Phase I trial of 17-allylamino-17-demethoxygeldanamycin in patients with advanced cancer. J Clin Oncol 2005;23:1078–1087.

92. Grem JL, Morrison G, Guo X-D, et al. Phase I and pharmacologic study of 17-(allylamino)-17-demethoxygeldanamycin in adult patients with solid tumors. J Clin Oncol 2005;23:1885 1893.

5 The Cancer Epigenome

Can it be Targeted for Therapy?

Sam Thiagalingam, PhD, and Douglas V. Faller, MD, PhD

SUMMARY

It has become increasingly clear in recent years that reversible alterations in the epigenome, which comprises the chromatin terrain and determines individual gene expression, has as great a role in defining the normal or malignant phenotype of a cell as does the more-familiar fixed genomic DNA sequence upon which it is superimposed. The "epigenomic code" is defined by DNA methylation patterns, unique combinations of post-translational modifications of histones and non-histone proteins, the nature of the remodeled chromatin structure, and the identity of nucleoprotein complexes assembled on the chromatin. The distinctive epigenomic code associated with each individual gene dictates the expression status of that gene, dependent upon its localization and the associated unique chromatin organization. For example, silencing of gene expression, including silencing of tumor suppressor genes, is associated with local deacetylation of histones, and often localized to genomic regions containing DNA methylation as well as methylation at lysine residues 9 and 27 of histone H3, and lysine residue 20 of histone H4. In contrast, activation of gene expression, including activation of oncogenes, is associated with locally acetylated histones and methylation at lysine residues 4, 36, and 79 of histone H3. Currently, there are two groups of drugs targeting the epigenome: inhibitors of DNA methyl transferases and inhibitors of histone deacetylases. These agents have been employed with some success in pre-clinical studies and in early limited clinical trials. In the long term, the development of therapeutic agents which can target with precision the activities of a wide array of specific chromatin-modifying enzymes may become useful in reversing the epigenomic alterations which define the cancer cell (the cancer epigenome) and provide a novel therapeutic approach to malignancy.

Key Words: Epigenome; DNA methylation; DNA methyl transferase (DNMT); histone acetyl transferase (HAT); histone deacetylase (HDAC); histone code; active epigenomic code (AEC); silenced epigenomic code (SEC); histone deacetylase inhibitor (HDACi); cancer therapy.

From: *Cancer Drug Discovery and Development: Molecular Targeting in Oncology*
Edited by: H. L. Kaufman, S. Wadler, and K. Antman © Humana Press, Totowa, NJ

1. INTRODUCTION

The major dilemma currently facing genomic researchers is correlating the existing genetic blueprint at the level of DNA sequence to the differential expression patterns of this library of genes in the form of RNA transcripts and functional proteins, at the level of individual cells, tissues, organs, and the whole organism. Not only do the cells making up different tissues express unique sets of genes that define their characteristics and functions, but we now understand that there are additional levels of control superimposed upon the genome, which are still not fully elucidated. A given genetic sequence or gene may or may not be expressed in a target tissue, depending upon maternal or paternal "imprinting," or on unique physical modifications of the DNA of the gene and its surrounding proteins (which together comprise *chromatin*). The resulting phenomenon is also referred to as altered *penetrance* by population geneticists to account for the absence, variability, and altered severity of disorders corresponding to specific gene abnormalities. Recently, it has become increasingly clear that the *epigenome*, comprised of the various proteins and RNA which associate with the DNA sequence to determine the nature and physical state of chromatin, plays a major role in defining gene expression patterns and hence the functionality and properties of any given cell. Therefore, following the enormous success of sequencing the human genomic DNA, the next major challenge to deciphering the molecular basis of various complex diseases including cancer is to understand the nature, functionality, and regulation of the human epigenome.

The basic structural unit of chromatin is the nucleosome core particle, which consists of 147 base pairs of DNA wrapped around an octamer of basic proteins known as histones. The DNA itself can be dynamically modified by methylation of cytosine residues, and the histone proteins can be modified by methylation, acetylation, or other changes. These reversible notations on the chromatin landscape resulting from differential DNA methylation patterns, and the unique modifications of histones comprising the nucleoprotein complexes, define the epigenome, and determine differential gene expression patterns under normal conditions as well as in disease states. The cellular epigenome is altered in diseases such as cancer, shifting the equilibrium of gene expression patterns from normal to pathological ones. Unlike the DNA sequence mutations which are characteristic of cancer and inherited diseases, and which alter gene expression but are fixed in the genome, epigenetic modifications are dynamic and reversible, implying that the epigenome of a disease state should be theoretically amenable to therapeutic perturbation in a defined and directed manner, with the aim of restoring normal, functional gene expression status. In order to achieve this therapeutic goal, it will be necessary to fully understand the unique and discrete combinations of epigenetic alterations and nucleoprotein associations specific to particular stages and types of cancer, as well as the actions of agents which precisely target epigenetic modifications or interaction. Here, we will discuss modification of genomic DNA by methylation, describe the modifications of the DNA-interacting histone proteins and the dynamics of chromatin, and then review the current understanding of the nature of cancer epigenome and progress on therapeutic efforts.

2. DNA METHYLATION

Approximately 3–5% of the cytosine residues in the genome of mammalian cells are modified as 5-methyl cytosine, and 70–80% of them are found in CpG residues *(1–3)*. Covalent methylation modification of cytosine residues (Fig. 1) represents a

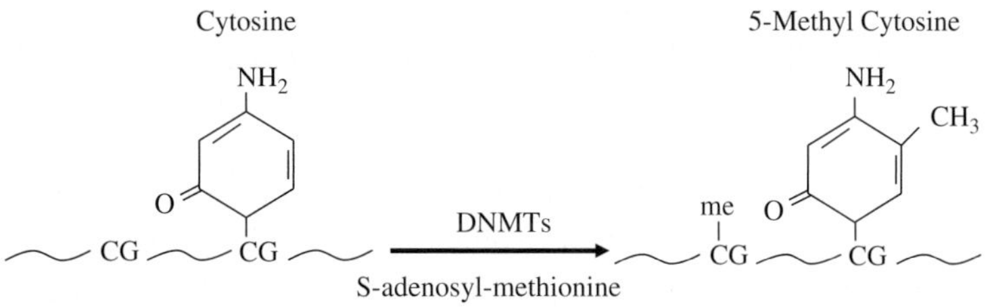

Fig. 1. The catalytic activity of the DNA methyl transferases (DNMTs). The DNMTs add a methyl group to the cytosine in a CpG island to create a reversible modification of the DNA, which then provides a platform for the assembly of unique nucleoprotein complexes initiated by the binding of methyl CpG-binding proteins. These initial complexes recruit histone deacetylases in addition to histone methyl transferases, such as the H3K9 methyl transferases. The deacetylated and methylated histones (e.g., H3K9Me) recruit transcriptional repressors (e.g., HP1) to "lock" the chromatin into a repressed state. See the text for additional details.

reversible but heritable change which can alter gene expression and is ultimately responsible for a diverse array of biological responses. The CpG residues are non-randomly distributed in the human genome, with the majority of the genome being CpG poor, and are often clustered in the promoter/regulatory regions of the genes as CpG islands. The CpG islands occur predominantly in the 5′ regions of genes, such as the promoter, the first exon, and sometimes in the first intron of housekeeping genes and many tissue-specific genes, and occasionally in the 3′ end of some tissue-specific genes *(4)*. DNA sequencing estimates suggest that there are approximately 29,000 CpG islands in the human genome *(5)*. In humans, DNA methyl transferases (DNMTs) add a methyl group preferentially to the 5′ carbon of a cytosine located adjacent to a guanine (5′-CpG-3′) (Fig. 1). Three active DNMTs (DNMT1, DNMT3a, and DNMT3b) have been identified in human and mouse *(6)*. While DNMT1 functions as a maintenance DNA methylase and copies the methylation patterns from the parental to daughter strands of DNA during replication (i.e., hemimethylated DNA), DNMT3a and DNMT3b can also perform additional roles as de novo DNA methylases *(7)*. Differential DNA methylation at CpG islands has been shown to be associated with regulation of gene expression and is essential for normal embryonic development, X chromosome inactivation, imprinting, chromatin modification, suppression of parasitic DNA sequences, and aberrant silencing of tumor suppressor genes or over-activation of oncogenes in cancer. Almost all DNA methylation is erased in the early morula stage of embryogenesis, and the basic pattern of CpG methylation is then re-established due to de novo methylation at the time of embryo implantation *(8)*. Methylation of CpGs in gene promoters is associated with decreased levels of transcription of these genes, when compared to the unmethylated genes. This "silencing" of gene expression by methylation could be mediated either by direct effects on the chromatin or through the transcription machinery.

Although DNA methylation and the associated unique chromatin modifications have been generally associated with silencing of gene expression, there has been no clear consensus as to which is the initiating event, that is, whether DNA methylation invariably precedes chromatin modifications, or vice versa *(9)*. While the marking of DNA by methylation of 5-methylcytosines (m^5C) could be envisioned to readily target specific genomic areas for more elaborate modification, including coating the genome with uniquely modified histones and other chromatin-binding proteins, it is also equally plausible that local modification of chromatin histones may occur first and then initiate recruitment of the DNA methyl transferases leading to methylation of cytosines. The lack of DNA methylation associated with silencing of genes such as *p16*, a tumor suppressor, in some cancers would argue that DNA methylation is not a pre-requisite for silencing of gene expression but may play a critical role in determining the nature of repressed state *(10)*. Overall, DNA methylation is associated with a more stable silenced state, due to a complex set of interactions involving the recruitment of methyl-binding domain (MBD) proteins such as MeCP2, MBD2, and MBD3 complexes and distinct histone deacetylases (HDACs) to the m^5C residues, which then induce unique methylation patterns on the histones in surrounding chromatin and facilitate interactions with transcriptional repressors such as Sin3 and/or Polycomb group protein repressive complexes such as PRC2 and PRC3 *(11–13)*.

3. HISTONE MODIFICATIONS AND THE HISTONE CODE

Changes in DNA methylation do not occur in isolation, but rather in the context of other complex epigenetic events. The second major mechanism of epigenetic regulation comprises the different types of histone modifications, including acetylation, methylation, phosphorylation, ubiquitination, and sumoylation, all of which can regulate local DNA transcription and replication (Fig. 2); *(14,15)*. Despite the long-standing recognition that histone modifications occur in chromatin, the first histone-modifying enzyme was only identified in 1996, with the elucidation of the histone acetyl transferase (HAT) in *Tetrahymena* as a homolog of the yeast GCN5 enzyme *(16)*.

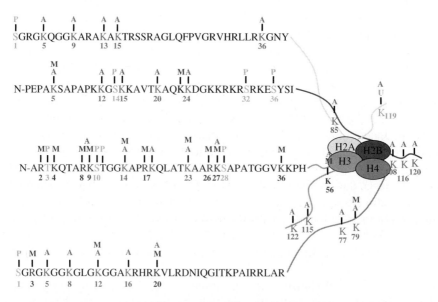

Fig. 2. A sub-set of the post-translational modifications responsible for the histone code. The N-terminal histone tails can become methylated (M) at lysine (K) or arginine (R) residues, phosphorylated (P) at serine (S) and threonine (T) residues, acetylated (A) and ubiquitylated (U) at lysines (K) in response to internal and external signaling events, converting chromatin locally to an "open" or "closed" configuration. See the text and the Fig. 3 legend for more details.

Nucleosome core particles are composed of a central tetramer of H3 and H4 histones with two peripheral heterodimers of H2A and H2B histones (or specialized natural variants of these proteins) *(17)*. Each core histone has a related globular C-terminal segment that mediates inter-histone interactions and a flexible basic amino acid-rich N-terminal tail, which is important for the formation of higher order structures of chromatin due to the extraordinary number of sites in the tail that are subject to post-translational modifications. Acetylation and methylation of lysine (K) and arginine (R) residues, ubiquitylation and sumoylation of lysines, and phosphorylation of serines (S) and threonines (T) and poly-ADP-ribosylation glutamic acid (E) are the key modifications that affect these histone tails *(17–19)* and thus the tertiary structure of chromatin. Additionally, the lysine residues are amenable to monomethylations, dimethylations, and trimethylations, and the arginines can be either monomethylated or dimethylated, adding to the complexity of histone modifications. Collectively, these modifications define the "histone code" and provide another layer of reversible, functional regulation superimposed upon the genetic code fixed in the DNA sequence *(20)*.

The acetylation of histones H3 and H4 mediated by the HATs leads to opening of the chromatin, providing local access of transcription factors and other proteins to the regulatory regions of genes and enabling the *active histone code (21)*. The HATs are composed of three super-families known as GNAT (Gcn5-related N-acetly transferase), MYST (MOZ, Ybf2/Sas3, Sas2, and TIP60), and p300/CBP and are distinguished by their mechanism of substrate binding and catalysis *(22)*. On the contrary, deacetylation of the lysine residues, mediated by HDACs, results in the compaction of chromatin, rendering it inaccessible to transcription factors and leading to the establishment of a *silenced histone code (21)*. The HDACs are grouped into three families, Class I, Class II, and the SIRT (sirtuin) enzymes, based on their sequence homologies *(21)*.

Methylation of lysines can occur in several lysine residues of histones H3 and H4, and unlike acetylation which signals active chromatin, histone methylation can signal either active or repressed chromatin. Methylation of the specific lysine residues of histones H3 and H4 are regarded as central to activated and repressed states of the chromatin and are mediated by histone methyl transferases, consisting of a SET (Su(var), Enhancer of Zeste and Trithorax) domain in conjunction with a chromo-domain adaptor protein *(23)*. While the H3K4, H3K36, and H3K79 methylations are associated with a transcriptionally active state, the H3K9, H3K27, and H4K20 methylations correspond to a repressed state. Up to three methyl groups per lysine residue can be progressively added to distinct states of the chromatin, resulting in both short-term and long-term imprints. For example, while monomethylated and dimethylated H3K9 can be present in active chromatin regions, trimethylated H3K9 can be present only in peri-centromeric heterochromatin.

Although histone methylation has been generally regarded as a very stable modification, the recent discovery of a newly emerging transcription factor family of histone lysine demethylases provides a molecular basis for the regulated reversion of methylation marks and suggests that these enzymes may play an important role in the differential maintenance of a dynamic chromatin structure in active or inactive states *(24,25)*. For example, lysine-specific demethylase 1, a nuclear amine oxidase, is a highly specific demethylase for H3K4me and H3K4me2; jumonji domain-containing histone demethylase 1 exhibits specificity for H3K36me2; and JMJD2A is a lysine trimethyl-specific demethylase specific to H3K9me3 and H3K36me3 *(26–28)*.

Histone arginine methylation and histone acetylation are often associated with functional synergy, leading to transcriptional activation of genes. Protein arginine methyl transferase 1 and co-activator-associated R-methyl transferase 1 methylate H4R3 and H3R2, respectively, and physically associate with HATs to activate transcription by NF-κB and p53 (29–32). Histone arginine methylation is often difficult to detect, as it often occurs transiently, concomitant with signaling events (33). Reversal of histone arginine methylation has been proposed to be mediated by peptidylarginine deiminases such as PADI4 (34).

The other major covalent histone or histone modifier modifications include phosphorylation, SUMOylation, and ubiquitination. Phosphorylation of histone H3 on at least two serine residues (Ser^{10} and Ser^{28}), and on Thr^{11}, is associated with chromosome condensation and segregation during mitosis and meiosis (35). H3S10 phosphorylation begins early in G2, with peak levels detected during metaphase, during which time Ser^{28} phosphorylation also becomes evident, ultimately followed by a general decrease in the amount of histone H3 phosphorylation during the progression through the cell cycle to telophase (36). Another component of the histone code involves the ubiquitination of C-terminal lysine residues of histones H2A and H2B. Although poly-ubiquitination targets proteins for degradation by the proteosome, mono-ubiquitination is a stable protein modification that does not affect the half-life of the protein. In yeast, H2B ubiquitination by the ubiquitin-conjugating enzyme RAD6, interacting with the ubiquitin ligase BRE1, is a pre-requisite for dimethylation of histone H3 lysine residues 4 and 79, which in turn is associated with increased gene activity. It is not clear, however, whether a similar trans-histone regulatory mechanism exists in the human (37). Conversely, extensive histone H2A ubiquitination is observed during meiotic prophase in mammalian cells (38). The small ubiquitin-related modifier (SUMO) can also reversibly modify transcription factors, cofactors, and chromatin-modifying enzymes such as HDACs. SUMOylation of HDAC1 increases both its deacetylase activity and its transcriptional repressor activity (39). Modification of MBD1 with either SUMO-2/3 or SUMO-1 mediates the interaction between MBD1 and MCAF1, suggesting a role for SUMOylation in the methylation of DNA (40). Furthermore, SUMOs are localized in MBD1-containing and MCAF1-containing heterochromatin regions that are enriched in trimethyl-H3-K9 and the heterochromatin proteins (HP) HP1ss and HP1 (40). These findings suggest a role for SUMOylation in the regulation of heterochromatin formation and gene silencing. Poly-ADP-ribose (PAR) polymerases, known as PARPs, maintain chromosome integrity at telomeres. The telomeric ankyrin-repeat-containing PARPs (tankyrases) bind to telomere-binding proteins, including TRF1, and are required for mitotic progression to anaphase, suggesting that poly-ADP-ribosylation is important for chromosome segregation (41). Several of the centromere proteins are also poly-ADP-ribosylated (42).

4. CHROMATIN REMODELING

Chromatin remodeling in response to signaling events in any given cell regulates the accessibility of gene-related DNA sequences to transcriptional regulatory proteins and plays a major role in determining the active or inactive states of the epigenome (43). Several enzyme complexes are known to modify the structure of chromatin in response to internal and external signaling events, rendering chromatin in a highly "fluid" state. All of the currently known ATP-dependent chromatin remodeling factors form multiprotein complexes containing nucleic acid-stimulated DEAD/H ATPases of the Swi2/Snf2 subfamily, which can be further classified into four major sub-families, the SWI/SNF, CHD1, ISWI, and INO80 proteins. These enzymes couple ATP hydrolysis to alterations of the chromatin structure at the level of the nucleosomal array, which can involve sliding of nucleosomes along the DNA, altered histone–DNA interactions, and removal or exchange of specific histones. Examples of human chromatin remodeling factors include two distinct Swi2/Snf2-like ATPase subunits, known as human brahma (hBRM) and BRG1 (brahma-related gene), and the NuRD (nucleosome remodeling HDAC).

5. ESTABLISHMENT OF THE CHROMATIN TERRAIN FOR UNIQUE MOLECULAR INTERACTIONS AND THE EPIGENOMIC CODE

Specific chromatin markings arising from differential DNA methylation status, covalent modifications of histones and association of various proteins, and the localized structure of the chromatin due to remodeling events generate the unique chromatin surface or "terrain" for interaction with the specific factors that regulate transcription of genes, genome replication and recombination, repair of DNA, and other functions concomitant with the physiological state of any given cell.

Certain histone modifications, such as acetylation of histone tails, are believed to play a role in chromatin dynamics by neutralizing the effects of the positive charges on the lysine residues. Alternatively, other specific histone modifications, such as methylation, can provide the chromatin landscape with binding sites for critical protein complexes to modulate replication or transcription or repair. There is evidence for recognition of acetylated lysines on histones by bromodomain-containing proteins and interactions of methylated lysines with chromodomain-containing, tudor domain-containing and WD-40-repeat domain-containing proteins *(44–47)*. For example, the H3K9 methylation creates a high-affinity binding site for the HP1, while other histone methylations create specific binding sites for other proteins, in a process that can lead to repressed/inactive chromatin *(48)*.

The complex and stepwise sets of protein–protein and protein–DNA/RNA interactions cause reshaping of the chromatin terrain, coincident with the formation of higher order complexes, and promote additional protein modifications or DNA methylation changes at each step in response to internal and external signaling events or by the microenvironmental changes in the immediately surrounding chromatin, ultimately generating the "epigenomic code." This epigenomic code, and its superimposed control over the genomic (DNA sequence) code, is then responsible for dictating local transcriptional events, which in turn determines the physiological state of the cell (Fig. 3).

6. EPIGENOMIC ALTERATIONS IN CANCER

Although fixed genetic alterations at the level of the DNA sequence (mutations) have been traditionally correlated to the genesis of cancer, recent studies suggest that the level of control superimposed upon the genetic code by the epigenetic code could prevent oncogenic or tumor suppressor gene mutations from causing cancer

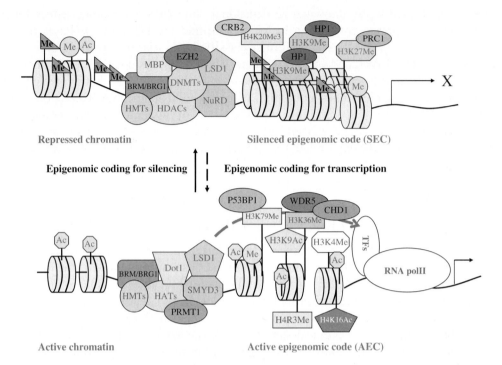

Fig. 3. A reversible epigenomic code dictates the functional status of the chromatin. Various modifications of the histone and non-histone proteins and the protein–protein and DNA–protein interactions that define the epigenome are responsible for the active and repressed epigenomic codes of any specific cell. The silencing of gene expression consisting of a "silenced epigenomic code" (SEC) is associated with local histone deacetylation and often found within regions of DNA methylation and methylation (Me) at the lysine (K) residues 9 and 27 of histone H3 and lysine residue 20 of histone H4. H3K9Me acts as a docking site for the heterochromatin protein 1 (HP1), facilitating maintenance of the repressed state of SEC. Conversely, the "active epigenomic code" (AEC), representing active chromatin supporting gene expression, is associated with local histone acetylation and methylation at lysine residues 4, 36, and 79 of histone H3. AEC can also be transiently associated with methylations of arginine residue 2 in histone H3 and arginine residue 4 in histone H4, or other histone arginine residues, resulting in transcriptional activation. Enzyme complexes, such as hBRM/BRG1 and NuRD, which modify the structure of chromatin, and enzymes that reverse the modified status of proteins and DNA render chromatin in a highly "fluid" state, with the ability to transition between SEC or AEC states in response to external or internal stimuli, thereby determining the gene expression profile, and ultimately the physiological state, of the cell. BRM/BRG1 (brahma/brahma-related gene 1), chromatin remodeling multiprotein complex with ATPase activity; CHD1, chromodomain helicase DNA binding protein 1; CRB2, Cut5 repeat binding protein 2; DNMT, DNA methyl transferase; Dot1, H3K79 methylase; EZH2, enhancer of zeste homolog 2 (H3K27 methylase); HATs, histone acetyltransferases; H3K4Me, H3K9Me, H3K27Me, H3K36Me, and H3K79Me indicate monomethylated histone H3 at lysines 4, t 9, 27, 36, and 79, respectively; H4K20Me3, trimethylated histone H4 at lysines 20; H4R3Me, methylated histone H4 at arginine 3; H3K9Ac, acetylated histone H3 at lysine 9; H4K16Ac, acetylated histone H4 at lysine 16; HDAC, histone deacetylase; HP1, heterochromatic protein 1; LSD1, lysine-specific demethylase 1; MBP, methyl CpG-binding proteins; Me in a triangle, CpG methylation of DNA; Me in a circle, methylation of histone/non-histone proteins; PRC1, polycomb repressive complex; p53BP1, p53 binding protein 1; PRMT1, protein arginine methyl transferase 1; RNA pol II, RNA polymerase II; SMYD3, H3K4 methylase; TFs, transcription factors; NuRD, nucleosome remodeling histone deacetylase; WDR5, WD40-repeat protein which binds to H3K4Me.

by suppressing their expression, or conversely could alter the expression of perfectly normal alleles to promote tumorigenesis. Alterations in the normal epigenomic code leading to cancer can occur at different levels, including altered DNA methylation patterns, aberrant histone modifications, replacement of core histones with specific histone variants, and abnormalities in the chromatin remodeling machinery, as well as alterations of other critical factors that determine the active or the repressed state of chromatin.

Although the bulk of the genome typically becomes hypomethylated in malignancy, a large body of data suggests that many CpG islands which are normally unmethylated can become hypermethylated, due to abnormal de novo methylation in cancer *(3,49,50)*. Hypomethylation is likely to contribute to cancer by activation of oncogenes, unmasking of normally repressed latent retrotransposons, or through increases in chromosomal instability *(51–55)*. Promoter hypermethylation has been shown to be as frequent as transcription-suppressing promoter mutations in disrupting the expression of established and candidate tumor suppressor genes in cancer *(56)*. Interestingly, nearly 50% of familial cancer-causing genes, such as the von Hippel–Lindau syndrome (*VHL*) gene, the breast cancer 1 (*BRCA1*) gene, the mutL homolog-1 (*MLH1*) gene, and the cyclin-dependent kinase inhibitor 2A ($p16^{INK4A/CDKN2A}$) gene, which are known to be frequently inactivated by mutations in various cancers, can instead sporadically undergo promoter methylation-associated silencing *(56)* in malignancy. Silencing of genes resulting from promoter methylation is also known to affect various essential molecular pathways that regulate normal cell cycle control, DNA repair, genomic stability, growth factor responses, apoptosis, and tumor cell invasion and metastasis (Table 1); *(56,57)*. It is widely accepted that the clonal evolution of tumor cells to more aggressive phenotypes is accompanied by the sequential selection and fixation of genetic changes that promote tumor progression *(58)*. There is now evidence for a similar phenomenon with epigenetic alterations, such as promoter DNA methylation. Recent analysis of NSCLC suggested that E-Cadherin (*ECAD*) and DAP-Kinase (DAPK) are targeted for methylation in the earliest stages of lung cancer, while DNA methylation silencing of *p16* and *hMGMT* are likely alterations that occur in

Table 1
Properties of Tumor Cells Affected by Silencing of Gene Promoters

Genes	Cancer	Property of tumor cells
BRCA1, MGMT, hMLH1, KIP2	Breast, colon, ovarian, prostate	Genomic instability
$p14^{ARF}$, $p15^{INK4b}$, $p16^{INK4a}$, RB, p73, TRβI TRβII, ER-α, AR, RARβ, SOS-1, 14-3-3σ	Breast, colon, esophagus, eye, lung, ovary, pancreas	Loss of growth control
DAPK, CASP8, BCL2, TMS-1, p73, HIC1	Breast, lung, lymph, head & neck	Loss of apoptosis
APC, VHL, ECAD, TIMP-3,CDH1 SMAD8, LKB1/STK11,THBS-1, RELN, RASSF1A,TWIST, GATA4, GATA5	Brain, breast, colon, esophagus, lung, pancreas	Invasion and metastasis

the later stages of cancer progression *(59)*. The recent development of assays which detect DNA methylation changes in specific gene promoters in small population of cells within a background of large number of unaffected cells in easily obtainable clinical samples now affords a new opportunity for efficient diagnosis and staging of cancer *(59,60)*.

Histone modification patterns of cancers of different types, and at different stages of progression, are highly variable, and an emerging array of data from ongoing studies of tumor banks may facilitate diagnosis as well as therapy in the near future. The loss of monoacetylation of lysine 16 (K16) and trimethylation of lysine 20 (K20) in the tail of histone H4 have been shown to be associated with malignant transformation *(61)*. Histone-modifying and chromatin-remodeling enzymes can also exhibit altered activity in cancer, due either to mutations in their genes or deregulation of their expression *(62)*. For example, mutations and chromosomal translocations of HATs, with associated aberrant over-activity, are often detected in both solid tumors and hematological malignancies *(62,63)*. Mutations in BRG1, a chromatin remodeling factor, have also been identified in various cancers *(64)*.

The identity of histone types comprising the nucleosomes may also be important in cancer etiology. For example, the gene for histone H2A.X localizes to 11q23.3, a region frequently deleted in cancer. Although histone H2A.X is apparently randomly incorporated into chromatin, it becomes rapidly phosphorylated during double-strand break repair at sites where protein factors for DNA repair are recruited *(65,66)*. Loss of H2A.X causes genomic stability and oncogenic transformation. Aurora B kinase, which phosphorylates serines in histone H3, can function as an oncogene in human cancers *(67)*. Increased ADP-ribosylation of histones and non-histones has been found in chromatin from oral cancers *(68)*. In summary, the evidence for altered functionality of chromatin structural proteins, as well as remodeling proteins, in cancer cells further substantiates the model that alterations of the normal epigenome are important contributors to cancer progression.

7. THE EPIGENOME AS A TARGET FOR CANCER THERAPY

The combinatorial nature of the factors involved in protein–protein as well as protein–nucleic acid interactions, as well as the reversible nature of the covalent modifications of the molecules that comprise the cancer epigenome, makes it a tantalizing target for therapeutic manipulation in the treatment of cancer. Additionally, because epigenetic changes are often observed in the pre-clinical stages of cancer, pharmacological approaches to modifying the epigenome may also be chemopreventive. At present, two groups of drugs are being utilized for epigenomic therapy. One class inhibits DNA methyl transferases, while the other class inhibits HDACs *(69)*. Whereas promoter DNA hypermethylation has been shown to be a critical mechanism for silencing of tumor suppressor genes, leading to loss-of-function, pharmacological reactivation of these genes using DNMT inhibitors (Table 1) would be predicted to result in suppression of tumorigenesis. Conversely, inhibition of HDACs, leading to the accumulation of acetylated histones and other transcription factors, may result in activation of aberrantly repressed genes that would normally initiate cell cycle checkpoints or apoptotic pathways, thereby promoting growth arrest or death in cancer cells.

The DNMT inhibitors are primarily analogs of the nucleoside deoxycytidine and are phosphorylated in the cells in vivo to generate deoxynucleotide triphosphates and then incorporated into the replicating DNA, where they act as inhibitors of DNMTs (Table 2). Various DNMT inhibitors have been extensively tested in in vitro cell culture systems and in animal models and have demonstrated restoration of the expression of aberrantly silenced genes and inhibition of tumor cell growth *(69,70)*. 5-aza-CR, decitabine and dihydro-5-5-azacytidine are administered by parenteral injections, and their limiting toxicity is on myelopoiesis, leading to cytopenias *(71)*. A newer cytosine analog, Zebularine, is less toxic and can be administered orally *(72)*. Newer, non-nucleoside inhibitors have been developed which bind directly to the catalytic domains of the DNMTs to inhibit their activity *(70,73)*. Most of these agents remain in the pre-clinical stages of development, although a few have entered clinical trials (Table 2).

Initial speculation might suggest that genome-wide inhibition of HDACs would have deleterious effects on key cellular functions. Fortunately, the discovery and potentially beneficial actions of certain naturally occurring HDAC inhibitors (HDACi) were realized before their activity on HDACs was understood. HDACi promote the acetylated state of histones and several non-histone proteins, including transcription factors, and cause inhibition of cell cycle progression, differentiation, and in some cases apoptosis of tumor cells *(21,74,75)*. In leukemias, HDACi induce tumor necrosis factor-related apoptosis inducing ligand and its receptor, death receptor 5, and FAS ligand and FAS, by acetylation and activation of transcription factors SP1 and SP2, resulting in apoptosis of leukemic cells *(75)*. An additional potential benefit of HDACi as cancer therapeutics is their ability to inhibit DNA repair responses, thus increasing the sensitivity of tumor cells to chemotherapy and radiotherapy *(76)*. Several studies have shown that normal cells are profoundly less sensitive to the growth arrest and pro-apoptotic activities of HDACi than are tumor cells, providing a relatively high therapeutic index for HDACi as anti-cancer agents. However, it is important to note that there may indeed be some potential undesirable effects resulting from global histone

Table 2
DNA Methyl Transferase Inhibitors

Drug	Dose range	Cancer
Nucleoside analogues		
5-Azacytidine (5-aza-CR)	μM	Leukemia, myelodysplasia
Decitabine (5-Aza-deoxycytidine; 5-aza-CdR)	μM	Leukemia, myelodysplasia, cervical, and NSCLC cancers
Dihydro-5-Azacytidine	μM	Lymphomas, ovarian cancer
Zebularine	μM–mM(Oral)	Bladder cancer (pre-clinical)
Non-nucleoside analogues		
Procaine	μM	Pre-clinical
Procainamide	μM	Pre-clinical
EGCG (gallate)	μM	Pre-clinical
Psammaplins	nM–μM	Pre-clinical
RG108	μM	Pre-clinical

acetylation, such as promotion of cell cycle progression and anti-apoptotic effects, due to general transcriptional activation of genes. For example, the retinoblastoma protein recruits HDACs to a subset of E2F target genes, and hence, HDACi might activate E2F target genes. While pre-clinical studies have suggested that in general HDACi produce less systemic toxicity compared to most other classes of chemotherapeutic agents, the specific dose-limiting toxicities of individual HDACi at the effective anti-tumor concentrations required for cancer therapy remain largely unknown until more extensive clinical studies have been carried out. There are at least six known structural classes of HDACi that are currently in clinical trials as potential cancer therapeutics (Table 3); *(21,69,75–81)*.

Table 3
Histone Deacetylase Inhibitors

Drug	Dose range	Cancer
Short-chain fatty acids		
Butyrate	mM	Hematological and colon tumors
AN-9 (Pivanex; pivaloyloxymethyl butyrate)	µM	Lung cancer, melanoma, and leukemia
Phenyl butyrate	mM	Leukemia and myelodysplasia
Valproic acid	mM	Leukemia, myelodysplasia, and cervical cancer
Hydroxamic acids		
Trichostatin A	nM	Pre-clinical
SAHA (Suberoylanilide hydroxamic acid)	µM	Hematological and solid tumors
Oxamflatin	µM	Pre-clinical
Scriptaid	µM	Pre-clinical
Benzamides		
MS-275	µM	Lymphoma, leukemia, and solid tumors
CI-994 (N-acetyl dinaline)	µM	Solid tumors
Epoxyketones		
Trapoxin	nM	Pre-clinical
2-amino-5-oxo-9, 10-epoxydeconic acid (AOE)	nM	Pre-clinical
Cyclic peptides		
Apicidin	nM	Pre-clinical
Depsipeptide (FK-228)	µM	T-cell lymphoma, leukemia, and solid tumors
Hybrid molecules		
CHAP31 (cyclic hydroxamic acid peptide)	nM	Pre-clinical
CHAP50	nM	Pre-clinical

8. FUTURE PERSPECTIVES

The "holy grail" of cancer therapy is the design and delivery of specific drugs to target cancer cells and the unique genetic and biochemical alterations inherent in these transformed cells, while sparing normal cells and the patient from undesirable side effects. The initial approach of the research community toward targeted cancer therapy has been to elucidate the precise genetic correlations connecting mutations at the level of the DNA sequence to the pathologic manifestation of the disease, under the assumption that this approach would hold all of the answers to molecular basis of malignancy and its cure. While the completion of the human genome sequencing project represented a critical first step, we now understand that the resulting huge body of information represents just the tip of the iceberg. The complex differential gene expression patterns that define cellular and tissue phenotypes in health and disease depend not only on the genomic sequence but also on the epigenome. To address this gap in our understanding, Human Epigenome Projects have been initiated in Europe and the United States *(82–84)*. These efforts will help to elucidate details of the various cancer epigenomes and are likely to assist the elucidation of the molecular basis of the disease progression of cancer using an integrated modeling approach such as the formulation of the multi-modular molecular network cancer progression models *(85)*. While we move toward achieving this goal, there will surely be parallel progress in developing therapeutic agents which will selectively target the aberrant epigenomic control of gene expression in cancer cells, with the specificity and precision to reap the maximum therapeutic benefit with minimal adverse effects on normal cells.

ACKNOWLEDGMENTS

Work in the authors' laboratories is supported by grants from the NIH and the Department of Defense, the Dolphin Trust Investigator Award from the Medical Foundation (S.T.), and the Karin Grunebaum Cancer Foundation (D.V.F.). The authors thank Panagiotis Papageorgis for help with the illustrations.

REFERENCES

1. Momparler RL, Bovenzi V. DNA methylation and cancer. J Cell Physiol 2000 183:145–54.
2. Robertson KD, Jones PA. DNA methylation: past, present and future directions. Carcinogenesis 2000 21: 461–7.
3. Herman JG, Baylin SB. Gene silencing in cancer in association with promoter hypermethylation. N Engl J Med 2003 349:2042–54.
4. Gardiner-Garden M, Frommer M. CpG islands in vertebrate genomes. J Mol Biol 1987 Jul 20; 196(2):261–82.
5. Lander ES, et al. Initial sequencing and analysis of the human genome. Nature 2001 Feb 15; 409(6822):860–921.
6. Okano M, Xie S, Li E Cloning and characterization of a family of novel mammalian DNA (cytosine-5) methyltransferases. Nat Genet 1998 Jul; 19(3):219–20.
7. Jones PA, Takai D. The role of DNA methylation in mammalian epigenetics. Science 2001 Aug 10; 293(5532):1068–70.
8. Brandeis M, Ariel M, Cedar H. Dynamics of DNA methylation during development. Bioessays 1993 Nov; 15(11):709–13.
9. Bird A. Molecular biology. Methylation talk between histones and DNA. Science 2001 Dec 7; 294(5549):2113–5.

10. Bachman KE, Park BH, Rhee I, Rajagopalan H, Herman JG, Baylin SB, Kinzler KW, Vogelstein B. Histone modifications and silencing prior to DNA methylation of a tumor suppressor gene. Cancer Cell 2003 Jan; 3(1):89–95.

11. Li E. The mojo of methylation. Nat Genet 1999 Sep; 23(1):5–6.

12. Fuks F, Hurd PJ, Wolf D, Nan X, Bird AP, Kouzarides T. The methyl-CpG-binding protein MeCP2 links DNA methylation to histone methylation. J Biol Chem 2003 Feb 7; 278(6):4035–40.

13. Kuzmichev A, Margueron R, Vaquero A, Preissner TS, Scher M, Kirmizis A, Ouyang X, Brockdorff N, Abate-Shen C, Farnham P, Reinberg D. Composition and histone substrates of polycomb repressive group complexes change during cellular differentiation. Proc Natl Acad Sci USA 2005 Feb 8; 102(6):1859–64.

14. Peterson CL, Laniel MA. Histones and histone modifications. Curr Biol 2004 Jul 27; 14(14):R546–51.

15. Mersfelder EL, Parthun MR. The tale beyond the tail: histone core domain modifications and the regulation of chromatin structure. Nucleic Acids Res 2006 May 19; 34(9):2653–62.

16. Brownell JE, Zhou J, Ranalli T, Kobayashi R, Edmondson DG, Roth SY, Allis CD. Tetrahymena histone acetyltransferase A: a homolog to yeast Gcn5p linking histone acetylation to gene activation. Cell 1996 Mar 22; 84(6):843–51.

17. Isenberg H. Histones. Annu Rev Biochem 1979; 48:159–91.

18. Wolffe AP. Chromatin: Structure and Function, 3rd Ed., 1999, Academic Press, Inc., San Diego, CA.

19. Strahl BD, Allis CD. The language of covalent histone modifications. Nature 2000 Jan 6; 403(6765):41–5.

20. Jenuwein T, Allis CD. Translating the histone code. Science 2001 Aug 10; 293(5532):1074–80.

21. Thiagalingam S, Cheng KH, Lee HJ, Mineva N, Thiagalingam A, Ponte JF. Histone deacetylases: unique players in shaping the epigenetic histone code. Ann N Y Acad Sci 2003 Mar; 983:84–100.

22. Gibbons RJ. Histone modifying and chromatin remodelling enzymes in cancer and dysplastic syndromes. Hum Mol Genet 2005 Apr 15; 14 Spec No 1:R85–92.

23. Zhang Y, Reinberg D. Transcription regulation by histone methylation: interplay between different covalent modifications of the core histone tails. Genes Dev 2001 Sep 15; 15(18):2343–60.

24. Bannister AJ, Kouzarides T. Reversing histone methylation. Nature 2005 Aug 25; 436(7054):1103–6.

25. Shi Y, Lan F, Matson C, Mulligan P, Whetstine JR, Cole PA, Casero RA, Shi Y. Histone demethylation mediated by the nuclear amine oxidase homolog LSD1. Cell 2004 Dec 29; 119(7):941–53.

26. Chen Z, Zang J, Whetstine J, Hong X, Davrazou F, Kutateladze TG, Simpson M, Mao Q, Pan CH, Dai S, Hagman J, Hansen K, Shi Y, Zhang G. Structural insights into histone demethylation by JMJD2 family members. Cell 2006 May 19; 125(4):691–702.

27. Tsukada Y, Fang J, Erdjument-Bromage H, Warren ME, Borchers CH, Tempst P, Zhang Y. Histone demethylation by a family of JmjC domain-containing proteins Nature 2006 Feb 16; 439(7078): 811–6.

28. Whetstine JR, Nottke A, Lan F, Huarte M, Smolikov S, Chen Z, Spooner E, Li E, Zhang G, Colaiacovo M, Shi Y. Reversal of histone lysine trimethylation by the JMJD2 family of histone demethylases. Cell 2006 May 5; 125(3):467–81.

29. Strahl BD, Briggs SD, Brame CJ, Caldwell JA, Koh SS, Ma H, Cook RG, Shabanowitz J, Hunt DF, Stallcup MR, Allis CD. Methylation of histone H4 at arginine 3 occurs in vivo and is mediated by the nuclear receptor coactivator PRMT1. Curr Biol 2001 Jun 26; 11(12):996–1000.

30. Wang H, Huang ZQ, Xia L, Feng Q, Erdjument-Bromage H, Strahl BD, Briggs SD, Allis CD, Wong J, Tempst P, Zhang Y. Methylation of histone H4 at arginine 3 facilitating transcriptional activation by nuclear hormone receptor. Science 2001 Aug 3; 293(5531):853–7.

31. An W, Kim J, Roeder RG. Ordered cooperative functions of PRMT1, p300, and CARM1 in transcriptional activation by p53. Cell 2004 June 11; 117(6):735–48.

32. Covic M, Hassa PO, Saccani S, Buerki C, Meier NI, Lombardi C, Imhof R, Bedford MT, Natoli G, Hottiger MO. Arginine methyltransferase CARM1 is a promoter-specific regulator of NF-kappaB-dependent gene expression. EMBO J 2005 Jan 12; 24(1):85–96.

33. Metivier R, Penot G, Hubner MR, Reid G, Brand H, Kos M, Gannon F. Estrogen receptor-alpha directs ordered, cyclical, and combinatorial recruitment of cofactors on a natural target promoter. Cell 2003 Dec 12; 115(6):751–63.

34. Cuthbert GL, Daujat S, Snowden AW, Erdjument-Bromage H, Hagiwara T, Yamada M, Schneider R, Gregory PD, Tempst P, Bannister AJ, Kouzarides T. Histone deimination antagonizes arginine methylation Cell 2004 Sep 3; 118(5):545–53.

35. Nowak SJ, Corces VG. Phosphorylation of histone H3: a balancing act between chromosome condensation and transcriptional activation. Trends Genet 2004 Apr; 20(4):214–20.

36. Goto H, Tomono Y, Ajiro K, Kosako H, Fujita M, Sakurai M, Okawa K, Iwamatsu A, Okigaki T, Takahashi T, Inagaki M. Identification of a novel phosphorylation site on histone H3 coupled with mitotic chromosome condensation. J Biol Chem 1999 274:25543–9.

37. Briggs SD, Xiao T, Sun ZW, Caldwell JA, Shabanowitz J, Hunt DF, Allis CD, Strahl BD. Gene silencing: trans-histone regulatory pathway in chromatin. Nature 2002 Aug 1; 418(6897):498.

38. Baarends WM, Hoogerbrugge JW, Roest HP, Ooms M, Vreeburg J, Hoeijmakers JH, Grootegoed JA. Histone ubiquitination and chromatin remodeling in mouse spermatogenesis Dev Biol 1999 Mar 15; 207(2):322–33.

39. Gill G. Something about SUMO inhibits transcription. Curr Opin Genet Dev 2005 Oct; 15(5):536–41.

40. Uchimura Y, Ichimura T, Uwada J, Tachibana T, Sugahara S, Nakao M, Saitoh H. Involvement of SUMO modification in MBD1- and MCAF1-mediated heterochromatin formation. J Biol Chem. 2006 Aug 11; 281(32): 23180–90.

41. Dynek JN, Smith S. Resolution of sister telomere association is required for progression through mitosis. Science 2004 Apr 2; 304(5667):97–100.

42. Earle E, Saxena A, MacDonald A, Hudson DF, Shaffer LG, Saffery R, Cancilla MR, Cutts SM, Howman E, Choo KH. Poly(ADP-ribose) polymerase at active centromeres and neocentromeres at metaphase. Hum Mol Genet 2000 Jan 22; 9(2):187–94.

43. Langst G, Becker PB. Nucleosome remodeling: one mechanism, many phenomena? Biochim Biophys Acta 2004 Mar 15; 1677(1–3):58–63.

44. Jacobson RH, Ladurner AG, King DS, Tjian R. Structure and function of a human TAFII250 double bromodomain module. Science 2000 May 26; 288(5470):1422–5.

45. Bannister AJ, Zegerman P, Partridge JF, Miska EA, Thomas JO, Allshire RC, Kouzarides T. Selective recognition of methylated lysine 9 on histone H3 by the HP1 chromo domain. Nature 2001 Mar 1; 410(6824):120–4.

46. Huyen Y, Zgheib O, Ditullio RA Jr, Gorgoulis VG, Zacharatos P, Petty TJ, Sheston EA, Mellert HS, Stavridi ES, Halazonetis TD. Methylated lysine 79 of histone H3 targets 53BP1 to DNA double-strand breaks. Nature 2004 Nov 18; 432(7015):406–11.

47. Wysocka J, Swigut T, Milne TA, Dou Y, Zhang X, Burlingame AL, Roeder RG, Brivanlou AH, Allis CD. WDR5 associates with histone H3 methylated at K4 and is essential for H3 K4 methylation and vertebrate development. Cell 2005 Jun 17; 121(6):859–72.

48. Lachner M, O'Carroll D, Rea S, Mechtler K, Jenuwein T. Methylation of histone H3 lysine 9 creates a binding site for HP1 proteins. Nature 2001 Mar 1; 410(6824):116–20.

49. Feinberg AP, Vogelstein B. Hypomethylation distinguishes genes of some human cancers from their normal counterparts. Nature 1983 Jan 6; 301(5895):89–92.

50. Jones PA, Laird PW. Cancer epigenetics comes of age. Nat Genet 1999 21:163–167.

51. Feinberg AP, Vogelstein B. Hypomethylation of ras oncogenes in primary human cancers. Biochem Biophys Res Commun 1983 Feb 28; 111(1):47–54.

52. Bestor TH, Tycko B. Creation of genomic methylation patterns. Nat Genet 1996 Apr; 12(4): 363–7.

53. De Smet C, De Backer O, Faraoni I, Lurquin C, Brasseur F, Boon T. The activation of human gene MAGE-1 in tumor cells is correlated with genome-wide demethylation. Proc Natl Acad Sci USA 1996 Jul 9; 93(14):7149–53.

54. Lengauer C, Kinzler KW, Vogelstein B. DNA methylation and genetic instability in colorectal cancer cells. Proc Natl Acad Sci USA 1997 Mar 18; 94(6):2545–50.

55. Gaudet F, Hodgson JG, Eden A, Jackson-Grusby L, Dausman J, Gray JW, Leonhardt H, Jaenisch R. Induction of tumors in mice by genomic hypomethylation. Science 2003 Apr 18; 300(5618): 489–92.

56. Jones PA, Baylin SB. The fundamental role of epigenetic events in cancer. Nat Rev Genet 2002 Jun; 3(6):415–28.

57. Esteller M, Corn PG, Baylin SB, Herman JG. A gene hypermethylation profile of human cancer. Cancer Res 2001 61:3225–9.

58. Kinzler KW, Vogelstein B. Lessons from hereditary colon cancer. Cell 1996 87:159–170.

59. Russo AL, Thiagalingam A, Pan H, Califano J, Cheng KH, Ponte JF, Chinnappan D, Nemani P, Sidransky D, Thiagalingam S. Differential DNA hypermethylation of critical genes mediate the stage

specific tobacco smoke induced neoplastic progression of lung cancer. Clin Cancer Res 2005 Apr 1; 11(7):2466–70.

60. Belinsky SA, Liechty KC, Gentry FD, Wolf HJ, Rogers J, Vu K, Haney J, Kennedy TC, Hirsch FR, Miller Y, Franklin WA, Herman JG, Baylin SB, Bunn PA, Byers T. Promoter hypermethylation of multiple genes in sputum precedes lung cancer incidence in a high-risk cohort. Cancer Res 2006 Mar 15; 66(6):3338–44.

61. Fraga MF, Ballestar E, Villar-Garea A, Boix-Chornet M, Espada J, Schotta G, Bonaldi T, Haydon C, Ropero S, Petrie K, Iyer NG, Perez-Rosado A, Calvo E, Lopez JA, Cano A, Calasanz MJ, Colomer D, Piris MA, Ahn N, Imhof A, Caldas C, Jenuwein T, Esteller M. Loss of acetylation at Lys16 and trimethylation at Lys20 of histone H4 is a common hallmark of human cancer. Nat Genet 2005 Apr; 37(4):391–400.

62. Esteller M. Epigenetics provides a new generation of oncogenes and tumour-suppressor genes. Br J Cancer 2006 Jan 30; 94(2):179–83.

63. Santos-Rosa H, Caldas C. Chromatin modifier enzymes, the histone code and cancer. Eur J Cancer 2005 Nov; 41(16):2381–402.

64. Wong AK, Shanahan F, Chen Y, Lian L, Ha P, Hendricks K, Ghaffari S, Iliev D, Penn B, Woodland AM, Smith R, Salada G, Carillo A, Laity K, Gupte J, Swedlund B, Tavtigian SV, Teng DH, Lees E. BRG1, a component of the SWI-SNF complex, is mutated in multiple human tumor cell lines. Cancer Res 2000 Nov 1; 60(21):6171–7.

65. Monni O, Knuutila S. 11q deletions in hematological malignancies. Leuk Lymphoma 2001 Jan; 40(3–4):259–66.

66. Redon C, Pilch D, Rogakou E, Sedelnikova O, Newrock K, Bonner W. Histone H2A variants H2AX and H2AZ. Curr Opin Genet Dev 2002 Apr; 12(2):162–9.

67. Giet R, Petretti C, Prigent C. Aurora kinases, aneuploidy and cancer, a coincidence or a real link? Trends Cell Biol 2005 May; 15(5):241–50.

68. Das BR. Increased ADP-ribosylation of histones in oral cancer. Cancer Lett 1993 Sep 15; 73(1): 29–34.

69. Yoo CB, Jones PA. Epigenetic therapy of cancer: past, present and future. Nat Rev Drug Discov 2006 Jan; 5(1):37–50.

70. Stresemann C, Brueckner B, Musch T, Stopper H, Lyko F. Functional diversity of DNA methyltransferase inhibitors in human cancer cell lines. Cancer Res 2006 Mar 1; 66(5):2794–800.

71. Gilbert J, Gore SD, Herman JG, Carducci MA. The clinical application of targeting cancer through histone acetylation and hypomethylation. Clin Cancer Res 2004 Jul 15; 10(14):4589–96.

72. Cheng JC, Matsen CB, Gonzales FA, Ye W, Greer S, Marquez VE, Jones PA, Selker EU. Inhibition of DNA methylation and reactivation of silenced genes by zebularine. J Natl Cancer Inst 2003 Mar 5; 95(5):399–409.

73. Lyko F, Brown R. DNA methyltransferase inhibitors and the development of epigenetic cancer therapies. J Natl Cancer Inst 2005 Oct 19; 97(20):1498–506.

74. Mork CN, Faller DV, Spanjaard RA. A mechanistic approach to anticancer therapy: targeting the cell cycle with histone deacetylase inhibitors. Curr Pharm Des 2005; 11(9):1091–104.

75. Minucci S, Pelicci PG. Histone deacetylase inhibitors and the promise of epigenetic (and more) treatments for cancer. Nat Rev Cancer 2006 Jan; 6(1):38–51.

76. Munshi A, Kurland JF, Nishikawa T, Tanaka T, Hobbs ML, Tucker SL, Ismail S, Stevens C, Meyn RE. Histone deacetylase inhibitors radiosensitize human melanoma cells by suppressing DNA repair activity. Clin Cancer Res 2005 Jul 1; 11(13):4912–22.

77. de Vos D. Epigenetic drugs: a longstanding story. Semin Oncol 2005 Oct; 32(5):437–42.

78. Inche AG, La Thangue NB. Chromatin control and cancer-drug discovery: realizing the promise. Drug Discov Today 2006 Feb; 11(3–4):97–109.

79. Furumai R, Komatsu Y, Nishino N, Khochbin S, Yoshida M, Horinouchi S. Potent histone deacetylase inhibitors built from trichostatin A and cyclic tetrapeptide antibiotics including trapoxin. Proc Natl Acad Sci USA 2001 Jan 2; 98(1):87–92.

80. Batova A, Shao LE, Diccianni MB, Yu AL, Tanaka T, Rephaeli A, Nudelman A, Yu J. The histone deacetylase inhibitor AN-9 has selective toxicity to acute leukemia and drug-resistant primary leukemia and cancer cell lines. Blood 2002 Nov 1; 100(9):3319–24.

81. Drummond DC, Noble CO, Kirpotin DB, Guo Z, Scott GK, Benz CC. Clinical development of histone deacetylase inhibitors as anticancer agents. Annu Rev Pharmacol Toxicol 2005; 45: 495–528.

82. Novik KL, Nimmrich I, Genc B, Maier S, Piepenbrock C, Olek A, Beck S. Epigenomics: genome-wide study of methylation phenomena. Curr Issues Mol Biol 2002 Oct; 4(4):111–28.
83. Bradbury J. Human epigenome project–up and running. PLoS Biol 2003 Dec; 1(3):E82.
84. Jones PA, Martienssen R. A blueprint for a Human Epigenome Project: the AACR Human Epigenome Workshop. Cancer Res 2005 Dec 15; 65(24):11241–6.
85. Thiagalingam S, A cascade of modules of a network defines cancer progression. Cancer Res. 2006 Aug 1; 66(15):7379–85.

II Molecular Targeting for Specific Disease Sites

6 Molecular Targeting in Upper Gastrointestinal Malignancies

Scott Wadler, MD

SUMMARY

Tumors of the upper gastrointestinal tract (GI) are not common but represent a significant clinical challenge. Treatment of locally advanced and metastatic tumors of the upper GI tract, liver and pancreas has met with limited success using standard cytotoxic agents. The development of agents that block signal transduction pathways, cell cycle proteins, proteosomal degredation, angiogenesis and immunologic-based approaches are being tested in this group of tumors. These studies have better defined the toxicity profiles of these agents but have resulted in limited successful clinical responses. The pre-clinical and preliminary clinical trial results will be reviewed and some of the problems with targeted therapy in the upper GI tract cancer population will be discussed.

Key Words: Hepatocellular carcinoma; Pancreatic cancer; Stomach cancer; Targeted therapy; Upper GI tract.

1. INTRODUCTION

Tumors of the upper gastrointestinal (GI) tract are uncommon, but not rare malignancies, comprised of neoplasms of the stomach (22,710), pancreas (31,860), and liver (18,920) *(1)*. Less common tumors, such as ampullary carcinomas, neuroendocrine tumors, and small bowel carcinomas will not be discussed here. Initial management of upper GI cancer is usually confined to dealing with locally advanced tumors. Eventually, these tumors progress, usually by metastasizing to the liver first. In addition to liver metastases, very far advanced tumors spread predominantly to the peritoneal cavity, but also to lung, bone, and lymph nodes. This chapter will restrict itself to covering the natural history of locally advanced, unresectable tumors and metastatic disease. Local recurrence will also be covered.

The mortality with these tumors is formidable. Only a very small percentage of patients will survive these tumors over a 2-year period after diagnosis *(1)*. In addition to the mortality, the morbidity of these tumors is considerable. For pancreatic cancer, this includes back pain, bowel obstruction, ascites, and thromboembolic events. For stomach cancer, bowel obstruction and ascites are common complications. For liver

From: *Cancer Drug Discovery and Development: Molecular Targeting in Oncology*
Edited by: H. L. Kaufman, S. Wadler, and K. Antman © Humana Press, Totowa, NJ

cancer, lumping together hepatocellular carcinoma, biliary duct, and gall bladder cancer, the most common complications are ascites, liver failure, fatigue, and symptoms of cirrhosis, such as skin findings and encephalopathy.

The epidemiology of these diseases is not well understood. Hepatocellular carcinoma may arise from cirrhotic livers, secondary to alcohol or viral hepatitis or other causes of chronic liver inflammation. Gall bladder carcinoma has been weakly associated with gall stones. Pancreatic cancer is weakly associated with smoking. The other tumors, however, have no clear environmental associations.

Treatment for these tumors using conventional cytotoxic agents has been unsuccessful. Commonly used treatments for stomach cancer include irinotecan and cisplatin (CP) administered on a weekly schedule or infusional fluorouracil, CP, and epirubicin (ECF; more generally used in Europe). Sometimes a taxane is added to CP. For patients with pancreatic carcinoma, standard of care is still gemcitabine, although the combination of gemcitabine and erlotinib (Tarceva), discussed in Section 3.2.1. has prolonged survival by a few weeks. Other combination chemotherapy regimens are used as well, with mixed success. Biliary and gall bladder tumors are treated like pancreatic cancer. For hepatocellular carcinoma, there is no standard treatment. Median survival for these diseases with conventional therapy is about 6–12 months depending on the disease, patient selection, and treatment used.

The following discussion will center around molecular targets and preliminary results with targeted therapies. The major goals of these studies are to attempt to duplicate the results achieved with chemotherapy with fewer side effects or, alternatively, to surpass the results achieved with chemotherapy with the same side effect profile. Most of the data presented here are preliminary, so it must be interpreted with caution. Nevertheless, the data are signposts to allow the reader to recognize emerging concepts in both the clinical and translational sciences.

2. EMERGING MOLECULAR TARGETS

2.1. Signal Transduction Pathways

Signal transduction pathways (STPs) are adequately covered in chapters 1–5. Certain facts require a brief review, however. STPs are responsible for communication of the extracellular environment with the cell nucleus, usually stimulating transcription of DNA. A variety of cell receptors bind ligands that influence the cell in multiple ways, but primarily to stimulate replication of DNA and transcription of S phase proteins. Manipulating these pathways may have profound cytostatic or even cytotoxic effects. Epidermal growth factor receptor (EGFR) pathways are an important target for anti-EGFR drugs (2). A phase III trial of gefitinib 250 versus 500 mg/day orally showed an 18% disease control rate with dose-related side effects of rash (25 versus 44%) and anorexia (8 versus 16%). Rojo and colleagues (3) found high levels of EGFR expression (64%) in patients with gastric cancer, but post-gefitinib treatment, the levels of pEGFR expression was markedly reduced, but only in patients with low p-akt levels prior to treatment.

2.2. Cell Cycle Proteins

Cell cycle proteins are necessary for cells to prepare for replication and actual replication. So far, efforts to manipulate expression, deactivation, or activation of these proteins have been ineffective. Therefore, clinical strategies to influence activity of these proteins have not taken off. Nevertheless, there is much preclinical research in this area, with a modest amount of early clinical investigation. Abnormalities in cell cycle regulators are involved in stomach carcinogenesis and cell proliferation *(4,5)*. The cyclin E gene is amplified in 15–20% of gastric cancers, and this correlates with aggressiveness of the tumors. Reduction in p27^{KIP1} expression correlates with deep invasion and lymph node metastases, especially in patients with cyclin E-expressing tumors *(6)*. Although over 200 articles have been published on the role of p53, it is controversial as to exactly what p53 does. Drugs that influence levels of cell cycle proteins will likely best be used in combination therapies for cancer.

2.3. Immunologic Therapies

Immune-based strategies, including newer, more potent vaccines, cell-based strategies, particularly dendritic cell, cytokine-based treatments, and combinations of the above, have had a resurgence recently. This is largely based on new research, which has revealed much more about immune function. The major challenges are deciphering which changes will provide the most effective therapies. For a formal review of this subject, see ref. *(7)*.

2.4. Proteasome-Based Research

The 26S proteasome is the cellular organelle responsible for digesting cellular proteins that are no longer needed for cell cycle, immune activation, or anti-apoptotic events. Velcade (PS-341) is the prototypical drug in this family. It has been approved by the FDA for the treatment of multiple myeloma, but more importantly is being studied in combination with cytotoxic drugs in a variety of solid tumors (Wadler, personal communication).

2.5. Anti-Angiogenesis Strategies

Tumors require an active blood supply in order to grow. Strategies designed to inhibit the growth of new vessels (anti-angiogenesis strategies) or destroy established blood vessels (anti-vascular strategies) can inhibit the growth of either established tumors or metastases. Bevacizumab (BV), the prototype drug, is a monoclonal antibody that has been approved for the treatment of metastatic colon cancer, based on a highly significant clinical demonstration of a nearly 5-month improvement in survival. Vascular endothelial growth factor (VEGF) expression has been associated with higher paracrine and autocrine activity associated with greater tumor aggressiveness *(8)*. Other drugs are in clinical trials, based on these promising early results.

2.6. Viral Therapies

A variety of viral species have been tested in the clinic with inconclusive results. The modified, chimeric, replication-competent adenovirus, ONYX-015, was withdrawn before conclusive findings were available. The type 3 recombinant Dearing strain reovirus, Reolysin, is currently entering clinical testing at this juncture. Preliminary results have been promising. Little is known about the latter virus; however, preliminary data suggest that it acts by cell necrosis rather than stimulating pro-apoptotic pathways. Phase I trials have been initiated.

2.7. Epigenetic Changes

Inactivation of chromatin by histone deacetylation is involved in transcriptional repression of multiple tumor suppressor genes, including p21$^{WAF1/CIP1}$. Hypoacetylation of histones H3 and H4 in the p21$^{WAF1/CIP1}$ promoter region is observed in >50% of gastric cancer tissues by chromatin immunoprecipitation *(9)*. The level of acetylated H4 is reduced 70% in patients with gastric cancer in comparison with non-neoplastic tissues; thus, low levels of global histone acetylation may be associated with high-grade malignancy. Trichostatin A, a histone deacetylator, induces growth arrest and apoptosis and suppresses the invasion of gastric cancer cells *(10)*. Multiple teams are working on ways to manipulate the methylation and acetylation status of DNA in tumor cells in order to either decrease or increase transcription of critical genes needed for growth inhibition or induction of apoptosis. Trials with new classes of these agents are currently underway without any definitive clinical results.

2.8. Microarray Analysis

Hippo and colleagues *(11)* identified 7 genes among 6800 genes in resected gastric cancer whose overexpression correlated with increased numbers of lymph node metastases. Inoue and colleagues *(12)* using a cDNA microarray showed that altered expression of 12 genes and 2 ESTs among 23,040 genes was associated with lymph node metastases. Hasegawa and colleagues developed a prognostic scoring system using a cDNA microarray. Seventy-eight genes were differentially expressed in patients with aggressive and non-aggressive gastric cancer *(12)*. Other investigators have set up a serial analysis through gene expression (SAGE) analysis to determine molecular prognosis and response to therapy among patients with gastric cancer *(13)*. The possibilities for personalized prognosis have become much more likely using this system.

3. NEW PARADIGMS IN THE TREATMENT OF UPPER GI MALIGNANCIES

The lack of success in treating malignancies of the upper GI tract has lead to the search for new paradigms. Specifically, empiric discoveries no longer seem to be the way to go in identifying new drugs. The combination of molecular biology and the ability to isolate, quantify, and identify specific proteins and nucleic acids and the clarification of the roles of various regulatory proteins in the cellular metabolism have all made it possible to move to a non-empiric level of drug discovery. This will hopefully bring a combination of more drug efficacy with less patient toxicity. The major questions to be asked, however, are whether the specificity of these newer agents is related to the disease in question or alternatively extends across disease borders. For example, Gleevec, which is a potent inhibitor of both c-kit and bcr-abl, is active against both GI stromal tumors and chronic myelogenous leukemia. A better example may be the EGFR inhibitors, which may have activity against any tumor that overexpresses EGFR. An alternative question is whether these drugs work specifically against a subset of tumors from a particular organ site. Take for example Iressa, which appears to work against subsets of lung cancers with one of several specific mutations, but does not work against non-small cell lung cancers (NSCLCs) which are wild-type for that specific gene. Thus, important questions remain about how best to use these drugs.

3.1. Stomach Cancer

An early study from Japan in 2003 suggested the possibility of personalized therapy for patients with gastric cancer *(14)*. In this case, the investigators quantified expression levels of dihydropyrimidine dehydrogenase, glutathione S-transferase-pi, beta-tubulin, O^6-methylguanine-DNA methyltransferase, multiple drug-resistant protein-1, NADPH/quinone oxidoreductase-1, and cytochrome p450 (P450) by reverse transcription-polymerase chain reaction analysis and constructed a flow chart by which to recommend specific therapies. They had a 42% response rate in the patients treated with this new paradigm versus 0% in a control group. A review of the molecular biology of stomach cancer from Italy the next year expanded on this panel of findings *(15)*. This laid the groundwork for future studies using microarray technology to predict outcome, optimize therapies, and discover new regimens for the treatment of gastric cancer.

At about the same time, based on early results from patients with breast cancer, c-erb2 expression and amplification were being investigated in Japan in patients with stomach cancer, and specifically the role of an antibody against the c-erbB-2 gene product, trastuzumab (Herceptin; Genentech, Inc., South San Francisco, CA) *(16)*. Unfortunately, overexpression of c-erbB-2 protein was found in only 29 (8.2%) of the 352 gastric carcinomas analyzed. Nevertheless, this suggested that targeted therapies for patients with gastric cancer were a possibility. Combination therapies of trastuzumab and conventional chemotherapy were tried against several cell lines. The combination of five consecutive days' treatment of trastuzumab with 1-day doxorubicin treatment showed significant growth inhibition only in YCC-2 and NCI-N87 gastric cancer cells *(17)*. After 1 day of trastuzumab treatment, the S-phase fraction was decreased by 52 and 70% in YCC-2 and SK-BR-3, respectively.

Measurements of EGFR were performed based on the promising preliminary results from studies in colorectal cancer *(18)* and of partial reversal of the malignant phenotype with anti-EGF antisense *(19)*. High levels of overexpression (2+ or 3+ staining) were found in 9 of 413 (2.2%) patients, whereas low levels of overexpression (1+) were found in 34 (8.2%) of the study cohort. Thus, as with trastuzumab, inhibition of EGFR was not an optimal target *(20)*. One caveat to this statement, however, is that EGF signaling is enhanced by CPT-11 *(21)*; therefore, inhibition of EGF with gefitinib (Iressa, ZD1839) may down-regulate EGF activity and prevent activation of the tumor cell in AGS gastric cancer cells. Furthermore, synergy studies were performed with gefitinib and either paclitaxel or oxaliplatin against SNU-1 gastric carcinoma cells *(22)*. Synergy using median effect analysis was observed with both cytotoxic agents. A phase I study of another EGFR inhibitor, EMD 72000, was initiated in 2004 in patients with elevated EGFR levels *(23)*; unfortunately, there were no responders among the 2/22 patients with upper GI malignancies. A follow-up study in patients with upper GI malignancies, including 2 cholangiocarcinomas and 1 gastric cancer, showed 7 objective responders among the 24 patients enrolled on the study *(24)*, demonstrating single-agent activity for the compound.

Flavopiridol (NSC 649890) is a synthetic flavone possessing significant antitumor activity in preclinical models. Flavopiridol is capable of inducing cell cycle arrest and apoptosis, presumably through its potent, specific inhibition of multiple cyclin-dependent kinases *(25,26)*. A phase I trial and pharmacokinetic study of flavopiridol given as a 72-h continuous intravenous infusion repeated every 2 weeks was performed. A total of 38 patients were treated at dose levels of 8, 16, 26.6, 40, 50 and

56 mg/m²/24 h. The maximum tolerated dose was determined to be 40 mg/m²/24 h. Of interest, a patient with metastatic gastric cancer at this dose level had a complete response and remained disease-free for more than 48 months after completing therapy *(25,26)*. This was very early evidence that targeted therapies could have potent activity against some subsets of refractory tumors. A phase II trial conducted at Memorial Sloan-Kettering enrolled 16 patients with no responders. Toxicities, specifically diarrhea, fatigue, and thromboembolic events, were greater than predicted. Thus, this particular regimen appeared to be less active and more toxic than expected.

3.2. Pancreatic Cancer

3.2.1. LABORATORY EXPERIENCE

Cetuximab (Erbitux; Merck) is an IgG1 monoclonal antibody that specifically targets the EGFR with high affinity. In early experiments in orthotopically placed tumors in nu/nu mice from M.D. Anderson Cancer Center in 2000, significant differences in microvessel density were observed 18 days after C225 or the combination of C225 and gemcitabine treatments (but not gemcitabine alone) in direct correlation with the difference in percentage of apoptotic endothelial cells, as visualized by double immunofluorescence microscopy *(27)*. These experiments indicate that therapeutic strategies targeting EGFR have a significant antitumor effect on human L3.6pl pancreatic carcinoma growing in nude mice, which is mediated in part by inhibition of tumor-induced angiogenesis, leading to tumor cell apoptosis and regression. Furthermore, this effect is potentiated in combination with gemcitabine. This was one of the first studies to show both a benefit and an anti-proliferative and an anti-angiogenic effect from C225. In vivo preclinical studies were performed against orthotopic pancreatic tumors in SCID mice at the Princess Margaret Hospital in Ontario, Canada *(28)*. They tested the EGFR inhibitor OSI-774 (Tarceva) alone and in combination with wortmannin and/or gemcitabine on downstream signaling molecules, as well as apoptosis in primary pancreatic cancer xenografts. The extent of apoptosis was significantly increased by twofold in OCIP#2 tumors treated with gemcitabine and wortmannin in combination; an additional twofold increase in apoptosis was evident in the presence of OSI-774.

In vitro studies from the University of Massachusetts *(29)* showed that for BxPC3 pancreatic cancer xenografts established in athymic nu/nu mice, weekly administration of 1 mg/kg PS-341 significantly inhibited tumor growth. Both cellular apoptosis and p21 protein levels were increased in PS-341-treated xenografts. Inhibition of tumor xenograft growth was greatest (89%) when PS-341 was combined with the tumoricidal agent CPT-11. Combined CPT-11/PS-341 therapy, but not single-agent therapy, yielded highly apoptotic tumors, significantly inhibited tumor cell proliferation, and blocked nuclear factor (NF)-κB activation indicating this systemic therapy was effective at the cancer cell level. Additional studies were performed at UC Davis in California *(30)*. Investigators found that gemcitabine followed by bortezomib induced the greatest induction of apoptosis and long-term inhibition of cell growth. Bortezomib treatment led to accumulation of p21 and p27 and decreased BCL-2; gemcitabine decreased p27, induced BCL-2, and had no effect on p21. A follow-up set of studies from M.D. Anderson *(31)* showed that inhibiting constitutive

NF-κB activity by expressing IκBaM suppresses liver metastasis, but not tumorigenesis, from the metastatic human pancreatic tumor cell line AsPc-1 in an orthotopic nude mouse model. Furthermore, inhibiting NF-κB activation by expressing IκBaM significantly reduced in vivo expression of a major proangiogenic molecule, VEGF, and, hence, decreased neoplastic angiogenesis. Inhibiting NF-κB activation by expressing IκBaM and using pharmacologic NF-κB inhibitor, PS-341, also significantly reduced cytokine-induced VEGF and interleukin-8 expression in AsPc-1 pancreatic cancer cells. In 2001, investigators studied the role of Velcade (PS-341), an anti-proteasome molecule, against MIA-PACA2 human pancreatic cell lines in vitro with or without gemcitabine *(32)*. PS-341 decreased BCL-2, without effect on BAX or BAK. The down-regulation of BCL-2 by PS-341 appears to be transcriptionally mediated. Xenograft growth was inhibited 59% by gemcitabine; the addition of PS-341 increased growth inhibition to 75%.

Pancreatic cancer cells, and particularly k-ras positive cells, express functional neurotensin (NT) receptors as well as other receptors that stimulate growth. Interruption of those pathways using an NT antagonist resulted in growth delay in an in vivo tumor model system, HPAF-II *(33)*. Furthermore, pancreatic cancer cell lines rely on EGFR and HER2 receptors to stimulate growth *(34)*. The EGFR and its ligands are expressed in 95% of pancreatic adenocarcinomas, resulting in constitutive activation that enhances cellular proliferation. Pancreatic head carcinomas overexpress HER3 while ampullary carcinomas overexpress HER2, and these immunohistochemical findings have been related to poor outcome. Furthermore, most pancreatic carcinomas harbor activating mutations of *K-ras*, the canonical downstream signaling intermediary of the EGFR family. The Her 1-2 inhibitor, lapatinib, GW572016, inhibited EGFR-dependent proliferation and anchorage-independent colony formation in pancreatic cancer cell lines through inhibition of MAPK and Akt pathways. Bruns et al. *(27)* have reported the results of a phase II trial of EGFR/HER2 targeted therapy against in vitro pancreatic cancer cell lines, and Bloom and colleagues *(35)* have reported that dual inhibition of FAK and insulin-like growth factor 1 receptor (IGR-IR) enhanced apoptosis through akt down-regulation in human pancreatic cell lines.

Fahy and colleagues investigated the role of bcl-2 expression in the aggressiveness of human pancreatic cell lines. BCL-2 expression varied both between and within tumor types; four of seven cell lines demonstrated high BCL-2 levels (MIA-PaCa-2, PC-3, Calu-1, and MCF-7), and no signaling pathway was uniformly responsible for overexpression of BCL-2. Inhibition of NF-κB activity decreased BCL-2 protein levels independently of the signaling pathway involved in transcriptional activation of the BCL-2 gene. The authors concluded that diverse signaling pathways variably regulate BCL-2 gene expression in a cell type-specific fashion. Therapy to decrease BCL-2 levels in various human cancers would be more broadly applicable if targeted to transcriptional activation rather than signal transduction cascades. Finally, the apoptotic efficacy of proteasome inhibition with bortezomib paralleled the ability to inhibit NF-κB activity and decrease BCL-2 levels *(36)*.

3.2.2. CLINICAL EXPERIENCE

Following the promising clinical data regarding anti-angiogenic strategies against human cancer, a phase III trial of one of the earliest angiogenic drugs, marimastat, given at 5, 10, and 25 mg/day versus gemcitabine was undertaken *(37)*. This was a multicenter trial with overall survival as an end-point. Overall survival was 20 and 19% for the highest dose of marimastat and gemcitabine, respectively; survival post-gemcitabine was better than for the lower doses of marimastat. The findings were inconclusive for a benefit for anti-angiogenic therapy. A more recent trial combined BV and gemcitabine for advanced pancreatic cancer led by Kindler at the University of Chicago (38). Preliminary results of that study were presented at ASCO in 2004 and showed encouraging results with response rates of 24% and overall survival of 12.4 months. Kindler is currently leading a phase II trial through the Chicago Phase II Consortium to test gemcitabine–BV + an EGFR inhibitor [either erbitux (C2250) or Tarceva], as well as a randomized trial through CALGB of gemcitabine ± BV. Furthermore, she recently led a study of SDX-102 in patients with pancreatic therapy who had failed gemcitabine. Results of this study are pending (39).

Another clinical trial tested an anti-growth factor strategy using antibodies against gastrin, a proven growth factor for pancreatic cancer (40). The end-point of the study was immunologic response; there were no clinical responders reported. Of interest, patients who responded to treatment had a longer survival time than the non-responders. Whether this was due to the fact that these patients were in better shape in general or to a legitimate improvement allowing a response is unclear from the study. A follow-up randomized study in 2005 by Shapiro and colleagues compared G17DT + gem with gem alone in 383 patients. There were intrinsic differences in the populations, but in the end, no difference in outcome between them (41). Again higher antibody titers and female sex suggested better outcomes.

In 2003, results of a multicenter clinical trial of the farnesyltransferase (FTase) inhibitor, R115777, which is a selective nonpeptidomimetic inhibitor of FTase and one of several enzymes responsible for posttranslational modification that is required for the function of ras (42). Twenty patients who had not received prior therapy for metastatic disease were treated with 300 mg of R115777 orally every 12 h for 21 of 28 days. Inhibition of FTase activity in peripheral blood mononuclear cells was measured using a lamin B C-terminus peptide as substrate. Western blot analysis was performed to monitor farnesylation status of the chaperone protein HDJ-2. No objective responses were seen, and median survival time was 19.7 weeks. FTase activity decreased by 49.8% ± 9.8% 4 h after treatment on day 1 and 36.1% ± 24.8% before treatment on day 15. HDJ-2 farnesylation (mean ± SD) decreased by 33.4% ± 19.8% on day 15. Although treatment with R115777 resulted in partial inhibition of FTase activity in mononuclear cells, it did not exhibit single-agent antitumor activity in patients with previously untreated metastatic pancreatic cancer. Nevertheless, these results were exciting for demonstrating the potential for performing translational research in this group of patients with difficult tumors.

In 2005, Moore and colleagues reported the results of a very important random assignment trial of erlotinib + gemcitabine versus gemcitabine alone in Canadian patients entered into an NCI-C study. EGFR status was not a criterion for study entry. No patients had had prior therapy except for either XBRT or XBRT + a sensitizer. There were 569 patients entered; 485 had died by the initial analysis. The toxicity profile in the initial submission was very favorable: there was a small increase in grade 1–2 rash, diarrhea, and hematologic toxicity among patients receiving the erlotinib. While the initial report showed strong evidence for a survival advantage, the report delivered at the ASCO 2005 meeting showed a borderline survival benefit. Along similar lines, Graeven et al. *(43)* have reported the results of a phase I study of the humanized EGFR-1 inhibitor, EMD 72000 + gemcitabine in pancreatic cancer. The early reports are preliminary.

Another randomized clinical trial was an NCIC-sponsored trial of the selective matrix metalloproteinase inhibitor BAY 12-9566 versus gemcitabine in patients with locally advanced or metastatic pancreatic cancer *(44)*. The study was closed early after the second planned interim analysis. Time to progression and survival were virtually identical; however, the toxicity profile was better for gemcitabine.

Viral therapy holds promise for the treatment of pancreatic cancer. A multicenter, industry-sponsored phase II clinical trial of the replicating adenovirus, ONYX-015, directly injected into the pancreatic tumor, failed to demonstrate responses, although it was well-tolerated *(45)*. A more sophisticated system involves a modified adenovirus that releases TNFa locally when it is radiated *(46)*. This system unfortunately requires endoscopic placement of the virus into the tumor; nevertheless, it is one of the more interesting ideas for modified viruses.

Several small clinical trials using drugs with poorly understood mechanisms of action failed to show a benefit, including a phase II trial of arsenic trioxide, which closed after 12 patients were enrolled *(47)*.

Two preliminary studies have attempted to combine external beam radiation therapy with either erlotinib + gemcitabine *(48)* or BV + capecitabine *(49)*. The results from these two trials were preliminary.

3.3. Hepatocellular Carcinoma

3.3.1. LABORATORY EXPERIENCE

Based on data in NSCLC that shows that gain-of-function EGFR and EGFR2 somatic mutations enhance response to gefitinib *(50)*, the investigators resequenced the flanking sequences, in order to confirm this in hepatocellular and bilary tumors. There were no somatic mutations among the flanking sequences in the biliary tumors and none in the EGFR sequences in the hepatocellular sequences. There were somatic H878Y mutations in exon 21 of the EGFR2 sequences that were predictive of response. In a separate study, investigators attempted to raise VEGF and platelet-derived growth factor levels with octreotide ± imatinib using double modulation of pathways; there were no beneficial effects noted *(51)*.

3.3.2. CLINICAL EXPERIENCE

Schwartz and colleagues *(52)* studied only low-risk patients with hepatocellular carcinoma (HCC) (no distant mets and no main portal vein invasion) with BV. Eleven of twelve patients who received BV tolerated therapy well, and some were advanced from 5 mg/kg to 10 mg/kg. Most of the patients remained stable during treatment. An additional study was reported by Britten et al. *(53)* of BV in patients with HCC undergoing transarterial chemoembolization (TACE) and thus a better group of patients. The rationale was that VEGF levels rise after TACE, providing rapid collateralization. Seven of ten patients were evaluable; however, results were not definitive.

Zhu and colleagues *(54)* studied gemcitabine, oxaliplatin + BV in a phase II study in patients with unresectable hepatocellular carcinoma. The drug administration was sequenced with BV administered first followed by gemcitabine and oxaliplatin.

Seventeen patients were enrolled. Response rates were not mentioned in the abstract; however, toxicities were relatively modest. Despite a bowel perforation and a grade 3 variceal bleed, the therapy was well tolerated.

Thomas and colleagues (55) enrolled patients to a stratified phase II trial of OSI-774, 150 mg/day based on low-risk or high-risk parameters, with 6 in the low-risk group and 19 in the high risk group, although all apparently received the same dose of OSI-774. There was 1 partial response (PR). Twenty of twenty-five patients were able to tolerate full-dose therapy and had a grade 1–2 rash. Median survival for all patients was 35 weeks. Philip and colleagues (56) also performed a phase II study of OSI-774 in patients with unresectable HCC. All patients received OSI-774 at 150 mg/day. Thirty-nine patients with biliary tumors and 28 patients with HCC were treated with OSI-774. Two of thirty-nine patients with biliary tumors and 3/28 patients with HCC had partial response with response durations of 3–12+ months.

Lin and colleagues (57) tested imatinib mesylate (Gleevec) at 300 mg/day with a 100-mg/week dose escalation in patients with unresectable or embolizable HCC. There were no grade 3–4 AEs reported. Sorafenib is a multikinase inhibitor with anti-angiogenic, pro-apoptotic and Raf kinase inhibitory activity. In 2007 the results of a multicenter, randomized, placebo-controlled phase III trial of sorafenib administered at 400 mg twice daily versus placebo was reported for 602 treatment naïve advanced HCC patients. A highly significant improvement in time to disease progression and overall survival was seen with a similar toxicity profile in both the sorafenib and placebo groups (58).

4. UNEXPECTED TOXICITIES WITH TARGETED THERAPIES

Treatment targeting the EGFR is associated with a high rate of rashes; there are significant cutaneous toxic effects, including follicular rashes (59), acneiform eruptions (60), nail bed changes, and seborrheic dermatitis-like eruptions. Follicular rashes occur in approximately one-third of patients (61). The probable explanation for these skin reactions may lie in the role of EGF in the development and mainte-nance of hair follicles. Failure of hair follicles to enter the catagen stage develops in transgenic mice expressing an EGFR dominant-negative mutation in the basal layer of the epidermis and the follicular outer root sheath, which causes severe inflammatory follicular necrosis and alopecia (62). Similarly, EGFR-null hair follicle buds grafted onto nude mice demonstrate an inability to progress from the anagen to telogen stages, resulting in inflammation and alopecia within weeks (63). These studies suggest that EGFR has a central role in follicular physiology and probably plays a role in the development of facial, chest, and back folliculitis.

A letter from Memorial Sloan Kettering revealed a high incidence of thromboembolic-based events among patients receiving BV (64). In the phase II trial of irinotecan, CP and BV, there were 25 patients enrolled on study; 6 (25%) had a thromboembolic event. Two of six remained on study having demonstrated a partial response. The remainder of the patients either came off study or were continued on without the BV.

5. PROBLEMS AND PITFALLS

One of the significant problems with these early studies is the small patient size. Therefore, it is difficult to actually establish a benefit for any of these agents. For example, one of the largest studies comparing gemcitabine ± erlotinib in a randomized trial for pancreatic cancer demonstrated a borderline survival benefit for patients receiving targeted therapies, but the differences were minor, at most a few weeks *(65)*, although originally reported as more significant. Therefore, it is hard to determine whether there is a benefit for the augmented therapy. The remainder of the randomized trials were negative except for the sorafenib HCC phase III trial.

A second problem is the underreporting of severe adverse events. The picture is confusing, especially when dealing with a sick population of patients and a relatively benign class of drugs. Exactly when is the decision made to change therapies? Clearly, patients having severe side effects need to come off treatment, but how about patients who are having a relatively benign event? How will this effect the reporting of further adverse events in this patient population? Furthermore, many of the events do not fall into standard criteria for adverse events, so are hard to grade and characterize.

A third problem is how to incorporate preclinical results into the design of clinical trials. This can be in the phase I–III setting. For example, drug synergy and modulation of molecular pathways is difficult to translate from the preclinical to clinical setting, largely because criteria are vague. Furthermore, with multiple preclinical tumor models, it is difficult to predict which human tumors are likely to be more responsive to combination therapy.

It is also more difficult to predict which combinations will add excessive toxicities, although this would presumably be covered in a phase I study.

6. CONCLUSIONS AND FUTURE DIRECTIONS

Targeted agents offer promises, but challenges in incorporating them into standard therapy. Some agents, such as BV, cetuximab (C225), erlotinib, and gefinitinib (Iressa) have already been approved. Some agents are waiting for approval. The side effect profiles are still being investigated. More PK and PD studies need to be performed to understand how best to incorporate these agents into standard regimens.

REFERENCES

1. Jemal, A., R.C. Tiwari, T. Murray, A. Ghafoor, A. Samuels, E. Ward, E.J. Feuer, and M.J. Thun, Cancer statistics. CA Cancer J Clin, 2004 54(1): p. 8–29.
2. Doi, T., W. Koizumi, S. Siena, et al., Efficacy, tolerability and pharmacokinetics of gefitinib (ZD 1839) in pretreated patients with metastatic gastric cancer. Proc ASCO, 2003, abstract 103.
3. Rojo, F., J. Tabernero, and E. Van Cutsem, Pharmacodynamic studies of tumor biopsy specimens from patients undergoing treatment with gefitinib. Proc ASCO, 2003, abstract 764.
4. Yasui, W., N. Oue, H. Kuniyasu, et al., Molecular diagnosis of gastric cancer: present and future. Gastric Cancer, 2001 4: p. 113–21.
5. Yasui, W., H. Yokozoki, F. Shimamato, et al., Molecular-pathologic diagnosis of gastrointestinal tissues and its contribution to cancer histopathology. Pathol Int, 1999 49: p. 763–74.
6. Xaingming, C., S. Natsugoe, S. Takao, et al., The cooperative role of p27 and cyclin E in the prognosis of advanced gastric cancer. Cancer, 2000 89: p. 1214–9.
7. Mosolits, S., G. Ullenhag, and H. Mellstedt, Therapeutic vaccination in patients with gastrointestinal malignancies: a review of immunological and clinical results. Ann Oncol, 2005 16: p. 847–62.

8. Zhang, H., J. Wu, L. Meng, and C. Shou, Expression of vascular endothelial growth factor and its receptors KDR and flt-1 in gastric cancer cells. World J Gastroenterol, 2002 8: p. 994–8.

9. Mitani, Y., N. Oue, Y. Hamai, et al., H3 acetylation is associated with reduced p21 WAF1/CIP1 expression in gastric carcinoma. J Pathol, 2005 205: p. 65–73.

10. Suzuki, T., H. Kuniyasu, K. Hayashi, et al., Effect of trichostatin A on cell growth and expression cell-cycle and apoptosis-related in human gastric and oral carcinoma cell lines. Int J Cancer, 2000 88: p. 992–7.

11. Hippo, Y., H. Taniguchi, S. Tsutsumi, et al., Global gene expression analysis of gastric cancer by oligonucleotide microarrays. Cancer Res, 2002 62: p. 233–40.

12. Hasegawa, S., Y. Furukawa, M. Li, et al., Genome-wide analysis of gene expression in intestinal-type gastric cancers using a complementary DNA microarray representing 23,040 genes. Cancer Res, 2002 62: p. 7012–7.

13. Yasui, W., N. Oue, N. Ito, et al., Search through new biomarkers of gastric cancer through serial analysis of gene expression and its clinical implications. Cancer Sci, 2004 95: p. 385–92.

14. Yoshida, K., K. Tanabe, H. Ueno, K. Ohta, J. Hihara, T. Toge, and M. Nishiyama, Future prospects of personalized chemotherapy in gastric cancer patients: results of a prospective randomized pilot study. Gastric Cancer, 2003 6 Suppl 1: p. 82–9.

15. Scartozzi, M., E. Galizia, F. Freddari, R. Berardi, R. Cellerino, and S. Cascinu, Molecular biology of sporadic gastric cancer: prognostic indicators and novel therapeutic approaches. Cancer Treat Rev, 2004 30(5): p. 451–9.

16. Takehana, T., K. Kunitomo, K. Kono, F. Kitahara, H. Iizuka, Y. Matsumoto, M.A. Fujino, and A. Ooi, Status of c-erbB-2 in gastric adenocarcinoma: a comparative study of immunohistochemistry, fluorescence in situ hybridization and enzyme-linked immuno-sorbent assay. Int J Cancer, 2002 98(6): p. 833–7.

17. Gong, S.J., C.J. Jin, S.Y. Rha, and H.C. Chung, Growth inhibitory effects of trastuzumab and chemotherapeutic drugs in gastric cancer cell lines. Cancer Lett, 2004 214(2): p. 215–24.

18. Takehana, T., K. Kunitomo, S. Suzuki, K. Kono, H. Fujii, Y. Matsumoto, and A. Ooi, Expression of epidermal growth factor receptor in gastric carcinomas. Clin Gastroenterol Hepatol, 2003. 1(6): p. 438–45.

19. Choudhury, A., J. Charo, S.K. Parapuram, R.C. Hunt, D.M. Hunt, B. Seliger, and R. Kiessling, Small interfering RNA (siRNA) inhibits the expression of the Her2/neu gene, upregulates HLA class I and induces apoptosis of Her2/neu positive tumor cell lines. Int J Cancer, 2004 108(1): p. 71–7.

20. Tanner, M., M. Hollmen, T.T. Junttila, A.I. Kapanen, S. Tommola, Y. Soini, H. Helin, J. Salo, H. Joensuu, E. Sihvo, K. Elenius, and J. Isola, Amplification of HER-2 in gastric carcinoma: association with Topoisomerase IIalpha gene amplification, intestinal type, poor prognosis and sensitivity to trastuzumab. Ann Oncol, 2005 16(2): p. 273–8.

21. Kishida, O., Y. Miyazaki, Y. Murayama, M. Ogasa, T. Miyazaki, T. Yamamoto, K. Watabe, S. Tsutsui, T. Kiyohara, I. Shimomura, and Y. Shinomura, Gefitinib ("Iressa", ZD1839) inhibits SN38-triggered EGF signals and IL-8 production in gastric cancer cells. Cancer Chemother Pharmacol, 2005 55(4): p. 393–403.

22. Park, J.K., S.H. Lee, J.H. Kang, K. Nishio, N. Saijo, and H.J. Kuh, Synergistic interaction between gefitinib (Iressa, ZD1839) and paclitaxel against human gastric carcinoma cells. Anticancer Drugs, 2004. 15(8): p. 809–18.

23. Vanhoefer, U., M. Tewes, F. Rojo, O. Dirsch, N. Schleucher, O. Rosen, J. Tillner, A. Kovar, A.H. Braun, T. Trarbach, S. Seeber, A. Harstrick, and J. Baselga, Phase I study of the humanized antiepidermal growth factor receptor monoclonal antibody EMD72000 in patients with advanced solid tumors that express the epidermal growth factor receptor. J Clin Oncol, 2004 22(1): p. 175–84.

24. Trarbach, T., T. Beyer, N. Schleucher, et al., A randomized phase I study of the humanized anti-epidermal growth factor monoclonal antibody EMD 72000 in subjects with advanced gastrointestinal cancers. J Clin Oncol, 2004 22(14S): p. 3018.

25. Schwartz, G.K., D. Ilson, L. Saltz, E. O'Reilly, W. Tong, P. Maslak, J. Werner, P. Perkins, M. Stoltz, and D. Kelsen, Phase II study of the cyclin-dependent kinase inhibitor flavopiridol administered to patients with advanced gastric carcinoma. J Clin Oncol, 2001 19(7): p. 1985–92.

26. Thomas, J.P., K.D. Tutsch, J.F. Cleary, H.H. Bailey, R. Arzoomanian, D. Alberti, K. Simon, C. Feierabend, K. Binger, R. Marnocha, A. Dresen, and G. Wilding, Phase I clinical and pharmacokinetic trial of the cyclin-dependent kinase inhibitor flavopiridol. Cancer Chemother Pharmacol, 2002 50(6): p. 465–72.

27. Bruns, C.J., M.T. Harbison, D.W. Davis, C.A. Portera, R. Tsan, D.J. McConkey, D.B. Evans, J.L. Abbruzzese, D.J. Hicklin, and R. Radinsky, Epidermal growth factor receptor blockade with C225 plus gemcitabine results in regression of human pancreatic carcinoma growing orthotopically in nude mice by antiangiogenic mechanisms. Clin Cancer Res, 2000 6(5): p. 1936–48.

28. Ng, S.S., M.S. Tsao, T. Nicklee, and D.W. Hedley, Effects of the epidermal growth factor receptor inhibitor OSI-774, Tarceva, on downstream signaling pathways and apoptosis in human pancreatic adenocarcinoma. Mol Cancer Ther, 2002 1(10): p. 777–83.

29. Shah, S.A., M.W. Potter, T.P. McDade, R. Ricciardi, R.A. Perugini, P.J. Elliott, J. Adams, and M.P. Callery, 26S proteasome inhibition induces apoptosis and limits growth of human pancreatic cancer. J Cell Biochem, 2001 82(1): p. 110–22.

30. Fahy, B.N., M.G. Schlieman, S. Virudachalam, and R.J. Bold, Schedule-dependent molecular effects of the proteasome inhibitor bortezomib and gemcitabine in pancreatic cancer. J Surg Res, 2003 113(1): p. 88–95.

31. Fujioka, S., G.M. Sclabas, C. Schmidt, W.A. Frederick, Q.G. Dong, J.L. Abbruzzese, D.B. Evans, C. Baker, and P.J. Chiao, Function of nuclear factor kappaB in pancreatic cancer metastasis. Clin Cancer Res, 2003 9(1): p. 346–54.

32. Bold, R.J., S. Virudachalam, and D.J. McConkey, Chemosensitization of pancreatic cancer by inhibition of the 26S proteasome. J Surg Res, 2001 100(1): p. 11–7.

33. Guha, S., K. Kisfalvi, G. Eibl, et al., [D Arg [1], DTrp [5,7,9], Leu [11]]SP: a novel, potent inhibitor of mitogenic signaling and growth in vitro and in vivo in human pancreatic cancer. Gastrointestinal Cancers Symposium, 2004.

34. Baerman, K., L. Caskey, F. Dasi, et al., EGFR/HER2 targeted therapy inhibits growth of pancreatic cancer cells. Gastrointestinal Cancers Symposium, 2005.

35. Bloom, D., W. Cance, E. Kurenova, et al., Dual inhibition of FAK and IGF-IR increases apoptosis through akt down-regulation in human pancreatic adenocarcinoma cells. Gastrointestinal Cancers Symposium, 2005.

36. Fahy, B.N., M.G. Schlieman, M.M. Mortenson, S. Virudachalam, and R.J. Bold, Targeting BCL-2 overexpression in various human malignancies through NF-kappaB inhibition by the proteasome inhibitor bortezomib. Cancer Chemother Pharmacol, 2005 56(1): p. 46–54.

37. Bramhall, S.R., A. Rosemurgy, P.D. Brown, C. Bowry, and J.A. Buckels, Marimastat as first-line therapy for patients with unresectable pancreatic cancer: a randomized trial. J Clin Oncol, 2001 19(15): p. 3447–55.

38. Kindler, H.L., G. Friberg, W. Stadler, et al. Bevacizumab plus gemcitabine is an active combination in patients with advanced pancreatic cancer: interim results of an ongoing phase II trial from the Univ of Chicago Phase II consortium. Gastrointestinal Cancers Symposium, 2004.

39. Kindler, H.L., M. Lobell, D. Smith, et al. SDX-102: a targeted therapy for previously treated, MTAP-deficient pancreatic cancer and gastrointestinal stromal tumors. Gastrointestinal Cancers Symposium, 2004.

40. Brett, B.T., S.C. Smith, C.V. Bouvier, D. Michaeli, D. Hochhauser, B.R. Davidson, T.R. Kurzaw-inski, A.F. Watkinson, N. Van Someren, R.E. Pounder, and M.E. Caplin, Phase II study of anti-gastrin-17 antibodies, raised to G17DT, in advanced pancreatic cancer. J Clin Oncol, 2002 20(20): p. 4225–31.

41. Shapiro, J., J. Marshall, P. Karasek, et al., G17DT + gemcitabine versus placebo + gemcitabine in untreated subjects with locally advanced, recurrent or metastatic adenocarcinoma of the pancreas: results of a randomized, double-blind, multinational, multicenter study. ASCO Ann Mtg-41st Ann Mtg, 2005. Orlando, Fl.

42. Cohen, S.J., L. Ho, S. Ranganathan, J.L. Abbruzzese, R.K. Alpaugh, M. Beard, N.L. Lewis, S. McLaughlin, A. Rogatko, J.J. Perez-Ruixo, A.M. Thistle, T. Verhaeghe, H. Wang, L.M. Weiner, J.J. Wright, G.R. Hudes, and N.J. Meropol, Phase II and pharmacodynamic study of the farnesyltrans-ferase inhibitor R115777 as initial therapy in patients with metastatic pancreatic adenocarcinoma. J Clin Oncol, 2003 21(7): p. 1301–6.

43. Graeven, U., I. Vogel, B. Killings, et al., Phase I study of humanized IgG1 anti-epidermal growth factor receptor monoclonal antibody EMD 72000 plus gemcitabine in advanced pancreatic cancer. J Clin Oncol, 2004 22(14S): p. 3061.

44. Moore, M.J., J. Hamm, J. Dancey, P.D. Eisenberg, M. Dagenais, A. Fields, K. Hagan, B. Greenberg, B. Colwell, B. Zee, D. Tu, J. Ottaway, R. Humphrey, and L. Seymour, Comparison of gemcitabine

versus the matrix metalloproteinase inhibitor BAY 12–9566 in patients with advanced or metastatic adenocarcinoma of the pancreas: a phase III trial of the National Cancer Institute of Canada Clinical Trials Group. J Clin Oncol, 2003 21(17): p. 3296–302.

45. Hecht, J.R., R. Bedford, J.L. Abbruzzese, S. Lahoti, T.R. Reid, R.M. Soetikno, D.H. Kirn, and S.M. Freeman, A phase I/II trial of intratumoral endoscopic ultrasound injection of ONYX-015 with intravenous gemcitabine in unresectable pancreatic carcinoma. Clin Cancer Res, 2003 9(2): p. 555–61.

46. Hanna, N., T. Chung, J. Hecht, et al., Safety and efficacy of TNFerade in unresectable locally advanced pancreatic cancer: results of the first two cohorts of a dose escalating study. Gastrointestinal Cancers Symposium, 2004.

47. Aklilu, M., H.L. Kindler, S. Nattam, et al., A phase II study of arsenic trioxide in patients with adenocarcinoma of the pancreas refractory to gemcitabine. Gastrointestinal Cancers Symposium, 2004.

48. Kortmansky, J., E. O'Reilly, B. Minsky, et al. A phase I trial of erlotinib, gemcitabine and radiation for patients with locally advanced, unresectable pancreatic cancer. Gastrointestinal Cancers Symposium, 2005.

49. Crane, C., L. Ellis, J. Abbruzzese, et al., Phase I trial of bevacizumab with concurrent radiotherapy and capecitabine in locally advanced pancreatic adenocarcinoma. ASCO Ann Mtg-41st Ann Mtg, 2005, Orlando, FL.

50. Bekaii-Saab, T., T. Sawada, N. Williams, et al. Intragenic EGFR and EGFR2 mutations in hepato-biliary tumors and potential role in predicting response to agents that target EGFR. ASCO Ann Mtg-41st Ann Mtg, 2005. Orlando, Fl.

51. Treiber, G., T. Wex, and P. Malfertheiner, Plasma PDGF and VEGF as sensitive biomarkers for monitoring of angiogenesis in hepatocellular cancer patients. ASCO Ann Mtg-41st Ann Mtg, 2005. Orlando, FL.

52. Schwartz, J.D., M. Schwartz, J. Goldman, et al., Bevacizumab in hepatocellular carcinoma for patients without metastasis and without invasion of the portal vein. Gastrointestinal Cancers Symposium, 2005.

53. Britten, C., R. Finn, A. Gomes, et al., A pilot study of IV bevacizumab in hepatocellular cancer patients undergoing chemoembolization. ASCO Ann Mtg-41st Ann Mtg, 2005. Orlando, FL.

54. Zhu, A., D. Sahani, A. Norden-Zfoni, et al. A phase II study of gemcitabine, oxaliplatin in combination with bevacizumab in patients with hepatocellular carcinoma. Gastrointestinal Cancers Symposium, 2005.

55. Thomas, M., A. Dutta, T. Brown, et al., A phase II open label study of OSI-774 in unresectable hepatocellular carcinoma. ASCO Ann Mtg-41st Ann Mtg, 2005. Orlando, FL.

56. Philip, P., M. Mahoney, J. Thomas, et al. Phase II trial of OSI-774 in patients with hepatocellular and biliary cancer. Gastrointestinal Cancers Symposium, 2004.

57. Lin, A., G. Fisher, S. So, et al., A phase II study of imatinib mesylate in patients with unresectable hepatocellular carcinoma. ASCO Ann Mtg-41st Ann Mtg, 2005. Orlando, FL.

58. Llovet, J., Ricci, S., Mazzaferro, V., et al. Sorafenib improves survival in advanced hepatocellular carcinoma (HCC): Results of a Phase III randomized placebo-controlled trial (SHARP trial). J Clin Oncol ASCO Proc, 2007 25: p LBA1

59. Mendelsohn, J., D. Shin, N. Donato, et al., A phase I study of chimerized anti-epidermal growth factor recombinant monoclonal antibody in combination with cisplatin in patients with recurrent head and neck squamous cell carcinoma. 35th Ann Mtg of the Am Soc Clin Oncol, 1999.

60. Baselga, J., D. Pfister, M. Cooper, et al., Phase I studies of anti-epidermal growth factor receptor chimeric C225 alone and in combination with cisplatin. J Clin Oncol, 2000 18: p. 904–14.

61. Kimyai-Asadi, A. and M.H. Jih, Follicular toxic effects of chimeric anti-epidermal growth factor receptor antibody cetuximab used to treat human solid tumors. Arch Dermatol, 2002 138(1): p. 129–31.

62. Murillas, R., F. Larcher, and F. Conti, Expression of a dominant negative mutant epidermal growth factor receptor in the epidermis of transgenic mice elicits striking alterations in hair follicle development and skin structure. EMBO J, 1995, 14: p. 5216–23.

63. Hansen, L., N. Alexander, M. Hogan, et al., Genetically null mice reveal a central role for epidermal growth factor receptor in the differentiation of the hair follicle and normal hair development. *Am J Pathol*, 1997, 160: p. 1969–75.

64. Shah, M., D. Ilson, and D. Kelson, Thromoembolic events in gastric cancer: high incidence in patients receiving irinotecan- or bevacizumab-based therapy. J Clin Oncol, 2005 23: p. 2574–6.
65. Moore, M.J., D. Goldstein, J. Hamm, et al., Erlotinib plus gemcitabine compared to gemcitabine alone in patients with advanced pancreatic cancer. A phase III trial of the National Cancer Institute of Canada Clinical Trials Group. ASCO Ann Mtg-41st Ann Mtg, 2005. Orlando, FL.

7 Molecular Targeting of Colorectal Cancer

An Idea Whose Time Has Come

Mark L. Sundermeyer, MD,
Alfonso Bellacosa, MD, PhD,
and Neal J. Meropol, MD

SUMMARY

Colorectal cancer is the fourth most common cancer worldwide and the fourth most common cause of cancer mortality, with approximately 529,000 deaths annually *(1)*. The concept of molecular targeting in colorectal cancer is not new. After all, 5-fluorouracil (5-FU), the standard bearer of "old school" treatment and continued mainstay of colon cancer systemic therapy, was developed as a "targeted" agent. In this case, the primary target is thymidylate synthase, a key enzyme in DNA synthesis, and the mechanism of action is competitive inhibition by a false substrate. Whereas 5-FU is clearly targeted, it lacks specificity, and the therapeutic window is therefore narrow. In the past decade, advances in understanding of the biology of colorectal cancer as well as the technology of drug development have permitted the identification of new targets and inhibitory pharmaceuticals with high specificity and favorable toxicity profiles. It is this specificity with regard to both target and tissue that characterizes the current generation of targeted therapeutics.

In contrast to other tumors that are driven by a single transforming molecular event, colorectal cancers are characterized by their genetic diversity. This diversity presents challenges for treatment and suggests that molecular profiling of individual patients and tumors will ultimately be required if we are to optimize the matching of patients and treatments. In this chapter, we will review the landscape of colorectal cancer treatment, with a focus on the most promising molecular targets in development. The cancer cell as well as surrounding stroma will be considered. In addition, we will review mechanisms of colorectal cancer pathogenesis and their implications for therapeutic intervention.

Key Words: Colorectal cancer; colon cancer; epidermal growth factor receptor (EGFR); vascular endothelial growth factor.

From: *Cancer Drug Discovery and Development: Molecular Targeting in Oncology*
Edited by: H. L. Kaufman, S. Wadler, and K. Antman © Humana Press, Totowa, NJ

1. THE COLORECTAL CANCER CELL AS TARGET

A great deal has been learned about those features that distinguish colorectal cancer cells from normal tissues, and a variety of efforts are under way to exploit these characteristics (see Fig. 1). We will begin with a discussion of cell surface targets and subsequently consider downstream intracellular events.

1.1. Surface Targets

1.1.1. Epidermal Growth Factor Receptor

Identification of the epidermal growth factor receptor (EGFR) as a therapeutic target has provided the first proof of concept that a highly specific molecularly targeted therapy can be of clinical benefit in patients with colorectal cancer. The EGFR is a 170-kDa transmembrane cell surface glycoprotein and was the first receptor protein to be recognized as a tyrosine-specific protein kinase *(2)*. This receptor consists of an extracellular portion that serves as a glycosylated ligand-binding domain, a transmembrane region, and an intracellular carboxy-terminal domain with tyrosine kinase (TK)

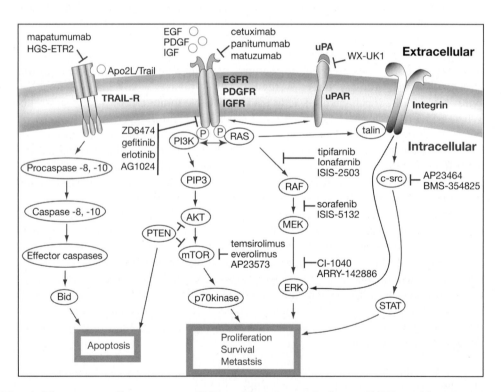

Fig. 1. The cancer cell as a target. EGF, epidermal growth factor; EGFR, epidermal growth factor receptor; IGF, insulin-like growth factor; IGFR, insulin-like growth factor receptor; mTOR, mammalian target of rapamycin; MEK, mitogen-activated protein kinase; ERK, extracellular signal-regulated protein kinase; PDGF, platelet-derived growth factor; PDGFR, platelet derived growth factor receptor; PI3K, phosphatidylinositol 3-kinase; PIP3, phosphatidylinositol *(3,4,5)* triphosphate; PTEN, phosphatase and tensin homologue deleted on chromosome ten; STAT, signal transducers and activators of transcription; TRAIL-R, tumor necrosis factor-related apoptosis-inducing ligand receptor; uPA, urokinase plasminogen activator; uPAR, urokinase plasminogen activator receptor.

activity. The EGFR is also referred to as c-erbB1 or human EGF receptor 1 (HER-1). The HER family of receptor TKs is the first structurally homologous receptor family to be characterized. Significant homology has been found in the TK portions for this receptor family that includes HER-2, HER-3, and HER-4 in addition to EGFR (2).

EGFR is activated through interaction of a ligand with the extracellular domain. The ligands that are capable of binding EGFR include EGF, heparin-binding EGF, transforming growth factor (TGF)-a, amphiregulin, betacellulin, and epiregulin. When these ligands bind, receptor dimerization occurs, and intrinsic TK activity autophosphorylates the tyrosine residues of the C-terminal tail. Intracellular signaling proteins with Src homology 2 (SH2) domains then bind to these phosphorylated tyrosine residues, subsequently activating intracellular signaling pathways (3). These signaling pathways include the ras-raf-MEK-ERK, PI3K, GTPases rac and rho, and others (2,4). How these pathways result in specific biologic responses continues to be studied, but successful transmission of the signal results in proliferation, differentiation, migration, and protection from apoptosis (5). In addition, EGFR can heterodimerize with other members of the HER family, with a wide variety of binding sites for proteins with the SH2 domain, with resultant downstream signaling (6).

While EGFR expression is a strictly controlled phenomenon in normal cells, EGFR overexpression favors receptor heterodimerization and activation of TK activity (7,8). EGFR overexpression and autocrine stimulation is common in colorectal adenocarcinomas, reported in 25–75% of cases (9,10). Overexpression of EGFR in colorectal cancer is associated with higher stage, poorer prognosis, and increased likelihood of metastasis (11,12). These observations led to the development of EGFR inhibitors in colorectal cancer.

Multiple strategies for targeting EGFR are under investigation. The two most extensively studied approaches are EGFR antibodies that block ligand binding to the EGFR extracellular domain and small molecules that bind intracellular domains and inhibit EGFR TK activity. Other approaches include immunoconjugates, multifunctional antibodies, anti-EGFR vaccines, and antisense oligonucleotides.

Monoclonal antibodies are highly specific and selective. In addition, chimeric, humanized, and human antibodies have prolonged half-lives that permit weekly or less frequent intravenous dosing schedules. Antibody binding to EGFR prevents ligand–receptor interaction. The antibody–receptor complex is internalized resulting in downregulation of EGFR on the cell surface (13). Potential drawbacks associated with therapeutic antibodies include the possibility of allergic reactions with those constructs containing murine components, and large molecular size, which may limit optimal tumor penetration. Furthermore, the specificity of monoclonal antibodies, while desirable in targeted therapy, may preclude interaction with EGFR variants in which the antibody-binding domain is altered.

In contrast to therapeutic antibodies, small molecule inhibitors of EGFR TK activity are orally administered, although shorter half-lives require more frequent dosing. Oral TK inhibitors have less target selectivity than antibodies and may interact with other members of the HER family, variants of EGFR, or other TKs. This promiscuity may result in unwanted toxicities, or perhaps augmented antitumor activity in contexts where multiple pathways are activated. Furthermore, their small size may facilitate greater tumor penetration. As described in section 2.1.5, in contrast to anti-EGFR antibodies, small molecule EGFR TKIs lack activity against colorectal cancer.

Preclinical evaluation of EGFR inhibitors against EGFR-expressing cell lines, whether in cell culture or xenograft models, yielded consistent biologic effects. EGFR blockade results in decreased expression of molecules associated with invasion and metastasis, such as matrix metalloproteinase 9 (MMP-9) and interleukin-8 (*14–16*). Angiogenesis is inhibited through downregulation of vascular endothelial growth factor (VEGF) (*17–19*). EGFR inhibition also results in decreased proliferation and promotion of apoptosis (*20,21*). Preclinical evidence also suggests that radiation effects and chemotherapy sensitivity may be enhanced (*22–24*). Based upon these findings, EGFR inhibitors entered clinical evaluation.

1.1.2. CLINICAL EXPERIENCE WITH EGFR ANTIBODIES IN PATIENTS WITH COLORECTAL CANCER

Several EGFR monoclonal antibodies are in clinical development. The three drugs furthest along in the clinic are discussed below.

1.1.2.1. Cetuximab. Cetuximab (Erbitux®, Bristol Myers Squibb) is an IgG1 chimeric counterpart of the murine monoclonal antibody M225 (*16*). Phase I studies identified skin rash as the predominant toxicity and saturable clearance consistent with receptor-antibody internalization serving as a primary elimination pathway. Selection of the phase II dose was based in part on the assumption that saturation of elimination represented full receptor occupancy. Phase II trials established the activity of cetuximab, both as a single agent, and in combination with chemotherapy when administered weekly to patients with metastatic colorectal cancer (see Table 1). Based on preclinical data suggesting augmented activity when used in combination (*25*), an initial phase II trial of cetuximab in combination with irinotecan was conducted in patients with metastatic colorectal cancer who had previously shown progressive disease during or shortly after treatment with irinotecan (*26*). In 120 patients, 27 (23%) experienced significant regression of their tumors ("partial responses"). The median survival of this irinotecan-refractory population was 7.6 months (*26*). In an effort to determine the single-agent activity of cetuximab, a study was conducted in a similar patient population. Among 57 irinotecan-refractory patients treated with cetuximab alone, a response rate of 9% (5/57) was obtained, with median survival 6.4 months (*27*). These data were confirmed in a randomized phase II trial in which 329 patients with refractory metastatic colorectal cancer were randomly assigned (2:1 randomization) to cetuximab plus irinotecan or cetuximab alone (*28*). In the cetuximab plus irinotecan arm, a 22.9% (50/218) partial response rate was reported. Single-agent cetuximab yielded a partial response rate of 10.8% (12/111). Median survival was 8.6 months in the combination group and 6.9 months in the single-agent cetuximab group (*28*). These data led to the licensure of cetuximab by the U.S. FDA either as a single agent or in combination with irinotecan for patients with metastatic colorectal cancer refractory to or intolerant of irinotecan. The activity of cetuximab was also confirmed in patients who were pretreated with both oxaliplatin and irinotecan. In a study of 350 patients, cetuximab monotherapy resulted in 28 (12%) antitumor responses (*29*).

It is notable that each of the studies described above required documentation of EGFR staining by immunohistochemistry. This selection criterion was based on the assumption that for a targeted therapy such as cetuximab to be effective, the target must be present. However, within these trials, the intensity of EGFR staining was not

Table 1
Clinical Activity of EGFR Antibodies in Patients with Colorectal Cancer

Study	Agents	Patient characteristics	Number of patients	Response rate (%)	Median survival (months)
Saltz et al. 2001 (26)	Cetuximab + irinotecan	5-FU, irinotecan-refractory	120	23.0	7.6
Cunningham et al. 2004,	Cetuximab	Irinotecan-refractory	111	10.8	6.9
randomized (28)	cetuximab + irinotecan		218	29.1	8.6
				p=0.007	p=0.48
Saltz et al. 2004 (27)	Cetuximab	Irinotecan-refractory	57	9.0	6.4
Lenz et al. 2004 (29)	Cetuximab	Oxaliplatin, irinotecan-refractory	350	12.0	NR
Malik et al. 2005 (32)	Panitumumab	Refractory to 5-FU and irinotectan, 5-FU and oxaliplatin, or both	148	10.0	9.4

5-FU, 5-flurouracil; NR, not reported.

associated with response *(26–28)*. Furthermore, several small series have indicated that response to cetuximab is possible in tumors without EGFR detectable by immunohistochemistry *(29,30)*. These observations do not indicate that the target of cetuximab is not the EGFR, but rather highlight the inadequacy of immunohistochemistry as a method for both EGFR detection and identification of those tumors driven by this signaling pathway *(31)*.

Activity in patients with refractory disease provided proof of concept and led to licensure of cetuximab. The current generation of clinical trials is seeking to characterize the activity of cetuximab (i) in combination with other chemotherapy regimens (e.g., oxaliplatin plus 5-FU) and molecularly targeted agents (e.g., bevacizumab), (ii) earlier in the course of metastatic disease, and (iii) in the adjuvant setting. In addition, as noted in Section 2.1.6., efforts are under way to identify predictors of response to EGFR inhibitory antibodies.

1.1.2.2. Panitumumab. Panitumumab (ABX-EGF, Abgenix, Inc.; Immunex Corp, a subsidiary of Amgen Inc.) is a fully human IgG2 monoclonal antibody against the EGFR. A phase II trial of weekly panitumumab has shown activity in metastatic colorectal cancer refractory to 5-FU plus irinotecan, oxaliplatin, or both *(32)*. This study enrolled two cohorts of patients based upon EGFR expression by immunohistochemistry. The first cohort of 105 patients was required to have 2–3+ EGFR staining in greater than 10% of tumor cells. The second cohort of 43 patients required at least 10% of tumor cells EGFR positive at any level *(1, 2,* or *3+)*, but with less than 10% of cells 2–3+. An overall response rate of 10% (15/148) was seen, with no differences between the cohorts. The median time to disease progression was 3.4 (cohort 1) and 2.1 (cohort 2) months, and median survival was 10.0 (cohort 1) and 9.4 (cohort 2) months *(32)* (Table 1). Ongoing development of panitumumab in patients with colorectal cancer includes (i) studies of less frequent dosing schedules *(33)*, (ii) studies in combination with chemotherapy (e.g., 5-FU/oxaliplatin) and bevacizumab, and (iii) a randomized study versus best supportive care in refractory disease.

1.1.2.3. Matuzumab. Matuzumab (EMD 72000, EMD Pharmaceuticals, Inc.) is a humanized monoclonal IgG1 antibody that binds to the extracellular domain of EGFR. A randomized phase I dose finding study of 24 patients that included 21 colorectal cancer patients with EGFR IHC-positive tumors was performed. Antitumor activity demonstrated a complete response in one patient, six patients with partial responses, and nine with stable disease *(34)*.

1.1.3. EGFR Antibody Toxicities

In general, treatment with anti-EGFR monoclonals is well tolerated. Whereas doselimiting toxicities were not observed with cetuximab and panitumumab in phase I trials, dose-limiting headache and fever were reported with matuzumab *(35)*. Acute allergic reactions occur in approximately 3% of patients treated with cetuximab *(36)*. The primary toxicity with these antibodies is an acne-like rash. This sterile, suppurative folliculitis occurs mainly on the scalp, face, and trunk and appears within the first few weeks of therapy. Peak intensity usually occurs within the first few weeks *(32,37)*. Interestingly, an association between skin rash intensity and response and survival has been reported with cetuximab *(27,28)*. This observation has led to the hypothesis that skin can serve as a surrogate pharmacodynamic marker for drug effect (vida infra). Furthermore, studies are under way to determine whether dose escalation of cetuximab based upon rash intensity can result in improved antitumor activity.

1.1.4. COMPARISON OF ANTI-EGFR ANTIBODIES

Each of the antibodies described above has clinical activity against colorectal cancer, and for those furthest along in development (cetuximab and panitumumab), the clinical activity appears comparable. In contrast to panitumumab, the IgG1 immunoglobulin backbones of cetuximab and matuzumab permit participation in antibody-dependent cellular cytotoxicity and other cell-mediated events (38). Whether this mechanism contributes to the clinical activity of these agents, or whether it may be exploited for clinical benefit, is as yet uncertain. The affinity of panitumumab for EGFR ($K_D = 5 \times 10^{-11}$) is higher than that of either cetuximab ($K_D = 20 \times 10^{-11}$) or matuzumab ($K_D = 34 \times 10^{-11}$M) (35,38). Again, the clinical significance of this potential biologic advantage is uncertain in view of the similarity in response rates observed with these agents. The half-lives of panitumumab and matuzumab permit dosing on an every 2-week or every 3-week schedule. If antitumor activity is preserved with less frequent dosing, this may provide an advantage for patients. Finally, those antibodies that contain murine components have been associated with infrequent but serious allergic reactions (27), in contrast to panitumumab which is a fully human product (32).

1.1.5. SMALL MOLECULE TYROSINE KINASE INHIBITORS OF EGFR

Small molecule EGFR TKIs have shown significant activity against non-small cell lung cancer and other malignancies. However, in contrast to EGFR antibodies, small molecule TKIs are inactive as single agents in patients with colorectal cancer.

Gefitinib (ZD1839, Iressa®, AstraZeneca) is a small molecule TKI that has been evaluated in several phase II trials in patients with refractory metastatic colorectal cancer. Overall, no objective responses were reported in 34 patients (39–41). When combined with 5-FU and oxaliplatin, responses were reported in 21/27 patients with previously untreated colorectal cancer and 8/22 patients with previously treated disease, raising the possibility of synergistic activity (42). Erlotinib (OSI-774, Tarceva®, OSI Pharmaceuticals, Genentech) is a small molecule TKI that has preclinical activity against colorectal cancer cell lines. However, as seen with other EGFR TKIs, no significant activity was identified in the clinical setting, with no objective responses in 31 patients with metastatic colorectal cancer (43).

1.1.6. PREDICTING RESPONSE TO EGFR INHIBITORS

Success has been achieved with the use of anti-EGFR monoclonal antibodies in patients with colorectal cancer. However, responses occur in only a minority of patients, and these responses tend to be short lived. It is clear that the mere presence of EGFR is not sufficient for therapeutic response; rather a subset of sensitive tumors are likely addicted to this pathway for survival (44). Identification of this subset has been challenging. Several studies have documented expected perturbations of downstream signaling in skin and tumors following EGFR inhibitor therapy (45–47); although there has been some suggestion that abrogation of AKT signaling is a prerequisite for clinical activity (47), for the most part these studies have failed to correlate these downstream effects (e.g., decreased activation of AKT and MEK/ERK pathways) with antitumor effect. These observations suggest that most colorectal cancers are able to circumvent the pathway blockade that originates at the cell surface.

Several studies have identified a subset of non-small cell lung cancers harboring somatic mutations in the TK domain, which are especially sensitive to EGFR TKIs (48–50). Notably, these early reports have not been uniformly corroborated (51). Colorectal cancers only rarely contain such mutations (52), perhaps explaining in part the lack of responsiveness of colorectal cancers to EGFR TK inhibitors.

A recent study of 31 patients treated with cetuximab or panitumumab has shown that EGFR gene copy number assessed by fluorescent in situ hybridization (FISH) was significantly elevated in responding colorectal tumors (53). This phenomenon of gene amplification predicting response has also been recently described in patients receiving gefitinib for non-small cell lung cancer (54). If this observation is confirmed, patient selection for anti-EGFR antibody therapy may ultimately be based upon prospective FISH analysis, similar to the practice of patient selection for breast cancer therapy with another HER family antibody, trastuzumab. Other efforts under way to help characterize tumors sensitive of EGFR inhibitory antibodies include (i) identification of a genomic signature predictive of response to cetuximab (55), (ii) describing the presence of germline polymorphisms in EGFR that alter expression (56) and measurement of variation in ligand expression.

The identification of tumors driven by the EGFR pathway could assist in the selection of patients most likely to benefit from EGFR blockade. However, for the vast majority of colorectal cancers, complexities related to HER family heterodimerization, other growth factor receptor interactions, downstream crosstalk, and ligand expression, to name a few, suggest that combination therapies selected based upon features of individual tumors will be required (31).

1.2. Cell Surface Targets Beyond the EGFR

1.2.1. INSULIN-LIKE GROWTH FACTOR RECEPTOR

The insulin-like growth factor receptor (IGFR), when stimulated by ligand binding, results in downstream signal activation of ras/raf/MEK/ERK, PI3K/AKT, and JAK/STAT (57). Circulating ligands, such as IGF1 and IGF2, are produced widely in the body, and when bound to IGFR, result in tumor growth and apoptosis inhibition (58,59). Overexpression of IGF2 has been documented in approximately one-third of colon cancers (60). There is almost no expression of IGF1R in normal colonic mucosa; however, it is present in over 90% of colon cancers (61). Preclinical evaluation of IGF1R blockade has demonstrated growth inhibition (62). Agents in development include small molecule inhibitors and monoclonal antibodies. One small molecule TKI targeting IGFR has recently entered clinical trials (e.g., AG1024).

1.2.2. TUMOR NECROSIS FACTOR-RELATED APOPTOSIS-INDUCING LIGAND AND ITS RECEPTORS

The targeting strategies described above interrupt aberrantly activated pathways that are pro-growth and anti-apoptotic. However, in addition to numerous anti-apoptotic networks, malignant cells also possess pro-apoptotic ligands, receptors, and pathways. Tumor necrosis factor-related apoptosis-inducing ligand (TRAIL) is a member of the

TNF ligand superfamily, and ligand binding results in apoptosis *(63)*. Four receptors have been identified. TRAIL-R1 and TRAIL-R2, when activated by TRAIL, mediate downstream signaling and subsequent apoptosis *(64)*. Current understanding of this ligand–receptor interaction suggests that binding recruits apoptosis-inducing caspases. Resultant activation of the pro-apoptotic proteins Bid and Bax leads to mitochondrial release of cytochrome c, loss of mitochondrial membrane potential, cell shrinkage, nuclear condensation, and apoptosis *(65)*. Evaluation of TRAIL-R3 and TRAIL-R4 suggests that these receptors have absent or non-functional death domains and may function as decoy receptors *(66)*. TRAIL receptors have been identified in the majority of human sporadic and inherited colorectal cancers *(67,68)*. Monoclonal antibodies against TRAIL-R1 (mapatumumab) and TRAIL-R2 (HGS-ETR2) as well as a recombinant ligand are under clinical investigation.

1.2.3. Urokinase Plasminogen Activator and Receptor

Urokinase plasminogen activator and receptor (UPAR) is a plasma membrane-bound glycoprotein receptor for uPA *(69)*. With receptor activation due to uPA binding, plasminogen is converted to plasmin. Plasmin activity has been shown to facilitate breakdown of the extracellular matrix (ECM) in coordination with MMPs *(70)*. Creation of a uPAR/integrin complex results in activation of the ERK pathway. Blockade of this complex resulted in decreased secretion of MMP-9 *(70)*. Recently, it has been suggested that uPAR can signal through EGFR to the ERK pathway without EGFR ligand binding *(71)*. Activity of uPA is elevated in colon adenocarcinomas and adenomatous polyps, but not in normal mucosa *(72)*. Elevated uPA levels in colorectal cancer specimens have been correlated with worse survival *(73)*. Inhibition of uPAR in colorectal cell lines results in decreased MEK/ERK activity and decreased cell invasion, migration, and adhesion *(70)*. uPAR antisense molecules result in decreased ECM degradation and lung metastases in mouse models *(74)*. Clinical use of uPA and uPAR inhibitors (e.g., WX-UK1) is in the earliest stages, but as a target they are attractive due to their dual role in signal transduction and shaping the ECM.

2. DIGGING BELOW THE SURFACE: INTRACELLULAR TARGETS

As described above, cell surface receptor targeting in colorectal cancer is based on the dependency of the cancers on the downstream events that follow receptor activation. Putative mechanisms of resistance to receptor targeting include the redundancy of signaling pathways and downstream pathway transactivation. Thus, the ability to selectively target signaling events more proximate to ultimate nuclear transcription has theoretical advantages. Those pathways that are shared by multiple surface receptors represent desirable targets. Two pathways that have gained the most interest to date are the Ras/Raf/MAPK and PI3K/AKT cascades, which conduct signals promoting proliferation, survival, angiogenesis, and metastasis (Fig. 1).

2.1. Ras

K-ras mutations are found in approximately 50% of colon cancers *(75)*. These mutations result in constitutive activation and signaling yielding increased proliferation *(76)*. p21ras protein activation requires posttranslational modifications, including farnesylation, that facilitate translocation to the cell membrane. Initial attempts at inhibiting p21ras function centered on the clinical development of farnesylation inhibitors, such as tipifarnib and lonafarnib. Unfortunately, this approach did not show clinical activity in patients with colorectal cancer *(77,78)*. The use of antisense oligonucleotides (e.g., ISIS-2503) also failed in this setting *(79)*. These negative results with farnesyl-transferase inhibitors may be due to lack of dependency of p21ras on farnesylation, or insufficient blockade of the target enzyme in colorectal cancers in vivo *(80)*. Subsequently, it has been found that p21ras can be activated by an alternate process, geranylgeranylation, which may have contributed to the lack of activity with farnesyl-transferase inhibitors *(81)*. Inhibitors of geranylgeranylation have shown some efficacy in preclinical models *(82)*.

2.2. Raf

Targets downstream of p21ras are also under investigation. Activated p21ras recruits raf to the cell membrane, binds it, resulting in raf kinase activity. Phosphorylation events secondary to raf kinase include activation of MEK1, MEK2, ERK1, and ERK2 resulting in growth and proliferation *(83,84)*. Of the three functional raf proteins, the B-raf isoform is the major protein linking p21ras to MEK signaling. Mutations of this protein have been described in approximately 10% of colon cancers in the absence of K-*ras* mutations *(84)*. Raf kinase targeting strategies, including antisense oligonucleotides and small molecule inhibitors, are under investigation. Recent clinical studies of these agents (e.g., sorafenib and ISIS 5132) in patients with metastatic colorectal cancer have demonstrated at best disease stabilization *(85,86)*. Preclinical studies have suggested additive benefit with cytotoxic chemotherapy, and combination studies are under way *(87,88)*.

2.3. MEK/ERK

Activated raf phosphorylates MEK *(89)*. Activated MEK facilitates the phospho-rylation of ERK1 and ERK2 [also referred to as mitogen-activated protein kinase (MAPK)]. Dimerization of phospo-ERK occurs, followed by nuclear translocation, transcription factor activation, and subsequent cell proliferation, differentiation, survival, invasion, and metastasis *(90)*. Constitutive activation of MAPK is seen with high frequency in colon cancers *(91,92)*. Additionally, its position late in the pathway suggests less potential for bypass through other activation pathways. The initial clinical approach to targeting this pathway involved CI-1040, an oral small molecule inhibitor of MEK1 and MEK2 with preclinical activity *(93)*. Phase I studies demonstrated the expected decreases in phospho-MAPK in tumor and surrogate tissues *(94)*, but phase II trials failed to demonstrate antitumor activity *(95)*. A second generation selective non-competitive inhibitor of MEK 1/2, ARRY-142886, is currently undergoing clinical evaluation *(96)*.

2.4. AKT/PI3K

The AKT/PI3K pathway is a signal transduction cascade that is primarily anti-apoptotic *(97)*. AKT phosphorylation and PI3K activation correlate with tumorigenic potential in colorectal cancer *(98)*. Inhibition of PI3K and AKT has shown preclinical activity *(99,100)*. Also, downstream targets of this pathway, such as mammalian target of rapamycin (mTOR) are being studied *(101,102)*. Activation of mTOR regulates protein translation through ribosomal S6 kinase, eukaryote initiation factor 4E protein 1, and p70 kinase *(103,104)*. Inhibition of mTOR may be most effective in tumors that lack PTEN expression, a tumor suppressor protein that inhibits mTOR activity *(105)*. While much initial research involved rapamycin, newer agents are under investigation. Temsirolimus (CCI-779), RAD001, and AP23573 have shown preclinical activity in colorectal cancer cell lines, and clinical trials have been initiated (*106–108*).

2.5. Src

Src is a non-receptor TK associated with the intracellular membrane. It functions as a mediator between growth factor receptors and downstream signaling through activation of Signal transducer and activator of transcriptions (STATs) for proliferation, differentiation, and survival *(109)*. In colon cancer, c-src associates with EGFR upon ligand binding *(110)*. Src activity in colorectal cancer has been associated with worse prognosis *(111)*. Clinical studies of src inhibitors such as AP23464 and BMS-354825 are ongoing *(112,113)*.

2.6. STAT

STAT is active or overexpressed in most malignancies, including colorectal cancer *(114)*. VEGF has been implicated as an activator of STAT3 signaling, which ultimately results in proliferation, survival, and metastasis of cancer cells *(115)*. In colorectal cell lines, inhibition of STAT3 results in downregulation of VEGF receptor 1 (VEGFR-1), neuropilin (NRP)-1, and NRP-2 and inhibition of malignant transformation *(115)*. STAT inhibitors have yet to enter clinical testing.

2.7. Aurora Kinase Inhibitors

The predominant function of aurora kinases is in mitosis. Aurora protein kinase-B is a component of a structure (the inner centromere protein) required for chromosome segregation and results in histone H3 phosphorylation *(116,117)*. The aurora2 kinase is amplified and oncogenic in colorectal cancer cell lines *(118)*. Preclinical evaluation of VX-680, a small molecule inhibitor of aurora kinases, has shown in vivo anti-tumor activity *(119)*. Another inhibitor, PH-739358, is in early clinical trials.

3. THE EXTRACELLULAR MATRIX AS TARGET

The ECM, or stroma, represents a promising target in colorectal cancer (Fig. 2). While acting as a barrier to invasion and metastasis, it also is a reservoir of growth factors and binding proteins that influence tumor behavior.

3.1. Angiogenesis

The development of vasculature is a critical component of cancer biology *(120)*. In 1971, Judah Folkman *(121)* proposed targeting angiogenesis as a method of cancer therapy. In general, tumors require a blood supply to grow beyond 1–2 mm^3 in size *(121)*. Angiogenesis allows for delivery of oxygen, nutrients, and hormones to the tumor cells *(122)*. Additionally, neovasculature allows migration of tumor cells resulting in metastasis *(122,123)*. In order for angiogenesis to occur, regulatory proteins

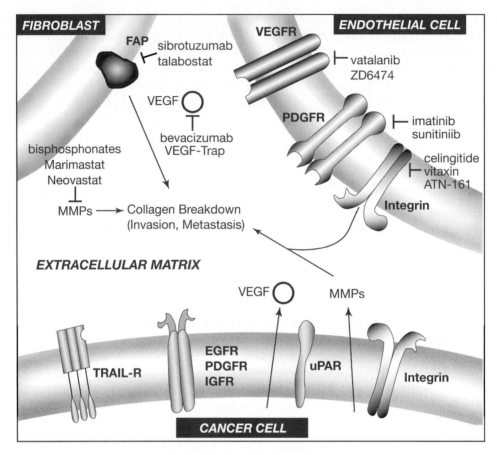

Fig. 2. The extracellular matrix as a target. EGFR, epidermal growth factor receptor; IGFR, insulin-like growth factor receptor; FAP, fibroblast activating protein; MMPs, matrix metallopro-teinases; PDGFR, platelet-derived growth factor receptor; TRAIL-R, tumor necrosis factor-related apoptosis-inducing ligand receptor; uPAR, urokinase plasminogen activator receptor; VEGF, vascular endothelial growth factor; VEGFR, vascular endothelial growth factor receptor..

favoring blood vessel formation must predominate over angiogenic inhibitory factors. This balance has been termed the "angiogenic switch" *(120)*.

Some factors known to facilitate angiogenesis include VEGF, fibroblast growth factor, platelet-derived growth factor (PDGF) and EGF. Those agents known to inhibit angiogenesis include thrombospondin-1 and thrombospondin-2, platelet factor-4, angiostatin, endostatin, canstatin, and turnstatin *(123,124)*. The "switch" to an angiogenic phenotype is considered one of the hallmarks of cancer and is regulated by oncogene activation, tumor suppressor gene mutations, tissue hypoxia, hyperglycemia, mechanical stress, and inflammation *(120,125,126)*.

3.1.1. VEGF/VEGFR as Therapeutic Targets in the Angiogenic Pathway

VEGFs and their associated receptors have become important molecular targets in the treatment of colorectal and other cancers. VEGFs include VEGF-A, VEGF-B,

VEGF-C, VEGF-D, and VEGF-E as well as placental growth factor-1 and placental growth factor-2 *(127,128)*. The receptor family includes VEGFR-1 *(129)*, VEGFR-2 *(130,131)*, and VEGFR-3 *(132,133)*. While VEGFR-3 functions in lymphangiogenesis, VEGFR-1 and VEGFR-2 are specific to angiogenesis *(134–137)*. In addition, several receptors such as NRP-1 and NRP-2 may act as co-stimulatory molecules *(138,139)*. Activation of the VEGF/VEGFR axis triggers multiple signaling networks. This results in prolonged endothelial cell survival *(127,140)*, increased proliferation *(141,142)*, migration, and invasion *(141,143,144)*. In addition, vascular permeability is enhanced *(145)*.

VEGF-A is expressed by vascular smooth muscle cells and cancer cells and is present in the stroma of colorectal cancer cell specimens *(128,146)*. Recently, the presence of VEGFR-1 has been identified on colorectal cancer cells themselves *(147)*. VEGF expression is seen in 40–60% of colorectal adenocarcinomas and may be higher in metastatic specimens *(148)*. Antibodies against VEGF showed antitumor activity in colorectal cancer xenograft models, leading to subsequent clinical development *(149)*.

Strategies to inhibit VEGF signaling include monoclonal antibodies against VEGF and VEGFR as well as TK inhibitors of the VEGFR. To date, the use of VEGF inhibitors as single agents has had limited success in patients with advanced colorectal cancer *(150)*. However, survival is improved when the VEGF inhibitor, bevacizumab, is combined with chemotherapy *(150,151)*.

3.1.1.1. Bevacizumab. Bevacizumab (Avastin® Genentech) is a humanized monoclonal antibody against VEGF-A. Phase I studies of bevacizumab did not identify dose-limiting toxicities *(152,153)*. In a small, randomized phase II trial (~90 patients) of initial therapy in patients with metastatic colorectal cancer, 5-FU plus leucovorin (5-FU/LV) was compared to 5-FU/LV with either 5 mg/kg or 10 mg/kg of bevacizumab every 2 weeks *(154)*. A suggestion of improved clinical outcome with bevacizumab led to further clinical development (Table 2) *(150,151,154–157)*.

A phase III trial of bevacizumab plus chemotherapy versus chemotherapy alone resulted in FDA approval of bevacizumab in combination with chemotherapy for use in patients with metastatic colorectal cancer *(151)*. In this trial, patients were randomized to receive either bevacizumab plus irinotecan, 5-FU, and leucovorin (IFL) (402 patients), or IFL plus placebo (411 patients). The addition of bevacizumab (5mg/kg) to IFL resulted in a survival advantage over IFL alone (20.3 months versus 15.6 months, $p < 0.0001(151)$

A third arm was initially included in this trial: 5-FU, leucovorin, plus bevacizumab. This arm was closed after approximately 300 patients were enrolled, when it was determined that the IFL/bevacizumab arm had an acceptable safety profile. Results of a comparison between the three original arms show similar efficacy of IFL to 5-FU, leucovorin and bevacizumab, suggesting a beneficial effect of bevacizumab when added to 5-FU/LV alone *(155)*. Further evidence of the benefit of bevacizumab when added to 5-FU/LV is provided by a randomized trial in patients felt by their physicians to not be suitable for therapy with IFL, based upon age or other clinical factors. Although not powered for survival as the primary endpoint, this study in 209 patients showed an improved progression-free survival with the addition of bevacizumab (9.2 versus 5.5 months, $p = 0.0002$) *(157)*.

Table 2
Clinical Activity of Bevacizumab in Patients with Colorectal Cancer

Study	Agents	Patient characteristics	Number of patients	Response rate (%)	Median survival (months)
Hurwitz et al. 2004 (151, 155)	IFL + bevacizumab (5 mg/kg)	Previously untreated metastatic disease	402	44.8	20.3
	IFL		411	34.8 $p = 0.004$	15.6 $p < 0.001$
Hurwitz et al. 2005 (155)	5-FU/LV + bevacizumab (5 mg/kg)	Previously untreated metastatic disease	110	40.0	18.3
	IFL		100	37.0 $p = 0.65$	15.1 $p = 0.25$
Kabbinavar et al. 2003 (154)	5-FU/LV + bevacizumab (10 mg/kg)	Previously untreated metastatic disease	33	24	16.1
	5-FU/LV + bevacizumab (5 mg/kg)		35	40	21.5
	5-FU/LV		36	17	13.8
Kabbinivar et al. 2005 (157)	5-FU/LV + bevacizumab (5 mg/kg)	Previously untreated metastatic disease, not irinotecan candidates	104	26.0	16.6
	5-FU/LV		105	15.2 $p = 0.055$	12.9 $p = 0.16$
Giantonio et al. 2005 (150)	FOLFOX + bevacizumab (10 mg/kg)	Prior therapy with 5-FU and irinotecan	290	21.8	12.9
	FOLFOX		289	9.2	10.8(12.9 vs. 10.8, $p = 0.0024$)
	bevacizumab (10 mg/kg)		243	3.0	10.2

5-FU/LV, 5-fluorouracil and leucovorin; FOLFOX, infusional and bolus 5-flurouracil, leucovorin and oxaliplatin; IFL, irinotecan, bolus of 5-flurouracil and leucovorin.

Bevacizumab also improves survival when added to infusional 5-FU, leucovorin, and oxaliplatin (FOLFOX) *(150)*. A North American cooperative group trial, ECOG 3200, evaluated the efficacy of bevacizumab in combination with FOLFOX in the second-line or third-line metastatic colorectal cancer setting. In this randomized phase III trial, patients received FOLFOX alone (289 patients), or FOLFOX plus bevacizumab (290 patients). A third arm, bevacizumab alone, was prematurely discontinued after enrollment of 243 patients, when the data safety monitoring board determined inferiority of bevacizumab monotherapy. Bevacizumab was administered at 10 mg/kg as opposed to the 5 mg/kg dose used in the IFL + bevacizumab study. The addition of bevacizumab to FOLFOX resulted in improved response rate (21.8 versus 9.2%, $p < 0.0001$) and median survial (12.9 versus 10.8 months, P=0.0024) (150). Although this study validated the addition of bevacizumab to an oxaliplatin-based regimen in the second-line or third-line setting, it did not address the important question of whether bevacizumab should be continued with subsequent chemotherapy in patients who progress on an initial bevacizumab-containing regimen, as none of the patients on this trial had received prior bevacizumab.

Based on the activity of these agents in colorectal cancer, and their pathway interrelationships, clinical trials are under way that combine bevacizumab and cetuximab. The BOND2 trial included the combination of bevacizumab and cetuximab alone or in conjunction with irinotecan in irinotecan-refractory patients with metastatic colorectal cancer *(158)*. This trial did not meet initial accrual goals, and data were presented for 81 patients. In the three-drug arm, there were 15/41 partial response (PRs) (37%), and time to progression (TTP) was 7.9 months. In the cetuximab/bevacizumab arm, there were 8/40 partial response (PRs) (20%) and time to progression (TTP) was 5.6 months *(158)*.

The studies described above clearly provide proof of concept that a VEGF inhibitor can improve survival when combined with chemotherapy in patients with metastatic colorectal cancer. To date, no clinical or biologic factors have been identified that predict for benefit from bevacizumab *(151)*. Furthermore, the mechanisms by which bevacizumab exerts its clinical effects in vivo are not entirely clear. It was initially presumed that VEGF inhibition would have antitumor effect by virtue of "choking off" the blood flow to tumors. However, it has also been postulated that normalization of vasculature, with decrease in interstitial pressures, may result in clinically meaningful improvement in delivery of cytotoxic agents *(159)*. These hypotheses have been validated by a clinical trial in which patients with rectal cancer received bevacizumab, with in vivo pharmacodynamic assessments confirming a decrease in tumoral interstitial fluid pressure, blood flow, and microvascular density *(160)*.

The toxicities of bevacizumab have been well documented in multiple trials. Common complications include hypertension and proteinuria *(151)*. Rare, but more serious, complications of bowel perforation, arterial thrombotic events, and bleeding have been noted *(150,151,154,157)*. Based on a pooled analysis of five bevacizumab trials, those patients most at risk for arterial thrombotic complications include age >65 years, prior history of arterial thrombotic events, and proteinuria >500mg/24 h *(161)*.

3.1.1.2. Vatalanib (PTK787/ZK 222584). Vatalanib (PTK787/ZK 222584, Novartis) is an oral TKI of VEGFR-1 and VEGFR-2. It also has inhibitory effects on c-kit and PDGF receptor (PDGFR)-β. Initial studies in patients with colorectal cancer established the safety of vatalanib in combination with infusional 5FU plus either oxaliplatin or irinotecan *(162,163)*. In addition, in vivo pharmacodynamic studies conducted in early phase clinical trials documented decreased tumor vascularity and permeability using dynamic contrast-enhanced magnetic resonance imaging *(164)*. These findings led to carry out two phase III trials combining vatalanib with chemotherapy in patients with metastatic colorectal cancer. In the CONFIRM1 trial, 1168 patients were randomized to receive FOLFOX or FOLFOX plus daily oral valatanib. Preliminary results failed to demonstrate a difference in progression-free survival between the study arms. Survival data are pending at the time of this chapter *(165)*. A similar study (CONFIRM2) with the same design is ongoing in the second-line metastatic setting. One plausible explanation for the discordant results with bevacizumab and vatalanib is that target inhibition with the oral TKI is not sustained as in the case of antibody treatment. In fact, the pharmacokinetic profile of vatalanib [half-life 3–6 h *(164)*] suggests that more frequent dosing than the once daily schedule employed in CONFIRM1 may be required *(166)*.

3.1.2. OTHER VEGF INHIBITORS IN CLINICAL DEVELOPMENT

ZD6474 (AstraZeneca) is an orally available small molecule TKI of VEGFR-1, VEGFR-2, VEGFR-3, and EGFR. It has shown preclinical efficacy and tolerability in the phase I setting. Side effects have included those seen with other VEGF targets agents (hypertension and proteinuria) and those seen with EGFR targeting agents (diarrhea and rash) and QTc prolongation *(167,168)*. Phase II studies are in progress.

The VEGF-Trap (NSC 724770 Sanafi-Aventis) is a unique fusion protein. It combines the Fc portion of human IgG1 with the principal extracellular ligand-binding domains of human VEGFR-1 and VEGFR-2. The VEGF Trap has the highest binding affinity for VEGF described to date ($K_D of 5 \times 10^{-12}$M) *(169)*. Phase I trials have been completed and phase II trials are anticipated *(170)*.

3.2. Platelet-Derived Growth Factor Receptor

PDGFs bind to TKα and TKβ receptors. These receptors, typically found on fibroblasts, smooth muscle and endothelial cells, autophosphorylate upon PDGF binding *(171)*. Downstream signaling results in increased migration, survival, and proliferation of stromal cells. In addition, pro-angiogenic features are present in activated PDGF pathways *(172)*. PDGF is associated with increased microvessel density and is expressed in approximately 83% of colon cancers, specifically on the stromal cells *(173)*. In preclinical models, PDGF stimulates colon cancer cell growth *(174)*.

There are many PDGF receptor TK inhibitors in development. Those furthest along include imatinib mesylate and sunitinib malate (SU11248). Both agents target the PDGFR as well as other receptor TKs such as the VEGF receptor and have shown preclinical evidence of efficacy against colon cancer cell lines *(175–177)*. Phase II clinical studies of sunitinib malate in colorectal cancer are currently under way.

3.3. Matrix Metalloproteinases

MMPs are a structurally related family of zinc-dependent endopeptidases that are capable of degrading components of the ECM, allowing tumor growth and invasion as well as angiogenesis *(178)*. Increased expression of MMP-9 in colorectal cancer has been associated with advanced stage and distant metastases *(179)*. Expression of

MMP-1 has been shown to be predictive of hematogenous spread in colorectal cancer as well *(180)*. MMP inhibitors have been effective in preclinical studies against malignant tumors *(181)*. Marimastat (BB-2516), a low molecular weight MMP inhibitor, was evaluated in a phase II trial of colorectal patients and showed evidence of a CEA response *(182)*. Neovastat (AE-941) is an extract from shark cartilage that has been shown to inhibit MMP-2, MMP-9, MMP-12, and MMP-13 *(178)*. Additionally, bisphosphonates have been identified as having some MMP inhibitor activity *(183)*. However, despite early signs of efficacy, no advantage has been identified in phase III trials of lung cancer patients *(184)*. Given lack of success of MMP inhibitors in phase III trials for other malignant diseases, trials in colorectal cancer are not anticipated.

3.4. Fibroblast-Activating Protein

Fibroblast-activating protein (FAP)-α is a 170 kDa cell membrane-associated serine protease, initially identified as an inducible antigen on reactive stromal cells *(185,186)*. FAP-α is not expressed in normal tissue, but is present in activated fibroblasts of malignant tissues *(187)*. The function of FAP-α has not been proven to date. It has been postulated to shape the ECM to facilitate tumor growth and metastasis, though emerging research suggests its primary function lies elsewhere *(188)*. The presence of FAP-α has been confirmed in tumor-associated fibroblasts in colorectal cancer *(189,190)*. In addition, FAP-α may work in conjunction with dipeptidyl peptidase IV to regulate tumor cell behavior *(191)*. Sibrotuzumab, a unconjugated monoclonal antibody, expected to rely on antibody-dependent cellular cytotoxicity against FAP-α, failed to show clinical activity in a phase II study of 25 patients with metastatic colorectal cancer *(192)*. This antibody may have a role as a radioimmunotherapeutic agent. Talabostat (PT-100) is a small molecule dipeptidyl peptidase inhibitor that has shown inhibitory activity in colon cancer xenografts *(193)*. A phase II trial in patients with colorectal cancer is under way.

3.5. **Integrins**

Integrins are adhesion receptors, expressed on a wide variety of cells, formed by the non-covalent association of α and β subunits. Integrins function in cell–cell adhesion as well as cell–ECM interaction *(194)*. Unlike most cellular receptors described above, integrins appear to require activation by intracellular ligands before the extracellular domain becomes active. Ras signaling pathways can activate integrins, most likely through an intermediary, talin *(194,195)*. Integrins are expressed on a variety of cell types, and in malignancy, their most important role may be in facilitating angiogenesis. Integrins link endothelial cells to the ECM and regulate cell survival, growth, and motility during angiogenesis *(196,197)*. Also, cooperation of MMP-1 and integrins has been described, thereby facilitating invasion and metastasis *(198)*. The small molecule integrin inhibitor, ATN-161, in combination with 5-FU has resulted in reduced metastasis and improved survival in a murine colorectal cancer model *(199)*. Other small molecule inhibitors and monoclonal antibodies against integrins (e.g., celingitide and vitaxin) are in early phase clinical investigation *(200,201)*.

4. COLORECTAL CANCER PATHOGENESIS: IMPLICATIONS FOR TARGETED THERAPY

All common adult epithelial cancers follow a typical incidence curve described by the formula $I = kt^{r-1}$, where the age-specific incidence (I) increases with time (t) according to an exponential that reflects a discrete number of mutations (r) according to a constant (k) *(202)*. Colorectal cancer is the archetype of multistep tumorigenesis, whereby several mutations in oncogenes and tumor suppressor genes, estimated to range from 4 to 7, accumulate in a normal cell and its progeny during the stepwise progression to early, intermediate, and late adenoma, and ultimately invasive and metastatic cancer *(203)*.

During multistep tumorigenesis, alterations of the function of several oncogenes and tumor suppressor genes is necessary to disrupt the complex genetic circuitry involving several interacting pathways that in the colonic crypt coordinately regulate proliferation, apoptosis, differentiation, and cell migration *(204)*. Understanding of these pathways suggests additional targets for therapeutic intervention.

4.1. Tumor Suppressor Genes in Colorectal Tumorigenesis

4.1.1. ADENOMATOUS POLYPOSIS COLI

While *K-ras* is the oncogene most frequently mutated in colorectal cancer, several tumor suppressor genes, and the pathways they critically control, are inactivated during colorectal tumorigenesis. These include adenomatous polyposis coli (*APC*) and the Wnt/β-catenin pathway; *SMAD2–SMAD4* in the TGF-β pathway; and *p53* and the p53-dependent checkpoint pathways regulating cell cycle and apoptosis.

APC has been dubbed the "gatekeeper" of cellular proliferation in the large intestine and is mutated in virtually all cases of colorectal cancer. In addition, germline mutations of *APC* cause familial adenomatous polyposis, an autosomal dominant syndrome predisposing to colorectal cancer that is becoming the testing ground of choice for new chemopreventive agents.

The APC protein appears to regulate the fine and directional balance of proliferation and differentiation along the crypt axis: proliferation normally occurs in the bottom third of the crypt, with differentiation in the upper crypt, where there is ultimately reduced cell–cell adhesion and shedding into the lumen. As a consequence of APC inactivation in early adenomas, this basic co-regulation of proliferation and differentiation is lost *(205)* along with the loss of positional cues along the crypt axis by adenomatous cells *(206)*.

How does APC integrate cell adhesion cues with regulation of proliferation– differentiation? APC is a critical component of the canonical Wnt/β-catenin pathway. In the absence of Wnt signals, the cell adhesion molecule β-catenin forms a complex with APC and axin that facilitates its phosphorylation by glycogen synthase kinase 3β. Phosphorylation of β-catenin results in its ubiquitin/proteasome-mediated degradation. However, when Wnt ligands bind frizzled receptors and activate Disheveled protein, β-catenin degradation is blocked; excess β-catenin enters the nucleus where in association with the transcription factor TCF4/LEF promotes the expression of several target genes, including *MYC* and cyclin D1 *(207–210)*. Nuclear accumulation of β-catenin also occurs when APC is mutated or when β-catenin point mutations in some colorectal cancer cases prevent its phosphorylation and degradation.

Based on these premises, preclinical therapeutic models include inhibiting the β-catenin–TCF4/LEF interaction, stimulating β-catenin degradation, and restraining β-catenin at the cellular membrane or in the cytoplasm *(211)*. In general, while there are numerous examples of chemopreventive agents whose pharmacodynamics is associated with modulation of the Wnt/β-catenin pathway, it remains to be seen whether inhibition of this pathway has merit in the context of cancer therapy.

4.1.2. Transforming Growth Factor-β

The TGF-β pathway is frequently mutated in colorectal cancer (CRC), by inactivating mutations of either the TGF-β receptor II or of the signal transducer and transcription factors SMAD2 and SMAD4, downstream of the activated receptor *(212)*. While the TGF-β pathway can initially inhibit proliferation and induce senescence or apoptosis, during late tumorigenesis it is associated with increased motility, invasiveness, and metastasis *(213)*. Thus, molecular targeting of TGF-β should ideally spare the downstream signaling associated with tumor suppression while blocking the pathways associated with tumor progression *(214)*. Numerous efforts focused on the development of anti-TGF-βtherapeutics are under way *(213)*.

4.1.3. p53

Point mutations or deletions of p53 are very frequent in malignant colorectal tumors, and their consequence is inactivation of the DNA damage checkpoints that normally cause cell cycle arrest at the G_1-S and G_2-M transitions and apoptosis. These functions are linked to the role of p53 as a sequence-specific transcription factor. One potential therapeutic approach involves the selective killing of p53-mutant cells using the engineered adenovirus ONYX-015 *(215)*. No maximum tolerated dose was identified in phase I trials. Liver directed therapy in conjunction with 5-FU in phase II colorectal cancer trials has been studied. Median survival in heavily pretreated patients was 10.7 months *(216)*. Another strategy in development for tumors bearing missense mutations of p53 is the use of small molecules that act as molecular "braces" forcing the mutated p53 protein to assume near-normal conformation and DNA binding *(217)*.

4.1.4. Mutator Phenotype in CRC

It is a matter of debate whether the several mutations required for colorectal cancer formation can occur at the normal mutation rate, that is, whether an increase in the mutation rate (or mutator phenotype) is necessary *(204)*. On the other hand, it is clear that a mutator phenotype *(218)*, also termed "genomic instability," becomes apparent during CRC tumorigenesis.

Two types of genomic instability have been described for colorectal cancer, chromosomal instability (CIN, 85% of the cases) and microsatellite instability (MIN, 15% of the cases, also called MSI) *(219)*. CIN tumors exhibit marked aneuploidy and frequent loss of heterozygosity, and the prognosis is often poor. The molecular basis of CIN is poorly understood and is likely a reflection of alterations in mitotic checkpoint genes *(220)*. MIN tumors are diploid or nearly diploid and typically have a good prognosis. In these tumors, a characteristic length instability of simple repetitive sequences (called microsatellites) takes place and is a consequence of a defective DNA mismatch repair (MMR) system. An intact MMR system repairs replication errors that result in mismatches, that is, non-Watson–Crick pairing of DNA bases, as well as slippage of microsatellite sequences (*221–224*). In MIN tumors, any of six MMR genes (*MSH2, MSH3, MSH6, MLH1, PMS2*, and *PMS1*) can be mutated. Germline MMR mutations cause hereditary non-polyposis colorectal cancer (HNPCC) or Lynch syndrome, an autosomal dominant disorder predisposing to CRC and extracolonic tumors (*225*). MMR deficiency also occurs in approximately 15% of sporadic tumors, frequently as a consequence of methylation and transcriptional silencing of the *MLH1* promoter (*226,227*).

MMR-defective tumors (either sporadic or in HNPCC individuals) are characterized by resistance to DNA-damaging agents (*228,229,230*). This apparently paradoxical effect is due to a role of MMR proteins distinct from their DNA repair function. Specifically, MMR proteins participate in the signal transduction cascade activated by DNA damage that normally engages the cell cycle and apoptosis checkpoints causing cell killing. In MMR-defective tumors, DNA damage accumulates but fails to activate these checkpoints. After many examples of this apparent paradox in MMR-deficient cell lines and model organisms, recent studies have shown that CRC with MMR defects may be resistant to treatment with 5-FU (*231*).

Promising strategies to overcome the resistance to DNA damage of MMR-defective cells are being explored. One approach involves the use of nucleotide analogs, such as gemcitabine (*232*) or iododeoxyuridine and bromodeoxyuridine (*233*) as radiosensitizing agents for selective killing of MMR-defective tumors. Selective killing of MMR-defective tumors has also been obtained with frameshift-inducing agents such as ICR191 (*234*).

5. CONCLUSION

Supported by expanded characterization of the malignant phenotype, and technological advances that facilitate the development of highly specific inhibitory reagents, the notion of molecular targeting in colorectal cancer has come of age. Clinical successes with inhibitors of EGFR and VEGF have validated the concept that drug development based on biology can bear fruit. However, this experience has also been sobering, insofar as these approaches have not been curative, and major benefits are achieved in a minority of patients. Markers predictive of response have not been fully elucidated. Furthermore, targeted agents such as cetuximab and bevacizumab require coadministration with traditional cytotoxics for maximal effect. Clearly, much remains to be learned from the standpoint of mechanisms of action as well as mechanisms of resistance.

The different sensitivity patterns of colorectal tumors to particular inhibitory strategies suggests that this is a heterogeneous disease in which specific pathway addictions, redundancies, interactions, and feedback mechanisms are operative. It is hoped that characterization of such networks of interrelated processes will lead to rationally designed combination strategies for the treatment of individual patients. Success will require that clinical investigators abandon some previously held tenets of drug development. First, combination strategies should no longer be selected based on the avoidance of overlapping toxicities, but rather an appreciation of colorectal cancer as a complex network. Second, in vivo pharmacodynamic assessment should be pursued early in clinical development to ensure target acquisition and define mechanism of action of new agents. Third, acceleration of drug development requires some acceptance of risk, with modification of the traditional sequence and design of phase I, II, and III trials. We must accept that toxicity evaluation will not be complete before phase III investigation. Finally, the importance of banking biologic material cannot be overstated. With a wealth of new agents to explore, redundancy in the clinical trials enterprise must be minimized, and the potential for missed opportunities is therefore increased. In the treatment of colorectal cancer, the commitment of laboratory investigators, clinical scientists, and patients has moved us from an era of relative nihilism to one of hopeful expectation.

REFERENCES

1. Parkin DM, Bray F, Ferlay J, Pisani P. Global cancer statistics, 2002. CA Cancer J Clin 2005; 55(2):74–108.
2. Wells A. EGF receptor. Int J Biochem Cell Biol 1999;31(6):637–43.
3. Davies DE, Chamberlin SG. Targeting the epidermal growth factor receptor for therapy of carcinomas. Biochem Pharmacol 1996;51(9):1101–10.
4. Mendelsohn J. Targeting the epidermal growth factor receptor for cancer therapy. J Clin Oncol 2002;20(18 Suppl):1S–13.
5. Hackel PO, Zwick E, Prenzel N, Ullrich A. Epidermal growth factor receptors: critical mediators of multiple receptor pathways. Curr Opin Cell Biol 1999;11(2):184–9.
6. Diermeier S, Horvath G, Knuechel-Clarke R, Hofstaedter F, Szollosi J, Brockhoff G. Epidermal growth factor receptor coexpression modulates susceptibility to Herceptin in HER2/neu overexpressing breast cancer cells via specific erbB-receptor interaction and activation. Exp Cell Res 2005;304(2):604–19.
7. Arteaga CL. The epidermal growth factor receptor: from mutant oncogene in nonhuman cancers to therapeutic target in human neoplasia. J Clin Oncol 2001;19(18 Suppl):32S–40S.
8. Lawrence DS, Niu J. Protein kinase inhibitors: the tyrosine-specific protein kinases. Pharmacol Ther 1998;77(2):81–114.
9. Goldstein NS, Armin M. Epidermal growth factor receptor immunohistochemical reactivity in patients with American Joint Committee on Cancer Stage IV colon adenocarcinoma: implications for a standardized scoring system. Cancer 2001;92(5):1331–46.
10. Salomon DS, Brandt R, Ciardiello F, Normanno N. Epidermal growth factor-related peptides and their receptors in human malignancies. Crit Rev Oncol Hematol 1995;19(3):183–232.
11. Mayer A, Takimoto M, Fritz E, Schellander G, Kofler K, Ludwig H. The prognostic significance of proliferating cell nuclear antigen, epidermal growth factor receptor, and mdr gene expression in colorectal cancer. Cancer 1993;71(8):2454–60.
12. Hemming AW, Davis NL, Kluftinger A, et al. Prognostic markers of colorectal cancer: an evaluation of DNA content, epidermal growth factor receptor, and Ki-67. J Surg Oncol 1992;51(3):147–52.
13. Ennis BW, Lippman ME, Dickson RB. The EGF receptor system as a target for antitumor therapy. Cancer Invest 1991;9(5):553–62.
14. Prenzel N, Zwick E, Daub H, et al. EGF receptor transactivation by G-protein-coupled receptors requires metalloproteinase cleavage of proHB-EGF. Nature 1999;402(6764):884–8.
15. Cox G, Jones JL, O'Byrne KJ. Matrix metalloproteinase 9 and the epidermal growth factor signal pathway in operable non-small cell lung cancer. Clin Cancer Res 2000;6(6):2349–55.
16. Naramura M, Gillies SD, Mendelsohn J, Reisfeld RA, Mueller BM. Therapeutic potential of chimeric and murine anti-(epidermal growth factor receptor) antibodies in a metastasis model for human melanoma. Cancer Immunol Immunother 1993;37(5):343–9.
17. Perrotte P, Matsumoto T, Inoue K, et al. Anti-epidermal growth factor receptor antibody C225 inhibits angiogenesis in human transitional cell carcinoma growing orthotopically in nude mice. Clin Cancer Res 1999;5(2):257–65.
18. Viloria-Petit A, Crombet T, Jothy S, et al. Acquired resistance to the antitumor effect of epidermal growth factor receptor-blocking antibodies in vivo: a role for altered tumor angiogenesis. Cancer Res 2001;61(13):5090–101.
19. Ciardiello F, Bianco R, Damiano V, et al. Antiangiogenic and antitumor activity of anti-epidermal growth factor receptor C225 monoclonal antibody in combination with vascular endothelial growth factor antisense oligonucleotide in human GEO colon cancer cells. Clin Cancer Res 2000;6(9): 3739–47.

20. Moyer JD, Barbacci EG, Iwata KK, et al. Induction of apoptosis and cell cycle arrest by CP-358,774, an inhibitor of epidermal growth factor receptor tyrosine kinase. Cancer Res 1997;57(21):4838–48.

21. Wu X, Fan Z, Masui H, Rosen N, Mendelsohn J. Apoptosis induced by an anti-epidermal growth factor receptor monoclonal antibody in a human colorectal carcinoma cell line and its delay by insulin. J Clin Invest 1995;95(4):1897–905.

22. Bruns CJ, Harbison MT, Davis DW, et al. Epidermal growth factor receptor blockade with C225 plus gemcitabine results in regression of human pancreatic carcinoma growing orthotopically in nude mice by antiangiogenic mechanisms. Clin Cancer Res 2000;6(5):1936–48.

23. Fan Z, Baselga J, Masui H, Mendelsohn J. Antitumor effect of anti-epidermal growth factor receptor monoclonal antibodies plus cis-diamminedichloroplatinum on well established A431 cell xenografts. Cancer Res 1993;53(19):4637–42.

24. Bonner JA, Raisch KP, Trummell HQ, et al. Enhanced apoptosis with combination C225/radiation treatment serves as the impetus for clinical investigation in head and neck cancers. J Clin Oncol 2000;18(21 Suppl):47S–53S.

25. Prewett MC, Hooper AT, Bassi R, Ellis LM, Waksal HW, Hicklin DJ. Enhanced antitumor activity of anti-epidermal growth factor receptor monoclonal antibody IMC-C225 in combination with irinotecan (CPT-11) against human colorectal tumor xenografts. Clin Cancer Res 2002;8(5): 994–1003.

26. Saltz L, Rubin M, Hochster H, et al. Cetuximab (IMC-C225) plus irinotecan (CPT-11) is active in CPT-11-refractory colorectal cancer (CRC) that expresses epidermal growth factor receptor (EGFR). Proc Am Soc Clin Oncol 2001;20:3.

27. Saltz LB, Meropol NJ, Loehrer PJ, Sr., Needle MN, Kopit J, Mayer RJ. Phase II trial of cetuximab in patients with refractory colorectal cancer that expresses the epidermal growth factor receptor. J Clin Oncol 2004;22(7):1201–8.

28. Cunningham D, Humblet Y, Siena S, et al. Cetuximab monotherapy and cetuximab plus irinotecan in irinotecan-refractory metastatic colorectal cancer. N Engl J Med 2004;351(4):337–45.

29. Lenz HJ, Mayer RJ, Gold PJ, et al. Activity of cetuximab in patients with colorectal cancer refractory to both irinotecan and oxaliplatin. Proc Am Soc Clin Oncol 2004;22:247.

30. Chung KY, Shia J, Kemeny NE, et al. Cetuximab shows activity in colorectal cancer patients with tumors that do not express the epidermal growth factor receptor by immunohistochemistry. J Clin Oncol 2005;23(9):1803–10.

31. Meropol NJ. Epidermal growth factor receptor inhibitors in colorectal cancer: it's time to get back on target. J Clin Oncol 2005;23(9):1791–3.

32. Malik I, Hecht JR, Patnaik A, et al. Safety and efficacy of panitumumab monotherapy in patients with metastatic colorectal cancer. Proc Am Soc Clin Oncol 2005;23:251.

33. Weiner LM, Belldegrun A, Rowinsky E, et al. Updated results from a dose and schedule study of Panitumumab (ABX-EGF) monotherapy in patients with advanced solid malignancies. Proc Am Soc Clin Oncol 2005;23:206.

34. Trarbach T, Beyer T, Schleucher N, et al. A randomized phase I study of the humanized anti-epidermal growth factor receptor (EGFR) monoclonal antibody EMD 72000 in subjects with advanced gastrointestinal cancers. Proc Am Soc Clin Oncol 2004;22:199.

35. Vanhoefer U, Tewes M, Rojo F, et al. Phase I study of the humanized antiepidermal growth factor receptor monoclonal antibody EMD72000 in patients with advanced solid tumors that express the epidermal growth factor receptor. J Clin Oncol 2004;22(1):175–84.

36. Erbitux, Package Insert. 2004. Accessed August 2, 2005, at http://www.erbitux.com.

37. Busam KJ, Capodieci P, Motzer R, Kiehn T, Phelan D, Halpern AC. Cutaneous side-effects in cancer patients treated with the antiepidermal growth factor receptor antibody C225. Br J Dermatol 2001;144(6):1169–76.

38. Damjanov N, Meropol NJ. Epidermal growth factor receptor inhibitors for the treatment of colorectal cancer: a promise fulfilled? Oncology (Williston Park) 2004;18(4):479–88; discussion 88, 93, 97 passim.

39. Mackenzie MJ, Hirte HW, Glenwood G, et al. A phase II trial of ZD1839 (Iressa) 750 mg per day, an oral epidermal growth factor receptor-tyrosine kinase inhibitor, in patients with metastatic colorectal cancer. Invest New Drugs 2005;23(2):165–70.

40. Goss G, Hirte H, Miller WH, Jr., et al. A phase I study of oral ZD 1839 given daily in patients with solid tumors: IND.122, a study of the Investigational New Drug Program of the National Cancer Institute of Canada Clinical Trials Group. Invest New Drugs 2005;23(2):147–55.

41. Dorligschaw O, Kegel T, Jordan K, Harba A, Grothey A, Schmoll HJ. ZD 1839 (Iressa)-based treatment as last-line therapy in patients with advanced colorectal cancer. Proc Am Soc Clin Oncol 2003.
42. Fisher GA, Kuo T, Cho CD, et al. A phase II study of gefitinib in combination with FOLFOX-4 in patients with metastatic colorectal cancer. Proc Am Soc Clin Oncol 2004;22:248.
43. Townsley C, Major P, Siu LL, et al. Phase II study of OSI-774 in patients with metastatic colorectal cancer. American Society of Clinical Oncology Gastrointestinal Symposium; 2004; San Francisco, California.
44. Weinstein IB. Cancer. Addiction to oncogenes–the Achilles heal of cancer. Science 2002; 297(5578):63–4.
45. Albanell J, Rojo F, Averbuch S, et al. Pharmacodynamic studies of the epidermal growth factor receptor inhibitor ZD1839 in skin from cancer patients: histopathologic and molecular consequences of receptor inhibition. J Clin Oncol 2002;20(1):110–24.
46. Malik SN, Siu LL, Rowinsky EK, et al. Pharmacodynamic evaluation of the epidermal growth factor receptor inhibitor OSI-774 in human epidermis of cancer patients. Clin Cancer Res 2003;9(7): 2478–86.
47. Tabernero J, Rojo F, Jimenez E, et al. A phase I PK and serial tumor and skin pharmacodynamic study of weekly, every 2-week or every 3-week 1-hour infusion EMD 72000, a humanized monoclonal anti-epidermal growth factor receptor antibody, in patients with advanced tumors. Proc Am Soc Clin Oncol 2004;22:69.
48. Lynch TJ, Bell DW, Sordella R, et al. Activating mutations in the epidermal growth factor receptor underlying responsiveness of non-small-cell lung cancer to gefitinib. N Engl J Med 2004;350(21):2129–39.
49. Paez JG, Janne PA, Lee JC, et al. EGFR mutations in lung cancer: correlation with clinical response to gefitinib therapy. Science 2004;304(5676):1497–500.
50. Pao W, Miller VA. Epidermal growth factor receptor mutations, small-molecule kinase inhibitors, and non-small-cell lung cancer: current knowledge and future directions. J Clin Oncol 2005;23(11):2556–68.
51. Tsao MS, Sakurada A, Cutz JC, et al. Erlotinib in lung cancer - molecular and clinical predictors of outcome. N Engl J Med 2005;353(2):133–44.
52. Barber TD, Vogelstein B, Kinzler KW, Velculescu VE. Somatic mutations of EGFR in colorectal cancers and glioblastomas. N Engl J Med 2004;351(27):2883.
53. Moroni M, Veronese S, Benvenuti S, et al. Gene copy number for epidermal growth factor receptor (EGFR) and clinical response to antiEGFR treatment in colorectal cancer: a cohort study. Lancet Oncol 2005;6(5):279–86.
54. Cappuzzo F, Hirsch FR, Rossi E, et al. Epidermal growth factor receptor gene and protein and gefitinib sensitivity in non-small-cell lung cancer. J Natl Cancer Inst 2005;97(9):643–55.
55. Garrett C, Takimoto M, Wojtowicz M, et al. Identification of a molecular signature of radiographic response to cetuximab in patients with advanced colorectal cancer. Proc Am Soc Clin Oncol 2005;23:277.
56. Gregorc V, Cusatis G, Spreafico A, et al. Association of germline mutations in EGFR and ABCG2 with gefitinib response in patients with non-small cell lung cancer. Proc Am Soc Clin Oncol 2005;23:197.
57. Petley T, Graff K, Jiang W, Yang H, Florini J. Variation among cell types in the signaling pathways by which IGF-I stimulates specific cellular responses. Horm Metab Res 1999;31(2–3):70–6.
58. D'Ercole AJ, Stiles AD, Underwood LE. Tissue concentrations of somatomedin C: further evidence for multiple sites of synthesis and paracrine or autocrine mechanisms of action. Proc Natl Acad Sci USA 1984;81(3):935–9.
59. Bahr C, Groner B. The IGF-1 receptor and its contributions to metastatic tumor growth-novel approaches to the inhibition of IGF-1R function. Growth Factors 2005;23(1):1–14.
60. Lambert S, Vivario J, Boniver J, Gol-Winkler R. Abnormal expression and structural modification of the insulin-like growth-factor-II gene in human colorectal tumors. Int J Cancer 1990;46(3): 405–10.
61. Weber MM, Fottner C, Liu SB, Jung MC, Engelhardt D, Baretton GB. Overexpression of the insulin-like growth factor I receptor in human colon carcinomas. Cancer 2002;95(10):2086–95.
62. Reinmuth N, Liu W, Fan F, et al. Blockade of insulin-like growth factor I receptor function inhibits growth and angiogenesis of colon cancer. Clin Cancer Res 2002;8(10):3259–69.

63. Wiley SR, Schooley K, Smolak PJ, et al. Identification and characterization of a new member of the TNF family that induces apoptosis. Immunity 1995;3(6):673–82.
64. MacFarlane M, Ahmad M, Srinivasula SM, Fernandes-Alnemri T, Cohen GM, Alnemri ES. Identification and molecular cloning of two novel receptors for the cytotoxic ligand TRAIL. J Biol Chem 1997;272(41):25417–20.
65. Li H, Zhu H, Xu CJ, Yuan J. Cleavage of BID by caspase 8 mediates the mitochondrial damage in the Fas pathway of apoptosis. Cell 1998;94(4):491–501.
66. Pan G, O'Rourke K, Chinnaiyan AM, et al. The receptor for the cytotoxic ligand TRAIL. Science 1997;276(5309):111–3.
67. Koornstra JJ, Jalving M, Rijcken FE, et al. Expression of tumour necrosis factor-related apoptosis-inducing ligand death receptors in sporadic and hereditary colorectal tumours: Potential targets for apoptosis induction. Eur J Cancer 2005;41(8):1195–202.
68. Kim YH, Park JW, Lee JY, Kwon TK. Sodium butyrate sensitizes TRAIL-mediated apoptosis by induction of transcription from the DR5 gene promoter through Sp1 sites in colon cancer cells. Carcinogenesis 2004;25(10):1813–20.
69. Chapman HA, Wei Y, Simon DI, Waltz DA. Role of urokinase receptor and caveolin in regulation of integrin signaling. Thromb Haemost 1999;82(2):291–7.
70. Ahmed N, Oliva K, Wang Y, Quinn M, Rice G. Downregulation of urokinase plasminogen activator receptor expression inhibits Erk signalling with concomitant suppression of invasiveness due to loss of uPAR-beta1 integrin complex in colon cancer cells. Br J Cancer 2003;89(2):374–84.
71. Liu D, Aguirre Ghiso J, Estrada Y, Ossowski L. EGFR is a transducer of the urokinase receptor initiated signal that is required for in vivo growth of a human carcinoma. Cancer Cell 2002;1(5):445–57.
72. de Bruin PA, Griffioen G, Verspaget HW, Verheijen JH, Lamers CB. Plasminogen activators and tumor development in the human colon: activity levels in normal mucosa, adenomatous polyps, and adenocarcinomas. Cancer Res 1987;47(17):4654–7.
73. Fujii T, Obara T, Tanno S, Ura H, Kohgo Y. Urokinase-type plasminogen activator and plasminogen activator inhibitor-1 as a prognostic factor in human colorectal carcinomas. Hepatogastroenterology 1999;46(28):2299–308.
74. Wang Y, Liang X, Wu S, Murrell GA, Doe WF. Inhibition of colon cancer metastasis by a 3´-end antisense urokinase receptor mRNA in a nude mouse model. Int J Cancer 2001;92(2):257–62.
75. Servomaa K, Kiuru A, Kosma VM, Hirvikoski P, Rytomaa T. p53 and K-ras gene mutations in carcinoma of the rectum among Finnish women. Mol Pathol 2000;53(1):24–30.
76. Bos JL. Ras oncogenes in human cancer: a review. Cancer Res 1989;49(17):4682–9.
77. Rao S, Cunningham D, de Gramont A, et al. Phase III double-blind placebo-controlled study of farnesyl transferase inhibitor R115777 in patients with refractory advanced colorectal cancer. J Clin Oncol 2004;22(19):3950–7.
78. Sharma S, Kemeny N, Kelsen DP, et al. A phase II trial of farnesyl protein transferase inhibitor SCH 66336, given by twice-daily oral administration, in patients with metastatic colorectal cancer refractory to 5-fluorouracil and irinotecan. Ann Oncol 2002;13(7):1067–71.
79. Saleh M, Posey J, Pleasant L, et al. A phase II trial of ISIS 2503, an antisense inhibitor of H-ras, as first line therapy for advanced colorectal cancer. Proc Am Soc Clin Oncol 2000;19.
80. Cohen SJ, Gallo J, Lewis NL, et al. Phase I and pharmacokinetic study of the farnesyltransferase inhibitor R115777 in combination with irinotecan in patients with advanced cancer. Cancer Chemother Pharmacol 2004;53(6):513–8.
81. Whyte DB, Kirschmeier P, Hockenberry TN, et al. K- and N-Ras are geranylgeranylated in cells treated with farnesyl protein transferase inhibitors. J Biol Chem 1997;272(22):14459–64.
82. Tilkin-Mariame AF, Cormary C, Ferro N, et al. Geranylgeranyl transferase inhibition stimulates antimelanoma immune response through MHC class I and costimulatory molecule expression. FASEB J 2005;19(11):1513–1515.
83. Wellbrock C, Karasarides M, Marais R. The RAF proteins take centre stage. Nat Rev Mol Cell Biol 2004;5(11):875–85.
84. Nagasaka T, Sasamoto H, Notohara K, et al. Colorectal cancer with mutation in BRAF, KRAS, and wild-type with respect to both oncogenes showing different patterns of DNA methylation. J Clin Oncol 2004;22(22):4584–94.

85. Cripps MC, Figueredo AT, Oza AM, et al. Phase II randomized study of ISIS 3521 and ISIS 5132 in patients with locally advanced or metastatic colorectal cancer: a National Cancer Institute of Canada clinical trials group study. Clin Cancer Res 2002;8(7):2188–92.

86. Wilhelm SM, Carter C, Tang L, et al. BAY 43–9006 exhibits broad spectrum oral antitumor activity and targets the RAF/MEK/ERK pathway and receptor tyrosine kinases involved in tumor progression and angiogenesis. Cancer Res 2004;64(19):7099–109.

87. Kupsch P, Passarge K, Richly H, et al. Results of a phase I trial of BAY 43–9006 in combination with oxaliplatin in patients with refractory solid tumors. Proc Am Soc Clin Oncol 2004;22:209.

88. Mross K, Steinbild S, Baas F, et al. Drug-drug interaction pharmacokinetic study with the Raf kinase inhibitor (RKI) BAY 43–9006 administered in combination with irinotecan (CPT-11) in patients with solid tumors. Int J Clin Pharmacol Ther 2003;41(12):618–9.

89. Cobb MH, Goldsmith EJ. How MAP kinases are regulated. J Biol Chem 1995;270(25):14843–6.

90. Sebolt-Leopold JS, Herrera R. Targeting the mitogen-activated protein kinase cascade to treat cancer. Nat Rev Cancer 2004;4(12):937–47.

91. Hoshino R, Chatani Y, Yamori T, et al. Constitutive activation of the 41-/43-kDa mitogen-activated protein kinase signaling pathway in human tumors. Oncogene 1999;18(3):813–22.

92. Sebolt-Leopold JS, Dudley DT, Herrera R, et al. Blockade of the MAP kinase pathway suppresses growth of colon tumors in vivo. Nat Med 1999;5(7):810–6.

93. Meng XW, Chandra J, Loegering D, et al. Central role of Fas-associated death domain protein in apoptosis induction by the mitogen-activated protein kinase kinase inhibitor CI-1040 (PD184352) in acute lymphocytic leukemia cells in vitro. J Biol Chem 2003;278(47):47326–39.

94. Lorusso PM, Adjei AA, Varterasian M, et al. Phase I and pharmacodynamic study of the oral MEK inhibitor CI-1040 in patients with advanced malignancies. J Clin Oncol 2005;23(24):5597–5604.

95. Rinehart J, Adjei AA, Lorusso PM, et al. Multicenter phase II study of the oral MEK inhibitor, CI-1040, in patients with advanced non-small-cell lung, breast, colon, and pancreatic cancer. J Clin Oncol 2004;22(22):4456–62.

96. Doyle MP, Yeh TC, Suzy B, et al. Validation and use of a biomarker of clinical development of the MEK1/2 inhibitor ARRY-142886 (AZD6244). Proc Am Soc Clin Oncol; 2005; Orlando, Florida.

97. Testa JR, Bellacosa A. AKT plays a central role in tumorigenesis. Proc Natl Acad Sci USA 2001;98(20):10983–5.

98. Semba S, Itoh N, Ito M, et al. Down-regulation of PIK3CG, a catalytic subunit of phosphatidylinositol 3-OH kinase, by CpG hypermethylation in human colorectal carcinoma. Clin Cancer Res 2002;8(12):3824–31.

99. Meuillet EJ, Ihle N, Baker AF, et al. In vivo molecular pharmacology and antitumor activity of the targeted Akt inhibitor PX-316. Oncol Res 2004;14(10):513–27.

100. Van Ummersen L, Binger K, Volkman J, et al. A phase I trial of perifosine (NSC 639966) on a loading dose/maintenance dose schedule in patients with advanced cancer. Clin Cancer Res 2004;10(22):7450–6.

101. Atkins MB, Hidalgo M, Stadler WM, et al. Randomized phase II study of multiple dose levels of CCI-779, a novel mammalian target of rapamycin kinase inhibitor, in patients with advanced refractory renal cell carcinoma. J Clin Oncol 2004;22(5):909–18.

102. Punt CJ, Boni J, Bruntsch U, Peters M, Thielert C. Phase I and pharmacokinetic study of CCI-779, a novel cytostatic cell-cycle inhibitor, in combination with 5-fluorouracil and leucovorin in patients with advanced solid tumors. Ann Oncol 2003;14(6):931–7.

103. Hara K, Yonezawa K, Kozlowski MT, et al. Regulation of eIF-4E BP1 phosphorylation by mTOR. J Biol Chem 1997;272(42):26457–63.

104. Kumar V, Sabatini D, Pandey P, et al. Regulation of the rapamycin and FKBP-target 1/mammalian target of rapamycin and cap-dependent initiation of translation by the c-Abl protein-tyrosine kinase. J Biol Chem 2000;275(15):10779–87.

105. DeGraffenried LA, Fulcher L, Friedrichs WE, Grunwald V, Ray RB, Hidalgo M. Reduced PTEN expression in breast cancer cells confers susceptibility to inhibitors of the PI3 kinase/Akt pathway. Ann Oncol 2004;15(10):1510–6.

106. Desai AA, Janisch L, Berk LR, et al. A phase I trial of a novel mTOR inhibitor AP23573 administered weekly in patients with refractory or advanced malignancies: a pharmacokinetic and pharmacodynamic analysis. Proc Am Soc Clin Oncol; 2005; Orlando, Florida.

107. Peralba JM, DeGraffenried L, Friedrichs W, et al. Pharmacodynamic evaluation of CCI-779, an inhibitor of mTOR, in cancer patients. Clin Cancer Res 2003;9(8):2887–92.

108. Beuvink I, Boulay A, Fumagalli S, et al. The mTOR inhibitor RAD001 sensitizes tumor cells to DNA-damaged induced apoptosis through inhibition of p21 translation. Cell 2005;120(6):747–59.

109. Cao X, Tay A, Guy GR, Tan YH. Activation and association of Stat3 with Src in v-Src-transformed cell lines. Mol Cell Biol 1996;16(4):1595–603.

110. Muthuswamy SK, Muller WJ. Direct and specific interaction of c-Src with Neu is involved in signaling by the epidermal growth factor receptor. Oncogene 1995;11(2):271–9.

111. Aligayer H, Boyd DD, Heiss MM, Abdalla EK, Curley SA, Gallick GE. Activation of Src kinase in primary colorectal carcinoma: an indicator of poor clinical prognosis. Cancer 2002;94(2):344–51.

112. Doggrell SA. BMS-354825: a novel drug with potential for the treatment of imatinib-resistant chronic myeloid leukaemia. Expert Opin Investig Drugs 2005;14(1):89–91.

113. Corbin AS, Demehri S, Griswold IJ, et al. In vitro and in vivo activity of ATP-based kinase inhibitors AP23464 and AP23848 against activation-loop mutants of Kit. Blood 2005;106(1):227–34.

114. Corvinus FM, Orth C, Moriggl R, et al. Persistent STAT3 activation in colon cancer is associated with enhanced cell proliferation and tumor growth. Neoplasia 2005;7(6):545–55.

115. Rivat C, Rodrigues S, Bruyneel E, et al. Implication of STAT3 signaling in human colonic cancer cells during intestinal trefoil factor 3 (TFF3) – and vascular endothelial growth factor-mediated cellular invasion and tumor growth. Cancer Res 2005;65(1):195–202.

116. Adams RR, Maiato H, Earnshaw WC, Carmena M. Essential roles of Drosophila inner centromere protein (INCENP) and aurora B in histone H3 phosphorylation, metaphase chromosome alignment, kinetochore disjunction, and chromosome segregation. J Cell Biol 2001;153(4):865–80.

117. Giet R, Glover DM. Drosophila aurora B kinase is required for histone H3 phosphorylation and condensin recruitment during chromosome condensation and to organize the central spindle during cytokinesis. J Cell Biol 2001;152(4):669–82.

118. Bischoff JR, Anderson L, Zhu Y, et al. A homologue of Drosophila aurora kinase is oncogenic and amplified in human colorectal cancers. EMBO J 1998;17(11):3052–65.

119. Harrington EA, Bebbington D, Moore J, et al. VX-680, a potent and selective small-molecule inhibitor of the Aurora kinases, suppresses tumor growth in vivo. Nat Med 2004;10(3):262–7.

120. Hanahan D, Weinberg RA. The hallmarks of cancer. Cell 2000;100(1):57–70.

121. Folkman J. Tumor angiogenesis: therapeutic implications. N Engl J Med 1971;285(21):1182–6.

122. Folkman J. What is the evidence that tumors are angiogenesis dependent? J Natl Cancer Inst 1990;82(1):4–6.

123. Bergers G, Benjamin LE. Tumorigenesis and the angiogenic switch. Nat Rev Cancer 2003;3(6): 401–10.

124. Longo R, Sarmiento R, Fanelli M, Capaccetti B, Gattuso D, Gasparini G. Anti-angiogenic therapy: rationale, challenges and clinical studies. Angiogenesis 2002;5(4):237–56.

125. Dor Y, Porat R, Keshet E. Vascular endothelial growth factor and vascular adjustments to perturbations in oxygen homeostasis. Am J Physiol Cell Physiol 2001;280(6):C1367–74.

126. Acker T, Plate KH. Hypoxia and hypoxia inducible factors (HIF) as important regulators of tumor physiology. Cancer Treat Res 2004;117:219–48.

127. Ferrara N, Gerber HP, LeCouter J. The biology of VEGF and its receptors. Nat Med 2003;9(6): 669–76.

128. Tischer E, Mitchell R, Hartman T, et al. The human gene for vascular endothelial growth factor. Multiple protein forms are encoded through alternative exon splicing. J Biol Chem 1991;266(18):11947–54.

129. Shibuya M, Yamaguchi S, Yamane A, et al. Nucleotide sequence and expression of a novel human receptor-type tyrosine kinase gene (flt) closely related to the fms family. Oncogene 1990;5(4): 519–24.

130. Terman BI, Dougher-Vermazen M, Carrion ME, et al. Identification of the KDR tyrosine kinase as a receptor for vascular endothelial cell growth factor. Biochem Biophys Res Commun 1992;187(3):1579–86.

131. Matthews W, Jordan CT, Gavin M, Jenkins NA, Copeland NG, Lemischka IR. A receptor tyrosine kinase cDNA isolated from a population of enriched primitive hematopoietic cells and exhibiting close genetic linkage to c-kit. Proc Natl Acad Sci USA 1991;88(20):9026–30.

132. Pajusola K, Aprelikova O, Armstrong E, Morris S, Alitalo K. Two human FLT4 receptor tyrosine kinase isoforms with distinct carboxy terminal tails are produced by alternative processing of primary transcripts. Oncogene 1993;8(11):2931–7.

133. Galland F, Karamysheva A, Pebusque MJ, et al. The FLT4 gene encodes a transmembrane tyrosine kinase related to the vascular endothelial growth factor receptor. Oncogene 1993;8(5):1233–40.
134. Hiratsuka S, Maru Y, Okada A, Seiki M, Noda T, Shibuya M. Involvement of Flt-1 tyrosine kinase (vascular endothelial growth factor receptor-1) in pathological angiogenesis. Cancer Res 2001;61(3):1207–13.
135. Dvorak HF. Vascular permeability factor/vascular endothelial growth factor: a critical cytokine in tumor angiogenesis and a potential target for diagnosis and therapy. J Clin Oncol 2002;20(21): 4368–80.
136. Zeng H, Dvorak HF, Mukhopadhyay D. Vascular permeability factor (VPF)/vascular endothelial growth factor (VEGF) peceptor-1 down-modulates VPF/VEGF receptor-2-mediated endothelial cell proliferation, but not migration, through phosphatidylinositol 3-kinase-dependent pathways. J Biol Chem 2001;276(29):26969–79.
137. Valtola R, Salven P, Heikkila P, et al. VEGFR-3 and its ligand VEGF-C are associated with angiogenesis in breast cancer. Am J Pathol 1999;154(5):1381–90.
138. Kitsukawa T, Shimono A, Kawakami A, Kondoh H, Fujisawa H. Overexpression of a membrane protein, neuropilin, in chimeric mice causes anomalies in the cardiovascular system, nervous system and limbs. Development 1995;121(12):4309–18.
139. Takashima S, Kitakaze M, Asakura M, et al. Targeting of both mouse neuropilin-1 and neuropilin-2 genes severely impairs developmental yolk sac and embryonic angiogenesis. Proc Natl Acad Sci USA 2002;99(6):3657–62.
140. Alon T, Hemo I, Itin A, Pe'er J, Stone J, Keshet E. Vascular endothelial growth factor acts as a survival factor for newly formed retinal vessels and has implications for retinopathy of prematurity. Nat Med 1995;1(10):1024–8.
141. Zachary I, Gliki G. Signaling transduction mechanisms mediating biological actions of the vascular endothelial growth factor family. Cardiovasc Res 2001;49(3):568–81.
142. Meadows KN, Bryant P, Pumiglia K. Vascular endothelial growth factor induction of the angiogenic phenotype requires Ras activation. J Biol Chem 2001;276(52):49289–98.
143. Ferrara N, Davis-Smyth T. The biology of vascular endothelial growth factor. Endocr Rev 1997;18(1):4–25.
144. Zachary I. Signaling mechanisms mediating vascular protective actions of vascular endothelial growth factor. Am J Physiol Cell Physiol 2001;280(6):C1375–86.
145. Dvorak HF, Brown LF, Detmar M, Dvorak AM. Vascular permeability factor/vascular endothelial growth factor, microvascular hyperpermeability, and angiogenesis. Am J Pathol 1995;146(5): 1029–39.
146. Takahashi Y, Kitadai Y, Bucana CD, Cleary KR, Ellis LM. Expression of vascular endothelial growth factor and its receptor, KDR, correlates with vascularity, metastasis, and proliferation of human colon cancer. Cancer Res 1995;55(18):3964–8.
147. Fan F, Wey JS, McCarty MF, et al. Expression and function of vascular endothelial growth factor receptor-1 on human colorectal cancer cells. Oncogene 2005;24(16):2647–53.
148. Lee JC, Chow NH, Wang ST, Huang SM. Prognostic value of vascular endothelial growth factor expression in colorectal cancer patients. Eur J Cancer 2000;36(6):748–53.
149. Kim KJ, Li B, Winer J, et al. Inhibition of vascular endothelial growth factor-induced angiogenesis suppresses tumour growth in vivo. Nature 1993;362(6423):841–4.
150. Giantonio BJ, Catalano PJ, Meropol NJ, et al. High-dose bevacizumab improves survival when combined with FOLFOX4 in patients with previously treated advanced colorectal cancer: results from the Eastern Cooperative Oncology Group (ECOG) Study E3200. Proc Am Soc Clin Oncol 2005;23:1.
151. Hurwitz H, Fehrenbacher L, Novotny W, et al. Bevacizumab plus irinotecan, fluorouracil, and leucovorin for metastatic colorectal cancer. N Engl J Med 2004;350(23):2335–42.
152. Margolin K, Gordon MS, Holmgren E, et al. Phase Ib trial of intravenous recombinant humanized monoclonal antibody to vascular endothelial growth factor in combination with chemotherapy in patients with advanced cancer: pharmacologic and long-term safety data. J Clin Oncol 2001;19(3):851–6.
153. Gordon MS, Margolin K, Talpaz M, et al. Phase I safety and pharmacokinetic study of recombinant human anti-vascular endothelial growth factor in patients with advanced cancer. J Clin Oncol 2001;19(3):843–50.

154. Kabbinavar F, Hurwitz HI, Fehrenbacher L, et al. Phase II, randomized trial comparing bevacizumab plus fluorouracil (FU)/leucovorin (LV) with FU/LV alone in patients with metastatic colorectal cancer. J Clin Oncol 2003;21(1):60–5.

155. Hurwitz HI, Fehrenbacher L, Hainsworth JD, et al. Bevacizumab in combination with fluorouracil and leucovorin: an active regimen for first-line metastatic colorectal cancer. J Clin Oncol 2005;23(15):3502–8.

156. Kabbinavar FF, Hambleton J, Mass RD, Hurwitz HI, Bergsland E, Sarkar S. Combined analysis of efficacy: the addition of bevacizumab to fluorouracil/leucovorin improves survival for patients with metastatic colorectal cancer. J Clin Oncol 2005;23(16):3706–12.

157. Kabbinavar FF, Schulz J, McCleod M, et al. Addition of bevacizumab to bolus fluorouracil and leucovorin in first-line metastatic colorectal cancer: results of a randomized phase II trial. J Clin Oncol 2005;23(16):3697–705.

158. Saltz LB, Lenz H, Kindler H, et al. Randomized phase II trial of cetuximab/bevacizumab/irinotecan (CBI) versus cetuximab/bevacizumab (CB) in irinotecan-refractory colorectal cancer. Proc Am Soc Clin Oncol 2005;23:16S.

159. Jain RK. Normalization of tumor vasculature: an emerging concept in antiangiogenic therapy. Science 2005;307(5706):58–62.

160. Willett CG, Boucher Y, di Tomaso E, et al. Direct evidence that the VEGF-specific antibody bevacizumab has antivascular effects in human rectal cancer. Nat Med 2004;10(2):145–7.

161. Skillings JR, Johnson DH, Miller K, et al. Arterial thromboembolic events in a pooled analysis of 5 randomized, controlled trials of bevacizumab with chemotherapy. J Clin Oncol 2005;23:16S.

162. Steward WP, Thomas A, Morgan B, et al. Expanded phase I/II study of PTK787/ZK 222584 (PTK/ZK), a novel, oral angiogenesis inhibitor, in combination with FOLFOX-4 as first-line treatment for patients with metastatic colorectal cancer. Proc Am Soc Clin Oncol 2004;22:259.

163. Schleucher N, Trarbach T, Junker U, et al. Phase I/II study of PTK787/ZK 222584 (PTK/ZK), a novel, oral angiogenesis inhibitor in combination with FOLFIRI as first-line treatment for patients with metastatic colorectal cancer. Proc Am Soc Clin Oncol 2004;22:259.

164. Morgan B, Thomas AL, Drevs J, et al. Dynamic contrast-enhanced magnetic resonance imaging as a biomarker for the pharmacological response of PTK787/ZK 222584, an inhibitor of the vascular endothelial growth factor receptor tyrosine kinases, in patients with advanced colorectal cancer and liver metastases: results from two phase I studies. J Clin Oncol 2003;21(21):3955–64.

165. Hecht JR, Trarbach T, Jaeger E, et al. A randomized, double-blind, placebo-controlled, phase III study in patients with metastatic adenocarcinoma of the colon or recturm receiving first-line chemotherapy with oxaliplatin/5-fluorouracil/leucovorin and PTK787/ZK 222584 or placebo (CONFIRM-1). Proc Am Soc Clin Oncol 2005;23:2.

166. Ellis LM. Anti-VEGF Therapy Goes Mainstream. Proc Am Soc Clin Oncol; 2005; Orlando, Florida.

167. Wedge SR, Ogilvie DJ, Dukes M, et al. ZD6474 inhibits vascular endothelial growth factor signaling, angiogenesis, and tumor growth following oral administration. Cancer Res 2002;62(16):4645–55.

168. Hurwitz H, Holden SN, Eckhardt SG, et al. Clinical evaluation of ZD6474, an orally active inhibitor of VEGF signaling, in patients with solid tumors. Proc Am Soc Clin Oncol 2002;21.

169. Holash J, Davis S, Papadopoulos N, et al. VEGF-Trap: a VEGF blocker with potent antitumor effects. Proc Natl Acad Sci USA 2002;99(17):11393–8.

170. Dupont J, Rothenberg ML, Spriggs DR, et al. Safety and pharmacokinetics of intravenous VEGF trap in a phase I clinical trial of patients with advanced solid tumors. Proc Am Soc Clin Oncol 2005;23:199.

171. Heldin CH, Westermark B. Mechanism of action and in vivo role of platelet-derived growth factor. Physiol Rev 1999;79(4):1283–316.

172. Bergers G, Song S, Meyer-Morse N, Bergsland E, Hanahan D. Benefits of targeting both pericytes and endothelial cells in the tumor vasculature with kinase inhibitors. J Clin Invest 2003;111(9): 1287–95.

173. Takahashi Y, Bucana CD, Liu W, et al. Platelet-derived endothelial cell growth factor in human colon cancer angiogenesis: role of infiltrating cells. J Natl Cancer Inst 1996;88(16):1146–51.

174. Hsu S, Huang F, Friedman E. Platelet-derived growth factor-B increases colon cancer cell growth in vivo by a paracrine effect. J Cell Physiol 1995;165(2):239–45.

175. Attoub S, Rivat C, Rodrigues S, et al. The c-kit tyrosine kinase inhibitor STI571 for colorectal cancer therapy. Cancer Res 2002;62(17):4879–83.

176. Zhou L, An N, Haydon RC, et al. Tyrosine kinase inhibitor STI-571/Gleevec down-regulates the beta-catenin signaling activity. Cancer Lett 2003;193(2):161–70.
177. Mendel DB, Laird AD, Xin X, et al. In vivo antitumor activity of SU11248, a novel tyrosine kinase inhibitor targeting vascular endothelial growth factor and platelet-derived growth factor receptors: determination of a pharmacokinetic/pharmacodynamic relationship. Clin Cancer Res 2003;9(1):327–37.
178. Vihinen P, Kahari VM. Matrix metalloproteinases in cancer: prognostic markers and therapeutic targets. Int J Cancer 2002;99(2):157–66.
179. Zeng ZS, Huang Y, Cohen AM, Guillem JG. Prediction of colorectal cancer relapse and survival via tissue RNA levels of matrix metalloproteinase-9. J Clin Oncol 1996;14(12):3133–40.
180. Sunami E, Tsuno N, Osada T, et al. MMP-1 is a prognostic marker for hematogenous metastasis of colorectal cancer. Oncologist 2000;5(2):108–14.
181. Nelson AR, Fingleton B, Rothenberg ML, Matrisian LM. Matrix metalloproteinases: biologic activity and clinical implications. J Clin Oncol 2000;18(5):1135–49.
182. Primrose JN, Bleiberg H, Daniel F, et al. Marimastat in recurrent colorectal cancer: exploratory evaluation of biological activity by measurement of carcinoembryonic antigen. Br J Cancer 1999;79(3–4):509–14.
183. Teronen O, Heikkila P, Konttinen YT, et al. MMP inhibition and downregulation by bisphosphonates. Ann N Y Acad Sci 1999;878:453–65.
184. Rigas JR, Denham CA, Rinaldi DA, et al. Randomized placebo-controlled trials of the matrix metalloproteinase inhibitor, BAY12–9566 as adjuvant therapy for patients with small cell and non-small cell lung cancer. Proc Am Soc Clin Oncol 2003;22:628.
185. Rettig WJ, Garin-Chesa P, Beresford HR, Oettgen HF, Melamed MR, Old LJ. Cell-surface glycoproteins of human sarcomas: differential expression in normal and malignant tissues and cultured cells. Proc Natl Acad Sci USA 1988;85(9):3110–4.
186. Aoyama A, Chen WT. A 170-kDa membrane-bound protease is associated with the expression of invasiveness by human malignant melanoma cells. Proc Natl Acad Sci USA 1990;87(21):8296–300.
187. Sappino AP, Skalli O, Jackson B, Schurch W, Gabbiani G. Smooth-muscle differentiation in stromal cells of malignant and non-malignant breast tissues. Int J Cancer 1988;41(5):707–12.
188. Cheng JD, Weiner LM. Tumors and their microenvironments: tilling the soil. Commentary re: A. M. Scott et al., A Phase I dose-escalation study of sibrotuzumab in patients with advanced or metastatic fibroblast activation protein-positive cancer. Clin Cancer Res, 9:1639–47, 2003. Clin Cancer Res 2003;9(5):1590–5.
189. Park JE, Lenter MC, Zimmermann RN, Garin-Chesa P, Old LJ, Rettig WJ. Fibroblast activation protein, a dual specificity serine protease expressed in reactive human tumor stromal fibroblasts. J Biol Chem 1999;274(51):36505–12.
190. Welt S, Divgi CR, Scott AM, et al. Antibody targeting in metastatic colon cancer: a phase I study of monoclonal antibody F19 against a cell-surface protein of reactive tumor stromal fibroblasts. J Clin Oncol 1994;12(6):1193–203.
191. Kelly T. Fibroblast activation protein-alpha and dipeptidyl peptidase IV (CD26): cell-surface proteases that activate cell signaling and are potential targets for cancer therapy. Drug Resist Updat 2005;8(1–2):51–8.
192. Hofheinz RD, al-Batran SE, Hartmann F, et al. Stromal antigen targeting by a humanised monoclonal antibody: an early phase II trial of sibrotuzumab in patients with metastatic colorectal cancer. Onkologie 2003;26(1):44–8.
193. Adams S, Miller GT, Jesson MI, Watanabe T, Jones B, Wallner BP. PT-100, a small molecule dipeptidyl peptidase inhibitor, has potent antitumor effects and augments antibody-mediated cytotoxicity via a novel immune mechanism. Cancer Res 2004;64(15):5471–80.
194. Calderwood DA. Integrin activation. J Cell Sci 2004;117(Pt 5):657–66.
195. Kinbara K, Goldfinger LE, Hansen M, Chou FL, Ginsberg MH. Ras GTPases: integrins' friends or foes? Nat Rev Mol Cell Biol 2003;4(10):767–76.
196. Varner JA. The role of vascular cell integrins alpha v beta 3 and alpha v beta 5 in angiogenesis. EXS 1997;79:361–90.
197. Yang JT, Rayburn H, Hynes RO. Embryonic mesodermal defects in alpha 5 integrin-deficient mice. Development 1993;119(4):1093–105.

198. Galvez BG, Matias-Roman S, Yanez-Mo M, Sanchez-Madrid F, Arroyo AG. ECM regulates MT1-MMP localization with beta1 or alphavbeta3 integrins at distinct cell compartments modulating its internalization and activity on human endothelial cells. J Cell Biol 2002;159(3):509–21.

199. Stoeltzing O, Liu W, Reinmuth N, et al. Inhibition of integrin alpha5beta1 function with a small peptide (ATN-161) plus continuous 5-FU infusion reduces colorectal liver metastases and improves survival in mice. Int J Cancer 2003;104(4):496–503.

200. Eskens FA, Dumez H, Hoekstra R, et al. Phase I and pharmacokinetic study of continuous twice weekly intravenous administration of Cilengitide (EMD 121974), a novel inhibitor of the integrins alphavbeta3 and alphavbeta5 in patients with advanced solid tumours. Eur J Cancer 2003;39(7): 917–26.

201. Posey JA, Khazaeli MB, DelGrosso A, et al. A pilot trial of Vitaxin, a humanized anti-vitronectin receptor (anti alpha v beta 3) antibody in patients with metastatic cancer. Cancer Biother Radiopharm 2001;16(2):125–32.

202. Armitage P, Doll R. The age distribution of cancer and a multistage theory of carcinogenesis. Br J Cancer 1954;8:1–12.

203. Fearon ER, Vogelstein B. A genetic model for colorectal tumorigenesis. Cell 1990;61:759–67.

204. Bellacosa A. Genetic hits and mutation rate in colorectal tumorigenesis: versatility of Knudson's theory and implications for cancer prevention. Genes Chromosomes Cancer 2003;38(4):382–8.

205. van de Wetering M, Sancho E, Verweij C, et al. The beta-catenin/TCF-4 complex imposes a crypt progenitor phenotype on colorectal cancer cells. Cell 2002;111(2):241–50.

206. Batlle E, Henderson JT, Beghtel H, et al. Beta-catenin and TCF mediate cell positioning in the intestinal epithelium by controlling the expression of EphB/ephrinB. Cell 2002;111(2): 251–63.

207. He TC, Sparks AB, Rago C, et al. Identification of c-MYC as a target of the APC pathway. Science 1998;281(5382):1509–12.

208. Korinek V, Barker N, Morin PJ, et al. Constitutive transcriptional activation by a beta-catenin-Tcf complex in APC-/- colon carcinoma. Science 1997;275(5307):1784–7.

209. Morin PJ, Sparks AB, Korinek V, et al. Activation of beta-catenin-Tcf signaling in colon cancer by mutations in beta-catenin or APC. Science 1997;275(5307):1787–90.

210. Rubinfeld B, Robbins P, El-Gamil M, Albert I, Porfiri E, Polakis P. Stabilization of beta-catenin by genetic defects in melanoma cell lines. Science 1997;275(5307):1790–2.

211. Clapper ML, Coudry J, Chang WC. Beta-catenin-mediated signaling: a molecular target for early chemopreventive intervention. Mutat Res 2004;555(1–2):97–105.

212. Fodde R, Smits R, Clevers H. APC, signal transduction and genetic instability in colorectal cancer. Nat Rev Cancer 2001;1(1):55–67.

213. Muraoka-Cook RS, Dumont N, Arteaga CL. Dual role of transforming growth factor beta in mammary tumorigenesis and metastatic progression. Clin Cancer Res 2005;11(2 Pt 2): 937s-43s.

214. Bhowmick NA, Ghiassi M, Bakin A, et al. Transforming growth factor-beta1 mediates epithelial to mesenchymal transdifferentiation through a RhoA-dependent mechanism. Mol Biol Cell 2001;12(1):27–36.

215. McCormick F. Cancer-specific viruses and the development of ONYX-015. Cancer Biol Ther 2003;2(4 Suppl 1):S157–60.

216. Reid TR, Freeman S, Post L, McCormick F, Sze DY. Effects of Onyx-015 among metastatic colorectal cancer patients that have failed prior treatment with 5-FU/leucovorin. Cancer Gene Ther 2005;12(8):673–81.

217. Foster BA, Coffey HA, Morin MJ, Rastinejad F. Pharmacological rescue of mutant p53 conformation and function. Science 1999;286(5449):2507–10.

218. Loeb LA. Mutator phenotype may be required for multistage carcinogenesis. Cancer Res 1991;51:3075–9.

219. Lengauer C, Kinzler KW, Vogelstein B. Genetic instability in colorectal cancers. Nature 1997;386:623–7.

220. Jallepalli PV, Lengauer C. Chromosome segregation and cancer: cutting through the mystery. Nat Rev Cancer 2001;1(2):109–17.

221. Fishel R. Signaling mismatch repair in cancer. Nat Med 1999;5(11):1239–41.

222. Kunkel TA, Erie DA. DNA mismatch repair. Annu Rev Biochem 2005;74:681–710.

223. Modrich P. Mismatch repair, genetic stability, and cancer. Science 1994;226:1959–60.

224. Kolodner RD, Marsischky GT. Eukaryotic DNA mismatch repair. Curr Opin Genet Dev 1999;9(1):89–96.
225. Lynch HT, de la Chapelle A. Hereditary colorectal cancer. N Engl J Med 2003;348(10):919–32.
226. Cunningham JM, Christensen ER, Tester DJ, et al. Hypermethylation of the hMLH1 promoter in colon cancer with microsatellite instability. Cancer Res 1998;58(15):3455–60.
227. Herman JG, Umar A, Polyak K, et al. Incidence and functional consequences of hMLH1 promoter hypermethylation in colorectal carcinoma. Proc Natl Acad Sci USA 1998;95(12):6870–5.
228. Bellacosa A. Functional interactions and signaling properties of mammalian DNA mismatch repair proteins. Cell Death Differ 2001;8:1076–92.
229. Karran P, Hampson R. Genomic instability and tolerance to alkylating agents. Cancer Surv 1996;28:69–85.
230. Stojic L, Brun R, Jiricny J. Mismatch repair and DNA damage signalling. DNA Repair (Amst) 2004;3(8–9):1091–101.
231. Ribic CM, Sargent DJ, Moore MJ, et al. Tumor microsatellite-instability as a predictor of benefit from fluorouracil-based adjuvant chemotherapy for colon cancer. N Engl J Med 2003;349: 247–57.
232. Robinson BW, Im MM, Ljungman M, Praz F, Shewach DS. Enhanced radiosensitization with gemcitabine in mismatch repair-deficient HCT116 cells. Cancer Res 2003;63(20):6935–41.
233. Berry SE, Kinsella TJ. Targeting DNA mismatch repair for radiosensitization. Semin Radiat Oncol 2001;11(4): 300–15.
234. Chen WD, Eshleman JR, Aminoshariae MR, et al. Cytotoxicity and mutagenicity of frameshift-inducing agent ICR191 in mismatch repair-deficient colon cancer cells. J Natl Cancer Inst 2000;92(6):480–5.

8

Molecular Targeting in Hepatocellular Carcinoma

Jonathan D. Schwartz, MD, and Josep M. Llovet, MD

SUMMARY

Globally, hepatocellular carcinoma (HCC) represents the fifth most common cause of cancer and is diagnosed in 500,000 patients annually. Recent molecular and epidemiologic analyses of tumor specimens suggest a complex and heterogeneous pathogenesis of HCC. Despite this hetergeneity, specific molecular pathways have been identified in the progression of HCC and there is increasing evidence that experimental approaches targeting these pathways have been initiated with a variety of targeted agents, which will be reviewed in this chapter. The potential of targeting therapeutics for HCC has been recently validated in a randomized Phase III clinical trial of sorafenib for advanced HCC. Some of the remaining challenges in using molecular targeted treatment for HCC will be discussed.

Key Words: Anti-angiogenesis therapy; Growth factors; Hepatocellular carcinoma; Molecular targeting.

1. THE MOLECULAR PATHOGENESIS OF HEPATOCELLULAR CARCINOMA

1.1. Overview, Epidemiology, and Natural History

Hepatocellular carcinoma (HCC) is a major global health problem. It is the fifth most common neoplasm in the world, accounting for more than 500,000 new cases per year *(1)*. In the USA, the incidence of HCC has risen in recent years, and this increase is expected to continue over the next two decades, equaling that currently experienced in Japan *(2,3)*. HCC is now the leading cause of death among cirrhotic patients *(4)*.

HCC develops in a cirrhotic liver in 80% of cases, and this pre-neoplastic condition is the strongest predisposing factor *(5)*. Hepatitis B virus (HBV) infection is the main risk factor in Asia and Africa *(6,7)*. Chronic carriers have a 100-fold relative risk for developing HCC, with an annual incidence rate of 2–6% in cirrhotic patients *(8)*. Aflatoxin B1 exposure further enhances the risk *(9)*. In Western countries and Japan, hepatitis C virus (HCV) infection is the main risk factor, together with other causes of

From: *Cancer Drug Discovery and Development: Molecular Targeting in Oncology*
Edited by: H. L. Kaufman, S. Wadler, and K. Antman © Humana Press, Totowa, NJ

cirrhosis including hemochromatosis *(10–12)*. Approximately 20–30% of the estimated 170 million HCV-infected individuals worldwide will develop cirrhosis. Once cirrhosis is established, the annual incidence of HCC is of 3–5%; one-third of all cirrhotic patients develop an HCC over their lifetime *(13)*. The multifactorial etiology of HCC may explain its complex molecular pathogenesis.

The prognosis of HCC was dismal two decades ago *(14)*. This remains the case in much of the developing world. Currently, at specialized centers, up to 40% of HCC patients may receive potentially curative treatments *(15,16)*. These treatments are believed to improve the natural history of the disease. The best survival outcomes are seen in patients with Child-Pugh class A liver function and solitary tumors; these patients experience 65% 3-year overall survival without treatment *(17)*. Radical therapy is believed to improve survival in this group to 70% at 5 years *(18–20)*. The natural course of advanced stage HCC has also been characterized, but varies considerably *(21,22)*. The 1- and 2-year survival rates of untreated patients followed in 25 randomized trials ranged between 10–72% and 8–50%, respectively *(23,24)*. Overall, two groups of patients with unresectable HCC have been identified. These include patients at intermediate stage (asymptomatic tumors) who experience a 3-year survival rate of 50%, compared with an 8% 3-year survival seen in patients with advanced stage HCC (impaired performance status, vascular invasion, or extrahepatic spread) *(21)*. Patients at terminal stages (Child-Pugh class C liver function, ECOG performance status 3–4) survive less than 6 months *(14,21)*.

1.2. Early Diagnosis and Potentially Curative Treatments

Screening is essential for the early diagnosis of HCC, enabling application of potentially curative treatments, which are expected to improve survival. However, reliable histologic differentiation of early HCC from pre-neoplastic lesions found in the setting of surveillance is difficult. Global genomic analysis can provide valuable help by assessing the biological markers differentially expressed in HCC versus pre-neoplastic lesions.

Resection and transplantation achieve the best outcomes in well-selected candidates with single tumors (5-year survival of 60–70%), and both are viable first options as assessed from an intention-to-treat perspective *(16,17,19)*. Percutaneous treatments provide good results but have not reliably demonstrated outcomes equivalent to surgery *(18)*. In unresectable HCC, transarterial chemoembolization (TACE) is the only therapy that has been proven to prolong survival in selected candidates with preserved performance status and hepatic function *(23)*. There is no proven, widely recognized first-line treatment option for patients with advanced HCC, and this is an area of intense clinical research. We recently reported the results of a Phase III placebo-controlled randomized trial demonstrating an improvement in overall survival with sorafenib as a first-line theray in patients with advanced HCC. It is hoped that these results and some of the newer molecularly targeted drugs currently under investigation will contribute to new therapeutic standards. These agents are discussed in Section 3.

1.2.1. RESECTION

The best candidates for resection are patients with single tumors without portal hypertension *(15,16,19)*. However, even with optimal selection of candidates, 5-year recurrence rates approach 70% and include both previously undetected metastases and de novo tumors *(19,25)*. Pathologic variables—including vascular invasion, poor histologic differentiation, and presence of satellite nodules—are predictive of metastases *(25)*. Biological markers that correlate with aggressive behavior are not well defined, although gene expression signature as a predictor of recurrence has been assessed in a recent study *(26)*. Adjuvant treatments reported to diminish recurrence include adoptive immunotherapy and retinoids *(27)*.

1.2.2. LIVER TRANSPLANTATION

Liver transplantation has changed the treatment strategy for HCC, because it may simultaneously cure the tumor and the underlying cirrhosis. The broad selection criteria applied in the early transplant experience led to poor results *(28)*. More recent, restrictive selection criteria (single HCC < 5 cm or up to 3 nodules < 3 cm) have enabled 70% 5-year overall survival with recurrence rates below 15% *(19,20)*. The major shortcoming of liver transplantation is the shortage of donors. Increased waiting time in recent years has led to the exclusion of approximately 20–30% of candidates because of tumor progression or death, worsening outcomes when evaluated on an intention-to-treat basis *(19,29)*. Identification of patients at high risk for dropout is the subject of current investigation *(30)*.

1.2.3. MOLECULAR PATHOGENESIS OF HCC

The molecular pathogenesis of HCC is complex. This neoplasm may arise in normal livers, in abnormal but non-cirrhotic livers, and most commonly, in cirrhotic livers. Furthermore, different risk factors are involved in HCC development. Each of these scenarios involves varied genetic and epigenetic alterations, chromosome aberrations, gene mutations, and altered molecular pathways (Table 1) *(31–38)*. Data from cDNA microarrays demonstrate disparate carcinogenetic pathways for HCC arising in cirrhotic versus non-cirrhotic livers *(36,39)*. Similarly, genomic aberrations and gene expression differ between HCC patients with HBV infection compared with HCV infection or with alcohol-induced disease *(38–42)*. In patients with HCV infection, HCC typically appears in the setting of cirrhosis after years of chronic inflammation, fibrosis, and proliferation. The direct oncogenic effect of the virus is unknown, although data in transgenic mice suggest that the core protein of HCV may have a direct carcinogenic role *(43)*.

Unfortunately, the current knowledge of HCC pathogenesis precludes defining the critical sequence of events occurring at different stages of liver carcinogenesis. Although HCV and HBV are key insults resulting in the development of HCC, the specific mechanism of initiation remains unknown. From the classical description of events in cancer, we can state that the first and second "hits" in HCC remain to be elucidated, at least for most cases. It is certain that aflatoxin, through p53 mutation, constitutes a clear first hit in Asian patients with HBV-related HCC. Similarly, it is clear that cirrhosis constitutes a pre-neoplastic condition, and in some cases is a condition sine qua non HCC cannot develop. However, the key molecular events that lead some of these damaged livers—and not others—to develop cancer is unknown. On the contrary, the main pathways involved in HCC progression and dissemination are not clearly defined. We have summarized the current knowledge regarding this issue subsequently. In principle, it is clear that the Wnt pathway is activated in one-third of cases, as also may be the case of the Ras/MAPKK pathway. A comprehensive approach that simultaneously tests the most relevant pathways in HCC progression is urgently needed.

1.2.4. THE HEPATOCARCINOGENIC PROCESS

The hepatocarcinogenic process involves different and incompletely understood oncogenic pathways. The most accepted hypothesis describes a step-by-step process through which external stimuli induce genetic alterations in mature hepatocytes leading to cell death and cellular proliferation (regeneration) *(31,32,34,38)*. In the progression of chronic inflammation to fibrosis and cirrhosis, the upregulation of mitogenic pathways leads to the production of monoclonal populations. These populations harbor dysplastic hepatocytes because of altered gene expression, telomere erosions, and even chromosomal aberrations. This process may develop over 10–30 years *(38)*. At this point, proliferation may be detected in isolated groups of cells, resulting in foci of dysplasia or, more frequently, dysplastic cells surrounded by a fibrotic ring resulting in low-grade dysplastic nodules (LGDN) or high-grade dysplastic nodules (HGDN) *(44–49)*. These are the major pre-neoplastic entities, although HCC may also arise from isolated small dysplastic cells, from non-conforming clear hepatic nodules, or even from progenitor cells, which may develop mixed cell-type tumors. Altered gene expression profiles are evident at these stages, as well as loss of heterozygosity (LOH); microsatellite instability is a marginal event *(34,38,50–56)*. HGDN are currently considered to be true pre-neoplastic lesions and may develop into malignant tumors in 30% of cases over 1–5 years *(46,47,49)*. There is no agreement on the gene expression profile that reflects the malignant phenotype, and differentiation of early well-differentiated carcinoma in situ from pre-neoplastic lesions continues to represent a real histopathological challenge *(45,57)*.

Some of the characteristics considered by Hanahan et al. to reflect the malignant phenotype are present at early stages of HCC *(58)*. Self-sufficiency in growth signals and resistance to apoptosis are recognized in early well-differentiated tumors of 1 cm, which exhibit high proliferative activity and may become less differentiated upon growth in the 1- to 1.5-cm range *(57)*. Unrestricted replicative potential is reflected in the high level of telomerase activation at these stages. Sustained angiogenesis, tissue invasion and metastasis (i.e., intrahepatic dissemination through portal veins) occur at early stages of the disease. This has been recently recognized both in histological studies and through gene expression profiling *(57,59)*. Cancer invasion and dissemination may occur in some tumors smaller than 2 cm, although a majority behaves in a manner more consistent with carcinoma in situ *(57)*. Kojiro et al. analyzed 106 resected HCC < 2 cm and distinguished a so-called indistinct type without local invasiveness (mean diameter: 12 mm) from a distinct nodular type that showed local invasiveness (mean diameter: 16 mm). In the latter type, local metastases surrounding the nodule were found in 10% of cases and microscopic portal invasion in up to 25%. The metastatic potential of early HCC has been confirmed by gene expression assessment through microarrays *(59)*.

1.2.5. GENE EXPRESSION PROFILES OF PRE-NEOPLASTIC LESIONS AND HCC

Genetic alterations in human cancer include quantitative changes in gene expression without known structural abnormalities, or structural genetic alterations, such as mutations and LOH. The genetic abnormalities of human cancers have been approached in two ways: (i) by analyzing gene expression profiles using reverse transcription-polymerase chain reaction (RT-PCR) technology or microarray, or (ii) by screening for chromosomal regions that may contain tumor suppressor genes or oncogenes, either by comparative genomic hybridization or by microsatellite genotyping *(60–65)*.

Table 1
Mutation Rate

Chromosome	Chromosomal aberration	LOH	Genes involved	Mutation Rate	Potential role
1	1p22 (del)	1p	L-myc		Oncogene
	1p32-p36 (del)	1p35-36	RIZ, p73		TSG
	1p36.13-23 (del)		WNT 9A		Wnt pathway
	1q21.1-q44 (gain)		FASL		Fas ligand
			RXR		Suppresses TGF-β
3	3p21		β-Catenin	30–40%	Wnt pathway
	3p21.3		RASSF1		TSG, Ras pathway
4	4q21-35 (del)	4p, 4q, 4q32	Caspase 3,6		Induction apoptosis
			Smad 1		TGF-β pathway
			AFP/Alb/α-FGF		
5	t5;9	5p			
		5q, 5q35-q-ter			
6	6(q13-q-ter)(del)	6q26-q27	APC	25%	
7	tl7;7 (p13-p14)		M6P/IGF-IIR		
8	8p21.2-p22 (del)	8q	FZ3, DLC-1		Wnt pathway
	inv 8(q10) (gain)		c-myc (8q24)		Oncogene
9	t5;9	p19 ARF (9p21)	p16 INK4A		Cell-cycle regulation
	inv 9p12:q12		&		
10		10q23	PTEN	4–10%	TSG
11	11p15 (gain)	11p	IGF-II		
	11(p13-p14) (del)	11p13-p15.1	CyclinD (11q13)		
	11p11 (del)				
	t11;22				

Chromosome	Aberration	Region	Gene	Marker/%	Function
13	13q14.1-22 (del)	13q	Rb-1, BRCA2	Rb 15%	TSG
		13q12			
		13q12-q31			
16	16q12.1-21 (del)	16p	Axin-1 (16p13.3)	6%	TSG
		16q22-24	E-cadherin (16q22.1)		TSG-Wnt pathway
				SOCS-1 (16p13.13)	TSG-Jak/Stat pathway
				TOP2A	DNA repair
17	17q21.2 (gain)	17p, 17p13	p53 (17p13)	30–50%	TSG
	t17;7 (p13;p14)	17p13-pter	HCCS1	35%	HCC-suppressor
	t17;18 (q25;q11)				
	t17:X				
	17(p12) (del)				
18	t17;18 (q25;q11)		SMAD2,4 (18q21)	10%	TGF-β pathway

HCC, Hepatocellular carcinoma; TGF, transforming growth factor; TSG, tumor suppressor gene.

There is an increasing interest in the characterization of gene alterations in pre-neoplastic lesions and HCC. In general, the majority of studies assess the expression of an individual candidate gene or a few genes in a selected group of patients *(66–79)*. Alternatively, microarray studies have attempted to discover new genes linked with hepatocarcinogenesis and prognosis of HCC *(39–42,59,80–85)*. Some studies published have compared tumor gene expression with that in adjacent cirrhotic tissue, while others compare tumors with normal liver tissue. The results obtained have been quite heterogeneous, which may reflect the complexity of the molecular pathways implicated in the initiation and progression of this neoplasm. However, other factors may explain heterogeneity as well. The selection of the tumors analyzed is of importance, as gene expression may vary according to the etiology of the underlying liver disease and the different evolutionary stages of the tumor. The most common target populations assessed to this point include patients with advanced HCC and portal vascular invasion or metastasis. However, more recent reports have evaluated patients with dysplastic nodules *(50,51)* and early HCC *(78,81)*.

1.2.6. GENE ALTERATIONS IN CIRRHOSIS AND PRE-NEOPLASTIC LESIONS

In patients with chronic HCV infection, increased transforming growth factor (TGF-α) and insulin-like growth factor-2 (IGF-2) contribute to accelerated hepatocyte proliferation *(38)*. Upregulation of these genes results from the combined action of cytokines released by inflammatory cells and from viral transactivation. Simultaneously, aberrant methylation occurs, and the genes involved in the process are clearly upregulated *(86)*. Structural changes such as allelic deletions may occur in 10% of cirrhotic livers *(38)*. Oxidative stress occurs in cirrhosis and may result in damage to genomic and mitochondrial DNA. All these changes result in a genetic and molecular portrait of HCV cirrhotic livers that differs significantly from that of the normal *(82)*. Gene expression is disrupted, affecting genes that are involved in remodeling matrix, cell–cell interactions and anti-apoptotic pathways.

Such is the carcinogenic field in which monoclonal hepatocyte populations develop. For a minority of cases, however, a stem cell origin has been postulated. There is a progressive telomere shortening from cirrhosis to HGDN, which likely limits the replicative lifespan of dysplastic hepatocytes *(52)*. However, telomerase activity progressively increases, differing significantly between LGDN and HGDN, the latter demonstrating a pattern similar to HCC. Telomerase dysfunction promotes chromosomal instability that drives early carcinogenesis *(75)*. This dysfunction may explain some of the structural changes already present in 30% of pre-neoplastic lesions *(38)*. Few allelic alterations are well characterized in pre-neoplastic lesions. Characteristically, LOH at 1p (1p36-p34) is associated with dysplastic nodules (10–15%) and small HCC (25%) *(53)*, a consistent finding for which a candidate gene has not been identified. Recently, LOH in M6F/IGF2R has been detected in more than half of patients with dysplastic nodules and early HCCs, both in the USA and Japan, and is postulated to be an early event in hepatocarcinogenesis *(54,69)*. Data regarding gene expression in dysplastic nodules are scarce. Two recent studies using quantitative RT-PCR and microarrays have described the gene expression profile of this pre-neoplastic state in comparison with non-neoplastic liver *(50,51)*. Genes involved in cell adhesion and invasion (caveolin-1, semaphorin, and collagen IV) and growth factors (IGF) were upregulated, whereas genes expressed in inflammatory cells [FMS-like tyrosine kinase-3 and lymphatic vessel endothelial hyaluronan receptor I (LYVE1)] were downregulated. Further investigations are needed to confirm these preliminary results.

1.2.7. Gene Alterations in HCC

Structural and genetic alterations are very heterogeneous in HCC. None of the well-characterized chromosomal gains, losses, mutations, LOH, or gene alterations have been detected in more than half of the cases *(38)*. This may be due to the heterogeneity of the populations studied, precluding homogeneous conclusions. Overall, more than 200 genes have been associated with neoplastic development; many of these genes contribute to cell-cycle regulation, cell adhesion, vascular invasion and metastasis.

1.2.7.1. Early HCC. Clear distinctions between cancer and dysplastic nodules are difficult to establish. It is estimated that only 30% of HGDN will ultimately develop the malignant phenotype, with the remaining nodules either disappearing or remaining stable for years *(46,47,49)*. Therefore, aspects of these lesions—including LOH, aberrant chromosomes, and mutations—may not be sufficient to drive hepatocarcinogenesis *(31–38)*. Gene expression profiles of early tumors have been infrequently assessed in HCV patients. HSP70 has been proposed as a molecular marker of early HCC, as it was clearly upregulated in early HCC compared with dysplastic nodules *(74)*. Glypican-3 has also been proposed as an early marker, but no comparison with dysplastic lesions has been provided *(73,78)*. LOH and mutations of KLF6 have been found in a majority of HCC patients, although the role of this tumor suppressor gene in early stages remains to be elucidated *(87)*. Smith and colleagues *(81)* have postulated a gene set for diagnosis of HCC, comprising 50 genes obtained by microarrays, including p53, members of the Ras oncogene family, TNF and STK6 among others. Serine/threonine kinase 15 (STK6) was postulated as the most relevant gene. If these genes or others are consistently upregulated in comparison with pre-neoplastic lesions, they may be considered for further evaluation as markers of early tumors.

1.2.7.2. Advanced HCC. In advanced HCC, allelic alterations have been widely described and likely involve all chromosomes. In a recent meta-analysis of comparative genomic hybridization studies that assessed genomic imbalances in 719 HCC patients, the most frequent gains of genomic material were in 1q (55%), 6p (22%), and 8q (45%), whereas losses were most prevalent in 4q (33%), 8p (36%), 16q (35%), and 17p (31%) *(88)*. Some of the alterations observed most likely correspond to non-specific phenomena secondary to neoplastic progression, as they are also observed widely in other cancers. The alterations most specific to HCC are found in chromosomes 1, 4, 8, 16, and 17, but with great variability and with none affecting more than 60% of cases *(31,36,38)*. Few studies have linked genetic abnormalities with gene expression, a topic that has been recently reviewed *(89)*. LOH at 16p and 17p occurs more frequently in advanced tumors and has been associated with loss in p53, SOCS-1, or axin, respectively. Loss of 6q and 9p coincides with loci of IGF-2R and p16, respectively. LOH of 6q and 9p were found to be independent predictors of survival in a series of 85 HCC patients undergoing resection *(56)*. A candidate tumor suppressor (SIAH) has been recently located in 16q, and two other tumor suppressor genes (Rb and BRCA2) are located in chromosome 13q (Table 1).

Several altered genes and pathways have been described in advanced HCC (some of these are detailed more extensively in the following sections). Among the most prevalent alterations, downregulation of p53—either by mutation or LOH—is present in 30–40% of cases *(31,35)*. This is the gene most widely studied in HCC. Large-scale studies throughout the world have revealed a high degree of heterogeneity in the prevalence of p53 mutations (higher in areas with high prevalence of HBV infection) and in their location along the p53 protein (for instance, a mutation at codon 249 is associated with aflatoxin AFB1 exposure). Similarly, alterations in the β-catenin pathways (including mutations of β-catenin and axin 1) involve 30% of liver tumors *(31)*. Epigenetic inactivation of p16 and SOCS-1 through promoter hypermethylation occurs in 65% of cases *(66,70)*. More recently, upregulation of PEG10 (associated with downregulation of SIAH-1 protein) has been implicated in HCC progression *(77)*. Serine protein kinase inhibitor-1 (SPINK-1) was the most upregulated gene in a series of resected HCC's, using microarray *(90)*. Other genes implicated include c-myc, cyclin D1, epidermal growth factor, and phosphatase and tensin homolog (PTEN) *(89)*. In advanced HCC, high telomerase activity is detected in 90% of cases and has been correlated with telomere elongation *(52,75)*. Genome-wide hypomethylation is also more evident in advanced HCC relative to pre-neoplastic lesions *(86)*. A well-characterized group of genes has been strongly related to tumor dissemination and metastasis. Among the most relevant candidate genes are nm23-H1 *(72)*, osteopontin (SPP1) *(59)*, ARHC (Rho C), KAI1 *(68)*, and MMP14 *(34,42)*. Osteopontin and Rho C were the most salient candidates identified by statistical analysis among a set of markers of dissemination *(59)*. The *nm23-H1* gene (a metastasis-suppressor gene) seems to influence progression and differentiation of tumor cells, and its downregulation is associated with a higher metastatic potential in HCC and other neoplasms *(72)*. KAI1 is a gene implicated in cell–cell interactions and cell–extracellular matrix interactions, and is significantly underexpressed in disseminated tumors. In a recent study with HBV-related HCC, a molecular signature related to metastasis and survival was identified; osteopontin (SPP1) overexpression was most highly correlated with an aggressive phenotype *(59)*. Genes regulating the extracellular matrix and cytoskeleton, such as MMP9, MMP14, osteonectin, and Rho A, have been implicated in HCC invasiveness *(42)*. Other genes that likely contribute to aggressive behavior include additional metalloproteases (responsible for degradation of basement membranes and extracellular matrices) and angiogenesis-related genes vascular endothelial growth factor (VEGF) and angiopoietin-2.

2. PRINCIPLES AND CAVEATS WITH RESPECT TO THE APPLICATION OF MOLECULARLY TARGETED THERAPY IN HCC

1. All HCCs may not be created equal. The potential for molecular heterogeneity in HCC is substantial. Although some authors have suggested that HCCs are relatively homogenous at a molecular level, these findings are preliminary, and many studies point to significant abnormalities in multiple cellular pathways *(91–93)*. HCC carcinogenesis is known to be a diverse process—depending on HBV, HCV, or non-viral

etiologies—and the possibility of marked genomic and molecular differences between tumors cannot be discounted *(94)*. Some investigators have described heterogeneity of hormone receptors with striking therapeutic ramifications *(95)*; it is entirely possible that individualized molecular profiling will be necessary to enable efficacious targeted therapy in the future.

2. There is overlap between angiogenic and tumor growth pathways. In experimental systems, mechanisms believed important for HCC cell growth (such as TGF-α and IGF-II) have also been shown to stimulate tumor angiogenesis *(96,97)*. Drugs currently in clinical investigation such as the raf-kinase inhibitor sorafenib have been associated with simultaneous cancer and blood vessel inhibition. Agents that inhibit matrix metalloproteases have potential to impair both local tumor invasion and tumor-related neoangiogenesis. Determining the actual mechanism of action of multi-targeted agents in human subjects presents many difficulties for cancer investigators and is mostly impossible in non-investigational clinical practice.

3. Molecular mechanisms essential to the development of pre-neoplastic lesions may be distinct from those associated with the transformation to carcinoma. In turn, the mechanisms responsible for carcinogenesis are not necessarily identical to those associated with growth of early- or late-stage cancer. Some agents may be efficacious as chemopreventants, anti-neoplastics, both, or neither. Agents which inhibit growth of more limited-stage HCC may not have comparable efficacy for advanced disease. Rigorous translational and clinical analysis represents the only means of determining appropriate, stage-specific therapy.

4. It is likely that inhibition of even well-chosen molecular pathways will result in modest, incremental benefit in HCC—a situation more analogous to the gradual improvement in survival for patients with advanced colorectal cancer than the marked alteration in prognosis seen with imatinib in chronic myelogenous leukemia (CML) or gastrointestinal (GI) stromal tumors. We hope for magic bullets but recognize that they arise rarely.

5. Reliance upon response rates as the most important preliminary indicator of efficacy must be abandoned. These traditional oncologic parameters were valuable in early anti-leukemia/lymphoma efforts but may be impediments in diseases in which meaningful, non-toxic cytoreduction is difficult. Long-term disease control represents a more logical and feasible goal for targeted therapy. Clinicians, investigators, and regulators must be willing to move beyond traditional paradigms; in advanced cancer, an agent which confers no response but 80% progression-free survival at 1 year is likely preferable to one with a 25% response rate and 50% 1-year PFS.

6. Cirrhosis frequently accompanies HCC (especially in HCV-infected populations). The potential of any anti-neoplastic agent to disturb hepatic function must be considered and incorporated into clinical trial design. Primum, non nocere.

3. MOLECULARLY TARGETED AGENTS IN CANCER THERAPY WITH RELEVANCE TO HCC

3.1. Anti-Angiogenic Agents

Drugs that inhibit the formation of blood vessels represent a highly promising direction in cancer therapy. Anti-angiogenic drugs include those which directly inhibit VEGF or the function of VEGF-receptor and associated tyrosine kinases, which are analogs of endogenous pro- or anti-angiogenic molecules, and which impair the function of cell-adhesion and matrix molecules essential for angiogenesis (integrins

Table 2
Circulating and Local Factors Associated with Angiogenesis

Pro-angiogenic factors	Anti-angiogenic factors
Angiopoetins	Angiostatin
Angiogenin	Anti-thrombin III
Angiotensin-2	Arrestin
Basic fibroblast growth factor-1/2	Canstatin
Endothelin-1	Endostatin
Hepatocyte growth factor	Fibronectin
Insulin-like growth factor-2	Interferon-α/β
Interleukin-4/8	Interleukin-12
Placental growth factor	Soluble VEGF-receptor 2 (sKDR)
Platelet-derived growth factor	Thrombospondin-1/2
Stromal derived factor-1α	Tissue inhibitors of metalloproteases
Tissue factor	Tumstatin
Transforming growth factor-α	Vasostatin
Vascular endothelial growth factor-A/B/C/E	

and matrix metalloproteases). Vascular targeting agents disrupt established vessel endothelium through inhibition of tubulin and other mechanisms. These are summarized in Table 2. Many other agents—ranging from cyclo-oxygenase (COX) inhibitors to cytotoxic chemotherapeutics given through alternate schedules—inhibit angiogenesis as a secondary mechanism (see Table 3).

3.1.1. Anti-VEGF Therapy

There is significant rationale for anti-VEGF therapy in HCC. VEGF is overexpressed in cirrhotic liver and HCC, and elevated tumor and circulating VEGF have been associated with the presence of other adverse prognostic features and more rapid clinical disease progression (98–103). Inhibition of VEGF-receptor binding to VEGF receptors 1 (Flt-1) and 2 (KDR/Flk-1) through investigational antibodies has been shown to decrease both HCC carcinogenesis and development of subsequent metastasis in murine HCC models (104,105). In a Morris rat hepatoma model, Graepler et al. (106) demonstrated a substantial reduction in HCC growth in cells genetically modified to express soluble VEGF-receptor 1 (sFlt-1); further modification resulting in constitutive expression of both soluble VEGF-receptor and endostatin markedly reduced tumor growth. Soluble VEGF-receptor- and endostatin-associated growth inhibition was not observed in vitro, suggesting a vital angiogenic contribution for in vivo HCC growth in this system.

Bevacizumab, a humanized anti-VEGF monoclonal antibody binds isoforms of VEGF-A and inhibits VEGF binding to VEGF receptors, with concomitant inhibition of endothelial cell proliferation and migration. Bevacizumab has been shown to improve response rates, time-to-progression, and survival in metastatic colorectal cancer when given in conjunction with cytotoxic chemotherapy (107). Recent preliminary results suggest a similar augmentation of tumor inhibition in advanced breast cancer and non-squamous non-small cell lung cancer. As monotherapy, bevacizumab (given at 10 mg/kg every 2 weeks) has been shown to prolong time-to-progression in refractory,

Table 3
Anti-Angiogenic Agents in Development and Clinical Practice

Type of agent	Drug	Mechanism	Targets	Phase of development
Anti-VEGF	Bevacizumab	Monoclonal antibody	VEGF-A	Approved (colorectal cancer)
	HuMV833	Monoclonal antibody (IgG4κ)	VEGF-A (-121 and -165 isomers)	Phase I
	VEGF-Trap	Soluble receptor	VEGF-A	Phase I
	VEGF-AS (Veglin)	Anti-sense oligonucleotide	VEGF-A, -C, -D	Phase I
Anti-VEGF receptor	CT322 (AdNectin)	Fibronectin protein VEGF receptor agonist	VEGF-R 2	Pre-clinical/ phase I
	CEP-7055	VEGF-R tyrosine kinase inhibitor	VEGF-R 1, 2, 3	Phase I
	CP547,632	VEGF-R tyrosine kinase inhibitor	VEGF-R 2	Phase II
	IMC-1C11	Monoclonal antibody	VEGF-receptor 2	Phase I
	PTK787 (vatalanib)	VEGF-R tyrosine kinase inhibitor	VEGF-R 1, 2, 3	Phase III
	ZD2171	VEGF-R tyrosine kinase inhibitor	VEGF-R 1, 2,	Phase II
	AE-941	Shark cartilage component	VEGF-receptor binding MMP-2, -9	Phase III
Vascular targeting agents	ADH-1 (exherin)	Cyclic pentapeptide	N-cadherin competitive inhibitor	Phase I
	AVE8062	Small molecule	Tubulin	Phase I
	Combrestatin A-4 Phosphate (CA4P)	Small molecule	Tubulin	Phase II
	DMXAA (AS1404)	Flavanoid	Actin (TNF induction)	Phase I-II
	ZD6126	Small molecule	Tubulin	Suspended

Category	Agent	Description	Mechanism	Phase
Endogenous anti-angiogenic factors	ABT-510	Mimetic	Thrombospondin	Phase II
	Endostatin	Recombinant human protein	–	Phase II
Other Anti-angiogenic pathways	MEDI-552	Monoclonal antibody	$\alpha_v\beta_3$ Integrin	Phase II
	Angiostatin	Recombinant human protein	–	Phase II
	Atrasentan	Oral endothelin-1 receptor antagonist	Endothelin-1 receptor	Phase II
	LY317615 (enzastaurin)	Acyclic bisindolyl maleimide	Protein kinase C-β isozyme	Phase I
	PI-88	Sulfonated mannose-P oligosaccharide	Heparanase inhibitor	Phase II
	TNP-470	Fumagillin analog	Multiple putative pathways	Phase II
	Thalidomide	Glutamic acid derivative	Downregulation of bFGF and VEGF; other potential mechanisms	Approved (multiple myeloma); Phase III study in HCC
MMP inhibitors	BMS-275291	Non hydroxamate small molecule	MMP	Phase III
	BAY12-9566		MMP	Phase III
	Marimastat		MMP	Phase III
Anti-integrin agents	CNTO 95	Monoclonal antibody	α_5 Integrins	Phase I
	E7820	Oral sulfonamide	α_2 Integrin inhibitor	Phase II
	EMD 121974 (cilengitide)	Peptide	$\alpha_v\beta_3$ and $\alpha_v\beta_5$ integrins	Phase II
	LM609	Monoclonal antibody	α_v Integrins	

TNF, tumor necrosis factor; VEGF, vascular endothelial growth factor.

metastatic renal cell cancer in a randomized phase II study; the improvement in median overall survival (4.8 versus 2.5 months, $p < 0.001$ log-rank test) occurred despite a modest response rate of 10% (4 of 39 patients) *(108)*. In rectal cancer, a single dose of bevacizumab monotherapy has been shown to reduce tumor interstitial pressure, tumor perfusion, and microvessel density *(109)*.

Preliminary results from our study of bevacizumab monotherapy in HCC patients without extrahepatic metastasis and without invasion of the main portal vein suggest disease-modifying activity. At time of publication, we have observed a 71% rate of disease control (2 partial responses and 10 patients with stable disease for at least 4 months of an initial 17 patients evaluable) with acceptable toxicity. This disease-control has been accompanied by a significant reduction in tumor arterial enhancement as measured by gadolinium-enhanced DCE-MRI. Enrollment has been limited by concerns of increased bleeding risk because of significant pulmonary hemorrhage in 4 of 13 (31%) non-small cell lung cancer patients treated with bevacizumab. HCC patients with cirrhosis are at high risk of bleeding esophageal varices. Our study (NCI 5611) has been modified to exclude patients with untreated esophageal varices; preliminary safety analyses suggest a low and acceptable risk of bleeding at both 5 and 10 mg/kg q 14 day doses. Additional studies are underway assessing bevacizumab in conjunction with transcatheter arterial chemoembolization (TACE).

3.1.2. OTHER ANTI-ANGIOGENIC THERAPIES: PRE-CLINICAL AND CLINICAL EXPERIENCE

Fumagillin, isolated from *Aspergillus fumigatus*, and its synthetic analog TNP-470 have been shown to impair tumor angiogenesis; postulated mechanisms include inhibition of endothelial retinoblastoma gene phosphorylation, cyclin-dependent kinase inactivation, and p53 activation *(110–112)*. TNP-470 has been studied in several in vivo HCC models and has been associated with decreased growth and metastasis of implanted human and investigational HCC cell lines *(113)*. Preliminary clinical development of TNP-470 suggests potential for growth inhibition in solid tumors, although there have been no specific trials in HCC.

The glutamic acid-derivative thalidomide is best known for teratogenicity reported after initial clinical experiences in the 1950s and 1960s *(114,115)*. Potential mechanisms of thalidomide's anti-cancer activity are several-fold and include downregulation of TNF-alpha and immunomodulatory properties *(116–119)*. Anti-angiogenic effects, well-demonstrated in vivo, are believed to result from metabolite-initiated inhibition of activity of the cytokines bFGF and VEGF *(120–122)*. Thalidomide is an approved therapy in multiple myeloma *(123–125)*. Limited disease-modifying activity has been demonstrated in renal cell carcinoma, Kaposi's sarcoma, and high-grade glioblastoma *(126–128)*. An initial report in 2000 described a durable clinical response in a patient with advanced HCC *(129)*. Several clinical trials utilizing thalidomide in advanced HCC have been published in 2003–2005. Response rates have uniformly been low (4–7%) with somewhat higher rates of disease stabilization, when reported (see Table 4) *(130–133)*. Toxicity has been significant in the reports from North-American centers, including an overall 21–35% incidence of severe fatigue or somnolence. In our experience, there was an 18% rate of disease-control (5% partial response, 13% stability), although 16% of patients discontinued therapy secondary to side effects, and 5% of patients had serious arteriothrombotic events that were attributed to this

Table 4
Clinical Trials of Thalidomide in Hepatocellular Carcinoma

Study	N	Efficacy			Toxicity		
		RR (%)	SD (%)	Disease (%) control	Median overall survival (months)	Grade 3/4 fatigue or somnolence (%)	Other serious AEs
Hsu	68	6	32	38	4.3	0	Not observed
Wang	99	6	N/R	N/R	0.8[a] 2.5	0	Not observed
Lin	27	4	7	11	4.1	26	30% neuropathy (gr1-2) 41% instability or dizziness (gr 1-2)
Patt	37	3	31	34	6.8	35	8% early grade 4 rash requiring discontinuation
Schwartz	38	5	13	18	5.5	21	5% grade 4 arteriothrombotic events (MI, CVA)

[a]Median OS in the study by Wang et al. was reported for two separate groups based on whether subjects had received cumulative thalidomide dose <5 g ($n = 22$) or $\geq$ 5g ($n = 77$)

agent (134). This collective experience indicates that thalidomide is not efficacious in HCC. The development of more potent and less toxic thalidomide analogs is underway, although no specific studies in HCC have been undertaken (135).

Ongoing questions remain as to the clinical mechanism of thalidomide, given the diverse pathways it has been postulated to affect. Angiogenic mechanisms have clearly been demonstrated in experimental systems. Hsu and colleagues (136) have reported that of 44 HCC patients evaluated with power Doppler ultrasonography, a significant reduction in tumor vascularity index was seen in 4 patients with disease-control versus minimal difference in the remaining 40 subjects for whom thalidomide did not confer disease control. We were able to assess circulating (plasma) VEGF levels before and following 8 weeks of therapy in 6 HCC patients receiving thalidomide. Although baseline VEGF was similar between 3 patients who had rapid disease progression and 3 who experienced disease-control, circulating VEGF was substantially elevated following therapy in the 3 subjects with progression; VEGF levels remained similar to baseline in the disease-control patients. The small number of subjects precludes meaningful statistical analysis. Although not conclusive, it is likely based on the findings above, that thalidomide acts as a weak anti-angiogenic agent in advanced HCC, albeit one that infrequently renders disease-control.

Angiotensin I and II, produced by angiotensin-converting enzyme (ACE), have been associated with in vivo angiogenic activity through VEGF and metalloprotease-related mechanisms. The angiotensin-converting enzyme inhibitor perindopril has been associated with reduction in hepatic fibrosis, carcinogenesis and HCC growth in HCC models *(137)*. Noguchi et al. *(138)* demonstrated that perindopril inhibited HCC carcinogenesis and growth in a murine HCC model both independently and in conjunction with IFN-beta. ACE-inhibitor-mediated HCC inhibition was associated with decreased VEGF-expression and angiogenesis.

3.1.3. RESISTANCE TO ANTI-ANGIOGENIC THERAPY: THE POTENTIAL CONTRIBUTION OF HYPOXIA-INDUCIBLE FACTOR AND POTENTIAL HIF-DIRECTED THERAPY

Despite the broad range of anti-angiogenic and vascular-targeted agents under development, it is unlikely that inhibition of any one pathway will provide indefinite disease control in a majority of patients. In advanced solid tumors, even studies demonstrating the most robust progression or survival advantages also indicate eventual resistance to anti-VEGF therapy. Given the large number of pro- and anti-angiogenic molecules known to impact on human cancer, the potential for acquired resistance to anti-angiogenic agents is marked *(139)*. One potential consequence of therapy impeding tumor arteriovascular supply is hypoxia, which results in upregulation of hypoxia-inducible factor (HIF)-1a; HIF-1a has been shown to activate genes associated with tumor growth, including pro-angiogenic factors, growth factors and their receptors, and extracellular proteases *(140,141)*. Lee and colleagues have demonstrated genetic upregulation of HIF-1a (and concomitant reduced expression of ENGL2, a negative HIF-1a regulator) in a subset of human HCCs associated with poor clinical outcomes *(142,143)*. Some cancer cells display in vitro resistance to hypoxia, which has been associated with in vivo resistance to anti-angiogenic therapy *(144)*. Anti-HIF agents represent an intriguing class of anti-cancer agents and include Hsp90-binding agents (geldanamycin and derivatives, such as 17AAG; see Chapter 4). Hsp90 is a chaperone protein that enables refolding and maturation of potential oncogenic proteins during cellular stress, preventing proteasomal degradation. Ansamycin antibiotics (such as geldanamycin) inactivate Hsp90, enabling HIF-1 degradation; these agents are in early-stage clinical trials *(145–147)*. Other agents, including the microtubule inhibitor 2ME2 and topoisomerase inhibitors block HIF-1 activity *(148,149)*. Agents that block signal transduction pathways (including EGF-receptors, RAF-kinase, mTOR, and MEK) have also been associated with reduction in HIF-1 activity and are discussed in the following sections *(150)*. Combined anti-angiogenic and anti-HIF therapy represents an appealing future direction for HCC research.

3.2. Integrin and Matrix Metalloprotease-Directed Therapy

Matrix metalloproteases contribute to liver remodeling in chronic hepatitis and cirrhosis, and are also significantly expressed in HCC. MMP-1, MMP-2, and MMP-9 are overexpressed in HCC, and MMP-2 and MMP-9 are associated with extracapsular tumor invasion and activation of matrix-bound VEGF *(151–158)*. Matrix metalloprotease inhibitors have been studied in human cancer with disappointing results; to date, several randomized pancreatic cancer trials involving MMP-inhibitors marimastat or BAY12-9566 indicated no advantage over conventional chemotherapy in patients with advanced disease *(159,160)*.

In HCC cell lines, marimastat inhibits hepatocyte growth factor (HGF)-induced MMP-3 activation and concomitant invasion as measured by an in vitro matrigel assay *(154)*. Korean investigators have demonstrated that Magnoliae cortex (an herbal preparation from the bark of Magnolia trees) inhibits HCC invasion in similar assays through diminished activation of MMP-2 and MMP-9 *(161)*. Caffeic acid and related compounds isolated from other plant species (*Euonymus alatus*) inhibit MMP-2 and MMP-9 expression in HepG2 cell lines; this inhibition was also evident in vivo and was associated with diminished tumor growth following implantation in nude mice *(162)*. The MMP inhibitor batimastat (BB-94) has been shown to reduce human HCC cell growth, invasion, and metastasis in an orthotopic nude mouse HCC model *(163)*.

HCC has also been characterized by altered expression and distribution of integrins, which are predominantly expressed on the cell–matrix interface in non-neoplastic hepatocytes (Grsh 100-2) *(164–166)*. Well-differentiated HCCs (and regenerative nodules) have been shown to overexpress α_1/β_1 and α_5/β_1 integrins *(167,168)*. Expression of α_5 integrins is reduced in more poorly differentiated HCCs with concomitant increase in α_2, α_3, and α_6; α_6 integrins have been specifically correlated with more aggressive tumor phenotype *(168,169)*.

Integrin inhibitors in early clinical trials include CNTO95, a human monoclonal antibody to α_5 integrin, and E7820, an oral α_2 integrin inhibitor. These agents inhibit in vitro vascular formation and endothelial activity, and tumor growth in animal xenografts. Preliminary efforts suggest that these agents can be administered safely, with some evidence of ongoing disease control in patients with refractory solid tumors *(170,171)*. Antibodies that inhibit α_1, α_2, α_6, and integrins have been shown to reduce growth factor (i.e., EGF, TGF-β, bFGF)-mediated migration of HCC cells over extracellular matrix membranes in vitro *(172,173)*.

Molecules that mediate adhesion between hepatocytes are also significantly dysregulated in HCC. Annexin-1 overexpression is common and is most marked in poorly differentiated HCCs. E-cadherin is overexpressed in more well-differentiated tumors but not in more aggressive HCCs *(93)*. ADH-1 is a competitive N-cadherin antagonist currently in early-stage clinical development. To date, disease-control with this agent has been observed exclusively in solid tumors that overexpress N-cadherin; use of this agent may be limited in HCC because N-cadherin is infrequently overexpressed in human HCC *(166,167,174)*.

3.3. Extracellular Growth Factors and Growth Factor Receptors

The cytokines bFGF, TGF-α, IGF-II, HGF, and TNF-alpha have been associated with carcinogenesis and pathogenesis in animal models and human HCC *(38,175,176)*. It is likely that the contribution of any individual factor diminishes as HCC becomes more advanced and that cytokines stimulate disease through both systemic and more local paracrine or autocrine mechanisms. Agents inhibiting these cytokine pathways include antibodies binding either growth factors or receptors, and small molecule receptor tyrosine kinase inhibitors. As detailed in Table 5, some agents inhibit a single receptor, some inhibit multiple receptors within a family (i.e., pan-EGF-receptor inhibitors), and others are multi-targeted inhibitors capable of blocking activation of multiple receptor families.

Agents that inactivate EGF or its receptor include cetuximab (C225), gefitinib, erlotinib, and others detailed in Table 5. Cetuximab has been shown to improve

Table 5
Multi-Targeted Tyrosine Kinase Inhibitors in Development for Cancer Therapy

Agent	Targets	Phase of development
ABT-869	VEGF-R 1, 2, 3 PDGF-R α, β Kit	Pre-clinical
AEE788	VEGF-R 2 ErbB- 1, 2	Phase I
AMG 706	VEGF-R 1, 2, 3 PDGF-R kit, Ret	Phase I-II
AG-013736	VEGF-R 1, 2 PDGF-R	Phase II
BIBF 1120	VEGF-R 1, 2, 3 PDGF-R bFGF-R SRC	Phase I
BMS-354825	PDGF-R kit, SRC BCR-ABL	Phase I
BMS-599626	ErbB-1, 2, 4	Phase I
CHIR-258	VEGF-R PDGF-R bFGF-R c-kit	Phase I
CI-1033	ErbB- 1, 2, 4	Phase I-II
GW786034	VEGF-R 1, 2, 3 PDGF-R α, β c-kit	Phase I
Lapatinib (GW5722016)	ErbB- 1, 2	Phase I-III
PKC412	VEGF-R 2 PDGF-R 1, 2 c-kit	Phase I-II
Sorafenib (BAY 43-9006)	VEGF-R 2 PDGF-R β raf	Phase III including HCC
SU5416	VEGF-R 2 c-kit c-met	Phase III
SU11248	VEGF-R 1, 2, 3 PDGF-R*beta* kit	Phase III
XL647	VEGF-R 2 ErbB-1, 2 EphB4	Phase I
ZD6474	VEGF-R 2, EGF-R (ErbB-1)	Phase II

response rates and time-to-progression in irinotecan-refractory advanced colorectal cancer. Erlotinib has been associated with disease-control in refractory, advanced non-small cell lung cancer and may also improve survival in conjunction with chemotherapy in advanced pancreatic cancer. Other agents such as GW572016, ABX-EGF, and EMD72000 are in various stages of clinical development.

3.3.1. INHIBITION OF EGF, TGF-α, AND THE EGF-RECEPTOR IN HCC

In transgenic mice engineered to overexpress TGF-α, TGF-α anti-sense transfection therapy inhibits hepatocarcinogenesis *(177–179)*. Antibodies to TGF-α have been shown to induce apoptosis in the TGF-α overexpressing OCUH-16 HCC cell line *(180)*. Jo et al. *(181)* have demonstrated that antibody blockade of TGF-α or its ligand EGF-receptor inhibited constitutive c-Met activation and growth in several HCC cell lines; c-Met activation is present in HCC lines in the absence of the c-Met ligand HGF, and this constitutive activation is not observed in non-neoplastic hepatocytes.

The EGF-receptor tyrosine kinase inhibitor gefitinib (ZD1839) diminished HCC carcinogenesis in a chemical-induced (diethylnitrosamine) rat model *(182)*. Gefitinib inhibited growth in human HCC cell lines by reduced mitogen-activated protein kinase (MAPK) ERK1/2 pathway activation, caspase activation, and suppression of Bcl-2- and Bcl-X(L)-mediated apoptosis *(183)*. Gefitinib has also been shown to inhibit in vitro HCC migration, adhesion, MMP-9 production, integrin α5 expression, and TNF-alpha-mediated growth. In mouse models, gefitinib inhibits growth of implanted HCC and intrahepatic metastases *(184,185)*. Cisplatin-mediated inhibition of HCC in murine models is also enhanced by gefitinib co-administration in a schedule-dependent manner *(186)*.

3.3.1.1. Clinical Trials of EGF-Receptor Inhibition in HCC.

Results of two phase II trials investigating the EGF-receptor tyrosine kinase inhibitor erlotinib (OSI-7*(44)*) in advanced HCC have indicated potential for efficacy. Philip and colleagues treated 41 HCC patients (44% with metastatic tumor, 68% with Child-Pugh Class A liver dysfunction) with oral erlotinib at a dose of 150 mg per day. Responses were infrequent (8%), although a 57% disease-control (response + stable disease) rate was reported. The rate of progression-free survival at 24 weeks was 29%. Frequent toxicities included rash (83%, 15% grade 3–4), diarrhea (53%, 7% grade 3–4), fatigue, and hypophosphatemia *(187)*. Results of an additional study by Thomas et al. *(188)* were slightly less robust and included 25% progression-free survival at 4 months. These differences were likely due to patient selection; the results nonetheless suggest that inhibition of EGF-receptor activity can confer disease control in advanced HCC.

3.3.2. INHIBITION OF OTHER EXTRACELLULAR GROWTH FACTORS IN HCC

Anti-sense inhibition of IGF-1 in a rat hepatoma model reduces cell growth, increases apoptosis, and results in tumor reduction when implanted in vivo *(189)*. A DNA-methylating sense oligonucleotide targeting the IGF-2 promoter reduced IFG-2 transcription in an HCC cell line and decreased tumor growth after implantation in nude mice *(190)*. Anti-sense inhibition of both IGF-1 and IGF-2 receptors has also been shown to reduce growth of an HCC cell line in culture and in vivo *(191)*. Hopfner et al. have recently demonstrated that a novel IGF-1 receptor tyrosine kinase inhibitor (NVP-AEW541) inhibits growth of HCC cell lines through caspase-3 activation and induction of apoptosis *(192)*. Other agents under consideration for clinical trials include CP-751,871, a human monoclonal antibody to the IGF-1 receptor that blocks phosphorylation in vitro *(193)*.

Inhibition of FGF-2 (bFGF) through anti-sense reduced the malignant phenotype of an HCC cell line and reduced tumor development in mice *(194)*. Recently, Shao and colleagues *(195)* have demonstrated that acyclic retinoid may exert tumor inhibition by FGF-receptor 3 downregulation and reduction in FGF signaling pathways. Retinoid mechanism and efficacy in HCC prevention and treatment are discussed more extensively in Section 3.9.

Hepatocyte growth factor (HGF) and its ligand c-Met have been associated with investigational and human HCC growth. Resveratrol, a polyphenolic anti-oxidant found in grape skin and associated with anti-carcinogenic effects in prostate carcinoma, has been shown to decrease HGF expression and invasive properties in the rat hepatoma cell line AH109A *(196)*. The HGF-antagonist NK4 inhibits growth, invasion, angiogenesis, and in vivo metastasis when administered through an adenoviral vector in HepG2 cells and xenograft models *(197)*. Small molecule inhibitors of the c-Met tyrosine kinase include PHA-665752 and the multi-targeted SU5416; these agents have inhibited in vitro growth of HCC lines through c-Met inhibition with concomitant reduction in phosphorylation of downstream signaling proteins ERK 1/2 and Akt *(198,199)*.

3.4. Intracellular Signaling Pathways: Inhibition and Relevance in HCC

Cytosolic signaling molecules, often activated by protein phosphorylation, represent potential targets in HCC therapy. These "downstream" pathways may be activated by overexpressed cytokines or membrane-bound tyrosine kinase receptors, or may be constitutively overexpressed or activated in the absence of extracellular membrane-mediated events. The Ras MAPK cascade involves downstream components c-Raf, MEK1/2, and ERK1/2; phosphorylation ultimately results in activation of transcriptional activators c-fos and c-jun and transcription of genes essential for cell proliferation [see Chapter 3 and *(200)*]. The phosphatidylinositol 3-kinase (PI3K) pathway involves components PIP-3, Akt, and mTOR (see Chapter 2). Activation of the PI3K pathway stimulates cell-cycle activation through phosphorylation of p70S6 kinase and other activators of cycle regulation.

3.4.1. Ras MAPK Pathway Inhibition in HCC

P21 ras protein is overexpressed in HCC and in high-risk dysplastic liver tissue, often in the absence of activating ras mutations. There is significant evidence from both human and investigational cell lines suggesting upregulation of the MAP kinase pathway in HCC, concomitant activation of nuclear transcription factors (c-myc, c-fos, c-jun), and increased cyclin D1 indicative of cell-cycle activation *(93,201–207)*.

In pre-clinical studies, agents that block ras-initiated MAPK can impair HCC growth. In HepG2 cells, the ras-farnesylation inhibitor manumycin inhibited ras binding to c-raf-1 and phosphorylation of ERK1/2 with concomitant reduction in cell growth *(208)*. In vitro HCC proliferation is inhibited by the MEK inhibitors PD98059 and U0126 and by ERK1/2 anti-sense oligonucleotide administration *(209,210)*. MEK and ERK1/2 growth inhibition is characterized by cell-cycle arrest and capase-3- and capase-7-mediated apoptosis *(201,209)*. Drugs in clinical practice with potential to impair MAPK-mediated signaling include COX-2 inhibitors. The COX-2 inhibitors NS398 and etodolac have been associated with in vitro HCC growth inhibition and cell-cycle arrest through variable inhibition of ERK1/2 phosphorylation, predominantly in conjunction with the MEK inhibitors PD98059 and U0126, and especially in the less-differentiated Hep3B cell line *(211,212)*. A phase I/II study of the COX-2 inhibitor celecoxib in conjunction with epirubicin chemotherapy in HCC is underway *(213)*.

Drugs in development targeting ras include ISIS 2503, an anti-sense oligodeoxynucleotide, which hybridizes to the 5′-untranslated region of H-ras RNA. Agents that inhibit protein farnesylation of ras include oral heterocyclic compounds R115777 (tipifarnib) and SCH66336, and intravenously administered BMS-214662 and L778,123; phase I–III assessment of these drugs is underway. Raf-kinase inhibitors include the c-Raf anti-sense ISIS 5132 and the multi-targeted BAY 43-9006 (sorafenib), which blocks Raf-kinase and VEGF-receptor activity *(214)*. Oral MEK inhibitors include PD184322 and PD0325901. PD0325901 inhibits both MEK-1 and MEK-2 activity and is in early-stage clinical development; dose-limiting toxicity to-date has been rash, although fatigue, diarrhea, nausea, and blurred vision were also observed *(215)*.

3.4.1.1. Preliminary Experience with Sorafenib (BAY 43-9006) in HCC.

A phase II study of sorafenib in HCC involved 137 patients (72% had Child-Pugh Class A liver dysfunction and the remainder were Child-Pugh Class B, 50% had ECOG performance status of 0, and 50% were PS = 1). Responses were infrequent (4%) but disease-control rate (response + stability) was 59%, with median time-to-progression of 5.5 months. Principal toxicities have been skin rash including hand-foot syndrome, diarrhea, and fatigue. Because of this evidence of disease-control, randomized studies are underway involving patients with advanced HCC (not eligible for resection or TACE) and preserved liver function and performance status. A randomized phase III trial involves randomization to sorafenib versus placebo and a second, smaller randomized study involves randomization to sorafenib versus doxorubicin. The randomized Phase III data was reported by our group at ASCO in 2007 and demonstrated that treatment-naïve, Child-Pugh status A patients who received 400 mg of sorafenib twice daily had a 44% improvement in median overall survival compared to placebo (10.7 versus 7.9 months), which was highly significant. The frequency of adverse events was similar between the treatment groups although more diarrhea and hand-foot syndrome was seen in sorafenib treated patients. This represents the first targeted therapy to show a benefit for first line therapy of HCC.

3.4.2. PI3K/Akt Pathway Inhibition in HCC

Human HCC has also been characterized by overexpression of phosphorylated downstream components of the PI3K pathway mTOR and p70 S6K (M10-70). The PI3K pathway is inhibited by the PTEN tumor suppression gene product, a lipid phosphatase that blocks PI3K phosphorylation of Akt. PTEN expression is absent or diminished in a significant percentage of human HCCs *(216)*. Inhibition of mTOR with rapamycin reduced p70 S6K phosphorylation and inhibited growth of HCC cell lines *(217)*. Drugs that inhibit this pathway include the PI3K inhibitor genistein-combined polysaccharide (GCP) and mTOR inhibitors rapamycin, everolimus (RAD001), AP23573, temsirolimus, and CCI-779 *(192,218–222)*. Because rapamycin also functions as an immunosuppressant, we have commenced studies at Mt. Sinai to investigate its utility following liver transplant, with specific assessment of effect on HCC recurrence versus more established immunosuppressive therapy.

3.4.3. Wnt/β-Catenin Pathway Inhibition in HCC

The Wnt/β-catenin pathway represents a compelling potential target for anti-HCC therapy. Extracellular Wnt family glycoprotein ligands inhibit membrane-bound Frizzled receptors, which in the absence of wnt-binding enable intracellular disheveled-mediated AXIN1/2-, APC-, and GSK3β-mediated phosphorylation of cytoplasmic β-catenin. Phosphorylation of β-catenin triggers its proteasomal degradation. In the absence of phosphorylation, β-catenin nuclear migration results in activation of transcription factors TCF and LEF, resulting in β-catenin target-gene transcription; target genes include c-myc, cell-cycle activators, growth factors, the growth factor receptor c-met, VEGF, and metalloproteases. β-Catenin mutations occur in 20–40% of HCCs and result in constitutive activation of target genes *(223–227)*. AXIN1/2 mutations resulting in similar activation have also been observed in a subset of HCCs *(228)*. Several drugs currently in development include small molecule inhibitors of β-catenin-mediated gene transcription (CGP049090 and PKF115-854) and anti-wnt monoclonal antibodies *(229–232)*. COX-1/2 inhibitors have also been associated with in vitro inhibition of Wnt/β-catenin pathway activation (M10-40-44) *(233–237)*.

3.5. Proteasome Inhibition in HCC

The ubiquitin-proteasome system accounts for the majority of eukaryotic cellular protein degradation and is essential to cellular homeostasis. Ubiquitin-conjugation is an essential precedent to degradation and is mediated by sequential activating (E1), conjugating (E2), and ligase (E3) enzymes. Poly-ubiquinated proteins are recognized and digested by the 26S proteasome into peptide fragments through caspase-like, trypsin-like, and chymotrypsin-like subunits *(238–240)*. Although proteasomes eliminate molecules essential for cancer growth (including EGF receptors, the p44/42 MAPK, and NF-kB), a large number of proteins that impede tumor cell proliferation and survival are degraded by this system, including tumor suppressors p53, cyclin-dependent kinase inhibitors p21 and p27, IKB, and the bax pro-apoptotic protein (prot1-5) *(241)*. Enhanced ubiquitin-proteasome activity has been demonstrated in many human and experimental malignancies including HCC and has been specifically associated with poor clinical outcomes in HCC *(142,242,243)*. Proteasome inhibition in HCC cell lines with the agent MG132 reduced growth and induced apoptosis through caspase-mediated fragmentation of β-catenin *(244)*.

Agents that inhibit ubiquitin-proteasome activity include peptidyl aldehydes MG132 (PSI), vinyl sulfones (NL-VS), beta lactones (lactacystin), epoxyketones (epoxomycin, eponemycin), cyclic peptides (TMC-95), and peptide boronic acids (bortezomib). Bortezomib (PS-341) inhibits proteasomal activity by forming reversible complexes within the chymotrypsin-like site. Bortezomib is an approved therapy in multiple myeloma and appears to confer disease-control both as monotherapy and in conjunction with cytotoxic chemotherapy. Phase II trials have also been conducted in other hematologic malignancies and in renal cell carcinoma. Toxicity in the renal cancer studies included significant neuropathy, fatigue, and myelosuppression. Although disease-control was observed in 34–49% of the patients evaluated, objective clinical responses were infrequent. Median time-to-progression was 1.4 months in one study; a second trial was stopped after enrollment of 21 patients because of lack of efficacy. Authors of both studies were not enthusiastic about further investigation in renal cell carcinoma (RCC) *(245,246)*. Preliminary results of a bortezomib trial in HCC suggest potential for disease-control, with stability lasting up to 8 months observed in 7 of 15 patients *(247)*. A North-American phase II study (ECOG-E6202) combining bortezomib and doxorubicin is currently underway *(248)*.

3.6. Inhibition of DNA Methylation and Histone Acetylation in HCC

In addition to upregulation of proliferation-inducing factors, HCC is also characterized by reduced expression of cellular components that exert a growth inhibitory effect on normal and neoplastic tissues. HCC has been characterized by reduced expression of tumor suppressor genes p53 and E-cadherin *(249)*; TGF-β receptors (TGF-BRI and TGF-BRII) are variably reduced in HCC cell lines and human HCC *(250,251)*. The mannose 6-phosphase/IGF-II receptor (M6P/IGF-2R), which inactivates bound IGF-II by enabling lysosomal degradation, is also significantly underexpressed in human HCC *(252)*. Cell-cycle inhibitors p16^{INK4A}, p21$^{WAF/CIP1}$, and p27^{KIP1} are frequently underexpressed in HCC *(253,254)*; reduced expression of these cell-cycle regulatory proteins and E-cadherin have been associated with more aggressive phenotype and poor clinical outcomes *(255,256)*. Additional suppressor genes potentially associated with allelic deletion in HCC include PTEN, AXIN1, STK11, and BAX *(92,257)*. Gene underexpression has been associated with DNA hypermethylation at CpG islands within promoter sites and with acetylation of related core histone N-terminal tails in human malignancies, including HCC *(258–263)*.

In HCC, altered methylation of the IGF-2 gene has also been associated with P3 promoter activation, increased IGF-2 expression, and concomitant in vitro hepatocyte proliferation and carcinogenesis *(264–268)*. DNA methyltransferases responsible for methylation of CpG sites are upregulated in HCC *(86,269)*. Drugs capable of altering expression of aberrant genes include DNA-demethylating agents 5-azacytidine and decitabine and histone deacetylase (HDAC) inhibitors.

HDAC inhibitors in clinical development include fatty acid derivatives phenylbutyrate and valproic acid, hydoxamic acids including trichostatin A (TSA), suberoylanilide hydroxamic acid (SAHA) and pyroximide, cyclic tetrapeptides such as depsipeptide (FK-228), and the benzamide MS-275 *(270,271)*. Other HDAC inhibitors in early clinical development are the hydroxamates PXD101 and LAQ824, and other agents CI-994 and LBH589B *(272–276)*.

Herold et al. *(277)* demonstrated that the HDAC inhibitor trichostatin-A inhibited growth and induced apoptosis in multiple chemotherapy-resistant human HCC lines. Yamashita and colleagues have recently demonstrated that trichostatin-A inhibited proliferation, induced differentiation and at times induced apoptosis in HCC cell lines *(278)*. HDI-mediated induction of apoptosis has been demonstrated in several chemotherapy-resistant cell lines. Investigators have recently demonstrated that HDIs trichostatin-A- and SAHA-enhanced chemotherapy-induced cytotoxicity in resistant cancer cell lines and have suggested that this effect may be most pronounced by combining HDIs with anti-cancer drugs targeting DNA (including cisplatin, doxorubicin, etoposide, fluoropyrimidines, and topoisomerase inhibitors) *(279,280)*.

Several groups have demonstrated that HDAC-inhibitor-mediated growth inhibition is accompanied by enhanced expression of cell-cycle regulatory genes (i.e., cyclin A, p21$^{WAF/CIP1}$), inducers of apoptosis (caspase-3, bax), growth factor-binding proteins (IGFBP-2 and IGFBP-3), and p53 gene products in HCC models *(277,281,282)*. Chiba and colleagues have recently evaluated the effect of trichostatin-A on gene regulation in HCC cell lines through cDNA microchip assay; they identified a number of genes highly induced by TSA therapy, including p21$^{WAF/CIP1}$ and extracellular matrix-signaling genes CYR61 and CTGF. Of the 57 genes whose expression was induced by histone deacetylation, only two were clearly associated with enhanced proliferation *(283)*.

Other pre-clinical investigation suggests that agents affecting DNA methylation can also impede in vitro and in vivo HCC growth. These include decitabine and a novel sense oligonucleotide designed to enhance methylation at the IGF-II P4 promoter site *(190,284)*. Decitabine is currently under early-stage evaluation in solid tumors *(285)*. MG98 is a novel anti-sense DNA methyltransferase inhibitor in early-phase clinical development *(286)*. Pre-clinical assessment also suggests a potent combined effect of DNA-demethylating drugs and HDAC inhibitors with respect to activation of hyper-methylated genes; combined therapy trials are underway in hematologic malignancies *(287,288)*.

3.7. Telomerase Inhibition in HCC

Neoplastic cells are characterized by an enhanced capacity for cell division enabled by preservation of telomere length *(289)*. Telomeres function as "chromosomal caps" and comprise repeat-sequences of the hexamere TTAGGG in conjunction with nuclear proteins. Intact, lengthy telomeres inhibit chromosomal degradation. Attrition of telomeres accompanies somatic cell aging, with resultant reduction in mitotic capacity. Telomere shortening beyond a threshold level results in chromosomal fusion, insta-bility, and cell death. In the majority of advanced human cancers (and germ line cells), telomere length is preserved by the activity of telomerase enzymes. Telomerases maintain telomeres through reverse transcription of an RNA template (hTR) by a critical catalytic subunit (hTERT), likely in conjunction with additional DNA-binding proteins and polymerases. In vitro telomerase activation has been shown to confer ongoing replicative capacity in several cell systems, including HCC lines *(290–293)*.

Hepatocarcinogenesis is characterized by telomeric stabilization and increased telomerase expression *(293,294)*. Increased telomerase activity has also been associated with more histologically aggressive HCC and more rapid recurrence following HCC resection *(295,296)*.

Several classes of agents have been studied in pre-clinical assessment of telom-erase inhibition for anti-cancer therapy. Reverse transcriptase inhibitors, including those used for anti-HIV therapy, appear limited secondary to resistance-development and toxicity considerations at doses required for adequate inhibition *(297,298)*. Other strategies include anti-sense compounds, inactivators of chaperone molecules (i.e., geldanamycin and analogs) that may inhibit assembly of telomerase complexes, inhibitors of telomerase phosphorylation sites necessary for activation, and perylene diimides capable of stabilizing telomeres *(299–305)*. An additional approach involves inhibition of tankyrase-1-mediated telomere accessibility. Tankyrase-1 ribosylates the telomere-binding protein TRF1 such that it disassociates from the telomere and enables telomerase activity; small molecule inhibitors of this have recently been identified *(306)*.

To date, telomerase inhibitory compounds have not yet been investigated in clinical situations, although development of agents for cancer therapy is proceeding *(307)*. The HDAC inhibitors trichostatin-A and sodium butyrate have been shown to reduce telomerase activity in several HCC cell lines and may represent an additional, clinically feasible means of impairing this aspect of HCC pathogenesis *(308)*.

3.8. Induction of Apoptosis in Human Cancer

Apoptosis, or programmed cell death, is essential for normal homeostasis in self-renewing tissues. Apoptosis is induced through distinct external (mediated by cell surface death receptors) and internal (caspase-mediated) pathways *(309,310)*. Defective apoptotic mechanisms contribute significantly to carcinogenesis and tumor pathogenesis in many human malignancies, including HCC. Cells resistant to apoptosis may proliferate in the absence of exogenous growth factors, in the presence of oxidative stress, hypoxia, DNA damage, and in the presence of other molecular alterations that derange normal proliferation, differentiation, and adhesive and vascular invasive mechanisms *(309,311,312)*. Aberrant oncogenes and proteins associated with cell proliferation, including c-myc and cyclin D1, have also been associated with impaired apoptotic mechanisms *(313)*.

There exists enormous potential in anti-cancer therapy for drugs that specifically promote apoptosis in cells with defective apoptotic mechanisms. The most extensive development of such agents to-date has been with the Bcl-2 anti-sense compound oblimersen sodium (G3139). There has been evidence of activity in hematologic malignancies and solid tumors, although the results have not been overwhelming. In advanced melanoma, the addition of oblimersen to dacarbazine resulted in higher response rates and time-to-progression in a phase III study; there were no differences in overall survival, and hematologic and gastrointestinal toxicities were higher in the patients receiving combined therapy *(314)*. A trial combining oblimersen with doxorubicin in HCC is currently underway *(315)*.

Additional Bcl-2-targeted agents currently under investigation include the small molecule Bcl-2 BH3 binding groove antagonist GX015-070 *(316–318)*. Potential targets stimulating extrinsic apoptotic pathways include tumor necrosis apoptosis-inducing ligand (TRAIL) receptors-1/-2; agonistic monoclonal antibodies under development are in early clinical trials and include HGS-ETR2, which is a high-affinity activator of the TRAIL receptor-2 *(319,320)*. Pre-clinical investigation with antibodies to alternate TRAIL receptors is also ongoing *(321)*.

3.8.1. PEROXISOME PROLIFERATOR-ACTIVATED RECEPTOR-γ-MEDIATED APOPTOSIS IN HCC

The nuclear transcription factor peroxisome proliferator-activated receptor (PPAR)-γ has recently been shown to stimulate in vitro growth in multiple cancer models, including HCC cell lines *(322–328)*. Inhibition of PPAR-γ reduces HCC growth in multiple cell lines through caspase-3-mediated apoptosis, which derives in part from altered integrin-mediated signaling *(329,330)*. PPAR-γ inhibitory ligands include the thiazolidenediones, currently approved for clinical use as anti-diabetic agents. Clinical trials utilizing these agents in prostate cancer, liposarcomas, and other tumors have demonstrated limited efficacy *(325,331,332)*. Recently, Schaefer and colleagues *(333)* demonstrated a more potent anti-adhesive and pro-apoptotic effect with the novel PPAR-γ inhibitor T0070907 in HepG2 cell lines. Agents such as T0070907 or GW9662 may prove more efficacious as anti-cancer agents, although these are not yet in clinical investigation. Combination therapy may also represent a worthwhile approach; PPAR-γ and retinoid receptor ligands appear to confer synergistic inhibitory properties in vitro; a phase 1 trial combining rostiglitazone with bexarotene in advanced cancers is underway *(334)*.

3.9. Retinoid Therapy in HCC

Retinoids are vitamin A metabolites that regulate physiologic epithelial growth and differentiation. Retinoids exert biologic effect through cellular binding proteins and nuclear retinoic acid receptors (RARs). Several related but distinct RAR subtypes have been characterized (α, β, and γ) each with two distinct promoter sites; these homodimeric receptors are expressed variably during development, accounting for the diverse functions attributed to this receptor family *(335)*. A related group of receptors (RXR) are activated by higher concentrations of retinoids and can mediate signaling both as hetero- and homodimers, conferring additional complexity to the potential physiologic and aberrant effects of retinoid–ligand interaction.

Retinoid activation of RAR/RXRs can induce differentiation and induce apoptosis in neoplastic hematologic and epithelial cell lines. All-trans-retinoic acid (ATRA) has become an efficacious standard therapy in promyelocytic leukemia characterized by the t*(15,17)* PML/RAR-α translocation; other retinoids have demonstrated efficacy in hematologic malignancies and in chemoprevention of epithelial skin and upper aerodigestive tract cancers *(336)*.

Muto and colleagues developed an acyclic retinoid that induced differentiation and apoptosis of HCC cell lines and inhibited in vivo growth in animal HCC models. In a randomized trial following potentially curative HCC resection, they demonstrated reduced development of subsequent HCC in patients receiving acyclic retinoid (versus placebo) with concomitant improvement in long-term survival *(337,338)*. Widespread clinical development of acyclic retinoid has been delayed in recent years for non-scientific reasons, although this agent may soon be available for more widespread clinical investigation in HCC.

Additional studies have indicated that retinoids, including acyclic retinoid, ATRA, and 9-cis retinoic acid, inhibit in vitro HCC growth. Several putative mechanisms have been suggested, including downregulation of TGF-α and FGF-receptor 3, telomerase inhibition, induction of caspase-3-mediated apoptosis, and induction of p21(CIP1) with concomitant cyclin D1 downregulation *(195,339–341)*. Matsushima-Nishiwaki and colleagues have demonstrated a potential mechanism for the increased anti-HCC activity of acyclic retinoid. Physiologic levels of 9-cis retinoic acid activate ERK by MAPK phosphatase-1 downregulation; ERK 1/2 signaling inactivates RXR-α and enables cell proliferation; acyclic retinoid inhibits ERK signaling and promotes RXR-α-mediated inhibition of cell growth *(342)*.

3.10. Arsenic Trioxide in HCC

Arsenic trioxide, first associated with anti-cancer effects by Chinese physicians who isolated it as an active component of a traditional herbal remedy, has been shown both in vitro and in clinical trials to induce apoptosis and remission in promyelocytic leukemia *(343,344)*. Arsenic has been associated with induction of apoptosis in several human solid tumors, including breast and gastric and esophageal cancers *(345–347)*. Arsenic has been shown to decrease cell growth through apoptosis induction in several HCC cell lines, including multidrug-resistant lines (R-HepG2) characterized by MDR1 and P-glycoprotein overexpression *(348,349)*. In vivo, arsenic inhibits tumor growth both in mouse xenograft and chemically induced rat hepatoma models *(350–353)*.

3.11. Vitamin K and Analogs

Vitamin K and analogs have also shown potential for inhibition of experimental and clinical HCC. Koike and colleagues *(354)* have observed diminished vitamin K activity in more aggressive HCC, as evidenced by increases in circulating des-gamma-carboxy-prothrombin (DCP), an abnormal prothrombin that circulates in vitamin K-deficient states. Other investigators have demonstrated reduced vitamin K in HCC lesions relative to concentrations in non-neoplastic liver *(355)*. Vitamins K1, K2 and K3, and analogs have been shown to reduce in vitro HCC growth in several cell systems. Vitamins K2 and K3 and the analog CPD 5 [2-(2-mercaptoethanol)-3-methyl-1,4-napthoquinone] have also inhibited in vivo HCC growth in rat hepatoma and nude mice models *(356–358)*. Proposed mechanisms of HCC inhibition include inhibition of EGF-receptor and ERK phosphorylation, downregulation of cyclin D1, Cdk4 and possibly other cell cycle-related proteins, and phosphorylation-mediated degradation of c-myc *(356–361)*. Preliminary results of a randomized trial involving HCC patients undergoing radio frequency ablation (RFA) or TACE suggest improved survival and diminished incidence of portal vein invasion for patients receiving post-procedure vitamin K2 versus no therapy *(362)*.

3.12. Hormone Receptor Antagonist Therapy in HCC

During the 1990s, several investigators postulated that the estrogen receptor antagonist tamoxifen conferred a clinical benefit in HCC. Investigations in hormonal therapy were undertaken in part because of the male predominance of the disease worldwide and the presence of estrogen receptors variably reported in HCC specimens. Although some initial studies reported a benefit from tamoxifen, more recent well-conducted, larger trials indicate no benefit, and these findings are corroborated by a recent meta-analysis *(23)*.

Villa et al. *(95)* identified a variant estrogen-receptor in approximately 33% of HCC patients. The variant receptor is characterized by an exon deletion that results in an altered receptor-binding domain and constitutive transcriptional activity. Tumors with variant receptors are associated with shorter doubling times, aggressive clinical behavior, and have recently been demonstrated as an independent predictor of limited survival *(363)*. Forty-five patients expressing variant receptors were randomized to receive either megestrol 160 mg per day or no therapy. Side effects were mild and infrequent. There were no responses. Megestrol was associated with significant increase in appetite and weight gain, and a significant survival benefit (median of 18 months versus 7 months for no therapy). These results have not yet been confirmed in additional study. They suggest nonetheless that even a non-specific hormonal agent may function as targeted therapy if employed in patients with relevant molecular abnormalities.

4. SUMMARY OF POTENTIAL ADVANCES IN HCC: MOLECULAR UNDERSTANDING AND THERAPEUTIC POTENTIAL

The first 5 years of the twenty-first century have been marked by enormous progress with respect to characterization of HCC at genetic, chromosomal, molecular, and cellular levels. The first targeted agent – sorafenib – has demonstrated a survival benefit in Phase III clinical trials for patients with advanced HCC. It is clear, however, that wide-range abnormalities contribute to the difference between neoplastic and non-neoplastic hepatic disease. As an example, a definitive 2001 book chapter devoted to defining these molecular abnormalities cited over 800 primary references (93). The current chapter details over 15 pharmacologic pathways for which therapy is under development, and over 100 agents in early-to-advanced stage trials for which a reasonable rationale exists in anti-HCC therapy. The possibilities for investigation—especially given the likely need for efficacious combination therapy—quickly become overwhelming. What follows is a primitive strategy as to how informed investigators might proceed in the coming years in the treatment of this devastating malignancy.

4.1. Targeted Therapy Is Not Necessarily Specific Therapy

Although anti-cancer agents are developed to exploit pathways unique to cancer cells, some of these agent classes nonetheless affect diverse cell populations with ramifications for toxicity and limitations on therapeutic efficacy. As an example, proteasomal degradation is an essential component of homeostasis in many normal cell systems. Drugs that inhibit proteasome function may be no more specific in their risk-benefit profile than traditional cytotoxic therapy. A recent randomized trial in multiple myeloma indicated improved efficacy for the proteasome inhibitor bortezomib, but this was accompanied by more frequent serious toxicity relative to the standard dexamethasone treatment (364). Other targeted agents, including more specific apoptosis-inducing strategies such as the bcl-2 anti-sense GD3139, have resulted in very modest benefit and enhanced toxicity in randomized melanoma trials (314). Tumor vasculature represents an appealing therapeutic target, but anti-vascular agents also have potential to impede the precious vascular supply of non-regenerative essential cerebral and coronary tissues. Development of some vascular targeting agents has been limited by arteriothrombotic events; despite life-prolonging benefits in colorectal cancer, the VEGF-inhibitor bevacizumab has been associated with additional risk of myocardial and cerebrovascular events (365).

4.2. Targeted Therapy as Determined by Tissue Analysis

As the preliminary investigation by Villa and colleagues illustrate, even relatively non-specific agents may be considered "targeted therapy" when employed in a subset of patients with a high likelihood of disease-control, as determined by clinical, immunohistochemical, or more sophisticated means of molecular assessment. It is highly likely that the 20–40% of HCC patients whose tumors display aberrant wnt/β-catenin signaling (or related AXIN-1/2 mutations) will benefit from therapies targeting this pathway than would a larger, more heterogeneous HCC population. Just as estrogen/progesterone and Her-2-receptor status factor significantly into the clinician's choices for breast cancer therapy, a range of clinically feasible assays may determine the appropriate therapeutic agents in HCC.

4.3. Selecting the Most Promising Targets and Agents from a Broad Range of Possibilities

Although sorafenib resulted in a modest improvement in survival, it remains unlikely that inhibition of a single pathway, or reversal of a single aberrant tumor suppressor gene/product, will result in sustained tumor control for a majority of HCC patients, especially those with substantial disease-burdens. Several groups of investigators have attempted to identify genes specifically identified with more aggressive HCC phenotypes. Although results of these gene-expression studies are inconsistent, it appears likely that genes associated with cellular proliferation, apoptosis inhibition, and protein ubiquitination are associated with a malignant phenotype and specifically with aggressive clinical disease profiles *(26,83,84,91,142,366,367)*. Although specific inhibitors of many of the proliferation-inducing gene products are not available, it is likely that reduction in aberrant histone acetylation and DNA methylation will result in a cellular environment favoring reduced proliferation *(283)*. Agents inhibiting proteasome function are also likely to exert meaningful growth control, especially if associated with specific degradation of HIF and factors enabling cell growth in the face of otherwise lethal damage; the potential for these agents to confer toxicity remains significant, given the reliance of many cell systems on proteasomal activity for homeostasis.

The contribution of extracellular growth factors—including IGF-II and TGF-α—to malignant transformation and pathogenesis makes it likely that successful treatment combinations will contain agents that can inhibit these pathways *(38)*. It remains to be seen whether inhibition of a single signaling pathway will be sufficient to confer meaningful disease control; it is more than likely that parallel inhibition (i.e., both ras/MAPK and PI3K/Akt) will provide more significant anti-tumor activity. Toxicity is likely to be more significant with use of multiple agents and also with drugs interfering with more downstream (intracellular) pathways. Additional genes and pathways are associated with invasion, metastasis, and poor clinical outcomes, and these include those promoting tumor neoangiogenesis. We have seen preliminary evidence of prolonged (but not indefinite) disease-control with anti-VEGF therapy, suggesting that this is a useful albeit not comprehensive clinical strategy. It is entirely conceivable that regimens comprising combinations of these compounds represent a real hope for future management of HCC, although this will become a reality even in the best of circumstances only through rigorous clinical investigation, careful patient selection, and willingness to persist despite potentially discouraging results of monotherapy trials.

REFERENCES

1. Parkin DM, Bray F, Ferlay J, Pisani P. Estimating the world cancer burden: Globocan 2000. *Int J Cancer* 2001;94(2):153–6.
2. Tanaka Y, Hanada K, Mizokami M, Yeo AE, Shih JW, Gojobori T, et al. A comparison of the molecular clock of hepatitis C virus in the United States and Japan predicts that hepatocellular carcinoma incidence in the United States will increase over the next two decades. *Proc Natl Acad Sci USA* 2002;99(24):15584–9.
3. El-Serag HB, Mason AC. Rising incidence of hepatocellular carcinoma in the united states. *N Engl J Med* 1999;340:745–750.
4. Fattovich G, Giustina G, Degos F, Tremolada F, Diodati G, Almasio P, et al. Morbidity and mortality in compensated cirrhosis type C: a retrospective follow-up study of 384 patients. *Gastroenterology* 1997;112(2):463–72.
5. Bosch FX, Ribes J, Borras J. Epidemiology of primary liver cancer. *Semin Liver Dis* 1999;19(3):271–85.
6. Brechot C, Thiers V, Kremsdorf D, Nalpas B, Pol S, Paterlini-Brechot P. Persistent hepatitis B virus infection in subjects without hepatitis B surface antigen: clinically significant or purely "occult." *Hepatology* 2001;34(1):194–203.
7. Okuda K. Hepatocellular carcinoma. *J Hepatol* 2000;32(1 Suppl):225–37.
8. Fattovich G, Giustina G, Schalm SW, Hadziyannis S, Sanchez-Tapias J, Almasio P, et al. Occurrence of hepatocellular carcinoma and decompensation in western European patients with cirrhosis type B. The EUROHEP Study Group on Hepatitis B Virus and Cirrhosis. *Hepatology* 1995;21(1):77–82.
9. Sun Z, Lu P, Gail MH, Pee D, Zhang Q, Ming L, et al. Increased risk of hepatocellular carcinoma in male hepatitis B surface antigen carriers with chronic hepatitis who have detectable urinary aflatoxin metabolite M1. *Hepatology* 1999;30(2):379–83.
10. Tsukuma H, Hiyama T, Tanaka S, Nakao M, Yabuuchi T, Kitamura T, et al. Risk factors for hepatocellular carcinoma among patients with chronic liver disease. *N Engl J Med* 1993;328(25):1797–801.
11. Colombo M, de Franchis R, Del Ninno E, Sangiovanni A, De Fazio C, Tommasini M, et al. Hepatocellular carcinoma in Italian patients with cirrhosis. *N Engl J Med* 1991;325(10):675–80.
12. Bruix J, Barrera JM, Calvet X, Ercilla G, Costa J, Sanchez-Tapias JM, et al. Prevalence of antibodies to hepatitis C virus in Spanish patients with hepatocellular carcinoma and hepatic cirrhosis. *Lancet* 1989;2(8670):1004–6.
13. WHO. Hepatitis C: global prevalence. *Wkly Epidemiol Rec* 1997;72:341–344.
14. Okuda K, Ohtsuki T, Obata H, Tomimatsu M, Okazaki N, Hasegawa H, et al. Natural history of hepatocellular carcinoma and prognosis in relation to treatment. Study of 850 patients. *Cancer* 1985;56(4):918–928.
15. Llovet JM, Burroughs A, Bruix J. Hepatocellular carcinoma. *Lancet* 2003;362(9399):1907–17.
16. Bruix J, Llovet JM. Prognostic prediction and treatment strategy in hepatocellular carcinoma. *Hepatology* 2002;35(3):519–24.
17. Barbara L, Benzi G, Gaiani S, Fusconi F, Zironi G, Siringo S, et al. Natural history of small untreated hepatocellular carcinoma in cirrhosis: a multivariate analysis of prognostic factors of tumor growth rate and patient survival. *Hepatology* 1992;16(1):132–7.
18. Arii S, Yamaoka Y, Futagawa S, Inoue K, Kobayashi K, Kojiro M, et al. Results of surgical and nonsurgical treatment for small-sized hepatocellular carcinomas: a retrospective and nationwide survey in Japan. The Liver Cancer Study Group of Japan. *Hepatology* 2000;32(6):1224–9.
19. Llovet JM, Fuster J, Bruix J. Intention-to-treat analysis of surgical treatment for early hepatocellular carcinoma: resection versus transplantation. *Hepatology* 1999;30(6):1434–1440.
20. Mazzaferro V, Regalia E, Doci R, Andreola S, Pulvirenti A, Bozzetti F, et al. Liver transplantation for the treatment of small hepatocellular carcinomas in patients with cirrhosis. *N Engl J Med* 1996;334(11):693–699.
21. Llovet JM, Bru C, Bruix J. Prognosis of hepatocellular carcinoma: the BCLC staging classification. *Semin Liver Dis* 1999;19(3):329–38.
22. Llovet J, Bustamente J, Castells A, Vilana R, Ayuso C, Sala M, et al. Natural history of untreated nonsurgical hepatocellular carcinoma: rational for the design and evaluation of therapeutic trials. *Hepatology* 1999;29:62–67.
23. Llovet JM, Bruix J. Systematic review of randomized trials for unresectable hepatocellular carcinoma: chemoembolization improves survival. *Hepatology* 2003;37:429–442.

24. Schwartz JD, Beutler AS. Therapy for unresectable hepatocellular carcinoma: Review of the randomized clinical trials – II: Systemic and local non-embolization based therapies in unresectable and advanced hepatocellular carcinoma. *Anticancer Drugs* 2004;15:439–452.

25. Nagasue N, Uchida M, Makino Y, Takemoto Y, Yamanoi A, Hayashi T, et al. Incidence and factors associated with intrahepatic recurrence following resection of hepatocellular carcinoma. *Gastroenterology* 1993;105(2):488–94.

26. Iizuka N, Oka M, Yamada-Okabe H, Nishida M, Maeda Y, Mori N, et al. Oligonucleotide microarray for prediction of early intrahepatic recurrence of hepatocellular carcinoma after curative resection. *Lancet* 2003;361(9361):923–929.

27. Schwartz JD, Schwartz M, Mandeli J, Sung M. Neoadjuvant and adjuvant therapy for resectable hepatocellular carcinoma: review of the randomised clinical trials. *Lancet Oncol* 2002;3(10): 593–603.

28. Ringe B, Pichlmayr R, Wittekind C, Tusch G. Surgical treatment of hepatocellular carcinoma: experience with liver resection and transplantation in 198 patients. *World J Surg* 1991;15(2):270–85.

29. United Network for Organ Sharing. Annual report. http://www.unos.org/data. 2000.

30. Llovet JM, Sala M, Fuster J, Navasa M, Pons F, Sole M, et al. Predictors of drop-out and survival of patients with hepatocellular carcinoma candidates for liver transplantation. *Hepatology* 2003;38:763A.

31. Buendia MA. Genetics of hepatocellular carcinoma. *Semin Cancer Biol* 2000;10(3):185–200.

32. Feitelson MA, Sun B, Satiroglu Tufan NL, Liu J, Pan J, Lian Z. Genetic mechanisms of hepato-carcinogenesis. *Oncogene* 2002;21(16):2593–604.

33. Kensler TW, Qian GS, Chen JG, Groopman JD. Translational strategies for cancer prevention in liver. *Nat Rev Cancer* 2003;3(5):321–9.

34. Kim JW, Wang XW. Gene expression profiling of preneoplastic liver disease and liver cancer: a new era for improved early detection and treatment of these deadly diseases. *Carcinogenesis* 2003;24(3):363–9.

35. Staib F, Hussain SP, Hofseth LJ, Wang XW, Harris CC. TP53 and liver carcinogenesis. *Hum Mutat* 2003;21(3):201–16.

36. Tannapfel A, Wittekind C. Genes involved in hepatocellular carcinoma: deregulation in cell cycling and apoptosis. *Virchows Arch* 2002;440(4):345–52.

37. Ozturk M, Cetin-Atalay R. Biology of hepatocellular cancer. In: Rustgi AK (Editor). *Gastrointestinal Cancers*. New York: Elsevier Science Ltd; pp 575–591, 2003.

38. Thorgeirsson SS, Grisham JW. Molecular pathogenesis of human hepatocellular carcinoma. *Nat Genet* 2002;31(4):339–346.

39. Iizuka N, Oka M, Yamada-Okabe H, Mori N, Tamesa T, Okada T, et al. Differential gene expression in distinct virologic types of hepatocellular carcinoma: association with liver cirrhosis. *Oncogene* 2003;22(19):3007–14.

40. Okabe H, Satoh S, Kato T, Kitahara O, Yanagawa R, Yamaoka Y, et al. Genome-wide analysis of gene expression in human hepatocellular carcinomas using cDNA microarray: identification of genes involved in viral carcinogenesis and tumor progression. *Cancer Res* 2001;61(5):2129–37.

41. Iizuka N, Oka M, Yamada-Okabe H, Mori N, Tamesa T, Okada T, et al. Comparison of gene expression profiles between hepatitis B virus- and hepatitis C virus-infected hepatocellular carcinoma by oligonucleotide microarray data on the basis of a supervised learning method. *Cancer Res* 2002;62(14):3939–44.

42. Delpuech O, Trabut JB, Carnot F, Feuillard J, Brechot C, Kremsdorf D. Identification, using cDNA macroarray analysis, of distinct gene expression profiles associated with pathological and virological features of hepatocellular carcinoma. *Oncogene* 2002;21(18):2926–37.

43. Moriya K, Fujie H, Shintani Y, Yotsuyanagi H, Tsutsumi T, Ishibashi K, et al. The core protein of hepatitis C virus induces hepatocellular carcinoma in transgenic mice. *Nat Med* 1998;4(9):1065–7.

44. Theise ND, Schwartz M, Miller C, Thung SN. Macroregenerative nodules and hepatocellular carcinoma in forty-four sequential adult liver explants with cirrhosis. *Hepatology* 1992;16(4):949–955.

45. Theise ND, Park YN, Kojiro M. Dysplastic nodules and hepatocarcinogenesis. *Clin Liver Dis* 2002;6(2):497–512.

46. Terasaki S, Kaneko S, Kobayashi K, Nonomura A, Nakanuma Y. Histological features predicting malignant transformation of nonmalignant hepatocellular nodules: a prospective study. *Gastroenterology* 1998;115(5):1216–22.

47. Seki S, Sakaguchi H, Kitada T, Tamori A, Takeda T, Kawada N, et al. Outcomes of dysplastic nodules in human cirrhotic liver: a clinicopathological study. *Clin Cancer Res* 2000;6(9):3469–73.
48. Maggioni M, Coggi G, Cassani B, Bianchi P, Romagnoli S, Mandelli A, et al. Molecular changes in hepatocellular dysplastic nodules on microdissected liver biopsies. *Hepatology* 2000;32(5):942–6.
49. Borzio M, Fargion S, Borzio F, Fracanzani AL, Croce AM, Stroffolini T, et al. Impact of large regenerative, low grade and high grade dysplastic nodules in hepatocellular carcinoma development. *J Hepatol* 2003;39(2):208–14.
50. Anders RA, Yerian LM, Tretiakova M, Davison JM, Quigg RJ, Domer PH, et al. cDNA microarray analysis of macroregenerative and dysplastic nodules in end-stage hepatitis C virus-induced cirrhosis. *Am J Pathol* 2003;162(3):991–1000.
51. Colombat M, Paradis V, Bieche I, Dargere D, Laurendeau I, Belghiti J, et al. Quantitative RT-PCR in cirrhotic nodules reveals gene expression changes associated with liver carcinogenesis. *J Pathol* 2003;201(2):260–7.
52. Oh BK, Jo Chae K, Park C, Kim K, Jung Lee W, Han KH, et al. Telomere shortening and telomerase reactivation in dysplastic nodules of human hepatocarcinogenesis. *J Hepatol* 2003;39(5):786–92.
53. Sun M, Eshleman JR, Ferrell LD, Jacobs G, Sudilovsky EC, Tuthill R, et al. An early lesion in hepatic carcinogenesis: loss of heterozygosity in human cirrhotic livers and dysplastic nodules at the 1p36-p34 region. *Hepatology* 2001;33(6):1415–24.
54. Yamada T, De Souza AT, Finkelstein S, Jirtle RL. Loss of the gene encoding mannose 6-phosphate/insulin-like growth factor II receptor is an early event in liver carcinogenesis. *Proc Natl Acad Sci USA* 1997;94(19):10351–5.
55. Oka Y, Waterland RA, Killian JK, Nolan CM, Jang HS, Tohara K, et al. M6P/IGF2R tumor suppressor gene mutated in hepatocellular carcinomas in Japan. *Hepatology* 2002;35(5):1153–63.
56. Laurent-Puig P, Legoix P, Bluteau O, Belghiti J, Franco D, Binot F, et al. Genetic alterations associated with hepatocellular carcinomas define distinct pathways of hepatocarcinogenesis. *Gastroenterology* 2001;120(7):1763–73.
57. Kojiro M. The evolution of pathologic features of hepatocellular carcinoma. In: Tabor E (Editor). *Viruses in Liver Cancer*. New York: Elsevier Science BV; pp 113–122, 2002.
58. Hanahan D, Weinberg RA. The hallmarks of cancer. *Cell* 2000;100(1):57–70.
59. Ye QH, Qin LX, Forgues M, He P, Kim JW, Peng AC, et al. Predicting hepatitis B virus-positive metastatic hepatocellular carcinomas using gene expression profiling and supervised machine learning. *Nat Med* 2003;9(4):416–23.
60. Bernard PS, Wittwer CT. Real-time PCR technology for cancer diagnostics. *Clin Chem* 2002;48(8):1178–85.
61. Alizadeh AA, Eisen MB, Davis RE, Ma C, Lossos IS, Rosenwald A, et al. Distinct types of diffuse large B-cell lymphoma identified by gene expression profiling. *Nature* 2000;403(6769):503–11.
62. Bittner M, Meltzer P, Chen Y, Jiang Y, Seftor E, Hendrix M, et al. Molecular classification of cutaneous malignant melanoma by gene expression profiling. *Nature* 2000;406(6795):536–40.
63. Dhanasekaran SM, Barrette TR, Ghosh D, Shah R, Varambally S, Kurachi K, et al. Delineation of prognostic biomarkers in prostate cancer. *Nature* 2001;412(6849):822–6.
64. Perou CM, Sorlie T, Eisen MB, van de Rijn M, Jeffrey SS, Rees CA, et al. Molecular portraits of human breast tumours. *Nature* 2000;406(6797):747–52.
65. van de Vijver MJ, He YD, van't Veer LJ, Dai H, Hart AA, Voskuil DW, et al. A gene-expression signature as a predictor of survival in breast cancer. *N Engl J Med* 2002;347(25):1999–2009.
66. Yoshikawa H, Matsubara K, Qian GS, Jackson P, Groopman JD, Manning JE, et al. SOCS-1, a negative regulator of the JAK/STAT pathway, is silenced by methylation in human hepatocellular carcinoma and shows growth-suppression activity. *Nat Genet* 2001;28(1):29–35.
67. Song BC, Chung YH, Kim JA, Choi WB, Suh DD, Pyo SI, et al. Transforming growth factor-beta1 as a useful serologic marker of small hepatocellular carcinoma. *Cancer* 2002;94(1):175–80.
68. Guo XZ, Friess H, Di Mola FF, Heinicke JM, Abou-Shady M, Graber HU, et al. KAI1, a new metastasis suppressor gene, is reduced in metastatic hepatocellular carcinoma. *Hepatology* 1998;28(6):1481–8.
69. De Souza AT, Hankins GR, Washington MK, Orton TC, Jirtle RL. M6P/IGF2R gene is mutated in human hepatocellular carcinomas with loss of heterozygosity. *Nat Genet* 1995;11(4):447–9.
70. Baek MJ, Piao Z, Kim NJ, Park C, Shin EC, Park JH, et al. p16 is a major inactivation target in hepatocellular carcinoma. *Cancer* 2000;89:60–68.

71. Armengol C, Boix L, Bachs O, Sole M, Fuster J, Sala M, et al. p27(Kip1) is an independent predictor of recurrence after surgical resection in patients with small hepatocellular carcinoma. *J Hepatol* 2003;38(5):591–7.
72. Boix L, Bruix J, Campo E, Sole M, Castells A, Fuster J, et al. nm23-H1 expression and disease recurrence after surgical resection of small hepatocellular carcinoma. *Gastroenterology* 1994;107(2): 486–91.
73. Capurro M, Wanless IR, Sherman M, Deboer G, Shi W, Miyoshi E, et al. Glypican-3: a novel serum and histochemical marker for hepatocellular carcinoma. *Gastroenterology* 2003;125(1):89–97.
74. Chuma M, Sakamoto M, Yamazaki K, Ohta T, Ohki M, Asaka M, et al. Expression profiling in multistage hepatocarcinogenesis: identification of HSP70 as a molecular marker of early hepatocellular carcinoma. *Hepatology* 2003;37(1):198–207.
75. Farazi PA, Glickman J, Jiang S, Yu A, Rudolph KL, DePinho RA. Differential impact of telomere dysfunction on initiation and progression of hepatocellular carcinoma. *Cancer Res* 2003;63(16):5021–7.
76. Iyoda K, Sasaki Y, Horimoto M, Toyama T, Yakushijin T, Sakakibara M, et al. Involvement of the p38 mitogen-activated protein kinase cascade in hepatocellular carcinoma. *Cancer* 2003;97(12):3017–26.
77. Okabe H, Satoh S, Furukawa Y, Kato T, Hasegawa S, Nakajima Y, et al. Involvement of PEG10 in human hepatocellular carcinogenesis through interaction with SIAH1. *Cancer Res* 2003;63(12):3043–8.
78. Sung YK, Hwang SY, Park MK, Farooq M, Han IS, Bae HI, et al. Glypican-3 is overexpressed in human hepatocellular carcinoma. *Cancer Sci* 2003;94(3):259–62.
79. Wang Y, Wu MC, Sham JS, Zhang W, Wu WQ, Guan XY. Prognostic significance of c-myc and AIB1 amplification in hepatocellular carcinoma. A broad survey using high-throughput tissue microarray. *Cancer* 2002;95(11):2346–52.
80. Xu XR, Huang J, Xu ZG, Qian BZ, Zhu ZD, Yan Q, et al. Insight into hepatocellular carcinogenesis at transcriptome level by comparing gene expression profiles of hepatocellular carcinoma with those of corresponding noncancerous liver. *Proc Natl Acad Sci USA* 2001;98(26):15089–94.
81. Smith MW, Yue ZN, Geiss GK, Sadovnikova NY, Carter VS, Boix L, et al. Identification of novel tumor markers in hepatitis C virus-associated hepatocellular carcinoma. *Cancer Res* 2003;63(4):859–64.
82. Smith MW, Yue ZN, Korth MJ, Do HA, Boix L, Fausto N, et al. Hepatitis C virus and liver disease: global transcriptional profiling and identification of potential markers. *Hepatology* 2003;38(6):1458–67.
83. Shirota Y, Kaneko S, Honda M, Kawai HF, Kobayashi K. Identification of differentially expressed genes in hepatocellular carcinoma with cDNA microarrays. *Hepatology* 2001;33:832–840.
84. Cheung ST, Chen X, Guan XY, Wong SY, Tai LS, Ng IO, et al. Identify metastasis-associated genes in hepatocellular carcinoma through clonality delineation for multinodular tumor. *Cancer Res* 2002;62(16):4711–4721.
85. Wurmbach E, Chen Y, Khitrov G, Zhang W, Roayaie S, Schwartz M, et al. Genomewide molecular profiles of HCV-induced dysplasia and hepatocellular carcinoma. *Hematology* 2007;45(4):938–947.
86. Lin CH, Hsieh SY, Sheen IS, Lee WC, Chen TC, Shyu WC, et al. Genome-wide hypomethylation in hepatocellular carcinogenesis. *Cancer Res* 2001;61(10):4238–4243.
87. Tal-Kremer S, Reeves H, Narla G, Thung SN, Schwartz M, Difeo A, et al. Frequent inactivation of the tumor suppressor Kruppel-like factor 6 (KLF6) in hepatocellular carcinoma. *Hepatology* 2004;40:1047–1052.
88. Moinzadeh P, Breuhahn K, Stutzer H, Schirmacher P. Chromosome alterations in human hepatocellular carcinomas correlate with aetiology and histologal grade – results of an explorative CGH meta-analysis. *Br J Cancer* 2005;92:935–41.
89. Lemmer ER, Friedman SL, Llovet JM. Molecular diagnosis of chronic liver disease and hepatocellular carcinoma: the potential of gene expression profiling. *Semin Liver Dis* 2006;26(4):373–834.
90. Ieta K, Ojima E, Tanaka F, Nakamura Y, Haraguchi N, Mimori K, et al. identification of overexpressed genes in hepatocellular carcinoma, with special reference to ubiquitin-conjugating enzyme E2C gene expression. *Int J Cancer* 2007;121(1):33–38.
91. Locker J. A new way to look at liver cancer. *Hepatology* 2004;40(3):521–523.
92. Thorgeirsson SS. Hunting for tumor suppressor genes in liver cancer. *Hepatology* 2003;37(4): 739–741.

93. Grisham JW. Molecular genetic alterations in primary hepatocellular neoplasms: hepatocellular adenoma, hepatocellular carcinoma, and hepatoblastoma. In: Coleman WB, Tsongalis GJ (Editors). *The Molecular Basis of Human Cancer*. Totowa, NJ: Humana Press Inc; pp 269–346, 2001.

94. Suriawinata A, Xu R. An update on the molecular genetics of hepatocellular carcinoma. *Semin Liver Dis* 2004;24(1):77–88.

95. Villa E, Ferretti I, Grottola A, Buttafoco P, Buono MG, Giannini F, et al. Hormonal therapy with megestrol in inoperable hepatocellular carcinoma characterized by variant oestrogen receptors. *Br J Cancer* 2001;84(7):881–5.

96. Kim KW, Bae SK, Lee OH, Bae MH, Lee MJ, Park BC. Insulin-like growth factor II induced by hypoxia may contribute to angiogenesis of human hepatocellular carcinoma. *Cancer Res* 1998;58(2):348–351.

97. Bae MH, Lee MJ, Bae SK, Lee OH, Lee YM, Park BC, et al. Insulin-like growth factor II (IGF-II) secreted from HepG2 human hepatocellular carcinoma cells shows angiogenic activity. *Cancer Lett* 1998;128(1):41–6.

98. El-Assal ON, Yamanoi A, Soda Y, Yamaguchi M, Igarashi M, Yamamoto A, et al. Clinical significance of microvessel density and vascular endothelial growth factor expression in hepatocellular carcinoma and surrounding liver: possible involvement of vascular endothelial growth factor in the angiogenesis of cirrhotic liver. *Hepatology* 1998;27(6):1554–1562.

99. Miura H, Miyazaki T, Kuroda M, Oka T, Machinami R, Kodama T, et al. Increased expression of vascular endothelial growth factor in human hepatocellular carcinoma. *J Hepatol* 1997;27(5): 854–861.

100. Shimoda K, Mori M, Shibuta K, Banner BF, Barnard GF. Vascular endothelial growth factor/vascular permeability factor mRNA expression in patients with chronic hepatitis C and hepatocellular carcinoma. *Int J Oncol* 1999;14(2):353–359.

101. Torimura T, Sata M, Ueno T, Kin M, Tsuji R, Suzaku K, et al. Increased expression of vascular endothelial growth factor is associated with tumor progression in hepatocellular carcinoma. *Hum Pathol* 1998;29(9):986–991.

102. Li XM, Tang ZY, Zhou G, Lui YK, Ye SL. Significance of vascular endothelial growth factor mRNA expression in invasion and metastasis of hepatocellular carcinoma. *J Exp Clin Cancer Res* 1998;17(1):13–17.

103. Chow NH, Hsu PI, Lin XZ, Yang HB, Chan SH, Cheng KS, et al. Expression of vascular endothelial growth factor in normal liver and hepatocellular carcinoma: an immunohistochemical study. *Hum Pathol* 1997;28(6):698–703.

104. Yoshiji H, Kuriyama S, Hicklin DJ, Huber J, Yoshii J, Miyamoto Y, et al. KDR/Flk-1 is a major regulator of vascular endothelial growth factor induced tumor development and angiogenesis in murine hepatocellular carcinoma cells. *Hepatology* 1999;30:1179–86.

105. Yoshiji H, Kuriyama S, Yoshii J, Ikenaka Y, Noguchi R, Hicklin DJ, et al. Halting the interaction between vascular endothelial growth factor and its receptors attenuates liver carcinogenesis in mice. *Hepatology* 2004;39(6):1517–24.

106. Graepler F, Verbeek B, Graeter T, Smirnow I, Kong HL, Schuppan D, et al. Combined endostatin/sFlt-1 antiangiogenic gene therapy is highly effective in a rat model of HCC. *Hepatology* 2005;41(4):879–86.

107. Hurwitz H, Fehrenbacher L, Novotny W, Cartwright T, Hainsworth J, Heim W, et al. Bevacizumab plus irinotecan, fluorouracil, and leucovorin for metastatic colorectal cancer. *N Engl J Med* 2004;350(23):2335–42.

108. Yang JC, Haworth L, Sherry RM, Hwu P, Schwartzentruber DJ, Topalian SL, et al. A randomized trial of bevacizumab, an anti-vascular endothelial growth factor antibody, for metastatic renal cancer. *N Engl J Med* 2003;349(5):427–34.

109. Willett CG, Boucher Y, di Tomaso E, Duda DG, Munn LL, Tong RT, et al. Direct evidence that the VEGF-specific antibody bevacizumab has antivascular effects in human rectal cancer. *Nat Med* 2004;10(2):145–147.

110. Abe J, Zhou W, Takuwa N, Taguchi J, Kurokawa K, Kumada M, et al. A fumagillin derivative angiogenesis inhibitor, AGM-1470, inhibits activation of cyclin-dependent kinases and phosphorylation of retinoblastoma gene product but not protein tyrosyl phosphorylation or protooncogene expression in vascular endothelial cells. *Cancer Res* 1994;54:3407–3412.

111. Yeh JR, Mohan R, Crews CM. The antiangiogenic agent TNP-470 requires p53 and p21CIP/WAF for endothelial growth arrest. *Proc Natl Acad Sci USA* 2000;97:12782–12787.

112. Sheen IS, Jeng KS, Jeng WJ, Jeng CJ, Wang YC, Gu SL, et al. Fumagillin treatment of hepatocellular carcinoma in rats: an in vivo study of antiangiogenesis. *World J Gastroenterol* 2005;11(6): 771–7.

113. Sun HC, Tang ZY. Angiogenesis in hepatocellular carcinoma: the retrospectives and perspectives. *J Cancer Res Clin Oncol* 2004;130(6):307–19.

114. Lenz W. Thalidomide and congenital abnormalities. *Lancet* 1962;1:45.

115. McBride WG. Thalidomide and congenital abnormalities. *Lancet* 1961;2:1358.

116. Moreira AL, Sampaio EP, Zmuidzinas A, Findt P, Smith KA, Kaplan G. Thalidomide exerts its inhibitory action on tumor necrosis factor alpha by enhancing mRNA degradation. *J Exp Med* 1993;177:1675–80.

117. Haslett PA, Corral LG, Albert M, Kaplan G. Thalidomide costimulates primary human T-lymphocytes, preferentially inducing proliferation, cytokine production, and cytotoxic responses in the CD8+ subset. *J Exp Med* 1998;187:1885–92.

118. McHugh SM, Rifkin IR, Deighton J. The immunosuppressive drug thalidomide induces T helper cell type 2 (Th2) and concomitantly inhibits Th1 cytokine production in mitogen- and antigen-stimulated human peripheral blood mononuclear cell cultures. *Clin Exp Immunol* 1995;99:160–67.

119. Greitz H, Handt S, Zwingenberger K. Thalidomide selectively modulates the density of cell surface molecules involved in the adhesion cascade. *Immunopharmacol* 1996;31:213–21.

120. Bauer KS, Dixon SC, Figg WD. Inhibition of angiogenesis by thalidomide requires metabolic activation, which is species-dependent. *Biochem Pharmacol* 1998;55:1827–1834.

121. Kruse FE, Joussen AM, Rohrschneider K, Becker MD, Volcker HE. Thalidomide inhibits corneal angiogenesis induced by vascular endothelial growth factor. *Graefes Arch Clin Exp Opthalmol* 1998;236:461–66.

122. D'Amato RJ, Loughnan MS, Flynn E, Folkman J. Thalidomide is an inhibitor of angiogenesis. *Proc Natl Acad Sci USA* 1994;91(9):4082–4085.

123. Weber D, Rankin K, Gavino M, Delasalle K, Alexanian R. Thalidomide alone or with dexamethasone for previously untreated multiple myeloma. *J Clin Oncol* 2003;21:16–19.

124. Dimopoulos MA, Zomas A, Viniou NA, Grigoraki V, Galani E, Matsouka C, et al. Treatment of Waldenstrom's macroglobulinemia with thalidomide. *J Clin Oncol* 2001;19:3596–601.

125. Raza A, Meyer P, Dutt D, Zorat F, Lisak L, Nascimben F, et al. Thalidomide produces transfusion independence in long-standing refractory anemias of patients with myelodysplastic syndromes. *Blood* 2001;98:958–65.

126. Fine H, Figg WD, Jaeckle K, Wen PY, Kyritsis AP, Loeffler JS, et al. Phase II trial of the antiangiogenic agent thalidomide in patients with recurrent high-grade gliomas. *J Clin Oncol* 2000;18:708–715.

127. Daliani DD, Papandreou CN, Thall PF, Wang X, Perez C, Oliva R, et al. A pilot study of thalidomide in patients with progressive metastatic renal cell carcinoma. *Cancer* 2002;95:758–65.

128. Little RF, Wyvill KM, Pluda JM, Welles L, Marshall V, Figg WD, et al. Activity of thalidomide in AIDS-related Kaposi's sarcoma. *J Clin Oncol* 2000;18:2593–602.

129. Patt YZ, Hassan MM, Lozano RD, Ellis LM, Petersen JA, Waugh K. Durable clinical response of refractory hepatocellular carcinoma to orally administered thalidomide. *Am J Clin Oncol* 2000;23:319–21.

130. Wang T-E, Kao C-R, Lin S-C, Chang W-H, Chu C-H, Lin J, et al. Salvage therapy for hepatocellular carcinoma with thalidomide. *World J Gastroenterol* 2004;10:649–53.

131. Hsu C, Chen C-N, Chen L-T, Wu C-Y, Yang P-M, Lai M-Y, et al. Low-dose thalidomide treatment for advanced hepatocellular carcinoma. *Oncology* 2003;65:242–9.

132. Lin AY, Brophy N, Fisher GA, So S, Biggs C, Yock T, et al. A phase II study of thalidomide in patients with unresectable hepatocellular carcinoma. *Cancer* 2005;103:119–125.

133. Patt YZ, Hassan MM, Lozano RD, Nooka AK, Schnirer II, Zeldis JB, et al. Thalidomide in the treatment of patients with hepatocellular carcinoma: a phase II trial. *Cancer* 2005;103(4):749–55.

134. Schwartz JD, Sung M, Schwartz M, Lehrer D, Mandeli J, Liebes L, et al. Thalidomide in advanced hepatocellular carcinoma with optional interferon alpha-2a upon progression. *The Oncologist* 2005;10(9):718–27.

135. Dredge K, Marriott JB, Macdonald CD, Man HW, Chen R, Muller GW, et al. Novel thalidomide analogues display anti-angiogenic activity independently of immunomodulatory effects. *Br J Cancer* 2002;87(10):1166–72.

136. Hsu C, Chen CN, Chen LT, Wu CY, Hsieh FJ, Cheng AL. Effect of thalidomide in hepatocellular carcinoma: assessment with power doppler US and analysis of circulating angiogenic factors. *Radiology* 2005;235(2):509–16.

137. Yoshiji H, Kuriyama S, Fukui H. Angiotensin-I-converting enzyme inhibitors may be an alternative anti-angiogenic strategy in the treatment of liver fibrosis and hepatocellular carcinoma. Possible role of vascular endothelial growth factor. *Tumour Biol* 2002;23(6):348–56.

138. Noguchi R, Yoshiji H, Kuriyama S, Yoshii J, Ikenaka Y, Yanase K, et al. Combination of interferon-beta and the angiotensin-converting enzyme inhibitor, perindopril, attenuates murine hepatocellular carcinoma development and angiogenesis. *Clin Cancer Res* 2003;9(16 Pt 1):6038–45.

139. Semela D, Dufour JF. Angiogenesis and hepatocellular carcinoma. *J Hepatol* 2004;41(5):864–80.

140. Semenza GL. Surviving ischemia: Adaptive responses mediated by hypoxia-inducible factor 1. *J Clin Invest* 2000;106:809–812.

141. Semenza GL. HIF-1 and tumor progression: pathophysiology and therapeutics. *Trends Mol Med* 2002;8:S62–S67.

142. Lee JS, Chu IS, Heo J, Calvisi DF, Sun Z, Roskams T, et al. Classification and prediction of survival in hepatocellular carcinoma by gene expression profiling. *Hepatology* 2004;40(3):667–76.

143. Bruick RK, McKnight SL. A conserved family of prolyl-4-hydroxylases that modify HIF. *Science* 2001;294:1337–1340.

144. Yu JL, Rak JW, Coomber BL, Hicklin DJ, Kerbel RS. Effect of p53 status on tumor response to antiangiogenic therapy. *Science* 2002;295:1526–1528.

145. Giaccia A, Siim BG, Johnson RS. HIF-1 as a target for drug development. *Nat Rev Drug Discov* 2003;2:803–811.

146. Ramanathan RK, Belani CP, Friedland D, Ramalingam S, Agarwala SS, Ivy P, et al. Phase I study (twice weekly schedule) of 17-allylamino-17 demethoxygeldanamycin (17AAG NSC-704057) in patients with advanced refractory tumors. *Proc Am Soc Clin Oncol* 2005;23(16S):204s.

147. Solit DB, Egorin M, Kopil C, Delacruz A, Shaffer D, Slovin S, et al. Phase 1 pharmacokinetic and pharmacodynamic trial of docetaxel and 17AAG (17-allylamino-17-demethoxygeldanamycin). *Proc Am Soc Clin Oncol* 2005;23(16S):204s.

148. Mabjeesh NJ, Escuin D, LaVallee TM, Pribluda VS, Swartz GM, Johnson MS, et al. 2ME2 inhibits tumor growth and angiogenesis by disrupting microtubules and dysregulating HIF. *Cancer Cell* 2003;3(4):363–75.

149. Rapisarda A, Uranchimeg B, Scudiero DA, Selby M, Sausville EA, Shoemaker RH, et al. Identification of small molecule inhibitors of hypoxia-inducible factor 1 transcriptional activation pathway. *Cancer Res* 2002;62(15):4316–24.

150. Semenza GL. Targeting HIF-1 for cancer therapy. *Nat Rev Cancer* 2003;3(10):721–32.

151. Grigioni WF, Garbisa S, D'Errico A, Baccarini P, Stetler-Stevenson WG, Liotta LA, et al. Evaluation of hepatocellular carcinoma aggressiveness by a panel of extracellular matrix antigens. *Am J Pathol* 1991;138(3):647–54.

152. Arii S, Mise M, Harada T, Furutani M, Ishigami S, Niwano M, et al. Overexpression of matrix metalloproteinase 9 gene in hepatocellular carcinoma with invasive potential. *Hepatology* 1996;24(2):316–22.

153. Okazaki I, Wada N, Nakano M, Saito A, Takasaki K, Doi M, et al. Difference in gene expression for matrix metalloproteinase-1 between early and advanced hepatocellular carcinomas. *Hepatology* 1997;25(3):580–4.

154. Monvoisin A, Bisson C, Si-Tayeb K, Balabaud C, Desmouliere A, Rosenbaum J. Involvement of matrix metalloproteinase type-3 in hepatocyte growth factor-induced invasion of human hepatocellular carcinoma cells. *Int J Cancer* 2002;97(2):157–62.

155. Martin DC, Sanchez-Sweatman OH, Ho AT, Inderdeo DS, Tsao MS, Khokha R. Transgenic TIMP-1 inhibits simian virus 40 T antigen-induced hepatocarcinogenesis by impairment of hepatocellular proliferation and tumor angiogenesis. *Lab Invest* 1999;79(2):225–34.

156. Kim JH, Kim TH, Jang JW, Jang YJ, Lee KH, Lee ST. Analysis of matrix metalloproteinase mRNAs expressed in hepatocellular carcinoma cell lines. *Mol Cells* 2001;12(1):32–40.

157. Giannelli G, Bergamini C, Fransvea E, Marinosci F, Quaranta V, Antonaci S. Human hepatocellular carcinoma (HCC) cells require both alpha3beta1 integrin and matrix metalloproteinases activity for migration and invasion. *Lab Invest* 2001;81(4):613–27.

158. Bergers G, Brekken R, McMahon G, Vu TH, Itoh T, Tamaki K, et al. Matrix metalloproteinase-9 triggers the angiogenic switch during carcinogenesis. *Nat Cell Biol* 2000;2(10):737–44.

159. Moore MJ, Hamm J, Dancey J, Eisenberg PD, Dagenais M, Fields A, et al. Comparison of gemcitabine versus the matrix metalloproteinase inhibitor BAY 12–9566 in patients with advanced or metastatic adenocarcinoma of the pancreas: a phase III trial of the National Cancer Institute of Canada Clinical Trials Group. *J Clin Oncol* 2003;21(17):3296–302.

160. Bramhall SR, Schulz J, Nemunaitis J, Brown PD, Baillet M, Buckels JA. A double-blind placebo-controlled, randomised study comparing gemcitabine and marimastat with gemcitabine and placebo as first line therapy in patients with advanced pancreatic cancer. *Br J Cancer* 2002;87(2):161–7.

161. Ha KT, Kim JK, Lee YC, Kim CH. Inhibitory effect of Daesungki-Tang on the invasiveness potential of hepatocellular carcinoma through inhibition of matrix metalloproteinase-2 and -9 activities. *Toxicol Appl Pharmacol* 2004;200(1):1–6.

162. Chung TW, Moon SK, Chang YC, Ko JH, Lee YC, Cho G, et al. Novel and therapeutic effect of caffeic acid and caffeic acid phenyl ester on hepatocarcinoma cells: complete regression of hepatoma growth and metastasis by dual mechanism. FASEB J 2004;18(14):1670–81.

163. Bu W, Tang ZY, Sun FX, Ye SL, Liu KD, Xue Q, et al. Effects of matrix metalloproteinase inhibitor BB-94 on liver cancer growth and metastasis in a patient-like orthotopic model LCI-D20. *Hepatogastroenterology* 1998;45(22):1056–61.

164. Couvelard A, Bringuier AF, Dauge MC, Nejjari M, Darai E, Benifla JL, et al. Expression of integrins during liver organogenesis in humans. *Hepatology* 1998;27(3):839–47.

165. Patriarca C, Roncalli M, Gambacorta M, Cominotti M, Coggi G, Viale G. Patterns of integrin common chain beta 1 and collagen IV immunoreactivity in hepatocellular carcinoma. Correlations with tumour growth rate, grade and size. *J Pathol* 1993;171(1):5–11.

166. Volpes R, van den Oord JJ, Desmet VJ. Integrins as differential cell lineage markers of primary liver tumors. *Am J Pathol* 1993;142(5):1483–92.

167. Scoazec JY, Flejou JF, D'Errico A, Fiorentino M, Zamparelli A, Bringuier AF, et al. Fibrolamellar carcinoma of the liver: composition of the extracellular matrix and expression of cell-matrix and cell-cell adhesion molecules. *Hepatology* 1996;24(5):1128–36.

168. Le Bail B, Faouzi S, Boussarie L, Balabaud C, Bioulac-Sage P, Rosenbaum J. Extracellular matrix composition and integrin expression in early hepatocarcinogenesis in human cirrhotic liver. *J Pathol* 1997;181(3):330–7.

169. Begum NA, Mori M, Matsumata T, Takenaka K, Sugimachi K, Barnard GF. Differential display and integrin alpha 6 messenger RNA overexpression in hepatocellular carcinoma. *Hepatology* 1995;22(5):1447–55.

170. Mita MM, Mita AC, Goldston M, Chu QS, Tolcher AW, Ricart A, et al. Pharmacokinetics (PK) and pharmacodynamics (PD) of E7820 - an oral sulfonamide with novel alpha-2 integrin mediated antiangiogenic properties: Results of a phase I study. *Proc Am Soc Clin Oncol* 2005;23(16S): 212s.

171. Jayson GC, Mullamitha S, Ton C, Valle J, Jackson A, Julyan P, et al. Phase I study of CNTO 95, a fully human monoclonal antibody (mAb) to alpha-v integrins, in patients with solid tumors. *Proc Am Soc Clin Oncol* 2005;23(16s):220s.

172. Nejjari M, Hafdi Z, Dumortier J, Bringuier AF, Feldmann G, Scoazec JY. alpha6beta1 integrin expression in hepatocarcinoma cells: regulation and role in cell adhesion and migration. *Int J Cancer* 1999;83(4):518–25.

173. Yang C, Zeisberg M, Lively JC, Nyberg P, Afdhal N, Kalluri R. Integrin alpha1beta1 and alpha2beta1 are the key regulators of hepatocarcinoma cell invasion across the fibrotic matrix microenvironment. *Cancer Res* 2003;63(23):8312–7.

174. Jonker DJ, Stewart DJ, Goel R, Avruch L, Goss G, Maroun J, et al. A phase I study of the novel molecularly targeted vascular targeting agent Exherin (ADH-1), shows activity in some patients with refractory solid tumors stratified according to N-cadherin expression. *Proc Am Soc Clin Oncol* 2005;23(16s):201s.

175. Kiss A, Wang NJ, Xie JP, Thorgeirsson SS. Analysis of transforming growth factor (TGF)-alpha/epidermal growth factor receptor, hepatocyte growth Factor/c-met, TGF-beta receptor type II, and p53 expression in human hepatocellular carcinomas. *Clin Cancer Res* 1997;3(7): 1059–66.

176. Mise M, Arii S, Higashituji H, Furutani M, Niwano M, Harada T, et al. Clinical significance of vascular endothelial growth factor and basic fibroblast growth factor gene expression in liver tumor. *Hepatology* 1996;23(3):455–464.

177. Lee GH, Merlino G, Fausto N. Development of liver tumors in transforming growth factor alpha transgenic mice. *Cancer Res* 1992;52(19):5162–5170.

178. Laird AD, Brown PI, Fausto N. Inhibition of tumor growth in liver epithelial cells transfected with a transforming growth factor alpha antisense gene. *Cancer Res* 1994;54(15):4224–4232.

179. Jhappan C, Stahle C, Harkins RN, Fausto N, Smith GH, Merlino GT. TGF alpha overexpression in transgenic mice induces liver neoplasia and abnormal development of the mammary gland and pancreas. *Cell* 1990;61(6):1137–1146.

180. Seki S, Sakai Y, Kitada T, Kawakita N, Yanai A, Tsutsui H, et al. Induction of apoptosis in a human hepatocellular carcinoma cell line by a neutralizing antibody to transforming growth factor-alpha. *Virchows Arch* 1997;430(1):29–35.

181. Jo M, Stolz DB, Esplen JE, Dorko K, Michalopoulos GK, Strom SC. Cross-talk between epidermal growth factor receptor and c-Met signal pathways in transformed cells. *J Biol Chem* 2000;275(12):8806–11.

182. Schiffer E, Housset C, Cacheux W, Wendum D, Desbois-Mouthon C, Rey C, et al. Gefitinib, an EGFR inhibitor, prevents hepatocellular carcinoma development in the rat liver with cirrhosis. *Hepatology* 2005;41(2):307–14.

183. Hopfner M, Sutter AP, Huether A, Schuppan D, Zeitz M, Scherubl H. Targeting the epidermal growth factor receptor by gefitinib for treatment of hepatocellular carcinoma. *J Hepatol* 2004;41(6):1008–16.

184. Matsuo M, Sakurai H, Saiki I. ZD1839, a selective epidermal growth factor receptor tyrosine kinase inhibitor, shows antimetastatic activity using a hepatocellular carcinoma model. *Mol Cancer Ther* 2003;2(6):557–61.

185. Ueno Y, Sakurai H, Matsuo M, Choo MK, Koizumi K, Saiki I. Selective inhibition of TNF-alpha-induced activation of mitogen-activated protein kinases and metastatic activities by gefitinib. *Br J Cancer* 2005;92(9):1690–5.

186. Zhu BD, Yuan SJ, Zhao QC, Li X, Li Y, Lu QY. Antitumor effect of Gefitinib, an epidermal growth factor receptor tyrosine kinase inhibitor, combined with cytotoxic agent on murine hepatocellular carcinoma. *World J Gastroenterol* 2005;11(9):1382–6.

187. Philip PA, Mahoney M, Thomas J, Pitot H, Donehower R, Kim G, et al. Phase II Trial of erlotinib (OSI-774) in patients with hepatocellular or biliary cancer. *Proc Am Soc Clin Onc* 2004;22:319s.

188. Thomas MB, Dutta A, Brown T, Charnsangavej C, Rashid A, Hoff PM, et al. A phase II open-label study of OSI-774 (NSC 718781) in unresectable hepatocellular carcinoma. *Proc Am Soc Clin Onc* 2005;23:317s.

189. Ellouk-Achard S, Djenabi S, De Oliveira GA, Desauty G, Duc HT, Zohair M, et al. Induction of apoptosis in rat hepatocarcinoma cells by expression of IGF-I antisense c-DNA. *J Hepatol* 1998;29(5):807–18.

190. Yao X, Hu JF, Daniels M, Shiran H, Zhou X, Yan H, et al. A methylated oligonucleotide inhibits IGF2 expression and enhances survival in a model of hepatocellular carcinoma. *J Clin Invest* 2003;111(2):265–273.

191. Yang JM, Chen WS, Liu ZP, Luo YH, Liu WW. Effects of insulin-like growth factors-IR and -IIR antisense gene transfection on the biological behaviors of SMMC-7721 human hepatoma cells. *J Gastroenterol Hepatol* 2003;18(3):296–301.

192. Di Cosimo S, Seoane J, Guzman M, Rojo F, Jimenez J, Anido J, et al. Combination of the mammalian target of rapamycin (mTOR) inhibitor everolimus (E) with the insulin like growth factor-1-receptor (IGF-1-R) inhibitor NVP-AEW-541: a mechanistic based anti-tumor strategy. *Proc Am Soc Clin Onc* 2005;23:219s.

193. Gualberto A, Alsina M, Lacy M, Poutney S, Birgin A, Littman B, et al. Inhibition of the insulin like growth factor 1 receptor by a specific monoclonal antibody in multiple myeloma. *Am Soc Clin Oncol Abstract Catalog* 2005;23:203s.

194. Maret A, Galy B, Arnaud E, Bayard F, Prats H. Inhibition of fibroblast growth factor 2 expression by antisense RNA induced a loss of the transformed phenotype in a human hepatoma cell line. *Cancer Res* 1995;55(21):5075–5079.

195. Shao RX, Otsuka M, Kato N, Taniguchi H, Hoshida Y, Moriyama M, et al. Acyclic retinoid inhibits human hepatoma cell growth by suppressing fibroblast growth factor-mediated signaling pathways. *Gastroenterology* 2005;128(1):86–95.

196. Miura D, Miura Y, Yagasaki K. Resveratrol inhibits hepatoma cell invasion by suppressing gene expression of hepatocyte growth factor via its reactive oxygen species-scavenging property. *Clin Exp Metastasis* 2004;21(5):445–51.

197. Heideman DA, Overmeer RM, van Beusechem VW, Lamers WH, Hakvoort TB, Snijders PJ, et al. Inhibition of angiogenesis and HGF-cMET-elicited malignant processes in human hepatocellular carcinoma cells using adenoviral vector-mediated NK4 gene therapy. *Cancer Gene Ther* 2005;12(12):954–62.

198. Christensen JG, Schreck R, Burrows J, Kuruganti P, Chan E, Le P, et al. A selective small molecule inhibitor of c-Met kinase inhibits c-Met-dependent phenotypes in vitro and exhibits cytoreductive antitumor activity in vivo. *Cancer Res* 2003;63(21):7345–55.

199. Wang SY, Chen B, Zhan YQ, Xu WX, Li CY, Yang RF, et al. SU5416 is a potent inhibitor of hepatocyte growth factor receptor (c-Met) and blocks HGF-induced invasiveness of human HepG2 hepatoma cells. *J Hepatol* 2004;41(2):267–73.

200. Adjei AA. Blocking oncogenic Ras signaling for cancer therapy. *J Natl Cancer Inst* 2001; 93(14):1062–74.

201. Huynh H, Nguyen TT, Chow KH, Tan PH, Soo KC, Tran E. Over-expression of the mitogen-activated protein kinase (MAPK) kinase (MEK)-MAPK in hepatocellular carcinoma: its role in tumor progression and apoptosis. *BMC Gastroenterol* 2003;3(1):19.

202. Jagirdar J, Nonomura A, Patil J, Thor A, Paronetto F. Ras oncogene p21 expression in hepatocellular carcinoma. *J Exp Pathol* 1989;4(1):37–46.

203. Nonomura A, Ohta G, Hayashi M, Izumi R, Watanabe K, Takayanagi N, et al. Immunohistochemical detection of ras oncogene p21 product in liver cirrhosis and hepatocellular carcinoma. *Am J Gastroenterol* 1987;82(6):512–8.

204. Tiniakos D, Spandidos DA, Yiagnisis M, Tiniakos G. Expression of ras and c-myc oncoproteins and hepatitis B surface antigen in human liver disease. *Hepatogastroenterology* 1993;40(1): 37–40.

205. Schmidt CM, McKillop IH, Cahill PA, Sitzmann JV. Increased MAPK expression and activity in primary human hepatocellular carcinoma. *Biochem Biophys Res Commun* 1997;236(1):54–8.

206. Ito Y, Sasaki Y, Horimoto M, Wada S, Tanaka Y, Kasahara A, et al. Activation of mitogen-activated protein kinases/extracellular signal-regulated kinases in human hepatocellular carcinoma. *Hepatology* 1998;27(4):951–8.

207. Tsou AP, Wu KM, Tsen TY, Chi CW, Chiu JH, Lui WY, et al. Parallel hybridization analysis of multiple protein kinase genes: identification of gene expression patterns characteristic of human hepatocellular carcinoma. *Genomics* 1998;50(3):331–40.

208. Zhou JM, Zhu XF, Pan QC, Liao DF, Li ZM, Liu ZC. Manumycin inhibits cell proliferation and the Ras signal transduction pathway in human hepatocellular carcinoma cells. *Int J Mol Med* 2003;11(6):767–71.

209. Wiesenauer CA, Yip-Schneider MT, Wang Y, Schmidt CM. Multiple anticancer effects of blocking MEK-ERK signaling in hepatocellular carcinoma. *J Am Coll Surg* 2004;198(3):410–21.

210. Tsukada Y, Miyazawa K, Kitamura N. High intensity ERK signal mediates hepatocyte growth factor-induced proliferation inhibition of the human hepatocellular carcinoma cell line HepG2. *J Biol Chem* 2001;276(44):40968–76.

211. Schmidt CM, Wang Y, Wiesenauer C. Novel combination of cyclooxygenase-2 and MEK inhibitors in human hepatocellular carcinoma provides a synergistic increase in apoptosis. *J Gastrointest Surg* 2003;7(8):1024–33.

212. Cheng J, Imanishi H, Liu W, Nakamura H, Morisaki T, Higashino K, et al. Involvement of cell cycle regulatory proteins and MAP kinase signaling pathway in growth inhibition and cell cycle arrest by a selective cyclooxygenase 2 inhibitor, etodolac, in human hepatocellular carcinoma cell lines. *Cancer Sci* 2004;95(8):666–73.

213. Phase I/II study of epirubicin and celecoxib in patients with hepatocellular carcinoma (NU-0216); 285669. In: DeVita VT (Editor). *Current Clinical Trials in Oncology: National Cancer Institute PDQ*. Yardley, PA: MediMedia; p I-46, 2005.

214. Strumberg D, Richly H, Hilger RA, Schleucher N, Korfee S, Tewes M, et al. Phase I clinical and pharmacokinetic study of the Novel Raf kinase and vascular endothelial growth factor receptor inhibitor BAY 43–9006 in patients with advanced refractory solid tumors. *J Clin Oncol* 2005;23(5):965–72.

215. Lorusso P, Krishnamurthi S, Rinehart JR, Nabell L, Croghan G, Varterasian M, et al. A phase 1–2 clinical study of a second generation oral MEK inhibitor, PD 0325901 in patients with advanced cancer. *Proc Am Soc Clin Onc* 2005;23:194s.

216. Horie Y, Suzuki A, Kataoka E, Sasaki T, Hamada K, Sasaki J, et al. Hepatocyte-specific Pten deficiency results in steatohepatitis and hepatocellular carcinomas. *J Clin Invest* 2004;113(12): 1774–1783.

217. Sahin F, Kannangai R, Adegbola O, Wang J, Su G, Torbenson M. mTOR and P70 S6 kinase expression in primary liver neoplasms. *Clin Cancer Res* 2004;10(24):8421–8425.

218. Yu K, Toral-Barza L, Discafani C, Zhang WG, Skotnicki J, Frost P, et al. mTOR, a novel target in breast cancer: the effect of CCI-779, an mTOR inhibitor, in preclinical models of breast cancer. *Endocr Relat Cancer* 2001;8(3):249–58.

219. Tabernero J, Rojo F, Burris H, Casado E, Macarulla T, Jones S, et al. A phase I study with tumor molecular pharmacodynamic (MPD) evaluation of dose and schedule of the oral mTOR-inhibitor Everolimus (RAD001) in patients (pts) with advanced solid tumors. *Proc Am Soc Clin Onc* 2005;23:193s.

220. Rivera VM, Kreisberg JI, Mita MM, Goldston M, Knowles HL, Herson J, et al. Pharmacodynamic study of skin biopsy specimens in patients (pts) with refractory or advanced malignancies following administration of AP23573, an MTOR inhibitor. *Proc Am Soc Clin Onc* 2005;23:200s.

221. Desai AA, Mita M, Fetterly GJ, Chang C, Netsch M, Knowles HL, et al. Development of a pharmacokinetic (PK) model and assessment of patient (pt) covariate effects on dose-dependent PK following different dosing schedules in two phase I trials of AP 23573 (AP), a mTOR inhibitor. *Proc Am Soc Clin Onc* 2005;23:202s.

222. Duran I, Le, L, Saltman D, Kortmansky J, Kocha W, Singh D, et al. A phase II trial of temsirolimus. *Proc Am Soc Clin Onc* 2005;23:215s.

223. Hsu HC, Jeng YM, Mao TL, Chu JS, Lai PL, Peng SY. Beta-catenin mutations are associated with a subset of low-stage hepatocellular carcinoma negative for hepatitis B virus and with favorable prognosis. *Am J Pathol* 2000;157(3):763–70.

224. Mao TL, Chu JS, Jeng YM, Lai PL, Hsu HC. Expression of mutant nuclear beta-catenin correlates with non-invasive hepatocellular carcinoma, absence of portal vein spread, and good prognosis. *J Pathol* 2001;193(1):95–101.

225. Terris B, Pineau P, Bregeaud L, Valla D, Belghiti J, Tiollais P, et al. Close correlation between beta-catenin gene alterations and nuclear accumulation of the protein in human hepatocellular carcinomas. *Oncogene* 1999;18(47):6583–8.

226. Wong CM, Fan ST, Ng IO. Beta-catenin mutation and overexpression in hepatocellular carcinoma: clinicopathologic and prognostic significance. *Cancer* 2001;92:136–145.

227. Nhieu JT, Renard CA, Wei Y, Cherqui D, Zafrani ES, Buendia MA. Nuclear accumulation of mutated beta-catenin in hepatocellular carcinoma is associated with increased cell proliferation. *Am J Pathol* 1999;155:703–710.

228. Giles RH, van Es JH, Clevers H. Caught up in a Wnt storm: Wnt signaling in cancer. *Biochim Biophys Acta* 2003;1653(1):1–24.

229. You L, He B, Xu Z, Uematsu K, Mazieres J, Fujii N, et al. An anti-Wnt-2 monoclonal antibody induces apoptosis in malignant melanoma cells and inhibits tumor growth. *Cancer Res* 2004;64(15):5385–5389.

230. You L, He B, Xu Z, Uematsu K, Mazieres J, Mikami I, et al. Inhibition of Wnt-2-mediated signaling induces programmed cell death in non-small-cell lung cancer cells. *Oncogene* 2004;23(36):6170–6174.

231. Lepourcelet M, Chen YN, France DS, Wang H, Crews P, Petersen F, et al. Small-molecule antagonists of the oncogenic Tcf/beta-catenin protein complex. *Cancer Cell* 2004;5(1):91–102.

232. Emami KH, Nguyen C, Ma H, Kim DH, Jeong KW, Eguchi M, et al. A small molecule inhibitor of beta-catenin/CREB-binding protein transcription. *Proc Natl Acad Sci USA* 2004;101(34): 12682–12687.

233. Yamada Y, Yoshimi N, Hirose Y, Hara A, Shimizu M, Kuno T, et al. Suppression of occurrence and advancement of beta-catenin-accumulated crypts, possible premalignant lesions of colon cancer, by selective cyclooxygenase-2 inhibitor, celecoxib. *Jpn J Cancer Res* 2001;92(6):617–623.

234. Williams JL, Nath N, Chen J, Hundley TR, Gao J, Kopelovich L, et al. Growth inhibition of human colon cancer cells by nitric oxide (NO)-donating aspirin is associated with cyclooxygenase-2 induction and beta-catenin/T-cell factor signaling, nuclear factor-kappaB, and NO synthase 2 inhibition: implications for chemoprevention. *Cancer Res* 2003;63(22):7613–7618.

235. Hawcroft G, D'Amico M, Albanese C, Markham AF, Pestell RG, Hull MA. Indomethacin induces differential expression of beta-catenin, gamma-catenin and T-cell factor target genes in human colorectal cancer cells. *Carcinogenesis* 2002;23(1):107–114.

236. Dihlmann S, Klein S, Doeberitz Mv MK. Reduction of beta-catenin/T-cell transcription factor signaling by aspirin and indomethacin is caused by an increased stabilization of phosphorylated beta-catenin. *Mol Cancer Ther* 2003;2(6):509–516.
237. Boon EM, Keller JJ, Wormhoudt TA, Giardiello FM, Offerhaus GJ, van der Neut R, et al. Sulindac targets nuclear beta-catenin accumulation and Wnt signalling in adenomas of patients with familial adenomatous polyposis and in human colorectal cancer cell lines. *Br J Cancer* 2004;90(1):224–229.
238. Voorhees PM, Dees EC, O'Neil B, Orlowski RZ. The proteasome as a target for cancer therapy. *Clin Cancer Res* 2003;9(17):6316–25.
239. Magill L, Walker B, Irvine AE. The proteasome: a novel therapeutic target in haematopoietic malignancy. *Hematology* 2003;8(5):275–83.
240. Adams J. The development of proteasome inhibitors as anticancer drugs. *Cancer Cell* 2004; 5(5): 417–21.
241. Adams J. The proteasome: structure, function, and role in the cell. *Cancer Treat Rev* 2003;29:3–9.
242. Pagano M, Benmaamar R. When protein destruction runs amok, malignancy is on the loose. *Cancer Cell* 2003;4(4):251–256.
243. Shirahashi H, Sakaida I, Terai S, Hironaka K, Kusano N, Okita K. Ubiquitin is a possible new predictive marker for the recurrence of human hepatocellular carcinoma. *Liver* 2002;22(5):413–418.
244. Cervello M, Giannitrapani L, La Rosa M, Notarbartolo M, Labbozzetta M, Poma P, et al. Induction of apoptosis by the proteasome inhibitor MG132 in human HCC cells: Possible correlation with specific caspase-dependent cleavage of beta-catenin and inhibition of beta-catenin-mediated trans-activation. *Int J Mol Med* 2004;13(5):741–8.
245. Davis NB, Taber DA, Ansari RH, Ryan CW, George C, Vokes EE, et al. Phase II trial of PS-341 in patients with renal cell cancer: a University of Chicago phase II consortium study. *J Clin Oncol* 2004;22(1):115–9.
246. Kondagunta GV, Drucker B, Schwartz L, Bacik J, Marion S, Russo P, et al. Phase II trial of bortezomib for patients with advanced renal cell carcinoma. *J Clin Oncol* 2004;22(18):3720–5.
247. Hegewisch-Becker S, Sterneck M, Schubert U, Rogiers X, Guerciolini R, Pierce JE, et al. Phase I/II trial of bortezomib in patients with unresectable hepatocellular carcinoma (HCC). *Proc Am Soc Clin Oncol* 2004;23:335s.
248. Berlin J, Chapman W. Phase II study of doxorubicin and bortezomib in patients with hepatocellular carcinoma (ECOG-E6202); 363801. In: DeVita VT (Editor). *Current Clinical Trials in Oncology: National Cancer Institute PDQ*. Yardley, PA: MediMedia; p. I-46, 2005.
249. Nishida N, Fukuda Y, Komeda T, Kita R, Sando T, Furukawa M, et al. Amplification and overexpression of the cyclin D1 gene in aggressive human hepatocellular carcinoma. *Cancer Res* 1994;54(12):3107–10.
250. Bedossa P, Peltier E, Terris B, Franco D, Poynard T. Transforming growth factor beta 1 (TGF-B1) and transforming growth factor beta 1 receptors in normal, cirrhotic and neoplastic human livers. *Hepatology* 1995;21:760–66.
251. Donaghy A, Ross R, Gimson A, Hughes SC, Holly J, Williams R. Growth hormone, insulin-like growth factor-I and insulin-like growth factor binding proteins 1 and 3 in chronic liver disease. *Hepatology* 1995;21:680–688.
252. Sue SR, Chari RS, Kong F-M, Mills JJ, Fine RL, Jirtle RL. Transforming growth factor beta receptors and mannose-6-phosphate/insulin-like growth factor-II receptor expression in human hepatocellular carcinoma. *Ann Surg* 1995;222:171–8.
253. Hui A-M, Sun L, Kanai Y, Sakamoto M, Tsuda H, Hirohashi S. Reduced P27Kip1 expression in hepatocellular carcinomas. *Cancer Lett* 1997;132:67–73.
254. Furutani M, Arii S, Tanaka H, Mise M, Niwano M, Harada T. Decreased expression of rare and somatic mutation of the CIP1/WAF1 gene in human hepatocellular carcinoma. *Cancer Lett* 1997;111:191–97.
255. Matsuda Y, Ichida T, Matsuzawa J, Sugimura K, Asakura H. P16-INK4 is inactivated by extensive CpG methylation in human hepatocellular carcinoma. *Gastroenterology* 1999;116:394–400.
256. Ito Y, Matsuura N, Sakon M, Miyoshi E, Noda K, Takeda T. Expression and prognostic roles of the G1-S modulators in hepatocellular carcinoma: P27 effectively predicts the recurrence. *Hepatology* 1999;30:90–99.
257. Pineau P, Marchio A, Nagamori S, Seki S, Tiollais P, Dejean A. Homozygous deletion scanning in hepatobiliary tumor cell lines reveals alternative pathways for liver carcinogenesis. *Hepatology* 2003;37(4):852–61.

258. Strahl BD, Allis CD. The language of covalent histone modifications. *Nature* 2000;403(6765):41–5.

259. Grunstein M. Histone acetylation in chromatin structure and transcription. *Nature* 1997;389(6649): 349–52.

260. Bernstein BE, Humphrey EL, Erlich RL, Schneider R, Bouman P, Liu JS, et al. Methylation of histone H3 Lys 4 in coding regions of active genes. *Proc Natl Acad Sci USA* 2002;99(13): 8695–700.

261. Kanai Y, Ushijima S, Hui AM, Ochiai A, Tsuda H, Sakamoto M, et al. The E-cadherin gene is silenced by CpG methylation in human hepatocellular carcinomas. *Int J Cancer* 1997;71(3): 355–9.

262. Yoshiura K, Kanai Y, Ochiai A, Shimoyama Y, Sugimura T, Hirohashi S. Silencing of the E-cadherin invasion-suppressor gene by CpG methylation in human carcinomas. *Proc Natl Acad Sci USA* 1995;92:7416–7419.

263. Herman JG, Baylin SB. Gene silencing in cancer in association with promoter hypermethylation. *N Engl J Med* 2003;349(21):2042–2054.

264. Schwienbacher C, Gramantieri L, Scelfo R, Veronese A, Calin GA, Bolondi L, et al. Gain of imprinting at chromosome 11p15: A pathogenetic mechanism identified in human hepatocarcinomas. *Proc Natl Acad Sci USA* 2000;97(10):5445–9.

265. Paradis V, Dargere D, Bonvoust F, Rubbia-Brandt L, Ba N, Bioulac-Sage P, et al. Clonal analysis of micronodules in virus C-induced liver cirrhosis using laser capture microdissection (LCM) and HUMARA assay. *Lab Invest* 2000;80(10):1553–1559.

266. Paradis V, Laurendeau I, Vidaud M, Bedossa P. Clonal analysis of macronodules in cirrhosis. *Hepatology* 1998;28:953–958.

267. Okuda T, Wakasa K, Kubo S, Hamada T, Fujita M, Enomoto T, et al. Clonal analysis of hepatocellular carcinoma and dysplastic nodule by methylation pattern of X-chromosome-linked human androgen receptor gene. *Cancer Lett* 2001;164(1):91–6.

268. Ochiai T, Urata Y, Yamano T, Yamagishi H, Ashihara T. Clonal expansion in evolution of chronic hepatitis to hepatocellular carcinoma as seen at an X-chromosome locus. *Hepatology* 2000;31(3):615–21.

269. Saito Y, Kanai Y, Sakamoto M, Saito H, Ishii H, Hirohashi S. Expression of mRNA for DNA methyltransferases and methyl-CpG-binding proteins and DNA methylation status on CpG islands and pericentromeric satellite regions during human hepatocarcinogenesis. *Hepatology* 2001;33(3):561–8.

270. Kelly WK, O'Connor OA, Krug LM, Chiao JH, Heaney M, Curley T, et al. Phase I study of an oral histone deacetylase inhibitor, suberoylanilide hydroxamic acid, in patients with advanced cancer. *J Clin Oncol* 2005;23(17):3923–31.

271. Ryan QC, Headlee D, Acharya M, Sparreboom A, Trepel JB, Ye J, et al. Phase I and pharmacokinetic study of MS-275, a histone deacetylase inhibitor, in patients with advanced and refractory solid tumors or lymphoma. *J Clin Oncol* 2005;23(17):3912–22.

272. Steele N, Vidal L, Plumb J, Attard G, Rasmussen A, Buhl-Jensen P, et al. A phase 1 pharmacokinetic (PK) and pharmacodynamic (PD) study of the histone deacetylase (HDAC) inhibitor PXD101 in patients (pts) with advanced solid tumours. *Proc Am Soc Clin Oncol* 2005;23:200s.

273. Donovan EA, Ryan Q, Acharya M, Chung E, Trepel J, Maynard K, et al. Phase I pharmacokinetic-pharmacodynamic trial of weekly MS-275, an oral histone deacetylase inhibitor. *Proc Am Soc Clin Oncol* 2005;:23.

274. Rowinsky EK, de Bono J, Deangelo DJ, van Oosterom A, Morganroth J, Laird GH, et al. Cardiac monitoring in phase I trials of a novel histone deacetylase (HDAC) inhibitor LAQ824 in patients with advanced solid tumors and hematologic malignancies. *Proc Am Soc Clin Oncol* 2005;:23.

275. Hansen M, Gimsing P, Rasmusen A, Buhl Jensen P, Meldgaard Knudsen L. A phase 1 study of the histone deacetylase (HDAC) inhibitor PXD101 in patients with advanced hematological tumors. *Proc Am Soc Clin Oncol* 2005;23:225s.

276. Beck J, Fischer T, George D, Huber C, Calvo E, Atadja P, et al. Phase I pharmacokinetic (PK) and pharmacodynamic (PD) study of ORAL LBH589B: A novel histone deacetylase (HDAC) inhibitor. *Proc Am Soc Clin Oncol* 2005;:23.

277. Herold C, Ganslmayer M, Ocker M, Hermann MAG, Hahn EG, et al. The histone deactylase inhibitor Trichostatin-A blocks proliferation and triggers apoptotic programs in hepatoma cells. *J Hepatol* 2002;36:233–240.

278. Yamashita Y, Shimada M, Harimoto N, Rikimaru T, Shirabe K, Tanaka S, et al. Histone deactylase inhibitor trichostatin A induces cell-cycle arrest/apoptosis and hepatocyte differentiation in human hepatoma cells. *Int J Cancer* 2003;103:572–76.

279. Kim MS, Blake M, Baek JH, Kohlhagen G, Pommier Y, Carrier F. Inhibition of histone deactylase increases cytotoxicity to anticancer drugs targeting DNA. *Cancer Res* 2003;63:7291–7300.

280. Ocker M, Alajati A, Ganslmayer M, Zopf S, Luders M, Neureiter D, et al. The histone-deacetylase inhibitor SAHA potentiates proapoptotic effects of 5-fluorouracil and irinotecan in hepatoma cells. *J Cancer Res Clin Oncol* 2005;131(6):385–394.

281. Choi HS, Lee JH, Park JG, Lee Trichostatin YIA. A histone deacetylase inhibitor, activates the IGFBP-3 promoter by upregulating Sp1 activity in hepatoma cells: alteration of the Sp1/Sp3/HDAC1 multiprotein complex. *Biochem Biophys Res Commun* 2002;296(4):1005–1012.

282. Chiba T, Yokosuka O, Fukai K, Kojima H, Tada M, Arai M, et al. Cell growth inhibition and gene expression induced by the histone deacetylase inhibitor, trichostatin A, on human hepatoma cells. *Oncology* 2004;66(6):481–491.

283. Chiba T, Yokosuka O, Arai M, Tada M, Fukai K, Imazeki F, et al. Identification of genes upregulated by histone deacetylase inhibition with cDNA microarray and exploration of epigenetic alterations on hepatoma cells. *J Hepatol* 2004;41(3):436–445.

284. Liu LH, Xiao WH, Liu WW. Effect of 5-Aza-2'-deoxycytidine on the P16 tumor suppressor gene in hepatocellular carcinoma cell line HepG2. *World J Gastroenterol* 2001;7(1):131–135.

285. Stewart DJ, Kurzrock R, Oki Y, Kehr K, Gupta S, Wistuba II, et al. Pharmacodynamics of decitabine 5 days/week x 2 weeks in advanced cancers. *Proc Am Soc Clin Oncol* 2005;23:219s.

286. Vidal L, Leslie M, Sludden J, Griffin MG, Plummer R, Judson I, et al. A phases I and pharmacodynamic study of a 7 day infusion schedule of the DNMT1 antisense compound MG98. *Proc Am Soc Clin Oncol* 2005;23:209s.

287. Gilbert J, Gore SD, Herman JG, Carducci MA. The clinical application of targeting cancer through histone acetylation and hypomethylation. *Clin Cancer Res* 2004;10(14):4589–96.

288. Zhu WG, Otterson GA. The interaction of histone deacetylase inhibitors and DNA methyltransferase inhibitors in the treatment of human cancer cells. *Curr Med Chem Anticancer Agents* 2003;3(3):187–99.

289. Helder MN, Wisman GB, van der Zee GJ. Telomerase and telomeres: from basic biology to cancer treatment. *Cancer Invest* 2002;20(1):82–101.

290. Bodnar AG, Ouellette M, Frolkis M, Holt SE, Chiu CP, Morin GB, et al. Extension of life-span by introduction of telomerase into normal human cells. *Science* 1998;279(5349):349–52.

291. Beattie TL, Zhou W, Robinson MO, Harrington L. Reconstitution of human telomerase activity in vitro. *Curr Biol* 1998;8(3):177–80.

292. Weinrich SL, Pruzan R, Ma L, Ouellette M, Tesmer VM, Holt SE, et al. Reconstitution of human telomerase with the template RNA component hTR and the catalytic protein subunit hTRT. *Nat Genet* 1997;17(4):498–502.

293. Nakayama J, Tahara H, Tahara E, Saito M, Ito K, Nakamura H, et al. Telomerase activation by hTRT in human normal fibroblasts and hepatocellular carcinomas. *Nat Genet* 1998;18(1):65–8.

294. Nagao K, Tomimatsu M, Endo H, Hisatomi H, Hikiji K. Telomerase reverse transcriptase mRNA expression and telomerase activity in hepatocellular carcinoma. *J Gastroenterol* 1999;34(1):83–87.

295. Kojima H, Yokosuka O, Imazeki F, Saisho H, Omata M. Telomerase activity and telomere length in hepatocellular carcinoma and chronic liver disease. *Gastroenterology* 1997;112:493–500.

296. Tahara H, Nakanishi T, Kitamoto M, Nakashio R, Shay JW, Tahara E. Telomerase activity in human liver tissue: comparison between chronic liver disease and hepatocellular carcinoma. *Cancer Res* 1995;55:2734–36.

297. Olivero OA, Poirier MC. Preferential incorporation of 3'-azido-2', 3'-dideoxythymidine into telomeric DNA and Z-DNA-containing regions of Chinese hamster ovary cells. *Mol Carcinog* 1993;8(2):81–8.

298. Arts EJ, Quinones-Mateu ME, Albright JL. Mechanisms of clinical resistance by HIV-1 variants to zidovudine and the paradox of reverse transcriptase sensitivity. *Drug Res Updates* 1998;1:21–28.

299. Prins J, De Vries EG, Mulder NH. Antisense of oligonucleotides and the inhibition of oncogene expression. *Clin Oncol (R Coll Radiol)* 1993;5:245–252.

300. Kondo Y, Koga S, Komata T, Kondo S. Treatment of prostate cancer in vitro and in vivo with 2–5A-anti-telomerase RNA component. *Oncogene* 2000;19(18):2205–11.

301. Kushner DM, Paranjape JM, Bandyopadhyay B, Cramer H, Leaman DW, Kennedy AW, et al. 2–5A antisense directed against telomerase RNA produces apoptosis in ovarian cancer cells. *Gynecol Oncol* 2000;76(2):183–92.

302. Kang SS, Kwon T, Kwon DY, Do SI. Akt protein kinase enhances human telomerase activity through phosphorylation of telomerase reverse transcriptase subunit. *J Biol Chem* 1999; 274(19):13085–90.

303. Han H, Bennett RJ, Hurley LH. Inhibition of unwinding of G-quadruplex structures by Sgs1 helicase in the presence of N,N'-bis[2-(1-piperidino)ethyl]-3,4,9,10-perylenetetracarboxylic diimide, a G-quadruplex-interactive ligand. *Biochemistry* 2000;39(31):9311–6.

304. Fedoroff OY, Salazar M, Han H, Chemeris VV, Kerwin SM, Hurley LH. NMR-Based model of a telomerase-inhibiting compound bound to G-quadruplex DNA. *Biochemistry* 1998;37(36):12367–74.

305. Tebes SJ, Johnson NC, Fiorica JV, Kruk PA. Inhibition of telomerase in ovarian cancer using siRNA technology. *Proc Am Soc Clin Onc* 2005;23:236s.

306. Seimiya H, Muramatsu Y, Ohishi T, Tsuruo T. Tankyrase 1 as a target for telomere-directed molecular cancer therapeutics. *Cancer Cell* 2005;7(1):25–37.

307. Tressler RJ, Chin AC, Gryaznov SM, Harley CB. Preclinical efficacy, safety, and ADME of GRN163l, a novel telomerase inhibitor developed for the treatment of cancer. *Proc Am Soc Clin Onc* 2005;23:233s.

308. Nakamura M, Saito H, Ebinuma H, Wakabayashi K, Saito Y, Takagi T, et al. Reduction of telomerase activity in human liver cancer cells by a histone deacetylase inhibitor. *J Cell Physiol* 2001;187(3):392–401.

309. Reed JC. Dysregulation of apoptosis in cancer. *J Clin Oncol* 1999;17(9):2941–53.

310. Reed JC. Mechanisms of apoptosis. *Am J Pathol* 2000;157(5):1415–30.

311. Makin G, Hickman JA. Apoptosis and cancer chemotherapy. *Cell Tissue Res* 2000;301(1):143–52.

312. Frisch SM, Screaton RA. Anoikis mechanisms. *Curr Opin Cell Biol* 2001;13(5):555–62.

313. Evan G, Littlewood T. A matter of life and cell death. *Science* 1998;281(5381):1317–22.

314. Millward MJ, Bedikian AY, Conry RM, Gore ME, Pehamberger HE, Sterry W, et al. Randomized multinational phase 3 trial of dacarbazine (DTIC) with or without Bcl-2 antisense (oblimersen sodium) in patients (pts) with advanced malignant melanoma (MM): Analysis of long-term survival. *Proc Am Soc Clin Oncol* 2004;22(14S):711s.

315. DeVita VT. Phase II study of oblimersen and doxorubicin in patients with advanced hepatocellular carcinoma (PMH-PHL-011 257565). In: DeVita VT (Editor). *Current Clinical Trials in Oncology: National Cancer Institute PDQ*. Yardley, PA: MediMedia; p I-46, 2005.

316. Olney HJ, Weng X, Watson M, Beauparlent P, Soulieres D, Viallet J, et al. Preclinical evaluation of apoptosis induction by the novel small molecule BCL-2 inhibitor, GX015–070, in ex vivo chronic lymphoid leukemia (CLL) cells. *Proc Am Soc Clin Oncol* 2005;23(16S):228s.

317. Castro JE, Prada CE, Kitada S, Contreras D, Viallet J, Reed JC, et al. GX-015–070MS, a synthetic small molecule induces apoptosis in vitro and in vivo in chronic lymphocytic leukemia. *Proc Am Soc Clin Oncol* 2005;23(16S):233s.

318. McGreivy JS, Marshall J, Cheson BD, Hwang J, Malik S, Lebowitz P, et al. Initial results from ongoing phase I trials of a novel pan bcl-2 family small molecule inhibitor. *Proc Am Soc Clin Oncol* 2005;23(16S):236s.

319. Ashkenazi A, Pai RC, Fong S, Leung S, Lawrence DA, Marsters SA, et al. Safety and antitumor activity of recombinant soluble Apo2 ligand. *J Clin Invest* 1999;104(2):155–62.

320. Pacey S, Plummer RE, Attard G, Bale C, Clavert AH, Blagden S, et al. Phase I and pharmacokinetic study of HGS-ETR2, a human monoclonal antibody to TRAIL R2, in patients with advanced solid malignancies. *Proc Am Soc Clin Oncol* 2005;23:205s.

321. Arafat WO, Buchsbaum DJ. TRAIL-mediated induction of apoptosis as a targeted therapy for prostate cancer. *Proc Am Soc Clin Oncol* 2005;23:237s.

322. Chang TH, Szabo E. Induction of differentiation and apoptosis by ligands of peroxisome proliferator-activated receptor gamma in non-small cell lung cancer. *Cancer Res* 2000;60(4):1129–38.

323. Elstner E, Muller C, Koshizuka K, Williamson EA, Park D, Asou H, et al. Ligands for peroxisome proliferator-activated receptorgamma and retinoic acid receptor inhibit growth and induce apoptosis of human breast cancer cells in vitro and in BNX mice. *Proc Natl Acad Sci USA* 1998;95(15): 8806–11.

324. Mueller E, Sarraf P, Tontonoz P, Evans RM, Martin KJ, Zhang M, et al. Terminal differentiation of human breast cancer through PPAR gamma. *Mol Cell* 1998;1(3):465–70.
325. Mueller E, Smith M, Sarraf P, Kroll T, Aiyer A, Kaufman DS, et al. Effects of ligand activation of peroxisome proliferator-activated receptor gamma in human prostate cancer. *Proc Natl Acad Sci USA* 2000;97(20):10990–5.
326. Rumi MAK, Sato H, Ishihara S, Kawashima K, Hamamoto S, Kazumori H, et al. Peroxisome proliferator-activated receptor gamma ligand-induced growth inhibition of human hepatocellular carcinoma. *Br J Cancer* 2001;84:1640–47.
327. Sarraf P, Mueller E, Jones D, King FJ, DeAngelo DJ, Partridge JB, et al. Differentiation and reversal of malignant changes in colon cancer through PPARgamma. *Nat Med* 1998;4(9):1046–52.
328. Tsubouchi Y, Sano H, Kawahito Y, Mukai S, Yamada R, Kohno M, et al. Inhibition of human lung cancer cell growth by the peroxisome proliferator-activated receptor-gamma agonists through induction of apoptosis. *Biochem Biophys Res Commun* 2000;270(2):400–5.
329. Toyoda M, Takagi H, Horiguchi N, Kakizaki S, Sato K, Takayama H, et al. A ligand for peroxisome proliferator activated receptor gamma inhibits cell growth and induces apoptosis in human liver cancer cells. *Gut* 2002;50(4):563–7.
330. Date M, Fukuchi K, Morita S, Takahashi H, Ohura K. 15-Deoxy-delta12, 14-prostaglandin J2, a ligand for peroxisome proliferators-activated receptor-gamma, induces apoptosis in human hepatoma cells. *Liver Int* 2003;23(6):460–6.
331. Debrock G, Vanhentenrijk V, Sciot R, Debiec-Rychter M, Oyen R, Van Oosterom A. A phase II trial with rosiglitazone in liposarcoma patients. *Br J Cancer* 2003;89(8):1409–12.
332. Kulke MH, Demetri GD, Sharpless NE, Ryan DP, Shivdasani R, Clark JS, et al. A phase II study of troglitazone, an activator of the PPARgamma receptor, in patients with chemotherapy-resistant metastatic colorectal cancer. *Cancer J* 2002;8(5):395–9.
333. Schaefer KL, Wada K, Takahashi H, Matsuhashi N, Ohnishi S, Wolfe MM, et al. Peroxisome proliferator-activated receptor gamma inhibition prevents adhesion to the extracellular matrix and induces anoikis in hepatocellular carcinoma cells. *Cancer Res* 2005;65(6):2251–9.
334. Read WL, Govindan R, James J, Picus J. Phase I study of bexarotene and rosiglitazone in patients with refractory cancers. *Proc Am Soc Clin Onc* 2005;23:232s.
335. Mangelsdorf DJ, Umesono K, Evans RM. The retinoid receptors. In: Sporn MB, Roberts AB, Goodman DS (Editors). *The Retinoids: Biology, Chemistry and Medicine*, 2nd Ed. New York: Raven Press; pp 319–349, 1994.
336. Miller WHJ. The emerging role of retinoids and retinoic acid metabolism blocking agents in the treatment of cancer. *Cancer* 1998;83:1471–1482.
337. Muto Y, Moriwaki H, Ninomiya M, Adachi S, Saito A, Takasaki KT, et al. Prevention of second primary tumors by an acyclic retinoid, polyprenoic acid, in patients with hepatocellular carcinoma. Hepatoma Prevention Study Group. *N Engl J Med* 1996;334(24):1561–1567.
338. Muto Y, Moriwaki H, Saito A. Prevention of second primary tumors by an acyclic retinoid in patients with hepatocellular carcinoma [letter; comment]. *N Engl J Med* 1999;340(13):1046–7.
339. Piao YF, Shi Y, Gao PJ. Inhibitory effect of all-trans retinoic acid on human hepatocellular carcinoma cell proliferation. *World J Gastroenterol* 2003;9(9):2117–2120.
340. Suzui M, Masuda M, Lim JT, Albanese C, Pestell RG, Weinstein IB. Growth inhibition of human hepatoma cells by acyclic retinoid is associated with induction of p21(CIP1) and inhibition of expression of cyclin D1. *Cancer Res* 2002;62(14):3997–4006.
341. Nakamura N, Shidoji Y, Moriwaki H, Muto Y. Apoptosis in human hepatoma cell line induced by 4, 5-didehydro geranylgeranoic acid (acyclic retinoid) via down-regulation of transforming growth factor-alpha. *Biochem Biophys Res Commun* 1996;219(1):100–4.
342. Matsushima-Nishiwaki R, Okuno M, Takano Y, Kojima S, Friedman SL, Moriwaki H. Molecular mechanism for growth suppression of human hepatocellular carcinoma cells by acyclic retinoid. *Carcinogenesis* 2003;24(8):1353–1359.
343. Mervis J. Ancient remedy performs new tricks. *Science* 1996;273:578.
344. Soignet SL, Maslak P, Wang ZG, Jhanwar S, Calleja E, Dardasthi LJ, et al. Complete remission after treatment of acute promyelocytic leukemia with arsenic trioxide. *N Engl J Med* 1998;339:1341–1348.
345. Shen ZY, Shen J, Cai WJ, Hong C, Zheng MH. The alteration of mitochondria is an early event of arsenic trioxide induced apoptosis in esophageal carcinoma cells. *Int J Mol Med* 2000;5:155–158.

346. Chow SK, Chan JY, Fung KP. Inhibition of cell proliferation and the action mechanisms of arsenic trioxide (As203) on human breast cancer cells. *J Cell Biochem* 2004;93:173–187.
347. Jiang XH, Wong BCY, Yuen ST, Jiang SH, Cho CH, Lai KC. Arsenic trioxide induces apoptosis in human gastric cancer cells through upregulation of p53 and activation of caspase 3. *Int J Cancer* 2001;93:173–179.
348. Chan JY, Siu KP, Fung KP. Effect of arsenic trioxide on multidrug resistant hepatocellular carcinoma cells. *Cancer Lett* 2005;236(2):250–8.
349. Oketani M, Kohara K, Tuvdendorj D, Ishitsuka K, Komorizono Y, Ishibashi K, et al. Inhibition by arsenic trioxide of human hepatoma cell growth. *Cancer Lett* 2002;183(2):147–53.
350. Kito M, Matsumoto K, Wada N, Sera K, Futatsugawa S, Naoe T, et al. Antitumor effect of arsenic trioxide in murine xenograft model. *Cancer Sci* 2003;94(11):1010–4.
351. Wang SS, Zhang T, Wang XL, Hong L, Qi QH. Effect of arsenic trioxide on rat hepatocellular carcinoma and its renal cytotoxicity. *World J Gastroenterol* 2003;9(5):930–5.
352. Xu HY, Yang YL, Liu SM, Bi L, Chen SX. Effect of arsenic trioxide on human hepatocarcinoma in nude mice. *World J Gastroenterol* 2004;10(24):3677–9.
353. Zhang T, Wang SS, Hong L, Wang XL, Qi QH. Arsenic trioxide induces apoptosis of rat hepatocellular carcinoma cells in vivo. *J Exp Clin Cancer Res* 2003;22(1):61–8.
354. Koike Y, Shiratori Y, Sato S, Obi S, Teratani T, Imamura M. Des-gamma-carboxy prothrombin as a useful predisposing factor for the development of portal venous invasion in patients with hepatocellular carcinoma: a prospective analysis of 227 patients. *Cancer* 2001;91:561–569.
355. Miyakawa T, Kajiwara Y, Shirahata A, Okamoto K, Itoh H, Ohsato K. Vitamin K contents in liver tissue of hepatocellular carcinoma patients. *Jpn J Cancer Res* 2000;91:68–74.
356. Hitomi M, Yokoyama F, Kita Y, Nonomura T, Masaki T, Yoshiji H, et al. Antitumor effects of vitamins K1, K2 and K3 on hepatocellular carcinoma in vitro and in vivo. *Int J Oncol* 2005;26(3):713–20.
357. Carr BI, Wang Z, Kar S. K vitamins, PTP antagonism, and cell growth arrest. *J Cell Physiol* 2002;193(3):263–74.
358. Ge L, Wang Z, Wang M, Kar S, Carr BI. Involvement of c-Myc in growth inhibition of Hep 3B human hepatoma cells by a vitamin K analog. *J Hepatol* 2004;41(5):823–9.
359. Markovits J, Wang Z, Carr BI, Sun TP, Mintz P, Le Bret M, et al. Differential effects of two growth inhibitory K vitamin analogs on cell cycle regulating proteins in human hepatoma cells. *Life Sci* 2003;72(24):2769–84.
360. Osada S, Carr BI. Mechanism of novel vitamin K analog induced growth inhibition in human hepatoma cell line. *J Hepatol* 2001;34(5):676–82.
361. Otsuka M, Kato N, Shao RX, Hoshida Y, Ijichi H, Koike Y, et al. Vitamin K2 inhibits the growth and invasiveness of hepatocellular carcinoma cells via protein kinase A activation. *Hepatology* 2004;40(1):243–51.
362. Koike Y, Shiratori Y, Shiina S, Teratani T, Obi S, Sato S, et al. Randomized prospective study of prevention from tumor invasion into portal vein in 120 patients with hepatocellular carcinoma by vitamin K. *Gastroenterology Suppl. Proc Ann Meeting AGA, AASL, GRG, SSAT, ASGE* 2002;122(4):P-43.
363. Villa E, Colantoni A, Camma C, Grottola A, Buttafoco P, Gelmini R, et al. Estrogen receptor classification for hepatocellular carcinoma: comparison with clinical staging systems. *J Clin Oncol* 2003;21:441–46.
364. Richardson PG, Sonneveld P, Schuster MW, Irwin D, Stadtmauer EA, Facon T, et al. Bortezomib or high-dose dexamethasone for relapsed multiple myeloma. *N Engl J Med* 2005;352:2487–2498.
365. Kozloff M, Cohn A, Christiansen N, Flynn P, Kabbinavar F, Robles R, et al. Safety of bevacizumab (BV) among patients (pts) receiving first-line chemotherapy (CT) for metastatic colorectal cancer (mCRC): preliminary results from a larger registry trial in the US. *Proc Am Soc Clin Oncol* 2005;23(16S):262s.
366. Li Y, Tang Y, Ye L, Liu B, Liu K, Chen J, et al. Establishment of a hepatocellular carcinoma cell line with unique metastatic characteristics through in vivo selection and screening for metastasis-related genes through cDNA microarray. *J Cancer Res Clin Oncol* 2003;129(1):43–51.
367. Kawai HF, Kaneko S, Honda M, Shirota Y, Kobayashi K. Alpha-fetoprotein producing hepatoma cell lines share common expression profiles of genes in various categories demonstrated by cDNA microarray analysis. *Hepatology* 2001;33:676–691.

9 Molecularly Targeted Therapy in Pancreatic Cancer

Anupama Goel, MD, *and Peter Kozuch,* MD

SUMMARY

Pancreatic cancer is predicted to remain as the fourth leading cause of cancer associated death in the United States in 2006. The annual death rate approximates the annual incidence rate because this disease, with rare exception, becomes locally advanced or metastatic and resistant to cytotoxic chemotherapy. Unfortunately, in eight randomized controlled clinical trials, the standard oncologic principle of combining drugs with demonstrated single agent activity and unique mechanism of action has not resulted in clinical benefit beyond gemcitabine alone. Molecular pathways that confer significant biological advantage to most human cancers have been identified over the past three decades. Drugs designed to specifically disrupt these essential pathways are currently in clinical development and hopefully will bear new standards of care for patients with pancreatic cancer.

Key Words: Pancreatic cancer; molecular therapies; bevacizumab; cetuximab; erlotinib.

1. MOLECULARLY TARGETED THERAPY IN PANCREATIC CANCER

Pancreatic cancer is the fourth leading cause of cancer death in the United States. In 2005, 32,180 new cases of pancreatic cancer were reported, associated with 31,800 pancreatic cancer-related deaths *(1)*. Pancreatic cancer is rarely curable; only 20% of patients have localized disease at presentation, which is potentially amenable to curative resection. The majority of patients have either locally advanced disease or metastatic disease with 5-year survival less than 2%.

Gemcitabine is currently the standard first line palliative therapy for patients with advanced or metastatic pancreatic cancer based on the results of a randomized trial comparing gemcitabine with bolus 5-fluorouracil (5-FU) *(2)*. This trial showed a significant improvement in clinical benefit response (23.8% versus 4.8%, $p = 0.0022$) and a statistically significant improvement in median survival (5.65 versus 4.41 months, $p = 0.0025$). The 1-year survival rate in the gemcitabine arm was 18% compared with 2% in the 5-FU arm.

From: *Cancer Drug Discovery and Development: Molecular Targeting in Oncology*
Edited by: H. L. Kaufman, S. Wadler, and K. Antman © Humana Press, Totowa, NJ

Gemcitabine has limited efficacy with overall response rates around 10% or less. Many different cytotoxic agents have been evaluated in combination with gemcitabine; unfortunately, none of these two drug combinations has significantly improved survival (Table 1). However, two recent meta-analyses presented at the 2005 American Society of Clinical Oncology annual meeting showed benefit from combination chemotherapy (11,12). One meta-analysis of trials in patients with inoperable pancreatic cancer showed a significant improvement for gemcitabine-based combinations with respect to 6-month survival rate [risk difference (RD) of 4%, $p = 0.02$], objective response rate (RD 5%, $p = 0.01$), and 6-month progression-free survival (RD 10%, $p < 0.0001$). However, there was only marginal improvement for gemcitabine-based combinations regarding 1-year survival rate (RD 3%, $p = 0.05$) and clinical benefit rate (RD 7%, $p = 0.06$). A second similar meta-analysis showed an overall survival benefit with gemcitabine combinations over

Table 1
Clinical Trials with Gemcitabine Alone and Gemcitabine Combinations in Patients with Advanced or Metastatic Pancreatic Cancer

Study	Evaluable patients	Treatment	Estimated 1-year survival	Median survival (months)	p value
Burris et al. (2)	126	Gemcitabine	18%	5.6	0.0025
		5-Fluorouracil (FU)	2%	4.4	
Berlin et al. (6)	322	Gemcitabine	<20%	5.4	0.09
		Gemcitabine/5-FU	<20%	6.7	
Rocha Lima et al. (7)	360	Gemcitabine	20%	6.6	0.789
		Gemcitabine/ Irinotecan	20%	6.3	
O'Riley et al. (8)	349	Gemcitabine		6.2	0.52
		Gemcitabine/ Exatecan		6.7	
Richards et al. (9)	365	Gemcitabine	20.1%	6.3	0.848
		Gemcitabine/ Pemetrexed	21.4%	6.2	
Louvet et al. (10)	313	Gemcitabine	8 months—45%	7.1	0.13
		Gemcitabine/ Oxaliplatin	8 months—56%	9.0	
Herrmann et al. (11)	319	Gemcitabine		7.3	0.314
		Gemcitabine/ Capecitabine		8.4	
Colucci et al. (12)	107	Gemcitabine		5 (20 weeks)	0.43
		Gemcitabine/ cisplatin		7.5 (30 weeks)	
Riess et al. (13)	466	Gemcitabine	22%	6.2	0.68
		Gemcitabine/ continuous infusion 5-FU/FA	21%	5.85	

gemcitabine alone (relative risk reduction of 9, 4, and 3% at 6, 12, and 18 months, respectively).

A small phase 3 trial compared a four drug combination of cisplatin and epirubicin, both at a dose of 40 mg/m^2 on day 1 combined with gemcitabine 600 mg/m^2 on days 1 and 8 plus continuous infusion 5-FU 200 mg/m^2/day on days 1 through 28 repeated every 4 weeks (PEFG) in comparison with single agent gemcitabine 1 g/m^2 over 30 min through typical weekly schedule *(13)*. Fifty patients assigned to the 4-drug combination and 48 patients assigned to single agent gemcitabine were evaluable. The response rate difference was statistically significant and favored the experimental arm 40% versus 8.5%. Progression-free survival at 4 months also favored PEFG 60% versus 28%, *p* value = 0.003. One-year overall survival trended in favor of the PEFG regimen, 38% versus 22%, *p* value = 0.06. The encouraging outcomes associated with the PEFG regimen aside, the theme of combining two currently available non-cross resistant cytotoxic drugs has not produced a new standard of care to replace single agent gemcitabine. Therefore, the current clinical research priority is to develop new treatment paradigms, particularly molecularly targeted therapy.

2. MOLECULARLY TARGETED THERAPIES

Molecularly targeted therapies are treatments directed against pathways that play an important role in carcinogenesis. For clinical purposes, oncologists should be familiar with six pathophysiological traits believed to be essential for cancer growth, the so-called hallmarks of cancer *(14)*. These characteristics, outlined in Table 2, network among themselves and seem to be shared by most human epithelial cancers. Therefore, drugs, particularly small molecules and monoclonal antibodies that target these hallmarks, are being developed in a broad spectrum of tumors including pancreatic cancer, reviewed herein.

Ideally, a molecular target should be unique and critical for development of malignant cells but not normal tissues. Molecules or molecular processes that have been targeted in pancreatic cancer include cell surface growth receptors, angiogenic pathways, farnesyltransferase inhibitors, matrix metalloproteinases (MMPs), and apoptotic pathways (Table 3).

3. EPIDERMAL GROWTH FACTOR RECEPTOR

Epidermal growth factor receptor (EGFR) is a cell membrane receptor that plays a key role in cancer development and progression. EGFR-signaling pathways control cell proliferation, apoptosis, and angiogenesis (Fig. 1). The EGFR family includes EGFR (erb-B1), Her-2 (or erbB-2), Her-3, and Her-4. EGF receptor erb-B1 can be targeted by either chimeric monoclonal antibodies such as cetuximab (IMC-C225) or by inhibition of downstream signaling pathways through tyrosine kinase inhibitors such as gefitinib (ZD 1839) and erlotinib (OSI-774).

Approximately 90% of pancreatic cancer patients have tumors that express EGFR *(15)*. It is suggested that the cytoplasmic overexpression of EGFR plays a significant role in progression of pancreatic adenocarcinoma, especially in invasiveness and acquisition of aggressive clinical behavior *(16)*.

Table 2
Molecular Hallmarks of Cancer

Cancer trait	Correlative molecular strategy	Target
1. Independence from growth signaling	1. Autocrine stimulation of growth factor	1. Cell surface receptors-epidermal growth factor receptor, HER2/neu
	2. Alteration of extracellular growth signals	2. Embedded tyrosine kinases
	3. Alteration of transcellular signal transduction	3. Ras–Raf–MAP kinase cascade
	4. Alteration of intracellular translational circuitry	
2. Invulnerability to anti-growth signals	1. Disruption of retinoblastoma protein (pRb) and its relative proteins p107 and p130. These proteins govern other factors (E2F transcription factors) that in turn control progression from G1 to S phase in the cell cycle	Transforming growth factor (TGF)-β
	2. Perturbation of cell adhesion molecules that normally send anti-growth signals	
3. Resistance to Apoptosis	1. Disturbance of apoptosis associated *sensors*	1. Ligand/receptors (apoptosis *sensor* components): insulin growth factor-1 and insulin growth factor-2, Fas ligand, tumor necrosis factor (TNF)-binding
	2. Disturbance of apoptosis associated *effectors*	2. p53 tumor suppressor gene and/or functionally inactivated p53 protein product
		3. Members of BCL-2 protein family
4. Immortalization (unlimited replicative potential)	Maintenance of telomere length	Telomerase (upregulated in 85–90% of cancers)
5. Angiogenesis	1. Angiogenesis-initiating signals	1. Vascular endothelial growth factor
	2. Downregulation of endogenous angiogenesis inhibitors	2. Acidic and basic fibroblast growth factors
		3. Platelet derived growth factors

| 6. Invasion and metastasis | 1. Cell–cell adhesion molecules, E-cadherin loss of function | 1. E-cadherin, β-catenin |
| | 2. Upregulation of protease genes, downregulation of protease inhibitor genes | 2. Matrix metalloproteinases |

From ref. *14*.

Table 3
Molecular Targets in Pancreatic Cancer

Molecular targets	Novel agents
Epidermal growth factor receptor (EGFR)	Monoclonal antibodies: cetuximab, trastuzumab Tyrosine kinase inhibitor: erlotinib, gefitinib
Vascular endothelial growth factor receptor (VEGF)	Bevacizumab
Farnesyltransferase (Ras protein)	Tipifarnib (R115777)
Matrix metalloproteinase (MMPs)	Marimastat (BB 2516) BAY 12-9566
Mammalian target of rapamycin (m-Tor)	CCI-779
Nuclear factor (NF)-κB	Bortezomib
Cyclooxygenase-2 (COX-2)	Celecoxib, rofecoxib

3.1. Erlotinib

Erlotinib (Tarceva, OSI pharmaceuticals, New York) is an orally administered, quinazolin-based agent that competes with adenosine triphosphate for binding with the intracellular catalytic domain of EGFR, tyrosine kinase. This action blocks downstream signaling pathways leading to inhibition of tumorogenic effects *(17)*. Currently, erlotinib is approved for treatment of locally advanced or metastatic non-small-cell lung cancer (NSCLC) refractory to at least one chemotherapy regimen *(18)*.

Moore et al. *(19)* conducted a randomized phase III trial comparing the combination of gemcitabine and erlotinib with gemcitabine and placebo in advanced pancreatic cancer. Patients were eligible if they had advanced, treatment-naïve pancreatic cancer. EGFR overexpression was not required. The primary endpoint of the study was overall survival. All patients were initially treated with gemcitabine at a dose of 1000mg/m^2 weekly for 7 weeks followed by 1 week of rest. In subsequent cycles, patients received weekly gemcitabine for 3 weeks followed by 1 week of rest. Patients were randomized to receive either erlotinib or placebo concurrently with initiation of gemcitabine. At the start of this study, a phase I dose-finding study of erlotinib with gemcitabine was enrolling patients at a 100-mg daily dose of erlotinib; therefore, patients were initially treated with 100 mg/day. As the 150-mg daily dose was established as the maximum tolerated dose (MTD) of single agent erlotinib, dose escalation of erlotinib to 150 mg was performed at selected centers in Canada. Two hundred and eighty-five patients received erlotinib, 237 patients received 100 mg daily, and 48 patients received 150 mg daily. Two hundred and eighty-four patients were treated with gemcitabine/placebo. Approximately 80% patients had 0–1 performance status, and approximately 75% had metastatic disease.

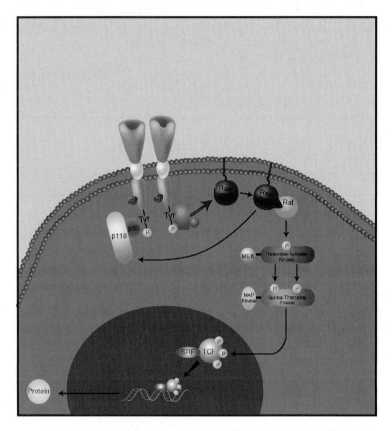

Fig. 1. Epidermal growth receptor pathway. (Content attributed to Harry Quon) (*14*).

The erlotinib with gemcitabine combination was associated with a statistically significant prolongation of median overall survival of 6.37 months compared with 5.91 months in the control arm, $p = 0.025$. The median time to progression (TTP) associated with gemcitabine/erlotinib was 3.75 versus 3.55 months attained with gemcitabine/placebo, $p = 0.003$. One-year survival was 24% with gemcitabine/erlotinib versus 17% with gemcitabine/placebo. A 19% relative risk reduction in death with addition of erlotinib to gemcitabine was reported. However, there was no difference in overall response rate between gemcitabine/erlotinib and gemcitabine/placebo, 9% versus 8%, respectively. The overall incidence of grade 3 or 4 toxicity was not increased, and there was no decline in global quality of life with the addition of erlotinib. Characteristic acne-like skin rash (all grades 72% versus 28% with grade 3/4 6% versus 1%) and grade 3–4 diarrhea (6% versus 2%) favored the placebo arm. Grade 3 or 4 neutropenia, infection, and fatigue were similar between the two treatment groups.

This trial is the first trial to demonstrate a survival benefit over gemcitabine alone in treatment of advanced pancreatic cancer. Notably, this is also the first trial to demonstrate survival benefit of oral tyrosine kinase inhibitors in combination with conventional chemotherapy. Phase III randomized trials of both erlotinib and gefitinib combined with platinum-based chemotherapy in NSCLC failed to show any survival benefit (*20–23*). The statistically significant outcome of this trial encourages further development of EGFR-targeted therapy, but the modest clinical benefit may lead to only minimal clinical application of gemcitabine with erlotinib. This contention is supported by the fact that randomized pivotal trials of single agent gemcitabine with or without cetuximab or bevacizumab continue to rapidly accrue patients.

3.2. Cetuximab

Cetuximab is a chimeric monoclonal antibody targeting the ligand-binding extracellular domain of EGFR and is currently approved for treatment of irinotecan refractory metastatic colon cancer (24). Xiong et al. (25) conducted a phase II multi-center clinical study in which patients with locally advanced and metastatic or recurrent pancreatic cancer were treated with a combination of gemcitabine and cetuximab. Patients were eligible if they had not received any prior chemotherapy and if their tumor expressed epidermal growth factor by immunohistochemical staining. Patients were treated with cetuximab at an initial dose of 400mg/m^2 followed by 250mg/m^2/week maintenance dose for 7 weeks. Gemcitabine 1000mg/m^2 was administered concurrently with cetuximab weekly for 7 weeks followed by 1 week of rest. After the first cycle, cetuximab 250mg/m^2 was administered weekly and gemcitabine 1000mg/m^2 was administered weekly for 3 weeks followed by 1 week of rest. Forty-one patients were enrolled in this trial. Five patients (12.2 %) achieved a partial response and 26 (63.4%) had stable disease. The median TTP was 3.8 months, and the median overall survival was 7.1 months. One-year progression-free survival and overall survival rates were 12 and 31.7%, respectively. The most frequently reported grade 3 or 4 adverse events were neutropenia (39%), asthenia (22%), abdominal pain (22%), and thrombocytopenia (17.1%). Almost all patients developed acne-like rashes (87.8%) with grade 3 in 5 (12.2%) patients. None of the patients discontinued therapy secondary to rash.

The encouraging response and survival outcomes prompted an ongoing phase III randomized trial, SWOG-S0205, in which patients with treatment-naïve locally advanced or metastatic pancreatic cancer are randomized to gemcitabine with or without cetuximab. Another phase II trial, ECOG-E8200, is underway to determine the efficacy of irinotecan and docetaxel with or without cetuximab in patients with metastatic pancreatic cancer.

3.3. Trastuzumab

HER2/neu is overexpressed in 20–30% of pancreatic adenocarcinomas. In a phase II study, herceptin with gemcitabine was associated with 22% (4/18) radiological partial response rate, and greater than 50% reduction in level of CA 19-9 in 50% (9/18) of patients (26).

4. VASCULAR ENDOTHELIAL GROWTH FACTOR RECEPTOR—ANTI-ANGIOGENIC THERAPY

Angiogenesis is the complex biologic process involved in the development and formation of new blood vessels (27). Angiogenesis is regulated by many growth factors secreted by tumor cells, including basic fibroblast growth factor, vascular endothelial growth factor (VEGF), platelet-derived epidermal growth factor, and transforming growth factor (TGF)-β. VEGF is known to be a potent angiogenic mitogen that plays an important role in maintaining these newly formed blood vessels *(28)*. VEGF and its receptor are overexpressed in pancreatic cancer and have prognostic importance. VEGF overexpression is positively correlated with local recurrence, metastatic potential, and overall survival in pancreatic cancer *(29–32)*. Thus, the VEGF pathway has important therapeutic potential in pancreatic cancer.

Kindler et al. *(33)* conducted a multi-center phase II trial of the anti-VGEF antibody bevacizumab plus gemcitabine in patients with advanced pancreatic cancer. Patients with advanced pancreatic cancer were eligible if they had not received prior chemotherapy except 5-FU given as a radiation sensitizer. Notably, patients were excluded if there was obvious tumor involvement of major blood vessels. Patients were treated with gemcitabine 1000 mg/m^2 given over 30 min on days 1, 8, and 15 of 28-day cycles. Bevacizumab 10 mg/kg was given after gemcitabine on days 1 and 15. Ten of 52 evaluable patients (19%) achieved a partial response (PR), and 25 (48%) had stable disease (SD), for an overall disease control rate of 67%. The combination of gemcitabine and bevacizumab produced a promising median survival of 8.7 months, median TTP of 5.8 months, and a 1-year survival rate of 29%. The combination was well tolerated, and complications of thrombosis and bleeding were not more frequent than expected for patients with advanced pancreatic cancer.

On the basis of the encouraging safety and efficacy of bevacizumab and gemcitabine from this phase II trial, the Cancer and Leukemia Group B (CALGB) has opened a phase III trial (CALGB 80303), which began accrual in June 2004. Patients are randomized to receive gemcitabine with either bevacizumab 10 mg/kg on days 1 and 15 or placebo, every 28 days. The primary endpoint of this trial is survival with safety and efficacy as secondary endpoints. Unlike the phase II trial, blood vessel involvement by tumor is not a contraindication to study entry.

The National Cancer Institute (NCI) is currently conducting a phase II trial of gemcitabine, capecitabine, and bevacizumab in patients with metastatic or unresectable pancreatic adenocarcinoma. Patients with newly diagnosed or previously treated metastatic cancer are eligible. Patients will receive bevacizumab on day 1, capecitabine orally, twice daily on days 1–14, and gemcitabine on days 1 and 8 of a 3-week cycle. The primary endpoint of this trial is progression-free survival.

4.1. Bevacizumab with Radiation Therapy for Locally Advanced and Resectable Pancreatic Cancer

In vivo studies have shown increased radiation sensitivity in association with VEGF blockade. A phase I study evaluating the feasibility and toxicity of concurrent bevacizumab and capecitabine showed no significant increase in acute toxicity with addition of bevacizumab *(34)*. Two ongoing trials are evaluating the efficacy of bevacizumab with concurrent radiation. RTOG 0411 is a phase II study of bevacizumab with concurrent capecitabine and radiation followed by maintenance gemcitabine and bevacizumab for patients with locally advanced pancreatic cancer unresectable by standard criteria. ACOSOG Z5041 is a phase II study evaluating preoperative gemcitabine and bevacizumab followed by surgery and adjuvant capecitabine, bevacizumab, and radiation.

Clinicians using bevacizumab should be familiar with important albeit rare side effects associated with this drug that include cerebrovascular and cardiovascular complications such as stroke and heart attack, gastrointestinal perforation, and hypertension (HTN) *(35)*.

5. COMBINED EGFR AND VEGF BLOCKADE—A RATIONALE FOR COMBINATION TREATMENT IN PANCREATIC CANCER

Although the EGFR pathway and angiogenesis are fundamental for progressive growth of human pancreatic carcinoma and have been independently evaluated as targets for therapy, there are a few observations that justify combining anti-EGFR and anti-VEGF drugs. EGFR and TGF appear to be strong stimulators for VEGF, *(36,37)*. Resistance to EGFR inhibition can be acquired in part by upregulation of angiogenic pathways, including VEGF, or by activation of one or more alternative proangiogenic growth factors *(38)*. Growth of EGFR-inhibitor-resistant tumor xenografts can be effectively inhibited with anti-VEGF therapy *(39)*. Dual EGFR/VEGF pathway inhibition with C225 monoclonal antibody and VEGF anti-sense has demonstrated additive anti-tumor activity *(40)*.

A phase I/II study in NSCLC and a phase II study in metastatic breast cancer have shown that bevacizumab plus erlotinib is well tolerated with no unexpected adverse events *(41,42)*.

The preclinical and clinical data for anti-EGFR pathway and anti-VEGF combinations have prompted a randomized phase II trial (NCI-6580), which will assess the effectiveness of bevacizumab plus gemcitabine with either cetuximab or erlotinib in advanced pancreatic adenocarcinoma. The primary objectives of this trial are response, survival, and safety outcomes. Patients in arm I will receive cetuximab on days 1, 8, 15, and 22, gemcitabine on days 1, 8, and 15 and bevacizumab on days 1 and 15. Patients in arm 2 will receive the same schedule of bevacizumab and gemcitabine as arm 1 plus erlotinib once daily on days 1–5, 8–12, and 15–26. In both arms, cycles will be repeated every 4 weeks.

There is an emerging hypothesis that the characteristic rash associated with the use of cetuximab and early HTN (defined as greater than grade 2 HTN during first 56 days of treatment) associated with the use of bevacizumab may be an early pharmacodynamic marker for survival. In the phase II trial of cetuximab with gemcitabine, median overall survival in patients with grade 1, 2, and 3 rash was 2.3 months, 5.7 months, and 13.9 months, respectively *(25)*. Similarly, a retrospective analysis of pancreatic cancer patients treated with bevacizumab and gemcitabine showed a median overall survival of 13.7 months in patients with early HTN compared with 8.7 months in patients without early HTN. If these observations hold true in prospective studies, rash and early HTN might become useful pharmacodynamic markers for survival *(43)*.

6. FARNESYLTRANSFERASE INHIBITORS

Ras proteins play a significant role in signal transduction pathways, which control various cellular processes including growth, differentiation, apoptosis, and cytoskeletal organization *(44)*. Ras proteins are encoded by three distinct proto-oncogenes: H-ras, K-ras, and N-ras *(45)*. Somatic mutations of *K-ras* gene are present in 85% of pancreatic tumor cells.

Ras is synthesized as biologically inactive propeptide, and its function depends on the addition of a 15-carbon farnesyl moiety, catalyzed by the enzyme farnesyltransferase

(46,47). The farnesyltransferase inhibitor tipifarnib (R115777) is a selective inhibitor of the farnesyltransferase enzyme. In a phase 1 trial conducted by Hudes and Schol *(48),* 300 mg given twice daily for 21 days of a 28-day cycle was found to be tolerable and safe. Two phase II trials with tipifarnib 300 mg twice daily have been conducted. In the first study by Cohen et al. *(49),* no objective responses were seen and median TTP was 4.9 weeks with a median survival of 19.7 weeks. The second trial accrued 47 patients and demonstrated a median survival of only 2.7 months *(50).* Grade 1–2 fatigue, nausea, vomiting, myelosuppression, and mild transaminitis were the main side effects seen in both trials.

Tipifarnib was also combined with gemcitabine in a randomized placebo-controlled phase III randomized trial *(51).* Patients with untreated advanced or metastatic pancreatic cancer were eligible. Treatment consisted of standard schedule gemcitabine with either tipifarnib 200 mg twice daily continuously or placebo. Six hundred and eighty-eight patients were enrolled in this trial. There was no statistical difference seen in terms of overall survival (193 versus 182 days) or median time to disease progression. Hematological toxicity was more frequent in the experimental arm with 40% grade 3–4 neutropenia in the gemcitabine/tipifarnib arm versus 30% in the gemcitabine/placebo arm.

These trials demonstrate that farnesyltransferase inhibitor tipifarnib has no single agent activity and it also does not improve survival with gemcitabine in patients with advanced and/or metastatic pancreatic adenocarcinoma despite encouraging in vitro activity. One hypothesis for this disappointing outcome is that farnesyltransferase inhibition alone is not sufficient to inhibit Ras function. It is possible that Ras may still undergo post-translational prenylation by other enzymes, such as geranylgeranyl transferase. An alternative hypothesis is that the anti-tumor activity of farnesyltransferase inhibition occurs irrespective of Ras mutational status. This hypothesis is supported by clinical activity of tipifarnib in acute and chronic myeloid leukemia, myelodysplastic syndrome, and breast cancer—malignancies in which Ras mutation plays a marginal role *(52,53).*

7. MMP INHIBITORS

Eighty percent of patients with pancreatic cancer have either locally advanced disease (i.e., invasion of visceral and vascular structures that contraindicates attempts at resection) or metastatic disease at presentation. This phenomenon requires breakdown of the surrounding extracellular matrix leading to invasion of tumor cells into vasculature and distant growth. MMPs are a family of zinc-dependent proteolytic enzymes capable of degrading the extracellular matrix. MMPs are upregulated in many different types of cancers (thyroid, prostate, head, and neck) and correlate with their invasive potential *(54,55).* Overexpressions of MMPs have also been demonstrated in pancreatic cancer *(56).*

In a preclinical study, the synthetic MMP inhibitor BB-94 was shown to inhibit pancreatic cancer cell lines through two mechanisms: prevention of pro-MMP activation and direct inhibition of catalytic site of activated MMP *(57).* Encouraging preclinical data led to phase II studies of oral synthetic MMP inhibitor BB-2516 (marimastat) and BAY 12-9566. In a dose-finding study by Rosemurgy et al. *(58),* marimastat doses of 5, 10, and 25 mg twice daily were identified as acceptable and safe with musculoskeletal pain, stiffness, and tenderness as major dose-limiting toxicities, especially at doses more than 50 mg twice daily.

In a phase II study of patients with advanced pancreatic cancer, marimastat administration was associated with a 30% serological response rate (unspecified decrease or stabilization of CA 19-9 level), 49% radiological response rate, and 51% clinical benefit response rate (decrease or stabilization of pain, mobility, and analgesia score) at a median follow up of 28 days. Patients who showed serological response had a significantly improved survival compared with non-responders (245 versus 128 days). No difference in survival was seen in patients with or without radiological response *(59)*. A large multi-institutional prospective randomized trial compared three different doses of marimastat with that of gemcitabine *(60)*. Four hundred and fourteen patients were randomized to receive marimastat (5, 10, and 25 mg twice daily) or gemcitabine (1000 mg/m^2). There was no significant difference in survival among the 4 groups ($p = 0.19$). One-year survival rates were 14, 14, 20, and 19% in 5, 10, or 25 mg marimastat and gemcitabine groups, respectively. Patients with non-metastatic disease had improved 1-year survival rates with marimastat compared with gemcitabine, but this difference was not statistically significant (30% versus 25%). Musculoskeletal side effects were seen in 44% of marimastat-treated patients and were found to be dose- and duration-dependent.

A double-blind placebo-controlled randomized study compared gemcitabine and marimastat 10 mg twice daily with gemcitabine and placebo as first line therapy in 239 patients with advanced pancreatic cancer *(61)*. There was no significant difference in either overall survival ($p = 0.95$) or 1-year survival between gemcitabine and marimastat versus gemcitabine and placebo. In a sub-group exploratory analysis, there was a trend toward better survival in marimastat-treated patients with disease confined only to the pancreas.

A phase III trial conducted by the NCI of Canada Clinical Trials Group compared gemcitabine with another metalloproteinase inhibitor, BAY 12-9566 in patients with advanced or metastatic pancreatic cancer *(62)*. Two hundred and seventy-seven patients were randomized to receive either gemcitabine (1000 mg/m^2) or BAY 12-9566 (800 mg twice daily). This study was closed after second interim analysis secondary to inferior median survival of BAY 12-9566 arm versus gemcitabine arm, 3.74 versus 6.59 months, respectively ($p < 0.001$).

The results of these trials indicate that marimastat or other MMPs, alone or in combination with gemcitabine, have no benefit in patients with advanced or metastatic pancreatic cancer. However, it is hypothesized that, because of its cytostatic effect, MMP inhibitor administration in patients with early stage pancreatic cancer could be beneficial.

8. OTHER TARGETED AGENTS

8.1. Nuclear factor-κB

Nuclear factor (NF)-κB plays an important role in cancer growth by developing resistance to apoptosis. NF-κB activity is increased in approximately 67% of pancreatic adenocarcinomas compared with normal pancreatic tissue *(63)*. In orthotopic xenograft models, blockade of NF-κB is associated with impaired angiogenesis in pancreatic cancer cells and is associated with retarded tumor growth and suppression of metastasis

(64,65). In vitro, inhibition of NF-κB with curcumin (diferuloylmethane) also showed marked growth inhibition and apoptosis of pancreatic cancer cells *(66)*. Bortezomib (velcade, PS-341), a proteasome inhibitor, is a potent inhibitor of NF-κB by blocking degradation of IκBκ. In an orthotopic xenograft model, bortezomib appeared to enhance anti-tumor activity of taxanes by enforcing cell growth arrest *(67)*.

However, these preclinical findings have not been confirmed in clinical trials. A randomized phase II study comparing bortezomib with or without gemcitabine did not show any benefit over what would be expected with gemcitabine alone. Median survival in association with single agent bortezomib was 2.5 and 4.8 months in association with bortezomib plus gemcitabine. Response rates were 0 and 10% for bortezomib and bortezomib plus gemcitabine, respectively *(68)*. Despite this negative result, NF-κB inhibition appears to have significant biological rationale for activity in pancreatic cancer and warrants further clinical research.

8.2. CCI-779

CCI-779, an inhibitor of mammalian target of rapamycin (m-TOR), inhibits synthesis of proteins required for cell cycle progression from G1 to S phase. Preclinical studies indicate that m-TOR is activated in pancreatic tumor cells and that CCI-779 is a potent inhibitor of some pancreatic cell lines especially those containing defective p53 *(69)*. Clinical evaluation of CCI-779 is underway.

8.3. Others

Cyclooxygenase-2 (COX-2) is involved in tumor cell growth, angiogenesis, and inhibition of apoptosis. COX-2 is overexpressed in 67–90% of pancreatic tumor cells *(70)*. A phase II study of gemcitabine with celecoxib reported a median survival duration of 6.2 months with acceptable toxicities *(71)*. Another phase II study demonstrated a 17% partial response rate *(72)*. Leukotriene B4 receptor antagonist LY293111 (LY) showed promising results in preclinical studies. However, a recent phase II study of LY293111 in combination with gemcitabine did not show any significant benefit in 6-month survival compared with gemcitabine alone *(73)*.

Gastrin is a trophic hormone and stimulates growth and proliferation of pancreatic cells *(74)*. G17DT (Aphton Corporation, Woodland, CA) is an immuno-conjugate of amino-terminal sequence of gastrin-17. It induces formation of antibodies, which neutralizes gastrin-17 and its precursor glycin-G-17, thereby causing growth inhibition. A phase II study randomized patients with advanced or metastatic pancreatic cancer to receive three intramuscular injections of either 100 or 250 mcg *(75)*. At 8 weeks, 67% of patients had an antibody response. Median survival from the day of first injection was statistically significant for immune responders compared with non-responders (217 versus 121 days, $p = 0.0023$). There was no statistically significant difference in median survival between the two doses, although the antibody response rate was higher with 250 mcg dose compared with 100 mcg, 82% versus 46%, respectively, $p = 0.018$. Treatment was generally well tolerated with mild local injection site reaction as a major side effect. Another multi-center trial randomized patients with advanced or metastatic pancreatic cancer, unsuitable or unwilling to take chemotherapy, to receive either G17DT or placebo. This trial showed statistically significant improvement in median survival (151 versus 82 days, $p = 0.03$) and time to deterioration in Karnofsky index (138 versus 78 days, $p = 0.038$) *(76)*.

Encouraging results from the above trials led to a phase III randomized, double-blind, multi-center trial in patients with advanced or metastatic pancreatic cancer *(77)*. Patients were randomized to receive gemcitabine plus G17DT, administered intramuscularly on weeks 0, 4, 8, and 24 weeks or placebo. Seventy-five percent of patients (131/175) developed an antibody response, with substantially weaker response in women. However, overall survival did not show any benefit with addition of G17DT (178 versus 201 days, $p = 0.10$).

9. CONCLUSIONS

The first and subsequent generations of molecularly targeted therapy will provide clinical scientists with promising research opportunities for many years. Research priorities should include identification and validation of predictors of response for specific drugs and development of new or surrogate endpoints with which to improve efficiency of drug development. Not only will such efforts, if successful, directly benefit patients but will permit more cost-effective application of this expensive technology. Lastly, practitioners will need to be ever vigilant for emergence of class-specific side effects as well as new drug–drug interactions so that patients may enjoy maximal benefit from these novel molecular therapies.

REFERENCES

1. American Cancer Society. *Cancer Facts and Figures 2005*. Atlanta, GA: American Cancer Society, 2005.
2. Burris HA, Moore MJ, Anderson J, et al. Improvements in survival and clinical benefit with gemcitabine as first-line therapy for patients with advanced pancreatic cancer: a randomized trial. *J Clin Oncol* 1997; 15: 2403–2413.
3. Liang H. Comparing gemcitabine-based combination chemotherapy with gemcitaine alone in inoperable pancreatic cancer: a meta-analysis. *Proc Am Soc Clin Oncol* 2005; Abstract 4110.
4. Banu E, Oudard S, et al. Cumulative meta-analysis of randomized trials comparing gemcitabine based-chemotherapy versus gemcitabine alone in patient with advanced or metastatic pancreatic cancer (PC). *Proc Am Soc Clin Oncol* 2005; Abstract 4101.
5. Reni M, Cordio S, et al. Final results of a phase III trial of gemcitabine versus PEFG regimen and stage IVA or metastatic pancreatic adenocarcinoma. *Proc Am Soc Clin Oncol* 2004; 22(14S): 4010 (post-meeting edition).
6. Berlin JD, Catalano P, et al. Phase III study of gemcitabine in combination with flourouracil versus gemcitabine alone in patients with advanced pancreatic carcinoma: Eastern Cooperative Oncology Group Trial E2297. *J Clin Oncol* 2002; 20: 3270–3275.
7. Rocha Lima CA, Green MR, et al. Irinotecan plus gemcitabine results in no survival advantage compared with gemcitabine monotherapy in patients with locally advanced or metastatic pancreatic cancer despite increase tumor response rate. *J Clin Oncol* 2004; 22: 3776–3783.
8. O'Riley EM, Abou-Alfa GK, et al. A randomized phase III trial of DX-8951f (exatecan mesylate; DX) and gemcitabine (GEM) vs gemcitabine alone in advanced pancreatic cancer. *J Clin Oncol* 2004; 22(14S): 4006.
9. Richards DA, Kindler HL, et al. A randomized phase III study comparing gemcitabine + pemetrexed versus gemcitabine in patients with locally advanced pancreatic cancer. *Proc Am Soc Clin Oncol* 2004; Abstract 4007.
10. Louvet C, Labianca R, et al. GemOx (gemcitabine + oxaliplatin) versus Gem (Gemcitabine) in non resectable pancreatic adenocarcinoma: final results of GERCOR/ GISCAD Intergroup phase III. *Proc Am Soc Clin Oncol* 2004; Abstract 4008.

11. Herrmann R, Bodoky G, et al. Gemcitabine (G) plus capecitabine (C) versus G alone in locally advanced or metastatic pancreatic cancer. A randomized phase III study of the Swiss Group for Clinical Cancer Research (SAKK) and Central European Cooperative Oncology Group (CECOG). *Proc Am Soc Clin Oncol* 2005; Abstract LBA40410.

12. Colucci G, Giuliani F, et al. Gemcitabine alone or with cisplatin for the treatment of patients with locally advanced and/or metastatic pancreatic carcinoma. *Cancer* 2002; 94: 902–910.

13. Riess H, Helm A, et al. A randomized, prospective multicenter, phase III trial of gemcitabine, 5-fluorouracil (5FU), folinic acid vs gemcitabine alone in patients with advanced pancreatic cancer. *Proc Am Soc Clin Oncol* 2005; Abstract LBA4009.

14. Hanahan D and Weinberg RA. The hallmarks of cancer. *Cell* 2000; 100(1): 57–70.

15. Abbruzzese JL, Rosenberg A, Xiong Q, et al. Phase II study of anti-epidermal growth factor receptor (EGFR) antibody cetuximab (IMC-C225) in combination with gemcitabine in patients with advanced pancreatic cancer. *Proc Am Soc Clin Oncol* 2001; 20: 130a (Abstract 518).

16. Ueda S, Ogata S, et al. The correlation between cytoplasmic overexpression of epidermal growth factor receptor and tumor aggressiveness: poor prognosis in patients with pancreatic ductal adenocarcinoma. *Pancreas* 2004; 29(1): e1–8.

17. Ng SS, Tsao MS, et al. Effects of the epidermal growth factor receptor inhibition OSI-774, Tarceva, on downstream signaling pathways and apoptosis in human pancreatic adenocarcinoma. *Mol Cancer Ther* 2002; 1(10): 777–783.

18. Shepherd FA, Pereira J, et al. A randomized placebo-controlled trial of erlotinib in patients with advanced non-small cell lung cancer (NSCLC) following failure of 1st line or 2nd line chemotherapy. A National Cancer Institute of Canada Clinical Trials Group (NCIC- CTG) trial. *J Clin Oncol* 2004; 22(Suppl 14): A-7022, 622s [Abstract].

19. Moore MJ, Goldstein D, et al. Erlotinib improves survival when added to gemcitabine in patients with advanced pancreatic cancer: a phase III trial of the National Cancer Institute of Canada Clinical Trials Group (NCIC-CTG). *Proc Am Soc Clin Oncol* 2005; Abstract 77.

20. Giaccone G, Herbst RS, et al. Gefitinib in combination with gemcitabine and cisplatin in advanced non-small cell lung cancer: a phase III trial- INTACT 1. *J Clin Oncol* 2004; 22: 777–784.

21. Herbst RS, Giaccone G, et al. Gefitinib in combination with paclitaxel and carboplatin in advanced non-small cell lung cancer: a phase III trail- INTACT 2. *J Clin Oncol* 2004; 22: 785–794.

22. Gatzemeier U, Pluzanska A, et al. Results of phase III trial of erlotinib (OSI-774) combined with cisplatin and gemcitabine chemotherapy in advanced non-small cell lung cancer (NSCLC). *Proc Am Soc Clin Oncol* 2004; 23: 617 (Abstract 7010).

23. Herbst RS, Prager D, et al. Tribute - a phase III trial of erlotinib HCL (OSI-774) combined with carboplatin and paclitaxel chemotherapy in advanced non-small cell lung cancer (NSCLC). *Proc Am Soc Clin Oncol* 2004; 23: 617 (Abstract 7011).

24. Cunningham D, Humblet Y, et al. Cetuximab monotherapy and cetuximab plus irinotecan in irinotecan-refractory metastatic colorectal cancer. *N Engl J Med* 2004; 351: 337–345.

25. Xiong HQ, Rosenberg A, et al. Cetuximab, a monoclonal antibody targeting the epidermal growth factor receptor, in combination with gemcitabine for advanced pancreatic cancer: A muticenter phase II trial. *J Clin Oncol* 2004; 22: 2610–2616.

26. Safran H, Ramanathan RK, et al. Herceptin and gemcitabine for metastatic pancreatic cancers that overexpress Her-2/neu. *Proc Am Soc Clin Oncol* 2001; Abstract 517.

27. Folkman J. What is the evidence that tumors are angiogenesis dependent? *J Natl Cancer Inst* 1990; 82: 4–6.

28. Leung DW, Cathians G, et al. Vascular endothelial growth factor is a secreted angiogenic mitogen. *Science* 1989; 246: 1305–1309.

29. Itakura J, Ishiwata T, et al. Concomitant over- expression of vascular endothelial growth factor and its receptors in pancreatic cancer. *Int J Cancer* 2000; 85: 27–34.

30. Ikeda N, Adachi M, et al. Prognostic significance of angiogenesis in human pancreatic cancer. *Br J Cancer* 1999; 79: 1553–1563.

31. Itakura J, Ishiwata T, et al. Enhanced expression of vascular endothelial growth factor in human pancreatic cancer correlates with local disease progression. *Clin Cancer Res* 1997; 3: 1309–1316.

32. Seo Y, Baba H, et al. High expression of vascular endothelial growth factor is associated with liver metastasis and a poor prognosis for patients with ductal pancreatic adenocarcinoma. *Cancer* 2000; 88: 2239–2245.

33. Kindler HL, Friberg G, et al. Bevacizumab (B) plus gemcitabine (G) in patients with advanced pancreatic cancer: updated results of a multi-center phase II trail. *Proc Am Soc Clin Oncol* 2004; 23: 314 (Abstract 4009).

34. Crane CH, Ellis LM, et al. Phase I trial of bevacizumab (BV) with concurrent radiotherapy (RT) and capecitabine (CAP) in locally advanced pancreatic carcinoma (PA). *Proc Am Soc Clin Oncol* 2005; Abstract 4033.

35. Olszewski AJ, Grossbard ML, Kozuch PS. The horizon of antiangiogenic therapy for colorectal cancer. *Oncology (Williston Park)* 2005; 19(3): 297–306.

36. Bruns CJ, Solorzano CC, et al. Blockade of the epidermal growth factor receptor signaling by a novel tyrosine kinase inhibitor leads to apoptosis of endothelial cells and therapy of human pancreatic carcinoma. *Cancer Res* 2000; 60(11): 2926–2935.

37. Petit AM, Rak J, et al. Neutralizing antibodies against epidermal growth factor and ErbB-2/neu receptor tyrosine kinases down-regulate vascular endothelial growth factor production by tumor cells in vitro and in vivo: angiogenic implications for signal transduction therapy of solid tumors. *Am J Pathol* 1997; 151(6): 1523–1530.

38. Viloria-Petit A, Crombet T, et al. Acquired resistance to the antitumor effect of epidermal growth factor receptor-blocking antibodies in vivo: a role for altered tumor angiogenesis. *Cancer Res* 2001; 61(13): 5090–5101.

39. Ciardiello F, Bianco R, et al. Antitumor activity of ZD6474, a vascular endothelial growth factor receptor tyrosine kinase inhibitor, in human cancer cells with acquired resistance to antiepidermal growth factor receptor therapy. *Clin Cancer Res* 2004; 10(2): 784–793.

40. Ciardiello F, Bianco R, et al. Antiangiogenic and antitumor activity of anti-epidermal growth factor receptor C225 monoclonal antibody in combination with vascular endothelial growth factor antisense oligonucleotide in human GEO colon cancer cells. *Clin Cancer Res* 2000; 6(9): 3739–3747.

41. Mininberg ED, Herbst RS, et al. PhaseI/II study of the recombinant humanized monoclonal anti-VEGF antibody bevacizumab and the EGFR-TK inhibitor erlotinib in patients with recurrent non-small cell lung cancer (NSCLC). *Proc Am Soc Clin Oncol* 2003; 22: 627 (Abstract 2521).

42. Dickler M, Rugo H, et al. Phase II trial of erlotinib (OSI-774), an epidermal growth factor receptor (EGFR)-tyrosine kinase inhibitor, and bevacizumab, a recombinant humanized monoclonal antibody to vascular endothelial growth factor (VEGF), in patients (pts) with metastatic breast cancer (MBC). *J Clin Oncol* 2004; 22: 14S (Abstract 2001).

43. Friberg G, Kasza K, et al. Early hypertension (HTN) as a potential pharmacodynamic (PD) marker for survival in pancreatic cancer (PC) patients (pts) treated with bevacizumab (B) and gemcitabine (G). *Proc Am Soc Clin Oncol* 2005; Abstract 3020.

44. Boguski MS, McCormik F: Proteins regulating Ras and its relatives. *Nature* 1993; 366: 643–654.

45. Barbacid M. Ras genes. *Annu Rev Biochem* 1987; 55: 779–827.

46. Rowinsky EK, Windle JJ, Von Hoff DD. Ras protein farnesyltransferase: a strategic target for anticancer therapeutic development. *J Clin Oncol* 1999; 17: 3631–3652.

47. Adjei AA. Blocking oncogenic ras signaling for cancer therapy. *J Natl Cancer Inst* 2001; 93: 1062–1074.

48. Hudes GR, Schol J. Phase I clinical and pharmacokinetic trial of the farnesyltransferase inhibitor R115777 on a 21-day dosing schedule. *Proc Am Soc Clin Oncol* 1999; 18: Abstract 601.

49. Cohen SJ, Ho L, et al. Phase II and pharmacodynamic study of the farnesyltransferase inhibitor R115777 as initial therapy in patients with metastatic pancreatic adenocarcinoma. *J Clin Oncol* 2003; 21: 1301–1306.

50. Macdonald JS, Chansky K, et al. A phase II study of farnesyltransferase inhibitor R115777 in pancreatic cancer: a Southwest Oncology Group (SWOG) study. *Proc Am Soc Clin Oncol* 2002; Abstract 548.

51. Van Cutsem E, van de Velde H, et al. Phase III trial of gemcitabine plus tipifarnib compared with gemcitabine plus placebo in advanced pancreatic cancer. *J Clin Oncol* 2004; 22: 1430–1438.

52. Cortes J, Albitar M, et al. Efficacy of the farnesyl transferase inhibitor R115777 in chronic myeloid leukemia and other hematological malignancies. *Blood* 2003; 101: 1692–1697.

53. Kurzrock R, Cortes J, et al. Clinical development of farnesyltransferase inhibitors in leukemia and myelodysplastic syndrome. *Semin Hematol* 2002; 39: 20–24.

54. Shapiro SD. Matrix metalloproteinase degradation of extracellular matrix: biological consequence. *Curr Opin Cell Biol* 1998; 10: 602–608.

55. Kahari VM and Saarialho-Kere U. Matrix metalloproteinases and their inhibitors in tumor growth and invasion. *Ann Med* 1999; 31: 34–45.

56. Bramhall SR, Neoptolemos JP, Stamp GW, and Lemoine NR. Imbalance of expression of matrix metalloproteinases (MMPs) ad tissue inhibitors of matrix metalloproteinases (TIMPs) in human pancreatic carcinoma. *J Pathol* 1997; 182: 347–355.

57. Zervos EE, Shafii AE, Haq M, and Rosemurgy AS. Matrix metalloproteinase inhibition suppresses MMP-2 activity and activation of PANC-1 cells in vitro. *J Surg Res* 1999; 84: 162–167.

58. Rosemurgy A, Harris J, et al. Marimastat in patients with advanced pancreatic cancer: a dose finding study. *Am J Clin Oncol* 1999; 22: 247–252.

59. Evans JD, Stark A, et al. A phase II trial of marimastat in advanced pancreatic cancer. *Br J Cancer* 2001; 85: 1865–1870.

60. Bramhall SR, Rosemurgy A, et al. Marimastat as first line therapy for patients with unresectable pancreatic cancer: a randomized trial. *J Clin Oncol* 2001; 19: 3447–3455.

61. Bramhall SR, Schultz J, et al. A double-blind placebo controlled, randomized study comparing gemcitabine and marimastat with gemcitabine and placebo as first line therapy in patients with advanced pancreatic cancer. *Br J Cancer* 2002; 87: 161–167.

62. Moore, MJ, Hamm J, et al. Comparison of gemcitabine versus the matrix metalloproteinase inhibitor BAY 12–9566 in patients with advanced or metastatic adenocarcinoma of pancreas: a phase III trial of the National cancer institute of Canada clinical trials group. *J Clin Oncol* 2003; 22: 3296–3302.

63. Wang W, Abbruzzese JL, et al. The Nuclear factor – *B RelA transcription factor is constitutively activated in human pancreatic adenocarcinoma cells. *Clin Cancer Res* 1999; 5: 119–127.

64. Xiong HQ, Abbruzzese JL, et al. NF-*B activity blockade impairs the angiogenic potential of human pancreatic cancer cells. *Int J Cancer* 2004; 108: 181–188.

65. Fujioka S, Sclabas GM, et al. Function of nuclear factor –*B in pancreatic cancer metastasis. *Clin Cancer Res* 2003; 9: 346–354.

66. Li L, Aggarwal BB, et al. Nuclear factor – *B and I*B kinase are constitutively active in human pancreatic cancer cells, and their down regulation by curcumin (diferuloylmethane) is associated with the suppression of proliferation and the induction of apoptosis. *Cancer* 2004; 101: 2351–2362.

67. Nawrocki ST, Sweeney-Gotsch B, Takamori R, McConkey DJ. The proteasome inhibitor bortezomib enhances the activity of docetaxel in orthotopic pancreatic tumor xenografts. *Mol Cancer Ther* 2004; 3(1): 59–70.

68. Alberts S, Gill S, et al. PS-341 and gemcitabine in patients with metastatic pancreatic adenocarcinoma (ACA): final results of a North Central Cancer Treatment Group (NCCTG) randomized phase II study. American Society of Clinical Oncology, Gastrointestinal Cancer Symposium 2005; Abstract 129.

69. Asano T, Yao Y, et al. The rapamycin analog CCI-779 is a potent inhibitor of pancreatic cancer cell proliferation. *Biochem Biophys Res Commun* 2005; 331(1):295–302.

70. Okami J, Yamamoto H, et al. Overexpression of cyclooxygenase-2 in carcinoma of the pancreas. *Clin Cancer Res* 1999; 5: 2018–2024.

71. Xiong HQ, Hess KR, et al. A phase II trial of gemcitabine and celecoxib for metastatic pancreatic cancer. *Proc Am Soc Clin Oncol* 2005; Abstract 4174.

72. Smith SE, Burris HA, et al. Preliminary report of a phase II trial of gemcitabine combined with celecoxib for advanced pancreatic carcinoma. *Proc Am Soc Clin Oncol* 2003; 22: Abstract 1502.

73. Richards DA, Oettle H, et al. Randomized double-blind phase II trial comparing gemcitabine (GEM) plus LY293111 vs GEM plus placebo in advanced adenocarcinoma of the pancreas. *Proc Am Soc Clin Oncol* 2005; Abstract 4092.

74. Smith JP, Shih A, et al. Gastrin regulates the growth of human pancreatic cancer in a tonic and autocrine fashion. *Am J Physiol* 1996; 270: R1078–R1084.

75. Brett BT, Smith SC, et al. Phase II study of anti-gastrin-17 antibodies, raised to G17DT, in advanced pancreatic cancer. *J Clin Oncol* 2002; 20(20): 4225–4231.

76. Gilliam AD, Topuzov, et al. Randomized, double blind, placebo controlled, multi-center, group-sequential trial of G17DT for patients with advanced pancreatic cancer unsuitable or unwilling to take chemotherapy. *Proc Am Soc Clin Oncol* 2004; Abstract 2511.

77. Shapiro J, Marshall J, et al. G17DT + gemcitabine [GEM] versus placebo + Gem in untreated subjects with locally advanced, recurrent, or metastatic adenocarcinoma of the pancreas: results of a randomized, double-blind, multinational, multi-center study. *Proc Am Soc Clin Oncol* 2005; Abstract LBA4012.

10

Untargeted Use of Targeted Therapy

A Dilemma in Non-Small Cell Lung Cancer

Cheryl Ho, MD, Angela M. Davies, MD, Primo N. Lara, MD, O. Gautschi, MD, P. C. Mack, MD, Paul H. Gumerlock, MD, and David R. Gandara, MD

SUMMARY

Lung cancer is the leading cause of cancer death in men and women and frequently presents as advanced disease. The majority of lung cancers are of the non-small cell type, for which chemotherapy has demonstrated modest survival benefits at all stages of disease. Clearly, more effective therapies are needed. Agents that alter critical molecular cell growth pathways, so-called targeted therapies, are a growing area of research and development. Targeted therapies including drugs directed at the epidermal growth factor receptor (EGFR) and vascular endothelial growth factor (VEGF) have recently established a role in the treatment of advanced stage non-small cell lung cancer (NSCLC). These drugs and many others are undergoing investigation, either as single agents or in combination with cytotoxics or other targeted therapies, with dual goals of improved efficacy and reduced toxicity. Although progress has been made in target identification for lung cancer treatment, the ability to select groups of NSCLC patients who benefit from these therapies based on predictive markers remains a challenge. Ongoing studies correlating potential predictive biomarkers with patient outcome are designed to refine the use of developing targeted therapies, that is, to provide a rational basis for "targeted use of targeted therapies." Despite some recent breakthroughs in identifying molecular signatures predictive of benefit, much remains to be learned. Using erlotinib and bevacizumab as prime examples, this chapter will review a dilemma currently facing both basic scientists and clinical investigators engaged in the study of NSCLC, namely how to develop and test paradigms for individualizing patient therapy.

Key Words: Non-small cell lung cancer; targeted therapy; EGFR; VEGF.

From: *Cancer Drug Discovery and Development: Molecular Targeting in Oncology*
Edited by: H. L. Kaufman, S. Wadler, and K. Antman © Humana Press, Totowa, NJ

1. INTRODUCTION

Lung cancer is the second most commonly diagnosed malignancy in North America and is the leading cause of cancer-related deaths. In 2005, an estimated 172,000 men and women were diagnosed with lung cancer in the USA *(1)*. Non-small cell lung cancer (NSCLC) accounts for 80–85% of lung cancer diagnoses. Despite recent data supporting the position that chemotherapy improves patient outcomes in every stage, NSCLC remains a fatal disease in the vast majority of patients.

1.1. Current Standards of Care

Previously, the standard of care for early stage lung cancer was surgery alone. However, four recent trials have established a survival benefit for adjuvant post-operative platinum-based chemotherapy, ranging from 4% to 15% absolute survival benefit at 5 years for resected stage IB–IIIA (Table 1) tumors *(2–5)*. Thus, current practice guidelines advocate the use of adjuvant platinum-based chemotherapy in good performance status patients with resected early-stage lung cancer *(6)*. In patients with locally advanced disease, who are not candidates for curative surgical resection (stage IIIA and IIIB), concurrent chemotherapy and radiation is now established as the standard, with chemotherapy playing a role not only in radiosensitization but also in the eradication of distant micrometastatic disease *(7,8)*.

In advanced NSCLC, palliative chemotherapy has demonstrated improved symptom control and improved survival *(9)*, with recently revised American Society of Clinical Oncology (ASCO) guidelines recommending a two-drug platinum-based regimen, with non-platinum-containing combinations as an alternative *(10)*. For patients who are elderly or have a poor performance status, single agent therapy can be considered. Food and Drug Administration (FDA)-approved second-line treatment options include chemotherapy, docetaxel, or pemetrexed, with the recent addition of one of the first molecular targeted agents to be approved in NSCLC, erlotinib.

1.2. Re-Shaping the Standard of Care

Current guidelines for the treatment of NSCLC are applicable to all subsets of disease. NSCLC, however, is a heterogenous group of histologies with different presentations, natural histories, clinical, and molecular characteristics. Clinical and radiographic findings can give clues to the tumor subtype based on patterns of behavior:

Table 1
Current Recommended Treatment for Non-Small Cell Lung Cancer (NSCLC)

Stage	Treatment
I, II, IIIA (N1)	Surgery + adjuvant chemotherapy
IIIA (N2)	Concurrent chemotherapy + radiotherapy +/- surgery
IIIB	Concurrent chemotherapy + radiotherapy
IV	Chemotherapy – platinum-based doublet
IV (PS 2, elderly)	Single agent chemotherapy
Relapsed/recurrent	Single agent chemotherapy or EGFR TKI

EGFR, epidermal growth factor receptor; EGFR TKI, EGFR tyrosine kinase inhibitor.

squamous cell carcinomas typically arise in the proximal portion of the tracheobronchial tree and present with hemoptysis, adenocarcinomas tend to be more peripherally situated, and large cell tumors frequently present as peripheral lesions with central cavitation. Molecular characteristics including mutations or abnormal expression of EGFR, HER2, thymidylate synthase, k-ras, and ki-67 proliferation index also vary according to histological subtype (11–13). Adding further complexity, microarray analyses have revealed a large number of additional differences at the gene, mRNA, and protein level, within each histopathological type of NSCLC. These biomolecular differences appear to affect prognosis and are likely relevant for treatment (14).

Given the heterogeneity between and within histological subtypes, employing targeted therapies in a rational manner is a tremendous challenge. It has become apparent that differences in tumor histology impact the efficacy of molecular therapies, for example, the increased sensitivity of adenocarcinoma and its subtype, bronchioloalveolar carcinoma, to EGFR tyrosine receptor kinases (TKIs) (15). These concepts may prove to be important not only for the application of new molecular targeted agents but also for conventional cytotoxic drugs, which are also utilized in an unselected fashion. Evidence suggests that the activity of certain chemotherapeutic drugs in NSCLC may be modulated by underlying tumor-specific or host-specific genetic factors, for example, tumor mRNA levels of ERCC1 (pertinent to platinum compounds) and/or protein levels of beta-tubulin III isoforms (pertinent to vinca alkaloids and taxanes), or polymorphisms in DNA repair enzymes observed in host genomic DNA (pertinent to platinum compounds), although these factors are not ready for utilization in clinical practice at this time (16,17).

Thus, there is an increasing need for elucidation of molecular biomarkers predictive for response or lack of response for all pharmacological agents active in lung cancer and other malignancies as well. While recent studies using erlotinib and bevacizumab have emphasized the emerging importance of histologic subtype of NSCLC (i.e., increased efficacy of erlotinib in adenocarcinoma and risk of hemorrhage in squamous histology) and inter-individual patient characteristics (female sex, performance status, degree of prior therapy), these studies have likewise demonstrated relevance for molecular profiling in predicting favorable outcomes (i.e., EGFR gene copy number and mutation status). To optimize the approach to individual patients with NSCLC, future therapeutic strategies must recognize and address these issues head on.

This chapter reviews current evidence in support of utilizing molecular targeted therapies in unselected NSCLC patient populations and suggests a paradigm shift toward incorporating patient-specific pharmacogenomic information and tumor-related molecular profiling into the treatment selection process.

2. THERAPEUTIC TARGETS

2.1. Angiogenesis and Tumor Hypoxia

2.1.1. VASCULAR ENDOTHELIAL GROWTH FACTOR: BEVACIZUMAB (AVASTIN®)

Despite multiple attempts to improve the overall survival of patients with advanced NSCLC, a plateau has seemingly been reached with currently available platinum-containing doublets. Furthermore, despite exhaustive study, there is no evidence that non-platinum chemotherapeutic combinations offer any benefit in either efficacy or toxicity compared with platinum-based therapy (10,18). Over the last decade, many new agents purportedly directed against important tumor-related molecular targets have been combined with front-line platinum-based therapy in the phase III setting without success, including the matrix metalloproteinases, tirapazimine (hypoxic cytoxin), LY90003 (protein kinase C alpha antisense oligonucleotide), bexarotene (retinoic acid receptor agonist), and gefitinib and erlotinib [EGFR tyrosine kinase inhibitors (EGFR TKIs)] (19–25). The negative results of these studies, and the tremendous patient resources consumed in these efforts, have raised serious questions about the validity of further testing the hypothesis that a targeted agent plus chemotherapy would improve survival compared with chemotherapy alone.

Bevacizumab, a recombinant, humanized monoclonal antibody that binds VEGF and inhibits angiogenesis, has proven to be an exception to this rule, being the first targeted therapy in NSCLC to improve clinical outcomes when given in combination with chemotherapy (Fig. 1). An Eastern Cooperative Oncology Group (ECOG) phase III study (ECOG 4599) was conducted in a clinically select group of patients comparing paclitaxel 200 mg/m^2 and carboplatin, area under the curve (AUC) 6, with or without bevacizumab 15 mg/kg every 3 weeks [paclitaxel/carboplatin (PC) versus paclitaxel carboplatin bevacizumab (PCB)] (26). Patients with chemotherapy naïve stage IIIB pleural effusion or IV NSCLC were eligible. Owing to hemorrhagic complications seen in the phase II study, including fatal and life-threatening pulmonary hemorrhages; patients with squamous histology, a history of thrombotic or hemorrhagic disorders, hemoptysis, or intracranial metastases were excluded. Thus, patients were selected for this study based on clinical features that predicted for risk of serious adverse events, not the drug target. Over 800 patients were enrolled in ECOG 4599, and although the response rate (RR) for PC alone was lower than would have been expected based on historical data (10%), the combination arm (PCB) demonstrated improvement in all measures of efficacy, including RR at 27% ($p < 0.0001$). More importantly, PCB increased median progression-free survival (4.5 versus 6.4 months, $p < 0.0001$) and

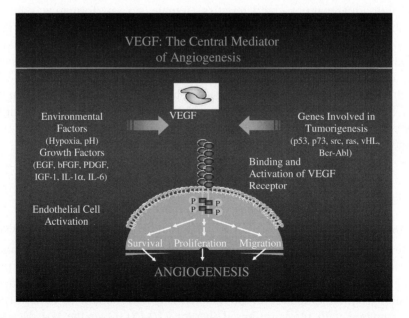

Fig. 1. The vascular endothelial growth factor receptor pathway.

median survival (10.2 versus 12.5 months, hazard ratio 0.77, $p = 0.007$). Treatment was generally well tolerated in both arms; however, there were higher rates of > grade 3 hemorrhage (PC 0.7%, PCB 4.5%, $p < 0.001$) and hypertension (0.7 versus 6%, $p < 0.001$) in the bevacizumab-containing arm. There were two treatment-related deaths in the PC arm and eight in PCB, seven of the latter were because of hemorrhagic events (six hemoptysis and two gastrointestinal bleeds). In addition, 1% of patients on PCB had > grade 3 central nervous system (CNS) hemorrhage, despite the requirement for a normal baseline brain computed tomography (CT) or magnetic resonance imaging (MRI). Thus, although the overall rates of toxicity with PCB were low, the association with severe or fatal hemorrhage was of concern in this selected population.

Unplanned sub-group analyses identified survival benefit across all subgroups (stage, weight loss, performance status, age, ethnicity, and prior radiotherapy treatment) except gender. No improvement in survival was observed in women receiving PCB, despite improved RRs and progression-free survival. The reason for this finding is unclear, although the authors speculate that unknown differences in baseline prognostic characteristics and the use of second- or third-line therapy may have contributed to this finding. Determining whether the survival benefit with PCB is gender-specific will require further studies. Although patients were not selected for participation in this trial based on molecular features, specimens were collected and correlative studies assessing VEGF, soluble E-selectin, soluble inter-cellular adhesion molecule 1 (ICAM), and basic fibroblast growth factor (bFGF) are being performed (27). However, to date, a measurable molecular marker predictive for benefit from bevacizumab has not been identified.

Prospective correlative studies in the randomized phase II trial of bevacizumab in renal cell carcinoma were conducted and did not demonstrate a correlation between response and baseline plasma vascular endothelial growth factor (VEGF) levels (28). In the metastatic colorectal cancer trial employing bevacizumab with irinotecan and 5-fluorouracil, retrospective analysis of k-ras, b-raf, and p53 expression did not demonstrate a correlation with tumor response (29). Thus, although ECOG 4599 is the first phase III trial to show significant improvement in overall survival by adding a molecular targeted agent to platinum-based chemotherapy in advanced NSCLC, enthusiasm must be tempered by the observation of increased severe and fatal hemorrhage, despite inclusion criteria designed to minimize this risk. Furthermore, these results were obtained in an unselected patient population: "untargeted use of targeted therapy," standing in stark contrast to trials in breast cancer with another targeted agent, trastuzumab, which limited patient accrual to those with HER2 positive tumors. At present, lack of known predictive markers prevents oncologists from selecting patients most likely to benefit from bevacizumab therapy, while minimizing risks to those least likely to benefit. To optimize the risk/benefit ratio of this and other anti-angiogenic therapies, establishing a reliable and reproducible set of predictive biomarkers is essential.

2.1.2. Hypoxic cell cytotoxin: Tirapazamine (Tirazone®)

Tirapazamine is a bioreductive drug that is cytotoxic only under hypoxic conditions (Fig. 2). In the setting of hypoxia, it is metabolized to an oxidizing radical; however, in the presence of oxygen, it reverts to the non-toxic parent compound (30). In pre-clinical

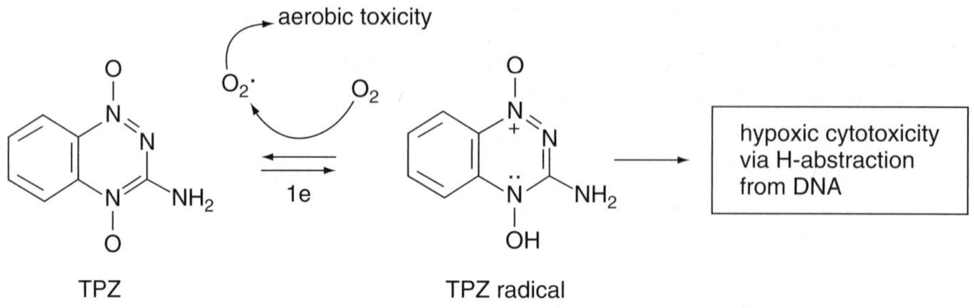

Fig. 2. Tirapazamine, the hypoxic cytotoxin.

studies, this property augments the cytotoxicity of various chemotherapeutic agents, facilitating cell death under low-oxygen conditions. Clinical proof of principle was provided by the CATAPULT I trial, in which cisplatin plus tirapazamine resulted in improved survival compared with cisplatin alone in advanced stage NSCLC *(21)*. CATAPULT II, however, did not demonstrate an improvement in time to progression or overall survival when tirapazamine was used to replace standard chemotherapy, etoposide, in combination with cisplatin *(25)*. The Southwest Oncology Group (SWOG) performed a phase III trial of carboplatin and paclitaxel with or without tirapazamine in advanced NSCLC patients (S0003) *(31)*. Unfortunately, the addition of tirapazamine to the platinum-based doublet did not improve survival, and the combination was significantly more toxic than anticipated. Twice as many patients in the tirapazamine arm were removed from treatment because of toxicity compared with chemotherapy alone (28% versus 14%, $p = 0.001$). Although the results were disappointing, studies examining hypoxia-induced genes, plasminogen activator inhibitor (PAI)-1, VEGF, and osteopontin, as prognostic and predictive factors demonstrated a potential correlation *(32,33)*. Early data from our group suggest that high osteopontin levels correlate with poor survival. Investigations are currently underway to evaluate the association between osteopontin and other markers as predictors of therapeutic response to facilitate use of tirapazamine in a selected population *(33)*.

2.2. Epidermal Growth Factor Receptor Pathways

2.2.1. EPIDERMAL GROWTH FACTOR RECEPTOR TKIS: GEFITINIB (IRESSA®)

Gefitinib was the first targeted drug to be evaluated in NSCLC. It is an oral, reversible TKI that blocks the epidermal growth factor receptor (EGFR) pathway, inhibiting proliferation, differentiation, and angiogenesis (Fig. 3). On the basis of an RR of 12–18% observed in phase II trials in patients with previously treated NSCLC, gefitinib received FDA-accelerated approval for third-line treatment, conditional on further investigations of this agent *(34,35)*. In a recent phase III trial (ISEL), gefitinib did not demonstrate a statistically significant survival advantage over placebo in an unselected NSCLC population *(36)*. As the RR to gefitinib in ISEL was similar to that seen with single agent erlotinib in the BR21 trial, the reasons why BR21 was positive for a survival benefit and ISEL showed only a non-significant trend remain unclear at this time. Although possibilities include increased potency of erlotinib and favorable pharmacodynamics, it appears more likely that differences in patient characteristics

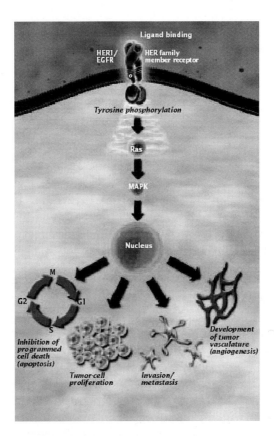

Fig. 3. The epidermal growth factor receptor pathway.

may be responsible. For example, BR21 included patients previously treated with chemotherapy, regardless of response status, whereas ISEL required progressive disease after chemotherapy. In support of this view, planned subgroup analysis of ISEL did demonstrate improved survival in never smokers ($n = 375$; hazard ratio (HR) 0.67; 8.9 versus 6.1 months, $p = 0.012$) and Asian patients ($n = 342$; HR 0.66; 9.5 versus 5.5 months, $p = 0.010$), indicating that patient selection may have been a factor in the results of this study.

Gefitinib was evaluated in two phase III trials for advanced NSCLC in combination with chemotherapy: INTACT 1 (cisplatin and gemcitabine) and INTACT 2 (carboplatin and paclitaxel). The addition of gefitinib did not show any benefit in RR, time to progression, or overall survival compared with chemotherapy alone *(37,38)*. More recently, gefitinib was also evaluated as maintenance therapy following definitive concurrent chemoradiation (cisplatin and etoposide) and consolidation docetaxel in the SWOG 0023 trial *(39)*. A preliminary analysis indicated that gefitinib did not improve survival when delivered in this clinical setting ($p = 0.0015$). In fact, overall survival from randomization was numerically inferior with maintenance gefitinib [19 months compared with 29 months with placebo ($p = 0.09$)]. At the present time, broad utilization of gefitinib cannot be recommended and the FDA is limiting the scope in which this drug can be utilized to patients who are currently or have previously taken gefitinib with evidence of benefit or in the clinical trial setting.

2.2.2. EGFR TKI: Erlotinib (Tarceva®)

Erlotinib is a small molecule EGFR TKI sharing a similar mechanism of action as gefitinib (Fig. 3). After phase II studies demonstrated promising efficacy and a favorable toxicity profile with single agent erlotinib in refractory NSCLC, large randomized controlled studies were conducted in combination with front-line chemotherapy. Erlotinib was combined with cisplatin/gemcitabine (TALENT) and carboplatin/paclitaxel (TRIBUTE) and compared with the platinum doublets alone: in an identical fashion to gefitinib, no improvement in RR, time to progression, or overall survival was observed *(40,41)*. Two hypotheses have been proposed to account for these negative results: (i) lack of patient selection for the EGFR target and (ii) potential negative interaction (i.e., antagonism) between chemotherapy and EGFR TKIs when delivered concurrently *(42)*. Each of these two hypotheses are currently being evaluated. To test the first hypothesis, clinical trials are ongoing with NSCLC populations enriched for clinical features (female, non-smoker, adenocarcinoma) or biomarkers (EGFR mutation, EGFR gene copy number) predictive of benefit from EGFR TKI therapy. For the second hypothesis: strategies being evaluated include sequential delivery of chemotherapy for several cycles followed by the EGFR TKI, as well as intermittent dosing schedules to achieve pharmacodynamic separation of chemotherapy and erlotinib *(43)*.

Single agent erlotinib was evaluated in a randomized, placebo-controlled trial (BR 21) in patients with previously treated advanced NSCLC *(44)*. The RR was 8.9% for erlotinib-treated patients, with improved survival, 6.7 versus 4.7 months in the placebo arm (hazard ratio 0.70, $p < 0.001$). Quality-of-life analysis also demonstrated a benefit with erlotinib with an improvement in the median time to deterioration of cough (4.9 versus 3.7 months, $p = 0.04$), dyspnea (4.7 versus 2.9 months, $p = 0.03$) and pain (2.8 versus 1.9 months, $p = 0.04$). This trial was the first to demonstrate a significant benefit in overall survival with an EGFR TKI and led to the approval of erlotinib for second-line treatment of advanced NSCLC.

2.2.3. Predictive Biomarkers for EGFR TKI Efficacy

Greater progress has been made in identifying predictive factors for benefit from EGFR TKIs than from VEGF-targeted therapies. In early trials with EGFR TKIs, greater efficacy was seen in specific clinical subgroups, namely non-smoking Asian females with adenocarcinoma (including bronchioloalveolar histology) *(34,35)*. Subsequent studies have revealed that many of these patients had mutations in the ATP-binding domain of EGFR *(45–47)*. Although higher RRs were seen in this specific population, an analysis from BR 21 (erlotinib versus placebo) indicated that EGFR mutation did not predict for survival endpoints. Instead, all subsets of patients benefited from therapy including smokers, men, and patients with squamous cell histology, groups known to have a low incidence of EGFR mutation *(44,48)*. In the BR 21 correlative studies, increased EGFR copy number assessed by fluorescent in situ hybridization (FISH) was predictive of response with erlotinib. In univariate analysis, survival was longer with erlotinib-treated patients with EGFR expression and high copy number of EGFR. These results are identical with those of the SWOG trial S0126, which also demonstrated that EGFR positivity predicted for improved response, progression-free survival, and overall survival, whereas EGFR mutation predicted only for response *(49)*. In addition to high EGFR gene copy number by FISH and EGFR mutation, EGFR protein expression by immunohistochemistry (IHC), k-ras mutations, and phosphory-lation of mitogen-activated protein kinase (MAPK) and AKT have all been reported to be relevant markers for activity of EGFR TKIs *(50)*. The latter findings indicate that activity of the overall signaling pathways, rather than the simple presence of the EGFR target, may be biologically and therapeutically important *(49,51)*. Although these molecular analyses have been helpful in explaining why certain patient character-istics predict for benefit with EGFR TKI treatment, none have yet to be prospectively validated as a screening tool, permitting incorporation into routine clinical practice.

2.3. Retinoic Acid Receptor Agonists

2.3.1. BEXAROTENE (TARGRETIN®)

Bexarotene selectively binds and activates retinoid X receptors (RXR) (Fig. 4), modulating downstream signaling pathways involved in cell cycle control, apoptosis and differentiation, and causing down-regulation of EGFR and cyclin D1. In NSCLC, high RXR-beta expression is reported to correlate with longer patient survival; therefore, it was hypothesized that bexarotene plus chemotherapy would be more effective than chemotherapy alone *(52)*. Two phase III trials of bexarotene with or without chemotherapy, cisplatin/vinorelbine (SPIRIT I) and carboplatin/paclitaxel (SPIRIT II), were performed *(23,24)*. Both trials were negative, with no overall improvement in PFS or OS when compared with chemotherapy alone. However, in subgroup analyses, severe hypertriglyceridemia in bexarotene-treated patients corre-lated with improved outcomes in both studies. In SPIRIT I and II, the median survival in patients with grade 3–4 triglycerides in comparison with grade 0–2 was extended by approximately 6 months. Thus, the investigators speculate that triglyceride levels may be a positive selection factor for therapy with bexarotene

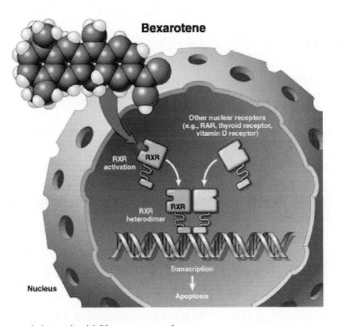

Fig. 4. Bexarotene and the retinoid X receptor pathway.

plus chemotherapy, whereby bexarotene-induced hypertriglyceridemia would be a requirement for continued treatment with the combination. Despite these intriguing correlative findings, at present there is no defined role for this agent in the therapy of NSCLC.

2.3.2. A Paradigm for Future Study of Targeted Therapies

As described section 2.2.2 and 2.1.1, clinical trials of the targeted therapies erlotinib (BR21) and bevacizumab (ECOG 4599) have shown improved survival in NSCLC, even when employed in patient populations unselected for the molecular target of interest. To optimize use of this and other emerging targeted therapies, prospective identification of patient subgroups who are most likely to benefit is essential. To achieve this goal, in 1999, the SWOG Lung Committee adopted a multi-tiered approach toward linking correlative science studies and clinical trials (Fig. 5) *(53)*. Recent recommendations from the National Cancer Institute provide a rationale, direction, and support for translational science efforts such as these *(52)*. The first step involves the selection and in vitro evaluation of potential biomarkers, based on current understanding of the drug target, signaling pathways, and downstream events. Observations derived from the initial laboratory investigations serve to refine the pre-clinical model and highlight assays of interest. Subsequent evaluation is performed in animal models and then tested in human tissues before prospective study in a clinical trial setting. In phase I trials, the goal is to further refine proposed biomarker assays and concepts, generating further hypotheses to be tested in subsequent larger studies. Within the phase II setting, assessment of proposed biomarkers in the context of patient outcomes facilitates further refinement and assists in differentiating prognostic versus predictive factors. Finally, correlative studies conducted within the context of phase III trials comparing targeted therapy with standard treatment allows for validation of the predictive value

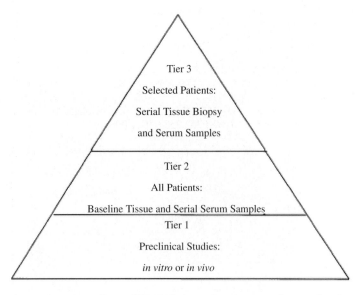

Fig. 5. The Southwest Oncology Group (SWOG) three-tiered approach to molecular–clinical correlative studies.

of biomarkers. This approach permits evaluation of both new and established research methodologies and rapid translation of new laboratory findings and technologies into clinical application. If successful, this approach provides the basis for defining a patient population most likely to benefit from the targeted therapy of interest and, equally important, those patients who should be excluded from such therapy because of lack of efficacy or high risk for toxicity.

For potential predictive markers to be of practical benefit within the clinic, several criteria should be considered. Firstly, expression should be altered in a significant number of patients and the marker should correlate with response or survival, as described above *(54)*. In addition, specimens required for analysis should be easily obtainable, that is, assays requiring tumor biopsy to obtain fresh tissue will have limited applicability compared with assays amenable to analysis of paraffin-embedded specimens already in hand from prior diagnostic biopsies. Methods for analysis should be generalizable to outside research centers, results must be available in a timely manner, and the assay must be cost effective.

3. CONCLUSIONS

Through a better understanding of the molecular pathways responsible for tumor growth, metastasis, and angiogenesis, drugs that are more cancer-specific, and potentially less toxic, are being developed, investigated, and integrated into NSCLC therapy. Although two targeted agents, erlotinib and bevacizumab, have demonstrated survival benefit in advanced NSCLC, at present they are being employed in an unselected patient population. Optimizing use of these agents and other targeted therapies now in development will require improved understanding of tumor biology, improved experimental methodologies, and close collaboration between basic and clinical scientists engaged in lung cancer research. Only through carefully designed and conducted translational and correlative science studies will predictive biomarkers be established that enable the practicing oncologist to employ targeted therapies in a truly targeted fashion.

REFERENCES

1. Society AC: Cancer Facts and Figures 2005, 2005.
2. Arriagada R, Bergman B, Dunant A, et al. Cisplatin-based adjuvant chemotherapy in patients with completely resected non-small-cell lung cancer. *N Engl J Med* 350:351–60, 2004.
3. Winton T, Livingston R, Johnson D, et al. Vinorelbine plus cisplatin vs. observation in resected non-small-cell lung cancer. *N Engl J Med* 352:2589–97, 2005.
4. Strauss GM, Herndon J, Maddaus MA, et al. Randomized clinical trial of adjuvant chemotherapy with paclitaxel and carboplatin following resection in stage IB non-small cell lung cancer (NSCLC): report of cancer and leukemia group B (CALGB) protocol 9633. *J Clin Oncol* 22:7019, 2004 (Meeting Abstracts).
5. Douillard J-Y, Rosell R, Delena M, et al. ANITA: Phase III adjuvant vinorelbine (N) and cisplatin (P) versus observation (OBS) in completely resected (stage I-III) non-small-cell lung cancer (NSCLC) patients (pts): Final results after 70-month median follow-up. On behalf of the Adjuvant Navelbine International Trialist Association. *J Clin Oncol* 23:7013, 2005 (Meeting Abstracts).
6. Ettinger DS, Akerley W, Bepler G, et al. NCCN clinical practice guidelines in oncology: non-small cell lung cancer, http://www.nccn.org/professionals/physician_gls/pdf/nscl.pdf. Accessed June 14, 2007.
7. Furuse K, Fukuoka M, Kawahara M, et al. Phase III study of concurrent versus sequential thoracic radiotherapy in combination with mitomycin, vindesine, and cisplatin in unresectable stage III non-small-cell lung cancer. *J Clin Oncol* 17:2692–9, 1999.

8. Sause W, Kolesar P, Taylor SI, et al. Final results of phase III trial in regionally advanced unresectable non-small cell lung cancer: Radiation Therapy Oncology Group, Eastern Cooperative Oncology Group, and Southwest Oncology Group. *Chest* 117:358–64, 2000.

9. Anonymous. Chemotherapy in non-small cell lung cancer: a meta-analysis using updated data on individual patients from 52 randomised clinical trials. Non-small Cell Lung Cancer Collaborative Group. *BMJ* 311:899–909, 1995.

10. Pfister DG, Johnson DH, Azzoli CG, et al. American Society of Clinical Oncology treatment of unresectable non-small-cell lung cancer guideline: update 2003. *J Clin Oncol* 22:330–53, 2004.

11. Eberhard DA, Johnson BE, Amler LC, et al. Mutations in the epidermal growth factor receptor and in KRAS are predictive and prognostic indicators in patients with non-small-cell lung cancer treated with chemotherapy alone and in combination with erlotinib. *J Clin Oncol* 23:5900–9, 2005.

12. Huang C, Liu D, Masuya D, et al. Clinical application of biological markers for treatments of resectable non-small-cell lung cancers. *Br J Cancer* 92:1231–9, 2005.

13. Nakamura H, Kawasaki N, Taguchi M, et al. Association of HER-2 overexpression with prognosis in nonsmall cell lung carcinoma: a metaanalysis. *Cancer* 103:1865–73, 2005.

14. Meyerson M, Carbone D. Genomic and proteomic profiling of lung cancers: lung cancer classification in the age of targeted therapy. *J Clin Oncol* 23:3219–26, 2005.

15. Miller VA, Herbst R, Prager D, et al. Long survival of never smoking non-small cell lung cancer (NSCLC) patients (pts) treated with erlotinib HCl (OSI-774) and chemotherapy: Sub-group analysis of TRIBUTE. *J Clin Oncol* 22:7061, 2004 (Meeting Abstracts).

16. Seve P, Isaac S, Tredan O, et al. Expression of class III {beta}-tubulin is predictive of patient outcome in patients with non-small cell lung cancer receiving vinorelbine-based chemotherapy. *Clin Cancer Res* 11:5481–6, 2005.

17. Seve P, Dumontet C. Chemoresistance in non-small cell lung cancer. *Curr Med Chem AntiCancer Agents* 5:73–88, 2005.

18. D'Addario G, Pintilie M, Leighl NB, et al. Platinum-based versus non-platinum-based chemotherapy in advanced non-small-cell lung cancer: a meta-analysis of the published literature. *J Clin Oncol* 23:2926–36, 2005.

19. Smylie M, Mercier R, Aboulafia D, et al. Phase III study of the matrix metalloprotease (MMP) inhibitor prinomastat in patients having advanced non-small cell lung cancer (NSCLC). ASCO 2001.

20. Leighl NB, Shepherd F, Paz-Ares L, et al. Randomized phase II-III study of matrix metalloproteinase inhibitor (MMPI) BMS-275291 in combination with paclitaxel (P) and carboplatin (C) in advanced non-small cell lung cancer (NSCLC): NCIC-CTG BR.18. *J Clin Oncol* 22:7038, 2004 (Meeting Abstracts).

21. von Pawel J, von Roemeling R, Gatzemeier U, et al. Tirapazamine plus cisplatin versus cisplatin in advanced non-small-cell lung cancer: a report of the international CATAPULT I study group. *J Clin Oncol* 18:1351–1359, 2000.

22. Lynch T RR, Lind M, Riviere A, Gatzemeier U, Drorr A, Holmlund J, Yuen A, Sikic B. Randomized phase III trial of chemotherapy and antisense oligonucleotide LY900003 (ISIS 3521) in patients with advanced NSCLC: Initial report, ASCO 2003.

23. Jassem J, Zatloukal P, Ramlau R, et al. A randomized phase III trial comparing bexarotene/cisplatin/vinorelbine versus cisplatin/vinorelbine in chemotherapy-naive patients with advanced or metastatic non-small cell lung cancer (NSCLC). *J Clin Oncol* 23:LBA7024, 2005 (Meeting Abstracts).

24. Blumenschein GR, Khuri F, Gatzemeier U, et al. A randomized phase III trial comparing bexarotene/carboplatin/paclitaxel versus carboplatin/paclitaxel in chemotherapy-naive patients with advanced or metastatic non-small cell lung cancer (NSCLC). *J Clin Oncol* 23:LBA7001, 2005 (Meeting Abstracts).

25. Shepherd F, Koschel G, von Pawel J, Gatzmeier U, Van Zandwiyk N, Woll P, Van Klavren R, Krasko P, Desimone P, Nicolson M, Pieters W, Bigelow R, Rey A, Viallet J, Loh E. Comparison of tirazone (tirapazamine) and cisplatin vs. etoposide and cisplatin in advanced non-small cell lung cancer (NSCLC): final results of the international Phase III CATAPULT II Trial. *Lung Cancer* 29:28, 2000.

26. Sandler AB, Gray R, Brahmer J, et al. Randomized phase II/III trial of paclitaxel (P) plus carboplatin (C) with or without bevacizumab (NSC #704865) in patients with advanced non-squamous non-small cell lung cancer (NSCLC): An Eastern Cooperative Oncology Group (ECOG) Trial - E4599. *J Clin Oncol* 23:LBA4, 2005 (Meeting Abstracts).

27. Sparano JA, Gray R, Giantonio B, et al. Evaluating antiangiogenesis agents in the clinic: the Eastern Cooperative Oncology Group Portfolio of Clinical Trials. *Clin Cancer Res* 10:1206–11, 2004.

28. Yang JC, Haworth L, Sherry RM, et al. A randomized trial of bevacizumab, an anti-vascular endothelial growth factor antibody, for metastatic renal cancer. *N Engl J Med* 349:427–434, 2003.

29. Ince WL, Jubb AM, Holden SN, et al. Association of k-ras, b-raf, and p53 status with the treatment effect of bevacizumab. *J Natl Cancer Inst* 97:981–989, 2005.

30. Gandara DR, Lara PN, Jr., Goldberg Z, et al. Tirapazamine: prototype for a novel class of therapeutic agents targeting tumor hypoxia. *Semin Oncol* 29:102–9, 2002.

31. Williamson S, Crowley JJ, Lara PN, Tucker RW, McCoy J, Lau DHM, Gandara DR. S0003: pacli-taxel/carboplatin (PC) v PC+tirapazamine (PCT) in advanced non small cell lung cancer (NSCLC) A phase III Southwest Oncology Group (SWOG) trial. *ASCO Chicago* 2003, pp 662.

32. Galvin IV, Lara PN, Le Q, et al. Hypoxia-related markers in the plasma of patients with advanced non-small cell lung cancer (NSCLC) and survival from chemotherapy: Southwest Oncology Group (SWOG) S0003. *J Clin Oncol* 22:7146, 2004 (Meeting Abstracts).

33. Mack PC, Gumerlock PH, Primo N. Lara Jr., Quynh-Thu Le, Jeffrey A. Longmate, Kari Chansky, John J. Crowley, David R. Gandara. Plasma markers of tumor hypoxia-angiogenesis as predictors of therapeutic response. AACR 2005.

34. Fukuoka M, Yano S, Giaccone G, et al. Multi-institutional randomized phase II trial of gefitinib for previously treated patients with advanced non-small-cell lung cancer. *J Clin Oncol* 21:2237–2246, 2003.

35. Kris MG, Natale RB, Herbst RS, et al. Efficacy of gefitinib, an inhibitor of the epidermal growth factor receptor tyrosine kinase, in symptomatic patients with non-small cell lung cancer: a randomized trial. JAMA 290:2149–58, 2003.

36. Thatcher N, Chang A, Parikh P, et al. ISEL: a phase III survival study comparing gefitinib (Iressa) plus best supportive care (BSC) with placebo plus BSC in patients with advanced non small cell lung cancer (NSCLC) who had received one or two prior chemotherapy regimens. *Lung Cancer* 49:S4, 2005.

37. Giaccone GJ, Johnson DH, Manegold C, Scagliotti GV, Rosell R, Wolf M, Rennie P, Ochs J, Averbuch S, Fandi A. A phase III clinical trials of ZD1839 ("Iressa") in combination with gemcitabine and cisplatin chemotherapy in chemotherapy-naive patients with advanced non-small cell lung cancer (INTACT-1). *Ann Oncol* 13:2, 2002.

38. Herbst RS, Giaccone G, Schiller JH, et al. Gefitinib in combination with paclitaxel and carboplatin in advanced non-small-cell lung cancer: a phase III trial–INTACT 2. *J Clin Oncol* 22:785–94, 2004.

39. Kelly K, Gaspar LE, Chansky K, et al. Low incidence of pneumonitis on SWOG 0023: A preliminary analysis of an ongoing phase III trial of concurrent chemoradiotherapy followed by consolidation docetaxel and gefitinib/placebo maintenance in patients with inoperable stage III non-small cell lung cancer. *J Clin Oncol* 23:7058, 2005 (Meeting Abstracts).

40. Gatzemeier U, Pluzanska A, Szczesna A, Kaukel E, Roubec J, Brennscheidt U, DeRosa F, Mueller B, von Pawel J. Results of a phase III trial of erlotinib (OSI-774) combined with cisplatin and gemcitabine chemotherapy in advanced non-small cell lung cancer. *ASCO New Orleans* 2004.

41. Herbst RS, Prager D, Hermann R, et al. TRIBUTE: a phase III trial of erlotinib hydrochloride (OSI-774) combined with carboplatin and paclitaxel chemotherapy in advanced non-small-cell lung cancer. *J Clin Oncol* 23:5892–5899, 2005.

42. Gandara DR, Gumerlock PH. EGFR tyrosine kinase inhibitors plus chemotherapy: Case closed or is the jury still out? *J Clin Oncol* in press, 2005.

43. Davies AM, Lara PN, Lau DH, et al. Intermittent erlotinib in combination with docetaxel (DOC): Phase I schedules designed to achieve pharmacodynamic separation. *J Clin Oncol* 23:7038, 2005 (Meeting Abstracts).

44. Shepherd FA, Rodrigues Pereira J, Ciuleanu T, et al. Erlotinib in previously treated non-small-cell lung cancer. *N Engl J Med* 353:123–32, 2005.

45. Paez JG, Janne PA, Lee JC, et al. EGFR Mutations in Lung Cancer: Correlation with Clinical Response to Gefitinib Therapy. *Science* 304:1497–1500, 2004.

46. Pao W, Miller V, Zakowski M, et al. EGF receptor gene mutations are common in lung cancers from "never smokers" and are associated with sensitivity of tumors to gefitinib and erlotinib. PNAS 101:13306–13311, 2004.

47. Lynch TJ, Bell DW, Sordella R, et al. Activating mutations in the epidermal growth factor receptor underlying responsiveness of non-small-cell lung cancer to gefitinib. *N Engl J Med* 350:2129–2139, 2004.

48. Tsao MS, Sakurada A, Cutz JC, et al. Erlotinib in lung cancer - molecular and clinical predictors of outcome. *N Engl J Med* 353:133–44, 2005.

49. Hirsch FR, Varella-Garcia M, McCoy J, et al. Increased epidermal growth factor receptor gene copy number detected by fluorescence in situ hybridization associates with increased sensitivity to gefitinib in patients with bronchioloalveolar carcinoma subtypes: a southwest oncology group study. *J Clin Oncol* 23:6838–6845, 2005.

50. Gandara DR, West H, Chansky K, et al. Bronchioloalveolar Carcinoma: A Model for Investigating the Biology of Epidermal Growth Factor Receptor Inhibition. *Clin Cancer Res* 10:4205S–4209, 2004.

51. Laskin J, Johnson DH. Receptor tyrosine kinase inhibitors, in Pass HI, Carbone DP, Muna JD, Johnson DH, Turrisi AT (eds): *Lung Cancer Principles and Practice*, 3rd edn, Lippincott Williams and Wilkins, Philadelphia 2005.

52. Brabender J, Metzger R, Salonga D, et al. Comprehensive expression analysis of retinoic acid receptors and retinoid X receptors in non-small cell lung cancer: implications for tumor development and prognosis. *Carcinogenesis* 26:525–30, 2005.

53. Gandara DR, Lara PN, Lau DH, et al. Molecular-clinical correlative studies in non-small cell lung cancer: application of a three-tiered approach. *Lung Cancer* 34(Suppl 3):S75–80, 2001.

54. Davies AM, Mack P, Lara P, Lau D, Danenberg K, Gumerlock P, Gandara D. Predictive molecular markers: Has the time come for routine use in lung cancer? JNCCN 2:125–134, 2004.

11

Renal Cell Carcinoma

Renal Cell Cancer

Olwen Hahn, MD,
and Walter Stadler, MD, FACP

SUMMARY

Renal cell carcinoma (RCC) accounts for 2–3% of all malignancies; however, its incidence has increased by 43% since 1973, with a 16% increase in the death rate. It is estimated that, in 2005, more than 36,000 new cases of kidney cancer will be diagnosed in the USA, and there will be 12,660 deaths. The recent Food and Drug Administration's (FDA) approval of two multi-tyrosine kinase inhibitors and encouraging results from a phase III trial of a mammalian target of rapamycin (mTOR) inhibitor represent significant advances in the treatment of metastatic renal cancer. An increased understanding of molecular oncology and cancer genetics has identified multiple genetic and cell-signaling defects in RCC that are prime targets for novel therapies.

Key Words: Renal cell carcinoma; molecular therapy.

1. INTRODUCTION

Renal cell carcinoma (RCC) accounts for 2–3% of all malignancies; however, its incidence has increased by 43% since 1973, with a 16% increase in the death rate *(1,2)*. It is estimated that, in 2005, more than 36,000 new cases of kidney cancer will be diagnosed in the USA, and there will be 12,660 deaths *(3)*. Approximately, 20–30% of patients will have metastatic disease at initial diagnosis, and 20–40% of patients with localized RCC who undergo nephrectomy will develop metastases *(4)*. The survival of patients with metastatic renal cancer is poor, with a median survival of less than 1 year *(5)*.

RCC is a collection of different neoplasms with distinct histologies and varying responses to therapy. The most common renal cell cancer is clear cell, accounting for approximately 75% of localized renal tumors and thought to originate from the proximal tubal epithelium *(6)*. Other histologic types include papillary (15%), chromophobe (5%), medullary (<2%), X-translocation tumors (<1%), and benign oncocytoma (5%). Molecular expression analyses suggest that additional subtypes may exist *(7)*.

From: *Cancer Drug Discovery and Development: Molecular Targeting in Oncology*
Edited by: H. L. Kaufman, S. Wadler, and K. Antman © Humana Press, Totowa, NJ

Surgery remains the mainstay of therapy for localized RCC *(8)*. Treatment of
metastatic RCC remains a challenge, as renal tumors are not radiosensitive and few
cytotoxic chemotherapies have shown consistent responses. Before the development
of tyrosine kinase inhibitors sorafenib and sunitinib, cytokine-based immunotherapy
with interferon (IFN)-α and interleukin (IL)-2 was the cornerstone of therapy for
metastatic RCC. In 1992, the FDA approved high-dose IL-2 for treatment of
metastatic RCC. The overall response rate for high-dose IL-2 in metastatic RCC is
15–20%, including some complete responders *(9)*. However, this therapy has considerable toxicities, including hypotension, respiratory distress, and renal impairment;
also, high-dose IL-2 requires administration in hospital ward with specialized nursing
care. A phase III trial compared high-dose IL-2 (720,000 U/kg) with low-dose IL-2
(72,000 U/kg) *(10)*. Whereas low-dose IL-2 had considerably fewer toxicities, high-
dose therapy had significant higher responses (21% versus 13%, $p = 0.048$). IFN-α
monotherapy has been studied in several trials and appears to be superior to placebo
with a survival advantage of 3.8 months and response rate of approximately 15% in
randomized phase III trials *(11)*.

Although there are considerable side effects to IFN-α, such as fever, malaise, and
flu-like symptoms, it is less toxic than high-dose IL-2 and can be administered at home
by the patient. Various combination IL-2 and IFN-α regimens have been described,
and recently, a phase III study of high-dose IL-2 versus IFN-α/IL-2 was reported *(12)*.
This trial reports a higher response rate with high-dose IL-2 in comparison with IFN-
α/IL-2 (23% versus 9.9%, $p = 0.018$); however, the median duration of response and
the median survival were not statistically significant. Chemoimmunotherapies have not
been shown to be any more effective than IL-2 and/or IFN-α.

It appears that low-dose immunotherapy has a modest antitumor and survival benefit,
whereas high-dose IL-2 leads to durable complete responses in a small proportion of
patients. It would therefore be useful to identify patients most likely to benefit from the
latter very toxic regimen. To this end, data strongly suggest that non-clear cell renal
carcinomas do not benefit from IL-2 or IFN-α. More recent data identify a subgroup
of well-differentiated clear cell cancer patients who are most likely to benefit from
high-dose IL-2, and there is accumulating evidence that high tumor expression of
carbonic anhydrase IX (CAIX) may help identify this subgroup *(13)*. Further research
in this area is clearly necessary. Additionally, with the emerging positive clinical data
with novel therapies for renal cancer, the role of immunotherapy will be diminished.

Therapy for metastatic renal carcinoma has thus remained a challenge for physicians.
An increased understanding of molecular oncology and cancer genetics has identified
multiple genetic and cell-signaling defects in RCC that are prime targets for novel
therapies. The use of novel therapies in renal cancer has gained momentum recently
with the FDA approval of two multi-tyrosine kinase inhibitors and encouraging results
from phase III trial of a mammalian target of rapamycin (mTOR) inhibitor. In the
subsequent sections, we will outline molecular abnormalities in renal cell cancer and
review novel therapies that are in development and that will likely change the approach
to this disease. Table 1 summarizes potential therapeutic targets and drugs discussed
in this chapter. Previously described cytotoxic and other immunotherapies have been
the subject of other reviews and will not be discussed further here *(14,15)*.

Table 1
Molecular Targets in Renal Cell Cancer

Target	Mechanism of action	Class of inhibitory drug	Examples of drug	Class side effects
EGFR	RTK, overexpressed	TKI	Gefitinib, Erlotinib	Rash, diarrhea, rarely interstitial lung disease
		Antibody	Cetuximab ABX-EGF	
c-Kit/PDGF	RTK	TKI	Imatinib	Fluid retention, muscle cramps
c-Met	RTK, (mutated in papillary RCC)	TKI	SU11274	Unknown
mTOR	Transcription factor	Inhibitor	Temsirolimus	Rash, mucositis
VEGFR	RTK, involved in angiogenesis	Antibody	Bevacizumab	Hemorrhage, thrombosis
		TKI	Vatalanib	
Proteasome Inhibitor	Degrade cellular proteins	Inhibitor	Bortezomib	Fatigue, weakness, thrombocytopenia
Multiple TKI				
Ras/Raf, VEGFR, PDGFR	RTK	TKI	Sorafenib	Hypertension, rash, hand foot syndrome
VEGF, PDGFR	RTK	TKI	Sunitinib AG13736	Fatigue, diarrhea, nausea, Hypertension

RTK, Receptor tyrosine kinase; TKI, tyrosine kinase inhibitor; RCC, renal cell carcinoma; mTOR, mammalian target of rapamycin; VEGF, vascular endothelial growth factor; VEGFR, vascular endothelial growth factor receptor.

243

2. HEREDITARY RCC

Renal cancer occurs sporadically and in a hereditary manner. Patients affected by the hereditary syndromes are often younger and have bilateral cancers. Four genes responsible for hereditary renal carcinoma have been identified to date: von Hippel Lindau (VHL), hereditary papillary renal carcinoma (HPRC), hereditary leiomyomatosis renal carcinoma, and Birt–Hogg–Dubé (BHD) syndrome. Additional families in which the relevant genetic lesion has not yet been identified have been described. Patients affected by tuberous sclerosis and polycystic kidney disease also have a higher incidence of renal carcinoma. Investigations of hereditary renal carcinomas have led to the identification of several renal cancer genes that are potential therapeutic targets.

VHL is a familial cancer syndrome in which affected patients have a predisposition to develop tumors in multiple organs, including the kidneys, central nervous system, adrenals, and eyes; 40% of patients with VHL develop multiple, bilateral renal tumors or cysts *(16)*. The *VHL* gene was identified by genetic linkage analysis and mapped to the 3p chromosomal arm; mutations of *VHL* gene have been detected in 100% of *VHL* patients and kindreds *(17,18)*. Interestingly, in one study with 108 tumor samples, the *VHL* gene was found to be mutated in 57% of sporadic clear cell renal cancers, with a loss of heterozygosity in 98% of the samples *(19)*. The *VHL* gene pathway will be discussed in Section 3.

Individuals affected by HPRC develop bilateral, multi-focal type 1 renal papillary carcinoma *(20)*. The oncogene linked to HPRC is *c-Met*, whose protein is a cell surface receptor for hepatocyte growth factor *(21)*. c-Met is a receptor tyrosine kinase (RTK) that is involved in cellular proliferation and motility *(22)*. The downstream effectors of c-Met include PI3K, Ras, p21GTPases, p85, and Gab1. HPRC kindreds carry activating mutations of the tyrosine kinase domain of *c-Met (23)*. c-Met mutations are very rare in sporadic papillary renal cancers *(23)*.

Clinical features of hereditary leiomyomatosis RCC (HLRC) include cutaneous nodules, uterine leiomyoma and leiomyosarcoma, and type 2 papillary renal cell cancer. The gene mutated in HLRC is fumarate hydratase, a Krebs cycle enzyme *(24)*. At this time, it is unclear how mutations of this gene lead to the development of papillary renal cell cancer.

BHD is an autosomal dominant hereditary syndrome that consists of fibrofolliculoma, pulmonary cysts, pneumothoracies, and multi-focal renal tumors *(25)*. The renal tumors in BHD can be chromophobe, clear cell carcinoma, or oncocytoma *(26)*. A *BHD* gene, which is mutated in a high percentage of BHD kindreds, was identified on chromosome 17q; it appears to be a tumor suppressor gene *(27)*. Further studies will hopefully elucidate how mutations of the *BHD* gene affect the development of renal cancer.

3. VHL

Mutations in the *VHL* gene are an early event in sporadic clear cell renal cancer. In the majority of clear cell RCC, both alleles of the *VHL* gene have been inactivated *(19)*. Typically, one allele is mutated by a deletion or a nonsense/missense point mutation; the second allele is deleted or promoted methylated. Gene mutations of VHL are not detected in non-clear cell carcinomas, such as papillary or chromophobe renal cancers. In vitro testing indicates that VHL is a key tumor suppressor gene, as tumor development in VHL$^{-/-}$ cell lines derived from clear cell human RCC was suppressed after reintroduction of wild-type (but not mutant) VHL *(28)*.

The VHL protein complexes with other proteins to target the hypoxia-inducible factors (HIF) for ubiquitin-mediated degradation. HIFα, which consists of two isomers, dimerizes with HIFβ to form a transcription factor that regulates the expression of multiple genes including *VEGF, GLUT-1, PDGF, EGFR,* and *TGF-α (29)*. Under normoxic condition, HIF1α and HIF2α are hydroxylated at a specific proline, bind to the VHL protein complex, and become quickly ubiquitinated and degraded *(30)*. Under hypoxic conditions, HIF1α and HIF2α do not bind to the VHL complex, are stabilized, and not degraded. Similarly, when VHL is mutated or silenced, HIF degradation is prevented *(31–33)*. The resulting upregulation of several angiogenesis-related genes explains the well-known vascular nature of clear cell renal cancer. Some data suggest that HIF2α is more relevant to clear cell RCC pathogenesis than HIF1α. Figure 1 provides a simplified diagram of VHL-signaling pathways.

HIF levels are also dependent on another pathway, namely, phosphatidyl-inosital-3 kinase (PI3K)/mitogen-activated protein kinase (MAPK) pathway. The PI3K family is involved in regulating motility, migration, adhesion, and proliferation of cells *(34)*. PI3K is activated by protein kinase C (PKC) and Ras; PI3K promotes HIF translation through the mTOR *(35)*.

There are a host of HIF-inducible proteins that are targets for antitumor therapy; these include proteins involved in angiogenesis such as vascular endothelial growth factors (VEGF), oncogenic growth factors such as platelet-derived growth factor (PDGF)-β, and tyrosine kinase receptors such as epidermal growth factor receptors (EGFR). Compounds targeting one or more of these factors or their downstream partners are

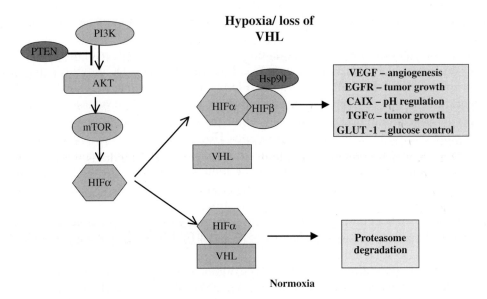

Fig. 1. von Hippel Lindau (VHL)-signaling pathway. A simplified schema of the pathways involved in VHL and hypoxia-inducible factor (HIF) protein complexes. HIF-α dimerizes with HIF-β to form a transcription factor that regulates the expression of multiple genes including *VEGF, GLUT-1, PDGF, EGFR,* and *TGF-α*.

being investigated. Additionally, direct HIF inhibitors have been described and are being explored as therapeutic agents.

3.1. Targets Downstream of Hypoxia-Inducible Factor

3.1.1. RTKs

Several molecules interacting with the HIF are oncogenes known as RTKs. The molecular structure of RTKs consists of an extracellular ligand-binding domain, a hydrophobic transmembrane domain, and an intracellular tyrosine kinase domain. When a ligand binds to the RTK receptor, the receptor dimerizes with another RTK; this dimerization induces autophosphorylation and the activation of downstream signaling molecules that play an important role in tumor progression and metastasis. The RTKs are further divided into groups based on structure and sequence homology. The subgroups include EGFR, VEGF, and PDGF.

3.1.2. EGFR

EGFR is overexpressed in the majority of renal cell cancer lines in comparison with normal renal tissues *(36,37)*. TGF-α is also upregulated by HIF and is a potent ligand for EGFR. In preclinical models, EGFR inhibition leads to decreased cell proliferation, increased apoptosis, and reduced angiogenesis *(38,39)*. Several strategies to target EGFR are in development and include inhibition of the RTK domain and monoclonal antibodies targeting the extracellular domain. The EGFR-signaling pathways are depicted in Fig. 2.

RTK inhibitors compete with ATP for an intracellular catalytic site. These inhibitors have been widely studied in non-small cell lung cancer. RTK inhibitors are administered orally and are well tolerated; common toxicities include an acneiform rash and diarrhea, presumably because of high levels of EGFR in the epidermis and intestines *(40)*. Less commonly, interstitial lung disease can occur with these agents.

Gefitinib (Iressa, ZD1839, AstraZeneca, London, UK) and erlotinib (OSI 774, Tarceva, Genen-tech, San Francisco, CA, USA) are two oral RTK inhibitors that have been approved for monotherapy by the FDA for use in metastatic non-small lung cancer. Several non-randomized, phase II studies with gefitinib in metastatic renal cell cancer have been reported *(41–44)*. No tumor responses were seen; the best response was stable disease in 23–48% of the patients. However, stable disease was prolonged for more than 6 months in several patients ($\sim$10%) *(41,43)*. Tumor samples of patients enrolled in this trial were not screened for mutations in the EGFR tyrosine kinase domain. Several studies with gefitinib and erlotinib in non-small lung cancer have described a positive relationship between the presence of mutations in EGFR's tyrosine kinase domain and a clinical response to RTK inhibitor *(45)*. It is possible that a subset of renal cell cancer patients may have mutations of EGFR tyrosine kinase domain that are responsive to RTK inhibitors.

Monoclonal antibodies against the extracellular domain of EGFR provide another avenue to inhibit the activity of EGFR. Cetuximab (C225, Erbitux, Bristol-Myers, New York, NY, US) is a human/mouse chimeric antibody that binds EGFR and competitively inhibits EGF RTK activation. The FDA approved cetuximab for patients with metastatic colorectal cancers that overexpress EGFR, after a clinical trial demonstrated that cetuximab has significant activity when given in combination with irinotecan

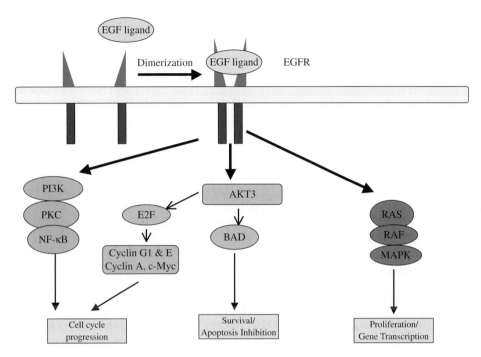

Fig. 2. The epidermal growth factor receptor (EGFR)-signaling pathway. A simplified schema of the pathways involved in EGFR signal transduction. After ligand activation, the EGFR dimerizes and is phosphorylated. Its activation results in the activation of molecules involved in apoptosis, the cell cycle, gene transcription, and proliferation. This figure has been adapted from N. Choong with permission.

in patients with irinotecan-refractory colorectal cancer *(46)*. A phase II trial tested cetuximab in 55 patients with metastatic RCC did not show a partial or complete response in any of the patients; additionally, the time to progression was not prolonged in comparison with historical controls *(47)*.

Another EGFR antibody is ABX-EGF (panitumumab, Abgenix, Fremont, CA, US), a fully humanized monoclonal antibody. A phase II trial evaluated four different doses of ABX-EGF in 88 patients and revealed three major responses (including one complete response) and two minor responses (total of 6%); 44 (55%) other patients had stable disease *(48)*. There was no relationship between the ABX-EGF dose given and the response.

Given the low activity of EGFR antibodies and EGFR tyrosine kinase inhibition in RCC, it seems unlikely that targeting EGFR alone is a viable approach. However, there remains interest in combining EGFR inhibitors with other targeted therapies.

3.1.3. ANGIOGENESIS: VEGF

Angiogenesis is an essential mechanism for tumor growth and metastasis, and its inhibition is an attractive target for investigators. VEGF and its tyrosine kinase receptor (VEGFR) are the principal molecules involved in endothelial cell proliferation, formation of new blood vessels, and vascular permeability *(49)*. VEGF is a HIF-1-regulated gene and is upregulated in clear cell tumors with mutated VHL. VEGF's role in creating a highly vascularized tumor makes it an attractive target. Inhibition of VEGF and its receptor are the primary focus of research targeting angiogenesis. Bevacizumab (Avastin, Genentech) is a humanized monoclonal antibody against VEGF that was recently approved as first line treatment in metastatic colon cancer. In renal cell cancer, a phase II trial randomized patients to receive low-dose (3 mg/kg) bevacizumab, high-dose (10 mg/kg) bevacizumab, or placebo. In the high-dose arm, 10% of patients achieved a partial response, with a significantly increased median time to progression of 4.8 months (versus 2.5 months in the placebo arm). Overall survival was not prolonged in the bevacizumab arms; however, the trial was not powered to determine a survival benefit. Bevacizumab was well tolerated; grade 3 toxicities were hypertension and proteinuria. Further study of bevacizumab combined with immunotherapy is being conducted. CALGB trial 90206 is a randomized phase III trial of IFN-α with or without bevacizumab as first line therapy in metastatic RCC that has recently completed accrual *(50)*.

Bevacizumab is currently being combined with other small molecules in the treatment of renal cell cancer. In one phase II trial, bevacizumab and erlotinib were combined to treat patients with metastatic renal cancer; preliminary results in 63 patients are promising with 15 (25%) patients achieved an objective response rate and 36 patients (61%) had stable disease after 8 weeks *(51)*. Progression-free survival at 1 year was 43%; median overall survival had not been reached at the time of the report. The combination was well tolerated with two patients discontinuing treatment secondary to adverse reactions. Another phase I/II trial combined bevacizumab, erlotinib, and imatinib; preliminary results with 15 patients in the phase I portion indicate that two patients had a partial response *(52)*. The combination of all three drugs was less well tolerated; reported side effects were rash, diarrhea, nausea, fatigue, and vomiting. A randomized, phase II trial evaluated bevacizumab with or without erlotinib in 104 patients *(53)*. Results were reported after a median follow-up of 9.8 months; although both treatment arms were well tolerated, adding erlotinib did not improve the progression-free survival (8.5 months in the bevacizamab arm versus 9.9 months in the combination arm).

VEGF receptor tyrosine kinase inhibitors inactivate VEGF signaling by inhibiting autophosphorylation of the receptor. Several molecules that inhibit VEGF are in development and are being tested including vatalanib (PTK787; Novartis, Basel, Switzerland) and ZD6474 (AstraZeneca). Vatalanib has been safely combined with standard chemotherapy regiments in metastatic colon cancer and inhibits all three isoforms of VEGFR: VEGFR1, VEGFR2, and VEGFR3 *(54)*. A phase I dose escalation study with vatalanib was conducted in 49 patients *(55)*. A substantial reduction in tumor size (>25%) was seen in 19% of patients, with 60% of patients maintaining stable disease. Agents targeting multiple tyrosine kinase receptors including VEGFR are discussed below in Section 3.1.5.

3.1.4. c-Kit/PDGF

c-Kit is a receptor tyrosine kinase with homology to the receptors for PDGF; c-Kit's ligand is mast cell growth factor or stem cell factor (SCF). The downstream pathways of c-Kit include molecules that prevent apoptosis, stimulate growth, and affect cell motility. Examples of these downstream effectors are the Ras/MAPK cascade, phophoinositide 3 kinase (PI3K), and Akt. In non-malignant cells, c-Kit signaling is critical for the development of early hematopoietic cells, mast cells, melanocytes, and germ cells *(56,57)*. Overexpression and aberrant activation of c-Kit has been identified in many malignancies including gastrointestinal stromal tumors and lung cancer. In kidney cancers, c-Kit overexpression has been reported in chromophobe RCC and sarcomatoid RCC *(58,59)*. However, c-Kit overexpression is infrequent in clear cell and papillary tumors *(60,61)*.

Imatinib (Gleevac; Novartis) is a tyrosine kinase inhibitor that was developed as a target against the BCR/ABL oncogene, a fusion protein found in chronic myelogenous leukemia. However, imatinib also targets c-Kit and platelet derived growth factor receptor (PDGFR), which lead to its use in gastrointestinal stromal tumors and other hematologic disorders. Imatinib's activity against PDGF prompted testing of this agent in renal cell cancer, as overexpression of PDGF-aa has been identified in grade 3 and 4 RCC and is associated with adverse outcomes *(62)*. A phase II trial with single agent imatinib 400 mg daily in 14 patients with metastatic RCC (12 with clear cell carcinoma) did not reveal any major responses, but four patients had stable disease after 1, 3, and 4 months *(61)*. A subsequent study by Polite et al. *(63)* combined imatinib 600 mg daily and IFN-α (9 million units subcutaneously three times a week) in 14 patients with metastatic renal cell cancer without prior therapy. One patient had a partial response (7%), four patients had stable disease (28%), and six patients had a progressive disease (43%); the median time to progression was 8 weeks. Significant toxicities were seen, with 50% of patients experiencing grade 3 adverse events including fatigue, anorexia, nausea, and vomiting; four patients withdrew from the trial secondary to toxicities. With a low response rate (7%) and significant toxicities, further investigation of imatinib in clear cell RCC is not recommended. However, therapy with imatinib may be a viable approach in sacromatoid or chromophobe RCC.

3.1.5. Other Multiple Targeted Tyrosine Kinase Inhibitors

Because of extensive homology between multiple tyrosine kinase receptors, especially in ATP-binding site targeted by all the currently studied small molecule inhibitors, several of these agents inhibit multiple kinases *(64)*. Manning et al. *(65)* identified 518 putative protein kinase genes in the human genome, of which 244 kinases map to known cancer amplicons. Tyrosine kinases from the VEGFR and EGFR families are the most pursued targets in oncologic drug development *(64)*. Small molecule inhibitors are profiled against a small subset of known kinases, as there are approximately 20 kinase targets associated with cancer indications. Thus, it is possible that the small molecule inhibitors target more kinases than are screened and that there are clinically relevant kinases in tumor pathogenesis that are not targeted by the inhibitors.

Sorafenib (Nexavar, Bayer, Momjton, NJ, US) is a kinase inhibitor that was initially developed to inhibit the Ras/Raf pathway; however, it was determined to inhibit multiple targets including VEGFR, PDGFR, and c-Kit *(66)*. Sorafenib was investigated in a randomized discontinuation phase II trial designed to elucidate the activity of a potential cytostatic agent *(67)*. Sorafenib was administered for 12 weeks; those patients who experienced >25% tumor reduction in bidimensional measurements were continued on the drug, whereas those patients with stable disease were randomized to sorafenib or placebo. Patients randomized to continued sorafenib had marked improvement in progression-free survival in comparison with patients who were discontinued. Preliminary results from an international, phase III trial were recently reported. Over 900 patients who failed first line systemic therapy were randomized to sorafenib 400 mg twice a day versus placebo (68). The preliminary results with 905 enrolled patients demonstrated that progression-free survival was significantly longer in the treatment arm, with those receiving sorafenib demonstrating a time to progression of 24 versus 12 weeks ($p < 0.000001$). The interim survival analysis of the phase III trial after 367 survival events revealed that sorafenib-treated patients experienced a prolonged overall survival of 19.3 versus 15.9 months for placebo arm ($p = 0.015$, hazard ratio 0.77) (69). These preliminary results did not meet the predetermined O'Brien-Fleming boundary for significance; final analysis is pending data maturation. Sorafenib is well tolerated with side effects that included hypertension (17%), hand foot syndrome (27%), rash (34%), cardiac events (3%), and fatigue (26%). In the frontline setting for patients with metastatic disease, a multi-center, randomized phase II trial of sorafenib versus IFN is ongoing; preliminary toxicity data in 188 patients again demonstrated that sorafenib was well tolerated, but efficacy data are not mature (70). Owing to sorafenib's presumed role in tumor perfusion and angiogenesis, there is currently interest in investigating its pharmacodynamic properties using dynamic contrast MRI and Doppler ultrasonography (70,71).

These promising phase II and III results led to the FDA's approval of sorafenib in December 2005 for the treatment of metastatic RCC. Currently, Sorafenib is undergoing evaluation in the adjuvant setting two large phase III trials in patients with locally advanced disease status post-nephrectomy; one trial sponsored by the US cooperative group, ECOG, will compare sorafenib versus sunitinib versus placebo for 1 year. The second trial, sponsored by investigators in the UK, will compare 1 versus 2 years of sorafenib versus placebo. Another phase II trial evaluated the combination of sorafenib and IFN as a first or second line therapy in metastatic RCC; preliminary results are encouraging (72).

Sunitinib (Sutent, Pfizer, New York, NY, US) is an oral multi-tyrosine kinase inhibitor that targets receptors for VEGF, PDGF, KIT, and FLT3. Activity of sunitinib was tested in two independent, single arm phase II trials with a total of 169 patients with cytokine refractory, metastatic RCC (73). Patients received 50 mg daily of sunitinib for 4 weeks, followed by a 2-week rest. The first trial with 63 patients completed accrual in July 2003; results reveal a 40% partial response rate with a median duration of response of 10+ months. Another 33% of patients had stable disease; the median survival was 16 months. The preliminary results of the second trial appear equally promising. The toxicity profile was acceptable, with grade 1–2 adverse events consisting of fatigue, nausea, diarrhea, and stomatitis; grade 3–4 adverse events included lymphopenia, elevated lipase, and amylase without clinical signs of pancreatitis. Sunitinib has received accelerated FDA approval for treatment of advanced RCC based on promising phase II data (74). A phase III trial of sunitinib versus IFN in treatment naïve metastatic clear cell RCC patients was recently reported (75). This study, which enrolled 750 patients with metastatic clear cell RCC, showed that the median progression-free survival was 47.3 weeks for sunitinib versus 24.9 weeks for IFN, with a hazard ratio 0.394 (95% CI 0.297, 0.521) ($p < 0.000001$). Overall survival has not yet been reached, as 632 patients (85%) are alive, with 49 deaths on sunitinib arm and 65 deaths on IFN arm.

AG-013736 (Axitinib, Pfizer) is an oral molecule that inhibits multiple tyrosine kinase receptors, including VEGFR and PDGFR-beta. Preliminary results from a phase II trial evaluating AG 013736 in 52 patients with metastatic RCC that failed one prior cytokine-based therapy were recently reported *(76)*. The agent had a substantial objective response rate, with a partial response reported in 21 patients (40%). After a median follow-up of 1 year, the median time to progression has not been reached and only one patient with a partial response relapsed after 232 days. Adverse reactions reported include hypertension (33%), fatigue (29%), nausea (29%), diarrhea (27%), and hoarseness (19%); three patients discontinued the three secondary to adverse reactions.

3.2. Carbonic Anhydrase IX

CAIX, a transmembrane enzyme, is a gene controlled by HIF-1 and plays a role in intracellular and extracellular pH regulation. It is possible that CAIX may allow tumors to proliferate in an acidic and hypoxic environment. Immunohistochemical studies of malignant and benign renal tissues indicate that CAIX is overexpressed in RCC but is not detected in normal renal or other tissues *(77)*. Analysis of tissue microarrays from 321 nephrectomy specimens reveals that 94% of clear cell RCCs express CAIX *(78)*. CAIX may be a molecular predictor for outcomes in RCC, as low CAIX staining was found to be an independent prognostic factor for poor survival in patients with metastatic RCC. One series analyzing outcomes and CAIX expression in 224 patients elucidated that patients with high CAIX expression had a median survival of 67 versus 22 months in patients with low CAIX expression ($p < 0.001$) *(79)*. Additionally, patients who respond to immunotherapy are more likely to have a high expression of CAIX *(12,78)*. Non-clear cell renal cancers express low levels of CAIX *(80)*.

The chimeric monoclonal antibody WX-G250 (Wilex AG, Munich, Germany) targets CAIX. A phase I dose escalation study with radiolabeled WX-G250 indicated that the antibody had specific and high accumulation in RCC lesions *(81)*. In vitro investigations of RCC cell lines reported that G250 mediates antibody-dependent cellular cytotoxicity *(82)*. An open-label, non-randomized phase II trial of G250 in 36 patients with metastatic RCC was completed *(83)*. Intravenous G250 was given weekly for 12 weeks; patients with stable disease or a response could receive an additional 8 weeks of therapy. This schedule was well tolerated with no grade 3 or 4 adverse events; 11 patients had stable disease, one patient had a complete response, and another had a partial response. The median survival was 15 months after therapy. These results were promising and suggested that G250 could modulate the natural history of RCC. Laboratory investigations by Liu et al. *(84)* in RCC cell lines reported that cytokines IFN-α and IL-2 upregulated antibody-dependent cellular cytotoxicity. Thus, phase I/II trials combining G250 and immunotherapy have been initiated. One trial combined G250 (50 mg weekly) with low-dose IL-2 (3 MIU daily) in 30 patients with metastatic RCC *(85)*. Early results after 32–39 weeks of therapy indicated that this regiment is well tolerated; eight patients had stable disease and an additional 3 had a partial response. Another trial with G250 (20 mg weekly) combined with IFN-α (3 MIU, three times a week) in 32 patients with metastatic RCC *(86)*. Preliminary results in 26 patients after 16 weeks of therapy revealed that 11.5% of patients had a partial response, 46% had stable disease, and 42% progressed. Drug-related adverse events consisted of constitutional symptoms and pain. A phase III trial with G250 in the adjuvant setting is slated to start in 2005 *(87)*.

Owing to CAIX specificity to renal cell tumors, it is a potential target for vaccine-based therapies. Early work with a fusion protein of G250 and granulocyte/macrophage colony-stimulating factor (GM-CSF) suggests that it is a potent immunostimulant that may activate a CD8-mediated antitumor response. Thus, a GM-CSF–CAIX fusion protein vaccine may be a viable strategy *(88,89)*.

4. TARGETS UPSTREAM OF HIF: PI3K/MTOR PATHWAY

mTOR activation increases *HIF-1*α gene expression; mTOR inhibition may prevent the increased angiogenesis seen in sporadic RCC and loss of VHL *(35,90)*. Additionally, the mTOR pathway is postulated to be involved in hereditary RCC seen in patients with tuberous sclerosis. Downstream signaling of the mTOR pathway is inhibited by a complex of tuberin and hamartin, the products of tuberous sclerosis complex genes *TSC-2* and *TSC-1 (91)*. In tuberous sclerosis, *TSC-1* and *TSC-2* are mutated, preventing mTOR-signaling inhibition and enabling cell proliferation.

Temsirolimus (CCI-779, Wyeth, Madison, NJ) is an mTOR inhibitor that has demonstrated promising results in patients with metastatic RCC. A phase II trial randomized 111 patients to three doses of weekly CCI-779 *(25, 75* and 250mg) *(92)*. An objective response rate of 7% was observed with one complete response and seven partial responses; another 26% of patients had minor responses. Tumor response rates did not differ significantly between the doses, even when patients were classified in good, intermediate, or poor risk groups. In light of these encouraging results, a phase I trial combining IFN-α and temsirolimus was initiated; an updated report with 71 patients established the maximum tolerated dose of 15 mg temsirolimus and 6 MU of IFN-α was selected *(93)*. Preliminary results were favorable with 13% with a partial response and 71% of patients with stable disease. Adverse events were tolerable and included leukopenia, hyperlidemia, and asthenia. Results of a phase III trial comparing IFN-α (up to 18 MU weekly) versus temsirolimus 25 mg weekly versus the combination (temsirolimus 15 mg + IFN 6 MU three times a week) for first line therapy in poor risk patients with RCC were recently reported. The trial enrolled 626 participants; patients treated with temsirolimus had a significantly longer overall survival than IFN alone (10.9 versus 7.3 months, $p = 0.0069$). The combination arm did not show any significant improvement compared with IFN alone. The temsirolimus arm was well tolerated with a lower incidence of adverse events including asthiam anemia, and dyspnea than the IFN arm *(94)*. In May 2006, the FDA approved temsirolimus for treatment in patients with kidney cancer.

5. OTHER ANGIOGENESIS TARGETS

5.1. Thalidomide

Thalidomide is a non-specific antiangiogenic drug; one mechanism of its actions is downregulating VEGF activation of endothelial cells. A series of phase II trials has been performed in patients with RCC; antitumor activity is seen in only a few patients, with response rates ranging between 0 and 10% *(95–98)*. Thalidomide has been combined with other anticancer agents, most notably IL-2 and IFN-α. A trial with low-dose thalidomide and IFN-α was conducted, but several patients experienced adverse neurologic side effects that led to the termination of the trial *(99)*.

However, further investigation with lower doses of IFN-α (0.9–1.2 MU three times a day) and thalidomide produced promising results, with a response rate of 20% *(100)*. Unfortunately, a phase III trial of low-dose IFN-α alone versus low-dose IFN-α and thalidomide failed to prove an advantage in the combination arm *(101)*.

CC-5013 (lenalidomide, Revlimid, Celgene, Summit, NJ) is an immunomodulatory drug with similar antiangiogenic properties to thalidomide but designed with a structural modification to avoid the toxicities associated with thalidomide. A phase II trial of CC-5013 enrolled 40 patients with metastatic RCC; preliminary results in 36 patients revealed a 7.5% partial response rate *(102)*.

5.2. ABT-510

ABT-510 (Abbott, Chicago, US) is a non-apeptide that inhibits angiogenesis by mimicking the activity of endogenous protein, thrombospondin-1, by competing for its cellular receptor, CD 36. After promising results were seen in 13 patients with RCC in a phase I trial, a randomized phase II study was undertaken with 103 previously untreated patients with advanced RCC *(103)*. Patients were randomized to receive ABT 510 at 10 or 100 mg twice daily by subcutaneous injection. Recently reported data for 88 patients revealed a 6-month progression-free survival rate of 24.6% (95% CI = 11.9–37.3); one partial response has been seen. Frequent adverse events included asthenia, pain, and injection site reaction. There were six serious adverse events reported that were possibly or probably related to the drug; these include hemoptysis, gastrointestinal bleeding, deep venous thrombosis, dehydration, pulmonary edema, and cardiomyopthy. Although these data are preliminary, ABT 510 alone does not appear to provide improved efficacy in comparison with historic controls.

6. TARGETING THE C-MET PATHWAY

As discussed earlier in Section 2, individuals affected by HPRC causing type 1 papillary renal carcinoma carry mutations of c-Met. C-Met is the receptor for hepatocyte growth factor and is involved in cellular migration and mitogenesis *(104)*. Germline mutations of c-Met in HPRC result in the constitutive activation of the kinase *(21)*. There are several proposed strategies for inhibition of the c-Met pathway, including kinase inhibitor, interfering with c-Met's binding to its ligand, and inhibition of c-Met's downstream pathways. The development of small molecules that inhibit c-Met is currently in preclinical stages. SU11274 (Sugen, New York, US) and PHA665752 (Sugen, New York, US) are two c-Met inhibitors under investigation. SU11274 specifically targets c-Met and Tpr/Met. Tpr/Met is a fusion oncoprotein that results in constitutive activation of c-Met *(105)*.

7. NEW CHEMOTHERAPEUTIC AGENTS

7.1. Epothilones

Epothilones are a new generation of agents that stabilize microtubules and inhibit cell growth. Preclinical studies indicate that epothilones induce microtubule polymerization by mechanism similar to paclitaxel *(106,107)*. Two epothilone analogs have been tested in patients with RCC: EPO906 (Novartis) and BMS-247550 (ixabepilone, Bristol Myers Squibb). A multi-institutional phase II trial with 53 patients with advanced renal cancer evaluated the safety and efficacy of EPO906 *(108)*. Patients received weekly infusions of EPO906 at 2.5 mg/m^2 for 3 weeks, followed by a week off. EPO906 was well tolerated, with one grade 4 adverse reaction (septic shock); grade 3 adverse events included diarrhea (8%), asthenia (4%), and anemia (4%). Seven patients discontinued EPO906 secondary to adverse reactions. Final evaluation of 52 patients revealed that 2 (4%) patients had a partial response of 3 and 5 months duration, and 24 (46%) patients had stable disease after 4 cycles of therapy. A phase II clinical trial BMS-247550 enrolled 39 patients to receive 6 mg/m^2/day for 5 days every 3 weeks *(109)*. The therapy was well tolerated, with 11 patients remaining on study at the time of the report. A partial response was seen in 4 (10%) patients, with five additional patients with a minor or mixed response. Correlative studies suggest that the post-translational modification of the a-tubulin (removal of C-terminal tyrosine, exposing glutamic acid as the C terminal residue) correlates with the extent of microtubule stability and pharmacodynamic effects of the agents. Another translational study of BMS 247550 assessed the drug's activity and target engagement in tumor cells *(110)*. Of the 67 patients enrolled, grade 3 adverse reactions were observed in two patients; partial responses were observed in 8 (12%). Twelve patients with accessible tumors underwent biopsies before treatment and after the fifth dose for correlative studies. Microtubular stabilization was compared in the pre- and post-treatment biopsies: increased glu-tubulin and acetylated tubulin levels were seen in 11/12 and 10/12 patients respectively. Thus, BMS 247550 is an active agent in RCC that warrants further investigation. Further investigation may determine whether VHL gene status predicts response to epothilone therapy.

7.2. Ubiquitin–Proteasome Pathway

Proteasomes are considered "cellular housekeepers" as they degrade cellular proteins. This function serves to regulate the cell cycle, apoptosis, and angiogenesis. Several proteins, such as cyclins, cyclin-dependent kinases, p53, and nuclear factor (NF)-κB, are degraded in this manner. Bortezomib (PS 341, Velcade, Millennium, Cambridge, MA) is a novel inhibitor of proteasomes; its antitumor activity has been demonstrated in several cell lines. Fleming et al. *(111)* demonstrated with DNA array analysis of multiple myeloma cells that bortezomib downregulates genes involved in growth pathways; it also upregulates genes involved in apoptosis and heat shock proteins. Bortezomib was approved by the FDA for therapy in relapsed multiple myeloma, after the CREST trial demonstrated a 30–38% response rate in relapsed multiple myeloma *(112)*. Bortezomib has been tested in patients with metastatic RCC in several trials. A phase II trial performed at the University of Chicago in 37 patients with metastatic RCC revealed a partial response in four patients (11%), with a response duration of 8–20+ months *(113)*. The toxicities of bortezomib were significant, as 53% of the patients reported grade 2 or 3 neuropathy. Another phase II trial with bortezomib in 25 patients conducted at Memorial Sloan Kettering Cancer Center revealed a similar response rate (9% with a partial response), but reported a lower rate of neuropathy of 28% *(114)*.

7.3. Heat Shock Proteins

Heat shock proteins (HSPs) are "molecular chaperones" that protect cellular proteins from degradation; they ensure proper folding of proteins after translation and refolding of denatured proteins. Proteins that are damaged are directed to proteasomes for repair; thus, HSPs and proteasomes have common substrates *(115)*. HSP regulate signaling proteins involved in proliferation and carcinogenesis (i.e., Akt, Her2, c-Met, Raf1), making them an attractive target for drug development *(116)*. Certain HSPs (HSP90, HSP70, and HSP27) are expressed in high levels in cancer cells. Specifically, Hsp90 acts as a chaperone for several proteins implicated in RCC tumorgenesis including HIF-1α, c-Kit, and c-Met. In vitro investigations suggest that Hsp90 inhibition may be a viable target in RCC, as inhibiting Hsp90 decrease HIF-dependent transcription *(117)*. An Hsp90 inhibitor in early stages of development includes 17-allyloaminogeldanamycin (17-AAG) *(118,119)*.

Another HSP, Hsp96, is under investigation as part of a vaccination strategy against RCC. This immunotherapy employs the HSP linked to tumor peptides from an individual's tumor (HSPPC96). When the HSPPC96 is administered, the antigenic tumor peptides are expressed on the surface of potent antigen-presenting cells such as dendritic cells, with a goal of stimulating a strong antitumor immune response *(120)*. A phase II trial tested the efficacy of HSPPC96 in 61 patients with metastatic RCC; IL-2 was added if the patients progressed on the vaccine alone *(121)*. The results were promising with one patient with a durable complete response lasting 2.5 years, and an additional two patients with partial responses. HSPPC-96 is currently in phase III trials in the adjuvant setting.

8. OTHER IMMUNOTHERAPIES

Although IFN-α and IL-2 have remained the cornerstones of immunotherapies against renal cell cancer, other targets are being investigated. CTLA4 is an inhibitory receptor on T lymphocytes; it is theorized that inhibition of CTLA4 may promote lymphocyte activation against tumor cells. MDX-010 (Bristol Myers Squib) is a human IgG antibody against CTLA4 that was tested in a phase II trial in 29 patients with metastatic renal cancer *(122)*. Although six patients (including three patients with prior IL-2) had an objective response that lasted 4–18 months, they all experienced autoimmune toxicities of the therapy (enteritis, meningitis, hypophysitis). There is interest in developing IL-12 and IL-18 as immunotherapies against renal cell cancer *(123,124)*.

9. CONCLUSION

Metastatic renal cancer is a challenging disease; however, significant strides in its clinical management have been made. Several molecular and genetic abnormalities have been identified in the various types of renal cancer, and multiple drugs are in development to target these pathways. Our understanding of the molecular pathways involved in tumorgenesis has increased in recent years. Although our ability to impact clinical outcomes has been relatively slower, the emerging data with tyrosine kinase inhibitors that target VEGF and mTOR inhibitors are encouraging. In addition to prolonging progression-free survival and overall survival, these therapies have fewer side effects than prior immune-based therapies. Hopefully, a better understanding of the molecular pathways involved in the development of renal cancers will pave the way to specific therapies that make an impact on cancer survival. Future challenges will be how to select appropriate novel therapies for clinical trials and for individual patients.

REFERENCES

1. Jemal A, Tiwari RC, Murray T, et al. Cancer statistics, 2004. *CA Cancer J Clin* 2004;54(1):8–29.
2. Ries LG KC, Hankey BF, Harras A, Edwards BK, eds. SEER *Cancer Statistics Review*. Bethesda, MD: National Cancer Institute [NIH Publication 1997;97(2789)].
3. American Cancer Society. *Cancer Facts and Figures 2005*. Available at http://wwwcancerorg/docroot/STT/stt_0asp (Accessed June 15, 2005).
4. Janzen NK, Kim HL, Figlin RA, Belldegrun AS. Surveillance after radical or partial nephrectomy for localized renal cell carcinoma and management of recurrent disease. *Urol Clin North Am* 2003;30(4):843–52.
5. Motzer RJ, Bander NH, Nanus DM. Renal-cell carcinoma. *N Engl J Med* 1996;335(12):865–75.
6. Linehan WM, Bates SE, Yang JC. Cancers of the genitourinary system. In: DeVita VTHS, Rosenberg SA, ed. *Cancer: Principles and Practice of Oncology*. Philadelphia, PA: Lippincott Williams & Wilkins; 2005.
7. Shi T, Seligson D, Belldegrun AS, Palotie A, Horvath S. Tumor classification by tissue microarray profiling: random forest clustering applied to renal cell carcinoma. *Mod Pathol* 2005;18(4):547–57.
8. Motzer RJ, Carducci MA, Fishman M, Hnacock SL, et al. The NCCN kidney cancer: clinical practice guidelines. *JNCCN* 2005;3(1):84–102.
9. Margolin KA. Interleukin-2 in the treatment of renal cancer. *Semin Oncol* 2000;27(2):194–203.
10. Yang JC, Sherry RM, Steinberg SM, et al. Randomized study of high-dose and low-dose interleukin-2 in patients with metastatic renal cancer. *J Clin Oncol* 2003;21(16):3127–32.
11. Coppin C, Porzsolt F, Awa A, Kumpf J, Coldman A, Wilt T. Immunotherapy for advanced renal cell cancer. *Cochrane Database Syst Rev* 2005(1):CD001425.
12. Atkins M MD, Mier J, Stanbridge E, et al. Carbonic anhydrase IX (CAIX) expression predicts for renal cell cancer patient response and survival to Il-2 therapy. *Proc Am Soc Clin Oncol* 2004;23(383).
13. Atkins M, Regan M, McDermott D, et al. Carbonic anhydrase IX expression predicts outcome of interleukin 2 therapy for renal cancer. *Clin Cancer Res* 2005;11(10):3714–21.
14. Jin JO, Stadler W. Chemotherapy for metastatic clear cell renal cell cancer. In: Bukowski RM, Novick AC, eds. *Clinical Management of Renal Tumors*. Humana Press (in press).
15. Stadler WM. Genitourinary cancer (Chapter 30). In: Govinden G, ed. *Cancer: Principles and Practices of Oncology Review*. Philadelphia, PA: Lippincott; 2005.
16. Linehan WM, Walther MM, Zbar B. The genetic basis of cancer of the kidney. *J Urol* 2003;170(6 Pt 1):2163–72.
17. Latif F, Tory K, Gnarra J, et al. Identification of the von Hippel-Lindau disease tumor suppressor gene. *Science* 1993;260(5112):1317–20.
18. Stolle C, Glenn G, Zbar B, et al. Improved detection of germline mutations in the von Hippel-Lindau disease tumor suppressor gene. *Hum Mutat* 1998;12(6):417–23.
19. Gnarra JR, Tory K, Weng Y, et al. Mutations of the VHL tumour suppressor gene in renal carcinoma. *Nat Genet* 1994;7(1):85–90.
20. Zbar B, Tory K, Merino M, et al. Hereditary papillary renal cell carcinoma. *J Urol* 1994;151(3):561–6.
21. Schmidt L, Duh FM, Chen F, et al. Germline and somatic mutations in the tyrosine kinase domain of the MET proto-oncogene in papillary renal carcinomas. *Nat Genet* 1997;16(1):68–73.
22. Ma PC, Maulik G, Christensen J, Salgia R. c-Met: structure, functions and potential for therapeutic inhibition. *Cancer Metastasis Rev* 2003;22(4):309–25.
23. Schmidt L, Junker K, Nakaigawa N, et al. Novel mutations of the MET proto-oncogene in papillary renal carcinomas. *Oncogene* 1999;18(14):2343–50.
24. Toro JR, Nickerson ML, Wei MH, et al. Mutations in the fumarate hydratase gene cause hereditary leiomyomatosis and renal cell cancer in families in North America. *Am J Hum Genet* 2003;73(1):95–106.
25. Zbar B, Alvord WG, Glenn G, et al. Risk of renal and colonic neoplasms and spontaneous pneumothorax in the Birt-Hogg-Dube syndrome. *Cancer Epidemiol Biomarkers Prev* 2002;11(4):393–400.
26. Pavlovich CP, Walther MM, Eyler RA, et al. Renal tumors in the Birt-Hogg-Dube syndrome. *Am J Surg Pathol* 2002;26(12):1542–52.

27. Nickerson ML, Warren MB, Toro JR, et al. Mutations in a novel gene lead to kidney tumors, lung wall defects, and benign tumors of the hair follicle in patients with the Birt-Hogg-Dube syndrome. *Cancer Cell* 2002;2(2):157–64.
28. Iliopoulos O, Kibel A, Gray S, Kaelin WG, Jr. Tumour suppression by the human von Hippel-Lindau gene product. *Nat Med* 1995;1(8):822–6.
29. Maxwell PH, Wiesener MS, Chang GW, et al. The tumour suppressor protein VHL targets hypoxia-inducible factors for oxygen-dependent proteolysis. Nature 1999;399(6733):271–5.
30. O'Rourke JF, Dachs GU, Gleadle JM, et al. Hypoxia response elements. *Oncol Res* 1997;9(6–7):327–32.
31. Gnarra J, Ahou S, Merrill MJ, et al. Post-transcriptional regulation of vascular endothelial growth factor mRNA by the VHL tumor suppressor gene product. Proc Nat Acad Sci USA 1996;93: 10589–94.
32. Iliopoulos O, Levy AP, Jiang C, et al. Negative regulation of hypoxia inducible genes by the von Hippel Lindau protein. *Proc Nat Acad Sci USA* 1996;93:10595–99.
33. Ohh M, Park CW, Ivan M, et al. Ubiquitination of hypoxia-inducible factor requires direct binding to the beta-domain of the von Hippel-Lindau protein. *Nat Cell Biol* 2000;2(7):423–7.
34. Carpenter CL, Cantley LC. Phosphoinositide kinases. *Curr Opin Cell Biol* 1996;8(2):153–8.
35. Hudson CC, Liu M, Chiang GG, et al. Regulation of hypoxia-inducible factor 1alpha expression and function by the mammalian target of rapamycin. *Mol Cell Biol* 2002;22(20):7004–14.
36. Asakuma J, Sumitomo M, Asano T, Hayakawa M. Modulation of tumor growth and tumor induced angiogenesis after epidermal growth factor receptor inhibition by ZD1839 in renal cell carcinoma. *J Urol* 2004;171(2 Pt 1):897–902.
37. Sumitomo M, Asano T, Asakuma J, Horiguchi A, Hayakawa M. ZD1839 modulates paclitaxel response in renal cancer by blocking paclitaxel-induced activation of the epidermal growth factor receptor-extracellular signal-regulated kinase pathway. *Clin Cancer Res* 2004;10(2):794–801.
38. Yarden Y. The EGFR family and its ligands in human cancer. Signalling mechanisms and therapeutic opportunities. *Eur J Cancer* 2001;37(Suppl 4):S3–8.
39. Sirotnak FM, Zakowski MF, Miller VA, Scher HI, Kris MG. Efficacy of cytotoxic agents against human tumor xenografts is markedly enhanced by coadministration of ZD1839 (Iressa), an inhibitor of EGFR tyrosine kinase. *Clin Cancer Res* 2000;6(12):4885–92.
40. Van Doorn R, Kirtschig G, Scheffer E, Stoof TJ, Giaccone G. Follicular and epidermal alterations in patients treated with ZD1839 (Iressa), an inhibitor of the epidermal growth factor receptor. *Br J Dermatol* 2002;147(3):598–601.
41. Jermann M, Joerger M, Pless M, et al. An open label phase II trial to evaluate the efficacy and safety of gefitinib in patients with locally advanced, relapsed or metastatic renal cell cancer. *Proc Am Soc Clin Oncol* 2003;22:418 (Abstract 1681).
42. Drucker BJ, Schwartz L, Marion S, et al. Phase II trial ZD1839 (iressa), an EGF receptor inhibitor, in patients with advanced renal cell carcinoma. *Proc Am Soc Clin Oncol* 2002; Abstract 720.
43. Beeram M, Rowinsky EK, Weiss GR, et al. Durable disease stabilization and antitumor activity with OSI-774 in renal cell carcinoma: a phase II, pharmacokinetic and biological correlative study with FDG-PET imaging. *Proc Am Soc Clin Oncol* 2004;22(14S):3050.
44. Dawson NA, Guo C, Zak R, et al. A phase II trial of gefitinib (Iressa, ZD1839) in stage IV and recurrent renal cell carcinoma. *Clin Cancer Res* 2004;10(23):7812–9.
45. Lynch TJ, Bell DW, Sordella R, et al. Activating mutations in the epidermal growth factor receptor underlying responsiveness of non-small-cell lung cancer to gefitinib. *N Engl J Med* 2004;350(21):2129–39.
46. Cunningham D, Humblet Y, Siena S, et al. Cetuximab monotherapy and cetuximab plus irinotecan in irinotecan-refractory metastatic colorectal cancer. *N Engl J Med* 2004;351(4):337–45.
47. Motzer RJ, Amato R, Todd M, et al. Phase II trial of antiepidermal growth factor receptor antibody C225 in patients with advanced renal cell carcinoma. *Invest New Drugs* 2003;21(1): 99–101.
48. Rowinsky EK, Schwartz GH, Gollob JA, et al. Safety, pharmacokinetics, and activity of ABX-EGF, a fully human anti-epidermal growth factor receptor monoclonal antibody in patients with metastatic renal cell cancer. *J Clin Oncol* 2004;22(15):3003–15.
49. Cox G, Jones JL, Walker RA, Steward WP, O'Byrne KJ. Angiogenesis and non-small cell lung cancer. *Lung Cancer* 2000;27(2):81–100.

50. Rini BI, Halabi S, Taylor J, Small EJ, Schilsky RL. Cancer and Leukemia Group B 90206: a randomized phase III trial of interferon-alpha or interferon-alpha plus anti-vascular endothelial growth factor antibody (bevacizumab) in metastatic renal cell carcinoma. *Clin Cancer Res* 2004;10(8):2584–6.

51. Hainsworth JD, Sosman JA, Spigel DR, Edwards DL, Baughman C, Greco A. Treatment of metastatic renal cell carcinoma with a combination of bevacizumab and erlotinib. *J Clin Oncol* 2005;23(31):7889–96.

52. Hainsworth JD, Sosman JA, Spigel DR, et al. Bevacizumab, erlotinib, and imatinib in the treatment of patients with advanced renal cell carcinoma: a Minnie Pearl Cancer Research Network phase I/II trial. *J Clin Oncol* 2005;23:388 (Abstract 4542).

53. Bukowski RM, Kabbinavar F, Figlin RA, et al. Bevacizumab with or without erlotinib in metastatic renal cell carcinoma (RCC). *J Clin Oncol* 2006;24:18S (Abstract 4523).

54. Early clinical data with small-molecule vascular endothelial growth factor tyrosine kinase receptor inhibitors. *Clin Lung Cancer* 2004;6(2):74–6.

55. George D, Michaelson D, Oh WK, et al. Phase I study of PTK787/ZK 222584 in metastatic renal cell carcinoma. *Proc Am Soc Clin Oncol* 2003;22:385 (Abstract 1548).

56. Sattler M, Salgia R. Molecular and cellular biology of small cell lung cancer. *Semin Oncol* 2003;30(1):57–71.

57. Park GH, Plummer HK, III, Krystal GW. Selective Sp1 binding is critical for maximal activity of the human c-kit promoter. Blood 1998;92(11):4138–49.

58. Yamazaki K, Sakamoto M, Ohta T, Kanai Y, Ohki M, Hirohashi S. Overexpression of KIT in chromophobe renal cell carcinoma. *Oncogene* 2003;22(6):847–52.

59. Castillo M, Petit A, Mellado B, Palacin A, Alcover JB, Mallofre C. C-kit expression in sarcomatoid renal cell carcinoma: potential therapy with imatinib. *J Urol* 2004;171(6 Pt 1):2176–80.

60. Zigeuner R, Ratschek M, Langner C. Kit (CD117) immunoreactivity is rare in renal cell and upper urinary tract transitional cell carcinomas. *BJU Int* 2005;95(3):315–8.

61. Vuky J, Fotoohi M, Isacson C, et al. Phase II trial of imatinib mesylate in patients with metastatic renal cell carcinoma. *Proc Am Soc Clin Oncol* 2003;22:416 (Abstract 1672).

62. Sulzbacher I, Birner P, Traxler M, Marberger M, Haitel A. Expression of platelet-derived growth factor-alpha alpha receptor is associated with tumor progression in clear cell renal cell carcinoma. *Am J Clin Pathol* 2003;120(1):107–12.

63. Polite BN, Desai AA, Peterson AC, Manchen B, Stadler WM. A phase II study of imatinib mesylate and interferon alpha in metastatic renal cell carcinoma. *J Clin Oncol* 2005; Abstract 4689.

64. Vieth M, Sutherland JJ, Robertson DH, Campbell RM. Kinomics: characterizing the therapeutically validated kinase space. *Drug Discov Today* 2005;10(12):839–46.

65. Manning G, Whyte DB, Martinez R, Hunter T, Sudarsanam S. The protein kinase complement of the human genome. *Science* 2002;298(5600):1912–34.

66. Wilhelm SM, Carter C, Tang L, et al. BAY 43–9006 exhibits broad spectrum oral antitumor activity and targets the RAF/MEK/ERK pathway and receptor tyrosine kinases involved in tumor progression and angiogenesis. *Cancer Res* 2004;64(19):7099–109.

67. Ratain MJ, Eisen T, Stadler WM, et al. Phase II placebo-controlled randomized discontinuation trial of sorafenib in patients with metastatic renal cell carcinoma. *J Clin Oncol* 2006;24(16):2505–12.

68. Escudier CS, Eisen T, Stadler WM, et al. Randomized Phase III trial of the Raf kinase and VEGFR inhibitor sorafenib (BAY 43–9006) in patients with advanced renal cell carcinoma (RCC). *J Clin Oncol* 2005; Abstract 4510.

69. Updated Overall Survival Analysis Presented on Nexavar Phase III Trial. http://www.news.boyer.com/BayNews.acessed 6/12/06.

70. Lamuraglia NL, Chami L, Jaziri S, Schwartz B, Leclere J, Escudier B. Doppler ultrasonography with perfusion software and contrast agent injection as a tool for early evaluation of metastatic renal cancers treated with the Raf kinase and VEGFR inhibitor: a prospective study. *J Clin Oncol* 2005; Abstract 3069.

71. O'Dwyer MR, Gallagher M, Schwartz B, Flaherty KT. Pharmacodynamic study of BAY 43–9006 in patients with metastatic renal cell carcinoma. *Proc Am Soc Clin Oncol* 2005; Abstract 3005.

72. Gollob J, Richmond T, Jones J, Rathmell WK, Grigson G, Watkins C, Peterson B, Wright J. Phase II trial of sorafenib plus interferon-alpha 2b (IFN-α2b) as first- or second-line therapy in patients (pts) with metastatic renal cell cancer (RCC). *J Clin Oncol* 2006; Abstract 4538.

73. Motzer RJ, Rini BI, Michaelson MD, et al. Phase 2 trials of SU 11248 show antitumor activity in second line therapy for patients with metastatic renal cell carcinoma. J Clin Oncol 2005; Abstract 4508.

74. Patel PH, Chaganti RS, Motzer RJ. Targeted therapy for metastatic renal cell carcinoma. *Br J Cancer* 2006;94(5):614–9.

75. Motzer RJ, Hutson TE, Tomczak P, et al. Phase III randomized trial of sunitinib malate (SU11248) versus interferon-alfa (IFN-α) as first-line systemic therapy for patients with metastatic renal cell carcinoma (mRCC). *J Clin Oncol* 2006; Abstract LBA3.

76. Rini B, Rixe O, Bukowski R, et al. AG-013736, a multi-target tyrosine kinase receptor inhibitor, demonstrates anti-tumor activity in a phase 2 study of cytokine refractory, metastatic renal cell cancer. *J Clin Oncol* 2005; Abstract 4509.

77. Uemura H, Nakagawa Y, Yoshida K, et al. MN/CA IX/G250 as a potential target for immunotherapy of renal cell carcinomas. *Br J Cancer* 1999;81(4):741–6.

78. Bui MH, Seligson D, Han KR, et al. Carbonic anhydrase IX is an independent predictor of survival in advanced renal clear cell carcinoma: implications for prognosis and therapy. *Clin Cancer Res* 2003;9(2):802–11.

79. Bui MH, Visapaa H, Seligson D, et al. Prognostic value of carbonic anhydrase IX and KI67 as predictors of survival for renal clear cell carcinoma. J Urol 2004;171(6 Pt 1):2461–6.

80. Motzer RJ, Bacik J, Mariani T, Russo P, Mazumdar M, Reuter V. Treatment outcome and survival associated with metastatic renal cell carcinoma of non-clear-cell histology. *J Clin Oncol* 2002;20(9):2376–81.

81. Oosterwijk E, Bander NH, Divgi CR, et al. Antibody localization in human renal cell carcinoma: a phase I study of monoclonal antibody G250. *J Clin Oncol* 1993;11(4):738–50.

82. Surfus JE, Hank JA, Oosterwijk E, et al. Anti-renal-cell carcinoma chimeric antibody G250 facilitates antibody-dependent cellular cytotoxicity with in vitro and in vivo interleukin-2-activated effectors. *J Immunother Emphasis Tumor Immunol* 1996;19(3):184–91.

83. Bleumer I, Knuth A, Oosterwijk E, et al. A phase II trial of chimeric monoclonal antibody G250 for advanced renal cell carcinoma patients. *Br J Cancer* 2004;90(5):985–90.

84. Liu Z, Smyth FE, Renner C, Lee FT, Oosterwijk E, Scott AM. Anti-renal cell carcinoma chimeric antibody G250: cytokine enhancement of in vitro antibody-dependent cellular cytotoxicity. *Cancer Immunol Immunother* 2002;51(3):171–7.

85. Ullrich S, Beck J, Mala C, et al. A phase I/II trial with monoclonal antibody WX-G250 in combination with low dose interleukin 2 in metastatic renal cell carcinoma. *Proc Am Soc Clin Oncol* 2003;22:173 (Abstract 692).

86. Bevan P, Mala C, Kindler M, et al. Results of a phase I/II study with monoclonal antibody CG250 in combination with IFN-alpha 2A in metastatic renal cell carcinoma patients. *Proc Am Soc Clin Oncol* 2004;22(14S):4606.

87. Lam JS, Leppert JT, Belldegrun AS, Figlin RA. Novel approaches in the therapy of metastatic renal cell carcinoma. *World J Urol* 2005;23(3):202–12.

88. Hernandez JM, Bui MH, Han KR, et al. Novel kidney cancer immunotherapy on the granulocyte-macrophage colony-stimulating factor and carbonic anhydrase IX fusion gene. *Clin Cancer Res* 2003;9(5):1906–16.

89. Tso CL, Zisman A, Pantuck A, et al. Induction of G250-targeted and T-cell-mediated antitumor activity against renal cell carcinoma using a chimeric fusion protein consisting of G250 and granulocyte/monocyte-colony stimulating factor. *Cancer Res* 2001;61(21):7925–33.

90. Turner KJ, Moore JW, Jones A, et al. Expression of hypoxia-inducible factors in human renal cancer: relationship to angiogenesis and to the von Hippel-Lindau gene mutation. *Cancer Res* 2002;62(10):2957–61.

91. Tee AR, Fingar DC, Manning BD, Kwiatkowski DJ, Cantley LC, Blenis J. Tuberous sclerosis complex-1 and -2 gene products function together to inhibit mammalian target of rapamycin (mTOR)-mediated downstream signaling. Proc Natl Acad Sci USA 2002;99(21):13571–6.

92. Atkins MB, Hidalgo M, Stadler WM, et al. Randomized phase II study of multiple dose levels of CCI-779, a novel mammalian target of rapamycin kinase inhibitor, in patients with advanced refractory renal cell carcinoma. *J Clin Oncol* 2004;22(5):909–18.

93. Smith JW, Ko Y-J, Dutcher J, et al. Update of a phase I study of intravenous CCI 779 given in combination with interferon-alpha to patients with advanced renal cell carcinoma. *Proc Am Soc Clin Oncol* 2004;22(14S):4513.

94. Hudes G, Carducci M, Tomczak P, et al. A phase 3, randomized, 3-arm study of temsirolimus (TEMSR) or interferon-alpha (IFN) or the combination of TEMSR + IFN in the treatment of first-line, poor-risk patients with advanced renal cell carcinoma (adv RCC). *J Clin Oncol* 2006; Abstract LBA4.

95. Daliani DD, Papandreou CN, Thall PF, et al. A pilot study of thalidomide in patients with progressive metastatic renal cell carcinoma. *Cancer* 2002;95(4):758–65.

96. Minor DR, Monroe D, Damico LA, Meng G, Suryadevara U, Elias L. A phase II study of thalidomide in advanced metastatic renal cell carcinoma. *Invest New Drugs* 2002;20(4):389–93.

97. Motzer RJ, Berg W, Ginsberg M, et al. Phase II trial of thalidomide for patients with advanced renal cell carcinoma. *J Clin Oncol* 2002;20(1):302–6.

98. Stebbing J, Benson C, Eisen T, et al. The treatment of advanced renal cell cancer with high-dose oral thalidomide. *Br J Cancer* 2001;85(7):953–8.

99. Nathan PD, Gore ME, Eisen TG. Unexpected toxicity of combination thalidomide and interferon alpha-2a treatment in metastatic renal cell carcinoma. *J Clin Oncol* 2002;20(5):1429–30.

100. Hernberg M, Virkkunen P, Bono P, Ahtinen H, Maenpaa H, Joensuu H. Interferon alfa-2b three times daily and thalidomide in the treatment of metastatic renal cell carcinoma. *J Clin Oncol* 2003;21(20):3770–6.

101. Tripathi R, Patel B, Heilbrun L, et al. Phase II study of interferon and thalidomide in metastatic renal cell carcinoma. *Proc Am Soc Clin Oncol* 2004;22(14S):4712.

102. Rawat A, Needle M, Miles B, Amato RJ. Phase II study of CC-5013 in patients with metastatic renal cell cancer. *J Clin Oncol* 2005; Abstract 4604.

103. Ebbinghaus SW, Hussain M, Tannir NM, et al. A randomized phase 2 study of the thrombospondin mimetic peptide ABT 510 in patients with previously untreated advanced renal cell carcinoma. *J Clin Oncol* 2005; Abstract 4607.

104. Michalopoulos GK, DeFrances MC. Liver regeneration. *Science* 1997;276(5309):60–6.

105. Sattler M, Pride YB, Ma P, et al. A novel small molecule met inhibitor induces apoptosis in cells transformed by the oncogenic TPR-MET tyrosine kinase. *Cancer Res* 2003;63(17):5462–9.

106. Kamath K, Jordan MA. Suppression of microtubule dynamics by epothilone B is associated with mitotic arrest. *Cancer Res* 2003;63(18):6026–31.

107. Kowalski RJ, Giannakakou P, Hamel E. Activities of the microtubule-stabilizing agents epothilones A and B with purified tubulin and in cells resistant to paclitaxel (Taxol(R)). *J Biol Chem* 1997;272(4):2534–41.

108. Thompson JA, Swerdloff J, Escudier B, et al. Phase II trial evaluating the safety and efficacy of EPO906 in patients with advanced renal cell cancer. *Proc Am Soc Clin Oncol* 2003;22(408):405 (Abstract 1640).

109. Zhuang SH, Menefee M, Kotz H, et al. A phase II clinical trial of BMS 247550 (ixabepilone), a microtubule-stabilizing agent in renal cell cancer. *Proc Am Soc Clin Oncol* 2004;22(14S):4550.

110. Fojo AT, Menefee ME, Poruchynsky M, et al. A translational study of ixabepilone (BMS 247550) in renal cell cancer: assessment of its activity and demonstration of target engagement in tumor cells. *J Clin Oncol* 2005; Abstract 4541.

111. Fleming JA, Lightcap ES, Sadis S, Thoroddsen V, Bulawa CE, Blackman RK. Complementary whole-genome technologies reveal the cellular response to proteasome inhibition by PS-341. *Proc Natl Acad Sci USA* 2002;99(3):1461–6.

112. Richardson P. Clinical update: proteasome inhibitors in hematologic malignancies. *Cancer Treat Rev* 2003;29(Suppl 1):33–9.

113. Davis NB, Taber DA, Ansari RH, et al. Phase II trial of PS-341 in patients with renal cell cancer: a University of Chicago phase II consortium study. *J Clin Oncol* 2004;22(1):115–9.

114. Drucker BJ, Schwartz L, Bacik J, et al. Phase II trial of PS 341 shows response in patients with advanced renal cell carcinoma. *Proc Am Soc Clin Oncol* 2003;22:386 (Abstract 1550).

115. Braun BC, Glickman M, Kraft R, et al. The base of the proteasome regulatory particle exhibits chaperone-like activity. *Nat Cell Biol* 1999;1(4):221–6.

116. Banerji U, Judson I, Workman P. The clinical applications of heat shock protein inhibitors in cancer – present and future. *Curr Cancer Drug Targets* 2003;3(5):385–90.

117. Isaacs JS, Jung YJ, Mimnaugh EG, Martinez A, Cuttitta F, Neckers LM. Hsp90 regulates a von Hippel Lindau-independent hypoxia-inducible factor-1 alpha-degradative pathway. *J Biol Chem* 2002;277(33):29936–44.

118. Munster PN, Tong W, Schwartz L, et al. Phase I trial of 17 (allylamino)-17-demethoxygeldanamycin (17-AAG) in patients with advanced solid tumors. *Proc Am Soc Clin Oncol* 2001;20(83a).
119. Wilson RH, Takimoto CH, Agnew EB, et al. Phase I pharmacologic study of 17 (allylamino)-17-demethooxygeldanamycin (17-AAG) in adult patients with advanced solid tumors. *Proc Am Soc Clin Oncol* 2001;20(82a).
120. Cancer vaccine–Antigenics. BioDrugs 2002;16(1):72–4.
121. Assikis VJ, Daliani D, Pagliaro L, et al. Phase II study of an autologous tumor derived heat shock protein-peptide complex vaccine (HSPPC-96) for patients with metastatic renal cell carcinoma. *Proc Am Soc Clin Oncol* 2003;22:386 (Abstract 1552).
122. Yang JC, Beck KE, Blansfied J, et al. Tumor regression in patients with metastatic renal cancer treated with a monoclonal antibody to CTLA4 (MDX-010). *J Clin Oncol* 2005; Abstract 2501.
123. Alatrash G, Hutson TE, Molto L, et al. Clinical and immunologic effects of subcutaneously administered interleukin-12 and interferon alfa-2b: phase I trial of patients with metastatic renal cell carcinoma or malignant melanoma. *J Clin Oncol* 2004;22(14):2891–900.
124. Hara I, Nagai H, Miyake H, et al. Effectiveness of cancer vaccine therapy using cells transduced with the interleukin-12 gene combined with systemic interleukin-18 administration. *Cancer Gene Ther* 2000;7(1):83–90.

12 Targeted Therapies for Prostate Cancer

Elisabeth I. Heath, MD, and Michael A. Carducci, MD

SUMMARY

Prostate cancer is the second leading cause of cancer deaths in males in the USA. Prostate cancer is a highly morbid disease, especially upon its progression toward metastatic disease. Hormonal therapy is the mainstay of treatment in patients with progressive disease, but, eventually most patients develop hormone-insensitive disease. Docetaxel-based chemotherapy increases overall survival in patients with metastatic hormone-insensitive prostate cancer. Other cytotoxic chemotherapies either in combination with docetaxel or as single agents are currently being evaluated in metastatic prostate cancer patients. There is a clear need for additional treatment options in this growing group of patients. Novel agents targeting specific aberrant molecular pathways in prostate cancer are actively being investigated in clinical trials. The development of novel agents requires thoughtful clinical trial design, selection of appropriate study endpoints and/or surrogate markers of efficacy, and close monitoring of adverse events. Novel agents discussed in this chapter include anti-tubulin agents, anti-mitotic agents, various signal transduction inhibitors, angiogenesis inhibitors, vascular targeting agents, and immunotherapy. These agents are being developed with a goal of maximizing tumor-specific cell death while minimizing adverse toxicities. Several novel agents are currently in phase III clinical trials whereas the majority remains in early clinical development. Our improved understanding of prostate tumorigenesis will undoubtedly contribute to further development of novel targeted therapies for prostate cancer.

Key Words: Prostate cancer; targeted therapies; signaling pathway inhibitors; angiogenesis inhibitors;immunotherapy.

1. INTRODUCTION

Approximately 218,890 men in the USA were diagnosed with prostate cancer in 2007 (1). Prostate cancer remains the second leading cause of cancer deaths with 27,050 men dying of the disease in 2007. Significant advances have been achieved in

From: *Cancer Drug Discovery and Development: Molecular Targeting in Oncology*
Edited by: H. L. Kaufman, S. Wadler, and K. Antman © Humana Press, Totowa, NJ

the early detection and diagnosis of prostate cancer, but treatment options for patients with metastatic disease are limited.

Androgen deprivation therapy (ADT) is often the front-line treatment offered to patients with hormone-sensitive prostate cancer. It is effective at slowing the progression of disease in most men and in general, well tolerated, with minimal toxicities compared with chemotherapy. Unfortunately, the period of response with ADT is on average 18–24 months, at which time the cancer develops hormone insensitivity (2). Additional hormonal manipulations including ketoconazole and anti-androgens such as bicalutamide, which are agents that inhibit adrenal production of androgens or block the androgen receptor, respectively, are often used. Second-line hormonal manipulation is initially effective, but after an average period of 4 months, disease progression continues. Recently, chemotherapy with docetaxel has demonstrated an increase in overall survival in men with metastatic hormone-insensitive prostate cancer. Unfortunately, docetaxel-based chemotherapy is, at best, palliative, and not curative. As with all palliative therapies, it is critical to recognize its potential impact on patient's quality of life. There is an urgent need for newer therapies that can offer patients meaningful survival benefit with less toxicity.

Toxicity from traditional chemotherapy occurs primarily because of its cytotoxic effects on cancerous tissues as well as on normal tissues. Recently, drug development in prostate cancer and in other solid tumors has focused on improving our understanding of aberrant tumor-specific pathways and developing agents that specifically attack them. In doing so, there is hope that systemic toxicities will be minimized.

There are over 200 novel agents in various stages of development and clinical trials for prostate cancer. It is beyond the scope of this chapter to discuss all agents in development. Instead, this chapter will focus on selected aberrant tumor-specific pathways in prostate cancer and the clinical trials of promising agents that target them. The agents discussed will include anti-tubulin compounds, anti-mitotic compounds, various signaling pathway inhibitors, angiogenesis inhibitors, immunotherapeutic agents, and radiopharmaceuticals.

2. ANTI-TUBULIN AGENTS

Mitoxantrone and prednisone therapy had been considered as front-line treatment in men with advanced, symptomatic, hormone-insensitive prostate cancer because of its effects on improving disease-related symptoms (3). Two recent, large, phase III trials evaluating docetaxel-based chemotherapy in men with metastatic hormone-insensitive prostate cancer have reported a survival benefit of 2–2.5 months over previous standard therapy of mitoxantrone and prednisone (4,5). In light of these findings, docetaxel-based combination regimens are now being tested in clinical trials. Taxanes act by inhibiting microtubule depolymerization and promotion of microtubule assembly, which eventually leads to increased cellular apoptosis.

Other novel formulations of taxanes, such as the nanoparticle albumin-bound paclitaxel formulation ABI-007 (abraxane) (Abraxis Oncology), have been evaluated in breast cancer but remain untested in prostate cancer. However, abraxane's novel formulation is one of emerging interest. Abraxane is the first biologically interactive nanoparticle composition exploiting the albumin receptor-mediated (gp60/caveolin-1) pathway, achieving high intratumor concentrations of the active ingredient paclitaxel (6). Abraxane may exploit secreted protein acidic rich in cysteine (SPARC or osteonectin) and caveolin-1 to deliver drug preferentially to tumors. Both SPARC and caveolin-1 are overexpressed in prostate cancer and are associated with poor prognosis *(7)*.

Additional agents that are known to disrupt the microtubule mechanism are the epothilones. Epothilones are macrolides from myxobacteria such as Myxococcus xanthus and Sorangium cellulosum. Both taxanes and epothilones bind to B-tubulin, stabilize the polymerized microtubule, cause cell cycle arrest in G2/M phase, initiate phosphorylation of bcl-2, and lead to apoptosis *(8,9)*. However, they are structurally different from each other and attach to different binding sites within the microtubule. These differences may account for the activity of epothilones in taxane-resistant cell lines. Preclinical studies support the cytotoxic effects of epothilones despite overexpression of the drug efflux protein, P-glycoprotein.

There are currently four epothilone analogs in clinical trials: BMS-247550 (epothilone B, ixabepilone) (Bristol Myers Squibb, NY, NY, USA), BMS-310705 (epothilone B) (Bristol Myers Squibb), EPO906 (epothilone B, patupilone) (Novartis, Cambridge, MA, USA), and KOS-862 (epothilone D) (Kosan Biosciences Inc, Hayward, CA, USA). Phase I trials of ixabepilone (Bristol Myers Squibb), a semi-synthetic analog of epothilone B, evaluating various drug schedules have been reported: every 21-day cycle, weekly, daily-times-five every 21 days, and daily-times-three every 21 days *(10–12)*. A phase II trial conducted in the Southwest Oncology Group (SWOG 0111) evaluated ixabepilone (Bristol Myers Squibb) as front-line therapy in 41 patients with metastatic hormone-insensitive prostate cancer *(13)*. Patients received ixabepilone (Bristol Myers Squibb) 40 mg/m^2 i.v. every 3 weeks. There was 34% confirmed prostate-specific antigen (PSA) decrease of 50% or greater and 15% partial response in patients with measurable disease. The median time to treatment failure was 3 months with progression-free survival of 6 months. As expected, with microtubule-stabilizing agents, hematologic and neurologic toxicities were the primary side effects. Seventeen percent of patients experienced grade 3/4 neutropenia, and 12% of patients experienced sensory neuropathy. Approximately 20% of patients terminated treatment secondary to neurotoxicity. These toxicities are less compared with those reported with docetaxel-based chemotherapy. Thirty-two percent of patients treated with every 3-week docetaxel and prednisone experienced grade 3/4 neutropenia compared with 22% of patients treated with mitoxantrone and prednisone *(4)*. Also, 30% of patients treated with every 3-week docetaxel and prednisone experienced grade 3 sensory neuropathy compared with 7% of patients treated with mitoxantrone and prednisone. A randomized phase II study is currently underway evaluating single agent ixabepilone (Bristol Myers Squibb) compared with mitoxantrone and prednisone in men with taxane-resistant metastatic hormone-insensitive prostate cancer.

Combination trials of ixabepilone (Bristol Myers Squibb) have also been reported. A phase I trial of front-line therapy of ixabepilone (Bristol Myers Squibb) in combination with estramustine in metastatic hormone-insensitive men with prostate cancer established the maximum tolerated dose of ixabepilone (Bristol Myers Squibb) as 35 mg/m^2 i.v. every 3 weeks *(14)*. Of the 12 evaluable patients, 11 experienced greater than 50% PSA decline. These promising results led to a multi-institutional, randomized, phase II trial comparing ixabepilone (Bristol Myers Squibb) versus ixabepilone (Bristol Myers Squibb) and estramustine *(15)*. Ixabepilone (Bristol Myers Squibb) 35 mg/m^2 i.v. was administered on day 2 with or without estramustine

280 mg orally thrice daily on days 1–5 every 3 weeks. Ninety-two patients were enrolled, with twenty-two patients continuing treatment. As anticipated, neutropenia and neuropathy were reported. Unexpectedly, grade 3/4 thrombotic events in 6% of patients on the ixabepilone (Bristol Myers Squibb)/estramustine arm were reported, despite administration of prophylactic warfarin. The combination ixabepilone (Bristol Myers Squibb)/estramustine arm resulted in 69% of patients who achieved greater than 50% PSA decline compared with 48% in ixabepilone (Bristol Myers Squibb) alone group. Given the relatively small additional benefits and excess toxicity to estramustine in combination with docetaxel, it is unlikely that the combination of ixabepilone (Bristol Myers Squibb)/estramustine will move forward.

A water soluble, semi-synthetic analog of another epothilone B, BMS-310705, has been evaluated in phase I trials. However, these trials did not include prostate cancer patients, and there are no reports to date that indicate the pursuit of this compound in patients with prostate cancer. Another analog of epothilone B being evaluated in prostate cancer is EPO906, patupilone (Novartis). Patupilone (Novartis) has been evaluated in a phase II trial of 37 patients with metastatic hormone-insensitive prostate cancer at a dose of 2.5 mg/m^2 i.v. once weekly for 3 weeks followed by 1-week rest (16). Unlike the trials with ixabepilone (Bristol Myers Squibb), this trial was conducted primarily in patients who have received prior cytotoxic chemotherapy. Sixty-four percent of patients received one prior chemotherapy regimen. The overall response rate was 16%, with seven of twenty-eight patients achieving a PSA response of 50% or greater. Interestingly, the most common adverse event was gastrointestinal in nature: diarrhea, vomiting, and abdominal pain. There was no grade 3/4 neutropenia or neuropathy reported. The gastrointestinal toxicities were not unexpected as diarrhea was the dose-limiting toxicity in phase I trials. The different toxicity profiles of these two epothilone analogs are not well understood. Patupilone (Novartis) is susceptible to inactivation by esterases whereas ixabepilone is not. This may contribute to the different toxicities observed. A phase I trial of patupilone (Novartis) in combination with estramustine reported a maximum tolerated dose of 2.5 mg i.v. on day 2 every 3 weeks followed by 1-week rest and 280 mg orally twice daily on days 1–3, respectively (17). Grade 3/4 toxicities included diarrhea, fatigue, and vomiting, with no observed neurotoxicity.

KOS-862 (Kosan Biosciences Inc), an epothilone D analog, has been evaluated in three phase I trials using various dosing schedules, including intravenous continuous infusion (9,18,19). One patient with prostate cancer experienced a 25% decline in PSA. The maximum tolerated dose is yet to be reported. Toxicities noted so far include sensory neuropathy, fatigue, nausea, and vomiting. Phase II trials are planned in patients with taxane-sensitive and taxane-resistant tumors.

The role of anti-tubulin agents, either as single or combination therapy, has yet to be defined in prostate cancer. Preclinical, phase I and II trials show promise for these agents. However, it is important to note that the toxicities associated with epothilones are similar to those reported with taxane-based therapies. Should further data support the initiation of larger, randomized, phase III trials, quality of life measures must be incorporated. Development of first-line or second-line therapy for treatment of advanced prostate cancer remains an active area of research.

3. ANTI-MITOTIC AGENTS

Additional agents that have the potential to target cellular proliferation, regulate centrosome function, impact G2/M phase, and induce apoptosis include polo-like kinase (Plk)-1 inhibitors, aurora kinase inhibitors, and kinesin spindle protein (KSP) inhibitors. HMN-214 (NS Pharma, Tokyo, Japan) is an oral Plk-1 inhibitor with potent anti-microtubular effects *(20)*. Several phase I trials evaluating various dosing schedules have been reported *(21–23)*. The maximum tolerated dose is 8 mg/m² orally once a day. Toxicities include hyperglycemia, myalgias, and bone pain. Although the phase I trials did not include prostate cancer patients, there is excellent biologic rationale for pursuing Plk-1 inhibitors in this patient population.

Aurora kinases are serine/threonine kinases that regulate mitotic spindle formation and centrosome maturation *(24,25)*. Inhibitors of aurora kinases disrupt cell cycle progression and, ultimately, induce apoptosis. VX-680 (Vertex Pharmaceuticals, Cambridge, MA, USA) is a small molecule inhibitor of the aurora kinases. Preclinical studies of VX-680 (Vertex Pharmaceuticals) showed inhibition of proliferation of solid tumors, including prostate cancer cells. Phase I trials are ongoing *(26)*.

KSP has an important role in the assembly process and function of the mitotic spindle. SB-715992 (Cytokinetics, San Francisco, CA, USA), an intravenous KSP inhibitor, is currently being evaluated in several phase I trials *(27,28)*. SB-715992 (Cytokinetics) is expected to disrupt mitotic spindle function, cause arrest of cell cycle in mitosis, and eventually cell death. The patient population in the phase I trials included those with colorectal, sarcoma, lung, and renal cell carcinoma; no prostate cancer patients were enrolled. However, based on the mechanism of action, anti-mitotic agents may offer additional alternatives for patients with prostate cancer.

4. SIGNALING PATHWAY INHIBITORS

4.1. Epidermal Growth Factor Receptor Family

The epidermal growth factor receptor (EGFR) family and its role in cellular signaling have emerged as one of the leading pathways integral in tumorigenesis of solid tumors *(29)*. The EGFR (also known as HER-1, c-erbB-1), a 170-kDa transmembrane glycoprotein, is one of four growth factor receptor proteins. The other three receptors are HER-2 (c-erbB-2), HER-3 (c-erbB3), and HER-4 (c-erbB-4). In addition to cellular signaling, the EGFR family also impacts cell proliferation, angiogenesis, and eventually apoptosis. A wide range of solid tumors have overexpression of EGFR, particularly HER-2. Targeting key receptors overexpressed in malignant cells and not in normal cells is appealing because of the potential to minimize treatment-related toxicities.

The EGFR inhibitors available are primarily monoclonal antibodies targeting the extracellular receptor domain of the EGFR and small molecule compounds disrupting intracellular tyrosine kinase activity. The monoclonal antibodies are administered intravenously, and small molecule compounds are available as oral agents. There are similar side effects associated with both types of inhibitors, with rash emerging as a common toxicity.

The role of the EGFR family in prostate tumorigenesis is not well defined. EGFR has been shown to be overexpressed in 40–90% of prostate cancer cells *(30)*. Unlike in breast cancer where HER-2 has been reported to be highly overexpressed, HER-2 is overexpressed in less than 50% of prostate cancer tissue. Inhibition of HER-2 pathway signaling may be due to reduced androgen receptor transcriptional activity instead of downstream signaling effects. Modulation of the androgen receptor activity may be primarily mediated by the HER-2/HER-3 pathway. There are also studies that report increased HER-2 expression in androgen-insensitive prostate cell lines as well as in metastatic prostate tissue. The clinical significance of EGFR and HER-2 overexpression in prostate cancer is actively being explored.

4.2. Monoclonal Antibodies

Monoclonal antibodies block natural ligands such as epidermal growth factor, transforming growth factor-a, amphiregulin, heparin-binding EGF, and betacellulin from binding to the extracellular receptor domain of the EGFR. This block prevents activation of the receptor tyrosine kinase. As a result, there is no receptor homo- or heterodimerization at the cell surface, no internalization of the dimerized receptor, and no autophosphorylation of the intracytoplasmic EGFR tyrosine kinase domains. Lack of phosphorylated tyrosine kinase results in deficient active binding sites for intra-cellular substrates necessary for initiation of signal transduction. The two recognized downstream pathways, which are impacted, include the Ras-Raf mitogen-activated protein kinase pathway and the phosphatidyl inositol 3′ kinase (PI3K) and Akt pathway. Therefore, monoclonal antibodies have a potential to shut down critical pathways in prostate tumorigenesis. However, one must be cognizant of the limitations posed by monoclonal antibodies; monoclonal antibodies may induce an immune-antibody response in which subsequent doses of antibody would be ineffective and mutated forms of EGFR may not be recognized by the antibodies.

Erbitux (IMC-C225, cetuximab) (ImClone Systems, NY, NY, USA) is a monoclonal antibody that has been studied extensively in preclinical and clinical trials of solid tumors, with FDA approval for use in colorectal cancer in February 2004. Erbitux (ImClone Systems) specifically targets EGFR and was effective as a single agent and in combination in reducing tumor growth and metastasis in preclinical studies in prostate cancer. Currently, there is one published abstract of a phase I/II trial of erbitux (ImClone Systems) in prostate cancer (31). Escalating doses of erbitux (ImClone Systems) combined with doxorubicin were administered to 22 men with hormone-insensitive prostate cancer. One patient had a greater than 50% decline in PSA. Sixty-four percent of patients had increased PSA.

Another monoclonal antibody against EGFR is ABX-EGF (Abgenix, Freemont, CA, USA), a fully humanized monoclonal antibody. ABX-EGF (Abgenix) blocks the ligand binding to EGFR, which inhibits tyrosine phosphorylation, and leads to internalization of EGFR (32). In vitro and in vivo studies of ABX-EGF (Abgenix) in prostate cancer cell lines and xenografts revealed significant growth inhibition (33). In a phase I clinical trial, 43 patients with solid tumors were treated with ABX-EGF (Abgenix) with doses ranging from 0.01 to 2.5 mg/kg i.v. weekly (34). One of thirteen patients with prostate cancer achieved a minor response. A phase II clinical trial is underway in patients with metastatic hormone-insensitive prostate cancer.

EMD 72000 (Matuzumab) (EMD Pharmaceuticals, Durham, NC, USA) is another EGFR-specific monoclonal antibody in early clinical development. Matuzumab (EMD Pharmaceuticals) has been evaluated in phase I trials with primarily patients with colorectal carcinoma (35). Clinical trials of Matuzumab (EMD Pharmaceuticals) in prostate cancer patients have not been reported.

Herceptin (trastuzumab) (Genentech, San Francisco, CA, USA) is a monoclonal antibody that specifically targets the HER-2 receptor. Herceptin (Genentech) has been studied alone and in combination with paclitaxel, docetaxel, and docetaxel/estramustine in patients with metastatic hormone-insensitive prostate cancer. Herceptin (Genentech) alone was administered in a phase II trial in 18 patients with metastatic hormone-insensitive prostate cancer at a loading dose of 8 mg/kg i.v. and maintenance doses of 4 mg/kg i.v. weekly (36). Only two patients experienced stable disease. Herceptin (Genentech) as a single agent showed poor efficacy in patients with advanced hormone-insensitive prostate cancer. Another study reported the combination of herceptin (Genentech) and paclitaxel in 23 patients with metastatic prostate cancer (37). The study aimed to assign patients to four treatment groups based on androgen-dependent status and HER-2 status. Interestingly, only 6 study patients were HER-2 positive. All patients experienced disease progression on the herceptin (Genentech) alone arm (4 mg/kg i.v. loading dose, 2 mg/kg i.v. maintenance dose). Three of fifteen patients receiving combination therapy (addition of weekly paclitaxel 100 mg/m^2 i.v.) experienced greater than 50% decline in the PSA levels. The study also revealed that significant proportions of HER-2 overexpression were found in metastatic tissue, not in primary prostate tumors. Therefore, HER-2 expression profiling was a challenge because of limitations in acquiring metastatic tissues for testing.

A phase II trial of herceptin (Genentech) in combination with docetaxel was planned in patients with metastatic hormone-insensitive prostate cancer (38). One hundred patients with this disease were screened for HER-2 overexpression by immunohistochemistry and fluorescent in situ hybridization. Shed HER-2 levels were also measured by enzyme-linked immunoadsorbent assay. The results revealed an overall HER-2 positivity rate of less than 20%. The trial was closed because of nonfeasibility, and therefore, the clinical efficacy of docetaxel and herceptin combination is at present unknown.

A phase I study evaluating the role of docetaxel, estramustine, and herceptin (Genentech) in a similar patient population was conducted in 13 patients (39). The triple combination appeared to be well tolerated with 69% of patients achieving a greater than 50% decline in PSA levels. Given the activity of docetaxel and estramustine, it is unclear whether there are any additional benefits of herceptin in this combination. To date, there have been no published reports of this combination in phase II trials.

Overall, the role of herceptin (Genentech) in prostate cancer has yet to be defined. The lack of efficacy of single agent herceptin (Genentech) and the difficulty in determining HER-2 overexpression, especially in metastatic tumors, must be considered. To obviate the requirement for HER-2 overexpression in prostate cancer, another monoclonal antibody against HER-2 was developed. Omnitarg (pertuzumab) (Genentech) is a humanized monoclonal antibody that acts by blocking ligand-associated heterodimerization of HER-2 with other EGFR family members (40). Lack of heterodimerization results in inhibition of intracellular signaling through the mitogen-activated protein kinase (MAPK) and the PI3K pathways. Both herceptin (Genentech) and omnitarg (Genentech) are monoclonal antibodies against HER-2, but they are different in their mechanism of action. Omnitarg (Genentech) should inhibit prostate cancers that do not overexpress HER-2 in either primary or metastatic tumors. Preclinical studies have supported this assumption in both androgen-sensitive and androgen-insensitive xenografts (41). Agus et al. (42) conducted a phase I trial of omnitarg (Genentech) in 21 patients with solid tumors at a dose range of 0.5–10.0 mg/kg. Of the five patients with prostate cancer, one patient achieved partial response and the rest experienced stable disease. A phase II study in taxane-resistant patients with prostate cancer is ongoing.

4.3. Small Molecule Inhibitors

Small molecule inhibitors are oral agents designed to competitively inhibit the binding of ATP to the tyrosine kinase domain of the EGF receptor. This inhibition should result in the inhibition of EGF autophosphorylation. The mechanism of action of small molecule inhibitors differs from that of monoclonal antibodies in that its action is based intracellularly, whereas monoclonal antibodies target the extracellular domain of the receptor. Many small molecule inhibitors have been evaluated in preclinical and clinical studies in solid tumors.

Iressa (gefitinib) (Astra Zeneca, Wilmington, DE, USA) is an oral low-molecular weight tyrosine kinase inhibitor with high affinity to EGFR tyrosine kinase. It noncompetitively disrupts EGFR ligand signaling. Multiple preclinical and clinical studies have been conducted in solid tumors, in particular lung cancer. FDA approval as third-line therapy in metastatic non-small cell lung cancer patients was granted in May of 2003 under the agency's accelerated approval program. The approval was granted because of data in phase II trials, which support significant anti-tumor activity in approximately 10% of patients. The study also reported longer survival in patients who are of Asian ethnicity. Unfortunately, in the large phase III confirmatory trial, tumor shrinkage did not translate to an overall survival benefit.

Preclinical studies of iressa (Astra Zeneca) on androgen-sensitive (LNCaP) and androgen-insensitive (PC3 and DU145) cell lines have shown inhibition of prostate cancer cell proliferation (43). Potential mechanisms of action include suppression of PI3K activation (44). CWR22 xenografts in nude mice treated with iressa (Astra Zeneca) and bicalutamide also showed inhibition of tumor growth (45). Thus, preclinical studies support the use of iressa (Astra Zeneca) in clinical studies. However, similar to the lung cancer experience, the results of iressa (Astra Zeneca) in clinical trials with prostate cancer patients are disappointing, but perhaps not surprising. Promising phase I clinical studies suggesting activity of iressa (Astra Zeneca) in patients with prostate cancer did not translate to successful phase II trial results. Two different randomized phase II studies evaluating Iressa (Astra Zeneca) at either the 250 mg/day or the 500 mg/day dose in patients with hormone-insensitive prostate cancer showed no significant differences in progression rates, time to progression, and overall survival between the two arms (46,47).

The apparent lack of efficacy of iressa (Astra Zeneca) as a single agent emphasizes our incomplete understanding of the biological mechanisms of cellular signal transduction. For example, the impact of race on the efficacy of iressa (Astra Zeneca) in patients with lung cancer has been recognized. Patients of Asian ethnicity have an apparent benefit with iressa (Astra Zeneca) compared with Caucasian patients with lung cancer. Recognizing these biologic differences may be important as they may potentially contribute to the success or failure of EGFR-targeted agents.

It is possible that the potency of iressa (Astra Zeneca) as a single agent in advanced prostate cancer is not adequate, and its role may be better suited in combination with other agents. From a biologic rationale standpoint, the combination of anti-EGFR antibody with a small molecule tyrosine kinase inhibitor and/or cytotoxic chemotherapy has tremendous potential. However, in a preclinical study, iressa (Astra Zeneca) and herceptin (Genentech), administered in DU145 androgen-insensitive prostate cancer cells resulted in less than additive effects on prostate cancer cell survival *(48)*. Therefore, treatment of androgen-insensitive prostate cancer in preclinical studies with two biologically targeted agents against EGFR did not result in the anticipated inhibitory effects.

In contrast, the combination of iressa (Astra Zeneca) and cytotoxic chemotherapy in preclinical studies does show additive and synergistic effects in prostate cancer cells. A dose-escalation clinical trial evaluating iressa (Astra Zeneca) in combination with docetaxel/estramustine or mitoxantrone/prednisone reported PSA and pain responses in patients with androgen-insensitive prostate cancer *(49)*. The combination appeared feasible with a reasonable toxicity profile. Larger confirmatory studies are certainly necessary but unlikely, given iressa's demise in lung cancer.

Tarceva (Genentech) is another oral agent that specifically inhibits tyrosine kinase activity of the intracellular portion of EGFR. Inhibition of tyrosine kinase activity appears to be reversible. Similar to iressa (Astra Zeneca), multiple preclinical and early clinical studies of tarceva (Genentech) have been conducted in patients with solid tumors. Encouraging results seen in patients with lung cancer led to phase II and III clinical trials. On the basis of a 731 patient phase III study conducted by the National Cancer Institute of Canada Clinical Trials Group (BR.21), tarceva (Genentech) administered as a second- or third-line agent compared with placebo resulted in increased overall survival and quality of life in patients with advanced nonsmall cell lung cancer *(50)*. On the basis of this study, the FDA approved tarceva (Genentech) in November of 2004 as a second-line agent for patients with progressive non-small cell lung cancer after failing one prior chemotherapy regimen. Improvement in overall survival with tarceva (Genentech) was 6.7 months compared with 4.7 months in the placebo arm. Rash and diarrhea were two common reported adverse events. Larger confirmatory post-marketing studies are required by the FDA.

Two randomized phase III trials evaluating tarceva (Genentech) and chemotherapy in patients with lung cancer have reported disappointing results. The primary endpoint of overall survival was not achieved when tarceva (Genentech) versus placebo in combination with gemcitabine/cisplatin (TALENT) and tarceva (Genentech) versus placebo in combination with paclitaxel/carboplatin (TRIBUTE) followed by maintenance with Tarceva (Genentech) or placebo was administered in patients with advanced nonsmall cell lung cancer as first-line treatment *(51,52)*. These results are indeed surprising as there is biologic rationale to combine cytotoxic chemotherapy with an agent targeting tyrosine kinase signaling pathways.

With regard to prostate cancer, the experience with single agent tarceva (Genentech) has been limited to phase I trials. To date, there are no published reports of tarceva (Genentech) as a single agent or in combination with cytotoxic agents in prostate cancer patients in phase II or III trials.

Iressa (Astra Zeneca) and tarceva (Genentech) are small molecules that target EGFR tyrosine kinase signaling. Additional oral agents have been developed to target more than one EGFR tyrosine kinase. PKI 166 (Novartis), GW 572016 (lapatinib) (Glaxo Welcome, Research Triangle, NC, USA), and EKB 569 (Wyeth, Madison, NJ, USA) are dual HER inhibitors, targeting both the EGFR and HER-2 receptors. CI-1033 (Pfizer, NY, NY, USA) is an irreversible pan-HER inhibitor. These agents are in early stages of clinical evaluation.

Administration of PKI 166 (Novartis) resulted in growth inhibitory effects on human prostate cancer xenografts *(53)*. Three phase I studies have been conducted in patients with advanced solid tumors but none in patients with advanced prostate cancer *(54–56)*. GW 572016 (Lapatinib) (Glaxo Welcome) is a reversible inhibitor of EGFR and HER-2 tyrosine kinases. Two phase I trials have been conducted in patients with solid tumors *(57,58)*. A phase II trial of lapatinib front-line therapy in patients with metastatic or recurrent prostate cancer is underway. EKB-569 (Wyeth), an irreversible inhibitor of EGFR and HER-2 tyrosine kinases, is being evaluated in early phase I trials *(59)*. No data are available regarding the toxicity of EKB-569 (Wyeth) in patients with advanced prostate cancer. Finally, CI-1033 (Pfizer) is an irreversible inhibitor of all tyrosine kinase domains in the HER family. Phase I trials have been reported, and these include a few patients with prostate cancer *(60)*.

In summary, further investigation is clearly warranted to study the potential disruptive effects of HER family signaling through extracellular monoclonal antibodies or small molecules inhibiting intracellular tyrosine kinases of HER receptors. The data from inhibition of targeted pathways from single agents are disappointing. However, combination studies with anti-androgens and/or cytotoxic chemotherapy appear warranted given supportive preclinical data and strong biologic rationale.

4.4. Vascular Endothelial Growth Factor Family

An important event in tumorigenesis and metastasis is angiogenesis. Angiogenesis occurs when new blood vessels are formed from existing vasculature. New blood vessel formation is critical for supply of oxygen, growth and other factors to sustain malignant growth, tumor invasion as well as spread of metastatic disease to distant sites. Vascular endothelial growth factor (VEGF) and its receptors play important roles in tumor growth and angiogenesis.

There are currently seven known VEGF molecules: VEGF-A, VEGF-B, VEGF-C, VEGF-D, VEGF-E, and placenta growth factor (PIGF)-1 and PIGF-2. These six growth factor ligands bind to VEGF receptors of which three are known receptors: VEGFR-1 [fms-like tyrosine kinase 1 (Flt-1)], VEGFR-2 (KDR, Flk-1), and VEGFR-3 (Flt-4). The VEGF receptors are primarily found on endothelial cells, hematopoietic stem cells, osteoblasts, and osteoclasts. Binding of the VEGF ligands to its various receptors (each receptor with its own affinity for different ligands) initiates the signaling cascade integral in angiogenesis and vasculogenesis. However, VEGF does not only function in initiation of the process but also in proliferation, invasion, migration, and other critical functions of the endothelial cells in angiogenesis and hematopoiesis. In addition, VEGF is regulated by multiple other factors, such as hypoxia, additional growth factors and cytokines, and other signaling pathways (including HER family).

Tissue and serum VEGF levels have been reported to be elevated in patients with prostate cancer *(61,62)*. Using serum samples from patients enrolled in CALGB 9480, increased VEGF levels were correlated with decreased survival in patients with hormone-insensitive prostate cancer. Increased immunohistochemical expression of VEGF in prostate cancer tissue was also correlated with increased tumor stage and grade. Similar to other solid tumors, VEGF appears to have a role in prostate cancer.

Avastin (bevacuzimab) (Genentech) is a humanized murine monoclonal antibody that targets VEGF. It was FDA approved in February of 2004 for use in patients with metastatic colon cancer in addition to 5-flouuracil-based chemotherapy. By inhibiting

VEGF from binding to its receptor, the angiogenic process is believed to be inhibited. Preclinical studies in prostate xenograft models showed that antibodies to VEGF in addition with chemotherapy inhibited tumor growth *(63)*. On the basis of encouraging preclinical studies and activity of Avastin (Genentech) with chemotherapy in colon cancer, CALGB initiated a trial of Avastin (Genentech) in combination with docetaxel and estramustine in patients with metastatic hormone-insensitive prostate cancer *(64)*. Avastin (Genentech) was administered at a dose of 15 mg/kg i.v. on day 2. Eighty-one percent of patients experienced PSA decrease of 50% or greater. However, excess thrombotic events were reported. This toxicity is most likely from estramustine (despite prophylactic warfarin). A phase III trial of avastin (Genentech) versus placebo in combination with docetaxel and prednisone is underway (CALGB 90401).

Two other VEGF-targeting agents are in early clinical development: 2C3 (monoclonal antibody against VEGF-A) (Peregrine, Tustin, CA, USA) and VEGF-trap (soluble hybrid receptor against VEGF-A and PlGF) (Bristol Myers Squibb). Additional agents in development are those that target primarily VEGF receptor tyrosine kinases. CP-547,632 (OSI Pharmaceuticals, Melville, NY, USA) is an ATP-competitive kinase inhibitor to tyrosine kinase domain of VEGFR-2. AZD2171 (Astra Zeneca) is an oral VEGFR-1 and VEGFR-2 inhibitor. PTK787/ZK 222584 (Novartis), ZD6474 (Astra Zeneca), and CEP-7055 (Antigenics, NY, NY, USA) are inhibitors of all three VEGF receptor tyrosine kinases. ZD6474 (Astra Zeneca) also targets the EGFR pathway similar to AEE788 (Novartis) compound. SU11248 (sutent) (Pfizer) and AG013736 (Pfizer) target VEGFR-1, VEGFR-2, and PDGFR pathways. Preclinical and early clinical studies support the inhibitory effects of these small molecule agents on tumor growth and angiogenesis in solid tumors. Phase II-specific studies in patients with prostate cancer have not been initiated with any of these compounds with the exception of SU5416 (Pfizer). SU5416 (Pfizer) is an intravenous agent inhibiting tyrosine kinase phosphorylation of VEGFR-2. Thirty-six chemotherapy naïve, hormone-insensitive patients with prostate cancer were randomized to treatment with SU5416 (Pfizer) (145 mg/m2 i.v. twice weekly) and dexamethasone premedication versus high dexamethasone dose alone *(65)*. There was no effect of SU5416 (Pfizer) on PSA secretion or time to progression. VEGF levels were not prognostic. Toxicities including headache, fatigue, hyperglycemia, hyponatremia, and other effects attributed to steroid use were noted. On the basis of these results, no further clinical development of SU5416 (Pfizer) will occur in prostate cancer . The experience with SU5416 (Pfizer) emphasizes the difficulty of translating promising molecular observations into successful treatment for prostate cancer patients. The difficulty in defining clinically meaningful, successful treatment for prostate cancer patients may explain why most of these newer agents are initially evaluated in other tumor types.

4.5. Platelet-Derived Growth Factor Family

Another tyrosine kinase receptor overexpressed in solid tumors is the PDGF receptor (PDGFR). Activation of the PDGFR initiates the paracrine and autocrine pathways crucial for tumor growth and bone metastasis. Platelet-derived growth factor (PDGF) is a 30-kDa protein that has four isoforms: AA, BB, AB, and CC. Initial publications have reported PDGFR expression in 88% of primary prostate cancer and 80% in androgen-insensitive metastatic lesions *(66)*. Preclinical studies support the use of PDGFR inhibition in prostate cancer *(67)*.

Two PDGFR inhibitors evaluated in clinical trials in prostate cancer are gleevec (Novartis) and SU101 (leflunomide) (Pfizer). Gleevec (Novartis) is an inhibitor of the Abl and BCR-Abl tyrosine kinases. It is currently being used for the treatment of patients with chronic myeloid leukemia and gastrointestinal stromal tumors with remarkable response rates. A phase II trial of gleevec (Novartis) in hormone-insensitive prostate cancer patients has been completed *(68)*. Patients received gleevec (Novartis) at a dose of 400 mg orally daily. Another trial evaluating the role of gleevec (Novartis) in patients with hormone-sensitive prostate cancer with PSA progression was conducted *(69)*. In this trial, gleevec (Novartis) was administered at a dose of 400 mg orally twice daily. Unfortunately, there was considerable toxicity reported with minimal drug activity. Administration of gleevec (Novartis) and zoledronic acid resulted in no PSA responses with no clinical or palliative benefit *(70)*. Combination phase I trial of gleevec (Novartis) with docetaxel in androgen-insensitive prostate cancer patients has been completed *(71)*. The trial evaluated 28 men with bony metastatic disease with a 30-day gleevec (Novartis) at 600 mg orally daily dose lead-in therapy. There was no meaningful clinical activity seen in the lead-in phase, and the investigators concluded that single agent gleevec (Novartis) in this patient population is ineffective. Combination therapy with weekly docetaxel for 4 of 6 weeks starting at 30 mg/m^2 was feasible and reasonably well tolerated. Greater than 50% of the patients received prior taxanes. Despite the negative early results and because of encouraging preclinical data, a randomized, placebo-controlled phase II study of docetaxel and gleevec (Novartis) in hormone-insensitive prostate cancer patients with bony metastases without prior taxane exposure is ongoing.

Leflunomide (Pfizer) has also been evaluated in a phase II trial of 39 evaluable patients with metastatic hormone-insensitive prostate cancer *(72)*. Unfortunately, the results showed very low response rates in PSA, although overall pain improvement was noted in 26% of patients. The results from single agent gleevec (Novartis) and leflunomide (Pfizer) to date have been less than spectacular. One possible reason is that the significance of PDGFR overexpression in prostate cancer is unknown. Although the receptor is overexpressed in cancer and the PDGFR inhibitors such as gleevec (Novartis) do specifically affect the target, the functional role of PDGFR in prostate cancer tumorigenesis has not been well established. Recent data would also suggest that PDGFR-B expression may not be increased in prostate cancer after all using expression array analysis *(73)*. On the basis of the available data, PDGFR inhibitors do not appear to be promising therapy for patients with advanced prostate cancer.

4.6. Endothelin Receptor Inhibitors

Three amino acid peptides, endothelin-1 (ET-1), endothelin-2 (ET-2), and endothelin-3 (ET-3), make up the family of endothelins (ETs). ETs bind to two different membrane receptors, ETA and ETB. Upon binding to its receptors, ETs trigger a pertussis-insensitive G protein that triggers downstream signal transduction pathways of adjacent receptors, such as EGFR. In addition to activating its own intracellular signaling pathway and potential cross-talking with other signaling pathways, ETs are involved in a multitude of other functions; vasoconstriction, autocrine and/or paracrine growth factor, angiogenesis, osteogenesis, and nociception *(74)*.

In the normal prostate gland, ET-1 is produced and is in high concentration in seminal fluid *(75)*. In prostate cancer tissue, ET-1 concentrations are increased along with decreased ETBs *(76)*. Plasma ET-1 levels are also higher in men with androgen-insensitive prostate cancer. Atrasentan (Abbott, Abbott park, IL) is a highly potent and selective antagonist to the ETA receptor. A phase II trial comparing atrasentan (Abbott) at 2.5 mg orally daily versus 10 mg orally daily versus placebo in patients with hormone-insensitive prostate cancer with asymptomatic metastasis was conducted *(77)*. The results showed that time to PSA progression was much longer at 155 days in the atrasentan (Abbott) 10 mg group compared with 71 days in the placebo group (p = 0.002). Another phase II trial similar in design was conducted in patients with hormone-insensitive prostate cancer with symptomatic metastasis. Patients in this trial reported improvement in pain rating while taking atrasentan (Abbott) when compared with placebo *(78)*.

Two large phase III trials evaluating the effects of atrasentan (Abbott) have been completed *(79)*. The M00-211 trial and the M00-244 trial completed patient enrollment in early 2003. The M00-211 trial evaluated atrasentan (Abbott) 10 mg orally daily in 809 men with hormone-insensitive metastatic disease. The primary endpoint of this trial was time to progression and secondary endpoints measuring drug effects on PSA and bone alkaline phosphatase levels. M00-211 showed no significant difference in the time to progression between the two arms based on an intent-to-treat analysis. However, upon review of progression events, atrasentan demonstrated a significant delay in time to onset of metastatic pain events. The M00-244 trial evaluated the atrasentan (Abbott) 10 mg orally daily in 941 patients with hormone-insensitive disease but without metastatic disease. Results from the M00-244 trial are not yet available given the anticipated lengthy natural history of prostate cancer in this situation; only 33% of patients developed bony metastatic disease at 2 years *(80)*.

The adverse events noted throughout all three phases of clinical trial testing included peripheral edema, rhinitis, and headache, all most likely because of the drug's vasodilatory effects. In the M00-211 trial, peripheral edema was reported in 40% of patients on atrasentan (Abbott) compared with 12% on placebo. Thirty-five percent of patients experienced rhinitis on the treatment arm compared with 14% on the placebo arm. In nearly 9% of patients on the treatment arm, these adverse events led to premature treatment termination. However significant these side effects were, it is important to note that overall quality of life measures in the M00-211 trial improved, especially scores measuring pain. The improvement in pain ratings, especially those caused by osteoblastic metastasis, is not surprising as endothelin receptor inhibitors are known to interfere with osteoblastic/osteoclastic interaction and interrupt ET-1-related pain pathways. The delay in pain onset and maintenance of quality of life summarize the clinically meaningful benefits of atrasentan (Abbott).

Although the primary objective of improved time to disease progression in the M00-211 study was not achieved, a pooled intent-to-treat meta-analysis of all patients in both the M00-211 and the M96-594 trials reported an improvement in the median time to disease progression, incidence of bone pain, and median time to bone pain in the treatment arm compared with placebo. With these supportive results to the clinical benefits of the individual studies above, Abbott Pharmaceutical has submitted an NDA application for atrasentan (Abbott) for FDA approval. SWOG recently activated a phase III study of docetaxel with or without atrasentan (Abbott) as front-line therapy for patients with metastatic, hormone-insensitive prostate cancer (DAHRT Study).

Atrasentan (Abbott) is the first oral agent in the endothelin receptor antagonist class to be studied in clinical trials in oncology. Bosentan is actually the first oral endothelin receptor antagonist studied in nononcologic clinical trials and is FDA approved for treatment of primary pulmonary hypertension. Another oral endothelin receptor inhibitor with promising potential is ZD4054 (Astra Zeneca), which is in phase III clinical trials *(81,82)*. Indeed, there is great potential of this class of agents to impact the biology of prostate cancer metastasis to bone.

4.7. Farnesyl Protein Transferase Inhibitor

Farnesyl protein transferase (FTI) is an enzyme that catalyzes a step in the post-translational addition of an isoprenoid side chain at the carboxyl terminus of many proteins including Ras protein. Ras is a critical protein that plays an integral role in cellular signaling. Because Ras is functional after post-translational modification, a group of agents known as FTIs were developed to inhibit this process. Four FTIs have been evaluated in clinical trials: Zarnestra (Johnson and Johnson, Rantan, NJ, USA) (oral agent), SCH-66336 (sarasar) (Schering Plough, Kenilworth, NJ, USA) (intravenous agent), L-778,123 (Merck) (intravenous agent), and BMS-214662 (Bristol Myers Squibb) (oral agent) *(83)*. Clinical development of L-778,123 (Merck, Whitehouse Station, NJ, USA) was halted because of prolonged QT.

Zarnestra (Johnson and Johnson) has been evaluated in a phase II trial in patients with hormone-insensitive prostate cancer *(84)*. Zarnestra (Johnson and Johnson) at a dose of 300 mg orally twice daily for 21 days every 28 days was administered to 15 patients with metastatic hormone-insensitive prostate cancer. Unfortunately, there was little anti-tumor activity seen as no patient experienced greater than 50% decline in the serum PSA. Phase III trials have been conducted with Zarnestra (Johnson and Johnson) in colorectal and pancreatic cancer, both trials showing no differences in overall survival.

Zarnestra (Johnson and Johnson) has also been studied in combination with various cytotoxic chemotherapy and other targeted therapies. None of these trials have shown the expected dramatic anti-tumor effects of inhibition of ras/raf/mapk pathway. FTIs were developed with the objective of targeting mutant Ras function in cancer cells. However, the evidence has shown that FTIs can exert inhibitory activity in cells regardless of the Ras status, implicating additional targets within the cell that were previously not recognized. The role of FTIs in treatment of patients with prostate cancer has yet to be defined.

4.8. Raf Kinase Inhibitor

BAY 43–9006 (sorafanib) (Bayer, Wayne, NJ, USA) is a novel, orally available small molecule inhibitor of c-Raf-1 and B-raf. The raf kinase family is important in regulating ras-signaling pathways, which impact tumor signal transduction and cellular proliferation. In addition, BAY 43–9006 (sorafanib) (Bayer) also inhibits tyrosine kinase phosphorylation of multiple receptors, including VEGFR-2, VEGFR-3, PDGFR-B, Flt3, and MAPK *(85)*. As a result, BAY 43-9006 (sorafanib) (Bayer) mechanism of action appears to impact both tumor cell proliferation and inhibition of angiogenesis. A phase I trial in patients with advanced refractory solid tumors was conducted in 69 patients *(86)*. The Maximum Tolerated Dose (MTD) was 400 mg orally twice daily.

The most common toxicities included diarrhea and skin toxicities, including rashes and hand-foot syndrome. Although none of the patients in the trial had prostate cancer, it is plausible that this agent may have activity in prostate cancer patients based on its mechanism of action.

4.9. Rapamycin Kinase Inhibitor (Mammalian Target of Rapamycin Pathway)

Another important signaling pathway in addition to the ras/raf/mapk pathway is the PI3K/Akt signal transduction pathway. The PI3K/Akt pathway is also regulated by PTEN (phosphatase and tensin homolog deleted on chromosome 10). PTEN is a lipid phosphatase involved in tumor-suppressive activities. PTEN cleaves the D3 phosphatase of the PIP3 and activates the PI3K/Akt pathway. Inactivation of PTEN has resulted in development of solid tumors, including prostate cancer. Loss of PTEN has also been associated with chemotherapy resistance. Mammalian target of rapamycin (mTOR) is a serine/threonine kinase downstream to Akt, which affect translational regulators p70 and 4EB-P1. In initial studies in PTEN-defective breast cancer cells, rapamycin, an inhibitor of mTOR, resulted in apoptosis and anti-proliferative effects. In preclinical studies in prostate cancer cell lines, PC-3 and DU-145, administration of rapamycin resulted in reversing chemotherapy resistance in PTEN-defective cells, most likely because of rapamycin's effects on the PI3K/Akt pathway (87). Currently, there are three mTOR inhibitors, derivatives of rapamycin, in clinical trials; CCI779 (temsirolimus) (Wyeth), RAD001 (everolimus) (Novartis), and AP-23573 (Ariad Pharmaceuticals, Cambridge, MA, USA). CCI779 (Wyeth) has completed phase I testing in solid tumors with several phase II trials in different tumor types underway, including renal cell carcinoma, nonsmall cell lung cancer, and breast cancer (88). RAD001 (Novartis) and AP-23573 (Ariad Pharmaceuticals), both in active development, may have a role in patients with PTEN-negative/Akt up-regulated, androgen-insensitive prostate cancer.

4.10. Proteasome Inhibitors

The proteasome is a large protein unit located in the cytoplasm and nucleus of all cells. Its function is to degrade ubiquinated proteins, including cyclin B1, p53 tumor-suppressor gene, p21 and p27 cyclin-dependent kinase inhibitors, inhibitor of NF-kB (IkB), and MAPK. The proteasome is important in cell cycle regulation, apoptosis, angiogenesis, and metastasis. Velcade (bortezomib) (Millenium, Cambridge, MA, USA) is a peptide boronic acid and the first proteasome inhibitor approved by the FDA for treatment of patients with refractory multiple myeloma. Velcade (Millenium) has shown activity in preclinical studies of prostate cancer cell lines and animals (89–91). A phase I/II trial of velcade (Millenium) was conducted in patients with solid tumors (although most patients had androgen-insensitive prostate cancer) (92). The MTD of weekly velcade (Millenium) was 1.6 mg/m² and diarrhea was the common toxicity. Two of the 24 patients treated at higher doses showed a greater than 50% decline in serum PSA. A combination Phase I/II study of velcade (Millenium) with docetaxel is ongoing in patients with advanced androgen-insensitive prostate cancer (93). The cohort receiving docetaxel 40 mg/m² i.v. on days 1 and 8 and velcade (Millenium) 1.6 mg/m² i.v. on days 2 and 9 is still accruing patients. Twenty-two patients are evaluable with 36% having a greater than 50% decline in PSA (with 5 of the 8 patients having 90% or greater decline in PSA). The effect of prior taxane treatment in this patient population will need to be further explored.

4.11. Histone Deacetylase Inhibitors

Genetic changes responsible in tumorigenesis include gene mutations and deletions. Changes in DNA other than mutations and deletions are epigenetic changes that cause cellular progression toward a malignant phenotype. Methylation of DNA nucleotides and changes in chromatin conformation by histone acetylation are two epigenetic events associated with transcriptional silencing. Histone acetylation is a post-translational modification of the core nucleosomal histones affecting chromatin structure and ultimately gene expression. The nucleosome inhibits transcription of genes by blocking transcriptional regulators from binding to the promoter regions. Histones are deacetylated by histone deacetylase (HDAC). Inhibitors of HDAC have potential to reverse epigenetic silencing. Phenylbutyrate is an HDAC inhibitor approved by the FDA for patients with urea cycle disorders. In a phase I trial of patients with solid tumors, phenylbutyrate actually resulted in an increase in serum PSA in greater than 90% of patients with advanced prostate cancer (94). Another HDAC inhibitor in early clinical trials is suberoylanilide hydroxamic acid (SAHA) in patients with advanced solid tumors. Intravenous SAHA resulted in objective tumor regression in two lymphoma and two bladder cancer patients (95). Although 32% of patients enrolled had prostate cancer, the effects of SAHA on serum PSA were not reported. Oral SAHA (Aton Pharma, Lawrenceville, NJ, USA) completed phase I testing in patients with advanced cancer (96). There were no reported responses in the seven prostate cancer patients in this trial.

CI-994 (Pfizer) is an oral HDAC inhibitor that has been tested in phase I and II studies in advanced solid tumors including renal cell carcinoma and nonsmall cell lung cancer patients. Additional agents in development and early clinical trials include PXD101 (CuraGen, Branford, CT, USA and TopoTarget, Rockaway, NJ, USA) and MS275 (97,98). Preclinical studies suggest enhanced gene expression in response to HDAC inhibitors may lead to anti-tumor effects with the use of retinoids or with DNA methyltransferase inhibitors (99). In a tumor in which serum PSA levels are significantly correlated to disease status, agents such as phenylbutyrate or other HDAC inhibitors that cause rising PSA must be further investigated.

5. MULTIPLE TARGETED PATHWAYS

5.1. Heat Shock Protein 90 Inhibitor

Heat shock proteins (HSP) are molecular chaperones that assist in general protein folding, prevent misfolding and aggregation, and facilitate refolding or degradation. There are five gene families identified thus far: Hsp100, Hsp90, Hsp70, Hsp60, and Hsp40. Members of each family are constitutively expressed, are inducible, and may target different cellular compartments. Hsp90 is one of the most abundant cellular chaperone proteins. Most of its client proteins are protein kinases or transcription factors involved in signal transduction including ligand-dependent steroid hormone receptors (androgen and estrogen receptors), tyrosine kinases (HER-2 and VEGFR2), and hypoxia-inducible factor-1a (HIF-1a). Additional proteins stabilized by Hsp90 include proteins in the MAP kinase pathway, the PI3K pathway, and NF-kB signaling pathway.

17-Allylamino-17-demethoxygeldanamycin (17-AAG) (Kosan, Hayward, CA, USA) is a small molecule that binds to Hsp90 and, as a result, alters its ability to properly chaperone multiple client proteins, including those involved in prostate cancer tumorigenesis *(100)*. 17-AAG (Kosan) has been studied in several single agent phase I trials in patients with advanced cancers, including prostate cancer *(101)*. A combination trial of docetaxel and 17-AAG (Kosan) has also been conducted in patients with advanced solid tumors *(102)*. The potential biologic effects of 17-AAG (Kosan) in patients with prostate cancer is tremendous *(103)*. Currently, a single agent phase II trial in patients with metastatic hormone-insensitive prostate cancer is underway.

5.2. Matrix Metalloproteinase Inhibitors

Matrix metalloproteinases (MMPIs) are a family of zinc- and calcium-dependent peptides. These proteins are involved in remodeling of extracellular matrix, angiogenesis, and metastasis. Overexpression of MMPIs has been correlated with malignant growth and potential. Inhibitors of MMPIs were evaluated in phase III clinical trials in patients with solid tumors *(104)*. Prinomastat, an oral MMPI, was administered in patients with metastatic hormone-insensitive prostate cancer in combination with mitoxantrone/prednisone. Unfortunately, the large trial showed no differences between prinomastat and placebo with regard to serum PSA and overall survival. As with the other MMP inhibitors in clinical trials showing no benefit in patients with various solid tumors, clinical development of previous oral MMPIs has been abandoned. However, the biology underlying MMPIs remains intriguing, especially in prostate cancer, a disease with a propensity for bone metastasis. A new MMPI, BMS-275291 (Bristol Myers Squibb) was evaluated in a randomized phase II trial of 1200 mg either orally daily or twice daily in men with metastatic hormone-insensitive prostate cancer *(105)*. The primary endpoint was 4-month progression-free survival. Eighty patients were accrued. There were no responders. The classical dose-limiting arthritis seen in other MMPI inhibitors was not reported with BMS-275291 (Bristol Myers Squibb). Grade 3 and 4 toxicities were less in the lower dose arm. Bone metabolism marker study results are still pending. Further understanding of MMPI inhibitors is required before undertaking another large phase III clinical trial.

6. ANTI-VASCULAR TARGETING AGENTS

6.1. Angiogenesis Inhibitors

New vascular proliferation surrounding tumors and their metastases helps regulate and maintain tumor growth. Angiogenesis occurs when new capillaries from existing vasculature develop to nourish tumor cells. As previously described in this chapter, VEGF family inhibitors are considered the first class of angiogenesis inhibitors. Another agent considered as an angiogenesis inhibitor is thalidomide. Thalidomide is a synthetic derivative of glutamic acid. It acts as an angiogenesis inhibitor by affecting multiple pathways, including inhibition of TNF-a and NF-kB, stimulation of CD8[+] T cells, and stimulation of interleukin (IL)-2. Thalidomide was initially developed in the late 1950s to treat morning sickness for pregnant women. However, teratogenic effects of thalidomide were seen in infants born to mothers taking the medication. Birth defects seen include severe congenital malformations including absent or hypoplastic limbs, malformations of internal organs, and congenital heart defects. Further development and use of thalidomide was appropriately halted at that time. However, a decade later, thalidomide use re-emerged, and the drug was shown to be effective in treating erythema nodosum leprosum, a painful, dermatological complication of leprosy. Currently, thalidomide is being studied in other medical conditions such as Behcet's disease, graft versus host disease, HIV disease, and cancer. Thalidomide has shown activity in treatment of multiple myeloma patients.

Access to thalidomide in the USA is still limited. Patients and physicians are extremely careful to not repeat the tragedy caused by thalidomide's teratogenic effects. Physicians are required to participate in the System for Thalidomide Education and Prescription Safety Program (STEPS) to prescribe thalidomide. Women of childbearing potential are required to undergo rigorous pregnancy testing before and during administration. Men are also required to practice abstinence or use a condom during sexual intercourse while on treatment.

Thalidomide in solid tumors, including prostate cancer, has been evaluated in clinical trials. The rationale to use thalidomide in the treatment of prostate cancer is primarily due to the important role of angiogenesis in prostate tumorigenesis and metastasis. A phase II randomized trial in metastatic hormone-insensitive prostate cancer patients was conducted evaluating two dose levels of thalidomide (106). Eighteen percent of patients taking 200 mg orally daily experienced greater than 50% reduction in the PSA. There were no PSA responses in the higher dose arm of up to 1200 mg orally daily. The side effects reported include sedation and fatigue. Another trial using thalidomide 100 mg orally daily in 20 patients reported a 15% response rate in PSA (107).

Combination trials of thalidomide with chemotherapy have been completed. Investigators at the National Cancer Institute reported results of a phase II trial evaluating docetaxel 30 mg/m^2/week i.v. for 3 of 4 weeks and thalidomide 200 mg orally daily versus docetaxel alone in 75 patients with metastatic hormone-insensitive prostate cancer (2:1 randomization) (108). Fifty-three percent of patients on the combination arm achieved a greater than 50% decline in PSA compared with 37% in the docetaxel arm. Partial responses in measurable soft tissue lesions were also higher in the combination arm. Overall, survival at 18 months was 68% in the combination arm versus 43% in the docetaxel arm. These results were not statistically significant because of the small number of patients on the trial. However, what was significant was the toxicity data in the patients treated with thalidomide and docetaxel. Nine of the first 43 patients experienced venous thrombosis and an additional three patients experienced transient ischemic attack or stroke. The remaining three patients on the combination arm were treated with enoxaparin, and no other thromboembolic events were reported. Neurotoxicity including depression, confusion, and fatigue were observed, but all were less than grade 3.

Additional trials of thalidomide in combination have been conducted. Preliminary results of thalidomide in combination with docetaxel and estramustine reported a greater than 50% decline in PSA in 6 of the 9 patients treated (109). A phase I trial of thalidomide and sargramostim in patients with metastatic hormone-insensitive prostate cancer resulted in a high PSA response rate in 9 evaluable patients (110). The true benefit of thalidomide as a single agent or in combination therapy in treatment of prostate cancer is yet to be determined. However significant the benefit may be, thalidomide remains an agent associated with a significant toxicity profile, especially when administered in higher doses.

In hopes of minimizing toxic side effects, new generation thalidomide analogs have been developed. These immunomodulatory drugs (IMiDs) are novel agents already in clinical trials in patients with multiple myeloma. Recently, Revlimid (CC-5013, lenalidomide) (Celgene, Summit, NJ, USA) was evaluated in a phase I study in patients with refractory solid tumors, primarily patients with hormone-insensitive prostate cancer (111). Another IMiD currently in phase II clinical trial is CC-4047 (actimid) (Celgene) at 2 mg orally daily (112). Undoubtedly, response as well as toxicity data will be equally important.

6.2. Vascular Targeting Agents

Vascular targeting agents are agents that destroy existing blood vessels in solid tumors. These agents are different and distinct from angiogenesis inhibitors because they destroy existing blood vessels whereas angiogenesis inhibitors inhibit new blood vessel formation. There are two types of vascular targeting agents being developed for use in cancer therapy: ligand-directed agents effecting tumor endothelium and small molecule agents indirectly effecting tumor endothelium. Ligand-directed agents including antibodies, peptides, and growth factors are primarily in preclinical development. Small molecules in early clinical development include combretastatin (Oxigene, Waltham, MA, USA), ZD6126 (Astra Zeneca), DMXAA (Antisoma PLC, London, England), and AVE8062A (Aventis, Pans, France). Phase I studies have been reported in combretastatin (Oxigene), ZD6126 (Astra Zeneca), and DMXAA (Antisoma PLC) (113–115). Three most common toxicities seen in all three compounds include cardiac toxicities, significant neurologic complications, and gastrointestinal toxicities. AVE8062A (Aventis) is currently being tested in phase I clinical trials. Phase II study of DMXAA (Antisoma PLC) in combination with docetaxel is being studied in a multicenter trial. The role of vascular targeted agents in treatment of prostate cancer patients is not yet established, but there is potential for this class of agents to be effective.

7. IMMUNOTHERAPY

7.1. Therapeutic Vaccines

Modulation of the immune system for purposes of enhancing tumor-specific deaths is a growing area of research in patients with advanced solid tumors. The immune system may be stimulated to target tumor-specific antigens with therapeutic vaccines or monoclonal antibodies. Therapeutic vaccines help the immune system to recognize foreign antigens on tumor cells, such as PSA, prostatic acid phosphatase (PAP), and prostate-specific membrane antigen (PSMA). Once recognized, the immune system initiates a response against the abnormal malignant cells. One area of difficulty in development of effective therapeutic vaccines has been the ability of the immune system to repeatedly recognize proteins as foreign antigens. To overcome immune tolerance and help improve recognition of foreign antigens, therapeutic vaccines have been prepared in conjunction with different modalities, such as viral vectors, and different adjuvants, such as cytokines.

To date, multiple therapeutic vaccines are in various stages of clinical development. Two of the vaccines, provenge (Dendreon, San Francisco, CA, USA) and GVAX (Cell Genesys, Seattle, WA, USA), are currently being tested in phase III clinical trials. Provenge (Dendreon) is an autologous $CD54^+$ dendritic cell vaccine with a recombinant fusion protein consisting of PAP and granulocyte macrophage-colony-stimulating factor (GM-CSF). It also includes macrophages, B cells, and T cells. Dendritic cells are antigen-presenting cells that can prime an immune response by T cells that have not been previously exposed to the antigen.

Encouraging phase I and II trial results led to a phase III randomized (2:1), placebo-controlled trial in asymptomatic men with androgen-insensitive prostate cancer *(116,117)*. Eligible patients had >25% of cancer cells positive for PAP by central pathology review. Preliminary findings in 127 patients revealed a trend toward a delay in disease progression, especially in patients with intermediate and lower grade disease compared with those on placebo arm *(118)*. Median time to progression was 16.0 weeks in the intermediate/lower grade arm compared with 9.0 weeks in the placebo arm. There appeared to be no benefit in patients with aggressive, high-grade disease. Provenge was delivered as an intravenous infusion every 2 weeks for a total of three infusions. Fevers and rigors were common adverse events. The current phase III trial of Provenge (Dendreon) is conducted in patients with intermediate and lower grade cancers (Gleason <7). Provenge (Dendreon) is being considered for fast track FDA approval. Provenge (Dendreon) is also being evaluated in combination with other biologic therapies, including bevacuzimab *(119)*.

Another therapeutic vaccine evaluated in phase III clinical trials is GVAX (Cell Genesys). GVAX (Cell Genesys) utilizes irradiated, allogeneic prostate cancer cell lines (PC-3 and LNCaP) genetically modified to secrete human GM-CSF. Phase I/II trials reported GVAX (Cell Genesys) to be reasonably well tolerated in patients with metastatic androgen-insensitive prostate cancer patients *(120)*. Currently, two phase III trials are planned in symptomatic and asymptomatic men with metastatic androgen-insensitive prostate cancer. Additional therapeutic vaccines in active investigation include prostvac-VF (Therion, Cambridge, MA, USA), BLP25 liposome (Biomira, Cranbury, NJ, USA), and recombinant soluble PSMA vaccine (Progenix, Tarrytown, NY, USA) *(121,122)*. ECOG will initiate a phase III trial (E1805) of Prostvac-VF versus placebo in patients with nonmetastatic, hormone-insensitive prostate cancer.

7.2. Monoclonal Antibodies

There are two primary roles of monoclonal antibodies in treatment of prostate cancer: to initiate antibody-dependent cell-mediated cytotoxicity using naked antibodies and to help direct cytotoxic agents such as chemotherapy, toxin, or radionuclide to specifically target and kill tumor cells. In so doing, the promise of monoclonal antibody therapy in increasing tumor death with minimal toxicities is similar with other biologic therapies. Development of monoclonal antibodies in treatment of advanced prostate cancer is actively ongoing. Several monoclonal antibodies are in phase I/II clinical trials; J591 (Millenium), MDX-070 (Medarex, Princeton, NJ, USA), MLN 2704 (Immunogen, Cambridge, MA, USA), MDX-010 (Medarex), and AMG162 (Amgen, Thousand Oaks, CA, USA). A frequent target of the antibodies is PSMA. PSMA expression on the cell membrane is increased with advancing prostate cancer. J591 (Millenium) is a humanized monoclonal antibody that binds to PSMA. Early trials of J591 (Millenium) administered as a radioconjugate, a chemoconjugate, or a naked antibody have reported encouraging results *(123)*. MDX-070 (Medarex) is an unconjugated monoclonal antibody targeting PSMA as well. MLN 2704 (Immunogen) is an immunoconjugate designed to deliver an anti-microtubule agent DM1 through PSMA-targeted monoclonal antibody MLN591 *(124)*. Phase I/II trial of MLN 2704 (Immunogen) reported PSA response of 50% or greater in 3 of 21 patients *(125)*. Four patients had a PSA decline between 25 and 50%.

MDX-010 (Medarex) is a human monoclonal antibody that binds to the CTLA-4 receptor on T cells, which stimulates cytotoxic T-lymphocyte immune response. In a phase I trial of MDX-010 (Medarex), fourteen patients were treated with one dose of 3 mg/kg intravenously *(126)*. Mild rashes and pruritis occurred in 4 patients. No other major adverse events were noted. A phase II clinical trial of MDX-010 (Medarex) in combination with docetaxel for patients with androgen-insensitive prostate cancer is ongoing.

AMG 162 (Amgen) is a fully humanized monoclonal antibody to receptor activator of NF-kB ligand currently being investigated in patients with cancer-related bone lesions. Although the phase I study evaluated patients with breast cancer and multiple myeloma, it is conceivable that this antibody may have activity in patients with prostate cancer *(127)*.

Prostate stem cell antigen (PSCA) is a surface-linked antigen expressed in approximately 80% of prostate cancers. AGS-PSCA (Agensys) is a fully human monoclonal antibody to PSCA. With encouraging preclinical data, a phase I trial of AGS-PSCA is anticipated *(128)*.

8. CONCLUSIONS

It is evident by the number of agents in clinical trials discussed in this chapter that treatment options for patients with advanced prostate cancer are increasing. We are fortunate to have discovered a survival advantage with docetaxel-based chemotherapy in the treatment of patients with metastatic androgen-insensitive disease. However, treatment with chemotherapy is associated with side effects, albeit minimal. Newer agents specifically targeting prostate cancer cells and not normal cells will hopefully minimize the toxicities while retaining maximum anti-tumor activity. Traditional endpoints of measuring treatment success with targeted biologic therapies are rapidly evolving. The role of chemotherapy and targeted agents will undoubtedly need to be investigated not only in the advanced hormone-insensitive prostate cancer patients but in the adjuvant and neoadjuvant setting. Signal transduction inhibitors, angiogenesis inhibitors and immunotherapeutic agents are promising emerging therapies for prostate cancer. Additional targeted agents not discussed in this chapter include radiopharmaceuticals (strontium and samarium), bisphosphonates, and nutritional agents (soy isoflavone and selenium). It is evident that no matter how exciting the agent, progress can only be achieved in a timely fashion with multidisciplinary collaboration and increasing patient enrollment in clinical trials. Although there is considerable work to be done in moving effective targeted biologic agents to the clinic, there is optimism that we are heading in the appropriate direction.

ACKNOWLEDGMENTS

Funding for some of the studies described in this article was provided by Abbott Laboratories or Sanofi-Aventis. Dr. Michael A. Carducci is a paid consultant to Abbott Laboratories and on the speaker's bureau for Sanofi-Aventis. The terms of this arrangement are being managed by the Johns Hopkins University in accordance with its conflict of interest policies.

REFERENCES

1. Jemal A, Siegel R, Ward E, et al. Cancer statistics, 2007. *CA Cancer J Clin* 2007; 57(1):43–66.
2. Rosenbaum E, Carducci MA. Pharmacotherapy of hormone refractory prostate cancer: new developments and challenges. *Expert Opin Pharmacother* 2003; 4(6):875–87.
3. Tannock IF, Osoba D, Stockler MR, et al. Chemotherapy with mitoxantrone plus prednisone or prednisone alone for symptomatic hormone-resistant prostate cancer: a Canadian randomized trial with palliative end points. *J Clin Oncol* 1996; 14(6):1756–64.
4. Tannock IF, de Wit R, Berry WR, et al. Docetaxel plus prednisone or mitoxantrone plus prednisone for advanced prostate cancer. *N Engl J Med* 2004; 351(15):1502–12.
5. Petrylak DP, Tangen CM, Hussain MH, et al. Docetaxel and estramustine compared with mitoxantrone and prednisone for advanced refractory prostate cancer. *N Engl J Med* 2004; 351(15): 1513–20.
6. Ibrahim NK, Desai N, Legha S, et al. Phase I and pharmacokinetic study of ABI-007, a cremophor-free, protein-stabilized, nanoparticle formulation of paclitaxel. *Clin Cancer Res* 2002; 8(5):1038–44.
7. Tahir SA, Ren C, Timme TL, et al. Development of an immunoassay for serum caveolin-1: a novel biomarker for prostate cancer. *Clin Cancer Res* 2003; 9(10 Pt 1):3653–9.
8. Goodin S, Kane MP, Rubin EH. Epothilones: mechanism of action and biologic activity. *J Clin Oncol* 2004; 22(10):2015–25.
9. Holen KD, Syed S, Hannah AL, et al. Phase I study using continuous intravenous (CI) KOS-862 (epothilone D) in patients with solid tumors. *Proc Am Soc Clin Oncol* 2004;22(14S):A2024.
10. Spriggs D, Soignet S, Bienvenu B, et al. Phase I first-in-man study of the epothilone analog of BMS-247550 in patients with advanced cancer. *Proc Am Soc Clin Oncol* 2001; 20(108a):428.
11. Mani S, McDaid H, Hamilton A, et al. Phase I pharmacokinetic and pharmacodynamic study of an epothilone B analog (BMS-247550) administered as a 1-hour infusion every 3 weeks: An update. *Proc Am Soc Clin Oncol* 2002; 21(103a):409.
12. Tripathi R, Gadgeel SM, Wozniak AJ, et al. Phase I clinical trial of BMS-247550 (epothilone B derivative) in adult patients with advanced solid tumors. *Proc Am Soc Clin Oncol* 2002; 21(102a):407.
13. Hussain M, Faulkner J, Vaishampayan U, et al. Epothilone B (Epo-B) analogue BMS-247550 (NSC#710428) administered every 21 days in patients (pts) with hormone refractory prostate cancer (HRPC). A southwest Oncology Group Study (S0111). *Proc Am Soc Clin Oncol* 2004; 23:4510.
14. Smaletz O, Galsky M, Scher HI, et al. Pilot study of epothilone B analog (BMS-247550) and estramustine phosphate in patients with progressive metastatic prostate cancer following castration. *Ann Oncol* 2003; 14(10):1518–24.
15. Galsky MD, Small EJ, Oh WK, et al. Multi-institutional randomized phase II trial of the epothilone B analog ixabepilone (BMS-247550) with or without estramustine phosphate in patients with progressive castrate metastatic prostate cancer. *J Clin Oncol* 2005; 23(7):1439–46.
16. Hussain A, Dipaola RS, Baron CS, et al. A phase IIa trial of weekly EPO906 in patients with hormone-refractory prostate cancer (HPRC). *Proc Am Soc Clin Oncol* 2004; 14S(4563).
17. Wojtowicz M, Rothermel JD, Anderson J, et al. Phase I dose-escalation trial investigating the safety and tolerability of EPO906 plus estramustine in patients with advanced cancer. *Proc Am Soc Clin Oncol* 2004; 14S(4623).
18. Rosen PJ, Rosen LS, Britten C, et al. KOS-862 (epothilone D): results of a phase I dose-escalating trial in patients with advanced malignancies. *Proc Am Soc Clin Oncol* 2002; 20(413).
19. Spriggs DR, Dupont J, Pezzulli S, et al. KOS-862 (Epothilone D): phase I dose escalating and pharmacokinetic (PK) study in patients with advanced malignancies. *Proc Am Soc Clin Oncol* 2003; 22(223):894.

20. Ahmad N. Polo-like kinase (Plk) 1: a novel target for the treatment of prostate cancer. *FASEB J* 2004; 18(1):5–7.
21. Von Hoff DD, Taylor C, Rubin S, et al. A phase I and pharmacokinetic study of HMN-214, a novel oral polo-like kinase inhibitor, in patients with advanced solid tumors. *Proc Am Soc Clin Oncol* 2004; 14S(3034).
22. Taylor C, Dragovich T, Simpson A, et al. A phase I and pharmacokinetic study of HMN-214 administered orally for 21 consecutive days, repeated every 28 days to patients with advanced solid tumors. *Proc Am Soc Clin Oncol* 2002; 21:419.
23. Patnaik A, Forero L, Goetz A, et al. HMN-214, a novel oral antimicrotubular agent and inhibitor of polo-like-and cyclin-dependent kinases: clinical, pharmacokinetic (PK) and pharmacodynamic (PD) relationships observed in a phase I trial of a daily x 5 schedule every 28 days. *Proc Am Soc Clin Oncol* 2003; 128(514).
24. Harrington EA, Bebbington D, Moore J, et al. VX-680, a potent and selective small-molecule inhibitor of the Aurora kinases, suppresses tumor growth in vivo. *Nat Med* 2004; 10(3):262–7.
25. Doggrell SA. Dawn of Aurora kinase inhibitors as anticancer drugs. *Expert Opin Investig Drugs* 2004; 13(9):1199–201.
26. Nair JS, Tse A, Keen N, et al. A novel Aurora B kinase inhibitor with potent anticancer activity either as a single agent or in combination with chemotherapy. *Proc Am Soc Clin Oncol* 2004; 14S(9568).
27. Chu QS, Holen KD, Rowinsky EK, et al. Phase I trial of novel kinesin spindle protein (KSP) inhibitor SB-715992 IV q 21 days. *Proc Am Soc Clin Oncol* 2004; 14S(2078).
28. Burris HA, Lorusso P, Jones S, et al. Phase I trial of novel kinesin spindle protein (KSP) inhibitor SB-715992 IV days 1,8,15 q 28 days. *Proc Am Soc Clin Oncol* 2004; 14S(2004).
29. Blackledge G. Growth factor receptor tyrosine kinase inhibitors; clinical development. *J Urol* 2003; 170(6 Pt 2):S77–83. discussion S.
30. Di Lorenzo G, Autorino R, De Laurentiis M, et al. HER-2/neu receptor in prostate cancer development and progression to androgen independence. *Tumori 2004*; 90(2):163–70.
31. Slovin SF, Kelly WK, Cohen R, et al. Epidermal growth factor receptor (EGFr) monoclonal antibody (MoAb) C225 and doxorubicin (DOC) in androgen independent (AI) prostate cancer (PC): results of a phase Ib/IIa study. *Proc Am Soc Clin Oncol* 1997;(1108).
32. Foon KA, Yang XD, Weiner LM, et al. Preclinical and clinical evaluations of ABX-EGF, a fully human anti-epidermal growth factor receptor antibody. *Int J Radiat Oncol Biol Phys* 2004; 58(3):984–90.
33. Yang X, Wang P, Fredlin P, et al. ABX-EGF, a fully human anti-EGF receptor monoclonal antibody: inhibition of prostate cancer in vitro and in vivo. *Proc Am Soc Clin Oncol* 2002; 21(2454).
34. Figlin RA, Belldegrun AS, Crawford J, et al. ABX-EGf, a fully human anti-epidermal growth factor receptor (EGFR) monoclonal antibody (mAb) in patients with advanced cancer: phase I clinical results. *Proc Am Soc Clin Oncol* 2002; 21(35).
35. Tewes M, Schleucher N, Dirsch O, et al. Results of a phase I trial of the humanized anti epidermal growth factor receptor (EGFR) monoclonal antibody EMD 72000 in patients with EGFR expressing solid tumors. *Proc Am Soc Clin Oncol* 2004; 21: 95a(378).
36. Ziada A, Barqawi A, Glode LM, et al. The use of trastuzumab in the treatment of hormone refractory prostate cancer; phase II trial. *Prostate* 2004; 60(4):332–7.
37. Morris MJ, Reuter VE, Kelly WK, et al. HER-2 profiling and targeting in prostate carcinoma. *Cancer* 2002; 94(4):980–6.
38. Lara PN Jr, Chee KG, Longmate J, et al. Trastuzumab plus docetaxel in HER-2/neu-positive prostate carcinoma: final results from the California Cancer Consortium Screening and Phase II Trial. *Cancer* 2004; 100(10):2125–31.
39. Small EJ, Bok R, Reese DM, Sudilovsky D, Frohlich M. Docetaxel, estramustine, plus trastuzumab in patients with metastatic androgen-independent prostate cancer. *Semin Oncol* 2001; 28(4 Suppl 15):71–6.
40. Franklin MC, Carey KD, Vajdos FF, Leahy DJ, de Vos AM, Sliwkowski MX. Insights into ErbB signaling from the structure of the ErbB2-pertuzumab complex. *Cancer Cell* 2004; 5(4):317–28.
41. Mendoza N, Phillips GL, Silva J, Schwall R, Wickramasinghe D. Inhibition of ligand-mediated HER2 activation in androgen-independent prostate cancer. *Cancer Res* 2002; 62(19):5485–8.
42. Agus DB, Gordon M, Taylor C, et al. Clinical activity in a phase I trial of HER-2 targeted rhuMab 2C4 (pertuzumab) in patients with advanced solid malignancies (AST). *Proc Am Soc Clin Oncol* 2003; 22(771).

43. Sgambato A, Camerini A, Faraglia B, et al. Targeted inhibition of the epidermal growth factor receptor-tyrosine kinase by ZD1839 ('Iressa') induces cell-cycle arrest and inhibits proliferation in prostate cancer cells. *J Cell Physiol* 2004; 201(1):97–105.

44. Bonaccorsi L, Marchiani S, Muratori M, Forti G, Baldi E. Gefitinib ('IRESSA', ZD1839) inhibits EGF-induced invasion in prostate cancer cells by suppressing PI3 K/AKT activation. *J Cancer Res Clin Oncol* 2004; 130(10):604–14.

45. Sirotnak FM, She Y, Lee F, Chen J, Scher HI. Studies with CWR22 xenografts in nude mice suggest that ZD1839 may have a role in the treatment of both androgen-dependent and androgen-independent human prostate cancer. *Clin Cancer Res* 2002; 8(12):3870–6.

46. Canil CM, Moore MJ, Winquist E, et al. Randomized phase II study of two doses of gefitinib in hormone-refractory prostate cancer: a trial of the National Cancer Institute of Canada-Clinical Trials Group. *J Clin Oncol* 2005; 23(3):455–60.

47. Schroeder FH, Wildhagen MF, et al. ZD1839 (gefitinib) and hormone resistant (HR) prostate cancer-final results of a double blind randomized placebo-controlled phase II study. *Proc Am Soc Clin Oncol* 2004; 14S(4698).

48. Formento P, Hannoun-Levi JM, Fischel JL, Magne N, Etienne-Grimaldi MC, Milano G. Dual HER 1–2 targeting of hormone-refractory prostate cancer by ZD1839 and trastuzumab. *Eur J Cancer* 2004; 40(18):2837–44.

49. Trump DL, Wilding G, Miller K, et al. Pilot trials of ZD1839 (Iressa), an orally active, selective epidermal growth factor receptor tyrosine kinase inhibitor, in combination with mitoxantrone/prednisone or docetaxel/estramustine in patients with hormone-refractory prostate cancer. *Am Urol Assoc* 2003; 945(105239).

50. Shepherd FA, Pereira J, Ciuleanu TE, et al. A randomized placebo-controlled of erlotinib in patients with advanced non-small cell lung cancer following failure of 1st or 2nd line chemotherapy. A National Cancer Institute of Canada Clinical Trials Group Trial. *Proc Am Soc Clin Oncol* 2004; 22(7022).

51. Gatzemeier U, Pluzanska A, Szczesna A, et al. Results of a phase III trial of erlotinib (OSI-774) combined with cisplatin and gemcitabine (GC) chemotherapy in non-small cell lung cancer. *Proc Am Soc Clin Oncol* 2004; 23(7010).

52. Herbst RS, Prager D, Hermann R, et al. TRIBUTE: a phase III trial of erlotinib HCL (OSI-774) combined with carboplatin and paclitaxel (CP) in advanced non-small cell lung cancer. *Proc Am Soc Clin Oncol* 2004; 23(7011).

53. Mellinghoff IK, Tran C, Sawyers CL. Growth inhibitory effects of the dual ErbB1/ErbB2 tyrosine kinase inhibitor PKI-166 on human prostate cancer xenografts. *Cancer Res* 2002; 62(18):5254–9.

54. Hoekstra R, Dumez H, van Oosterom AT, et al. A phase I and pharmacological study of PKI166, an epidermal growth factor receptor (EGFR) tyrosine kinase inhibitor, administered orally in a two weeks on, two weeks off scheme to patients with advanced cancer. *Proc Am Soc Clin Oncol* 2002; 21(340).

55. Dumez H, Hoekstra R, Eskens F, et al. A phase I and pharmacological study of PKI166, an epidermal growth factor receptor (EGFR) tyrosine kinase inhibitor, administered orally 3 times a week to patients with advanced cancer. *Proc Am Soc Clin Oncol* 2002; 21(341).

56. Murren JR, Papadimitrakopoulou VA, Sizer KC, et al. A phase I dose-escalating study to evaluate the biological activity and pharmacokinetics of PKI166, a novel tyrosine kinase inhibitor, in patients with advanced cancers. *Proc Am Soc Clin Oncol* 2002; 21(377).

57. Dees EC, Burris H, Hurwitz H, et al. Clinical summary of 67 heavily pretreated patients with metastatic carcinomas treated with GW572016 in a phase Ib study. *Proc Am Soc Clin Oncol* 2004; 14S(3188).

58. Minami H, Nakagawa K, Kawada K, et al. A phase I study of GW572016 in patients with solid tumors. *Proc Am Soc Clin Oncol* 2004; 14S(3048).

59. Hidalgo M, Erlichman C, Rowinsky EK, et al. Phase I trial of EKB-569, an irreversible inhibitor of the epidermal growth factor receptor (EGFR), in patients with advanced solid tumors. *Proc Am Soc Clin Oncol* 2002; 22(65).

60. Calvo E, Tolcher AW, Hammond LA, et al. Administration of CI-1033, an irreversible pan-erbB tyrosine kinase inhibitor, is feasible on a 7-day on, 7-day off schedule: a phase I pharmacokinetic and food effect study. *Clin Cancer Res* 2004; 10(21):7112–20.

61. Strohmeyer D, Strauss F, Rossing C, et al. Expression of bFGF, VEGF and c-met and their correlation with microvessel density and progression in prostate carcinoma. *Anticancer Res* 2004; 24(3a):1797–804.

62. Duque JL, Loughlin KR, Adam RM, Kantoff PW, Zurakowski D, Freeman MR. Plasma levels of vascular endothelial growth factor are increased in patients with metastatic prostate cancer. *Urology* 1999; 54(3):523–7.

63. Retter AS, Figg WD, Dahut WL. The combination of antiangiogenic and cytotoxic agents in the treatment of prostate cancer. *Clin Prostate Cancer* 2003; 2(3):153–9.

64. Picus J, Halabi S, Rini B, et al. The use of bevacizumab (B) with docetaxel (D) and estramustine (E) in hormone refractory prostate cancer (HPRC): initial results of CALGB 90006. *Proc Am Soc Clin Oncol* 2003; 22(1578).

65. Stadler WM, Cao D, Vogelzang NJ, et al. A randomized Phase II trial of the antiangiogenic agent SU5416 in hormone-refractory prostate cancer. *Clin Cancer Res* 2004; 10(10):3365–70.

66. Chott A, Sun Z, Morganstern D, et al. Tyrosine kinases expressed in vivo by human prostate cancer bone marrow metastases and loss of the type 1 insulin-like growth factor receptor. *Am J Pathol* 1999; 155(4):1271–9.

67. Kubler HR, Randenborgh HV, Treiber U, et al. In vitro cytotoxic effects of imatinib in combination with anticancer drugs in human prostate cancer cell lines. *Prostate* 2004; 63(4):385–94.

68. George D. Targeting PDGF receptors in cancer–rationales and proof of concept clinical trials. *Adv Exp Med Biol* 2003; 532:141–51.

69. Rao KV, Goodin S, Capanna T, et al. A phase II trial of imatinib mesylate in patients with PSA progression after local therapy for prostate cancer. *Proc Am Soc Clin Oncol* 2003; 22(1645).

70. Tiffany NM, Wersinger EM, Garzotto M, Beer TM. Imatinib mesylate and zoledronic acid in androgen-independent prostate cancer. *Urology* 2004; 63(5):934–9.

71. Mathew P, Fidler IJ, Logothetis CJ. Combination docetaxel and platelet-derived growth factor receptor inhibition with imatinib mesylate in prostate cancer. *Semin Oncol* 2004; 31(2 Suppl 6): 24–9.

72. Ko YJ, Small EJ, Kabbinavar F, et al. A multi-institutional phase ii study of SU101, a platelet-derived growth factor receptor inhibitor, for patients with hormone-refractory prostate cancer. *Clin Cancer Res* 2001; 7(4):800–5.

73. Hofer MD, Fecko A, Shen R, et al. Expression of the platelet-derived growth factor receptor in prostate cancer and treatment implications with tyrosine kinase inhibitors. *Neoplasia* 2004; 6(5):503–12.

74. Jimeno A, Carducci M. Atrasentan: targeting the endothelin axis in prostate cancer. *Expert Opin Investig Drugs* 2004; 13(12):1631–40.

75. Langenstroer P, Tang R, Shapiro E, Divish B, Opgenorth T, Lepor H. Endothelin-1 in the human prostate: tissue levels, source of production and isometric tension studies. *J Urol* 1993; 150(2 Pt 1):495–9.

76. Nelson JB, Chan-Tack K, Hedican SP, et al. Endothelin-1 production and decreased endothelin B receptor expression in advanced prostate cancer. *Cancer Res* 1996; 56(4):663–8.

77. Carducci MA, Padley RJ, Breul J, et al. Effect of endothelin-A receptor blockade with atrasentan on tumor progression in men with hormone-refractory prostate cancer: a randomized, phase II, placebo-controlled trial. *J Clin Oncol* 2003; 21(4):679–89.

78. Carducci MA, Nelson JB, Padley RJ, et al. The endothelin-1 receptor antagonist atrasentan (ABT-627) delays clinical progression in hormone refractory prostate cancer: a multinational, randomized, double-blind, placebo-controlled trial. *Proc Am Soc Clin Oncol* 2001; 20(694).

79. Carducci MA, Nelson JB, Saad F, et al. Effects of atrasentan on disease progression and biological markers in men with metastatic hormone-refractory prostate cancer: phase 3 study. *Proc Am Soc Clin Oncol* 2004; 23(4508).

80. Smith MR, Kabbinavar F, Saad F, et al. Natural history of rising serum prostate-specific antigen in men with castrate nonmetastatic prostate cancer. *J Clin Oncol* 2005; 23(13):2918–25.

81. Yuyama H, Noguchi Y, Fujimori A, et al. Superiority of YM598 over atrasentan as a selective endothelin ETA receptor antagonist. *Eur J Pharmacol* 2004; 498(1–3):171–7.

82. Dreicer R, Curtis N, Morris C, et al. ZD4054 specifically inhibits endothelin A receptor-mediated effects, but not endothelin B receptor-mediated effects. *ASCO Prostate Cancer Symp* 2005; Abstract 237.

83. Mazieres J, Pradines A, Favre G. Perspectives on farnesyl transferase inhibitors in cancer therapy. *Cancer Lett* 2004; 206(2):159–67.
84. Haas N, Peereboom D, Ranganathan S, et al. Phase II trial of R115777, an inhibitor of farnesyl-transferase, in patients with hormone refractory prostate cancer. *Proc Am Soc Clin Oncol* 2002; 21(721).
85. Wilhelm SM, Carter C, Tang L, et al. BAY 43–9006 exhibits broad spectrum oral antitumor activity and targets the RAF/MEK/ERK pathway and receptor tyrosine kinases involved in tumor progression and angiogenesis. *Cancer Res* 2004; 64(19):7099–109.
86. Strumberg D, Richly H, Hilger RA, et al. Phase I clinical and pharmacokinetic study of the Novel Raf kinase and vascular endothelial growth factor receptor inhibitor BAY 43–9006 in patients with advanced refractory solid tumors. *J Clin Oncol* 2005; 23(5):965–72.
87. Grunwald V, DeGraffenried L, Russel D, Friedrichs WE, Ray RB, Hidalgo M. Inhibitors of mTOR reverse doxorubicin resistance conferred by PTEN status in prostate cancer cells. Cancer Res 2002; 62(21):6141–5.
88. Dancey JE. Inhibitors of the mammalian target of rapamycin. *Expert Opin Investig Drugs* 2005; 14(3):313–28.
89. Williams S, Pettaway C, Song R, Papandreou C, Logothetis C, McConkey DJ. Differential effects of the proteasome inhibitor bortezomib on apoptosis and angiogenesis in human prostate tumor xenografts. *Mol Cancer Ther* 2003; 2(9):835–43.
90. Ikezoe T, Yang Y, Saito T, Koeffler HP, Taguchi H. Proteasome inhibitor PS-341 down-regulates prostate-specific antigen (PSA) and induces growth arrest and apoptosis of androgen-dependent human prostate cancer LNCaP cells. *Cancer Sci* 2004; 95(3):271–5.
91. Williams SA, Papandreou C, McConkey D, et al. Preclinical effects of proteasome inhibitor PS-341 in combination chemotherapy for prostate cancer. *Proc Am Soc Clin Oncol* 2001; 20(2427).
92. Papandreou CN, Daliani DD, Nix D, et al. Phase I trial of the proteasome inhibitor bortezomib in patients with advanced solid tumors with observations in androgen-independent prostate cancer. *J Clin Oncol* 2004; 22(11):2108–21.
93. Dreicer R, Roth B, Petrylak D, et al. Phase I/II trial of bortezomib plus docetaxel in patients with advanced androgen-independent prostate cancer. *Proc Am Soc Clin Oncol* 2004; 14S(4654).
94. Carducci MA, Gilbert J, Bowling MK, et al. A Phase I clinical and pharmacological evaluation of sodium phenylbutyrate on an 120-h infusion schedule. *Clin Cancer Res* 2001; 7(10):3047–55.
95. Kelly WK, Richon VM, O'Connor O, et al. Phase I clinical trial of histone deacetylase inhibitor: suberoylanilide hydroxamic acid administered intravenously. *Clin Cancer Res* 2003; 9(10 Pt 1):3578–88.
96. Kelly WK, O'Connor OA, Krug LM, et al. Phase I study of an oral histone deacetylase inhibitor, suberoylanilide hydroxamic acid, in patients with advanced cancer. *J Clin Oncol* 2005; 23(17): 3923–31.
97. Qian DZ, Wang X, Kachhap SK, et al. The histone deacetylase inhibitor NVP-LAQ824 inhibits angiogenesis and has a greater antitumor effect in combination with the vascular endothelial growth factor receptor tyrosine kinase inhibitor PTK787/ZK222584. *Cancer Res* 2004; 64(18):6626–34.
98. Plumb JA, Finn PW, Williams RJ, et al. Pharmacodynamic response and inhibition of growth of human tumor xenografts by the novel histone deacetylase inhibitor PXD101. *Mol Cancer Ther* 2003; 2(8):721–8.
99. Qian DZ, Ren M, Wei Y, et al. In vivo imaging of retinoic acid receptor beta2 transcriptional activation by the histone deacetylase inhibitor MS-275 in retinoid-resistant prostate cancer cells. *Prostate* 2005; 64(1):20–8.
100. Neckers L. Heat shock protein 90 inhibition by 17-allylamino-17- demethoxygeldanamycin: a novel therapeutic approach for treating hormone-refractory prostate cancer. *Clin Cancer Res* 2002; 8(5):962–6.
101. Grem JL, Morrison G, Guo XD, et al. Phase I and pharmacologic study of 17-(allylamino)-17-demethoxygeldanamycin in adult patients with solid tumors. *J Clin Oncol* 2005; 23(9):1885–93.
102. Solit DB, Egorin M, Valentin G, et al. Phase I pharmacokinetic and pharmacodynamic trial of docetaxel and 17AAG (17-allylamino-17-demethoxygeldanamycin). *Proc Am Soc Clin Oncol* 2004; 14S(3032).
103. Solit DB, Scher HI, Rosen N. Hsp90 as a therapeutic target in prostate cancer. *Semin Oncol* 2003; 30(5):709–16.

104. Zucker S, Cao J, Chen WT. Critical appraisal of the use of matrix metalloproteinase inhibitors in cancer treatment. *Oncogene* 2000; 19(56):6642–50.

105. Lara PN, Longmate J, Stadler W, et al. Angiogenesis inhibition in metastatic hormone refractory prostate cancer (HPRC): A randomized phase II trial of two doses of the matrix metalloproteinase inhibitor (MMPI) BMS-275291. *Proc Am Soc Clin Oncol* 2004; 14S(4647).

106. Figg WD, Dahut W, Duray P, et al. A randomized phase II trial of thalidomide, an angiogenesis inhibitor, in patients with androgen-independent prostate cancer. *Clin Cancer Res* 2001; 7(7): 1888–93.

107. Drake MJ, Robson W, Mehta P, Schofield I, Neal DE, Leung HY. An open-label phase II study of low-dose thalidomide in androgen-independent prostate cancer. *Br J Cancer* 2003; 88(6):822–7.

108. Dahut WL, Gulley JL, Arlen PM, et al. Randomized phase II trial of docetaxel plus thalidomide in androgen-independent prostate cancer. *J Clin Oncol* 2004; 22(13):2532–9.

109. Frank RC, Coscia A, Versea L, et al. Low dose docetaxel, estramustine and thalidomide followed by maintenance thalidomide for the treatment of hormone refractory prostate cancer (HPRC): a phase II community based trial. *Proc Am Soc Clin Oncol* 2004; 23(4681).

110. Lilly M, Rowsell EH, Gurrola R, et al. Phase I trial of sargramostim and thalidomide for treatment of hormone-refractory prostate cancer. *Proc Am Soc Clin Oncol* 2004; 23(4690).

111. Tohnya TM, Ng SS, Dahut WL, et al. A phase I study of oral CC-5013 (lenalidomide, Revlimid), a thalidomide derivative, in patients with refractory metastatic cancer. *Clin Prostate Cancer* 2004; 2(4):241–3.

112. Sison B, Bone T, Amato RJ, et al. Phase II study of CC-4047 in patients with metastatic hormone-refractory prostate cancer (HPRCa). *Proc Am Soc Clin Oncol* 2004; 23(4701).

113. Rustin GJ, Galbraith SM, Anderson H, et al. Phase I clinical trial of weekly combretastatin A4 phosphate: clinical and pharmacokinetic results. *J Clin Oncol* 2003; 21(15):2815–22.

114. Evelhoch JL, LoRusso PM, He Z, et al. Magnetic resonance imaging measurements of the response of murine and human tumors to the vascular-targeting agent ZD6126. *Clin Cancer Res* 2004; 10(11):3650–7.

115. Thorpe PE. Vascular targeting agents as cancer therapeutics. *Clin Cancer Res* 2004; 10(2):415–27.

116. Small EJ, Fratesi P, Reese DM, et al. Immunotherapy of hormone-refractory prostate cancer with antigen-loaded dendritic cells. *J Clin Oncol* 2000; 18(23):3894–903.

117. Burch PA, Croghan GA, Gastineau DA, et al. Immunotherapy (APC8015, Provenge) targeting prostatic acid phosphatase can induce durable remission of metastatic androgen-independent prostate cancer: a Phase 2 trial. *Prostate* 2004; 60(3):197–204.

118. Small EJ, Rini B, Higano C, et al. A randomized, placebo-controlled phase III trial of APC8015 in patients with androgen-independent prostate cancer (AiPCa). *Proc Am Soc Clin Oncol* 2003; 22(1534).

119. Rini BI, Weinberg V, Bok RA, et al. A phase 2 study of prostatic acid phosphatase-pulsed dendritic cells (APC8015; provenge) in combination with bevacizumab in patients with serologic progression of prostate cancer after local therapy. *Proc Am Soc Clin Oncol* 2003; 22(699).

120. Simons J, Higano C, Corman J, et al. A phase I/II study of high dose allogeneic GM-CSF gene-transduced prostate cancer cell line vaccine in patients with metastatic hormone-refractory prostate cancer. *Proc Am Soc Clin Oncol* 2003; 22(667).

121. Sanda MG, Smith DC, Charles LG. Recombinant vaccinia-PSA (PROSTVAC) can induce a prostate-specific immune response in androgen-modulated human prostate cancer. *Urology* 1999; 53(2):260–6.

122. Cavacini LA, Duval M, Eder JP, Posner MR. Evidence of determinant spreading in the antibody responses to prostate cell surface antigens in patients immunized with prostate-specific antigen. *Clin Cancer Res* 2002; 8(2):368–73.

123. Milowsky MI, Nanus DM, Kostakoglu L, Shoehan CE, Vallabhajosula S, Goldsmith SJ, Ross JS, Bander NH. Vascular tracked therapy with anti prostate specific membarane. Antigen monoclonal antibody JS91 in advanced solid tumors. J Clin Oncol 2007; Feb 10; 25(5):540–7.

124. Henry MD, Wen S, Silva MD, Chandra S, Milton M, Worland PJ. A prostate-specific membrane antigen-targeted monoclonal antibody-chemotherapeutic conjugate designed for the treatment of prostate cancer. *Cancer Res* 2004; 64(21):7995–8001.

125. Galsky MD, Eisenberger M, Moore-Cooper S, et al. Phase I trail of MLN2704 in patients with castrate-metastatic prostate cancer (CMPC). *Proc Am Soc Clin Oncol* 2004; 14S(4592).

126. Davis TA, Tchekmedyian S, Korman A, et al. MDX-010 (human anti-CTLA4): a phase I trial in hormone refractory prostate carcinoma (HPRC). *Proc Am Soc Clin Oncol* 2002; 21(74).
127. Peterson MC, Martin SW, Stouch B, et al. Pharmacokinetics (PK) and pharmacodynamics (PD) of AMG 162, a fully human monoclonal antibody to Receptor Activator of NF kappa B Ligand (RANKL), following a single subcutaneous dose to patients with cancer-related bone lesions. *Proc Am Soc Clin Oncol* 2004; 14S(8106).
128. Lam JS, Yamashiro J, Shintaku IP, et al. Prostate stem cell antigen is overexpressed in prostate cancer metastases. *Clin Cancer Res* 2005; 11(7):2591–6.

13

Molecular Targets in Ovarian Cancer and Endometrial Cancer

Nilofer S. Azad, MD, Gisele Sarosy, MD, and Elise C. Kohn, MD

SUMMARY

Molecular targets represent the cumulative and ongoing dissection of the dynamic process of malignant progression and the hope for improved survival for the future. Identification of a precursor lesion, molecular, or physiologic makes molecular dissection and therapeutic targeting more direct. Thus, clinical and laboratory investigators are using the role of common molecular targets, such as those underlying angiogenesis or general proliferation, survival, and invasion from which to design therapeutic strategies. New therapies directed at molecular targets are being actively evaluated in many malignancies, including ovarian and endometrial cancers. Increasing knowledge about the molecular basis of disease has allowed dissection of the events that underlie tumor development and progression as well as the supportive changes that occur in the microenvironment of the tumor and its metastases. Taking the next step to validate the regulation of the targets in the patient, in the tumor, and in the local microenvironment may disprove the specific target hypothesis but will enrich our understanding of the next steps. Early detection has changed the face of endometrial cancer, and molecular targeting will further improve that landscape. This chapter details the major molecular pathways that are known to be important in ovarian and endometrial cancer and their potential as therapeutic targets.

Key Words: Molecular targeting; ovarian cancer; endometrial cancer.

1. INTRODUCTION

Gynecologic malignancies are a major cause of morbidity and mortality in women. More than 25,000 women receive a diagnosis of ovarian cancer every year. Most of these women have late-stage disease and will ultimately die. Endometrial cancer is the most common gynecologic malignancy in women. Many women present with early-stage disease and histology that can be well managed surgically. Nevertheless, more than 16,000 women with ovarian cancer and more than 7000 women with endometrial cancer died of their disease in 2005 *(1)*. Surgery and cytotoxic chemotherapy are

From: *Cancer Drug Discovery and Development: Molecular Targeting in Oncology*
Edited by: H. L. Kaufman, S. Wadler, and K. Antman © Humana Press, Totowa, NJ

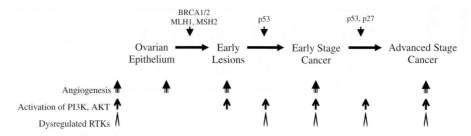

Fig. 1. Signal pathway and gene regulation in the development and progression of ovarian and endometrial cancers. Growth dysregulation may occur for many reasons including genetic disruption through mutation or other insults or downstream changes. Progressive cellular changes occur with augmentation of local angiogenesis, activation of survival signals and through dysregulation of normally balanced signaling events. All can participate in the progression of malignancy and therefore be points for therapeutic interruption.

the mainstay of treatment for advanced disease but ultimately fail to control these malignancies in many patients.

New therapies directed at molecular targets are being actively evaluated in many malignancies, including ovarian and endometrial cancers. Increasing knowledge about the molecular basis of disease has allowed dissection of the events that underlie tumor development and progression as well as the supportive changes that occur in the microenvironment of the tumor and its metastases (Fig. 1). Chemotherapy and radiation therapy have a relatively broad spectrum of injury, affecting both tumor and normal cell populations. Many currently available agents injure the very systems that can contain the tumor, the local stroma, and the immune system. Harnessing the molecular knowledge and using it to focus therapeutic strategies to the sites where they are needed most logically would be more successful. Effective drug development is now coming into its own with the plethora of small molecules and antibodies that are more selective in their target and function.

2. TYROSINE KINASES

Tyrosine kinases, enzymes that phosphorylate selected tyrosine moieties, are essential initiators and messengers in many cell-signaling pathways. Sequencing of the human genome reveals more than 500 tyrosine kinases *(2)*. The complex interactions between tyrosine kinases and their intracellular partner proteins play a major role in normal cell homeostasis. Knowledge in this area has increased exponentially in the past decade, providing new insight into the pathophysiology of many human malignancies and identifying new targets against which novel anticancer strategies can be developed.

Tyrosine kinases exist in two forms: receptor and non-receptor. Receptor tyrosine kinases (RTK) consist of three parts: an extracellular ligand-binding domain, the transmembrane domain, and the intracellular cytoplasmic domain, which contains the kinase region and partner protein-binding sites. Activation of the RTK, through either homodimerization or heterodimerization, results in auto- or transphosphorylation, which then initiates intracellular signaling, critical for normal and malignant cell functioning (Fig. 2). Activated tyrosine kinases initiate multiple and diverse cell-signaling pathways.

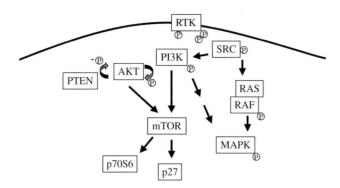

Fig. 2. Receptor tyrosine kinase (RTK) signaling. RTK signaling may propagate through several different downstream pathways. Pathways commonly associated with malignancy and targets for molecular intervention for ovarian and endometrial cancers include the PI3K and mitogen-activated protein kinase (MAPK) pathways. AKT, a major survival protein, is a major downstream effector of PI3K, which activates it through phosphorylation as well as production of a lipid activator, phosphatidylinositol 3′ phosphate. The normal function of PTEN is dephosphorylation of this lipid, putting it into an opposing role. Overexpression or overactivation of PI3K or mutational silencing of PTEN have the same ultimate function of AKT activation, driving survival of tumor, endothelial, and stromal cells.

Perturbation of tyrosine kinase function can occur through several mechanisms. Normal homeostasis results from autocrine, paracrine, and other signals regulating the activity of RTKs. If these are interrupted, unopposed stimulation of RTKs (or suppression of inhibitors) can lead to abnormal intracellular signaling, resulting in malignancy (Fig. 1). Another mechanism for overexpression of RTK is gene amplification (e.g., Her-2-neu-positive breast cancers) (3). Genomic rearrangements also can alter tyrosine kinase activity. In chronic myelogenous leukemia (CML), the fusion protein resulting from p210[bcr/abl] has abnormally active tyrosine kinases. Finally, mutation of the kinase moiety can confer constitutive activation, for example, in c-kit in a subset of gastrointestinal stromal tumors (4) and in the epidermal growth factor receptor (EGFR) in a subset of non-small-cell lung tumors (5). Several families of RTKs have been extensively studied, including EGFR, vascular endothelial growth factor receptor (VEGFR), platelet-derived growth factor receptor (PDGFR), and insulin growth factor receptor (IGFR). These have been shown to be important in the process of ovarian and endometrial cancer progression at the tumor and stromal levels. All have been the focus of development of therapeutic inhibitors. Although sharing fundamental similarities, each possesses unique characteristics, which have been or could be exploited to develop targeted anticancer strategies.

3. EGFR

The EGFR family consists of four members: ErbB1 (EGFR or HER1), ErbB2 (HER2/neu), ErbB3, and ErbB4. Ligands that bind to this family of EGFR can be separated into three groups based on the receptor with which they bind. The first group, EGF, transforming growth factor alpha (TGF-α), and amphiregulin, bind only to EGFR. The second group, betacellulin, heparin-binding EGF, and epiregulin, bind to both EGFR and ErbB3. The last group, neuroregulin, consists of two subtypes, one of which binds to both ErbB3 and ERrbB4 and the other that binds only to ErbB4. No ligand has been identified that preferentially binds to ErbB2. However, ErbB2 participates as a preferred heterodimerization partner with other EGFRs and can amplify their ligand response *(6)*. As such, it plays a pivotal role in the signaling expressed by all receptors in this group *(7)*. Overlap in the signaling cascade provoked by the activation of ErbBs is extensive, occurring in ovarian and endometrial cancer as well as other carcinomas. Two major downstream pathways activated include the mitogen-activated protein kinase (MAPK) pathway and the phosphatidylinositol 3′ kinase (PI3K)-AKT pathway (Fig. 2). Other important pathways include signal transducer and activator of transcription proteins (STATs), src tyrosine kinase, and mammalian target of rapamycin (mTOR) *(8)*.

Amplification and overexpression of ErbB2 is found in approximately 25% of patients with early-stage breast cancer and, when amplified, is associated with a worse prognosis *(9–11)*. Trastuzumab is a humanized monoclonal antibody that binds to and inhibits the ErbB2 receptor and has been shown to have activity in ErbB2-overexpressing breast cancer *(12–14)*. ErbB2 is overexpressed in ovarian cancer, where it is reported to be associated with a poorer prognosis *(15,16)*. However, the frequency of overexpression and the magnitude of overexpression of ErbB2 in ovarian cancer are substantially less than those in breast cancer. In a Gynecologic Oncology Group (GOG) study in patients with measurable recurrent epithelial ovarian or primary peritoneal carcinoma, ErbB2 overexpression was documented in only 95 of the 837 samples tested (11.4%). Forty-one of the 45 patients treated with trastuzumab were eligible for response; the response rate was 7.4%, with a median progression-free interval of 2 months *(17)*.

ErbB2 is overexpressed in a substantial number of patients with endometrial carcinoma where it is also associated with worse prognosis *(18)*. Because overexpression appears more prevalent in uterine papillary serous endometrial carcinoma *(19)*, trastuzumab may be an appropriate therapy for patients with this an aggressive subtype *(20)*.

4. VEGFR

VEGF plays a critical role in tumor angiogenesis, stimulating new blood vessel formation through increased vascular cell survival, proliferation, and vessel remodeling *(2,21)*. VEGF is encoded by a single gene that is expressed by many cell types in the cancer microenvironment, including tumor cells, immune cells, stromal cells, and endothelial cells themselves *(22)*. VEGF was initially identified in 1983 as vascular permeability factor in a mouse ovarian cancer ascites model *(23)*. Expression of VEGFs is induced by hypoxia, autocrine, and paracrine release of other growth factors, inflammatory cytokines, and activation of other RTK pathways *(24)*. VEGF-A plays a pivotal role in angiogenesis, whereas VEGF-C and VEGF-D stimulate lymphatic proliferation *(6)*; all have been implicated in ovarian and endometrial cancer. The biologic activity of VEGF is further refined by the RTK to which it binds. Two tyrosine kinase receptors are found primarily in blood vessel endothelial cells, VEGFR-1 and VEGFR-2, each coded by a separate gene. VEGFR-2 has been the most successfully targeted RTK to date. Monoclonal antibodies and several small molecule inhibitors have been developed with several clinical trials showing promising results. A third tyrosine kinase receptor,

VEGFR-3, is found primarily in lymphatic endothelium; therapeutic application of VEGFR-3 has not been demonstrated to date *(25)*.

VEGF underpins the development of human malignancies by supporting new blood vessel growth *(2)*. High levels of VEGF are found in many malignancies, including ovarian and uterine cancers *(26)*. VEGF levels and microvessel density are related to aggressiveness of disease or poor prognosis in many tumor types, including breast, non-small-cell lung, prostate, head and neck, ovary, and endometrial cancers. VEGF expression correlates variably with prognosis in ovarian and breast cancers *(27)*. Increased VEGF expression even in early-stage epithelial ovarian cancer is a poor prognostic indicator *(28)*, whereas recent studies of endometrial cancer have not validated VEGF expression as a useful prognostic indicator *(29)*.

A number of strategies have been undertaken to block VEGF and its signaling. One approach, disrupting the interaction of VEGF with its RTKs, can be accomplished through competitive removal of VEGF with a VEGF antibody (e.g., bevacizumab) or a VEGF-Trap *(30)*. Another approach is to target VEGFR with neutralizing antibodies or small molecule inhibitors. In addition, interrupting the autocrine and paracrine signals that either induce VEGF or propagate its message can suppress the pathway.

Bevacizumab is a humanized monoclonal antibody directed against VEGF, not specific against the tumor per se. Preclinical evaluation confirmed the antitumor activity of bevacizumab in vivo *(26)*. Phase I clinical trials indicated that bevacizumab was reasonably well tolerated; side effects included hypertension, proteinuria, thrombosis, and hemorrhage. In 2003, a phase III trial demonstrated prolonged survival in previously untreated patients with colorectal cancer treated with cytotoxic chemotherapy and bevacizumab compared with those treated with cytotoxic chemotherapy alone *(31)*. On the basis of these exciting data, the FDA approved bevacizumab for this indication.

The role of bevacizumab in the treatment of ovarian and endometrial cancer is a subject of active clinical investigation. In a GOG phase II study of bevacizumab in more than 50 patients with persistent or recurrent epithelial ovarian cancer or primary peritoneal cancer, 17% responded, with stabilization of disease in an additional 30% *(32)*. A phase III trial of bevacizumab in combination with cytotoxic chemotherapy is planned for newly diagnosed patients with advanced ovarian cancer. The benefits observed are consistent with the predictive and prognostic correlations of expression of VEGF in ovarian cancer.

Other agents that inhibit the VEGF pathway have been or are being tested in patients with gynecologic malignancies, including SU5416, AZ2171, and Bay 43-9006 (sorafenib). Sorafenib was originally described as an inhibitor of B- and c-raf kinase isoforms based on in vitro kinase assays. It also inhibits VEGFR-2 with high affinity and perturbs the Ras/Raf/MAK pathway, c-kit, and PDGFR-ß *(33,34)*. Toxicities of this oral agent include hand-foot syndrome, dermatitis, hypertension, fatigue, anorexia, nausea, and diarrhea *(35–37)*. Single-agent clinical trials in ovarian cancer are underway.

Other antiangiogenic therapy has been evaluated in patients with epithelial ovarian cancer with limited evidence of measurable tumor response, but demonstrated tumor stabilization. Carboxyamidotriazole (CAI), a cytostatic inhibitor of calcium channel-mediated signaling, yielded one partial response and 11 patients (31%) with disease stabilization of 6 months or longer. Its side effects were minimal, supporting the hypothesis that prolongation of progression-free survival in the setting of minimal side effects through dysregulation of signaling in the microenvironment might be a reasonable endpoint (38).

Data regarding the efficacy of thalidomide is limited in patients with endometrial or ovarian carcinoma. Abramson et al. reported that thalidomide was well tolerated in eight patients with ovarian or peritoneal papillary serous carcinoma; one had an objective response and two experienced a decrease in CA-125. A cooperative group trial of thalidomide versus tamoxifen is underway in patients with a first recurrence of ovarian cancer associated with a rise in CA125 but without measurable disease. VEGF and basic fibroblast growth factor (FGF-β) levels will be measured to determine any association with recurrence free survival.

5. PDGFR AND c-KIT

Platelet-derived growth factor, initially described more than 30 years ago, is also an RTK ligand (39,40). Activated PDGFR complexes induce similar biologic effects in endothelial cells to those effects described for VEGFR, including mitogenesis, chemotaxis, stimulation of wound healing, and interstitial fluid pressure. However, other stromal cells and also carcinoma cells express PDGFR, making inhibition of this target a broader sweep. This nuanced activation of PDGF is further complicated by the crosstalk of PDGFR with other RTKs, further amplifying or refining the number of downstream signaling pathways. The biologic effect can be modulated not only by the type of ligand initiating the response but also by the amount and type of each of the component compounds in the signaling pathways (41).

C-Kit and the abl cytoplasmic-tyrosine kinase are members of the PDGFR family. Kit is activated by steel factor, also known as stem cell factor or kit-ligand. Heterogeneity is substantial among malignancies in the expression of PDGFR and/or c-kit and in the manner and degree to which they are activated.

Increased PDGFRα expression in epithelial ovarian cancer is associated with a worse prognosis (42). Of 45 malignant tumors, 16 were positive for PDGFRα, whereas only one of seven borderline tumors and none of the 16 benign tumors were positive; none of these samples were positive for PDGFRβ.

Schmandt et al. (43) found that 81% of 52 serous ovarian cancer samples expressed PDGFRβ as well as 93% of 14 samples of normal ovarian tissue. In a second study by Dabrow et al., of 21 ovarian cancer samples from patients undergoing surgical debulking for Stage III serous cystadenocarcinoma, only one was positive for PDGFR-α, whereas 8 of the 21 were positive for PDGFRß. PDGFRß positivity correlated with a favorable prognosis in this study. The median time to progression was 42 compared with 20 months for those with no observed expression (44). Further studies are required to dissect this dichotomy.

Tonary et al. studied the expression of c-kit and stem cell factor (SCF) in normal human ovary, cultured ovarian surface epithelium, and samples from patients with epithelial ovarian cancer. Ovarian surface epithelium did not express c-kit. In contrast, 38 of 50 ovarian cancer samples (76%) expressed c-kit. C-Kit expression was higher in patients with benign tumors or tumors of low malignant potential. In this study, lowered c-kit expression was associated with worse prognosis, suggesting that c-kit expression decreases with the development of malignancy (45).

An additional study looked at the expression of the three targets of imatinib: PDGFR, c-kit, and abl kinase. C-Kit protein was seen in 26% of high-grade serous carcinomas but not detected in 14 samples of normal ovarian epithelium or in 21 low-grade serous tumors. PDGFR-β was observed in 42 of 52 samples from patients with serous carcinoma and was more likely seen in high-grade tumors. In contrast, c-abl was expressed in all 13 samples of normal ovarian surface epithelium and in 37 of 52 serous carcinomas, where it seems to be more highly expressed in low-grade tumors. These data suggest that a substantial subset of patients with serous carcinomas express one or more tyrosine kinases that are targeted by imatinib mesylate *(43)*.

Except for the trastuzumab study where target expression was required for enrollment, no other studies in ovarian cancer to date have required target expression. However, several have addressed expression of target within the treated population. Several clinical studies have evaluated the activity of imatinib mesylate, targeting abl kinase, and PDGFR and c-kit kinases, in gynecologic malignancies. In a phase II trial of imatinib mesylate in advanced recurrent epithelial ovarian, fallopian tube, or primary peritoneal cancer at the NCI, core-needle biopsies of index tumor lesions were collected from patients before and at 4 weeks of therapy. Modulation of c-kit activation in response to imatinib therapy was shown by decreased phosphorylation of tyrosine 721, the autophosphorylation site of c-kit in six of eight patients for whom matched samples were available *(46)*.

Endometrial carcinomas also express the tyrosine kinases targeted by imatinib mesylate. Newly diagnosed endometrioid and papillary serous endometrial cancer specimens had a high frequency of abl and PDGFR expression by immunohistochem- istry; none expressed c-kit. The presence of PDGFR appeared to be favorable, appearing more commonly in lower grade tumors *(47)*. In a similar analysis of paraffin blocks from 38 patients with gynecologic sarcomas by Caudell et al., all 21 malignant mixed muellerian tumors were positive for PDGFRβ. Abl was expressed in fewer than 50% of the samples and c-kit even less frequently *(48)*. These results have been translated to the clinic with an ongoing clinical trial of imatinib mesylate for women with uterine sarcomas; target expression is not required for eligibility.

6. TYROSINE KINASE INHIBITORS AND COMBINATORIAL THERAPY

Several challenges face the development of tyrosine kinase inhibitors. Although tyrosine kinase inhibitors tested clinically show promise, the clinical benefit does not appear to date to parallel the biologic effect of tyrosine kinase inhibition seen preclinically. One reason is that the efficacy and target selectivity determined in kinase assays may have no bearing on the activity in vivo. If so, we need better models and methods to validate target regulation in the patient. Furthermore, optimal dose may not be maximally tolerated dose (used in cytotoxic chemotherapy) but rather a biologic effective dose, a concept that remains clinically elusive.

Obtaining appropriate tissue with which to assess suppression of the targeted signaling pathway(s) remains a challenge. Using accessible tissue (e.g., blood cells) to assess biologic endpoints may be misleading because effects on signaling pathways may differ among tumor types and organ systems. Approaches to study vascular regulation in vivo include new forms of imaging. Dynamic contrast-enhanced magnetic resonance imaging (DCE-MRI) measures change in contrast flow over time and can serve as a surrogate for measurement of blood flow *(49,50)*. Additionally, the optimal clinical

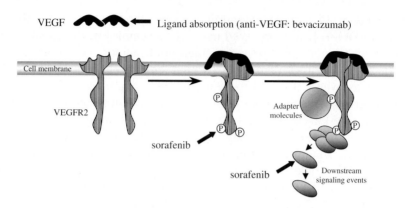

Fig. 3. Combinatorial therapy: multiple hits to the vascular endothelial growth factor (VEGF)/VEGFR pathway. Targeting multiple points within a signaling cascade may improve outcome and reduce the required drug dose. We tested a combination of bevacizumab, to adsorb circulating VEGF, with sorafenib, an inhibitor of VEGFR2 and downstream kinases, such as Raf. This combination is active in ovarian cancer patients with partial responses and disease stabilization of up to 15+ months in heavily pretreated patients. A phase II study is pending.

endpoint must be addressed. Stabilization of disease and increased time to progression may be more useful clinical outcomes than measurable decreases in tumor volume. These questions are important in determining the next steps of development of the use of these new agents.

Combinatorial signal inhibition therapy is a promising area of investigation. Whereas some drugs, such as bevacizumab, are selective in their target, other drugs are promiscuous and can effect signaling at multiple points, including non-target sites. An example of combinatorial therapy is the combination of bevacizumab and sorafenib, under investigation at the national cancer institute (NCI) (Fig. 3). Sorafenib can reduce production of VEGF because of downregulation of its target signaling pathways as an indirect result of inhibition of its kinase targets (discussed in section 4: VEGF). The combination of these two agents may synergistically suppress VEGF and its RTK pathway activation, producing increased activity as well as increased toxicity. In a phase I study of this combination in 12 patients with epithelial ovarian, renal, and colon cancers and melanoma two patients had responses (both with ovarian cancer), and eight had disease stabilization >4 months. Grade 3 dose-limiting toxicity was hypertension (one patient) and proteinuria (two patients). Common toxicities included hand-foot syndrome, weight loss, and anorexia and increasing blood pressure requiring medical management *(51)*.

Other investigators are combining angiogenesis signal inhibitors with tyrosine kinase inhibitors. Combinations under investigation in solid tumors include bevacizumab with EGFR inhibitors, imatinib, or chemotherapy. Several such trials are underway and patients with these cancers are encouraged to participate in phase I combination studies.

7. ONCOGENES

7.1. PI3K/AKT

The PI3K pathway plays an important antiapoptotic role in early and advanced ovarian cancer (Fig. 1). PI3K is a protein and lipid kinase that promotes cell survival through phosphorylation of various downstream proteins. The most well-studied class of PI3K has two subunits, the regulatory p85 with two SH2 domains and the catalytic p110 subunit *(52)*. PIK3CA, the gene for the p110 subunit of PI3K, is amplified in a substantial proportion of ovarian cancers and cell lines and is the kinase that activates the pathway causing phosphorylation of the prosurvival protein, AKT *(53)*. PI3K is activated by binding to the phosphorylated, active states of multiple RTKs, thus making it an important downstream target of RTK inhibition. This was confirmed by remarkable activity of the direct PI3K inhibitor, LY294002, in preclinical models, including its ability to increase ovarian cell line responsiveness to paclitaxel *(54)*. LY294002, unfortunately, is toxic and thus inappropriate for clinical development. Alternative agents that directly inhibit PI3K are under preclinical development.

AKT promotes cell survival and prevents apoptosis by multiple mechanisms. It activates mTOR that, in turn, activates the translational initiation factor 4E-binding protein 1 (4E-BP1) and p70S6 kinase (Fig. 2). These two proteins stimulate synthesis of ribosomes and mRNA transcription for proteins needed for cell-cycle progression *(55)*. In addition, AKT phosphorylates and inhibits function of pro-apoptotic Bcl-associated death promoter (BAD), preventing it from binding to its usual partner, Bcl-xL, leaving Bcl-xL free to exert its antiapoptotic effects *(56)*. Expression of BCL-xl is increased in ovarian cancer *(57)*.

The PI3K/AKT pathways are important therapeutic targets in ovarian and endometrial cancer through both direct and indirect inhibitors. The mTOR inhibitor, rapamycin *(58)*, has activity in vitro and in vivo in animal models but is too toxic for general use. However, two rapamycin analogs, CCI-779 and RAD-001, are in single agent and combination trials and show promising activity in various solid tumors, including ovarian and endometrial cancers *(59–62)*. AKT activation may also be responsible for chemotherapy resistance. In cell line studies, endometrial cells expressing AKT2 and AKT3 are resistant to platinum. Accordingly, in a clinical setting, AKT expression and activation may be a discriminating biomarker for treatment directions and itself a target *(63)*. A number of inhibitors of AKT are under development, heading to clinical trial.

7.2. Ras/Raf/MAP Kinase Pathway

The Ras FORMAT cell-signaling pathway is a pivotal potentiator of RTK activation. Ras is one of 10 proteins in a large family of membrane-bound small GTP-binding proteins. RTK activation forces inactive RAS to release GDP and preferentially bind GTP, which activates RAS to function as a GTPase. Harvey (h)-RAS has many downstream effectors, including PI3K, MEK kinase 1 (MEKK1), protein kinase C (PKC), and the raf family of serine/threonine kinases *(64–66)*. Activated RAS binds to RAF, a target of sorafenib, bringing the protein complex to the cell membrane for further signaling events. Once activated, Raf propagates its signal through the MEK-ERK-MAPK pathway (Fig. 2). ERK activation increases DNA synthesis and cell proliferation through increasing cyclin D1 expression and activation of the ETS family of transcription factors *(67)*.

K-ras, a member of the Ras family, is mutated in the development of endometrioid endometrial carcinoma and can help diagnostically differentiate between types of endometrial cancer. K-ras undergoes mutation as an early event in endometrioid carcinoma but not in serous endometrial carcinoma *(68)*. In various series, up to 20% of epithelial ovarian carcinomas have aberrant K-ras expression or mutation. Mutations are more common in borderline ovarian tumors and mucinous ovarian cancer where they are closer to 50% *(69)*. This was modeled in a transgenic system in which crossing a K-ras transgenic mouse with a p53$^{-/-}$ strain resulted in the formation of endometrioid ovarian cancer in the animals *(70)*.

7.3. APC/Wnt/ß-Catenin

Mutation of the APC tumor suppressor gene has been extensively described as a transforming event in colon cancer but also has an important role in endometrioid endometrial and ovarian cancers. APC gene products phosphorylate, and inactivate, ß-catenin. In the presence of mutated APC, non-phosphorylated ß-catenin accumulates and travels to the nucleus as a transcription factor for cyclin D1, leading to cell-cycle progression *(71)*. Wnt modulates the activity of this pathway by destabilizing the APC/ß-catenin complex, effecting causing higher levels of active ß-catenin *(72)*. In addition, some malignancies have a mutated and more active form of the ß-catenin protein itself. Mutated ß-catenin is seen in 40% of endometrioid ovarian cancer and 15–47% of endometrioid endometrial cancer *(73)*. From one-third to one-half of endometrioid endometrial carcinoma have increased levels of non-mutated ß-catenin *(74)*, especially in hereditary non-polyposis colon cancer-related tumors, suggesting another pathway for ß-catenin modulation in these cancers and a potential importance as a putative therapeutic target. APC inexpression or ß-catenin overaccumulation was not associated in higher risk of recurrence in a case–control study of Stage 1 endometrial cancer *(75)*. Diagnostically, immunostaining for ß-catenin can differentiate between two similar histologies, ovarian mucinous carcinoma and colon cancer, the latter being more likely to stain for ß-catenin *(76)*.

7.4. The Lyosphosphatidic Acid Pathway

Lyosphosphatidic acid (LPA) is a phospholipid produced by many cells including epithelial ovarian cancer cells. LPA has diverse cellular effects, from cellular proliferation, chemotaxis, and increasing endothelial permeability to affecting cellular differentiation, neovascularization, and wound healing *(77)*. LPA functions by activating at least three different G-protein-coupled transmembrane receptors *(78,79)*. Crosstalk occurs between the LPA receptors and multiple RTKs, including EGFR and PDGFR *(80)*. Activation by LPA stimulates a multitude of downstream pathways including PI3K/AKT, mTOR, and the MAPK pathways *(81)*. Thus targeting LPA and its immediate events is logical for therapeutic intervention and also for combinatorial therapy.

Production and circulation of LPA also has potential as a diagnostic marker, but studies to date have been controversial. LPA is produced by ovarian epithelial carcinoma, but not normal ovarian epithelial tissue, and can function as an autocrine factor *(82)*. One study showed 90% of ovarian cancer patients had increased LPA plasma concentrations *(83)*. However, patients with other gynecologic malignancies, breast cancer, and those on dialysis also had increased LPA concentrations, making it sensitive but not specific to ovarian cancer. Other studies have failed to show an increase in LPA plasma concentrations in ovarian cancer patients *(84)*.

G-protein receptors, the category into which the LPA receptor falls, are amongst the most druggable receptors. Examples of successful G-protein-coupled receptor antagonists include ß adrenergic receptor blockers and serotonin receptor antagonists. In addition to targeting the LPA receptors themselves, there are other LPA regulatory steps. Bioactive LPA is produced extracellularly by a series of ecto-enzymes, most importantly autotaxin/lysophospholipase D (ATX/lysoPLD) that is overproduced in ovarian cancer *(82)*. Inactivators of ATX/lysoPLD are being evaluated as potential targets for therapy by decreasing the production of bioactive LPA.

8. TUMOR SUPPRESSOR GENES

8.1. PTEN

PTEN, also known as MMAC1, is a tumor suppressor gene that is responsible for the rare autosomal-dominant Cowden's syndrome, a disease associated with benign hamartomas, breast cancer, and thyroid cancer *(85)*. *PTEN* is the most commonly mutated gene found in endometrioid endometrial cancer. The *PTEN* gene encodes a tyrosine phosphatase that inhibits the AKT/PI3K pathway by dephosphorylating the activated lipid product of PI3K (Fig. 2). When mutated, PTEN loses its lipid phosphatase activity, thus losing its tumor proliferation and invasion suppressor function *(86,87)*.

PTEN may be a useful diagnostic marker in endometrial biopsy specimens *(88)*. Sporadic endometrial cancer has PTEN mutations in 30–50% of cases, with most mutations resulting in a truncated protein. Wild-type PTEN expression is associated with a favorable clinical behavior, early stage, and endometrioid histology *(89)*, whereas lack of PTEN is associated with increased metastases in endometrial cancer *(90)*. PTEN can be mutated in both preinvasive endometrial carcinoma and normal appearing endometrial glands before the development of overt hyperplasia, which suggests that PTEN mutations are an early and potentially carcinogenic event *(91)*.

PTEN is rarely found mutated in papillary serous ovarian cancer (<7% in one series of tumor samples) *(92)*. The AKT survival pathway is frequently already overactive in these ovarian cancers through PI3K activation rather than inhibition of the PTEN inhibitor (Fig. 2). However, PTEN mutations have been found in approximately 15% endometrioid ovarian carcinomas *(93)*. PTEN could have prognostic value, as wild-type PTEN upregulation correlates with slow progression in ovarian cancer, whereas PTEN mutation correlates with rapid progression *(94)*.

Numerous preclinical models have shown the possible therapeutic importance of PTEN/AKT in endometrial cancer *(95)*. A small molecule inhibitor of the PTEN/AKT pathway, API-59CJ-Ome, had increased activity in endometrial cancer cell lines containing PTEN mutations and no activity in wild-type endometrial cancer cell lines *(96)*. Gene therapy to upregulate PTEN expression in endometrial cancer with PTEN mutations could re-establish cell-cycle regulation *(97)* and has been tried preclinically in endometrial and ovarian cancer cell lines *(98,99)*. Targeting mutant PTEN may thus be an alternative method to inhibit overactive AKT.

8.2. P53

P53 mutations are the most common genetic abnormality found in ovarian cancer *(100)*. Normal P53 binds to DNA and induces proapoptotic gene transcription

Table 1
Potential Molecular Targets and Effective Agents that Are Approved or in Testing

Target	Tumor	Agents	Trial status
EGFR	EOC	Cetuximab	In combination with carboplatin (27% RR) in platinum-sensitive patients
		Gefitinib	Minimal activity in phase II trial (RR 4%)
		Erlotinib	Marginal activity in phase II trial (6% RR)
		Erlotinib	In phase I/II trial in combination with docetaxel/carboplatin
		Erlotinib	In phase I/II trial in combination with paclitaxel/carboplatin
		Erlotinib	In phase II testing in combination with bevacizumab (two studies)
		Erlotinib	In phase III testing as adjuvant therapy
	EEC	Gefitinib	Minimal activity in phase II trial
	EEC/EPSC	Lapatinib	In phase II trial
	GS		No agents in testing
Her-2-neu	EOC	Trastuzumab	Minimal activity in phase II trial
		Trastuzumab	In phase II trial in combination with paclitaxel/carboplatin
		Lapatinib	In phase II testing
	EPSC/EEC	Trastuzumab	Minimal activity in phase II trial
PDGFR/kit	EOC	Imatinib	No RR in phase II trial
	GS	Imatinib	In phase II testing
		Sunitinib	Not being tested in these tumors presently
	EEC		No agents in testing
VEGF	EOC	Bevacizumab	17% RR in single agent phase II trial
		Bevacizumab	PIn phase III testing in untreated EOC patients in combination with carboplatin/paclitaxel (2 studies)
		Bevacizumab	In combination with erlotinib (see above)
		VEGF trap	In phase II trial
	GS		No agents in testing
VEGFR	EOC	Sorafenib	In phase II trial
		Sorafenib	In phase II trial in combination with paclitaxel/carboplatin
		AZD2171	In phase II trial (two studies)
	GS	Sorafenib	In phase II trial
		Sunitinib	No current study in these tumors
		AZ6474	No current study in these tumors
mTor	EOC	Rapamycin	No current study in these tumors
	EPSC	CCI-779	No current study in these tumors
	EEC	RAD-001	In phase II trial

EOC, epithelial ovarian cancer; EEC, endometrioid endometrial cancer; EPSC, endometrial papillary serous cancer; GS, gynecologic sarcoma.

and p21, an inhibitor of cyclin-dependent kinase phosphorylation, halting the cell cycle *(101)*. P53 promotes apoptosis through BAX upregulation and decreased expression of the antiapoptotic protein, BCL-2. The frequency of p53 overexpression in ovarian and endometrial cancers generally increases with stage. Approximately 10–15% of early stage cancers overexpress p53 compared with overexpression of 40–50% of advanced sage tumors *(102)*. P53 mutation is more likely an early event in serous ovarian cancer in which the frequency of mutation is similar to that of advanced stage disease. P53 overexpression is more common in serous papillary and endometrioid carcinoma than in clear-cell carcinoma *(103)* and is rarely overexpressed in borderline cancers.

Overexpression portends a poorer prognosis with a 6-fold higher mortality *(104)*. P53 expression in ovarian carcinoma was thought to confer resistance to chemotherapy, but study results have been conflicting *(105)*. p53 mutation may render disease resistant to taxane therapy *(106,107)*.

P53 was considered an important therapeutic target for ovarian cancer. p53 gene transfer was successful in human subjects using an adenoviral vector *(108)* and was tested in phase I and II trials. Ad-p53 gene transfer with SCH58500 was shown to upregulate BAX and p21 as proof of concept *(109)*; however, a randomized trial showed no benefit over gemcitabine chemotherapy alone in patients with recurrent ovarian cancer *(108)*.

9. CONCLUSIONS

Molecular targets are more than the oncologic focus of the decade. They represent the cumulative dissection of the dynamic process of malignant progression and the hope for improved survival for the future. Identification of a molecular or physiologic precursor lesion has provided a potential therapeutic target for some cancers. However, specific targets are unknown in epithelial ovarian cancer and just developing in endometrial cancer. Thus, clinical and laboratory investigators are evaluating common molecular targets, such as those underlying angiogenesis or general proliferation, survival, and invasion to design therapeutic strategies (Table 1). The reliance endometrial and ovarian cancers on the local microenvironment (e.g., angiogenesis and stromal activation) makes tyrosine kinase inhibitors and ligand-neutralizing monoclonal antibodies logical initial therapeutic choices. Evaluating regulation of the targets in the tumor, in the local microenvironment, and in the patient may not support the specific target hypothesis but will suggest which steps to take next.

Early detection has changed the prognosis of patients with endometrial cancer and molecular targeting will further improve their treatment. Until we have reliable early detection for ovarian cancer, focused, effective, minimally toxic therapies are needed. Survival with ovarian cancer has improved greatly; we now need to improve survival from ovarian cancer.

ACKNOWLEDGMENT

This work is supported by the Intramural Research Program of the NIH, National Cancer Institute, Center for Cancer Research.

REFERENCES

1. Jemal A, Murray T, Ward E, et al. Cancer statistics, 2005. *CA Cancer J Clin* 2005;55(1):10–30.
2. Neufeld G, Cohen T, Gengrinovitch S, Poltorak Z. Vascular endothelial growth factor (VEGF) and its receptors. *FASEB J* 1999;13(1):9–22.
3. Monk BJ, Choi DC, Pugmire G, Burger RA. Activity of bevacizumab (rhuMAB VEGF) in advanced refractory epithelial ovarian cancer. *Gynecol Oncol* 2005;96(3):902–5.
4. Hirota S, Isozaki K, Moriyama Y, et al. Gain-of-function mutations of c-kit in human gastrointestinal stromal tumors. *Science* 1998;279(5350):577–80.
5. Lynch TJ, Bell DW, Sordella R, et al. Activating mutations in the epidermal growth factor receptor underlying responsiveness of non-small-cell lung cancer to gefitinib. *N Engl J Med* 2004;350(21):2129–39.
6. Karkkainen MJ, Alitalo K. Lymphatic endothelial regulation, lymphoedema, and lymph node metastasis. *Semin Cell Dev Biol* 2002;13(1):9–18.
7. Tammela T, Enholm B, Alitalo K, Paavonen K. The biology of vascular endothelial growth factors. *Cardiovasc Res* 2005;65(3):550–63.
8. Mendelsohn J, Baselga J. The EGF receptor family as targets for cancer therapy. *Oncogene* 2000;19(56):6550–65.
9. Toikkanen S, Helin H, Isola J, Joensuu H. Prognostic significance of HER-2 oncoprotein expression in breast cancer: a 30-year follow-up. *J Clin Oncol* 1992;10(7):1044–8.
10. Slamon DJ, Clark GM, Wong SG, Levin WJ, Ullrich A, McGuire WL. Human breast cancer: correlation of relapse and survival with amplification of the HER-2/neu oncogene. *Science* 1987;235(4785):177–82.
11. Pegram MD, Pauletti G, Slamon DJ. HER-2/neu as a predictive marker of response to breast cancer therapy. *Breast Cancer Res Treat* 1998;52(1–3):65–77.
12. Slamon DJ, Leyland-Jones B, Shak S, et al. Use of chemotherapy plus a monoclonal antibody against HER2 for metastatic breast cancer that overexpresses HER2. *N Engl J Med* 2001;344(11):783–92.
13. Cobleigh MA, Vogel CL, Tripathy D, et al. Multinational study of the efficacy and safety of humanized anti-HER2 monoclonal antibody in women who have HER2-overexpressing metastatic breast cancer that has progressed after chemotherapy for metastatic disease. *J Clin Oncol* 1999;17(9):2639–48.
14. Carter P, Presta L, Gorman CM, et al. Humanization of an anti-p185HER2 antibody for human cancer therapy. *Proc Natl Acad Sci USA* 1992;89(10):4285–9.
15. Meden H, Kuhn W. Overexpression of the oncogene c-erbB-2 (HER2/neu) in ovarian cancer: a new prognostic factor. *Eur J Obstet Gynecol Reprod Biol* 1997;71(2):173–9.
16. Afify AM, Werness BA, Mark HF. HER-2/neu oncogene amplification in stage I and stage III ovarian papillary serous carcinoma. *Exp Mol Pathol* 1999;66(2):163–9.
17. Bookman MA, Darcy KM, Clarke-Pearson D, Boothby RA, Horowitz IR. Evaluation of monoclonal humanized anti-HER2 antibody, trastuzumab, in patients with recurrent or refractory ovarian or primary peritoneal carcinoma with overexpression of HER2: a phase II trial of the Gynecologic Oncology Group. *J Clin Oncol* 2003;21(2):283–90.
18. Saffari B, Jones LA, el-Naggar A, Felix JC, George J, Press MF. Amplification and overexpression of HER-2/neu (c-erbB2) in endometrial cancers: correlation with overall survival. *Cancer Res* 1995;55(23):5693–8.
19. Prat J, Oliva E, Lerma E, Vaquero M, Matias-Guiu X. Uterine papillary serous adenocarcinoma. A 10-case study of p53 and c-erbB-2 expression and DNA content. *Cancer* 1994;74(6):1778–83.
20. Slomovitz BM, Broaddus RR, Burke TW, et al. Her-2/neu overexpression and amplification in uterine papillary serous carcinoma. *J Clin Oncol* 2004;22(15):3126–32.
21. Conejo-Garcia JR, Benencia F, Courreges MC, et al. Tumor-infiltrating dendritic cell precursors recruited by a beta-defensin contribute to vasculogenesis under the influence of Vegf-A. *Nat Med* 2004;10(9):950–8.
22. Houck KA, Ferrara N, Winer J, Cachianes G, Li B, Leung DW. The vascular endothelial growth factor family: identification of a fourth molecular species and characterization of alternative splicing of RNA. *Mol Endocrinol* 1991;5(12):1806–14.
23. Senger DR, Perruzzi CA, Feder J, Dvorak HF. A highly conserved vascular permeability factor secreted by a variety of human and rodent tumor cell lines. *Cancer Res* 1986;46(11):5629–32.

24. Ferrara N, Gerber HP, LeCouter J. The biology of VEGF and its receptors. *Nat Med* 2003;9(6): 669–76.
25. Huang K, Andersson C, Roomans GM, Ito N, Claesson-Welsh L. Signaling properties of VEGF receptor-1 and -2 homo- and heterodimers. *Int J Biochem Cell Biol* 2001;33(4):315–24.
26. Ferrara N, Hillan KJ, Gerber HP, Novotny W. Discovery and development of bevacizumab, an anti-VEGF antibody for treating cancer. *Nat Rev Drug Discov* 2004;3(5):391–400.
27. Fox SB GG, and Harris AL. Antiogenesis: pathological, prognostic, and growth-factor pathways and their link to trial design and anticancer drugs. *Lancet Oncol* 2001;2:278–89.
28. Paley PJ, Staskus KA, Gebhard K, et al. Vascular endothelial growth factor expression in early stage ovarian carcinoma. *Cancer* 1997;80(1):98–106.
29. Talvensaari-Mattila A, Soini Y, Santala M. VEGF and its receptors (flt-1 and KDR/flk-1) as prognostic indicators in endometrial carcinoma. *Tumour Biol* 2005;26(2):81–7.
30. Holash J, Davis S, Papadopoulos N, et al. VEGF-Trap: a VEGF blocker with potent antitumor effects. *Proc Natl Acad Sci USA* 2002;99(17):11393–8.
31. Hurwitz H, Fehrenbacher L, Novotny W, et al. Bevacizumab plus irinotecan, fluorouracil, and leucovorin for metastatic colorectal cancer. *N Engl J Med* 2004;350(23):2335–42.
32. Burger RA SM. Monk BJ, Greer B, Sorosky J. Phase II trial of bevacizumab in persistent or recurrent epithelial ovarian cancer (EOC) or primary peritoneal cancer (PPC): a Gynecologic Oncology Group (GOG) study. *Am Soc Clin Oncol Annu Meet* 2005:5009.
33. Wilhelm SM, Carter C, Tang L, et al. BAY 43–9006 exhibits broad spectrum oral antitumor activity and targets the RAF/MEK/ERK pathway and receptor tyrosine kinases involved in tumor progression and angiogenesis. *Cancer Res* 2004;64(19):7099–109.
34. Wilhelm S, Chien DS. BAY 43–9006: preclinical data. *Curr Pharm Des* 2002;8(25):2255–7.
35. Strumberg D, Richly H, Hilger RA, et al. Phase I clinical and pharmacokinetic study of the novel Raf kinase and vascular endothelial growth factor receptor inhibitor BAY 43–9006 in patients with advanced refractory solid tumors. *J Clin Oncol* 2005;23(5):965–72.
36. Hotte SJ, Hirte HW. BAY 43–9006: early clinical data in patients with advanced solid malignancies. *Curr Pharm Des* 2002;8(25):2249–53.
37. Ahmad T, Eisen T. Kinase inhibition with BAY 43–9006 in renal cell carcinoma. *Clin Cancer Res* 2004;10(18 Pt 2):6388S–92S.
38. Hussain MM, Kotz H, Minasian L, et al. Phase II trial of carboxyamidotriazole in patients with relapsed epithelial ovarian cancer. *J Clin Oncol* 2003;21(23):4356–63.
39. Ross R, Glomset J, Kariya B, Harker L. A platelet-dependent serum factor that stimulates the proliferation of arterial smooth muscle cells in vitro. *Proc Natl Acad Sci USA* 1974;71(4): 1207–10.
40. Fletcher JA. Role of KIT and platelet-derived growth factor receptors as oncoproteins. *Semin Oncol* 2004;31(2 Suppl 6):4–11.
41. Heldin CH, Ostman A, Ronnstrand L. Signal transduction via platelet-derived growth factor receptors. *Biochim Biophys Acta* 1998;1378(1):F79–113.
42. Henriksen R, Funa K, Wilander E, Backstrom T, Ridderheim M, Oberg K. Expression and prognostic significance of platelet-derived growth factor and its receptors in epithelial ovarian neoplasms. *Cancer Res* 1993;53(19):4550–4.
43. Schmandt RE, Broaddus R, Lu KH, et al. Expression of c-ABL, c-KIT, and platelet-derived growth factor receptor-beta in ovarian serous carcinoma and normal ovarian surface epithelium. *Cancer* 2003;98(4):758–64.
44. Dabrow MB, Francesco MR, McBrearty FX, Caradonna S. The effects of platelet-derived growth factor and receptor on normal and neoplastic human ovarian surface epithelium. *Gynecol Oncol* 1998;71(1):29–37.
45. Tonary AM, Macdonald EA, Faught W, Senterman MK, Vanderhyden BC. Lack of expression of c-KIT in ovarian cancers is associated with poor prognosis. *Int J Cancer* 2000;89(3):242–50.
46. Posadas EM, Kwitkowski V, Kotz HL, et al. A prospective analysis of imatinib-induced c-kit modulation in ovarian cancer: a phase II clinical study with proteomic profiling. Cancer 2007; 110 (2): 309–317.
47. Slomovitz BM, Broaddus RR, Schmandt R, et al. Expression of imatinib mesylate-targeted kinases in endometrial carcinoma. *Gynecol Oncol* 2004;95(1):32–6.
48. Caudell JJ, Deavers MT, Slomovitz BM, et al. Imatinib mesylate (gleevec)–targeted kinases are expressed in uterine sarcomas. *Appl Immunohistochem Mol Morphol* 2005;13(2):167–70.

49. Rehman S, Jayson GC. Molecular imaging of antiangiogenic agents. *Oncologist* 2005;10(2):92–103.

50. Rudisch A, Kremser C, Judmaier W, Zunterer H, DeVries AF. Dynamic contrast-enhanced magnetic resonance imaging: a non-invasive method to evaluate significant differences between malignant and normal tissue. *Eur J Radiol* 2005;53(3):514–9.

51. Azad NS, Annunziata CM, Barrett T, et al. Dual targeting of vascular endothelial growth factor (VEGF) with sorafenib and bevacizumab: clinical and translational results. Presented at Annual Meeting American Society of Clinical Oncology, Chicago, IL; 2007.

52. Osaki M, Oshimura M, Ito H. PI3K-Akt pathway: its functions and alterations in human cancer. *Apoptosis* 2004;9(6):667–76.

53. Shayesteh L, Lu Y, Kuo WL, et al. PIK3CA is implicated as an oncogene in ovarian cancer. *Nat Genet* 1999;21(1):99–102.

54. Hu L, Hofmann J, Lu Y, Mills GB, Jaffe RB. Inhibition of phosphatidylinositol 3´-kinase increases efficacy of paclitaxel in in vitro and in vivo ovarian cancer models. *Cancer Res* 2002;62(4):1087–92.

55. Schmelzle T, Hall MN. TOR, a central controller of cell growth. *Cell* 2000;103(2):253–62.

56. Soini Y, Paakko P, Lehto VP. Histopathological evaluation of apoptosis in cancer. *Am J Pathol* 1998;153(4):1041–53.

57. Marone M, Scambia G, Mozzetti S, et al. Bcl-2, bax, bcl-XL, and bcl-XS expression in normal and neoplastic ovarian tissues. *Clin Cancer Res* 1998;4(2):517–24.

58. Altomare DA, Wang HQ, Skele KL, et al. AKT and mTOR phosphorylation is frequently detected in ovarian cancer and can be targeted to disrupt ovarian tumor cell growth. *Oncogene* 2004;23(34):5853–7.

59. Atkins MB, Hidalgo M, Stadler WM, et al. Randomized phase II study of multiple dose levels of CCI-779, a novel mammalian target of rapamycin kinase inhibitor, in patients with advanced refractory renal cell carcinoma. *J Clin Oncol* 2004;22(5):909–18.

60. Chan S. Targeting the mammalian target of rapamycin (mTOR): a new approach to treating cancer. *Br J Cancer* 2004;91(8):1420–4.

61. Raymond E, Alexandre J, Faivre S, et al. Safety and pharmacokinetics of escalated doses of weekly intravenous infusion of CCI-779, a novel mTOR inhibitor, in patients with cancer. *J Clin Oncol* 2004;22(12):2336–47.

62. Vignot S, Faivre S, Aguirre D, Raymond E. mTOR-targeted therapy of cancer with rapamycin derivatives. *Ann Oncol* 2005;16(4):525–37.

63. Gagnon V, Mathieu I, Sexton E, Leblanc K, Asselin E. AKT involvement in cisplatin chemoresistance of human uterine cancer cells. *Gynecol Oncol* 2004;94(3):785–95.

64. Santos E, Nebreda AR. Structural and functional properties of ras proteins. *FASEB J* 1989;3(10): 2151–63.

65. Campbell SL, Khosravi-Far R, Rossman KL, Clark GJ, Der CJ. Increasing complexity of Ras signaling. *Oncogene* 1998;17(11 Reviews):1395–413.

66. Cox AD, Der CJ. The dark side of Ras: regulation of apoptosis. *Oncogene* 2003;22(56):8999–9006.

67. Chang F, Lee JT, Navolanic PM, et al. Involvement of PI3K/Akt pathway in cell cycle progression, apoptosis, and neoplastic transformation: a target for cancer chemotherapy. *Leukemia* 2003;17(3):590–603.

68. Lax SF. Molecular genetic pathways in various types of endometrial carcinoma: from a phenotypical to a molecular-based classification. *Virchows Arch* 2004;444(3):213–23.

69. Mammas IN, Zafiropoulos A, Spandidos DA. Involvement of the ras genes in female genital tract cancer. *Int J Oncol* 2005;26(5):1241–55.

70. Orsulic S, Li Y, Soslow RA, Vitale-Cross LA, Gutkind JS, Varmus HE. Induction of ovarian cancer by defined multiple genetic changes in a mouse model system. *Cancer Cell* 2002;1(1):53–62.

71. Saegusa M, Hashimura M, Kuwata T, Hamano M, Okayasu I. Beta-catenin simultaneously induces activation of the p53-p21WAF1 pathway and overexpression of cyclin D1 during squamous differentiation of endometrial carcinoma cells. *Am J Pathol* 2004;164(5):1739–49.

72. Rask K, Nilsson A, Brannstrom M, et al. Wnt-signalling pathway in ovarian epithelial tumours: increased expression of beta-catenin and GSK3beta. *Br J Cancer* 2003;89(7):1298–304.

73. Schlosshauer PW, Ellenson LH, Soslow RA. Beta-catenin and E-cadherin expression patterns in high-grade endometrial carcinoma are associated with histological subtype. *Mod Pathol* 2002;15(10):1032–7.

74. Kariola R, Abdel-Rahman WM, Ollikainen M, Butzow R, Peltomaki P, Nystrom M. APC and beta-catenin protein expression patterns in HNPCC-related endometrial and colorectal cancers short communication. *Fam Cancer* 2005;4(2):187–90.

75. Pijnenborg JM, Kisters N, van Engeland M, et al. APC, beta-catenin, and E-cadherin and the development of recurrent endometrial carcinoma. *Int J Gynecol Cancer* 2004;14(5):947–56.

76. Chou YY, Jeng YM, Kao HL, Chen T, Mao TL, Lin MC. Differentiation of ovarian mucinous carcinoma and metastatic colorectal adenocarcinoma by immunostaining with beta-catenin. *Histopathology* 2003;43(2):151–6.

77. Mills GB, Moolenaar WH. The emerging role of lysophosphatidic acid in cancer. *Nat Rev Cancer* 2003;3(8):582–91.

78. Goetzl EJ, Dolezalova H, Kong Y, et al. Distinctive expression and functions of the type 4 endothelial differentiation gene-encoded G protein-coupled receptor for lysophosphatidic acid in ovarian cancer. *Cancer Res* 1999;59(20):5370–5.

79. Noguchi K, Ishii S, Shimizu T. Identification of p2y9/GPR23 as a novel G protein-coupled receptor for lysophosphatidic acid, structurally distant from the Edg family. *J Biol Chem* 2003;278(28):25600–6.

80. Daub H, Weiss FU, Wallasch C, Ullrich A. Role of transactivation of the EGF receptor in signalling by G-protein-coupled receptors. *Nature* 1996;379(6565):557–60.

81. Fang X, Gaudette D, Furui T, et al. Lysophospholipid growth factors in the initiation, progression, metastases, and management of ovarian cancer. *Ann N Y Acad Sci* 2000;905:188–208.

82. Tokumura A, Majima E, Kariya Y, et al. Identification of human plasma lysophospholipase D, a lysophosphatidic acid-producing enzyme, as autotaxin, a multifunctional phosphodiesterase. *J Biol Chem* 2002;277(42):39436–42.

83. Xu Y, Shen Z, Wiper DW, et al. Lysophosphatidic acid as a potential biomarker for ovarian and other gynecologic cancers. JAMA 1998;280(8):719–23.

84. Baker DL, Morrison P, Miller B, et al. Plasma lysophosphatidic acid concentration and ovarian cancer. JAMA 2002;287(23):3081–2.

85. Eng C. PTEN: one gene, many syndromes. *Hum Mutat* 2003;22(3):183–98.

86. Chu EC, Tarnawski AS. PTEN regulatory functions in tumor suppression and cell biology. *Med Sci Monit* 2004;10(10):RA235–41.

87. Sansal I, Sellers WR. The biology and clinical relevance of the PTEN tumor suppressor pathway. *J Clin Oncol* 2004;22(14):2954–63.

88. Taranger-Charpin C, Carpentier S, Dales JP, et al. [Immunohistochemical expression of PTEN antigen: a new tool for diagnosis of early endometrial neoplasia]. *Bull Acad Natl Med* 2004;188(3): 415–27; discussion 27–9.

89. Kanamori Y, Kigawa J, Itamochi H, et al. PTEN expression is associated with prognosis for patients with advanced endometrial carcinoma undergoing postoperative chemotherapy. *Int J Cancer* 2002;100(6):686–9.

90. Salvesen HB, Stefansson I, Kalvenes MB, Das S, Akslen LA. Loss of PTEN expression is associated with metastatic disease in patients with endometrial carcinoma. *Cancer* 2002;94(8):2185–91.

91. Mutter GL, Lin MC, Fitzgerald JT, et al. Altered PTEN expression as a diagnostic marker for the earliest endometrial precancers. *J Natl Cancer Inst* 2000;92(11):924–30.

92. Chen Y, Zheng H, Yang X, Sun L, Xin Y. Effects of mutation and expression of PTEN gene mRNA on tumorigenesis and progression of epithelial ovarian cancer. *Chin Med Sci J* 2004;19(1):25–30.

93. Catasus L, Bussaglia E, Rodrguez I, et al. Molecular genetic alterations in endometrioid carcinomas of the ovary: similar frequency of beta-catenin abnormalities but lower rate of microsatellite instability and PTEN alterations than in uterine endometrioid carcinomas. *Hum Pathol* 2004;35(11):1360–8.

94. Schondorf T, Gohring UJ, Roth G, et al. Time to progression is dependent on the expression of the tumour suppressor PTEN in ovarian cancer patients. *Eur J Clin Invest* 2003;33(3):256–60.

95. Lilja JF, Wu D, Reynolds RK, Lin J. Growth suppression activity of the PTEN tumor suppressor gene in human endometrial cancer cells. *Anticancer Res* 2001;21(3B):1969–74.

96. Jin X, Gossett DR, Wang S, et al. Inhibition of AKT survival pathway by a small molecule inhibitor in human endometrial cancer cells. *Br J Cancer* 2004;91(10):1808–12.

97. Russell W. Adenovirus gene therapy for ovarian cancer. *J Natl Cancer Inst* 2002;94(10):706–7.

98. Minaguchi T, Mori T, Kanamori Y, et al. Growth suppression of human ovarian cancer cells by adenovirus-mediated transfer of the PTEN gene. *Cancer Res* 1999;59(24):6063–7.

99. Sakurada A, Hamada H, Fukushige S, et al. Adenovirus-mediated delivery of the PTEN gene inhibits cell growth by induction of apoptosis in endometrial cancer. *Int J Oncol* 1999;15(6):1069–74.

100. Marks JR, Davidoff AM, Kerns BJ, et al. Overexpression and mutation of p53 in epithelial ovarian cancer. *Cancer Res* 1991;51(11):2979–84.
101. Harris SL, Levine AJ. The p53 pathway: positive and negative feedback loops. *Oncogene* 2005;24(17):2899–908.
102. Berchuck A, Kohler MF, Marks JR, Wiseman R, Boyd J, Bast RC, Jr. The p53 tumor suppressor gene frequently is altered in gynecologic cancers. *Am J Obstet Gynecol* 1994;170(1 Pt 1):246–52.
103. Ho ES, Lai CR, Hsieh YT, et al. p53 mutation is infrequent in clear cell carcinoma of the ovary. *Gynecol Oncol* 2001;80(2):189–93.
104. Gershenson DM, Deavers M, Diaz S, et al. Prognostic significance of p53 expression in advanced-stage ovarian serous borderline tumors. *Clin Cancer Res* 1999;5(12):4053–8.
105. Debernardis D, Sire EG, De Feudis P, et al. p53 status does not affect sensitivity of human ovarian cancer cell lines to paclitaxel. *Cancer Res* 1997;57(5):870–4.
106. Vasey PA. Resistance to chemotherapy in advanced ovarian cancer: mechanisms and current strategies. *Br J Cancer* 2003;89(Suppl 3):S23–8.
107. Cassinelli G, Supino R, Perego P, et al. A role for loss of p53 function in sensitivity of ovarian carcinoma cells to taxanes. *Int J Cancer* 2001;92(5):738–47.
108. Wen SF, Mahavni V, Quijano E, et al. Assessment of p53 gene transfer and biological activities in a clinical study of adenovirus-p53 gene therapy for recurrent ovarian cancer. *Cancer Gene Ther* 2003;10(3):224–38.
109. Buller RE, Runnebaum IB, Karlan BY, et al. A phase I/II trial of rAd/p53 (SCH 58500) gene replacement in recurrent ovarian cancer. *Cancer Gene Ther* 2002;9(7):553–66.

14

Targeted Therapy For Breast Cancer

Deena M. Atieh, MD,
and Linda T. Vahdat, MD

SUMMARY

With the completion of human genome project, the confluence of information on drivers of cell proliferation and the emergence of technology to define and produce therapeutics, molecularly targeted therapies are beginning to be integrated into the overall treatment of patients with cancer. This chapter focuses on the development, mechanism of action, and clinical utility of these agents in breast cancer. Monoclonal antibodies (trastuzumab, bevacizumab, and cetiximab), small molecule tyrosine kinase and farnesyl transferase inhibitors (gefitinib, erlotinib, lapatinib, and tipifarnib), mammalian target of rapamycin and Raf kinase inhibitors (temsirolimus, everolimus, and sorafenib), and other novel agents (ZD 6474, SU 11248) are discussed. In addition, the use of molecular taxonomy and microarray analysis in predicting outcomes and response to therapy are reviewed.

Key Words: Breast cancer; targeted therapy; trastuzumab; HER-2; microarray; bevacizumab; VEGF; lapatinib; monoclonal antibody; tyrosine kinase inhibitor.

1. INTRODUCTION

Molecularly targeted therapy for the molecularly defined patient. Ten years ago, this statement was conjecture, but today this statement is reality. With the completion of the human genome project, the confluence of information on drivers of cell proliferation and the emergence of technology to define and produce therapeutics, molecularly targeted therapies are beginning to be integrated into the overall treatment of patients with breast cancer.

While breast cancer mortality continues to decline, the incidence continues to rise *(1)*. In the year 2005, more than one million women worldwide were diagnosed with breast cancer with almost one-quarter (over 200,000 women) diagnosed in the United States alone. Furthermore, this disease continues to be the most common cause of death in women between 35 and 50 years of age. Annually, 40,870 women die of breast cancer

From: *Cancer Drug Discovery and Development: Molecular Targeting in Oncology*
Edited by: H. L. Kaufman, S. Wadler, and K. Antman © Humana Press, Totowa, NJ

in the United States, and more than 400,000 deaths are reported around the world *(2)*. Although tangible, incremental advances have been made in the chemotherapeutic and hormonal treatments of breast cancer, there is still much room for improvement in the prevention, diagnosis, and treatment. Targeted therapy holds the promise for tailored prevention and treatment strategies.

2. GENOMIC DEFINITION OF BREAST CANCER

Historically, breast cancer classification/staging, prognosis, and treatment allocation has been determined by anatomic and histopathologic criteria (i.e., lymph node and hormone receptor status, tumor size, and grade). With elucidation of the human genome project and the advent of new technologies, a more complex and heterogeneous, molecular breast cancer taxonomy has emerged. DNA and tissue microarray analyses of archival tumor specimens have uncovered molecular signatures, characteristic patterns of gene expression that cluster together, which define distinct subtypes of breast cancer. In locally advanced breast cancer, these subtypes [luminal A, luminal B, basal-type, human epidermal receptor (HER)-2 or ERB-B2, and normal-like] are reproducibly identified in different cohorts of patients and are associated with specific, varied, and predictable clinical outcomes*(3–5)*. The gene expression profiles identified in these subgroups are thought to be a sensitive and pure indicator of tumor biology and behavior. Others have identified groups of genes that can prognosticate early stage breast cancers. Furthermore, even when compared to lymph node status in multivariate analyses, these genetic profiles are a strong and independent predictor of prognosis and survival *(6,7)*.

2.1. Molecular Breast Tumor Subtypes

Initial gene cluster analyses on a small number of breast tumor specimens and mammary carcinoma cell lines identified two definitive patterns, which correlated with proliferation and specific cell signaling cascades *(8)*. To determine if this "molecular portrait" idea was valid, gene expression patterns of pre-chemotherapy and post-chemotherapy tumor specimens were examined from a cohort of women receiving neo-adjuvant doxorubicin *(5)*. While distinct patterns from different tumors were identified, very little variation was seen from samples derived from the same individual at different time points. In addition, some patients in the study had specimens retrieved from nodal metastases, and conservation of the molecular profile was observed even in these specimens. This provided further evidence that a pre-determined molecular composition existed. Genes identified within these patterns correlated with tumor proliferative index, signaling activation, and also cell type(s) of which the tumor was composed. Based on established immunohistochemical data (keratin staining profile), malignant epithelial cells could be identified as basal or luminal types found in normal mammary gland epithelium. Next, to systematically characterize tumors based on gene expression, genes with the highest inter-tumor variability were compiled into an "intrinsic gene subset," and a cluster analysis utilizing this group of genes was applied to the cohort specimens. The results yielded four molecular categories: luminal/estrogen receptor (ER) positive, basal, HER-2, and normal breast. When this "intrinsic" cluster analysis was performed in an expanded cohort, two sub-groups of the luminal/ER positive category emerged: luminal Type A with high ER gene expression and luminal Type B with moderate to low ER gene expression. The HER-2 or ERB-B2 group highly expressed HER-2 and other nearby genes on chromosome 17 *(3)*. The normal-like cluster is predominantly comprised of non-epithelial and adipose cell genes and was seen most frequently in patients who had a response to neo-adjuvant chemotherapy *(3,5)*. The basal cell subtype is discussed below. In a prospective analysis, each of the molecular subclasses correlated with specific patient outcomes. Inferior overall and relapse-free survival was seen in the basal and HER-2 subtypes. Larger, independent

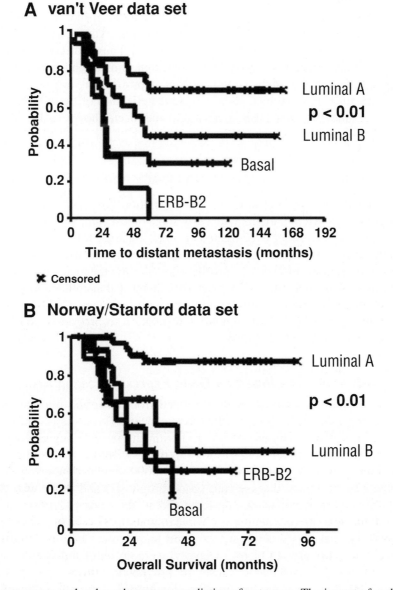

Fig. 1. Breast cancer molecular subtypes are predictive of outcomes. The impact of molecular subtype classification on prognosis and development of metastasis is represented in this Kaplan-Meier Plot. Copyright 2003 National Academy of Sciences, USA *(138)*.

data sets have been similarly probed, and molecular subtype classification, distribution, and impact on prognosis are preserved (see Fig. 1; *4*).

The basal cell subtype is also known as the "triple negative" variant because ER, progesterone receptor (PR), and HER-2 overexpression is absent. These cells do, however, overexpress basal or mammary stem cell markers such as cytokeratins 5, 6, and 17 as well as the *EGFR/HER-1* gene. Prevalence of this sub-type is estimated to be between 15 and 20%, but may be higher in high-risk and/or poor prognosis populations such as women with BRCA-1 mutations and pre-menopausal African Americans *(4,9)*. While basal cell sub-type does correlate with tumor grade, there is no association with lymph node involvement, suggesting an alternative mechanism of spread for these tumors. Sensitivity of these tumors and basal-like mammary cell lines to cytotoxic chemotherapy and targeted anti-epidermal growth factor receptor (EGFR) agents, respectively, have been demonstrated *(10,11)*. Thus, correct identification of basal-like breast cancer subtypes may have important clinical and treatment ramifications. Although gene array analysis and molecular profiling is not readily or widely available at this time, immunohistochemical characterization of these tumors (CK 5/6, HER-1 positive and ER, and HER-2 negative) has been shown to be accurate and specific in identifying this subtype *(12)*.

3. GENE EXPRESSION SIGNATURES AND PROGNOSIS

Cytotoxic chemotherapy and hormonal manipulation have considerably improved risk of disease recurrence in patients with early stage breast cancer *(13)*. However, there are still patients who are either over-treated or under-treated for their disease. The need to better tailor therapies to patients that will derive the most benefit has motivated the search for more accurate methods to profile clinical outcomes and thereby determine appropriate treatment allocation. Emerging data suggest that molecular signatures of gene expression within breast tumors correlates with prognosis *(5,14,15)*. In a series of studies, tumor gene expression signatures in relation to prognosis and response to therapy have been piloted *(6,7,16)*.

3.1. Prognosis Based on Gene Expression Signatures

A 70-gene expression signature was generated from an initial microarray analysis carried out on approximately 100 sporadic breast cancers. These snap-frozen specimens were derived from young women (<55 years) with lymph node negative disease who were followed for outcomes (median follow-up of 8.7 years) *(7)*. The initial unsupervised cluster analysis of this cohort revealed over 5000 regulated genes that correlated with patients who remained disease free (good prognosis) and those who developed metastases (poor prognosis). Genes up-regulated in the poor prognosis group were indicative of an aggressive phenotype. For example, genes involved in cell cycle control, invasion, metastasis, angiogenesis, and signal transduction (cyclins, matrix metalloproteinases, and growth factor receptors) were up-regulated. Supervised classification (to identify a manageable subset of prognostic reporter genes) ultimately yielded the 70 marker genes comprising the signature. In a small cohort of patients, this signature proved to be valid, reproducible, and accurate in predicting development of metastatic disease (odds ratio of 15, $p = 4 \times 10^{-6}$). Compared to currently available clinical and histopathological predictors of recurrent disease, this signature demonstrated superior predictive capacity *(7)*.

To authenticate this gene expression signature as a predictor of survival and outcomes in an independent, larger population, the Netherlands Cancer Institute Group evaluated the gene prognosis profile in about 300 young, untreated patients with early stage (I or II) breast cancer *(6)*. Fresh frozen tissue was obtained and all patients had definitive surgery with axillary lymph node dissection, radiation therapy (if appropriate) followed by adjuvant systemic therapy (chemotherapy, hormonal therapy, or both). Patients were then followed at least annually for development of metastatic disease and overall survival. The median follow-up for the entire cohort was about 7 years. Again, the outcome for both disease-free and overall survival was markedly different between those with a "good prognosis" gene expression signature compared to a "poor prognosis" gene expression signature (hazard ratios of 5.1 and 8.6, respectively, $p < 0.001$). In multivariate analysis, the poor prognosis signature was the strongest predictor for likelihood of developing distant metastases (hazard ratio of 4.6, $p < 0.001$), and this effect was independent of lymph node involvement. Furthermore, when this data set was applied to and stratified by risk using other established prognostic classification schemes (St. Gallen or NIH Consensus Criteria), the Dutch gene expression profile was more predictive of outcome than either of these two models *(6)*.

3.2. Likelihood of Response to Tamoxifen for Node Negative Breast Cancer Patients

Along the same lines, a NSABP group developed a 21-gene assay of tumor-related genes and correlated it to prognosis and response to tamoxifen therapy in patients with lymph node negative, hormone receptor positive breast cancer *(16)*. After developing a practical and novel method to measure gene expression [reverse transcriptase polymerase chain reaction (RT-PCR)] in paraffin-embedded tissue specimens, 250 candidate predictor genes were compiled. Subsequently, candidate gene expression was tested in three independent clinical studies (447 specimens) *(17–19)*. Based on their predicative ability and reproducibility, 16 representative, cancer-related genes and 5 reference genes were selected for the assay. Genes were grouped based on function (proliferation, HER-2, ER, or invasion), and their unique expression pattern in tumors was then folded into a recurrence score (range 1–100: low risk <18, intermediate risk 18–30, and high risk ≥31) for individual tumors. To test the ability of this genetic model to predict likelihood of distant recurrence and overall survival, a prospective evaluation of just under 700 paraffin block specimens from women treated with tamoxifen alone in the NSABP B-14 trial was performed. The number of patients without distant recurrence at more than 10 years after surgery was significantly greater in the low-risk versus the high-risk group (93 versus 70%, $p < 0.001$). In fact, the risk of recurrence in the high-risk group (approximately 30%) was similar to that observed in patients with positive lymph nodes. Recurrence scores generated significantly correlated with both disease-free and overall survival. Subgroup analysis identified the recurrence score as a strong, independent predictor of outcome (hazard ratio of 3.21, $p = 0.001$). It is important to note that these findings may not solely be indicative of the natural history of breast cancer, but may also represent a phenotype likely to respond to tamoxifen *(16)*. Therefore, at this time, treatment decisions based strictly on this assay are not recommended.

Differences between this analysis and the Dutch study include pre-selection of a candidate gene set derived from the literature (versus collection from unsupervised microarray analysis), use of real-time, RT-PCR on paraffin-embedded, readily available tumor blocks (versus DNA array analysis on snap-frozen tissue which is limited in availability), and evaluation of a totally independent data set (versus analysis including a subset of patients upon which the initial predictor genes were derived).

4. MOLECULARLY TARGETED THERAPY OF BREAST CANCER—TARGET: HER-2

HER-2, also known as HER-2/*neu* or the ERB-B2 receptor, is a member of the EGFR super-family of transmembrane tyrosine kinase receptors. The *HER-2* gene encodes a 185-kilodalton protein (also referred to as p185[HER2]). The HER-2 receptor is one of four known members of the HER family of transmembrane tyrosine kinase receptors. These growth factor receptors are implicated in tumor cell growth, survival,

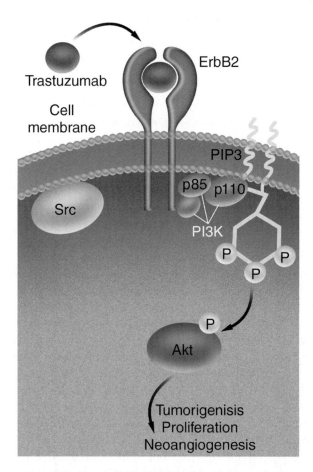

Fig. 2. ERB-B2 Receptor Activation and Signalling favors Cell Survival. The activated receptor triggers AKT (pictured here) and various other downstream pathways such as Ras, Raf, MAPK, PI3K, PKC, STAT to promote tumorigenesis, proliferation and neo-angiogenesis *(25)*.

metastasis, invasiveness, as well as angiogenesis (20). The HER-2 receptor has no known specific ligand. Instead, it signals through dimerization (heterodimerizes or homodimerizes) with other receptors of the receptor family (HER-1 or EGFR, HER-3, HER-4). Subsequently, there is auto-phosphorylation of the HER-2 receptor and initiation of downstream signaling (21–23). The activated receptor triggers various downstream pathways such as Ras, Raf, mitogen-activated protein kinase (MAPK), phosphatidylinositol-3-OH kinase (PI3K), protein kinase C, signal transducer and activator of transcription, and AKT family of serine/threonine specific protein kinase family. Collectively, these pathways favor cell survival and cell proliferation (see Fig. 2; 24,25)

4.1. HER-2 and Breast Cancer

A series of experiments established that 20–30% of breast cancers show evidence of HER-2 gene overamplification (26). Overexpression of this gene is associated with aggressive features and poor prognosis across all subtypes of breast cancer. This includes invasive breast cancer both with and without lymph node involvement and also ductal carcinoma in situ (26–28). Furthermore, *HER-2* gene overamplification correlates with increased levels of the HER-2 protein, which is sufficient (but not essential) to induce transformation and tumorigenesis of breast cancer both in vitro and in vivo (29). Because of its preferential expression in malignant breast tissue, the HER-2 receptor represents a logical target in breast cancer.

5. TRASTUZUMAB

Trastuzumab serves as a model for development of novel, rational targeted therapies. The steps involved, which culminated in FDA approval for advanced breast cancer, reflect at least 10 years of research and encompass the discovery of a functional target in cancer cell lines, development of an antibody with specificity for the receptor in pre-clinical models, and finally well-designed clinical trials leading to its incorporation into clinical practice. As the first monoclonal antibody approved for use in metastatic breast cancer, trastuzumab is the prototype for targeted, antibody therapy in solid tumors (22).

5.1. Development

Trastuzumab represents the integration of genetic engineering and translational research. It is a humanized monoclonal antibody with specificity for the extracellular domain of the HER-2 receptor. The construct was created from a murine monoclonal antibody, 4D5, derived from mice immunized with HER-2 overexpressing cell lines. 4D5 exhibits anti-proliferative effects in cell lines and xenografts that specifically overexpress HER-2 (30–32). As murine antibodies cannot be utilized for repetitive dosing in humans due to their immunogenicity [human anti-mouse antibodies production], a humanized form of the antibody was synthesized. Trastuzumab, the humanized, recombinant form of the 4D5 murine monoclonal antibody, allows repeated administration and increased stability (22). It is comprised of the murine complimentary determining region (i.e., hypervariable or specific binding region) of 4D5 fused onto a backbone of human IgG. In preclinical models, this monoclonal, humanized antibody showed similar efficacy to the murine antibody (33).

5.2. Mechanism of Action

The anti-tumor effect of trastuzumab is multi-factorial. Trastuzumab has effects on signal transduction, the innate immune response, and angiogenesis. Monoclonal antibody-receptor binding accelerates endocytosis and internalization of the HER-2 receptor (34–36). This in turn inhibits dimerization of the receptor with other HER family members, and thereby decreases HER-2-mediated potentiation of ERB-B signaling (37). Trastuzumab also directly inhibits the AKT survival signaling pathway, promoting early cell death and apoptosis (38). Immunomodulatory effects include augmentation of both antibody-dependent cellular cytotoxicity and the T-cell response (39–41). There are also effects on neo-angiogenesis. HER-2 overexpression promotes new vessel formation through a variety of mechanisms, including increased expression and production of vascular endothelial growth factor (VEGF). In vitro experiments demonstrate that trastuzumab binding inhibits the HER-2-mediated production of VEGF (42). Modulation of other angiogenic factors such as thrombospondin-1, transforming growth factor-α (TGF-α), angiopoietin-1, and plasminogen activator inhibitor-1 are also observed (43). Trastuzumab impairs DNA repair and cellular proliferation through its effects on p21 and promotion of DNA strand breaks (44,45). It also potentiates effects of tumor suppressor genes and proteins such as the programmed cell death 4 gene and lipid phosphatase and tensin homolog (PTEN) (46). This latter effect is implicated in trastuzumab resistance, as patients with PTEN-deficient tumors have an inferior response to trastuzumab therapy (47).

Trastuzumab's effects on apoptosis, DNA repair, cell signaling, and angiogenesis are synergistic and/or additive with the anti-proliferative effects of chemotherapy (48–50). These findings have guided the design of clinical trials combining trastuzumab with chemotherapeutic and other biologic agents (51). Synergistic cell kill of breast cancer cell lines that overexpress HER-2 is observed when trastuzumab is used in conjunction with various agents including carboplatin, docetaxel, cyclophosphamide, vinorelbine, and etoposide. Also, additive effects are seen with anthracyclines, paclitaxel, and methotrexate (50,52). Trastuzumab's effect on malignant neo-vasculature probably allows for more efficient delivery of chemotherapy to the tumor bed. In mouse models, it restores vessel structure to a more normal appearance, by decreasing the diameter and permeability of tumor infiltrating vessels (43).

5.3. Pharmacokinetics

Phase I clinical trials established the pharmacokinetic and safety profiles of trastuzumab (53,54). Weekly administration of trastuzumab in the phase II and III studies resulted in target serum levels greater than 10 μg/mL, a concentration associated with optimal growth inhibition in vitro (33,53). Although initial studies employed fixed dosing, current practice is with weight-based dosing to ensure adequate serum levels. Weekly administration (4 mg/kg loading dose followed by 2 mg/kg weekly infusions) is based on initial pharmacokinetic analysis of the pivotal trials, in which a one-compartment distribution of the drug was assumed. This model suggests that trastuzumab obeys a dose-dependent, non-linear profile with an estimated half-life of approximately 6 days (53). Hence, weekly dosing is accepted as an appropriate dosing interval.

Recently, additional pharmacokinetic studies suggest that a once every 3-week schedule is also feasible from a pharmacokinetic standpoint. Re-evaluation of the initial pharmacokinetic data suggests an alternative, two-compartment drug distribution, which results in a longer half-life of nearly 30 days. This allows a prolonged dosing interval to once every 3 weeks while still maintaining adequate serum target levels *(55)*. The every 3-week dosing requires a higher loading dose of 8 mg/kg followed by maintenance doses of 6 mg/kg every 3 weeks. Trastuzumab administration every 3 weeks as monotherapy and/or in combination with chemotherapy demonstrates similar pharmacokinetics, response rates, and toxicity as the weekly schedule *(55–57)*. There appears to be no excess adverse effects (cardiac, infectious, or infusional) attributable to the prolonged dosing regimen *(56,58)*. The majority of these trials include prospective cardiac monitoring, and clinical congestive heart failure is rarely observed. As a result, the current National Comprehensive Cancer Network (NCCN) practice guidelines include both weekly and once every 3-week dosing intervals *(59)*.

Concurrent cytotoxic chemotherapy, tumor burden, and plasma levels of the extra-cellular domain of HER-2 do not have any consistent effect on the pharmacokinetics of trastuzumab, regardless of the dosing schedule employed. Thus, dose adjustment based on these factors is not recommended *(52,55)*.

5.4. Predicting Response and Measurement of HER-2 Overexpression

Current practice guidelines recommend HER-2 testing for all invasive breast cancers utilizing immunohistochemistry (IHC) for protein overexpression and/or fluorescent in-situ hybridization (FISH) for gene amplification *(59,60)*. FISH testing is more sensitive and has emerged as a gold standard in determining HER-2 overexpression and response to trastuzumab. Three to four percent of IHC-negative tumors will be positive by FISH *(61,62)*. Discordance is most common with the IHC 2+ score, and only approximately 30% of IHC 2+ specimens demonstrate gene overamplification by FISH *(62,63)*. Therefore, confirmation by FISH of IHC 2+ specimens is recommended *(59, 64,65)*. Concordance between FISH and IHC testing has been prospectively examined in certain adjuvant trastuzumab clinical trials (NSABP B-31 and NCCTG N 9831). IHC testing is more accurate in experienced reference labs, as community testing facilities have shown up to a 20% false-positive rate *(66)*. In addition, there is frequent, substantial discordance (15% for FISH and 20% for IHC) between local and central laboratories *(67,68)*. Overall, FISH testing is superior to IHC in determining whether a tumor is HER-2 positive or not. Furthermore, only patients with tumors that are IHC 3+ and/or FISH positive will derive a benefit from trastuzumab therapy *(23,69–71)*.

Novel methods for predicting response and/or resistance to trastuzumab are currently under evaluation and include quantifying circulating levels of the HER-2 extracellular domain and HER-2 receptor internalization *(72–77)*.

5.5. Clinical Trials

Limitations of the early clinical studies included inadequate identification of patients most likely to benefit from trastuzumab (many IHC 2+ patients were enrolled) and lack of prospective cardiac monitoring. Nonetheless, these trials demonstrate improved survival and response rates with trastuzumab treatment.

Similar to other biological agents, trastuzumab is cytostatic in preclinical models *(78,79)*. As a result, the benefit derived includes not only regression of disease, but stabilization as well. Thus, the clinical benefit rate, a collective term encompassing complete, partial, or minor responses, as well as stable disease lasting ≥6 months becomes clinically relevant *(69)*. Documented clinical benefit of trastuzumab is associated with durable response duration even in heavily pre-treated patients with predominantly visceral disease. Clinical benefit rates range from 35 to 45% as monotherapy and 60–80% in combination with cytotoxic chemotherapy (see Table 1).

5.5.1. PHASE II SINGLE-AGENT CLINICAL TRIALS

As shown in Table 1, trastuzumab monotherapy produces overall response rates of 26% in the first-line setting and 15% in second or third line refractory disease *(69,70)*. The median time to progression in both of these cohorts is in the range of just over 3 months. These are rates comparable to traditional chemotherapeutics employed in metastatic breast cancer. With the important exception of cardiac toxicity, trastuzumab is generally well tolerated. Traditional side effects associated with cytotoxic chemotherapy such as myelosuppression, alopecia, nausea, and vomiting are rare. Toxicity occurring in more than 10% of patients includes mild to moderate chills, fever, asthenia, and pain. Severe adverse events are rare and occur in only 9% of patients. Although cardiac toxicity occurs with trastuzumab monotherapy, the incidence is higher when trastuzumab is used in combination with cytotoxic chemotherapy.

5.5.2. PHASE II CLINICAL TRIALS IN COMBINATION WITH CHEMOTHERAPY

Trastuzumab has been combined with various chemotherapeutic agents (Table 1). In general, the combinations are well tolerated with the most notable adverse events being a reduction in left ventricular ejection fraction and clinical congestive heart failure in some, but not all trials.

Doublets incorporating the taxanes (docetaxel and paclitaxel), vinorelbine, gemcitabine, and capecitabine in combination with trastuzumab appear to have, at a minimum, similar response rates when compared to similar single-agent studies. With the exception of cardiac complications, common side effects with combinations were almost exclusively chemotherapy-type side effects.

Triplets including a platinum salt, taxane, and trastuzumab were developed as a result of preclinical data demonstrating synergistic effects on DNA repair and receptor-enhanced chemosensitivity *(50,94)*. Studies of these three drugs in combination generate response rates between 56 and 79% with median time to progression of 10–12 months. Toxicities attributable to the platinum salt are common, but tolerable *(92)*. Activity of this triplet is advantageous because it does not contain an anthracycline, which is known to augment the cardiac toxicity of trastuzumab.

5.5.3. USE WITH THE ANTHRACYCLINES

Tumors which overexpress HER-2 are hypothesized to have an inherent sensitivity to anthracyclines *(28,95,96)*. Caution regarding overlapping cardiac toxicity, however, limits the use of trastuzumab in combination with traditional anthracyclines. For example, single agent doxorubicin itself has a 3–18% incidence of cardiac toxicity depending on total cumulative dose. Alternative anthracycline derivatives with

Table 1

Trials of Trastuzumab +/- Chemotherapy in Metastatic Breast Cancer

Author/study	N	Line Tx	Tx	RR (%)	CBR (%)[a]	TTP (months)	All CD (%)	Symp CD (%)
Cobleigh et al. (70)	222	2nd or 3rd	T alone	15	29[a]	3	5[b]	4[b]
Vogel et al. (69)	114	1st	T alone	26	38	4	3[b]	1–3[b]
Baselga et al. (57)	105	1st	T q3 alone	19	33	3	15	1
Seidman et al. (80)	95	1st, 2nd or 3rd	T + P	57	87	7	7	3
John/FAKT et al. (81)	109	1st or 2nd	T + P	75	89[a]	NR	NR	0
Leyland-Jones et al. (56)	32	1st or 2nd	T q3 + P q3	59	81	12	28	3
Tedesco et al. (82)	26	1st or 2nd	T + D	50	81[a]	12	8	NR
Esteva et al. (83)	30	1st or 2nd	T + D	63	83	9	26	3
Burstein et al. (84)	40	1st, 2nd or 3rd	T + V	75	80	9 (1st), 4 (2nd/3rd)	20	0
Burstein et al. (85)	54	1st	T + V	68	85	6	13	2
Jahanzeb et al. (86)	40	1st	T + V	72	NR	17	2.5	0
O'Shaughnessy (87,88)	64	2nd or 3rd	T + G	38	74[a]	6	8	0
Xu et al. (89)	41	1st	T + X	77	NR	NR	NR	0
Sledge (90)	46	1st	T + G + P q3	63	61[a]	10	NR	2
Perez/N98-32-52 et al. (91)	43	1st	T + P + C q3	65	NR	9	NR	0
Perez/N98-32-52 et al. (91)	53	1st	T + P + C	71	NR	13	NR	0
BCIRG 101 et al. (92)	62	1st	T + D + CDDP	79	81[a]	10	17	2
BCIRG 102/ULCA-ORN et al. (92)	62	1st, 2nd or 3rd	T + D + C	56	84[a]	13	13	2

All CD, asymptomatic and symptomatic cardiac dysfunction [Grades 0–4 by the CTCAE criteria (93)]; C, carboplatin; CDDP, cisplatin; CBR, clinical benefit rate (complete responses + partial responses + stable disease for ≥ 6 months); D, docetaxel; G, gemcitabine; N, number of patients; NR, not reported; P, paclitaxel; RR, response rate; Symp CD, symptomatic cardiac dysfunction [Grades 3–4 by the CTCAE criteria (93)]; T, trastuzumab; TTP, time to progression; TX, treatment (weekly unless noted otherwise); V, vinorelbine; X, capecitabine.

[a] Duration of stable disease not mentioned or <6 months.

[b] Retrospective review of cardiac toxicity.

q3 = every 3 weeks

similar efficacy but less cardiac toxicity have been safely and effectively administered concomitantly with trastuzumab *(97)*. Liposomal doxorubicin shows an overall response rate of 58% when combined with trastuzumab. In patients who had received <240 mg/m^2 cumulative dose of prior anthracycline, the rates of cardiac toxicity were comparable to other doublet, trastuzumab-containing regimens *(98)*. Other studies have included liposomal doxorubicin in triplet regimens along with trastuzumab and a taxane *(99–101)*. Small phase I/II studies combining epirubicin with trastuzumab yield comparable response rates and acceptable cardiac toxicity *(102,103)*.

5.5.4. RANDOMIZED CLINICAL TRIALS

Randomized studies of chemotherapy alone compared to chemotherapy plus trastuzumab in patients with advanced breast cancer unequivocally show a benefit for combination therapy with respect to response rate, time to progression, and overall survival *(54,104)*. A multi-national phase III trial evaluated chemotherapy alone (paclitaxel or doxorubicin + cyclophosphamide) versus chemotherapy plus trastuzumab. Approximately 470 patients received standard doxorubicin and cyclophosphamide or paclitaxel every 3 weeks for six cycles depending on which prior agents they had received. The intention to treat analysis at 31 months follow-up favored the combined therapy arms. Chemotherapy plus trastuzumab provided a 20% relative risk reduction for death (11% absolute benefit). Crossover to the treatment arm was permitted and frequent, thus the benefit of combination therapy on survival is probably underestimated. Cardiac toxicity in this trial was much greater than in the phase II trials; attributed primarily to the concurrent (and prior) use of anthracyclines and trastuzumab. These findings prompted future studies to employ prospective monitoring of left ventricular ejection fraction (see Table 2; *54)*. Docetaxel (every 3 weeks) with or without trastuzumab also reveals significant improvement in response rate, time to progression, and survival for the combination arm with a median follow-up of 2 years. The incidence of symptomatic cardiac dysfunction (CD) was acceptable at 1% *(104)*.

5.5.5. ADJUVANT CLINICAL TRIALS

Phase III (NSABP B31, NCCTG N 9831, and HERA) trials have identified a role for trastuzumab in the adjuvant setting. When compared to adjuvant chemotherapy alone, trastuzumab administered concurrently or after the completion of adjuvant chemotherapy shows a highly statistically significant improvement in disease-free survival, time to distant disease progression, and even with short follow-up, an overall survival advantage. Eligible patients in these trials included those with HER-2 overexpression (IHC 3+ and/or FISH+ by central or reference laboratories), a high risk of recurrence (lymph node positive disease or node negative with tumors >1 cm and poor prognostic features), and normal cardiac function. Patient demographics and results of these trials are shown in Table 3.

The US-based multi-center trials, NSABP B31 and NCCTG N 9831, randomized high-risk women to conventionally dosed doxorubicin and cyclophosphamide followed by paclitaxel (every week or every 3 weeks) with or without weekly, concurrent trastuzumab. A pooling of the trial results and subsequent joint analysis of these trials demonstrates a 52% relative reduction in risk of breast cancer recurrence and a 33% relative mortality reduction for women who receive trastuzumab as part of their adjuvant chemotherapy regimen over those who do not. At a median follow-up of 2

Table 2

Randomized Trials of Chemotherapy +/- Trastuzumab in Metastatic Breast Cancer

Author/study	N	TX		RR (%)		TTP (months)		OS (months)		All CD (%)		Symp CD (%)	
		T + CTX	CTX	T + CTX	CTX	T + CTX	CTX	T + CTX	CTX	T + CTX	CTX	T + CTX	CTX
Slamon et al. (54)	469	T + AC or T + P	AC or P	**50**	**32**	**7**	**5**	**25**	**20**	26 (AC)[a] 13 (P)	8 (AC)[a] 1 (P)	16 (AC)[a] 2 (P)	3 (AC)[a] 1 (P)
Marty/M77001 et al. (104)	186	T + D q3w	D q3w	**61**	**34**	**12**	**6**	**31**	**23**	17	8	1	0
Gasparini et al. (105)	85	T + P	P	78	60	13	7	NR	NR	0	0	0	0

All CD, asymptomatic and symptomatic cardiac dysfunction [Grades 0–4 by the CTCAE criteria (93)]; C, cyclophosphamide; CTX, chemotherapy; N, number of patients; OS, overall survival; P, paclitaxel; RR, response rate; Symp CD, symptomatic cardiac dysfunction [Grades 3–4 by the CTCAE criteria (93)]; T, trastuzumab; TTP, time to progression; TX, Treatment (weekly unless noted otherwise).

[a]Retrospective review of cardiac toxicity.

Statistically significant values are represented in bold.

A = Adriamycin

q3w = Every 3 weeks

Table 3
Trials of Adjuvant Chemotherapy +/- Trastuzumab in High Risk Breast Cancer

	Joint analysis (106,107)	HERA (108)
N[a]	3351	3387
Control arm	AC→P→Obv	Chemo→Obv
Experimental arm(s)	AC→P + T (qwk) × 1 year	Chemo→T (q3wk) × 1 yearChemo→T (q3wk) × 2 years
Tumor Size		
<2 cm	39%	NR
>2 cm	61%	NR
Positive lymph nodes		
None	6%	32%[b]
<4	52%	28%
≥4	41%	28%
ER/PR positive	50%	50%
Adjuvant chemotherapy		
Non-anthracycline, non-Taxane	0%	6%
Anthracycline	100%	67%
Anthracycline + taxane	100%	25%
Median follow-up	2 years	1 year
DFS, HR	**0.48**[c]	**0.54**[c]
OS, HR	**0.67**	0.76
CHF	3.3–4%(0–0.5% in control arm)	0.5%(0% in control arm)

AC, doxorubicin 60 mg/m^2 and cyclophosphamide 600 mg/m^2 every 3 weeks × 4; Chemo, ≥4 cycles of accepted adjuvant or neo-adjuvant chemotherapy; CHF, NYHA Class III–IV congestive heart failure (109); DFS, disease-free survival; ER/PR, estrogen receptor/progesterone receptor HR, hormone receptor; HR, hazard ratio; Joint analysis, NSABP-B-31 and NCCTG N 9831; P, Paclitaxel 175 mg/m^2 every 3 weeks × 4 or 80 mg/m^2 every week × 12; Obv, observation; OS, overall survival; T, trastuzumab (qwk, 4 mg/m^2 × 1 followed by 2 mg/m^2 weekly; q3wk, 8 mg/m^2 × 1 followed by 6 mg/m^2 every 3 weeks).

[a] N, Number of patients analyzed, of note: the sequential trastuzumab arm of the NCCTG N 9831 and the 2-year trastuzumab arm of the HERA trials were not reported and/or analyzed.

[b] Approximately 10% of patients received neo-adjuvant therapy and nodal status was just noted as "any" and not quantified in the report.

[c] $p < 3 \times 10^{-12}$.

[d] $p < 0.0001$

$p \leq 0.015$ values are represented in bold.

years, there is a highly statistically significant improvement in disease-free survival ($p < 3 \times 10^{-12}$), time to first distant recurrence ($p = 8 \times 10^{-10}$), as well as overall survival ($p = 0.015$). These benefits are consistent across all subgroups, except those with lymph node negative disease because the number of patients in this subset is quite small (6% of the joint analysis). To address the effect of schedule on trastuzumab benefit in the adjuvant setting, a third arm in the NCCTG N 9831 also randomized patients to a sequential regimen of trastuzumab following adjuvant chemotherapy (106). An unplanned analysis of the sequential versus the concomitant trastuzumab arm of NCCTG N 9831 suggests an advantage of concomitant over sequential trastuzumab;

however, the number of events (25% of total needed) observed at this time precludes definitive conclusions *(107)*.

The HERA Trial is a multi-national (39 countries, 478 medical centers) EORTC-led study of more than 5000 women, which also examined the role of trastuzumab in the adjuvant setting. In addition to investigating the overall benefit of trastuzumab, this trial also attempts to evaluate the optimal duration of treatment with the monoclonal antibody as patients were randomized to no trastuzumab, 1 year of trastuzumab or 2 years of trastuzumab. Important differences between this trial and the US-led trials are that trastuzumab was administered *after* the completion of chemotherapy, on an every 3-week schedule and to more patients with lymph node negative disease (1/3 of all patients). Results from the first planned efficacy interim analysis are currently available only for the 1-year trastuzumab arm. At a median follow-up of 1 year, the addition of trastuzumab shows a benefit over the control groups in all sub-groups analyzed. Similar to the US-led trials, patients receiving trastuzumab after chemotherapy were 49% less likely to have a breast cancer recurrence. A trend toward overall survival improvement was observed (hazard ratio of 0.76), but this was not statistically significant. However, the distant disease-free survival (often used as a surrogate marker for overall survival as it represents progression to metastatic disease) was significantly improved in the trastuzumab arm, with a hazard ratio of 0.51 ($p < 0.001$) *(108)*.

The risk of cardiac toxicity was carefully followed in these trials. Pre-treatment and prospective cardiac evaluation with serial echocardiograms was employed. Conservative cardiac enrollment criteria, discontinuation, and dose adjustment parameters were applied as well. Despite careful surveillance, an excess of cardiac toxicity was observed in the treatment arms. Incidence of clinical congestive heart failure was approximately 3% with concurrent chemotherapy and trastuzumab (joint analysis of NSABP B31 and NCCTG N 9831) and a half percent with trastuzumab administered after chemotherapy (HERA) *(106–108)*. Early data suggest that this reduction in the left ventricular ejection fraction is reversible with cessation of trastuzumab use *(110)*.

5.5.6. NEO-ADJUVANT CLINICAL TRIALS

Neo-adjuvant trastuzumab-containing regimens demonstrate, at a minimum, similar to superior pathological complete remission rates as compared to historical controls of non-trastuzumab, anthracycline–taxane regimens (see Table 4). One randomized study demonstrates a significantly higher pathological complete response rate with a paclitaxel–epirubicin–trastuzumab-containing regimen when compared to chemotherapy alone (66 versus 25%). Prospective, serial cardiac monitoring in this trial shows that trastuzumab administered with concurrent epirubicin did not result in a higher number of cardiac events compared to the non-trastuzumab arm *(111)*. This poses the question as to whether it is the specific type of anthracycline used with trastuzumab versus an anthracycline class effect, responsible for the relatively high incidence of CD observed with the doxorubicin–trastuzumab combination.

5.6. Non-Cardiac Toxicity Profile

Trastuzumab is generally very well tolerated. Minor to moderate infusional, hypersensitivity reactions while seen in a number of initial infusions (up to 40%) are rare thereafter and readily managed with diphenhydramine, acetaminophen, and/or

Table 4

Trials of Neo-Adjuvant Trastuzumab + Chemotherapy in Locally Advanced Breast Cancer

Author	N	Patient stage	Neoadj CTX	Length neoadj TX (months)	Adj CTX	PCR (%)	OCRR (%)	CD G1–2 (%)	CD G3–4 (%)
Buzdar[a] (111)	42	II–IIIa	P q3[b]→FEC75 ± T	6	None	**65 (+T) versus 26 (−T)**	NR	30 (+T) versus 26 (−T)	0
Burstein (112)	40	II/III	P q3 + T	3	AC	18	75	10	0
Hurley (113)	34	II–IV	D q3 + CDDP + T + GCSF	3	None	26	100	38	6
Bines (114)	33	III	D qwk + T	3.5	NR	12	70	NR	NR
Coudert (115)	30	II/III	D q3wk + T	4.5	NR	47[c]	96	3	0
Limentani (116)	31	II–IV	D q2wk + V q2wk + T + GCSF	3	AC	39	94	NR	NR
Harris (117)	42	II/III	V + T	3	AC	19	87	5	0
Wenzel (118)	14	II/III	Ed + D qwk + T	1.5	NR	7	NR	NR	0

A, doxorubicin; C, cyclophosphamide; CD, cardiac dysfunction [Grades 0–4 by the CTCAE criteria (93)]; CDDP, cisplatin; CTX, chemotherapy; D, docetaxel; E, epirubicin; Ed, epidoxorubicin; F, fluorouracil; N, number of patients; NR, not reported; PCR, pathological complete response rate (no evidence of residual invasive cancer in breast and axilla following surgical intervention); OCRR, objective clinical response rate (clinical complete and partial responders); P, paclitaxel; T, trastuzumab; V, vinorelbine.

[a] Prospective, randomized, phase III study, median follow-up 20 months.
[b] Twenty-four-hour continuous intravenous infusion of paclitaxel 225 mg/m².
[c] Alternative classification of PCR.

p < 0.02 values are represented in bold.
TX = treatment
AC = adriamycin/cyclophosphamide
q3wk = every three weeks
GCSF = granulocyte stimulating factor
q2wk = every two weeks,
qwk = every week.

meperidine. Other side effects include myelosuppression, pain (especially at tumor site), diarrhea, and asthenia *(54,69,70)*. In addition, rare but severe adverse events such as anaphylactic reactions with pulmonary injury and nephrotoxicity have been reported *(113)*.

5.7. Cardiac Toxicity Profile

The pivotal trials of trastuzumab in the metastatic setting revealed CD manifested by a dilated cardiomyopathy and symptomatic congestive heart failure as an unexpected adverse event. As compared to trastuzumab monotherapy, more events occurred in patients receiving trastuzumab in combination with taxanes and anthracyclines (monotherapy, 3–7%; taxane combination, 13%; anthracycline combination, 26%) *(54,69,70)*. Subsequently, the Cardiac Review and Evaluation Committee (CREC) was established and retrospectively evaluated the incidence, risk factors, and clinical course of trastuzumab-associated CD *(114)*. In contrast to anthracycline CD, the majority of trastuzumab-associated cardiac events are symptomatic, reversible, and not dose dependent. Furthermore, in most cases, the CD is easily managed with medical therapy *(68,115)*. Age greater than 60 years and prior exposure to anthracyclines are confirmed risk factors; however, prior chest wall irradiation and pre-existing cardiac disease may also contribute to the development of symptomatic CD *(68,114,116)*.

The etiology of trastuzumab-associated CD remains elusive. Theories include direct effects on the HER-2 cardiac receptor (which is essential for prevention of dilated cardiomyopathy in experimental models), immunologic destruction of myocardial tissue, and synergism with anthracycline-induced myocardial disarray *(117–119)*.

5.8. Future Directions

Clinical trials have demonstrated an unequivocal benefit for the addition of trastuzumab in the metastatic, adjuvant, and neo-adjuvant settings. In general, with the important exception of cardiac toxicity, the treatment is well tolerated. Ongoing research areas include CD prevention strategies, prediction of response, and optimal treatment strategies (trastuzumab alone or combined with other agents) in all clinical settings.

6. MOLECULARLY TARGETED THERAPY OF BREAST CANCER—TARGET: VASCULAR ENDOTHELIAL GROWTH FACTOR

Neo-angiogenesis is a fundamental and necessary event in embryogenesis as well as malignant processes. Angiogenesis supports the proliferation, propagation, and metastasis of tumors. Therefore, disruption of this phenomenon has been targeted for therapeutic intervention. *(120,121)*. Multiple lines of evidence have established VEGF production as the critical, rate-limiting step of the complex pro-angiogenic process in solid tumors. Among its many functions, VEGF promotes growth, migration, and proliferation of endothelial cells in vitro. VEGF also augments vascular permeability, hence its original name of VPF or vascular permeability factor *(122–124)*. The increase in vascular permeability leads to extravasation of plasma proteins and subsequent alteration of the extracellular matrix that ultimately catalyzes new blood vessel formation. VEGF gene expression is induced by a number of factors including nitric oxide, reduced oxygen tension, circulating growth factors [interleukins 1 and 6, EGFR, TGF-α and TGF-β, insulin-like growth factor-1, and platelet-derived growth factor (PDGF)] and hormones (TSH and ACTH) *(125–131)*.

6.1. VEGF and Breast Cancer

VEGF is a highly conserved heparin-binding, homo-dimeric 45-kilodalton glyco-protein. Multiple isoforms of VEGF have been characterized. $VEGF_{165}$, the most abundant isoform, binds with high affinity to VEGF receptor-1 (VEGFR-1, Flt-1) and VEGFR-2 (KDR/Flk-1) *(120)*. Clinical studies demonstrate elevated serum VEGF levels in invasive breast cancers. Furthermore, $VEGF_{165}$ content inversely correlates with disease-free and overall survival in primary breast cancer. Thus, VEGF plays a pivotal role in breast cancer progression and carcinogenesis *(132–134)*.

6.2. Bevacizumab

Bevacizumab is a recombinant, humanized monoclonal antibody with specificity for VEGF-A. The majority of the amino acid sequence (93%) is derived from human immunoglobulin, and the remainder is murine in origin *(135)*. The antibody has a molecular mass of 149,000 Daltons and has a high affinity for all VEGF isoforms. Thus, upon binding, bevacizumab neutralizes VEGF-mediated neo-angiogenesis.

A phase I/II trial established single-agent activity of bevacizumab in advanced breast cancer. Patients received escalating doses (3–20 mg/kg) of drug intravenously every other week *(136)*. The response rate was just over 9%, with a median time to progression of 5.5 months (range, 2.3–13.7 months). Responses were durable. Twelve of the 75 patients (16%) had stable disease or an ongoing response at the day 154 evaluation. In this trial, the incidence of any hypertension was 22%, and the optimal dose of bevacizumab was 10 mg/kg every other week. As shown in Table 5, bevacizumab has single-agent activity in refractory breast cancer. Response rates double when this agent is combined with chemotherapy.

In patients with advanced breast cancer, bevacizumab has been combined with several chemotherapeutic and biological agents in a variety of clinical scenarios (see Table 5). Overall, the combinations are well tolerated. When combined with paclitaxel as first-line therapy for advanced breast cancer, bevacizumab doubles the time to treatment failure.

Preclinical data support the use of bevacizumab in combination with trastuzumab (a humanized, monoclonal antibody to HER-2). A series of experiments demonstrates a 4.5-fold increase in VEGF expression in breast cancer cell lines which overex-press HER-2 (MCF-7/HER-2) versus control breast cancer cells which do not (MCF-7/controls) *(137)*. Furthermore, HER-2 activation results in up-regulation of VEGF expression at the protein level. Based on this observation, it is hypothesized that up-regulation of VEGF in HER-2 overexpressing breast cancers contributes to the aggressive phenotype observed in these tumors *(138,139,140)*. Consequently, Pegram et al. initiated a phase I/II trial of a fixed dose of trastuzumab with dose escalation of bevacizumab in patients with HER-2 overexpressing/amplified, advanced breast cancer. In the first 9 patients, five responses (one complete and four partial) were observed. In addition, two patients had stable disease, while only two progressed. Overall, the combination was well tolerated and side effects were mild (diarrhea and

Table 5
Phase I and II Trials of Bevacizumab in Advanced Breast Cancer

Author	N	prior TX	Additional TX	Bev dose	RR (%)	TTP (months)
Cobleigh et al. (136)	75	>1	None	3–20 mg/kg	9.3	5.5
Ramaswamy and Shapiro (141)	21	≤1	Docetaxel[a]	10 mg/kg q2w	56	8
Rugo (142)	55	≤2	Vinorelbine[b]	10 mg/kg q2w	31	NR
Pegram et al. (138)	9	No limit	Trastuzumab	10 mg/kg[c]	57	Ongoing
Dickler et al. (143)	18	≤2	Erlotinib	15 mg/kg q3w	6	Ongoing

BEV, bevacizumab; N, number of patients; NR, not reported; RR, response rate; TTP, time to progression; TX, treatment.

[a]35 mg/m^2/week (days 1, 8, 15 every 28 days).

[b]25 mg/m^2/week.

[c]Dose escalation of bevacizumab for first six patients enrolled.

q2wk = every two weeks, q3wk = every three weeks.

fatigue). The phase I dose of bevacizumab is 10 mg/kg every 2 weeks, and the phase II component of this trial is ongoing (138). Taken together, bevacizumab has emerged as an important biologic in the treatment of breast cancer with its optimal utilization still to be determined.

Two phase III trials of chemotherapy with or without bevacizumab have been performed (see Table 6). In the first trial, 462 patients were randomly assigned to receive capecitabine (2500 mg/m^2) twice daily on days 1 through 14 every 3 weeks, alone or in combination with bevacizumab (15 mg/kg) on day 1 of an every 21-day cycle. The response rate favored the combination arm; however, this did not result in a longer progression-free (hazard ratio of 0.98) or overall (15.1 versus 14.5 months) survival (144). The second trial was performed in patients receiving first-line chemotherapy for metastatic breast cancer. Patients were randomized to either paclitaxel or paclitaxel plus bevacizumab every 2 weeks (10 mg/kg). The response rate was nearly double in the combination arm, and the time to progression was increased as well (hazard ratio of 0.5, $p < 0.01$). Clinically relevant Grade 3 and 4 side effects of paclitaxel versus paclitaxel and bevacizumab included hypertension (0 versus 1.6%), fatigue (2.7 versus 5%), and neurotoxicity (14.2 versus 20.5%). The incidence of thromboembolism was equivalent (1.2%). Clinical trials in the neo-adjuvant setting

Table 6
Phase III Trials of Chemotherapy ± Bevacizumab in Metastatic Breast Cancer

Author	N	CTX	Bev dose	RR (%)	TTP (months)
Miller et al. (144)	462	Capecitabine	15 mg/kg q3w	**20 versus 9**	5 versus 4
Miller (145)	715	Paclitaxel	10 mg/kg q2w	**28 versus 14**	**6 versus 11**

BEV, bevacizumab; CTX, chemotherapy; N, number of patients; RR, response rate; TTP, time to progression.

Statistically significant differences are represented in bold.

are ongoing. The combination appears to be safe and well tolerated with toxicity rates similar to those observed in other studies of advanced disease *(146)*.

7. MOLECULARLY TARGETED THERAPY OF BREAST CANCER—TARGET: HER-1 AND HER-2

ERB-B1 or HER-1 (EGFR) and ERB-B2 or HER-2 promote tumor growth and survival in a variety of tumor models including breast cancer *(147,148)*. These transmembrane growth factor receptors contain multiple tyrosine kinase phosphorylation sites which ultimately link to downstream cell proliferation and cell survival pathways including MAPK and PI3K*(149–154)*.

8. LAPATINIB (GW572016)

Lapatinib (GW572016) is a selective, oral, reversible small molecule inhibitor of both ERB-B1 and ERB-B2 tyrosine kinases (so-called dual kinase inhibitor). Downstream effects include blocking phosphorylation and activation of ERK 1/2 and AKT in both tumor cell lines and xenograft models *(155–158)*.

8.1. Phase I and II Single-Agent Clinical Trials

As shown in Table 7, the response rate of lapatinib ranges from 9 to 37%. As first-line treatment for HER-2 positive breast cancer (by FISH), the response rate is just over 37%, and a 33% stable disease rate is observed. Side effects are mild with no evidence of Grade 3 or 4 toxicity. Grade 1 and 2 events include pruritus (38%), diarrhea (25%), skin changes (<15%), and dyspepsia (10%) *(159)*. Overall, it is quite active with an acceptable side effect profile.

8.2. Phase I and II Clinical Trials in Combination with Trastuzumab

An in vitro model of HER-2 overexpressing/amplified breast cancer (MDA-MB 361 cell lines) suggests the combination of trastuzumab and lapatinib is synergistic *(160)*. A phase I clinical trial of lapatinib and trastuzumab in advanced breast cancer reveals

Table 7
Phase I and II Trials of Lapatinib in Advanced Breast Cancer

Author	Line TX	N	Lapatinib daily dose (mg)	RR (%)	SD (%)	Toxicity G3/4 $\geq 10\%$
Gomez et al. (159)	1st	20	1000	37	42	0
Gomez (159)	1st	20	1500	37	26.5	0
Blackwell et al. (162)	No limit	81	Varied	9	20	NR
Storniolo et al. (161)	No limit	40	1000 + T	30	32.5	Diarrhea, fatigue

N, number of patients; NR, not reported; RR, response rate; SD, stable disease; T, trastuzumab; TX, treatment.

a response rate of 15%. The recommended phase II dose of lapatinib is 1000 mg daily. Grade 3 adverse events include diarrhea and fatigue *(161)*.

9. MOLECULARLY TARGETED THERAPY OF BREAST CANCER—TARGET: FARNESYL TRANSFERASE

Ras proteins are guanine nucleotide–binding proteins that play pivotal roles in the control of normal and transformed cell growth. After stimulation by various growth factors and cytokines, Ras activates a number of downstream effectors and initiates proliferation, cell survival, and stress kinase pathway signaling *(163,164)*. Although Ras is not usually mutated in breast cancer, signaling through this protien is common. Ras undergoes several post-translational modifications, which facilitate its attachment to the inner surface of the plasma membrane. The first and most critical of these modifications is acquisition of a farnesyl-isoprenoid moiety in a reaction catalyzed by the enzyme farnesyl transferase. Therefore, inhibition of this enzyme prevents Ras from maturing into its biologically active form.

9.1. Tipifarnib

Tipifarnib (R115777) is an oral, non-peptidomimetic farnesyl transferase inhibitor *(165)*. Preclinical data demonstrate activity against human tumor cell lines in vitro *(166, 167)*. A phase I trial of tipifarnib given for 5 consecutive days every 2 weeks identified a maximum tolerated dose of 500 mg twice daily. Nausea, vomiting, headache, fatigue, anemia, and hypotension were observed dose-limiting events *(168)*.

Tipifarnib has single-agent activity as demonstrated by a phase II trial in advanced breast cancer. Johnston et al. reported the results of a phase II trial of tipifarnib in 76 patients with metastatic breast cancer who failed second-line endocrine therapy or had ER-negative disease. Response rates ranged from 10 to 15%, and an additional 9–15% of patients sustained stable disease lasting at least 6 months depending on schedule of administration (continuous versus intermittent). Intermittent dosing was better tolerated with lower rates of myelosuppression and peripheral neuropathy *(169)*.

9.2. Phase I and II Clinical Trials in Combination with Chemotherapy

Tipifarnib has been combined with multiple chemotherapy agents including docetaxel, capecitabine, and doxorubicin/cyclophosphamide *(170–172)*. In a phase I/II trial conducted by the Montefiore Phase II consortium, tipifarnib was safely administered with standard dose doxorubicin and cyclophosphamide administered every 2 weeks in conjunction with growth factor support. In this trial, tipifarnib was administered concurrently with chemotherapy twice a day (200 mg) for 1 week. For patients who received therapy in the neo-adjuvant portion of the study, the pathologic complete response rate observed was quite high (33%) *(172)*. Preclinical data with BMS 214662 and SCH 66336 also suggest activity in breast cancer, although clinical trial results are not yet available *(173,174)*.

10. MOLECULARLY TARGETED THERAPY OF BREAST CANCER—TARGET: EPIDERMAL GROWTH FACTOR RECEPTOR

The EGFR plays a major role in the proliferation and malignant growth of breast cancer cells in both preclinical models and clinical scenarios *(175–178)*. A variety of agents have been designed to target EGFR. These include antibodies (cetixumab and EMD 72000) and tyrosine kinase inhibitors (TKIs, i.e., gefitinib and erlotinib). As shown in Table 8, while stable disease is observed in some monotherapy trials, the response rate is quite low. However, when combined with other agents, such as anastrazole, the response rate appears to increase substantially. This highlights the challenge with many agents in this class.

11. MOLECULARLY TARGETED THERAPY OF BREAST CANCER—TARGET: MAMMALIAN TARGET OF RAPAMYCIN

The mammalian target of rapamycin (mTOR) is a downstream effector of the PI3K and AKT (protein kinase B) signaling pathway *(185)*. Because this pathway also mediates cell survival and cell proliferation, it is a logical target for oncologic drug development. By targeting mTOR, there is inhibition of the signals required for cell cycle progression, cell growth, and cell proliferation. Highly specific clinical inhibitors of mTOR such as rapamycin and rapamycin analogs (CCI-779, RAD 001, and AP23573) are in clinical development.

11.1. Temsirolimus (CCI-779)

Temsirolimus produced a response rate of approximately 9% (all partial responders) as a single agent in heavily pretreated breast cancer. In that trial, 109 patients were randomized to receive temsirolimus weekly (75 or 250 mg intravenous infusion) *(186)*. Median time to tumor progression was 3 months. Efficacy was similar for both dose levels but toxicity was more common with the higher doses, especially Grade 3/4 depression (10% of patients at the 250-mg dose level versus 0% at the 75-mg dose level). In addition, a relatively high incidence of Grade 3/4 mucositis was observed (9%).

11.2. Everolimus (RAD-001)

Everolimus is an orally bioavailable inhibitor of rapamycin. Clinical trials are underway with this drug combined with an aromatase inhibitor *(187)*.

11.3. AP23573

AP23573 is an intravenously administered agent of this subclass. A phase I study of this drug in patients with advanced cancer was presented at the Annual meeting of the American Society of Clinical Oncology. Data are maturing, thus assessment of its activity in breast cancer is premature *(188)*.

Table 8
Clinical Trials of EGFR Inhibitors in Advanced Breast Cancer

Author	Line TX	TX	N	Daily dose (mg)	RR (%)	SD (%)	TTP (days)	Toxicity G 3/4 =10%
Baselga et al. (179)	1st or 2nd	G	31	500	0	39	55	Diarrhea, rash
von Minckwitz et al. (180)	1st	G	57	500	1.7	0	61	None
Albain et al. (181)	No limit	G	63	500	4.8	9.5	57	Diarrhea, nausea, rash
Robertson et al. (182)	No limit	G	22	500	9	46	60	Diarrhea, nausea, rash
Tan et al. (183)	No limit	E	18	150	0	0	29	Diarrhea
Polychronis et al. (184)	Neoadj	G + A	27	250	50	NR	NA	None
Polychronis et al. (184)	Neoadj	G	29	250	55	NR	NA	None

A, anastrazole; E, erlotinib; EGFR, epidermal growth factor receptor; G, gefitinib; N, number of patients; NA, not applicable; Neoadj, neo-adjuvant therapy; NR, not reported; RR, response rate; SD, stable disease; TTP, time to progression; TX, treatment.

12. MOLECULARLY TARGETED THERAPY OF BREAST CANCER—TARGET: RAF-1-KINASE

The RAS/RAF signal transduction pathway is an important mediator of tumor cell proliferation. It regulates several pathways that synergistically induce cellular transformation *(189)*. RAF kinase is a second-messenger, serine/threonine kinase that functions as a downstream effector molecule of RAS. The RAF kinase family is composed of three members, A-RAF, B-RAF, and RAF-1 (also known as C-RAF), which are variably mutated in a variety of human cancers including breast cancer *(190)*.

Sorafenib (formerly BAY 43-9006) is an orally bioavailable bi-aryl urea that is a potent inhibitor of RAF-1 as well as the wildtype and mutant forms of B-RAF *(191)*. In vitro, sorafenib reduces MEK and ERK phosphorylation without directly affecting kinase activity *(191,192)*. In addition, sorafenib also demonstrates significant inhibition of receptor tyrosine kinases involved in neo-angiogenesis including VEGFR-2 and VEGFR3 and PDGF receptor-β in both tumor cell lines and xenograft models *(192,193)*.

A phase I trial of sorafenib was conducted in 69 patients with refractory, advanced cancer, four (6%) of whom had breast cancer *(194)*. Stable disease for at least 6 months was observed in 18% of the patients. Diarrhea was the most frequent adverse event (55%, predominantly Grades 1 and 2, followed by rash (26%) and hand-foot syndrome (23%). A Phase II Trial by the North Central Cancer Treatment Group (NCCTG) is underway.

13. MISCELLANEOUS TARGETED AGENTS

A number of agents have multiple distinct targets. For example, JNJ-17029259 (VEGF-R/PDGF TKI), Bay 57-9352, BMS 582664 (VEGFR1/FGF1), and BMS 354825 (src/abl kinase inhibitor) are currently in preclinical or phase I clinical trials for advanced cancers. Whether there is activity in breast cancer is yet to be determined. SU11248 is a very interesting molecule, and activity in heavily pre-treated breast cancer has been observed *(194)*. Matrix metalloproteinase inhibitors have been

Table 9
Clinical Trials of Novel Targeted Agents in Breast Cancer

Agent	Class	N	Efficacy			Toxicity
			RR	SD	POD	
BMS-27521 *(195)*	MMPI[a]	72	NA			Arthralgia (35%)
Marimastat *(196)*	MMPI[a]	63	NA			Arthralgia
ZD6474 *(197)*	VEGFR-R2	44	0	1	43	Diarrhea (dose related)
SU11248 *(194)*	TKI[b]	24	4	5	14	Diarrhea (32% any grade)

MMPI, matrix metalloproteinase inhibitor; N, number of patients; NA, not applicable; POD, progression of disease; RR, response rate; SD, stable disease; TKI, tyrosine kinase inhibitor; VEGFR-R2, vascular endothelial growth factor receptor 2 inhibitor.

[a] adjuvant setting.

[b] TIK of VEGF, flt-3, PDGF and c-kit.

evaluated in breast cancer patients in the adjuvant setting and also as a "maintenance therapy" after the attainment of a complete response in the advanced disease setting. Toxicity has precluded further development of these compounds (see Table 9).

14. PREDICTION OF RESPONSE BASED ON ASSESSMENT OF THE BIOLOGIC TARGET

A plethora of anti-tumor, targeted agents are now available. The breast cancer clinical trials' experience with trastuzumab and hormonal therapies highlights the need for accurate identification of appropriate and practical biomarkers in order to predict response to targeted, biological therapeutics *(23,66,69–71,201–202)*. While many candidate biomarkers exist, none predict response to therapy with 100% accuracy *(162,172,179)*. This suggests that redundant pathways for cell growth and survival must be targeted in order to eradicate disease. This is an area of ongoing and active research.

15. CONCLUSION

Undoubtedly the trend is to molecularly define the patient in order to prescribe a molecularly defined therapy. The median survival for advanced breast cancer has almost doubled over the past 20 years, and the addition of trastuzumab has improved survival for HER-2 overexpressing breast cancer. The future probably includes not only targeting the tumor, but also targeting the surrounding stromal tissues and other components of the tumor vasculature that allow tumors to grow and metastasize. Elucidating the tropism of individual tumors will allow the administration an individualized "cocktail" to treat patients with breast cancer and ultimately to prevent disease recurrence.

REFERENCES

1. Jemal A, Murray T, Ward E, et al. Cancer statistics, 2005. CA Cancer J Clin 2005; 55:10–30.
2. Cancer IUA. Global Action Against Cancer: Cancer Statistics. http://www.iucc.org, 2005.
3. Sorlie T, Perou CM, Tibshirani R, et al. Gene expression patterns of breast carcinomas distinguish tumor subclasses with clinical implications. Proc Natl Acad Sci USA 2001; 98:10869–74.
4. Sorlie T, Tibshirani R, Parker J, et al. Repeated observation of breast tumor subtypes in independent gene expression data sets. Proc Natl Acad Sci USA 2003; 100:8418–23.
5. Perou CM, Sorlie T, Eisen MB, et al. Molecular portraits of human breast tumours. Nature 2000; 406:747–52.
6. van de Vijver MJ, He YD, van't Veer LJ, et al. A gene-expression signature as a predictor of survival in breast cancer. N Engl J Med 2002; 347:1999–2009.
7. van't Veer LJ, Dai H, van de Vijver MJ, et al. Gene expression profiling predicts clinical outcome of breast cancer. Nature 2002; 415:530–6.
8. Perou CM, Jeffrey SS, van de Rijn M, et al. Distinctive gene expression patterns in human mammary epithelial cells and breast cancers. Proc Natl Acad Sci USA 1999; 96:9212–7.
9. Carey LA, Perou CM, Dressler LG, et al. Race and poor prognosis basal breast tumor (BBT) phenotype in the population-based Caroline Breast Cancer Study (CBCS). Proc Am Soc Clin Oncol 2004; 23:833: Abstract 9510.
10. Rouzier R, Anderson K, Hess KR, et al. Basal and luminal types of breast cancer defined by gene expression patterns respond differenty to neoadjuvant chemotherapy. Breast Cancer Res Treat 2004; 88:S24: Abstract 201.

11. Sartor CI, Zhou H, Perou CM, Ethier SP. Basal-like breast tumor-derrived cell lines are growth inhibited and radiosensitized by epidermal growth factor receptor (EGFR) tyrosine kinase inhibitors. Breast Cancer Res Treat 2004; 88:S34: Abstract 311.
12. Nielsen TO, Hsu FD, Jensen K, et al. Immunohistochemical and clinical characterization of the basal-like subtype of invasive breast carcinoma. Clin Cancer Res 2004; 10:5367–74.
13. Early Breast Cancer Trialists' Collaborative Group (EBCTCG). Effects of chemotherapy and hormonal therapy for early breast cancer on recurrence and 15-year survival: an overview of the randomised trials. Lancet 2005; 365:1687–717.
14. Golub TR, Slonim DK, Tamayo P, et al. Molecular classification of cancer: class discovery and class prediction by gene expression monitoring. Science 1999; 286:531–7.
15. Davis RE, Staudt LM. Molecular diagnosis of lymphoid malignancies by gene expression profiling. Curr Opin Hematol 2002; 9:333–8.
16. Paik S, Shak S, Tang G, et al. A multigene assay to predict recurrence of tamoxifen-treated, node-negative breast cancer. N Engl J Med 2004; 351:2817–26.
17. Paik S, Shak S, Tang G, et al. Multi-gene RT-PCR assay for predicting recurrence in node negative breast cancer patients - NSABP studies B-20 and B-14. Breast Cancer Res Treat 2003; 82:S10: Abstract 16.
18. Esteban J, Baker J, Cronin M, et al. Tumor gene expression and prognosis in breast cancer: multi-gene RT-PCR assay of paraffin-embedded tissue. Proc Am Soc Clin Oncol 2003; 22:850: Abstract 3416.
19. Cobleigh MA, Bitterman P, Baker J, et al. Tumor gene expression predicts distant disease-free survival (DDFS) in breast cancer patients with 10 or more positive nodes: high throughout RT-PCR assay of paraffin-embedded tumor tissues. Proc Am Soc Clin Oncol 2003; 22:850: Abstract 3415.
20. Laskin JJ, Sandler AB. Epidermal growth factor receptor: a promising target in solid tumours. Cancer Treat Rev 2004; 30:1–17.
21. Riese DJ, II, Stern DF. Specificity within the EGF family/ErbB receptor family signaling network. Bioessays 1998; 20:41–8.
22. Harris M. Monoclonal antibodies as therapeutic agents for cancer. Lancet Oncol 2004; 5:292–302.
23. Emens LA, Davidson NE. Trastuzumab in breast cancer. Oncology (Huntingt) 2004; 18:1117–28; discussion 1131–2, 1137–8.
24. Karunagaran D, Tzahar E, Beerli RR, et al. ErbB-2 is a common auxiliary subunit of NDF and EGF receptors: implications for breast cancer. EMBO J 1996; 15:254–64.
25. Graves TA. Trastuzumab. Macintosh-PhotoShop, 2005.
26. Slamon DJ, Clark GM, Wong SG, Levin WJ, Ullrich A, McGuire WL. Human breast cancer: correlation of relapse and survival with amplification of the HER-2/neu oncogene. Science 1987; 235:177–82.
27. Volpi A, Nanni O, De Paola F, et al. HER-2 expression and cell proliferation: prognostic markers in patients with node-negative breast cancer. J Clin Oncol 2003; 21:2708–12.
28. Ross JS, Fletcher JA. The HER-2/neu oncogene in breast cancer: prognostic factor, predictive factor, and target for therapy. Stem Cells 1998; 16:413–28.
29. Hudziak RM, Schlessinger J, Ullrich A. Increased expression of the putative growth factor receptor p185HER2 causes transformation and tumorigenesis of NIH 3T3 cells. Proc Natl Acad Sci USA 1987; 84:7159–63.
30. Chazin VR, Kaleko M, Miller AD, Slamon DJ. Transformation mediated by the human HER-2 gene independent of the epidermal growth factor receptor. Oncogene 1992; 7:1859–66.
31. Fendly BM, Winget M, Hudziak RM, Lipari MT, Napier MA, Ullrich A. Characterization of murine monoclonal antibodies reactive to either the human epidermal growth factor receptor or HER2/neu gene product. Cancer Res 1990; 50:1550–8.
32. Hudziak RM, Lewis GD, Winget M, Fendly BM, Shepard HM, Ullrich A. p185HER2 monoclonal antibody has antiproliferative effects in vitro and sensitizes human breast tumor cells to tumor necrosis factor. Mol Cell Biol 1989; 9:1165–72.
33. Carter P, Presta L, Gorman CM, et al. Humanization of an anti-p185HER2 antibody for human cancer therapy. Proc Natl Acad Sci USA 1992; 89:4285–9.
34. Sliwkowski MX, Lofgren JA, Lewis GD, Hotaling TE, Fendly BM, Fox JA. Nonclinical studies addressing the mechanism of action of trastuzumab (Herceptin). Semin Oncol 1999; 26:60–70.
35. Nahta R, Esteva FJ. HER-2-targeted therapy: lessons learned and future directions. Clin Cancer Res 2003; 9:5078–84.

36. Baselga J, Albanell J, Molina MA, Arribas J. Mechanism of action of trastuzumab and scientific update. Semin Oncol 2001; 28:4–11.

37. Arteaga CL. Trastuzumab, an appropriate first-line single-agent therapy for HER2-overexpressing metastatic breast cancer. Breast Cancer Res 2003; 5:96–100.

38. Yakes FM, Chinratanalab W, Ritter CA, King W, Seelig S, Arteaga CL. Herceptin-induced inhibition of phosphatidylinositol-3 kinase and Akt1 is required for antibody-mediated effects on p27, cyclin D1, and antitumor action. Cancer Res 2002; 62:4132–41.

39. Gennari R, Menard S, Fagnoni F, et al. Pilot study of the mechanism of action of preoperative trastuzumab in patients with primary operable breast tumors overexpressing HER2. Clin Cancer Res 2004; 10:5650–5.

40. zum Buschenfelde CM, Hermann C, Schmidt B, Peschel C, Bernhard H. Antihuman epidermal growth factor receptor 2 (HER2) monoclonal antibody trastuzumab enhances cytolytic activity of class I-restricted HER2-specific T lymphocytes against HER2-overexpressing tumor cells. Cancer Res 2002; 62:2244–7.

41. Clynes RA, Towers TL, Presta LG, Ravetch JV. Inhibitory Fc receptors modulate in vivo cytoxicity against tumor targets. Nat Med 2000; 6:443–6.

42. Kumar R, Yarmand-Bagheri R. The role of HER2 in angiogenesis. Semin Oncol 2001; 28:27–32.

43. Izumi Y, Xu L, di Tomaso E, Fukumura D, Jain RK. Tumour biology: herceptin acts as an anti-angiogenic cocktail. Nature 2002; 416:279–80.

44. Mayfield S, Vaughn JP, Kute TE. DNA strand breaks and cell cycle perturbation in herceptin treated breast cancer cell lines. Breast Cancer Res Treat 2001; 70:123–9.

45. Pietras RJ, Poen JC, Gallardo D, Wongvipat PN, Lee HJ, Slamon DJ. Monoclonal antibody to HER-2/neureceptor modulates repair of radiation-induced DNA damage and enhances radiosensitivity of human breast cancer cells overexpressing this oncogene. Cancer Res 1999; 59:1347–55.

46. Nagata Y, Lan KH, Zhou X, et al. PTEN activation contributes to tumor inhibition by trastuzumab, and loss of PTEN predicts trastuzumab resistance in patients. Cancer Cell 2004; 6:117–27.

47. Pandolfi PP. Breast cancer – loss of PTEN predicts resistance to treatment. N Engl J Med 2004; 351:2337–8.

48. Barnes DM, Bartkova J, Camplejohn RS, Gullick WJ, Smith PJ, Millis RR. Overexpression of the c-erbB-2 oncoprotein: why does this occur more frequently in ductal carcinoma in situ than in invasive mammary carcinoma and is this of prognostic significance? Eur J Cancer 1992; 28:644–8.

49. Mohsin SK, Weiss HL, Gutierrez MC, et al. Neoadjuvant trastuzumab induces apoptosis in primary breast cancers. J Clin Oncol 2005; 23:2460–8.

50. Pegram MD, Konecny GE, O'Callaghan C, Beryt M, Pietras R, Slamon DJ. Rational combinations of trastuzumab with chemotherapeutic drugs used in the treatment of breast cancer. J Natl Cancer Inst 2004; 96:739–49.

51. Slamon D, Pegram M. Rationale for trastuzumab (Herceptin) in adjuvant breast cancer trials. Semin Oncol 2001; 28:13–9.

52. Pegram M, Hsu S, Lewis G, et al. Inhibitory effects of combinations of HER-2/neu antibody and chemotherapeutic agents used for treatment of human breast cancers. Oncogene 1999; 18:2241–51.

53. Baselga J, Tripathy D, Mendelsohn J, et al. Phase II study of weekly intravenous recombinant humanized anti-p185HER2 monoclonal antibody in patients with HER2/neu-overexpressing metastatic breast cancer. J Clin Oncol 1996; 14:737–44.

54. Slamon DJ, Leyland-Jones B, Shak S, et al. Use of chemotherapy plus a monoclonal antibody against HER2 for metastatic breast cancer that overexpresses HER2. N Engl J Med 2001; 344:783–92.

55. Harris KA, Carla B Washington GL, Jian-feng Lu, Robert Mass, Rene Bruno. A population pharmacokinetic (PK) model for trastuzumab (Herceptin) and implications for clinical dosing. Proc Am Soc Clin Oncol 2002; 21:123: Abstract 488.

56. Leyland-Jones B, Gelmon K, Ayoub JP, et al. Pharmacokinetics, safety, and efficacy of trastuzumab administered every three weeks in combination with paclitaxel. J Clin Oncol 2003; 21:3965–71.

57. Baselga J, Carbonell X, Castaneda-Soto NJ, et al. Phase II study of efficacy, safety, and pharmacokinetics of trastuzumab monotherapy administered on a 3-weekly schedule. J Clin Oncol 2005; 23:2162–71.

58. Stewart J, Fehrenbacher L, Blanchard R, Rodriguez G, Vogel C, Anavekar P. Phase II trial of trastuzumab and paclitaxel or docetaxel administered every 3 weeks to patients receiving a first treatment for HER2+ metastatic breast cancer. Proc Am Soc Clin Oncol 2004; 22:78: Abstract 806.

59. NCCN. Breast Cancer. Clinical Practice Guidelines in Oncology: Version 2005.

60. Bast RC, Jr., Ravdin P, Hayes DF, et al. 2000 update of recommendations for the use of tumor markers in breast and colorectal cancer: clinical practice guidelines of the American Society of Clinical Oncology. J Clin Oncol 2001; 19:1865–78.

61. O'Malley F, Thomson T, Julian J, et al. HER2 status: A Canadian experience of concordance between central and local testing laboratories. Breast Cancer Res Treat 2003; 82:S70: Abstract 305.

62. Anderson S, Reddy JC, Rai S, Lieberman G, Klein P. Concordance between central and local lab IHC and FISH HER2 testing in a community-based trial of first line trastuzumab plus a taxane in HER2+ metastatic breast cancer (MBC). Proc Am Soc Clin Oncol 2004; 23:851: Abstract 9580.

63. Tubbs RR, Pettay JD, Roche PC, Stoler MH, Jenkins RB, Grogan TM. Discrepancies in clinical laboratory testing of eligibility for trastuzumab therapy: apparent immunohistochemical false-positives do not get the message. J Clin Oncol 2001; 19:2714–21.

64. Bilous M, Dowsett M, Hanna W, et al. Current perspectives on HER2 testing: a review of national testing guidelines. Mod Pathol 2003; 16:173–82.

65. Zarbo RJ, Hammond ME. Conference summary, Strategic Science Symposium. Her-2/neu testing of breast cancer patients in clinical practice. Arch Pathol Lab Med 2003; 127:549–53.

66. Paik S, Bryant J, Tan-Chiu E, et al. Real-world performance of HER2 testing – National Surgical Adjuvant Breast and Bowel Project experience. J Natl Cancer Inst 2002; 94:852–4.

67. Roche PC, Suman VJ, Jenkins RB, et al. Concordance between local and central laboratory HER2 testing in the breast intergroup trial N9831. J Natl Cancer Inst 2002; 94:855–7.

68. Perez EA, Suman VJ, Davidson NE, et al. Effect of doxorubicin plus cyclophosphamide on left ventricular ejection fraction in patients with breast cancer in the North Central Cancer Treatment Group N9831 Intergroup Adjuvant Trial. J Clin Oncol 2004; 22:3700–4.

69. Vogel CL, Cobleigh MA, Tripathy D, et al. Efficacy and safety of trastuzumab as a single agent in first-line treatment of HER2-overexpressing metastatic breast cancer. J Clin Oncol 2002; 20:719–26.

70. Cobleigh MA, Vogel CL, Tripathy D, et al. Multinational study of the efficacy and safety of humanized anti-HER2 monoclonal antibody in women who have HER2-overexpressing metastatic breast cancer that has progressed after chemotherapy for metastatic disease. J Clin Oncol 1999; 17:2639–48.

71. Mass RD, Press M, Anderson S, Murphy M, Slamon D. Improved survival benefit from herceptin (trastuzumab) in patients selected by fluorescence in situ hybridization (FISH). Proc Am Soc Clin Oncol 2001; 20:22: Abstract 85.

72. Lueftner DI, Schaller G, Henschke P, et al. Complex formation between the shed antigen of HER-2/neu and the monoclonal anti-HER-2/neu antibody trastuzumab. Breast Cancer Res Treat 2003; 82:S49: Abstract 222.

73. Ghahramani P, Baselga J, Leyland-Jones B, Gianni L, Gatzemeier U. Meta-analysis of the relationship between serum trastuzumab concentrations and shed HER2 extracellular domain (ECD). Breast Cancer Res Treat 2002; 76:S31: Abstract 10.

74. Ross JS, Fletcher JA, Linette GP, et al. The Her-2/neu gene and protein in breast cancer 2003: biomarker and target of therapy. Oncologist 2003; 8:307–25.

75. Seidman AD, Broadwater G, Carney W, et al. Serum HER2 extracellular domain (ECD) levels and efficacy of weekly (W) or every 3-weekly (q3W) paclitaxel (P) with or without trastuzumab (T) in patients (pts) with metastatic breast cancer (MBC): CALGB 150002/9840. Proc Am Soc Clin Oncol 2005; 23:18s: Abstract 558.

76. Kallab V, Benz CC, Kirpotin D, Marks JD, Park JW. HER2/EGFR internalization: a novel biomarker for ErbB-targeted theraputics. Breast Cancer Res Treat 2004; 88:126–7: Abstract 3044.

77. Luftner DI, Possinger K, Henschke P, et al. Longitudinal HER-2/neu measurements during treatment with herceptin, epirubicin plus cyclophosphamide (HEC): interim serum results of a phase ii study in patients with metastatic breast cancer. Breast Cancer Res Treat 2002; 76:S109: Abstract 423.

78. Argiris A, Wang CX, Whalen SG, DiGiovanna MP. Synergistic interactions between tamoxifen and trastuzumab (Herceptin). Clin Cancer Res 2004; 10:1409–20.

79. Raymond E, Faivre S, Armand JP. Epidermal growth factor receptor tyrosine kinase as a target for anticancer therapy. Drugs 2000; 60 Suppl 1:15–23; discussion 41–2.

80. Seidman AD, Fornier MN, Esteva FJ, et al. Weekly trastuzumab and paclitaxel therapy for metastatic breast cancer with analysis of efficacy by HER2 immunophenotype and gene amplification. J Clin Oncol 2001; 19:2587–95.

81. John M, Kriebel-Schmitt R, Stauch M, Wolf H, Mohr B, Klare P, Hinke R, Schlosser J. Weekly paclitaxel plus trastuzumab shows promising efficacy in advanced breast cancer. Breast Cancer Res Treat 2003; 82:S49: Abstract 221.

82. Tedesco KL, Thor AD, Johnson DH, et al. Docetaxel combined with trastuzumab is an active regimen in HER-2 3+ overexpressing and fluorescent in situ hybridization-positive metastatic breast cancer: a multi-institutional phase II trial. J Clin Oncol 2004; 22:1071–7.

83. Esteva FJ, Valero V, Booser D, et al. Phase II study of weekly docetaxel and trastuzumab for patients with HER-2-overexpressing metastatic breast cancer. J Clin Oncol 2002; 20:1800–8.

84. Burstein HJ, Kuter I, Campos SM, et al. Clinical activity of trastuzumab and vinorelbine in women with HER2-overexpressing metastatic breast cancer. J Clin Oncol 2001; 19:2722–30.

85. Burstein HJ, Harris LN, Marcom PK, et al. Trastuzumab and vinorelbine as first-line therapy for HER2-overexpressing metastatic breast cancer: multicenter phase II trial with clinical outcomes, analysis of serum tumor markers as predictive factors, and cardiac surveillance algorithm. J Clin Oncol 2003; 21:2889–95.

86. Jahanzeb M, Mortimer JE, Yunus F, et al. Phase II trial of weekly vinorelbine and trastuzumab as first-line therapy in patients with HER2(+) metastatic breast cancer. Oncologist 2002; 7:410–7.

87. O'Shaughnessy J. Gemcitabine and trastuzumab in metastatic breast cancer. Semin Oncol 2003; 30:22–6.

88. O'Shaughnessy JA, Vukelja S, Marsland T, Kimmel G, Ratnam S, Pippen JE. Phase II study of trastuzumab plus gemcitabine in chemotherapy-pretreated patients with metastatic breast cancer. Clin Breast Cancer 2004; 5:142–7.

89. Xu L, Song S, Zhu J, et al. Capecitabine (Xeloda) combined with trastuzumab (Herceptin) as first-line therapy in patients with HER2-overexpressing metastatic breast cancer (MBC): an interim analysis. Breast Cancer Res Treat 2004; 88:S128: Abstract 3049.

90. Sledge GW, Jr. Gemcitabine combined with paclitaxel or paclitaxel/trastuzumab in metastatic breast cancer. Semin Oncol 2003; 30:19–21.

91. Perez EA, Rowland KM, Suman VJ, et al. N98–32–52: efficacy and tolerability of two schedules of paclitaxel, carboplatin and trastuzumab in women with HER2 positive metastatic breast cancer: a North Central Cancer Treatment Group Randomized phase II trial. Breast Cancer Res Treat 2003; 82:S47:Abstract 216.

92. Pegram MD, Pienkowski T, Northfelt DW, et al. Results of two open-label, multicenter phase II studies of docetaxel, platinum salts, and trastuzumab in HER2-positive advanced breast cancer. J Natl Cancer Inst 2004; 96:759–69.

93. Common Terminology Criteria for Adverse Events, Version 3.0, DCTD, NCI, NIH, DHHS. 2003: 7–10.

94. Pegram MD, Lipton A, Hayes DF, et al. Phase II study of receptor-enhanced chemosensitivity using recombinant humanized anti-p185HER2/neu monoclonal antibody plus cisplatin in patients with HER2/neu-overexpressing metastatic breast cancer refractory to chemotherapy treatment. J Clin Oncol 1998; 16:2659–71.

95. Piccart M, Lohrisch C, Di Leo A, Larsimont D. The predictive value of HER2 in breast cancer. Oncology 2001; 61 Suppl 2:73–82.

96. Yamauchi H, Stearns V, Hayes DF. When is a tumor marker ready for prime time? A case study of c-erbB-2 as a predictive factor in breast cancer. J Clin Oncol 2001; 19:2334–56.

97. Theodoulou M, Hudis C. Cardiac profiles of liposomal anthracyclines: greater cardiac safety versus conventional doxorubicin? Cancer 2004; 100:2052–63.

98. Theodoulou M, Campos SM, Batist G, Winer E, Norton L, Hudis C, Welles L. TLC D99 (D, Myocet) and Herceptin (H) is safe in advanced breast cancer (ABC): final cardiac safety and efficacy analysis. Proc Am Soc Clin Oncol 2002; 21:55: Abstract 216.

99. Wolff AC, Bonetti M, Sparano JA, Wang M, Davidson NE. Cardiac safety of trastuzumab (H) in combination with pegylated liposomal doxorubicin (D) and docetaxel (T) in HER2-positive metastatic breast cancer (MBC): preliminary results of the Eastern Cooperative Oncology Group trial E3198. Proc Am Soc Clin Oncol 2003; 22:18: Abstract 70.

100. Trigo J, Climent MA, Lluch A, Gascon P, Hornedo J, Gil M, Cirera L, Guillem V, Regueiro P, Baselga J. Liposomal doxorubicin Myocet in combination with Herceptin and paclitaxel is active and well tolerated in patients with HER2-positive locally advanced or metastatic breast cancer: a phase II study. Breast Cancer Res Treat 2003; 82:S83: Abstract 351.

101. Cortes J, Climent M, Lluch A, et al. Updated results of a phase II study (M77035) of Myocet combined with weekly Herceptin and paclitaxel in patients with HER2-positive locally advanced or metastatic breast cancer (LABC/MBC). Breast Cancer Res Treat 2004; 88:S125–6: Abstract 3041.

102. Thomssen CH, Eidtmann H, Untch M, et al. Cardiac saftey of epirubicin/cyclophosphamide (EC) alone and in combination with herceptin in women with metastatic breast cancer (MBC). Breast Cancer Res Treat 2002; 76:S111: Abstract 430.

103. Untch M, Eidtmann H, du Bois A, et al. Cardiac safety of trastuzumab in combination with epirubicin and cyclophosphamide in women with metastatic breast cancer: results of a phase I trial. Eur J Cancer 2004; 40:988–97.

104. Marty M, Cognetti F, Maraninchi D, et al. Efficacy and safety of trastuzumab combined with docetaxel in patients with human epidermal growth factor receptor 2-positive metastatic breast cancer administered as first-line treatment: results of a randomized phase II trial of the efficacy and sapiey of trastuzumab comsined with docetaxel in patients with HER2 positive metastatic breast cancer administration as first line treatment: the M77001 study group. J Clin Oncol 2005;19:19: 7/1/2005:4265–4274.

105. Gasparini G, Morabito A, Sio LD, et al. Preliminary clinical results of a randomized phase IIb study of weekly paclitaxel (PCT) +/– trastuzumab (T) as a first line therapy of patients (pts) with HER-2/neu positive metastatic breast cancer (MBC). Breast Cancer Res Treat 2003; 82:S51: Abstract 227.

106. Romond E. Joint Analysis of NSABP-B-31 and NCCTG-N9831. American Society of Clinical Oncololgy Annual Meeting, Orlando, FL, 2005.

107. Perez E. Further analysis of NCCTG-N9831. American Society of Clinical Oncology Annual Meeting, Orlando, FL, 2005.

108. Piccart-Gebhart M. HERA Trial. American Society of Clinical Oncology Annual Meeting, Orlando, FL, 2005.

109. Hunt SA; American College of Cardiology; American Heart Association Task Force on Practice Guidelines (Writing Committee to Update the 2001 Guidelines for the Evaluation and Management of Heart Failure). ACC/AHA 2005 guideline update for the diagnosis and management of chronic heart failure in the adult: a report of the American College of Cardiology/American Heart Association Task Force on Practice Guidelines (Writing Committee to Update the 2001 Guidelines for the Evaluation and Management of Heart Failure). J Am Coll Cardiol. 2005 Sep 20;46(6):e1–82.

110. Perez EA, Suman VJ, Davidson NE, et al. Interim cardiac safety analysis of NCCTG N9831 Intergroup adjuvant trastuzumab trial. Proc Am Soc Clin Oncol 2005; 23:17s: Abstract 556.

111. Buzdar AU, Ibrahim NK, Francis D, et al. Significantly higher pathologic complete remission rate after neoadjuvant therapy with trastuzumab, paclitaxel, and epirubicin chemotherapy: results of a randomized trial in human epidermal growth factor receptor 2-positive operable breast cancer. J Clin Oncol 2005; 23:3676–85.

112. Burstein HJ, Harris LN, Gelman R, Lester SC, Nunes RA, Kaelin CM, Parker LM, Ellisen LW, Kuter I, Gadd MA, Christian RL, Kennedy PR, Borges VF, Bunnell CA, Younger J, Smith BL, Winer EP. Preoperative therapy with trastuzumab and paclitaxel followed by sequential adjuvant doxorubicin/cyclophosphamide for HER2 overexpressing stage II or III breast cancer: a pilot study. J Clin Oncol. 2003 Jan 1;21(1):46–53.

113. Hurley J, Doliny P, Silva O et al. Neoadjuvant herceptin/taxotere/cisplatin in the treatment of locally advanced and inflammatory breast cancer. Proc Am Soc Clin Oncol 2002; 21: 50a (Abstr 196).

114. Bines J, Murad A, Lago S et al. Multicenter Brazilian study of weekly docetaxel and trastuzumab as primary therapy in stage III, HER-2 overexpressing breast cancer. Proc Am Soc Clin Oncol 2003; 22: 67 (Abstr 268).

115. Coudert BP, Arnould L, Moreau L, Chollet P, Weber B, Vanlemmens L, Moluçon C, Tubiana N, Causeret S, Misset JL, Feutray S, Mery-Mignard D, Garnier J, Fumoleau P. Pre-operative systemic (neo-adjuvant) therapy with trastuzumab and docetaxel for HER2-overexpressing stage II or III breast cancer: results of a multicenter phase II trial. Ann Oncol. 2006 Mar;17(3):409–14.

116. Limentani SA, Brufsky AM, Erban JK, Jahanzeb M, Lewis D.Phase II study of neoadjuvant docetaxel, vinorelbine, and trastuzumab followed by surgery and adjuvant doxorubicin plus cyclophosphamide in women with human epidermal growth factor receptor 2-overexpressing locally advanced breast cancer. J Clin Oncol. 2007 Apr 1;25(10):1232–8.

117. Harris LN, You F, Schnitt SJ, Witkiewicz A, Lu X, Sgroi D, Ryan PD, Come SE, Burstein HJ, Lesnikoski BA, Kamma M, Friedman PN, Gelman R, Iglehart JD, Winer EP. Predictors of resistance to preoperative trastuzumab and vinorelbine for HER2-positive early breast cancer. Clin Cancer Res. 2007 Feb 15;13(4):1198–207.
118. Wenzel C, Hussain D, Bartsch R et al. Preoperative therapy with epidoxorubicin and docetaxel plus trastuzumab in patients with primary breast cancer: a pilot study. J Cancer Res Clin Oncol 2004; 130: 400–404.
119. Crone SA, Zhao YY, Fan L, et al. ErbB2 is essential in the prevention of dilated cardiomyopathy. Nat Med 2002; 8:459–65.
120. Ferrara N. Vascular endothelial growth factor: basic science and clinical progress. Endocr Rev 2004; 25:581–611.
121. Folkman J. Tumor angiogenesis: therapeutic implications. N Engl J Med 1971; 285:1182–6.
122. Carmeliet P. Angiogenesis in health and disease. Nat Med 2003; 9:653–60.
123. Jain RK. Molecular regulation of vessel maturation. Nat Med 2003; 9:685–93.
124. Yancopoulos GD, Davis S, Gale NW, Rudge JS, Wiegand SJ, Holash J. Vascular-specific growth factors and blood vessel formation. Nature 2000; 407:242–8.
125. Liu Y, Cox SR, Morita T, Kourembanas S. Hypoxia regulates vascular endothelial growth factor gene expression in endothelial cells. Identification of a 5′ enhancer. Circ Res 1995; 77:638–43.
126. Levy AP, Levy NS, Wegner S, Goldberg MA. Transcriptional regulation of the rat vascular endothelial growth factor gene by hypoxia. J Biol Chem 1995; 270:13333–40.
127. Frank S, Hubner G, Breier G, Longaker MT, Greenhalgh DG, Werner S. Regulation of vascular endothelial growth factor expression in cultured keratinocytes. Implications for normal and impaired wound healing. J Biol Chem 1995; 270:12607–13.
128. Pertovaara L, Kaipainen A, Mustonen T, et al. Vascular endothelial growth factor is induced in response to transforming growth factor-beta in fibroblastic and epithelial cells. J Biol Chem 1994; 269:6271–4.
129. Warren RS, Yuan H, Matli MR, Ferrara N, Donner DB. Induction of vascular endothelial growth factor by insulin-like growth factor 1 in colorectal carcinoma. J Biol Chem 1996; 271:29483–8.
130. Soh EY, Sobhi SA, Wong MG, et al. Thyroid-stimulating hormone promotes the secretion of vascular endothelial growth factor in thyroid cancer cell lines. Surgery 1996; 120:944–7.
131. Shifren JL, Mesiano S, Taylor RN, Ferrara N, Jaffe RB. Corticotropin regulates vascular endothelial growth factor expression in human fetal adrenal cortical cells. J Clin Endocrinol Metab 1998; 83:1342–7.
132. Konecny GE, Meng YG, Untch M, et al. Association between HER-2/neu and vascular endothelial growth factor expression predicts clinical outcome in primary breast cancer patients. Clin Cancer Res 2004; 10:1706–16.
133. Meunier-Carpentier S, Dales JP, Djemli A, et al. Comparison of the prognosis indication of VEGFR-1 and VEGFR-2 and Tie2 receptor expression in breast carcinoma. Int J Oncol 2005; 26:977–84.
134. Bando H, Weich HA, Brokelmann M, et al. Association between intratumoral free and total VEGF, soluble VEGFR-1, VEGFR-2 and prognosis in breast cancer. Br J Cancer 2005; 92:553–61.
135. Ignoffo RJ. Overview of bevacizumab: a new cancer therapeutic strategy targeting vascular endothelial growth factor. Am J Health Syst Pharm 2004; 61:S21–6.
136. Cobleigh MA, Langmuir VK, Sledge GW, et al. A phase I/II dose-escalation trial of bevacizumab in previously treated metastatic breast cancer. Semin Oncol 2003; 30:117–24.
137. Pegram MD, Reese DM. Combined biological therapy of breast cancer using monoclonal antibodies directed against HER2/neu protein and vascular endothelial growth factor. Semin Oncol 2002; 29:29–37.
138. Pegram M, Yeon C, Durna L, et al. Phase I combined biologic therapy of breast cancer using two humanized monoclonal antibodies directed against HER2 proto-oncogene and vascular endothelial growth factor. Breast Cancer Res Treat 2004; 88:S124: Abstract 3039.
139. Petit AM, Rak J, Hung MC, et al. Neutralizing antibodies against epidermal growth factor and ErbB-2/neu receptor tyrosine kinases down-regulate vascular endothelial growth factor production by tumor cells in vitro and in vivo: angiogenic implications for signal transduction therapy of solid tumors. Am J Pathol 1997; 151:1523–30.
140. Borgstrom P, Hillan KJ, Sriramarao P, Ferrara N. Complete inhibition of angiogenesis and growth of microtumors by anti-vascular endothelial growth factor neutralizing antibody: novel concepts of angiostatic therapy from intravital videomicroscopy. Cancer Res 1996; 56:4032–9.

141. Ramaswamy B, Shapiro CL. Phase II trial of bevacizumab in combination with docetaxel in women with advanced breast cancer. Clin Breast Cancer 2003; 4:292–4.

142. Rugo HS. Bevacizumab in the treatment of breast cancer: rationale and current data. Oncologist 2004; 9 Suppl 1:43–9.

143. Dickler M, Rugo H, Caravelli J, et al. Phase II trial of erlotinib (OSI-774), an epidermal growth factor receptor (EGFR)-tyrosine kinase inhibitor, and bevacizumab, a recombinant humanized monoclonal antibody to vascular endothelial growth factor (VEGF) in patients with metastatic breast cancer. Proc Am Soc Clin Oncol 2004; 22:14S (July 15 supplement).

144. Miller KD, Chap LI, Holmes FA, et al. Randomized phase III trial of capecitabine compared with bevacizumab plus capecitabine in patients with previously treated metastatic breast cancer. J Clin Oncol 2005; 23:792–9.

145. Miller K. ECOG 2100. Scientific symposium of the American Society of Clinical Oncology, Orlando, Florida, 2005.

146. Overmoyer B, Silverman P, Leeming R, et al. Phase II trial of neoadjuvant docetaxel with or without bevacizumab in patients with locally advanced breast cancer. Proc Amer Soc Clin Oncol 2004; 22:15S: Abstract 727.

147. Bacus SS, Zelnick CR, Plowman G, Yarden Y. Expression of the erbB-2 family of growth factor receptors and their ligands in breast cancers. Implication for tumor biology and clinical behavior. Am J Clin Pathol 1994; 102:S13–24.

148. Slamon DJ. Studies of the HER-2/neu proto-oncogene in human breast cancer. Cancer Invest 1990; 8:253.

149. Olayioye MA, Graus-Porta D, Beerli RR, Rohrer J, Gay B, Hynes NE. ErbB-1 and ErbB-2 acquire distinct signaling properties dependent upon their dimerization partner. Mol Cell Biol 1998; 18:5042–51.

150. Fukazawa T, Miyake S, Band V, Band H. Tyrosine phosphorylation of Cbl upon epidermal growth factor (EGF) stimulation and its association with EGF receptor and downstream signaling proteins. J Biol Chem 1996; 271:14554–9.

151. Hackel PO, Zwick E, Prenzel N, Ullrich A. Epidermal growth factor receptors: critical mediators of multiple receptor pathways. Curr Opin Cell Biol 1999; 11:184–9.

152. Tzahar E, Waterman H, Chen X, et al. A hierarchical network of interreceptor interactions determines signal transduction by Neu differentiation factor/neuregulin and epidermal growth factor. Mol Cell Biol 1996; 16:5276–87.

153. Lange CA, Richer JK, Shen T, Horwitz KB. Convergence of progesterone and epidermal growth factor signaling in breast cancer. Potentiation of mitogen-activated protein kinase pathways. J Biol Chem 1998; 273:31308–16.

154. Bacus SS, Altomare DA, Lyass L, et al. AKT2 is frequently upregulated in HER-2/neu-positive breast cancers and may contribute to tumor aggressiveness by enhancing cell survival. Oncogene 2002; 21:3532–40.

155. Cockerill S, Stubberfield C, Stables J, et al. Indazolylamino quinazolines and pyridopyrimidines as inhibitors of the EGFr and C-erbB-2. Bioorg Med Chem Lett 2001; 11:1401–5.

156. Rusnak DW, Lackey K, Affleck K, et al. The effects of the novel, reversible epidermal growth factor receptor/ErbB-2 tyrosine kinase inhibitor, GW2016, on the growth of human normal and tumor-derived cell lines in vitro and in vivo. Mol Cancer Ther 2001; 1:85–94.

157. Rusnak DW, Affleck K, Cockerill SG, et al. The characterization of novel, dual ErbB-2/EGFR, tyrosine kinase inhibitors: potential therapy for cancer. Cancer Res 2001; 61:7196–203.

158. Xia W, Mullin RJ, Keith BR, et al. Anti-tumor activity of GW572016: a dual tyrosine kinase inhibitor blocks EGF activation of EGFR/erbB2 and downstream Erk1/2 and AKT pathways. Oncogene 2002; 21:6255–63.

159. Gomez H, Chavez M, Doval D, et al. A phase II, randomized trial using the small molecule tyrosine kinase inhibitor lapatinib as a fist-line treatment in patients with FISH positive advanced or metastatic breast cancer. Proc Am Soc Clin Oncol 2005; 24: Abstract 3046.

160. Konecny G, Venkatesan N, Beryt M, et al. Therapeutic advantage of a dual tyrosine kinase inhibitor (GW2016) in combination with chemotherapy drugs or trastuzumab against human breast cancer cells with HER2 overexpression. Proc Am Assoc Cancer Res 2002; 43:1003: Abstract 4974.

161. Storniolo A, Burris H, Pegram M, et al. A Phase I, open-label study of the safely, tolerability, and pharmacokinetics of lapatinib (GW572016) in combination with trastuzumab. Proc Am Soc Clin Oncol 2005; 44: Abstract 559.

162. Blackwell KL, Burnstein H, Pegram M, et al. Determining relevant biomarkers from tissue and serum that may predict response to single agent lapatinib in trastuzumab refractory metastatic breast cancer. Proc Am Soc Clin Oncol 2005: Abstract 3004.

163. Kato K, Cox AD, Hisaka MM, Graham SM, Buss JE, Der CJ. Isoprenoid addition to Ras protein is the critical modification for its membrane association and transforming activity. Proc Natl Acad Sci USA 1992; 89:6403–7.

164. Clark GJ, Der CJ. Aberrant function of the Ras signal transduction pathway in human breast cancer. Breast Cancer Res Treat 1995; 35:133–44.

165. Venet M, End D, Angibaud P. Farnesyl protein transferase inhibitor ZARNESTRA R115777– history of a discovery. Curr Top Med Chem 2003; 3:1095–102.

166. End DW, Smets G, Todd AV, et al. Characterization of the antitumor effects of the selective farnesyl protein transferase inhibitor R115777 in vivo and in vitro. Cancer Res 2001; 61:131–7.

167. Kelland LR, Smith V, Valenti M, et al. Preclinical antitumor activity and pharmacodynamic studies with the farnesyl protein transferase inhibitor R115777 in human breast cancer. Clin Cancer Res 2001; 7:3544–50.

168. Zujewski J, Horak ID, Bol CJ, et al. Phase I and pharmacokinetic study of farnesyl protein transferase inhibitor R115777 in advanced cancer. J Clin Oncol 2000; 18:927–41.

169. Johnston SR, Hickish T, Ellis P, et al. Phase II study of the efficacy and tolerability of two dosing regimens of the farnesyl transferase inhibitor, R115777, in advanced breast cancer. J Clin Oncol 2003; 21:2492–9.

170. Holden S, Eckhardt S, Fisher S, et al. A phase I pharmacokinetic and biological study of the farnesyl transferse inhibitor R 115777 and capecitabine in patients with advanced solid malignancies. Proc Am Soc Clin Oncol 2001; 40:80a: Abstract 316.

171. Piccart-Gebhart M, Branle F, Valeriola D, et al. A phase I, clinical anad pharmacokinetic trial of the farnesyltransferase inhibitor R1155777 + docetaxel: a promising combination in patients with solid tumors. Proc Am Soc Clin Oncol 2001; 40:80a: Abstract 316.

172. Sparano JA, Vahdat L, Moulder S, Kazi A, Sebti S. Phase I-II trial of tipifarnib plus cyclophosphamide and doxorubicin in patients with metastatic and locally advanced breast cancer: clinical and molecular effects. Breast Cancer Res Treat 2004; 88:S64: Abstract 1067.

173. Long B, Liu G, CH M. Combining the farnesyl transferase inhibitor lonafarnib (SCH66336) with antiestrogens and aromatase inhibitors results in enhanced growth inhibition of hormondependent human breast cancer cells and tumor xenografts. Proc Am Assoc Cancer Res 2004; 45: Abstract 3868.

174. Lee F, Camuso M, Clark J, et al. The FT inhibitor BMS214662 selectively targets the nonproliferating cell subpopulation in solid tumors-implications for a synergistic therapeutic strategy. Proc Am Assoc Cancer Res 2001; 42:260: Abstract 1402.

175. Salomon DS, Brandt R, Ciardiello F, Normanno N. Epidermal growth factor-related peptides and their receptors in human malignancies. Crit Rev Oncol Hematol 1995; 19:183–232.

176. Fox SB, Harris AL. The epidermal growth factor receptor in breast cancer. J Mammary Gland Biol Neoplasia 1997; 2:131–41.

177. Klijn JG, Berns PM, Schmitz PI, Foekens JA. The clinical significance of epidermal growth factor receptor (EGF-R) in human breast cancer: a review on 5232 patients. Endocr Rev 1992; 13:3–17.

178. Massarweh S, Shou J, Dipietro M. Targeting the epidermal growth factor receptor pathway improves the anti-tumor effect of tamoxifen and delays acquired resistance in a xenograft model of breast cancer. Breast Cancer Res Treat 2002; 76:S33: Abstract 18.

179. Baselga J, Albanell J, Ruiz A, et al. Phase II and tumor pharmacodynamic study of gefitinib in patients with advanced breast cancer. J Clin Oncol 2005; 23:1–11.

180. von Minckwitz G, Jonat W, Fasching P, et al. A multicentre phase II study on gefitinib in taxane- and anthracycline-pretreated metastatic breast cancer. Breast Cancer Res Treat 2005; 89:165–72.

181. Albain K, Elledge R, Gradishar W, et al. Open label, phase II, multicenter trial of ZD 1839(Iressa) in patients with advanced breast cancer. Breast Cancer Res Treat 2002; 76:S33:December 2002.

182. Robertson J, Gutteridge E, Cheung K, Owens R, Koehler M, Hamilton L. A Phase II study of ZD1849 (Iressa) in tamoxifen-resistant ER positive and endocrine-insensitive breast cancer. Breast Cancer Res Treat 2002; 76:S96: Abstract 57 December 2002.

183. Tan AR, Yang X, Hewitt SM, et al. Evaluation of biologic end points and pharmacokinetics in patients with metastatic breast cancer after treatment with erlotinib, an epidermal growth factor receptor tyrosine kinase inhibitor. J Clin Oncol 2004; 22:3080–90.

184. Polychronis A, Sinnett HD, Hadjiminas D, et al. Preoperative gefitinib versus gefitinib and anastrozole in postmenopausal patients with oestrogen-receptor positive and epidermal-growth-factor-receptor-positive primary breast cancer: a double-blind placebo-controlled phase II randomised trial. Lancet Oncol 2005; 6:383–91.

185. Mita MM, Mita A, Rowinsky EK. Mammalian target of rapamycin: a new molecular target for breast cancer. Clin Breast Cancer 2003; 4:126–37.

186. Chan S, Scheulen ME, Johnston S, et al. Phase II study of Temsirolimus (CCI-779), a novel inhibitor of mTOR, in heavily pretreated patients with locally advanced or metastatic breast cancer. J Clin Oncol 2005; 23:23: 8/10/05 5314-5322.

187. Rudoff J, Boulay A, Zumstein-Meckerr S. The mTOR pathway in estrogen response: a potential for combining the rapamycin derivative RAD001 with the aromatase inhibitor letrozole (Femara) in breast carcinoma. Proc Am Assoc Cancer Res 2004; 45: Abstract 5619.

188. Rivera V, Kreisberg J, Mita M, et al. Pharmacodynamic study of skin biopsy specimens in patients with refractory or advanced malignancies following administration of AO23573, an mTOR inhibitor. Proc Am Soc Clin Oncol 2005; 44: Abstract 3033.

189. Herrera R, Sebolt-Leopold JS. Unraveling the complexities of the Raf/MAP kinase pathway for pharmacological intervention. Trends Mol Med 2002; 8:S27–31.

190. Davies H, Bignell GR, Cox C, et al. Mutations of the BRAF gene in human cancer. Nature 2002; 417:949–54.

191. Wilhelm SM, Carter C, Tang L, et al. BAY 43–9006 exhibits broad spectrum oral antitumor activity and targets the RAF/MEK/ERK pathway and receptor tyrosine kinases involved in tumor progression and angiogenesis. Cancer Res 2004; 64:7099–109.

192. Wilhelm S, Chien DS. BAY 43–9006: preclinical data. Curr Pharm Des 2002; 8:2255–7.

193. Lyons JF, Wilhelm S, Hibner B, Bollag G. Discovery of a novel Raf kinase inhibitor. Endocr Relat Cancer 2001; 8:219–25.

194. Strumberg D, Voliotis D, Moeller JG, et al. Results of phase I pharmacokinetic and pharmacodynamic studies of the Raf kinase inhibitor BAY 43–9006 in patients with solid tumors. Int J Clin Pharmacol Ther 2002; 40:580–1.

195. Miller K, Burstein H, Elias A, et al. Phase II study of SU11248, a multitargeted receptor tyrosine kinase inhibitor, in patients with previously treated metastatic breast cancer . Proc Am Soc Clin Oncol 2005; 24: Abstract 563.

196. Miller KD, Saphner TJ, Waterhouse DM, et al. A randomized phase II feasibility trial of BMS-275291 in patients with early stage breast cancer. Clin Cancer Res 2004; 10:1971–5.

197. Miller KD, Gradishar W, Schuchter L, et al. A randomized phase II pilot trial of adjuvant marimastat in patients with early-stage breast cancer. Ann Oncol 2002; 13:1220–4.

198. Miller KD, Trigo JM, Wheeler C, et al. A multicenter phase II trial of ZD6474, a vascular endothelial growth factor receptor-2 and epidermal growth factor receptor tyrosine kinase inhibitor, in patients with previously treated metastatic breast cancer. Clin Cancer Res 2005; 11:3369–76.

199. De Placido S, De Laurentiis M, Carlomagno C, et al. Twenty-year results of the Naples GUN randomized trial: predictive factors of adjuvant tamoxifen efficacy in early breast cancer. Clin Cancer Res 2003; 9:1039–46.

200. Ellis MJ, Coop A, Singh B, et al. Letrozole is more effective neoadjuvant endocrine therapy than tamoxifen for ErbB-1- and/or ErbB-2-positive, estrogen receptor-positive primary breast cancer: evidence from a phase III randomized trial. J Clin Oncol 2001; 19:3808–16.

201. Schiff R, Massarweh S, Shou J, Osborne CK. Breast cancer endocrine resistance: how growth factor signaling and estrogen receptor coregulators modulate response. Clin Cancer Res 2003; 9:447S–54S.

202. Schmid P, Wischnewsky MB, Sezer O, Bohm R, Possinger K. Prediction of response to hormonal treatment in metastatic breast cancer. Oncology 2002; 63:309–16.

15   Melanoma

Akshay Gupta, MD,
and John M. Kirkwood, MD

SUMMARY

Melanoma is the fastest rising form of cancer among men and the second fastest rising form of cancer among women and has become an important public health hazard. There has been a 3–7% worldwide annual age standardized incidence increase over the past 5 decades. The current lifetime risk is 1.9% for men and 1.37% for women. Once the disease progresses in regional or distant sites, it is very difficult to treat—and when disseminated it is a devastating illness. Despite an epic number of clinical trials to test a wide variety of anticancer strategies, ranging from surgery to immuno-, radio- and chemotherapy, the average survival rate for patients with metastatic melanoma is still 6-to-10 months. The only 2 FDA approved agents for advanced melanoma are Decarbazine (DTIC) and interleukin-2 (IL-2). The overall response rate (RR) is 15% with the former and 15–20% with the latter. There have been several advances made in understanding the molecular pathways playing important roles in the pathophysiology and chemoresistance of melanoma. Based on this ongoing research, several current novel management strategies, such as immunomodulators, inhibitors of the Raf kinase signal transduction cascade, proapop-totic agents etc., have evolved targeting the molecular pathways critical for survival and progression of melanoma. This chapter reviews the current concepts in the molecular pathogenesis of melanoma, the current medical treatment alternatives and outlines the future directions of further improvement.

Key Words: Melanoma; molecular; pathogenesis; targeting; interferon; STAT; antisense Bc12; BRaf; apoptosis; cell cycle.

1. INTRODUCTION

Melanoma has become an important public health hazard owing to its rising incidence as this has been well documented over the past 50 years. There has been a 3–7% worldwide annual age standardized incidence increase over the past five decades. The current lifetime risk is 1.9% for men and 1.37% for women *(1)*. Melanoma is the fastest rising form of cancer among men and the second fastest rising form of cancer among women. The overall mortality from melanoma has increased because

From: *Cancer Drug Discovery and Development: Molecular Targeting in Oncology*
Edited by: H. L. Kaufman, S. Wadler, and K. Antman © Humana Press, Totowa, NJ

of the increase in incidence; however, because the case fatality rates have decreased, today 89% of patients are alive 5 years after their diagnosis. If detected at an early stage, cutaneous melanoma is curable with complete surgical excision in most patients. However, once the disease progresses in regional or distant sites, it is more difficult to treat—and when disseminated, it is a devastating illness. Despite an epic number of clinical trials to test a wide variety of anticancer strategies ranging from surgery to immunotherapy, radiotherapy, and chemotherapy, the median survival for patients with metastatic melanoma (MM) is still 6–10 months *(2,3)*. Because melanoma affects young- and middle-aged individuals, the number of years of life lost to this malignancy is high and exceeds all other adult solid tumors except testicular carcinoma. Unfortunately, only two agents in current use have been approved by the FDA for advanced melanoma which are dacarbazine (DTIC) and interleukin-2 (IL-2). The overall response rate (RR) has fallen over the years from 15 to 7% in phase III trials of the former agent, whereas it is 15% with the latter in the pivotal collection of phase II trials that led to approval of IL-2. Less than 5% of patients achieve complete remission (CR) with IL-2 and less than 2% with DTIC. Until recently, no single drug or combination therapy has proven to be superior to the single agent DTIC. There have been several advances made in understanding the molecular pathways that play important roles in the pathophysiology and chemotherapy resistance of melanoma. On the basis of this ongoing research, several current novel management strategies have evolved targeting the molecular pathways critical for progression of melanoma. This chapter reviews the current concepts in the molecular pathogenesis of melanoma and the current novel medical treatment alternatives and outlines the future directions of further improvement.

2. MOLECULAR PATHOGENESIS

Melanoma arises as a result of transformation of melanocytes, either located in pre-existing nevi or de novo, arising from the single melanocytes that are located at the junction of the epidermis. Melanocytes are derived from neural crest cells, which after differentiating into melanoblasts follow tightly controlled migration routes to reach the skin. They are normally embedded in the basal layer of the epidermis and have contacts with keratinocytes. Stimulated by environmental stimulants such as ultraviolet radiation (UVR), the melanocytes synthesize melanin, using the enzyme tyrosinase, which is distributed through their dendritic processes. Despite the constant growth, differentiation, and vertical migration of keratinocytes, melanocyte proliferation is rarely observed under physiological conditions. Its regulation is mediated by keratinocytes through direct cell–cell contact or soluble growth factors *(4,5)*. The molecular events associated with the development of melanoma represent an interaction between host and environmental factors. Upon malignant transformation, melanoma cells initially have a superficial spreading, or radial growth phase, followed by a vertical growth phase in which the tumor cells invade the dermis. This is followed by a more aggressive metastatic phase, where the tumor cells can metastasize lymphatically or hematologically to regional and distant sites in the human body.

2.1. Ultraviolet Radiation

Perhaps the best recognized environmental risk factor for melanoma is exposure to UVR. UVR has direct and indirect effects on melanocyte homeostasis and function. Epidemiological studies implicating a role of UVR in melanoma development have been reviewed by the International Agency for Research on Cancer, which definitively concluded that "there is sufficient evidence in humans for carcinogenicity of solar radiation in causing melanoma" *(6)*. Using c-DNA microarray technology, both UVA and UVB have been shown to differentially affect gene expression in melanocytes, inducing various transcription factors, cell cycle regulators, and proteins involved in cell stress and apoptosis *(7,8)*. The result of the activation of these pathways may be DNA damage, apoptosis, or cell cycle arrest, depending upon the radiation dosage, melanin content, and the inherent capacity to repair the UVR-induced DNA damage *(9)*. Recently, prospects for elucidating this relationship have brightened considerably through the development of UV-responsive experimental animal models of melanoma *(10,11)*. In the skin, absorption of UV photons by the DNA of epidermal cells and the rearrangement of electrons leads to the formation of photoproducts at adjacent pyrimidine sites, and unrepaired damage can lead to specific gene mutations, which are usually C to T or CC to TT, termed the "UV molecular signature." Recent studies using laboratory animals have identified components of the retinoblastoma (Rb) pathway (divided into two genetically distinct pathways: "p16^{INK4A}/Rb" and "p14^{ARF}/p53") as major target(s) of UV in early stages of melanoma *(12)*. Mutations affecting either of these pathways can lead to loss of cell cycle control following UVR-induced DNA damage. The locus, which encodes the proteins p16^{INK4A} and p14^{ARF}, has also been implicated in familial forms of melanoma (discussed in Section 2.2.1.).

Melanocytes are also affected by UVR-induced changes in the surrounding tissue manifested by production of pro-inflammatory cytokines (including IL-1α and β, tumor necrosis factor, growth factors, and neuropeptides by keratinocytes and other cell types) *(13)*. This paracrine mechanism affects melanocyte cell cycle and melanin synthesis and survival *(14)*. For example, endothelin-1, a peptide produced by keratinocytes in response to UVR, promotes survival of melanocytes through the endothelin B receptor and mitogen-activated protein kinase (MAPK) *(15)* and downregulates E-cadherin expression on melanocytes with the potential of creating a subset of melanocytes capable of escaping the epidermis *(16)*.

2.2. Genetic Alterations

2.2.1. CELL CYCLE CONTROL PROTEINS

Studies of familial melanoma have identified two genes predisposing to melanoma, *CDKN2A* (INK4a/ARF) and *CDK4* located at 9p21 and 12q13, respectively *(17,18)*. The *CDKN2A* locus is also altered in 10–60% of sporadic melanomas *(19)*. In humans, this locus codes for two distinct tumor suppressor proteins, the inhibitor of kinase 4A, p16^{INK4A} and p14^{ARF}, as a result of alternative splicing and translation of two different reading frames *(20)*. P16 is a cyclin dependent kinase 4/6 (CDK 4/6) inhibitor which, when activated, binds to *CDK4* and maintains the Rb protein in its nonphosphorylated (active) state, causing cell cycle arrest in the G1 phase. If p16^{INK4A} is inactivated through missense mutation, deletion, or methylation, Rb protein is no longer maintained in its active form, and cell replication goes unchecked. The *CDK4* mutations that have been identified in melanoma prone families produce a defective protein that interferes with its binding to p16. The *CDK4* is, therefore, constantly active promoting Rb phosphorylation and subsequent cell division *(21)*.

The *p14^{ARF}* gene is a principal regulator of MDM2, an E3 ubiquitin ligase regulating p53 degradation and stability *(22)*. When activated, it prevents the interaction between p53 and MDM2 leading to elevation of the level of p53. This inhibits Rb protein phosphorylation, leading to cell cycle arrest in G1 and G2 phases. If the *ARF* gene is mutated, the above events are disrupted, leading to unchecked cell division promoting melanoma development. A recent analysis of p53 transcription levels and allelic imbalance in cutaneous and mucosal melanoma has detected altered p53 expression levels in both, indicating a primary disturbance in p53 expression *(23)*.

Other types of aberrations in cell cycle regulation known in melanoma are cyclin D1 gene amplification *(24)*, mutations of the M-phase cell cycle inhibitor p34^{CDC2L} *(25)*, and increased accumulation of the cell cycle inhibitors p27^{KIP-1} and p21$^{WAF-1/SDI-1/Cip-1}$ *(26)*.

The above changes in cell cycle control mechanisms render the normally non- proliferating melanocytes into a rapidly proliferating cells which opens up the possibility of development of more aggressive malignant cell types.

2.2.2. RAS/RAF/MAPK SIGNAL TRANSDUCTION PATHWAY

Despite their important roles in melanoma predisposition, mutations of *CDKN2A* and *CDK4* account for only a small proportion of sporadic and familial melanomas, indicating that additional genes relevant to melanoma must exist. The MAPK signaling cascade is activated through sequential phosphorylation of a number of kinases to rapidly and reciprocally alter cellular behavior in response to diverse environmental cues *(27)*. The extracellular-signal-regulated kinases (ERK1 and ERK2) belong to one branch of this cascade that is responsible for sensing extracellular stimuli, including UV light. Such stimuli activate the RAS family of proto-oncoproteins (NRAS, HRAS, and KRAS), which in turn activate the RAF family of serine/threonine kinases (c-RAF1, BRAF, and ARAF), known as MAPK kinase kinases (MKKK). RAF then phosphorylates the MAPK kinase MEK (MKK), which subsequently phosphorylates and activates the MAPKs ERK1 and ERK2. In addition to activation by upstream receptor tyrosine kinases (RTKs), ERK phosphorylation has been linked to G-protein-coupled receptor (GPCR) signaling through an as-yet poorly understood crosstalk mechanism *(28)*. Considering the fact that melanocyte proliferation and survival is tightly regulated by the paracrine control of surrounding cells through several growth factors that signal through GPCRs and RTKs, it can be speculated that activating mutations in RAS or RAF can mimic these mitogenic signals, leading to uncontrolled melanocyte proliferation.

Recent genome-based high-throughput sequencing efforts have identified activating *BRAF* mutations in as many as 60% of human melanoma samples and cell lines *(29)*. Importantly, these point mutations clustered in specific regions of biochemical importance, and 80% of them resulted in a single phosphomimetic substitution in the kinase-activation domain (V599E) that is known to confer constitutive activation of *BRAF*. More recently, *BRAF* mutations have also been shown to be common in benign and dysplastic nevi, which supports the observation that activation of ERKs is an early event in melanoma progression *(30)*. Activating mutations in RAS have also been reported. Recent studies have reported that as many as 33% of primary melanomas and 26% of MM samples harbor activating *NRAS* point mutations *(31)*. Transgenic studies in mice have shown that activated *HRAS* mutations in melanocytes can lead to their aberrant proliferation and transformation, particularly in cooperation with inactivating mutations in tumor suppressors such as *CDKN2A (32)*.

2.2.3. TELOMERASE ACTIVATION

Telomeres, repetitive DNA sequences for stabilizing the replicating chromosomes, normally slightly shorten with each cell division until they reach a critical length, which leads to cell cycle arrest. It has been proposed that the rarely observed phenomena of spontaneous regression of melanoma could be attributed to progressive telomere dysfunction, leading to accumulating DNA breaks, ultimately triggering cell death, termed "crisis" *(33)*. During this period, however, new clones may arise with enhanced ability to stabilize their telomeres, through activation of the telomerase gene, leading to unchecked cell division *(34)* and to the development of more aggressive forms of cancer.

2.2.4. RESISTANCE TO APOPTOSIS

Impaired ability to undergo programmed cell death in response to a wide range of external stimuli provides melanoma cells a selective advantage for progression and metastasis as well as their notorious resistance to therapy. Melanocytes themselves are inherently resistant to apoptosis, as evidenced by their ability to become activated in response to UVR, secrete melanin, and protect surrounding keratinocytes and other epidermal cells from further damage. Fibroblasts and keratinocytes, through paracrine stimulation, promote the survival of melanocytes. For example, keratinocytes promote expression of bcl-2, an antiapoptotic factor, in melanocytes by secreting neuronal growth factor (NGF) and stem cell factor (SCF) *(35)*. The dependence of normal melanocytes on Bcl-2 for survival is illustrated by the depigmentation and loss of melanocytes of mice deficient in Bcl-2 *(36)*.

Apoptotic indices are typically low in melanoma tumors, particularly at advanced stages *(37)*. The role of acquired apoptotic defects during melanoma progression can be grouped into three categories: (i) activation of antiapoptotic factors, (ii) inactivation of proapoptotic effectors, and (iii) reinforcement of survival signals.

2.2.4.1. Antiapoptotic Factors and Melanoma. Studies on viruses that block apoptosis in mammalian cells to favor infection led to the identification of two groups of apoptotic "breaks": inhibitors of apoptosis (IAPs) and FLICE inhibitory proteins (FLIPs), later found to be overexpressed in multiple tumor types. In melanoma, two members of the IAP family (survivin and ML-IAP) and FLIP have been associated with tumor progression, as they become detectable in melanocytic nevi and are further overexpressed in invasive and MMs *(38)*. At the beginning of mitosis, survivin associates with microtubules of the mitotic spindle apparently preventing the activation of caspase 3 in response to abnormal cell division *(39)*. The potential impact of survivin overexpression on melanoma progression is illustrated in xenograft studies, where a dominant-negative mutant of survivin (Thr34-Ala) reduced the tumorigenicity of melanoma cells injected in immunosuppressed mice *(40)*. ML-IAP is also upregulated in melanoma cell lines but absent in normal melanocytes *(41)*. The impact of FLIP in melanoma resistance to chemotherapy is controversial. Although overexpression of FLIP increases the resistance of melanoma cells to both TRAIL and FasL *(42)*, the endogenous levels of FLIP do not necessarily correlate with drug response in patients *(43)*.

2.2.4.2. Inactivation of Proapoptotic Factors. Unlike many other tumors, p53 mutations have not been associated with melanoma chemoresistance to any significant extent. However, many disruptions in other components of the pathway, which amount to functional p53 deficiency, have been described. Mutations in p14ARF, a principal regulator of HDM2, which lead to loss of p53 activity, have already been discussed in Section 2.2.1. Apap–1 and caspase-9 are essential downstream effectors of p53 induced apoptosis, which if mutated or downregulated could lead to unchecked cell division. Apaf-1 protein and mRNA expression are frequently downregulated in metastatic cell lines and tumor specimens. Restoring physiological levels of Apaf-1 through gene transfer or 5aza2dC treatment enhances chemosensitivity, alleviating cell death defects associated with reduced Apaf-1 expression *(44)*. These results raise the possibility that restoring Apaf-1 regulation to some melanomas could have therapeutic benefit.

2.2.4.3. Survival Signaling in Melanoma. Activation of several survival pathways such as phosphatidylinositol 3-kinase (PI3K)/AKT/PTEN, nuclear factor-κB (NFκB), and Ras/Raf/MAPK pathways has been shown to contribute to melanoma aggressiveness and resistance to chemotherapy. The PI3K/AKT/PTEN pathway is activated by several mitogens binding to cell surface receptor kinases. Once activated, PI3K converts the lipid PIP2 into PIP3. PIP3 activates the protein kinase B (PKB)/AKT, which, in turn, targets multiple factors involved in cell proliferation, migration, and survival. PTEN is a phosphatase that targets PIP3 and prevents the activation of AKT. One-third of primary melanomas and about 50% of MM cell lines showed reduced expression of PTEN as a result of allelic deletion, mutation, or transcriptional silencing *(45,46)*. NFκB is a transcription factor with plieotropic effects, functioning as a modulator in inflammation, angiogenesis, and cell death, survival, adhesion and migration. In melanoma cells, the NFκB pathway can be altered by upregulation of the NFκB subunits p50 and Rel A *(47,48)* and downregulation of the NFκB inhibitor IκB, which amounts to upregulation of its downstream targets in melanoma. The role of Ras/Raf/MAPK pathway in melanoma pathogenesis has already been discussed in Section 2.2.2.

2.2.5. Low-Penetrance Genes: MC1R Polymorphism

Melanocortin-1 receptor *(MC1R)*, a GPCR, mediates the effects of the melanocyte mitogen a-melanocyte-stimulating hormone (α-MSH) by upregulating cyclic AMP (cAMP). cAMP in turn activates the PKA, which translocates to the nucleus and activates the transcription of several genes including microphthalmia-associated transcription factor *(MITF)*, a helix-loop-helix transcription factor which is one of the most crucial for regulation of pigmentary genes *(49)*. *MC1R* is highly polymorphic in human populations—accounting in large part for the variations in pigmentation phenotypes and skin phenotypes. Several distinct variants of *MC1R* have been associated with the red hair color phenotype (RHC), an independent risk factor for melanoma. These variants shift the balance of melanin synthesis from eumelanin to pheomelanin. In addition to its diminished UV-light protective capacity, pheomelanin produces metabolites that are believed to be mutagenic and cytotoxic *(50)*, which could further contribute to increased cancer risk. Indeed, cell-culture-based studies have shown that primary human melanocytes harboring RHC variants of *MC1R* show a pronounced increase in sensitivity to UV-light-induced cytotoxicity *(51)*. Besides affecting the pigment type, some variants of *MC1R* may also potentiate the effects of other mutations, such as increasing the penetrance of the mutations at the *CDKN2A* locus *(52)*.

2.2.6. STAT SIGNAL TRANSDUCTION PATHWAY

The Janus kinase-signal transducer and activator of transcription (JAK-STAT) pathway transmits information received from extracellular polypeptide signals, through transmembrane receptors, directly to target gene promoters in the nucleus, providing a mechanism for transcriptional regulation without second messengers *(53)*. In mammals, seven STAT genes have been described including *STAT1, STAT2, STAT3, STAT4, STAT5A, STAT5B*, and *STAT6*. STAT proteins are present in an inactive monomer form in the cytoplasm. On ligand engagement at the receptor, they are recruited with the help of receptor associated tyrosine kinases and converted into activated dimers. STAT dimers are rapidly transported from the cytoplasm to the nucleus and are competent for DNA binding. Once the activated STAT dimer recognizes a target promoter, the transcription rate from this promoter is dramatically increased. STATs are involved in regulating many genes in normal cells that control fundamental biological processes — such as cell proliferation, apoptosis, angiogenesis, and immune responses. These STAT proteins, particularly STAT3 and the STAT5 proteins, are frequently overactivated in various human solid tumors and blood malignancies *(54)*. This continuous activation promotes growth and survival of tumor cells, induces tumor angiogenesis, and suppresses host antitumor immune responses *(55)*. In the case of melanoma, there appears to be distinct roles of STAT1 and STAT3 in progression and pathogenesis. STAT1 has been shown to be important in the intracellular signal transduction transmitting the antitumor response of interferon-α (IFNα), the only agent proven to be useful as adjuvant therapy for high-risk melanoma patients. Hence, defective STAT1 pathway signaling is thought to be associated with melanoma resistance to IFN *(56)*. In contrast to STAT1, constitutively activated STAT3 DNA-binding activity has been observed in the majority of human melanoma cell lines and primary tumors tested but not in matched normal skin specimens from the same patients *(27)*. Of the tested melanoma cell lines, abrogation of Src phosphorylation resulted in growth inhibition and cell death of melanoma cells exhibiting constitutive STAT3 activation *(57)*.

Targeting STAT3 in melanoma tumor models induces tumor cell death/tumor regression *(55,57)*, inhibits angiogenesis *(58)*, prevents metastasis *(59)*, and inactivates antitumor immune responses *(60)*. As the majority of melanoma cell lines and tumor specimens display constitutively-activated STAT3, targeting STAT3 is expected to affect a significant population of melanoma patients *(61)*.

2.3. Changes in Adhesion Molecules

Besides the genetic changes in the several signal transduction pathways, changes in cell surface molecules, leading to altered interaction with neighboring cells, are important in melanoma pathogenesis. Disruption of normal contacts between keratinocytes and melanocytes occurs early in melanoma progression and is reflected in the expression of cell surface cell adhesion molecules (CAMs). There is a switch of homotypic cell–cell CAMs (i.e., loss of E-cadherin and gain of N-cadherin expression) *(62)*, increased expression of heterophilic CAMs (i.e., melanocytes CAMs

MelCAM/MUC18, L1CAM, and activated leukocyte CAM) *(63)*, integrins *(64)*, and junctions (i.e., connexin 26) *(65)*. These changes result in changes in interaction partners from the normally observed basal keratinocytes to other non-epithelial cells such as melanocytes themselves, vascular endothelial cells, smooth muscle, and activated T cells, promoting cell migration.

2.4. Angiogenesis

Angiogenesis is a prominent feature in melanoma progression and metastasis, similar to other solid tumors. The pathways for angiogenesis and vasculogenesis are guided through the cooperation of fibroblasts and melanoma cells perpetuated by the dominance of the MM cells *(66)*. Evidence of angiogenesis can be detected as early as during the radial growth phase *(67)* in thin lesions *(68)* and is correlated with melanoma progression *(69)* and may be inversely correlated with disease-free survival (DFS) and overall survival (OS) *(70)*.

Melanoma cells also secrete various angiogenic factors, such as vascular endothelial growth factor (VEGF), bFGF, IL-8, and platelet like growth factor *(71)*. Gene expression profiling of uveal melanomas with liver metastasis has shown that aggressive tumor cells express genes associated with pluripotent embryonic-like phenotypes *(72)*, implying that tumor cell plasticity (dedifferentiation or transdifferentiation) is important in melanoma progression. Moreover, unique to melanoma and pancreatic cell carcinoma, highly aggressive melanoma cells are capable of expressing endothelium-associated genes, and form matrix-rich networks de novo, when cultured on a three-dimensional matrix, thus mimicking embryonic vasculogenesis, a phenomenon called "vasculogenic mimicry" *(73)*. Melanoma cells are also known to express the $\alpha_V\beta_3$ integrin, a protein that has been shown to be essential for endothelial cell proliferation, maturation, and survival *(74)*. Experiments in SCID mice have shown an inhibition of human melanoma cells in human skin explants by anti-$\alpha_V\beta_3$ monoclonal antibody *(75)*. Other molecular determinants of this phenomenon, currently being delineated, may help in design of more efficient antivascular treatments.

2.5. Immune Evasion

Human melanoma has been considered to represent a neoplasm of high immunogenicity, in which powerful mechanisms of immunological evasion or tolerogenesis have developed. In fact, in a fraction of patients, T cells directed against several different tumor-associated antigens (TAAs) have been documented in both peripheral blood and tumor tissues, providing support for the notion of immunogenicity *(76)*. Beyond this, melanoma has been shown to be relatively susceptible to nonspecific immunotherapy and multiple agents ranging from BCG and *Cryptosporidium parvum* to defined cytokines, and IFNs have shown RRs of ~15% in patients with melanoma *(77)*. Moreover, melanoma cells also secrete chemoattractive cytokines, such as IL-8, granulocyte–monocyte colony-stimulating factors (GM-CSF) and monocyte chemoattractive protein-1 (MCP-1). The presence of tumor-infiltrating lymphocytes in primary melanoma has been associated with improved clinical prognosis *(78)*. At the same time, most mechanisms of tumor escape from immune surveillance have also been observed during melanoma progression *(76)*. Melanoma has been shown to constitutively produce STAT3, for instance, which has been correlated with the induction of IL-10 and VEGF, and conspires to suppress immune response (section 3.1.1.2). However, the fact that melanoma has been a prototype for mechanisms of immune escape and immunosuppression is, in a sense, a proof of principal—suggesting that tumor progression can only occur in the setting that one or more of these mechanisms of evasion has developed. The mechanisms demonstrated to be associated with tumor progression following immunotherapy of melanoma include loss/downregulation of HLA, of adhesion molecules, and TAAs, as well as the production of immunosuppressive factors, and factors capable of killing activated T cells or inducing T-cell receptor (TCR) signaling defects, or inhibiting pro-inflammatory signals. At a level prior to the T cell, melanoma has been associated with impaired maturation of dendritic cells (DCs) and defective $CD4^+$ T-cell help *(79)*. In addition to these classical examples of tumor escape mechanisms, new pathways of immune resistance are being continuously discovered in human melanoma, such as those based on tryptophan degradation, leading to impaired T-cell proliferation *(80)*, increased resistance to cell death through triggering of CCR10 *(81)*, and inhibition of DC maturation through production of gangliosides *(82)*.

Many of these immune evasion mechanisms have been correlated with clinical aggressiveness of the disease, as exemplified by the association between tumor stage and loss/down-modulation of HLA class I antigens *(83,84)* and of HLA class I-processing molecules *(85)*. Even in patients with infiltrating $CD8^+$ lymphocytes, the maturation of these $CD8^+$ cells seems to be arrested at the final stage, significantly hampering their cytotoxic activity *(86)*. A subset of $CD4^+$ T cells responsive to the TAA LAGE1, expressed in melanoma cells, has been shown to have immunosuppressive ability ($CD4^+T_{reg}$), opening up another potential mechanism of TAA-induced immune suppression *(87)*.

The evolution of immune escape mechanisms in advanced disease, despite development of immunity, may help to explain the lack of relationship between evidence of immunity in peripheral blood or at the tumor site, and effective tumor destruction in the lesions. Attempts to generate antitumor immunity have so far resulted in short-term, low-titer, antigen-specific T cells necessitating design of treatment options that promote more durable and efficacious antitumor immunity *(78)*.

3. CURRENT MOLECULAR TARGETING

Based on these advances in the understanding of the molecular pathways playing important roles in the pathophysiology and chemoresistance of melanoma, several current novel management strategies (Table 1) have evolved targeting the molecular pathways critical for survival and progression of melanoma.

3.1. Localized Non-Metastatic Disease

Adequate surgical resection is curative for most early melanomas. The risk of local/regional recurrence is dictated mostly by the depth of invasion, as measured in millimeters. The limited impact of surgical or medical therapy on the prognosis of patients with MM has prompted the search for effective adjuvant therapy options to prevent a relapse after adequate surgical excision.

Table 1
Current Novel Management Strategies for Melanoma, Targeting the Molecular Pathways
Critical for Its Survival and Progression

Molecular lesion	Targeted intervention
Antiapoptotic processes: e.g., Bcl-2 overexpression → resistance to apoptosis	Oblimersan, antisense oligonucleotide directed against Bcl-2 messenger RNA
MAPK pathway activation: e.g., activating *BRAF* mutation	BAY 43-9006, small molecule inhibitor of Raf-1, VEGFR-2, wild-type BRAF and the V599E mutant
	AZD6244 is a potent, selective, orally active MEK inhibitor
	CHIR-258, an orally bioavailable, potent inhibitor of class III-V receptor tyrosine kinases
Defective antigen presentation	MDX-010, human anti-CTLA-4 antibody, enhances T cell-dependent immunity
Angiogenesis, expression of the $\alpha_V\beta_3$ integrin	MEDI-522, a humanized monoclonal antibody to $\alpha_V\beta_3$ integrin directed against a conformational epitope of the $\alpha_V\beta_3$ integrin
Immune tolerance	Interferon a2b, activates the host immune system to generate a better antitumor response
Loss of MHC-I expression DC immaturity Loss of TAP I expression Defective CD4+ T-cell help	IL-2

CTLA-4, Cytotoxic T-cell-associated antigen-4; IL-2, interleukin-2, MAPK, mitogen-activated protein kinase; MHC, major histocompatibility complex.

3.1.1. ADJUVANT MEDICAL THERAPY

According to the current 6th Edition American Joint Committee of Cancer (AJCC) staging system, Stage I and II disease represent localized cutaneous melanoma of increasing thickness, and Stage III disease corresponds to regional lymph node involvement. On the basis of data from the Sydney Melanoma Unit and the University of Alabama at Birmingham, patients with AJCC Stage IIB (Breslow depth >4 mm) have a 5-year survival of 56.3% and patients with Stage III melanoma have a 5-year survival of 10–46%, depending on the size and number of involved nodes *(88)*. It is this group of patients who has been the target of most adjuvant therapy trials. A complete review of all the major trials using several combinations of chemo/immuno/biotherapeutic agents to be used in adjuvant settings is beyond the scope of this chapter. We would instead direct our focus on IFNs, the only adjuvant therapy demonstrated in multiple trials to be useful in high-risk melanoma patients.

3.1.1.1. Adjuvant Interferon Therapy. The IFNs are a group of complex proteins first identified by Isaacs and Lindeman in the 1950s. They possess diverse functions and can be broadly divided into two types: type I (IFNα and IFNβ) and type II (IFNγ). Recombinant DNA technology has allowed the production of virtually unlimited quantities of purified IFN, which has facilitated their testing in both the laboratory and the clinic. High-dose IFN (HDI) therapy using IFNα was the first form of medical therapy to be approved by the FDA for use in melanoma in the adjuvant setting. The Eastern Cooperative Oncology Group (ECOG) and intergroups have conducted several trials over the last 15 years, to observe the effect of this therapy on the overall and relapse-free survival of patients (89).

E1684 was the first randomized comparison of HDI versus observation. Treatment was given with IFNα2b at 20 million units (MU)/m²/day IV 5 days a week for 4 weeks, then 10 MU/m²/day SC 3 days a week for the next 48 weeks for a full year's therapy. The results of this trial were first reported in 1996 with a median follow-up interval of 6.9 years (90). The median relapse-free survival (RFS) was 1.72 years in the HDI arm versus 0.98 year in the Obs arm [stratified log-rank one-sided P value = 0.0023], and the median OS was 3.82 versus 2.78 years ($P = 0.0237$), respectively. E1690 attempted to replicate the findings of E1684 by HDI and low-dose IFN (LDI) versus observation. At a follow-up of 4.3 years, there was a statistically significant improvement in the RFS of the HDI-treated patients [Hazard Ratio (HR) = 1.28; $P_1 = 0.025$]. LDI was not associated with any RFS benefit, and neither HDI nor LDI had any benefit in terms of OS (91). However, a retrospective analysis of salvage therapy suggested a disproportionate crossover of patients from the observation arm to HDI therapy off protocol at the time of regional recurrence (stage IIB patients in this trial were not required to undergo lymphadenectomy), which may have confounded the survival analysis. The intergroup trial E1694 was designed to test whether the ganglioside GM2/keyhole limpet hemocyanin vaccine (GMK) was superior to HDI. This trial was prematurely terminated as the GMK vaccine proved to be significantly inferior to HDI, both for relapse and for mortality endpoints, on interim analysis. Among eligible patients in this trial, HDI provided a statistically significant RFS benefit (HR = 1.47; $P_1 = 0.0015$) and OS benefit (HR = 1.52; $P_1 = 0.009$) compared with GMK (92).

In an updated analysis of E1684, E1690, and E1694, the clinical benefit of HDI in terms of RFS was still evident in E1684 and E1690, at a mean follow-up of 12.6 and 2.1 years, respectively (89). In E1690, the RFS showed a trend toward statistical significance ($P = 0.09$). The OS benefit was maintained in the E1694 ($P = 0.04$) but was diminished in E1684 ($P = 0.18$) and not present in E1690. In a pooled analysis of patients in E1684 and E1690, the RFS benefit was maintained ($P < 0.006$) but no OS benefit was seen ($P = 0.42$). This discrepancy between durable RFS benefits and eroding OS benefits on long-term follow-up is hitherto unexplained.

HDI is the standard of care for high-risk melanoma patients in the adjuvant setting. HDI therapy is however associated with significant toxicity, with the incidence and severity of these adverse events clearly dose-related. In the pivotal trial E1684, 66% of patients treated with HDI had at least one grade 3 (ECOG toxicity scale) adverse event and 14% of patients had a grade 4 event (93). Consequently, there has been a great deal of interest in intermediate- and low-dose regimens administered through subcutaneous injection. However, none of the trials using intermediate or LDI dosing so far have been able to demonstrate any reliable benefit in terms of RFS or OS. For patients at intermediate-risk (T3, ≥1.5 mm primary melanoma, stage IIA), the US Intergroup, together with the NCI-Canada and selected Australian sites, has joined the E1697 trial in which the initial intravenous induction component of HDI regimen therapy (one month of IV HDI) is compared to observation alone.

3.1.1.2. Mechanism of Action of Interferon. IFNα, a member of the type I IFN family, is classically known to be produced by cells in response to exposure to viruses and double-stranded RNAs. Their role as an antitumor agent has emerged recently, and many aspects of the mechanism behind its antitumor efficacy against a few select neoplasms including melanoma are still unknown. Type I IFN-dependant signaling *(94)* requires recruitment of two IFN receptor chains, IFNAR1 (human type I IFN receptor chain 1) and IFNAR2 (human type I IFN receptor chain 2). Binding of type I IFNs induces the assembly of these receptor chains, which leads to the phosphorylation of tyrosine residues located in the intracellular domain of each receptor chain, thought to be carried out by the Janus kinases TYK2 and JAK1. This assembly further leads to the recruitment, phosphorylation, and dimerization of STAT proteins – classically STAT1 and STAT 2. Homo- and heterodimers of STAT proteins then translocate into the nucleus and bind to GAS regulatory elements of the IFN-activated genes, whereby inducing the expression of a large number of genes affecting important biological responses, including antiviral, antiproliferative, and immunomodulatory activities. The pleiotropic effects of IFN on host immunity leading to activation of antitumor immune mechanisms were reviewed recently *(95)*.

Over the last decade, several other transcription factors have been shown to be associated with IFN-mediated signaling. STAT3, for example, has been shown to be activated in a human leukemia cell line in response to IFNα *(96)*. Activated STAT3, in turn, activates the PI3K, acting as an adaptor to couple another signaling pathway to the IFN receptor *(97)*. Also, STAT5 has been demonstrated to be activated by IFNα in lymphoma cells *(98)*.

Among the many postulated mechanisms of the antitumor activity of IFNα, induction of apoptosis has been shown to be important in vitro, if not in vivo. IFNα can induce apoptosis in transformed cell lines as well as primary tumor cells *(99)*. Although a detailed molecular mechanism of IFN-induced apoptosis remains to be elucidated, it has recently been shown in hematopoetic tumor cell lines to involve the loss of mitochondrial membrane potential, cytochrome c release, caspase 9 activation, and activation of proapoptotic members of the bcl-2 family of proteins Bak and Bax *(100,101)*. Apart from STATs, type I IFNs have been also shown to directly activate enzymatic isoforms of MAPK, such as Jun kinase 1 (JNK1) and p38 kinase, which can mediate antiproliferative and apoptotic stimuli *(102)*. Besides modulating gene expression, IFNα has also been recently shown to affect protein synthesis pathways. One such molecular target is protein kinase-dependent dsRNA (PKR), whose activation induced by the cytokine regulates translational and transcriptional pathways resulting in the expression of selected proteins (p53, bax, fas, etc.) that trigger cell death *(103)*.

Most of the data relating to the molecular impact of IFN therapy on melanoma comes either from animal studies or from cell culture-based experiments, as discussed earlier. As it is used in an adjuvant setting, that is, weeks after the surgical excision of the tumor, tumor tissue has not been available to assess the actual mechanism of action on human melanoma in vivo, which may help in precisely defining the molecular processes that form the basis of its reliable effect on prolonging the RFS. More recently, an alternate approach has been taken to enable the direct study of tumor tissue at the University of Pittsburgh Cancer Institute: in this trial UPCI 00-008, melanoma patients with stage III nodal disease are being treated with "neoadjuvant" IFN for I month (20 MU/m²/day of IFNα2b 5 days a week for 4 weeks) before definitive surgical lymphadenectomy.

This treatment design has the advantage of earlier delivery of the systemic agent for the control of distant disease and has enabled the direct evaluation of the effects of high-dose IFNα2b on the tumor cell, its vasculature, and its lymphoid and DC content. Improved efficacy of adjuvant therapies has been observed when treatment has been administered before definitive surgery, in a "neoadjuvant" setting for a number of solid tumors, including breast adenocarcinoma, esophageal carcinoma, soft tissue sarcoma, and non-small-cell carcinoma of the lung. This study has demonstrated the feasibility and usefulness of neoadjuvant IFN therapy and outlines a strategy that may help to augment the mechanistic analysis as well as to improve the impact of this therapy, while also potentially helping to define intermediate endpoints of analysis that will accelerate progress and select subjects most likely to derive benefit from this therapy *(104)*.

Preliminary findings from this study have yielded valuable insights into the molecular mechanism of action of IFN. Tumor regression was observed in 11 of 20 patients with stage IIIB-IIIC disease or recurrent nodal involvement, when IFN was used in the neoadjuvant setting. Analyzing the molecular alterations caused by IFN, we found that the STAT1/STAT3 expression ratios rose in association with IFN treatment. The clinical effects of IFNα2b in human melanoma are inversely related to STAT3 expression. No changes were observed in tumor tissue expression of vascular markers, markers of apoptosis, Tap1, or major histocompatibility complex I (MHCI) or MHCII molecules *(105)*.

3.2. Metastatic Disease

Current approaches for the treatment of MM include chemotherapy, immunotherapy, and biochemotherapy. DTIC is one of the only two agents in current usage that are approved for the treatment of MM, producing overall RRs in the range of 7–15% and a CR rate of 2–5%. Temozolomide, which is an oral agent that spontaneously gives rise to the same active metabolite as DTIC, has the additional benefit of oral administration and good penetration to the central nervous system (CNS). All novel chemotherapeutic agents are compared against the RRs and treatment results of DTIC. High-dose IL-2, a form of nonspecific immunotherapy, is the other agent considered as a standard of care in patients with MM, producing overall RRs in the range of 15% and a CR rate of 5–6%. Despite an epic number of clinical trials evaluating combinations of available chemotherapeutic hormonal and bio/immunotherapeutic agents for the management of MM, only a few strategies focusing on targeted therapy of specific signaling pathways have yielded promising results to date. A complete review of the various chemo-immuno-biotherapy trials is beyond the scope of this chapter. We would however focus on the specific molecular targeting strategies that have recently emerged.

3.2.1. RAS/RAF/MAPK Signal Transduction Pathway Inhibitors

Recent genome-based high-throughput sequencing efforts have identified activating *BRAF* mutations in 60% of human melanoma samples and cell lines as well as in melanocytes of benign and dysplastic nevi (discussed in Section 2.2.2.). This finding points to a direct and early role of the Ras/Raf/MAPK signal transduction pathway in melanomapathogenesis. It has been shown that the melanoma cell lines harboring activating *BRAF* mutations have constitutively elevated ERK (the final common effector of the MAPK signal transduction pathway). Inhibition of *BRAF* activity using RNA interference or the RAF kinase inhibitor BAY 43-9006 inhibits DNA synthesis and induces apoptosis in these cell lines. BAY 43-9006 also induces growth delay in melanoma xenografts *(106,107)*.

BAY 43-9006 is an orally active small molecule inhibitor of Raf-1, VEGFR-2, wild-type *BRAF* and the V599E mutant, in addition to a number of other pro-angiogenic RTKs *(108)*. It is currently in phase II–III trials in various solid tumors, including some that do not have *BRAF* mutations. Used as a single agent, it only has modest activity against advanced melanoma *(107)*. A phase II trial of BAY 43-9006 combined with carboplatin and paclitaxel in 35 patients with advanced melanoma with predominantly M1c disease (68%), the majority of whom had received previous treatment, reported a RR of 31% *(109)*. A phase III clinical trial of carboplatin, paclitaxel ± BAY 43-9006, as first-line therapy in patients with unresectable locally advanced or stage IV melanoma is in progress in the US Cooperative Groups (E2603).

A few other inhibitors of this pathway have recently entered clinical trials. MEK is a critical enzyme at the intersection of several biological pathways, which regulates cell proliferation and survival as part of the Ras/Raf/MEK/ERK pathway. AZD6244 (Astra-Zeneca, Inc) is a potent, selective MEK inhibitor that is orally active and is being evaluated in a randomized phase II study that will compare AZD6244 to temozolomide in the treatment of unresectable stage III/IV melanoma. The antiproliferative potential of AZD6244 has been studied in cell lines harboring Ras and *BRAF* mutations and in various human tumor xenograft models. RTK-258 (Novartis-Chiron), an orally bioavailable, potent inhibitor of class III-V RTKs, has been studied in animal models of solid and hematological malignancies with promising results *(110)*. A phase I study of CHIR-258 in patients with locally advanced or MM is underway.

3.2.2. Antisense bcl-2 (Oblimersan)

Oblimersan is a phosphorothioate antisense oligonucleotide directed against the first six codons of the Bcl-2 messenger RNA. Binding of the drug to the mRNA recruits RNAse H, resulting in cleavage of the mRNA. As a result, further translation is halted and intracellular protein concentrations of Bcl-2 decrease with time. The dependence of normal melanocytes on Bcl-2 expression for survival has already been discussed in Section 2.2.4. Melanoma cell lines having bcl-2 overexpression have been shown to enhance activity of metastasis-related proteinases, in vitro cell invasion, and in vivo tumor growth *(111)*. Many in vitro studies have demonstrated increased sensitivity of melanoma cells to chemotherapy when combined with antisense bcl-2 therapy *(112)*. These data prompted the start of numerous clinical trials evaluating the addition of oblimersan to chemotherapy in various solid tumors, including melanoma. Updated analysis from a randomized phase III trial *(113)*, comparing DTIC combined with oblimersan, with DTIC alone in 771 patients with Stage IV or unresectable Stage III melanoma who had not previously received chemotherapy has shown a RR of 12.4% in the former compared with 6.8% in the latter group (P = .007). Median progression-free survival for the oblimersan group was 2.4 months as compared with 1.6 months for the DTIC group, with a relative risk reduction of 27% (P = .0003). The median survival was increased from 7.8 months in the DTIC arm to 9 months in the oblimersan arm with a P value of .077, which became significant when the patients with normal baseline LDH were analyzed (.02). In terms of toxicity, no new or unexpected adverse events were observed in this study, which had not been seen with DTIC alone.

3.2.3. ANTIANGIOGENIC AGENTS

The ability of a tumor to grow and metastasize is dependent on its ability to develop and acquire new blood supply, through a complex process called angiogenesis. Angiogenesis requires the orchestration of vascular basement membrane degradation, endothelial cell migration, endothelial cell proliferation, capillary tube formation, and finally differentiation into a mature vessel. Melanoma metastases tend to be very vascular making melanoma a possible candidate for antiangiogenic therapy (pathogenesis discussed in Section 1.5.).

Thalidomide has antiangiogenic and immunomodulatory properties and has been used successfully in the treatment of Kaposi's sarcoma, myeloma, and renal cell cancer *(114)*. Some recent trials in melanoma that added thalidomide to temozolomide have reported improved RRs when this combination is compared with temozolomide alone, but others in the cooperative groups have not *(115,116)*.

Melanoma cells are also known to express the $\alpha_{v\beta3}$ integrin, a protein that has been shown to be essential for angiogenesis (Section 1.5.). MEDI-522 is a humanized monoclonal antibody to $\alpha_{v\beta3}$ integrin engineered from the murine monoclonal LM609, an antibody directed against a conformational epitope of the $\alpha_{v\beta3}$ integrin *(117)*. A phase II, randomized, open-label study evaluating the antitumor activity of MEDI-522 ± DTIC in patients with MM has been conducted at the University of Pittsburgh Cancer Institute. MEDI-522 with or without DTIC appears well tolerated. The preliminary OS results in both arms suggest potential clinical activity of MEDI-522 ± DTIC in MM. However, MEDI 522 as antitumor agent alone did not show any compelling activity.

Other antiangiogenic agents currently under investigation as potential therapeutic agents against melanoma include Lenalidomide, a potent analog of thalidomide *(118)*, Semaxanib, a selective inhibitor of VEGF receptor 2 *(119)*, and Bevacizumab, a monoclonal antibody against VEGF *(120)*.

3.2.4. IMMUNE MODULATORS AND VACCINES

The role of IFNs in the adjuvant therapy of melanoma has been reviewed earlier (Section 3.1.2.1.). This section outlines the newer nonspecific and specific immune modulators that have been under investigation for melanoma treatment.

In vitro experiments have shown that addition of histamine to IL-2 increases its antitumor efficacy in lymphoma and melanoma cells *(121)*. Pilot studies in small samples of patients with MM also have shown improved RRs when histamine was added to IL-2 *(122)*. A phase III trial randomized patients with melanoma to histamine dihydrochloride combined with IL-2 versus IL-2 alone in patients with stage IV disease. Although there was a trend toward an improvement in median survival for the combination therapy (9.1 vs. 8.2 months, P-value not significant) and subgroup analysis of patients with liver metastases showed a significant improvement in median survival for the IL-2/histamine dihydrochloride combination (9.4 vs. 5.1 months, P = 0.008) *(123)*, two subsequent larger randomized phase III trials comparing IL-2 + histamine (Maxim Pharmaceuticals Press Conference on phase III trial for advanced melanoma fails to meet its primary endpoint September 20, 2004) and IL-2 + IFN + histamine versus DTIC *(124)* in patients with MM showed no survival advantage.

3.2.4.1. Vaccines. Ultimately, the most selective therapies may be those with immunological specificity for melanoma. Over the last decade, dramatic improvements in our understanding of tumor immunology have occurred, and several different vaccine-based therapeutic approaches have been developed and tested for melanoma in the clinic, both for MM and in the adjuvant setting. There are several factors that make melanoma an attractive potential target for active specific immunotherapy, such as expression of developmental and tumor-specific antigens on tumor tissue, the frequent infiltration of lymphocytes at the primary site of the cancer and its potential prognostic significance, the evidence for objective tumor regressions with cytokines such as IFNα and IL-2, and the isolation of functional T cells that recognize melanoma antigens from melanoma patients (125). However, despite the several different approaches and trials, use of vaccines in melanoma management still remains investigational. In this section, we have presented the different approaches through representative samples as a detailed discussion of all the trials is beyond the scope of this review.

Antibody responses detected in the sera of patients with melanoma have identified the importance of the gangliosides, which are a series of glycolipids, i.e. GM1, GM2, GM3, and GD2, GD3. These molecules have been recognized as major constituents of melanoma cells, but in general, the gangliosides have not been found to be immunogenic. An exception to this rule is the ganglioside GM2, to which antibody responses have been detected in up to five patients studied in the untreated setting and 10% of patients immunized with whole-cell tumor vaccines. A trial of the ganglioside vaccine GM2 showed an improvement in RFS in the subset of patients that developed antibodies to the vaccine when compared with Bacillus Calmette-Guerin (BCG) (126). However, on direct evaluation of the GM2 vaccine in comparison with HDI (in the intergroup trial E1694), there was no evidence of benefit from the vaccine and HDI proved to be significantly more effective in preventing relapse as well as death from melanoma (127).

A number of recent trials have examined the benefit of peptide-based vaccination with antigens derived from proteins present within melanoma cells that are bound to the MHC (such as melanoma-associated antigen E (MAGE), as well as pigmentation associated antigens such as Melan-A/MART-1, gp100, and tyrosinase), with or without an immune adjuvant (128), with some clinical responses. It is hoped that the use of multiple peptides, epitopes, or proteins will generate multi-specific cytotoxic T lymphocytes (CTLs), translating into a better clinical response. A recently completed study by ECOG using multiepitope vaccine targeting multiple CD8 T-cell epitopes of lineage antigens expressed in melanoma ± GM-CSF and IFNα-2b showed that an immune response to the tyrosinase antigen appeared to correlate with progression-free survival (129). Polyvalent vaccines derived from cell lysates of melanoma cell lines, Melacine (130) and Cancervax (131), have shown various clinical responses. A phase III trial indicated a survival benefit for Melacine in the subset of melanoma patients who express the HLA class I antigens A2 and/or HLA-C3. This finding will now require prospective confirmation, although transitions in the ownership of this product make it unlikely that this will ever occur (132). Both for stage III resectable and stage IV resectable disease, large prospective randomized controlled trials have been closed for futility analyses in the past 1–2 years, indicating that Cancervax (C-vax) has no significant benefit in the adjuvant treatment of melanoma. An autologous tumor-derived heat-shock protein gp-96 peptide complex has shown activity in a phase II study (133) with phase III studies underway.

A number of gene therapy-based vaccine strategies are under investigation, most commonly using viral or plasmid DNA delivery systems. Studies have examined in vitro transfection of autologous cells with genes encoding GM-CSF *(134)*, melanoma-derived CD8+ T-cell epitopes, or melanoma-associated antigens *(135)*.

DCs are central to the induction of immune responses and may be used as adjuvants for tumor vaccine therapy. DCs can be pulsed in vitro with many antigens before being adoptively transferred back into the patient. Vaccines developed in this way tend to generate immunity with acceptable toxicity and some tumor responses in melanoma patients *(136)*. A randomized study comparing DTIC with autologous DCs pulsed with peptide antigens however failed to show any benefit associated with the vaccination strategy *(137)*. Adoptive transfer of highly selected tumor-reactive T cells directed against overexpressed self-derived differentiation antigens after a nonmyeloablative conditioning regimen to patients with MM is another novel and promising approach shown to result in regression of the patients' MM as well as to the onset of autoimmune melanocyte destruction *(138)*.

3.2.4.2. Newer Immune Modulators. Cytotoxic T-cell-associated antigen-4 (CTLA-4) is a critical immunomodulatory molecule. It is expressed on activated T cells and some other regulatory T cells and is capable of down-regulating T-cell activation. Blockade of CTLA-4 can potentially enhance T-cell-dependent immunity and so increase response to vaccine therapies and a number of other treatment approaches. In murine systems, the administration of antibodies that block CTLA-4 function inhibits the growth of moderately immunogenic tumors *(139)* and, in combination with cancer vaccines, increases the rejection of poorly immunogenic tumors, albeit with a loss of tolerance to normal differentiation antigens *(140)*. In a recent study of melanoma patients, serial IV administration of a fully human anti-CTLA-4 antibody (MDX-010) in conjunction with sc vaccination with two modified HLA-A*0201-restricted peptides from the gp100 melanoma-associated antigen, induced objective cancer regression in 21% of the patients, at the cost of grade III/IV autoimmune manifestations in 43% of the patients *(141)*. Another group of immunomodulators under investigation are activators of human toll-like receptors (TLRs), such as Imiquimod and Resiquimod, as vaccine adjuvants and nonspecific immune stimulants *(142)*. Human TLRs are crucial for the recognition of invading pathogens and for the activation of both innate and adaptive immunity. TLRs are preferentially expressed on cells of the innate immune system. Clinical trials investigating their efficacy in melanoma therapy are underway.

Another immunotherapy employed for MM is IL-2. Although there are no studies of IL-2 that have demonstrated a specific molecular mechanism of action that would serve to guide more rational combinations based on this modality, it is hoped that future work on IL-2 will permit an understanding of its mechanism.

REFERENCES

1. de Vries, E., Bray, F. I., Coebergh, J. W. W., and Parkin, D. M. Changing epidemiology of malignant cutaneous melanoma in Europe 1953–1997: rising trends in incidence and mortality but recent stabilizations on western Europe and decreases in Scandinavia. *Int. J. Cancer*, *107*: 119–126, 2003.
2. Manola, J., Atkins, M., Ibrahim, J., and Kirkwood, J. Prognostic factors in metastatic melanoma: a pooled analysis of eastern cooperative oncology group trials. *J. Clin. Oncol.*, *18*: 3782–3793, 2000.

3. Jemal, A., Thomas, A., Murray, T., and Thun, T. Cancer Statistics, 2002. *CA Cancer J. Clin.*, *52*: 23–47, 2002.

4. Li, G., Satyamoorthy, K., and Herlyn, M. Dynamics of cell interactions and communications during melanoma development. *Crit. Rev. Oral Biol. Med.*, *13*: 62–70, 2002.

5. Halaban, R. The regulation of normal melanocyte proliferation. *Pig. Cell Res.*, *13*: 4–14, 2000.

6. IARC. *Solar and Ultraviolet Radiation. International Agency for Research on Cancer Monographs on the Evaluation of Carcinogenic Risk to Humans*, Vol. 55. Lyon, France: International Agency for Research on Cancer, 1992.

7. Valery, C., Grob, J. J., and Verrando, P. Identification by cDNA microarray technology of genes modulated by artificial ultraviolet radiation in normal human melanocytes: relation to melanocarcinogenesis. *J. Invest. Dermatol.*, *117*: 1482, 2001.

8. Jean, S., Bideau, C., Bellon, L., Halimi, G., De Méo, M., Orsière, T., Dumenil, G., Bergé-Lefranc, J.-L., and Botta, A. The expression of genes induced in melanocytes by exposure to 35-nm UVA: study by cDNA arrays and real-time quantitative RT-PCR. *Biochim. Biophys. Acta*, *1522*: 89–96, 2001.

9. Kadekaro, A. L., Kavanagh, R. J., Wakamatsu, K., Ito, S., Pipitone, M. A., and Abdel-Malek, Z. A. Cutaneous photobiology. The melanocyte vs. the sun: who will win the final round? *Pig. Cell Res.*, *16*: 434–447, 2003.

10. Noonan, F. P., Recio, J. A., and Takayama, H. Neonatal sunburn and melanoma in mice. *Nature*, *413*: 271–272, 2001.

11. Jhappan, C., Noonan, F. P., and Merlino, G. Ultraviolet radiation and cutaneous malignant melanoma. *Oncogene*, *22*: 3099–3112, 2003.

12. Kannan, K., Sharpless, N. E., Xu, J., O'Hagan, R. C., Bosenberg, M., and Chin, L. Components of the Rb pathway are critical targets of UV mutagenesis in a murine melanoma model. *Proc. Natl. Acad. Sci. U. S. A.*, *100*: 1221–1225, 2003.

13. Dazard, J.-E., Gal, H., Amariglio, N., Rechavi, G., Domany, E., and Givol, D. Genome-wide comparison of human keratinocyte and squamous cell carcinoma responses to UVB irradiation: implications for skin and epithelial cancer. *Oncogene*, *22*: 2993–3006, 2003.

14. Clydesdale, G. J., Dandie, G. W., and Muller, H. K. Ultraviolet light induced injury: immunological and inflammatory effects. *Immunol. Cell Biol.*, *79*: 547–568, 2001.

15. Tada, A., Pereira, E., Beitner-Johnson, D., Kavanagh, R., and Abdel-Malek, Z. A. Mitogen- and ultraviolet-B-induced signaling pathways in normal human melanocytes. *J. Invest. Dermatol.*, *118*: 316–322, 2002.

16. Jamal, S. and Schneider, R. UV-induction of keratinocyte endothelin-1 down regulates E-cadherin in melanocytes and melanoma cells. *J. Clin. Invest.*, *110*: 443–452, 2002.

17. Hussussian, C. J., Struewing, J. P., Goldstein, A. M., Higgins, P. A., Ally, D. S., Sheahan, M. D., Clark, W. H., Tucker, M. A., and Dracopoli, N. C. Germline p16 mutations in familial melanoma. *Nat. Genet.*, *8*: 15–21, 1994.

18. Zuo, L., Weger, J., Yang, Z., et al. Germline mutations in the p16INK4a binding domain of CDK4 in familial melanoma. *Nat. Genet.*, *12*: 97–99, 1996.

19. Castellano, M. and Parmiani, G. Genes involved in melanoma: an overview of INK4a and other loci. *Melanoma Res.*, *9*: 421–432, 1999.

20. Quelle, D. E., Zindy, F., Ashnum, R. A., and Sherr, C. J. Alternative reading frames of the INK4a tumor suppressor gene encode two unrelated proteins capable of inducing cell cycle arrest. *Cell*, *83*: 993–1000, 1995.

21. Wolfel, T., Hauer, M., Schneider, J., Serrano, M., Wolfel, C., Klehmann-Hieb, E., DePlaen, E., Hankeln, T., Meyer zum Buschenfelde, K. H., and Beach, D. A p161NK4a-insensitive CDK4 mutant targeted by cytolytic T lymphocytes in a human melanoma. *Science*, *269*: 1281–1284, 1995.

22. Honda, R. and Yasuda, H. Association of p19^{ARF} with Mdm2 inhibits ubiquitin ligase activity of Mdm2 for tumor suppressor p53. *EMBO* J, 18: 22–27, 1999.

23. Gwosdz, C., Scheckenbach, K., Lieven, O., Reifenberger, J., Knopf, A., Bier, H., and Balz, V. Comprehensive analysis of the p53 status in mucosal and cutaneous melanomas. *Int. J. Cancer*, 118:577–582, 2006.

24. Sauter, E. R., Yeo, U.-C., von Stemm, A., Zhu, W., Litwin, S., Tichansky, D. S., Pistritto, G., Nesbit, M., Pinkel, D., Herlyn, M., and Bastian, B. C. Cyclin D1 is a candidate oncogene in cutaneous melanoma. *Cancer Res.*, *62*: 3200–3206, 2002.

25. Nelson, M. A., Ariza, M. E., Yang, J. M., Thompson, F. H., Taetle, R., Trent, J. M., Wymer, J., Massey-Brown, K., Broome-Powel, M., Easton, J., Lahti, J. M., and Kidd, V. J. Abnormalities in the p34^{cdc2}-related PITSLRE protein kinase gene complex (CDC2L) on chromosome band 1p36 in melanoma. *Cancer Genet. Cytogenet.*, *108*: 91–99, 1999.

26. Bales, E. S., Dietrich, C., Bandyopadhyay, D., Schwahn, D. J., Weidong, X., Didenko, V., Leiss, P., Conrad, N., Pereira-Smith, O., Orengo, I., and Medrano, E. E. High levels of expression of p27^{KIP1} and cyclin E in invasive primary malignant melanomas. *J. Invest. Dermatol.*, *113*: 1039–1046, 1999.

27. Johnson, G. R. and Lapadat, R. Mitogen-activated protein kinase pathways mediated by ERK, JNK, and p38 protein kinase. *Science*, *298*: 1911–1912, 2002.

28. Busca, R., Abbe, P., Mantoux, F., Aberdam, E., Peyssonnaux, C., Eychène, A., Ortonne, J.-P., and Ballotti, R. Ras mediates the cAMP-dependent activation of extracellular signal-regulated kinases (ERFs) in melanocytes. *EMBO J.*, 19: 2900–2910, 2000.

29. Davies, H., Bignell, G., Cox, C., Stephens, P., Edkins, S., Clegg, S., Teague, J., Woffendin, H., Garnett, M., Bottomley, W., Davis, N., Dicks, E., Ewing, R., Floyd, Y., Gray, K., Hall, S., Hawes, R., Hughes, J., Kosmidou, V., Menzies, A., Mould, C., Parker, A., Stevens, C., Watt, S., Hooper, S., Wilson, R., Jayatilake, H., Gusterson, B., Cooper, C., Shipley, J., Hargrave, D., Pritchard-Jones, K., Maitland, N., Chenevix-Trench, G., Riggins, G., Bigner, D., Palmieri, G., Cossu, A., Flanagan, A., Nicholson, A., Ho, J., Leung, S., Yuen, S., Weber, B., Seigler, H., Darrow, T., Paterson, H., Marais, R., Marshall, C., Wooster, R., Stratton, M., and Futreal, P. Mutations of the BRAF gene in human cancer. *Nature*, *417*: 949–954, 2002.

30. Pollock, P. M., Harper, L., Hansen, K. S., Yudt, M. S., Robbins, C. R., Moses, T. Y., Galen, H., Wagner, U., Kakareka, J., Salem, G., Pohida, T., Heenan, P., Duray, P., Kallioniemi, O., Hayward, N. K., Trent, J. M., and Meltzer, P. S. High frequency of BRAF mutations in nevi. *Nat. Genet.*, *33*: 19–20, 2003.

31. Demunter, A., Stas, M., Degreef, H., De Wolf-Peeters, C., and van den Oord, J. J. Analysis of N- and K-ras mutations in the distinctive tumor progression phases of melanoma. *J. Invest. Dermatol.*, *117*: 1483–1489, 2001.

32. Powell, M. B., Hyman, P., Bell, O. D., Balmain, A., Brown, K., Alberts, D., and Bowden, G. T. Hyperpigmentation and melanocytic hyperplasia in transgenic mice expressing the human T24 Ha-ras gene regulated by a mouse tyrosinase promoter. *Mol Carcinog.*, *12*: 82–90, 1995.

33. Bastian, B. C. Hypothesis: A role for telomere crisis in spontaneous regression of melanoma. *Arch. Dermatol.*, *139*: 667–668, 2003.

34. Ramirez, R. D., D'Atri, S., Pagani, E., Faraggiana, T., Lacal, P. M., Taylor, R. S., and Shay, J. W. Progressive increase in telomerase activity from benign melanocytic conditions to malignant melanoma. *Neoplasia (New York)*, *1*: 42–49, 1999.

35. Zhai, S., Yaar, M., Doyle, S. M., and Gilchrest, B. A. Nerve growth factor rescues pigment cells from ultraviolet-induced apoptosis by upregulating BCL-2 levels. *Exp. Cell Res.*, *224*: 335–343, 1996.

36. Hodgkinson, C. A., Moore, K. J., Nakayama, A., Steingrimsson, E., Copeland, N. G., Jenkins, N. A., and Arnheiter, H. Mutations at the mouse microphthalmia locus are associated with defects in a gene encoding a novel basic-helix-loop-helix-zipper protein. *Cell*, *74*: 395–404, 1993.

37. Glinsky, G. V., Glinsky, V. V., Ivanova, A. B., and Hueser, C. J. Apoptosis and metastasis: increased apoptosis resistance of metastatic cancer cells is associated with the profound deficiency of apoptosis execution mechanisms. *Cancer Lett.*, *115*: 185–193, 1997.

38. Irmler, M., Thome, M., Hahne, M., Schneider, P., Hofmann, K., Steiner, V., Bodmer, J.-L., Schröter, M., Burns, K., Mattmann, C., Rimoldi, D., French, L. E., and Tschopp, J. Inhibition of death receptor signals by cellular FLIP. *Nature*, *388*: 190–195, 1997.

39. Li, F., Ambrosini, G., Chu, E. Y., Plescia, J., Tognin, S., Marchisio, P. C., and Altieri, D. C. Control of apoptosis and mitotic spindle checkpoint by survivin. *Nature*, *396*: 580–584, 1998.

40. Ambrosini, G., Adida, C., Sirugo, G., and Altieri, D. C. Induction of apoptosis and inhibition of cell proliferation by *survivin* gene targeting. *J. Biol. Chem.*, *273*: 11177–11182, 1998.

41. Vucic, D., Deshayes, K., Ackerly, H., Pisabarro, M. T., Kadkhokayan, S., Fairbrother, W. J., and Dixit, V. M. SMAC negatively regulates the anti-apoptotic activity of melanoma inhibitor of apoptosis (ML-IAP)*. *J. Biol. Chem.*, *277*: 12275–12279, 2002.

42. Bullani, R. R., Huard, B., Viard-Leveugle, I., Byers, H. R., Irmler, M., Saurat, J.-H., Tschopp, J., and French, L. E. Selective expression of FLIP in malignant melanocytic skin lesions. *J. Invest. Dermatol.*, *117*: 360–364, 2001.

43. Ugurel, S., Seiter, S., Rappl, G., Stark, A., Tilgen, W., and Reinhold, U. Heterogenous susceptibility to CD95-induced apoptosis in melanoma cells correlates with bcl-2 and bcl-x expression and is sensitive to modulation by interferon-y. *Int. J. Cancer*, *82*: 727–736, 1999.

44. Soengas, M. S., Capodieci, P., Polsky, D., Mora, J., Esteller, M., Opitz-Araya, X., McCombie, R., Herman, J. G., Gerald, W. L., Lazebnik, Y. A., Cordón-Cardó, C., and Lowe, S. W. Inactivation of the apoptosis effector *Apaf-1* in malignant melanoma. *Nature*, *409*: 207–211, 2001.

45. Birck, A., Ahrenkiel, V., Zeuthen, J., Hou-Jensen, K., and Guldberg, P. Mutation and allelic loss of the PTEN/MMAC1 gene in primary and metastatic melanoma biopsies. *J. Invest. Dermatol.*, *114*: 277–280, 2000.

46. Zhou, X.-P., Gimm, O., Hampel, H., Niemann, T., Walker, M. J., and Eng, C. Epigenetic PTEN silencing in malignant melanomas without PTEN mutation. *Am. J. Pathol.*, *157*: 1123–1128, 2000.

47. McNulty, S. E., Del Rosario, R., Cen, D., Meyskens, F. L., and Yang, S. Comparative expression of NF_{KB} proteins in melanocytes of normal skin vs. benign intradermal naevus and human metastatic melanoma biopsies. *Pig. Cell Res.*, *17*: 173–180, 2004.

48. McNulty, S. E., Tohidian, N. B., and Meyskens, F. L. RelA, p50 and Inhibitor of kappa B alpha are elevated in human metastatic melanoma cells and respond aberrantly to ultraviolet light B. *Pig. Cell Res.*, *14*: 456–465, 2001.

49. Buscà, R. and Ballotti, R. Cyclic AMP a key messenger in the regulation of skin pigmentation. *Pig. Cell Res.*, *13*: 60–69, 2000.

50. Harsanyi, Z. P., Post, P. W., Brinkmann, J. P., Chedekel, M. R., and Deibel, R. M. Mutagenicity of melanin from human red hair. *Experientia*, *36*: 291–292, 1980.

51. Scot, M. C., Wakamatsu, K., Shosuke, I., Kadedaro, A. L., Kobayashi, N., Groden, J., Kavanagh, R., Takauwa, T., Virador, V., Hearing, V. J., and Abdel-Malek, Z. A. Human melanocortin 1 receptor variants, receptor function and melanocyte response to UV radiation. *J. Cell Sci.*, *115*: 2349–2355, 2002.

52. Box, N. F., Duffy, D. L., Chen, W., Stark, M., Martin, N. G., Sturm, R. A., and Hayward, N. K. MC1R genotype modifies risk of melanoma in families segregating CDKN2A mutations. *Am. J. Hum. Genet.*, *69*: 765–773, 2001.

53. Aaronson, D. S. and Horvath, C. M. A road map for those who don't know JAK-STAT. *Science*, *296*: 1653–1655, 2002.

54. Yu, H. and Jove, R. The stats of cancer - new molecular targets come of age. *Nat. Rev.*, *4*: 97–105, 2004.

55. Yu, H. and Jove, R. The STATs of cancer–new molecular targets come of age. *Nat Rev Cancer*, *4*: 97–105, 2004.

56. Wong, L. H., Krauer, K. G., Hatzinisiriou, I., Estcourt, M. J., Hersey, P., Tam, N. D., Edmondson, S., Devenish, R. J., and Ralph, S. J. Interferon-resistant human melanoma cells are deficient in ISGF3 components, STAT1, STAT2, and p48-ISGF3γ. *J. Biol. Chem.*, *272*: 28779–28785, 1997.

57. Niu, G., Bowman, T., Huang, M., Shivers, S., Reintgen, D., Daud, A., Chang, A., Kraker, A., Jove, R., and Yu, H. Roles of activated Src and Stat3 signaling in melanoma tumor cell growth. *Oncogene*, *21*: 7001–7010, 2002.

58. Xu, Q., Briggs, J., Park, S., Niu, G., Kortylewski, M., Zhang, S., Gritsko, T., Turkson, J., Kay, H., Semenza, G. L., Cheng, J. Q., Jove, R., and Yu, H. Targeting Stat3 blicks both HIF-1 and VEGF expression induced by multiple oncogenic growth signaling pathways. *Oncogene*, *24*: 5552–5560, 2005.

59. Xie, T.-X., Wei, D., Liu, M., Gao, A. C., Ali-Osman, F., Sawaya, R., and Huang, S. Stat3 activation regulates the expression of matrix metalloproteinase-2 and tumor invasion and metastasis. *Oncogene*, *23*: 3550–3560, 2004.

60. Wang, T., Niu, G., Kortylewski, M., Burdelya, L., Shain, K., Zhang, S., Bhattacharya, R., Gabrilovich, D., Heller, R., Coppola, D., Dalton, W., Jove, R., Pardoll, D., and Yu, H. Regulation of the innate and adaptive immune responses by Stat-3 signaling in tumor cells. *Nat. Med.*, *10*: 48–54, 2004.

61. Kortylewski, M., Jove, R., and Yu, H. Targeting STAT3 affects melanoma on multiple fronts. *Cancer Metastasis Rev.*, *24*: 315–327, 2005.

62. Hsu, M. Y., Wheelock, M. J., Johnson, K. R., and Herlyn, M. Shifts in cadherin profiles between human normal melanocytes and melanomas. *J. Investig. Dermatol. Symp. Proc.*, *1*: 188–194, 1996.

63. Xie, S., Luca, M., Huang, S., Gutman, M., Reich, R., Johnson, J. P., and Bar-Eli, M. Expression of MCAM/MUC18 by human melanoma cells leads to increased tumor growth and metastasis. *Cancer Res.*, *57*: 2295–2303, 1997.

64. Voura, E. B., Ramjeesingh, R. A., Montgomery, A. M. P., and Siu, C.-H. Involvement of integrin α_vβ3 and cell adhesion molecule L1 in transendothelial migration of melanoma cells. *Mol. Biol. Cell*, *12*: 2699–2710, 2001.

65. Ito, A., Katoh, F., Kataoka, T. R., Okada, M., Tsubota, N., Asada, H., Yoshikawa, K., Maeda, S., Kitamura, Y., Yamasaki, H., and Nojima, H. A role for heterologous gap junctions between melanoma and endothelial cells in metastasis. *J. Clin. Invest.*, *105*: 1189–1197, 2000.

66. Velazquez, O. C. and Herlyn, M. The vascular phenotype of melanoma metastasis. *Clin. Exp. Metastasis*, *20*: 229–235, 2003.

67. Barnhill, R. L., Fandrey, K., Levy, M. A., Mihm, M. C., Jr., and Hyman, B. Angiogenesis and tumor progression of melanoma quantification of vascularity in melanocytic nevi and cutaneous malignant melanoma. *Lab. Invest.*, *67*: 331–337, 1992.

68. Srivastava, A., Laidler, P., Hughes, L. E., Woodcock, J., and Shedden, E. J. Neovascularization in human cutaneous melanoma: a quantitative morphological and Doppler ultrasound study. *Eur. J. Cancer Clin. Oncol.*, *22*: 1205–1209, 2003.

69. Massi, D., Franchi, A., Borgognoni, L., Paglierani, M., Reali, U. M., and Santucci, M. Tumor angiogenesis as a prognostic factor in thick cutaneous malignant melanoma. A quantitative morphologic analysis. *Virchows Arch.*, *440*: 22–28, 2002.

70. Kashani-Sabet, M., Sagebiel, R. W., Carlow, M., Ferreira, M., and Miller, J. R. Tumor vascularity in the prognostic assessment of primary cutaneous melanoma. *J Clin Oncol*, *20*: 1826–1831, 2002.

71. Streit, M. and Detmar, M. Angiogenesis, lymphangiogenesis, and melanoma metastasis. *Oncogene*, *22*: 3172–3179, 2003.

72. Seftor, E. A., Meltzer, P. S., Kirschmann, D. A., Peer, J., Maniotis, A. J., Trent, J. M., Folberg, R., and Hendrix, M. J. C. Molecular determinants of human uveal melanoma invasion and metastasis. *Clin. Exp. Metastasis*, *19*: 233–246, 2002.

73. Hendrix, M. J. C., Seftor, E. A., Hess, A. R., and Seftor, R. E. B. Vasculogenic mimicry and tumour-cell plasticity: lessons from melanoma. *Nat. Rev. Cancer*, *3*: 411–421, 2003.

74. Van Belle, P. A., Elenitsas, R., Satyamoorthy, K., Wolfe, J. T., Guerry, D., Schuchter, L., Van Belle, T. J., Albelda, S., Tahin, P., Herlyn, M., and Elder, D. E. Progression-related expression of beta3 integrin in melanomas and nevi. *Hum. Pathol.*, *30*: 562–567, 1999.

75. Mitjans, F., Meyer, T., Fittschen, C., Goodman, S., Jonczyk, A., Marshall, J. F., Reyes, G., and Piulats, J. In vivo therapy of malignant melanoma by means of antagonists of αv integrins. *Int. J. Cancer*, *87*: 716–723, 2000.

76. Anichini, A., Vegetti, C., and Mortarini, R. The paradox of T cell-mediated antitumor immunity in spite of poor clinical outcome in human melanoma. *Cancer Immunol. Immunother.*, *53*: 855–864, 2004.

77. Agarwala, S. S. and Kirkwood, J. M. Adjuvant therapy of melanoma. *Semin. Surg. Oncol.*, *14*: 302–310, 1998.

78. Clemente, C. G., Mihm, M. C., Jr., Bufalino, R., Zurrida, S., Collini, P., and Cascinelli, N. Prognostic value of tumor infiltrating lymphocytes in the vertical growth phase of primary cutaneous melanoma. *Cancer*, *77*: 1303–1310, 1996.

79. Marincola, F. M., Wang, E., Herlyn, M., Seliger, B., and Ferrone, S. Tumors as elusive targets of T-cell-based active immunotherapy. *Trend Immunol.*, *24*: 334–341, 2003.

80. Uyttenhove, C., Pilotte, L., Théate, I., Stroobant, V., Colau, D., Parmentier, N., Boon, T., and Van den Eynde, B. J. Evidence for a tumoral immune resistance mechanism based on tryptophan degradation by indoleamine 2,3-dioxygenase. *Nat. Med.*, *9*: 1269–1274, 2005.

81. Murakami, T., Cardones, A. R., Finkelstein, S. E., Restifo, N. P., Klaunberg, B. A., Nestle, F. O., Castillo, S. S., Dennis, P. A., and Hwang, S. T. Immune evasion by murine melanoma mediated through CC chenokine recptor-10. *J. Exp. Med.*, *198*: 1337–1347, 2003.

82. Péguet-Navarro, J., Sportouch, M., Popa, I., Berthier, O., Schmitt, D., and Portoukalian, J. Gangliosides from human melanoma tumors impair dendritic cell differentiation from monocytes and induce their apoptosis. *J. Immunol.*, *170*: 3488–3494, 2003.

83. Geertsen, R. C., Hofbauer, G. F. L., Yue, F.-Y., Manolio, S., Berg, G., and Dummer, R. Higher frequency of selective losses of HLA-A and -B allospecificities in metastasis than in primary melanoma lesions. *J. Invest. Dermatol.*, *111*: 497–502, 1998.

84. Kageshita, T., Hirai, S., Ono, T., Hicklin, D. J., and Ferrone, S. Down-regulation of HLA class I antigen-processing molecules in malignant melanoma. *Am. J. Pathol.*, *154*: 745–754, 1999.

85. Dissemond, J., Kothen, T., Mörs, J., Weinann, T. K., Lindeke, A., Goos, M., and Wagner, S. N. Downregulation of tapasin expression in progressive human malignant melanoma. *Arch. Dermatol. Res.*, *295*: 43–49, 2003.

86. Mortarini, R., Piris, A., Maurichi, A., Molla, A., Bersani, I., Bono, A., Bartoli, C., Santinami, M., Lombardo, C., Ravagnani, F., Cascinelli, N., Parmiani, G., and Anichini, A. Lack of terminally differentiated tumor-specific CD8$^+$ T cells at tumor site in spite of antitumor immunity to self-antigens in human metastatic melanoma. *Cancer Res.*, *63*: 2535–2545, 2003.

87. Wang, H. Y., Lee, D. A., Peng, G., Guo, Z., Li, Y., Kiniwa, Y., Shevach, E. M., and Wang, R.-F. Tumor-specific human CD4$^+$ regulatory T cells and their ligands: implications for immunotherapy. *Immunity*, *20*: 107–118, 2004.

88. Buzaid, A. C., Ross, M. I., Balch, C. M., Soong, S.-J., McCarthy, W. H., Tinoco, L., Mansfield, P., Lee, J. E., Bedikian, A., Eton, O., Plager, C., Papadopoulos, N., Legha, S. S., and Benjamin, R. S. Critical analysis of the current American Joint Committee on Cancer staging system for cutaneous melanoma and proposal of a new staging system. *J. Clin. Oncol.*, *15*: 1039–1051, 1997.

89. Kirkwood, J. M., Manola, J., Ibrahim, J., Sondak, V., Ernstoff, M. S., and Rao, U. A pooled analysis of Eastern Cooperative Oncology Group and intergroup trials of adjuvant high-dose interferon for melanoma. *Clin.Cancer Res.*, *10*: 1670–1677, 2004.

90. Kirkwood, J. M., Strawderman, M. H., Ernstoff, M. S., Smith, T. J., Borden, E. C., and Blum, R. H. Interferon alfa-2b adjuvant therapy of high-risk resected cutaneous melanoma: the Eastern Cooperative Oncology Group Trial EST 1684. *J. Clin. Oncol.*, *14*: 7–17, 1996.

91. Kirkwood, J. M., Ibrahim, J. G., Sondak, V. K., Richards, J., Flaherty, L. E., Ernstoff, M. S., Smith, T. J., Rao, U. N. M., Steele, M., and Blum, R. H. High- and low-dose interferon alfa-2b in high-risk melanoma: first analysis of intergroup trial E1690/S9111/C9190. *J. Clin. Oncol.*, *18*: 2444–2458, 2000.

92. Kirkwood, J. M., Ibrahim, J., Lawson, D. H., Atkins, M. B., Agarwala, S. S., Collins, K., Mascari, R., Morrissey, D. M., and Chapman, P. B. High-dose interferon alfa-2b does not diminish antibody response to GM2 vaccination in patients with resected melanoma: results of the multicenter Eastern Cooperative Oncology Group phase II trial E2696. *J. Clin. Oncol.*, *19*: 1430–1436, 2001.

93. Kirkwood, J. M., Bender, C., Agarwala, S. S., Tarhini, A., Shipe-Spotloe, J., Smith, D., Smelko, B., Donnelly, S., and Stover, L. Mechanisms and management of toxicities associated with high-dose interferon alfa-2b therapy. *J. Clin. Oncol.*, *20*: 3703–3718, 2002.

94. Caraglia, M., Marra, M., Pelaia, G., Maselli, R., Caputi, M., Marsico, S. A., and Abbruzzese, A. Alpha-interferon and its effects on signal transduction pathways. *J. Cell. Physiol.*, *202*: 323–335, 2005.

95. Belardelli, F., Ferrantini, M., Proietti, E., and Kirkwood, J. M. Interferon-alpha in tumor immunity and immunotherapy. *Cytokine Growth Factor Rev.*, *13*: 119–134, 2002.

96. Beadling, C., Guschin, D., Witthuhn, B. A., Ziemiecki, A., Ihle, J. N., Kerr, I. M., and Cantrell, D. A. Activation of JAK kinases and STAT proteins by interleukin-2 and interferon α, but not the T cell antigen receptor, in human T lymphocytes. EMBO J., 13: 5605–5615, 1994.

97. Yang, C.-H., Murti, A., and Pfeffer, L. M. STAT3 complements defects in an interferon-resistant cell line: Evidence for an essential role for STAT3 in interferon signaling and biological activities. *Proc. Natl. Acad. Sci. U. S. A.*, *95*: 5568–5572, 1998.

98. Fish, E. N., Uddin, S., Korkmaz, M., Majchrzak, B., Druker, B. J., and Platanias, L. C. Activation of a CrkL-stat5 signaling complex by type I interferons. *J. Biol. Chem.*, *274*: 571–573, 1999.

99. Sangfelt, O., Erickson, S., Castro, J., Heiden, T., Einhorn, S., and Grander, D. Induction of apoptosis and inhibition of cell growth are independent responses to interferon-alpha in hematopoietic cell lines. *Cell Growth Differ.*, *8*: 343–352, 1997.

100. Thyrell, L., Erickson, S., Zhivotovsky, B., Pokrovskaja, K., Sangfelt, O., Castro, J., Einhorn, S., and Grander, D. Mechanisms of Interferon-alpha induced apoptosis in malignant cells. *Oncogene*, *21*: 1251–1262, 2002.

101. Panaretakis, T., Pokrovskaja, K., Shoshan, M., and Grandér, D. Interferon-α-induced apoptosis in U266 cells is associated with activation of the proapoptotic Bcl-2 family members Bak and Bax. *Oncogene*, *22*: 4543–4556, 2003.

102. David, M., Petricoin, E., III, Benjamin, C., Pine, R., Weber, M. J., and Larner, A. C. Requirement for MAP kinase (ERK2) activity in interferon alpha- and interferon beta-stimulated gene expression through STAT proteins. *Science*, *269*: 1721–1723, 1995.

103. Gil, J. and Esteban, M. Induction of apoptosis by the dsRNA-dependent protein kinase (PKR): mechanism of action. *Apoptosis*, 5: 107–114, 2000.

104. Moschos, S. J., Edington, H. D., Land, S. R., Rao, U. N., Jukic, D., Shipe-Spotloe, J., and Kirkwood, J. M. Neoadjuvant treatment of regional stage IIIB melanoma with high-dose interferon alfa-2b induces objective tumor regression in association with modulation of tumor infiltrating host cellular immune responses. *J. Clin. Oncol.*, 24: 3164–3171, 2006.

105. Wang, W., Edington, H., Rao, U., Jukic, D., Moschos, S., Mascari, R., Sander, C., Becker, D., Ferrone, S., and Kirkwood, J. M. Effects of neoadjuvant high-dose interferon (IFNα2β) upon STAT signaling, IFNαRβ, MHC and Tap expression in lymph node metastatic melanoma (UPCI 008). *Proc. Am. Assoc. Cancer Res.*, 13: 1523–31, 2005.

106. Collisson, E. A., De, A., Suzuki, H., Gambhir, S. S., and Kolodney, M. S. Treatment of metastatic melanoma with an orally available inhibitor of the Ras-Raf-MAPK cascade. *Cancer Res.*, 63: 5669–5673, 2003.

107. Karasarides, M., Chiloeches, A., Hayward, R., Niculescu-Duvaz, D., Scanlon, I., Friedlos, F., Ogilvie, L., Hedley, D., Martin, J., Marshall, C. J., Springer, C. J., and Marais, R. B-RAF is a therapeutic target in melanoma. *Oncogene*, 23: 6292–6298, 2004.

108. Ahmad, T., Marais, R., Pyle, L., James, M., Schwartz, B., Gore, M., and Eiden, T. BAY 43–9006 in patients with advanced melanoma: the Royal Marsden experience. *Proc. Am. Soc. Clin. Oncol.*, 22(14S): 711s, 2004.

109. Flaherty, K. T., Brose, M., Schuchter, L., Tuveson, D., Lee, R., Schwartz, B., Lathia, C., Weber, B., and O'Dwyer, P. Phase I/II trial of BAY 43–9006, carboplatin (C) and paclitaxel (P) demonstrates preliminary antitumor activity in the expansion cohort of patients with metastatic melanoma. *Proc. Am. Soc. Clin. Oncol.*, 22(14S): 711s, 2004.

110. Morgan, G. J., Krishnan, B., Jenner, M., and Davies, F. E. Advances in oral therapy for multiple myeloma. *Lancet Oncol.*, 7: 316–325, 2006.

111. Trisciuoglio, D., Desideri, M., Ciuffreda, L., Mottolese, M., Ribatti, D., Vacca, A., Del Rosso, M., Larcocci, L., Zupi, G., and Del Bufalo, D. Bcl-2 overexpression in melanoma cells increases tumor progression-associated properties and in vivo tumor growth. *J. Cell. Physiol.*, 205: 141–421, 2005.

112. Jansen, B., Wacheck, V., Heere-Ress, E., Schlagbauer-Wadl, H., Hoeller, C., Lucas, T., Hoermann, M., Hollenstein, U., Wolff, K., and Pehamberger, H. Chemosensitisation of malignant melanoma by BCL2 antisense therapy. *Lancet*, 356: 1728–1733, 2000.

113. Kirkwood, J. M., Bedikian, A. Y., Millward, M. J., Conry, R. M., Gore, M. E., Pehamberger, H. E., Sterry, W., Pavlick, A. C., DeConti, R. C., and Itri, L. M. Long-term survival results of a randomized multinational phase 3 trial of dacarbazine (DTIC) with or without Bcl-2 antisense (oblimersen sodium) in patients (pts) with advanced malignant melanoma (MM). *Proc. Am. Soc. Clin. Oncol.*, 23(16S): 711s, 2005.

114. Eisen, T., Boshoff, C., Mak, I., Sapunar, F., Vaughan, M. M., Pyle, L., Johnston, S. R. D., Ahern, R., Smith, I. E., and Gore, M. E. Continuous low dose thalidomide: a phase II study in advanced melanoma, renal cell, ovarian and breast cancer. *Br. J. Cancer*, 82: 812–817, 2000.

115. Danson, S., Lorigan, P., Arance, A., Clamp, A., Ranson, M., Hodgetts, J., Lomax, L., Ashcroft, L., Thatcher, N., and Middleton, M. R. Randomized phase II study of temozolomide given every 8 hours or daily with either interferon alfa-2b or thalidomide in metastatic malignant melanoma. *J. Clin. Oncol.*, 21: 2551–2557, 2003.

116. Hwu, W.-J., Krown, S. E., Menell, J. H., Panageas, K. S., Merrell, J., Lamb, L. A., Williams, L. J., Quinn, C. J., Foster, T., Chapman, P. B., Livingston, P. O., Wolchok, J. D., and Houghton, A. N. Phase II study of temozolomide plus thalidomide for the treatment of metastatic melanoma. *J. Clin.Oncol.*, 21: 3351–3356, 2003.

117. Wu, H., Beuerlein, G., Nie, Y., Smith, H., Lee, B. A., Hensler, M., Huse, W. D., and Watkins, J. D. Stepwise in vitro affinity maturation of Vitaxin, an $\alpha_{v\beta3}$-specific humanized mAb. *Proc. Natl. Acad. Sci. U. S. A.*, 95: 6037–6042, 1998.

118. Bartlett, J. B., Michael, A., Clarke, I. A., Dredge, K., Nicholson, S., Kristeleit, H., Polychronis, A., Pandha, H., Muller, G. W., Stirling, D. I., Zeldis, J., and Dalgleish, A. G. Phase I study to determine the safety, tolerability and immunostimulatory activity of thalidomide analogue CC-5013 in patients with metastatic malignant melanoma and other advanced cancers. *Br. J. Cancer*, 90: 955–961, 2004.

119. Peterson, A. C., Swiger, S., Stadler, W. M., Medved, M., Karczmar, G., Gajewski, T. F. Phase II study of the Flk-1 TK inhibitor SU5416 in patients with advanced melanoma. *Proc. Am. Soc. Clin. Oncol.*, 22: 712, 2003.

120. Carson, W. E., Biber, J., Shah, N., Reddy, K., Kefauver, C., Leming, P. D., Kendre, K., and Walker, M. A phase 2 trial of a recombinant humanized monoclonal anti-vascular endothelial growth factor (VEGF) antibody in patients with malignant melanoma. *Proc. Am. Soc. Clin. Oncol.*, *22*: 715, 2003.

121. Asea, A., Hermodsson, S., and Hellstrand, K. Histaminergic regulation of natural killer cell-mediated clearance of tumor cells in mice. *Scand. J. Immunol.*, *43*: 9–15, 1996.

122. Hellstrand, K., Naredi, P., Lindner, P., Lundholm, K., Rudenstam, C. M., Hermodsson, S., Asztely, M., and Hafstrom, L. Histamine in immunotherapy of advanced melanoma: a pilot study. *Cancer Immunol. Immunother.*, *39*: 416–419, 1994.

123. Agarwala, S. S., Glaspy, J., O'Day, S. J., Mitchell, M., Gutheil, J., Whitman, E., Gonzalez, R., Hersh, E., Feun, L., Belt, R., Meyskens, F., Hellstrand, K., Wood, D., Kirkwood, J. M., Gehlsen, K. R., and Naredi, P. Results from a randomized phase III study comparing combined treatment with histamine dihydrochloride plus interleukin-2 versus interleukin-2 alone in patients with metastatic melanoma. *J. Clin. Oncol.*, *20*: 125–133, 2002.

124. Hauschild, A. First analysis of international M-02 trial: histamine, interferon alpha-2b (IFN), interleukin (IL)-2 vs dacarbazine (DTIC). Proceedings of perspectives in Melanoma Management, Tampa, Florida, 2003 (abstr.).

125. Kawakami, Y., Eliyahu, S., Delgado, C. H., Robbins, P. F., Rivoltini, L., Topalian, S. L., Miki, T., and Rosenberg, S. A. Cloning of the gene coding for a shared human melanoma antigen recognized by autologous T cells infiltrating into tumor. *Proc. Natl. Acad. Sci. U. S. A.*, *91*: 3515–3519, 1994.

126. Livingston, P. O., Wong, G. Y. C., Adluri, S., Tao, Y., Padavan, M., Parente, R., Hanlon, C., Calves, M. J., Helling, F., Ritter, G., Oettgen, H. F., and Old, L. J. Improved survival in stage III melanoma patients with gm2 antibodies: a randomized trial of adjuvant vaccination with GM2 ganglioside. *J. Clin. Oncol.*, *12*: 1036–1044, 1994.

127. Kirkwood, J. M., Ibrahim, J., Sosman, J. A., Sondak, V. K., Agarwala, S. S., Ernstoff, M. S., and Rao, U. High-dose interferon alfa-2b significantly prolongs relapse-free and overall survival compared with the GM2-KLH/QS-21 vaccine in patients with resected stage IIB-III melanoma: results of intergroup trial E1694/S9512/C509801. *J. Clin. Oncol.*, *19*: 2370–2380, 2001.

128. Cebon, J., Jäger, E., Shackleton, M. J., Gibbs, P., Davis, I. D., Hopkins, W., Gibbs, S., Chen, Q., Karbach, J., Jackson, H., MacGregor, D. P., Sturrock, S., Vaughan, H., Maraskovisky, E., Neumann, A., Hoffman, E., Sherman, M., and Knuth, A. Two phase I studies of low dose recombinant human IL-12 with Melan-A and influenza peptides in subjects with advanced malignant melanoma. *Cancer Immun.*, *3*: 7, 2003.

129. Kirkwood, J. M., Lee, S., Land, S., Sander, C., Mascari, R., Weiner, L. M., and Whiteside, T. L. E1696: Final analysis of the clinical and immunological results of a multicenter ECOG phase II trial of multi-epitope peptide vaccination for stage IV melanoma with MART-1 (27–35), gp100 (209–217, 210M), and tyrosinase (368–376, 370D) (MGT) =/- IFNα2b and GM-CSF. *Proc. Am. Soc. Clin. Oncol.*, *22(14S)*: 145. 2004.

130. Sosman, J. A., Weeraratna, A. T., and Sondak, V. K. When will melanoma vaccines be proven effective. *J. Clin. Oncol.*, *22*: 387–389, 2004.

131. Chan, A. D. and Morton, D. L. Active immunotherapy with allogeneic tumor cell vaccines: present status. *Semin. Oncol.*, *25*: 611–622, 1998.

132. Sondak, V. K. and Sosman, J. A. Results of clinical trials with an allogenic melanoma tumor cell lysate vaccine: Melacine®. *Semin. Cancer Biol.*, *13*: 409–415, 2003.

133. Belli, F., Testori, A., Rivoltini, L., Maio, M., Andreola, G., Sertoli, M., Gallino, G., Piris, A., Cattelan, A., Lazzari, I., Carrabba, M., Scita, G., Santantonio, C., Pilla, L., Tragni, G., Lombardo, C., Arienti, F., Marchiano, A., Queirolo, P., Bertolini, F., Cova, A., Lamaj, E., Ascani, L., Camerini, R., Corsi, M., Cascinelli, N., Lewis, J., Srivastava, P., and Parmiani, G. Vaccination of metastatic melanoma patients with autologous tumor-derived heat shock protein gp96-peptide complexes: clinical and immunologic findings. *J. Clin. Oncol.*, *20*: 4169–4180, 2002.

134. Soiffer, R., Hodi, F. S., Haluska, F., Jung, K., Gillessen, S., Singer, S., Tanabe, K., Duda, R., Mentzer, S., Jaklitsch, M., Bueno, R., Clift, S., Hardy, S., Neuberg, D., Mulligan, R., Webb, I., Mihm, M., and Dranoff, G. Vaccination with irradiated, autologous melanoma cells engineered to secrete granulocyte-macrophage colony-stimulating factor by adenoviral-mediated gene transfer augments antitumor immunity in patients with metastatic melanoma. *J. Clin. Oncol.*, *21*: 3343–3350, 2003.

135. Smith, C. L. Results of a phase I study evaluating "prime-boost" therapeutic vaccination strategies using a string of melanoma-derived CD8+ T cell epitopes in stage II/III/IV melanoma patients. *Proc. Am. Soc. Clin. Oncol.*, *22*, 175. 2003.

136. Bedrosian, I., Mick, R., Xu, S., Nisenbaum, H., Faries, M., Zhang, P., Cohen, P. A., Koski, G., and Czerniecki, B. J. Intranodal administration of peptide-pulsed mature dendritic cell vaccines results in superior CD8+ T-cell function in melanoma patients. *J. Clin. Oncol.*, *21*: 3826–3835, 2003.

137. Schadendorf, D., Nestle, F. O., Broecker, E.-B., Enk, A., Grabbe, S., Ugurel, S., Edler, L., Schuler, G., and DeCPG-DC Stidu Group. Dacarbacine (DTIC) versus vaccination with autologous peptide-pulsed dendritic cells (DC) as first-line treatment of patients with metastatic melanoma: results of a prospective-randomized phase III study. *Proc. Am. Soc. Clin. Oncol.*, *22*(14S). 2005.

138. Dudley, M., Wunderlich, J., Robbins, P., Yang, J., Hwu, P., Schwartzentruber, D., Topalian, S., Sherry, R., Restifo, N., Hubicki, A., Robinson, M., Raffeld, M., Duray, P., Seipp, C., Rogers-Freezer, L., Morton, K., Mavroukakis, S., White, D., and Rosenberg, S. Cancer regression and autoimmunity in patients after clonal repopulation with antitumor lymphocytes. *Science*, *289*: 850–854, 2002.

139. Leach, D. R., Krummel, M. F., and Allison, J. P. Enhancement of antitumor immunity by CTLA-4 blockade. *Science*, *271*: 1734–1736, 1996.

140. van Elsas, A., Hurwitz, A. A., and Allison, J. P. Combination immunotherapy of B16 melanoma using anti-cytotoxic T lymphocyte-associated antigen 4 (CTLA-4) and granulocyte/macrophage colony-stimulating factor (GM-CSF)-producing vaccines induces rejection of subcutaneous and metastatic tumors accompanied by autoimmune depigmentation. *J. Exp. Med.*, *190*: 355–366, 1999.

141. Phan, G. Q., Yang, J. C., Sherry, R. M., Hwu, P., Topalian, S. L., Schwartzentruber, D. J., Restifo, N. P., Haworth, L. R., Seipp, C. A., Freezer, L. J., Morton, K. E., Mavroukakis, S. A., Duray, P. H., Steinberg, S. M., Allison, J. P., and Davis, T. A. Cancer regression and autoimmunity induced by cytotoxic T lymphocyte-associated antigen 4 blockade in patients with metastatic melanoma. *Proc. Natl. Acad. Sci. U. S. A.*, *100*: 8372–8377, 2003.

142. Schneeberger, A., Wagner, C., Zemann, A., Lührs, P., Kutil, R., Goos, M., Stingl, G., and Wagner, S. N. CpG motifs are efficient adjuvants for DNA cancer vaccines. *J. Invest. Dermatol.*, *123*: 371–379, 2004.

III Classes of Drugs for Molecular Targeting in Oncology

16 Antibody Therapy of Cancer

Hossein Borghaei, DO, MS,
Liat Binyamin, PhD,
Igor Astsaturov, MD, PhD,
and Louis M. Weiner, MD

SUMMARY

Antibody-based therapy has emerged as an integral part of effective therapies for a number of malignancies. Antibodies raised against CD20 on the surface of B-cells (e.g., rituximab) have emerged as major components of lymphoma treatment *(1)*. Exciting and clinically important antibody therapies are widely used to treat HER2/*neu* overexpressing breast cancer *(2)*, epidermal growth factor receptor (EGFR) overexpressing malignancies *(3,4)*, and cancers that are driven by vascular endothelial growth factor (VEGF)-driven tumor angiogenesis *(5)*. This chapter reviews some of the pertinent data regarding antibodies, their mechanisms of actions, and some of the available clinical data for the more clinically relevant monoclonal antibodies.

Key Words: Monoclonal antibodies; antibody therapy; Antibody Dependent Cellular Cytotoxicity (ADCC); Complement Dependent Cytotoxicity (CDC); immunoconjugates.

1. INTRODUCTION

An expanding understanding regarding the role of the immune system in cancer, coupled with the success of immune-based treatments has mapped the road for monoclonal antibodies to become major therapeutic vehicles in the treatment of malignant and nonmalignant diseases.

The antibody therapy concept was first illustrated a century ago by Paul Ehrlich as the "magic bullet" hypothesis, describing selectively target malignant cells based on the unique expressed determinants profile of the disease. The development of hybridoma technology by Kohler and Milstein provided practical skill to produce monoclonal antibody (mAb) with highly specific associations to their targeted antigens *(6,7)*. Since then, extensive efforts have been taken to wisely apply and generate mAb for cancer therapy. Many mAbs that are currently in clinical use or under evaluation were originally derived from hybridoma-derived murine antibodies *(8)*. Considering the induction of human anti-mouse antibody (HAMA) responses and the inefficient interaction of

From: *Cancer Drug Discovery and Development: Molecular Targeting in Oncology*
Edited by: H. L. Kaufman, S. Wadler, and K. Antman © Humana Press, Totowa, NJ

the murine origin constant region of the antibodies (Fc) with human immune-accessory cells, a second generation of engineered antibody was employed *(9)*. Chimeric and humanized antibodies were generated by incorporating portions of the murine variable regions into the human immunoglobulin G (IgG) framework *(10)*. Grafting either the entire murine variable regions (chimeric mAb) or the murine complementary-determining regions (humanized mAb) to create mAbs that contain human Fc domains and retain targeting specificity was accomplished *(11)*. Moreover, using molecular-engineered techniques, critical human heavy-chain backbone sequences were grafted onto the xenogeneic murine antibody structure, reducing the immunogenicity and introducing important human origin structures (e.g., Fc) to the resulting antibodies *(12)*. To further improve the power of mAb and to design antibodies that specifically target and subsequently eliminate cancer cells, large combinatorial antibody libraries of murine, human, or synthetic origin were constructed. Effective in vitro screening systems now allow bypassing immunization and selecting recombinant antibodies of defined specificity without the need for hybridoma production *(13,14)*. In addition, advances in molecular biology have to a great extent facilitated the genetic manipulation, recombinant production, and conjugation of antibody fragments *(15,16)*. New forms of antibody modules with different size, flexibility, and valency suited for in vivo imaging and therapy were created. A major breakthrough in the technology of antibody engineering was the derivation of single-chain molecules (scFv). These molecules were obtained by joining the heavy and light variable domains (VH and VL) from a mAb with a flexible linker, which allowed the reconstitution of the original antigen-binding fragment association *(17)*. A number of multivalent scFv-based structures have been engineered, including bispecific and bivalent antibodies such as diabodies and minibodies *(18–21)*. This has led to the development of a variety of engineered mAb molecules for research, diagnosis, and therapy. As will be discussed, therapeutic mAbs are most notably being used in unconjugated form, as drug/toxin conjugates and as radiolabeled, fragmented or genetically recombinant modified versions. Each format possesses advantages and disadvantages with respect to their potential mechanisms of function and clinical applications. The desire to improve efficacy while decreasing treatment toxicities has accelerated the development of more specifically targeted therapies. This can be deduced from a review of ongoing research and development efforts and by the increasing number of mAbs which have gained approval from the Food and Drug Administration (FDA) for use in cancer therapy *(11)*. At present, seven cancer-directed mAbs have been approved for use in the clinic and five of them are directed against hematological tumors. Their anti-tumor effects have been achieved through multiple mechanisms *(22)*.

2. MECHANISMS OF mAbs ACTION

Most therapeutic mAbs that are in use for therapy of cancer possess human backbones of the IgG1 isotype. IgG1 effectively mediates Fc domain-based functions such as antibody-dependent cellular cytotoxicity (ADCC) and complement fixation *(23)*. The therapeutic effects of other mAbs can be achieved directly by binding to the specific antigen and thus excluding the natural ligand from binding to its receptor, leading to signal transduction alteration *(24,25)*. As "naked" antibodies, anti-idiotypic antibodies were raised and provided a proof of concept, taking advantage of differentially expressed tumor-associated antigens (e.g., cell surface membrane immunoglobulin present on human B-cell lymphomas) *(26)*. These antibodies were designed to bind unique idiotypes associated with cell surface membrane immunoglobulin, leading to perturbation of downstream signaling and enhanced programmed cell death. The next generation of mAbs currently under development incorporates additional beneficial modifications to combine and to increase the efficiency of each mechanism. For example, introducing alterations of Fc domain glycosylation and sequences that enhance ADCC or modification in size and antigen-binding affinity that increase the ability of the mAbs to penetrate solid tumors *(27)*. mAbs directed against human cancer-associated antigens also has been used to selectively deliver radionuclides to malignant cell populations *(28)*. In the past 40 years, although studies have confirmed this concept of using labeled antibodies for cancer diagnosis and therapy, progress in this area was obstructed by methodological limitations in the characterization and production of antibody, as well as in labeling and imaging. In addition to radioisotopes, mAbs have been used extensively in clinical trials to target cytotoxic agents to tumor cells. These agents include catalytic toxins, drugs, and enzymes *(28)*. Also, synergistic effects can be achieved combining mAb therapy with traditional chemotherapy agents, attacking tumors through complementary mechanisms of action.

3. ANTIBODY-BASED CELLULAR CYTOTOXICITY

Clinically promising mAbs specific for tumor antigens may mediate their effects in part through ADCC. ADCC is a well-recognized immune effector mechanism in which antigen-specific antibodies direct immune effector cells of the innate immune system to the killing of the antigen-expressing cancer cells *(29)*. This property is dependent on interactions between cellular Fc receptors (FcRs) on immune accessory cells and the antibody Fc domains (effector) *(30)*. FcRs for IgG were identified over 40 years ago with the observation that IgG antibodies could be directly cytophilic for macrophages when presented as opsonized red blood cells *(31)*. The binding property of IgG antibodies was independent of the F(ab) region of the antibody and required only Fc interactions. The binding cross-links FcR on the effector cells, and as a result, the FcR-bearing effector cells become activated and trigger their function. For example, NK cells kill cancer cells and also release cytokines and chemokines *(22)*. NK cell-secreted interferon-γ (IFN-γ) inhibits cell proliferation, increases cell surface expression of MHC and antigens, and is anti-angiogenic. This is an example of cooperation between the innate and adaptive immune systems. The effector cells that may mediate ADCC include NK cells, monocyte macrophages, and neutrophils. Of the above, NK cells comprise the principal ADCC effector cells. They bear low-affinity type II (FcγIIc; CD32c) and type IIIA (FcγIIIa; CD16a) FcRs on their surface *(32–34)*. In ADCC, NK cells generally kill their target cells by releasing cytotoxic granules, that is, perforin, granulysin, and granzymes. CD16a plays a dominant role in NK cell-mediated ADCC, thereby it has been known as the ADCC receptor *(35)*. ADCC-mediated elimination of tumor cells was demonstrated in vitro in the presence of NK cells and tumor-specific antibodies of appropriate IgG isotypes *(36)*. Clynes and Ravetch studied the magnitude of FcR interaction in vivo by examining the anti-tumor activities of clinically effective mAb against human tumor xenografts growing in either wild-type mice or in murine FcγRII/III knock-out mice *(37)*. Anti-tumor activity was reduced in the Fcγ receptor knock-out mice and was conserved when only the inhibitory Fcγ receptor isoform was removed *(37)*. The role of FcγR in the anti-tumor effects of rituximab, the first FDA-approved chimeric mAb for lymphoma treatment, was further supported with the finding that CD16 polymorphisms predict responses to rituximab in patients suffering from follicular lymphoma *(38)*. These findings indicate that interactions between the antibody Fc domain and the FcR underlie at least some of the clinical benefit of some mAbs and imply the importance of ADCC. Indeed, ongoing research projects are focused on strategies to design and test new mAbs with an improved capacity to mediate ADCC. The approaches include manipulation of the mAb Fc region, which directly participates in activating complement through the classical pathway and in recruiting FcγRIII on immune effector cells to mediate ADCC. Combined computational and experimental methods have identified mutations within the Fc domain of mAbs to selectively tune the affinity for FcγRIII and other Fcγ receptors *(39)*. An alternative strategy to enhance ADCC by mAb is to modify Fc glycosylation by the producing cell line *(40)*. Modification of the Fc region to interact with activating or with inhibitory FcRs could enhance antigen presentation by dendritic cells and can be biased to promote or to inhibit the generation of cytotoxic T-cell responses against the targeted antigen, indicating that induction of ADCC can lead to adaptive immune responses and finally to the elimination of tumor cells *(27,41)*.

4. COMPLEMENT-DEPENDENT CYTOTOXICITY

Most mAbs that mediate ADCC also activate the complement system *(42)*. Notably, chimerized or humanized mouse mAb containing the IgG1 Fc region trigger both ADCC and complement activation. The classical activation cascade of complement, the "complement-dependent cytotoxicity" (CDC), involves direct killing of tumor cells by forming a "membrane attack complex" (MAC). Following antibody binding to antigens on the target cell, C1q-binding sites on the antibodies Fc region become available. Multiple C1q binding then changes the low-affinity interactions of the single C1q–IgG interaction to high-avidity interactions. This leads to the release of C3a and C5a that will attract effector cells to the target cells and activate them, resulting in improvement of ADCC. The complement proteins will also create MAC that cause pore formation in the target cell membrane, leading to elimination of the tumor cell. Several clinically approved mAbs activate complement on tumor cells in vitro; however, the clinical relevance of CDC has been difficult to convincingly demonstrate *(43,44)*. Improving mAb ability to activate CDC has the potential to increase clinical efficacy and could be additive to the existing effector mechanism of such antibodies. For example, it is known that C1q high-binding avidity requires the dimerization of two IgGs; however, occasionally, the targeted antigens are present in low density, thus enabling the formation of IgG dimers. A recent study has suggested that multiple, different epitope-targeted mAbs directed to HER2/neu antigen could be used to increase the number and the density of the mAbs on the target cell *(45)*. To increase the density of bound antibodies, a secondary antibody could be also used against either the anti-tumor mAb *(46)* or against iC3b deposited on tumor cells by the primary mAb *(47)*. Another approach to enhance complement-mediated effector mechanisms is to conjugate an anti-tumor mAb to a complement activation protein, such as cobra venom factor (CVF) or

C3b *(48,49)*. Complement receptor 3-dependent cellular cytotoxicity (CR3-DCC) could be activated through the complement C3 activation product iC3b deposited on the tumor cell. Although this is mainly an anti-yeast and fungal infection defense mechanism that requires the presence of the cell-wall β-glucan, it was shown that administration of soluble yeast β-glucan can serve as adjuvant that greatly promoted the tumor regression activity of mAbs that activated complement *(50)*. Down-regulating the expression of mCRPs has a potential to improve mAb-mediated complement activation *(23)*. In this regard, the use of various cytokines for in vitro studies has been reported *(51)*. Furthermore, mAbs that block the function of mCRPs could enhance the complement susceptibility of tumor cells. Limitation of this approach is that mCRP is widely expressed on normal tissue. Bispecific antibodies that can recognize both a tumor-related antigen and a mCRP can selectively target tumor cells and enhance their susceptibility to complement deposition and lysis *(52)*.

5. MANIPULATION OF SIGNAL TRANSDUCTION

The signaling events leading to cellular proliferation are mainly triggered through the interaction of extracellular ligands with cell surface receptors. These interactions can be specifically targeted by antibodies *(53,54)*. The concept of mAb binding to the ligand and prevention of its interaction with the cell surface receptor has been used in the case of vascular endothelial growth factor (VEGF) *(54)*. The inhibition of angiogenesis and the prevention of the development of the tumor neovasculature is the process which deprives the tumor of new blood vessels and inhibits growth beyond a minimal size. The vasculature is directly accessible to antibodies, and vascular damage might affect many tumor cells depending on each capillary. A mAb, bevacizumab (Avastin; Genentech), binds to VEGF and blocks the interaction of VEGF with its receptor *(55)*. Another exploited growth factor receptor family with respect to interference and inhibition by mAb for therapeutic purposes is the ErbB or epidermal growth factor receptor (EGFR). The activation of the EGFR is controlled by binding of specific ligands, and this induces the formation of heterodimers and activation of the intrinsic kinase domain (reviewed in ref.56). It has been shown that cancer patients with tumor cells expressing high levels of ErbB1 and ErbB2 have a more aggressive disease and an unfavorable prognosis. For this reason, ErbB receptors are attractive therapeutic targets, and many different approaches to inhibit the receptors have been tried *(57)*. Cetuximab, a humanized mAb, recognizes the EGFR ectodomain and competes for ligand binding to the receptor resulting in the inhibition of mitogenesis *(58)*. Others, such as pertuzumab (2C4), allow ligand binding to occur but sterically inhibit the subsequent receptor heterodimerization required for signal transduction *(59)*. Subsequently, in vitro and in vivo growth of breast and prostate tumor cells is inhibited by 2C4. The therapeutic benefit of 2C4 could complement that of other antibodies, as the prevention of receptor heterodimerization appears to be a promising novel approach which could possibly complement the use of trastuzumab and tyrosine kinase inhibitors *(60)*. As overexpression of growth factor receptors is essential for the maintenance of tumor cells, therapeutic agents that have the effect of reducing the density of target antigen expression are of interest. Of these, antagonistic antibodies, small molecular weight kinase inhibitors, compounds causing ErbB2 degradation, and scFv-mediated inactivation of ErbB2 through its retention in the endoplasmic reticulum have all been employed and interfere with ErbB2 function and demonstrate strong anti-proliferative effects *(61)*.

6. DELIVERY OF CYTOTOXIC
COMPOUNDS—IMMUNOCONJUGATES

Immunoconjugates are bifunctional molecules that consist of a "targeting" domain that localizes in tumors coupled to a therapeutic moiety. Immunoconjugates, in the broadest definition, may utilize mAb, mAb fragments, hormones, peptides, or growth factors to selectively localize cytotoxic drugs, plant and bacterial toxins, enzymes, radionuclides, or cytokines to antigens presented on tumor cells or on cells of the tumor neovasculature (11).

For radioimmunotherapy, the central issue is balancing the dose delivered to tumor balanced against exposure of normal organs and tissue to radiation. Two radiolabeled mAbs have been licensed to date. These are ibritumomab tiuxetan (Zevalin) and tositumomab (Bexxar), which are mAbs conjugated to the β-particle-emitting radioisotopes, ^{90}Y and ^{131}I, respectively (62). These agents are used for the therapy of non-Hodgkin's lymphoma (NHL) and are directed to the CD20 antigen (63). In patients with solid tumors, response rates to radio immunotherapy agents are modest. In general, after intravenous injection, mAb accumulates in solid tumors comparatively slowly, and <0.1% per gram of the injected dose typically is localized per gram of tumor. Such inefficient accumulation has been attributed to a number of physiological barriers between the blood circulation and the tumor cell surface (64). These barriers include the vascular endothelium, size-dependent diffusion properties of the antibodies, hydrostatic pressure within the tumor, and long transport distances in the tumor tissue. Thus, for these less radiosensitive tumors, more effective targeting of tumor with mAb is required. To enhance the efficacy of immunotherapy, several strategies have been developed, such as the use of mAb fragments, the use of high affinity mAbs, the use of labeling techniques that are stable in vivo, active removal of the radiolabeled mAb from the circulation, and pretargeting strategies (63,65,66). Also, radioimmunotherapy has been combined with other agents or modalities, such as cytokine administration or hyperthermia, to increase antigen expression and tumor uptake. Chemotherapy has also been used to enhance radiosensitivity (67,68). An important application of radionuclide-labeled mAb is for imaging and localization of the tumor and prediction of clinical outcome following treatment. For in vivo imaging, contrast (signal: noise ratio) is the key parameter for success. Biodistribution studies demonstrate that smaller fragments, such as diabodies and minibodies, reach their maximal tumor uptakes within 1–6 h of administration in xenograft-bearing mice. Because of rapid blood clearance, tumor-blood ratios increase steadily over time and reach high values (>20:1) by 24 h, making these fragments prime candidates for imaging (69). Much recent work has focused on positron emission tomography (PET) imaging for evaluating, targeting, and distribution of drugs and tracers because of the higher sensitivity, resolution, and quantification afforded by this imaging mode (70,71). Achieving high tumor to nontumor ratios with the slow process of antibody uptake into the solid tumor are major challenges in immunoconjugate application (72).

In pretargeting, the radionuclide is administered separately from the tumor targeting mAb. In the first step the unlabeled anti-tumor mAb is administered and allowed to accumulate in the tumor. In the later phase, preferably when the mAb has been cleared from the circulation, the radionuclide is administered as a rapidly clearing agent with high affinity to the unlabeled molecule that was injected in the first phase (63,73).

Rather than radionuclides, second phase administration of cytotoxic drugs or enzymes specifically designed to bind the mAb could be applied. This has been referred to as ADEPT (Antibody-Directed Enzyme Prodrug Therapy). As with radiolabeled mAb for cancer therapy, the tumor to nontumor conjugate ratio is a critical parameter in ADEPT that directly affects the amount of prodrug that can be safely administered (74–76).

An example of a drug conjugate mAb, FDA approved for use in relapsing acute myeloid leukemia (AML), is gemtuzumab ozogamicin (Mylotarg), a humanized antibody directed to the MUC1 antigen, linked to calicheamicin, an antibiotic that cleaves DNA (77). The specificity of this very toxic anti-tumor compound is assured by the recognition of CD33 antigen on the cell surface; through internalization of the conjugated mAb, the calicheamicin portion can selectively exert its tumor effects. (62). Studies involving doxorubicin as the payload drug also have been reported (78–80). It is only in the past few years that the critical parameters for optimization have been identified and begun to be addressed. This includes the physiological barriers to mAb extravasation and intratumoral penetration, mAb immunogenicity, normal tissue expression of the targeted antigen, low-drug potency, inefficient drug release from the mAb, and difficulties in releasing drugs in their active states (28,72,81,82).

Antibody toxins, also termed immunotoxins, are hybrid molecules derived by coupling bacterial, plant, or fungi toxins to monoclonal antibodies specific for molecules on the surface of tumor cells. The elucidation of the molecular structure of bacterial toxins such as *Pseudomonas aeruginosa* exotoxin A and the development of recombinant antibody technologies have allowed the minimization of the size of these molecules through recombinant DNA techniques and their production as single polypeptides in large quantities and consistent quality in bacteria (83). The principal function of these immunotoxins is to inhibit protein synthesis after internalization, leading to death of the targeted cell. Accordingly, a mAb that has no intrinsic cell-elimination function in an unconjugated format still might be useful for immunotoxin design. An advantage of immunotoxins is that an exceedingly small amount of the protein is actually required to kill both resting and dividing cells. Part of this property derives from the catalytic properties of the fusion proteins; one molecule can attack multiple intracellular targets (84,85). Over the years, a number of clinical trails have been conducted with immunotoxins and fusion toxins (83,84,86,87). These studies defined a number of pharmacologic and toxicologic barriers that needed to be overcome. The original rationale for the production and testing of these reagents was that they had a different mechanism of action than DNA or cell-division-damaging therapeutics and thus might be effective either alone or in combination in patients with chemotherapy-resistant malignancies.

We will now discuss some of the clinically relevant monoclonal antibodies and their targets in detail.

6.1. EGFR Structure and Mechanism of EGF Receptor Signaling

The EGFR initially was implicated in cancer by virtue of its tyrosine kinase activity and the discovery of the v-erbB oncogene, a truncated EGF receptor, in avian erythroblastosis virus (88–90). Two decades of in-depth studies of EGF signaling have recently yielded a new family of antibody and small molecule therapeutics that are rapidly changing our views of how to treat many types of cancer with head and neck squamous cell carcinoma (HNSCC) in the vanguard.

Table 1
Erb Family of Receptors and Their Ligands

Erb family receptor	Ligand	Dimerization partner
ErbB1/EGFR	EGF, TGF-α, amphiregulin, epigenin, epiregulin, betacellulin, and HB-EGF	EGFR/Erb2
ErbB2/Her2	None	Erb2/Her2 and ErbB3
ErbB3 and ErbB4	Neuregulins 1–4 and heregulin	ErbB1–4
EGFRvIII, de2-7	Much lower affinity	

The EGFR has four family members (Erb1-4 or alternatively, Her1-4) of which the EGFR (Her1) and Her2 (EGFR2) are the most fully characterized. The members of EGFR family, their principal ligands, and nomenclature are summarized in Table 1. All EGFR family members consist of a heavily glycosylated extracellular region containing 11 potential glycosylation sites spanning approximately 620 amino acids. A single transmembrane domain (23 residues) is flanked by the so-called juxtamembrane regulatory domain (~40 amino acids) followed by a tyrosine kinase domain (~260 amino acids) and a C-terminal regulatory region of 232 amino acids.

Ligands for the EGFR include EGF itself, transforming growth factor-α (TGF-α) amphiregulin, epigene, epiregulin, betacellulin, and heparin-binding EGF (HB-EGF)-like growth factor. EGF and TGF-α, often coexpressed with EGFR *(91)*, are produced by normal and tumor tissue epithelial cells, with higher levels found in neoplastic tissue and stroma than in surrounding normal mucosa. EGFR2 (erbB2/Her2/neu) has no known ligands. Neuroregulins (NRG) serve as ligands for ErbB3 and ErbB4. NRG1, in particular, is implicated in pathogenesis of breast cancer, whereas the biologic functions of the other three members of the family remain poorly understood *(92)*.

Four distinct protein domains comprise the EGFR extracellular region. Domains II and IV (also known as CR1 and CR2, respectively) are cysteine rich with a number of disulfide bonds: eight disulfide modules in domain II and seven disulfide modules in domain IV. Domains I and III are leucine rich and are binding sites for the growth factor ligands. In studies of different mutant forms of EGFR, it became evident that cooperation between domains I and III was necessary for high-affinity binding of EGF *(93)*. In 1997, Lemmon et al. *(94)* suggested a model in which "one EGF monomer binds to one sEGFR monomer, and that receptor dimerization involves subsequent association of two monomeric (1:1) EGF-sEGFR complexes." The dimer form of EGFR possessed biologically significant tyrosine kinase activity (through the Akt pathway) and was shown to possess transforming activity in keratinocytes *(95)*.

7. ROLE OF EGFR AND MECHANISMS OF RADIORESISTANCE

The EGFR is universally overexpressed in squamous cell cancer of the head and neck as well as in a variety of less common head and neck cancers (e.g., salivary gland and thyroid). EGFR signaling exerts its cellular effects through the PI3 kinase and Akt kinase pathways leading to radiation and apoptosis resistance and cellular proliferation. In the following sections, we will focus on therapeutic antibodies that alter EGFR signaling and review how these drugs are being used alone and in combination with other therapeutics to increase susceptibility of head and neck cancers to radiation and chemotherapy and improving patient outcome.

Ang et al. *(96)* recently published correlative data on EGFR expression and outcomes in patients with head and neck squamous cell carcinoma (HNSCC) treated with radiation. Across the board, patients with tumors with high EGFR expression fared significantly worse for overall ($p = 0.0006$) and disease-free survival ($p = 0.0016$). EGFR level did not correlate with T- and N-stage or rate of distant metastases implying a very specific role of this protein in tumor responsiveness to radiotherapy (RT). Multivariate analysis showed that EGFR expression was an independent determinant of survival and a robust independent predictor of locoregional relapse *(96)*.

Ionizing radiation exposure of cancer cells triggers repair processes that involve signaling through EGFR. It was proposed that radiation induces autocrine production of EGFR ligands, such as EGF and TGF-α. In vitro cancer cell lines increase their proliferative rate and up-regulate the levels of EGFR and increase the activity of downstream signaling mediators including Raf-1, mitogen-activated protein kinase, and levels of inositol-triphosphate in response to radiation *(97)*. Akimoto et al. *(98)* in a murine model of diverse epithelial carcinomas demonstrated a great difference in their radiosensitivity and susceptibility to radiation-induced apoptosis. Not surprisingly, the magnitude of EGFR expression varied as much as 21-fold and correlated positively with increased tumor radioresistance. In contrast to tumors with wild-type p53, this correlation was not seen in tumors with p53 mutation.

The recognition that EGFR activity correlates with radioresistance sparked intense interest in interrupting EGFR signaling for therapeutic benefit as well as a desire to understand the mechanism underlying the anti-tumor effects of anti-EGFR therapy. Human A431 tumor transfectants with high levels of EGFR expression showed decreased tumor radiocurability *(99)*. Radiation activated the EGFR and its downstream signaling pathways in radioresistant but not in radiosensitive tumors *(99)*. The proliferative response to radiation can be blocked by tyrphostine (AG1478), a specific inhibitor of the EGFR tyrosine kinase *(98,99)*. Treatment of human tumor xenografts with cetuximab (C225) markedly enhanced the tumor response to RT, as assessed by both tumor growth delay and cure rate *(100)*. Thus, Huang et al. *(101)* showed that the anti-EGFR chimeric monoclonal antibody cetuximab (C225) enhanced radiation-induced apoptosis in vitro and increased the proportion of squamous cell cancer cells in the radiosensitive G1 phase of the cell cycle. Bonner et al. *(102)* determined in preclinical studies that the combination of C225 with radiation resulted in a greater decrement in cellular proliferation than either treatment alone. These findings correlated with reduction of EGFR and signal transducer and activator of transcription-3 (STAT-3) protein tyrosine phosphorylation *(102)* and increased levels of the cell cycle inhibitor p27^{KIP1}, hypophosphorylated Rb, and proapoptotic protein Bax. Milas et al. *(103)* studied the effects of cetuximab on radiation responsiveness of A431 tumor xenografts. This model provided the first evidence that serial injection of the antibody in conjunction with irradiation produced suppression of tumor growth lasting as long as 4 weeks. Many tumors in these experiments had central necrosis and heavy infiltration with granulocytes, suggesting angiogenesis inhibition and antibody-mediated cellular cytotoxicity as additional mechanisms of the anti-tumor effects of C225.

8. ANTI-EGFR MONOCLONAL ANTIBODIES

At least five antibodies targeting different epitopes of EGFR have entered clinical trials. Chimeric and humanized antibodies (e.g., cetuximab or C225 and matuzumab or EMD72000) were originally raised in mice. To reduce immunogenicity of these antibodies, their "backbones" have been exchanged with those of human immunoglobulin. In contrast, fully human, IgG kappa antibodies have been derived employing XenoMouse™ technology. The latter is the result of insertion of large portion of the human immunoglobulin gene locus into the mouse genome. By contrast to xenogeneic (i.e., murine or chimerized) antibodies, human or humanized antibodies are less likely to elicit host immune responses that may limit therapy or cause side effects.

There are several proposed mechanisms by which anti-EGFR monoclonal antibodies can elicit an anti-tumor response:

1. The binding of the EGFR on cell surface triggers receptor internalization and degradation through the lysosomal pathway, prevents nuclear translocation of DNA-PK, and thereby inhibits DNA break repair induced by radiation and DNA-avid chemotherapeutic agents.
2. Interference with EGFR signaling by preventing ligand binding and unfolding of the tethered EGFR to the extended, more active conformation. This is the likely scenario with domain III targeting anti-EGFR antibodies such as Cetuximab, Matuzumab, or panitumumab. Antibodies specific to epitopes in domain II of EGFR (the dimerization arm) will abrogate EGFR signaling by preventing its dimerization.
3. ADCC, or antibody-dependent cellular cytotoxicity, as discussed above.
4. Another mechanism that is less well understood invokes broad interference with signaling events through antibody disruption of lipid rafts. The EFGR is intimately associated with signaling microplatforms on the cellular membrane *(104)*. High concentration of EGFR ligands or, perhaps, cross-linking with an antibody may shuttle EGFR internalization from the clathrin-mediated pathway of recycling to the

Table 2
Monoclonal Antibodies Targeting the EGFR Receptor

Antibody	Origin	Phase of Development
C225/cetuximab	Humanized mAb	Approved as salvage therapy for CRC, combined therapy with RT in head and neck cancers
EMD72000/matuzumab	Humanized mAb	II
ABX-EGF/panitumumab	Human mAb	III
ICR62	Rat mAb	I
806 (EGFRvIII)	Humanized mAb	Imaging pilot trials
hR3/nimotuzumab	Humanized mAb	I/II
MDX-447	Humanized bispecific EGFR/CDε	I/II

CRC, colorectal cancer; EGFR, epidermal growth factor receptor; mAB, monoclonal antibody; RT, radiotherapy.

caveolin-associated route leading to receptor ubiquitination and degradation *(105)*. Table 2 lists some of the anti-EGFR antibodies that are currently either approved or under investigation.

9. EGFR-TARGETING MONOCLONAL ANTIBODIES IN HEAD AND NECK CANCER

9.1. Cetuximab

Cetuximab (Erbitux™), a chimeric IgG1 antibody, binds domain III of EGFR, thereby interfering with ligand binding. It is the most extensively studied anti-EGFR monoclonal antibody and received FDA approval in 2004 for the treatment of patients with metastatic colorectal cancer who are not suitable for or are refractory to irinotecan. Cetuximab binding to EGFR prevents the receptor from adopting an extended conformation, thereby inhibiting EGFR activation *(106)*.

HNSCC poses a significant health risk worldwide with approximately 38,530 newly diagnosed cases and over 11,000 deaths per year from the disease in 2004 in the USA alone as projected by the American Cancer Society *(107)*. The incidence is rising likely due to the role of environmental exposures that include tobacco and alcohol as well as emerging evidence for a role for viruses such as human papilloma virus (HPV) *(108)*.

Although the mainstay of treatment for stage 1 and 2 disease remains surgical excision, the majority of patients with HNSCC unfortunately present with large tumors (greater than T2), nodal involvement, or metastatic disease. With combined therapy, cure is still possible for many of these patients but the side effects of treatment and resultant long-term morbidity remain formidable, leaving numerous opportunities for treatment innovations. HNSCC is radiosensitive, and radiation therapy plays an integral role in treatment, yielding results comparable to surgery in early stage disease (T1 and T2). RT is less effective as a solo modality for intermediate size tumors and is typically used in an adjuvant fashion following surgery or as a component of induction chemoradiotherapy for more advanced stages of disease *(109,110)*.

Patients with stage 3 and 4 disease continue to represent a significant unmet medical need. Current state-of-the-art radiation, surgery, and chemotherapy cure only 40–50% of stage 3 and fewer than 20% of stage 4 HNSCC patients. Of note, many patients continue to succumb to cancer that persists or recurs above the clavicles; death from distant metastases occurs in less than one fifth of patients *(108)*. There have been significant advances in concomitant chemoradiotherapy *(111–113)* over the past three decades, and to date, this is the only approach for locally advanced disease that consistently shows improved locoregional control and a survival advantage (approximately 4% at 5 years) as shown in a recent meta-analysis *(114)*. The improved locoregional control achieved with cisplatin-based chemoradiotherapy regimens is likely attributable to the drug's ability to damage irreversibly DNA of resting clonogenic cells thereby sensitizing them to the effects of RT.

Despite these advances, an unacceptable proportion of patients continue to experience locoregional disease progression or recurrence leading to death. Therefore, a better understanding of the disease process and newer treatments are needed.

10. CLINICAL RESULTS OF COMBINATION OF CETUXIMAB AND RT FOR LOCO-REGIONALLY ADVANCED HNSCC

The use of cetuximab in this clinical scenario has resulted in a survival benefit, the first time in a randomized clinical trial that a statistically significant survival benefit has been conferred by the use of any EGFR antagonist. The pivotal trial performed by Bonner et al. *(115)* was preceded by a smaller phase I trial *(116)*, in which the safety was established of coadministration of cetuximab with a course of definitive radiation delivered once or twice daily over 7 weeks. In this study, all patients were given a loading dose of antibody 1 week prior to initiation of RT. The purpose of the loading dose was to saturate rapidly all EGFR binding sites, including those in normal liver and skin. After the loading dose, patients received a weekly maintenance dose of 200–250 mg/m^2 during radiation therapy. The weekly maintenance dose range was based on an earlier PK study showing a pattern of nonlinear pharmacokinetics consistent with saturation of clearance and relatively stable antibody levels in blood with this pattern of dosing *(116,117)*. The authors estimated the $T_{1/2}$ of cetuximab at around 90 h.

The 16 patients in that study had unresectable squamous cell carcinomas of the oropharynx and oral cavity *(12)*, larynx, and hypopharynx *(4)*. The three evaluated dose levels of cetuximab were 100, 200, and 250 mg/m^2 weekly; three subsequent cohorts received loading doses of 400 or 500 mg/m^2. The most common toxicities attributable to cetuximab were desquamating skin reactions in the radiation field (grade 3 in five patients) and infusion reactions in four patients (one resulting in discontinuation of protocol therapy). None of these was considered dose limiting. Mucositis and odynophagia occurred in most of the patients as expected and at the expected intensity with RT alone. Fourteen of 15 patients evaluable for response had a PR or CR, many of which have been durable over a prolonged period of follow-up *(116)*.

The pivotal multicenter trial *(115,118–123)* enrolled 424 patients who underwent randomization to definitive RT alone *(124)* or the combination of definitive RT and weekly cetuximab at a loading dose of 400 mg/m^2 followed by 250 mg/m^2. Patients were stratified by performance status, tumor stage, nodal involvement, and RT fractionation regimen. The majority of patients in this study were men *(119)* (80%) in relatively good health (68% had a Karnofsky score of *(90–100)*. Study entry did not require tumor testing for EGFR expression. Tumor sites were oropharyngeal (60%), larynx (25%), and hypopharynx (15%). The authors reported doubling of median duration of locoregional control from 14.9 to 24.4 months (log rank $p = 0.005$). This result, in particular, validated preclinical models in which improvement of regional control in the radiation field was anticipated based on the radiosensitizing and apoptosis-promoting effects of cetuximab. EGFR blockade with RT reduced the risk of locoregional failure by 32% [hazard ratio (HR) = 0.68] and the risk of death by 26% (HR = 0.74, $p = 0.03$). There was a lack of substantial improvement in locoregional control of carcinomas of larynx and hypopharynx (median locoregional control lasting in range of 10.3–12.9 months). Similarly, overall survival of laryngeal cancer patients improved only marginally from median of 31.6 to 32.8 months, and hypopharyngeal cancers uniformly showed poor outcome with no change in short median survival (13.5 versus 13.7% with Cetuximab). At the same time, the cancers of oropharynx showed an outstanding doubling in median duration of locoregional control with cetuximab (23 months versus 49 months) and staggering improvement in overall survival that in the cetuximab arm did not even reach the median at 3 years. The 3-year survival rates were 54 and 45% in favor of cetuximab and RT, largely because of the success of the investigational treatment in the oropharyngeal subgroup. Updated survival and toxicity data have been presented showing the durability of the locoregional control and survival data *(116)*. Of interest, there was no influence of EGFR expression on patient outcome.

11. COMBINATION OF CISPLATIN, CETUXIMAB, AND RT IN LOCALLY ADVANCED HNSCC

Platinum compounds are established potentiators of radiation-induced DNA damage in tumor cells. To test the clinical hypothesis that two radiosensitizing agents such as cisplatin and cetuximab would act synergistically, investigators from Memorial Sloan Kettering Cancer Center reported a phase 1/2 clinical study exploring the feasibility of cisplatin, cetuximab, and RT in locally advanced HNSCC *(125)*. All patients received 70 Gy RT in twice-daily fractions with concurrent cetuximab and high-dose cisplatin (100 mg/m^2 on weeks 1 and 4) during radiation. The study was closed early due to apparent grade 4 and 5 toxicity of the treatment regimen with two deaths (one pneumonia and one cause unknown) and one occurrence each of myocardial infarction, bacteremia, and atrial fibrillation. With median follow-up of 44 months, and all survivors followed for minimum 31 months, 3-year overall survival was 76%, and 3-year progression-free survival of 59% was reported. This regimen was selected as the experimental arm of the recently activated follow-up study to the Bonner trial.

12. CLINICAL RESULTS WITH CETUXIMAB IN COLORECTAL CANCER

Cetuximab has been evaluated both as a single agent and in combination with irinotecan in phase II clinical trials. In a phase II trial of patients whose tumors exhibited EGFR expression and had failed irinotecan, 5 of 57 patients (9%) achieved a partial response, with 21 additional patients having stable disease or minor responses *(126)*. Toxicities included rash (18% with grade 3) and allergic reaction that required drug discontinuation in 3.5% of patients. In a phase II trial of cetuximab versus cetuximab plus irinotecan, in patients whose disease had progressed on irinotecan alone, single-agent cetuximab produced a similar rate of partial response, 10.8%, whereas the irinotecan–cetuximab combination produced a partial response rate of 22.9% *(4)*. The combination also resulted in a statistically significant increase in time-to-progression (4.1 versus 1.5 months, $p < 0.001$) and a trend toward improved median survival. Building on these results, a phase II trial has evaluated the combination of cetuximab and bevacizumab with or without irinotecan in irinotecan-refractory colon cancer *(127)*. Cetuximab has also been tested with 5-fluorouracil, leucovorin, and irinotecan in the first-line setting in phase I/II trials, with response rates of 48–74% *(128–130)*. The combination of cetuximab and FOLFOX-4 is currently being evaluated in phase III trials in the first-line setting.

Numerous clinical trials are ongoing assessing the addition of cetuximab to a variety of commonly used regimens in metastatic and stage III colorectal cancer. Each of these studies was preceded by phase 1/2 data demonstrating that the addition of cetuximab to various chemotherapy regimens was feasible and safe *(131–137)*. A phase

II randomized clinical trial conducted by Saltz et al. *(127)* in irinotecan-refractory metastatic colorectal cancer patients showed similar objective response rates whether or not patients were treated with a cytotoxic drug (in that case irinotecan) added to combination of cetuximab and bevacizumab. Interesting in this context is the report from Kerbel's group at Sunnybrook Hospital in Toronto *(138)*, which showed increased level of VEGF production by the A431 xenografts selected for resistance to EGFR blocking antibodies. Treatment with anti-VEGF antibody resulted in normalization of anomalous vessels in the tumor. Clinical exploration of bevacizumab and EGFR antagonists has been initiated in at least seven clinical trials registered with NCI.

Cetuximab received accelerated approval from the US FDA in 2004 on the basis of data showing single-agent activity *(126)* and enhanced activity in combination with irinotecan in patients with metatstatic colorectal cancer (CRC) who had failed or were unsuitable for irinotecan-based therapy *(4)*. The benefits of Cetuximab in the metastatic, chemotherapy-refractory CRC setting are modest, conferring only 1.7 months of improvement in survival and a doubling of the partial response rate when given in combination with irinotecan compared with irinotecan alone *(4)*. Given the fact that this study enrolled heavily pretreated patients who were refractory to 5-FU and irinotecan, however, the result was considered highly encouraging, especially as the addition of cetuximab to irinotecan did not appear to intensify chemotherapy side effects. Other than acneiform rash and the occasional infusion reaction, the regimen was generally well tolerated.

13. INDICATORS OF RESPONSE TO CETUXIMAB

An acneiform or maculopapular rash is a characteristic side effect of anti-EGFR therapies. EGFR plays a role in maintaining the integrity of the skin *(139)* and is expressed in epidermal and follicular keratinocytes, sebaceous and eccrine epithelia, dendritic antigen-presenting cells, and connective tissue cells. After histologic analysis of rash biopsies from ten patients receiving cetuximab, Busam et al. concluded that rash is characterized by lymphocytic perifolliculitis or suppurative superficial folliculitis but without an infectious component *(140)*. Some studies of rash associated with EGFR inhibitors state that rash is sterile *(141,142)*, whereas others report that micro-organisms are present *(140)*. The sebaceous glands are not affected *(142,143)*, leading to the conclusion reached at a recent EGFR inhibitor rash management forum *(144)* that the rash is not acne vulgaris and does not appear to have an acne-like etiology, although the exact etiology is unclear.

Skin toxicity has been shown to be significantly associated with response and overall survival in metastatic colorectal cancer receiving cetuximab. Most patients in these studies were EGFR positive. In a phase II study comparing treatment with cetuximab, cetuximab plus irinotecan in irinotecan-refractory metastatic colorectal cancer, the response rates in patients with skin reactions after cetuximab treatment were higher than those in patients without skin reactions [25.8% versus 6.3% in the combination-therapy group ($p = 0.005$) and 13.0% versus 0% in the monotherapy group] *(4)*.

Rash was also not predictive of response among Chung's group of patients whose tumors were EGFR negative by immunohistochemistry (IHC). Of seven patients with tumor response (partial response and stable disease), five had a grade 1–2 rash. However, three patients who progressed on cetuximab/irinotecan also developed rashes

(145). At this time we view rash as a pharmacodynamic marker indicating the presence of significant levels of antibody in the body; such levels are sufficient to mediate anti-tumor effects. However, the mere attainment of potentially inhibitory antibody levels in tumor or surrogate tissues such as skin does not assure clinical response. Clinical response will require the presence of the relevant target, and possibly gene amplification, and a permissive tumor and host environment. Much work remains to be done to define the precise determinants of response to EGFR-targeted monoclonal antibody therapy.

14. EGFR EXPRESSION AND RESPONSE TO CETUXIMAB

Approximately 70–75% of human colorectal carcinomas express EGFR when assayed by IHC. The results of clinical studies suggest that EGFR expression is neither sufficient nor necessary for tumor response to cetuximab. Early clinical trials of cetuximab required EGFR positivity by IHC for study entry. Within those studies, no relationship between intensity of EGFR expression and clinical activity was demonstrated *(4,126)*. This led many oncologists to not exclude EGFR-negative colorectal patients from standard off-protocol treatment on the basis of EGFR status alone. Reporting on the experience of sixteen such EGFR-negative irinotecan-refractory patients treated with cetuximab/irinotecan at MSKCC, Chung et al. *(145)* noted that seven achieved a degree of tumor control, with four partial responses, two minor responses, and one patient with a more than 50% drop in carcinoembryonic antigen (CEA). This study provides confirmation that tumor response can be seen in EGFR-negative (by IHC) patients. Further, the 25% response rate in this small study is comparable to the 23–45% response rates seen in two cetuximab-plus-irinotecan clinical trials in EGFR-positive patients *(4,126)*.

Vallbohmer et al. *(146)* analyzed mRNA levels of enzymes involved in the EGFR signaling pathway to determine whether activation of these pathways correlates with clinical response to cetuximab. Thirty-nine patients with metastatic CRC refractory to both irinotecan and oxaliplatin were treated with single-agent cetuximab, with intra-tumoral mRNA levels of CCND1, Cox-2, EGFR, interleukin-8 (IL-8), and VEGF assessed from paraffin-embedded tissue samples. All patients had IHC evidence of EGFR expression. Only two patients had partial responses, limiting the robustness of their data, although 21 additional patients had stable disease. In this study, only low intratumoral gene expression of VEGF was associated with response to cetuximab therapy, independent of skin toxicity. Ciardiello et al. *(147)* demonstrated a dose-dependent inhibition of VEGF with cetuximab therapy. These findings suggest a possible role of intratumoral VEGF levels in determining response to cetuximab. They are consistent with the findings in other studies *(138)* of increased expression and secretion of VEGF in tumor cells with acquired resistance to cetuximab.

In an important recent study, Moroni et al. *(148)* evaluated nine colorectal cancer patients with EGFR-expressing tumors who had a response to treatment with cetuximab or panitumumab. They found the mutational status of the EGFR catalytic domain (exons 18, 19, and 21) and its immediate downstream effectors IK3CA, KRAS, and BRAF that did not correlate with disease response. However, eight of nine patients with objective clinical responses were found to have an EGFR copy number of three or more as determined by fluorescent in situ hybridization (FISH) performed on tumor samples. By contrast, only 1 of 20 nonresponders had an increased EGFR copy number. These findings suggest that, as in breast cancers sensitive to trastuzumab, anti-EGFR monoclonal antibodies are more likely to work against amplified rather than mutated targets. They do not explain the response of EGFR-negative colorectal tumors to anti-EGFR mAbs described by Chung, however. Further studies will be required to resolve this apparent contradiction. However, the gene amplification explanation is scientifically plausible and offers the possibility that target populations can be enriched to maximize the odds that antibody therapy will be useful.

The question of why EGFR-negative tumors would respond to an anti-EGFR mAb remains unanswered for now. Some studies have reported variability in EGFR staining depending on how the tissue has been processed or stored. For example, in one study, a decreasing EGFR staining intensity was seen with increasing storage time of tissue samples *(149)*. Thus, a tumor could appear falsely negative for EGFR expression. Because DNA is a generally more stable substance than protein, EGFR copy number assays may have predictive value even in cases in which tumor expression of EGFR was thought to be negative.

15. PANITUMUMAB

Panitumumab (ABX-EGF) is a fully human IgG2 monoclonal antibody to EGFR that, similarly to cetuximab, blocks ligand binding by the EGFR and receptor activation. Note that IgG2 class of antibodies lack ability to induce activation of immune system cell through Fc-receptor mechanism, the latter deemed important component in overall effect of antibody targeting of cancer cells.

In recently released updated results from a phase I clinical trial, ninety-six patients, including 39 colorectal cancer patients, with at least 1+ expression of EGFR, were treated with ABX-EGF. Grade 3 skin-related toxicities occurred in 7% of patients but no MTD was reached and no infusion-related reactions were observed. Partial responses were observed in 5 of 39 patients with colorectal cancer, and stable disease was observed in additional 18 patients *(69)*. A phase II trial is currently investigating ABX-EGF monotherapy in subjects with metastatic colorectal cancer whose tumors express low or negative EGFR levels following treatment with fluoropyrimidine, irinotecan, and oxaliplatin chemotherapy.

16. ANTIBODY FOR THERAPY OF HER2-POSITIVE MALIGNANCIES

The ErbB2/Her2/neu protein is amplified in a number of malignancies, including about 30% of breast cancers. It confers significantly worse prognosis and identifies a specific subset of the breast adenocarcinomas distinct from basal-like, normal breast-like, and luminal epithelial/ER+ *(150)*. The Her2 overexpressing breast carcinomas tend to have higher proliferative markers, metastasize early, and are more responsive to taxane chemotherapy *(151)*. Recent ground-breaking basic research led to better understanding of biologic function of Her2 and development of strategies of interference with its role in the oncogenic phenotype.

Her2 is a transmembrane protein of the Erb family and is active in a homodimeric or heterodimeric form as was discussed earlier. The preferred dimerization partners of Her2 are Her3/EbrB3, Her4/ErbB4, and EGFR/ErbB1 (Table 1). In normal cells, few HER2 molecules exist at the cell surface emanating only limited background "noise" of growth signals. Quite strikingly, when HER2 is overexpressed, multiple HER2 heterodimers are formed and cell signaling is stronger, resulting in enhanced responsiveness to growth factors and malignant growth. The unique stimulatory properties of Her2 are conferred by its constitutive exposure of the dimerization arm of the second domain of the molecule. This fixed conformation of HER2 resembles a ligand-activated state, making HER2 poised to interact with other ErbB receptors in the absence of direct ligand binding *(152)*.

Besides breast cancer, a number of other epithelial malignancies have been shown to overexpress HER2 protein *(153)*.

17. TRASTUZUMAB

17.1. Metastatic Breast Cancer

The recombinant humanized anti-HER2 monoclonal antibody (rhuMAb-HER2, trastuzumab, and Herceptin) induces rapid removal of HER2 from the cell surface, thereby reducing its availability to heterodimers and reducing oncogenicity *(154)*. Herceptin binds to the juxtamembrane region of HER2, identifying this site as a target for anti-cancer therapies. Its clinical use in metastatic breast cancer patients revealed modest activity as a single agent with response rate of 11.6% reported in a pilot study of 46 heavily pretreated patients *(155)*. In vitro and xenograft studies of MCF-7 breast carcinoma overexpressing HER2 revealed synergistic interaction of trastuzumab with alkylating agents, platinum analogs, and topoisomerase II inhibitors and additive interaction with taxanes, anthracyclines, and some anti-metabolites *(156)*. These rational combinations laid the foundation of human clinical trials of trastuzumab *(2,151)*.

In the pivotal randomized clinical trial, patients with immunohistochemically 2+/3+ HER2-positive metastatic breast cancer were assigned to receive six cycles of chemotherapy with or without weekly trastuzumab. This study excluded patients with untreated brain or osteoblastic bone metsastases or those who had pleural effusions or ascites as the only evidence for disease. The choice of chemotherapy was based on prior exposure to anthracyclines, that is, anthracycline-naïve patients were treated with cyclophosphamide/doxorubicin or epirubicin combination (143 in the trastuzumab arm and 138 in chemotherapy only arm), whereas anthracycline-exposed patients received paclitaxel chemotherapy alone (96 women) or with trastuzumab (92 women). Regardless of chemotherapy treatment, trastuzumab led to nearly doubling of time to progression (TTP) (median 7.4 versus 4.6 months; $p < 0.001$), higher rate of objective response (50 versus 36%), and better median survival by almost 5 months (25.1 versus 20.3 months; $p = 0.046$). Cardiotoxicity as expected was a quite prominent side effect, with New York Heart Association class III/IV severity cardiac failure seen in 16% of anthracycline/cyclophosphamide/trastuzumab arm and only in 1–3% in the remaining three study groups. Treatment with trastuzumab was otherwise well tolerated and was safely completed to 80% of planned in 92% of patients indicating excellent compliance rate. After achieving the end point of the study, which was time to disease progression, many patients switched to combinations of trastuzumab plus chemotherapy. This trial design may have masked an even greater advantage for survival derived from trastuzumab.

The second trastuzumab clinical trial was conducted in Europe. Patients (a total of 186) were randomly assigned to six cycles of docetaxel 100 mg/m^2 every

3 weeks, with or without trastuzumab 4 mg/kg loading dose followed by 2 mg/kg weekly until disease progression. This study included 97% of patients having Her2/neu gene amplification confirmed by FISH or 3+ reactivity by IHC. Trastuzumab plus docetaxel was significantly superior to docetaxel alone in terms of overall response rate (ORR) (61 versus 34%; $p = 0.0002$), overall survival (median 31.2 versus 22.7 months; $p = 0.0325$), time to disease progression (median 11.7 versus 6.1 months; $p = 0.0001$), time to treatment failure (median 9.8 versus 5.3 months; $p = 0.0001$), and duration of response (median 11.7 versus 5.7 months; $p = 0.009$). Investigators in these trials conducted close monitoring of their patients' cardiac performance by determining ejection fraction every third cycle of treatment. With these precautions, only one patient in the combination arm experienced symptomatic heart failure (1%), whereas two other patients died of progressive disease, but cardiac failure could not be excluded *(157)*.

Attempts to incorporate trastuzumab into chemotherapeutic regimens for treatment of lung cancer, prostate, or colorectal cancer have not yielded positive results, primarily due to the low incidence of Her2/neu gene amplification in these studies. A phase II randomized trial conducted in Germany included 103 eligible patients to receive gemcitabine–cisplatin with or without trastuzumab. Efficacy was similar in the trastuzumab and control arms *(158)*. The ECOG 2598 phase II clinical trial was also negative for any advantage of adding trastuzumab to carboplatin/paclitaxel combo *(159)*.

17.2. Adjuvant Therapy

In those women who undergo curative surgery, chemotherapy is given to eradicate micrometastatic disease that is responsible for distant recurrences of breast cancer. Several major clinical trials reported exciting results *(160,161)* for the European herceptin adjuvant trial (HERA) trial and combined results of NSABP B-31 and North Central Cancer Treatment Group trial N9831. All three trials included patients operated for breast cancer that tested 3+ by IHC or showed Her2 gene amplification by FISH. The North American trials included 94% node-positive patients, whereas in the HERA trial, 30% of participants were node negative. After a brief observation of 1–2.5 years, the absolute benefit for disease-free survival was 8% in the North American trial and 6% in the HERA trial at 2 years. This benefit would project to 18% at 4 years. Analysis of disease-free survival curves indicates an early separation in favor of trastuzumab. These findings indicate that trastuzumab therapy dramatically alters the natural history of this disease in appropriately selected patients.

Docetaxel and the platinum salts are logical candidates to be combined with trastuzumab as these agents exhibit potent synergy with the antibody in preclinical experiments. Furthermore, the two phase II clinical trials conducted by Breast Cancer International Research Group and the University of California at Los Angeles-Oncology Research Network using the TCH (docetaxel/platinum/trastuzumab) regimen suggest this combination has significant activity with response rates in these two studies reported as 59 and 78%. The BCIRG006 trial is a three-arm adjuvant study comparing doxorubicin/cyclophosphamide followed by docetaxel, the same regimen with trastuzumab administered with docetaxel (TH), and TCH in 3150 women with node-positive or high-risk node-negative, HER2-positive breast cancer. BCIRG 007 compares TH and TCH as first-line therapy in patients with HER2-positive metastatic breast cancer. In both trials, entry is restricted to patients whose tumors are positive for HER2 gene amplification as determined by fluorescence in situ hybridization.

The results of first interim analysis of BCIRG006 adjuvant trial were presented at the 2005 San Antonio Breast Cancer Symposium. Over 1000 women enrolled in each of three arms of this trial with median follow-up of 23 months and 322 disease-free survival events recorded (84 deaths). Twenty-nine percent of patients were node-negative (comparable with HERA trial population). At 4 years, the DFS projects to 73% in AC-T, 80% in TCH, and 84% in AC-TH arms. The number of events in TCH and AC-TH arms (98 and 77, respectively) did not reach a statistically significant difference ($p = 0.16$) indicating equivalency of the two regimens. One of the hypotheses tested in this trial was the possibility of reduction of cardiac events with nonanthracycline-containing TCH regimen compared to AC-TH. The total number of grade 3 and 4 cardiac events, including arrhythmias, was higher in AC-TH ($n = 25$) than in TCH ($n = 14$) but not statistically significant ($p = 0.11$). This clinical trial also provided deeper insight in the genetic events associated with Her2/neu amplification. Locus 17q21.2, the region of chromosome 17, contains the gene for Topo II. Patients who did not have coamplification of this gene with Her2/neu locus (65%) fared significantly worse overall and showed poor survival when treated with or without trastuzumab.

Future clinical research employing trastuzumab/chemotherapy combinations likely will be directed to identification of genetic profiles and chromosomal rearrangements associated with poor response to HER2 blockade and identification of the "escape mechanisms" that cause treatment resistance. The minimum necessary duration of trastuzumab treatment has not been defined. The HERA trial unambiguously showed equivalency of 1 and 2 years of therapy. The data from several neoadjuvant trials *(162–164)* as well as a recently presented Finnish adjuvant study *(165)* indicated that even a short course of trastuzumab confers a significant treatment benefit for Her2-positive breast cancer. In the latter trial *(165)*, adjuvant 9-week trastuzumab was effective in preventing any recurrence (HR = 0.46, $p = 0.0078$). Three-year distant disease-free survival of patients who received trastuzumab was 93% and that of patients who did not receive trastuzumab was 76% ($p = 0.0078$, 11 of 115 versus 26 of 116 events, HR = 0.43). There is a paucity of data to suggest any indication for trastuzumab therapy in highly curable node-negative breast cancer. The subset analysis of the HERA and BCIRG006 trials may shed some light on this issue as the follow-up data mature.

18. PERTUZUMAB

In many human cancers, Her2 is detectable on the cell surface without gene amplification. The molecular structure of Her2 allows it to exhibit strong tyrosine kinase activity in the absence of any extracellular ligand. Its dimerization arm of domain II (also known as CR1) is constitutively exposed and "ready" to dimerize with any activated isoform of an ErbB family member in physical proximity. Her2 is thus capable of retaining EGFR on the cell surface and augmenting the signaling cascade initiated by EGFR activation *(166,167)*. Moreover, an interaction between insulin-like growth factor type I receptor (IGF-IR) and Her2 has been described by Nahta et al. *(168)*. The investigators demonstrated reversal of trastuzumab resistance in the SKBR3 cell line by combining trastuzumab with either antibody to IGF-IR or a novel antibody, pertuzumab, that targets the Her2 dimerization arm. Interestingly, Her2 blockade with trastuzumab in the sensitive cell lines is associated with recruitment of the inhibitory molecule PTEN, which in turn results in reduced Akt phosphorylation *(167)*. Clearly, the mechanism of Her2-driven resistance to apoptosis and proliferation is highly complex and involves interaction with other growth factor tyrosine kinase receptors and complex trafficking of Her2 to and from the cellular membrane.

Pertuzumab, a "dimerization inhibitor," was demonstrated in preclinical studies to have a growth-inhibitory effect on breast, prostate, and nonsmall cell lung cancer cell lines, expressing varying levels of Her2. In phase I clinical trials, pertuzumab showed activity in a number of human cancers. Agus et al. *(169)* conducted a dose escalation phase I clinical trial of pertuzumab in 21 patients of whom 19 completed at least two cycles. Pertuzumab was well tolerated. The pharmacokinetics of pertuzumab were similar to other humanized IgG antibodies, supporting a 3-week dosing regimen. Trough plasma concentrations were in excess of target concentrations at doses >5 mg/kg. Two patients, one with ovarian cancer (5.0 mg/kg) and one with pancreatic islet cell carcinoma (15.0 mg/kg), achieved a partial response after 1.5 and 6 months of pertuzumab therapy and lasted for 11 and 10 months, respectively. Stable disease lasting for more than 2.5 months was observed in six patients. With good tolerance and favorable pharmacokinetics allowing 3-week dosing, pertuzumab is clinically active and proves the concept that inhibition of dimerization can be an effective anti-cancer strategy.

19. BEVACIZUMAB

Malignant tumor progression requires a blood supply capable of delivering essential nutrients and oxygen to the cells. Blood vessels consist of endothelial cells, smooth muscle cells, and other supporting cells and are responsible for the transport of these nutrients. Numerous researchers have investigated the process of tumor angiogenesis with a goal of manipulating the host vascular response to tumors based on the pioneering work of Folkman and others *(170)*. Blockade of tumor vessels deprives malignant cells of essential nutrients and causes tumor destruction. As one vessel supplies a vast number of cancer cells, this approach can amplify the consequences of an initial cytotoxic insult. Moreover, this strategy does not require the penetration of other cells or molecules into the tumor parenchyma.

Bevacizumab is a monoclonal antibody targeting VEGF. Of the identified angiogenic factors, VEGF is the most potent and specific regulator of both normal and pathologic angiogenesis *(171)*. VEGF produces a number of biologic effects, including endothelial cell mitogenesis and migration, induction of proteinases, leading to remodeling of the extracellular matrix, increased vascular permeability, and maintenance of survival for newly formed blood vessels *(171)*. The biologic effects of VEGF are mediated through binding and stimulation of two receptors on the surface of endothelial cells: Flt-1 (fms-like tyrosine kinase) and kinase domain region (KDR) *(171,172)*.

Increased expression of VEGF has been demonstrated in most human tumors examined to date, including tumors of the lung, breast, thyroid, gastrointestinal tract, kidney, bladder, ovary, and cervix, as well as angiosarcomas and glioblastomas *(171)*. Inhibition of VEGF by using an anti-VEGF monoclonal antibody blocks the growth of a number of human cancer cell lines in nude mice *(171)*. In addition, the combination of anti-VEGF antibody and chemotherapy in nude mice injected with human cancer xenografts results in an increased anti-tumor effect compared with antibody or chemotherapy treatment alone *(173)*.

Bevacizumab, previously known as rhuMAb (VEGF), is a recombinant humanized version of a murine anti-human VEGF monoclonal antibody *(174)*. Approximately 93% of the amino acid sequence, including most of the antibody framework, is derived from human IgG, and 7% of the sequence is derived from the murine antibody. Bevacizumab has been studied in at least 3500 patients in a number of phase I, II, and III clinical trials. These clinical trials have included patients with a number of tumor types, including colorectal, breast, lung, and renal carcinoma *(5,175–177)*.

19.1. Metastatic Colorectal Cancer

In a large phase III study (AVF2107g) in patients with metastatic colorectal cancer, the addition of bevacizumab to irinotecan/5-fluorouracil/leucovorin (IFL) chemotherapy resulted in a clinically and statistically significant increase in duration of survival, with a HR of death of 0.660 (median survival 15.6 versus 20.3 months; $p < 0.0001$). Similar increases were seen in progression-free survival (6.2 versus 10.6 months; $p < 0.0001$), ORR (35 versus 45%; $p < 0.0029$), and duration of response (7.1 versus 10.4 months; $p < 0.0014$) for the combination arm versus the chemotherapy only arm *(5)*. Based on this survival advantage, bevacizumab was designated for priority review and was approved in the USA for first-line treatment in combination with IV 5-FU-based chemotherapy for patients with metastatic CRC *(5)*.

One of the pivotal studies that addressed the nature of biologic effects of bevacizumab in patients with rectal cancer was a reported by Willet et al. *(178)*. The authors provided evidence of rapid changes in metabolic activity and reduction of vascularization, microvascular density, interstitial fluid pressure, and the number of viable, circulating endothelial, and progenitor cells. This study also paved the way for the rational combination of bevacizumab with chemotherapy as the blockade of VEGF led to increases in the fraction of vessels with pericyte coverage. This "normalization" of leaky vessels in the tumor bed reduces interstitial oncotic pressure and hence should improve diffusion of the chemotherapeutic drugs within tumor masses.

Preliminary results from a large, randomized clinical trial (ECOG 3200) for patients with advanced colorectal cancer who had previously received treatment were released subsequently. This trial offered a different choice of chemotherapy. Patients with previously treated metastatic colorectal cancer who received bevacizumab in combination with FOLFOX4 (a regimen of oxaliplatin, 5-fluorouracil, and leucovorin) had a median overall survival of 12.5 months that was significantly better than the 10.7-month median overall survival in patients treated with FOLFOX4 alone. There was a 26% reduction in the risk of death for patients in this study who received bevacizumab plus FOLFOX4 compared with those who received FOLFOX4 alone.

19.2. Metastatic Non-Small Cell Lung Cancer

In non–small cell lung cancer, bevacizumab plus carboplatin and paclitaxel was tested in a phase II trial in patients with advanced or recurrent disease *(179)*. In the three arms of this study, 99 patients were randomly assigned to bevacizumab 7.5 ($n = 32$) or 15 mg/kg ($n = 35$) plus carboplatin and paclitaxel every 3 weeks or carboplatin and paclitaxel alone ($n = 32$). Compared with the control arm, addition of bevacizumab resulted in a dose-dependent increase in the response rate (31.3 versus 40% as per independent review), longer median TTP (7.4 versus 4.2 months), and a modest increase in survival (17.7 versus 14.9 months). Unexpectedly, patients treated with lower dose of bevacizumab had worse response rates, TTP, and survival than in the control arm. These differences can be attributed to the small number of patients in the study and the unusually long median survival of patients in the control arm exceeding 1 year (14.9 months). Also, 19 of the 32 control patients crossed over to single-agent bevacizumab at 15 mg/kg at the time of disease progression, and five maintained stable disease for more than 6 months. Bleeding was the most prominent adverse event and was manifested in two distinct clinical patterns: minor mucocutaneous hemorrhage and major hemoptysis. Major hemoptysis was associated with squamous cell histology, tumor necrosis and cavitation, and disease location close to major blood vessels. Patients with centrally located tumors and squamous cell histology were excluded from the phase III trial due to tendency of these cancers to bleed.

The results of a randomized phase II/III trial (E4599) of paclitaxel and carboplatin with or without bevacizumab in patients with advanced nonsquamous nonsmall cell lung cancer were reported by the ECOG at the 2005 ASCO Annual Meeting (180). About 878 patients were randomized to receive paclitaxel plus carboplatin or the same chemotherapy plus bevacizumab (15 mg/kg) every 3 weeks. Chemotherapy was continued up to six cycles; patients in the experimental arm received single-agent bevacizumab after the six cycles of chemotherapy until progressive disease or intolerable toxicity. Patients in the chemotherapy arm alone were not allowed to cross over to bevacizumab. The results of the second interim analysis were reported after 469 (72.2%) of the 650 deaths required for final analysis had occurred. There was a significant advantage for patients in the bevacizumab arm in terms of median survival (12.5 versus 10.2 months; $p = 0.0075$). In addition, patients treated with bevacizumab had a significantly higher response rate (27 versus 10%; $p < 0.0001$) and a significantly longer progression-free survival time (6.4 versus 4.5 months; $p < 0.0001$). Both regimens were well tolerated; a higher incidence of bleeding was associated with bevacizumab administration (4.5 versus 0.7%). Five of ten treatment-related deaths occurred as a result of hemoptysis, all in the experimental arm.

19.3. Breast Cancer

The safety and efficacy of bevacizumab in patients with previously treated metastatic breast cancer was evaluated in a phase I/II trial (181). Seventy-five patients were treated with escalating doses of bevacizumab ranging from 3 mg/kg to 20 mg/kg administered intravenously every other week. Tumor response was assessed before the 6th (70 days) and 12th (154 days) doses. Safety was evaluated during every cycle. Eighteen patients were treated at 3 mg/kg, 41 at 10 mg/kg, and 16 at 20 mg/kg. Four patients discontinued study treatment because of an adverse event. Hypertension was reported as an adverse event in 17 patients (22%). This study demonstrated that bevacizumab as a single agent has minimal activity: the ORR was 9.3% (confirmed response rate, 6.7%). The optimal dose of bevacizumab in this trial was 10 mg/kg every other week, and toxicity was acceptable.

Miller and colleagues (182) reported the results of a randomized phase III trial of capecitabine/bevacizumab versus capecitabine alone in 462 anthracycline and taxane pretreated metastatic breast cancer patients. The primary end point of this study, progression-free survival, did not show a statistically significant difference (4.86 versus 4.17 months; HR = 0.98). Combination therapy significantly increased the response rates (19.8 versus 9.1%; $p = .001$). Overall survival (15.1 versus 14.5 months) and time to deterioration in quality of life were comparable in both treatment groups.

19.4. Toxicity

Four major bevacizumab-associated toxicities have been identified: hypertension, proteinuria, thromboembolic (TE) events, and hemorrhage.

1. Proteinuria, ranging from asymptomatic and transient events detected on routine dipstick urinalysis to nephrotic syndrome, has been seen in all clinical trials to date. The majority of proteinuria events have been grade 1 or 2. In the phase III pivotal trial in metastatic CRC, the rate of grade 3 or greater proteinuria (NCI CTAEC Scale v3) was < 1% in both treatment arms.
2. Venous and arterial TE events, ranging in severity from catheter-associated phlebitis to fatal, have been reported in patients treated with bevacizumab in the colorectal cancer trials and, to a lesser extent, in patients treated with bevacizumab in non–small cell lung cancer (NSCLC) and breast cancer trials. In the phase III pivotal trial in metastatic CRC, there was a slightly higher rate of venous TE events that was not statistically significant in the patients on the bevacizumab treatment arm (16 versus 19%). There was also a higher rate of arterial TE events (1 versus 3%) such as myocardial infarction, transient ischemia attack, cerebrovascular accident/stroke, and angina/unstable angina. A pooled analysis of the rate of arterial TE events from five randomized studies (1745 patients) showed that treatment with bevacizumab increased the risk of having an arterial TE event from 1.9 to 4.4%. Furthermore, certain baseline characteristics conferred additional risk, specifically age >65 years and a history of a prior arterial TE event.
3. Gastrointestinal perforation was not seen in the bevacizumab phase I or II clinical trials; however, in the phase III colorectal trial, gastrointestinal perforation and wound dehiscence, complicated by intra-abdominal abscesses, occurred at an increased incidence in patients receiving bevacizumab. These events varied in type and severity, ranging from free air seen on kidney, ureter and bladder X-ray (KUB), which resolved without treatment, to a colonic perforation with abdominal abscess, which was fatal.
4. In a phase II NSCLC trial, 6 of 66 bevacizumab-treated patients experienced life-threatening hemoptysis or hematemesis. Four of these events were fatal. Centrally located lesions that were necrotic or cavitary and squamous cell histology were identified as possible risk factors for bleeding.
5. Bevacizumab increases the risk of CHF in patients with a history of or concurrent anthracycline exposure. Prior radiation therapy to the chest wall may also increase the risk of CHF in these patients. There are no data to suggest that bevacizumab increases the risk of CHF in patients without exposure to anthracyclines.

20. HEMATOLOGIC MALIGNANCIES

20.1. Rituximab and CD20

The first successful application of a monoclonal antibody was in the treatment of low-grade NHLs. The anti-CD20 monoclonal antibody, rituximab (Rituxan), is the first mAb to be approved by the FDA in 1997, for use in human malignancy *(183,184)*. Treatment with this chimeric anti-CD20 antibody led to impressive clinical responses and dramatically changed the available treatment options for patients with NHL. Rituximab is humanized and multiple doses can be safely administered.

In vitro studies have demonstrated multiple mechanisms by which anti-CD20 antibodies lead to cell death *(185)*. CD20 is a cell surface marker that is expressed on all normal and most of the malignant B-cells *(186)*. It has a low rate of internalization and therefore is an attractive target for antibody-based therapeutics *(187)*. The CD20 protein has 297 amino acids, has intracellular termini, spans the plasma membrane four times, and has a single nonglycosylated extracellular loop of 43 residues *(186)*. Its functions remain unknown, but there is evidence that it participates in calcium influx *(188,189)*. Also, there is evidence that after binding of rituximab to CD20, in addition to ADCC and complement, apoptosis could play a role as a direct cytotoxic effect of rituximab on B-cells. There is evidence of apoptosis occurring in B-cells isolated from patients with chronic lymphocytic lymphoma (CLL) who were treated with rituximab and thus supporting a role for direct CD20 signaling *(190)*.

In the phase I study to determine the maximum tolerated dose, patients with relapsed low-grade and intermediate/high-grade NHL received four weekly infusions of rituximab *(184)*. Thrombocytopenia and B-cell lymphocytopenia were observed. The lymphocytopenia persisted for 3–6 months. Thirty three percent of patients with low-grade lymphomas had a partial response to the treatment. Phase II studies confirmed the efficacy of this therapy, demonstrating response rates of 46% and 48% in two separate studies *(1,191)*. A phase II trial evaluated rituximab in relapsing or refractory diffuse large B-cell lymphoma (DLBCL), mantle cell lymphoma, or other intermediate or high-grade B-cell NHLs *(192)*. The study randomized 54 patients to either eight weekly treatments of 375 mg/m^2 intravenous rituximab or 375 mg/m^2 in week 1 followed by seven weekly intravenous infusions of 500 mg/m^2. Five complete responses and 12 partial responses were observed, for an ORR of 31 % with a 14% CR rate and a median TTP of 246 days in responding patients. There was no evidence of superiority of either treatment regimens. Patients with refractory disease and those with histologies other than DLBCL appeared to have lower response rates.

In some patients with circulating B-cells, treatment with rituximab has induced an infusion-related syndrome characterized by fever, rigors, thrombocytopenia, tumor lysis, bronchospasm, and hypoxemia, requiring discontinuation of the antibody infusion. Symptoms typically resolve with supportive care and patients may continue further therapy without sequelae *(193)*. Circulating CD20-positive cells, including lymphoma cells, may affect the efficacy of rituximab. Peak levels of circulating antibody inversely correlate with pretreatment B-cell counts as well as the bulk of tumor *(194,195)*. Greater numbers of peripheral lymphocytes and/or tumor bulk serve as an antigen sink, removing antibody from the circulation. For patients with bulky disease, a higher antibody dose or a greater number of cycles may be warranted, as patients with lower serum rituximab concentrations have had statistically significant lower response rates.

Despite these exciting outcomes, most patients eventually relapsed and became resistant to therapy. In an attempt to overcome these obstacles and to take advantage of the multiple postulated mechanisms of actions of this monoclonal antibody, combination studies with rituximab and chemotherapy or cytokines were undertaken. Czuczman et al. *(196)* reported the first successful rituximab–chemotherapy combination study. In this study, which now has been updated, forty patients with low-grade or follicular B-cell NHL were enrolled in the study. Thirty-five patients received all six planned cycles of Cyclophosphamide, Adriamycin, Vincristine and Prednisone (CHOP) every 21 days, with six infusions of rituximab at a dose of 375 mg/m^2 given before, during, and after chemotherapy. Five patients were removed from the study for a variety of reasons. The ORR was 95% (38/40), with 55% complete responses and 40% partial responses. Fewer complete responses were noted in patients with bulky disease. Median TTP was 63.6 months. Seven of 8 patients who had initially been positive for the bcl-2 translocation became negative for the translocation by PCR assay after therapy; this has not been seen with CHOP chemotherapy alone *(197)*.

A phase II study reported by Vose and colleagues *(198)* demonstrated that the combination of CHOP plus rituximab in patients with DLBCL was safe and efficacious. Subsequently, a phase III randomized study in elderly patients with diffuse large-cell lymphoma comparing standard CHOP chemotherapy to CHOP with rituximab demonstrated a 76% complete response rate in the combination arm compared with 60% complete response rate in the chemotherapy arm ($p = 0.005$), without significant differences in toxicity between the two groups *(199)*. After a median follow up of 2 years, both overall (70 versus 57%) and event-free survival were significantly better in the combination arm ($p = 0.007$ and $p < 0.001$, respectively). This combination is now the standard treatment regimen for patients with DLBCL.

Emerging data also indicate that low-grade B-cell lymphoma patients possessing the (158v/v) polymorphism in FcγRIII experience superior response rates and outcomes when treated with this antibody. *(38,200)* These findings indicate that antibody Fc domain : FcR interactions underlie at least some of the clinical benefit of rituximab, and indicate a possible role for ADCC, which depends on such interactions.

20.1.1. COMBINATION WITH CYTOKINES

The immunomodulatory effects of cytokines have been exploited in a number of early studies to augment the observed activity of rituximab in patients with NHL. IL-2 is a lymphokine produced by T-lymphocytes that has direct effect on B-cells, NK cells, and monocytes. Several studies have shown that patients treated with IL-2 can have greatly increased NK cell concentrations *(201,202)*, and as these cells are important effectors of ADCC, this can lead to improved cell killing in combination with rituximab. A phase I trial conducted to test this hypothesis reported an ORR of 55% in rituximab-naïve patients with follicular lymphomas *(203)*. Patients were treated with the standard doses of rituximab on a weekly basis for four treatments in addition to daily subcutaneous injections of 1.2 million units/m^2 of IL-2. This combination seems to be feasible, and further work is needed to determine the risks and benefits of the addition of IL-2.

Similarly, IL-12 has been studied in combination with rituximab *(204)*. Escalating doses of IL-12 were used in a phase I trial in combination with standard doses of rituximab in 43 patients with NHL of various histologies. At the maximum tolerated dose of 300 ng/kg of IL-12, dose-limiting toxicities were liver function elevations and hematotoxicity. A response rate of 69% was reported, with a 25% CR rate across all histologies.

Type I IFN not only has immunomodulatory effects on T-cells and NK cells, it also has been shown to up-regulate the expression of CD20 on B-cells *(205,206)*. Two trials have examined the safety and feasibility of combination therapy with

IFN and rituximab *(207,208)*. Davis and colleagues *(207)* conducted a single-arm, multicenter, phase II trial to assess the safety and efficacy of combination therapy with rituximab and IFN-α-2a in 38 patients with relapsed or refractory, low-grade or follicular, B-cell NHL. IFN-α-2a [2.5 or 5 million units (MIU)] was administered s.c., three times weekly for 12 weeks. Starting on the fifth week of treatment, rituximab was administered by i.v. infusion (375 mg/m^2) weekly for four doses. All patients received four complete infusions of rituximab and were evaluable for efficacy, but 11 patients did not receive all 36 injections of IFN. The study treatment was reasonably well tolerated with no unexpected toxicities stemming from the combination therapy. The ORR was 45% (17 of 38 patients); 11% had a complete response and 34% had a partial response. The median response duration and the median TTP in responders were 22.3 and 25.2 months, respectively. Further follow-up is needed to determine whether this treatment combination leads to a significantly longer TTP than single-agent treatment with rituximab.

In a study reported by Sacchi et al., 64 patients with various lymphomas were treated with standard dose rituximab in addition to daily subcutaneous injection of IFN starting with 1.5 million units/day on week 1 and escalating to 6 million units on weeks 4 and 5. The OR was 70% with 33% CR and a median duration of remission of 19 months *(208)*. These results need to be confirmed in subsequent trials to better define the role of combination immunotherapy in patients with NHL.

20.2. New Anti-CD20 Antibodies

The pharmacokinetic analysis in the first pivotal trial of rituximab indicated that patients with higher and more prolonged blood level of active drug had a better chance of responding *(191)*. This not only led to investigations into the role of maintenance therapy with rituximab, it also sparked an interest in developing second generation anti-CD20 antibodies with more favorable pharmacokinetics and improved efficacy. Several of these antibodies are in clinical development *(209,210)*. The HuMax-CD20 antibody is being studied in several phase I/II trials. Hagenbeek and colleagues *(211)* have reported the results of a study involving 40 patients with follicular lymphoma. Patients in this study were treated with four weekly infusions of 300–1000 mg of HuMax antibody after premedication. Treatment led to the depletion of circulating B-cells and no dose-limiting toxicities were encountered. Twelve of 24 patients in this study had an objective response including four CR.

Coiffier et al. *(212)* recently reported the early results of a phase I/II trial with HuMax-CD20 antibody in patients with CLL. Thirty-three patients in three cohorts with refractory or relapsed CLL received four weekly infusions of the antibody in escalating doses up to 2000 mg. The maximally tolerated dose was not identified. Adverse events were noted mainly on the days of infusions and were related to cytokine release. Other adverse events included hepatic cytolysis, herpes zoster, neutropenia and one death from pneumonia. A response rate of 52% was observed at week 11. Treatment also led to significant depletion of CD19$^+$CD5$^+$ cells. These results suggest that HuMax-CD20 has a favorable toxicity profile and activity in low-grade lymphomas.

20.3. Alemtuzumab

Alemtuzumab is a humanized anti-CD52 monoclonal antibody that has been approved by the US FDAfor the treatment of fludarabine-refractory CLL. CD52 is expressed in almost all human lymphocytes *(124)*. The rat IgM and IgG antibodies were lytic with complement, but only IgG2b had ADCC activity. The function of this antigen remains unknown. It is known, however, that CD52 is expressed on all lymphocytes at various stages of differentiation, as well as monocytes, macrophages, and eosinophils *(214)*. The only other site of expression is the male reproductive tract *(215)*. Hematopoietic stem cells, erythrocytes, and platelets do not express this antigen and are thus spared from a direct antibody effect. The highest levels of expression is on T-prolymphocytic leukemia (PLL) cells, followed by B-cell CLL (B-CLL), with the lowest levels on normal B-cells *(216)*.

The first CD52 antibodies were isolated in the 1980s while researchers sought antibodies that would kill T-cells by activating human complement. An IgM antibody was initially selected because of the efficiency with which it activated complement and because it was cytotoxic to T-cells in vitro. Prior iterations included Campath-1G, a murine derivative *(213)*, which demonstrated clinical activity in refractory CLL even in patients who had experienced treatment failure with Campath-1M. Possible mechanisms of action of this agent include antibody-dependent cellular toxicity (ADCC), CDC, and induction of apoptosis. CAMPATH has been tested as a therapeutic agent in patients with CLL, low-grade NHL, and as a mean to deplete T-cells in bone marrow transplantation.

In one study, half of the patients with fludarabine-resistant chronic lymphocytic leukemia or B-prolymphocytic leukemia exhibited clinical responses to CAMPATH-1 *(29)*. A larger phase II study reported a 42% response rate in patients with relapsed or refractory chronic lymphocytic leukemia but at the cost of an increase in opportunistic infections and septicemia. CAMPATH-1 has also been evaluated as first-line therapy for patients with chronic lymphocytic leukemia. Loss of peripheral blood malignant lymphocytes was seen in all patients treated with this antibody. However, patients with involvement of lymph nodes and/or spleen were less likely to respond completely. There was evidence of reactivation of cytomegalovirus infections. Subcutaneous administration of the antibody was found to be safe and effective.

In a phase II multicenter study of CAMPATH-1H in previously treated patients with low-grade NHLs, 50 patients with relapsed or refractory disease were treated with 30 mg of CAMPATH-1H three times weekly for up to 12 weeks *(30)*. Infection, anemia, and thrombocytopenia were common, and myocardial infarction occurred in one patient with a prior history of angina and congestive heart failure. The ORR was 20% (16% partial response and 4% complete response). Responses were short in duration, with a median TTP of 4 months. Patients with mycosis fungoides responded more frequently and had a longer TTP (10 months) than did patients with low-grade NHL (4 months). Treatment was associated with reactivation of herpes simplex, oral candidiasis, *pneumocystis carinii* pneumonia, cytomegalovirus pneumonitis, pulmonary aspergillosis, disseminated tuberculosis, and seven cases of pneumonia and septicemia.

Alemtuzumab is also capable of inducing a minimal residual disease (MRD)-negative remission in patients with relapsed CLL, refractory to fludarabine *(217,218)*. One study has examined the relationship between MRD-negative state and patient survival. Moreton and colleagues treated 91 patients with refractory CLL with standard dose of Alemtuzumab (84 patients received IV dosing and the rest subcutaneously) three times a week until a maximum response was achieved with a goal of achieving an MRD-negative remission as determined by four-color flow cytometry *(219)*. Fifty-three percent of patients in this study responded to treatment, and the responses were clearly correlated with the degree of adenopathy such that patients with less adenopathy had better responses. Also, patients with minimal adenopathy had higher rates of MRD negativity. Overall, 18 patients achieved MRD-negative state. None had bulky disease, and 72% had no lymphadenopathy prior to initiating treatment although over 40% of them had failed prior treatments. Two patients in this group died of opportunistic infections, but 88% were alive at a median follow up time of 36 months. The most common adverse event reported for this agent in the trial conducted by Morton was infusion-related toxicities including rigors and fever. Other nonhematologic toxicities reported included fatigue, dyspnea, and bronchospasm. The most frequent hematologic toxicity reported was neutropenia *(219)*. There were three deaths due to fungal infections and one due to CMV reactivation. Overall, over 40% of patients developed an infection either during therapy or within a month of finishing their treatments.

20.4. Radioimmunotherapy with Labeled Antibodies

Radioimmunotherapy is a novel treatment approach, which combines the targeting capability of monoclonal antibodies with the additional cytotoxic effects of radiation. This form of therapy is particularly useful in B-cell lymphomas as these malignancies are exquisitely radiosensitive. Two labeled antibodies have been approved for the treatment of patients with NHL. Ibritumomab tiuxetan ^{90}Y is a radiolabeled monoclonal antibody that has been shown to produce clinically significant responses in patients with NHL and is the first radiolabeled monoclonal antibody approved for therapeutic use in the treatment of lymphoma.

20.5. Ibritumomab Tiuxetan ^{90}Y

Ibritumomab tiuxetan ^{90}Y is indicated for the treatment of patients with relapsed or refractory low-grade, follicular, or transformed B-cell NHL including patients with rituximab-refractory follicular NHL. Yttrium-90 ibritumomab tiuxetan is a murine IgG1 monoclonal antibody and is the parent molecule from which the chimeric monoclonal antibody, rituximab, was derived. It is covalently bound to the linker-chelator tiuxetan, which forms a strong bond with the radionuclide. Tiuxetan provides a high-affinity chelation site for either indium-111 or yttrium-90. Indium-111 is a gamma emitter that permits imaging and dosimetry, whereas yttrium-90 is a pure beta emitter and is used to deliver the therapeutic radiation payload. Being a pure beta emitter, the radiation emitted by ^{90}Y does not escape the body; this facilitates its use for therapeutic purposes in the outpatient setting *(220)*.

^{90}Y has a half-life of 64 h, a maximum energy of 2.3 MeV, and a mean path length of 5 mm in soft tissue. Its high energy and long path length makes it potentially well suited for the treatment of bulky or poorly vascularized tumors. Several clinical trials have established the efficacy of treatment with ^{90}Y ibritumomab tiuxetan in patients with NHL (Table 3).

One hundred forty three patients with histologically proven relapsed or refractory follicular, low-grade or transformed NHL were prospectively randomized, in a phase III trial, to receive either the ^{90}Y ibritumomab regimen (0.4 mCi/kg) or four weekly doses of rituximab (375 mg/m^2) *(222)*. The primary end point of this study was the ORR (ORR). The ORR for the ^{90}Y ibritumomab group was 80 versus 56% for the

Table 3
Early trials with rituximab

	Phase	Patient population	Number of patients	TTP (months)	CR (%)	ORR (%)
Witzig (221)	Phase I/II	Various histologies	51	12.9	26	67
Witzig (222)	Phase III	Randomized and no prior rituximab	143	11.2	34	80
Witzig (223)	Phase II	Rituximab refractory	57	6.8	15	74
Wiseman (224)	Phase II	Thrombocytopenia	30	9.4	37	83

rituximab arm ($p = 0.002$). This response rate for the rituximab arm was comparable to the previously published results with this monoclonal antibody (225). The complete response rate was 30% in the yttrium-90 ibritumomab tiuxetan arm and 16% in the rituximab arm ($p = 0.04$). However, there were no significant differences between the two treatment groups in median duration of response (14.2 and 12.1 months, $p = 0.6$) or the median TTP for the two groups (11.2 months for yttrium-90 ibritumomab tiuxetan versus 10.1 months for rituximab, $p = 0.17$) (222). There was a trend toward longer duration of response in patients who received the radiolabeled antibody and achieved a complete response. This study established yttrium-90 ibritumomab tiuxetan as a viable therapy for patients with relapsed/refractory NHL.

Owing to concerns for bone marrow toxicity, in the clinical studies conducted thus far with this agent, patients with more than a 25% marrow involvement with lymphoma have been excluded. Also, for patients with mild thrombocytopenia ($100–149 \times 10^9$ platelets/L) treatment at a lower dose (0.3 mCi/kg) is recommended based on a phase II study reported by Wiseman and coinvestigators (224). Several factors seem to predict for a lower response rate and duration of response. These include the number of prior therapies, presence of bulky disease, and histologic subtypes other than follicular lymphoma. Factors such as age and stage do not seem to affect treatment outcomes (226,227). This data has recently been reviewed elsewhere (228).

20.6. I-131 Tositumomab (Bexxar)

The second radioimmunotherapy agent to be approved by the FDA was I-131 tositumomab. This agent has been studied in several phase II trials. The safety and efficacy of therapy with this agent was first demonstrated by Kaminski and coworkers in phase I and II studies. In these trials, an ORR of 71% was achieved in patients with relapsed low-grade and transformed NHL. This included complete responses in 34% of patients who participated in these trials (229). Subsequently, two multicenter trials have confirmed the efficacy of this treatment modality (230,231). Vose and colleagues enrolled 47 patients with CD20-positive low-grade or transformed NHL. All patients had at least one prior therapy, with a median of four prior treatments, and all had failed or relapsed within a year of their last qualifying chemotherapy. In this trial, 57% of patients had an objective response and 32% of patients achieved complete responses. The median duration of response in patients who had a complete response was 19.9 months (230). The second multicenter trial involved 60 patients, nearly 40% of whom had transformed NHL. Again this patient population was heavily pretreated with a median of four prior treatments. The ORR was 65% with a 20% complete response rate. It was shown that both the response rate and duration of response was longer for patients treated with this agent than their last chemotherapy regimen *(231)*.

This treatment has also been evaluated in patients with newly diagnosed follicular lymphoma *(232)*. Overall, 95% of patients responded, with a complete response rate of 75%. BCR gene rearrangement studies as assessed by PCR showed molecular responses in 80% of patients. The actuarial 5-year progression-free survival for all patients was 59%, with a median progression-free survival of 6.1 years. Hematologic toxicities were acceptable, and there have not been any reports of myelodysplastic syndrome or secondary leukemias. This study has established 131-I tositumumab as a potential, frontline therapeutic option for patients with advanced follicular lymphoma.

20.7. Vascular Targeting Antibodies

Tumor-associated blood vessels offer numerous tumor-specific targets for therapy. Several markers have been identified by various groups, and these markers provide numerous opportunities for targeted therapy strategies. However, it is likely that great care must be taken when such markers are chosen for targeted therapy. If the target is also expressed by the endothelial cells or by other cells in a vital organ and if the targeting agent is not consumed by targeting the tumor vascular target, the results could be catastrophic. Therefore, regulation of the cytotoxic therapy to avoid excessive host toxicity is important.

Vascular targeting has been validated by a number of groups as a viable and effective method of treating solid tumors in mice. One of the earlier vascular-targeting agents, (although not an antibody) Combretastatin A-4, had shown significant activity in a number of animal studies *(233,234)*. Combretastatin A-4 (CA4P) is a tubulin-binding agent that inhibits tubulin polymerization *(235)*. This agent has limited water solubility leading to the development of a water soluble prodrug. In experimental tumors, CA4P causes rapid and extensive vascular destruction. This effect is highly selective and leads to a hemorrhagic necrosis of the tumors and a significant reduction in tumor perfusion within 1 h of treatment *(233)*. This general reduction in perfusion observed at 1 h continued at 3 and 6 h with residual areas of perfusion at the periphery of the tumors but complete vascular shut down in the tumor center *(233)*. By 24 h, perfusion in the peripheral zone had increased, but the center of the tumor remained unperfused. Phase I studies of this agent with different treatment schedules and in combination with chemotherapy lead to minimal responses although a reduction in tumor perfusion was indirectly shown for some of the patients in these trials *(236,237)*. Although toxicity profile of this agent was reasonable, responses were limited. The clinical results of this strategy have been a bit disappointing, leading to approaches that provide more targeting specificity. For example, significant treatment efficacy has been observed in large solid tumors using either toxin-linked or tissue-factor-linked vascular targeting *(238,239)*. A single-chain antibody fragment against the ED-B domain of fibronectin (scFv L19) has shown specific tumor targeting in a murine tumor model *(240)*. This potential target is expressed in the extracellular matrix around the newly formed blood vessels in a number of solid tumors. Fusion of this single chain with IFN-γ and IL-2 has led to impressive but inconsistent results in therapeutic studies in animal models *(241,242)*. A monoclonal antibody (3G4) directed against anionic phospholipids on the external membrane of hydrogen peroxide-treated endothelial cells localizes to the tumor vascular endothelium in *scid* mice bearing MDA-MB-435 breast cancer tumors *(243)*. An average of 40 ± 10% of the vessels were stained with 3G4 in this tumor model. Interestingly, the binding of 3G4 by ELISA to anionic phospholipids requires the presence of β2-glycoprotein I. Another intriguing finding from these studies is the demonstration that 3G4 causes monocytes to bind tumor blood vessels and macrophages to infiltrate tumors. It is possible that binding of 3G4 to exposed anionic phospholipids on tumor vessels can stimulate monocyte and macrophage binding through Fcγ receptors. A potential limitation of this antibody is that the target is only expressed under hypoxic and other oxidative distress conditions. Thus, the concept of antibody-directed vascular targeting has been validated, but new targets are needed to facilitate optimal clinical development of this strategy.

21. CONCLUSIONS

Antibodies have emerged as important therapeutic vehicles in a number of cancers. Future investigations are likely to define additional therapeutic targets and will exploit a variety of anti-tumor mechanisms. These efforts will be abetted by advances in antibody engineering. Although virtually all currently approved antibodies require the presence of the antibody target on the cell surface, in the tumor stroma or in the circulating blood, it is likely that continued progress in the area of immunotoxins will lead to important new directions for antibody therapy.

REFERENCES

1. McLaughlin P, Grillo-Lopez AJ, Link BK, et al. Rituximab chimeric anti-CD20 monoclonal antibody therapy for relapsed indolent lymphoma: half of patients respond to a four-dose treatment program. *J Clin Oncol* 1998;16(8):2825–33.
2. Slamon DJ, Leyland-Jones B, Shak S, et al. Use of chemotherapy plus a monoclonal antibody against HER2 for metastatic breast cancer that overexpresses HER2. *N Engl J Med* 2001;344(11):783–92.
3. Rowinsky EK, Schwartz GH, Gollob JA, et al. Safety, pharmacokinetics, and activity of ABX-EGF, a fully human anti-epidermal growth factor receptor monoclonal antibody in patients with metastatic renal cell cancer. *J Clin Oncol* 2004;22(15):3003–15.
4. Cunningham D, Humblet Y, Siena S, et al. Cetuximab monotherapy and cetuximab plus irinotecan in irinotecan-refractory metastatic colorectal cancer. *N Engl J Med* 2004;351(4):337–45.
5. Hurwitz H, Fehrenbacher L, Novotny W, et al. Bevacizumab plus irinotecan, fluorouracil, and leucovorin for metastatic colorectal cancer. *N Engl J Med* 2004;350(23):2335–42.
6. Kohler G, Milstein C. Derivation of specific antibody-producing tissue culture and tumor lines by cell fusion. *Eur J Immunol* 1976;6(7):511–9.
7. Kohler G, Milstein C. Continuous cultures of fused cells secreting antibody of predefined specificity. *Nature* 1975;256(5517):495–7.
8. Brekke OH, Sandlie I. Therapeutic antibodies for human diseases at the dawn of the twenty-first century. *Nat Rev Drug Discov* 2003;2(1):52–62.
9. Mirick GR, Bradt BM, Denardo SJ, Denardo GL. A review of human anti-globulin antibody (HAGA, HAMA, HACA, HAHA) responses to monoclonal antibodies. Not four letter words. *Q J Nucl Med Mol Imaging* 2004;48(4):251–7.
10. Kim JA. Targeted therapies for the treatment of cancer. *Am J Surg* 2003;186(3):264–8.
11. Trail PA, King HD, Dubowchik GM. Monoclonal antibody drug immunoconjugates for targeted treatment of cancer. *Cancer Immunol Immunother* 2003;52(5):328–37.
12. Khazaeli MB, Conry RM, LoBuglio AF. Human immune response to monoclonal antibodies. *J Immunother* 1994;15(1):42–52.
13. Benhar I, Azriel R, Nahary L, et al. Highly efficient selection of phage antibodies mediated by display of antigen as Lpp-OmpA' fusions on live bacteria. *J Mol Biol* 2000;301(4):893–904.

14. Hoogenboom HR. Overview of antibody phage-display technology and its applications. *Methods Mol Biol* 2002;178:1–37.

15. Worn A, Pluckthun A. Stability engineering of antibody single-chain Fv fragments. *J Mol Biol* 2001;305(5):989–1010.

16. Irving RA, Coia G, Roberts A, Nuttall SD, Hudson PJ. Ribosome display and affinity maturation: from antibodies to single V-domains and steps towards cancer therapeutics. *J Immunol Methods* 2001;248(1–2):31–45.

17. Bird RE, Hardman KD, Jacobson JW, et al. Single-chain antigen-binding proteins. *Science* 1988;242(4877):423–6.

18. Reiter Y, Brinkmann U, Lee B, Pastan I. Engineering antibody Fv fragments for cancer detection and therapy: disulfide-stabilized Fv fragments. *Nat Biotechnol* 1996;14(10):1239–45.

19. Todorovska A, Roovers RC, Dolezal O, Kortt AA, Hoogenboom HR, Hudson PJ. Design and application of diabodies, triabodies and tetrabodies for cancer targeting. *J Immunol Methods* 2001;248(1–2):47–66.

20. Pluckthun A, Pack P. New protein engineering approaches to multivalent and bispecific antibody fragments. *Immunotechnology* 1997;3(2):83–105.

21. Le Gall F, Kipriyanov SM, Moldenhauer G, Little M. Di-, tri- and tetrameric single chain Fv antibody fragments against human CD19: effect of valency on cell binding. *FEBS Lett* 1999;453(1–2):164–8.

22. Iannello A, Ahmad A. Role of antibody-dependent cell-mediated cytotoxicity in the efficacy of therapeutic anti-cancer monoclonal antibodies. *Cancer Metastasis Rev* 2005;24(4):487–99.

23. Gelderman KA, Tomlinson S, Ross GD, Gorter A. Complement function in mAb-mediated cancer immunotherapy. *Trends Immunol* 2004;25(3):158–64.

24. Schmidt KV, Wood BA. Trends in cancer therapy: role of monoclonal antibodies. *Semin Oncol Nurs* 2003;19(3):169–79.

25. Farah RA, Clinchy B, Herrera L, Vitetta ES. The development of monoclonal antibodies for the therapy of cancer. *Crit Rev Eukaryot Gene Expr* 1998;8(3–4):321–56.

26. Goldenberg DM. Challenges to the therapy of cancer with monoclonal antibodies. *J Natl Cancer Inst* 1991;83(2):78–9.

27. Weiner LM, Carter P. Tunable antibodies. *Nat Biotechnol* 2005;23(5):556–7.

28. Wu AM, Senter PD. Arming antibodies: prospects and challenges for immunoconjugates. *Nat Biotechnol* 2005;23(9):1137–46.

29. Ahmad A, Menezes J. Antibody-dependent cellular cytotoxicity in HIV infections. FASEB J 1996;10(2):258–66.

30. Steplewski Z, Lubeck MD, Koprowski H. Human macrophages armed with murine immunoglobulin G2a antibodies to tumors destroy human cancer cells. *Science* 1983;221(4613):865–7.

31. Berken A, Benacerraf B. Properties of antibodies cytophilic for macrophages. *J Exp Med* 1966;123(1):119–44.

32. Ernst LK, Metes D, Herberman RB, Morel PA. Allelic polymorphisms in the FcgammaRIIC gene can influence its function on normal human natural killer cells. *J Mol Med* 2002;80(4):248–57.

33. Metes D, Gambotto AA, Nellis J, et al. Identification of the CD32/FcgammaRIIc-Q13/STP13 polymorphism using an allele-specific restriction enzyme digestion assay. *J Immunol Methods* 2001;258(1–2):85–95.

34. Morel PA, Ernst LK, Metes D. Functional CD32 molecules on human NK cells. *Leuk Lymphoma* 1999;35(1–2):47–56.

35. O'Hanlon LH. Natural born killers: NK cells drafted into the cancer fight. *J Natl Cancer Inst* 2004;96(9):651–3.

36. Whiteside TL, Herberman RB. The role of natural killer cells in immune surveillance of cancer. *Curr Opin Immunol* 1995;7(5):704–10.

37. Clynes RA, Towers TL, Presta LG, Ravetch JV. Inhibitory Fc receptors modulate in vivo cytoxicity against tumor targets. *Nat Med* 2000;6(4):443–6.

38. Weng WK, Levy R. Two immunoglobulin G fragment C receptor polymorphisms independently predict response to rituximab in patients with follicular lymphoma. *J Clin Oncol* 2003;21(21):3940–7.

39. Hayes RJ, Bentzien J, Ary ML, et al. Combining computational and experimental screening for rapid optimization of protein properties. *Proc Natl Acad Sci USA* 2002;99(25):15926–31.

40. Umana P, Jean-Mairet J, Moudry R, Amstutz H, Bailey JE. Engineered glycoforms of an antineu-roblastoma IgG1 with optimized antibody-dependent cellular cytotoxic activity. *Nat Biotechnol* 1999;17(2):176–80.

41. Rafiq K, Bergtold A, Clynes R. Immune complex-mediated antigen presentation induces tumor immunity. *J Clin Invest* 2002;110(1):71–9.

42. Gorter A, Meri S. Immune evasion of tumor cells using membrane-bound complement regulatory proteins. *Immunol Today* 1999;20(12):576–82.

43. Trikha M, Yan L, Nakada MT. Monoclonal antibodies as therapeutics in oncology. *Curr Opin Biotechnol* 2002;13(6):609–14.

44. Velders MP, Litvinov SV, Warnaar SO, et al. New chimeric anti-pancarcinoma monoclonal antibody with superior cytotoxicity-mediating potency. *Cancer Res* 1994;54(7):1753–9.

45. Spiridon CI, Ghetie MA, Uhr J, et al. Targeting multiple Her-2 epitopes with monoclonal antibodies results in improved antigrowth activity of a human breast cancer cell line in vitro and in vivo. *Clin Cancer Res* 2002;8(6):1720–30.

46. Kroesen BJ, McLaughlin PM, Schuilenga-Hut PH, et al. Tumor-targeted immune complex formation: effects on myeloid cell activation and tumor-directed immune cell migration. *Int J Cancer* 2002;98(6):857–63.

47. Sokoloff MH, Nardin A, Solga MD, et al. Targeting of cancer cells with monoclonal antibodies specific for C3b(i). *Cancer Immunol Immunother* 2000;49(10):551–62.

48. Reiter Y, Fishelson Z. Targeting of complement to tumor cells by heteroconjugates composed of antibodies and of the complement component C3b. *J Immunol* 1989;142(8):2771–7.

49. Juhl H, Petrella EC, Cheung NK, Bredehorst R, Vogel CW. Complement killing of human neurob-lastoma cells: a cytotoxic monoclonal antibody and its F(ab´)2-cobra venom factor conjugate are equally cytotoxic. *Mol Immunol* 1990;27(10):957–64.

50. Cheung NK, Modak S, Vickers A, Knuckles B. Orally administered beta-glucans enhance anti-tumor effects of monoclonal antibodies. *Cancer Immunol Immunother* 2002;51(10):557–64.

51. Blok VT, Gelderman KA, Tijsma OH, Daha MR, Gorter A. Cytokines affect resistance of human renal tumour cells to complement-mediated injury. *Scand J Immunol* 2003;57(6):591–9.

52. Blok VT, Daha MR, Tijsma O, et al. A bispecific monoclonal antibody directed against both the membrane-bound complement regulator CD55 and the renal tumor-associated antigen G250 enhances C3 deposition and tumor cell lysis by complement. *J Immunol* 1998;160(7):3437–43.

53. Mendelsohn J, Baselga J. The EGF receptor family as targets for cancer therapy. *Oncogene* 2000;19(56):6550–65.

54. Olayioye MA, Neve RM, Lane HA, Hynes NE. The ErbB signaling network: receptor heterodimer-ization in development and cancer. EMBO J 2000;19(13):3159–67.

55. Rosen LS. Clinical experience with angiogenesis signaling inhibitors: focus on vascular endothelial growth factor (VEGF) blockers. *Cancer Control* 2002;9(2 Suppl):36–44.

56. Holbro T, Civenni G, Hynes NE. The ErbB receptors and their role in cancer progression. *Exp Cell Res* 2003;284(1):99–110.

57. Groner B, Hartmann C, Wels W. Therapeutic antibodies. *Curr Mol Med* 2004;4(5):539–47.

58. Mendelsohn J. Antibody-mediated EGF receptor blockade as an anticancer therapy: from the laboratory to the clinic. *Cancer Immunol Immunother* 2003;52(5):342–6.

59. Franklin MC, Carey KD, Vajdos FF, Leahy DJ, de Vos AM, Sliwkowski MX. Insights into ErbB signaling from the structure of the ErbB2-pertuzumab complex. *Cancer Cell* 2004;5(4):317–28.

60. Agus DB, Akita RW, Fox WD, et al. Targeting ligand-activated ErbB2 signaling inhibits breast and prostate tumor growth. *Cancer Cell* 2002;2(2):127–37.

61. Yarden Y. The EGFR family and its ligands in human cancer. Signalling mechanisms and thera-peutic opportunities. *Eur J Cancer* 2001;37(Suppl 4):S3–8.

62. Gatto B. Monoclonal antibodies in cancer therapy. *Curr Med Chem Anticancer Agents* 2004;4(5):411–4.

63. Boerman OC, van Schaijk FG, Oyen WJ, Corstens FH. Pretargeted radioimmunotherapy of cancer: progress step by step. *J Nucl Med* 2003;44(3):400–11.

64. DeNardo SJ, DeNardo GL, Brush J, Carter P. Phage library-derived human anti-TETA and anti-DOTA ScFv for pretargeting RIT. *Hybridoma* 1999;18(1):13–21.

65. Milenic DE, Brechbiel MW. Targeting of radio-isotopes for cancer therapy. *Cancer Biol Ther* 2004;3(4):361–70.

66. Goldenberg DM. Advancing role of radiolabeled antibodies in the therapy of cancer. *Cancer Immunol Immunother* 2003;52(5):281–96.

67. Wong JY, Shibata S, Williams LE, et al. A Phase I trial of 90Y-anti-carcinoembryonic antigen chimeric T84.66 radioimmunotherapy with 5-fluorouracil in patients with metastatic colorectal cancer. *Clin Cancer Res* 2003;9(16 Pt 1):5842–52.

68. Sharkey RM, Hajjar G, Yeldell D, et al. A phase I trial combining high-dose 90Y-labeled humanized anti-CEA monoclonal antibody with doxorubicin and peripheral blood stem cell rescue in advanced medullary thyroid cancer. *J Nucl Med* 2005;46(4):620–33.

69. Wu AM, Yazaki PJ. Designer genes: recombinant antibody fragments for biological imaging. *Q J Nucl Med* 2000;44(3):268–83.

70. Begent RH, Verhaar MJ, Chester KA, et al. Clinical evidence of efficient tumor targeting based on single-chain Fv antibody selected from a combinatorial library. *Nat Med* 1996;2(9):979–84.

71. Larson SM, El-Shirbiny AM, Divgi CR, et al. Single chain antigen binding protein (sFv CC49): first human studies in colorectal carcinoma metastatic to liver. *Cancer* 1997;80(12 Suppl):2458–68.

72. Jain RK. Tumor physiology and antibody delivery. *Front Radiat Ther Oncol* 1990;24:32–46; discussion 64–8.

73. Sharkey RM, Karacay H, Cardillo TM, et al. Improving the delivery of radionuclides for imaging and therapy of cancer using pretargeting methods. *Clin Cancer Res* 2005;11(19 Pt 2):7109s–21s.

74. Senter PD, Springer CJ. Selective activation of anticancer prodrugs by monoclonal antibody-enzyme conjugates. *Adv Drug Deliv Rev* 2001;53(3):247–64.

75. Bagshawe KD, Sharma SK, Begent RH. Antibody-directed enzyme prodrug therapy (ADEPT) for cancer. *Expert Opin Biol Ther* 2004;4(11):1777–89.

76. Sharma SK, Bagshawe KD, Begent RH. Advances in antibody-directed enzyme prodrug therapy. *Curr Opin Investig Drugs* 2005;6(6):611–5.

77. Zein N, Sinha AM, McGahren WJ, Ellestad GA. Calicheamicin gamma 1I: an antitumor antibiotic that cleaves double-stranded DNA site specifically. *Science* 1988;240(4856):1198–201.

78. Trail PA, Willner D, Knipe J, et al. Effect of linker variation on the stability, potency, and efficacy of carcinoma-reactive BR64-doxorubicin immunoconjugates. *Cancer Res* 1997;57(1):100–5.

79. Trail PA, Willner D, Lasch SJ, et al. Cure of xenografted human carcinomas by BR96-doxorubicin immunoconjugates. *Science* 1993;261(5118):212–5.

80. Mosure KW, Henderson AJ, Klunk LJ, Knipe JO. Disposition of conjugate-bound and free doxorubicin in tumor-bearing mice following administration of a BR96-doxorubicin immunoconjugate (BMS 182248). *Cancer Chemother Pharmacol* 1997;40(3):251–8.

81. Payne G. Progress in immunoconjugate cancer therapeutics. *Cancer Cell* 2003;3(3):207–12.

82. Dubowchik GM, Walker MA. Receptor-mediated and enzyme-dependent targeting of cytotoxic anticancer drugs. *Pharmacol Ther* 1999;83(2):67–123.

83. Kreitman RJ. Recombinant toxins for the treatment of cancer. *Curr Opin Mol Ther* 2003;5(1):44–51.

84. Pastan I, Kreitman RJ. Immunotoxins in cancer therapy. *Curr Opin Investig Drugs* 2002;3(7):1089–91.

85. Pastan I, Beers R, Bera TK. Recombinant immunotoxins in the treatment of cancer. *Methods Mol Biol* 2004;248:503–18.

86. Uckun FM. Immunotoxins for the treatment of leukaemia. *Br J Haematol* 1993;85(3):435–8.

87. Kreitman RJ. Immunotoxins in cancer therapy. *Curr Opin Immunol* 1999;11(5):570–8.

88. Ushiro H, Cohen S. Identification of phosphotyrosine as a product of epidermal growth factor-activated protein kinase in A-431 cell membranes. *J Biol Chem* 1980;255(18):8363–5.

89. Cohen S, Carpenter G, King L, Jr. Epidermal growth factor-receptor-protein kinase interactions. Co-purification of receptor and epidermal growth factor-enhanced phosphorylation activity. *J Biol Chem* 1980;255(10):4834–42.

90. Downward J, Yarden Y, Mayes E, et al. Close similarity of epidermal growth factor receptor and v-erb-B oncogene protein sequences. *Nature* 1984;307(5951):521–7.

91. Messa C, Russo F, Caruso MG, Di Leo A. EGF, TGF-alpha, and EGF-R in human colorectal adenocarcinoma. *Acta Oncol* 1998;37(3):285–9.

92. Falls DL. Neuregulins: functions, forms, and signaling strategies. *Exp Cell Res* 2003;284(1):14–30.

93. Lax I, Fischer R, Ng C, et al. Noncontiguous regions in the extracellular domain of EGF receptor define ligand-binding specificity. *Cell Regul* 1991;2(5):337–45.

94. Lemmon MA, Bu Z, Ladbury JE, et al. Two EGF molecules contribute additively to stabilization of the EGFR dimer. EMBO J 1997;16(2):281–94.

95. Sibilia M, Fleischmann A, Behrens A, et al. The EGF receptor provides an essential survival signal for SOS-dependent skin tumor development. *Cell* 2000;102(2):211–20.

96. Ang KK, Berkey BA, Tu X, et al. Impact of epidermal growth factor receptor expression on survival and pattern of relapse in patients with advanced head and neck carcinoma. *Cancer Res* 2002;62(24):7350–6.

97. Schmidt-Ullrich RK, Valerie KC, Chan W, McWilliams D. Altered expression of epidermal growth factor receptor and estrogen receptor in MCF-7 cells after single and repeated radiation exposures. *Int J Radiat Oncol Biol Phys* 1994;29(4):813–9.

98. Akimoto T, Hunter NR, Buchmiller L, Mason K, Ang KK, Milas L. Inverse relationship between epidermal growth factor receptor expression and radiocurability of murine carcinomas. *Clin Cancer Res* 1999;5(10):2884–90.

99. Schmidt-Ullrich RK, Mikkelsen RB, Dent P, et al. Radiation-induced proliferation of the human A431 squamous carcinoma cells is dependent on EGFR tyrosine phosphorylation. *Oncogene* 1997;15(10):1191–7.

100. Milas L, Fan Z, Andratschke NH, Ang KK. Epidermal growth factor receptor and tumor response to radiation: in vivo preclinical studies. *Int J Radiat Oncol Biol Phys* 2004;58(3):966–71.

101. Huang SM, Bock JM, Harari PM. Epidermal growth factor receptor blockade with C225 modulates proliferation, apoptosis, and radiosensitivity in squamous cell carcinomas of the head and neck. *Cancer Res* 1999;59(8):1935–40.

102. Bonner JA, Raisch KP, Trummell HQ, et al. Enhanced apoptosis with combination C225/radiation treatment serves as the impetus for clinical investigation in head and neck cancers. *J Clin Oncol* 2000;18(21 Suppl):47S–53S.

103. Milas L, Mason K, Hunter N, et al. In vivo enhancement of tumor radioresponse by C225 antiepidermal growth factor receptor antibody. *Clin Cancer Res* 2000;6(2):701–8.

104. Puri C, Tosoni D, Comai R, et al. Relationships between EGFR signaling-competent and endocytosis-competent membrane microdomains. *Mol Biol Cell* 2005;16(6):2704–18.

105. Sigismund S, Woelk T, Puri C, et al. Clathrin-independent endocytosis of ubiquitinated cargos. *Proc Natl Acad Sci USA* 2005;102(8):2760–5.

106. Li S, Schmitz KR, Jeffrey PD, Wiltzius JJ, Kussie P, Ferguson KM. Structural basis for inhibition of the epidermal growth factor receptor by cetuximab. *Cancer Cell* 2005;7(4):301–11.

107. Jemal A, Tiwari RC, Murray T, et al. Cancer statistics, 2004. *CA Cancer J Clin* 2004;54(1):8–29.

108. Kufe D, Holland J, Frei E, Society AC. *Cancer medicine 6*, 6th ed, vol. 2. Hamilton, ON: BC Decker; 2003.

109. Porter EH. The statistics of dose/cure relationships for irradiated tumours. Part II. *Br J Radiol* 1980;53(628):336–45.

110. Porter EH. The statistics of dose/cure relationships for irradiated tumours. Part I. *Br J Radiol* 1980;53(627):210–27.

111. Ohnishi K, Ota I, Takahashi A, Yane K, Matsumoto H, Ohnishi T. Transfection of mutant p53 gene depresses X-ray- or CDDP-induced apoptosis in a human squamous cell carcinoma of the head and neck. *Apoptosis* 2002;7(4):367–72.

112. Kojima H, Endo K, Moriyama H, et al. Abrogation of mitochondrial cytochrome c release and caspase-3 activation in acquired multidrug resistance. *J Biol Chem* 1998;273(27):16647–50.

113. Toyozumi Y, Arima N, Izumaru S, Kato S, Morimatsu M, Nakashima T. Loss of caspase-8 activation pathway is a possible mechanism for CDDP resistance in human laryngeal squamous cell carcinoma, HEp-2 cells. *Int J Oncol* 2004;25(3):721–8.

114. Pignon JP, Bourhis J, Domenge C, Designe L. Chemotherapy added to locoregional treatment for head and neck squamous-cell carcinoma: three meta-analyses of updated individual data. MACH-NC Collaborative Group. Meta-Analysis of Chemotherapy on Head and Neck Cancer. *Lancet* 2000;355(9208):949–55.

115. Bonner JA, Harari PM, Giralt J, et al. Radiotherapy plus Cetuximab for squamous-cell carcinoma of the head and neck. *N Engl J Med* 2006;354(6):567–78.

116. Robert F, Ezekiel MP, Spencer SA, et al. Phase I study of anti-epidermal growth factor receptor antibody cetuximab in combination with radiation therapy in patients with advanced head and neck cancer. *J Clin Oncol* 2001;19(13):3234–43.

117. Baselga J, Pfister D, Cooper MR, et al. Phase I studies of anti-epidermal growth factor receptor chimeric antibody C225 alone and in combination with cisplatin. *J Clin Oncol* 2000;18(4):904–14.

118. Bonner JA, et al. Phase III evaluation of radiation with and without cetuximab for locoregionally advanced head and neck cancer. *Int J Radiat Oncol Biol Phys* 2004;60(Suppl 1):S147–8.

119. Bonner JA, et al. Cetuximab improves locoregional control and survival of locoregionally advanced head and neck cancer: independent review of mature data with a median followup of 45 months. AACR-NCI-EORTC International Conference Molecular Targets and Cancer Therapeutics. *Clin Cancer Res* 2005;11:9058s.

120. Bonner JA, Giralt J, Harari PM, et al. Phase III evaluation of radiation with and without cetuximab for locoregionally advanced head and neck cancer. *Int J Radiat Oncol Biol Phys* 2004;60(Suppl 1):S147-S8.

121. Bonner JA, Harari PM, Giralt J, et al. Cetuximab improves locoregional control and survival of locoregionally advanced head and neck cancer: independent review of mature data with a median followup of 45 months. AACR-NCI-EORTC International Conference Molecular Targets and Cancer Therapeutics. *Clin Cancer Res* 2005;11(24 part 2):9058s.

122. Bonner JA, Harari PM, Giralt J, et al. *Improved Preservation of Larynx with the Addition of Cetuximab to Radiation for Cancers of the Larynx and Hypopharynx*. Abstract 5533. ASCO Annual Meeting Abstract 5533, 2005.

123. Bonner JA, Giralt J, Harari PM, et al. *Cetuximab Prolongs Survival in Patients with Locoregionally Advanced Squamous Cell Carcinoma of Head and Neck: A Phase III Study of High Dose Radiation Therapy with or Without Cetuximab*. ASCO 2004 Annual Meeting Abstract 5507, 2004.

124. Dyer MJ, Hale G, Hayhoe FG, Waldmann H. Effects of CAMPATH-1 antibodies in vivo in patients with lymphoid malignancies: influence of antibody isotype. *Blood* 1989;73(6):1431–9.

125. Su Y, et al. *Concurrent Cetuximab, Cisplatin, and Radiotherapy (RT) for Locoregionally Advanced Squamous Cell Carcinoma of the Head and Neck (SCCHN): Updated Results of a Novel Combined Modality Paradigm*. ASCO Annual Meeting, 2005.

126. Saltz LB, Meropol NJ, Loehrer PJ, Needle MN, Kopit J, Mayer RJ. Phase II trial of cetuximab in patients with refractory colorectal cancer that expresses the epidermal growth factor receptor. *J Clin Oncol* 2004;22(7):1201–8.

127. Saltz LB, Lenz H, Hochster H, Wadler S, Hoff P, Kemeny N, Hollywood E, Gonen M, Wetherbee S, Chen H. *Randomized Phase II Trial of Cetuximab/Bevacizumab/Irinotecan (CBI) Versus Cetuximab/Bevacizumab (CB) in Irinotecan-Refractory Colorectal Cancer*. Abstract 3508, ASCO Annual Meeting, 2005.

128. Lutz M, Schoffski P, Folprecht G, et al. A phase I/II study of cetuximab (C225) plus irinotecan (CPT-11) and 24h infusional 5FU/folinic acid (FA) in the treatment of metastatic colorectal cancer expressing the epidermal growth factor receptor. *Ann Oncol* 2002;13 (Suppl 5):73.

129. Rosenberg A, Loehrer PJ, Needle MN, et al. Erbitux (IMC-c225) plus weekly iriniotecan (CPT-11), fluorouracil (5FU) and leucovorin (LV) in colorectal cancer (CRC that expresses the epidermal growth factor receptor (EGFr). *Proc Am Soc Clin Oncol* 2002;21:135a.

130. Van Laethem J-L, Raoul J-L, Mitry E, et al. Cetuximab (C225) in combination with bi-weekly irinotecan (CPT-11), infusional 5-fluorouracil (5-FU), and folinic acid (FA) in patients (pts) with metastatic colorectal cancer (CRC) expressing the epidermal growth factor receptor (EGFR). Preliminary safety and efficacy results. *Proc Am Soc Clin Oncol* 2003;22:264.

131. Burtness B. The role of cetuximab in the treatment of squamous cell cancer of the head and neck. *Expert Opin Biol Ther* 2005;5(8):1085–93.

132. Herbst RS, Arquette M, Shin DM, et al. Phase II multicenter study of the epidermal growth factor receptor antibody cetuximab and cisplatin for recurrent and refractory squamous cell carcinoma of the head and neck. *J Clin Oncol* 2005;23(24):5578–87.

133. Baselga J, Trigo JM, Bourhis J, et al. Phase II multicenter study of the antiepidermal growth factor receptor monoclonal antibody cetuximab in combination with platinum-based chemotherapy in patients with platinum-refractory metastatic and/or recurrent squamous cell carcinoma of the head and neck. *J Clin Oncol* 2005;23(24):5568–77.

134. Chan ATC, Hsu M-M, Goh BC, et al. Multicenter, phase II study of cetuximab in combination with carboplatin in patients with recurrent or metastatic nasopharyngeal carcinoma. *J Clin Oncol* 2005;23(15):3568–76.

135. Vermorken JB, Bourhis J, Trigo J, et al. Cetuximab in recurrent/metastatic (R&M) squamous cell carcinoma of the head and neck (SCCHN) refractory to first-line platinum-based therapies. *J Clin Oncol* (Meeting Abstracts) 2005;23(Suppl 16):5505.

136. Govindan R. Cetuximab in advanced non-small cell lung cancer. *Clin Cancer Res* 2004;10 (12 Pt 2):4241s–4s.

137. Humblet Y, Vega-Villegas E, Mesia R, et al. Phase I study of cetuximab in combination with cisplatin or carboplatin and 5-fluorouracil (5-FU) in patients (pts) with recurrent and/or metastatic squamous cell carcinoma of the head and neck (SCCHN). *J Clin Oncol* (Meeting Abstracts) 2004;22(Suppl 14):5513.

138. Viloria-Petit A, Crombet T, Jothy S, et al. Acquired resistance to the antitumor effect of epidermal growth factor receptor-blocking antibodies in vivo: a role for altered tumor angiogenesis. *Cancer Res* 2001;61(13):5090–101.

139. Jost M, Kari C, Rodeck U. The EGF receptor - an essential regulator of multiple epidermal functions. *Eur J Dermatol* 2000;10(7):505–10.

140. Busam KJ, Capodieci P, Motzer R, Kiehn T, Phelan D, Halpern AC. Cutaneous side-effects in cancer patients treated with the antiepidermal growth factor receptor antibody C225. *Br J Dermatol* 2001;144(6):1169–76.

141. Kimyai-Asadi A, Jih MH. Follicular toxic effects of chimeric anti-epidermal growth factor receptor antibody cetuximab used to treat human solid tumors. *Arch Dermatol* 2002;138(1):129–31.

142. Van Doorn R, Kirtschig G, Scheffer E, Stoof TJ, Giaccone G. Follicular and epidermal alterations in patients treated with ZD1839 (Iressa), an inhibitor of the epidermal growth factor receptor. *Br J Dermatol* 2002;147(3):598–601.

143. Baselga J, Rischin D, Ranson M, et al. Phase I safety, pharmacokinetic, and pharmacodynamic trial of ZD1839, a selective oral epidermal growth factor receptor tyrosine kinase inhibitor, in patients with five selected solid tumor types. *J Clin Oncol* 2002;20(21):4292–302.

144. Perez-Soler R, Delord JP, Halpern A, et al. HER1/EGFR inhibitor-associated rash: future directions for management and investigation outcomes from the HER1/EGFR inhibitor rash management forum. *Oncologist* 2005;10(5):345–56.

145. Chung KY, Shia J, Kemeny NE, et al. Cetuximab shows activity in colorectal cancer patients with tumors that do not express the epidermal growth factor receptor by immunohistochemistry. *J Clin Oncol* 2005;23(9):1803–10.

146. Vallbohmer D, Zhang W, Gordon M, et al. Molecular determinants of cetuximab efficacy. *J Clin Oncol* 2005;23(15):3536–44.

147. Ciardiello F, Bianco R, Damiano V, et al. Antiangiogenic and antitumor activity of anti-epidermal growth factor receptor C225 monoclonal antibody in combination with vascular endothelial growth factor antisense oligonucleotide in human GEO colon cancer cells. *Clin Cancer Res* 2000;6(9): 3739–47.

148. Moroni M, Veronese S, Benvenuti S, et al. Gene copy number for epidermal growth factor receptor (EGFR) and clinical response to anti-EGFR treatment in colorectal cancer: a cohort study. *Lancet Oncol* 2005;6(5):279–86.

149. Atkins D, Reiffen KA, Tegtmeier CL, Winther H, Bonato MS, Storkel S. Immunohistochemical detection of EGFR in paraffin-embedded tumor tissues: variation in staining intensity due to choice of fixative and storage time of tissue sections. *J Histochem Cytochem* 2004;52(7):893–901.

150. Perou CM, Sorlie T, Eisen MB, et al. Molecular portraits of human breast tumours. *Nature* 2000;406(6797):747–52.

151. Baselga J, Seidman AD, Rosen PP, Norton L. HER2 overexpression and paclitaxel sensitivity in breast cancer: therapeutic implications. *Oncology* (Williston Park) 1997;11(3 Suppl 2):43–8.

152. Cho HS, Mason K, Ramyar KX, et al. Structure of the extracellular region of HER2 alone and in complex with the Herceptin Fab. *Nature* 2003;421(6924):756–60.

153. Scholl S, Beuzeboc P, Pouillart P. Targeting HER2 in other tumor types. *Ann Oncol* 2001;12 (Suppl 1):S81–7.

154. Rubin I, Yarden Y. The basic biology of HER2. *Ann Oncol* 2001;12(Suppl 1):S3–8.

155. Baselga J, Tripathy D, Mendelsohn J, et al. Phase II study of weekly intravenous recombinant humanized anti-p185HER2 monoclonal antibody in patients with HER2/neu-overexpressing metastatic breast cancer. *J Clin Oncol* 1996;14(3):737–44.

156. Pegram M, Hsu S, Lewis G, et al. Inhibitory effects of combinations of HER-2/neu antibody and chemotherapeutic agents used for treatment of human breast cancers. *Oncogene* 1999;18(13): 2241–51.

157. Marty M, Cognetti F, Maraninchi D, et al. Randomized phase II trial of the efficacy and safety of trastuzumab combined with docetaxel in patients with human epidermal growth factor receptor

2-positive metastatic breast cancer administered as first-line treatment: the M77001 study group. *J Clin Oncol* 2005;23(19):4265–74.

158. Gatzemeier U, Groth G, Butts C, et al. Randomized phase II trial of gemcitabine-cisplatin with or without trastuzumab in HER2-positive non-small-cell lung cancer. *Ann Oncol* 2004;15(1):19–27.

159. Langer CJ, Stephenson P, Thor A, Vangel M, Johnson DH. Trastuzumab in the treatment of advanced non-small-cell lung cancer: is there a role? Focus on Eastern Cooperative Oncology Group study 2598. *J Clin Oncol* 2004;22(7):1180–7.

160. Romond EH, Perez EA, Bryant J, et al. Trastuzumab plus adjuvant chemotherapy for operable HER2-positive breast cancer. *N Engl J Med* 2005;353(16):1673–84.

161. Piccart-Gebhart MJ, Procter M, Leyland-Jones B, et al. Trastuzumab after adjuvant chemotherapy in HER2-positive breast cancer. *N Engl J Med* 2005;353(16):1659–72.

162. Buzdar AU, Ibrahim NK, Francis D, et al. Significantly higher pathologic complete remission rate after neoadjuvant therapy with trastuzumab, paclitaxel, and epirubicin chemotherapy: results of a randomized trial in human epidermal growth factor receptor 2-positive operable breast cancer. *J Clin Oncol* 2005;23(16):3676–85.

163. Mohsin SK, Weiss HL, Gutierrez MC, et al. Neoadjuvant trastuzumab induces apoptosis in primary breast cancers. *J Clin Oncol* 2005;23(11):2460–8.

164. Kostler WJ, Steger GG, Soleiman A, et al. Monitoring of serum Her-2/neu predicts histopatho-logical response to neoadjuvant trastuzumab-based therapy for breast cancer. *Anticancer Res* 2004;24(2C):1127–30.

165. Joensuu H, Kellokumpu-Lehtinen P-L, Bono P, et al. Trastuzumab in combination with docetaxel or vinorelbine as adjuvant treatment of breast cancer: the FinHer Trial. In: *San Antonio Breast Cancer Symposium*, Abstract #2. San Antonio, TX; 2005.

166. Haslekas C, Breen K, Pedersen KW, Johannessen LE, Stang E, Madshus IH. The inhibitory effect of ErbB2 on epidermal growth factor-induced formation of clathrin-coated pits correlates with retention of epidermal growth factor receptor-ErbB2 oligomeric complexes at the plasma membrane. *Mol Biol Cell* 2005;16(12):5832–42.

167. Longva KE, Pedersen NM, Haslekas C, Stang E, Madshus IH. Herceptin-induced inhibition of ErbB2 signaling involves reduced phosphorylation of Akt but not endocytic down-regulation of ErbB2. *Int J Cancer* 2005;116(3):359–67.

168. Nahta R, Yuan LX, Zhang B, Kobayashi R, Esteva FJ. Insulin-like growth factor-I receptor/human epidermal growth factor receptor 2 heterodimerization contributes to trastuzumab resistance of breast cancer cells. *Cancer Res* 2005;65(23):11118–28.

169. Agus DB, Gordon MS, Taylor C, et al. Phase I clinical study of pertuzumab, a novel HER dimerization inhibitor, in patients with advanced cancer. *J Clin Oncol* 2005;23(11):2534–43.

170. O'Reilly MS, Holmgren L, Shing Y, et al. Angiostatin: a novel angiogenesis inhibitor that mediates the suppression of metastases by a Lewis lung carcinoma. *Cell* 1994;79(2):315–28.

171. Ferrara N, Davis-Smyth T. The biology of vascular endothelial growth factor. *Endocr Rev* 1997;18(1):4–25.

172. Davis-Smyth T, Chen H, Park J, Presta LG, Ferrara N. The second immunoglobulin-like domain of the VEGF tyrosine kinase receptor Flt-1 determines ligand binding and may initiate a signal transduction cascade. EMBO J 1996;15(18):4919–27.

173. Borgstrom P, Gold DP, Hillan KJ, Ferrara N. Importance of VEGF for breast cancer angiogenesis in vivo: implications from intravital microscopy of combination treatments with an anti-VEGF neutralizing monoclonal antibody and doxorubicin. *Anticancer Res* 1999;19(5B):4203–14.

174. Presta LG, Chen H, O'Connor SJ, et al. Humanization of an anti-vascular endothelial growth factor monoclonal antibody for the therapy of solid tumors and other disorders. *Cancer Res* 1997;57(20):4593–9.

175. Hurwitz HI, Fehrenbacher L, Hainsworth JD, et al. Bevacizumab in combination with fluorouracil and leucovorin: an active regimen for first-line metastatic colorectal cancer. *J Clin Oncol* 2005;23(15):3502–8.

176. Miller KD, Chap LI, Holmes FA, et al. Randomized phase III trial of capecitabine compared with bevacizumab plus capecitabine in patients with previously treated metastatic breast cancer. *J Clin Oncol* 2005;23(4):792–9.

177. D'Adamo DR, Anderson SE, Albritton K, et al. Phase II study of doxorubicin and bevacizumab for patients with metastatic soft-tissue sarcomas. *J Clin Oncol* 2005;23(28):7135–42.

178. Willett CG, Boucher Y, di Tomaso E, et al. Direct evidence that the VEGF-specific antibody bevacizumab has antivascular effects in human rectal cancer. *Nat Med* 2004;10(2):145–7.

179. Johnson DH, Fehrenbacher L, Novotny WF, et al. Randomized phase II trial comparing bevacizumab plus carboplatin and paclitaxel with carboplatin and paclitaxel alone in previously untreated locally advanced or metastatic non-small-cell lung cancer. *J Clin Oncol* 2004;22(11):2184–91.

180. Sandler AB, Gray R, Brahmer J, Dowlati A, Schiller JH, Perry MC, Johnson DH. *Randomized Phase II/III Trial of Paclitaxel (P) Plus Carboplatin (C) with or Without Bevacizumab (NSC # 704865) in Patients with Advanced Non-Squamous Non-Small Cell Lung Cancer (NSCLC): An Eastern Cooperative Oncology Group (ECOG) Trial - E4599.* ASCO, 2005, Orlando, FL.

181. Cobleigh MA, Langmuir VK, Sledge GW, et al. A phase I/II dose-escalation trial of bevacizumab in previously treated metastatic breast cancer. *Semin Oncol* 2003;30(5 Suppl 16):117–24.

182. Miller KD, Chap LI, Holmes FA, et al. Randomized phase III trial of capecitabine compared with bevacizumab plus capecitabine in patients with previously treated metastatic breast cancer. *J Clin Oncol* 2005;23(4):792–9.

183. Maloney DG, Grillo-Lopez AJ, Bodkin DJ, et al. IDEC-C2B8: results of a phase I multiple-dose trial in patients with relapsed non-Hodgkin's lymphoma. *J Clin Oncol* 1997;15(10):3266–74.

184. Maloney DG, Grillo-Lopez AJ, White CA, et al. IDEC-C2B8 (Rituximab) anti-CD20 monoclonal antibody therapy in patients with relapsed low-grade non-Hodgkin's lymphoma. *Blood* 1997;90(6):2188–95.

185. Shan D, Ledbetter JA, Press OW. Signaling events involved in anti-CD20-induced apoptosis of malignant human B cells. *Cancer Immunol Immunother* 2000;48(12):673–83.

186. Deans JP, Li H, Polyak MJ. CD20-mediated apoptosis: signalling through lipid rafts. *Immunology* 2002;107(2):176–82.

187. Press OW, Howell-Clark J, Anderson S, Bernstein I. Retention of B-cell-specific monoclonal antibodies by human lymphoma cells. *Blood* 1994;83(5):1390–7.

188. Bubien JK, Zhou LJ, Bell PD, Frizzell RA, Tedder TF. Transfection of the CD20 cell surface molecule into ectopic cell types generates a Ca2+ conductance found constitutively in B lymphocytes. *J Cell Biol* 1993;121(5):1121–32.

189. Kanzaki M, Shibata H, Mogami H, Kojima I. Expression of calcium-permeable cation channel CD20 accelerates progression through the G1 phase in Balb/c 3T3 cells. *J Biol Chem* 1995;270(22):13099–104.

190. Byrd JC, Kitada S, Flinn IW, et al. The mechanism of tumor cell clearance by rituximab in vivo in patients with B-cell chronic lymphocytic leukemia: evidence of caspase activation and apoptosis induction. *Blood* 2002;99(3):1038–43.

191. Berinstein NL, Grillo-Lopez AJ, White CA, et al. Association of serum Rituximab (IDEC-C2B8) concentration and anti-tumor response in the treatment of recurrent low-grade or follicular non-Hodgkin's lymphoma. *Ann Oncol* 1998;9(9):995–1001.

192. Coiffier B, Haioun C, Ketterer N, et al. Rituximab (anti-CD20 monoclonal antibody) for the treatment of patients with relapsing or refractory aggressive lymphoma: a multicenter phase II study. *Blood* 1998;92(6):1927–32.

193. Byrd JC, Waselenko JK, Maneatis TJ, et al. Rituximab therapy in hematologic malignancy patients with circulating blood tumor cells: association with increased infusion-related side effects and rapid blood tumor clearance. *J Clin Oncol* 1999;17(3):791–5.

194. McLaughlin P, Grillo-Lopez A, Link B, et al. Rituximab chimeric anti-CD20 monoclonal antibody therapy for relapsed indolent lymphoma: half of patients respond to a four-dose treatment program. *J Clin Oncol* 1998;16:2825–33.

195. Berinstein N, Grillo-Lopez A, White C, et al. Association of serum Rituximab (IDEC-C2B8) concentration and anti-tumor response in the treatment of recurrent low-grade or follicular non-Hodgkin's lymphoma. *Annals of Oncology* 1988;9:995–1001.

196. Czuczman MS, Grillo-Lopez AJ, White CA, et al. Treatment of patients with low-grade B-cell lymphoma with the combination of chimeric anti-CD20 monoclonal antibody and CHOP chemotherapy. *J Clin Oncol* 1999;17(1):268–76.

197. Gribben JG, Freedman A, Woo SD, et al. All advanced stage non-Hodgkin's lymphomas with a polymerase chain reaction amplifiable breakpoint of bcl-2 have residual cells containing the bcl-2 rearrangement at evaluation and after treatment. *Blood* 1991;78(12):3275–80.

198. Vose JM, Link BK, Grossbard ML, et al. Phase II study of rituximab in combination with chop chemotherapy in patients with previously untreated, aggressive non-Hodgkin's lymphoma. *J Clin Oncol* 2001;19(2):389–97.

199. Coiffier B, Lepage E, Briere J, et al. CHOP chemotherapy plus rituximab compared with CHOP alone in elderly patients with diffuse large-B-cell lymphoma. *N Engl J Med* 2002;346(4):235–42.

200. Cartron G, Dacheux L, Salles G, et al. Therapeutic activity of humanized anti-CD20 monoclonal antibody and polymorphism in IgG Fc receptor FcgammaRIIIa gene. *Blood* 2002;99(3):754–8.

201. Meropol NJ, Porter M, Blumenson LE, et al. Daily subcutaneous injection of low-dose interleukin 2 expands natural killer cells in vivo without significant toxicity. *Clin Cancer Res* 1996;2(4):669–77.

202. Caligiuri MA, Murray C, Robertson MJ, et al. Selective modulation of human natural killer cells in vivo after prolonged infusion of low dose recombinant interleukin 2. *J Clin Invest* 1993;91(1):123–32.

203. Friedberg JW, Neuberg D, Gribben JG, et al. Combination immunotherapy with rituximab and interleukin 2 in patients with relapsed or refractory follicular non-Hodgkin's lymphoma. *Br J Haematol* 2002;117(4):828–34.

204. Ansell SM, Witzig TE, Kurtin PJ, et al. Phase 1 study of interleukin-12 in combination with rituximab in patients with B-cell non-Hodgkin lymphoma. *Blood* 2002;99(1):67–74.

205. Sivaraman S, Venugopal P, Ranganathan R, Deshpande CG, Huang X, Jajeh A, Gregory SA, O'Brien T, Preisler HD. Effect of interferon-alpha on CD20 antigen expression of B-cell chronic lymphocytic leukemia. Cytokines Cell Mol Ther. 2000 Jun; 6(2):81–7.

206. Herberman RB. Effect of alpha-interferons on immune function. *Semin Oncol* 1997;24(3 Suppl 9):S9–78–80.

207. Davis TA, Maloney DG, Grillo-Lopez AJ, et al. Combination immunotherapy of relapsed or refractory low-grade or follicular non-Hodgkin's lymphoma with rituximab and interferon-alpha-2a. *Clin Cancer Res* 2000;6(7):2644–52.

208. Sacchi S, Federico M, Vitolo U, et al. Clinical activity and safety of combination immunotherapy with IFN-alpha 2a and Rituximab in patients with relapsed low grade non-Hodgkin's lymphoma. *Haematologica* 2001;86(9):951–8.

209. Teeling JL, French RR, Cragg MS, et al. Characterization of new human CD20 monoclonal antibodies with potent cytolytic activity against non-Hodgkin lymphomas. *Blood* 2004;104(6): 1793–800.

210. Stein R, Qu Z, Chen S, et al. Characterization of a new humanized anti-CD20 monoclonal antibody, IMMU-106, and Its use in combination with the humanized anti-CD22 antibody, epratuzumab, for the therapy of non-Hodgkin's lymphoma. *Clin Cancer Res* 2004;10(8):2868–78.

211. Hagenbeek A, Plesner T, Walewski A. A novel Fully Human anti CD20 monoclonal antibody, first clinical results from an ongoing phase I/II trial in patients with follicular non-Hodgkin's lymphoma. Abstract # 114. *Ann Oncol* 2005;16:S5.

212. Coiffier B, Tilly H, Pederson H, et al. HuMax CD20 human monoclonal antibody in chronic lymphocytic leukemia. Early results from an ongoing phase I/II clinical trial. Abstract # 448. *Blood* 2005;106:11.

213. Hale G, Xia MQ, Tighe HP, Dyer MJ, Waldmann H. The CAMPATH-1 antigen (CDw52). *Tissue Antigens* 1990;35(3):118–27.

214. Mavromatis B, Cheson BD. Monoclonal antibody therapy of chronic lymphocytic leukemia. *J Clin Oncol* 2003;21(9):1874–81.

215. Ginaldi L, De Martinis M, Matutes E, et al. Levels of expression of CD52 in normal and leukemic B and T cells: correlation with in vivo therapeutic responses to Campath-1H. *Leuk Res* 1998;22(2):185–91.

216. Rawstron AC, Kennedy B, Evans PA, et al. Quantitation of minimal disease levels in chronic lymphocytic leukemia using a sensitive flow cytometric assay improves the prediction of outcome and can be used to optimize therapy. *Blood* 2001;98(1):29–35.

217. Keating MJ, Flinn I, Jain V, et al. Therapeutic role of alemtuzumab (Campath-1H) in patients who have failed fludarabine: results of a large international study. *Blood* 2002;99(10):3554–61.

218. Moreton P, Kennedy B, Lucas G, et al. Eradication of minimal residual disease in B-cell chronic lymphocytic leukemia after alemtuzumab therapy is associated with prolonged survival. *J Clin Oncol* 2005;23(13):2971–9.

219. Wagner HN, Jr., Wiseman GA, Marcus CS, et al. Administration guidelines for radioimmunotherapy of non-Hodgkin's lymphoma with (90)Y-labeled anti-CD20 monoclonal antibody. *J Nucl Med* 2002;43(2):267–72.

220. Witzig TE, White CA, Wiseman GA, et al. Phase I/II trial of IDEC-Y2B8 radioimmunotherapy for treatment of relapsed or refractory CD20(+) B-cell non-Hodgkin's lymphoma. *J Clin Oncol* 1999;17(12):3793–803.

221. Witzig TE, Gordon LI, Cabanillas F, et al. Randomized controlled trial of yttrium-90-labeled ibritumomab tiuxetan radioimmunotherapy versus rituximab immunotherapy for patients with relapsed or refractory low-grade, follicular, or transformed B-cell non-Hodgkin's lymphoma. *J Clin Oncol* 2002;20(10):2453–63.

222. Witzig TE, Flinn IW, Gordon LI, et al. Treatment with ibritumomab tiuxetan radioimmunotherapy in patients with rituximab-refractory follicular non-Hodgkin's lymphoma. *J Clin Oncol* 2002;20(15):3262–9.

223. Wiseman GA, Gordon LI, Multani PS, et al. Ibritumomab tiuxetan radioimmunotherapy for patients with relapsed or refractory non-Hodgkin lymphoma and mild thrombocytopenia: a phase II multicenter trial. *Blood* 2002;99(12):4336–42.

224. McLaughlin P, Grillo-Lopez A, Link B, et al. Rituximab chimeric anti-CD20 monoclonal antibody therapy for relapsed indolent lymphoma: half of patients respond to a four-dose treatment program. *J Clin Oncol* 1998;16(8):2825–33.

225. Emmanouilides C WT, Molina A, et al. Improved safety and efficacy of yttrium-90 ibritumomab tiuxetan radioimmunotherapy when administered as 2nd or 3rd line therapy for relapsed low-grade, follicular and transformed B-cell non-Hodgkin's lymphoma (NHL). Abstract 2392. *Proc Am Soc Clin Oncol* 2003;22:595.

226. Emmanoulidies C, Gordon LI. *Zevalin Radioimmunotherapy (RIT) is Safe and Effective in Geriatric Patients with Low Grade, Follicular or cd20+ Transformed (LG/F/T) Non-Hodgkin Lymphoma (NHL)*. Abstract 1143, ASCO, 2001.

227. Borghaei H, Wallace SG, Schilder RJ. Factors associated with toxicity and response to yttrium 90-labeled ibritumomab tiuxetan in patients with indolent non-Hodgkin's lymphoma. *Clin Lymphoma* 2004;5(Suppl 1):S16–21.

228. Kaminski MS, Estes J, Zasadny KR, et al. Radioimmunotherapy with iodine (131)I tositumomab for relapsed or refractory B-cell non-Hodgkin lymphoma: updated results and long-term follow-up of the University of Michigan experience. *Blood* 2000;96(4):1259–66.

229. Vose JM, Wahl RL, Saleh M, et al. Multicenter phase II study of iodine-131 tositumomab for chemotherapy-relapsed/refractory low-grade and transformed low-grade B-cell non-Hodgkin's lymphomas. *J Clin Oncol* 2000;18(6):1316–23.

230. Kaminski MS, Zelenetz AD, Press OW, et al. Pivotal study of iodine I 131 tositumomab for chemotherapy-refractory low-grade or transformed low-grade B-cell non-Hodgkin's lymphomas. *J Clin Oncol* 2001;19(19):3918–28.

231. Kaminski MS, Tuck M, Estes J, et al. 131I-tositumomab therapy as initial treatment for follicular lymphoma. *N Engl J Med* 2005;352(5):441–9.

232. Pedley RB, Hill SA, Boxer GM, et al. Eradication of colorectal xenografts by combined radioimmunotherapy and combretastatin a-4 3-O-phosphate. *Cancer Res* 2001;61(12):4716–22.

233. Dark GG, Hill SA, Prise VE, Tozer GM, Pettit GR, Chaplin DJ. Combretastatin A-4, an agent that displays potent and selective toxicity toward tumor vasculature. *Cancer Res* 1997;57(10):1829–34.

234. Thorpe PE. Vascular targeting agents as cancer therapeutics. *Clin Cancer Res* 2004;10(2):415–27.

235. Galbraith SM, Maxwell RJ, Lodge MA, et al. Combretastatin A4 phosphate has tumor antivascular activity in rat and man as demonstrated by dynamic magnetic resonance imaging. *J Clin Oncol* 2003;21(15):2831–42.

236. Rustin GJ, Galbraith SM, Anderson H, et al. Phase I clinical trial of weekly combretastatin A4 phosphate: clinical and pharmacokinetic results. *J Clin Oncol* 2003;21(15):2815–22.

237. Burrows FJ, Thorpe PE. Eradication of large solid tumors in mice with an immunotoxin directed against tumor vasculature. *Proc Natl Acad Sci USA* 1993;90(19):8996–9000.

238. Huang X, Molema G, King S, Watkins L, Edgington TS, Thorpe PE. Tumor infarction in mice by antibody-directed targeting of tissue factor to tumor vasculature. *Science* 1997;275(5299):547–50.

239. Tarli L, Balza E, Viti F, et al. A high-affinity human antibody that targets tumoral blood vessels. *Blood* 1999;94(1):192–8.

240. Ebbinghaus C, Ronca R, Kaspar M, et al. Engineered vascular-targeting antibody-interferon-gamma fusion protein for cancer therapy. *Int J Cancer* 2005;116(2):304–313.

241. Carnemolla B, Borsi L, Balza E, et al. Enhancement of the antitumor properties of interleukin-2 by its targeted delivery to the tumor blood vessel extracellular matrix. *Blood* 2002;99(5):1659–65.
242. Ran S, He J, Huang X, Soares M, Scothorn D, Thorpe PE. Antitumor effects of a monoclonal antibody that binds anionic phospholipids on the surface of tumor blood vessels in mice. *Clin Cancer Res* 2005;11(4):1551–62.

17 Nucleic Acid Therapies for Cancer Treatment

Dan T. Vogl, MD, and Alan M. Gewirtz, MD

SUMMARY

A better understanding of the biochemical pathways mediating tumor growth and progression has provided a new set of targets for therapeutic intervention. The ability to target highly specific segments of genetic material using oligonucleotide probes has been the subject of nucleic acid based therapy and is being applied to a variety of human diseases, including cancer. This chapter will discuss the various strategies utilizing nucleic acids for inhibiting gene transcription and translation with a major focus on RNAi and antisense oligodeoxynucleotides. Specific targets will be described and results of preliminary clinical trials reported. Future challenges in the clinical translation of nucleic acid treatments will also be described.

Key Words: Antisense oligonucleotides; cancer therapy; DNA; RNAi.

1. INTRODUCTION

Advances in knowledge of the biochemical pathways that mediate cancer biology have identified many of the genes that promote cell growth, survival, invasion, and metastasis. The recent development of small molecules and monoclonal antibodies that block the activity of specific gene products has demonstrated the utility of targeted approaches in treating human cancer. Potentially more useful, however, would be inhibiting transcription and translation of pathologic genes by using sequence-specific oligonucleotides. Such a strategy should improve the specificity of cancer therapy and, at the same time, decrease toxicity to normal tissues as only pathological processes are being targeted.

Strategies for oligonucleotide cancer therapies include single- and double-stranded DNA and RNA oligonucleotides, in many cases chemically modified to optimize delivery, pharmacokinetics, and the ability to inhibit gene expression. Theoretic mechanisms of action still being evaluated in laboratory studies include transcription inhibition by homologous recombination, triple-helix formation, and promoter-sequence decoys, as well as translation inhibition by RNA decoys, antisense oligodeoxynucleotides, and antisense RNA and DNA enzymes. Recent advances in

From: *Cancer Drug Discovery and Development: Molecular Targeting in Oncology*
Edited by: H. L. Kaufman, S. Wadler, and K. Antman © Humana Press, Totowa, NJ

understanding RNA interference (RNAi) have created tremendous interest in using this mechanism for therapeutic benefit.

2. INHIBITION OF TRANSCRIPTION

Homologous recombination events occur during DNA replication, and this rare but natural process can be harnessed to alter somatic DNA structure and suppress transcription of a specific gene. This technique requires a viral vector capable of infecting target cells, into which a sequence from the target gene is cloned for introduction into the target cell. During subsequent DNA replication, the DNA from the vector integrates into the cellular genome through recombination events, altering the structure of the gene and interfering with its function *(1,2)*. The technique is potentially powerful but limited by the rare occurrence of homologous recombination events and difficulties in vector construction and delivery. Its use has been restricted to cell lines and animal models, and it has not been applied in human beings yet.

Some oligonucleotides can bind non-covalently in the major and minor grooves of double-stranded DNA in a sequence-specific manner and can inhibit transcription of a target gene. These molecules, known as triple-helix-forming oligonucleotides (TFOs), typically are 10–30 nucleotides in length and bind to areas of DNA with runs of purines on one strand and pyrimidines on the other *(3,4)*. This technique has been used in cell line and animal models to suppress transcription of the human c-myc oncogene. Polyamide molecules can also bind in the major groove of double-stranded DNA to inhibit transcription, which is an intriguing strategy for sequence-specific transcription inhibition *(5–7)*. Neither of these techniques has yet been explored for therapeutic potential. More recent developments have suggested that natural RNA may interact directly with genomic DNA to regulate gene expression *(8)*, an intriguing possibility for future development.

Short double-stranded oligodeoxynucleotides that mimic transcription factor-binding sequences can compete with genomic DNA for available intracellular transcription factors and thereby decrease transcription *(9)*. Intracellular nucleases can rapidly degrade short double-stranded oligodeoxynucleotides, and several structural modifications, including αβ-anomeric, phosphorothioate, and circular dumbbell oligonucleotides, have been used to increase stability. The major barrier to this technique remains gene specificity, and these molecules are still in early development.

3. INHIBITION OF TRANSLATION

Although inhibition of gene transcription is currently restricted to laboratory use, strategies for preventing translation of already transcribed mRNA are being actively investigated for the treatment of human malignancies. Antisense oligonucleotides to many target genes are in clinical development, including randomized trials, and other techniques are showing promising advances, including the use of RNA decoys, nucleotide enzymes (ribozymes and DNAzymes), and the intrinsic RNAi pathway.

3.1. Decoys and Enzymes

Short strands of single-stranded RNA can mimic mRNA sequences and compete for proteins that act as translational activators and stabilize mRNA. Although the use of these RNA decoys is potentially attractive, their therapeutic use remains theoretic.

Both ribonucleic acid and deoxyribonucleic acid molecules can possess catalytic activity, binding to an RNA substrate through sequence-specific Watson–Crick base pairing and cleaving the target transcript. Ribonucleic acid enzymes (ribozymes) recognize and cleave specific trinucleotide sequences, either GUX (where X is C, U, or A) or NUX (where N is any nucleotide). Five major catalytic motifs derived from naturally occurring ribozymes have been well studied, including hammerhead, hairpin, group I intron, ribonuclease P, and hepatitis delta virus ribozyme, of which hairpin and hammerhead are especially attractive for therapeutic use, because of relative simplicity, small size, and conserved catalytic activity with various flanking sequences (10,11). Catalytic nucleic acids are susceptible to degradation by endogenous nucleases, and chemical modifications at the 2′ ribosyl moiety enhance stability without diminishing catalytic ability. Deoxyribonucleic acid enzymes (DNAzymes) are potentially more resistant to endogenous nucleases and easier to synthesize (12), but their development as therapeutic agents is still in early stages of development.

3.2. RNAi

Recent developments have taken advantage of a cellular process that degrades mRNA at sequences homologous to endogenously produced short double-stranded oligoribonucleotides. This process functions as a natural method of transcription repression in yeast (where it is termed quelling), plants (post-transcriptional gene silencing or co-suppression), single-cell organisms, and invertebrates and mammals (known as RNAi). Naturally occurring double-stranded oligoribonucleotides are known as micro-RNA (miRNA), which are expressed in the nucleus as parts of long primary miRNA transcripts (pri-miRNA) that have 5′ caps and 3′ poly(A) tails. These longer transcripts form hairpin structures around the miRNA sequence, and the hairpin structures likely act as signals for digestion by a double-stranded ribonuclease (known as Drosha) to produce the precursor miRNA (pre-miRNA). The export of pre-miRNA from the nucleus is mediated by exportin-5, and a cytoplasmic double-stranded RNA nuclease (known as Dicer) cleaves the pre-miRNA leaving 21–27 nucleotide strands with 1–4 nucleotide 3′ overhangs (13–15). The single-stranded mature miRNA associates with a protein complex known as the RNA-induced silencing complex (RISC), and the miRNA/RISC complex recognizes the target mRNA through complementarity between the 5′ region of the antisense miRNA strand and the 3′ untranslated region of the target and then cleaves the target at a position 10 base pairs upstream of the 5′ end of the miRNA (14–16). This cleavage step silences any further translation of the target mRNA, which is rapidly degraded. The guide strand of miRNA that has been incorporated into RISC appears to remain intact and can therefore act as a catalyst for further cleavage of target mRNA.

Investigations have identified several natural examples of gene regulation through short double-stranded RNA. The ability of RNA to suppress gene expression in a sequence-specific manner was first noted in plants, through unexpected results of experiments aimed at overexpressing pigmentation genes in petunias (17,18). The natural occurrence of this process was first identified in *Caenorhabditis elegans*, where genes such as lin-4 and let-7 encode miRNA and were identified through loss-of-function mutations causing developmental abnormalities (19,20). Similar investigations led to the discovery of miRNA genes *bantam* and *mir-14* in *Drosophila* (21,22).

Later experiments showed that exogenously produced double-stranded RNA molecules could inhibit gene expression in *C. elegans (23)*. These molecules, known as siRNA (small interfering RNA), appear to require similar processing by Dicer and incorporation in RISC for functioning and are now widely used in cellular and animal experiments. Interest in siRNA increased as studies showed that the RNAi mechanism is operative in mammalian cells *(24)*. Many siRNA genes are available commercially for gene knockout experiments, to be transfected or introduced into viral vectors for integration into the genome. They offer several experimental advantages, including the ability to pass intact through generations, the ability to knock down expression of several homologous genes with redundant function, and the ability to target specific splicing products of a single gene *(25)*.

Although early experiments suggested that the RNAi system suppressed translation in a highly sequence-specific manner, more recent work has questioned that specificity. In some experimental systems, binding of siRNA to mRNA has not depended on precise matching, raising the possibility of one siRNA affecting multiple target genes *(14,15)*. In addition, the existence of non-specific cellular responses to double-stranded RNA, mediated through dsRNA protein kinase (PKR) or RNase L, complicates the identification of sequence-specific effects of siRNA *(15)*. These non-specific responses are presumably defense mechanisms against viral RNA and are clearly activated by long double-stranded RNA molecules (longer than 35 base pairs), leading to a general suppression of protein synthesis. Some siRNA or shRNA can activate PKR responses, which means that pro-apoptotic effects in cell models can be due to these non-specific reactions rather than homology of the siRNA to the mRNA target *(15)*. The specificity of siRNA effects is also limited by the theoretic possibility that siRNA could saturate cellular RNAi machinery, leading to inhibition of endogenous RNAi gene regulation and off-target effects.

3.3. Targets Under Development for Human RNAi Treatment

Utilizing the cellular RNAi apparatus to treat human disease has immediate appeal, given the potential specificity and efficient mechanism of inhibition. Although comparisons are limited by the nature of experimental conditions, siRNA appear to inhibit expression with 100–1000 times more potency than optimally modified oligodeoxynucleotides (see section 4) directed against the same target *(26)*. Animal models of fulminant hepatitis, viral infections, sepsis, and ocular neovascularization have validated the concept of systemically administered siRNA *(25)*. Recent applications have included attempts to use siRNA against respiratory syncytial virus (RSV) *(27)*, HIV, and hepatitis viruses *(28)*. Targeting vascular growth through inhibition of production of vascular endothelial growth factor (VEGF) is being investigated for the treatment of age-related macular degeneration (AMD) and diabetic retinopathy. Ideal targets for siRNA inhibition in cancer include oncogenes that are overexpressed in cancer cells, oncogenes active through point mutations, and fusion products of chromosomal translocations.

Overexpressed oncogenes that have been targeted in preclinical models include bcl-2, telomerase, WT1, c-myc, NF-κB, and c-kit. A "Stealth" siRNA (Invitrogen, Carlsbad, CA, USA), modified for intracellular stability, can specifically suppress bcl-2 expression in a prostate cancer cell line *(29)*. Another siRNA (D6, Dharmacon,

Lafayette, CO, USA) can suppress bcl-2 expression in Waldenström's macroglobulinemia cells *(30)*. An siRNA against the catalytic subunit of telomerase (hTERT)
inhibited expression by 99–100% in an ovarian cancer cell line, better than a corresponding antisense oligonucleotide *(31)*. In leukemic blast cell lines, an siRNA against
Wilms' tumor gene-1 (WT-1) decreases protein expression and cell proliferation, and
increases apoptosis *(32)*. An siRNA is able to reduce *c-myc* expression by 95% in
platinum-resistant ovarian cancer cells, thereby inducing platinum sensitivity *(33)*. In
human colon cancer cells, siRNA to the p65 subunit of NF-κB was able to reduce
NF-κB expression and increase susceptibility to irinotecan *(34)*. In human malignant
neuroepithelial cells, siRNA was able to reduce expression of c-kit by up to 42% *(35)*.

siRNA can selectively inhibit mutated versions of oncogenes, even when the
difference from the wild-type counterpart is limited to a point mutation. For example,
an siRNA to mutant K-Ras^{V12} can suppress expression of the mutant gene in pancreatic
carcinoma cells, while leaving intact expression of wild-type K-Ras *(36)*. Similarly, an
siRNA to mutant p53 can restore expression of wild-type p53 in cells expressing both
forms.

siRNA against cancer-specific fusion products associated with chromosomal translocations have been investigated in cell line models. These have included BCR-ABL in
chronic myeloid leukemia *(37–39)* and AML1/MTG8 in acute myeloid leukemia with
t(8;21) translocation *(40)*. In addition, siRNA to *lyn* kinase, which forms a signaling
complex with BCR-ABL, can decrease *lyn* expression and inhibit growth of BCR-ABL
positive leukemic blast cells, including those resistant to imatinib, while not affecting
the growth of normal bone marrow blasts *(41)*. In anaplastic large cell lymphoma
cells containing the t(2;5) translocation, an siRNA to the characteristic NPM–ALK
fusion product decreased cell proliferation and increased apoptosis *(42)*. Undergoing
laboratory investigation are several siRNA molecules for the EWS–FLI1 fusion product
of the t(11;22) translocation characteristic of Ewing's sarcoma family tumors *(43)*.
Appropriate siRNA sequences for additional fusion products have been suggested *(44)*.

The in vivo demonstration of antitumor efficacy of siRNA inhibition is limited
to a few experiments. Intratumoral injection of an atelocollagen-complexed siRNA
to FGF-4 in a mouse xenograft model of testicular cancer resulted in inhibition of
both FGF-4 expression and tumor growth *(45)*. Intravenous injection of atelocollagen-
complexed siRNA to EZH2 (enhancer of zestehomolog 2) and the phosphoinositide
3′-hydroxykinase p110-α subunit in a mouse model of prostate cancer bone metastasis resulted in delivery of siRNA to the tumor cells and inhibition of metastatic
growth *(46)*. An siRNA against Skp-2, delivered in an adenovirus vector by intratumoral injection in a mouse model of human small cell lung carcinoma, inhibited
both Skp-2 expression and tumor growth *(47)*. Systemic high-pressure infusion of an
siRNA to CEACAM6 (carcinoembryonic antigen-associated cell-adhesion molecule 6)
suppressed tumor growth in a mouse xenograft model of human pancreatic carcinoma
(48). Another experiment used a mouse model of intracranial human glioblastoma and
intratumoral injection of plasmids encoding siRNA to cathepsin B (a serine protease),
a matrix metalloproteinase (MMP)-9, or both, with evidence of synergy in antitumor
efficacy *(49)*.

3.4. Delivery of siRNA

The main limitation to application of RNAi to the treatment of human diseases has been the lack of reliable delivery techniques. Double-stranded RNA has an extremely short half-life in blood and is poorly taken up by mammalian cells. Hydrodynamic delivery (high-pressure intravenous injection) has been used successfully in rodents with experimental hepatitis (*50–53*) but is not suitable for use in human beings. Viral vectors can efficiently deliver DNA or RNA to cells, but have not been fully developed in humans, in part because of concerns of potential toxicity. Modifications to the oligonucleotide backbone can enhance stability without sacrificing function (*54*), and the recent observation of therapeutic effect after tail-vein injection of a modified siRNA (in a mouse model of hypercholesterolemia) has been encouraging (*55*). However, the feasibility of intravenous delivery has yet to be evaluated in larger animals. Other options for delivery have included intravenous injection in complex with liposomes or atelocollagen (Section 3.3), or local delivery enhanced by electric stimulation (*56*).

4. ANTISENSE OLIGONUCLEOTIDES

Among nucleic acid therapies, antisense oligodeoxynucleotides have been furthest developed for the treatment of human cancers. Although pre-clinical models have shown antitumor activity for some of these molecules and early human studies have suggested in vivo activity, the few available comparative studies have shown only limited efficacy and then only in selected sub-populations of patients. This might however be expected, as experience with at least some antibody therapies, such as those directed to human epidermal growth factor receptor 2 (HER2), have also been found useful only in selected sub-populations.

Antisense oligonucleotides are relatively short (17–26 base) single strands of either DNA or RNA that are complementary to specific mRNA sequences. Endogenous nucleases rapidly degrade unmodified phosphodiester oligonucleotides, limiting bioavailability; a process that can be slowed by modifications of the nucleic acid structure, the most common of which is the use of a phosphorothioate backbone (replacing the non-bridging phosphoryl oxygen of each nucleotide with sulfur). Theoretically, the therapeutic mechanism of action involves binding to mRNA through Watson–Crick pairing; the presence of double-stranded RNA then activates RNase-H-mediated cleavage of both the target mRNA and the antisense molecule. Other proposed mechanisms of action include prevention of mRNA transport, modulation or inhibition of splicing, and translational arrest.

The use of exogenous antisense oligonucleotides to modify gene expression was first used in a cell-free system in 1977 (*57*), followed shortly afterwards by the demonstration that a 13 base DNA strand that was antisense to the Rous sarcoma virus could inhibit viral replication in culture (*58*). Later, demonstration of the role of naturally occurring antisense RNA in gene regulation (*59–61*) encouraged further investigation into the possibility of therapeutic manipulation of gene transcription.

4.1. Bcl-2 and Bcl-x$_L$

BCL2 is a mitochondrial membrane protein that heterodimerizes with BAX and other pro-apoptotic regulators, inhibiting both the release of cytochrome *c* from mitochondria and the subsequent activation of the apoptotic cascade. The *bcl-2* gene was originally identified in follicular lymphoma through the characteristic t(14;18) translocation that results in its overexpression, but bcl-2 is overexpressed in a wide variety of tumor types and is associated with increased resistance to chemotherapy and radiation. Oblimersen (G3139 or Genasense) is an 18-base phosphorothioate oligonucleotide developed by Genta, Berkeley Heights, NJ as an antisense to the first six codons of bcl-2. In cell line and animal models, oblimersen has been able to decrease tumor bcl-2 expression, promote apoptosis, and induce tumor responses.

A preliminary dose-finding study in 9 patients with non-Hodgkin's lymphoma and positive immunohistochemical staining for bcl-2, used a 2-week subcutaneous infusion of single-agent oblimersen at doses ranging from 4.6 to 73.6 $mg/m^2/day$ produced one complete response and three partial responses, with evidence by flow cytometry of decreased bcl-2 expression in lymphoma cells of 2 patients *(62)*. Subsequent dose-finding studies have investigated the combination of oblimersen with cytotoxic chemotherapy agents, including gemcitabine *(63)*, taxanes *(64,65)*, and mitoxantrone, in general without finding toxicities beyond that expected with chemotherapy alone. Single-arm studies have explored the efficacy of oblimersen in treating various malignancies, including myeloma [in combination with dexamethasone *(66)*], B-cell lymphoma [with rituximab *(67)*], acute myeloid leukemia [with daunorubicin and cytarabine *(68)*], small cell lung cancer [with paclitaxel *(69)*, carboplatin *(70)*, or cisplatin and etoposide *(71)*], hormone-refractory prostate cancer [with mitoxantrone *(72)* or docetaxel *(73)*], metastatic renal cell carcinoma [with interferon-α *(74)*], gastric or esophageal cancer [with cisplatin and infusional fluorouracil *(75)*], and metastatic melanoma [with dacarbazine *(76)*].

These dose-finding and single-arm trials in general confirmed the safety of combining oblimersen with chemotherapy, with the exception of the combination with fluorouracil, leucovorin, and oxaliplatin in patients with metastatic colorectal cancer, which was not tolerated because of prolonged bone marrow suppression *(77)*. These trials also confirmed the ability of prolonged infusions of oblimersen to decrease expression of bcl-2 in peripheral blood leukocytes and, to a lesser degree, in repeated samples of tumor cells. However, correlation of decreased expression with clinical response has been more elusive, with only one of the above trials, in older patients with previously untreated acute myeloid leukemia, showing that bcl-2 mRNA levels were higher before therapy in completely responding patients than in non-responding patients, mRNA levels of bcl-2 (but not other cell-cycle proteins) decreased in completely responding patients whereas increasing in non-responding patients, and bcl-2 protein levels decreased in completely responding patients but remained unchanged in non-responding patients *(68)*.

Three comparative trials evaluating the efficacy of oblimersen for treatment of melanoma, chronic lymphocytic leukemia, and multiple myeloma have been reported. For advanced untreated melanoma, an international randomized trial used a 5-day continuous intravenous pretreatment regimen of 7 mg/kg/day followed by DTIC at 1000 mg/m^2. The intent-to-treat analysis reported a median survival of 9.1 months in patients receiving oblimersen plus dacarbazine compared with 7.9 months for dacarbazine alone ($p = 0.184$). Progression-free survival was significantly longer (median 78 versus 49 days, $p < 0.001$), response rates were higher (11.7 versus 6.8%, $p = 0.019$), and a per-protocol analysis of 480 patients who completed at least 12 months of follow-up showed significantly longer overall survival (10.1 versus 8.1 months, $p = 0.035$), but a recent evaluation by the US Food and Drug Administration did not result in approval for marketing *(78)*. For treatment of refractory chronic lymphocytic leukemia, a randomized trial compared oblimersen (3 mg/kg/day continuous intravenous infusion on days 1–7) in combination with fludarabine and cyclophosphamide with chemotherapy alone in 120 patients. Despite a lower dose than in other trials, the primary endpoint, complete response plus nodal partial response, was increased in the group receiving oblimersen (16 versus 7%, $p = 0.039$). However, the overall response rate was similar when partial responders were included *(79)*. For patients with refractory multiple myeloma, a randomized trial compared oblimersen (7 mg/kg/day) combined with high-dose dexamethasone with steroid treatment alone. In 224 patients, the primary endpoint of time to progression showed no benefit from the addition of oblimersen, nor was any benefit observed in response rate, toxicity, or overall survival *(80)*.

Bcl-x_L is another antiapoptotic gene that is similar to bcl-2, overexpressed in some tumor types, and may be important for tumor cell survival. Antisense oligonucleotides to bcl-x_L can induce apoptosis in various cells lines and sensitize tumor cells to chemotherapy *(81,82,83)* and in combination with bcl-2 antisense can synergistically enhance chemotherapy sensitivity *(84)*. An antisense oligonucleotide complementary to similar specific regions of both bcl-2 and bcl-x_L can inhibit expression of both proteins and is a potent inducer of apoptosis in tumor cell lines *(85)*.

4.2. PKC-α

Protein kinase C (PKC) is an important mediator of intracellular proliferative signals and is frequently overexpressed in human cancers. Aprinocarsen (Affinitak, also known as ISIS 3521 or LY900003) is a 20-base phosphorothioate antisense oligonucleotide to PKC-α mRNA. In animal models and cell lines, this compound is able to decrease PKC-α expression and induce apoptosis *(86,87)*. Dose-finding studies in patients with advanced cancer identified only mild toxicities on a thrice-weekly schedule *(88)* and dose-limiting toxicities of thrombocytopenia and fatigue at a dose of 3mg/kg/day continuous infusion for 21 days *(89)*. A phase II study in patients with hormone-refractory prostate cancer failed to identify any clinically significant single-agent activity *(90)*, but three studies in non-small cell lung cancer showed encouraging response and survival rates in combination with gemcitabine and cisplatin *(91)*, carboplatin and paclitaxel *(92)*, and single-agent docetaxel *(93)*.

On the basis of the promising phase II results, two large randomized trials examined the utility of aprinocarsen in combination with chemotherapy for the treatment of newly diagnosed metastatic non-small cell lung cancer. In one trial, 1000 patients with inoperable non-small cell lung cancer were randomized to gemcitabine (1250 mg/m^2 on days 1 and 8) and cisplatin (80 mg/m^2 on day 8) or the same chemotherapy combined with aprinocarsen (2mg/kg/day continuous infusion on days 1–15). The addition of the antisense therapy resulted in no improvement in response or survival, but did increase toxicity, with more discontinuations because of adverse events in the experimental arm *(94)*. In another trial, 600 patients with metastatic NSCLC were randomized to 21-day cycles of carboplatin (AUC 6 on day 1) and paclitaxel (175mg/m^2 on day 1) or chemotherapy combined with a 15-day continuous infusion of aprinocarsen (2mg/kg/day starting 3 days before chemotherapy), with similarly disappointing results: no difference in response rate, time to progression, or overall survival between groups *(95)*.

4.3. Clusterin

Clusterin encodes a chaperone protein that promotes cell survival and is expressed in various cancers; in prostate cancer, clusterin expression increases after androgen ablation, which in pre-clinical models confers a resistant phenotype. A clusterin antisense 21-base phosphorothioate oligonucleotide has been developed, with 2´-*o*-methoxy-ethyl modifications to the four bases at either end of the molecule (OGX-011, Oncogenex Technologies, Vancouver, BC, Canada). In a dose-escalation trial, clusterin antisense was administered in conjunction with antiandrogen therapy to 20 patients with localized prostate cancer before prostatectomy, with toxicity limited to grade 1–2 infusion reactions and transaminase elevations, and produced a dose-dependent decrease in clusterin expression in tumor samples by immunohistochemistry and in situ hybridization *(96)*. In a dose-escalation trial in combination with docetaxel (30 mg/m^2 for 5 of 6 weeks or 75 mg/m^2 every 3 weeks) in 26 patients with various carcinomas, clusterin antisense produced one partial response and at the highest dose level decreased serum clusterin levels by a mean of 52% *(97)*. Using a dose of 640 mg over 2 h, phase II studies in combination with chemotherapy have begun in patients with prostate, breast, and lung cancers.

4.4. Inhibitor of Apoptosis Family (Survivin and XIAP)

Survivin is a member of the inhibitor of apoptosis (IAP) gene family, has an important role in both cell division and apoptosis inhibition, at least in part through inhibition of caspases, and is expressed at a high level in a wide range of human cancer types, including lung, colon, pancreas, breast, and prostate cancers *(98–101)*, but not in most normal tissues. Survivin expression correlates with a lower apoptotic index in tumor cells and poorer prognosis in cancer patients, and overexpression of survivin in tumor cells inhibits chemotherapy-induced apoptosis. Conversely, expression of dominant-negative mutants of survivin induces apoptosis in tumor cell lines *(102)*. A 2´-MOE antisense oligonucleotide (LY2181308, ISIS Pharmaceuticals, Carlsbad, CA, USA) potently and specifically downregulates survivin expression in a broad range of human cancer cells, resulting in caspase-3-dependent apoptosis, cell-cycle arrest in the G2/M phase, the formation of multinucleated cells, and the sensitization of tumor cells to chemotherapy-induced apoptosis *(103–105)*. LY2181308 also possesses potent antitumor activity against a broad range of human cancers in xenograft models, activity that is sequence specific and associated with reduced survivin levels in tumors. On the basis of these promising preclinical results, LY2181308 is in clinical development in phase I studies encompassing a broad range of cancers *(78)*.

X-linked IAP (XIAP) overexpression inhibits apoptosis arising from chemotherapy, radiation, and growth-factor deprivation through inhibition of caspase activity *(106,107)*. A negative regulator of XIAP is underexpressed in tumor cell lines *(108)*, and XIAP is overexpressed in human cancers, including glioblastoma, prostate, pancreatic, gastric, and colorectal tumors, as well as in acute myeloid leukemia, in which overexpression has been associated with poor clinical outcome *(109)*. A 2´-*o*-methyl phosphorothioate 19-base antisense oligonucleotide to XIAP (AEG35156/GEM640, Aegera

Therapeutics, Montreal, QB, Canada) inhibits XIAP protein expression and enhances chemotherapy activity in xenograft models *(110,111)*. A dose-escalation study as a 7-day continuous intravenous infusion of single-agent AEG35156/GEM640 in patients with advanced tumors is currently accruing patients, and another dose-escalation trial in combination with docetaxel is planned *(78)*.

4.5. BCR–ABL

The BCR–ABL fusion protein is the hallmark of chronic myelogenous leukemia, produced by the t(9;22) translocation resulting in the Philadelphia chromosome. A junction-specific 26 nucleotide antisense to BCR–ABL has been used to purge autologous peripheral stem cell harvests before high-dose chemotherapy, with acceptable engraftment and reasonable duration of response *(112)*.

4.6. STAT3

The signal transducer and activator of transcription (STAT) factors function as downstream effectors of cytokine and growth-factor receptors, including Janus tyrosine kinases and SRC family members, and regulate gene expression through binding to DNA-regulatory elements. Preclinical studies implicate persistent STAT3 signaling in malignant transformation, through increased expression of genes associated with proliferation, cell survival, and angiogenesis *(113,114)*. A 2′-MOE antisense oligonucleotide against STAT3 (ISIS 345794) results in reduced STAT3 levels in cell lines and xenograft models of a range of tumor types, including multiple myeloma, melanoma, lymphoma, and prostate cancer *(115,116)*. Phase I studies are planned in multiple myeloma, lymphoma, and other cancers *(78)*.

4.7. p53

p53 is a classic oncogene, overexpressed in various solid tumors. EL625, a phosphorothioate antisense to p53, induces RNase-H-dependent cleavage of p53 transcripts resulting in loss of p53 production. In 11 patients with primary refractory acute myeloid leukemia, EL625 (0.1 mg/kg/h over 4 days with idarubicin 12 mg/m^2 on days 2–4 and either low- or high-dose cytarabine) produced four complete remissions and two morphologic remissions *(117)*.

4.8. Transforming Growth Factor-β2

Transforming growth factor (TGF)-β2 is an inflammatory cytokine that induces proliferation, invasion, metastasis, angiogenesis, and immunosuppression, and is highly over-expressed in malignant glioma and pancreatic cancer. A phosphorothioate antisense to TGF-β2 (AP12009) has been administered as an intratumoral infusion in patients with high-grade glioma; an ongoing randomized trial is comparing two doses of intratumoral AP12009 to single-agent systemic temozolomide or combination procarbazine, carmustine, and vincristine *(118)*. In pancreatic cancer cell lines and xenograft models, AP12009 inhibits TGF-β2 secretion and tumor cell proliferation and migration, and a systemic-use dose escalation trial in patients with metastatic pancreatic carcinoma is in progress *(119)*.

4.9. Raf

Raf kinases play key roles in integrating the K-Ras and mitogen-activated protein kinase (MAPK) intracellular signaling cascades, which are important regulators of tumor cell proliferation. An antisense to c-raf mRNA (LErafAON-ETU), formulated in liposomes and therefore requiring phosphorothioate modification only of the 3′ and 5′ nucleotides, is being examined in a dose-escalation trial *(120)*.

4.10. DNA methyltransferase

Suppression of pro-apoptotic and antiproliferative genes through methylation of DNA is an important component of tumor progression. An antisense to DNA methyltransferase (MG98) has been evaluated in a dose-escalation trial in 23 patients with advanced solid tumors, with consistent decreases in DNA methyltransferase expression in peripheral blood lymphocytes (by 6–69% on cycle 1 and 34–85% on cycle 2) and stable disease in 2 patients *(121)*.

4.11. Ribonucleotide Reductase

Ribonucleotide reductase, or RNR, is a cell-cycle controlled enzyme required for deoxyribonucleotide synthesis. An Antisense to the R2 subunit (GTI 2040, Lorus Therapeutics, Toronto, ON, Canada) has been administered in dose-finding studies as a single agent to patients with advanced cancer *(122)* and in combination with docetaxel to patients with previously treated non-small cell lung cancer *(123)*. Another antisense to RNR (GTI-2501, Lorus) is also in early trials in cancer patients.

4.12. c-Myb

The proto-oncogene c-myb encodes a nuclear binding protein that plays a major role in cell-cycle regulation in hematopoietic cells. An antisense oligonucleotide to c-myb (formerly LR3001, now G4460, Genta) has been used to purge ex vivo the bone marrow stem cells of patients with chronic myeloid leukemia undergoing autologous transplantation and was able to suppress c-myb mRNA levels in approximately half of the stem cell samples, leading to major and complete cytogenetic complete remissions after transplantation *(124)*. Further trials of systemic administration in advanced hematologic malignancies are planned.

4.13. Other Targets

Many more gene targets are in preclinical development, including eIF4E (LY2275796, Eli Lilly and Company, Indianapolis, IN, USA and ISIS Pharmaceuticals, Carlsbad, CA, USA), MDM2 (GEM240, Hybridon), IGFBP2 and IGFBP5 (OGX-225, Oncogenex), HSP27 (OGX-427, Oncogenex), MCL1 (ISIS 20408), androgen receptor (as750/15), and PKA (GEM 231, Hybridon, Inc., Cambridge, MA, USA).

5. CONSIDERATIONS FOR FURTHER DEVELOPMENT

The ability to selectively suppress expression of a single gene product remains a tantalizingly appealing modality for the treatment of cancers whose molecular basis is increasingly clear. However, the rush to develop antisense nucleotides led to disappointing clinical results and carries lessons in the danger of implementing clinical trials without a clear understanding of pharmacodynamics and demonstration of target suppression in malignant tissue. The failure of antisense oligonucleotides to produce significant clinical benefits as single agents or in combination with cytotoxic chemotherapy suggests several areas of possible improvement in the development of future sequence-specific oligonucleotide agents, both antisense molecules and siRNA. The success of these agents as cancer therapeutics will depend on

1. The development of delivery systems that achieve meaningful inhibitory concentrations in malignant cells.
2. The selection of gene targets on which malignant processes depend, such that suppression of expression of a single gene will lead to clinical effects.
3. Choice of a patient population with high expression of the target gene in malignant cells.
4. Accurate measurement in early phase clinical trials of expression of the target gene in malignant cells before, during, and after treatment, to identify a biologically optimal dose.

With these challenges in mind, development of nucleic acid agents continues. Ongoing laboratory work continues to identify new targets and confirm the utility of targeted gene suppression as cancer therapy, with the promise of exciting applications in the treatment of human disease.

REFERENCES

1. Melton DW. Gene targeting in the mouse. *Bioessays* 1994;16:633–638.
2. Stasiak A. Getting down to the core of homologous recombination. *Science* 1996;272:828–829.
3. Helene C. Control of oncogene expression by antisense nucleic acids. *Eur J Cancer* 1994;30A: 1721–1726.
4. Knauert MP, Glazer PM. Triplex forming oligonucleotides: sequence-specific tools for gene targeting. *Hum Mol Genet* 2001;10:2243–2251.
5. Kielkopf CL, Bremer RE, White S, et al. Structural effects of DNA sequence on T.A recognition by hydroxypyrrole/pyrrole pairs in the minor groove. *J Mol Biol* 2000;295:557–567.
6. Kielkopf CL, Baird EE, Dervan PB, Rees DC. Structural basis for G.C recognition in the DNA minor groove. *Nat Struct Biol* 1998;5:*104–109*.
7. Kielkopf CL, White S, Szewczyk JW, et al. A structural basis for recognition of A.T and T.A base pairs in the minor groove of B-DNA. *Science* 1998;282:111–115.
8. Corey DR. Regulating mammalian transcription with RNA. *Trends Biochem Sci* 2005;9.
9. Sharma HW, Perez JR, Higgins-Sochaski K, Hsiao R, Narayanan R. Transcription factor decoy approach to decipher the role of NF-kappa B in oncogenesis. *Anticancer Res* 1996;16:61–69.
10. Earnshaw DJ, Gait MJ. Progress toward the structure and therapeutic use of the hairpin ribozyme. *Antisense Nucleic Acid Drug Dev* 1997;7:403–411.
11. Hampel A. The hairpin ribozyme: discovery, two-dimensional model, and development for gene therapy. *Prog Nucleic Acid Res Mol Biol* 1998;58:1–39.
12. Santoro SW, Joyce GF. A general purpose RNA-cleaving DNA enzyme. *Proc Natl Acad Sci USA* 1997;94:4262–4266.
13. Elbashir SM, Lendeckel W, Tuschl T. RNA interference is mediated by 21- and 22-nucleotide RNAs. *Genes Dev* 2001;15:188–200.
14. Ambros V. The functions of animal microRNAs. *Nature* 2004;431:350–355.
15. Hannon GJ, Rossi JJ. Unlocking the potential of the human genome with RNA interference. *Nature* 2004;431:371–378.
16. Elbashir SM, Martinez J, Patkaniowska A, Lendeckel W, Tuschl T. Functional anatomy of siRNAs for mediating efficient RNAi in Drosophila melanogaster embryo lysate. *EMBO J* 2001;20: 6877–6888.

17. Napoli C, Lemieux C, Jorgensen R. Introduction of a chimeric chalcone synthase gene into petunia results in reversible co-suppression of homologous genes in trans. *Plant Cell* 1990;2:279–289.

18. van der Krol AR, Mur LA, Beld M, Mol JNM, Stuitje AR. Flavonoid genes in petunia: addition of a limited number of gene copies may lead to a suppression of gene expression. *Plant Cell* 1990;2:291–299.

19. Lee RC, Feinbaum RL, Ambros V. The C. elegans heterochronic gene lin-4 encodes small RNAs with antisense complementarity to lin-14. *Cell* 1993;75:843–854.

20. Reinhart BJ, Slack FJ, Basson M, et al. The 21-nucleotide let-7 RNA regulates developmental timing in Caenorhabditis elegans. *Nature* 2000;403:901.

21. Brennecke J, Hipfner DR, Stark A, Russell RB, Cohen SM. Bantam encodes a developmentally regulated microRNA that controls cell proliferation and regulates the proapoptotic gene hid in Drosophila. *Cell* 2003;113:36.

22. Xu P, Vernooy SY, Guo M, Hay BA. The Drosophila microRNA Mir-14 suppresses cell death and is required for normal fat metabolism. *Curr Biol* 2003;13:795.

23. Fire A, Xu S, Montgomery MK, Kostas SA, Driver SE, Mello CC. Potent and specific genetic interference by double-stranded RNA in Caenorhabditis elegans. *Nature* 1998;391:806–811.

24. Elbashir SM, Harborth J, Lendeckel W, Yalcin A, Weber K, Tuschl T. Duplexes of 21-nucleotide RNAs mediate RNA interference in cultured mammalian cells. *Nature* 2001;411:498.

25. Dorsett Y, Tuschl T. siRNAS: applications in functional genomics and potential as therapeutics. *Nat Rev Drug Discov* 2004;3:318.

26. Bertrand J-R, Pottier M, Vekris A, Opolon P, Maksimenko A, Malvy C. Comparison of antisense oligonucleotides and siRNAs in cell culture and in vivo. *Biochem Biophys Res Commun* 2002;296:1004.

27. Barik S. Development of gene-specific double-stranded RNA drugs. *Ann Med* 2004;36:540–551.

28. Seo MY, Abrignani S, Houghton M, Han JH. Small interfering RNA-mediated inhibition of hepatitis C virus replication in the human hepatoma cell line Huh-7. *J Virol* 2003;77:810–812.

29. Mack PC, Burich RA, Axentiev P, Gandara DR, Devere White RW. Inhibition of BCL-2 by Stealth siRNA results in growth suppression of LNCaP cells. *Am Soc Clin Oncol* 2005; 867s.

30. Nichols GL, Benimetskaya L, Stein CA. Inhibition of Bcl-2 by anti-sense oligonucleotide and siRNA in specimens from patients with Waldenström's macroglobulinemia (WM). *Am Soc Hematol* 2004.

31. Tebes SJ, Johnson NC, Fiorica JV, Kruk PA. Inhibition of Telomerase in Ovarian Cancer using siRNA Technology. *Am Soc Clin Oncol* 2005; 236s.

32. Glienke W, Seil I, Bauer N, Bergmann L. siRNA mediated silencing of Wilms tumor gene-1 (WT1) in leukemia cell lines. *Am Soc Hematol* 2005.

33. Elahi A, Martino MA, Cragun J, et al. Silencing pathways that underlie platinum resistance in ovarian cancer cells using small interfering RNA. *Am Soc Hematol* 2005; 483s.

34. Becerra CR, Verma U, Guo J, Gaynor RB. Enhanced chemosensitivity of CPT-11 in colon cancer by RNA interference of NF-kB p65 subunit. *Am Soc Clin Oncol (Gastroint Cancers Symp)* 2004; 344s.

35. Henningson CT, Jr., Demir G, Robbins M, Kalota A, Gewirtz AM. Inhibition of c-Kit receptor expression in malignant human neuroepithelial cells by RNA interference. *Am Soc Clin Oncol* 2002.

36. Brummelkamp TR, Bernards R, Agami R. Stable suppression of tumorigenicity by virus-mediated RNA interference. *Cancer Cell* 2002;2:247.

37. Scherr M, Battmer K, Winkler T, Heidenreich O, Ganser A, Eder M. Specific inhibition of bcr-abl gene expression by small interfering RNA. *Blood* 2003;101:1566–1569.

38. Wohlbold L, van der Kuip H, Moehring A, et al. Repeated application of sequence-specific siRNA molecules leads to an effective downmodulation of all clinically relevant bcr-abl gene variants. *Am Soc Hematol* 2004.

39. Wilda M, Fuchs U, Wössmann W, Borkhardt A. Killing of leukemic cells with a BCR/ABL fusion gene by RNA interference (RNAi). *Oncogene* 2002;21:5716–5724.

40. Heidenreich O, Krauter J, Riehle H, et al. AML1/MTG8 oncogene suppression by small interfering RNAs supports myeloid differentiation of t(8;21)-positive leukemic cells. *Blood* 2003;101: 3157–3163.

41. Ptasznik A, Nakata Y, Kalota A, Emerson SG, Gewirtz AM. Short interfering RNA (siRNA) targeting the Lyn kinase induces apoptosis in primary, and drug-resistant, BCR-ABL1(+) leukemia cells. *Nat Med* 2004;10:1189.

42. Hsu FY, Anderson WF, Johnston PB. Targeted siRNA inhibition of NPM-ALK in anaplastic large cell lymphoma causes disease specific growth inhibition which augments chemotherapeutic agents. *Am Soc Hematol* 2005.

43. Kovar H, Ban J, Pospisilova S. Potentials for RNAi in sarcoma research and therapy: Ewing's sarcoma as a model. *Semin Cancer Biol* 2003;13:281.

44. Damm-Welk C, Fuchs U, Wossmann W, Borkhardt A. Targeting oncogenic fusion genes in leukemias and lymphomas by RNA interference. *Semin Cancer Biol* 2003;13:283–292.

45. Minakuchi Y, Takeshita F, Kosaka N, et al. Atelocollagen-mediated synthetic small interfering RNA delivery for effective gene silencing in vitro and in vivo. *Nucleic Acids Res* 2004;32:e109.

46. Takeshita F, Minakuchi Y, Nagahara S, et al. Efficient delivery of small interfering RNA to bone-metastatic tumors by using atelocollagen in vivo. *PNAS* 2005;102:12177–12182.

47. Sumimoto H, Yamagata S, Shimizu A, et al. Gene therapy for human small-cell lung carcinoma by inactivation of Skp-2 with virally mediated RNA interference. *Gene Ther* 2004;12:100.

48. Duxbury MS, Matros E, Ito H, Zinner MJ, Ashley SW, Whang EE. Systemic siRNA-Mediated Gene Silencing: A New Approach to Targeted Therapy of Cancer. *Ann Surg* 2004;240:667–676.

49. Lakka SS, Gondi CS, Yanamandra N, et al. Inhibition of cathepsin B and MMP-9 gene expression in glioblastoma cell line via RNA interference reduces tumor cell invasion, tumor growth and angiogenesis. *Oncogene* 2004;23:4681–4689.

50. Song E, Lee S-K, Wang J, et al. RNA interference targeting Fas protects mice from fulminant hepatitis. *Nat Med* 2003;9:351.

51. Giladi H, Ketzinel-Gilad M, Rivkin L, Felig Y, Nussbaum O, Galun E. Small interfering RNA Inhibits Hepatitis B virus replication in mice. *Mol Ther* 2003;8:776.

52. Klein C, Bock CT, Wedemeyer H, et al. Inhibition of hepatitis B virus replication in vivo by nucleoside analogues and siRNA. *Gastroenterology* 2003;125:18.

53. McCaffrey AP, Nakai H, Pandey K, et al. Inhibition of hepatitis B virus in mice by RNA interference. *Nat Biotechnol* 2003;21:644.

54. Chiu Y-L, Rana TM. siRNA function in RNAi: A chemical modification analysis. *RNA* 2003;9:1034–1048.

55. Soutschek J, Akinc A, Bramlage B, et al. Therapeutic silencing of an endogenous gene by systemic administration of modified siRNAs. *Nature* 2004;432:173–178.

56. Dykxhoorn DM, Lieberman J. The silent revolution: RNA interference as basic biology, research tool, and therapeutic. *Annu Rev Med* 2005;56:401–423.

57. Paterson BM, Roberts BE, Kuff EL. Structural gene identification and mapping by DNA-mRNA hybrid-arrested cell-free translation. *Proc Natl Acad Sci USA* 1977;74:4370–4374.

58. Zamecnik PC, Stephenson ML. Inhibition of Rous sarcoma virus replication and cell transformation by a specific oligodeoxynucleotide. *Proc Natl Acad Sci USA* 1978;75:280–284.

59. Simons RW, Kleckner N. Translational control of IS10 transposition. *Cell* 1983;34:683–691.

60. Izant JG, Weintraub H. Inhibition of thymidine kinase gene expression by anti-sense RNA: a molecular approach to genetic analysis. *Cell* 1984;36:1007–1015.

61. Mizuno T, Chou MY, Inouye M. A unique mechanism regulating gene expression: translational inhibition by a complementary RNA transcript (micRNA). *Proc Natl Acad Sci USA* 1984;81:1966–1970.

62. Webb A, Cunningham D, Cotter F, et al. BCL-2 antisense therapy in patients with non-Hodgkin lymphoma. *Lancet* 1997;349:1137–1141.

63. Fisher G, Advani R, Wakelee H, et al. A Phase I trial of Oblimersen and Gemcitabine in refractory and advanced malignancies. *Am Soc Clin Oncol* 2005:3174.

64. Marshall J, Chen H, Yang D, et al. A phase I trial of a Bcl-2 antisense (G3139) and weekly docetaxel in patients with advanced breast cancer and other solid tumors. *Ann Oncol* 2004;15:1274–1283.

65. Morris MJ, Cordon-Cardo C, Kelly WK, et al. Safety and biologic activity of intravenous BCL-2 antisense oligonucleotide (G3139) and taxane chemotherapy in patients with advanced cancer. *Appl Immunohistochem Mol Morphol* 2005;13:6–13.

66. Badros AZ, Goloubeva O, Rapoport AP, et al. Phase II Study of G3139, a Bcl-2 antisense oligonucleotide, in combination with dexamethasone and thalidomide in relapsed multiple myeloma patients. *J Clin Oncol* 2005;23:4089–4099.

67. Pro B, Smith MR, Younes A, et al. Oblimersen sodium (Bcl-2 antisense) plus rituximab in patients with recurrent B-cell non-Hodgkin's lymphoma: preliminary phase II results. *Am Soc Clin Oncol* 2004:6572.

68. Marcucci G, Stock W, Dai G, et al. Phase I study of oblimersen sodium, an antisense to bcl-2, in untreated older patients with acute myeloid leukemia: pharmacokinetics, pharmacodynamics, and clinical activity. *J Clin Oncol* 2005;23:3404–3411.

69. Rudin CM, Otterson GA, Mauer AM, et al. A pilot trial of G3139, a bcl-2 antisense oligonucleotide, and paclitaxel in patients with chemorefractory small-cell lung cancer. *Ann Oncol* 2002;13:539–545.

70. Rudin CM, Kozloff M, Hoffman PC, et al. Phase I study of G3139, a bcl-2 antisense oligonucleotide, combined with carboplatin and etoposide in patients with small-cell lung cancer. *J Clin Oncol* 2004;22:1110–1117.

71. Rudin CM, Salgia R, Wang XF, Green MR, Vokes EE. CALGB 30103: A randomized phase II study of carboplatin and etoposide (CE) with or without G3139 in patients with extensive stage small cell lung cancer (ES-SCLC). *Am Soc Clin Oncol* 2005:7168.

72. Chi KN, Gleave ME, Klasa R, et al. A phase I dose-finding study of combined treatment with an antisense Bcl-2 oligonucleotide (Genasense) and mitoxantrone in patients with metastatic hormone-refractory prostate cancer. *Clin Cancer Res* 2001;7:3920–3927.

73. Tolcher AW, Kuhn J, Schwartz G, et al. A phase I pharmacokinetic and biological correlative study of oblimersen sodium (Genasense, G3139), an antisense oligonucleotide to the bcl-2 mRNA, and of docetaxel in patients with hormone-refractory prostate cancer. *Clin Cancer Res* 2004;10:5048–5057.

74. Margolin KA, Lara P, Quinn D, et al. G3139 plus a-Interferon (IFN) in metastatic renal cancer (RCC): a phase II study of the California Cancer Consortium. *Am Soc Clin Oncol* 2005:4694.

75. Kaubisch A, Wu Y, Wadler S. Phase I/II study of genasense (G3139) in combination with cisplatin (cis) and fluorouracil (FU) in patients with advanced esophageal, gastro-esophageal junction and gastric cancer (P5385). *Am Soc Clin Oncol* 2005; 884s.

76. Jansen B, Wacheck V, Heere-Ress E, et al. Chemosensitisation of malignant melanoma by BCL2 antisense therapy. *Lancet* 2000;356:1728–1733.

77. Mays TA, Mita AC, Takimoto C, et al. Bcl-2 biomodulation with oblimersen sodium in combination with FOLFOX4 chemotherapy: a phase I study in metastatic colon carcinoma. *Am Soc Clin Oncol* 2005:3158.

78. Gleave ME, Monia BP. Antisense therapy for cancer. *Nat Rev Cancer* 2005;5:468.

79. Rai KR, Moore JO, Boyd TE, et al. Phase 3 randomized trial of fludarabine/cyclophosphamide chemotherapy with or without oblimersen sodium (Bcl-2 antisense; genasense; G3139) for patients with relapsed or refractory chronic lymphocytic leukemia (CLL). *ASH Annu Meet Abstr* 2004;104:338.

80. Chanan-Khan AA, Niesvizky R, Hohl RJ, et al. Randomized multicenter phase 3 trial of high-dose dexamethasone (dex) with or without oblimersen sodium (G3139; Bcl-2 antisense; genasense) for patients with advanced multiple myeloma (MM). *ASH Annu Meet Abstr* 2004;104:1477.

81. Leech SH, Olie RA, Gautschi O, et al. Induction of apoptosis in lung-cancer cells following bcl-xL anti-sense treatment. *Int J Cancer* 2000;86:570–576.

82. Simões-Wüst AP, Olie RA, Gautschi O, et al. Bcl-xl antisense treatment induces apoptosis in breast carcinoma cells. *Int J Cancer* 2000;87:582–590.

83. Lebedeva I, Rando R, Ojwang J, Cossum P, Stein CA. Bcl-xl in prostate cancer cells: effects of overexpression and down-regulation on chemosensitivity. *Cancer Res* 2000;60:6052–6060.

84. Miyake H, Monia BP, Gleave ME. Inhibition of progression to androgen-independence by combined adjuvant treatment with antisense BCL-XL and antisense Bcl-2 oligonucleotides plus taxol after castration in the Shionogi tumor model. *Int J Cancer* 2004;86:855–862.

85. Gautschi O, Tschopp S, Olie RA, et al. Activity of a novel bcl-2/bcl-xl-bispecific antisense oligonucleotide against tumors of diverse histologic origins. *J Natl Cancer Inst* 2001;93:463–471.

86. Shen L, Dean NM, Glazer RI. Induction of p53-dependent, insulin-like growth factor-binding protein-3-mediated apoptosis in glioblastoma multiforme cells by a protein kinase calpha antisense oligonucleotide. *Mol Pharmacol* 1999;55:396–402.

87. Wang X-Y, Repasky E, Liu H-T. Antisense inhibition of protein kinase C[alpha] reverses the transformed phenotype in human lung carcinoma cells. *Exp Cell Res* 1999;250:253–263.

88. Nemunaitis J, Holmlund JT, Kraynak M, et al. Phase I evaluation of ISIS 3521, an antisense oligodeoxynucleotide to protein kinase C-alpha, in patients with advanced cancer. *J Clin Oncol* 1999;17:3586–3595.

89. Yuen AR, Halsey J, Fisher GA, et al. Phase I study of an antisense oligonucleotide to protein kinase C-{{alpha}} (ISIS 3521/CGP 64128A) in patients with cancer. *Clin Cancer Res* 1999;5:3357–3363.

90. Tolcher AW, Reyno L, Venner PM, et al. A randomized phase II and pharmacokinetic study of the antisense oligonucleotides ISIS 3521 and ISIS 5132 in patients with hormone-refractory prostate cancer. *Clin Cancer Res* 2002;8:2530–2535.

91. Ritch PS, Belt R, George S, et al. Phase I/II trial of ISIS 3521/LY900003, an antinsense inhibitor of PKC-alpha with cisplatin and gemcitabine in advanced non-small cell lung cancer (NSCLC). *Am Soc Clin Oncol* 2002.

92. Yuen A, Halsey J, Fisher G, et al. Phase I/II trial of ISIS 3521, an antisense inhibitor of PKC-alpha, with carboplatin and paclitaxel in non-small cell lung cancer. *Am Soc Clin Oncol* 2001.

93. Moore MR, Saleh M, Jones CM, et al. Phase II trial of ISIS 3521/LY900003, an antisense inhibitor of PKC-alpha, with docetaxel in non-small cell lung cancer (NSCLC). *Am Soc Clin Oncol* 2002.

94. Paz-Ares L, Douillard J, Koralewski P, et al. Randomized phase III trial of gemcitabine/cisplatin (GC) and protein kinase C a (PKCa) antisense oligonucleotide aprinocarsen in patients (pts) with advanced stage non-small cell lung cancer (NSCLC). *Am Soc Clin Oncol* 2005:7053.

95. Lynch TJ, Raju R, Lind M, et al. Randomized phase III trial of chemotherapy and antisense oligonucleotide LY900003 (ISIS 3521) in patients with advanced NSCLC: Initial report. *Am Soc Clin Oncol* 2003:623.

96. Chi KN, Eisenhauer E, Fazli L, et al. A phase I pharmacokinetic (PK) and pharmacodynamic (PD) study of OGX-011, a 2'-methoxyethyl phosphorothioate antisense to clusterin, in patients with prostate cancer prior to radical prostatectomy. *Am Soc Clin Oncol* 2004:3033.

97. Chi KN, Eisenhauer E, Siu L, et al. A phase I study of a second generation antisense oligonucleotide to clusterin (OGX-011) in combination with docetaxel: NCIC CTG IND.154. *Am Soc Clin Oncol* 2005:3085.

98. Li F, Ambrosini G, Chu EY, et al. Control of apoptosis and mitotic spindle checkpoint by survivin. *Nature* 1998;396:580–584.

99. Ambrosini G, Adida C, Altieri DC. A novel anti-apoptosis gene, survivin, expressed in cancer and lymphoma. *Nat Med* 1997;3:917–921.

100. Lu CD, Altieri DC, Tanigawa N. Expression of a novel antiapoptosis gene, survivin, correlated with tumor cell apoptosis and p53 accumulation in gastric carcinomas. *Cancer Res* 1998;58:1808–1812.

101. Tamm I, Wang Y, Sausville E, et al. IAP-family protein survivin inhibits caspase activity and apoptosis induced by Fas (CD95), Bax, caspases, and anticancer drugs. *Cancer Res* 1998;58: 5315–5320.

102. Altieri DC. Survivin, versatile modulation of cell division and apoptosis in cancer. *Oncogene* 2003;22:8581–8589.

103. Li F, Ackermann EJ, Bennett CF, et al. Pleiotropic cell-division defects and apoptosis induced by interference with survivin function. *Nat Cell Biol* 1999;1:461–466.

104. Chen J, Wu W, Tahir SK, et al. Down-regulation of survivin by antisense oligonucleotides increases apoptosis, inhibits cytokinesis and anchorage-independent growth. *Neoplasia* 2000;2:235–241.

105. Ansell SM, Arendt BK, Grote DM, et al. Inhibition of survivin expression suppresses the growth of aggressive non-Hodgkin's lymphoma. *Leukemia* 2004;18:616–623.

106. Deveraux QL, Takahashi R, Salvesen GS, Reed JC. X-linked IAP is a direct inhibitor of cell-death proteases. *Nature* 1997;388:300–304.

107. Silke J, Hawkins CJ, Ekert PG, et al. The anti-apoptotic activity of XIAP is retained upon mutation of both the caspase 3- and caspase 9-interacting sites. *J Cell Biol* 2002;157:115–124.

108. Fong WG, Liston P, Rajcan-Separovic E, St Jean M, Craig C, Korneluk RG. Expression and genetic analysis of XIAP-associated factor 1 (XAF1) in cancer cell lines. *Genomics* 2000;70:113–122.

109. Tamm I, Kornblau SM, Segall H, et al. Expression and prognostic significance of IAP-family genes in human cancers and myeloid leukemias. *Clin Cancer Res* 2000;6:1796–1803.

110. Hu Y, Cherton-Horvat G, Dragowska V, et al. Antisense oligonucleotides targeting XIAP induce apoptosis and enhance chemotherapeutic activity against human lung cancer cells in vitro and in vivo. *Clin Cancer Res* 2003;9:2826–2836.

111. Sasaki H, Sheng Y, Kotsuji F, Tsang BK. Down-regulation of X-linked inhibitor of apoptosis protein induces apoptosis in chemoresistant human ovarian cancer cells. *Cancer Res* 2000;60:5659–5666.

112. de Fabritiis P, Petti MC, Montefusco E, et al. BCR-ABL antisense oligodeoxynucleotide in vitro purging and autologous bone marrow transplantation for patients with chronic myelogenous leukemia in advanced phase. *Blood* 1998;91:3156–3162.

113. Yu CL, Meyer DJ, Campbell GS, et al. Enhanced DNA-binding activity of a Stat3-related protein in cells transformed by the Src oncoprotein. *Science* 1995;269:81–83.

114. Buettner R, Mora LB, Jove R. Activated STAT signaling in human tumors provides novel molecular targets for therapeutic intervention. *Clin Cancer Res* 2002;8:945–954.

115. Mora LB, Buettner R, Seigne J, et al. Constitutive activation of Stat3 in human prostate tumors and cell lines: direct inhibition of Stat3 signaling induces apoptosis of prostate cancer cells. *Cancer Res* 2002;62:6659–6666.

116. Epling-Burnette PK, Liu JH, Catlett-Falcone R, et al. Inhibition of STAT3 signaling leads to apoptosis of leukemic large granular lymphocytes and decreased Mcl-1 expression. *J Clin Invest* 2001;107:351–362.

117. Freireich EJ, Kantarjian H, Garcia-Manero G, et al. Phase II Study of EL625, a p53 antisense oligonucleotide, and chemotherapy in refractory and relapsed acute myelogenous leukemia (AML). *Am Soc Clin Oncol* 2005:6617.

118. Hau P, Kunst M, Pichler J, et al. Targeted downregulation of TGF-beta2 as immunotherapy for high-grade glioma: a phase IIb study. *Am Soc Clin Oncol* 2005:1537.

119. Schlingensiepen K, Bischof A, Egger T, et al. Targeted down regulation of TGF-beta2 in pancreatic carcinoma: A phase I/II dose escalation study to evaluate the safety and tolerability of the antisense oligonucleotide AP 12009. *Am Soc Clin Oncol* 2005; 123s.

120. Steinberg JL, Mendelson DS, Block H, et al. Phase I study of LErafAON-ETU, an easy-to-use formulation of Liposome Entrapped c-raf Antisense Oligonucleotide, in advanced cancer patients. *Am Soc Clin Oncol* 2005; 244s.

121. Vidal L, Leslie M, Sludden J, et al. A phase I and pharmacodynamic study of a 7 day infusion schedule of the DNMT1 antisense compound MG98. *Am Soc Clin Oncol* 2005; 209s.

122. Janisch LA, Schilsky RL, Vogelzang NJ, et al. Phase I study of GTI-2040 given by continuous intravenous infusion (CVI) in patients with advanced cancer. *Am Soc Clin Oncol* 2001.

123. Leighl NB, Laurie SA, Knox JJ, et al. Phase I/II study of GTI-2040 plus docetaxel as 2nd-line treatment in non-small cell lung cancer (NSCLC) and other solid tumors. *Am Soc Clin Oncol* 2005.

124. Luger SM, O'Brien SG, Ratajczak J, et al. Oligodeoxynucleotide-mediated inhibition of c-myb gene expression in autografted bone marrow: a pilot study. *Blood* 2002;99:1150–1158; 683s.

18

Engineering Oncolytic Measles Viruses for Targeted Cancer Therapy

Takafumi Nakamura, PhD,
and Stephen J. Russell, MD, PhD

SUMMARY

Many viruses are capable of destructive propagation in tumors. The goal of cancer virotherapy is to harness this destructive power to selectively destroy tumors without causing damage to normal tissues. Compared with normal tissues, tumors are more highly permissive for virus propagation because they fail to shut down protein synthesis in response to virus infection and do not readily undergo apoptosis. Additional oncolytic specificity can be achieved by engineering viruses such that their life cycles become dependent on a factor or factors supplied exclusively by tumor cells. One such strategy is transductional targeting whereby the virus is modified such that its attachment and entry are redirected through a receptor unique to the tumor cells. The two key components of transductional targeting are incorporation of functional polypeptide ligands into the virus coat and ablation of natural receptor tropisms. Available polypeptide ligands include short peptides, single-domain growth factors and cytokines, and two-domain single-chain antibodies that have higher affinities and more versatile binding specificities than short peptide ligands but more stringent folding requirements such that they can be displayed only as fusions to the surface glycoproteins of enveloped viruses. Unfortunately, for many of the viruses that have been tested, retargeted attachment fails to mediate efficient virus entry through the targeted receptor. Oncolytic measles virus (MV) provides a notable exception to this rule—not only can this virus tolerate the insertion of a wide variety of polypeptide ligands as C-terminal extensions of its attachment glycoprotein, but retargeted virus attachment usually leads to efficient virus entry through the targeted receptor. A versatile system has therefore been developed for the construction, rescue, and amplification of fully retargeted oncolytic MVs, and studies are currently underway to determine the value of transductional targeting using these agents in a variety of cancer therapy models.

Key Words: Target; oncolytic; measles virus; cancer; virotherapy.

From: *Cancer Drug Discovery and Development: Molecular Targeting in Oncology*
Edited by: H. L. Kaufman, S. Wadler, and K. Antman © Humana Press, Totowa, NJ

1. INTRODUCTION

New strategies using biological agents are being developed to treat cancer. Live viruses are among these new agents. Wild-type viruses infect normal cells and tissues, replicate in them, induce cell death, release progeny viral particles, and spread through the body. The key to virotherapy is to generate an oncolytic virus, which selectively infects and efficiently replicates within tumor cells but not in normal cells. In addition, oncolytic viruses kill tumor cells by several unique mechanisms. Viruses infect and replicate within target cells, directly lysing and killing them. Viruses also can kill cells by inducing expression of toxic proteins that induce both inflammatory cytokines and T-cell-mediated immunity. Because of these different modes of action, cross-resistance, which occurs after standard chemotherapy or radiotherapy, is much less likely to develop with virotherapy.

Advances in the molecular biology and genetics of viruses have led to a fundamental understanding of their pathogenicity and replicative mechanisms. These advances have also enabled virus engineering to increase the safety or antitumor potency of thera-peutic viruses. Adenoviruses and herpesviruses have been modified genetically to replicate selectively within tumor cells *(1)*. Other viruses, such as reovirus, autonomous parvoviruses, Newcastle disease virus, measles virus (MV), and vesicular stomatitis virus (VSV), have inherent tumor selectivity *(2)*. Each of these viruses has demonstrated promising tumor selectivity and antitumor potency after administration by intratumoral, intraperitoneal, or intravenous routes in preclinical and clinical studies.

We recently discovered that attenuated Edmonston lineage MVs (MV-Edm) are selectively oncolytic for a broad spectrum of human tumors, including ovarian cancer, lymphoma, glioma, multiple myeloma, and pancreatic cancer *(3–6)*, and we are currently testing their oncolytic activity in a phase I clinical trial in patients with relapsed ovarian cancer. The virus targets and destroys tumor cells through CD46, a membrane regulator of complement activation that is known to be over-expressed on many human malignancies *(7–18)*. CD46 is one of the two cellular receptors used by MV-Edm *(19,20)*, mediating both virus entry and subsequent cell killing through cell/cell fusion. The second receptor is signaling lymphocyte activation molecule (SLAM) that is expressed only on activated T cells, B cells, and monocytes *(21–23)*. The cytopathic effect of MV-Edm increases exponentially as the density of CD46 receptors on target cells increases and is therefore dramatic at high CD46 receptor densities (malignancy) but minimal at low densities (normal tissues) *(24)*.

When an oncolytic virus is used for human cancer therapy, its native tropisms have the potential to cause unwanted damage to normal tissues. On the contrary, a prerequisite to gene therapy of metastatic cancer is that the virus must gain access to disseminated tumor cells while avoiding depletion through infection of nontarget cells through its native receptors. Virus attachment specificity is determined by specific coat proteins that have evolved for this purpose. Attachment specificity can be retargeted by engineering new binding domains into these coat proteins or by the use of soluble bifunctional crosslinkers that bind to the viral coat and to a cell-surface receptor *(25)*. For oncolytic virus targeting, genetic modification of the coat protein is preferred. However, the range of polypeptide ligands that can be incorporated into the viral coat is highly variable between different virus families. Thus, the adenovirus fiber that folds in the cytoplasmic compartment cannot tolerate the insertion of complex polypeptides such as single-chain antibodies (scFvs) *(26)*. Efforts to reengineer adenovirus tropism are therefore restricted by the availability of short peptide ligands recognizing the desired targets. Peptides with the potential to target viruses to tumor blood vessels have been isolated by in vivo panning of phage display libraries *(27,28)*. However, the targeting properties of these peptides have been less impressive when they have been transplanted from filamentous phages to adenoviral vectors *(29,30)*. As a logical extension of these studies, investigators have therefore displayed peptide libraries on the surface of gene therapy vectors, such as adeno-associated virus (AAV), which can then be subjected to ex vivo or in vivo panning *(31,32)*. Enveloped viruses, in contrast to adenovirus and AAV, are able to tolerate the insertion of large domains such as growth factors or single-chain antibodies at specific sites in their membrane proteins that fold in the more favorable environment of the endoplasmic reticulum *(33,34)*. However, the limitation here is that, at least in the case of retroviral and herpes virus vectors, targeted attachment to alternative cell-surface receptors is not associated with efficient cell entry *(35–37)*.

Oncolytic measles virotherapy may lead to unwanted damage to normal tissues, and toxicity, because of broad tropism through its two native receptors CD46 *(19,20)* and SLAM *(21–23)*. To prevent infection of non-target normal tissues, we sought to ablate the native viral tropisms and to add new tropisms thereby retargeting the virus for more exclusive interaction with tumor cells. However, ablation of the native receptor interactions was not compatible with virus growth. This article reviews the development of retargeted oncolytic MVs, which are more selective and potentially safer oncolytic agents than their untargeted counterparts. Approaches of viral modification to overcome the current limitations are summarized. This article also argues that, at least in the case of MV, receptor choice is not a significant limitation for targeting oncolytic virotherapy. Finally, the article summarizes strategies for further improvements to current MV-Edm vectors for future clinical trials.

2. MEASLES VIRUS

2.1. Biology of MV

MV, a member of the genus *Morbillivirus* in the Paramyxoviridae family, is an enveloped RNA virus that contains a single-strand, negative-sense, nonsegmented genome *(38)*. The 15,894-kb MV genome encodes eight proteins from six nonoverlapping cistrons arranged 3′-N-P-M-F-H-L-5′. As shown in Fig. 1, its genome is associated with a nucleocapsid (N) protein, a helical ribonucleoprotein complex that serves as the template for transcription and replication, and is packaged into progeny virions *(38)*. The P cistron specifies three polypeptides: P, C, and V. The phospho (P) protein acts as a chaperone that interacts with and regulates the cellular localization of the N protein and assists in assembly of the ribonucleoprotein complex. The C and V polypeptides are nonstructural proteins that are translated from P mRNAs through the use of alternative reading frames; C protein is synthesized from a downstream translation start signal, whereas V protein is translated from an edited mRNA that contains an extra G residue *(38)*. The viral envelope consists of the matrix (M) protein and two transmembrane glycoproteins, the hemagglutinin (H) and fusion protein (F). The M protein is also involved in viral budding *(38)*. The two MV envelope glycoproteins, H and F, are the mediators of virus–cell membrane fusion during infection *(39)*.

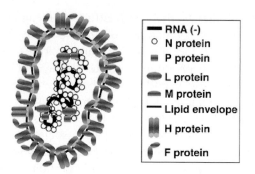

Fig. 1. Structure of MV.

Attachment to target cells is mediated by H and is followed by membrane fusion, mediated by F *(40)*. The tropism of MV is determined by binding of the H protein to one of the two possible cellular receptors, CD46 or SLAM. CD46 is a ubiquitous regulator of complement activation and is found on all human nucleated cells *(19,20)*. SLAM is expressed on activated T and B cells, dendritic cells, and macrophages *(21–23)*. Finally, the L protein is a multifunctional catalytic subunit of the RNA-dependent RNA polymerase *(38)*.

2.2. Oncolytic Properties of MV

MV is a serious human pathogen, responsible for the deaths of approximately 1 million children annually worldwide *(41)*. In contrast to wild-type MV that can cause potentially serious disease, vaccine strains of MV have an excellent safety record. The administration of millions of vaccine doses has substantially decreased measles incidence, morbidity, and mortality worldwide *(42)*. The wild-type virus propagates efficiently on Epstein–Barr virus-transformed B-cell lines, which may help explain how a large, untreated retro-orbital Burkitt's lymphoma regressed completely during concomitant MV infection in an 8-year-old child *(43)*. However, wild-type MV cannot propagate in SLAM-negative cell lines derived from human sarcomas or epithelial malignancies.

In contrast to wild-type MV, certain members of the Edmonston vaccine lineage are potently and selectively oncolytic against a broad spectrum of lymphoid and nonlymphoid human malignancies in vitro and in vivo. Replication-competent, attenuated MV-Edm strains have demonstrated potent antitumor activity against xenograft models of human multiple myeloma *(3)*, ovarian cancer *(4)*, lymphoma *(6)*, and glioma *(5)*. The virus is selectively oncolytic in human tumor cells, inducing extensive cell/cell fusion, which finally leads to apoptotic cell death in both cancer cell lines and primary cancer cells from patients *(3,4)*.

The differences between wild-type MV and MV-Edm are in part related to their receptor usage: wild-type MV enters cells predominantly through the SLAM receptor, and MV-Edm enters cells more efficiently through the CD46 receptor *(44)*. Indeed, MV strains with wild-type H protein use SLAM more efficiently than does MV with the H protein of Edm strain *(45)*. In contrast, the efficiency of cell entry through CD46 was considerably lower for MV with wild-type H protein than for MV with H protein of the Edm strain *(45)*. These differences are seen clearly at the cell fusion stage because wild-type MV fails to fuse CD46-positive cells *(21,44,46)*. It is well established that the broadened host range of attenuated MV-Edm is a consequence of mutations in the viral attachment glycoprotein, which enhance its ability to interact with CD46, which is expressed at high levels on the surface of most human tumor cells *(7–18)*. Kinetic rates for the MV-Edm H protein binding to CD46 are considerably higher than those for binding to SLAM *(47)*. Thus, tissue culture-adapted strains of MV-Edm have altered receptor specificity, have attenuated pathogenicity, and are coincidentally, potently and selectively oncolytic.

In contrast to the oncolytic potency of attenuated MV-Edm in human tumor cells, the virus causes only minor cytopathic effects in nontransformed human cells, such as normal ovarian surface epithelial cells, primary mesothelial cells, primary normal human dermal fibroblasts, primary coronary artery smooth muscle cells, and peripheral blood lymphocytes *(3,4)*. The basis for the enhanced infection of tumor cells by MV-Edm has recently been determined. The high CD46 receptor density on tumor cells leads to preferential infection and killing by oncolytic MV *(24)*. This dramatic difference in cytopathic effects was not due to a lack of virus entry or viral gene expression in the normal cells. Instead, it is the cell-surface expression levels of CD46 receptors that play a pivotal role in modulating the extent and size of syncytia at a given level of H and F expression. Using a panel of engineered Chinese hamster ovary (CHO) cells expressing a range of CD46 receptors, it was determined that although virus entry increased progressively with increase in receptor density, cell fusion was dependent on a threshold level of CD46 receptors, below which cell fusion was minimal and above which fusion was extensive. In line with this finding, tumor cells have been found to express more CD46 receptors on their surfaces compared with their normal counterparts, possibly as a mechanism to resist lysis by human complement proteins. As a result of the higher density of CD46 receptors on tumor cells, they are more susceptible to the cytopathic effects of MV-Edm-induced cell fusion and are able to recruit more neighboring cells into the syncytium thereby further amplifying the bystander killing.

It is possible that there are additional factors contributing to the tumor specificity of attenuated MV besides the increased CD46 receptor density on tumor cells. Upon infection, the primary innate response of a virally infected cell leads to the inhibition of viral protein synthesis, a response that is coordinated through the double-strand RNA-dependent protein kinase (PKR) and interferon-$\alpha\beta$ *(48)*. Since tumor cells frequently have defective PKR signaling and defective interferon response pathways, they may be relatively permissive substrates for RNA viruses such as VSV *(49,50)*, reovirus *(51,52)*, herpes simplex virus *(53,54)*, and possibly MV.

2.3. Approaches to Monitoring Virus Replication and/or Enhancing Anti-Tumor Effects

When administered as a therapeutic agent, MV-Edm has been shown to mediate regression of large, established myeloma, lymphoma, ovarian cancer, and glioma tumor xenografts *(3–6)*. However, not all human tumor xenografts were equally responsive to measles virotherapy even though all the human tumor cell lines tested were highly susceptible to the virus in tissue culture. To further elucidate the basis for treatment failure in resistant human xenografts, MV-Edm was engineered to encode a soluble marker peptide whose concentration could be monitored noninvasively in the serum of treated animals (55).

Recombinant viruses encoding the soluble extracellular domain of carcinoembryonic antigen (CEA) or β-chain of human chorionic gonadotropin (β-HCG) were shown to propagate almost as efficiently as the unmodified MV-Edm from which they were derived. Analysis of the gene expression profiles of MV-CEA-treated tumor xenografts indicated that resistance to therapy (when it occurred) could be due to either failure of virus uptake by the tumor cells or failure of tumor cell killing in the presence of persistent intratumoral virus infection. On the basis of its demonstrated potency against an intraperitoneal ovarian tumor xenograft (4), the MV-CEA virus is currently being tested in a phase I clinical trial. The analysis of the kinetics of the MV-CEA infection will be straightforward in this trial because it is convenient and inexpensive to regularly monitor CEA levels in blood. Subsequently, this information will be useful for optimizing the dosing and administration schedule for MV-CEA without the need for histologic analysis of tissue.

MV-Edm has been demonstrated to have oncolytic potency in various human tumor xenograft models. However, some tumors are resistant to MV oncolysis despite repeated virus injections (56). These data point to the need for methods to increase the oncolytic potency of MV-Edm. There are several strategies to enhance antitumor activity. The simplest approach is to use a higher multiplicity of infection. However, very high viral doses may not be feasible since it is difficult to manufacture sufficient quantities of the virus.

Another approach is to incorporate a therapeutic gene with a potent bystander effect or that induces immunity to the tumor. The human sodium iodide symporter (hNIS), the thyroidal protein responsible for concentrating iodide in the thyroid gland, has been used as a therapeutic gene in cancer (57,58). ^{131}I is efficiently taken up by tumor cells expressing hNIS, and in some cases, the tumor can be eliminated by such irradiation. The electron emitted by ^{131}I has a path length of approximately 0.4 mm (59), and therefore MV-Edm expressing hNIS (in combination with ^{131}I) was able to mediate complete regression of a human myeloma tumor xenograft, which was resistant to virotherapy with oncolytic MV-Edm (56). The hNIS gene also has another advantage. As described above, it is important to have a convenient, noninvasive method to track whether oncolytic MV replicates and spreads in the treated tumors. With the hNIS method, in vivo tracking of MV-NIS can be achieved by serial gamma camera imaging of ^{123}I uptake (56).

The antitumor activity of MV-Edm may also be enhanced by administering it in combination with immunotherapy, which may stimulate host defenses to recognize and destroy the tumors. Various viral vectors have been used for in vivo delivery of genes encoding immunostimulatory proteins such as cytokines (60–62) and chemokines (63). This type of in vivo vaccination has been applied clinically with promising results (64). Oncolytic viruses such as herpes simplex virus expressing IL-12 (65) and MV-Edm expressing granulocyte macrophage colony-stimulating factor (GM-CSF) (66) had superior in vivo antitumor activity against a squamous cell carcinoma and a lymphoid tumor, respectively. These oncolytic viruses infect tumor cells and not only induce tumor lysis but also produce cytokines locally, which may then amplify the tumor-specific host-immune response.

3. RETARGETED MV

3.1. Modification of H Protein for Targeted Cell Fusion

Fusion of measles-infected cells is mediated by the viral hemagglutinin (H) and fusion (F) proteins that together form a fusogenic membrane glycoprotein complex. The H protein mediates attachment to either one of the viral receptors, CD46 *(19,20)* or SLAM *(21)*, on the cell surface and signals to the F protein to trigger cell fusion *(67)*. As shown in Fig. 2, the steps required to retarget this cell fusion reaction are ablation of H-mediated CD46 and SLAM receptor recognition and introduction of a new binding specificity in the H glycoprotein while preserving its ability to trigger conformational changes in the F protein that lead to membrane fusion. To retarget H attachment, we fused an anti-CD38 single-chain antibody (scFv) to its C terminus and mutated residues involved in binding to CD46 (451 and 481) and SLAM (529 and 533) (Fig. 2). Receptor-specific fusion support by the chimeric H expression plasmids was determined after F-plasmid co-transfection in CHO cells expressing CD46 or SLAM or CD38. Syncytial cytopathic effect was scored by counting syncytia. Paired mutations at positions 451 and 529 or 481 and 533 supported fusion through the targeted CD38 receptor but not through CD46 or SLAM. These data proved conclusively that antibody-targeted cell fusion can be achieved. However, the fusion support activity of the fully retargeted H chimeras on CHO-CD38 cells was considerably reduced compared with the original nonablated chimeric protein displaying scFv against CD38. To address the issue of suboptimal fusion support by fully retargeted chimeric H glycoproteins, we focused on residue 481 as it has been reported that amino acid substitutions at this position can have a strong effect on fusion triggering activity *(68)*. We therefore generated additional H chimeras mutated as before at residue 533 (R533A) but with different substitutions at position 481 (Y481M, Y481Q, and Y481A) in place of Y481N. Interestingly, all the new 481-substituted H protein chimeras retained the fully retargeted phenotype but demonstrated higher fusion support activity than the original Y481N mutant on CHO-CD38 cells. Finally, we concluded that the Y481A and R533A mutations on H protein (H_{AA}) provide the optimal platform for the generation of fully retargeted H proteins by scFv display *(69)*.

3.2. Rescue and Propagation of Retargeted MV

Previously, nontargeted MVs have been rescued on genetically modified 293 cells and amplified on Vero cells *(70)*. However, ablation of the native receptor interactions was not compatible with virus rescue and growth on 293 or Vero cells as the retargeted virus no longer bind to the native measles receptors. We therefore rescued and propagated retargeted MVs carrying CD38-displayed H_{AA} protein on Vero cell transfectants stably expressing human CD38 *(71)*. Both the CD38-targeted viruses and the untargeted parental viruses grew efficiently on Vero-CD38 cells. When the tropism of the targeted viruses was evaluated in CHO cells transfected with each of the respective receptors, the untargeted virus infected CHO cells expressing native CD46 or SLAM, whereas the retargeted viruses did not. Fully CD38-retargeted viruses showed the expected host-range properties (CD46 and SLAM blind but efficiently entering CHO cells through CD38). On the contrary, like other RNA viruses, MV has a high mutation rate, estimated at 9×10^{-5} per base per replication giving a genomic mutation rate of 1.43 per replication *(72)*. Therefore, a stock of MVs derived from a single infectious

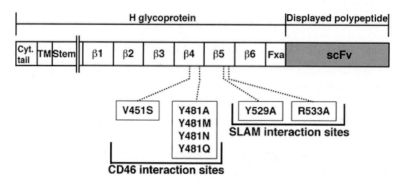

Fig. 2. Schematic representation of chimeric H protein of MV-Edm showing H residues mediating CD46 or SLAM interactions. The single-chain antibody is displayed as a C-terminal extension of H glycoprotein. cyt., cytoplasmic; TM, transmembrane; Fxa, factor Xa cleavage site. Abbreviations for the amino acid residues are as follows: A, Ala; M, Met; N, Asn; Q, Gln; R, Arg; S, Ser; V, Val; Y, Tyr.

unit is not clonal but consists of a diverse population of genetic microvariants known as quasispecies. It is therefore expected that a stock of "retargeted" MVs may contain occasional revertants capable of using the original virus receptors, CD46 or SLAM. To determine whether receptor revertants have a selective advantage, we subjected the retargeted viruses to serial passage on Vero-CD38 cells that express abundant native MV receptors (CD46) as well as the retargeted receptors (CD38). Even after multiple serial passages on Vero-CD38 cells, the host-range properties of the retargeted viruses did not change. They did not lose the ability to utilize their targeted receptors nor did they reacquire the ability to enter cells through CD46.

3.3. Versatile Six-Histidine Tagging and Retargeting System for Retargeted MVs

More recently, a versatile method was developed to rescue and propagate retargeted viruses without having to go through the slow process of generating an antigen-specific virus rescue cell line. We generated Vero-αHis cells expressing membrane-anchored single-chain antibody that recognizes a six-histidine (H6) peptide. Viruses incorporating H6 peptide at the C terminus of their ablated H proteins could be rescued and propagated on Vero-αHis cells under conditions where the interaction between H and its native receptor CD46 was absent (Fig. 3). We named this rescue system: six-histidine tagging and retargeting (STAR) (73). Tumor-selective scFvs against CD38, epidermal growth factor receptor (EGFR), or the tumor-associated mutant form of EGFR, EGFRvIII, were inserted as *Sfi*I/*Not*I fragments into a shuttle plasmid for display as C-terminal extensions of a doubly ablated H protein. Subsequently, a *Pac*I/*Spe*I-digested fragment encoding the targeted *H* gene was inserted into the corresponding sites of a full-length infectious measles clone. For rescue of retargeted virus, a well-established system (70) was modified by using Vero-αHis cells. 293–3–46 cells, which stably express phage T7 RNA polymerase and measles N and P proteins, were transfected with each recombinant MV genomic cDNA plasmid and measles L-encoding plasmid. Seventy-two hours after transfection, the cells were harvested and overlaid on Vero-αHis cells. MV antigenomic RNAs were transcribed through the T7 promoter, and the interaction between

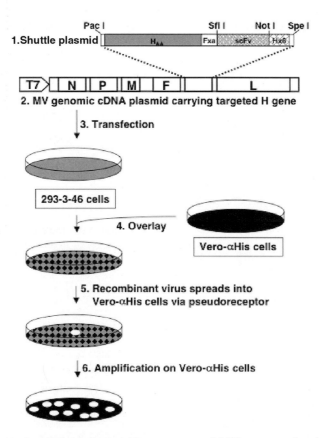

Fig. 3. Schematic representation of recombinant retargeted MV genome showing the mutated H protein and STAR system. The single-chain antibody, which is flanked by *Sfi*I (AAQPA)/*Not*I (AAA) restriction sites, is displayed as a C-terminal extension of the ablated H glycoprotein. White triangle and circle represent Y481A and R533A mutations in H protein. Fxa, factor Xa cleavage site (IEGR); H6, six-histidine peptide (HHHHHH). T7 indicates T7 promoter.

the transcribed negative-strand RNAs and N, P, and L proteins led to virus rescue in the transfected 293–3–46 cells. The rescued viruses were able to cause syncytia in the Vero-αHis cells through His tag binding to its pseudoreceptor. When individual syncytia were seen in the overlays, they were picked up and used to infect new Vero-αHis cells for amplification. All viruses grew efficiently on Vero-αHis cells, achieving approximately the same titer as the parental MV with unmodified H. In addition, all the fully retargeted viruses showed the expected host-range properties (CD46 and SLAM blind but efficiently entering cells through their respective targeted receptors). More importantly, the targeted viruses demonstrated specific receptor-mediated anti-tumor activities that were comparable with the parental MV when administered intratumorally or intravenously to mice bearing human tumor xenografts. Thus, recombinant MV-Edms rescued using the STAR system demonstrated efficient and specific infection and oncolytic properties even when receptor usage was fully retargeted by displaying different kinds of scFvs at the C terminus of a receptor blind H_{AA} protein. The STAR system therefore provides a versatile platform for targeting oncolytic MVs to any desired receptor.

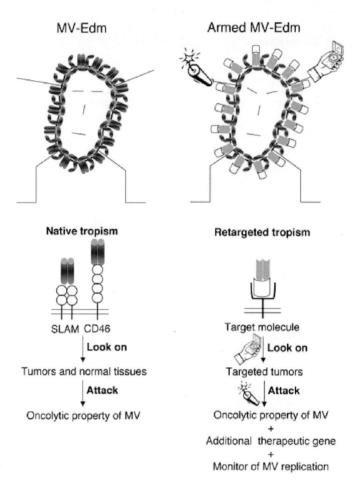

Fig. 4. Concept of targeted, armed, and trackable MV-Edm for cancer virotherapy. MV-Edm containing H protein binds to cells through CD46 and SLAM native receptors. The high CD46 receptor density on tumor cells leads to preferential infection and killing by oncolytic MV-Edm. In case of the targeted, armed, and trackable MV-Edm, incorporation of targeting polypeptide ligand in the C terminus of the blind H protein permits binding to a novel target receptor on the cell surface. The receptor-mediated oncolytic activity is enhanced by additional therapeutic genes such as *NIS* and cytokine genes. Viral replication and distribution can be noninvasively monitored through CEA or NIS incorporation into the virus genome.

3.4. Potential Targeting Ligands for Construction of Fully Retargeted MVs

Polypeptide ligands that have been successfully displayed on the H protein of MVs are listed in Table. 1 These tumor-targeting ligands belong to widely differing receptor families. Echistatin is a naturally occurring $\alpha V\beta 3$ integrin-binding snake venom peptide and shows 100-fold higher binding affinity for $\alpha V\beta 3$ than short arginine-glycine-aspartate (RGD)-containing peptides *(74,75)*. The single-domain growth factors EGF and insulin-like growth factor 1 (IGF1) bind to cell-surface tyrosine kinase receptors EGFR and type I IGF receptor (IGFR), respectively *(76)*. So far, six different single-chain antibodies have been displayed on recombinant MVs, each are recognizing a distinct target. CD20 is a cell-surface protein containing four membrane-spanning

Table 1
Ligands and Target Molecules for Retargeted Measles Viruses

Ligand	Class	Size (aa)	Target	H platform	Reference
His tag	Short	6	αHis scFv	H, H$_{AA}$	Nakamura et al. (73)
Echistatin	peptide	49	α$_v$β$_3$ integrin	H	Hallak et al. (83)
EGF	Growth	53	EGFR	H	Schneider et al. (84)
IGF1	factor	70	IGFR	H	Schneider et al. (84)
αCD20	Single-chain		CD20	H	Bucheit et al. (85)
αCD38	antibody		CD38	H, H$_{AA}$	Peng (86)/
	(scFv)				Nakamura (73)
αCEA		−250	CEA	H, H$_{AA}$	Nakamura (69)/
					Hammond (88)
αEGFR			EGFR	H$_{AA}$	Nakamura (73)/
					Hadac (71)
αEGFRvIII			EGFRvIII	H$_{AA}$	Nakamura et al. (73)
2C-TCR	scTCR	−250	SIYRYYGL/	H	Peng et al. (87)
			mouse K(b)		

regions, with N- and C-terminal cytoplasmic domains (77). CD38 is a 45-kDa type II transmembrane glycoprotein with NAD(P)+glycohydrolase and cell-signaling activity (78). CEA is a type II cell-surface glycoprotein and is a member of larger immunoglobulin supergene family (79). EGFR is a type I membrane glycoprotein that binds EGF (80), and the mutant EGFRvIII is a deletion mutation of EGFR lacking a portion of the extracellular domain, resulting in a ligand-independent and constitutive tyrosine kinase activity with transforming activity (81). In contrast to monoclonal antibodies, T-cell receptors (TCRs) determine the specificity of T-cell recognition by binding to peptide fragments of intracellular proteins presented at the cell surface in association with molecules of the major histocompatibility complex (MHC) (82).

4. CONCLUSION

The STAR virus rescue system enables the rescue and propagation of retargeted oncolytic MVs capable of entering and killing tumor cells through a broad array of cell-surface antigens. These engineered tropisms are stably maintained during multiple serial virus passages without reversion to native receptor usage. Thus, the available data suggest that receptor choice is not a significant limitation for targeting oncolytic measles virotherapy, and the use of single-chain antibodies or polypeptides for targeting potentially allows us to redirect the virus against any chosen cellular receptor. This flexibility of genetically modified MV for ligand-directed targeting is superior to other viral vectors (25).

Attenuated measles has a number of additional advantages compared with other oncolytic viruses. For example, oncolytic MVs kill their target cells by the induction of cell fusion, which is a distinct killing mechanism compared with most other oncolytic viruses. Also, the genomes of recombinant MVs can be easily expanded to incorporate additional therapeutic genes and/or genes that can be used for non-invasive imaging.

The expanded recombinant genomes are very stable during virus growth and do not easily rearrange or delete their foreign transgenes.

With regard to the immune system, most people have anti-measles antibodies, and the vaccination status of medical personnel is routinely tested such that careers are not at risk from environmental exposure to the therapeutic agent. There is therefore an urgent need for studies that evaluate the oncolytic potential of MV in the presence of an intact immune system and, perhaps more importantly, in the presence of pre-existing antiviral immunity. Immune-mediated destruction of virus-infected cells may lead to enhancement of the antitumoral effect. Conversely, neutralization of free virus by circulating antibody and rapid elimination of virus-infected cells may negate the potential therapeutic benefit.

The first phase I clinical trial of an oncolytic MV expressing a foreign transgene (MV-CEA) has been opened at the Mayo Clinic, Rochester, MN. Meanwhile, new and improved (targeted, armed, and trackable) MVs are being developed and tested in the laboratory. During the next few years, we expect to learn a great deal about the activity and toxicity profiles of these promising novel agents and how best to use them in the never-ending war on cancer.

ACKNOWLEDGMENTS

This work is supported by NIH grants CA100634 and HL66958 and the Harold W. Siebens Foundation.

REFERENCES

1. Kirn D, Martuza RL, Zwiebel J. Replication-selective virotherapy for cancer: biological principles, risk management and future directions. *Nat Med* 2001;7(7):781–7.
2. Russell SJ. RNA viruses as virotherapy agents. *Cancer Gene Ther* 2002;9(12):961–6.
3. Peng KW, Ahmann GJ, Pham L, Greipp PR, Cattaneo R, Russell SJ. Systemic therapy of myeloma xenografts by an attenuated measles virus. *Blood* 2001;98(7):2002–7.
4. Peng KW, Teneyck CJ, Galanis E, Kalli KR, Hartmann LC, Russell SJ. Intraperitoneal therapy of ovarian cancer using an engineered measles virus. *Cancer Res* 2002;62(16):4656–62.
5. Phuong LK, Allen C, Peng KW, et al. Use of a vaccine strain of measles virus genetically engineered to produce carcinoembryonic antigen as a novel therapeutic agent against glioblastoma multiforme. *Cancer Res* 2003;63(10):2462–9.
6. Grote D, Russell SJ, Cornu TI, et al. Live attenuated measles virus induces regression of human lymphoma xenografts in immunodeficient mice. *Blood* 2001;97(12):3746–54.
7. Hara T, Suzuki Y, Semba T, Hatanaka M, Matsumoto M, Seya T. High expression of membrane cofactor protein of complement (CD46) in human leukaemia cell lines: implication of an alternatively spliced form containing the STA domain in CD46 up-regulation. *Scand J Immunol* 1995;42(6): 581–90.
8. Seya T, Hara T, Matsumoto M, Akedo H. Quantitative analysis of membrane cofactor protein (MCP) of complement. High expression of MCP on human leukemia cell lines, which is down-regulated during cell differentiation. *J Immunol* 1990;145(1):238–45.
9. Yamakawa M, Yamada K, Tsuge T, et al. Protection of thyroid cancer cells by complement-regulatory factors. *Cancer* 1994;73(11):2808–17.
10. Thorsteinsson L, O'Dowd GM, Harrington PM, Johnson PM. The complement regulatory proteins CD46 and CD59, but not CD55, are highly expressed by glandular epithelium of human breast and colorectal tumour tissues. *APMIS* 1998;106(9):869–78.
11. Gorter A, Blok VT, Haasnoot WH, Ensink NG, Daha MR, Fleuren GJ. Expression of CD46, CD55, and CD59 on renal tumor cell lines and their role in preventing complement-mediated tumor cell lysis. *Lab Invest* 1996;74(6):1039–49.

12. Juhl H, Helmig F, Baltzer K, Kalthoff H, Henne-Bruns D, Kremer B. Frequent expression of complement resistance factors CD46, CD55, and CD59 on gastrointestinal cancer cells limits the therapeutic potential of monoclonal antibody 17–1A. *J Surg Oncol* 1997;64(3):222–30.

13. Kinugasa N, Higashi T, Nouso K, et al. Expression of membrane cofactor protein (MCP, CD46) in human liver diseases. *Br J Cancer* 1999;80(11):1820–5.

14. Murray KP, Mathure S, Kaul R, et al. Expression of complement regulatory proteins-CD 35, CD 46, CD 55, and CD 59-in benign and malignant endometrial tissue. *Gynecol Oncol* 2000;76(2):176–82.

15. Simpson KL, Jones A, Norman S, Holmes CH. Expression of the complement regulatory proteins decay accelerating factor (DAF, CD55), membrane cofactor protein (MCP, CD46) and CD59 in the normal human uterine cervix and in premalignant and malignant cervical disease. *Am J Pathol* 1997;151(5):1455–67.

16. Bjorge L, Hakulinen J, Wahlstrom T, Matre R, Meri S. Complement-regulatory proteins in ovarian malignancies. *Int J Cancer* 1997;70(1):14–25.

17. Blok VT, Daha MR, Tijsma OM, Weissglas MG, van den Broek LJ, Gorter A. A possible role of CD46 for the protection in vivo of human renal tumor cells from complement-mediated damage. *Lab Invest* 2000;80(3):335–44.

18. Fishelson Z, Donin N, Zell S, Schultz S, Kirschfink M. Obstacles to cancer immunotherapy: expression of membrane complement regulatory proteins (mCRPs) in tumors. *Mol Immunol* 2003;40(2–4):109–23.

19. Dorig RE, Marcil A, Chopra A, Richardson CD. The human CD46 molecule is a receptor for measles virus (Edmonston strain). *Cell* 1993;75(2):295–305.

20. Naniche D, Varior-Krishnan G, Cervoni F, et al. Human membrane cofactor protein (CD46) acts as a cellular receptor for measles virus. *J Virol* 1993;67(10):6025–32.

21. Tatsuo H, Ono N, Tanaka K, Yanagi Y. SLAM (CDw150) is a cellular receptor for measles virus. *Nature* 2000;406(6798):893–7.

22. Hsu EC, Iorio C, Sarangi F, Khine AA, Richardson CD. CDw150(SLAM) is a receptor for a lymphotropic strain of measles virus and may account for the immunosuppressive properties of this virus. *Virology* 2001;279(1):9–21.

23. Minagawa H, Tanaka K, Ono N, Tatsuo H, Yanagi Y. Induction of the measles virus receptor SLAM (CD150) on monocytes. *J Gen Virol* 2001;82(Pt 12):2913–7.

24. Anderson BD, Nakamura T, Russell SJ, Peng KW. High CD46 receptor density determines preferential killing of tumor cells by oncolytic measles virus. *Cancer Res* 2004;64(14):4919–26.

25. Wickham TJ. Ligand-directed targeting of genes to the site of disease. *Nat Med* 2003;9(1):135–9.

26. Mizuguchi H, Hayakawa T. Targeted adenovirus vectors. *Hum Gene Ther* 2004;15(11):1034–44.

27. Ruoslahti E. Vascular zip codes in angiogenesis and metastasis. *Biochem Soc Trans* 2004;32 (Pt 3):397–402.

28. Zurita AJ, Arap W, Pasqualini R. Mapping tumor vascular diversity by screening phage display libraries. *J Control Release* 2003;91(1–2):183–6.

29. Takahashi S, Mok H, Parrott MB, et al. Selection of chronic lymphocytic leukemia binding peptides. *Cancer Res* 2003;63(17):5213–7.

30. Work LM, Nicklin SA, Brain NJ, et al. Development of efficient viral vectors selective for vascular smooth muscle cells. *Mol Ther* 2004;9(2):198–208.

31. Muller OJ, Kaul F, Weitzman MD, et al. Random peptide libraries displayed on adeno-associated virus to select for targeted gene therapy vectors. *Nat Biotechnol* 2003;21(9):1040–6.

32. Perabo L, Buning H, Kofler DM, et al. In vitro selection of viral vectors with modified tropism: the adeno-associated virus display. *Mol Ther* 2003;8(1):151–7.

33. Weber E, Anderson WF, Kasahara N. Recent advances in retrovirus vector-mediated gene therapy: teaching an old vector new tricks. *Curr Opin Mol Ther* 2001;3(5):439–53.

34. Lavillette D, Russell SJ, Cosset FL. Retargeting gene delivery using surface-engineered retroviral vector particles. *Curr Opin Biotechnol* 2001;12(5):461–6.

35. Peng KW, Russell SJ. Viral vector targeting. *Curr Opin Biotechnol* 1999;10(5):454–7.

36. Tai CK, Logg CR, Park JM, Anderson WF, Press MF, Kasahara N. Antibody-mediated targeting of replication-competent retroviral vectors. *Hum Gene Ther* 2003;14(8):789–802.

37. Laquerre S, Anderson DB, Stolz DB, Glorioso JC. Recombinant herpes simplex virus type 1 engineered for targeted binding to erythropoietin receptor-bearing cells. *J Virol* 1998;72(12):9683–97.

38. Griffin DE. Measles virus. In: Knipe DM, Howley PM, eds. *Fields Virology*. Philadelphia, PA: Lippincott Williams & Wilkins; 2001:1402–42.

39. Cathomen T, Naim HY, Cattaneo R. Measles viruses with altered envelope protein cytoplasmic tails gain cell fusion competence. *J Virol* 1998;72(2):1224–34.
40. Wild TF, Fayolle J, Beauverger P, Buckland R. Measles virus fusion: role of the cysteine-rich region of the fusion glycoprotein. *J Virol* 1994;68(11):7546–8.
41. Oldstone MB. Measles virus. In: Oldstone MB, ed. *Viruses, Plagues and History*. New York: Oxford Press; 1998.
42. Cutts FT, Markowitz LE. Successes and failures in measles control. *J Infect Dis* 1994;170(Suppl 1): S32–41.
43. Bluming AZ, Ziegler JL. Regression of Burkitt's lymphoma in association with measles infection. *Lancet* 1971;2(7715):105–6.
44. Yanagi Y. The cellular receptor for measles virus–elusive no more. *Rev Med Virol* 2001;11(3): 149–56.
45. Schneider U, von Messling V, Devaux P, Cattaneo R. Efficiency of measles virus entry and dissemination through different receptors. *J Virol* 2002;76(15):7460–7.
46. Hsu EC, Sarangi F, Iorio C, et al. A single amino acid change in the hemagglutinin protein of measles virus determines its ability to bind CD46 and reveals another receptor on marmoset B cells. *J Virol* 1998;72(4):2905–16.
47. Santiago C, Bjorling E, Stehle T, Casasnovas JM. Distinct kinetics for binding of the CD46 and SLAM receptors to overlapping sites in the measles virus hemagglutinin protein. *J Biol Chem* 2002;277(35):32294–301.
48. Katze MG, He Y, Gale M, Jr. Viruses and interferon: a fight for supremacy. *Nat Rev Immunol* 2002;2(9):675–87.
49. Stojdl DF, Lichty BD, Tenoever BR, et al. VSV strains with defects in their ability to shutdown innate immunity are potent systemic anti-cancer agents. *Cancer Cell* 2003;4(4):263–75.
50. Balachandran S, Barber GN. Vesicular stomatitis virus (VSV) therapy of tumors. *IUBMB Life* 2000;50(2):135–8.
51. Strong JE, Coffey MC, Tang D, Sabinin P, Lee PW. The molecular basis of viral oncolysis: usurpation of the Ras signaling pathway by reovirus. *EMBO J* 1998;17(12):3351–62.
52. Norman KL, Lee PW. Reovirus as a novel oncolytic agent. *J Clin Invest* 2000;105(8):1035–8.
53. Farassati F, Yang AD, Lee PW. Oncogenes in Ras signalling pathway dictate host-cell permissiveness to herpes simplex virus 1. *Nat Cell Biol* 2001;3(8):745–50.
54. Varghese S, Rabkin SD. Oncolytic herpes simplex virus vectors for cancer virotherapy. *Cancer Gene Ther* 2002;9(12):967–78.
55. Peng KW, Facteau S, Wegman T, O'Kane D, Russell SJ. Non-invasive in vivo monitoring of trackable viruses expressing soluble marker peptides. *Nat Med* 2002;8(5):527–31.
56. Dingli D, Peng KW, Harvey ME, et al. Image-guided radiovirotherapy for multiple myeloma using a recombinant measles virus expressing the thyroidal sodium iodide symporter. *Blood* 2004;103(5):1641–6.
57. Spitzweg C, Zhang S, Bergert ER, et al. Prostate-specific antigen (PSA) promoter-driven androgen-inducible expression of sodium iodide symporter in prostate cancer cell lines. *Cancer Res* 1999;59(9):2136–41.
58. Cho JY, Shen DH, Yang W, et al. In vivo imaging and radioiodine therapy following sodium iodide symporter gene transfer in animal model of intracerebral gliomas. *Gene Ther* 2002;9(17):1139–45.
59. Simpkin DJ, Mackie TR. EGS4 Monte Carlo determination of the beta dose kernel in water. *Med Phys* 1990;17(2):179–86.
60. Bramson JL, Hitt M, Addison CL, Muller WJ, Gauldie J, Graham FL. Direct intratumoral injection of an adenovirus expressing interleukin-12 induces regression and long-lasting immunity that is associated with highly localized expression of interleukin-12. *Hum Gene Ther* 1996;7(16):1995–2002.
61. Andreansky S, He B, van Cott J, et al. Treatment of intracranial gliomas in immunocompetent mice using herpes simplex viruses that express murine interleukins. *Gene Ther* 1998;5(1):121–30.
62. D'Angelica M, Karpoff H, Halterman M, et al. In vivo interleukin-2 gene therapy of established tumors with herpes simplex amplicon vectors. *Cancer Immunol Immunother* 1999;47(5):265–71.
63. Kutubuddin M, Federoff HJ, Challita-Eid PM, et al. Eradication of pre-established lymphoma using herpes simplex virus amplicon vectors. *Blood* 1999;93(2):643–54.
64. Trudel S, Trachtenberg J, Toi A, et al. A phase I trial of adenovector-mediated delivery of interleukin-2 (AdIL-2) in high-risk localized prostate cancer. *Cancer Gene Ther* 2003;10(10):755–63.

65. Wong RJ, Patel SG, Kim S, et al. Cytokine gene transfer enhances herpes oncolytic therapy in murine squamous cell carcinoma. *Hum Gene Ther* 2001;12(3):253–65.

66. Grote D, Cattaneo R, Fielding AK. Neutrophils contribute to the measles virus-induced antitumor effect: enhancement by granulocyte macrophage colony-stimulating factor expression. *Cancer Res* 2003;63(19):6463–8.

67. von Messling V, Zimmer G, Herrler G, Haas L, Cattaneo R. The hemagglutinin of canine distemper virus determines tropism and cytopathogenicity. *J Virol* 2001;75(14):6418–27.

68. Xie M, Tanaka K, Ono N, Minagawa H, Yanagi Y. Amino acid substitutions at position 481 differently affect the ability of the measles virus hemagglutinin to induce cell fusion in monkey and marmoset cells co-expressing the fusion protein. *Arch Virol* 1999;144(9):1689–99.

69. Nakamura T, Peng KW, Vongpunsawad S, et al. Antibody-targeted cell fusion. *Nat Biotechnol* 2004;22(3):331–6.

70. Radecke F, Spielhofer P, Schneider H, et al. Rescue of measles viruses from cloned DNA. *EMBO J* 1995;14(23):5773–84.

71. Hadac EM, Peng KW, Nakamura T, Russell SJ. Reengineering paramyxovirus tropism. *Virology* 2004;329(2):217–25.

72. Schrag SJ, Rota PA, Bellini WJ. Spontaneous mutation rate of measles virus: direct estimation based on mutations conferring monoclonal antibody resistance. *J Virol* 1999;73(1):51–4.

73. Nakamura T, Peng KW, Harvey M, et al. Rescue and propagation of fully retargeted oncolytic measles viruses. *Nat Biotechnol* 2005;23(2):209–14.

74. Thibault G. Sodium dodecyl sulfate-stable complexes of echistatin and RGD-dependent integrins: a novel approach to study integrins. *Mol Pharmacol* 2000;58(5):1137–45.

75. Kumar CC, Nie H, Rogers CP, et al. Biochemical characterization of the binding of echistatin to integrin alphavbeta3 receptor. *J Pharmacol Exp Ther* 1997;283(2):843–53.

76. Ward CW, Garrett TP. Structural relationships between the insulin receptor and epidermal growth factor receptor families and other proteins. *Curr Opin Drug Discov Dev* 2004;7(5):630–8.

77. Einfeld DA, Brown JP, Valentine MA, Clark EA, Ledbetter JA. Molecular cloning of the human B cell CD20 receptor predicts a hydrophobic protein with multiple transmembrane domains. *EMBO J* 1988;7(3):711–7.

78. Mehta K, Shahid U, Malavasi F. Human CD38, a cell-surface protein with multiple functions. *FASEB J* 1996;10(12):1408–17.

79. Thomas P, Toth CA, Saini KS, Jessup JM, Steele G, Jr. The structure, metabolism and function of the carcinoembryonic antigen gene family. *Biochim Biophys Acta* 1990;1032(2–3):177–89.

80. Carpenter G. Receptor tyrosine kinase substrates: src homology domains and signal transduction. *FASEB J* 1992;6(14):3283–9.

81. Kuan CT, Wikstrand CJ, Bigner DD. EGF mutant receptor vIII as a molecular target in cancer therapy. *Endocr Relat Cancer* 2001;8(2):83–96.

82. Germain RN. The T cell receptor for antigen: signaling and ligand discrimination. *J Biol Chem* 2001;276(38):35223–6.

83. Hallak LK, Merchan JR, Storgard CM, Loftus JC, Russell SJ. Targeted measles virus vector displaying echistatin infects endothelial cells via alpha(v)beta3 and leads to tumor regression. *Cancer Res* 2005;65(12):5292–300.

84. Schneider U, Bullough F, Vongpunsawad S, Russell SJ, Cattaneo R. Recombinant measles viruses efficiently entering cells through targeted receptors. *J Virol* 2000;74(21):9928–36.

85. Bucheit AD, Kumar S, Grote DM, et al. An oncolytic measles virus engineered to enter cells through the CD20 antigen. *Mol Ther* 2003;7(1):62–72.

86. Peng KW, Donovan KA, Schneider U, Cattaneo R, Lust JA, Russell SJ. Oncolytic measles viruses displaying a single-chain antibody against CD38, a myeloma cell marker. *Blood* 2003;101(7): 2557–62.

87. Peng KW, Frenzke M, Myers R, et al. Biodistribution of oncolytic measles virus after intraperitoneal administration into Ifnar-CD46Ge transgenic mice. *Hum Gene Ther* 2003;14(16):1565–77.

88. Hammond AL, Plemper RK, Zhang J, Schneider U, Russell SJ, Cattaneo R. Single-chain antibody displayed on a recombinant measles virus confers entry through the tumor-associated carcinoembryonic antigen. *J Virol* 2001;75(5):2087–96.

19 Vaccines as Targeted Cancer Therapy

Miguel-Angel Perales, MD,
Jedd D. Wolchok, MD, PhD,
and Howard L. Kaufman, MD

SUMMARY

The identification of antigens on tumor cells has led to significant contributions to the field of immunotherapy. One of the most active areas under investigation in cancer immunotherapy is the development of vaccines against melanoma antigens. Induction of immunity against tumor antigens can follow multiple routes using different mechanisms. Crucial to the development of active immunization and other immunotherapies are the discovery and understanding of the molecular identity of tumor antigens and the mechanisms involved in tumor immunity. In this chapter, we will discuss strategies to induce active immunity against tumors, using melanoma as a paradigm for targeted vaccine therapy.

Key Words: Tumor; vaccine; melanoma; clinical trials.

1. INTRODUCTION

Cancer poses a difficult problem for immunotherapy, because it arises from the host's own tissues *(1)*. Many of the target antigens are tissue-specific molecules shared by cancer cells and normal cells. Thus, antigens are usually weakly immunogenic and do not typically elicit immunity because of immunologic ignorance and/or tolerance. In addition, tumors have several features that help them avoid recognition and destruction by the immune system. Despite these obstacles, several strategies for developing effective tumor immunity have been developed, including vaccines targeted to specific antigens with or without other adjuvants that can manipulate the type and quality of immune response. Crucial to these approaches is the discovery and understanding of the molecular identity of tumor-associated antigens and the mechanisms involved in tumor immunity. In this chapter, we will review attempts to induce active immunity against melanoma cells using vaccines targeted to specific antigens and will contrast these with vaccines using undefined tumor antigens (Table 1). Although this chapter

From: *Cancer Drug Discovery and Development: Molecular Targeting in Oncology*
Edited by: H. L. Kaufman, S. Wadler, and K. Antman © Humana Press, Totowa, NJ

Table 1
Types of Vaccines

Type of vaccine	Relative advantage(s)	Relative disadvantage(s)
Allogeneic cellular	Simple to prepare and broad spectrum of potential antigens	Irrelevant "allo" antigens, difficult to precisely characterize components, and requires adjuvant
Autologous cellular	Patient-specific unique antigens and presents numerous antigens	Custom-made individual vaccine production and requires adjuvant
Autologous heat shock proteins	Patient-specific unique antigens and presents numerous antigens	Custom-made individual vaccine production, production can be difficult
Purified protein or carbohydrate	Well-defined components, safety, and immunogenicity established (carbohydrates) in mature clinical trials	Production can be difficult and requires adjuvant
Peptide	Simple to prepare and safety established in early trials	Single epitope, HLA-restricted, and requires adjuvant
Dendritic cell	Inherently immunogenic and potentially numerous epitopes	Production can be difficult and limited epitopes and HLA restriction when used with peptides
Recombinant virus	Inherently immunogenic and numerous epitopes	Neutralizing immunity to vector
DNA	Simple to prepare, numerous epitopes, and immuno-stimulatory sequences in vector	Little clinical data to date

will focus on melanoma, the knowledge gained from studies using targeted melanoma vaccines are generally applicable to tumor vaccines developed against most other cancers as well.

2. UNDEFINED ANTIGEN TARGETS

2.1. Allogeneic Vaccines

Some of the earliest attempts at vaccination against melanoma used allogeneic cultured melanoma cells as vaccines based on the premise that relevant tumor-rejection antigens would be present among the thousands of different molecules injected. This strategy included whole irradiated cells (2–10), cell lysates (11–14), and shed antigens isolated from tissue culture supernatants (15–19). In addition to possible batch-to-batch variability and induction of immunity directed to irrelevant allogeneic antigens, one of the difficulties in developing these vaccines has been the inability to monitor a relevant immune response other than measuring responses against specific defined antigens. Immune responses have been noted with these vaccines against a number of melanoma antigens, such as members of the tyrosinase family of melanosomal antigens as well as the GM2 ganglioside *(19,20)*. A number of early non-randomized trials of allogeneic vaccines also suggested clinical benefit *(21,22)*. In a small randomized, placebo-controlled study of a shed antigen vaccine in 38 patients with resected high-risk melanoma, an improvement in time to progression was observed although there was no statistical improvement in overall survival *(18)*.

Canvaxin, an allogeneic whole-cell vaccine developed from three melanoma cell lines and administered with the Bacillus of Calmette and Guerin (BCG) as adjuvant, was shown to induce humoral and cellular immunity *(23)*, as well as improved overall survival in a non-randomized case–control study of 263 patients with resected melanoma *(10)*. However, two double-blind, placebo-controlled, randomized phase III trials in patients with resected American Joint Committee on Cancer (AJCC) stage III and IV melanoma were recently terminated after interim analyses failed to demonstrate a benefit.

Melacine, another allogeneic cellular vaccine, has been approved for use in Canada and Australia. The vaccine was compared to the Dartmouth chemotherapy regimen (cisplatin, carmustine, dacarbazine, and tamoxifen) in patients with metastatic melanoma. Response rates in both arms were low (13% in the Dartmouth arm versus 9% in the Melacine arm), and there was no difference in median survival. Treatment with Melacine resulted in far less toxicity than the combination chemotherapy *(24)*. Melacine has also been investigated in a phase III trial as adjuvant therapy in patients with stage II melanoma *(7)*. There was no difference in disease-free survival between vaccine and placebo groups. However, in a subset of patients who were Human leukocyte antigen (HLA)-A2 and/or HLA-C3—two of the allelotypes expressed by the tumor cells making up the vaccine—there was a statistically significant improvement in relapse-free survival (RFS) for vaccinated patients *(8)*. These results suggest that melanoma peptides expressed in Melacine may be presented by (HLA)-A2 and HLA-C3, and the immune response induced may prevent relapse in immunized patients. A second phase III trial is planned in which stage II patients who are free of disease after surgical resection and who are HLA-A2 or HLA-C3 positive will be randomized to receive vaccine or observation alone. This trial should serve to test directly the hypothesis that this vaccine can improve RFS after complete surgical resection among stage II patients expressing either HLA-A2 or HLA-C3.

In an attempt to increase the immunogenicity of allogeneic cellular vaccines, investigators transfected tumor cells with cytokine genes *(25,26)* or infected them with a cytolytic virus (vaccinia) and used the viral lysate to immunize. The rationale for the latter strategy is that highly immunogenic vaccinia proteins will function as an immunological adjuvant. Two randomized trials of this type of vaccine, in which AJCC stage III melanoma patients received a vaccinia oncolysate, did not demonstrate a significant survival benefit *(13,27)*. In the latter trial, there was a trend in favor of improved RFS, but it did not reach statistical significance *(13)*. The concept of using viruses as melanoma vaccine delivery systems remains appealing and is described below in the context of defined antigen systems.

2.2. Autologous Vaccines

Autologous tumor vaccines have several potential advantages over allogeneic vaccines. They are more likely to contain antigens of specific immunologic importance for the individual patient including unique antigens resulting from tumor cell-specific mutations and do not contain irrelevant allogeneic antigens. This approach requires a relatively large amount of tumor tissue for preparation of the customized vaccines, which restricts the eligible patient population and skews the results by including patients who have a relatively higher burden of disease.

To increase the immunogenicity of autologous tumor vaccines, Berd and colleagues have developed haptenated autologous vaccines for melanoma (M-Vax) as well as other epithelial malignancies (O-Vax). This approach involves conjugation of the highly immunogenic hapten dinitrophenyl, to proteins on autologous tumor cells and injection with BCG. Although there is no data from randomized trials, the haptenated autologous vaccines are intriguing because of their ability to mediate inflammation at tumor sites distant from the point of injection *(28)*. Interestingly, development of a delayed-type hypersensitivity (DTH) response to the autologous melanoma following immunization was associated with an increase in overall survival *(29)*.

As described for allogeneic vaccines, another strategy to enhance the immuno-genicity of tumor cells is to introduce genes encoding cytokines or chemokines. In syngeneic animal models, expression of a variety of cytokines seems to enhance tumor rejection. Special interest has been focused on granulocyte-macrophage colony-stimulating factor, largely due to work by Dranoff et al. *(30)* who showed in the mouse B16 melanoma model that vaccination with syngeneic melanoma cells secreting GM-CSF stimulated a more potent and long-lasting anti-tumor immunity compared with vaccines that secreted other cytokines. Initial clinical results using autologous vaccines expressing GM-CSF have demonstrated the ability of the vaccine to induce an inflammatory response at the injection site as well as in distant metastatic lesions in patients *(31–36)*. In other studies, investigators have immunized patients with autol-ogous melanoma expressing IL-2 and IL-4 *(37,38)*. Although clinical responses have been reported, interpretation is difficult because of relatively small numbers of patients.

2.3. Heat Shock Proteins

Another recent development in the field of autologous cell vaccines has been the use of heat shock proteins (HSP) derived from patient tumors. HSPs are a family of proteins, highly conserved through evolution, which are produced by cells in response to physical, chemical, or immunological stress. HSPs have been found to function as intracellular peptide carriers. There is evidence that HSP–peptide complexes are readily taken up by dendritic cells (DCs) for presentation to naive T-cells *(39–42)*. HSP–peptide complexes are easily purified from individual tumors and theoretically represent the total set of processed peptides from that population of tumor cells. Pilot trials immunizing patients with autologous HSP–peptide complexes are currently underway for several malignancies, including melanoma *(43–45)*. Although immune responses and, in some cases, clinical responses have been observed, a number of patients who were enrolled in studies after surgical resection did not complete the treatment because insufficient amounts of vaccine were produced or they progressed prior to being treated.

3. DEFINED ANTIGEN TARGETS

In an attempt to identify antigens that could be recognized by the human immune system, Old and colleagues *(46)* used autologous serum to identify antigens on melanoma cells. Using this technique, it was possible to identify both unique antigens as well as antigens shared by more than one patient's tumor. These observations

Table 2
Antigens on Melanoma Recognized by the Immune System

Category of antigen	Example
Mutations of genes or atypical gene products	Mutated cdk4, p53, β-catenin, ras, and CDC27
Cancer-testis antigens	MAGE-1, MAGE-3, BAGE, GAGE-1, GAGE-2, NY-ESO-1, and PRAME
Differentiation antigens	Gangliosides (GM2, GD2, and GD3) Tyrosinase, tyrosinase-related protein-1 (gp75), tyrosinase-related protein-2 (TRP-2/DCT), gp100/pmel17, and Melan A/MART-1

supported the general concept that the human immune system could recognize shared antigens on tumor cells. The molecular characterization of cancer antigens recognized by autologous antibodies and T-cells has proceeded rapidly over the last decade and has had a major impact on the field of cancer immunology. Tumor antigens can be divided into three categories: (i) antigens encoded by genetic alterations and alternative transcripts, (ii) cancer-testes antigens, and (iii) differentiation antigens (Table 2).

3.1. Mutations of Genes or Atypical Gene Products

Cancer is thought to arise as a result of an accumulation of genetic mutations; however, the immune system is not always able to detect genetic change. In other cases, mutations can lead to increased immunogenicity of an epitope due to enhanced stability of peptide–MHC interactions or improved intracellular trafficking of MHC-restricted peptides. For example, point mutations in the genes encoding p53, the p16/INK4a target cyclin-dependent kinase 4 (cdk4), β-catenin, and ras can be recognized by the immune system *(47–51)*. Boon and colleagues were the first to show that unique antigens could be created by nucleotide point mutations resulting in individual amino acid changes *(52)*. Recognition of mutated epitopes appears to be possible for both CD8[+] and CD4[+] T-cells in melanoma patients. A mutation within the glycolytic enzyme triosephosphate isomerase creating a heteroclitic epitope (antigens with higher biologic potency than the native antigen), has been recognized by CD4[+] T-cells *(53)*. Another MHC class II-restricted mutated epitope came from a cell-cycle regulator, CDC27, which is part of the anaphase-promoting cell division complex *(54)*. In this case, rather than creating a new epitope, the mutation affected a post-translational modification (phosphorylation site), which led to the mutated protein being directed for degradation in the endocytic pathway where peptides could be loaded onto MHC class II molecules. Thus, mutations may not only create new epitopes for binding to MHC or for recognition by T-cell receptors, may also promote more effective processing of degradation products for loading MHC molecules.

The immune system can also see products from alternative transcripts, including those from cryptic start sites and alternative reading frames, as well as pseudogenes *(55)* and antisense strands of DNA *(56)*. The frequency of these mutations in the melanoma population, however, is not completely defined. This general category of tumor antigens represents the only truly tumor-specific targets, as expression is limited to the particular cell clones which possess the mutation. However, because these mutated proteins seem to be present only in sporadic patients, no trials using mutated tumor-rejection proteins have been conducted thus far.

3.2. Cancer-Testis Antigens

This class of antigens is not unique to cancer cells but rather is shared with germ line cells that do not express MHC molecules. These normally silent antigens are sometimes expressed in cancer cells, perhaps due to changes in transcriptional regulation, as in abnormal DNA methylation. The MAGE-1 protein, which was the first human gene product recognized by CD8$^+$ T-cells identified in a patient with cancer, is considered the prototype cancer-testis antigen *(57)*. MAGE-1 is present on approximately 40% of melanoma tumor samples. The MAGE family, and related GAGE and BAGE families, as well as the NY-ESO-1 antigen *(58)* and PRAME (Preferentially Expressed Antigen of Melanoma) *(59)* are the main cancer-testis antigens defined so far. One of the interesting aspects of NY-ESO-1 is that it was identified using the SEREX technique, a method of autologous tissue typing in which the tumor cells' total cDNA is expressed in viral clones and probed with autologous serum *(60–63)*. In trying to identify antigens expressed by tumor cells that can be recognized by autologous antibodies, SEREX has several distinct advantages such as allowing the entire cDNA repertoire to be probed and permitting rapid identification of the antigen. A growing number of antigens have been identified by SEREX on melanoma and other tumor types, including antigens previously identified using T-cell clones, such as MAGE-1 and tyrosinase. Many of the antigens identified by SEREX are also recognized by T-cells, demonstrating that SEREX may be able to identify relevant T-cell antigens as well.

The attraction of the cancer-testis antigens is that they have relatively restricted normal tissue expression, which may lead to less common or less severe autoimmune sequelae when used as targets in vaccination strategies.

3.3. Differentiation Antigens

Differentiation antigens are antigens shared by cancer cells and their normal cell counterparts *(64)*. Melanoma and normal melanocytes share the melanosomal differentiation antigens *(65)*. The two prototypes are carbohydrate antigens, particularly gangliosides, and the melanosomal membrane proteins. Melanocytes express a variety of gangliosides including GM3, GD3, GD2, GM2, and *O*-acetyl GD3. The melanosomal membrane glycoproteins include tyrosinase (product of the mouse *albino* locus), TRP-1/gp75/tyrp1 (*brown* locus), TRP-2/DCT (*slaty* locus), gp100/pmel17 (*silver* locus), and Melan A/MART-1. The melanosome membrane glycoproteins are not only recognized by antibodies but also by CD4$^+$ and CD8$^+$ T-cells *(65–68)*. In fact, differentiation antigens are the most commonly recognized and studied tumor antigens. These antigens are often expressed at different times during melanocyte differentiation, and this feature suggest that these antigens are appropriate targets for vaccine development as the antigens are absent from fully differentiated normal melanocytes.

4. VACCINES DIRECTED AT DEFINED MELANOMA ANTIGENS

4.1. Ganglioside Antigens

Gangliosides are acidic glycolipids that consist of a hydrophobic ceramide moiety that anchors the molecule into the plasma membrane exposing the immunogenic sugars. The diversity of gangliosides is expressed by the composition of the glycosidic portion of the molecule, which consists of both neutral sugars and sialic acids. GM3 and GD3 are the gangliosides expressed most abundantly on melanoma, but expression of GM3 is ubiquitous, making it a less attractive target for immunotherapy. Efforts to develop ganglioside cancer vaccines have focused on the following gangliosides, which have much more limited expression on normal tissue.

4.1.1. GM2

GM2 appears to be one of the most immunogenic gangliosides. GM2 mixed with BCG can induce anti-GM2 IgM antibodies in over 85% of patients treated. In a trial of 122 patients with resected AJCC stage III melanoma, patients were randomized to receive GM2 plus BCG or BCG alone *(69)*. Using an intent-to-treat analysis, there was no statistically significant difference in RFS between the two groups. However, excluding the six patients who had antibodies against GM2 prior to entering the trial, there was a clear survival advantage for patients immunized with GM2/BCG ($p = 0.02$). This was the first randomized data suggesting that anti-ganglioside antibodies might have an impact on the natural history of melanoma.

Newer vaccines, in which GM2 is conjugated to keyhole limpet hemocyanin (KLH) and mixed with the adjuvant QS-21 (GM2-KLH), have resulted in antibodies to GM2 in nearly 100% of patients, with almost all of them developing IgG antibodies *(70,71)*. GMK vaccine has been tested in an adjuvant phase III randomized multicenter trial (ECOG 1694), in which 880 patients were randomized to receive either high-dose interferon-α-2b or GMK vaccine *(72)*. At a median follow-up of 16 months, high-dose interferon-α-2b produced a significant increase in relapse-free and overall survival. Further follow-up will be required to determine the durability of this benefit as the results of previous studies indicating an early benefit of interferon-α adjuvant therapy have lost statistical significance over time. As the original randomized trial using GM2/BCG suggested a benefit in RFS after 12 months and there is evidence that interferon-α-2b can have an early impact on RFS, a trial was conducted combining the two *(73)*. The primary endpoint of this trial was the anti-GM2 antibody response. Interferon-a-2b did not affect the peak titers or the percent of patients responding to GM2. These results show that interferon-α-2b and the GMK vaccine can be combined without diminishing the immunogenicity of the vaccine. Combinations of interferon-α-2b and other experimental vaccines are currently being considered for randomized trials.

4.1.2. GD2

Although GD2 is less immunogenic than GM2, immunizing patients with GD2 conjugated to KLH and mixed with QS21 (GD2-KLH/QS21) has been shown to induce anti-GD2 antibodies in majority of patients *(74)*. Although the severe pain syndrome noted with the infusion of monoclonal antibody (mAb) against GM2 was not observed after immunization against GD2, it remains to be seen whether vaccination against GD2 can have an impact on the natural history of tumors expressing GD2.

4.1.3. GD3

GD3 is one of the most abundantly expressed gangliosides on melanoma cells and is expressed on virtually all melanoma tumors *(75)*. This makes it a particularly attractive immunological target. However, compared with GM2 and GD2, GD3 is far less immunogenic. Immunization of patients with melanoma cells, GD3, or congeners of GD3 mixed with BCG failed to induce antibodies against GD3 *(76)*. Nevertheless, Livingston has shown that it is possible to induce antibodies against GD3 in four of six patients immunized with GD3-lactone conjugated to KLH and mixed with the adjuvant QS-21 *(77)*.

4.1.4. ANTI-IDIOTYPIC mAb MIMICKING GANGLIOSIDE ANTIGENS

Given the limited immunogenicity of GD2 and GD3, other approaches were sought to immunize against these gangliosides. One strategy is to develop anti-idiotypic mAb vaccines. In this approach, mice are immunized with a mAb (designated Ab1) against the antigen of interest and a secondary mAb specifically against Ab1. This secondary mAb is an anti-idiotypic mAb (also designated Ab2). It is possible to identify an Ab2 antibody that binds to the antigen-binding site of Ab1 and functionally mimics the original antigen—in this case either GD2 or GD3. That is, an Ab2 can sometimes be used as a surrogate antigen in place of the original antigen. An anti-idiotypic mAb vaccine may be more immunogenic than a ganglioside, as it is a xenogeneic protein rather than a carbohydrate self-antigen. An anti-idiotypic mAb vaccine may also contain helper T-cell epitopes capable of inducing class switching to IgG antibody responses. Several investigators have developed anti-idiotypic mAb that mimic the GD2 ganglioside and can induce anti-GD2 immune responses in animals *(78–80)*. Immunization of patients with one of these anti-idiotypic mAb vaccines, designated 1A7, can induce detectable antibodies against GD2 *(81)*.

We have developed an anti-idiotypic mAb that mimics GD3, designated BEC2. In approximately 20–30% of patients immunized to date, it has been possible to detect circulating antibodies against GD3 *(82,83)*. In a recent trial of 50 patients with melanoma at high risk for recurrence who were immunized with BEC2, all patients developed detectable IgG titers against BEC2 except for one patient at the lowest BEC2 dose level. Six patients developed detectable antibody responses to GD3 *(84)*. In a second trial, melanoma patients were randomized to be immunized with either BEC2 followed by GD3-L-KLH or in the opposite order *(85)*. Overall, 10 of 24 patients (42%) developed anti-GD3 antibodies detectable by ELISA. All antibody responses were in response to the GD3-L-KLH vaccine. This study confirmed that GD3-L-KLH vaccine induces anti-GD3 antibodies but did not confirm our previous finding that BEC2 is immunogenic.

In addition, because small cell lung carcinoma (SCLC) expresses GD3, we conducted a pilot study of the BEC2 vaccine in 15 patients with SCLC (eight patients with extensive disease and seven patients with limited disease) who had a complete or partial response after initial chemoradiation therapy *(86)*. Only 33% of the patients developed detectable anti-GD3 antibodies, a proportion similar to that seen previously in melanoma patients. However, after a median follow-up of 4 years, only one of the seven patients with limited disease has relapsed with SCLC. A multicenter randomized EORTC trial was recently completed in limited disease SCLC patients *(87)*. Although no survival benefit was observed in vaccinated patients, a trend toward prolonged survival was observed in those patients (one third) who developed a humoral response.

Once immunogenic formulations have been established for vaccines against GD2 and GD3, it will be possible to test bivalent or trivalent ganglioside vaccines *(88)*.

4.2. Protein Antigens

It is thought that full T-cell tolerance to certain protein antigens does not always occur during development because these antigens are not expressed in the thymus. Therefore, some reactive T-cells are not completely deleted, although in general, it is likely that T-cells with high affinity are deleted. Thus, T-cells subject to peripheral regulation of tolerance may be better targets for vaccine development. The goal of protein and peptide vaccination is to induce activation of such T-cells in patients with cancer.

4.2.1. PROTEIN VACCINES

Although the majority of studies targeting protein antigens have focused on peptide vaccines, recent studies have investigated the use of full-length protein vaccines *(89–91)*. These studies have shown that both antibody and T-cell responses ($CD4^+$ and $CD8^+$) can be induced by these vaccines, but strong adjuvants are required (see section 4.2.2.2.).

4.2.2. PEPTIDE VACCINES

Numerous clinical trials have been reported in which patients with melanoma were immunized with peptides derived from either cancer-testis antigens *(92–94)* or differentiation antigens *(95–117)*. These studies have highlighted several important issues: (i) immunity to peptide vaccines can be enhanced by modifications in the peptide sequence and the use of adjuvants, (ii) there is significant variability in the assays used for immune monitoring, and (iii) immune responses do not always correlate with clinical responses.

4.2.2.1. Heteroclitic Peptides.
Understanding how peptides bind to HLA molecules has led to the observation that amino acids can be substituted to increase the affinity of peptide binding. Using modified or "heteroclitic" peptides to immunize can result in enhanced immunogenicity toward the native peptide *(118,119)*. One example is the gp100 209–217 peptide in which substituting methionine for threonine at position 210 results in a much more immunogenic peptide *(120)*. The modified gp100 peptide (210M) has been used in a number of the peptide vaccine trials, with T-cell responses ranging from 52–96% *(99–101, 108–110, 113,114,117)*.

4.2.2.2. Adjuvants.
Simply injecting synthetic peptides without an immunological adjuvant rarely stimulates an immune response and can result in tolerance. Some of the most potent adjuvants utilized in animal models are thought to be too toxic for routine use in human clinical trials *(121)*. A number of adjuvants, however, are commonly used in peptide vaccine trials in cancer patients. These include alum, incomplete Freund's

adjuvant (IFA), QS-21, and GM-CSF *(122)*. We recently reported the results of a randomized peptide vaccine trial comparing three adjuvants: IFA, QS-21, and GM-CSF. Immune responses, assessed by ELISPOT assays, showed that GM-CSF and QS-21 were superior to IFA in stimulating $CD8^+$ T-cell responses against a tyrosinase peptide *(106)*. GM-CSF has also been demonstrated to be an effective adjuvant by other groups *(97,105,107,108,113)*. GM-CSF is produced by monocytes/macrophages and activated T-cells *(123)*. The ability of GM-CSF to act as a growth factor to stimulate and recruit DCs *(124)*, thus augmenting the survival and density of these antigen-presenting cells (APCs), may explain its potent adjuvant role *(122)*.

In a number of studies, interleukin-2 (IL-2) *(113,114,117)* and IL-12 *(100,101,111, 112)* have been administered in conjunction with peptide vaccines. Interestingly, in a study in which the timing of low-dose IL-2 was assessed, the use of IL-2 early during the course of immunization resulted in decreased T-cell responses, suggesting that IL-2 may be inducing activation-induced cell death in reactive T-cells. This may also explain the negative results of a recently reported CALGB phase II study of a gp100 peptide vaccine in conjunction with IL-2 *(117)*. IL-12, a cytokine that acts directly on T-cells, has also been used to enhance responses to peptide vaccines although clinical results are preliminary *(100,101,111,112)*.

Finally, another approach to enhance the immunogenicity of peptide vaccines is to administer peptides bound to MHC molecules on APCs such as DCs (see section 4.2.3.).

4.2.2.3. Monitoring of Responses to Peptide Vaccines.

A number of different assays are available to monitor T-cell responses to peptide vaccines *(125–127)*. Historically, assays to measure T-cell responses were based on the lymphoproliferation assay and the limited dilution assay of cytotoxic T-lymphocytes. These assays, as well as the measurement of cytokine release by ELISA, rely on in vitro re-stimulation of T-cells. More recently, several newer assays have been developed that can detect antigen-specific responses without the need for in vitro stimulation. Three assays assess $CD8^+$ T-cell precursor frequency: the ELISPOT *(128)*, the intracellular cytokine assay, and the tetramer assay *(129)*. The first two provide functional information by detecting antigen-specific cytokine production by $CD8^+$ T-cells. The latter assay directly measures T-cells that react with particular peptide:MHC complexes, including cells that may have no functional activity. When analyzing data from different trials, it is important to note which assay is being used and whether the cells were assayed ex vivo or after one or more rounds of in vitro stimulation.

Another issue in peptide vaccine trials has been that T-cell responses do not always correlate with clinical responses *(92,96,100,101)*. In the trial reported by Marchand et al. *(92)*, 39 patients were immunized with a MAGE-3 peptide. Two patients with cutaneous metastases underwent a complete remission. However, no T-cell responses were observed in any of the patients. In contrast, in the study by Rosenberg et al. *(99)*, patients who received a gp100 peptide had a significant T-cell response (10 of 11 patients) despite the absence of clinical responses. In fact, in the same report, the patients who were treated with the gp100 peptide vaccine and high-dose IL-2 had a 42% clinical response rate in the absence of documented T-cell responses. The lack of correlation between T-cell responses and clinical responses may be due to limitations in the T-cell assays, reflect mechanisms of immune evasion, or be related to the fact that most vaccines only target a limited number of epitopes.

Peptide vaccines have been investigated in patients with melanoma for the past several years. These clinical trials have used either cancer-testes antigens or differentiation antigens. A number of studies have demonstrated the generation of T-cell responses following immunization, but several issues remain unanswered: (i) What is the optimal dose and schedule of immunization? (ii) Which is the best adjuvant? (iii) Does the use of multiple peptides lead to a broader response or is there competition between peptides? (iv) Should vaccines target CD4$^+$ as well as CD8$^+$ T-cells? (v) What are the best assays to monitor relevant immune responses (that correlate with clinical responses)? Furthermore, despite the relative simplicity of producing peptide vaccines, their use is restricted to patients with a MHC haplotype that can present the specific peptide. Many of the peptides identified to date have been those binding to the HLA-A2.1 molecule. Although this is the most common haplotype among melanoma patients (expressed in ~40% of patients), more than half of patients are ineligible for current peptide vaccine studies because of MHC restriction. Another drawback of peptide vaccines is the presentation of only a limited number of epitopes. Future studies will need to examine these questions and potentially broaden the immune response as well as the population for which vaccines are available.

4.2.3. DENDRITIC CELLS

DCs are bone marrow-derived leaucocytes that are the most potent cells for the initiation of T-cell mediated immunity (130). They are unique among APCs in that they can prime T-cells to both class I and class II MHC-restricted antigens in vivo without additional adjuvants (131–134). Various approaches to immunization against antigens present on melanoma cells have been devised to take advantage of the unique properties of DCs. Techniques have been developed to generate large quantities of DCs in vitro, which are then "loaded" with peptides of interest and re-infused into patients (132, 135–140). In addition, several approaches have been used to introduce full-length sequences containing multiple possible epitopes. In situations where MHC-restricted peptides are not defined, DCs can be exposed to tumor lysates before injection as a means of generating a response to tumor-associated antigens, or they can be transfected with total tumor RNA, allowing for endogenous production and processing of antigens. Both the lysate and RNA methods run the risk of inducing potentially toxic autoimmune responses to unknown antigens in the lysate or RNA pool. Alternative approaches have utilized transduction of DCs with retroviruses, poxviruses, or adenoviruses encoding specific antigens of interest.

A number of clinical trials using DC-based vaccination in melanoma patients have recently been reported (132,135, 138–143). One trial used immature DCs pulsed with multiple peptides or tumor lysates and showed objective tumor responses in 5 of 16 patients (135). Mature DCs pulsed with MAGE-3A1 peptide were used in a second trial in which clinical responses were seen in 6 of 11 patients, including complete resolution of individual metastases in skin, liver, lung, and lymph nodes (132). In a study comparing different routes of administration, intranodal injection of peptide-loaded DCs was more effective in inducing CD8$^+$ T-cell responses than intravenous or intradermal injections (138). In a recent phase III study comparing peptide-loaded DCs to DTIC in patients with stage IV melanoma, low clinical response rates were observed in both arms (140).

Additional studies have examined DCs loaded with autologous tumor or melanoma lysates in conjunction with low-dose IL-2 and have reported immune responses, but few clinical responses were noted *(141–143)*. Although the immune responses observed were encouraging, they must be interpreted with caution, as many of the clinical responses were mixed, and the groups of patients were relatively small.

Ex vivo expansion and loading of DCs is a labor-intensive and expensive approach. We and others are exploring in vivo strategies to exploit the power of DCs by developing means of DC recruitment to sites of vaccination. One approach is the use of GM-CSF, in either the protein or DNA formulation. Intradermal or subcutaneous delivery of GM-CSF results in infiltration of DCs to the site of application *(122,144)*. After several days, the same site is used to administer the vaccine (peptide, DNA, virus, etc), and the hypothesis is that the locally concentrated DCs will present antigen efficiently and induce an immune response to the vaccine. As described above, this strategy has been used in several pilot trials of peptide vaccines with GM-CSF protein. The use of the GM-CSF gene, however, may allow for fewer days of GM-CSF injections because of persistent local production of the cytokine, in comparison to the necessity of repeated protein injections. Preclinical animal studies have demonstrated that administration of the murine GM-CSF gene results in recruitment of epidermal DCs and acts as a potent adjuvant for both peptide and DNA vaccines *(145–147)*. Furthermore, in a recent study of a DNA vaccine targeting tyrosinase in dogs with advanced melanoma (see section 4.2.5), animals that received human GM-CSF DNA in conjunction with the vaccine had a higher overall survival than those that received the vaccine alone *(148)*.

The CC chemokine CCL21 (also known as secondary lymphoid chemokine, SLC) induces the migration of naïve T-cells and mature DCs. The constitutive expression of CCL21 within the high endothelial venules of secondary lymphoid tissue induces the co-localization of antigen-presenting DCs and T-cells and promotes T-cell activation under physiologic conditions *(149)*. When recombinant CCL21 is injected directly into established tumors, DCs and T-cells are recruited to the tumor site and significant therapeutic anti-tumor effects have been seen in murine models *(150,151)*. The use of chemokines, such as CCL21, represents another approach for manipulating the migration of DCs and activation of tumor-specific T-cells in vivo.

4.2.4. Recombinant Viral Vaccines

The use of recombinant viruses encoding melanoma-associated antigens is also an area of active investigation. In murine models, recombinant vaccinia viruses (rVVs) have been used in several studies to induce immunity to self-antigens. Vaccination with rVV encoding murine TYRP1 resulted in tumor-protective immunity and autoimmunity, whereas plasmid DNA encoding this self-antigen failed to induce an immune response or skin depigmentation *(152)*. Vaccination of B16 melanoma-bearing mice with rVV encoding murine TRP-2 led to tumor rejection in 50% of the animals *(153)*. In contrast to immunity induced by rVV encoding TYRP1 or TRP-2, CTL against murine gp100 could not be generated by immunization of mice with rVV encoding murine gp100 *(154)*. One drawback to the use of vaccinia in humans is the high prevalence of neutralizing antibodies in the adult population that has received vaccinia immunization for prevention of smallpox. This can result in immune responses to vaccinia in the absence of any response to the melanoma antigen included in the vaccine *(155)*. To circumvent this problem, the latest generation of recombinant viral vaccines uses non-replicating poxviruses (i.e., fowlpox, canarypox, and modified vaccinia Ankara) that do not elicit the same degree of neutralizing anti-viral antibody titers *(156,157)*.

Recombinant adenovirus is also being used as a vector for delivery of melanoma-associated antigens. Immunization of mice with an adenoviral vector encoding murine TYRP1 induced an immune response to a B16 melanoma challenge that was enhanced by administration of IL-2 *(158)*. Despite encouraging results in pre-clinical mouse models, immunization of patients with recombinant adenovirus expressing either human MART-1 or gp100 have so far failed to induce an immune response, even with the addition of IL-2 *(159)*.

Finally, another approach using viral vectors has been the direct transfection of tumors with a vector expressing B7.1 and/or cytokine genes, which would lead to in vivo generation of a "vaccine" *(160–162)*. The intralesional administration of a vaccinia virus expressing B7.1 in patients with melanoma resulted in elevated expression of CD8, IFN-γ, and IL-10 in stable or regressing lesions determined by quantitative real-time polymerase chain reaction (RT–PCR) compared with growing lesions *(162)*. This suggested that rVV could be safely injected into established tumors in patients, and CD8$^+$ T-cells were able to mediate tumor regression following local vaccination.

4.2.5. DNA Vaccination

With DNA vaccination, cDNA encoding the antigen of interest is cloned into a bacterial expression plasmid with a constitutively active promoter. The plasmid is injected into the skin or muscle where it is taken up by professional APCs, particularly DCs. One proposed mechanism for the activity of DNA vaccines is direct transfection of DCs by the plasmid DNA *(163)*. Despite the fact that DCs represent a very small percent of the cells in the skin or muscle, the greatly enhanced ability of DCs to present antigen allows this mechanism to be practical. An alternative mechanism of presentation has been termed cross-priming *(164,165)*. This involves transcription and translation of the antigen by non-APCs, such as keratinocytes or myocytes, and release of mature protein antigen through either secretion or cell death. The pre-formed antigen is then captured by APCs and presented to naïve T-cells in local-draining lymph nodes. Most likely, both mechanisms (direct transfection of APC and cross-priming) are operative during successful DNA immunization.

DNA immunization offers several potential advantages: (i) The presence of the full-length cDNA provides multiple potential epitopes, thus alleviating the limitations of MHC restriction, (ii) Bacterial plasmid DNA itself contains immunogenic unmethylated CpG motifs (immunostimulatory sequences) that may act as a potent immunological adjuvant *(166,167)*, and (iii) DNA is relatively simple to purify in large quantities.

Immunization of mice with the xenogeneic (human) DNA-encoding tyrosinase, TYRP1, TRP-2, or gp100 results in protection from syngeneic tumor challenge with B16 mouse melanoma as well as rapid and extensive depigmentation of coat *(145, 168–172)*. The depigmentation is a graphic demonstration of the ability of the immune system to recognize "self" proteins in mouse melanocytes.

Based on pre-clinical murine data with xenogeneic DNA immunization, clinical trials in dogs with advanced canine malignant melanoma (CMM) were conducted. Three cohorts of three dogs each with advanced (WHO stage II, III, or IV) CMM received four biweekly IM. injections (dose levels 100, 500, or 1500 µg, respectively/vaccination) of human tyrosinase plasmid DNA. Mild local reactions at injection sites were the only toxicities observed, with no signs of autoimmunity. One dog with stage IV disease had a complete clinical response in multiple lung metastases for 329 days. The Kaplan–Meier median survival time for all nine dogs was 389 days, which is significantly longer than that seen in canine historical controls *(173)*. On the basis of these studies in dogs with melanoma, human tyrosinase DNA has been conditionally licensed by the USDA for treatment of canine melanoma.

DNA vaccines have also been investigated in patients. In clinical trials for infectious disease, DNA immunization has been shown to be safe and effective in developing immune responses to malaria and human immunodeficiency virus*(174–176)*. Recently, the first human trials of DNA vaccines in patients with melanoma have also been reported *(177–180)*. In two of these studies, no immune responses to the vaccine were observed *(177,180)*. In the other two studies, immune responses were generated following intranodal injection of tyrosinase epitopes DNA *(178)* or following a prime-boost strategy in which patients were immunized with DNA for Melan-A followed by a viral vaccine expressing the same antigen *(179)*.

5. CONCLUSION

A major achievement of the field of tumor immunology in the 1980s and 1990s was the identification of a rapidly growing set of genes that encode antigens recognized by the immune system. These include mutations and even products from genomic regions that have been assumed to be inert. Another class of cancer antigens is the cancer-testes antigens that are expressed by germ cells, normally silenced in somatic cells but re-expressed by a variety of cancers. Differentiation antigens represent a third class of cancer antigens and are the most commonly recognized and studied tumor antigens. The discovery of the molecular identity of antigens and the dissection of mechanisms involved in tumor immunity and escape from immunity are crucial to the development of rational active immunization and immunotherapy. A number of vaccine strategies are currently under investigation, and some early results have been promising.

Prior to identification of specific antigens, vaccine approaches included autologous and allogeneic whole tumor cells, although no benefit of these vaccines has been seen in randomized phase III clinical trials. The availability of defined tumor antigens has led to several strategies for vaccine development, including peptide vaccines, DCs loaded with peptides or tumor-derived RNA, recombinant viral vaccines, and DNA immunization. The advances in melanoma are being replicated in patients with other types of cancer as more antigenic targets have been isolated. The ultimate role of targeted vaccines in patients with cancer awaits completion of larger, randomized clinical trials. Additional studies of vaccines in combination with other immunologic approaches, especially treatments aimed at blocking immune suppression, and with other standard cancer therapeutics are also promising areas of future investigation.

REFERENCES

1. Perales MA, Blachere NE, Engelhorn ME, et al. Strategies to overcome immune ignorance and tolerance. *Semin Cancer Biol* 2002; 12: 63–71.
2. Livingston PO, Takeyama H, Pollack MS, et al. Serological responses of melanoma patients to vaccines derived from allogeneic cultured melanoma cells. *Int J Cancer* 1983; 31: 567–575.

3. Gercovich FG, Gutterman JU, Mavligit GM, Hersh EM. Active specific immunization in malignant melanoma. *Med Pediatr Oncol* 1975; 1: 277–287.
4. Morton DL, Foshag LJ, Nizze JA, et al. Active specific immunotherapy in malignant melanoma. *Semin Surg Oncol* 1989; 5: 420–425.
5. Morton DL, Foshag LJ, Hoon DS, et al. Prolongation of survival in metastatic melanoma after active specific immunotherapy with a new polyvalent melanoma vaccine. *Ann Surg* 1992; 216: 463–482.
6. Slingluff CL, Seigler HF. Immunotherapy for malignant melanoma with a tumor cell vaccine. *Ann Plast Surg* 1992; 28: 104–107.
7. Sondak VK, Liu PY, Tuthill RJ, et al. Adjuvant immunotherapy of resected, intermediate-thickness, node-negative melanoma with an allogeneic tumor vaccine: overall results of a randomized trial of the Southwest Oncology Group. *J Clin Oncol* 2002; 20: 2058–2066.
8. Sosman JA, Unger JM, Liu PY, et al. Adjuvant immunotherapy of resected, intermediate-thickness, node-negative melanoma with an allogeneic tumor vaccine: impact of HLA class I antigen expression on outcome. *J Clin Oncol* 2002; 20: 2067–2075.
9. Morton DL, Hsueh EC, Essner R, et al. Prolonged survival of patients receiving active immunotherapy with Canvaxin therapeutic polyvalent vaccine after complete resection of melanoma metastatic to regional lymph nodes. *Ann Surg* 2002; 236: 438–448.
10. Hsueh EC, Essner R, Foshag LJ, et al. Prolonged survival after complete resection of disseminated melanoma and active immunotherapy with a therapeutic cancer vaccine. *J Clin Oncol* 2002; 20: 4549–4554.
11. Haigh PI, Difronzo LA, Gammon G, Morton DL. Vaccine therapy for patients with melanoma. *Oncology (Huntingt)* 1999; 13: 1561–1574; discussion 1574 passim.
12. Chan AD, Morton DL. Active immunotherapy with allogeneic tumor cell vaccines: present status. *Semin Oncol* 1998; 25: 611–622.
13. Hersey P, Coates AS, McCarthy WH, et al. Adjuvant immunotherapy of patients with high-risk melanoma using vaccinia viral lysates of melanoma: results of a randomized trial. *J Clin Oncol* 2002; 20: 4181–4190.
14. Mitchell MS, Kan-Mitchell J, Morrow PR, et al. Phase I trial of large multivalent immunogen derived from melanoma lysates in patients with disseminated melanoma. *Clin Cancer Res* 2004; 10: 76–83.
15. Bystryn JC, Henn M, Li J, Shroba S. Identification of immunogenic human melanoma antigens in a polyvalent melanoma vaccine. *Cancer Res* 1992; 52: 5948–5953.
16. Bystryn JC. Immunogenicity and clinical activity of a polyvalent melanoma antigen vaccine prepared from shed antigens. *Ann N Y Acad Sci* 1993; 690: 190–203.
17. Bystryn JC. Clinical activity of a polyvalent melanoma antigen vaccine. *Recent Results Cancer Res* 1995; 139: 337–348.
18. Bystryn JC, Zeleniuch-Jacquotte A, Oratz R, et al. Double-blind trial of a polyvalent, shed-antigen, melanoma vaccine. *Clin Cancer Res* 2001; 7: 1882–1887.
19. Reynolds SR, Zeleniuch-Jacquotte A, Shapiro RL, et al. Vaccine-induced CD8+ T-cell responses to MAGE-3 correlate with clinical outcome in patients with melanoma. *Clin Cancer Res* 2003; 9: 657–662.
20. Okamoto T, Irie RF, Fujii S, et al. Anti-tyrosinase-related protein-2 immune response in vitiligo patients and melanoma patients receiving active-specific immunotherapy. *J Invest Dermatol* 1998; 111: 1034–1039.
21. Huang SK, Okamoto T, Morton DL, Hoon DS. Antibody responses to melanoma/melanocyte autoantigens in melanoma patients. *J Invest Dermatol* 1998; 111: 662–667.
22. Reynolds SR, Celis E, Sette A, et al. HLA-independent heterogeneity of CD8+ T cell responses to MAGE-3, Melan-A/MART-1, gp100, tyrosinase, MC1R, and TRP-2 in vaccine-treated melanoma patients. *J Immunol* 1998; 161: 6970–6976.
23. Hsueh EC, Gupta RK, Qi K, Morton DL. Correlation of specific immune responses with survival in melanoma patients with distant metastases receiving polyvalent melanoma cell vaccine. *J Clin Oncol* 1998; 16: 2913–2920.
24. Mitchell MS. Perspective on allogeneic melanoma lysates in active specific immunotherapy. *Semin Oncol* 1998; 25: 623–635.
25. Osanto S, Schiphorst PP, Weijl NI, et al. Vaccination of melanoma patients with an allogeneic, genetically modified interleukin 2-producing melanoma cell line. *Hum Gene Ther* 2000; 11: 739–750.

26. Arienti F, Belli F, Napolitano F, et al. Vaccination of melanoma patients with interleukin 4 gene-transduced allogeneic melanoma cells. *Hum Gene Ther* 1999; 10: 2907–2916.

27. Wallack MK, Sivanandham M, Balch CM, et al. Surgical adjuvant active specific immunotherapy for patients with stage III melanoma: the final analysis of data from a phase III, randomized, double-blind, multicenter vaccinia melanoma oncolysate trial. *J Am Coll Surg* 1998; 187: 69–77.

28. Berd D, Murphy G, Maguire HC, Mastrangelo MJ. Immunization with haptenized, autologous tumor cells induces inflammation of human melanoma metastases. *Cancer Res* 1991; 51: 2731–2734.

29. Berd D, Sato T, Maguire HC, Jr. et al. Immunopharmacologic analysis of an autologous, hapten-modified human melanoma vaccine. *J Clin Oncol* 2004; 22: 403–415.

30. Dranoff G, Jaffee E, Lazenby A, et al. Vaccination with irradiated tumor cells engineered to secrete murine granulocyte-macrophage colony-stimulating factor stimulates potent, specific, and long-lasting anti-tumor immunity. *Proc Natl Acad Sci USA* 1993; 90: 3539–3543.

31. Soiffer R, Lynch T, Mihm M, et al. Vaccination with irradiated autologous melanoma cells engineered to secrete human granulocyte-macrophage colony-stimulating factor generates potent antitumor immunity in patients with metastatic melanoma. *Proc Natl Acad Sci USA* 1998; 95: 13141–13146.

32. Ellem KA, O'Rourke MG, Johnson GR, et al. A case report: immune responses and clinical course of the first human use of granulocyte/macrophage-colony-stimulating-factor-transduced autologous melanoma cells for immunotherapy. *Cancer Immunol Immunother* 1997; 44: 10–20.

33. Dranoff G, Soiffer R, Lynch T, et al. A phase I study of vaccination with autologous, irradiated melanoma cells engineered to secrete human granulocyte-macrophage colony stimulating factor. *Hum Gene Ther* 1997; 8: 111–123.

34. Chang AE, Li Q, Bishop DK, et al. Immunogenetic therapy of human melanoma utilizing autologous tumor cells transduced to secrete granulocyte-macrophage colony-stimulating factor. *Hum Gene Ther* 2000; 11: 839–850.

35. Soiffer R, Hodi FS, Haluska F, et al. Vaccination with irradiated, autologous melanoma cells engineered to secrete granulocyte-macrophage colony-stimulating factor by adenoviral-mediated gene transfer augments antitumor immunity in patients with metastatic melanoma. *J Clin Oncol* 2003; 21: 3343–3350.

36. Luiten RM, Kueter EW, Mooi W, et al. Immunogenicity, including vitiligo, and feasibility of vaccination with autologous GM-CSF-transduced tumor cells in metastatic melanoma patients. *J Clin Oncol* 2005; 23: 8978–8991.

37. Arienti F, Belli F, Napolitano F, et al. Vaccination of melanoma patients with interleukin 4 gene-transduced allogeneic melanoma cells. *Hum Gene Ther* 1999; 10: 2907–2916.

38. Maio M, Fonsatti E, Lamaj E, et al. Vaccination of stage IV patients with allogeneic IL-4- or IL-2-gene-transduced melanoma cells generates functional antibodies against vaccinating and autologous melanoma cells. *Cancer Immunol Immunother* 2002; 51: 9–14.

39. Przepiorka D, Srivastava PK. Heat shock protein—peptide complexes as immunotherapy for human cancer. *Mol Med Today* 1998; 4: 478–484.

40. Srivastava PK, Udono H. Heat shock protein-peptide complexes in cancer immunotherapy. *Curr Opin Immunol* 1994; 6: 728–732.

41. Srivastava PK. Purification of heat shock protein-peptide complexes for use in vaccination against cancers and intracellular pathogens. *Methods* 1997; 12: 165–171.

42. Blachere NE, Srivastava PK. Heat shock protein-based cancer vaccines and related thoughts on immunogenicity of human tumors. *Semin Cancer Biol* 1995; 6: 349–355.

43. Eton O, East MJ, Ross MI, et al. Autologous tumor-derived heat-shock protein peptide complex-96 in patients with metastatic melanoma. In *American Association for Cancer Research*. San Francisco, CA: 2000, 543 (Abstract) meeting proceedings.

44. Belli F, Testori A, Rivoltini L, et al. Vaccination of metastatic melanoma patients with autologous tumor-derived heat shock protein gp96-peptide complexes: clinical and immunologic findings. *J Clin Oncol* 2002; 20: 4169–4180.

45. Pilla L, Patuzzo R, Rivoltini L, et al. A phase II trial of vaccination with autologous, tumor-derived heat-shock protein peptide complexes Gp96, in combination with GM-CSF and interferon-alpha in metastatic melanoma patients. *Cancer Immunol Immunother* 2006; 55: 958–968.

46. Old and colleagues LJ. Cancer immunology: the search for specificity - G. H. A. Clowes Menorial lecture. *Cancer Res* 1981; 41: 361–375.
47. Wolfel T, Hauer M, Schneider J, et al. A p16INK4a-insensitive CDK4 mutant targeted by cytolytic T lymphocytes in a human melanoma. *Science* 1995; 269: 1281–1284.
48. Robbins PF, El-Gamil M, Li YF, et al. A mutated beta-catenin gene encodes a melanoma-specific antigen recognized by tumor infiltrating lymphocytes. *J Exp Med* 1996; 183: 1185–1192.
49. Chiari R, Foury F, De Plaen E, et al. Two antigens recognized by autologous cytolytic T lymphocytes on a melanoma result from a single point mutation in an essential housekeeping gene. *Cancer Res* 1999; 59: 5785–5792.
50. Coulie PG, Lehmann F, Lethe B, et al. A mutated intron sequence codes for an antigenic peptide recognized by cytolytic T lymphocytes on a human melanoma. *Proc Natl Acad Sci USA* 1995; 92: 7976–7980.
51. Linard B, Bezieau S, Benlalam H, et al. A ras-mutated peptide targeted by CTL infiltrating a human melanoma lesion. *J Immunol* 2002; 168: 4802–4808.
52. Lurquin C, Van Pel A, Mariame B, et al. Structure of the gene of tum- transplantation antigen P91A: the mutated exon encodes a peptide recognized with Ld by cytolytic T cells. *Cell* 1989; 58: 293–303.
53. Pieper R, Christian RE, Gonzales MI, et al. Biochemical identification of a mutated human melanoma antigen recognized by CD4(+) T cells. *J Exp Med* 1999; 189: 757–766.
54. Wang RF, Wang X, Atwood AC, et al. Cloning genes encoding MHC class II-restricted antigens: mutated CDC27 as a tumor antigen. *Science* 1999; 284: 1351–1354.
55. Moreau-Aubry A, Le Guiner S, Labarriere N, et al. A processed pseudogene codes for a new antigen recognized by a CD8(+) T cell clone on melanoma. *J Exp Med* 2000; 191: 1617–1624.
56. Van Den Eynde BJ, Gaugler B, Probst-Kepper M, et al. A new antigen recognized by cytolytic T lymphocytes on a human kidney tumor results from reverse strand transcription. *J Exp Med* 1999; 190: 1793–1800.
57. van der Bruggen P, Traversari C, Chomez P, et al. A gene encoding an antigen recognized by cytolytic T lymphocytes on a human melanoma. *Science* 1991; 254: 1643–1647.
58. Chen YT, Scanlan MJ, Sahin U, et al. A testicular antigen aberrantly expressed in human cancers detected by autologous antibody screening. *Proc Natl Acad Sci USA* 1997; 94: 1914–1918.
59. Ikeda H, Lethe B, Lehmann F, et al. Characterization of an antigen that is recognized on a melanoma showing partial HLA loss by CTL expressing an NK inhibitory receptor. *Immunity* 1997; 6: 199–208.
60. Old LJ, Chen YT. New paths in human cancer serology. *J Exp Med* 1998; 187: 1163–1167.
61. Jager E, Chen YT, Drijfhout JW, et al. Simultaneous humoral and cellular immune response against cancer-testis antigen NY-ESO-1: definition of human histocompatibility leukocyte antigen (HLA)-A2-binding peptide epitopes. *J Exp Med* 1998; 187: 265–270.
62. Jager E, Maeurer M, Hohn H, et al. Clonal expansion of Melan A-specific cytotoxic T lymphocytes in a melanoma patient responding to continued immunization with melanoma-associated peptides. *Int J Cancer* 2000; 86: 538–547.
63. Segal NH, Blachere NE, Guevara-Patino JA, et al. Identification of cancer-testis genes expressed by melanoma and soft tissue sarcoma using bioinformatics. *Cancer Immun* 2005; 5: 2.
64. Boyse EA, Old LJ. Some aspects of normal and abnormal cell surface genetics. *Ann Rev Genet* 1969; 3: 269–290.
65. Houghton AN, Eisinger M, Albino AP, et al. Surface antigens of melanocytes and melanomas. Markers of melanocyte differentiation and melanoma subsets. *J Exp Med* 1982; 156: 1755–1766.
66. Vijayasaradhi S, Bouchard B, Houghton AN. The melanoma antigen gp75 is the human homologue of the mouse b (brown) locus gene product. *J Exp Med* 1990; 171: 1375–1380.
67. Brichard V, Van Pel A, Wolfel T, et al. The tyrosinase gene codes for an antigen recognized by autologous cytolytic T lymphocytes on HLA-A2 melanomas. *J Exp Med* 1993; 178: 489–495.
68. Wang RF, Appella E, Kawakami Y, et al. Identification of TRP-2 as a human tumor antigen recognized by cytotoxic T lymphocytes. *J Exp Med* 1996; 184: 2207–2216.
69. Livingston P, Wong G, Adluri S, et al. Improved survival in stage III melanoma patients with GM2 antibodies: a randomized trial of adjuvant vaccination with GM2 ganglioside. *J Clin Oncol* 1994; 12: 1036–1044.

70. Helling F, Zhang S, Shang A, et al. GM2-KLH conjugate vaccine: increased immunogenicity in melanoma patients after administration with immunological adjuvant QS-21. *Cancer Res* 1995; 55: 2783–2788.

71. Chapman PB, Morrissey DM, Panageas KS, et al. Induction of antibodies against GM2 ganglioside by immunizing melanoma patients using GM2-keyhole limpet hemocyanin + QS21 vaccine: a dose-response study. *Clin Cancer Res* 2000; 6: 874–879.

72. Kirkwood JM, Ibrahim JG, Sosman JA, et al. High-dose interferon alfa-2b significantly prolongs relapse-free and overall survival compared with the GM2-KLH/QS-21 vaccine in patients with resected stage IIB-III melanoma: results of intergroup trial E1694/S9512/C509801. *J Clin Oncol* 2001; 19: 2370–2380.

73. Kirkwood JM, Ibrahim J, Lawson DH, et al. High-dose interferon alfa-2b does not diminish antibody response to GM2 vaccination in patients with resected melanoma: results of the Multicenter Eastern Cooperative Oncology Group Phase II Trial E2696. *J Clin Oncol* 2001; 19: 1430–1436.

74. Ragupathi G, Livingston PO, Hood C, et al. Consistent antibody response against ganglioside GD2 induced in patients with melanoma by a GD2 lactone-keyhole limpet hemocyanin conjugate vaccine plus immunological adjuvant QS-21. *Clin Cancer Res* 2003; 9: 5214–5220.

75. Hamilton WB, Helling F, Lloyd KO, Livingston PO. Ganglioside expression on human malignant melanoma assessed by quantitative immune thin-layer chromatography. *Int J Cancer* 1993; 53: 566–573.

76. Ritter G, Boosfeld E, Adluri R, et al. Antibody response to immunization with ganglioside GD3 and GD3 congeners (lactones, amide and gangliosidol) in patients with malignant melanoma. *Int J Cancer* 1991; 48: 379–385.

77. Ragupathi G, Meyers M, Adluri S, et al. Induction of antibodies against GD3 ganglioside in melanoma patients by vaccination with GD3-lactone-KLH conjugate plus immunological adjuvant QS-21. *Int J Cancer* 2000; 85: 659–666.

78. Cheung NK, Canete A, Cheung IY, et al. Disialoganglioside GD2 anti-idiotypic monoclonal antibodies. *Int J Cancer* 1993; 54: 499–505.

79. Saleh MN, Stapleton JD, Khazaeli MB, LoBuglio AF. Generation of a human anti-idiotypic antibody that mimics the GD2 antigen. *J Immunol* 1993; 151: 3390–3398.

80. Sen G, Chakraborty M, Foon KA, et al. Preclinical evaluation in nonhuman primates of murine monoclonal anti-idiotype antibody that mimics the disialoganglioside GD2. *Clin Cancer Res* 1997; 3: 1969–1976.

81. Foon KA, Lutzky J, Baral RN, et al. Clinical and immune responses in advanced melanoma patients immunized with an anti-idiotype antibody mimicking disialoganglioside GD2. *J Clin Oncol* 2000; 18: 376–384.

82. Yao TJ, Meyers M, Livingston PO, et al. Immunization of melanoma patients with BEC2-keyhole limpet hemocyanin plus BCG intradermally followed by intravenous booster immunizations with BEC2 to induce anti-GD3 ganglioside antibodies. *Clin Cancer Res* 1999; 5: 77–81.

83. McCaffery M, Yao TJ, Williams L, et al. Immunization of melanoma patients with BEC2 anti-idiotypic monoclonal antibody that mimics GD3 ganglioside: enhanced immunogenicity when combined with adjuvant. *Clin Cancer Res* 1996; 2: 679–686.

84. Chapman PB, Williams L, Salibi N, et al. A phase II trial comparing five dose levels of BEC2 anti-idiotypic monoclonal antibody vaccine that mimics GD3 ganglioside. *Vaccine* 2004; 22: 2904–2909.

85. Chapman PB, Wu D, Ragupathi G, et al. Sequential immunization of melanoma patients with GD3 ganglioside vaccine and anti-idiotypic monoclonal antibody that mimics GD3 ganglioside. *Clin Cancer Res* 2004; 10: 4717–4723.

86. Grant SC, Kris MG, Houghton AN, Chapman PB. Long survival of patients with small cell lung cancer after adjuvant treatment with the anti-idiotypic antibody BEC2 plus Bacillus Calmette-Guerin. *Clin Cancer Res* 1999; 5: 1319–1323.

87. Giaccone G, Debruyne C, Felip E, et al. Phase III study of adjuvant vaccination with Bec2/bacille Calmette-Guerin in responding patients with limited-disease small-cell lung cancer (European Organisation for Research and Treatment of Cancer 08971–08971B; Silva Study). *J Clin Oncol* 2005; 23: 6854–6864.

88. Chapman PB, Morrisey D, Panageas KS, et al. Vaccination with a bivalent G(M2) and G(D2) ganglioside conjugate vaccine: a trial comparing doses of G(D2)-keyhole limpet hemocyanin. *Clin Cancer Res* 2000; 6: 4658–4662.

89. Marchand M, Punt CJ, Aamdal S, et al. Immunisation of metastatic cancer patients with MAGE-3 protein combined with adjuvant SBAS-2: a clinical report. *Eur J Cancer* 2003; 39: 70–77.
90. Vantomme V, Dantinne C, Amrani N, et al. Immunologic analysis of a phase I/II study of vaccination with MAGE-3 protein combined with the AS02B adjuvant in patients with MAGE-3-positive tumors. *J Immunother* 2004; 27: 124–135.
91. Davis ID, Chen W, Jackson H, et al. Recombinant NY-ESO-1 protein with ISCOMATRIX adjuvant induces broad integrated antibody and CD4(+) and CD8(+) T cell responses in humans. *Proc Natl Acad Sci USA* 2004; 101: 10697–10702.
92. Marchand M, van Baren N, Weynants P, et al. Tumor regressions observed in patients with metastatic melanoma treated with an antigenic peptide encoded by gene MAGE-3 and presented by HLA-A1. *Int J Cancer* 1999; 80: 219–230.
93. Weber JS, Hua FL, Spears L, et al. A phase I trial of an HLA-A1 restricted MAGE-3 epitope peptide with incomplete Freund's adjuvant in patients with resected high-risk melanoma. *J Immunother* 1999; 22: 431–440.
94. Chianese-Bullock KA, Pressley J, Garbee C, et al. MAGE-A1-, MAGE-A10-, and gp100-derived peptides are immunogenic when combined with granulocyte-macrophage colony-stimulating factor and montanide ISA-51 adjuvant and administered as part of a multipeptide vaccine for melanoma. *J Immunol* 2005; 174: 3080–3086.
95. Salgaller ML, Marincola FM, Cormier JN, Rosenberg SA. Immunization against epitopes in the human melanoma antigen gp100 following patient immunization with synthetic peptides. *Cancer Res* 1996; 56: 4749–4757.
96. Jaeger E, Bernhard H, Romero P, et al. Generation of cytotoxic T-cell responses with synthetic melanoma-associated peptides in vivo: implications for tumor vaccines with melanoma-associated antigens. *Int J Cancer* 1996; 66: 162–169.
97. Jager E, Ringhoffer M, Dienes HP, et al. Granulocyte-macrophage-colony-stimulating factor enhances immune responses to melanoma-associated peptides in vivo. *Int J Cancer* 1996; 67: 54–62.
98. Cormier JN, Salgaller ML, Prevette T, et al. Enhancement of cellular immunity in melanoma patients immunized with a peptide from MART-1/Melan A. *Cancer J Sci Am* 1997; 3: 37–44.
99. Rosenberg SA, Yang JC, Schwartzentruber DJ, et al. Immunologic and therapeutic evaluation of a synthetic peptide vaccine for the treatment of patients with metastatic melanoma. *Nat Med* 1998; 4: 321–327.
100. Lee KH, Wang E, Nielsen MB, et al. Increased vaccine-specific T cell frequency after peptide-based vaccination correlates with increased susceptibility to in vitro stimulation but does not lead to tumor regression. *J Immunol* 1999; 163: 6292–6300.
101. Lee P, Wang F, Kuniyoshi J, et al. Effects of interleukin-12 on the immune response to a multi-peptide vaccine for resected metastatic melanoma. *J Clin Oncol* 2001; 19: 3836–3847.
102. Wang F, Bade E, Kuniyoshi C, et al. Phase I trial of a MART-1 peptide vaccine with incomplete Freund's adjuvant for resected high-risk melanoma. *Clin Cancer Res* 1999; 5: 2756–2765.
103. Zarour HM, Kirkwood JM, Kierstead LS, et al. Melan-A/MART-1(51–73) represents an immunogenic HLA-DR4-restricted epitope recognized by melanoma-reactive CD4(+) T cells. *Proc Natl Acad Sci USA* 2000; 97: 400–405.
104. Lewis JJ, Janetzki S, Schaed S, et al. Evaluation of CD8(+) T-cell frequencies by the Elispot assay in healthy individuals and in patients with metastatic melanoma immunized with tyrosinase peptide. *Int J Cancer* 2000; 87: 391–398.
105. Scheibenbogen C, Schmittel A, Keilholz U, et al. Phase 2 trial of vaccination with tyrosinase peptides and granulocyte-macrophage colony-stimulating factor in patients with metastatic melanoma. *J Immunother* 2000; 23: 275–281.
106. Schaed SG, Klimek VM, Panageas KS, et al. T-cell responses against tyrosinase 368–376(370D) peptide in HLA(*)A0201(+) melanoma patients: randomized trial comparing incomplete Freund's adjuvant, granulocyte macrophage colony-stimulating factor, and QS-21 as immunological adjuvants. *Clin Cancer Res* 2002; 8: 967–972.
107. Scheibenbogen C, Schadendorf D, Bechrakis NE, et al. Effects of granulocyte-macrophage colony-stimulating factor and foreign helper protein as immunologic adjuvants on the T-cell response to vaccination with tyrosinase peptides. *Int J Cancer* 2003; 104: 188–194.

108. Weber J, Sondak VK, Scotland R, et al. Granulocyte-macrophage-colony-stimulating factor added to a multipeptide vaccine for resected Stage II melanoma. *Cancer* 2003; 97: 186–200.

109. Smith JW, 2*nd*, Walker EB, Fox BA, et al. Adjuvant immunization of HLA-A2-positive melanoma patients with a modified gp100 peptide induces peptide-specific CD8+ T-cell responses. *J Clin Oncol* 2003; 21: 1562–1573.

110. Slingluff CL, Jr., Yamshchikov G, Neese P, et al. Phase I trial of a melanoma vaccine with gp100(280–288) peptide and tetanus helper peptide in adjuvant: immunologic and clinical outcomes. *Clin Cancer Res* 2001; 7: 3012–3024.

111. Peterson AC, Harlin H, Gajewski TF. Immunization with Melan-A peptide-pulsed peripheral blood mononuclear cells plus recombinant human interleukin-12 induces clinical activity and T-cell responses in advanced melanoma. *J Clin Oncol* 2003; 21: 2342–2348.

112. Cebon J, Jager E, Shackleton MJ, et al. Two phase I studies of low dose recombinant human IL-12 with Melan-A and influenza peptides in subjects with advanced malignant melanoma. *Cancer Immun* 2003; 3: 7.

113. Slingluff CL, Jr., Petroni GR, Yamshchikov GV, et al. Clinical and immunologic results of a randomized phase II trial of vaccination using four melanoma peptides either administered in granulocyte-macrophage colony-stimulating factor in adjuvant or pulsed on dendritic cells. *J Clin Oncol* 2003; 21: 4016–4026.

114. Slingluff CL, Jr., Petroni GR, Yamshchikov GV, et al. Immunologic and clinical outcomes of vaccination with a multiepitope melanoma peptide vaccine plus low-dose interleukin-2 administered either concurrently or on a delayed schedule. *J Clin Oncol* 2004; 22: 4474–4485.

115. Lienard D, Rimoldi D, Marchand M, et al. Ex vivo detectable activation of Melan-A-specific T cells correlating with inflammatory skin reactions in melanoma patients vaccinated with peptides in IFA. *Cancer Immun* 2004; 4: 4.

116. Speiser DE, Lienard D, Rufer N, et al. Rapid and strong human CD8+ T cell responses to vaccination with peptide, IFA, and CpG oligodeoxynucleotide 7909. *J Clin Invest* 2005; 115: 739–746.

117. Roberts JD, Niedzwiecki D, Carson WE, et al. Phase 2 study of the g209–2M melanoma peptide vaccine and low-dose interleukin-2 in advanced melanoma: Cancer and Leukemia Group B 509901. *J Immunother* 2006; 29: 95–101.

118. Dyall R, Bowne WB, Weber LW, et al. Heteroclitic immunization induces tumor immunity. *J Exp Med* 1998; 188: 1553–1561.

119. Slansky EJ, Rattis MF, Boyd FL, et al. Enhanced antigen-specific antitumor immunity with altered peptide ligands that stabilize the MHC-peptide-TCR complex. *Immunity* 2000; 13: 529–538.

120. Parkhurst MR, Salgaller ML, Southwood S, et al. Improved induction of melanoma-reactive CTL with peptides from the melanoma antigen gp100 modified at HLA-A*0201-binding residues. *J Immunol* 1996; 157: 2539–2548.

121. Weeratna RD, McCluskie MJ, Xu Y, Davis HL. CpG DNA induces stronger immune responses with less toxicity than other adjuvants. *Vaccine* 2000; 18: 1755–1762.

122. Chang DZ, Lomazow W, Joy Somberg C, et al. Granulocyte-macrophage colony stimulating factor: an adjuvant for cancer vaccines. *Hematology* 2004; 9: 207–215.

123. Armitage JO. Emerging applications of recombinant human granulocyte-macrophage colony-stimulating factor. *Blood* 1998; 92: 4491–4508.

124. Nasi ML, Lieberman P, Busam KJ, et al. Intradermal injection of granulocyte-macrophage colony-stimulating factor (GM-CSF) in patients with metastatic melanoma recruits dendritic cells. *Cytokines Cell Mol Ther* 1999; 5: 139–144.

125. Clay TM, Hobeika AC, Mosca PJ, et al. Assays for Monitoring Cellular Immune Responses to Active Immunotherapy of Cancer. *Clin Cancer Res* 2001; 7: 1127–1135.

126. Keilholz U, Weber J, Finke JH, et al. Immunologic monitoring of cancer vaccine therapy: results of a workshop sponsored by the Society for Biological Therapy. *J Immunother* 2002; 25: 97–138.

127. Wolchok JD, Chapman PB. How can we tell when cancer vaccines vaccinate? *J Clin Oncol* 2003; 21: 586–587.

128. Scheibenbogen C, Lee KH, Stevanovic S, et al. Analysis of the T cell response to tumor and viral peptide antigens by an IFNgamma-ELISPOT assay. *Int J Cancer* 1997; 71: 932–936.

129. Klenerman P, Cerundolo V, Dunbar PR. Tracking T cells with tetramers: new tales from new tools. *Nat Rev Immunol* 2002; 2: 263–272.

130. Banchereau J, Steinman RM. Dendritic cells and the control of immunity. *Nature* 1998; 392: 245–252.

131. Dhodapkar MV, Steinman RM, Sapp M, et al. Rapid generation of broad T-cell immunity in humans after a single injection of mature dendritic cells. *J Clin Invest* 1999; 104: 173–180.

132. Thurner B, Haendle I, Roder C, et al. Vaccination with mage-3A1 peptide-pulsed mature, monocyte-derived dendritic cells expands specific cytotoxic T cells and induces regression of some metastases in advanced stage IV melanoma. *J Exp Med* 1999; 190: 1669–1678.

133. Inaba K, Metlay JP, Crowley MT, Steinman RM. Dendritic cells pulsed with protein antigens in vitro can prime antigen-specific, MHC-restricted T cells in situ. *J Exp Med* 1990; 172: 631–640.

134. Porgador A, Gilboa E. Bone marrow-generated dendritic cells pulsed with a class I-restricted peptide are potent inducers of cytotoxic T lymphocytes. *J Exp Med* 1995; 182: 255–260.

135. Nestle FO, Alijagic S, Gilliet M, et al. Vaccination of melanoma patients with peptide- or tumor lysate-pulsed dendritic cells. *Nat Med* 1998; 4: 328–332.

136. Mackensen A, Herbst B, Chen JL, et al. Phase I study in melanoma patients of a vaccine with peptide-pulsed dendritic cells generated in vitro from CD34(+) hematopoietic progenitor cells. *Int J Cancer* 2000; 86: 385–392.

137. Schuler-Thurner B, Dieckmann D, Keikavoussi P, et al. Mage-3 and influenza-matrix peptide-specific cytotoxic T cells are inducible in terminal stage HLA-A2.1+ melanoma patients by mature monocyte-derived dendritic cells. *J Immunol* 2000; 165: 3492–3496.

138. Bedrosian I, Mick R, Xu S, et al. Intranodal administration of peptide-pulsed mature dendritic cell vaccines results in superior CD8+ T-cell function in melanoma patients. *J Clin Oncol* 2003; 21: 3826–3835.

139. Fay JW, Palucka AK, Paczesny S, et al. Long-term outcomes in patients with metastatic melanoma vaccinated with melanoma peptide-pulsed CD34(+) progenitor-derived dendritic cells. *Cancer Immunol Immunother* 2006; 55: 1207–1218.

140. Schadendorf D, Ugurel S, Schuler-Thurner B, et al. Dacarbazine (DTIC) versus vaccination with autologous peptide-pulsed dendritic cells (DC) in first-line treatment of patients with metastatic melanoma: a randomized phase III trial of the DC study group of the DeCOG. *Ann Oncol* 2006.

141. Haenssle HA, Krause SW, Emmert S, et al. Hybrid cell vaccination in metastatic melanoma: clinical and immunologic results of a phase I/II study. *J Immunother* 2004; 27: 147–155.

142. Nagayama H, Sato K, Morishita M, et al. Results of a phase I clinical study using autologous tumour lysate-pulsed monocyte-derived mature dendritic cell vaccinations for stage IV malignant melanoma patients combined with low dose interleukin-2. *Melanoma Res* 2003; 13: 521–530.

143. Escobar A, Lopez M, Serrano A, et al. Dendritic cell immunizations alone or combined with low doses of interleukin-2 induce specific immune responses in melanoma patients. *Clin Exp Immunol* 2005; 142: 555–568.

144. Kaplan G, Walsh G, Guido LS, et al. Novel responses of human skin to intradermal recombinant granulocyte/macrophage-colony-stimulating factor: Langerhans cell recruitment, keratinocyte growth, and enhanced wound healing. *J Exp Med* 1992; 175: 1717–1728.

145. Bowne WB, Srinivasan R, Wolchok JD, et al. Coupling and uncoupling of tumor immunity and autoimmunity. *J Exp Med* 1999; 190: 1717–1722.

146. Bowne WB, Wolchok JD, Hawkins WG, et al. Injection of DNA encoding granulocyte-macrophage colony-stimulating factor recruits dendritic cells for immune adjuvant effects. *Cytokines Cell Mol Ther* 1999; 5: 217–225.

147. Perales MA, Fantuzzi G, Goldberg SM, et al. GM-CSF DNA induces specific patterns of cytokines and chemokines in the skin: implications for DNA vaccines. *Cytokines Cell Mol Ther* 2002; 7: 125–133.

148. Bergman PJ, Camps-Palau MA, McKnight JA, et al. Development of a xenogeneic DNA vaccine program for canine malignant melanoma at the Animal Medical Center. *Vaccine* 2006; 24: 4582–4585.

149. Flanagan K, Moroziewicz D, Kwak H, et al. The lymphoid chemokine CCL21 costimulates naive T cell expansion and Th1 polarization of non-regulatory CD4+ T cells. *Cell Immunol* 2004; 231: 75–84.

150. Kirk CJ, Hartigan-O'Connor D, Nickoloff BJ, et al. T cell-dependent antitumor immunity mediated by secondary lymphoid tissue chemokine: augmentation of dendritic cell-based immunotherapy. *Cancer Res* 2001; 61: 2062–2070.

151. Flanagan K, Glover RT, Horig H, et al. Local delivery of recombinant vaccinia virus expressing secondary lymphoid chemokine (SLC) results in a CD4 T-cell dependent antitumor response. *Vaccine* 2004; 22: 2894–2903.

152. Overwijk WW, Lee DS, Surman DR, et al. Vaccination with a recombinant vaccinia virus encoding a "self" antigen induces autoimmune vitiligo and tumor cell destruction in mice: Requirement for CD4(+) T lymphocytes. *Proc Natl Acad Sci USA* 1999; 96: 2982–2987.

153. Bronte V, Apolloni E, Ronca R, et al. Genetic vaccination with "self" tyrosinase-related protein 2 causes melanoma eradication but not vitiligo. *Cancer Res* 2000; 60: 253–258.

154. Irvine KR, Chamberlain RS, Shulman EP, et al. Route of immunization and the therapeutic impact of recombinant anticancer vaccines. *J Natl Cancer Inst* 1997; 89: 390–392.

155. Meyer RG, Britten CM, Siepmann U, et al. A phase I vaccination study with tyrosinase in patients with stage II melanoma using recombinant modified vaccinia virus Ankara (MVA-hTyr). *Cancer Immunol Immunother* 2005; 54: 453–467.

156. Rosenberg SA, Yang JC, Schwartzentruber DJ, et al. Recombinant fowlpox viruses encoding the anchor-modified gp100 melanoma antigen can generate antitumor immune responses in patients with metastatic melanoma. *Clin Cancer Res* 2003; 9: 2973–2980.

157. van Baren N, Bonnet MC, Dreno B, et al. Tumoral and immunologic response after vaccination of melanoma patients with an ALVAC virus encoding MAGE antigens recognized by T cells. *J Clin Oncol* 2005; 23: 9008–9021.

158. Hirschowitz EA, Leonard S, Song W, et al. Adenovirus-mediated expression of melanoma antigen gp75 as immunotherapy for metastatic melanoma. *Gene Ther* 1998; 5: 975–983.

159. Rosenberg SA, Zhai Y, Yang JC, et al. Immunizing patients with metastatic melanoma using recombinant adenoviruses encoding MART-1 or gp100 melanoma antigens. *J Natl Cancer Inst* 1998; 90: 1894–1900.

160. Triozzi PL, Allen KO, Carlisle RR, et al. Phase I study of the intratumoral administration of recombinant canarypox viruses expressing B7.1 and interleukin 12 in patients with metastatic melanoma. *Clin Cancer Res* 2005; 11: 4168–4175.

161. Triozzi PL, Strong TV, Bucy RP, et al. Intratumoral administration of a recombinant canarypox virus expressing interleukin 12 in patients with metastatic melanoma. *Hum Gene Ther* 2005; 16: 91–100.

162. Kaufman HL, Deraffele G, Mitcham J, et al. Targeting the local tumor microenvironment with vaccinia virus expressing B7.1 for the treatment of melanoma. *J Clin Invest* 2005; 115: 1903–1912.

163. Porgador A, Irvine KR, Iwasaki A, et al. Predominant role for directly transfected dendritic cells in antigen presentation to CD8+ T cells after gene gun immunization. *J Exp Med* 1998; 188: 1075–1082.

164. Casares S, Inaba K, Brumeanu TD, et al. Antigen presentation by dendritic cells after immunization with DNA encoding a major histocompatibility complex class II-restricted viral epitope. *J Exp Med* 1997; 186: 1481–1486.

165. Akbari O, Panjwani N, Garcia S, et al. DNA vaccination: transfection and activation of dendritic cells as key events for immunity. *J Exp Med* 1999; 189: 169–178.

166. Klinman DM, Yi AK, Beaucage SL, et al. CpG motifs present in bacteria DNA rapidly induce lymphocytes to secrete interleukin 6, interleukin 12, and interferon gamma. *Proc Natl Acad Sci USA* 1996; 93: 2879–2883.

167. Sato Y, Roman M, Tighe H, et al. Immunostimulatory DNA sequences necessary for effective intradermal gene immunization. *Science* 1996; 273: 352–354.

168. Weber LW, Bowne WB, Wolchok JD, et al. Tumor immunity and autoimmunity induced by immunization with homologous DNA. *J Clin Invest* 1998; 102: 1258–1264.

169. Hawkins WG, Gold JS, Blachere NE, et al. Xenogeneic DNA immunization in melanoma models for minimal residual disease. *J Surg Res* 2002; 102: 137–143.

170. Hawkins WG, Gold JS, Dyall R, et al. Immunization with DNA coding for gp100 results in CD4 T-cell independent antitumor immunity. *Surgery* 2000; 128: 273–280.

171. Gold JS, Ferrone CR, Guevara-Patino JA, et al. A single heteroclitic epitope determines cancer immunity after xenogeneic DNA immunization against a tumor differentiation antigen. *J Immunol* 2003; 170: 5188–5194.

172. Goldberg SM, Bartido SM, Gardner JP, et al. Comparison of two cancer vaccines targeting tyrosinase: plasmid DNA and recombinant alphavirus replicon particles. *Clin Cancer Res* 2005; 11: 8114–8121.

173. Bergman PJ, McKnight J, Novosad A, et al. Long-term survival of dogs with advanced malignant melanoma after DNA vaccination with xenogeneic human tyrosinase: a phase I trial. *Clin Cancer Res* 2003; 9: 1284–1290.

174. Wang R, Doolan DL, Le TP, et al. Induction of antigen-specific cytotoxic T lymphocytes in humans by a malaria DNA vaccine. *Science* 1998; 282: 476–480.
175. Boyer JD, Cohen AD, Vogt S, et al. Vaccination of seronegative volunteers with a human immunodeficiency virus type 1 env/rev DNA vaccine induces antigen-specific proliferation and lymphocyte production of beta-chemokines. *J Infect Dis* 2000; 181: 476–483.
176. Ugen KE, Nyland SB, Boyer JD, et al. DNA vaccination with HIV-1 expressing constructs elicits immune responses in humans. *Vaccine* 1998; 16: 1818–1821.
177. Rosenberg SA, Yang JC, Sherry RM, et al. Inability to immunize patients with metastatic melanoma using plasmid DNA encoding the gp100 melanoma-melanocyte antigen. *Hum Gene Ther* 2003; 14: 709–714.
178. Tagawa ST, Lee P, Snively J, et al. Phase I study of intranodal delivery of a plasmid DNA vaccine for patients with stage IV melanoma. *Cancer* 2003; 98: 144–154.
179. Smith CL, Dunbar PR, Mirza F, et al. Recombinant modified vaccinia Ankara primes functionally activated CTL specific for a melanoma tumor antigen epitope in melanoma patients with a high risk of disease recurrence. *Int J Cancer* 2005; 113: 259–266.
180. Triozzi PL, Aldrich W, Allen KO, et al. Phase I study of a plasmid DNA vaccine encoding MART-1 in patients with resected melanoma at risk for relapse. *J Immunother* 2005; 28: 382–388.

20 Cytokine-Based Therapy for Cancer

Henry B. Koon, MD,
and Michael B. Atkins, MD

SUMMARY

Cytokine therapy has been extensively investigated in the treatment of malignancies. However, only a few agents, such as interferon and interleukin-2, have proven to have sufficient clinical benefit to justify their more widespread use. This chapter reviews the biology and clinical data for cytokine-based therapies that have been approved for clinical use, as well as cytokines that are currently under investigation.

Key Words: Cancer; cytokine; interferon; interleukin; immunotherapy; resistance.

1. INTRODUCTION

Cytokines play a critical role in the recognition of malignancy by the immune system. Mice that are deficient in interferon-γ (IFN-γ), the type I or type II IFN receptors, or portions of their downstream signal transduction intermediates have a higher frequency of tumors compared with control mice *(1–5)*. These data demonstrate that cytokines play a role in immunosurveillance and also suggest that cytokines would be useful as cancer immunotherapies. The development of recombinant DNA technology allowed for production of cytokines in sufficient quantities to enable their utility as anti-tumor agents to be tested in the clinic.

Although multiple approaches aimed at enhancing the immune system's ability to recognize and eradicate tumors have been examined, cytokine-based therapies are the most widely used at this time. This chapter reviews the basic biology of cytokines and the role of cytokine therapy for malignancy. The focus will be on cytokines that have proven to be effective as cancer therapy or are still undergoing clinical testing. In addition, this chapter will discuss predictors of cytokine response, mechanisms of tumor resistance, and strategies for overcoming such resistance.

2. CYTOKINE BIOLOGY AND CYTOKINE RECEPTORS

Cytokines are secreted proteins that have pleiotropic effects including regulation of innate immunity, adaptive immunity, and hematopoiesis. Distinct cytokines often have overlapping effects providing a level of redundancy to the immune system. The

From: *Cancer Drug Discovery and Development: Molecular Targeting in Oncology*
Edited by: H. L. Kaufman, S. Wadler, and K. Antman © Humana Press, Totowa, NJ

Table 1
Functional Classification of Cytokines

Innate	Adaptive	Hematopoiesis
Interferon-α	Interferon-γ	Interleukin-3
Interferon-β	Interleukin-2	Interleukin-7
Interleukin-1	Interleukin-4	Interleukin-9
Interleukin-6	Interleukin-5	Interleukin-11
Interleukin-10	Interleukin-13	Colony-stimulating factors
Interleukin-12	Interleukin-16	
Interleukin-15	Interleukin-17	
Interleukin-18	Lymphotoxin	
Tumor necrosis factor alpha		

first cytokines identified were the IFNs. The name IFN was adopted based on the ability of these agents to "interfere" with viral infection of cells. Subsequently, characterized cytokines were referred to as interleukins (ILs) because they were produced by and acted on leucocytes. Older nomenclature may refer to ILs as monokines, cytokines produced by monocytes, or lymphokines, cytokines produced by lymphocytes. Although the term, IL, does not accurately reflect the biologic properties of all such cytokines, some of which have been shown to be produced by cells other than leucocytes, it has been adopted as standard nomenclature.

Because of their pleiotropic effects, multiple classification systems for cytokines have been devised. A functional classification of cytokines has been proposed, which segregates cytokines based on whether they effect innate immunity, adaptive immunity, or hematopoiesis (Table 1) *(6)*. However, given that a number of cytokines effect both the

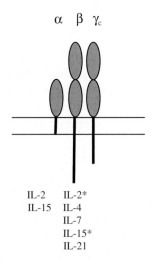

Fig. 1. The interleukin-2 (IL-2) receptor family is the prototypic type I cytokine receptors. All members of the family share the common gamma chain (γc). IL-2 and IL-15 share a common β-signaling subunit (*) but have distinct α-subunit that determines their affinity for the receptor.

innate and adaptive immune systems, a more practical classification system is based on homology of their cognate receptors—the type I cytokine receptors, the type II cytokine receptors, the immunoglobulin superfamily receptors, tumor necrosis factor (TNF) receptors, and G-protein-coupled receptors. The type I cytokines receptors are characterized by a common signaling subunit that complexes with a cytokine-specific subunit or units to initiate a signal. The prototypic type I cytokine receptor is the IL-2 receptor family in which the common γ-chain (γ_c) is shared by the IL-2, IL-4, IL-7, IL-9, IL-15, and IL-21 receptors (Fig. 1). Additional type I cytokine receptor subgroups include the granulocyte/monocyte colony-stimulating factor (GM-CSF) and IL-6 receptor families. IL-6, IL-11, and IL-12 share gp130 as a common subunit. IL-3, IL-5, and GM-CSF are members of the GM-CSF receptor subfamily and share a common β-chain that complexes with cytokine-specific α-chain. The effects of IFN-α, IFN-β, IFN-γ, and IL-10 are mediated by type II cytokine receptors, which are composed of a signaling chain and a ligand-binding chain. The immunoglobulin super-family receptors contain extracellular immunoglobulin domains and include the receptors for IL-1, IL-18, stem cell factor, and GM-CSF. The TNF receptor family is characterized by conserved cysteine-rich domains and appear to trimerize upon receptor binding.

3. INTERFERONS

3.1. Biology

The IFNs may be grouped based on their ability to bind specific IFN receptors. The type I and type II IFN receptors are a subset of the type II cytokine receptors (7–9). IFN-α and IFN-β are predominantly involved in responses of cellular immunity to viral infections (7,10,11). Both IFN-α and IFN-β activate the type I IFN receptor and are referred to as the type I IFNs (7–9). IFN-γ, the only type II IFN, is important in responses of the cellular immune system and activates the type II IFN receptor (7, 9–11).

The type I IFNs are the most clinically useful of all cytokines for treatment of malignancy. The type I IFNs consist of at least five classes in humans of which IFN-α and IFN-β have been used the most clinically (7). Both IFN-α and IFN-β have a number of effects that make them attractive as immunotherapies. They up-regulate major histocompatibility complex (MHC) class I molecules and induce maturation of a subset of dendritic cells (DC) (12,13). Type I IFNs also activate cytotoxic T-cell lymphocytes (CTLs), natural killer (NK) cells, and macrophages (14–16). In addition to their immunologic effects, the type I IFNs can have a cytostatic effect on tumors cells and may be proapoptotic (17,18). They also can have anti-angiogenic effects on the tumor vasculature (19,20). Mice with targeted deletion of the type I IFN receptor have a higher rate of carcinogen-induced cancer compared with controls and have enhanced tumor development in transplantable tumor models supporting the hypothesis that the type I IFNs are important in immunosurveillance (2,21). These multiple mechanisms of action explain why the type I IFNs are effective in such a board array of malignancies.

IFN-α, initially referred to as leucocyte IFN, is comprised of a group of at least twelve distinct proteins (7). Recombinant IFN-α-2a, IFN-α-2b, and IFN-α-2c differ by one to two amino acids and are the forms of IFN-α the have been tested clinically (7). In the USA, IFN-α-2a is sold under the trade name Roferon (Hoffmann-La Roche, Nutley, NJ) and IFN-α-2b is available as Intron A (Shering, Kenilworth, NJ). IFN-α-2c is available in Europe as Berofor (Bender, Vienna, Austria). These three compounds have never been compared in a randomized fashion; however, their spectrum of activity is likely to be similar. The approved indications for these agents include treatment of viral related diseases such as hepatitis C and Kaposi's sarcoma (KS) as well as treatment of cancers such as melanoma and chronic myelogenous leukemia (CML) *(22–24)*. Recently, IFN-α conjugated to polymer polyethylene glycol (PEG-IFN) has been introduced. PEG-IFN was designed to increase the half-life, thus allowing for longer dosing intervals and long exposure times *(25)*. Pegylated IFN-α-2a (Pegasys; Hoffmann-La Roche) and pegylated IFN-α-2b (Peg-Intron; Shering) are the two forms of PEG-IFN available in the USA *(26,27)*. These agents are widely used in combination with ribavirin in the treatment of hepatitis C. The role of the PEG-IFNs as monotherapies for cancer is still under study *(28,29)*.

As IFN-α and IFN-β signal through the same receptor, they would be expected to have similar biologic effects and have overlapping indications. However, this is not always the case. Although both IFN-α and IFN-β have activity against gliomas, one small study suggests that IFN-β has a higher response rate compared in contrast to IFN-α *(30)*. In contrast to IFN-α, IFN-β has been reported to have no clinical activity against CML and no responses were seen in a phase I trial of 35 patients with metastatic solid tumors *(31,32)*. Two forms of IFN-β, also known as fibroblast IFN, have been approved for relapsing multiple sclerosis: IFN-β-1a (Avonex; Biogen Idec, Cambridge, MA) and IFN-β-1b (Betaseron; Berlex, Montville, NJ). Their use in treatment of malignancy is currently limited to clinical trials.

IFN-γ, also known as immune IFN, is the only type II IFN and has effects on the innate and adaptive immune system. IFN-γ is secreted by NK cells, natural killer T-cells (NKT), Th1 CD4+ T-cells, CD8+ T-cells, antigen-presenting cells (APCs), and B-cells *(33–36)*. IFN-γ activates macrophages and stimulates up-regulation of MHC class I, MHC class II, and co-stimulatory molecules on APCs *(13–39)*. Additionally, IFN-γ induces changes in the proteosome to enhance antigen presentation *(40–42)*. It promotes Th1 differentiation of CD4+ T-cells and blocks IL-4-dependent isotype switching in B-cells *(37–44)*. Mice with targeted deletion of IFN-γ or the type II IFN receptor have an increased risk of spontaneous and chemically induced tumors compared with controls *(1–45)*. IFN-γ is cytotoxic to some malignant cells and has anti-angiogenic activity *(46–50)*.

The anti-tumor effects of IFN-γ suggested it would be effective against a wide spectrum of malignancies; however, IFN-γ has demonstrated limited clinical utility in cancer *(51–53)*. Actimmune (Intermune; Brisbane, CA) is an IFN-γ preparation that has been approved for the treatment of chronic granulomatous disease *(54)*. Although IFN-γ likely plays a critical role in mediating the *in vivo* effects of other cytokines, clinically significant benefit in treatment of malignancies has been largely restricted to type I IFN.

3.2. Indications

3.2.1. HEMATOLOGIC MALIGNANCIES: HAIRY CELL LEUKEMIA

In clinical usage, the type I IFNs have had their most success against two hematologic malignancies: hairy cell leukemia (HCL) and CML. A regimen of IFN-α-2b 2 million units/m^2 subcutaneously three times a week for 52 weeks produced an overall response rate of 77% with a complete response rate of 5% in patients with HCL *(55)*. The vast majority of these patients (61 out of 66) had undergone splenectomy but were otherwise untreated *(55)*. Subsequent studies demonstrated complete responses in 25–35% of patients who had not had splenectomies leading to regulatory approval for IFN in this patient population *(56)*. Although IFN has a significant response rate and improves survival in HCL, the majority of patients relapse after discontinuation of therapy *(57)*. Subsequent studies demonstrated that 80% of patients who relapsed would respond to another course of IFN *(57)*. It remains unclear whether IFN's effect in HCL is mediated by immune mechanisms or direct effects on the leukemic cells *(59–60)*. Although IFN was once considered first-line therapy, the introduction of the nucleoside analogs which have a greater than 90% complete response (CR) rate has limited the use of IFN therapy to patients who have disease that is refractory to nucleosides or with contraindications to these agents *(61,62)*.

3.2.2. HEMATOLOGIC MALIGNANCIES: CHRONIC MYELOGENOUS LEUKEMIA

Initial trials of IFN-α in CML suggested that as a single agent, IFN produced complete hematologic responses in over 50% of patients and a complete cytogenic response in up to 25% of patients *(63,64)*. Follow-up randomized studies demonstrated that IFN was superior to hydroxyurea or busulfan or both (Table 2) *(65–69)*. Four of

Table 2
Interferon Trials in CML

Study	Number of patients	Treatment arms	Response rate Hematologic (CR/PR)	Cytogenetic (CR/MR)	Median survival (months)
Broustet	30	IFN-α-2b (5 MIU/m^2/day)	53/–	–/–	NA
	28	Hydroxyurea	82/–	–/–	NA
Hehlmanm	133	IFN-α-2a (5 MIU/m^2/day)	31/52	5/6	58
	186	Hydroxyurea	39/51	0/0.5	46
	194	Busulfan	23/69	0.5/1	48
Italian Cooperative Group	218	IFN-α-2a (9 MIU/day)	62 (CR + PR)	10/21	72
	104	Hydroxyurea/ busulfan	53 (CR + PR)	0/1	52 $p = 0.002$
Allan	293	IFN-α-n1 (12 MIU/day)	69/18	6/11	61
	294	Hydroxyurea/ busulfan	NA	0/3	41 $p = 0.009$
Ohnishi	80	IFN-α-2a (9–18 MIU/day)	39/39	9/8	60+
	79	Busulfan	54/43	2/2	50 $p = 0.03$

MIU, million international units.

these studies demonstrated a improved overall survival (OS) for the IFN-treated patients *(65,66,68–71)*. A meta-analysis of the randomized trials demonstrated an improvement in the 5-year survival in the IFN-treated group of 12% over hydroxyurea and 20% over busulfan-treated patients *(72)*. Additionally, the meta-analysis showed the benefit extended to all risk groups. All three commercially available IFN-α were used in the CML trials, and although not formally compared, their activity in CML appeared similar.

The mechanism of response of CML to IFN has been extensively investigated. Reports that human leucocyte antigen (HLA) type and development of an immune response to BCR-Abl correlate with a complete response suggest that IFN works through an immune mechanism in patients with CML *(73)*. Further evidence supporting an immune mechanism of response in CML is the observation that patients who obtain a complete response correct abnormalities in the secretion of Th1 cytokines *(74)*. However, IFN also exerts a direct anti-proliferative effect in CML through inhibition of DNA polymerase *(75)*. These data suggest that the mechanism of action of IFN in CML is multifactorial.

In an effort to enhance the efficacy of IFN in CML, a number of trials were conducted with IFN combined with chemotherapy *(72)*. The combination of IFN and low-dose (LD) ara-C was shown to improve the number of cytogenetic remissions compared with IFN alone. However, the beneficial impact of the combination on OS is small and was achieved with a substantial increase in toxicity *(76–78)*. Although largely supplanted as first-line therapy by kinase inhibitors, IFN and IFN-containing regimens remain a valid second-line therapeutic option for patients with CML *(79–82)*.

3.2.3. NON-HODGKIN'S LYMPHOMA

Early studies on IFN-α monotherapy in follicular lymphomas demonstrated a response rate of over 50% *(83–85)*. Subsequently, a number of trials combined IFN with chemotherapy in an induction regimen or as maintenance therapy after induction. The results of these trials were mixed in terms of OS benefit. The Groupe d'Etude des Lymphoma Folliculaires (GELF) study demonstrated an advantage in response rate (85 versus 69%, $p < 0.001$) and OS (34 versus 19 months, $p = 0.02$) for chemotherapy plus IFN-α 5 million international units (MIU) thrice weekly for 18 months compared with chemotherapy alone using an anthracycline-based regimen *(86,87)*. These results were a major impetus for the approval of IFN-α for the treatment of follicular lymphoma. A meta-analysis of the IFN trials supports a survival advantage for intensive chemotherapy regimens containing IFN *(88)*. Interestingly, a large SWOG trial did not show any survival advantage for IFN-α at 2 MIU thrice weekly for 24 months versus observation *(89)*. These data suggest that the dose of IFN-α used may be critical to the beneficial effect in patients with follicular lymphomas. IFN is approved for treatment of follicular lymphoma but its use is limited because of its associated toxicities and the activity of a variety of other agents.

3.2.4. MELANOMA

The natural history of melanoma suggests that it is an immune responsive tumor. Up to 25% of primary cutaneous melanomas show histologic regression at the time of biopsy *(90)*. Approximately 5% of all melanoma cases that present as metastatic melanoma have no known primary *(91–94)*. It is thought that the majority of these patients had a primary that was completely eliminated by the immune system, but the immune system was unable to eliminate the micrometastatic disease *(92,95,96)*. Once a patient has metastatic disease, the rate of spontaneous remission is estimated to be 0.2–0.3% *(97)*.

All three IFN-αs have been investigated in patients with metastatic (stage IV) melanoma. Multiple dose levels and schedules have been tested, and the overall response rate for single agent IFN in patients with metastatic melanoma is approximately 15% *(98–106)*. There is no clear best regimen in terms of response rate; however, IFN administered on thrice weekly schedule is the most widely used schedule, because it has a significantly better toxicity profile than a daily administration schedule with no diminution in response rates *(107,108)*. It is unknown whether IFN provides a survival advantage, because there are no randomized trials in metastatic melanoma comparing IFN to either cytotoxic chemotherapy or best supportive care *(109)*. IFN works best in patients with low metastatic tumor burden, perhaps presaging its clinical activity in the adjuvant setting *(110)*.

IFN has proved most useful in the management of melanoma in the adjuvant setting. Multiple IFN regimens have been used in the adjuvant setting for patients with intermediate and high-risk melanoma (Table 3) *(111–121)*. IFN was approved in Europe based on studies that used lower dose of regimens. Two trials of LD IFN-α-2a given for 12–18 months demonstrated a benefit in relapse-free survival (RFS) in patients with melanomas >1.5 mm or locoregional disease, but neither trial showed an OS benefit *(112,116)*. Subsequent trials using LD IFN-α-2a and IFN-α-2b have failed to demonstrate a durable RFS or an OS benefit (Table 3) *(111,113–115,117,118)*.

In the USA, IFN was approved for adjuvant therapy in patients with high-risk melanoma based on the ECOG 1684 trial (Table 3B) *(122)*. In this trial, patients with high-risk melanoma defined as primary tumors >4 mm or pathologic or clinical regional lymph node involvement who had undergone lymphadenectomy were treated with 1 year of high-dose IFN-α-2b (HDI) *(122)*. The HD IFN regimen consists of 20 million units/m^2/day 5 days per week for 4 weeks followed by 10 million units/m^2/day thrice weekly for 48 weeks *(122)*. This initial trial demonstrated an overall improvement in median RFS from 1 to 1.7 years and median OS from 2.8 to 3.8 years *(122)*. In addition, there was a significant (42%) reduction in the risk of relapse. In a subsequent intergroup trial, ECOG 1690, an improvement in median and overall RFS was seen in the HDI arm compared with observation, but there was no difference in OS (Table 3B) *(120)*. The reason for the lack of an OS advantage for the HDI in this trial appeared to be related to improved survival in patients on the observation arm following relapse (6 years verses 2.8 years in ECOG 1684). *(120)*. Multiple explanations for this have been postulated for this observation. In contrast to E1684, patients on E1690 were not required to undergo elective node dissection prior to enrollment on study *(120,122)*. Consequently, many patients were enrolled with >4 mm thick primary tumors who had no evaluation of their regional nodal basin *(120)*. Additionally, IFN-α received FDA approval in 1996, while E1690 was ongoing. Consequently, many patients on the observation arm who relapsed in regional nodes received off protocol adjuvant IFN following therapeutic node dissection, perhaps contributing to their better than anticipated survival while obscuring the survival benefit related to upfront IFN administration *(120)*.

No other trials have compared HDI to observation. However, ECOG 1694, a trial that compared HDI to the ganglioside GM2/keyhole limpet hemocyanin vaccine (GMK),

Table 3A
Adjuvant Trials of Low-Dose to Intermediate-Dose IFN for Melanoma

Trial	Number of patients	Dose	Population	RFS	OS
WHO-16	218	3 MIU SC three times a week for 3 years	>1.5 mm		
	208	Observation	Clinically node negative	NS	NS
NCCTG 83-7052	131	20 MIU/m² IM three times a week for 3 months	>1.5 mm and/or resected regional disease		
	131	Observation		NS	NS
French Cooperative Group	244	3 MIU SC three times a week for 18 months	>1.5mm		
	245	Observation	Clinically node negative	$p = 0.04$	NS
Austrian Melanoma Cooperative Group	154	3 MIU/day for 3 weeks then 3 MIU SQ/week for 1 year	>1.5 mm		
	157	Observation	Clinically node negative	$p = 0.02$	NS
EORTC 18871	244	1 MIU qOD SQ for 1 year	>3.0 mm and/or resected regional disease		
	240	IFN-γ for 1 year		NS	NS
	244	Observation			
Scottish Melanoma Group	46	3 MIU SC three times a week for 6 months	>3.0mm and/or resected regional disease		
	49	Observation		NS	NS
EORTC 18952	553	10 MIU/day for 4 weeks then 10 MIU SQ three times a week for 1 year	>4.0mm and/or resected regional disease	NS	NS
	556	5 MIU SQ three times a week for 2 years		NS	NS
	279	Observation			
AIM HIGH	338	3MIU SQ three times a week for 2 years	>4.0mm and/or resected regional disease		
	336	Observation		NS	NS

IFN, interferon; MIU, million international units; OS, overall survival; qOD, every otherday; and RFS, relapse-free survival.

Table 3B
Adjuvant Trials of High-Dose IFN for Melanoma

Trial	Number of patients	Dose	Population	RFS	OS
ECOG 1684	287	20 MIU/m^2 five times for a week for 1 month, then 10 MIU/m^2 three times a week for 48 weeks	>4.0 mm and/or resected regional disease	$p = 0.004$	$p = 0.046$
ECOG 1690	642	20 MIU/m^2 five times for a week for 1 month, then 10 MIU/m^2 three times a week for 48 weeks	>4.0 mm and/or resected regional disease	$p = 0.05$	NS
		3MIU SC three times a week for 3 years Observation		$p = 0.17$	NS
ECOG 1694	385	20 MIU/m^2 five times a week for 1 month, then 10 MIU/m^2 three times a week for 48 weeks	>4.0mm and/or resected regional disease		
	389	GM2-KLH/QS-21 vaccine		$p = 0.0027$	$p = 0.0147$

IFN, interferon; MIU, million international units; OS, overall survival; and qOD, every other day RFS, relapse-free survival.

showed an improvement in both RFS and OS for patients receiving HDI (Table 3B) *(119)*. A long-term follow-up of the RFS and OS of ECOG 1684, 1690, and 1694 has recently been published (Figs 2 and 3) *(123)*. The HDI arm of ECOG 1684 continues to demonstrate a persistent improvement in RFS (HR = 1.38, $p = 0.02$) at a median follow-up of 12.6 years (Fig. 2) *(123)*. The OS benefit also persisted (HR = 1.22, $p = 0.18$) but was no longer statistically significant (Fig. 3) *(123)*. Given that the median age of the subjects is now greater than 60 years, competing causes of death may be blunting the survival benefit. Analysis of the pooled data from both 1684 and 1690 (median follow-up 7.2 years) demonstrates a RFS benefit (HR = 1.30, $p<0.006$), but there was no OS benefit, which is not surprising as there was no OS benefit in the larger 1690 trial *(123)*.

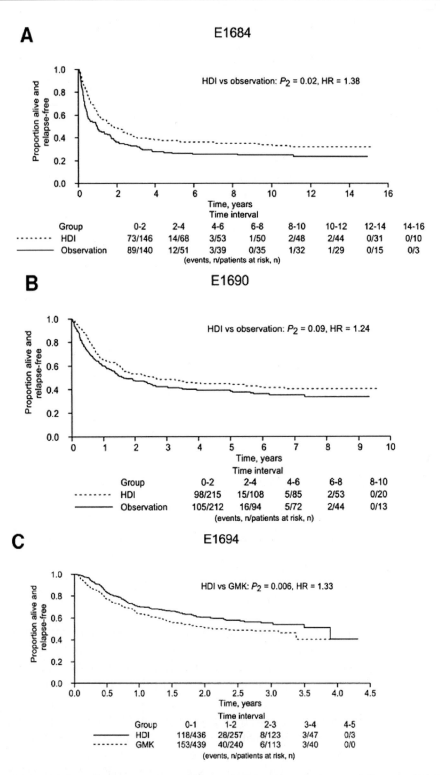

Fig. 2. Kaplan–Meier estimates of relapse-free survival based on long-term follow-up. (**A**) ECOG 1684 (median follow-up 12.6 years), (**B**) ECOG 1690 (median follow-up 6.6 years), and (**C**) ECOG 1694 (median follow-up 2.1 years).

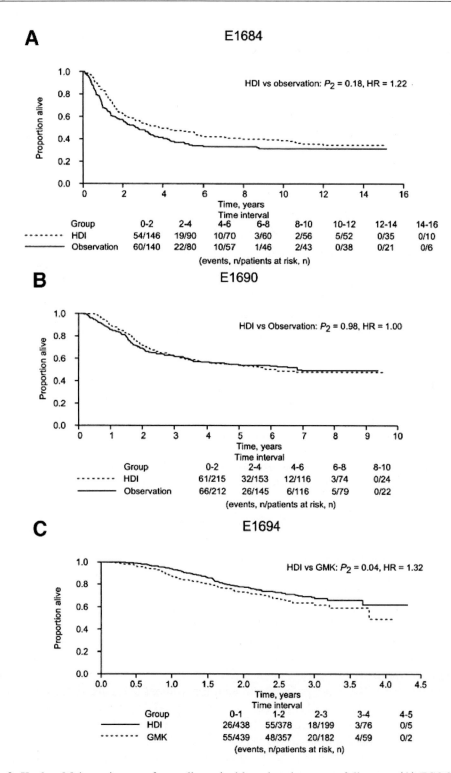

Fig. 3. Kaplan–Meier estimates of overall survival based on long-term follow-up. (**A**) ECOG 1684 (median follow-up 12.6 years), (**B**) ECOG 1690 (median follow-up 6.6 years), and (**C**) ECOG 1694 (median follow-up 2.1 years).

The ECOG adjuvant trials were stratified based on stage to ensure balance between the arms. Although not powered for subset analysis, these data were reviewed to determine whether any populations disproportionately benefited from adjuvant IFN (120,122). In ECOG 1684, the patients with microscopic involvement of their lymph nodes (stage IIIa T4pN1) had the greatest improvement in their hazard ratio when treated with IFN (120,122). As patients were not required to have lymphadenectomies, the group of patients who had microscopic disease were mixed in with the T4N0 subgroup in the E1690 trial (120,122). In both studies, there was an improvement in node positive disease that was proportionate to the patient's risk, whereas on E1694, the patients who benefited the most were those with no evidence of nodal involvement. Taken together, the benefit of IFN appears to be proportionate to risk, with a 20–30% reduction in risk of relapse and 10–20% reduction in risk of death regardless of the patients underlying risk of relapse and death. Consideration of IFN therapy should take into account the patient's risk of relapse and co-morbidities and potential side effects. Generally, IFN treatment should be considered in otherwise healthy patients whose risk of relapse is greater than 30% (120,122).

As with CML and NHL, combinations of IFN with chemotherapies and other cytokines have been studied in the metastatic melanoma setting in hopes of improving clinical benefit. One early single institution randomized phase II trial reported that combined IFN and dacarbazine demonstrated an improved response rate (53% versus 20%) and improved OS (18 months versus 10 months) relative to dacarbazine alone (124). Unfortunately, this benefit could not be confirmed in a larger randomized phase III trial (E3690) (109,125,126). In addition to chemotherapy, IFN has been combined with other cytokines. A trial testing the combination of a type I IFN combined with IFN-γ did not demonstrate any substantial benefit over type I IFN alone (127). Despite some controversy, IFN-α remains the standard of care for adjuvant treatment of patients with high-risk melanoma, whereas its role in metastatic melanoma has been limited to its use in combination with other drugs in various clinical trials.

3.3. Renal Cell Carcinoma

Like melanoma, renal cell carcinoma (RCC) has spontaneous regression response rate of approximately 0.3% in untreated patients and may be as high as approximately 6% in patients who have undergone cytoreductive nephrectomies. Multiple immunotherapies have been examined including the IFNs. Trials of IFN-α monotherapy in RCC demonstrate a response rate of 10–15% (128–138). All three IFN-α preparations have been used in RCC with no substantial differences in efficacy. IFN-α appears to have a dose–response curve in RCC that plateaus at 10 MIU per day (133,139). IFN-α appears to have a modest survival benefit in patients with advanced RCC (138). A randomized trial of IFN-α-2a versus medroxyprogesterone demonstrated an increased response rate for IFN-treated patients and an improvement in median OS from 6 versus 8.5 months (138). Additionally, patients who had cytoreductive nephrectomy have an improved survival when treated with IFN when compared with patients treated with IFN with their primary tumor still in place, suggesting that nephrectomy improves the beneficial effects of cytokine therapy (140). Because of the reported survival advantage, IFN has been widely used as monotherapy for RCC.

Despite the evidence of a survival benefit in patients with advanced disease, IFN has shown no benefit when studied in the adjuvant setting in patients with high risk of relapse *(141)*. Although some early trials suggested that combining IFN with chemotherapy may improve response rates relative to chemotherapy alone, no study has shown benefit for the combination of any agent, vinblastine, 13-cis retinoic acid IFN-γ or IL-2 to IFN relative to IFN alone. Nonetheless, because of its proven survival benefit, defined toxicities, and familiarity to most oncologists, it has served as an excellent agent to combine with novel targeted and anti-angiogenic agents in renal cancer and as the control arm in phase III trials. *(142–147)*.

3.3.1. Kaposi's Sarcoma

AIDS-related KS is a multifocal vascular proliferative disease associated with HIV and KS herpesvirus (KSHV)/human herpes virus-8 (HHV-8) co-infection *(148)*. Histologically, these lesions are composed of clusters of spindle-shaped cells (KS spindle cells) with prominent microvasculature. This angiogenic lesion is driven by autocrine and paracrine cytokine loops *(149,150)*. Because of the anti-angiogenic activity IFN demonstrated in treating hemangiomas, it was tested in patients with KS *(151–153)*.

As a single agent for the treatment of KS, IFN has a response rate of 30–40% that appears to be dose dependent *(23,154–156)*. When combined with anti-retroviral therapy, the response rate appears to be over 40% *(157,158)*. Although IFN is useful in KS, initiation of effective anti-retroviral therapy, cytotoxic chemotherapy or local therapy are currently the first-line treatments *(159)*.

3.4. Toxicities

The enthusiasm for IFN use is tempered by its side effects. Nonetheless, the side effects are typically dose related, and most resolve quickly with discontinuation of treatment. The toxicities can be broken down into five major categories—constitutional, neuropsychiatric, gastrointestinal, hematologic and autoimmune.

Constitutional symptoms are the most common with more than 80% of the patients in the HD IFN trials reporting fever and fatigue *(160)*. Additionally, more than half of patients report headache and myalgias *(160)*. The majority of these symptoms can be controlled with acetaminophen or NSAIDs; however, severe fatigue often requires a break from therapy with a subsequent dose reduction for amelioration.

Neuropsychiatric issues are not as common but are potentially life threatening. As many as 10% of patients complain of confusion and rarely (<1%) patients develop mania *(160,161)*. In some studies, up to 45% of patients reported depression, and suicides were occasionally reported *(162,163)*. In one small double blind placebo controlled trial in patients receiving HD IFN for high-risk melanoma, prophylactic use of anti-depressants significantly reduced the risk of depression from 45% to 11% after 12 weeks *(162)*. These data suggest that at a minimum, patients with a history of depression should be treated with anti-depressants if they are not currently taking them at the time of IFN initiation. All other patients should be monitored closely and anti-depressant therapy instituted at the earliest sign of depression.

Gastrointestinal side effects are common with up one third of patients having diarrhea, which is usually well controlled with over-the-counter anti-diarrheal medications *(160)*. Two thirds of patients have problems with nausea and anorexia. Anti-emetics often alleviate the nausea; however, the combination of nausea and anorexia can lead to significant weight loss *(160)*. Additionally, IFN can produce significant hepatic toxicity, which requires serial monitoring of liver function tests. In the early trials, some patients had fatal hepatic failure. Usually, a drug holiday until the liver function improves followed by dose reduction allows the majority of patients with liver toxicity to continue treatment.

IFN can affect all of the hematopoietic lineages. Thrombocytopenia, leucopenia, and neutropenia are common and are typically managed with dose reductions *(160)*. Anemia, if not hemolytic, can be treated with transfusions or dose reductions. Rarely thrombotic thrombocytopenic purpura (TTP) has been reported in association with IFN *(164–167)*. Hemolytic anemia and TTP require permanent drug discontinuation *(168–171)*.

In addition to autoimmune hemolytic anemia and thrombocytopenia, other manifestations of immune dysfunction can also be observed. Thyroid dysfunction in the form of hyperthyroidism or hypothyroidism occurs in about 15% of patients, and therefore, thyroid function tests should be routinely monitored in patients receiving IFN therapy *(172)*. The hyperthyroidism often presents as fatigue, restlessness, and/or significant weight loss and may be attributed to other causes if thyroid function tests are not checked. Sarcoid can occur in patients receiving IFN and can also present a diagnostic dilemma. It can present as skin lesions masquerading as subcutaneous metastases or as flurodeoxyglucose (FDG)-avid lymph nodes on PET scan *(173,174)*. Vitiligo, lupus, rheumatoid arthritis, polymyalgia rheumatica, and psoriasis are among the other autoimmune disorders that have been observed *(175,176)*. Of interest, patients who develop vitiligo or autoantibodies such as anti-thyroid and anti-nuclear antibodies during adjuvant IFN therapy for high-risk melanoma appear to have an improved relapse-free and OS relative to the total IFN-treated population, perhaps suggesting that at least in patients with melanoma, IFN mediates its anti-tumor effect through an autoimmune mechanism *(177)*. The clinician using IFN should be aware that a change in symptomatology of a patient on long-term IFN might herald the development of an autoimmune disease.

4. INTERLEUKINS

The ILs have pleiotropic effects on the innate and cellular immunity as well as hematopoiesis. Studies of cytokines in animal tumor models suggested that they would have broad anti-tumor activity, which led to intense clinical study a large number of cytokines as cancer therapy. Unfortunately, only IL-2 has shown sufficient activity to obtain FDA approval.

4.1. Interleukin-2

4.1.1. Biology

IL-2's effects are mediated by the IL-2 receptor which is a class I cytokine receptor *(178)*. The IL-2 receptor is composed of an α-chain, β-chain, and γ_c. The β-chain and γ_c are involved in signaling, whereas the α-chain is only involved in cytokine binding *(178)*. These subunits form a high-affinity, intermediate-affinity, or low-affinity receptor depending on which of the chains are in the receptor complex (Fig. 1). The high-affinity receptor is a complex of all three subunits with a Ka of 10^{-11} M and the intermediate affinity receptor is composed on the β-chain and γ_c with a Ka 10^{-9} M *(178)*. The α-chain (CD25) is the low-affinity receptor with a Ka of 10^{-8} M; however, this receptor does not initiate intracellular signaling *(178)*. Although the β-chain and

γ_c are expressed on T-cells, B-cells, and NK cells, the α-chain (CD25) is inducible, and its expression is restricted to T-cells *(178)*.

IL-2 has a myriad of effects on the immune system. When T-cells are stimulated with antigen, IL-2 is produced resulting in autocrine and paracrine effects on T-cells. Additionally, IL-2 stimulates release of other cytokines by T-cells. NK-cells express the intermediate affinity IL-2 receptor *(179)*. Exposure of NK cells to IL-2 results in their proliferation, enhanced cytolytic activity, and secretion of other cytokines *(179)*. B-cells also express intermediate affinity IL-2 receptor and IL-2 in co-operation with other cytokines results in B-cell proliferation and differentiation *(180,181)*.

Recent reports suggest that IL-2 also plays a critical role in suppressing immune responses. A subpopulation of CD4+ T-lymphocytes that co-express CD25 function as T-regulatory (Treg) cells, and these cells suppress self-reactive T-cells *(182)*. Depletion of CD4+CD25+ Tregs breaks tolerance to self-antigens and can lead to increased autoimmunity in animal models *(182)*. Additionally, depletion of CD4+CD25+ Tregs enhances tumor rejection and improves response to cancer vaccines by promoting the function of CD8+ CTLs due to lack of inhibition by CD4+CD25+ lymphocytes *(183)*. The mechanism by which CD4+CD25+ lymphocytes inhibit the function of CD8+ CTLs is poorly understood. Mice with targeted deletion of IL-2 and the IL-2 receptor develop a generalized inflammatory syndrome and often die of autoimmune colitis *(184–187)*. These data suggest that IL-2 not only activates immune responses but also participates in a negative feedback loop to limit immune responses.

4.2. Indications

4.2.1. RENAL CELL CARCINOMA

HD IL-2, either alone or in combination with adoptive transfer of lymphokine-activated killer (LAK) cells, has produced durable remissions in patients with advanced renal cell cancer. Randomized studies have suggested that the addition of LAK cells was not required for therapeutic benefit *(188–190)*. Although multiple IL-2 regimens have been examined including continuous infusion, LD bolus, and subcutaneous administration, none have shown superior anti-tumor activity relative to the HD bolus regimen. This regimen involves IL-2 (600,000 or 720,000 IU/kg intravenously every 8 hours days 1–5 and 15–19 of an 8–12 week course) *(191,192)*. This regimen has produced an overall response rate of 15–23% with 7–10% complete responses *(191,192)*. Responses have been relatively durable with the median duration of response ranging from 24 to 54 months and over 70% of complete responders being long-term free from relapse *(191,192)*. Based on this data bolus HD IL-2 received regulatory approved for RCC in 1992.

Other regimens of IL-2 have been tested in an attempt to reduce the toxicity of therapy while maintaining the clinical benefit. A three-arm randomized trial compared IV bolus regimens of HD IL-2 and LD IL-2 (~10% of the total dose in HD IL-2) given over two 5-day periods separated by a 9-day break with subcutaneous IL-2 given over 6 weeks *(193)*. The response rate in the HD arm was twice that of both LD arms (21% versus 11% and 10%). Although there was no significant difference in OS, the patients who achieved a CR on HD IL-2 had a significantly greater RFS than those who achieved a CR on the LD IL-2 arm suggesting that the trade off for reduced toxicity is reduced durability of the response *(193)*.

Combinations of IL-2 and IFN with and without 5FU-based chemotherapy have shown apparent improvement in response rates relative to IFN alone *(194–197)*.

However, a randomized phase III study comparing HD IL-2 to one of the more active LD IL-2 and IFN regimens showed a clear advantage for HD IL-2 in terms of response rate and percentage of durable CRs and a survival advantage in prestratified subsets of patients *(198)*. These results suggest that HD-IL-2 therapy should be considered the standard of care for selected patients with access to such treatment.

4.2.2. MELANOMA

HD IL-2 received regulatory approval for patients with advanced melanoma in 1998. This approval was also largely based on its demonstrated ability to produce durable complete responses in a minority of patients *(199)*. Data collected from multiple phase II studies showed a response rate of 16% with 6% of patients achieving a CR and 10% a partial response (PR) *(199)*. The median response duration was 11.2 months for all responders and exceeded 59 months for patients with a CR (Fig. 4) *(199,200)*. No patient achieving a response lasting in excess of 30 months has relapsed. Given that follow-up on many of these patients exceeds 15 years, this remarkable durability suggests that some if not all of these patients may actually be cured.

A variety IL-2-containing regimens have been tested in patients with melanoma in an effort to improve the effectiveness of general applicability of IL-2. A study from the NCI Surgery Branch suggested that the addition of a gp100 peptide vaccine to HD IL-2 might increase the response rate to 40% *(201,202)*. A randomized phase II study of 131 patients conducted by the Cytokine Working Group was designed to verify these results and the optimal vaccine and IL-2 schedule *(203)*. Although this study was not designed to compare IL-2 to IL-2 plus vaccine, the fact that the best response rate seen in any arm was 19% suggests that the addition of vaccine to HD IL-2 therapy is of little clinical benefit *(203)*. However, a definitive assessment must await the completion of a currently ongoing randomized phase III trial comparing HD IL-2 + gp100 vaccine to HD IL-2 alone. Data from phase II studies also suggested that the addition of IFN to IL-2 enhanced the response rate *(204,205)*; however, a randomized phase III study showed only a modest improvement in the overall response rate from 5% in the IL-2 alone group to 10% in the combination therapy group *(206)*.

In the 1990s, biochemotherapy regimens were developed under the hypothesis that the combination of chemotherapy and cytokines would increase response rates and lead to an increase in the number of durable responses. The phase II data suggested that these regimens had response rates in excess of 40% with up to 10% of patients achieving a durable response *(207–211)*. Unfortunately, four recent randomized trials showed no survival advantage to biochemotherapy compared with chemotherapy alone *(212–215)*. A study from the John Wayne Cancer Center administered IL-2/GM-CSF consolidation therapy to patients who were stable or responding following biochemotherapy in an effort to obtain more durable responses *(216,217)*. A recent multicenter study using this regimen reported an overall response rate to biochemotherapy of 44% with 8% CRs *(217)*. Interestingly, three patients with PRs after biochemotherapy went on to CR while on maintenance IL-2/GM-CSF *(217)*. However, almost 40% of the patients had CNS disease as their first or only site of progression. Although the data with maintenance IL-2 and GM-CSF are encouraging, the results from the randomized studies of biochemotherapy alone suggest that it has a limited role outside of clinical trials.

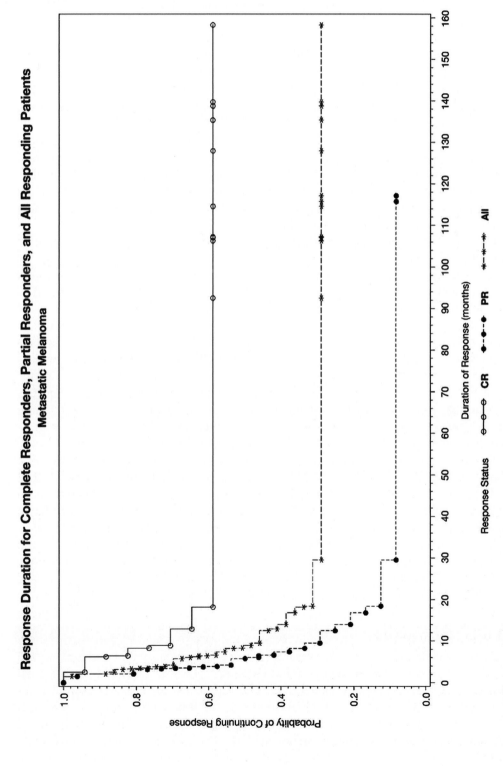

Fig. 4. Kaplan–Meier estimates of overall survival of responders to high-dose IL-2 therapy (median follow-up >7 years)

4.3. Toxicity

IL-2 is associated with a myriad of side effects, and owing to its toxicity, HD IL-2 cannot be given outside the hospital setting. IL-2 can cause constitutional symptoms such as fever, chill, and fatigue *(218)*. Gastrointestinal side effects such as nausea, vomiting, anorexia, transaminitis, cholestasis, and diarrhea are common *(218)*. IL-2 administration leads to increased vascular permeability, which can manifest as fluid retention including pleural effusions and occasionally pulmonary edema. Cardiovascular toxicity is often dose limiting and may present as cardiac arrhythmias, or hypotension, requiring vasopressor support *(218)*. Reversible renal and hepatic dysfunction is common *(218)*. Hematopoietic toxicity can manifest as thrombocytopenia, anemia, or coagulopathy *(218)*. Because IL-2 reversibly inhibits neutrophil chemotaxis, patients are usually placed on antibiotic prophylaxis to prevent catheter-related bacteremia that occurred commonly, sometimes with fatal consequences in the early trials *(219)*. Almost all of the side effects of IL-2 resolve rapidly with holding of the drug and are therefore manageable.

There are three treatment side effects that can worsen or persist for a period of time after drug discontinuation: autoimmunity, neurotoxicity, and myocarditis *(218)*. As with IFN therapy, patients develop autoimmune disorders such as thyroid dysfunction, which may take 6–10 months to resolve, and vitiligo, which is often progressive. IL-2 neurotoxicity can be subtle presenting as lethargy and irritability or it can present as florid psychosis *(218)*. Neurotoxicity appears to peak 24 h after the last dose and requires vigilance on the part of the physician and staff to recognize the symptoms of neurotoxicity early. Some patients develop a myocarditis, which typically develops on day 6 of the first cycle of therapy and is often only manifested by a rise of the cardiac enzymes. Although this typically resolves within a few days without sequelae, it occasionally can be associated with reversible cardiac dysfunction and ventricular ectopy *(218)*. The severity of the side effects observed with HD IL-2 require it to be administered as an inpatient by an experienced team of physicians and nurses.

IL-2 probably mediates its toxic effects through other compounds such as nitric oxide, IL-1, TNF, and IFN *(220,221)*. In hopes of reducing the toxicity of IL-2 while preserving the therapeutic benefit, trials of toxicity-modifying agents have been conducted. Inhibitors of the TNF and IL-1 pathways have been used with no significant change in the toxicity of the IL-2 *(222–224)*. A phase I trial of *N*-mono-methyl-L-arginine, which inhibits nitric oxide, improved hypotension in the setting of IL-2 given by continuous infusion, but there was no improvement in the ability to deliver therapy *(225,226)*. No inhibitor of IL-2 toxicity has shown sufficient ability to dissociate toxicity from anti-tumor activity to merit widespread use.

4.3.1. INTERLEUKIN-12

IL-12 is a heterodimer that consists of a 35-kDa and 40-kDa subunit that is expressed by activated mononuclear phagocytes and DCs *(227)*. LPS, intracellular bacteria, and viruses stimulate IL-12 expression *(228–231)*. This early innate response enhances the activity of NK cells *(232)*. IL-12 also stimulates IFN-γ production by NK cells and T-lymphocytes that in turn activates macrophages *(232,233)*. Additionally, IL-12 stimulates the CD4+ cells to differentiate into Th1 CD4+ cells and enhances the cytoxic activity of CD8+ CTLs *(232–235)*. Thus, IL-12 not only serves to activate the innate immune response, but it also helps initiate an adaptive immune response to infection or foreign antigen.

IL-12 has demonstrated anti-tumor activity in murine models of melanoma, colon carcinoma, mammary carcinoma, and sarcoma *(236–239)*. Investigations of the mechanism of IL-12 activity using mice with molecularly targeted defects suggest that different branches of the immune system mediate its anti-tumor effects. Studies using the B16 melanoma model demonstrate a role for NK cells in mediating the anti-tumor effect of IL-12 when administered at high doses *(240,241)*. In contrast, anti-tumor responses at low doses of IL-12 appear to be mediated by NKT cells *(242)*. In addition to its immune effects, IL-12 has anti-angiogenic effects that are mediated by IFN-γ and IP-10 *(243)*. The anti-tumor activity of IL-12 exhibited in animal models encouraged its clinical investigation.

A peculiar schedule dependency associated with IL-12 in which a single "test dose" increased tolerance to subsequent therapy has limited its clinical development. In the initial phase I trials, a single dose, "test dose," was given intravenously followed 2 weeks later by once daily IV bolus injections for 5 days every 3 weeks. When the phase II trial was initiated at the maximum-tolerated dose (MTD) of 500 ng/kg from the phase I trial, the test dose was dropped from the schedule resulting in unexpected significant toxicity. Animal studies demonstrated that the test dose markedly decreased IFN-γ production and the toxicity associated with IL-12 administration. After this discovery, intravenous and subcutaneous dosage schedules were developed that omitted the "test dose."

The results from these trials suggest that the response rate of IL-12 monotherapy in RCC and melanoma is less than 5% *(206)*. A trial of IL-12 in ovarian cancer resulted in stabilization of disease in 13 of 26 patients; however, only one patient had a PR. Although IL-12 had a low response rate, it was noted that patients who responded had sustained IFN-γ, IL-15, and IL-18 production after treatment. These data suggested that if IFN-γ production could be sustained the response rate might be improved. In a phase I setting the combination of LD IL-2 and IL-12 produced sustained IFN-γ production as well as an expansion of NK cells; however, only one patient achieved a PR *(244)*. Trials involving IL-12 as an adjuvant for vaccine therapy are ongoing; however, toxicity and low efficacy have dimmed enthusiasm for its use in the advanced disease setting.

4.3.2. INTERLEUKIN-18

IL-18 was initially identified as IFN-γ-inducing factor and is structurally related to IL-1 *(245)*. Like IL-1β, IL-18 is expressed as a precursor that requires processing by IL-1β converting enzyme to its 18-kDa active form. IL-18 synergizes with IL-12 and thus has many overlapping activities *(36,245,246)*. IL-18 stimulates IFN-γ production by NK and CD8+ T-cells as well as enhancing their cytotoxicity *(247–249)*. It also activates macrophages and promotes the development of Th1 helper cells that secrete IL-2, IFN-γ, and GM-CSF *(36,250)*. Additionally, IL-18 up-regulates FasL on NK, CD8, and CD4 cells *(247,251,252)*. As with IL-12, IL-18 has been shown to have anti-angiogenic effects in some systems *(253,254)*. Interestingly, IL-18 induces IL-18-binding protein (IL-18 BP) that can bind to and neutralize the activity of IL-18 in the circulation *(255,256)*.

Phase I studies of IL-18 have shown it to be tolerable as an IV regimen once daily for 5 days every 28 days or as once daily for 14 days as a subcutaneous regimen *(257–259)*. IL-18 treatment leads to increases in IFN-γ, GM-CSF, and IL-18BP as well as up-regulation of FasL on NK, CD8, and CD4 cells as predicted from the murine models *(257–259)*. To date the clinical effects have been modest with 2 of 26 patients experiencing tumor responses and three of nine patients with stable disease in phase I studies *(257–259)*. Phase II testing is under way, and the role of IL-18 combined with other cytokines has yet to be investigated.

4.3.3. INTERLEUKIN-21

IL-21 is a member of the IL-2 cytokine family that was isolated using a ligand–receptor pairing method. IL-21 receptor is a heterodimer consisting of the common IL-2-γ_c and the IL-21 receptor *(260)*. IL-21 is produced by CD4+ T-cells and like other members of the γc-dependent cytokines has pleiotropic effects *(260,261)*. IL-21 mediates proliferation CD4 and CD8 T-cells and also enhances CD8 and NK cell cytotoxicity *(260)*. Although the role of IL21 in Th1/Th2 differentiation is unclear, it is required for normal humoral responses *(260–264)*. IL-21 demonstrated activity in murine tumor models of thymoma, melanoma, sarcoma, lymphoma, and adenocarcinoma *(265–271)*. IL-21 has been developed for clinical use, and a phase I trial is currently ongoing *(272)*.

4.3.4. GM-CSF

GM-CSF was initially felt to be critical for hematopoiesis and was approved for clinical use in chemotherapy-related neutropenia. Investigations into the use of cytokines as adjuvants for vaccines revealed the unexpected immunotherapeutic potential of GM-CSF *(273)*. In a murine melanoma model, injection of irradiated melanoma cells expressing GM-CSF provided protection to subsequent tumor challenge in over 90% of mice *(273)*. Administration of irradiated melanoma cells expressing GM-CSF in mice with established tumors improved the survival by 40–60% depending on the initial tumor inoculum *(273)*. These initial data were validated in other animal model systems using various vaccination strategies. The anti-tumor activity of GM-CSF appears to be related to its ability to activate macrophages and DCs. GM-CSF-activated macrophages are cytotoxic to melanoma cells *(274)*. GM-CSF also matures DCs leading to up-regulation of co-stimulatory molecules and CD1d receptors *(275–278)*. Initial studies suggested that CD4+ and CD8+ T-cells mediated GM-CSF tumor immunity, but recent models using CD1d-deficient mice support a critical role for NKT cells in GM-CSF anti-tumor immune responses *(279)*.

Clinical trials of GM-CSF have suggested activity as monotherapy in patients with melanoma when injected intralesionally *(280–282)*. Additionally, multiple trials using autologous tumor vaccines engineered to secrete GM-CSF have shown biologic activity although few clinical responses have been observed *(283)*. Data from administration of GM-CSF in an adjuvant setting in patients with melanoma stage IV disease resected to NED suggested that GM-CSF prolonged survival-relative historical controls *(284)*. Attempts to validate this observation in a phase III intergroup trial comparing GM-CSF to placebo are ongoing. GM-CSF when used in combination with IL-2 has been shown to have a CR rate of 15% in patients who had previously obtained SD or better on biochemotherapy *(216)*. These data as an aggregate demonstrate that GM-CSF has potential anti-tumor activity; however, its role as an immunotherapeutic agent remains to be evaluated.

5. STRATEGIES FOR OVERCOMING RESISTANCE TO IMMUNOTHERAPY

The immunosurveillance hypothesis suggests that tumors are constantly under attack from the immune system. The natural extension of this hypothesis, "immunoediting," suggests that tumors are constantly under selective pressure that favors clones that can escape the onslaught of the immune system (285). Thus, even when immune responses are seen in patients, they often do not correlate with clinical benefit as the tumors may have already undergone "immune escape."

Tumors utilize two broad strategies to escape immune surveillance: altered antigen presentation/T-cell response and immunosuppression. Tumor cells have down-regulated MHC class I molecules, CD80 (B7-1), CD86 (B7-2), and ICAM-1 that are important for antigen presentation and activation of CD8+ CTLs (286). In addition to tumor-specific defects, down-regulation of the T-cell signaling components, zeta chain and Lck, is seen in a number of malignancies (287,288). Of interest, some cytokines appear to reverse the immune defects seen in the setting of malignancy. The IFNs up-regulate MHC I and MHC II, and IL-2 has been shown to restore zeta chain and Lck expression on T-cells (287,288).

Malignancies induce an immunosuppressive microenvironment by multiple mechanisms (Fig. 5). Tumors secret a number of cytokines that are potentially immunosuppressive such as IL-10, transforming growth factor beta (TGF-β), and IL-6 (289–293). Also nutrient catabolizing enzymes such as indoleamine 2,3-dioxygenase (IDO) and arginase contribute to an immunosuppressive environment (294). Tumors cells are protected by mechanisms designed to prevent autoimmune disease. Expression of CLTA-4 on activated T-cells limits the immune response. Invariant NKT (iNKT) cells, CD4+CD25+ T-regulatory (Tregs), Th3 cells, and type I regulatory cells are all immunoregulatory T-cells that are part of the natural course of an immune response. The fact that the induction of a tumor-focused immune response does not correlate with clinical benefit suggests that some of these of immune escape mechanisms are operant in vivo.

iNKT cells are unique in that they may be immunostimulatory or immunosuppressive (295). Under normal conditions, iNKT cells augment the initial immune response and stimulate IFN-γ production (295). However, cancer patients have been shown to have decreased number of iNKT cells, and the iNKT cells that are present have a Th2 phenotype (296–299). Tregs, Th3 cells and type I regulatory cells are CD4+ cells that are hypothesized to down-regulate normal immune responses. These regulatory cells are increased in regional lymph nodes draining primary tumors and limit the ability of CTLs to attack the tumor cells (300–302). Lymphocytes with a Treg phenotype increase in number during HD IL-2 therapy; however, these cells do not appear to have suppressive function. In one study objective response in patients with melanoma and RCC correlated with a subsequent decrease in Tregs four weeks after HD IL-2 therapy 303. These data support the hypothesis that regulatory cells may inhibit responses to immunotherapy.

A number of strategies have emerged for selectively targeting immunoregulatory cells with the aim of enhancing the effects cytokine therapy. IFN-γ secretion by iNKT cells is induced in response to activation with α-galactosylceramide (α-GalCer) *(304,305).* A trial of a-GalCer has recently been completed, which demonstrated increased IFN-γ and IL-12 secretion at the tumor site *(305).* In vitro data show that stimulation of iNKT cells with DC loaded with α-GalCer reverses the immunosuppressive phenotype seen in cancer patients *(305,306).* Although α-GalCer is still early in clinical development, it would be reasonable to combine it with cytokines in hopes of enhancing their efficacy.

Efforts to deplete Tregs, Th3 cells, and type I regulatory T-cells have included the use of the anti-CD25 agent, denileukin diftitox (Ontak), and lymphodepletion. A single dose of Ontak has been shown to deplete circulating regulatory T-cell populations and enhances the anti-tumor response to DC-based vaccines *(307,308).* The NCI Surgery Branch piloted a strategy of lymphodepletion using fludarabine and cyclophosphamide to deplete immunoregulatory cells followed by adoptive transfer of T-cells and IL-2 therapy. This regimen produces a response rate of 50% in patients who were previously resistant to IL-2-based immunotherapy *(309,310).* A phase II study of lymphodepletion followed by HD IL-2 that omits the adoptive transfer of ex vivo-selected T-cells is underway within the Cytokine Working Group. The use of antibodies to block the activation of CTLA-4 has resulted in dramatic responses in early clinical trials *(311–314).* The addition of this agent to HD IL-2 therapy in a phase Ib/II study resulted in a higher response than would have been expected with either agent alone but did not appear to be synergistic *(315).* These early results suggest that strategies to disrupt

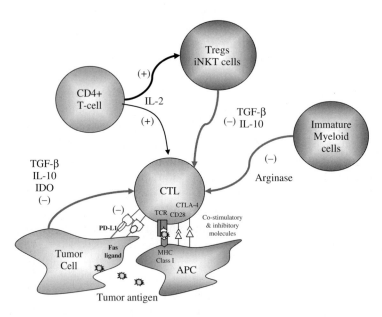

Fig. 5. Anti-tumor responses are inhibited directly by the tumor cells through Fas and CTLA-4 pathways. Tumors create an immunosuppressive microenvironment by secreting cytokines such as interleukin-10 (IL-10) and transforming growth factor beta (TGF-β) and by inducing/recruiting Tregs, Th3 cells, and type 1 regulatory cells. Production of indoleamine 2,3-dioxygenase (IDO) and arginase contribute to the inhibition of the immune response. Targeting these immunosuppressive networks may improve the efficacy of cytokine-based therapy.

the immunosuppressive microenvironment of the tumor may significantly enhance the efficacy of cytokines in the treatment of cancer.

6. PREDICTORS OF RESPONSE

Cytokine therapy has produced durable responses in melanoma and RCC. The majority of patients, however, are exposed to substantial toxicity with only a small percentage obtaining clinical benefit. Potential predictive markers have been studied in an effort to prospectively identify the subset of patients most likely to benefit from IL-2. In a blinded retrospective pathology analysis of 236 cases of RCC, clear cell histology with alveolar and granular but no papillary features predicted for a response rate to IL-2 therapy of 41% (316). Additionally, high expression of carbonic anhydrase IX (CA IX) by RCC tissue has been shown to predict for both response and survival in patients receiving IL-2 therapy. In these reports, durable response was limited to the high CA IX-expressing group. (317,318). A prospective trial is currently being planned to validate these promising biomarkers in RCC.

In melanoma, patients with cutaneous metastases and HLA Cw7 have been reported to be more likely to respond IL-2-containing therapy (110,319,320). Response to HD IL-2 for melanoma also has been reported to be higher in patients with autoimmunity phenomenon such as thyroid dysfunction or vitiligo, low pretreatment IL-6 levels, and low pretreatment C-reactive protein (CRP) levels (320). The development of autoimmunity during adjuvant IFN treatment for melanoma is associated with a dramatic improvement in survival (177). The correlation of autoimmunity with response and the potential role of immunosuppression in resistance to immunotherapy has led to the investigation of polymorphisms of immune pathways as predictors of response.

Single-nucleotide polymorphisms (SNPs) may serve as prognostic or predictive markers due to their linkage to variable expression of critical genes. For example, SNPs that reportedly alter IL-10 and IFN-γ expression are associated with response to biochemotherapy (321). Perhaps paradoxically, SNPs related to increased IL-10 and decreased IFN-γ expression have been associated with good prognosis, calling into question the mechanism of this linkage or the validity of these results (321–324). Clearly, more research is needed to sort out the influence of genotype variations on the anti-tumor effects of various cytokine-based immunotherapies.

7. CONCLUSIONS

Preclinical data suggested that cytokine-based therapies would have efficacy against a broad spectrum of malignancies. Unfortunately, only IFN and IL-2 have found a place in the therapeutic armamentarium against cancer. IL-2 remains the only agent that reproducibly produces long-term remissions in patients with melanoma or RCC. Efforts focused on improved patient selection for IL-2 therapy have recently made significant progress and will allow patients with low likelihood of response to pursue other treatment options while avoiding potentially significant toxicity. Strategies for overcoming resistance to immunotherapy are under intense investigation and hold the promise of extending the clinical benefits of cytokines to larger populations of patients. Although the success of cytokines has been modest to date, immunoreactive cytokines will likely remain a critical component of any curative strategy for the treatment of advanced or high-risk malignancies.

REFERENCES

1. Kaplan, D.H., et al., Demonstration of an interferon gamma-dependent tumor surveillance system in immunocompetent mice. *Proc Natl Acad Sci USA*, 1998. 95(13): p. 7556–61.
2. Picaud, S., et al., Enhanced tumor development in mice lacking a functional type I interferon receptor. *J Interferon Cytokine Res*, 2002. 22(4): p. 457–62.
3. Shankaran, V., et al., IFNgamma and lymphocytes prevent primary tumour development and shape tumour immunogenicity. *Nature*, 2001. 410(6832): p. 1107–11.
4. Street, S.E., E. Cretney, and M.J. Smyth, Perforin and interferon-gamma activities independently control tumor initiation, growth, and metastasis. *Blood*, 2001. 97(1): p. 192–7.
5. Street, S.E., et al., Suppression of lymphoma and epithelial malignancies effected by interferon gamma. *J Exp Med*, 2002. 196(1): p. 129–34.
6. Abbas, A.K., A.H. Lichtman, and J.S. Pober. *Cellular and Molecular Immunology*. W.B. Saunders company; Philadelphia, PA 19103 4th edition (May 15, 2000).
7. Pestka, S., C.D. Krause, and M.R. Walter, Interferons, interferon-like cytokines, and their receptors. *Immunol Rev*, 2004. 202: p. 8–32.
8. Pestka, S., et al., Interferons and their actions. *Annu Rev Biochem*, 1987. 56: p. 727–77.
9. Stewart, W.E., 2nd, Interferon nomenclature recommendations. *J Infect Dis*, 1980. 142(4): p. 643.
10. Isaacs, A. and J. Lindenmann, Virus interference. I. The interferon. *J Interferon Res*, 1987. 7(5): p. 429–38.
11. Muller, U., et al., Functional role of type I and type II interferons in antiviral defense. *Science*, 1994. 264(5167): p. 1918–21.
12. Basham, T.Y., et al., Interferon increases HLA synthesis in melanoma cells: interferon-resistant and -sensitive cell lines. *Proc Natl Acad Sci USA*, 1982. 79(10): p. 3265–9.
13. Dolei, A., M.R. Capobianchi, and F. Ameglio, Human interferon-gamma enhances the expression of class I and class II major histocompatibility complex products in neoplastic cells more effectively than interferon-alpha and interferon-beta. *Infect Immun*, 1983. 40(1): p. 172–6.
14. Herberman, R.B., et al., Effect of human recombinant interferon on cytotoxic activity of natural killer (NK) cells and monocytes. *Cell Immunol*, 1982. 67(1): p. 160–7.
15. Ortaldo, J.R., et al., Effects of several species of human leukocyte interferon on cytotoxic activity of NK cells and monocytes. *Int J Cancer*, 1983. 31(3): p. 285–9.
16. Ortaldo, J.R., et al., Effects of recombinant and hybrid recombinant human leukocyte interferons on cytotoxic activity of natural killer cells. *J Biol Chem*, 1983. 258(24): p. 15011–5.
17. Wagner, T.C., et al., Interferon receptor expression regulates the antiproliferative effects of interferons on cancer cells and solid tumors. *Int J Cancer*, 2004. 111(1): p. 32–42.
18. Clemens, M.J., Interferons and apoptosis. *J Interferon Cytokine Res*, 2003. 23(6): p. 277–92.
19. Sidky, Y.A. and E.C. Borden, Inhibition of angiogenesis by interferons: effects on tumor- and lymphocyte-induced vascular responses. *Cancer Res*, 1987. 47(19): p. 5155–61.
20. Tsuruoka, N., et al., Inhibition of in vitro angiogenesis by lymphotoxin and interferon-gamma. *Biochem Biophys Res Commun*, 1988. 155(1): p. 429–35.
21. Dunn, G.P., et al., A critical function for type I interferons in cancer immunoediting. *Nat Immunol*, 2005. 6(7): p. 722–9.
22. Greenberg, H.B., et al., Effect of human leukocyte interferon on hepatitis B virus infection in patients with chronic active hepatitis. *N Engl J Med*, 1976. 295(10): p. 517–22.
23. Krown, S.E., et al., Preliminary observations on the effect of recombinant leukocyte A interferon in homosexual men with Kaposi's sarcoma. *N Engl J Med*, 1983. 308(18): p. 1071–6.
24. Talpaz, M., et al., Chronic myelogenous leukaemia: haematological remissions with alpha interferon. *Br J Haematol*, 1986. 64(1): p. 87–95.
25. Harris, J.M., N.E. Martin, and M. Modi, Pegylation: a novel process for modifying pharmacokinetics. *Clin Pharmacokinet*, 2001. 40(7): p. 539–51.
26. Reddy, K.R., Development and pharmacokinetics and pharmacodynamics of pegylated interferon alfa-2a (40 kD). *Semin Liver Dis*, 2004. 24(Suppl 2): p. 33–8.
27. Youngster, S., et al., Structure, biology, and therapeutic implications of pegylated interferon alpha-2b. *Curr Pharm Des*, 2002. 8(24): p. 2139–57.
28. Motzer, R.J., et al., Phase II trial of branched peginterferon-alpha 2a (40 kDa) for patients with advanced renal cell carcinoma. *Ann Oncol*, 2002. 13(11): p. 1799–805.

29. Talpaz, M., et al., Phase 1 study of polyethylene glycol formulation of interferon alpha-2B (Schering 54031) in Philadelphia chromosome-positive chronic myelogenous leukemia. *Blood*, 2001. 98(6): p. 1708–13.

30. Nagai, M. and T. Arai, Clinical effect of interferon in malignant brain tumours. *Neurosurg Rev*, 1984. 7(1): p. 55–64.

31. Bukowski, R.M., et al., Phase I trial of natural human interferon beta in metastatic malignancy. *Cancer Res*, 1991. 51(3): p. 836–40.

32. Aulitzky, W.E., et al., Divergent in vivo and in vitro antileukemic activity of recombinant interferon beta in patients with chronic-phase chronic myelogenous leukemia. *Ann Hematol*, 1993. 67(5): p. 205–11.

33. Carnaud, C., et al., Cutting edge: cross-talk between cells of the innate immune system: NKT cells rapidly activate NK cells. *J Immunol*, 1999. 163(9): p. 4647–50.

34. Frucht, D.M., et al., IFN-gamma production by antigen-presenting cells: mechanisms emerge. *Trends Immunol*, 2001. 22(10): p. 556–60.

35. Harris, D.P., et al., Reciprocal regulation of polarized cytokine production by effector B and T cells. *Nat Immunol*, 2000. 1(6): p. 475–82.

36. Yoshimoto, T., et al., IL-12 up-regulates IL-18 receptor expression on T cells, Th1 cells, and B cells: synergism with IL-18 for IFN-gamma production. *J Immunol*, 1998. 161(7): p. 3400–7.

37. Boehm, U., et al., Cellular responses to interferon-gamma. *Annu Rev Immunol*, 1997. 15: p. 749–95.

38. Freedman, A.S., et al., Selective induction of B7/BB-1 on interferon-gamma stimulated monocytes: a potential mechanism for amplification of T cell activation through the CD28 pathway. *Cell Immunol*, 1991. 137(2): p. 429–37.

39. Wallach, D., M. Fellous, and M. Revel, Preferential effect of gamma interferon on the synthesis of HLA antigens and their mRNAs in human cells. *Nature*, 1982. 299(5886): p. 833–6.

40. Groettrup, M., et al., Interferon-gamma inducible exchanges of 20S proteasome active site subunits: why? *Biochimie*, 2001. 83(3–4): p. 367–72.

41. Groettrup, M., et al., A third interferon-gamma-induced subunit exchange in the 20S proteasome. *Eur J Immunol*, 1996. 26(4): p. 863–9.

42. Hisamatsu, H., et al., Newly identified pair of proteasomal subunits regulated reciprocally by interferon gamma. *J Exp Med*, 1996. 183(4): p. 1807–16.

43. Gajewski, T.F. and F.W. Fitch, Anti-proliferative effect of IFN-gamma in immune regulation. I. IFN-gamma inhibits the proliferation of Th2 but not Th1 murine helper T lymphocyte clones. *J Immunol*, 1988. 140(12): p. 4245–52.

44. Snapper, C.M. and W.E. Paul, Interferon-gamma and B cell stimulatory factor-1 reciprocally regulate Ig isotype production. *Science*, 1987. 236(4804): p. 944–7.

45. Dighe, A.S., et al., Enhanced in vivo growth and resistance to rejection of tumor cells expressing dominant negative IFN gamma receptors. *Immunity*, 1994. 1(6): p. 447–56.

46. Coughlin, C.M., et al., Tumor cell responses to IFNgamma affect tumorigenicity and response to IL-12 therapy and antiangiogenesis. *Immunity*, 1998. 9(1): p. 25–34.

47. Friesel, R., A. Komoriya, and T. Maciag, Inhibition of endothelial cell proliferation by gamma-interferon. *J Cell Biol*, 1987. 104(3): p. 689–96.

48. Pfizenmaier, K., et al., Differential gamma-interferon response of human colon carcinoma cells: inhibition of proliferation and modulation of immunogenicity as independent effects of gamma-interferon on tumor cell growth. *Cancer Res*, 1985. 45(8): p. 3503–9.

49. Ratliff, T.L., et al., Inhibition of mouse bladder tumor proliferation by murine interferon-gamma and its synergism with interferon-beta. *Cancer Res*, 1984. 44(10): p. 4377–81.

50. Rubin, B.Y., V. Sekar, and W.A. Martimucci, Comparative antiproliferative efficacies of human alpha and gamma interferons. *J Gen Virol*, 1983. 64(Pt 8): p. 1743–8.

51. Elhilali, M.M., et al., Placebo-associated remissions in a multicentre, randomized, double-blind trial of interferon gamma-1b for the treatment of metastatic renal cell carcinoma. The Canadian Urologic Oncology Group. *BJU Int*, 2000. 86(6): p. 613–8.

52. Koziner, B., et al., Double-blind prospective randomized comparison of interferon gamma-1b versus placebo after autologous stem cell transplantation. *Acta Haematol*, 2002. 108(2): p. 66–73.

53. Small, E.J., et al., The treatment of metastatic renal cell carcinoma patients with recombinant human gamma interferon. *Cancer J Sci Am*, 1998. 4(3): p. 162–7.

54. Todd, P.A. and K.L. Goa, Interferon gamma-1b. A review of its pharmacology and therapeutic potential in chronic granulomatous disease. *Drugs*, 1992. 43(1): p. 111–22.

55. Golomb, H.M., et al., Alpha-2 interferon therapy of hairy-cell leukemia: a multicenter study of 64 patients. *J Clin Oncol*, 1986. 4(6): p. 900–5.

56. Quesada, J.R., et al., Treatment of hairy cell leukemia with recombinant alpha-interferon. *Blood*, 1986. 68(2): p. 493–7.

57. Golomb, H.M., et al., Interferon treatment for hairy cell leukemia: an update on a cohort of 69 patients treated from 1983–1986. *Leukemia*, 1992. 6(11): p. 1177–80.

58. Dadmarz, R., et al., The mechanism of action of interferon-alpha (IFN-alpha) in hairy-cell leukaemia; Hu-IFN-alpha 2 receptor expression by hairy cells and other normal and leukaemic cell types. *Leuk Res*, 1986. 10(11): p. 1279–85.

59. Paganelli, K.A., et al., B cell growth factor-induced proliferation of hairy cell lymphocytes and inhibition by type I interferon in vitro. *Blood*, 1986. 67(4): p. 937–42.

60. Huber, C., et al., Studies on the optimal dose and the mode of action of alpha-interferon in the treatment of hairy cell leukemia. *Leukemia*, 1987. 1(4): p. 355–7.

61. Goodman, G.R., et al., Extended follow-up of patients with hairy cell leukemia after treatment with cladribine. *J Clin Oncol*, 2003. 21(5): p. 891–6.

62. Seymour, J.F., et al., Response to interferon-alpha in patients with hairy cell leukemia relapsing after treatment with 2-chlorodeoxyadenosine. *Leukemia*, 1995. 9(5): p. 929–32.

63. Talpaz, M., et al., Hematologic remission and cytogenetic improvement induced by recombinant human interferon alpha A in chronic myelogenous leukemia. *N Engl J Med*, 1986. 314(17): p. 1065–9.

64. Talpaz, M., et al., Leukocyte interferon-induced myeloid cytoreduction in chronic myelogenous leukemia. *Blood*, 1983. 62(3): p. 689–92.

65. The Italian Cooperative Study Group on Chronic Myeloid Leukemia, Interferon alfa-2a as compared with conventional chemotherapy for the treatment of chronic myeloid leukemia. *N Engl J Med*, 1994. 330(12): p. 820–5.

66. Allan, N.C., S.M. Richards, and P.C. Shepherd, UK Medical Research Council randomised, multi-centre trial of interferon-alpha n1 for chronic myeloid leukaemia: improved survival irrespective of cytogenetic response. The UK Medical Research Council's Working Parties for Therapeutic Trials in Adult Leukaemia. *Lancet*, 1995. 345(8962): p. 1392–7.

67. Broustet, A., et al., Hydroxyurea versus interferon alfa-2b in chronic myelogenous leukaemia: preliminary results of an open French multicentre randomized study. *Eur J Cancer*, 1991. 27(Suppl 4): p. S18–21.

68. Hehlmann, R., et al., Randomized comparison of interferon-alpha with busulfan and hydroxyurea in chronic myelogenous leukemia. The German CML Study Group. *Blood*, 1994. 84(12): p. 4064–77.

69. Ohnishi, K., et al., A randomized trial comparing interferon-alpha with busulfan for newly diagnosed chronic myelogenous leukemia in chronic phase. *Blood*, 1995. 86(3): p. 906–16.

70. The Italian Cooperative Study Group on Chronic Myeloid Leukemia, Long-term follow-Up of the italian trial of interferon-alpha versus conventional chemotherapy in chronic myeloid leukemia. *Blood*, 1998. 92(5): p. 1541–8.

71. Hehlmann, R., et al., Randomized comparison of interferon alpha and hydroxyurea with hydroxyurea monotherapy in chronic myeloid leukemia (CML-study II): prolongation of survival by the combination of interferon alpha and hydroxyurea. *Leukemia*, 2003. 17(8): p. 1529–37.

72. Chronic Myeloid Leukemia Trialists' Collaborative Group, Interferon alfa versus chemotherapy for chronic myeloid leukemia: a meta-analysis of seven randomized trials. *J Natl Cancer Inst*, 1997. 89(21): p. 1616–20.

73. Yasukawa, M., et al., CD4(+) cytotoxic T-cell clones specific for bcr-abl b3a2 fusion peptide augment colony formation by chronic myelogenous leukemia cells in a b3a2-specific and HLA-DR-restricted manner. *Blood*, 1998. 92(9): p. 3355–61.

74. Aswald, J.M., J.H. Lipton, and H.A. Messner, Intracellular cytokine analysis of interferon-gamma in T cells of patients with chronic myeloid leukemia. *Cytokines Cell Mol Ther*, 2002. 7(2): p. 75–82.

75. Nicolson, N.L., M. Talpaz, and G.L. Nicolson, Interferon-alpha directly inhibits DNA polymerase activity in isolated chromatin nucleoprotein complexes: correlation with IFN-alpha treatment outcome in patients with chronic myelogenous leukemia. *Gene*, 1995. 159(1): p. 105–11.

76. Kuhr, T., et al., A randomized study comparing interferon (IFN alpha) plus low-dose cytarabine and interferon plus hydroxyurea (HU) in early chronic-phase chronic myeloid leukemia (CML). *Leuk Res*, 2003. 27(5): p. 405–11.

77. Giles, F.J., et al., A prospective randomized study of alpha-2b interferon plus hydroxyurea or cytarabine for patients with early chronic phase chronic myelogenous leukemia: the International Oncology Study Group CML1 study. *Leuk Lymphoma*, 2000. 37(3–4): p. 367–77.

78. Guilhot, F., et al., Interferon alfa-2b combined with cytarabine versus interferon alone in chronic myelogenous leukemia. French Chronic Myeloid Leukemia Study Group. *N Engl J Med*, 1997. 337(4): p. 223–9.

79. Anstrom, K.J., et al., Long-term survival estimates for imatinib versus interferon-alpha plus low-dose cytarabine for patients with newly diagnosed chronic-phase chronic myeloid leukemia. *Cancer*, 2004. 101(11): p. 2584–92.

80. O'Brien, S.G. and M.W. Deininger, Imatinib in patients with newly diagnosed chronic-phase chronic myeloid leukemia. *Semin Hematol*, 2003. 40(2 Suppl 2): p. 26–30.

81. Kantarjian, H.M., et al., Imatinib mesylate therapy improves survival in patients with newly diagnosed Philadelphia chromosome-positive chronic myelogenous leukemia in the chronic phase: comparison with historic data. *Cancer*, 2003. 98(12): p. 2636–42.

82. Branford, S., et al., Imatinib produces significantly superior molecular responses compared to interferon alfa plus cytarabine in patients with newly diagnosed chronic myeloid leukemia in chronic phase. *Leukemia*, 2003. 17(12): p. 2401–9.

83. Foon, K.A., et al., Treatment of advanced non-Hodgkin's lymphoma with recombinant leukocyte A interferon. *N Engl J Med*, 1984. 311(18): p. 1148–52.

84. Siegert, W., et al., Treatment of non-hodgkin's lymphoma of low-grade malignancy with human fibroblast interferon. *Anticancer Res*, 1982. 2(4): p. 193–8.

85. Louie, A.C., et al., Follow-up observations on the effect of human leukocyte interferon in non-Hodgkin's lymphoma. *Blood*, 1981. 58(4): p. 712–8.

86. Solal-Celigny, P., et al., Doxorubicin-containing regimen with or without interferon alfa-2b for advanced follicular lymphomas: final analysis of survival and toxicity in the Groupe d'Etude des Lymphomes Folliculaires 86 Trial. *J Clin Oncol*, 1998. 16(7): p. 2332–8.

87. Solal-Celigny, P., et al., Recombinant interferon alfa-2b combined with a regimen containing doxorubicin in patients with advanced follicular lymphoma. Groupe d'Etude des Lymphomes de l'Adulte. *N Engl J Med*, 1993. 329(22): p. 1608–14.

88. Rohatiner, A.Z., et al., Meta-analysis to evaluate the role of interferon in follicular lymphoma. *J Clin Oncol*, 2005. 23(10): p. 2215–23.

89. Fisher, R.I., et al., Interferon alpha consolidation after intensive chemotherapy does not prolong the progression-free survival of patients with low-grade non-Hodgkin's lymphoma: results of the Southwest Oncology Group randomized phase III study 8809. *J Clin Oncol*, 2000. 18(10): p. 2010–6.

90. Barnetson, R.S. and G.M. Halliday, Regression in skin tumours: a common phenomenon. *Australas J Dermatol*, 1997. 38(Suppl 1): p. S63–5.

91. Chang, P. and W.H. Knapper, Metastatic melanoma of unknown primary. *Cancer*, 1982. 49(6): p. 1106–11.

92. Panagopoulos, E. and D. Murray, Metastatic malignant melanoma of unknown primary origin: a study of 30 cases. *J Surg Oncol*, 1983. 23(1): p. 8–10.

93. Norman, J., et al., Metastatic melanoma with an unknown primary. *Ann Plast Surg*, 1992. 28(1): p. 81–4.

94. Reintgen, D.S., et al., Metastatic malignant melanoma with an unknown primary. *Surg Gynecol Obstet*, 1983. 156(3): p. 335–40.

95. Gromet, M.A., W.L. Epstein, and M.S. Blois, The regressing thin malignant melanoma: a distinctive lesion with metastatic potential. *Cancer*, 1978. 42(5): p. 2282–92.

96. Baab, G.H. and C.M. McBride, Malignant melanoma: the patient with an unknown site of primary origin. *Arch Surg*, 1975. 110(8): p. 896–900.

97. King, M., D. Spooner, and D.C. Rowlands, Spontaneous regression of metastatic malignant melanoma of the parotid gland and neck lymph nodes: a case report and a review of the literature. *Clin Oncol (R Coll Radiol)*, 2001. 13(6): p. 466–9.

98. Coates, A., et al., Phase-II study of recombinant alpha 2-interferon in advanced malignant melanoma. *J Interferon Res*, 1986. 6(1): p. 1–4.

99. Creagan, E.T., et al., Phase II study of recombinant leukocyte A interferon (rIFN-alpha A) in disseminated malignant melanoma. *Cancer*, 1984. 54(12): p. 2844–9.

100. Dorval, T., et al., Clinical phase II trial of recombinant DNA interferon (interferon alpha 2b) in patients with metastatic malignant melanoma. *Cancer*, 1986. 58(2): p. 215–8.
101. Hersey, P., et al., Effects of recombinant leukocyte interferon (rIFN-alpha A) on tumour growth and immune responses in patients with metastatic melanoma. *Br J Cancer*, 1985. 51(6): p. 815–26.
102. Mughal, T.I., et al., Role of recombinant interferon alpha 2 and cimetidine in patients with advanced malignant melanoma. *J Cancer Res Clin Oncol*, 1988. 114(1): p. 108–9.
103. Neefe, J.R., et al., Phase II study of recombinant alpha-interferon in malignant melanoma. *Am J Clin Oncol*, 1990. 13(6): p. 472–6.
104. Robinson, W.A., et al., Treatment of metastatic malignant melanoma with recombinant interferon alpha 2. *Immunobiology*, 1986. 172(3–5): p. 275–82.
105. Sertoli, M.R., et al., Phase II trial of recombinant alpha-2b interferon in the treatment of metastatic skin melanoma. *Oncology*, 1989. 46(2): p. 96–8.
106. Steiner, A., C. Wolf, and H. Pehamberger, Comparison of the effects of three different treatment regimens of recombinant interferons (r-IFN alpha, r-IFN gamma, and r-IFN alpha + cimetidine) in disseminated malignant melanoma. *J Cancer Res Clin Oncol*, 1987. 113(5): p. 459–65.
107. Decatris, M., S. Santhanam, and K. O'Byrne, Potential of interferon-alpha in solid tumours: part 1. *BioDrugs*, 2002. 16(4): p. 261–81.
108. Legha, S.S., et al., Clinical evaluation of recombinant interferon alfa-2a (Roferon-A) in metastatic melanoma using two different schedules. *J Clin Oncol*, 1987. 5(8): p. 1240–6.
109. Falkson, C.I., et al., Phase III trial of dacarbazine versus dacarbazine with interferon alpha-2b versus dacarbazine with tamoxifen versus dacarbazine with interferon alpha-2b and tamoxifen in patients with metastatic malignant melanoma: an Eastern Cooperative Oncology Group study. *J Clin Oncol*, 1998. 16(5): p. 1743–51.
110. Creagan, E.T., et al., Recombinant leukocyte A interferon (rIFN-alpha A) in the treatment of disseminated malignant melanoma. Analysis of complete and long-term responding patients. *Cancer*, 1986. 58(12): p. 2576–8.
111. Eggermont, A.M., et al., Post-surgery adjuvant therapy with intermediate doses of interferon alfa 2b versus observation in patients with stage IIb/III melanoma (EORTC 18952): randomised controlled trial. *Lancet*, 2005. 366(9492): p. 1189–96.
112. Grob, J.J., et al., Randomised trial of interferon alpha-2a as adjuvant therapy in resected primary melanoma thicker than 1.5 mm without clinically detectable node metastases. French Cooperative Group on Melanoma. *Lancet*, 1998. 351(9120): p. 1905–10.
113. Castello, G., et al., Immunological and clinical effects of intramuscular rIFN alpha-2a and low dose subcutaneous rIL-2 in patients with advanced malignant melanoma. *Melanoma Res*, 1993. 3(1): p. 43–9.
114. Creagan, E.T., et al., Randomized, surgical adjuvant clinical trial of recombinant interferon alfa-2a in selected patients with malignant melanoma. *J Clin Oncol*, 1995. 13(11): p. 2776–83.
115. Kleeberg, U.R., et al., Final results of the EORTC 18871/DKG 80–1 randomised phase III trial. rIFN-alpha2b versus rIFN-gamma versus ISCADOR M versus observation after surgery in melanoma patients with either high-risk primary (thickness >3 mm) or regional lymph node metastasis. *Eur J Cancer*, 2004. 40(3): p. 390–402.
116. Pehamberger, H., et al., Adjuvant interferon alfa-2a treatment in resected primary stage II cutaneous melanoma. Austrian Malignant Melanoma Cooperative Group. *J Clin Oncol*, 1998. 16(4): p. 1425–9.
117. Cameron, D.A., et al., Adjuvant interferon alpha 2b in high risk melanoma - the Scottish study. *Br J Cancer*, 2001. 84(9): p. 1146–9.
118. Hancock, B.W., et al., Adjuvant interferon in high-risk melanoma: the AIM HIGH Study–United Kingdom Coordinating Committee on Cancer Research randomized study of adjuvant low-dose extended-duration interferon Alfa-2a in high-risk resected malignant melanoma. *J Clin Oncol*, 2004. 22(1): p. 53–61.
119. Kirkwood, J.M., et al., High-dose interferon alfa-2b significantly prolongs relapse-free and overall survival compared with the GM2-KLH/QS-21 vaccine in patients with resected stage IIB-III melanoma: results of intergroup trial E1694/S9512/C509801. *J Clin Oncol*, 2001. 19(9): p. 2370–80.
120. Kirkwood, J.M., et al., High- and low-dose interferon alfa-2b in high-risk melanoma: first analysis of intergroup trial E1690/S9111/C9190. *J Clin Oncol*, 2000. 18(12): p. 2444–58.

121. Hillner, B.E., et al., Economic analysis of adjuvant interferon alfa-2b in high-risk melanoma based on projections from Eastern Cooperative Oncology Group 1684. *J Clin Oncol*, 1997. 15(6): p. 2351–8.

122. Kirkwood, J.M., et al., Interferon alfa-2b adjuvant therapy of high-risk resected cutaneous melanoma: the Eastern Cooperative Oncology Group Trial EST 1684. *J Clin Oncol*, 1996. 14(1): p. 7–17.

123. Kirkwood, J.M., et al., A pooled analysis of eastern cooperative oncology group and intergroup trials of adjuvant high-dose interferon for melanoma. *Clin Cancer Res*, 2004. 10(5): p. 1670–7.

124. Falkson, C.I., G. Falkson, and H.C. Falkson, Improved results with the addition of interferon alfa-2b to dacarbazine in the treatment of patients with metastatic malignant melanoma. *J Clin Oncol*, 1991. 9(8): p. 1403–8.

125. Thomson, D.B., et al., Interferon-alpha 2a does not improve response or survival when combined with dacarbazine in metastatic malignant melanoma: results of a multi-institutional Australian randomized trial. *Melanoma Res*, 1993. 3(2): p. 133–8.

126. Bajetta, E., et al., Multicenter randomized trial of dacarbazine alone or in combination with two different doses and schedules of interferon alfa-2a in the treatment of advanced melanoma. *J Clin Oncol*, 1994. 12(4): p. 806–11.

127. Creagan, E.T., et al., A phase I-II trial of the combination of recombinant leukocyte A interferon and recombinant human interferon-gamma in patients with metastatic malignant melanoma. *Cancer*, 1988. 62(12): p. 2472–4.

128. deKernion, J.B., et al., The treatment of renal cell carcinoma with human leukocyte alpha-interferon. *J Urol*, 1983. 130(6): p. 1063–6.

129. Quesada, J.R., D.A. Swanson, and J.U. Gutterman, Phase II study of interferon alpha in metastatic renal-cell carcinoma: a progress report. *J Clin Oncol*, 1985. 3(8): p. 1086–92.

130. Quesada, J.R., et al., Antitumor activity of recombinant-derived interferon alpha in metastatic renal cell carcinoma. *J Clin Oncol*, 1985. 3(11): p. 1522–8.

131. Kempf, R.A., et al., Recombinant interferon alpha-2 (INTRON A) in a phase II study of renal cell carcinoma. *J Biol Response Mod*, 1986. 5(1): p. 27–35.

132. Umeda, T. and T. Niijima, Phase II study of alpha interferon on renal cell carcinoma. Summary of three collaborative trials. *Cancer*, 1986. 58(6): p. 1231–5.

133. Trump, D.L., et al., High-dose lymphoblastoid interferon in advanced renal cell carcinoma: an Eastern Cooperative Oncology Group Study. *Cancer Treat Rep*, 1987. 71(2): p. 165–9.

134. Muss, H.B., et al., Recombinant alfa interferon in renal cell carcinoma: a randomized trial of two routes of administration. *J Clin Oncol*, 1987. 5(2): p. 286–91.

135. Porzsolt, F., et al., Treatment of advanced renal cell cancer with recombinant interferon alpha as a single agent and in combination with medroxyprogesterone acetate. A randomized multicenter trial. *J Cancer Res Clin Oncol*, 1988. 114(1): p. 95–100.

136. Steineck, G., et al., Recombinant leukocyte interferon alpha-2a and medroxyprogesterone in advanced renal cell carcinoma. A randomized trial. *Acta Oncol*, 1990. 29(2): p. 155–62.

137. Minasian, L.M., et al., Interferon alfa-2a in advanced renal cell carcinoma: treatment results and survival in 159 patients with long-term follow-up. *J Clin Oncol*, 1993. 11(7): p. 1368–75.

138. Interferon-alpha and survival in metastatic renal carcinoma: early results of a randomised controlled trial. Medical Research Council Renal Cancer Collaborators. *Lancet*, 1999. 353(9146): p. 14–7.

139. Kirkwood, J.M., et al., A randomized study of low and high doses of leukocyte alpha-interferon in metastatic renal cell carcinoma: the American Cancer Society collaborative trial. *Cancer Res*, 1985. 45(2): p. 863–71.

140. Flanigan, R.C., et al., Nephrectomy followed by interferon alfa-2b compared with interferon alfa-2b alone for metastatic renal-cell cancer. *N Engl J Med*, 2001. 345(23): p. 1655–9.

141. Pizzocaro, G., et al., Interferon adjuvant to radical nephrectomy in Robson stages II and III renal cell carcinoma: a multicentric randomized study. *J Clin Oncol*, 2001. 19(2): p. 425–31.

142. Kriegmair, M., R. Oberneder, and A. Hofstetter, Interferon alfa and vinblastine versus medroxyprogesterone acetate in the treatment of metastatic renal cell carcinoma. *Urology*, 1995. 45(5): p. 758–62.

143. Fossa, S.D., et al., Recombinant interferon alfa-2a with or without vinblastine in metastatic renal cell carcinoma: results of a European multi-center phase III study. *Ann Oncol*, 1992. 3(4): p. 301–5.

144. Nanus, D.M., et al., Interaction of retinoic acid and interferon in renal cancer cell lines. *J Interferon Cytokine Res*, 2000. 20(9): p. 787–94.

145. Motzer, R.J., et al., Phase III trial of interferon alfa-2a with or without 13-cis-retinoic acid for patients with advanced renal cell carcinoma. *J Clin Oncol*, 2000. 18(16): p. 2972–80.
146. Foon, K., et al., A prospective randomized trial of alpha 2B-interferon/gamma-interferon or the combination in advanced metastatic renal cell carcinoma. *J Biol Response Mod*, 1988. 7(6): p. 540–5.
147. De Mulder, P.H., et al., EORTC (30885) randomised phase III study with recombinant interferon alpha and recombinant interferon alpha and gamma in patients with advanced renal cell carcinoma. The EORTC Genitourinary Group. *Br J Cancer*, 1995. 71(2): p. 371–5.
148. Schalling, M., et al., A role for a new herpes virus (KSHV) in different forms of Kaposi's sarcoma. *Nat Med*, 1995. 1(7): p. 707–8.
149. Sinkovics, J.G., Kaposi's sarcoma: its 'oncogenes' and growth factors. *Crit Rev Oncol Hematol*, 1991. 11(2): p. 87–107.
150. Ensoli, B., et al., AIDS-Kaposi's sarcoma-derived cells express cytokines with autocrine and paracrine growth effects. *Science*, 1989. 243(4888): p. 223–6.
151. Folkman, J., Successful treatment of an angiogenic disease. *N Engl J Med*, 1989. 320(18): p. 1211–2.
152. White, C.W., et al., Treatment of pulmonary hemangiomatosis with recombinant interferon alfa-2a. *N Engl J Med*, 1989. 320(18): p. 1197–200.
153. Ezekowitz, A., J. Mulliken, and J. Folkman, Interferon alpha therapy of haemangiomas in newborns and infants. *Br J Haematol*, 1991. 79(Suppl 1): p. 67–8.
154. Real, F.X., H.F. Oettgen, and S.E. Krown, Kaposi's sarcoma and the acquired immunodeficiency syndrome: treatment with high and low doses of recombinant leukocyte A interferon. *J Clin Oncol*, 1986. 4(4): p. 544–51.
155. Gelmann, E.P., et al., Human lymphoblastoid interferon treatment of Kaposi's sarcoma in the acquired immune deficiency syndrome. Clinical response and prognostic parameters. *Am J Med*, 1985. 78(5): p. 737–41.
156. Groopman, J.E., et al., Recombinant alpha-2 interferon therapy for Kaposi's sarcoma associated with the acquired immunodeficiency syndrome. *Ann Intern Med*, 1984. 100(5): p. 671–6.
157. Mauss, S. and H. Jablonowski, Efficacy, safety, and tolerance of low-dose, long-term interferon-alpha 2b and zidovudine in early-stage AIDS-associated Kaposi's sarcoma. *J Acquir Immune Defic Syndr Hum Retrovirol*, 1995. 10(2): p. 157–62.
158. Krown, S.E., et al., Interferon-alpha with zidovudine: safety, tolerance, and clinical and virologic effects in patients with Kaposi sarcoma associated with the acquired immunodeficiency syndrome (AIDS). *Ann Intern Med*, 1990. 112(11): p. 812–21.
159. Dezube, B.J., L. Pantanowitz, and D.M. Aboulafia, Management of AIDS-related Kaposi sarcoma: advances in target discovery and treatment. *AIDS Read*, 2004. 14(5): p. 236–8, 243–4, 251–3.
160. Jonasch, E. and F.G. Haluska, Interferon in oncological practice: review of interferon biology, clinical applications, and toxicities. *Oncologist*, 2001. 6(1): p. 34–55.
161. Greenberg, D.B., et al., Adjuvant therapy of melanoma with interferon-alpha-2b is associated with mania and bipolar syndromes. *Cancer*, 2000. 89(2): p. 356–62.
162. Musselman, D.L., et al., Paroxetine for the prevention of depression induced by high-dose interferon alfa. *N Engl J Med*, 2001. 344(13): p. 961–6.
163. Jonasch, E., et al., Adjuvant high-dose interferon alfa-2b in patients with high-risk melanoma. *Cancer J*, 2000. 6(3): p. 139–45.
164. Lacotte, L., et al., Thrombotic thrombocytopenic purpura during interferon alpha treatment for chronic myelogenous leukemia. *Acta Haematol*, 2000. 102(3): p. 160–2.
165. Ravandi-Kashani, F., et al., Thrombotic microangiopathy associated with interferon therapy for patients with chronic myelogenous leukemia: coincidence or true side effect? *Cancer*, 1999. 85(12): p. 2583–8.
166. Rachmani, R., et al., Thrombotic thrombocytopenic purpura complicating chronic myelogenous leukemia treated with interferon-alpha. A report of two successfully treated patients. *Acta Haematol*, 1998. 100(4): p. 204–6.
167. Iyoda, K., et al., Thrombotic thrombocytopenic purpura developed suddenly during interferon treatment for chronic hepatitis C. *J Gastroenterol*, 1998. 33(4): p. 588–92.
168. Gentile, I., et al., Hemolytic anemia during pegylated IFN-alpha2b plus ribavirin treatment for chronic hepatitis C: ribavirin is not always the culprit. *J Interferon Cytokine Res*, 2005. 25(5): p. 283–5.

169. Braathen, L.R. and P. Stavem, Autoimmune haemolytic anaemia associated with interferon alfa-2a in a patient with mycosis fungoides. *BMJ*, 1989. 298(6689): p. 1713.

170. Pangalis, G.A. and E. Griva, Recombinant alfa-2b-interferon therapy in untreated, stages A and B chronic lymphocytic leukemia. A preliminary report. *Cancer*, 1988. 61(5): p. 869–72.

171. Akard, L.P., et al., Alpha-interferon and immune hemolytic anemia. *Ann Intern Med*, 1986. 105(2): p. 306.

172. Jones, T.H., S. Wadler, and K.H. Hupart, Endocrine-mediated mechanisms of fatigue during treatment with interferon-alpha. *Semin Oncol*, 1998. 25(1 Suppl 1): p. 54–63.

173. Brudin, L.H., et al., Fluorine-18 deoxyglucose uptake in sarcoidosis measured with positron emission tomography. *Eur J Nucl Med*, 1994. 21(4): p. 297–305.

174. Lewis, P.J. and A. Salama, Uptake of fluorine-18-fluorodeoxyglucose in sarcoidosis. *J Nucl Med*, 1994. 35(10): p. 1647–9.

175. Brenard, R., Practical management of patients treated with alpha interferon. *Acta Gastroenterol Belg*, 1997. 60(3): p. 211–3.

176. Dalekos, G.N., et al., A prospective evaluation of dermatological side-effects during alpha-interferon therapy for chronic viral hepatitis. *Eur J Gastroenterol Hepatol*, 1998. 10(11): p. 933–9.

177. Gogas, H., et al., Prognostic significance of autoimmunity during treatment of melanoma with interferon. *N Engl J Med*, 2006. 354(7): p. 709–18.

178. Waldmann, T.A. and M. Tsudo, Interleukin-2 receptors: biology and therapeutic potentials. *Hosp Pract (Off Ed)*, 1987. 22(1): p. 77–84, 93–4.

179. Chan, W.C., et al., Large granular lymphocyte proliferation: an analysis of T-cell receptor gene arrangement and expression and the effect of in vitro culture with inducing agents. *Blood*, 1988. 71(1): p. 52–8.

180. Begley, C.G., et al., Human B lymphocytes express the p75 component of the interleukin 2 receptor. *Leuk Res*, 1990. 14(3): p. 263–71.

181. Mingari, M.C., et al., Human interleukin-2 promotes proliferation of activated B cells via surface receptors similar to those of activated T cells. *Nature*, 1984. 312(5995): p. 641–3.

182. Sakaguchi, S., et al., Immunologic self-tolerance maintained by activated T cells expressing IL-2 receptor alpha-chains (CD25). Breakdown of a single mechanism of self-tolerance causes various autoimmune diseases. *J Immunol*, 1995. 155(3): p. 1151–64.

183. Golgher, D., et al., Depletion of CD25+ regulatory cells uncovers immune responses to shared murine tumor rejection antigens. *Eur J Immunol*, 2002. 32(11): p. 3267–75.

184. Barmeyer, C., et al., The interleukin-2-deficient mouse model. *Pathobiology*, 2002. 70(3): p. 139–42.

185. Baumgart, D.C., et al., Mechanisms of intestinal epithelial cell injury and colitis in interleukin 2 (IL2)-deficient mice. *Cell Immunol*, 1998. 187(1): p. 52–66.

186. Garrelds, I.M., et al., Interleukin-2-Deficient mice: effect on cytokines and inflammatory cells in chronic colonic disease. *Dig Dis Sci*, 2002. 47(3): p. 503–10.

187. Kung, J.T., D. Beller, and S.T. Ju, Lymphokine regulation of activation-induced apoptosis in T cells of IL-2 and IL-2R beta knockout mice. *Cell Immunol*, 1998. 185(2): p. 158–63.

188. Negrier, S., et al., Interleukin-2 with or without LAK cells in metastatic renal cell carcinoma: a report of a European multicentre study. *Eur J Cancer Clin Oncol*, 1989. 25(Suppl 3): p. S21–8.

189. Topalian, S.L., et al., Immunotherapy of patients with advanced cancer using tumor-infiltrating lymphocytes and recombinant interleukin-2: a pilot study. *J Clin Oncol*, 1988. 6(5): p. 839–53.

190. Rosenberg, S.A., et al., Observations on the systemic administration of autologous lymphokine-activated killer cells and recombinant interleukin-2 to patients with metastatic cancer. *N Engl J Med*, 1985. 313(23): p. 1485–92.

191. Fisher, R.I., S.A. Rosenberg, and G. Fyfe, Long-term survival update for high-dose recombinant interleukin-2 in patients with renal cell carcinoma. *Cancer J Sci Am*, 2000. 6(Suppl 1): p. S55–7.

192. Fyfe, G.A., et al., Long-term response data for 255 patients with metastatic renal cell carcinoma treated with high-dose recombinant interleukin-2 therapy. *J Clin Oncol*, 1996. 14(8): p. 2410–1.

193. Yang, J.C., et al., Randomized study of high-dose and low-dose interleukin-2 in patients with metastatic renal cancer. *J Clin Oncol*, 2003. 21(16): p. 3127–32.

194. Atzpodien, J., et al., Treatment of metastatic renal cell cancer patients with recombinant subcutaneous human interleukin-2 and interferon-alpha. *Ann Oncol*, 1990. 1(5): p. 377–8.

195. Figlin, R.A., et al., Concomitant administration of recombinant human interleukin-2 and recombinant interferon alfa-2A: an active outpatient regimen in metastatic renal cell carcinoma. *J Clin Oncol*, 1992. 10(3): p. 414–21.

196. Veelken, H., et al., Combination of interleukin-2 and interferon-alpha in renal cell carcinoma and malignant melanoma: a phase II clinical trial. *Biotechnol Ther*, 1992. 3(1–2): p. 1–14.

197. Atzpodien, J., et al., Interleukin-2- and interferon alfa-2a-based immunochemotherapy in advanced renal cell carcinoma: a Prospectively Randomized Trial of the German Cooperative Renal Carcinoma Chemoimmunotherapy Group (DGCIN). *J Clin Oncol*, 2004. 22(7): p. 1188–94.

198. McDermott, D.F., et al., Randomized phase III trial of high-dose interleukin-2 versus subcutaneous interleukin-2 and interferon in patients with metastatic renal cell carcinoma. *J Clin Oncol*, 2005. 23(1): p. 133–41.

199. Atkins, M.B., et al., High-dose recombinant interleukin-2 therapy in patients with metastatic melanoma: long-term survival update. *Cancer J Sci Am*, 2000. 6(Suppl 1): p. S11–4.

200. Atkins, M.B., Cytokine-based and biochemotherapy for advanced melanoma. *Clin Cancer Res*, 2006. 12(7 Pt 2): p. 2353s–8s.

201. Rosenberg, S.A., et al., Immunologic and therapeutic evaluation of a synthetic peptide vaccine for the treatment of patients with metastatic melanoma. *Nat Med*, 1998. 4(3): p. 321–7.

202. Rosenberg, S.A., et al., Impact of cytokine administration on the generation of antitumor reactivity in patients with metastatic melanoma receiving a peptide vaccine. *J Immunol*, 1999. 163(3): p. 1690–5.

203. Gollob, J., L. Flaherty, and J. Smith. A Cytokine Working Group (CWG) phase II trial of a modified gp100 melanoma peptide (gp100 (209M)) and high dose interleukin-2 (HD IL-2) administered q3 weeks in patients with stage IV melanoma: limited antitumor activity. *Prog Proc Am Soc Clin Oncol*, 2001, abstr 1423.

204. Rosenberg, S.A., et al., Combination therapy with interleukin-2 and alpha-interferon for the treatment of patients with advanced cancer. *J Clin Oncol*, 1989. 7(12): p. 1863–74.

205. Kruit, W.H., et al., Dose efficacy study of two schedules of high-dose bolus administration of interleukin 2 and interferon alpha in metastatic melanoma. *Br J Cancer*, 1996. 74(6): p. 951–5.

206. Sparano, J.A., et al., Randomized phase III trial of treatment with high-dose interleukin-2 either alone or in combination with interferon alfa-2a in patients with advanced melanoma. *J Clin Oncol*, 1993. 11(10): p. 1969–77.

207. Feun, L., et al., Cyclosporine A, alpha-Interferon and interleukin-2 following chemotherapy with BCNU, DTIC, cisplatin, and tamoxifen: a phase II study in advanced melanoma. *Cancer Invest*, 2005. 23(1): p. 3–8.

208. Atkins, M.B., et al., A phase II pilot trial of concurrent biochemotherapy with cisplatin, vinblastine, temozolomide, interleukin 2, and IFN-alpha 2B in patients with metastatic melanoma. *Clin Cancer Res*, 2002. 8(10): p. 3075–81.

209. McDermott, D.F., et al., A phase II pilot trial of concurrent biochemotherapy with cisplatin, vinblastine, dacarbazine, interleukin 2, and interferon alpha-2B in patients with metastatic melanoma. *Clin Cancer Res*, 2000. 6(6): p. 2201–8.

210. Gibbs, P., et al., A phase II study of biochemotherapy for the treatment of metastatic malignant melanoma. *Melanoma Res*, 2000. 10(2): p. 171–9.

211. Johnston, S.R., et al., Randomized phase II trial of BCDT [carmustine (BCNU), cisplatin, dacarbazine (DTIC) and tamoxifen] with or without interferon alpha (IFN-alpha) and interleukin (IL-2) in patients with metastatic melanoma. *Br J Cancer*, 1998. 77(8): p. 1280–6.

212. Eton, O., et al., Sequential biochemotherapy versus chemotherapy for metastatic melanoma: results from a phase III randomized trial. *J Clin Oncol*, 2002. 20(8): p. 2045–52.

213. Hauschild, A., et al., Dacarbazine and interferon alpha with or without interleukin 2 in metastatic melanoma: a randomized phase III multicentre trial of the Dermatologic Cooperative Oncology Group (DeCOG). *Br J Cancer*, 2001. 84(8): p. 1036–42.

214. Rosenberg, S.A., et al., Prospective randomized trial of the treatment of patients with metastatic melanoma using chemotherapy with cisplatin, dacarbazine, and tamoxifen alone or in combination with interleukin-2 and interferon alfa-2b. *J Clin Oncol*, 1999. 17(3): p. 968–75.

215. Atkins, M.B., et al., A prospective randomized phase III trial of concurrent biochemotherapy (BCT) with cisplatin, vinblastine, dacarbazine (CVD), IL-2 and interferon alpha-2b (IFN) versus CVD alone in patients with metastatic melanoma (E3695): An ECOG-coordinated intergroup trial. *ASCO Annual Meeting Proceedings*, 2003: p. 2847.

216. O'Day, S.J., et al., Maintenance biotherapy for metastatic melanoma with interleukin-2 and granulocyte macrophage-colony stimulating factor improves survival for patients responding to induction concurrent biochemotherapy. *Clin Cancer Res*, 2002. 8(9): p. 2775–81.

217. O'Day, S., et al., A phase II multi-center trial of maintenance biotherapy (MBT) after induction concurrent biochemotherapy (BCT) for patients (Pts) with metastatic melanoma (MM). *ASCO Annual Meetings Proceedings*, 2005. 23(Suppl 16): p. 7503.

218. Schwartz, R.N., L. Stover, and J. Dutcher, Managing toxicities of high-dose interleukin-2. *Oncology*, 2002. 16(11 Suppl 13): p. 11–20.

219. Klempner, M.S., et al., An acquired chemotactic defect in neutrophils from patients receiving interleukin-2 immunotherapy. *N Engl J Med*, 1990. 322(14): p. 959–65.

220. Tilg, H., et al., Induction of circulating soluble tumour necrosis factor receptor and interleukin 1 receptor antagonist following interleukin 1 alpha infusion in humans. *Cytokine*, 1994. 6(2): p. 215–9.

221. Tilg, H., et al., Induction of circulating and erythrocyte-bound IL-8 by IL-2 immunotherapy and suppression of its in vitro production by IL-1 receptor antagonist and soluble tumor necrosis factor receptor (p75) chimera. *J Immunol*, 1993. 151(6): p. 3299–307.

222. Margolin, K., et al., Prospective randomized trial of lisofylline for the prevention of toxicities of high-dose interleukin 2 therapy in advanced renal cancer and malignant melanoma. *Clin Cancer Res*, 1997. 3(4): p. 565–72.

223. Du Bois, J.S., et al., Randomized placebo-controlled clinical trial of high-dose interleukin-2 in combination with a soluble p75 tumor necrosis factor receptor immunoglobulin G chimera in patients with advanced melanoma and renal cell carcinoma. *J Clin Oncol*, 1997. 15(3): p. 1052–62.

224. Atkins, M.B., et al., A phase I study of CNI-1493, an inhibitor of cytokine release, in combination with high-dose interleukin-2 in patients with renal cancer and melanoma. *Clin Cancer Res*, 2001. 7(3): p. 486–92.

225. Kilbourn, R.G., et al., Strategies to reduce side effects of interleukin-2: evaluation of the antihypotensive agent NG-monomethyl-L-arginine. *Cancer J Sci Am*, 2000. 6(Suppl 1): p. S21–30.

226. Kilbourn, R.G., et al., NG-methyl-L-arginine, an inhibitor of nitric oxide synthase, reverses interleukin-2-induced hypotension. *Crit Care Med*, 1995. 23(6): p. 1018–24.

227. Trinchieri, G., Interleukin-12 and the regulation of innate resistance and adaptive immunity. *Nat Rev Immunol*, 2003. 3(2): p. 133–46.

228. Bermudez, L.E., M. Wu, and L.S. Young, Interleukin-12-stimulated natural killer cells can activate human macrophages to inhibit growth of Mycobacterium avium. *Infect Immun*, 1995. 63(10): p. 4099–104.

229. Kaufmann, S.H., C.H. Ladel, and I.E. Flesch, T cells and cytokines in intracellular bacterial infections: experiences with Mycobacterium bovis BCG. *Ciba Found Symp*, 1995. 195: p. 123–32; discussion 132–6.

230. Reis e Sousa, C., et al., In vivo microbial stimulation induces rapid CD40 ligand-independent production of interleukin 12 by dendritic cells and their redistribution to T cell areas. *J Exp Med*, 1997. 186(11): p. 1819–29.

231. Coutelier, J.P., J. Van Broeck, and S.F. Wolf, Interleukin-12 gene expression after viral infection in the mouse. *J Virol*, 1995. 69(3): p. 1955–8.

232. Kobayashi, M., et al., Identification and purification of natural killer cell stimulatory factor (NKSF), a cytokine with multiple biologic effects on human lymphocytes. *J Exp Med*, 1989. 170(3): p. 827–45.

233. Kubin, M., M. Kamoun, and G. Trinchieri, Interleukin 12 synergizes with B7/CD28 interaction in inducing efficient proliferation and cytokine production of human T cells. *J Exp Med*, 1994. 180(1): p. 211–22.

234. Perussia, B., et al., Natural killer (NK) cell stimulatory factor or IL-12 has differential effects on the proliferation of TCR-alpha beta+, TCR-gamma delta+ T lymphocytes, and NK cells. *J Immunol*, 1992. 149(11): p. 3495–502.

235. Manetti, R., et al., Natural killer cell stimulatory factor (interleukin 12 [IL-12]) induces T helper type 1 (Th1)-specific immune responses and inhibits the development of IL-4-producing Th cells. *J Exp Med*, 1993. 177(4): p. 1199–204.

236. Satoh, Y., et al., Local administration of IL-12-transfected dendritic cells induces antitumor immune responses to colon adenocarcinoma in the liver in mice. *J Exp Ther Oncol*, 2002. 2(6): p. 337–49.

237. Mazzolini, G., et al., Regression of colon cancer and induction of antitumor immunity by intratumoral injection of adenovirus expressing interleukin-12. *Cancer Gene Ther*, 1999. 6(6): p. 514–22.

238. Gao, J.Q., et al., A single intratumoral injection of a fiber-mutant adenoviral vector encoding interleukin 12 induces remarkable anti-tumor and anti-metastatic activity in mice with Meth-A fibrosarcoma. *Biochem Biophys Res Commun*, 2005. 328(4): p. 1043–50.

239. Bramson, J.L., et al., Direct intratumoral injection of an adenovirus expressing interleukin-12 induces regression and long-lasting immunity that is associated with highly localized expression of interleukin-12. *Hum Gene Ther*, 1996. 7(16): p. 1995–2002.

240. Kodama, T., et al., Perforin-dependent NK cell cytotoxicity is sufficient for anti-metastatic effect of IL-12. *Eur J Immunol*, 1999. 29(4): p. 1390–6.

241. Smyth, M.J., et al., NKG2D recognition and perforin effector function mediate effective cytokine immunotherapy of cancer. *J Exp Med*, 2004. 200(10): p. 1325–35.

242. Kawamura, T., et al., Critical role of NK1+ T cells in IL-12-induced immune responses in vivo. *J Immunol*, 1998. 160(1): p. 16–9.

243. Boggio, K., et al., Interleukin 12-mediated prevention of spontaneous mammary adenocarcinomas in two lines of Her-2/neu transgenic mice. *J Exp Med*, 1998. 188(3): p. 589–96.

244. Gollob, J.A., et al., Phase I trial of twice-weekly intravenous interleukin 12 in patients with metastatic renal cell cancer or malignant melanoma: ability to maintain IFN-gamma induction is associated with clinical response. *Clin Cancer Res*, 2000. 6(5): p. 1678–92.

245. Okamura, H., et al., Cloning of a new cytokine that induces IFN-gamma production by T cells. *Nature*, 1995. 378(6552): p. 88–91.

246. Micallef, M.J., et al., Interferon-gamma-inducing factor enhances T helper 1 cytokine production by stimulated human T cells: synergism with interleukin-12 for interferon-gamma production. *Eur J Immunol*, 1996. 26(7): p. 1647–51.

247. Tsutsui, H., et al., IL-18 accounts for both TNF-alpha- and Fas ligand-mediated hepatotoxic pathways in endotoxin-induced liver injury in mice. *J Immunol*, 1997. 159(8): p. 3961–7.

248. Tomura, M., et al., A critical role for IL-18 in the proliferation and activation of NK1.1+ CD3-cells. *J Immunol*, 1998. 160(10): p. 4738–46.

249. Hunter, C.A., et al., Comparison of the effects of interleukin-1 alpha, interleukin-1 beta and interferon-gamma-inducing factor on the production of interferon-gamma by natural killer. *Eur J Immunol*, 1997. 27(11): p. 2787–92.

250. Hoshino, K., et al., The absence of interleukin 1 receptor-related T1/ST2 does not affect T helper cell type 2 development and its effector function. *J Exp Med*, 1999. 190(10): p. 1541–8.

251. Tsutsui, H., et al., IFN-gamma-inducing factor up-regulates Fas ligand-mediated cytotoxic activity of murine natural killer cell clones. *J Immunol*, 1996. 157(9): p. 3967–73.

252. Dao, T., et al., Interferon-gamma-inducing factor, a novel cytokine, enhances Fas ligand-mediated cytotoxicity of murine T helper 1 cells. *Cell Immunol*, 1996. 173(2): p. 230–5.

253. Park, C.C., et al., Evidence of IL-18 as a novel angiogenic mediator. *J Immunol*, 2001. 167(3): p. 1644–53.

254. Coughlin, C.M., et al., Interleukin-12 and interleukin-18 synergistically induce murine tumor regression which involves inhibition of angiogenesis. *J Clin Invest*, 1998. 101(6): p. 1441–52.

255. Paulukat, J., et al., Expression and release of IL-18 binding protein in response to IFN-gamma. *J Immunol*, 2001. 167(12): p. 7038–43.

256. Novick, D., et al., Interleukin-18 binding protein: a novel modulator of the Th1 cytokine response. *Immunity*, 1999. 10(1): p. 127–36.

257. Robertson, M.J., et al., Phase I study of recombinant human IL-18 (rhIL-18) administered as five daily intravenous infusions every 28 days in patients with solid tumors. *ASCO Annual Meeting Proceedings*, 2005: p. 2513.

258. Koch, K.M., et al., PK and PD of recombinant human IL-18 (rhIL-18) administered IV in repeated cycles to patients with solid tumors. *ASCO Annual Meeting Proceedings*, 2005: p. 2535.

259. Lewis, N., et al., Phase I dose escalation study to assess tolerability and pharmacokinetics of recombinant human IL-18 (rhIL-18) administered as fourteen daily subcutaneous injections in patients with solid tumors. *ASCO Annual Meeting Proceedings*, 2004: p. 2591.

260. Parrish-Novak, J., et al., Interleukin 21 and its receptor are involved in NK cell expansion and regulation of lymphocyte function. *Nature*, 2000. 408(6808): p. 57–63.

261. Brandt, K., et al., Interleukin-21 inhibits dendritic cell activation and maturation. *Blood*, 2003. 102(12): p. 4090–8.

262. Suto, A., et al., Interleukin 21 prevents antigen-induced IgE production by inhibiting germ line C(epsilon) transcription of IL-4-stimulated B cells. *Blood*, 2002. 100(13): p. 4565–73.

263. Pene, J., et al., Cutting edge: IL-21 is a switch factor for the production of IgG1 and IgG3 by human B cells. *J Immunol*, 2004. 172(9): p. 5154–7.

264. Habib, T., A. Nelson, and K. Kaushansky, IL-21: a novel IL-2-family lymphokine that modulates B, T, and natural killer cell responses. *J Allergy Clin Immunol*, 2003. 112(6): p. 1033–45.

265. Wang, G., et al., In vivo antitumor activity of interleukin 21 mediated by natural killer cells. *Cancer Res*, 2003. 63(24): p. 9016–22.

266. Ugai, S., et al., Transduction of the IL-21 and IL-23 genes in human pancreatic carcinoma cells produces natural killer cell-dependent and -independent antitumor effects. *Cancer Gene Ther*, 2003. 10(10): p. 771–8.

267. Ugai, S., et al., Expression of the interleukin-21 gene in murine colon carcinoma cells generates systemic immunity in the inoculated hosts. *Cancer Gene Ther*, 2003. 10(3): p. 187–92.

268. Takaki, R., et al., IL-21 enhances tumor rejection through a NKG2D-dependent mechanism. *J Immunol*, 2005. 175(4): p. 2167–73.

269. Moroz, A., et al., IL-21 enhances and sustains CD8+ T cell responses to achieve durable tumor immunity: comparative evaluation of IL-2, IL-15, and IL-21. *J Immunol*, 2004. 173(2): p. 900–9.

270. Ma, H.L., et al., IL-21 activates both innate and adaptive immunity to generate potent antitumor responses that require perforin but are independent of IFN-gamma. *J Immunol*, 2003. 171(2): p. 608–15.

271. Kishida, T., et al., Interleukin (IL)-21 and IL-15 genetic transfer synergistically augments therapeutic antitumor immunity and promotes regression of metastatic lymphoma. *Mol Ther*, 2003. 8(4): p. 552–8.

272. Curti, B.D., et al., Preliminary tolerability and anti-tumor activity of intravenous recombinant human Interleukin-21 (IL-21) in patients with metastatic melanoma and metastatic renal cell carcinoma. *ASCO Annual Meeting Proceedings*, 2005: p. 2502.

273. Dranoff, G., et al., Vaccination with irradiated tumor cells engineered to secrete murine granulocyte-macrophage colony-stimulating factor stimulates potent, specific, and long-lasting anti-tumor immunity. *Proc Natl Acad Sci USA*, 1993. 90(8): p. 3539–43.

274. Grabstein, K.H., et al., Induction of macrophage tumoricidal activity by granulocyte-macrophage colony-stimulating factor. *Science*, 1986. 232(4749): p. 506–8.

275. Yamasaki, S., et al., Presentation of synthetic peptide antigen encoded by the MAGE-1 gene by granulocyte/macrophage-colony-stimulating-factor-cultured macrophages from HLA-A1 melanoma patients. *Cancer Immunol Immunother*, 1995. 40(4): p. 268–71.

276. Hanada, K., R. Tsunoda, and H. Hamada, GM-CSF-induced in vivo expansion of splenic dendritic cells and their strong costimulation activity. *J Leukoc Biol*, 1996. 60(2): p. 181–90.

277. Armstrong, C.A., et al., Antitumor effects of granulocyte-macrophage colony-stimulating factor production by melanoma cells. *Cancer Res*, 1996. 56(9): p. 2191–8.

278. Mach, N., et al., Differences in dendritic cells stimulated in vivo by tumors engineered to secrete granulocyte-macrophage colony-stimulating factor or Flt3-ligand. *Cancer Res*, 2000. 60(12): p. 3239–46.

279. Hu, H.M., et al., Divergent roles for CD4+ T cells in the priming and effector/memory phases of adoptive immunotherapy. *J Immunol*, 2000. 165(8): p. 4246–53.

280. Ridolfi, L. and R. Ridolfi, Preliminary experiences of intralesional immunotherapy in cutaneous metastatic melanoma. *Hepatogastroenterology*, 2002. 49(44): p. 335–9.

281. Vaquerano, J.E., et al., Regression of in-transit melanoma of the scalp with intralesional recombinant human granulocyte-macrophage colony-stimulating factor. *Arch Dermatol*, 1999. 135(10): p. 1276–7.

282. Si, Z., P. Hersey, and A.S. Coates, Clinical responses and lymphoid infiltrates in metastatic melanoma following treatment with intralesional GM-CSF. *Melanoma Res*, 1996. 6(3): p. 247–55.

283. Dranoff, G., GM-CSF-secreting melanoma vaccines. *Oncogene*, 2003. 22(20): p. 3188–92.

284. Spitler, L.E., et al., Adjuvant therapy of stage III and IV malignant melanoma using granulocyte-macrophage colony-stimulating factor. *J Clin Oncol*, 2000. 18(8): p. 1614–21.

285. Dunn, G.P., L.J. Old, and R.D. Schreiber, The immunobiology of cancer immunosurveillance and immunoediting. *Immunity*, 2004. 21(2): p. 137–48.

286. Rivoltini, L., et al., Immunity to cancer: attack and escape in T lymphocyte-tumor cell interaction. *Immunol Rev*, 2002. 188: p. 97–113.

287. De Paola, F., et al., Restored T-cell activation mechanisms in human tumour-infiltrating lymphocytes from melanomas and colorectal carcinomas after exposure to interleukin-2. *Br J Cancer*, 2003. 88(2): p. 320–6.

288. Bukowski, R.M., et al., Signal transduction abnormalities in T lymphocytes from patients with advanced renal carcinoma: clinical relevance and effects of cytokine therapy. *Clin Cancer Res*, 1998. 4(10): p. 2337–47.

289. Piancatelli, D., et al., Local expression of cytokines in human colorectal carcinoma: evidence of specific interleukin-6 gene expression. *J Immunother*, 1999. 22(1): p. 25–32.

290. Lissoni, P., et al., Relation between macrophage and T helper-2 lymphocyte functions in human neoplasms: neopterin, interleukin-10 and interleukin-6 blood levels in early or advanced solid tumors. *J Biol Regul Homeost Agents*, 1995. 9(4): p. 146–9.

291. Chen, C.K., et al., T lymphocytes and cytokine production in ascitic fluid of ovarian malignancies. *J Formos Med Assoc*, 1999. 98(1): p. 24–30.

292. Nemunaitis, J., et al., Comparison of serum interleukin-10 (IL-10) levels between normal volunteers and patients with advanced melanoma. *Cancer Invest*, 2001. 19(3): p. 239–47.

293. Chen, Q., et al., Production of IL-10 by melanoma cells: examination of its role in immunosuppression mediated by melanoma. *Int J Cancer*, 1994. 56(5): p. 755–60.

294. Kim, R., et al., Tumor-driven evolution of immunosuppressive networks during malignant progression. *Cancer Res*, 2006. 66(11): p. 5527–36.

295. Smyth, M.J. and D.I. Godfrey, NKT cells and tumor immunity–a double-edged sword. *Nat Immunol*, 2000. 1(6): p. 459–60.

296. Yanagisawa, K., et al., Impaired proliferative response of V alpha 24 NKT cells from cancer patients against alpha-galactosylceramide. *J Immunol*, 2002. 168(12): p. 6494–9.

297. van der Vliet, H.J., et al., Polarization of Valpha24+ Vbeta11+ natural killer T cells of healthy volunteers and cancer patients using alpha-galactosylceramide-loaded and environmentally instructed dendritic cells. *Cancer Res*, 2003. 63(14): p. 4101–6.

298. Tahir, S.M., et al., Loss of IFN-gamma production by invariant NK T cells in advanced cancer. *J Immunol*, 2001. 167(7): p. 4046–50.

299. Dhodapkar, M.V., et al., A reversible defect in natural killer T cell function characterizes the progression of premalignant to malignant multiple myeloma. *J Exp Med*, 2003. 197(12): p. 1667–76.

300. Viguier, M., et al., Foxp3 expressing CD4+CD25(high) regulatory T cells are overrepresented in human metastatic melanoma lymph nodes and inhibit the function of infiltrating T cells. *J Immunol*, 2004. 173(2): p. 1444–53.

301. Marshall, N.A., et al., Immunosuppressive regulatory T cells are abundant in the reactive lymphocytes of Hodgkin lymphoma. *Blood*, 2004. 103(5): p. 1755–62.

302. Akasaki, Y., et al., Induction of a CD4+ T regulatory type 1 response by cyclooxygenase-2-overexpressing glioma. *J Immunol*, 2004. 173(7): p. 4352–9.

303. Cesana, G.C., et al., Characterization of CD4+CD25+ regulatory T cells in patients treated with high-dose interleukin-2 for metastatic melanoma or renal cell carcinoma. *J Clin Oncol*, 2006. 24(7): p. 1169–77.

304. Tomura, M., et al., A novel function of Valpha14+CD4+NKT cells: stimulation of IL-12 production by antigen-presenting cells in the innate immune system. *J Immunol*, 1999. 163(1): p. 93–101.

305. Kitamura, H., et al., The natural killer T (NKT) cell ligand alpha-galactosylceramide demonstrates its immunopotentiating effect by inducing interleukin (IL)- 12 production by dendritic cells and IL-12 receptor expression on NKT cells. *J Exp Med*, 1999. 189(7): p. 1121–8.

306. Chang, D.H., et al., Sustained expansion of NKT cells and antigen-specific T cells after injection of alpha-galactosyl-ceramide loaded mature dendritic cells in cancer patients. *J Exp Med*, 2005. 201(9): p. 1503–17.

307. Fujii, S., et al., Prolonged IFN-gamma-producing NKT response induced with alpha-galactosylceramide-loaded DCs. *Nat Immunol*, 2002. 3(9): p. 867–74.

308. Barnett, B., et al. Depleting CD4+ CD25+ Regulatory T-cells improves immunity in cancer-bearing patients. *Proceedings of AACR*, 2004.

309. Vieweg, J., Z. Su, and J. Dannuli. Enhancement of antitumor immunity following depletion of CD+CD25+ regulatory T-cells. *Proceedings of ASCO*, 2004.

310. Dudley, M.E., et al., A phase I study of nonmyeloablative chemotherapy and adoptive transfer of autologous tumor antigen-specific T lymphocytes in patients with metastatic melanoma. *J Immunother*, 2002. 25(3): p. 243–51.

311. Dudley, M.E., et al., Cancer regression and autoimmunity in patients after clonal repopulation with antitumor lymphocytes. *Science*, 2002. 298(5594): p. 850–4.

312. Phan, G.Q., et al., Cancer regression and autoimmunity induced by cytotoxic T lymphocyte-associated antigen 4 blockade in patients with metastatic melanoma. *Proc Natl Acad Sci USA*, 2003. 100(14): p. 8372–7.

313. Reuben, J.M., et al., Biologic and immunomodulatory events after CTLA-4 blockade with ticilimumab in patients with advanced malignant melanoma. *Cancer*, 2006. 106(11): p. 2437–44.

314. Ribas, A., et al., Antitumor activity in melanoma and anti-self responses in a phase I trial with the anti-cytotoxic T lymphocyte-associated antigen 4 monoclonal antibody CP-675,206. *J Clin Oncol*, 2005. 23(35): p. 8968–77.

315. Maker, A.V., et al., Tumor regression and autoimmunity in patients treated with cytotoxic T lymphocyte-associated antigen 4 blockade and interleukin 2: a phase I/II study. *Ann Surg Oncol*, 2005. 12(12): p. 1005–16.

316. Upton, M.P., et al., Histologic predictors of renal cell carcinoma (RCC) response to interleukin-2-based therapy. *ASCO Annual Meeting Proceedings*, 2004: p. 3420.

317. Bui, M.H., et al., Carbonic anhydrase IX is an independent predictor of survival in advanced renal clear cell carcinoma: implications for prognosis and therapy. *Clin Cancer Res*, 2004. 9(2): p. 802–11.

318. Atkins, M., et al., Carbonic Anhydrase IX (CAIX) expression predicts for renal cell cancer (RCC) patient response and survival to IL-2 therapy. *ASCO Annual Meeting Proceedings*, 2004. p. 4512.

319. Scheibenbogen, C., et al., HLA class I alleles and responsiveness of melanoma to immunotherapy with interferon-alpha (IFN-alpha) and interleukin-2 (IL-2). *Melanoma Res*, 1994. 4(3): p. 191–4.

320. Tartour, E., et al., Predictors of clinical response to interleukin-2–based immunotherapy in melanoma patients: a French multiinstitutional study. *J Clin Oncol*, 1996. 14(5): p. 1697–703.

321. Liu, D., et al., Impact of gene polymorphisms on clinical outcome for stage IV melanoma patients treated with biochemotherapy: an exploratory study. *Clin Cancer Res*, 2005. 11(3): p. 1237–46.

322. Garcia-Hernandez, M.L., et al., Interleukin-10 promotes B16-melanoma growth by inhibition of macrophage functions and induction of tumour and vascular cell proliferation. *Immunology*, 2002. 105(2): p. 231–43.

323. Huang, S., et al., Interleukin 10 suppresses tumor growth and metastasis of human melanoma cells: potential inhibition of angiogenesis. *Clin Cancer Res*, 1996. 2(12): p. 1969–79.

324. Howell, W.M., et al., IL-10 promoter polymorphisms influence tumour development in cutaneous malignant melanoma. *Genes Immun*, 2001. 2(1): p. 25–31.

21

Cyclooxygenase-2 as a Target for Cancer Prevention and Treatment

Monica Bertagnolli, MD,
Jaye L. Viner, MD, MA,
and Ernest T. Hawk, MD, MPH

SUMMARY

This chapter reviews our current understanding of the relationship between Cox-2 expression and activity and tumor promotion. In addition, this chapter reviews the status of clinical trails of Cox-2 inhibition in both pre-malignant and cancer treatment settings.

Key Words: Cyclooxygenase-2; inflammation; prostaglandins; tumorigenesis.

1. INTRODUCTION: CANCER AND INFLAMMATION

The disease states of cancer and inflammation have been linked in medical knowledge from the earliest recorded times. The term "tumor" comes from Latin, tumere, meaning "to swell." In his 1837 text, *Surgical Observations on Tumours (1)* the Boston surgeon John C. Warren noted, "A third supposable mode of the production of tumours is chronic inflammation of a natural texture, ...most frequently in parts disposed to inflame." A crucial mechanistic link between inflammation and cancer was discovered in the early 1990s with the identification of cyclooxygenase-2 (Cox-2). Researchers studying the acute inflammatory response identified a Cox-related gene product that was induced in response to serum and was inhibited by glucocorticoids *(2)*. Structural studies confirmed the presence of a second form of cyclooxygenase, termed Cox-2, produced in response to inflammatory mediators and mitogens (reviewed in ref. *3*). The mechanisms of Cox-2 activity and their relationship to cancer have been the subject of intense investigation in recent years. This work has been facilitated by the development of selective Cox-2 inhibitors, which have now been tested in human cancer clinical trials.

This chapter will review the mechanisms by which Cox-2 and related substances promote and maintain tumors. It will also describe evidence for the anti-cancer effects

From: *Cancer Drug Discovery and Development: Molecular Targeting in Oncology*
Edited by: H. L. Kaufman, S. Wadler, and K. Antman © Humana Press, Totowa, NJ

of Cox inhibitors, including the selective Cox-2 inhibitors and other non-steroidal anti-inflammatory drugs (NSAIDs).

2. THE BIOLOGY OF COX-2

2.1. Arachidonic Acid Metabolism

The inflammatory response is mediated by a cascade of bioactive substances that are produced in response to trauma or other stimuli. This response begins with the release of arachidonic acid from the cell membrane, followed by its metabolism through a series of tissue-specific reactions. The products of arachidonic acid metabolism exert a vast range of downstream effects on cell-signaling pathways. In a highly complex network of cell signaling, these mediators influence many different systems, including those governing cell proliferation and differentiation [e.g., mitogen-activated protein kinase (MAPK) and peroxisome proliferator-activated receptors (PPARs)], cytoskeletal dynamics (e.g., Rho GTPases), apoptosis (e.g., Akt and PI_3K), and ion transport (e.g., Ca^{2+} channels).

In a resting cell, arachidonic acid is stored by esterification to glycerol in membrane phospholipids, especially phosphatidylethanolamine, phosphatidylcholine, and the phosphatidylinositides. Arachidonic acid is released from the cell membrane by phospholipase A2 (PLA2), in response to local trauma or activation of a G-protein-coupled receptor by a growth factor or cytokine. Free arachidonate is metabolized to bioactive substances known as eicosanoids by three distinct enzyme pathways, defined by the activities of Coxs, lipoxygenases, and cytochrome P450. The prefix *eicosa-* (from the Greek for twenty) denotes the number of carbon atoms in arachidonic acid. The term "eicosanoids" is used as a collective name for molecules derived from 20-carbon fatty acids, including the prostanoids and leukotrienes, as well as several other classes such as the isoprostanes, lipoxins, and epoxyeicosatrienoic acids (EETs). Prostanoids are the subset of eicosanoids that are produced by Cox activity and include prostaglandins (PGs), prostacyclin, and thromboxanes (Txs). Leukotrienes are a family of active hydroperoxy derivatives resulting from metabolism of arachidonic acid by lipoxygenases. The numbering of eicosanoids is used to denote the number of double bonds in each molecule. The arachidonic acid-derived prostanoids [e.g., prostaglandin E_2 (PGE_2)] contain two double bonds, whereas the leukotrienes (e.g., LTB_4) have four (reviewed in ref. *4*).

Prostanoids are produced from arachidonic acid through sequential metabolism by Cox to the intermediate, short-lived prostaglandins PGG_2 and PGH_2, respectively *(5)*. PGH_2 is then converted by tissue-specific isomerases to other PGs [PGD_2, PGE_2, PGF_{2d}, and prostaglandin I_2 (PGI_2)] as well as Tx and prostacyclins (Fig. 1). Cell-specific profiles of arachidonic metabolites exist because of differential expression of both downstream metabolizing enzymes and receptor isoforms. For example, epithelial cells contain PG synthetase, leading to the production of PGE_2, platelets contain Tx synthetase and therefore produce thromboxane A_2 (TxA_2), and endothelial cells produce PGI_2, also known as prostacyclin, through the activity of prostacyclin synthase. Prostanoids mediate their local effects on cells by binding to G-protein-linked receptors, which are also present in a tissue-specific distribution. There are at least nine known PG receptor forms, conveying an additional level of tissue specificity to PG-mediated activities. Four of the receptor subtypes bind PGE_2 (EP_1–EP_4), two bind PDG_2 (DP_1 and

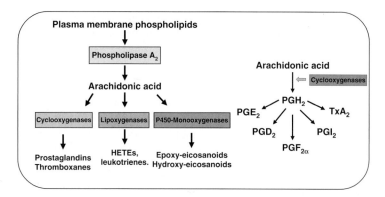

Fig. 1. Arachidonic acid metabolism.

DP_2), and separate receptors bind $PGF_{2\alpha}$ (FP), PGI_2 (IP), and TxA_2 (TP). These receptors are transmembrane G-protein-coupled proteins linked to a number of different signaling pathways *(6)*. In complex tissues, receptors for a wide variety of PGs are present on the surface of various components, such as epithelial cells, stromal fibroblasts, stromal endothelial cells, and inflammatory cells (Fig. 2).

Prostanoids are critical lipid mediators of many normal physiologic processes including platelet aggregation, renal function, gastric mucosal protection, reproduction, inflammation, and vascular integrity and tone. In response to inflammatory stimuli, prostanoids mediate the local symptoms associated with inflammation, including pain, vasoconstriction or vasodilation, coagulation, and fever. PGs, particularly PGE_2, also modulate cell behavior is ways that support tumor formation. For example, PGE_2 binds to specific G-protein-coupled receptors on the epithelial cell surface, initiating signaling cascades that promote cell growth and motility *(6)*. In epithelial cell lines, PGE_2 suppresses apoptosis by increasing expression of Bcl-2 and also increases expression of activated MAPK, promotes cell migration/invasiveness and activates epidermal

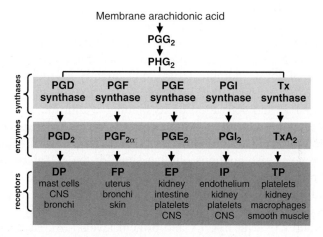

Fig. 2. Tissue specificity of arachidonic acid products. Specificity of prostaglandin signaling is governed by a tissue-specific distribution of enzymes and receptors.

growth factor receptor (EGFR) *(7–10)*. In addition, PGE_2 induces angiogenesis, thereby providing a mechanism for growth of both primary and metastatic disease *(11,12)*.

2.2. Distinct Functions of Cox-2

Until 1991, only one form of Cox was recognized. This family of enzymes is now known to contain at least two forms, Cox-1 and Cox-2, that share similar enzymatic activities and substrate but have distinct functional properties and expression patterns *(13,14)*. Cox-1 is constitutively expressed in the gastrointestinal mucosa, kidneys, platelets, and vascular endothelium and is responsible for maintenance of the normal physiologic function of these tissues. Cox-2 is minimally present in normal tissues and instead is a short-lived product of an intermediate-early response gene whose tissue expression increases 20-fold in response to growth factors, cytokines, tumor promoters, and oncogenic mutations (reviewed in ref. *15,16*). Cox-2 is not found in significant quantities in the absence of stimulation, which explains why it remained undetected as a distinct molecule for 20 years.

2.3. Regulation of Cox-2 Activity

Cox-2 gene expression is influenced by a wide variety of inflammatory mediators and tumor promoters. These signals include the oncogenes Ha-*ras*, v-*src*, *Her-2/neu*, and *wnt1* and the cytokines transforming growth factor-beta 1 (TGF-β1), TGF-α, tumor necrosis factor-α, interferon-γ (IFN-γ), and interleukins *(7,16–22)*. Transcriptional up-regulation by these diverse stimuli occurs through multiple transcription factor-binding sites within the Cox-2 gene promoter (Fig. 3). These transcriptional response elements include binding sites for cAMP, *Myb*, nuclear factor-interleukin 6 (NF-IL-6), CCAAT/enhancer binding proteins (C/EBPs), NF-kB, polyoma virus enhancer activator 3 (PEA3), and activator protein 1 (AP-1) (reviewed in ref.*23*). Several growth factors and other mitogenic stimuli that activate the ERK and JNK MAPK cellular-signaling pathways induce Cox-2 production through the cAMP response element of *Cox-2 (24,25)*. Cox-2 up-regulation by TNF is mediated through the NF-IL-6 and NF-kB sites *(26)*.

Based upon both animal and human studies, Cox-2 appears to be particularly important in promotion of colorectal tumorigenesis. Studies in zebrafish and human colorectal cancer (CRC) cell lines show that mutation of the *APC* gene, which is a characteristic of most sporadic CRC, may directly or indirectly induce Cox-2 expression *(27)*. In one model, this increased Cox-2 expression occurs through up-regulation of C/EBP-β and activation of its binding within the *Cox-2* promoter. The increased Cox-2 activity induced by APC mutation through C/EBP-β is suppressed by treatment with retinoic acid, a compound that decreases C/EBP-β expression *(27)*.

The Cox-2 promoter also contains binding sites for the c-*Myb* transcription factor, and activation through these sites stimulates Cox-2 transcription *(28)*. Myb transcription factors are broadly expressed and likely play an important role in cell growth *(29)*. Colorectal adenomas, cancer cell lines, and cancer tissue arrays all express high levels of c-Myb and Cox-2 at mRNA and protein levels *(28)*, suggesting that c-Myb contributes to Cox-2 induction early in colorectal carcinogenesis. Based on these data, c-Myb may be an important molecular target for the prevention of CRC, as well as other

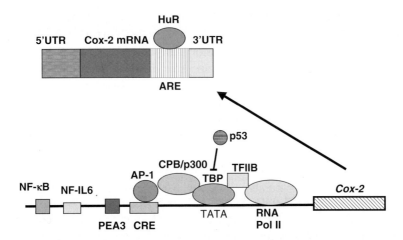

Fig. 3. Regulation of Cox-2 expression. Functional components of the *Cox-2* promoter include binding sites for a variety of transcription factors whose binding is induced in response to mitogen-activated protein kinase (MAPK) activation, including nuclear factor κB (NF-κB), nuclear factor interleukin-6 (NF-IL6), polyoma virus enhancer activator 3 (PEA3), activator protein 1 (AP-1), and nuclear factor of activated T cells (NFAT). AP-1-mediated activation of Cox-2 transcription involves the histone deacetylase activity of the CREB-binding protein/p300 co-activator complex *(23)*.Wild-type p53 suppresses Cox-2 transcription by competing with TATA-binding protein for binding to the TATA box *(31)*. Message stability is modified by Shaw–Kamen sequences located in the 3'-untranslated region (UTR) of Cox-2 mRNA. Association of the RNA-binding protein, HuR, with the Shaw–Kamen sequences is associated with Cox-2 message stability in colon cancer *(33)*.

malignancies in which it is commonly overexpressed (e.g., myeloid leukemia, lung cancer, breast cancer, neuroblastoma, osteogenic sarcoma, and melanoma) *(20,30)*.

Wild-type p53 suppresses Cox-2 transcription by competing with TATA-binding protein for binding to the TATA box *(31)*. Cox-2 protein expression is also regulated by post-transcriptional and post-translational events that are active during pathophysiologic processes such as inflammation and carcinogenesis. These mechanisms involve modulation of transcript stability as well as control of the rate of protein synthesis and degradation. Multiple regulatory elements, known as Shaw–Kamen sequences, have been identified in the Cox-2 gene 3'-untranslated region (UTR) *(32–34)*. Binding of regulatory proteins to these regions regulates message stability and translational efficiency. For example, IL-1β increases the half-life of Cox-2 mRNA *(32,35)*, and this effect is associated with induction of RNA-binding proteins that interact with sequences in the 3'-UTR of Cox-2. In addition, association of the RNA-binding protein, HuR, with the Shaw–Kamen sequences is associated with Cox-2 message stability in human colon cancer *(33)*.

2.4. Cancer-Promoting Activities of Cox-2

Over the last 10 years, increased Cox-2 activity has been associated with a broad range of malignancies (reviewed in ref. *36*). Cox-2 overexpression is found in human carcinogenesis, across the spectrum from pre-invasive to metastatic disease. Nearly all human epithelial neoplasias contain levels of Cox-2 mRNA, Cox-2 protein, or both that are elevated relative to normal tissues. For example, human tissue studies document twofold to 50-fold increases in Cox-2 mRNA and protein levels in 40% of pre-malignant adenomas and 80–90% of CRCs *(37)*. High Cox-2 expression is also characteristic of more aggressive tumor subtypes. High Cox-2 levels in gliomas correlate with increasing histological grade and predict poor survival *(38)*. Cox-2 overexpression also correlates with a poor prognosis in colorectal *(39–41)*, breast *(42)*, and cervical cancer *(43)*.

Evidence for a causal relationship between Cox-2 and tumor formation comes from many different sources. The first direct link was provided by studies involving targeted disruption of the murine Cox-2 gene (*Pghs2*) in mice prone to spontaneous tumor formation due to a truncated *Apc* gene (Apc$^{\Delta 716}$) *(44)*. The number and size of intestinal polyps driven by the Apc$^{\Delta 716}$ genetic background were substantially reduced in *Pghs2$^{-/-}$* mice compared with *Pghs2* wild-type mice. This animal study demonstrated a prominent gene-dosing effect, as *Pghs2$^{+/-}$* and *Pghs2$^{-/-}$* mice developed 68% and 86% fewer polyps, respectively, than *Pghs2* wild-type mice. In another study, targeted deletion of Cox-2 but not Cox-1 slowed the growth of Lewis lung carcinoma cells subcutaneously injected into mice *(45)*. However, it is important to note that this effect may not be entirely Cox-2 specific. For example, knockout of the Cox-1 gene in Apc^{Min} mice, an animal that carries a germline nonsense mutation of *Apc*, reduced tumor multiplicity to the same extent as Cox-2 gene deletion *(46)*. These results highlight the essential role of prostanoids, whether derived from Cox-1 or Cox-2, in intestinal polyp formation in the Apc^{Min} mouse model.

Complementary experiments support a specific role for Cox-2 in extracolonic carcinogenesis, such as in lung, mammary gland, and skin. Studies in animal models for these tumors show that overexpression of the Cox-2 gene alone is sufficient for tumorigenesis. Compared with age-matched controls, mammary gland carcinomas were increased in multiparous mice transgenic for a Cox-2 gene driven by the mouse mammary tumor virus promoter *(47)*. Cox-2 overexpression in basal keratinocytes in mice transgenic for the Cox-2 gene driven by the keratin 5 promoter also results in hyperplastic and dysplastic cutaneous changes consistent with neoplastic development *(48)*.

Emerging insights into Cox molecular pathways now show that PGE$_2$ is an important downstream mediator of epithelial carcinogenesis, possibly through effects that enhance cell survival and/or cell–cell adhesion (Figs 4 and 5). Treatment of Apc^{Min} mice with PGE$_2$ significantly increases adenoma formation *(49)*, and exogenous administration of PGE$_2$ inhibits NSAID-mediated suppression of these tumors *(50)*. Blockade of PGE$_2$ signaling by targeted disruption of EP$_2$ and EP$_4$ and by treatment with EP$_1$-selective and EP$_4$-selective antagonists significantly reduces tumors in animal intestinal tumor models *(51–53)*. Other studies show that EP$_3$ may mediate angiogenesis in response to PGE$_2$ *(54)*. Mechanisms of tumor formation through Cox-2/PGE$_2$-mediated signaling and will be described in the following sections.

2.4.1. MODULATION OF LOCAL IMMUNE RESPONSE

The earliest work on the link between PGs and tumor formation focused upon the relationship between PGE$_2$ and immune function. Studies performed in the 1980s showed that epithelial tumors contained increased levels of PGE$_2$ compared with normal tissues *(55,56)*. At this time, the leading hypothesis explaining this result was a lack of anti-tumor immunity produced by the immunosuppressive effects of PGE$_2$ *(55)*. In CRC, PGE$_2$ production is associated with immune suppression

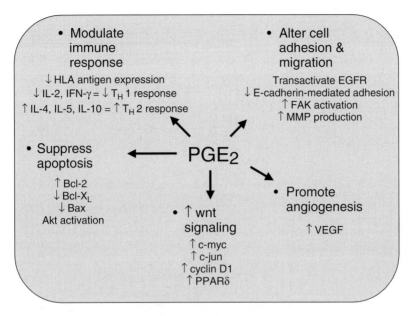

Fig. 4. Tumor-promoting activities of prostaglandin E_2 (PGE_2).

and loss of human leukocyte antigen (HLA) antigens *(57)*. Later studies described widespread immunomodulatory effects of PGE_2 through enhanced production of cytokines including IL-4, IL-5, and IL-10 (reviewed in ref. *58*). PGE_2 also inhibits the production of IL-2 and IFN-γ, thereby enhancing T helper 2 (Th2) response (resulting in increased antibody production) and inhibiting Th1 response (producing decreased cytotoxic T cell and macrophage killing efficacy). At the present time, the relationship of this effect to epithelial tumor formation and progression is unclear.

2.4.2. Effects on Apoptosis

In normal cellular physiology, apoptosis prevents uncontrolled proliferation by eliminating cells that are senescent or those that have become molecularly impaired. This function is particularly important in epithelial tissues, where continual tissue renewal involves a cycle of cell proliferation, followed by terminal differentiation, senescence, and cell death. Several studies suggest that Cox-2 and PGE_2 may be critical mediators of epithelial cell apoptosis. A variety of *in vitro* studies showed that Cox-2 enhances both cellular proliferation and resistance to apoptosis, whereas Cox-2 suppression restores apoptosis. For example, PGE_2 treatment of a CRC cell line decreased basal apoptotic rates and increased levels of the apoptosis suppressor protein, Bcl-2 *(7)*. Similarly, Cox-2 induction by mitogen-induced cancer cells is associated with decreased apoptosis *(59)*. The MAPK/extracellular signal-regulated protein kinase (MEK) pathway plays an important role in growth and differentiation of epithelial cells, and MEK activation modulates apoptosis by governing expression and activity of Bcl family members *(60)*. In intestinal epithelial cells, MEK activation causes up-regulation of Cox-2, an effect associated with anti-apoptotic response, including increased Bcl-X_L, phosphorylated Bad, and decreased Bak *(61)*. This work supports earlier findings in rat intestinal epithelial cells, where experimentally induced Cox-2 overexpression

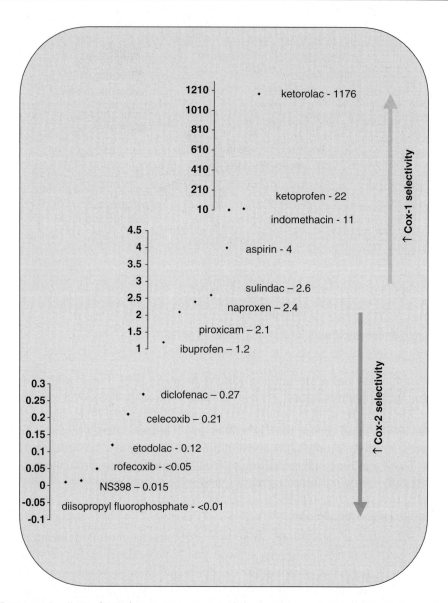

Fig. 5. Prostaglandin E$_2$ (PGE$_2$) signaling network. (A) PGE$_2$ transactivates epidermal growth factor receptor (EGFR) through binding to specific G-protein-coupled receptors in the cell membrane; downstream events include activation of PI$_3$K/Akt pathway signaling, resulting in suppression of apoptosis and increased cell survival *(9)*. (B) Stimulation of β-catenin/Tcf-4 survival signaling is induced by PGE$_2$ through EP$_2$-mediated G-protein signaling, a process that prevents regulatory phosphorylation of β-catenin by GSK3β *(65)*. (C) Activation of EGFR can alter cadherin-mediated cell–cell adhesion through phosphorylation of β-catenin, causing its dissociation from the intercellular adhesion complex *(69,70)*. (D) In the Apc^{Min} mouse, PGE$_2$-associated activation of EGFR produces cytoskeletal changes through RhoA that alter cell migration and intercellular adhesion *(72)*. (E) EGFR and integrin-mediated signaling networks are linked, providing necessary cross-talk between growth factor and adhesion-mediated survival signaling.

decreased expression of the pro-apoptotic protein, Bax, increased Bcl-2, and produced resistance to sodium butyrate-induced apoptosis *(62)*.

2.4.3. ALTERATIONS IN CELL SURVIVAL SIGNALING

Events mediated by Cox-2 clearly contribute to cell proliferation, differentiation, and survival although our understanding of the network of interactions involved is still incomplete. Prostanoids such as PGE_2 modulate the Wnt-signaling pathway, a major regulator of cell survival in epithelial cells. In normal tissues, Wnt signaling is activated in the proliferative zone of a renewable epithelium and suppressed during cell differentiation and maturation. Inappropriately activated Wnt signaling is found in >90% of human CRC due to loss of APC protein function. Cells with this defect are unable to degrade free intracellular β-catenin that is elaborated during cell migration and in response to growth factors. Excess β-catenin is transported to the nucleus, where it binds to Tcf-4, creating a transcription factor that activates production of gene products supporting cell proliferation, including *c-myc, c-jun, cyclin D1*, and *PPAR-δ (63)*. Evidence of activated Wnt signaling, in the form of nuclear-localized β-catenin, is a common characteristic of may epithelial tumors, present throughout the range of disease from pre-invasive to metastatic *(64)*. In addition to constitutive activation in response to APC loss, Wnt signaling can be directly induced by PGE_2. Stimulation of the EP_2 receptor by PGE_2 causes the G-protein, Gs, to release its regulatory subunits, Gβγ and Gα. Downstream events involve Gβγ-induced activation of Akt and Gα-mediated release of GSK3β from the β-catenin regulatory complex. The end result of this process is activation of cell proliferation through β-catenin/Tcf-4 *(65)*.

2.4.4. PROMOTION OF ANGIOGENESIS

Angiogenesis is a process required both for establishing new tumor colonies and for supporting growth of existing tumor masses. Cox-2 and PGs (i.e., PGE_2 and PGI_2) are potentially important factors in tumor angiogenesis. Cox-2 is consistently expressed in the neoangiogenic vasculature within tumors and pre-existing vasculature adjacent to tumors in human breast, lung, pancreas, prostate, bladder, and colon cancers (reviewed in ref. *66*). The pro-angiogenic effects of Cox-2 may be due to increased expression of vascular endothelial growth factor (VEGF). For example, in fibroblasts of Cox-$2^{-/-}$ mice, VEGF protein levels were reduced by 94% compared with wild-type or Cox-$1^{-/-}$mice *(45)*. A 30% decrease in vascular density and decreased VEGF mRNA expression was observed in Lewis lung tumors grown in these mice. A separate study of human colorectal tumor samples found a significant correlation between Cox-2 and VEGF expression and between the expression of both genes and microvessel density *(67)*. Although the molecular basis for these findings is as yet unknown, these studies establish a link between Cox-2 and VEGF gene regulation.

2.4.5. EFFECTS ON CELL MIGRATION AND INVASION

The processes of cell survival and migration are tightly linked in epithelial cells. PGE_2 can transactivate EGFR by both direct and indirect mechanisms and in doing so activates pathways governing both cell survival and migration *(9,68,69)*. Events downstream of this interaction are mediated through the Akt/PI$_3$K-signaling pathway, producing both modification of the actin cytoskeleton and suppression of apoptosis *(70)*.

A variety of *in vitro* assays document a link between EGFR activation and integrin-mediated adhesion of epithelial cells to the extracellular matrix *(71,72)*. The Cox-2 and EGFR-signaling pathways demonstrate a high degree of cross-talk. For example, EGFR transactivation stimulated AP-1-mediated induction of Cox-2 expression *(73)*. Cox-2 may also influence cell invasiveness directly. For example, transfection with a Cox-2 expression vector increases the invasiveness of human colon cancer cells (Caco-2). Induction of Cox-2 expression in this assay activates matrix metalloproteinase-2 (MMP-2) and increases RNA levels for the membrane-type MMPs *(74)*. Other evidence that Cox-2 promotes tumor invasiveness comes from discovery of nuclear factor of activated T cell (NFAT)-binding sites in the Cox-2 promoter. NFAT denotes a family of transcription factors first identified in hematopoietic cells whose activity promotes breast and colon cancer cell invasiveness *(75)*. NFAT recognition sequences are present adjacent to AP-1-binding sites in the Cox-2 promoter, and deletion studies show that these sequences are important for Cox-2 transcriptional activation. Activation of NFAT increases Cox-2 and PGE_2 expression and promotes invasion of tumor cells through Matrigel *(75)*.

2.4.6. Cox-2 and DNA Damage

Although the principal tumorigenic effects of Cox-2 involve PGE_2 generation, Cox-2 also exhibits peroxidase activity and as a result may potentiate formation of DNA mutagens in susceptible tissues. For example, isomerization of the Cox-2 product, PGH_2, leads to the formation of malondialdehyde, a potent mutagen that may cause DNA frame shifts and base-pair substitutions *(76)*. Depending on the tissue environment, carcinogens may be formed by the peroxidase activity of Cox acting upon a variety of substrates including aromatic amines, heterocyclic amines, and derivatives of polycyclic hydrocarbons *(77)*. This effect may be particularly important for smokers, where a link between exposure to carcinogenic tobacco products and Cox-2 expression has been identified *(78)*.

3. PHARMACOLOGIC INHIBITION OF COX ACTIVITY

Derivatives of aspirin have been used for centuries to treat pain and fever. Beginning in the 1930s, a variety of other NSAIDs were developed, primarily for management of severe arthritis. Aspirin and most NSAIDs inhibit the activity of both Cox-1 and Cox-2. In the case of aspirin, the drug covalently transfers an acetyl group from the aspirin to a serine residue (Ser 530) that lies in the arachidonic acid-binding channel, preventing binding of the substrate to the active site for its oxygenation. Although aspirin acetylates both Cox forms, its inhibitory activity is 10 to 100 times more potent against Cox-1 than Cox-2 *(79)*. Most other NSAIDs do not bind covalently to the channel and therefore exert a transient blockade of Cox activity.

Following the recognition that Cox-2 was primarily responsible for the consequences of inflammation, Cox-2-selective inhibitors were synthesized to preferentially target the function of Cox-2. Selective Cox-2 inhibitors were developed as potentially safer alternatives to non-selective NSAIDs for patients who required long-term use, such as those with severe arthritis or those with bleeding, gastrointestinal or renal intolerance to non-selective NSAIDs. The first selective Cox-2 inhibitors developed for clinical use belong to the diarylheterocycle class of compounds and include celecoxib, rofecoxib, valdecoxib, and etoricoxib. Structure-function studies show that a *cis*-stilbene moiety containing a 4-methylsulfonyl or sulfonamide in one of the phenyl rings of these compounds is required for Cox-2 selectivity *(80)*. A second class of selective Cox-2 inhibitors was developed by modification of non-selective NSAIDs, such as indomethacin, through esterification/amidation of their carboxylic acid moiety. This process produced compounds with ~500-fold selectivity for Cox-2 compared with Cox-1 *(81)*.

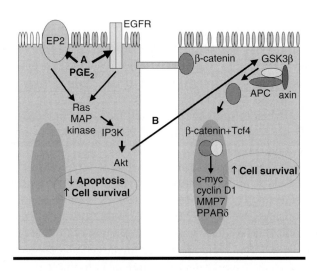

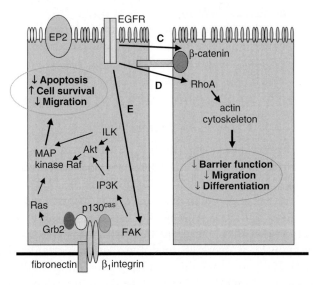

Fig. 6. Relative Cox-1/Cox-2 selectivity of non-steroidal anti-inflammatory drugs (NSAIDs). Values obtained using whole blood assay to determine IC_{80} ratio = dose able to suppress Cox-1 activity by 80% divided by dose able to suppress Cox-2 activity by 80%; aspirin could not be evaluated by this method because of its instability in whole blood, therefore aspirin value was obtained using the William Harvey Human Modified Whole Blood assay and showed an approximately fourfold greater selectivity for Cox-1 versus Cox-2 *(82)*.

NSAID activities range from those that are highly selective for either Cox-1 or Cox-2 to those with almost equal activity against the two isoforms. The relative degree of Cox-1 versus Cox-2 selectivity varies depending upon the assay used to measure this activity. In an extensive comparison using the human whole blood assay with inhibition of prostanoid function as a read-out, Warner et al. *(82)* compared the relative Cox-1 and Cox-2 selectivity of a variety of NSAIDs (Fig. 6). NSAID toxicities generally arise from impaired PG homeostasis in non-target tissues. Emerging data show that NSAID toxicities correlate with the relative degree of Cox-1 versus Cox-2 inhibition. In general, the greater the selectivity for Cox-1, the higher the risk of severe gastrointestinal toxicity, such as gastric ulceration and bleeding. Among the many NSAID side effects, gastrointestinal toxicities are probably the most significant, accounting for an estimated 100,000 hospitalizations and 16,500 deaths per year in the USA *(83)*. In addition, approximately 10–20% of NSAID users develop dyspepsia *(84)*, and yet another 1% have serious gastrointestinal complications *(85)*. These agents are also associated with an increased risk of bleeding, edema, hypertension, and renal insufficiency *(86,87)*. Most of these toxicities tend to occur in the context of increasing age, higher NSAID dose, concomitant use of corticosteroids or anti-coagulants, and certain co-morbid conditions *(88)*.

Selective Cox-2 inhibitors demonstrate minimal, if any, significant gastrointestinal toxicity or bleeding risk. It is likely, however, that selective Cox-2 inhibitors promote hypertension and also increase the risk of serious cardiovascular events such as myocardial infarction and stroke *(89,90)*. Not originally recognized in the relatively short term, active-controlled studies used for assessment of efficacy in treating arthritis, these cardiovascular toxicities were identified in large, placebo-controlled colorectal adenoma prevention trials using rofecoxib and celecoxib. It is unclear whether or not this cardiovascular risk extends to the non-selective NSAIDs. Case–control and cohort studies do suggest that increased cardiovascular risk may extend to non-selective NSAIDs *(91,92)*; however, there are no placebo-controlled randomized clinical trials of these agents that adequately address this issue.

4. EVIDENCE FOR ANTI-CANCER EFFECTS OF NSAIDS

4.1. Clues from Cancer Epidemiology

A wide range of observational data links Cox inhibitors to a reduced risk for all stages of colorectal neoplasia (adenomas, cancers, and cancer-associated mortality) (Fig. 7). A large case–control study from Australia was the first to report a protective effect against CRC among individuals who used aspirin, with a relative risk (RR) of 0.60 (95% CI = 0.44–0.82) *(93)*. Among the American Cancer Society prospective cohort study, an effort that involved more than 630,000 subjects, use of aspirin was associated with a significant decrease in colon cancer mortality (RR = 0.63; 95% CI = 0.44–0.89) *(94,95)*. Numerous other prospective cohort studies confirmed these results, with reported RRs ranging from 0.51 to 0.68, and a dose dependence based on frequency of use *(96–98)*. As detailed in recent reviews *(36,99,100)*, more than 40 retrospective and prospective epidemiological studies have found that regular aspirin or NSAID use reduces the risk of colorectal adenoma, carcinoma, and/or carcinoma-related mortality by approximately 40–50%. Only two observational studies showed no risk reduction for CRC in aspirin users *(101,102)*.

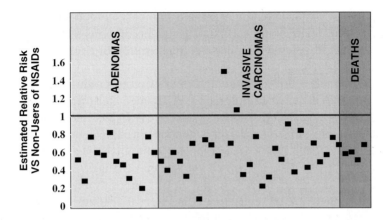

Fig. 7. Non-steroidal anti-inflammatory drug (NSAID) cancer epidemiology. Pictorial summary of results from 48 case–control and cohort studies of the relationship between NSAID use and the incidence of colorectal adenomas, colorectal cancer, and death due to colorectal cancer (reviewed in ref. *36*).

Although the data are less consistent and less compelling for extracolonic organs, in aggregate, they show important associations between NSAID use and reductions in tumor incidence. In a large population-based case–control study, Farrow et al. determined that frequent aspirin users showed reduced rates of esophageal cancer (RR = 0.37–0.49) and non-cardia gastric adenocarcinoma (RR = 0.46) *(103)*. This protection was not significant for cancers of the gastric cardia (RR = 0.8). A case–control study examining development of invasive breast cancer found that frequent users of NSAIDs were less likely to develop large or metastatic tumors *(104)*. Other studies show NSAID-associated reductions in bladder *(105)*, stomach *(106,107)*, and lung *(108,109)* cancer risks.

4.2. Anti-Cancer Mechanisms of Cox-2 Inhibitors: Cox-Dependent Effects

Evidence for Cox-2 selective anti-tumor effects comes from animal models and studies using selective Cox-2 inhibitors. As described above, either targeted deletion of the murine Cox-2 gene or treatment with selective Cox-2 inhibitors achieved dramatic reduction in tumor number in animals of tumorigenesis driven by loss of APC *(44,110)*. In UV-exposed skin, topical application with the Cox-2 inhibitor celecoxib effectively reduced edema, dermal neutrophil infiltration and activation, PGE_2 levels, and the production of sunburn cells *(111)*. Thus, by blocking inflammation, Cox-2 inhibitors may be effective in preventing UVB-induced skin tumor development.

In tissue culture, selective Cox-2 inhibitors induce apoptosis in cancer cells of the colon *(7,112)*, stomach *(113)*, prostate *(114,115)*, pancreas *(116)*, esophagus *(117)*, lung *(118)*, and head and neck squamous cell carcinomas *(119)*. Although Cox-2 inhibitors likely promote apoptosis by reducing PGE_2 levels, data also suggest that certain NSAIDs may directly affect the concentrations of proteins that regulate apoptosis, such as BAX and $Bcl-X_L$. Another intriguing mechanism of Cox-2 inhibitor activity involves promotion of tumor necrosis factor-related apoptosis-inducing ligand (TRAIL)-mediated apoptosis. In human CRC cell lines, treatment with Cox-2 inhibitors causes clustering of the TRAIL receptor, DR5, at the cell surface. This is associated with localization of death-inducing signaling complex components, including DR5, fas associated with death domain (FADD), and procaspase-8 in caveolae, together with activation of acid sphingomyelinase and generation of ceramide within the outer plasma membrane *(120)*.

NSAIDs alter signaling through growth factor receptors, producing changes that affect both cell survival and cell adhesion/migration. PGE_2-mediated transactivation of EGFR is strongly inhibited by both selective and non-selective NSAIDs, and resulting downstream effects include suppression of survival signaling through activated MAPK and the PI_3K/Akt pathway *(121)*. EGFR activation also diminishes cadherin-mediated cell–cell adhesion, a property associated with cytoskeletal changes favoring focal adhesion formation and cell contraction *(72)*. In the Apc^{Min} mouse model, all of these effects can be reversed upon treatment with either NSAIDs or EGFR inhibitors *(69,72,121)*.

The anti-angiogenic effects of Cox inhibitors include reversal of both invasiveness and MMP activation. In contrast to a Cox-1-selective inhibitor, a Cox-2-selective inhibitor blocked experimental colon and lung tumor growth and basic fibroblast growth factor (bFGF)-induced corneal neoangiogenesis *(122)*. Other anti-angiogenic effects of Cox-2 inhibitors include their ability to decrease tumor blood vessel and capillary formation and profoundly inhibit expression of angiogenic peptides such as VEGF and bFGF. In one study, a selective Cox-2 inhibitor reduced VEGF production by 92% *(45)*. Further, Cox-2 inhibitors reduce tumor cell adhesion and suppress endothelial cell growth. Cox-2 inhibitors may target integrin $\alpha V\beta 3$, an adhesion receptor critical for tumor neoangiogenesis *(123)*. Inhibition of endothelial-cell Cox-2 by NSAIDs suppressed $\alpha V\beta 3$-dependent activation of the small GTPases Cdc42 and Rac, resulting in inhibition of endothelial-cell spreading and migration *in vitro* and suppression of FGF-2-induced neoangiogenesis *in vivo*. These results establish a novel functional link between Cox-2, $\alpha V\beta 3$, and Cdc42-/Rac-dependent endothelial-cell migration.

Cox-2 inhibition may also prevent tumor formation by blocking metabolic activation of carcinogens. Evidence for this is particularly strong for the lung. Cox, through its peroxidase activity, catalyzes the oxidation of a wide range of substances such as aromatic amines and phenols. Although Cox-2-selective inhibitors do not inhibit the peroxidase function of Cox, they reduce substrate by blocking the production of PGG_2. For example, the Cox-2-selective inhibitor NS-398 prevents peroxidase-mediated bioactivation of the tobacco-specific carcinogen NNK in a microsomal preparation of lung tissue *(124)*. Cox-2-selective inhibitors may also block the formation of endogenous carcinogens from prostanoids. This result is produced by the activity of the selective Cox-2 inhibitor, NS-398, which inhibits the formation of malondialdehyde, a mutagenic by-product of PGH_2 breakdown, in cultured human colon cells *(125)*.

4.3. Anti-Cancer Mechanisms of Cox-2 Inhibitors: Cox-Independent Effects

In certain model systems, NSAIDs produce anti-tumor effects that are not dependent upon suppression of PG activity. In particular, gene deletion studies support a role for alternative mechanisms underlying the efficacy of Cox inhibitors. For example, sensitivity to NSAID-induced cell death mediated by NS-398 was comparable among transformed fibroblasts derived from wild-type, Cox-1null, Cox-2null, or Cox-1/Cox-2 null mice *(126)*. Additional study showed that NSAIDs may directly increase the BAX to $Bcl-X_L$ ratio and thereby restore apoptosis in cancer cells *(127)*. In this particular study, cultured human CRC cells null for the BAX gene were resistant to NSAID-induced apoptosis.

Studies *in vitro* also showed that the selective Cox-2 inhibitor, celecoxib, inhibited Akt signaling by directly blocking Akt phosphorylation and that overexpression of constitutively active Akt protected cells from celecoxib-induced apoptosis *(114)*. In addition, celecoxib induced apoptosis in HT-29 colon cancer cells, which lack Cox-2 activity *(128)*. The underlying mechanism in this model involved inhibition of 3-phosphoinositide-dependent kinase PDK1, an Akt/PKB upstream kinase. In aggregate, these data suggest that celecoxib induces apoptosis by blocking the anti-apoptotic PDK1/Akt/PKB pathway through targets other than Cox-2. This theory is supported by data showing that p70^{S6} kinase, a downstream substrate for PDK1, is responsible for site-specific phosphorylation of bcl-z-associated death promoter (BAD), which inactivates the proapoptotic molecule, thereby enhancing cell survival *(129)*.

Cox-2 inhibitors, like other NSAIDs, may also target PPARs. Members of the nuclear hormone-receptor superfamily, PPARs, are a family of ligand-activated transcription factors. Modulation of certain PPARs (i.e., PPAR-δ repression, and PPAR-α and PPAR-γ activation) has been associated with chemopreventive potential *(130)*. Some NSAIDs are weak PPAR-γ agonists and receptor ligands and also activate PPAR-α. For example, indomethacin can bind to and transcriptionally activate PPAR-α *(131)*. Inhibition of PPAR-δ may play a role in apoptosis and other chemopreventive effects (i.e., anti-inflammation) of sulindac. PPAR-δ expression is normally down-regulated by the APC tumor suppressor gene, and up-regulated by non-functional APC mutations. Sulindac was shown to block the DNA-binding activity of PPAR-δ, and PPAR-δ overexpression prevented sulindac-induced apoptosis *(131)*. These findings suggest that PPAR-δ may be a direct target of sulindac. This model is likely overly simplistic, however, as data also showed that PPAR-δ gene ablation does not alter the *in vitro* sensitivity of CRC cells to sulindac-induced apoptosis *(132)*.

Certain NSAIDs, such as aspirin and sulindac, may directly block signaling through NF-κB. NF-κB controls the expression of genes important for cellular survival and proliferation, and constitutive activation of NF-kB is associated with increased expression of anti-apoptotic proteins. Aspirin and sulindac inhibit IκB kinase-β (IκB) *(133,134)*, which phosphorylates the inhibitory subunit of NF-κB, targeting it for destruction. Although a likely contributing factor, *in vivo* studies suggest that effects on NF-κB signaling are neither necessary nor sufficient for suppression of intestinal polyposis by NSAIDs. For example, unlike aspirin and sulindac, indomethacin has no effect on NF-κB signaling *in vitro*, yet all three agents are efficacious against intestinal polyposis *in vivo*.

4.4. Studies in Animal Models

Animal spontaneous and carcinogen-induced tumor models clearly demonstrate a role for Cox-2 in tumor formation. In keeping with this, rodent tumor models show consistent reductions in tumor incidence, multiplicity and size, and increases in latency upon administration of selective and non-selective NSAIDs. Although CRC models have been studied most extensively, *in vivo* anti-tumor responses to NSAIDs have been demonstrated in intestine, skin, lung, mammary, oral, esophagus, and bladder carcinogenesis. A number of studies show that NSAIDs are protective even when given 14 weeks after carcinogen administration, indicating that these agents may be effective either early or late in neoplastic development *(135,136)*. Interestingly, these studies also suggest a gradient of efficacy among the non-selective NSAIDs commonly tested, in which piroxicam has the greatest effects, followed by sulindac, ASA, and finally ibuprofen. A variety of selective Cox-2 inhibitors, including JTE-522, NS-398, MF tricyclic, nimesulide, celecoxib, and rofecoxib were effective in preventing colorectal tumors in carcinogen-induced and/or genetically-induced rodent models *(44–140)*. Preventive effects have also been reported with selective Cox-2 inhibitors in animal models of mammary *(40,141)*, skin *(142,143)*, bladder *(144,145)*, esophagus *(146)*, lung *(124)*, and oral cavity *(147)* cancer.

NSAIDs inhibit tumor growth in studies involving a broad range of tumor xenografts, justifying research into the anti-tumor potential of this class of agents in established malignancies. Used alone or in combination with chemotherapy or radiotherapy in various animal models, NSAIDs generally delayed or reduced tumor growth, as opposed to inducing tumor regression, indicating a cytostatic rather than cytotoxic effect *(148–152)*.

5. COX-2 INHIBITION IN CANCER PREVENTION CLINICAL TRIALS

Randomized clinical trials using development of cancer as an endpoint require many thousands of patients and a decade or more of follow-up in many instances. As a result, NSAID cancer prevention trials have most often relied upon intermediate endpoints that are closely associated with cancer development, such as pre-invasive neoplastic lesions. The most significant tests of NSAID-mediated cancer prevention thus far have involved studies evaluating NSAIDs' effects against colorectal adenomas.

5.1. Studies in Familial Adenomatous Polyposis

In the 1980s, a surgeon named William Waddell prescribed indomethacin for management of desmoid tumors in a patient with familial adenomatous polyposis (FAP). At the time, desmoids, also known as "fibromatoses," were thought to represent an abnormal tissue reaction to local inflammation. After the patient developed side effects to indomethacin, Dr. Waddell switched to sulindac. Following this change, he observed a dramatic regression of the patient's colorectal adenomas. In 1983, Dr. Waddell published the first case series, including this patient, reporting results in four patients in whom there was almost complete adenoma regression following treatment with sulindac for 4–12 months *(153)*. Subsequent randomized, placebo-controlled studies *(154–156)* confirmed significant reductions in colorectal adenoma size and number with sulindac treatment. Independent case series also reported adenoma regression with sulindac sulfone, a metabolite of sulindac with no anti-PG activity *(157)* and with intrarectal indomethacin treatment *(158,159)*. These studies suggest that the anti-cancer effects of NSAIDs in FAP are class wide.

For several reasons, sulindac is not an optimal agent of cancer prevention for FAP patients. Responses to sulindac generally occurred within a few months but were not durable, with adenomas recurring soon after drug cessation. Because of gastrointestinal side effects, this drug is poorly tolerated in some patients. Complete adenoma regression with sulindac is rare, and cases of CRC have been reported in patients taking sulindac *(160,161)*. In contrast to effects against lower gastrointestinal tract tumors, effects of sulindac against duodenal neoplasia in patients with FAP have been far less striking and reproducible *(155,162,163)*. Giardiello et al. *(164)* recently investigated the chemopreventive potential of sulindac against early colorectal adenoma formation in 41 pre-phenotypic, genotype-positive patients with FAP. Although compliance was excellent, as documented by significant reductions in rectal mucosal prostanoid concentrations, incident adenomas were not significantly reduced. These results are the first to suggest that sulindac may be less effective against early stages of adenoma development in patients with FAP. Low expression of Cox-2 in small adenomas with less dysplasia has been reported by several groups *(37,165)* and may account for the study's results. Some patients develop adenomas that are resistant to NSAID inhibition. A recent study noted lower Cox-2 expression in sulindac-resistant adenomas excised from sulindac-treated FAP patients as compared with Cox-2 levels in adenomas excised prior to treatment *(166)*.

Cox-2-selective inhibitors have shown considerable promise in FAP. Although one small case series of nimesulide administered over 10 weeks found no effect on rectal adenoma burden in seven FAP patients *(167)*, celecoxib administered over 6 months to 83 FAP subjects significantly reduced colorectal adenoma number and size *(168)*. In this relatively brief randomized, placebo-controlled trial, no side effects attributable to celecoxib treatment were observed. Celecoxib improved the endoscopic appearance of both the colorectum and the duodenum of FAP patients, suggesting that it may reduce cancer risk in both organs *(169)*. Based on these findings and a commitment to further investigations, the FDA granted accelerated approval for the use of celecoxib in FAP patients as a complement to standard care that includes endoscopic surveillance and prophylactic surgery. Additional data in FAP patients suggest that NSAID-associated chemopreventive effects may not be entirely dependent upon anti-Cox-2 activity. One study of patients with FAP and measurable colonic disease examined the efficacy of exisulind, a metabolite of sulindac that induces tumor cell apoptosis *in vitro* but lacks anti-PG activity *(170)*. In 281 patients randomized to receive one of two doses of exisulind or placebo, a modest reduction in median polyp size was observed following 12 months at the high drug dose ($p = 0.03$). This favorable response, however, was associated with significant toxicity in the form of increased liver enzymes and abdominal pain.

Pre-clinical studies show synergistic efficacy when NSAIDs are combined with compounds directed against shared signaling pathways. For example, pre-clinical studies of NSAIDs in combination with inhibitors of inducible nitric oxide synthase *(171)*, EGFR *(172)*, or ornithine decarboxylase *(173)* achieve anti-tumor effects that exceed those of either agent alone.

5.2. Prevention of Sporadic Colorectal Adenomas

Small phase II biomarker clinical studies in patients with sporadic colorectal neoplasia showed that PG synthesis can be inhibited in the target tissue by oral administration of NSAIDs. By comparison of pre-treatment and post-treatment rectal biopsies, short-term to moderate-term administration of piroxicam, ibuprofen, or aspirin achieved significant reductions in mucosal prostanoids *(174–177)*. Aspirin doses as low as 81 mg daily reliably reduced mucosal prostanoid concentrations although these effects lasted no more than 3 months.

Several important randomized trials examined the effect of aspirin in the prevention of sporadic colorectal adenomas. These studies yielded mixed results. Initially, Baron et al. studied the effects of aspirin at 81 or 325 mg daily compared with placebo administered for 48 months in 1121 patients at moderate risk for CRC based on prior adenoma history *(178)*. A perplexing inverse dose–response was observed, with 19% and 4% reductions in recurrent adenomas and 40% and 19% reductions in recurrent advanced adenomas among patients using 81 and 325 mg per day, respectively. In a placebo-controlled study of patients who were successfully treated for a primary CRC, Sandler et al. showed that post-treatment administration of 325 mg aspirin daily produced a 35% reduction in adenoma detection over a median post-surgery observation period of 12.8 months *(179)*. Benamouzig et al. recently presented the results of a 4-year study that randomized adenoma patients to either placebo or lysine acetylsalicylate, a form of aspirin with increased solubility. Early results following 1 year of treatment indicated a 37% reduction in recurrent adenomas among those taking aspirin compared with those taking placebo; the result was of borderline statistical significance *(180)*. Long-term follow-up at 4 years demonstrated only a reduction in adenoma multiplicity *(181)*. Finally, Logan et al. recently reported the results of a placebo-controlled study that randomized 945 patients to enteric aspirin 300 mg daily with or without folate 0.5 mg daily in a 2 × 2 factorial design. The adenoma detection rate over 3 years was reduced by 29% (not statistically significant) for aspirin users, with a 41% reduction (statistically significant) in advanced adenomas *(182)*. The results from the arms containing folate have yet to be reported.

Recent placebo-controlled studies showed that selective Cox-2 inhibition produces more significant reductions in sporadic colorectal adenoma formation. In the Adenoma Prevention with Celecoxib (APC) Trial, 2035 patients with a history of large or multiple adenomas were randomized to receive either celecoxib 200 mg twice daily, celecoxib 400 mg twice daily, or placebo. Colonoscopies were performed after 1 and 3 years of study drug use, and efficacy analyses considered adenomas discovered at any point after randomization as a treatment failure. In this study, celecoxib produced a 33% reduction in adenoma detection at the 200 mg bid dose, and a 45% reduction in adenoma detection at the 400 mg bid dose *(183)*. The incidence of advanced adenomas in study participants was reduced by 57% and 66% for the 200 mg bid and 400 mg bid celecoxib arms, respectively. Unfortunately, although celecoxib was well tolerated in all other categories, serious adverse events in the cardiovascular system were significantly more common among celecoxib users. For a combined serious cardiovascular event endpoint that included myocardial infarction, stroke, congestive heart failure, and death due to cardiovascular causes, the RR of an event was 2.3 for the 200 mg bid arm and 3.4 for the 400 mg bid arm. A separate randomized trial by Arber et al. compared celecoxib at a dose of 400 mg once daily to placebo, with similar colonoscopic follow-up intervals of 1 and 3 years *(184)*. Reduction in patients with adenomas was 32% for 400 mg celecoxib daily. This regimen was associated with fewer serious cardiovascular adverse events (RR = 1.30). Comparisons with the APC Trial suggest that, although once daily lower dose celecoxib still carries a risk of cardiovascular toxicity, this regimen may be safer than twice daily. A third study, Adenomatous Polyp Prevention with Rofecoxib (APPROVe), found a 25% reduction in sporadic adenoma detection in a similar patient cohort, randomized to either rofecoxib 25 mg daily or placebo. Like the APC Trial, the APPROVe Trial found a threefold increased risk of serious cardiovascular complications in patients using rofecoxib *(89)* . Based upon these results, the maker of rofecoxib, Merck Inc., voluntarily withdrew the drug from the market in late 2004. Celecoxib continues to be used for management of arthritis and pain although its use is cautioned in those with risk factors for cardiovascular disease.

The efficacy of NSAIDs in regressing sporadic colon polyps is less well established than in FAP-associated tumors. Three small studies have reported modest effects of NSAIDs in this setting *(185–187)*. In one study, sporadic adenoma regression occurred in one of seven patients treated with either sulindac or piroxicam for 6 months *(185)*. Sulindac administered for 4 months regressed 13 of 20 polyps in a total of 15 patients in a second open label study *(186)*. In a double-blind placebo-controlled randomized trial of 44 subjects, 4 months of sulindac treatment did not result in a clinically significant regression of sporadic colonic polyps *(187)*.

5.3. Prevention of Other Epithelial Tumors

The chemopreventive efficacy of selective Cox-2 inhibitors has been demonstrated in a wide range of pre-clinical models, including mammary *(141)*, skin *(142,188)*, bladder *(144,145)*, esophagus *(189)*, lung *(124)*, and oral cavity *(190)* cancer. Based on strong pre-clinical data in a variety of epithelial tumors, phase II prevention trials of selective Cox-2 inhibitors, alone or in combination with other agents, were initiated in cohorts at risk for skin, cervical, bladder, breast, lung, oral cavity, and esophageal tumors. Most of these studies have yet to report results. One completed study examined 267 patients living in Linxian, China, who were at risk for esophageal carcinoma based on the presence of mild or moderate esophageal squamous dysplasia at a baseline endoscopy *(191)*. This study randomized patients using a 2 × 2 factorial design to a 10-month intervention of selenomethionine 200 mg daily and/or celecoxib 200 mg twice daily versus placebo. This study demonstrated a trend toward increased dysplasia regression (43% versus 32%) and decreased dysplasia progression (14% versus 19%) in the patients receiving selenomethionine but no effect with celecoxib use.

6. COX-2 INHIBITION IN CANCER TREATMENT

6.1. Cox-2 Expression and Cancer Prognosis

The presence of high levels of Cox-2 within tumor cells may indicate either more aggressive tumor behavior or a lack of response to standard treatment. In retrospective analyses, increased Cox-2 expression by immunostain was associated with decreased survival following surgical resection for patients with adenocarcinoma of the duodenal ampulla *(192)*, esophagus *(193)*, prostate *(194)*, bladder *(195)*, and colon *(196)*. Tumor Cox-2 expression was also associated with a reduced chemotherapy response in a broad variety of cancers. For example, in patients with stage IV CRC, intratumoral Cox-2 gene expression measured by RT–PCR indicated reduced overall survival in response to fluoropyrimidine-based chemotherapy *(197)*. The presence of Cox-2 by immunostain predicted poor response to cisplatin-based neoadjuvant chemotherapy for patients with locally advanced cervical cancer *(198)*, and a study of 87 primary ovarian carcinoma patients found that the degree of tumor Cox-2 expression was higher in patients resistant to chemotherapy with a cisplatin regimen *(199)*. In addition, overexpression of Cox-2 in tumor cells was associated with reduced survival following chemotherapy for multiple myeloma *(200)*, metastatic renal cell carcinoma *(201)*, and ovarian cancer *(202)*.

The mechanisms of Cox-2-associated chemotherapy resistance are only partially known. In human breast cancer, Cox-2 expression correlates with the presence of MDR-1 P-glycoprotein (MDR1/Pgp170), a molecule responsible for some forms of chemotherapy resistance *(203)*. In studies of non-small cell lung cancer (NSCLC), preoperative treatment with taxanes increased intratumor expression of Cox-2 and PGE_2 *(204)*. *In vitro* studies suggest that this effect is due to a taxane-associated increase in AP-1-mediated transcription of the Cox-2 gene *(204,205)*. Another possibility is induction of an intratumoral inflammatory response by chemotherapy. Concomitant treatment with taxanes and a selective Cox-2 inhibitor abrogated chemotherapy-associated increases in intratumor PGE_2 concentrations *(204)*.

6.2. Pre-Clinical Studies of Cox-2 Inhibitors in Combination with Chemotherapy or Radiation

Selective Cox-2 inhibitors demonstrate effects against tumor cells, such as apoptosis induction and angiogenesis inhibition, indicating that they may improve the effectiveness of cytotoxic chemotherapy. A variety of *in vitro* studies support this hypothesis. For example, cell culture studies show that combinations of NSAIDs and various chemotherapeutic drugs, including cisplatin, paclitaxel, docitaxel and VP-16, achieve a synergistic effect against human NSCLC, colorectal, pancreas, leukemia, and CNS cell lines *(149–210)*. In the Lewis lung carcinoma model, combination of a Cox-2 inhibitor with cisplatin or cyclophosphamide delayed the growth of primary tumors and decreased the number of lung metastases *(211)*. In dogs with carcinogen-induced transitional cell carcinoma of the urinary bladder, enhanced ability to achieve tumor remission was observed with a regimen of cisplatin plus piroxicam *(150)*.

NSAIDs may also potentiate the anti-tumor effects of radiotherapy. Ibuprofen, a non-selective inhibitor of both Cox isoforms, increases the sensitivity of prostate cancer cells to radiation *in vitro (212)*. Similarly, indomethacin potentiates the radiosensitivity of tumors with little effect on normal tissue in murine systems *(213–215)*. Radiation response was significantly enhanced when Cox-2 inhibitors were used to treat Cox-2-expressing tumor cells *in vitro (216)*. Xenograft studies show enhanced tumor growth delay when a Cox-2 inhibitor was combined with radiotherapy *(211,217,218)*. The combined effect of Cox-2 inhibition and radiation is more than additive, suggesting a synergistic or radiosensitizing effect. The mechanism of radiosensitization by NSAIDs is not fully characterized. A reasonable hypothesis is that enhanced anti-tumor response is produced by NSAID-mediated inhibition of angiogenesis, a response known to promote radiosensitization *(219,220)*. In one animal model, assessment of vascular function by direct contrast enhancement-magnetic resonance imaging showed that selective Cox-2 inhibition with celecoxib enhanced vascular permeability during radiation therapy *(221)*.

6.3. Clinical Trials in Cancer Treatment

The first use of NSAIDs in cancer patients was directed toward palliation of cancer symptoms. The efficacy of NSAIDs in the management of cancer pain has been well established in case series and randomized clinical trials *(222)*. NSAIDs are most often inadequate as monotherapy for cancer pain although a trial of cancer patients treated with sublingual piroxicam found that 7 of 21 patients treated achieved complete pain relief *(223)*. The mechanisms underlying NSAID-induced analgesia have yet to be fully defined *(224)*. Interestingly, NSAIDs administered with palliative intent may improve patient survival. In a study of 135 solid tumor patients with malignancy-associated malnutrition, indomethacin reduced both pain and the consumption of additional analgesics *(225)*. In addition, patients treated with indomethacin maintained their performance status compared with those on placebo, and median survival doubled from 250 days to 510 days ($p < 0.05$). Patients taking prednisolone also had a significantly prolonged survival, suggesting that anti-inflammatory agents may not only palliate disease but also improve prognosis.

The efficacy of chemotherapy in combination with selective Cox-2 inhibitors in first-line cancer treatment has been examined in trials of pre-operative chemotherapy for NSCLC *(226)*, breast cancer *(227,228)*, and esophageal cancer *(229)*. Patients with resectable NSCLC treated pre-operatively with paclitaxel, carboplatin, and celecoxib achieved a high response rate of 65%, with a pathological complete response rate of 17% *(226)*. In a small three-arm study of patients with hormone-sensitive post-menopausal breast cancer, subjects were randomized to receive letrozole, exemestane, or exemestane with celecoxib *(228)*. Preliminary results showed complete clinical responses only in patients on the celecoxib-containing arm. Finally, celecoxib was added to cisplatin, 5-fluorouracil, and radiation therapy in a phase II study of patients with potentially resectable esophageal cancer *(229)*. Of 31 patients treated, five (22%) achieved a pathological complete response, and no increase in toxicity due to the addition of celecoxib was noted.

Only a few studies examined selective Cox-2 inhibition in combination with chemotherapy in first-line treatment of advanced disease. A phase I study of stage IV CRC patients involved addition of celecoxib and glutamine to irinotecan, 5-fluorouracil, and leucovorin (IFL) *(230)*. The aim of this study was to decrease IFL-related gastrointestinal toxicity with glutamine and to increase anti-tumor efficacy with celecoxib. The addition of celecoxib and glutamine did not significantly alter the efficacy or toxicity of IFL in these patients with metastatic CRC, and anti-tumor response did not correlate with pre-treatment tumor expression of Cox-2.

The impact of selective Cox-2 inhibitors in second-line chemotherapy has been studied in a variety of tumors including pancreatic carcinoma *(231)*, multiple myeloma *(232)*, metastatic differentiated thyroid carcinoma *(233)*, and glioblastoma *(234)*. These phase I and II studies all reported a small number of patients who responded to combination treatment. Chemotherapy in combination with celecoxib has been more thoroughly examined in phase II trials of patients with recurrent NSCLC. In one study, patients with advanced NSCLC who progressed after platinum-based chemotherapy received docetaxel in combination with celecoxib. These patients achieved a response rate of 10.2% and an overall survival of 11.3 months, with toxicity comparable to docetaxel alone *(235)*. A similar study in refractory NSCLC patients examined the effect of the same regimen on urinary excretion of the PGE_2 metabolite, PGE-M *(236)*. This study confirmed a celecoxib-related reduction in PGE-M in NSCLC patients and also demonstrated anti-tumor efficacy, with a treatment response in 11% and a 14.8-month survival in the subset of patients with the greatest extent of drug-associated PGE-M reduction. A multicenter phase II study involving 58 patients in Italy also examined response to weekly paclitaxel and celecoxib as second-line therapy for NSCLC *(237)*. Objective responses occurred in 24.1%, with disease stabilization in

41.3%. Interestingly, treatment-associated decrease in expression of circulating VEGF correlated with anti-tumor response, indicating that the anti-tumor efficacy resulted from inhibition of angiogenesis.

Patients with stage IV CRC who progressed following treatment with oxiliplatin-based first-line chemotherapy were treated with continuous infusion of 5-fluorouracil, irinotecan, and rofecoxib *(238)*. Partial responses were observed in 48.5% of the 48 patients treated, with 30.3% achieving stable disease. The toxicity profile of this combination was consistent with that seen with irinotecan and 5-fluorouracil without rofecoxib. Phase II trials in advanced pancreatic cancer combined celecoxib and gemcitabine *(239)* or celecoxib, gemcitabine, and cisplatin *(240)*. Results from these studies were mixed, with little improvement on the expected median survival of 5–9 months.

A few studies examined combinations of selective Cox-2 inhibitors and targeted agents. The combination of gefitinib and celecoxib in patients with advanced squamous cell carcinoma of the head and neck produced partial responses in 4 of 18 patients *(241)*. In patients with metastatic HER-2/neu overexpressing breast cancer who progressed during treatment with trastuzumab, however, addition of celecoxib to trastuzumab failed to achieve an anti-tumor response *(242)*.

7. THE FUTURE: OPTIMIZING CLINICAL RESPONSE TO COX-2 INHIBITORS

The hypothesis that Cox-2 inhibition will improve cancer treatment and prevention is based upon a substantial body of pre-clinical data, demonstrating conclusively that inflammation promotes and maintains tumor formation (Fig. 8). Clinical trials available to date confirm the importance of Cox-2 and PGE_2 to tumor formation and demonstrate that Cox-2 inhibition achieves a significant anti-tumor response throughout all stages of tumorigenesis. Results from early studies of selective Cox-2 inhibitors in combination with chemotherapy and/or radiation therapy for invasive cancers are promising. In the

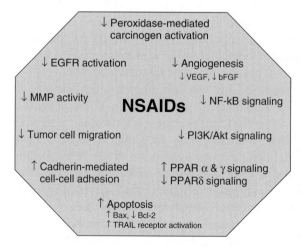

Fig. 8. Anti-tumor effects of Cox inhibition.

next few years, additional significant studies testing the efficacy of selective Cox-2 inhibitions in multimodality cancer treatment will reach maturity.

Despite an impressive demonstration of anti-tumor efficacy in human colorectal adenoma prevention trials, successful application of NSAIDs to cancer prevention is presently limited by toxicity, which unfortunately was not eliminated upon development of the selective Cox-2 inhibitiors. Two avenues must be pursued to carry this promising field forward. First, new research must provide a better understanding of the risks of all NSAIDs, both selective and non-selective, in terms of cardiovascular toxicity. Many questions central to this issue remain, including the relative contribution of Cox-1 versus Cox-2 to NSAID side effects, the impact of dosing regimen and drug metabolism, and the patient-specific factors that predict adverse drug effects. These studies are crucial not only for the field of oncology but for improving our management of important co-morbid conditions such as arthritis and pain. The second key area of new research is the study of combination regimens. Drug development efforts in oncology have produced numerous new compounds that specifically target components of Cox-2 and PGE_2-mediated signaling. Some of these, such as inhibitors of EGFR and VEGF, have advanced through phase III studies and found application in treatment of established solid tumors. A wide range of other new agents are beginning human trials, including inhibitors of prostanoid receptors, src kinase, FAK, components of PI_3K signaling, and inducible nitric oxide synthase, to name only a few. Owing to the tissue-specific nature of these signaling pathways, it is possible that combination therapy will decrease the toxicity of NSAIDs, yet maintain anti-tumor efficacy in the target tissue.

REFERENCES

1. Warren JC. *Surgical Observations on Tumours with Cases and Operations*. Boston, MA: Crocker and Brewster; 1837.
2. O'Banion MK, Sadowski HB, Winn V, Young DA. A serum- and glucocorticoid-regulated 4-kilobase mRNA encodes a cyclooxygenase-related protein. *J Biol Chem* 1991;266(34):23261–23267.
3. Herschman HR. Function and regulation of prostaglandin synthase 2. *Adv Exp Med Biol* 1999;469:3–8.
4. Bertagnolli MM. Eicosanoids. In Shepro D, ed. *Microvascular Research: Biology and Pathology*. Elsevier North America; 2006;727–734.
5. Needleman P, Turk J, Jakschik BA, Morrison AR, Lefkowith JB. Arachidonic acid metabolism. *Annu Rev Biochem* 1986;55:69–102.
6. Coleman RA, Smith WL, Narumya S. IUP classification of prostanoid receptors: properties, distribution, and structure of the receptors and their subtypes. *Pharm Rev* 1994;46:205–229.
7. Sheng H, Shao J, Morrow JD, Beauchamp RD, DuBois RN. Modulation of apoptosis and Bcl-2 expression by prostaglandin E2 in human colon cancer cells. *Cancer Res* 1998;58:362–366.
8. Hixon LJ, Alberts DX, Krutzsch M, et al. Antiproliferative effect of nonsteroidal anti-inflammatory drugs against human colon cancer cells. *Cancer Epidemiol Biomarkers Prev* 1994;3:433–438.
9. Pai R, Sorgehan B, Szabo IL, Pavelka M, Baatar D, Tarnawski AS. Prostaglandin E2 transactivates EGF receptor: a novel mechanism for promoting colon cancer growth and gastrointestinal hypertrophy. *Nat Med* 2002;8(3):289–293.
10. Pai R, Nakamura T, Moon WS, Tarnawski AS. Prostaglandins promote colon cancer cell invasion; signaling by cross-talk between two distinct growth factor receptors. *FASEB J* 2003;17(12):1640–1647.
11. Form DM, Auerbach R. PGE2 and angiogenesis. *Proc Soc Exp Biol Med* 1983;172:214–218.
12. Gately S, Kerbel R. Therapeutic potential of selective cyclooxygenase-2 inhibitors in the management of tumor angiogenesis. *Prog Exp Tumor Res* 2003;37:179–192.

13. Smith WL, Garavito RM, DeWitt DL. Prostaglandin endoperoxide H synthases (cyclooxygenases)-1 and -2. *J Biol Chem* 1996;271(52):33157–33160.

14. Williams CS, DuBois RN. Prostaglandin endoperoxide synthase: why two isoforms? *Am J Physiol* 1996;270(3 Pt 1):G393–G400.

15. Herschman HR. Prostaglandin synthase 2. *Biochim Biophys Acta* 1996;1299(1):125–140.

16. Howe LR, Subbaramaiah K, Brown AM, Dannenberg AJ. Cyclooxygenase-2: a target for the prevention and treatment of breast cancer. *Endocr Relat Cancer* 2001;8(2):97–114.

17. Xie W, Herschman HR. v-src induces prostaglandin synthase 2 gene expression by activation of the c-Jun N-terminal kinase and the c-Jun transcription factor. *J Biol Chem* 1995;270(46):27622–27628.

18. Subbaramaiah K, Norton L, Gerald W, Dannenberg AJ. Increased expression of cyclooxygenase-2 in HER-2-overexpressing human breast cancer cells. *NCI 7th SPORE Investigators' Workshop*; 1999.

19. Vadlamudi R, Mandal M, Adam L, Steinbach G, Mendelsohn J, Kumar R. Regulation of cyclooxygenase-2 pathway by HER2 receptor. *Oncogene* 1999;18(2):305–314.

20. Chen G, Wilson R, McKillop JH, Walker JJ. The role of cytokines in the production of prostacyclin and thromboxane in human mononuclear cells. *Immunol Invest* 1994;23(4–5):269–279.

21. Subbaramaiah K, Telang N, Ramonetti JT, et al. Transcription of cyclooxygenase-2 is enhanced in transformed mammary epithelial cells. *Cancer Res* 1996;56(19):4424–4429.

22. Subbarayan V, Sabichi AL, Llansa N, Lippman SM, Menter DG. Differential expression of cyclooxygenase-2 and its regulation by tumor necrosis factor-alpha in normal and malignant prostate cells. *Cancer Res* 2001;61(6):2720–2726.

23. Subbaramaiah K, Dannenberg AJ. Cyclooxygenase 2: a molecular target for cancer prevention and treatment. *Trends Pharmacol Sci* 2003;24:296–102.

24. Xie W, Herschman HR. Transcriptional regulation of prostaglandin synthase 2 gene expression by platelet-derived growth factor and serum. *J Biol Chem* 1996;271(49):31742–31748.

25. Subbaramaiah K, Chung WJ, Dannenberg AJ. Ceramide regulates the transcription of cyclooygenase-2. Evidence for involvement of extracellular signal-regulated kinase/c-Jun N-terminal kinase and p38 mitogen-activated protein kinase pathways. *J Biol Chem* 1998;273(49):32943–32949.

26. Chen C-C, Sun Y-T, Chen J-J, Chang Y-J. Tumor necrosis factor-alpha-induced cyclooxygenase-2 expression via sequential activation of ceramide-dependent mitogen-activated protein kinases, and I-kappaB kinase 1/2 in human alveolar epithelial cells. *Mol Pharmacol* 2001;59(3):493–500.

27. Eisinger AL, Nadauld LD, Shelton DN, et al. The APC tumor suppressor gene regulates expression of cycloosygenase-2 by a mechanism that involves retinoic acid. *J Biol Chem* 2006;281(29): 20474–20482.

28. Ramsay RG, Friend A, Vizantios Y, et al. Cyclooxygenase-2, a colorectal cancer nonsteroidal anti-inflammatory drug target, is regulated by c-MYB. *Cancer Res.* 2000;60(7):1805–1809.

29. Raschella G, Cesi V, Amendolo R, et al. Expression of B-myb in neuroblastoma tumors is a poor prognostic factor independent from MYCN amplification. *Cancer Res* 1999;59:3365–3368.

30. Lutz PG, Houzel-Charavel A, Moog-Lutz C, Cayre YE. Myeloblastin is an Myb targer gene: mechanisms of regulation in myeloid leukemia cells growth-arrested by retinoic adic. *Blood* 2001;97:2449–2456.

31. Subbaramaiah K, Altorki N, Chung WJ, Mestre JR, Sampat A, Dannenberg AJ. Inhibition of cyclooxygenase-2 gene expression by p53. *J Biol Chem* 1999;274:10911–10915.

32. Cok SJ, Morrison AR. The 3′-untranslated region of murine cyclooxygenase-2 contains multiple regulatory elements that alter message stability and translational efficiency. *J Biol Chem* 2001;276(25):23179–23185.

33. Dixon DA, Tolley ND, King PH, et al. Altered expression of the mRNA stability factor HuR promotes cyclooxygenase-2 expression in colon cancer cells. *J Clin Invest* 2001;108(11): 1657–1665.

34. Sheng H, Shao J, Dixon DA, et al. Transforming growth factor-beta1 enhances Ha-ras-induced expression of cyclooxygenase-2 in intestinal epithelial cells via stabilization of mRNA. *J Biol Chem* 2000;275(9):6628–6635.

35. Srivastava SK, Tetsuka T, Daphna-Iken D, Morrison AR. IL-1 beta stabilizes COX II mRNA in renal mesangial cells: role of 3′-untranslated region. *Am J Physiol* 1994;267(3 Pt 2):F504–8.

36. Anderson WF, Umar A, Viner JL, Hawk ET. The role of cyclooxygenase inhibitors in cancer prevention. *Curr Pharm Des* 2002;8:99–110.

37. Eberhart CE, Coffey RJ, Radhika A, Giardiello FM, Ferrenbach S, DuBois RN. Up-regulation of cyclooxygenase 2 gene expression in human colorectal adenomas and adenocarcinomas. *Gastroenterology* 1994;107(4):1183–1188.

38. Shono T, Tofilon PJ, Bruner JM, Owolabi O, Lang FF. Cyclooxygenase-2 expression in human gliomas: prognostic significance and molecular correlations. *Cancer Res* 2001;61(11): 4375–4381.

39. Sheehan KM, Sheahan K, O'Donoghue DP, et al. The relationship between cyclooxygenase-2 expression and colorectal cancer. *JAMA* 1999;282(13):1254–1257.

40. Masunaga R, Kohno H, Dhar DK, et al. Cyclooxygenase-2 expression correlates with tumor neovascularization and prognosis in human colorectal carcinoma patients. *Clin Cancer Res* 2000;6(10):4064–4068.

41. Tomozawa S, Tsuno NH, Sunami E, et al. Cyclooxygenase-2 overexpression correlates with tumour recurrence, especially haematogenous metastasis, of colorectal cancer. *Br J Cancer* 2000;83(3): 324–328.

42. Ristimaki A, Sivula A, Lundin J, et al. Prognostic significance of elevated cyclooxygenase-2 expression in breast cancer. *Cancer Res* 2002;62(3):632–635.

43. Gaffney DK, Holden J, Davis M, Zempolich K, Murphy KJ, Dodson M. Elevated cyclooxygenase-2 expression correlates with diminished survival in carcinoma of the cervix treated with radiotherapy. *Int J Radiat Oncol Biol Phys* 2001;49(5):1213–1217.

44. Oshima M, Dinchuk JE, Kargman SL, Oshima H, Hancock B, Kwong E. Suppression of intestinal polyposis in Apc delta knockout mice by inhibition of cyclooxygenase 2 (COX-2). *Cell* 1996;87:803–809.

45. Williams CS, Tsujii M, Reese J, Dey SK, DuBois RN. Host cyclooxygenase-2 modulates carcinoma growth. *J Clin Invest* 2000;105(11):1589–1594.

46. Chulada PC, Thompson MB, Mahler JF, et al. Genetic disruption of Ptgs-1, as well as Ptgs-2, reduces intestinal tumorigenesis in Min mice. *Cancer Res* 2000;60(17):4705–4708.

47. Liu CH, Chang SH, Narko K, et al. Overexpression of cyclooxygenase-2 is sufficient to induce tumorigenesis in transgenic mice. *J Biol Chem* 2001;276(21):18563–18569.

48. Neufang G, Furstenberger G, Heidt M, et al. Abnormal differentiation of epidermis in transgenic mice constitutively expressing cyclooxygenase-2 in skin. *Proc Natl Acad Sci USA* 2001;98(13):7629–7634.

49. Wang D, Wang H, Shi Q, et al. Prostaglandin E(2) promotes colorectal adenoma growth via transactivation of the nuclear peroxisome proliferator-activated receptor delta. *Cancer Cell* 2004;6: 285–295.

50. Hansen-Petrik MB, McEntee MF, Jull B, Shi H, Zemel MB, Whelan J. Prostaglandin E2 protects intestinal tumors from nonsteroidal anti-inflammatory drug-induced regression in ApcMin/+ Mice. *Cancer Res* 2002;62(2):403–408.

51. Watanabe K, Kawamori T, Nakatsugi S, et al. Role of the prostaglandin E receptor subtype EP1 in colon carcinogenesis. *Cancer Res* 1999;59(20):5093–5096.

52. Mutoh M, Watanabe K, Kitamura T, et al. Involvement of prostaglandin E receptor subtype EP4 in colon carcinogenesis. *Cancer Res* 2002;62:28–32.

53. Sonoshita M, Takaku K, Sasaki N, et al. Acceleration of intestinal polyposis through prostaglandin receptor EP2 in Apc(Delta 716) knockout mice. *Nat Med* 2001;7(9):1048–1051.

54. Amano H, Hayashi I, Endo H, et al. Host prostaglandin E(2)-EP3 signaling regulates tumor-associated antiogenesis and tumor growth. *J Exp Med* 2003;197:221–232.

55. Balch CM, Dougherty PA, Tilden AB. Excessive prostaglandin E2 production by suppressor monocytes in head and neck cancer patients. *Ann Surg* 1982;196:645–650.

56. Bennett A, Tacca MD, Stamford IF, Zebro T. Prostaglandins from tumour of human large bowel. *Br J Cancer* 1977;6:881–884.

57. McDougall CJ, Ngoi SS, Goldman IS, et al. Reduced expression of HLA class I and II antigens in colon cancer. *Cancer Res* 1990;50(24):8023–8027.

58. Harris SG, Padilla J, Koumas L, Ray D, Phipps RP. Prostaglandins as modulators of immunity. *Tends Immunol* 2002;23:144–150.

59. Grossman EM, Longo WE, Panesar N, Mazuski JE, Kaminski DL. The role of cyclooxygenase enzymes in the growth of human gallbladder cancer cells. *Carcinogenesis* 2000;21(7):1403–1409.

60. Boucher M-J, Morisset J, Vachon PH, Reed JC, Laine J, Rivard N. MEK/ERK signaling pathway regulates the expression of Bcl2, Bcl-XL, and Mcl-1 and promotes survival of human pancreatic cancer cells. *J Cell Biochem* 2000;79:355–369.

61. Komatsu K, Buchanan FG, Katkuri S, et al. Oncogenic potential of MEK1 in rat intestinal epithelial cells is mediated via cyclooxygenase-2. *Gastroenterology* 2005;129:577–590.

62. Tsujii M, DuBois RN. Alterations in cellular adhesion and apoptosis in epithelial cells overexpressing prostaglandin endoperoxide synthase 2. *Cell* 1995;83(3):493–501.

63. Kikuchi A, Kishida S, Yamamoto H. Regulation of Wnt signaling by protein-protein interaction and post-translational modifications. *Exp Mol Med* 2006;36(1):1–10.

64. Shao J, Jung C, Liu C, Sheng H. Prostaglandin E2 stimulates beta-catenin/T cell factor-dependent transcription in colon cancer. *J Biol Chem* 2005;280:26565–26572.

65. Castellone MD, Teramoto H, Williams BO, Druey KM, Gutkind JS. Prostaglandin E2 promotes colon cancer cell growth through a novel Gs-axin-{β}-catenin signaling axis. *Science* 2005;310:1504–1510.

66. Koki AT, Leahy KM, Masferrer JL. Potential utility of COX-2 inhibitors in chemoprevention and chemotherapy. *Expert Opin Investig Drugs* 1999;8(10):1623–1638.

67. Cianchi F, Cortesini C, Bechi P, et al. Up-regulation of cyclooxygenase-2 gene expression correlates with tumor angiogenesis in human colorectal cancer. *Gastroenterology* 2001;121(6):1339–1347.

68. Shao J, Lee SB, Guo H, et al. Prostaglandin E2 stiumlates the growth of colon cancer cells via induction of amphiregulin. *Cancer Res* 2003;63:5218–5223.

69. Buchanan FG, Wang D, Bargiacchi F, et al. Prostaglandin E2 regulates cell migration via the intracellular activation of the epidermal growth factor receptor. *J Biol Chem* 2003;278: 35451–35457.

70. Sheng H, Shao J, Washington MK, DuBois RN. Prostaglandin E2 increases growth and motility of colorectal carcinoma cells. *J Biol Chem* 2001;276(21):18075–18081.

71. Bill HM, Knudsen B, Moores SL, et al. Epidermal growth factor receptor-dependent regulation of integrin-mediated signaling and cell cycle entry in epithelial cells. *Mol Cell Biol* 2004;24(19): 8586–8599.

72. Carothers AM, Javid SH, Moran AE, Hunt DM, Redston M, Bertagnolli MM. Deficient E-cadherin adhesion in C57BL/6J-Min/+ mice is associated with increased tyrosine kinase activity and RhoA-dependent actomyosin contractility. *Exp Cell Res* 2006;312:387–400.

73. Dannenberg AJ, Lippman SM, Mann JR, Subbaramaiah K, DuBois RN. Cyclooxygenase-2 and epidermal growth factor receptor: pharmacologic targets for chemoprevention. *J Clin Oncol* 2005;23(2):254–266.

74. Tsujii M, Kawano S, DuBois RN. Cyclooxygenase-2 expression in human colon cancer cells increases metastatic potential. *Proc Natl Acad Sci USA* 1997;94(7):3336–3340.

75. Yiu GK, Toker A. NFAT induces breast cancer cell invasion by promoting the induction of cyclooxygenase-2. *J Biol Chem* 2006;281(18):12210–12217.

76. Marnett LJ. Aspirin and related nonsteroidal anti-inflammatory drugs as chemopreventive agents against colon cancer. *Prev Med* 1995;24(2):103–106.

77. Wiese FW, Thompson PA, Kadlubar FF. Carcinogen substrate specificity of human COX-1 and COX-2. *Carcinogenesis* 2001;22(1):5–10.

78. Moraitis D, Du B, DeLorenzo MS, et al. Levels of cyclooxygenase-2 are increased in the oral mucosa of smokers: evidence for the role of epidermal growth factor receptor and its ligands. *Cancer Res* 2005;65(2):664–670.

79. Kalgutkar AS, Crews BC, Rowlinson SW, Farner C, Seibert K, Marnett LJ. Aspirin-like molecules that covalently inactivate cyclooxygenase-2. *Science* 1998;280:1268–1270.

80. Marnett LJ, Kalgutkar AS. Cyclooxygenase 2 inhibitors: discovery, selectivity and the future. *TiPS* 1999;20:465–469.

81. Kalgutkar AS, Crews BC, Saleh S, Prudhomme D, Marnett LJ. Indolyl esters and amides related to indomethacin are selective COX-2 inhibitors. *Bioorg Med Chem* 2005;13:6810–6822.

82. Warner TD, Giuliano F, Vojnovic I, Bukasa A, Mitchell JA, Vane JR. Nonsteroid drug selectivities for cyclo-oxygenase-1 rather than cyclo-oxygenase-2 are associated with human gastrointestinal toxicity: a full *in vitro* analysis. *Proc Natl Acad Sci USA* 1999;96:7563–7568.

83. Singh AK, Trotman BW. Use and safety of aspirin in the chemoprevention of colorectal cancer. *J Assoc Acad Minor Phys* 1998;9:40–44.

84. Larkai EN, Smith JL, Lidsky MD, Graham DY. Gastroduodenal mucosa and dyspeptic symptoms in arthritic patients during chronic nonsteroidal anti-inflammatory drug use. *Am J Gastroenterol* 1987;82(11):1153–1158.

85. Singh G, Triadafilopoulos G. Epidemiology of NSAID induced gastrointestinal complications. *J Rheumatol* 1999;26(Suppl 56):18–24.
86. Schafer AI. Effects of nonsteroidal anti-inflammatory therapy on platelets. *Am J Med* 1999;106(5B):25S–36S.
87. Whelton A. Renal and related cardiovascular effects of conventional and Cox-2-specific NSAIDs and non-NSAID analgesics. *Am J Ther* 2000;7:63–74.
88. Bjorkman DJ. Current status of nonsteroidal anti-inflammatory drug (NSAID) use in the United States: risk factors and frequency of complications. *Am J Med* 1999;107(6A):3S–8S; discussion 8S–10S.
89. Bresalier RS, Sandler RS, Quan H, et al. Cardiovascular events associated with rofecoxib in a colorectal adenoma chemoprevention trial. *N Engl J Med* 2005;352(11):1092–1102.
90. Solomon SD, McMurray JJ, Pfeffer MA, et al. Cardiovascular risk associated with celecoxib in a clinical trial for colorectal adenoma prevention. *N Engl J Med* 2005;352(11):1071–1080.
91. Hawkey CJ, Hawkey GM, Everitt S, Skelly MM, Stack WA, Gray D. Increased risk of myocardial infarction as first manifestation of ischaemic heart disease and nonselective nonsteroidal anti-inflammatory drugs. *Br J Clin Pharmacol* 2006;61(6):730–737.
92. Weir MR, Sperling RS, Reicin A, Gertz BJ. Selective Cox-2 inhibition and cardiovascular effects: a review of the rofecoxib development program. *Am Heart J* 2003;146:591–604.
93. Kune GA, Kune S, Watson LF. Colorectal cancer risk, chronic illnesses, operations, and medications: case control results from the Melbourne Colorectal Cancer Study. *Cancer Res* 1988;48(15):4399–4404.
94. Thun MJ, Namboodiri MM, Heath CW, Jr. Aspirin use and reduced risk of fatal colon cancer. *N Engl J Med* 1991;325(23):1593–1596.
95. Thun MJ, Namboodiri MM, Calle EE, Flanders WD, Heath CW, Jr. Aspirin use and risk of fatal cancer. *Cancer Res* 1993;53(6):1322–1327.
96. Giovannucci E, Rimm EB, Stampfer MJ, Colditz GA, Ascherio A, Willett WC. Aspirin use and the risk for colorectal cancer and adenoma in male health professionals. *Ann Intern Med* 1994;121(4):241–246.
97. Garcia-Rodriguez LA, Huerta-Alvarez C. Reduced incidence of colorectal adenoma among long-term users of nonsteroidal antiinflammatory drugs: a pooled analysis of published studies and a new population-based study. *Epidemiology* 2000;11(4):376–381.
98. Garcia-Rodriguez LA, Heuerta-Alvarez C. Reduced risk of colorectal cancer among long-term users of aspirin and nonaspirin nonsteroidal antiinflammatory drugs. *Epidemiology* 2001;12:88–93.
99. Giovannucci E. The prevention of colorectal cancer by aspirin use. *Biomed Pharmacother* 1999;53(7):303–308.
100. Thun MJ, Henley SJ, Patrono C. Nonsteroidal anti-inflammatory drugs as anticancer agents: mechanistic, pharmacologic, and clinical issues. *J Natl Cancer Inst* 2002;94(4):252–266.
101. Paganini-Hill A. Aspirin and colorectal cancer: the Leisure World cohort revisited. *Prev Med* 1995;24(2):113–115.
102. Sturmer T, Glynn RJ, Lee IM, Manson JE, Buring JE, Hennekens CH. Aspirin use and colorectal cancer: post-trial follow-up data from the Physicians' Health Study. *Ann Intern Med* 1998;128(9):713–720.
103. Farrow DC, Vaughan TL, Hansten PD, et al. Use of aspirin and other nonsteroidal antiinflammatory durgs and risk of esophageal and gastric cancer. *Cancer Epidemiol Biomarkers Prev* 1998;7:97–102.
104. Sharpe CR, Collet JP, McNutt M, Belzile E, Boivin JF, Hanley JA. Nested case-control study of the effects of non-steroidal anti- inflammatory drugs on breast cancer risk and stage. *Br J Cancer* 2000;83(1):112–120.
105. Castelao JE, Yuan JM, Gago-Dominguez M, Yu MC, Ross RK. Non-steroidal anti-inflammatory drugs and bladder cancer prevention. *Br J Cancer* 2000;82(7):1364–1369.
106. Langman MJ, Cheng KK, Gilman EA, Lancashire RJ. Effect of anti-inflammatory drugs on overall risk of common cancer: case-control study in general practice research database. *BMJ* 2000;320(7250):1642–1646.
107. Gridley G, McLaughlin JK, Ekbom A, et al. Incidence of cancer among patients with rheumatoid arthritis. *J Natl Cancer Inst* 1993;85(4):307–311.
108. Bucher C, Jordan P, Nickeleit V, Torhorst J, Mihatsch MJ. Relative risk of malignant tumors in analgesic abusers. Effects of long- term intake of aspirin. *Clin Nephrol* 1999;51(2):67–72.
109. Schreinemachers DM, Everson RB. Aspirin use and lung, colon, and breast cancer incidence in a prospective study . *Epidemiology* 1994;5(2):138–146.

110. Jacoby RF, Seibert K, Cole CE, Kelloff G, Lubet RA. The cyclooxygenase-2 inhibitor celecoxib is a potent preventive and therapeutic agent in the Min mouse model of adenomatous polyposis. *Cancer Res* 2000;60(18):5040–5044.

111. Wilgus TA, Ross MS, Parrett ML, Oberyszyn TM. Topical application of a selective cyclooxygenase inhibitor suppresses UVB mediated cutaneous inflammation. *Prostaglandins Other Lipid Mediat* 2000;62(4):367–384.

112. Hara A, Yoshimi N, Niwa M, Ino N, Mori H. Apoptosis induced by NS-398, a selective cyclooxygenase-2 inhibitor, in human colorectal cancer cell lines. *Jpn J Cancer Res* 1997;88(6): 600–604.

113. Sawaoka H, Kawano S, Tsuji S, et al. Cyclooxygenase-2 inhibitors suppress the growth of gastric cancer xenografts via induction of apoptosis in nude mice. *Am J Physiol* 1998;274(6 Pt 1): G1061–G1067.

114. Hsu AL, Ching TT, Wang DS, Song X, Rangnekar VM, Chen CS. The cyclooxygenase-2 inhibitor celecoxib induces apoptosis by blocking Akt activation in human prostate cancer cells independently of Bcl-2. *J Biol Chem* 2000;275(15):11397–11403.

115. Liu XH, Yao S, Kirschenbaum A, Levine AC. NS398, a selective cyclooxygenase-2 inhibitor, induces apoptosis and down-regulates bcl-2 expression in LNCaP cells. *Cancer Res* 1998;58(19):4245–4249.

116. Ding XZ, Tong WG, Adrian TE. Blockade of cyclooxygenase-2 inhibits proliferation and induces apoptosis in human pancreatic cancer cells. *Anticancer Res* 2000;20(4):2625–2631.

117. Zimmermann KC, Sarbia M, Weber AA, Borchard F, Gabbert HE, Schror K. Cyclooxygenase-2 expression in human esophageal carcinoma. *Cancer Res* 1999;59(1):198–204.

118. Yao R, Rioux N, Castonguay A, You M. Inhibition of COX-2 and induction of apoptosis: two determinants of nonsteroidal anti-inflammatory drugs' chemopreventive efficacies in mouse lung tumorigenesis. *Exp Lung Res* 2000;26(8):731–742.

119. Nishimura G, Yanoma S, Mizuno H, Kawakami K, Tsukuda M. A selective cyclooxygenase-2 inhibitor suppresses tumor growth in nude mouse xenografted with human head and neck squamous carcinoma cells. *Jpn J Cancer Res* 1999;90(10):1152–1162.

120. Martin S, Phillips DC, Szekely-Szucs K, Elghazi L, Desmots F, Houghton JA. Cyclooxygenase-2 inhibition sensitizes human colon carcinoma cells to TRAIL-induced apoptosis through clustering of DR5 and concentrating death-inducing signaling complex components into ceramide-enriched caveolae. *Cancer Res* 2005;65(24):11447–11458.

121. Moran AE, Hunt DH, Javid SH, Redston M, Carothers AM, Bertagnolli MM. Apc deficiency is associated with increased Egfr activity in the intestinal enterocytes and adenomas of C57BL/6J-Min/+ mice. *J Biol Chem* 2004;279(4):43261–43272.

122. Masferrer JL, Leahy KM, Koki AT, et al. Antiangiogenic and antitumor activities of cyclooxygenase-2 inhibitors. *Cancer Res* 2000;60(5):1306–1311.

123. Dormond O, Foletti A, Paroz C, Ruegg C. NSAIDs inhibit alpha V beta 3 integrin-mediated and Cdc42/Rac-dependent endothelial-cell spreading, migration and angiogenesis. *Nat Med* 2001;7(9):1041–1047.

124. Rioux N, Castonguay A. Prevention of NNK-induced lung tumorigenesis in A/J mice by acetylsalicylic acid and NS-398. *Cancer Res* 1998;58(23):5354–5360.

125. Sharma RA, Gescher A, Plastaras JP, et al. Cyclooxygenase-2, malondialdehyde and pyrimidopurinone adducts of deoxyguanosine in human colon cells. *Carcinogenesis* 2001;22(9):1557–1560.

126. Zhang X, Morham SG, Langenbach R, Young DA. Malignant transformation and antineoplastic actions of nonsteroidal antiinflammatory drugs (NSAIDs) on cyclooxygenase-null embryo fibroblasts. *J Exp Med* 1999;190(4):451–459.

127. Zhang L, Yu J, Park BH, Kinzler KW, Vogelstein B. Role of BAX in the apoptotic response to anticancer agents. *Science* 2000;290(5493):989–992.

128. Arico S, Pattingre S, Bauvy C, et al. Celecoxib induces apoptosis by inhibiting 3-phosphoinositide-dependent protein kinase-1 activity in the human colon cancer HT-29 cell line. *J Biol Chem* 2002;277:27613–27621.

129. Harada H, Andersen JS, Mann M, Terada N, Korsmeyer SJ. p170S6 kinase signals cell survival as well as growth, inactivating the pro-apoptotic molecule BAD. *Proc Natl Acad Sci USA* 2001;98:9666–9670.

130. Kopelovich L, Fay JR, Glazer RI, Crowell JA. Peroxisome proliferator-activated receptor modulators as potential chemopreventive agents. *Mol Cancer Ther* 2002;1(5):357–363.

131. He TC, Chan TA, Vogelstein B, Kinzler KW. PPARdelta is an APC-regulated target of nonsteroidal anti-inflammatory drugs. *Cell* 1999;99(3):335–345.

132. Park BH, Vogelstein B, Kinzler KW. Genetic disruption of PPARdelta decreases the tumorigenicity of human colon cancer cells. *Proc Natl Acad Sci USA* 2001;98(5):2598–2603.

133. Goppelt-Struebe M. Molecular mechanisms involved in the regulation of prostaglandin biosynthesis by glucocorticoids. *Biochem Pharmacol* 1997;53(10):1389–1395.

134. Yamamoto Y, Yin MJ, Lin KM, Gaynor RB. Sulindac inhibits activation of the NF-kappaB pathway. *J Biol Chem* 1999;274(38):27307–27314.

135. Sun BC, Zhao XL, Zhang SW, Liu YX, Wang L, Wang X. Sulindac induces apoptosis and protects against colon carcinoma in mice. *World J Gastroenterol* 2005;11(18):2822–2826.

136. Reddy BS, Hirose Y, Lubet R, et al. Chemoprevention of colon cancer by specific cyclooxygenase-2 inhibitor, celecoxib, administered during different stages of carcinogenesis. *Cancer Res* 2000;60(2):293–297.

137. Reddy BS, Rao CV, Seibert K. Evaluation of cyclooxygenase-2 inhibitor for potential chemopreventive properties in colon carcinogenesis. *Cancer Res* 1996;56(20):4566–4569.

138. Kawamori T, Rao CV, Seibert K, Reddy BS. Chemopreventive activity of celecoxib, a specific cyclooxygenase-2 inhibitor, against colon carcinogenesis. *Cancer Res* 1998;58(3):409–412.

139. Sasai H, Masaki M, Wakitani K. Suppression of polypogenesis in a new mouse strain with a truncated Apc(Delta474) by a novel COX-2 inhibitor, JTE-522. *Carcinogenesis* 2000;21(5):953–958.

140. Lal G, Ash C, Hay K, et al. Suppression of intestinal polyps in Msh2-deficient and non-Msh2-deficient multiple intestinal neoplasia mice by a specific cyclooxygenase-2 inhibitor and by a dual cyclooxygenase-1/2 inhibitor. *Cancer Res* 2001;61(16):6131–6136.

141. Harris RE, Alshafie GA, Abou-Issa H, Seibert K. Chemoprevention of breast cancer in rats by celecoxib, a cyclooxygenase 2 inhibitor. *Cancer Res* 2000;60(8):2101–2103.

142. Fischer SM, Lo HH, Gordon GB, et al. Chemopreventive activity of celecoxib, a specific cyclooxygenase-2 inhibitor, and indomethacin against ultraviolet light-induced skin carcinogenesis. *Mol Carcinog* 1999;25(4):231–240.

143. Pentland AP, Schoggins JW, Scott GA, Khan KN, Han R. Reduction of UV-induced skin tumors in hairless mice by selective COX-2 inhibition. *Carcinogenesis* 1999;20(10):1939–1944.

144. Grubbs CJ, Lubet RA, Koki AT, et al. Celecoxib inhibits N-butyl-N-(4-hydroxybutyl)-nitrosamine-induced urinary bladder cancers in male B6D2F1 mice and female Fischer-344 rats. *Cancer Res* 2000;60(20):5599–5602.

145. Okajima E, Denda A, Ozono S, et al. Chemopreventive effects of nimesulide, a selective cyclooxygenase-2 inhibitor, on the development of rat urinary bladder carcinomas initiated by N-butyl-N-(4-hydroxybutyl)nitrosamine. *Cancer Res* 1998;58(14):3028–3031.

146. Li Z, Shimada Y, Kawabe A, et al. Suppression of N-nitrosomethylbenzylamine (NMBA)-induced esophageal tumorigenesis in F344 rats by JTE-522, a selective COX-2 inhibitor. *Carcinogenesis* 2001;22(4):547–551.

147. Shiotani H, Denda A, Yamamoto K, et al. Increased expression of cyclooxygenase-2 protein in 4-nitroquinoline-1-oxide-induced rat tongue carcinomas and chemopreventive efficacy of a specific inhibitor, nimesulide. *Cancer Res* 2001;61(4):1451–1456.

148. Alshafie GA, Abou-Issa HM, Seibert K, Harris RE. Chemotherapeutic evaluation of Celecoxib, a cyclooxygenase-2 inhibitor, in a rat mammary tumor model. *Oncol Rep* 2000;7(6):1377–1381.

149. Duffy CP, Elliott CJ, O'Connor RA, et al. Enhancement of chemotherapeutic drug toxicity to human tumour cells *in vitro* by a subset of non-steroidal anti-inflammatory drugs (NSAIDs). *Eur J Cancer* 1998;34(8):1250–1259.

150. Knapp DW, Glickman NW, Widmer WR, et al. Cisplatin versus cisplatin combined with piroxicam in a canine model of human invasive urinary bladder cancer. *Cancer Chemother Pharmacol* 2000;46(3):221–226.

151. Petersen C, Petersen S, Milas L, Lang FF, Tofilon PJ. Enhancement of intrinsic tumor cell radiosensitivity induced by a selective cyclooxygenase-2 inhibitor. *Clin Cancer Res* 2000;6(6):2513–2520.

152. Kobayashi S, Okada S, Hasumi T, Sato N, Fujimura S. The marked anticancer effect of combined VCR, MTX, and indomethacin against drug-resistant recurrent small cell lung carcinoma after conventional chemotherapy: report of a case. *Surg Today* 1999;29(7):666–669.

153. Waddell WR, Loughry RW. Sulindac for polyposis of the colon. *J Surg Oncol* 1983;24(1):83–87.

154. Labayle D, Fischer D, Vielh P, et al. Sulindac causes regression of rectal polyps in familial adenomatous polyposis. *Gastroenterology* 1991;101(3):635–639.
155. Nugent KP, Farmer KC, Spigelman AD, Williams CB, Phillips RK. Randomized controlled trial of the effect of sulindac on duodenal and rectal polyposis and cell proliferation in patients with familial adenomatous polyposis. *Br J Surg* 1993;80(12):1618–1619.
156. Giardiello FM, Hamilton SR, Krush AJ, et al. Treatment of colonic and rectal adenomas with sulindac in familial adenomatous polyposis. *N Engl J Med* 1993;328(18):1313–1316.
157. van Stolk R, Stoner G, Hayton WL, et al. Phase I trial of exisulind (sulindac sulfone, FGN-1) as a chemopreventive agent in patients with familial adenomatous polyposis. *Clin Cancer Res* 2000;6(1):78–89.
158. Hirata K, Itoh H, Ohsato K. Regression of rectal polyps by indomethacin suppository in familial adenomatous polyposis. Report of two cases. *Dis Colon Rectum* 1994;37(9): 943–946.
159. Hirota C, Iida M, Aoyagi K, et al. Effect of indomethacin suppositories on rectal polyposis in patients with familial adenomatous polyposis. *Cancer* 1996;78(8):1660–1665.
160. Lynch HT, Thorson AG, Smyrk T. Rectal cancer after prolonged sulindac chemoprevention. A case report. *Cancer* 1995;75(4):936–938.
161. Thorson AG, Lynch HT, Smyrk TC. Rectal cancer in FAP patient after sulindac. *Lancet* 1994;343(8890):180.
162. Seow-Choen F, Vijayan V, Keng V. Prospective randomized study of sulindac versus calcium and calciferol for upper gastrointestinal polyps in familial adenomatous polyposis. *Br J Surg* 1996;83(12):1763–1766.
163. Richard CS, Berk T, Bapat BV, Haber G, Cohen Z, Gallinger S. Sulindac for periampullary polyps in FAP patients. *Int J Colorectal Dis* 1997;12(1):14–18.
164. Giardiello FM, Yang VW, Hylind LM, et al. Primary chemoprevention of familial adenomatous polyposis with sulindac. *N Engl J Med* 2002;346(14):1054–1059.
165. Fujita M, Fukui H, Kusaka T, et al. Relationship between cyclooxygenase-2 expression and K-ras gene mutation in colorectal adenomas. *J Gastroenterol Hepatol* 2000;15(11):1277–1281.
166. Keller JJ, Offerhaus GJ, Drillenburg P, et al. Molecular analysis of sulindac-resistant adenomas in familial adenomatous polyposis. *Clin Cancer Res* 2001;7(12):4000–4007.
167. Dolara P, Caderni G, Tonelli F. Nimesulide, a selective anti-inflammatory cyclooxygenase-2 inhibitor, does not affect polyp number and mucosal proliferation in familial adenomatous polyposis. *Scand J Gastroenterol* 1999;34(11):1168.
168. Steinbach G, Lynch PM, Phillips RK, et al. The effect of celecoxib, a cyclooxygenase-2 inhibitor, in familial adenomatous polyposis. *N Engl J Med* 2000;342(26):1946–1952.
169. Phillips RK, Wallace MH, Lynch P, et al. A randomised, double-blind, placebo-controlled study of celecoxib, a selective cyclooxygenase-2 inhibitor, on duodenal polyposis in familial adenomatous polyposis. *Gut* 2002;50(6):857–860.
170. Arber N, Kuwada S, Leshno M, Sjodahl R, Hultcrantz R, Rex D. Sporadic adenomatous polyp regression with exisulind is effective but toxic: a randomized, double blind, placebo controlled, dose-response study. *Gut* 2006;55(3):367–373.
171. Rao CV, Indranie, C. Simi B, Manning PT, Connor JR, Reddy BS. Chemopreventive properties of a selective inducible nitric oxide synthase inhibitor in colon carcinogenesis, administered alone or in combination with celecoxib, a selective cyclooygenase-2 inhibitors. *Cancer Res* 2002;62:165–170.
172. Torrance CJ, Jackson PE, Montgomery E, et al. Combinatorial chemoprevention of intestinal neoplasia. *Nat Med.* 2000;6(9):1024–1028.
173. Li H, Schut HA, Conran P, et al. Prevention by aspirin and its combination with alpha-difluoromethylornithine of azoxymethane-induced tumors, aberrant crypt foci and prostaglandin E2 levels in rat colon. *Carcinogenesis* 1999;20:425–420.
174. Chow HH, Earnest DL, Clark D, et al. Effect of subacute ibuprofen dosing on rectal mucosal prostaglandin E2 levels in healthy subjects with a history of resected polyps. *Cancer Epidemiol Biomarkers Prev* 2000;9(4):351–356.
175. Barnes CJ, Hamby-Mason RL, Hardman WE, Cameron IL, Speeg KV, Lee M. Effect of aspirin on prostaglandin E2 formation and transforming growth factor alpha expression in human rectal mucosa from individuals with a history of adenomatous polyps of the colon. *Cancer Epidemiol Biomarkers Prev* 1999;8(4 Pt 1):311–315.

176. Ruffin MT, Krishnan K, Rock CL, et al. Suppression of human colorectal mucosal prostaglandins: determining the lowest effective aspirin dose. *J Natl Cancer Inst* 1997;89(15):1152–1160.

177. Calaluce R, Earnest DL, Heddens D, et al. Effects of piroxicam on prostaglandin E2 levels in rectal mucosa of adenomatous polyp patients: a randomized phase IIb trial. *Cancer Epidemiol Biomarkers Prev* 2000;9(12):1287–1292.

178. Baron JA, Cole BF, Sandler RS, et al. A randomized trial of aspirin to prevent colorectal adenomas. *N Engl J Med* 2003;348(10):891–899.

179. Sandler RS, Halabi S, Baron JA, et al. A randomized trial of aspirin to prevent colorectal adenomas in patients with previous colorectal cancer. *N Engl J Med* 2003;348(10):1939.

180. Benamouzig R, Deyra J, Martin A, et al. Daily soluble aspirin and prevention of colorectal adenoma recurrence: one-year results of the APACC trial. *Gastroenterology* 2003;125(2):612–614.

181. Benamouzig R, Deyra J, Martin A, et al. Daily soluble aspirin and prevention of colorectal adenoma recurrence: four years results of the APACC trial. *American Gastroenterological Society Annal Meeting Abstracts* 2006;689(A101).

182. Logan RE, Muir KR, Grainge MJ, Armitage NC, Shepherd VC, Group UT. Aspirin for the prevention of recurrent colorectal adenomas – results of the UKCAP trial. *American Gastroenterological Society Meeting Abstracts* 2006;438:A64.

183. Bertagnolli MM, Eagle CJ, Zauber AG, et al. A celecoxib for the prevention of sporadic colorectal adenomas. *N Engl J Med* 2006;355(9):873–884.

184. Arber N, Eagle C, Spicak J et al. Celecoxib for prevention of colorectal adenomatons polyps. *N Engl J Med* 2006;355(9):885–889.

185. Hixson LJ, Earnest DL, Fennerty MB, Sampliner RE. NSAID effect on sporadic colon polyps . *Am J Gastroenterol* 1993;88(10):1652–1656.

186. Matsuhashi N, Nakajima A, Fukushima Y, Yazaki Y, Oka T. Effects of sulindac on sporadic colorectal adenomatous polyps. *Gut* 1997;40(3):344–349.

187. Ladenheim J, Garcia G, Titzer D, et al. Effect of sulindac on sporadic colonic polyps. *Gastroenterology* 1995;108(4):1083–1087.

188. Muller-Decker K, Kopp-Schneider A, Marks F, Seibert K, Furstenberger G. Localization of prostaglandin H synthase isoenzymes in murine epidermal tumors: suppression of skin tumor promotion by inhibition of prostaglandin H synthase-2. *Mol Carcinog* 1998;23(1):36–44.

189. Li M, Wu X, Xu XC. Induction of apoptosis in colon cancer cells by cyclooxygenase-2 inhibitor NS398 through a cytochrome c-dependent pathway. *Clin Cancer Res* 2001;7:1010–1016.

190. Shiotani H, Denda A, Yamamoto K, et al. Increased expression of cyclooxygenase-2 protein in 4-nitroquinoline-1-oxide-induced rat tongue carcinomas and chemopreventive efficacy of a specific inhibitor, nimesulide. *Cancer Res* 2001;61:1451–1456.

191. Limburg PJ, Wei W, Ahnen DJ, et al. Randomized, placebo-controlled, esophageal squamous cell cancer chemoprevention trial of selenomethionine and celecoxib. *Gastroenterology* 2005;129(3):863–873.

192. Santini D, Vincenzi B, Tonini G, et al. Cyclooxygenase-2 overexpression is associated with a poor outcome in resected ampullary cancer patients. *Clin Cancer Res* 2005;11(10):3784–3789.

193. Bhandari P, Bateman AC, Mehta RL, et al. Prognostic significance of cyclooxygenase-2 (COX-2) expression in patients with surgically resectable adenocarcinoma of the oesophagus. *BMC Cancer* 2006;6(1):134.

194. Cohen BL, Gomez P, Omori Y, et al. Cyclooxygenase-2 (cox-2) expression is an independent predictor of prostate cancer recurrence. *Int J Cancer* 2006;119(5):1082–1087.

195. Diamantopoulou K, Laxaris A, Mylona E, et al. Cyclooxygenase-2 protein expression in relation to apoptotic potential and its prognostic significance in bladder urothelial carcinoma. *Anticancer Res* 2005;25:4543–4549.

196. Soumaoro LT, Uetake H, Takagi Y, et al. Coexpression of VEGF-C and Cox-2 in human colorectal cancer and its association with lymph node metastasis. *Dis Colon Rectum* 2006;49(3):392–398.

197. Uchida K, Schneider S, Yochim JM, et al. Intratumoral COX-2 gene expression is a predictive factor for colorectal cancer response to fluoropyrimidine-based chemotherapy. *Clin Cancer Res* 2005;11(9):3363–3368.

198. Ferrandina G, Lauriola L, Distefano MG, et al. Increased cyclooxygenase-2 expression is associated with chemotherapy resistance and poor survival in cervical cancer patients. *J Clin Oncol* 2002;20:973–981.

199. Ferrandina G, Lauriola L, Zannoni GF, et al. Increased cyclooxygenase-2 (COX-2) expression is associated with chemotherapy resistance and outcome in ovarian cancer patients. *Ann Oncol* 2002;13:1205–1211.

200. Cetin M, Buyukberber S, Demir MS, et al. Overexpression of cyclooxygenase-2 in multiple myeloma: association with reduced survival. *Am J Hematol* 2005;80(3):169–173.

201. Rini BI, Weinberg V, Dunlap S, et al. Maximal COX-2 immunostaining and clinical response to celecoxib and interferon alpha therapy in metastatic renal cell carcinoma. *Cancer* 2006;106(3): 566–575.

202. Raspollini MR, Amunni G, Villanucci A, Boddi V, Taddei GL. Cox-2 and preoperative CA-125 level are strongly correlated with survival and clinical responsiveness to chemotherapy in ovarian cancer. *Acta Obstet Gynecol Scand* 2006;85(4):493–498.

203. Ratnasinghe D, Daschner PJ, Anver MR, et al. Cyclooxygenase-2 P-glycoprotein-170 and drug resistance: is chemoprevention against multidrug resistance possible? *Anticancer Res* 2001;21: 2141–2147.

204. Altorki NK, Port JL, Zhang F, et al. Chemotherapy induces the expression of cyclooxygenase-2 in non-small cell lung cancer. *Clin Cancer Res* 2005;11:114191–114197.

205. Subbaramaiah K, Hart JC, Norton L, Dannenberg AJ. Microtubule-interfering agents stiumlate the transcription of cyclooxygenase-2. Evidence for involvement of ERK1/2 and p38 mitogen-activated protein kinase pathways. *J Biol Chem* 2000;275:14838–14845.

206. Mann M, Sheng H, Shao J, et al. Targeting cyclooxygenase 2 and HER-2/neu pathways inhibits colorectal carcinoma growth. *Gastroenterology* 2001;120(7):1713–1719.

207. Yip-Schneider MT, Sweeney CJ, Jung SH, Crowell PL, Marshall MS. Cell cycle effects of nonsteroidal anti-inflammatory drugs and enhanced growth inhibition in combination with gemcitabine in pancreatic carcinoma cells. *J Pharmacol Exp Ther* 2001;298(3):976–985.

208. Roller A, Bahr OR, Streffer J, et al. Selective potentiation of drug cytotoxicity by NSAID in human glioma cells: the role of COX-1 and MRP. *Biochem Biophys Res Commun* 1999;259(3):600–605.

209. Soriano AF, Helfrich B, Chan DC, et al. Synergistic effects of new chemopreventive agents and conventional cytotoxic agents against human lung cancer cell lines. *Cancer Res* 1999;59:6178–6184.

210. Hida T, Kozaki K, Muramatsu H, et al. Cyclooxygenase-2 inhibitor induces apoptosis and enhances cytotoxicity of various anticancer agents in non-small cell lung cancer cell lines. *Clin Cancer Res* 2000;6:2006–2011.

211. Teicher BA, Korbut TT, Meon K, et al. Cyclooxygenase and lipoxygenase inhibitors as modulators of cancer therapies. *Cancer Chemother Pharmacol* 1994;33:515–522.

212. Palayoor ST, Bump EA, Calderwood SK, Bartol S, Coleman CN. Combined antitumor effect of radiation and ibuprofen in human prostate carcinoma cells. *Clin Cancer Res* 1998;4:763–771.

213. Furuta Y, Hunter N, Barkley TJ, Hall E, Milas L. Increase in radioresponse of murine tumors by treatment with indomethacin. *Cancer Res* 1988;48:3008–3013.

214. Milas L, Furuta Y, Hunter N, Nishiguchi I, Runkel S. Dependence of indomethacin-induced potentiation of murine tumor radioresponse on tumor host immunocompetence. *Cancer Res* 1990;50: 4473–4477.

215. Milas L, Ito H, Nakayama T, Hunter N. Improvement in therapeutic ratio of radiotherapy for a murine sarcoma by indomethacin plus misonidazole. *Cancer Res* 1991;51:3639–3642.

216. Pyo H, Choy H, Amorino GP, et al. A selective cyclooxygenase-2 inhibitor, NS-398, enhances the effect of radiation *in vitro* and *in vivo* preferentially on the cells that express cyclooxygenase-2. *Clin Cancer Res* 2001;7:2998–3005.

217. Milas L, Kishi K, Hunter N, Mason K, Masferrer JL, Tofilon PJ. Enhancement of tumor response to gamma-radiation by an inhibitor of cyclooxygenase-2 enzyme. *J Natl Cancer Inst* 1999;91: 1501–1504.

218. Kishi K, Petersen S, Petersen C, et al. Preferential enhancement of tumor radioresponse by a cyclooxygenase-2 inhibitor. *Cancer Res* 2000;60:1326–1331.

219. Mauceri HJ, Hanna NN, Beckett MA, et al. Combined effects of angiostatin and ionizing radiation in antitumour therapy. *Nature* 1998;394:287–291.

220. Masferrer JL, Koki A, Seibert K. COX-2 inhibitors. A new class of antiangiogenic agents. *Ann N Y Acad Sci* 1999;889:84–86.

221. Davis TW, O'Neal JM, Pagel MD, et al. Synergy between celecoxib and radiotherapy results from inhibition of cyclooxygenase-2-derived prostaglandin E2, a survival factor for tumor and associated vasculature. *Cancer Res* 2004;64:279–285.

222. Jenkins CA, Bruera E. Nonsteroidal anti-inflammatory drugs as adjuvant analgesics in cancer patients. *Palliat Med* 1999;13(3):183–196.
223. Yalcin S, Altundag K, Asil M, Tekuzman G. Sublingual piroxicam for cancer pain. *Med Oncol* 1998;15(2):137–139.
224. Mercadante S. The use of anti-inflammatory drugs in cancer pain. *Cancer Treat Rev* 2001;27(1): 51–61.
225. Lundholm K, Gelin J, Hyltander A, et al. Anti-inflammatory treatment may prolong survival in undernourished patients with metastatic solid tumors. *Cancer Res* 1994;54(21):5602–5606.
226. Altorki NK, Keresztes RS, Port JL, et al. Celecoxib, a selective cyclooxygenase-2 inhibitor, enhances the response to preoperative paclitaxel/carboplatin in early-stage non-small cell lung cancer. *J Clin Oncol* 2003;21:2645–2650.
227. Chow LW, Loo WT, Wai CC, Lui EL, Zhu L, Toi M. Study of Cox-2, Ki67, and p53 expression to predict effectiveness of 5-fluorouracil, epirubicin and cyclophosphamide with celecoxib in treatment of breast cancer patients. *Biomed Pharmacother* 2005;59(Suppl 2):S298–S301.
228. Chow LW, Wong JL, Toi ML. Celecoxib anti-aromatase neoadjuvant trial for locally advanced breast cancer: preliminary report. *J Steroid Biochem Mol Biol* 2003;86:443–447.
229. Govindan R, McLeod H, Mantravadi P, et al. Cisplatin, fluorouracil, celecoxib, and RT in resectable esophageal cancer: preliminary results. *Oncology* 2004;14(Suppl 14):18–21.
230. Pan CX, Loehrer P, Seitz D, et al. A phase II trial of irinotecan, 5-fluorouracil and leucovorin combined with celecoxib and glutamine as first-line therapy for advanced colorectal cancer. *Oncology* 2005;69(1):63–70.
231. Milella M, Gelibter A, DiCosimo S, et al. Pilot study of celecoxib and infusional 5-fluorouracil as second-line treatment for advanced pancreatic carcinoma. *Cancer* 2004;101(1):133–138.
232. Prince HM, Mileshkin L, Roberts A, et al. A multicenter phase II trial of thalidomide and celecoxib for patients with relapsed and refractory multiple myeloma. *Clin Cancer Res* 2005;11(15): 5504–5514.
233. Mrozek E, Kloos RT, Ringel MD, et al. Phase II study of celecoxib in metastatic differentiated thyroid carcinoma. *J Clin Endocrinol Metab* 2006;91(6):2201–2204.
234. Reardon DA, Quinn JA, Vredenburgh J, et al. Phase II trial of irinotecan plus celecoxib in adults with recurrent malignant glioma. *Cancer* 2005;103(2):329–338.
235. Nugent FW, Mertens WC, Graziano S, et al. Docetaxel and cyclooxygenase-2 inhibition with celecoxib for advanced non-small cell lung cancer progressing after patinum-based chemotherapy: a multicenter phase II trial. *Lung Cancer* 2005;48(2):267–273.
236. Csiki I, Morrow JD, Sandler A, et al. Targeting cyclooxygenase-2 in recurrent non-small cell lung cancer: a phase II trial of celecoxib and docetaxel. *Clin Cancer Res* 2005;11(18):6634–6640.
237. Gasparini G, Meo S, Comella G, et al. The combination of the selective cyclooxygenase-2 inhibitor celecoxib with weekly paclitaxel is a safe and active second-line therapy for non-small cell lung cancer: a phase II study with biological correlates. *Cancer J* 2005;11(3):209–216.
238. Gasparini G, Gattuso D, Morabito A, et al. Combined therapy with weekly irinotecan, infusional 5-fluorouracil and the selective Cox-2 inhibitor rofecoxib is a safe and effective second-line treatment in metastatic colorectal cancer. *Oncologist* 2005;10:710–717.
239. Ferrari V, Valcamonico F, Amoroso V, et al. Gemcitabine plus celecoxib (GECO) in advanced pancreatic cancer: a phase II trial. *Cancer Chemother Pharmacol* 2006;57(2):185–190.
240. El-Rayes BF, Zalupski MM, Shields AF, et al. A phase II study of celecoxib, gemcitabine, and cisplatin in advanced pancreatic cancer. *Invest New Drugs* 2005;23(6):583–590.
241. Wirth LJ, Haddad RI, Lindeman NI, et al. Phase I study of gefitinib plus celecoxib in recurrent or metastatic squamous cell carcinoma of the head and neck. *J Clin Oncol* 2005;23(28):6976–6981.
242. Dang CT, Dannenberg AJ, Subbaramaiah K, et al. Phase II study of celecoxib and trastuzumab in metastatic breast cancer patients who have progresses after prior trastuzumab-based treatments. *Clin Cancer Res* 2004;10(12):4062–4067.

IV   SPECIFIC DRUGS FOR MOLECULAR TARGETING IN ONCOLOGY

22

Imatinib Mesylate (Gleevec®) and the Emergence of Chemotherapeutic Drug-Resistant Mutations

Gerald V. Denis, PhD

SUMMARY

Imatinib mesylate (Gleevec®; Novartis) is an important, new, molecularly targeted, anti-cancer agent with clinical efficacy in chronic myelogenous leukemia (CML) and gastrointestinal stromal tumor (GIST). These malignancies develop after constitutive activation of Abelson (Abl) or c-kit (CD117) tyrosine kinases, respectively; Imatinib specifically inhibits such kinase activity. Many CML and GIST patients have relapsed while on imatinib treatment, however. The emergence of resistance to imatinib chemotherapeutic intervention is in retrospect neither surprising nor insoluble. Principles previously used to develop combination chemotherapy to avoid the development of multidrug resistance in leukemias and lymphomas (and later used in developing combinations for treatment of human immunodeficiency virus infections) may prove useful in the approach to the next generation of targeted molecular therapeutics for CML. In general, multidrug protocols and agents targeted to mutation sites simultaneously are likely to have a greater chance of success than single-agent therapy.

Key Words: Imatinib mesylate; Gleevec; STI-571; drug resistance; cancer chemotherapy; chronic myelogenous leukemia; gastrointestinal stromal tumor; BCR-ABL; c-kit; stem cell factor; structural biology.

1. INTRODUCTION

Unquestionably, the most exciting success story of new targeted therapeutics highlights the selective tyrosine kinase inhibitor imatinib mesylate. This therapeutic agent, formerly called STI-571, was first developed as an inhibitor of the platelet-derived growth factor (PDGF) receptor and was found to target the Bcr-Abl kinase in chronic myelogenous leukemia (CML) and the overexpressed c-kit protein in gastrointestinal stromal tumor (GIST). Imatinib mesylate originated 25 years ago at the

From: *Cancer Drug Discovery and Development: Molecular Targeting in Oncology*
Edited by: H. L. Kaufman, S. Wadler, and K. Antman © Humana Press, Totowa, NJ

pharmaceutical company Ciba Geigy, now Novartis, as a candidate inhibitor of protein kinases. Gradual refinements in structure restricted its specificity first to tyrosine kinases and then to a very limited set of tyrosine kinases that include primarily the Abl kinase family (v-Abl, c-Abl, $p185^{BCR-ABL}$, and $p210^{BCR-ABL}$, which are important in CML) and others with binding-site architecture similar to Abl, including c-kit (Kit, CD117) and PDGF receptor α-form and β-form, which are important in GIST. The excitement surrounding this agent arose in 1998 when phase 1 clinical trials in CML patients showed dramatic improvements and very good tolerance of the drug. Phase 2 and 3 trials confirmed these results. Scientific reports have been numerous (2724 Medline citations as of April 2006) including 655 excellent and comprehensive reviews to date. However, despite the early success of imatinib mesylate treatment, resistance to the drug began to be reported quickly, which is not surprising, given the basic principles of mutation selection in single-agent therapy, and the race is on for new derivatives that will overcome the problem of resistance. We can now reflect on basic principles of chemoresistance and plan rational strategies to continue the development of this agent and its analogs or derivatives.

2. CML

CML is a relatively common adult hematologic neoplasm that occurs rarely in children. Based on incidence rates from 2000–2002, approximately one in 619 men and women will be diagnosed with CML during their lifetime. Five-year relative survival rates by race and sex are 37.6% for White men, 41.2% for White women, 33.9% for Black men, and 35.3% for Black women *(1)*. The only well-characterized risk factor is exposure to ionizing radiation *(2)*. CML has three phases, and disease progression is well understood. The malignancy typically presents with a long "chronic phase" that can last years, with mild symptoms. By definition, in this phase, 5% or fewer of the cells in the peripheral blood or bone marrow are blasts (immature cells of the myeloid lineage). The chronic phase is followed by an "accelerated phase," in which these compartments are populated with 6–30% blasts and then a terminal "blast phase," wherein the fraction of blasts exceeds 30%. If additional clinical signs are present, such as splenomegaly or fever, the phase is termed "blast crisis." Untreated blast crisis is fatal.

Increasing severity of the disease and deteriorating prognosis are associated with the appearance of cells of the leukemic clone that characteristically contain a reciprocal chromosomal translocation involving the p-arms of chromosomes 9 and 22, called the Philadelphia (Ph) chromosome. Measures of disease progression or responses to therapy therefore consider both molecular and cytogenetic characteristics, as well as clinical signs. Molecular assessments typically include reverse transcriptase (RT) treatment of peripheral blood cell RNA followed by amplification of the transcribed BCR-ABL message by polymerase chain reaction (PCR). Typical detection limits are one cell in 10^5 *(3)*. Cytogenetic assessments require viable bone marrow cells or more than 10% blasts in the peripheral blood to visualize metaphases. Fluorescence in situ hybridization of the t(9;22) translocation junction has become an important diagnostic tool *(4,5)*.

The t(9;22) reciprocal translocation creates a chromosomal fusion between the *BCR* gene, which stands for "break point cluster," and the *ABL* gene, termed Ph$^+$, which leads to localization of the resultant protein to the cytoskeleton and unfettered tyrosine kinase activity in the Abl protein kinase domain. The fusion protein has many unregulated functions; most potently, an elevated and constitutive protein tyrosine kinase activity but also aberrant initiation of mitogenic signaling cascades that lead to uncontrolled growth in CML and recruitment of downstream effectors of cell survival. Reduced apoptotic signaling is a uniquely important contributor to CML *(6)*. The key role of the Bcr-Abl tyrosine kinase in CML etiology makes Bcr-Abl an appealing target for rational drug design. Nevertheless, for a CML patient subpopulation that does not show evidence of Ph^+ abnormality, no molecular mechanism of leukemogenesis is known. The early Ph^- stage of CML initiates a specific kind of genetic instability that may involve the Ataxia Telangiectasia and Rad 3-related (ATR) protein *(7)*, leaving the *BCR* and *ABL* genes especially prone to translocation. Such genetic instability may promote the occurrence of imatinib-resistant Abl mutations even before exposure to imatinib in some CML cases *(8)*. Epistatic factors probably also contribute to the individual-level variation in this instability stage. Once the chromosomal translocation event has occurred, however, the abnormality is irreversible and progression is extremely frequent.

The advent of combination chemotherapy has extended the lifespan for many hematologic malignancies. CML was considered incurable and fatal until the 1980s. Traditional therapies for CML have improved overall survival; high-dose chemotherapy (especially hydroxyurea), donor-lymphocyte infusion, stem cell transplant, and biologic therapy, such as interferon-α (sometimes in combination with cytarabine), remain important therapeutic avenues and have been thoroughly reviewed elsewhere *(9)*.

The active site of the Abl protein kinase is conserved among related protein tyrosine kinases and has been well mapped and understood in structural studies *(10,11)*. The kinase shifts between active and inactive conformations with the movement of a three-dimensionally unstructured "activation loop," dependent on its phosphorylation state *(12)*. Tyr^{393} is the site of phosphorylation within the activation loop of Abl *(10)*, and Tyr^{823} is the site of phosphorylation within the activation loop of Kit *(11)*. Phosphorylation appears to stabilize the active conformer of the activation loop. Imatinib, with its structure based on a 2-phenylaminopyrimidine core (Fig. 1), functions as a competitive inhibitor of ATP and is able to bind only to the inactive conformation, freezing movement of the loop and thus interrupting the catalytic cycle. The drug provides a high degree of specific inhibition while being essentially inactive against serine threonine protein kinases and most other tyrosine kinases. The 6-methyl substituent of the phenyl aniline moiety (Fig. 1) forms a hydrogen bond with Thr^{315} in the Abl active site *(13)*, as does the nearby secondary amine *(10)*. The important 6-methyl residue seems to be a primary determinant of the specificity of imatinib for the Abl-related family of kinases, whereas a benzamide group at the phenyl ring is a determinant of activity against the PDGF receptor *(2)*. Several other kinases for which imatinib has weak inhibition constants harbor a bulky nonpolar amino acid at this position instead of threonine, which probably excludes the imatinib molecule due to steric hindrance in the binding pocket. Hydrogen bonding between imatinib and Thr^{315} clearly identifies a central requirement for imatinib's ability to inhibit Abl *(10)*. Indeed, as discussed in Section 2.2, the substitution of isoleucine for this threonine accounts for a plurality of the reported imatinib-resistant point mutations in the Abl active site.

2.1. Bcr-Abl Signal Transduction

A prominent feature of both Bcr-Abl and Kit signaling is the number of proliferative and anti-apoptotic effector molecules that are mobilized in both cases. The proliferative component of Bcr-Abl signal transduction involves the canonical ras

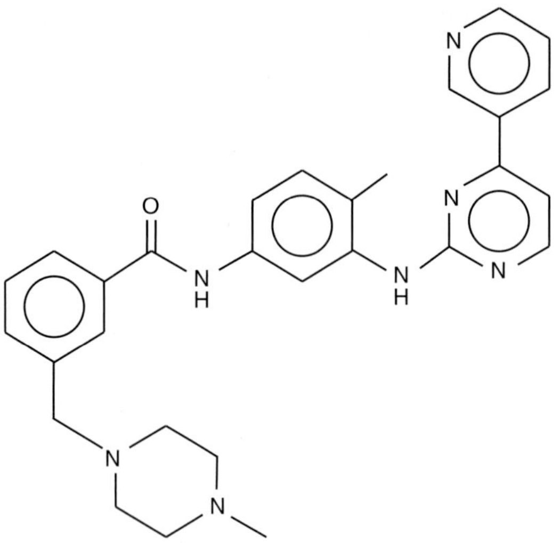

Fig. 1. The chemical structure of imatinib mesylate. Note the predicted highly hydrophobic property of the molecule.

pathway of the Raf/MEK/ERK1/2 kinase cascade, c-Jun NH$_2$-terminal kinase (JNK), and a Jak-Stat pathway, including Stat-1, Stat-3, Stat-5, and Stat-6 *(14)*. Anti-apoptotic signaling involves phosphatidylinositol 3-kinase (PI3K) and Akt/protein kinase B (PKB) signaling. The PI3K pathway has been identified as an essential signaling mechanism in CML *(15)*. PI3K inhibitors also synergize with imatinib to increase apoptosis in CML chronic phase and blast crisis patient cells *(16)*. Bcr-Abl appears also to induce transcription of the *BCL2* gene, the anti-apoptotic function of which has been shown to be essential for Bcr-Abl-mediated transformation *(17)*.

Elsewhere, I have discussed the oncogenic properties of fusion proteins arising from reciprocal chromosomal translocations *(18)*. These include not only Bcr-Abl oncoprotein, but Tel-Aml1, Mll-Cbp, NUP98-HOXC13, and RARα-PML among others. Not only do such proteins possess intrinsic transforming ability through improper transcription, but their communication with signal transduction pathways is corrupted *(18)*. Aberrant signaling, either proliferative or anti-apoptotic, drives the cell-cycle deregulation or extended survival that is characteristic of malignancy. Therefore, therapeutic approaches should consider signaling perturbations, particularly perturbations which, through secondary genetic changes, have become independent of the initiating oncoprotein. Such independence is a likely cause of some types of imatinib chemoresistance.

Imatinib mesylate was FDA-approved to treat CML in May 2001 based on the remarkable results of three clinical trials. Brian J. Druker, MD, of Oregon Health Sciences University, deserves the lion's share of the credit for proof-of-concept and for shepherding imatinib through the approval process. Imatinib is approved for treatment of patients who have progressed from the manageable chronic phase of the disease to the acute "blast crisis" phase, which is frequently fatal and is useful in early stages of CML, as well as in GISTs (section 4). Clinical trials of imatinib in combination with other agents are ongoing. Complete hematological response is defined as the normalization of the blood counts and the white cell differential, and the alleviation of all clinical signs. Complete cytogenetic response is defined as no detection of Ph$^+$metaphases. Major cytogenetic response is defined as the detection of <35% Ph$^+$metaphases. Molecular remission is defined as no BCR-ABL mRNA detectable by RT, coupled to PCR amplification *(2)*. These negative definitions obviously require appropriate statistical power and controls, lest failure to detect derives from failure of the assay.

2.2. Chemoresistance in CML

In clinical trials, most CML patients treated with imatinib responded, but CML relapse developed in about 80% of individuals successfully brought into remission. In every relapsed patient studied, the level of Bcr-Abl kinase activity was elevated to pre-treatment values. Most interestingly, the common T315I point mutation in the active site eliminated imatinib binding but did not compromise kinase activity *(10)*. Mutations of Y253, E255, T315, and M351 in Bcr-Abl account for approximately 60% of those detected at the time of relapse *(19)*. Goldman and Melo *(20)* have reported that in a sample of 179 imatinib-resistant patients, 114 mutations were detected, and some patients had more than one mutation in the resistant CML clone; most of these were in the tyrosine kinase domain of Abl.

The mechanisms of genetic instability in CML clonal expansion *(21)* are not well understood, but probably involve rates of point mutation that are much higher than the background rate. Such instability therefore provides a central and essential factor for the rapid emergence of chemoresistance. Of particular seriousness, Fabarius et al. *(22)* used centrosome immunostaining and conventional cytogenetics to reveal that imatinib treatment of normal fibroblasts (from human dermis, Chinese hamster embryo, or Indian muntjak) causes dose-dependent centrosome and chromosomal aberrations, independently of species. Thus, imatinib treatment per se is likely to exacerbate the accumulation of genetic lesions.

3. HISTORICAL PERSPECTIVE

The phenomenon of biological resistance to chemical agents has been widely reported in fields as diverse as insect control with pesticides and antibiotic control of microorganisms, especially tuberculosis. Primary resistance refers to innate or natural ability to resist an agent and is of marginal interest here. Secondary or acquired resistance arises from the biological processes of selection under the pressure of exposure to an agent and is a significant medical problem. Mutation of specific genes within a microorganism that are responsible for the transport or metabolism of the drug, or the signaling environment within the organism, enables the acquisition of drug resistance, often with dire consequences for the health and survival of a human host infected with that microorganism *(23)*.

The elements required for the appearance of stable resistant clones of CML cells are a mechanism for the introduction of frequent mutations, DNA replication to "stabilize" and perpetuate the mutations, the possibility that adventitious mutations exist, and selective pressure to provide a proliferative advantage to the cells that harbor the adventitious mutations *(24)*. Each of these elements is in place in the setting of CML under the conditions of imatinib single-agent therapy.

The well-known anti-metabolite methotrexate has been used for many years to treat acute leukemia and other neoplasms *(25)*; it is often prescribed in combination with other anti-metabolites such as 6-mercaptopurine or 6-thioguanine. Methotrexate came to prominence in 1956 when it was used successfully to achieve the first cure of a metastatic malignancy, a choriocarcinoma. However, resistance to this first-line agent soon became a problem for successful therapy. The five main mechanisms of resistance to methotrexate treatment are as follows: (i) amplification of the gene that encodes dihydrofolate reductase (*DHFR*), the protein product of which methotrexate is a competitive inhibitor; (ii) increased cellular export of methotrexate by the multidrug resistance transporter or P-glycoprotein; (iii) decreased cellular import of methotrexate; (iv) mutation in the active site of dihydrofolate reductase to better discriminate between methotrexate and folic acid, the natural substrate of dihydrofolate reductase; and (v) decreased polyglutamination of methotrexate, which causes reduced cellular retention of methotrexate. Each of these general classes of resistance mechanism but the last has now been identified in connection with resistance to imatinib: (i) amplification the *BCR-ABL* gene *(26)* is a frequent mechanism of resistance; (ii) the P-glycoprotein *(27)* and multidrug resistance transporter *(28)* are implicated in imatinib resistance and RNAi against the P-glycoprotein can confer imatinib sensitivity to resistant CML cells *(29)*. (iii) Furthermore, variable expression of influx transporters such as hOCT1 are also involved in resistance *(30,31)*. (iv) (in Section 2.2), point mutations that disrupt the binding of imatinib to Bcr-Abl protein are numerous *(32)* and are among the most widely reported of resistance phenomena. In short, the well-known mechanisms of chemotherapy resistance discovered over many years of methotrexate therapy now plague imatinib therapy; molecular remission in CML almost never occurs with imatinib treatment alone *(32)*. The only consistently successful curative treatment of CML has been high-dose chemotherapy followed by allogeneic marrow or stem cell transplantation *(9,33)*.

In imatinib-resistant CML, repopulation of the marrow and peripheral blood with Ph[+] clones is almost inevitable. Deininger and Druker *(2)* have reported that imatinib's selective pressure favors the outgrowth of pre-existing resistant clones, similar to a bacterial culture treated with a single antibiotic. However, the pace of basic research into drug design for the next generation of chemical inhibitors of Abl suggests that a respectable arsenal of agents will be available for physician choice, such as AP23464 (Ariad Pharmaceuticals), BMS-354825 (Bristol-Myers Squibb) *(34,35)*, SKI606 (Wyeth), PD180970 (Parke Davis), CGP76030, AMN107 (Novartis) *(36)*, and VX-680 (Vertex Pharmaceuticals/Merck) *(37)*. Other targeted agents such as SU11248 (Pfizer) have value against Kit and PDGF receptor α-form *(38)*.

The most productive places in the Bcr-Abl molecule to design new targeted therapeutics, to be used in combination with imatinib, are probably outside the active site and are likely to involve inhibition of movement of the activation loop. The novel investigational agent BMS-354825 (dasatinib) is effective against some resistant forms of Bcr-Abl *(39,40)* but not the T315I point mutant form *(41)*. On the other hand, VX-680 has been successful in the treatment of imatinib-resistant patients who harbor the T315I mutation, in which cases, unlike imatinib, VX-680 binds to the active form of the kinase. VX-680 inhibits the kinase activity of both wild-type (Ki 68 nM) and T315I Abl (Ki 114 nM), but imatinib inhibits the activity of only wild-type Abl *(41)*. VX-680 and similar novel agents that bind independently of imatinib, yet cooperate with imatinib to stabilize the Abl activation loop, might be good examples of a suitable first-line combination therapy and may work exceptionally well together to minimize chemoresistance.

Our molecular understanding of the Abl-active site is still evolving, and the principles by which mutations are selected are not yet completely understood. However, certain combinations of drugs may challenge the Abl protein with an insoluble problem: it may be impossible for Abl to mutate to overcome combination drug inhibition because of structural constraints within the protein. This feature made possible the success of curative combination chemotherapy for acute lymphocytic leukemia and acute myelogenous leukemia.

4. GISTS

The most frequently occurring mesenchymal tumor of the gastrointestinal tract humans is GIST *(42)*; up to 6000 new cases are diagnosed annually in the USA *(34,43)*. The tumors are found primarily in the stomach but also in the small intestine and elsewhere in the gastrointestinal tract. Rarely, GISTs are found in the pancreas *(35)*, gallbladder *(44)*, and appendix *(45)*. GISTs have traditionally been difficult to diagnose, and immunohistochemical techniques are essential in differential diagnosis true of smooth muscle tumors (leiomyomas and leiomyosarcomas), schwannoma, inflammatory fibroid polyp, and desmoid fibromatosis *(46)*.

The cell of origin for GIST is thought to be the Kit^+ interstitial cells of Cajal, which are the "pacemaker cells" for the gastrointestinal tract *(41)*. The pacemaker mechanism for adjacent smooth muscle cells uses "slow wave" calcium channel signaling *(47)* although Kit expression is probably independent of pacemaker function. GISTs derive primarily from activating mutations in the *KIT* gene *(48)*, which encodes a receptor tyrosine kinase. Kit was originally named stem cell factor (SCF) receptor *(49)*. This receptor dimerizes upon binding SCF, autophosphorylates on specific tyrosines, and then engages downstream effector pathways similar to other growth factor receptors to promote proliferation *(50)*, tumorigenesis, adhesion, and differentiation *(36)*. The activating mutations induce constitutive, ligand-independent activation of Kit and confer constitutive signaling, which is a frequent hallmark of growth factor receptor-driven malignancy. *KIT* contains 21 exons, but mutations tend to cluster within only four exons: exon 9 encoding the extracellular transmembrane domain, exon 11 encoding the intracellular juxtamembrane domain, exon 13 encoding the first portion of the split kinase domain, and exon 17 encoding the kinase-activation loop *(46)*. Most GISTs harbor *KIT* mutations, but 3% harbor mutations in the PDGF receptor α-form *(51)*, the active site structure of which is closely related to Kit and is also inhibited with imatinib. Treatment with imatinib has clearly been of great value for GIST patients, given their poor prognosis: median survival, including patients with unresectable or metastatic tumors, is 15 months without imatinib *(52)*, and response rates for standard chemotherapy have been reported to be as low as 10% *(53)*.

Interestingly, Kit loss-of-function mutations in knockout mouse models are correlated with depletion of the interstitial cells of Cajal *(54)* (among other abnormalities) but not nearby smooth muscle cells, which strongly support the view that Kit signaling is an essential requirement for survival or proliferation of the interstitial cells of Cajal. It follows that targeted chemotherapeutic inhibition of Kit signaling with imatinib might reasonably be expected to interfere with survival or growth of the GIST cells, providing that secondary genetic changes have not rendered the tumor independent of Kit signaling.

4.1. Kit Signal Transduction

Similar to Bcr-Abl signal transduction, Kit mobilizes several proliferative and anti-apoptotic signaling pathways. The proliferative component of Kit signal transduction involves ras activation and the canonical ras pathway of the Raf/MEK/ERK1/2 kinase cascade, which is mobilized through the Grb2/Sos effector system *(55)*. Other Kit proliferative pathways include the Src-Rac1-JNK cascade and a Jak-Stat pathway, specifically Jak2 mobilization of Stat-1α, Stat-3, Stat-5a, and Stat-5b transcription factors *(55)*. Anti-apoptotic activity is also conferred through PI3K and Akt/PKB signaling *(56)*. Imatinib treatment inhibits SCF-dependent phosphorylation of Kit *(57–62)*, ERK, and MAP kinase phosphorylation *(57–60)* (an expected functional outcome of blockade of ras signaling) and also inhibits Akt signaling *(58–60)*, without altering total protein levels of Kit, MAP kinase, or Akt *(59,60)*. Imatinib treatment also blocks SCF-dependent cell proliferation in vivo and in vitro *(60)*. In one in vitro system, HMC-1 cells transfected with Kit receptor that harbored an activating mutation showed strong survival dependence on the mutant protein; 85–95% of cells exposed to 1–10 μM imatinib underwent apoptosis in 48–72 h *(59)*. Moreover, the importance of the PI3K and Akt/PKB survival arm is apparent from the observation that GIST cells die in culture when treated with a PI3K inhibitor but not with a MEK inhibitor *(63)*. Imatinib therefore appears to block both the proliferative and anti-apoptotic arms of Kit signal transduction.

In summary, we might hypothesize that mutation in pluripotent signaling molecules like Bcr-Abl and Kit has a greater capacity to distort cellular metabolism than mutation in a tighter, more "directed" signaling molecule, such as an individual cytokine receptor. Imatinib is initially effective as a single molecularly targeted agent because it inhibits multiple effector pathways. Unfortunately, simple interventions rarely solve complex problems, and early responsiveness frequently gives way to chemoresistance.

4.2. Chemoresistance in GIST

Given the arguments above and long clinical experience, it is not surprising that some GIST patients develop resistance to imatinib. Secondary mutations in Kit after treatment of GIST with imatinib have now been reported; one study categorized up to four newly acquired *KIT* mutations in 14 patients (43.8%) *(64)*. In addition, 62.6% of GISTs that exhibit activating mutations in the gene that encodes the PDGF receptor α-form (*PDGFRA*) show point mutations associated with resistance to imatinib (181 out of 289 cases) *(65)*. In another study of 31 GIST patients *(66)*, who were treated with imatinib and then surgical resection, 13 patients were nonresistant to imatinib, 3 exhibited primary resistance, and 15 exhibited acquired resistance after initial benefit. There were no secondary mutations in *KIT* or *PDGFRA* in the nonresistant or primary resistance groups. In contrast, secondary mutations were found in 7 of 15 (46%) patients with acquired resistance, each of whom had a primary mutation in *KIT* exon 11. Most secondary mutations were located in *KIT* exon 17. Other investigators have reported that imatinib resistance in GISTs involves missense mutation in the Kit kinase domain, including T670I, Y823D, and V654A *(66–68)*. The T670I mutation is thought to be a structural analog of the T315I mutation in Bcr-Abl *(69)* discussed above. Functional studies have shown that Kit protein harboring the T670I mutation is insensitive to imatinib and that, if introduced in a Kit receptor that responds to imatinib, T670I ablates its sensitivity *(69)*. The V654A and T670I mutations have been reported repeatedly in the context of imatinib resistance *(70)*.

Resistance mechanisms encountered in CML are also observed in GIST, such as multidrug resistance. GIST cell lines are likely to alter influx and efflux transporters under imatinib selection *(71)*. One study of 21 GIST patients reported that all the GISTs were positive for multidrug resistance transporters and exhibit elevated expression of either P-glycoprotein (86% of cases) or MRP1 (62%) *(72)*. Analogously to *DHFR* amplification in methotrexate resistance, amplification of *KIT* or *PDGFRA* has also been reported *(70)*. On the other hand, the observation in GIST cases that there was never more than one new mutation in the same sample from patients treated with imatinib may bode well for such multiagent therapy that is targeted to a single protein *(64)*. Nevertheless, imatinib resistance will pose an ongoing problem, best solved through combination chemotherapy studies already underway.

5. PERSPECTIVE

Hirota and Isozaki *(48)* have pointed out that both CML and GIST appear to be special cases, because a single genetic activating event, such as translocation to produce *BCR-ABL* in the former case and mutation in *KIT* or *PDFGRA* in the latter, is necessary and sufficient to drive carcinogenesis. Progressive malignancy requires additional mutations, and involvement of the effector pathways seems most likely, such as abrogation of pro-apoptotic mechanisms. Multiple benign GISTs have been detected in the context of mutated Kit *(73)*, but malignancy appears to require additional factors *(74)*. Interestingly, two cases have been reported of progressive GISTs that have lost Kit protein expression *(70)*, implying that in some cases certain secondary genetic changes are sufficient to confer imatinib resistance and enable GIST progression in the absence of Kit signaling per se.

However, relatively simple genetic lesions that increase the risk for neoplastic transformation are rare and do not characterize the cancers to which the majority of morbidity and mortality in the USA may be attributed, such as breast cancer. Genetic lesions in the *BRCA1* and *BRCA2* loci increase the risk of breast cancer to about 80% for individuals with a family history but account for only about a tenth of the cases that occur sporadically. Yet, sporadic breast cancer is diagnosed in about 190,000 women in the USA annually, with a mortality rate approaching 20%. Multiple genetic lesions, including loss of tumor suppressor function, cytogenetic abnormalities, and epigenetic factors almost certainly cooperate to create the tumorigenic environment within breast ductal tissue. Furthermore, another diverse set of genes and epistatic factors is likely to control the invasiveness or metastatic potential of the primary tumor. A single-chemical agent has not proven to be sufficiently robust to inhibit reliably such a complex and multifactorial process, no matter how successful single agents may be in special cases. (To date, methotrexate has been curative for some choriocarcinomas and cyclophosphamide for some Burkitt's lymphomas; in both cases single-agent therapy is not considered optimal current therapy.) It is with the aforementioned principles in mind that new research into the next generation of chemical inhibitors should be undertaken and clinical trials designed for multiagent combination chemotherapy of CML and GIST.

ACKNOWLEDGMENTS

Supported by grants from the National Cancer Institute and the American Cancer Society.

REFERENCES

1. Ries LAG, Eisner MP, Kosary CL, Hankey BF, Miller BA, Clegg L, Mariotto A, Feuer EJ, Edwards BK (Eds). *SEER Cancer Statistics Review, 1975–2002*. National Cancer Institute: Bethesda, MD.
2. Deininger MW, Druker BJ. Specific targeted therapy of chronic myelogenous leukemia with imatinib. *Pharmacol Rev* 2003;55:401–423.
3. Nashed AL, Rao KW, Gulley ML. Clinical applications of BCR-ABL molecular testing in acute leukemia. *J Mol Diagn* 2003;5:63–72.
4. Pelz AF, Kröning H, Franke A, Wieacker P, Stumm M. High reliability and sensitivity of the bcr/ABL1 D-FISH test for the detection of BCR/ABL rearrangements. *Ann Hematol* 2002;81: 147–153.
5. Chase A, Grand F, Zhang J-G, Blackett N, Goldman M. Factors influencing false positive and negative rates of BCR-ABL fluorescence in situ hybridization. *Genes Chromosomes Cancer* 1997;18:246–253.
6. Bedi A, Zehnbauer BA, Barber JP, Sharkis SJ, Jones RJ. Inhibition of apoptosis by BCR-ABL in chronic myeloid leukemia. *Blood* 1994;83:2038–2044.
7. Dierov JK, Dierova R, Carroll M. BCR/ABL translocates to the nucleus and disrupts an ATR-dependent intra-s phase check point. *Cancer Cell* 2004;5:275–285.
8. Shah N, Nicoll J, Nagar B, et al. Multiple BCR-ABL kinase domain mutations confer polyclonal resistance to the tyrosine kinase inhibitor imatinib (STI571) in chronic phase and blast crisis chronic myeloid leukemia. *Cancer Cell* 2002;2:117–125.
9. Savage DG, Antman KH. Imtinib mesylate – a new oral targeted therapy. *N Engl J Med* 2002;346:683–693.
10. Schindler T, Bornmann W, Pellicena P, Miller WT, Clarkson B, Kuriyan J. Structural mechanism for STI-571 inhibition of Abelson tyrosine kinase. *Science* 2000;289:1938–1942.
11. Roskoski, R. Jr. Structure and regulation of Kit protein-tyrosine kinase—The stem cell factor receptor. *Biochem Biophys Res Commun* 2005;338:1307–1315.
12. Johnson LN, Noble ME, Owen DJ. Active and inactive protein kinases: structural basis for regulation. *Cell* 1996;85:149–158.
13. Noble MEM, Endicott JA, Johnson LN. Protein kinase inhibitors: Insights into drug design from structure. *Science* 2004;303:1800–1805.
14. Zou X, Calame K. Signaling pathways activated by oncogenic forms of Abl tyrosine kinase. *J Biol Chem* 1999;274:18141–18144.
15. Kharas MG, Fruman DA. ABL oncogenes and phosphoinositide 3-kinase: mechanism of activation and downstream effectors. *Cancer Res* 2005;65:2047–2053.
16. Klejman A, Rushen L, Morrione A, et al. Phosphatidylinositol-3 kinase inhibitors enhance the anti-leukemia effect of STI571. *Oncogene* 2002;21:5868–5876.
17. Sanchez-Garcia I, Martin-Zanca D. Regulation of Bcl-2 gene expression by BCR-ABL is mediated by Ras. *J Mol Biol* 1997;267:225–228.
18. Denis GV. Bromodomain motifs and "scaffolding"? *Front Biosci* 2001;6:D1065–D1068.
19. Chu S, Xu H, Shah NP, Snyder DS, Forman SJ, Sawyers CL, Bhatia R. Detection of BCR-ABL kinase mutations in CD34+ cells from chronic myelogenous leukemia patients in complete cytogenetic remission on imatinib mesylate treatment. *Blood* 2005;105:2093–2098.
20. Goldman JM, Melo JV. Chronic myeloid leukemia–advances in biology and new approaches to treatment. *N Engl J Med* 2003;349:1451–1464.

21. Cortes J, O'Dwyer ME. Clonal evolution in chronic myelogenous leukemia. *Hematol Oncol Clin North Am* 2004;18:671–684.
22. Fabarius A, Giehl M, Frank O, Duesberg P, Hochhaus A, Hehlmann R, Seifarth W. Induction of centrosome and chromosome aberrations by imatinib in vitro. *Leukemia* 2005;19:1573–1578.
23. Sharma SK, Mohan A. Multidrug-resistant tuberculosis. *Indian J Med Res* 2004;120:354–376.
24. Mayr E. *The Growth of Biological Thought*. Cambridge, MA: Belknap, 1982;Ch. 12:535–570.
25. Ward JR. Historical perspective on the use of methotrexate for the treatment of rheumatoid arthritis. *J Rheumatol* 1985;12(suppl 12):3–6.
26. Yoshida C, Melo JV. Biology of chronic myeloid leukemia and possible therapeutic approaches to imatinib-resistant disease. *Int J Hematol* 2004;79:420–433.
27. Illmer T, Schaich M, Platzbecker U, Freiberg-Richter J, Oelschlagel U, von Bonin M, Pursche S, Bergemann T, Ehninger G, Schleyer E. P-glycoprotein-mediated drug efflux is a resistance mechanism of chronic myelogenous leukemia cells to treatment with imatinib mesylate. *Leukemia* 2004;18:401–408.
28. Mahon FX, Belloc F, Lagarde V, Chollet C, Moreau-Gaudry F, Reiffers J, Goldman JM, Melo JV. MDR1 gene overexpression confers resistance to imatinib mesylate in leukemia cell line models. *Blood* 2003;101:2368–2373.
29. Rumpold H, Wolf AM, Gruenewald K, Gastl G, Gunsilius E, Wolf D. RNAi-mediated knockdown of P-glycoprotein using a transposon-based vector system durably restores imatinib sensitivity in imatinib-resistant CML cell lines. *Exp Hematol* 2005;33:767–775.
30. Thomas J, Wang L, Clark RE, Pirmohamed M. Active transport of imatinib into and out of cells: implications for drug resistance. *Blood* 2004;104:3739–3745.
31. Crossman LC, Druker BJ, Deininger MW, Pirmohamed M, Wang L, Clark RE. hOCT 1 and resistance to imatinib. *Blood* 2005;106:1133–1134.
32. Nardi V, Azam M, Daley GQ. Mechanisms and implications of imatinib resistance mutations in BCR-ABL. *Curr Opin Hematol* 2004;11:35–43.
33. Gratwohl A, Hermans J. Allogeneic bone marrow transplantation for chronic myeloid leukemia. Working Party Chronic Leukemia of the European Group for Blood and Marrow Transplantation (EBMT). *Bone Marrow Transplant* 1996;17(Suppl 3):S7–S9.
34. Fletcher CD, Berman JJ, Corless C, et al. Diagnosis of gastrointestinal stromal tumors: a consensus approach. *Hum Pathol* 2002;33:459–465.
35. Daum O, Klecka J, Ferda J, et al. Gastrointestinal stromal tumor of the pancreas: case report with documentation of KIT gene mutation. *Virchows Arch* 2005;446:470–472.
36. O'Hare T, Walters DK, Stoffregen EP, Jia T, Manley PW, Mestan J, Cowan-Jacob SW, Lee FY, Heinrich MC, Deininger MW, Druker BJ. In vitro activity of Bcr-Abl inhibitors AMN107 and BMS-354825 against clinically relevant imatinib-resistant Abl kinase domain mutants. *Cancer Res* 2005;65:4500–4505.
37. Harrington EA, Bebbington D, Moore J, Rasmussen RK, Ajose-Adeogun AO, Nakayama T, Graham JA, Demur C, Hercend T, Diu-Hercend A, Su M, Golec JM, Miller KM. VX-680, a potent and selective small-molecule inhibitor of the Aurora kinases, suppresses tumor growth in vivo. *Nat Med* 2004;10:262–267.
38. van der Zwan SM, DeMatteo RP. Gastrointestinal stromal tumor: 5 years later. *Cancer* 2005;104:1781–1788.
39. Shah NP, Tran C, Lee FY, Chen P, Norris D, Sawyers CL. Overriding imatinib resistance with a novel ABL kinase inhibitor. *Science* 2004;305:399–401.
40. Burgess MR, Skaggs BJ, Shah NP, Lee FY, Sawyers CL. Comparative analysis of two clinically active BCR-ABL kinase inhibitors reveals the role of conformation-specific binding in resistance. *Proc Natl Acad Sci USA* 2005;102:3395–400.
41. Young MA, Shah NP, Chao LH, Seeliger M, Milanov ZV, Biggs WH 3rd, Treiber DK, Patel HK, Zarrinkar PP, Lockhart DJ, Sawyers CL, Kuriyan J. Structure of the kinase domain of an imatinib-resistant Abl mutant in complex with the Aurora kinase inhibitor VX-680. *Cancer Res* 2006;66: 1007–1014.
42. Fishman AP. Gastrointestinal tract. In: Rosai J, Ed. *Ackerman's Surgical Pathology*. 8th ed. St. Louis: Mosby, 1996. 645–647.
43. Kindblom LG, Remotti HE, Aldenborg F, Meis-Kindblom JM. Gastrointestinal pacemaker cell tumor (GIPACT): gastrointestinal stromal tumors show phenotypic characteristics of the interstitial cells of Cajal. *Am J Pathol* 1998;152:1259–1269.

44. Ortiz-Hidalgo C, de Leon Bojorge B, Albores-Saavedra J. Stromal tumor of the gallbladder with phenotype of interstitial cells of Cajal: a previously unrecognized neoplasm. *Am J Surg Pathol* 2000;24:1420–1423.

45. Miettinen M, Sobin LH. Gastrointestinal stromal tumors in the appendix: a clinicopathologic and immunohistochemical study of four cases. *Am J Surg Pathol* 2001;25:1433–1437.

46. Rubin BP. Gastrointestinal stromal tumours: an update. *Histopathology* 2006;48:83–96.

47. Harhun MI, Pucovsky V, Povstyan OV, Gordienko DV, Bolton TB. Interstitial cells in the vasculature. *J Cell Mol Med* 2005;9:232–243.

48. Hirota S, Isozaki K, Moriyama Y, et al. Gain-of-function mutations of c-KIT in human gastrointestinal stromal tumors. *Science* 1998;279:577–580.

49. Williams DE, Eisenman J, Baird A, et al. Identification of a ligand for the c-kit proto-oncogene. *Cell* 1990;63:167–174.

50. Ullrich A, Schlessinger J. Signal transduction by receptors with tyrosine kinase activity. *Cell* 1990;61:203–212.

51. Heinrich MC, Corless CL, Duensing A, et al. PDGFRA activating mutations in gastrointestinal stromal tumors. *Science* 2003;299:708–710.

52. Mudan SS, Conlon KC, Woodruff J, Lewis J, Brennan MF. Salvage surgery in recurrent gastrointestinal sarcoma: prognostic factors to guide patient selection. *Cancer* 2000;88:51–58.

53. Edmonson J, Marks R, Buckner J, Mahoney M. Contrast of response to D-MAP + Sargramostim between patients with advance malignant gastrointestinal stromal tumors and patients with other advanced leiomyosarcomas. *Proc Am Soc Clin Oncol* 1999;18:541a.

54. Maeda H, Yamagata A, Nishikawa S, Yoshinaga K, Kobayashi S, Nishi K, Nishikawa S. Requirement of c-kit for development of intestinal pacemaker system. *Development* 1992;116:369–375.

55. Reber L, Da Silva CA, Frossard N. Stem cell factor and its receptor c-Kit as targets for inflammatory diseases. *Eur J Pharmacol* 2006;533:327–340.

56. Kitamura Y, Hirota S. Kit as a human oncogenic tyrosine kinase. *Cell Mol Life Sci* 2004;61:2924–2931.

57. Buchdunger E, Cioffi CL, Law N, Stover D, Ohno-Jones S, Druker BJ, Lydon NB. Abl protein-tyrosine kinase inhibitor STI571 inhibits in vitro signal transduction mediated by c-KIT and platelet-derived growth factor receptors. *J Pharmacol Exp Ther* 2000;295:139–145.

58. Krystal GW, Honsawek S, Litz J, Buchdunger E. The selective tyrosine kinase inhibitor STI571 inhibits small cell lung cancer growth. *Clin Cancer Res* 2000;6:3319–3326.

59. Heinrich MC, Griffith DJ, Druker BJ, Wait CL, Ott KA, Zigler AJ. Inhibition of c-kit receptor tyrosine kinase activity by STI 571, a selective tyrosine kinase inhibitor. *Blood* 2000;96:925–932.

60. Chen H, Isozaki K, Kinoshita K, Ohashi A, Shinomura Y, Matsuzawa Y, Kitamura Y, Hirota S. Imatinib inhibits various types of activating mutant kit found in gastrointestinal stromal tumors. *Int J Cancer* 2003;105:130–135.

61. Tuveson DA, Willis NA, Jacks T, Griffin JD, Singer S, Fletcher C-DM, Fletcher JA, Demetri GD. STI571 inactivation of the gastrointestinal stromal tumor c-KIT oncoprotein: biological and clinical implications. *Oncogene* 2001;20:5054–5058.

62. Ma Y, Zeng S, Metcalfe DD, Akin C, Dimitrijevic S, Butterfield JH, McMahon G, Longley BJ. The c-KIT mutation causing human mastocytosis is resistant to STI571 and other KIT kinase inhibitors: kinases with enzymatic site mutations show different inhibitor sensitivity profiles than wild-type kinases and those with regulatory-type mutations. *Blood* 2002;99:1741–1744.

63. Duensing A, Medeiros F, McConarty B, et al. Mechanisms of oncogenic KIT signal transduction in primary gastrointestinal stromal tumors (GISTs). *Oncogene* 2004;23:3999–4006.

64. Wardelmann E, Merkelbach-Bruse S, Pauls K, Thomas N, Schildhaus HU, Heinicke T, Speidel N, Pietsch T, Buettner R, Pink D, Reichardt P, Hohenberger P. Polyclonal evolution of multiple secondary KIT mutations in gastrointestinal stromal tumors under treatment with imatinib mesylate. *Clin Cancer Res* 2006;12:1743–1749.

65. Corless CL, Schroeder A, Griffith D, Town A, McGreevey L, Harrell P, Shiraga S, Bainbridge T, Morich J, Heinrich MC. PDGFRA mutations in gastrointestinal stromal tumors: frequency, spectrum and in vitro sensitivity to imatinib. *J Clin Oncol* 2005;23:5357–5364.

66. Antonescu CR, Besmer P, Guo T, Arkun K, Hom G, Koryotowski B, Leversha MA, Jeffrey PD, Desantis D, Singer S, Brennan MF, Maki RG, DeMatteo RP. Acquired resistance to imatinib in gastrointestinal stromal tumor occurs through secondary gene mutation. *Clin Cancer Res* 2005;11:4182–4190.

67. Chen LL, Trent JC, Wu EF, Fuller GN, Ramdas L, Zhang W, Raymond AK, Prieto VG, Oyedeji CO, Hunt KK, Pollock RE, Feig BW, Hayes KJ, Choi H, Macapinlac HA, Hittelman W, Velasco MA, Patel S, Burgess MA, Benjamin RS, Frazier ML. A missense mutation in KIT kinase domain 1 correlates with imatinib resistance in gastrointestinal stromal tumors. *Cancer Res* 2004;64: 5913–5919.
68. Chen LL, Sabripour M, Andtbacka RH, Patel SR, Feig BW, Macapinlac HA, Choi H, Wu EF, Frazier ML, Benjamin RS. Imatinib resistance in gastrointestinal stromal tumors. *Curr Oncol Rep* 2005;7:293–299.
69. Tamborini E, Bonadiman L, Greco A, Albertini V, Negri T, Gronchi A, Bertulli R, Colecchia M, Casali PG, Pierotti MA, Pilotti S. A new mutation in the KIT ATP pocket causes acquired resistance to imatinib in a gastrointestinal stromal tumor patient. *Gastroenterology* 2004;127:294–299.
70. Debiec-Rychter M, Cools J, Dumez H, Sciot R, Stul M, Mentens N, Vranckx H, Wasag B, Prenen H, Roesel J, Hagemeijer A, Van Oosterom A, Marynen P. Mechanisms of resistance to imatinib mesylate in gastrointestinal stromal tumors and activity of the PKC412 inhibitor against imatinib-resistant mutants. *Gastroenterology* 2005;128:270–279.
71. Prenen H, Guetens G, de Boeck G, Debiec-Rychter M, Manley P, Schoffski P, van Oosterom AT, de Bruijn E. Cellular uptake of the tyrosine kinase inhibitors imatinib and AMN107 in gastrointestinal stromal tumor cell lines. *Pharmacology* 2006;77:11–16.
72. Theou N, Gil S, Devocelle A, Julie C, Lavergne-Slove A, Beauchet A, Callard P, Farinotti R, Le Cesne A, Lemoine A, Faivre-Bonhomme L, Emile JF. Multidrug resistance proteins in gastrointestinal stromal tumors: site-dependent expression and initial response to imatinib. *Clin Cancer Res* 2005;11:7593–7598.
73. Nishida T, Hirota S, Taniguchi M, Hashimoto K, Isozaki K, Nakamura H, Kanakura Y, Tanaka T, Takabayashi A, Matsuda H, Kitamura Y. Familial gastrointestinal stromal tumours with germline mutation of the KIT gene. *Nat Genet* 1998;19:323–324.
74. Kitamura Y, Hirota S, Nishida T. Gastrointestinal stromal tumors (GIST): a model for molecule-based diagnosis and treatment of solid tumors. *Cancer Sci* 2003;94:315–320.

23

Development of a Targeted Treatment for Cancer

The Example of C225 (Cetuximab)

John Mendelsohn, MD

SUMMARY

This is a review of the discovery and development of C225/Cetuximab, a novel monoclonal antibody (mAb) treatment for cancer. Cetuximab was the first anticancer agent that successfully targeted a receptor for a growth factor and a protein tyrosine kinase. Blocking the signaling activity of the epidermal growth factor (EGF) receptor represented a new targeted approach to cancer therapy. Preclinical studies with human cancer xenografts suggested that C225 worked best in combination with chemotherapy or radiation. Many possible mechanisms of action have been uncovered, including inhibition of each of the six characteristics of a cancer cell described by Hanahan and Weinberg *(1)*. The research on C225 over a period of two decades has involved dozens of academic collaborations, numerous grants from the National Cancer Institute, and the work of four pharmaceutical/biotech companies. There was a setback in the Food and Drug Administration (FDA) review process, but approval for clinical use in advanced refractory colorectal cancer was obtained in 2004. Much additional research is needed (and is ongoing) to discover markers that predict clinical responsiveness and to determine how and for whom to use this therapy most effectively.

Key Words: Epidermal growth factor; epidermal growth factor receptor; Cetuximab (Erbitux™); tyrosine kinase; monoclonal antibody C225.

1. INTRODUCTION

This chapter will review a novel approach to cancer treatment that was proposed in 1983 by Drs John Mendelsohn and Gordon Sato. They were the first to demonstrate that inhibition of a growth factor receptor on the cell surface and that inhibition of a protein tyrosine kinase could prevent cell proliferation and serve as a potential anticancer therapy.

From: *Cancer Drug Discovery and Development: Molecular Targeting in Oncology*
Edited by: H. L. Kaufman, S. Wadler, and K. Antman © Humana Press, Totowa, NJ

2. HYPOTHESIS

The hypothesis behind this research, which was initiated in 1981, was that monoclonal antibodies (mAbs) that bind to EGF receptors and block receptor access to ligands may prevent cell proliferation by inhibiting activation of the EGF receptor tyrosine kinase (2).

3. RATIONALE

What led us to this hypothesis First, the backgrounds of the investigators were conducive. Dr. Sato had spent over a decade systematically determining which components of serum were required for cell growth in culture. His goal was to replace serum with defined culture medium. The work culminated in an understanding of the critical role of growth factors, especially EGF, insulin, and transferrin, as well as less well-defined components such as certain lipids (3). Dr. Mendelsohn also had an interest in defined culture medium (4), in activation of lymphocyte proliferation by ligands such as tuberculin (unpublished) and phytohemagglutin (5), and in ligand-triggered signal transduction in lymphocytes (6). Both scientists shared an interest in mAbs and in the study of human tumor xenografts in nude mice, which were new technologies. Core capabilities for these approaches were made available by Dr. Sato in the UCSD Cancer Center, which Dr. Mendelsohn directed.

And second, breaking scientific discoveries relevant to this hypothesis were accumulating in rapid fire just prior to the time these investigations were initiated. Autocrine stimulation of EGF receptors by transforming growth factor alpha (TGF-α) had been described, and the autocrine stimulation of cancer cell growth was postulated (7). A novel kinase that phosphorylated tyrosine was discovered, and among the very first protein tyrosine kinases identified were the *src* oncogene and the EGF receptor (8–10). Overexpression of EGF receptors on human cancer cells was found to be a common occurrence, often correlating with a worse clinical outcome (11).

Finally, the investigators were aware that circulating antireceptor antibodies can cause stable physiologic change in people—in myasthenia gravis, rare forms of diabetes and thyroid disease (12). This rich accumulation of information provided the rationale for embarking on experiments to create a mAb against the EGF receptor that could block binding of EGF or TGF-α.

4. INITIAL RESULTS FROM AN ACADEMIC RESEARCH LABORATORY

Largely through the hard work of two postdoctoral fellows, Tomo Kawamoto and J. Denry Sato, immunizations of mice were performed with A431 cells that are rich in EGF receptors, and hybridomas were created. The screen for hybridomas producing the desired murine mAb involved measuring reduction of P^{32} incorporation into A431 cell lysates—a difficult and time-consuming assay. Two such mAbs were identified. Reports were published in 1983 and 1984 demonstrating that blockade of human EGF receptors by murine mAb 225 or 528 inhibits proliferation of cultured human cancer cells (which secrete TGF-α) and their human tumor xenografts (2,13,14). Additionally, blockade of EGF receptors by either of these mAbs competitively inhibits activation of receptor tyrosine kinase by TGF-α or EGF (15).

When nontransformed human cells were cultured under serum-free conditions, they required addition of TGF-α or EGF to the medium, and proliferation was inhibited by the antireceptor mAbs *(13)*. Further studies in this investigator's laboratory explored the mechanism of inhibiting cell proliferation by blocking EGF receptor function *(16–23)*. Among these findings was the demonstration that the autocrine activation of the EGF receptor by TGF-α occurred at the cell surface by externalized ligand. It also was observed that not all cultured cells bearing EGF receptors responded to antireceptor mAb and responding cells could have high or normal levels of receptors *(24)*.

A study of [III]In-225 in xenografted nude mice provided important preclinical data. With A431 cell xenografts, which contain high EGF receptor levels, labeled mAb uptake was preferential, and imaging was obtained using a gamma camera. With MCF7 cells, containing low receptor levels, tumor uptake was not preferential and there was no image *(25)*.

5. FIRST CLINICAL TRIAL SPONSORED BY A BIOTECH COMPANY

Hybritech Inc. licensed mAb 225 from the University of California and carried out scale up, formulation, toxicology, and preclinical pharmacokinetic studies. Based on the published preclinical observations with mAb 225, FDA approval was obtained for a phase I clinical trial with tracer-labeled [III]In-mAb 225 *(26)*. This was the first ever clinical trial with an inhibitor of a growth factor receptor and an inhibitor of a tyrosine kinase. It is important to emphasize that initially there were great concerns over the potential for toxicity to normal tissues, as EGF receptors are widely expressed in epithelial cells.

The study was carried out in patients with advanced lung cancer during 1989–1990, and the results were gratifying:

1. Single doses of murine mAb 225 were escalated from 1 mg to 300 mg with excellent tolerance. At that point, the supply of mAb was depleted.
2. There were no serious toxicities.
3. At ≥40 mg dose, tumors were imaged. There was also high uptake in the liver.
4. All metastases observed by CT scan that were >1 cm in diameter were imaged.
5. At ≥120 mg doses, serum levels of mAb 225 were maintained at >40 μg/ml after 3 days. This is a receptor-saturating concentration.
6. Tumor uptake with the 120 mg dose was 3.4%.
7. Human antibodies against murine mAb 225 (HAMA) were observed in all patients after 2 weeks.

Why did this investigator and Hybritech select murine mAb 225 IgG1, instead of mAb 528 IgG2a, for the major preclinical studies and, therefore, for the initial clinical trial? The IgG1 murine mAb 225 had far less immunological capacity than the IgG2a mAb 528, and the goal in this research was to test this mAb-mediated therapy as an inhibitor of a tyrosine kinase not as a form of immunotherapy. The F(ab')$_2$ fragment of mAb 225 had been shown to be active against xenografts, demonstrating that an immune attack was not the primary mechanism of the antitumor activity observed in mice, and this observation was subsequently published *(27)*.

Table 1
Properties of C225

- IgG$_1$ (chimerized antibody), complement binding
- Binds with EGF receptor with high affinity ($K_d = 0.2\,nM$)
- Competes with growth factor binding to receptor
- Inhibits activation of receptor tyrosine kinase
- Stimulates receptor internalization

6. NCI PARTICIPATION IN THE RESEARCH

The first research proposal to the National Cancer Institute (NCI) for producing anti-EGF receptor mAbs was turned down by the study section because it was felt that mAbs against the EGF receptor could not be produced. Funding was awarded upon reapplication, after the initial positive results were obtained. Thereafter, for a period of 23 years, the NCI continuously funded research in this investigator's laboratory and with collaborators through the R01, P01, and National Cooperative Drug Discovery Group (NCDDG) mechanisms. The successful approval of anti-EGF receptor mAb 225 for clinical use is an example of success for the NCDDG program.

Based on the preclinical studies and the results of the phase I trial with murine mAb 225, the NCI Biologics Decision Network Committee agreed to contract with a biotech company to convert murine mAb 225 IgG1 into a human : murine chimeric mAb C225. It also was agreed that the chimeric mAb would have the human IgG1 isotope—which was subsequently shown (as predicted) to have the capacity to mediate immunological antitumor activity through complement or by antibody-dependent cellular cytotoxicity (22). The binding properties of chimeric mAb C225 were found to be similar to those of the murine mAb 225, except for greater affinity (28). Its properties are listed in Table. 1

7. IMPASS

At this point, further clinical trials hit a roadblock, lasting 5 years until clinical protocols were again activated in 1995 under a new company. What accounted for the delay? First, and most importantly, there was great skepticism that mAbs would be successful therapeutic agents in 1991. Reasons for this sentiment included their high molecular weight, inhibiting distribution, the requirement for intravenous administration, the expense and difficulty of production, the potential for an allergic reaction, and the lack of a track record—there were no successes with mAbs in the clinic that had led to FDA approval. As a result, there was far less incentive than today for pharmaceutical companies to pursue mAbs as therapies for human diseases such as cancer.

At that time, most researchers in academia and pharmaceutical companies felt that low molecular weight, soluble inhibitors targeting the ATP-binding site and the kinase domain were far more likely to succeed, and nearly a dozen pharmaceutical companies have pursued this approach (29). In addition, a few did pursue the antireceptor mAb approach. Subsequent to the early reports with mAb 225, Genentech developed Herceptin against the HER-2 receptor.

Hybritech and its new parent company did not pursue clinical testing of mAb C225, and the University of California relicensed the mAb to ImClone Systems in 1993. ImClone carried out preclinical, scale up, formulation, toxicology and clinical studies for a decade that culminated in FDA approval in 2004 (see Sections 8–12).

8. FURTHER MECHANISTIC RESEARCH IN ACADEMIC LABORATORIES

The original hypothesis of this research was premised on cell-cycle inhibition as a mechanism of tumor inhibition. Further studies in this investigator's laboratory clearly demonstrated that tumor growth inhibition occurred in G1 phase of the cell cycle, mediated by increased levels of p27^{KIP1}, which resulted in hyperphosphorylation of the Rb protein *(30–33)*. This mechanism was subsequently demonstrated to explain the growth inhibition by the low-molecular weight inhibitors of EGF receptor tyrosine kinase and by mAb Herceptin against the HER-2 receptor.

Preclinical studies by this investigator and by large number of collaborators identified additional mechanisms that could be contributing to the antitumor activity of mAb C225. In response to blockade of EGF receptors, human tumor cells were found to produce markedly reduced levels of factors promoting angiogenesis, including vascular endothelial growth factor (VEGF), interleukin 8 (IL-8), and basic Fibroblast Growth Factor (FGF). New blood vessel formation was inhibited in human tumor xenografts, and endothelial cell apoptosis was observed, suggesting that for endothelial cells in the neovasculature of cancers, these growth factors may act as survival factors *(34,35)* (Fig. 1).

In other studies with human tumor xenografts, EGF receptor inhibition by C225 was found to promote apoptosis through a number of pathways, including involvement of the Bcl family of proteins and the caspases *(36,37)*. In addition, metastasis was inhibited by treatment of established xenografts with mAb C225 *(34)*. It was formally demonstrated that downstream signaling through mitogen-activated protein (MAP) kinase and Akt was inhibited by C225 treatment, in cultured cells and in vivo *(38)*.

When the published data from many laboratories were analyzed collectively, it was apparent that all of the cancer-promoting properties of tumor cells listed by Hanahan and Weinberg *(1)* were blocked/reversed when EGF receptors were inhibited with mAb C225 (Table 2). Possible contributing mechanisms in addition to those in their list include inhibition of DNA repair and the potential for immune activity mediated by the human : murine chimeric mAb.

Following up a report from the laboratory of Michael Sela that the antitumor activity of another anti-EGF receptor mAb was enhanced against xenografts by concurrent administration of chemotherapy *(39)*, this investigator and many colleagues pursued combination therapy of human tumor xenografts with mAb C225 plus chemotherapy. Positive results were obtained against a variety of tumor types, including both adenocarcinomas and squamous carcinomas, using doxorubicin, cyclophosphamide, paclitaxel, and topoisomerase inhibitors *(40–44)*. In addition to enhancement of growth inhibition, combination therapy enhanced apoptosis, antiangiogenesis, and antimetastatic effects compared with either therapy alone.

CD31(texas red)/TUNEL (fitc)

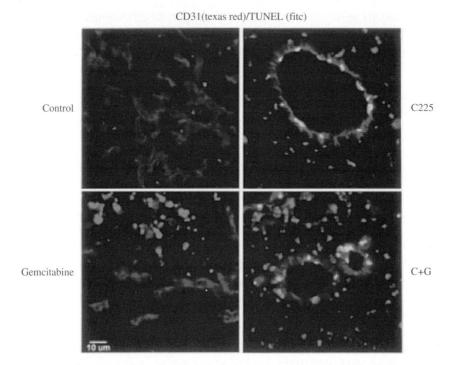

Fig. 1. Immunoflorescence double-staining for CD31 (endothelial cells) and TUNEL (apoptotic cells) in L3.6pl human pancreatic tumors after 18 days of therapy. Frozen tissue sections were fixed, treated with a rat anti-CD31 antibody, and then incubated with goat antirat IgG conjucated to Texas Red. After the sections were washed, TUNEL was performed using a commercial kit with modifications (ref. 35). Immunofluorescence microscopy was performed using ×400. Endothelial cells were identified by red fluorescence, and DNA fragmentation was detected by localized green and yellow fluorescence within the nucleus of apoptotic cells.

Subsequently, experiments were reported by this investigator and collaborators and by others, demonstrating that the combination of radiation therapy and mAb C225 produced enhanced activity against human tumor xenografts *(45,46)*.

Table 2
Acquired Capabilities of Cancer Cells *(38)*

Characteristics	Increased by EGFR stimulation	Decreased by EGFR inhibition
Self-sufficiency in growth signals	Yes	Yes
Insensitivity to antigrowth signals	Yes	Yes
Evading apoptosis	Yes	Yes
Limitless replicative potential	Yes	Yes
Sustained angiogenesis	Yes	Yes
Tissue invasion and metastasis	Yes	Yes

9. CLINICAL TRIALS WITH C225 SPONSORED BY A SECOND COMPANY

ImClone Systems Inc. was the second biotech company that licensed C225 from the University of California for clinical purposes. The funds necessary for scaled up production, formulation, and toxicology studies were raised by the company from investors and through collaborations with Merck (GMBH) and, subsequently, Bristol-Myers Squibb. ImClone sponsored a series of phase I and II trials that established the safety and pharmacokinetics of C225 (47). A schedule of 400 mg/m^2 loading dose followed by 250 mg/m^2 weekly doses was found to produce the desired serum levels of antibody that will maintain saturating concentrations. The only adverse event that reached grade 3 or 4 levels was an acneaform rash. Allergic sensitivity to C225 was observed in 2% of patients, and anaphylactic reactions (treated with standard therapy) occurred in 1% (ImClone data, 2002).

A novel clinical trial design was submitted by ImClone to the FDA for accelerated approval. Patients with advanced colorectal cancer were treated with the best single-agent chemotherapy, Irinotecan, until there was evidence of disease progression, and at that point C225 therapy was added, with continuation of Irinotecan treatment. A number of factors led to this trial design. In general, the earlier series of clinical trials involving about 2000 patients showed poor objective response rates (<10%) to C225 alone against a variety of tumors. The early results of a phase II trial of cisplatin plus C225 in head and neck cancer showed responses in 6 of 9 evaluable patients, and three had previously been treated with cisplatin with failure of this chemotherapy. These results, and preliminary results of combination treatment of lung cancer with C225 plus chemotherapy, suggested that C225 could act to potentiate chemotherapy in patients, as had been consistently observed in the preclinical xenograft studies. In addition, the previous experience with Herceptin against HER-2-positive breast cancer in the FDA registration trial had demonstrated that this mAb, against a closely related receptor, had greater clinical activity when combined with paclitaxel. A trial of C225 alone against a similar patient population was also initiated as a separate study.

The results of the phase II trial adding C225 to the treatment of patients with advanced colorectal cancer failing on Irinotecan therapy were reported at the American Society of Clinical Oncology meeting in 2001 (48) and are shown in Table 3. The investigators reported a 22.5% partial response rate, with an additional 26.7% of patients achieving stable disease. Also shown in the table are the results of treatment

Table 3
Cetuximab in Colorectal Cancer for Patients with Progression
on Irinotecan Therapy (Two Separate Clinical Trials)

	Irinotecan + cetuximab (48)	Cetuximab (49)
No. of patients	121	57
PR	22.5%	10.5%
SD	26.7%	35.1%
Median duration of response	186 days	164 days

with C225 alone in a similar population of colorectal cancer patients who had demon-strated progressing disease on Irinotecan (which was discontinued), reported a year later *(49,50)*. A partial response rate of 10.5% was observed.

10. ROLE OF THE FOOD AND DRUG ADMINISTRATION

When the combination therapy trial was presented to the FDA, the agency turned down the application for accelerated approval based on a number of issues:

1. The data records were felt to be incomplete.
2. The radiographic documentation of progression on Irinotecan treatment alone was felt to be inadequate.
3. The agency demanded that the company perform a randomized trial to determine whether continuation of Irinotecan was necessary to achieve responses to C225.

The fact that this was not anticipated by the company underscores the importance of clarity, consistency, and careful listening on the part of the drug producer and the FDA in the complex sequence of meetings that lead up to registration trials of new cancer therapies.

11. FURTHER CLINICAL TRIALS INVOLVING TWO ADDITIONAL COMPANIES

By this time, ImClone had a collaboration with Merck GMB in Germany and that company's randomized clinical trial with colorectal cancer patients failing Irinotecan treatment was expanded to meet the criteria demanded by the FDA. The results of this clinical trial were reported at the American Society of Clinical Oncology meeting in 2003 and are shown in Table 4. The partial response rate was 22.9% with combined therapy and 10.8% with C225 alone, and the difference was highly significant *(51,52)*. The median survival times in the table cannot be compared because patients on monotherapy were allowed to cross over and receive the combination if monotherapy failed. On the basis of these results, the FDA gave accelerated approval for C225 plus Irinotecan for the treatment of advanced colorectal cancer that had failed treatment with Irinotecan and for treatment with C225 alone if Irinotecan was not tolerated.

The necessary phase III randomized studies comparing chemotherapy with or without C225 in advanced colorectal cancer are ongoing. Clinical trials are also ongoing investigating C225 with or without concurrent chemotherapy in patients with head and neck, pancreatic, ovarian, and lung cancer.

Table 4
Cetuximab in Colorectal Cancer Randomized European Trial (PR + SD) *(51,52)*

	Combination *(n = 218) (95% Cl)*	*Monotherapy* *(n = 111) (95% Cl)*	*p-value*
PR	22.9% (17.5–29.1)	10.8% (5.7–18.1)	0.0074
Disease control	55.5% (48.6–62.2)	32.4% (23.9–42.0)	0.0001
Median TTP	4.1 months	1.5 months	<0.0001
Median Survival Time	8.6 months	6.9 months	0.48

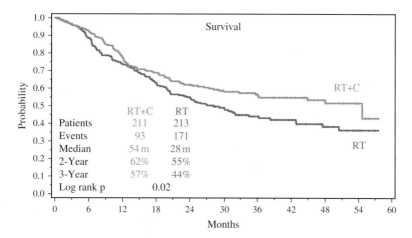

Fig. 2. Survival graph of patients with inoperable head and neck cancers treated with radiation therapy or radiation therapy and C225 (ASCO 2004).

As noted, synergistic interaction has been observed between C225 and radiotherapy in xenograft models. These observations led to a phase II clinical trial of radiation with C225 in patients with advanced head and neck cancer. The response rate was 100%, including 12 out of 14 complete responses *(53)*. These findings were followed up with a phase III clinical trial of radiation, with or without C225, in patients with inoperable head and neck cancer. The results, reported at ASCO in 2004, are shown in Fig. 2. Addition of C225 extended median survival by nearly 100% *(54,55)*. These findings were submitted to the FDA, resulting in approval in March 2006 for the combination of cisplatin and C225, as well as C225 alone, for the treatment of head and neck cancer.

It is not the purpose of this review to provide an accounting of research with the soluble, low-molecular weight inhibitors of EGF receptor tyrosine kinase that act by binding intracellularly to the receptor's ATP-binding site. Suffice it to say that large numbers of patients have been studied with these agents. Iressa (alone) received accelerated approval for advanced, chemotherapy-refractory lung cancer in 2003 based on an 11–18% response rate, but a recent randomized clinical trial was unable to show a survival benefit of Iressa over placebo (AstraZeneca communication). Tarceva was approved for advanced lung cancer after failure of chemotherapy in late 2004 based on an improvement in survival of 2 months compared with placebo.

For all agents active against EGF receptors, the major toxicity is an acneaform rash that can become grade 3 or 4 in less than 20% of patients. Although the level of EGF receptor expression in the cancer did not predict responders to C225, almost all responders had skin rashes *(48)*. The oral antityrosine kinases, but not Cetuximab, cause dose-limiting gastrointestinal toxicity. Other differences between C225 and the oral agents include differences in activity against specific types of cancer. For example, C225, but not Iressa or Tarceva, is active, alone, against colon cancer. Biological and molecular explanations for these differences must be sought to better understand the antitumor mechanisms of these agents and to enable selection of appropriate therapy for individual patients.

12. UNRESOLVED ISSUES

The most burning issue is the need for markers to identify these patients who are likely to respond to therapy with C225. In the case of Iressa, the presence of mutations in the EGF receptor identify lung cancer patients (a minority) likely to respond to treatment. To date, no such marker has been identified for responsiveness to C225. A number of laboratories are pursuing this question. The lesson from the results of dozen of trials with a variety of EGF receptor inhibitors is clear: it is important to identify markers that can predict patients likely to respond to a new therapy earlier in its development—even at the preclinical research stage. In the case of treatments targeting EGF receptors, over 60,000 patients received treatment on clinical trials before systematic attempts to discover markers for sensitivity were undertaken.

Because a patient's cancer, with few exceptions, has multiple genetic and molecular abnormalities, it is not surprising that treatment with a new therapy targeting one of these abnormalities, administered as a single agent, has a low response rate. For clinical trials of new targeted agents, a way to improve chances for detecting antitumor activity would be to allow and encourage combination therapies of two (or more) targeted agents, which may lead to detection of antitumor activity that either agent alone would fail to exhibit.

For C225, there will need to be extensive clinical trials with and without concurrent chemotherapy (or radiation) in a wide variety of cancer types at early stages of disease, along with detailed pharmacodynamic studies to determine markers for which patients are responsive to EGF receptor blockade. Combination therapy appears to be critical for best results with this agent, and current clinical trials include agents that target other molecular processes such as angiogenesis or apoptosis.

It is important to stress that the mechanisms explaining the antitumor activity of C225, alone or with chemotherapy/radiation, still remain unknown. The possibilities outlined in this article are numerous. This investigator favors pro-apoptotic and antiangiogenesis mechanisms that result from inhibition of EGF receptor tyrosine kinase activity as the likely candidates, but other mechanisms discovered in preclinical studies that have been discussed may be relevant. This may include recruitment of an immune response by the antibody.

Finally, it must be noted that there were 22 years between the initial hypothesis that inhibition of EGF receptor tyrosine kinase activity might be an effective cancer therapy and FDA approval of C225. Reasons for this prolongation of effort have been described. The timeline for developing a new cancer therapy today is still not as rapid as would be desirable, for both patients and for commercial enterprises. Shortening this timeline for new drug development is a challenge that will require increased open dialog and true collaboration between academia, drug companies, and governmental agencies.

REFERENCES

1. Hanahan D, Weinberg RA. The hallmarks of cancer. *Cell* 2000;100:57–70.
2. Kawamoto T, Sato JD, Le A, Polikoff J, Sato GH, Mendelsohn J. Growth stimulation of A431 cells by EGF: identification of high affinity receptors for epidermal growth factor by an anti-receptor monoclonal antibody. *Proc Natl Acad Sci USA* 1983;80:1337–1341.
3. Barnes D, Sato G. Serum-free cell culture: a unifying approach. *Cell* 1980;22:649–655.
4. Paul J. The nutrition of animal cells in vitro. *Proc Nutr Soc* 1960;19:45–50 (Reference to unpublished observations).

5. Bernheim JL, Dorian R, Mendelsohn J. DNA synthesis and proliferation of human lymphocytes in vitro: II. Cell kinetics of response to phytohemagglutinin. *J Immunol* 1978;120:955–962.

6. Mendelsohn J, Nordberg J. Adenylate cyclase in thymus-derived and bone marrow-derived lymphocytes from normal donors and patients with chronic lymphocytic leukemia. *J Clin Invest* 1979;63:1124–1132.

7. Sporn MB, Todaro GJ. Autocrine secretion and malignant transformation of cells. *N Engl J Med* 1980;303:878–880.

8. Cooper JA, Hunter T. Similarities and differences between the effect of epidermal growth factor Rous sarcoma virus. *J Cell Biol* 1981;91:878–883.

9. Chinkers M, Cohen S. Purified EGF receptor-kinase interacts specifically with antibodies to Rous sarcoma virus transforming protein. *Nature (Lond.)* 1981;290:516–519.

10. Erickson E, Shealy DJ, Erickson RL. Evidence that viral transforming gene products and epidermal growth factor stimulate phosphorylation of the same cellular protein with similar specificity. *J Biol Chem* 1981;25:11381–11384.

11. Ozanne B, Richards CS, Hendler F, Burns D, Gusterson B. Over-expression of the EGF receptor is a hallmark of squamous cell carcinomas. *J Pathol* 1986;149:9–14.

12. Mendelsohn J. Epidermal growth factor receptor inhibition by a monoclonal antibody as anticancer therapy. *Clin Cancer Res* 1997;3:2703–2707.

13. Sato JD, Kawamoto T, Le AD, Mendelsohn J, Polikoff J, Sato GH. Biological effects in vitro of monoclonal antibodies to human EGF receptors. *Mol Biol Med* 1983;1:511–529.

14. Masui H, Kawamoto T, Sato JD, Wolf B, Sato GH, Mendelsohn J. Growth inhibition of human tumor cells in athymic mice by anti-EGF receptor monoclonal antibodies. *Cancer Res* 1984;44:1002–1007.

15. Gill GN, Kawamoto T, Cochet C, Le A, Sato JD, Masui H, McLeod C, Mendelsohn J. Monoclonal anti-epidermal growth factor receptor antibodies which are inhibitors of epidermal growth factor binding and antagonists of epidermal growth factor-stimulated tyrosine protein kinase activity. *J Biol Chem* 1984;259:7755–7760.

16. Sunada H, Magun B, Mendelsohn J, MacLeod CL. Monoclonal antibody against EGF receptor is internalized without stimulating receptor phosphorylation. *Proc Natl Acad Sci USA* 1986;83: 3825–3829.

17. Masui H, Morayama T, Mendelsohn J. Mechanism of antitumor activity in mice for anti-EGF receptor monoclonal antibodies with different isotypes. *Cancer Res* 1986;46:5592–5598.

18. Sunada H, Yu P, Peacock JS, Mendelsohn J. Modulation of tyrosine serine and threonine phosphorylation and intracellular processing of the epidermal growth factor receptor by anti-receptor monoclonal antibody. *J Cell Physiol* 1990;142:284–292.

19. Van de Vijver M, Kumar R, Mendelsohn J. Ligand-induced activation of A431 cell EGF receptors occurs primarily by an autocrine pathway that acts upon receptors on the surface rather than intracellularly. *J Biol Chem* 1991;266:7503–7508.

20. Masui H, Castro L, Mendelsohn J. Consumption of epidermal growth factor by A431 cells: Evidence for receptor recycling. *J Cell Biol* 1983;120:85–93.

21. Fan Z, Masui H, Atlas I, Mendelsohn J. Blockade of epidermal growth factor (EGF) receptor function by bivalent and monovalent fragments of 225 anti-EGF receptor monoclonal antibody. *Cancer Res* 1993;53:4322–4328.

22. Naramura M, Gillies SD, Mendelsohn J, Reisfeld RA, Mueller BM. Therapeutic potential of chimeric and murine anti-(epidermal growth factor receptor) antibodies in a metastasis model for human melanoma. *Cancer Immunol Immunother* 1993;37:343–349.

23. Fan Z, Lu Y, Wu X, Mendelsohn J. Antibody-induced epidermal growth factor receptor dimerization mediates inhibition of autocrine proliferation of A431 squamous carcinoma cells. *J Biol Chem* 1994;269:27595–27602.

24. Mendelsohn J, Masui H, Sunada H, and MacLeod C. Monoclonal antibodies against the receptor for epidermal growth factor as potential anticancer agents. In *Cellular and Molecular Biology of Tumors and Potential Clinical Applications*, pp. 307–312. New York: Alan R. Liss Inc., 1988.

25. Goldenberg A, Masui H, Divgi C, Kamrath H, Pentlow K, Mendelsohn J. EGF receptor overexpression and localization of nude mouse xenografts using [111]Indium labeled anti-EGF receptor monoclonal antibody. *J Natl Cancer Inst* 1989;81:1616–1625.

26. Divgi CR, Welt C, Kris M, Real FX, Yeh SDJ, Gralla R, Merchant B, Schweighart S, Unger M, Larson SM, Mendelsohn J. Phase I and imaging trial of indium-111 labeled anti-EGF receptor monoclonal antibody 225 in patients with squamous cell lung carcinoma. *J Natl Cancer Inst* 1991;83:97–104.

27. Fan Z, Masui H, Atlas I, Mendelsohn J. Blockade of epidermal growth factor (EGF) receptor function by bivalent and monovalent fragments of 225 anti-EGF receptor monoclonal antibody. *Cancer Res* 1993;53:4322–4328.

28. Goldstein NI, Prewett M, Zuklys K, Rockwell P, Mendelsohn J. Biological efficacy of a chimeric antibody to the epidermal growth factor receptor in a human tumor xenograft model. *Clin Cancer Res* 1995;1:1311–1318.

29. Mendelsohn J, Baselga J. Status of EGF-receptor antagonists in the biology and treatment of cancer. *J Clin Oncol* (Biology of Neoplasia series) 2003;21:2787–2799.

30. Wu X, Fan Z, Masui H, Rosen N, Mendelsohn J. Apoptosis induced by an anti-epidermal growth factor receptor monoclonal antibody in a human colorectal carcinoma cell line and its delay by insulin. *J Clin Invest* 1995;95:1897–1905.

31. Soos T, Kiyokawa H, Yan JS, Rubin MS, Giordano A, DeBlasio A, Bottega S, Wong B, Mendelsohn J, Koff A. Formation of p27-CDK complexes during the human mitotic cell cycle. *Cell Growth Differ* 1996;7:135–146.

32. Wu X, Rubin M, Fan Z, DeBlasio T, Soos T, Koff A, Mendelsohn J. Involvement of p27^{KIP1} in G$_1$ arrest mediated by an anti-epidermal growth factor receptor monoclonal antibody. *Oncogene* 1996;12:1397–1403.

33. Peng D, Fan Z, Lu Y, DeBlasio T, Scher H, Mendelsohn J. Anti-epidermal growth factor receptor monoclonal antibody 225 upregulates p27^{Kip1} and induces G1 arrest in prostatic cancer cell line DU145. [Advances in Brief] *Cancer Res* 1996;56:3666–3669.

34. Perrotte P, Matsumoto T, Inoue K, Kuniyasu H, Eve BY, Hicklin DJ, Radinsky R, Dinney CP. Anti-epidermal growth factor receptor antibody C225 inhibits angiogenesis in human transitional cell carcinoma growing orthotopically in nude mice. *Clin Cancer Res* 1999;5:257–265.

35. Bruns CJ, Harbison MT, Davis DW, Portera CA, Tsan R, McConkey DJ, Evans DB, Abbruzzese JL, Hicklin DJ, Radinsky R. Epidermal growth factor receptor blockade with C225 plus gemcitabine results in regression of human pancreatic carcinoma growing orthotopically in nude mice by anti-angiogenic mechanisms. *Clin Cancer Res* 2000;6:1936–1948.

36. Mandal M, Adam L, Mendelsohn J, Kumar R. Nuclear targeting of Bax during epidermal growth factor receptor-induced apoptosis in colorectal cancer cells. *Oncogene* 1998;17:999–1007.

37. Liu B, Fang M, Schmidt M, Lu Y, Mendelsohn J, Fan Z. Induction of apoptosis and activation of the caspase cascade by anti-EGF receptor monoclonal antibodies in DiFi human colon cancer cells do not involve the c-*jun* N-terminal kinase activity. *Br J Cancer* 2000;82:1991–1999.

38. Albanell J, Codony-Servat J, Rojo F, Del Campo J, Sauleda S, Anido J, Raspall G, Giralt J, Rosello J, Nicholson R, Mendelsohn J, Baselga J. Activated extracellular signal-regulated kinases: association with epidermal growth factor receptor/transforming growth factor alpha expression in head and neck squamous carcinoma and inhibition by anti-EGF receptor treatments. *Cancer Res* 2001;61(17):6500–6510.

39. Aboud-Pirak E, Hurwitz E, Pirak ME, Bellot F, Schlessinger J, Sela M. Efficacy of antibodies to epidermal growth factor receptor against KB carcinoma in vitro and in nude mice. *J Natl Cancer Inst* 1988;80:1605–1611.

40. Baselga J, Norton L, Masui H, Pandiella A, Coplan K, Miller WH, Jr, Mendelsohn J. Antitumor effects of doxorubicin in combination with anti-epidermal growth factor receptor monoclonal antibodies. *J Natl Cancer Inst* 1993;85:1327–1333.

41. Fan Z, Baselga J, Masui H, Mendelsohn J. Antitumor effect of anti-epidermal growth factor receptor monoclonal antibodies plus *cis*-diamminedichloroplatinum on well established A431 cell xenografts. *Cancer Res* 1993;53:4637–4642.

42. Baselga J, Norton L, Coplan K, Shalaby R, Mendelsohn J. Antitumor activity of paclitaxel in combination with anti-growth factor receptor monoclonal antibodies in breast cancer xenografts. *Proc Annu Meet Am Assoc Cancer Res* 1994;35:A2262–A2380.

43. Ciardiello F, Bianco R, Damiano V, De Lorenzo S, Pepe S, De Placido S, Fan Z, Mendelsohn J, Bianco AR, Tortora G. Antitumor activity of sequential treatment with Topotecan and anti-epidermal growth factor receptor monoclonal antibody C225. *Clin Cancer Res* 1999;5:909–916.

44. Prewett MC, Hooper AT, Bassi R, Ellis LM, Waksal H, Hicklin DJ. Enhanced antitumor activity of anti-epidermal growth factor receptor monoclonal antibody IMC-C225 in combination with irinotecan (CPT-11) against human colerectal tumor xenografts. *Clin Cancer Res* 2002;8:994–1003.

45. Huang SM, Bock JM, Harari PM. Epidermal growth factor receptor blockade with C225 modulates proliferation, apoptosis, and radiosensitivity in squamous cell carcinomas of the head and neck. *Cancer Res* 1999;59:1935–1940.

46. Milas L, Mason K, Hunter N, Petersen S, Yamakawa M, Ang K, Mendelsohn J, Fan Z. *In vivo* enhancement of tumor radioresponse by C225 antiepidermal growth factor receptor antibody. *Clin Cancer Res* 2000;6:701–708.

47. Baselga J, Pfister D, Cooper MR, Cohen R, Burtness B, Bos M, D'Andrea G, Seidman A, Norton L, Gunnet K, Anderson V, Waksal H, Mendelsohn J. Phase I studies of anti-epidermal growth factor receptor chimeric antibody C225 alone and in combination with cisplatin. *J Clin Oncol* 2000;18: 904–914.

48. Saltz L, Rubin M, Hochster H, Tchekmevdian NS, Waksal H, Needle M, LoBuglio AF. Cetuximab (IMC-225) plus irinotecan (CPT-11) is active in CPT-11-refractory colorectal cancer (CRC) that expresses epidermal growth factor receptor (EGFR). *Proc Am Soc Clin Oncol* 2001;20:3a (abstract 7).

49. Saltz L, Meropol NJ, Loehrer PJ, Waksal H, Needle MN, Mayer RJ. Single agent IMC-C225 (Erbitux™) has activity in CPT-11-refractory colorectal cancer (CRC) that expresses the epidermal growth factor receptor (EGFR). *Proc Annu Meet Am Soc Clin Oncol* 2002;21:127a (abstract 504).

50. Saltz LB, Meropol NJ, Lochrer PJ Sr, Needle MN, Kopit J, Mayer RJ. Phase II trial of cetuximab in patients with refractory colorectal cancer that expresses the epidermal growth factor receptor. *J Clin Oncol* 2004;22(7):1177–1179.

51. Cunningham D, Humblet Y, Siena S, Khayat D, Bleiberg H, Santoro A, Bets D, Mueser M, Harstrick Y, Van Cutsem E. Cetuximab (C225) alone or in combination with irinotecan (CPT-11) in patients with epidermal growth factor receptor (EGFR)-positive, irinotecan-refractory metastatic colorectal cancer (MCRC). *Proc Annu Meet Am Soc Clin Oncol* 2003;22:252 (abstract 1012).

52. Cunningham D, Humblet Y, Siena S, Khayat D, Bleiberg H, Santor AD, Mueser M, Harstrick A, Verslype C, Chau I, Van Cutsem E. Cetuximab monotherapy and cetuximab plus irinotecan in irinotecan-refractory metastatic colorectal cancer. *N Engl J Med* 2004;22;351(4):337–345.

53. Robert F, Ezekiel MP, Spencer SA, Meredith RF, Bonner JA, Khazaeli MB, Saleh MN, Carey D, LoBuglio AF, Wheeler RH, Cooper MR, Waksal HW. Phase I study of anti-epidermal growth factor receptor antibody cetuximab in combination with radiation therapy in patients with advanced head and neck cancer. *J Clin Oncol* 2001;19:3234–3243.

54. Bonner JA, Harari PM, Giralt J, Azarnia N, Cohen RB, Raben D, Jones C, Kies MS, Baselga J, Ang KK. Cetuximab prolongs survival in patients with locoregionally advanced squamous cell carcinoma of head and neck: a phase III study of high dose radiation therapy with our without cetuximab. *Proc Annu Meet Am Soc Clin Oncol* 2004;23(14S):489s (abstract 5507).

55. Bonner JA, Harari PM, Giralt J, Azarnia N, Shin DM, Cohen RB, Jones CU, Sur R, Raben D, Jassem J, Ove R, Kies MS, Baselga J, Youssoufian H, Amellal N, Rowinsky EK, and Ang K. Radiotherapy plus cetuximab for squamous-cell carcinoma in head and neck. *N Engl J Med* 2006;354:567–578.

24

VEGF Inhibition
for Cancer Therapy

Shermini Saini, MD,
and Herbert Hurwitz, MD

SUMMARY

Drugs targeting vasculoendothelial cell growth factor (VEGF) are now among the most commonly used anti-cancer agents. The agent with the greatest current clinical experience is bevacizumab, a monoclonal antibody to VEGF which has been studied primarily in combination regimens. Clinical benefit with bevacizumab has been seen across multiple tumor types, including colon cancer, non–small cell lung cancer, breast cancer, renal cell cancer, among others. Two other small molecule inhibitors of the VEGF axis, sunitinib and sorafenib, have shown benefit in renal cell cancer, as well as other tumor types. The efficacy and toxicity of these and other VEGF inhibitors is reviewed.

Key Words: Cancer; VEGF; anti-angiogenesis; tyrosine kinase inhibition; bevacizumab; sorafenib; BAY 43–9006; sunitinib malate; SU11248; vatalilnib.

1. INTRODUCTION AND HISTORY

The clinical importance of tumor angiogenesis has recently been validated with the success of several anti-angiogenesis agents across a number of clinical settings. This success builds upon decades of preclinical research. The earliest reported observations that tumors could induce the growth of new blood vessels first appeared over a century ago *(1)*. Subsequently in the early twentieth century numerous investigators furthered this concept by describing the vascular architecture of various tumor types *(2)*, demonstrating by direct visualization that tumor growth can be accompanied by rapid and extensive vascular formation *(3,4)*. Algire and colleagues noted that in contrast to normal tissues, transplanted tumors developed significantly greater vascularity and that vascular growth appeared to temporally precede the phase of rapid tumor growth. Their observations led the authors to conclude that "the rapid growth of tumor explants is dependent on the development of a rich vascular supply" and that a tumor's ability to induce vasculogenesis may be one of the critical features distinguishing malignant tumors from non-malignant tissues *(4)*.

From: *Cancer Drug Discovery and Development: Molecular Targeting in Oncology*
Edited by: H. L. Kaufman, S. Wadler, and K. Antman © Humana Press, Totowa, NJ

The presence of a chemical mediator or diffusible growth factor responsible for the rapid growth of blood vessels in growing tumors was suggested by several investigators *(3,4)*. In 1971, Judah Folkman and his collaborators were among the first to isolate what they referred to as tumor angiogenesis factor (TAF) from human and animal tumors *(5)*. Folkman proposed the novel, and at the time highly controversial, idea that inhibition of these angiogenic factors would inhibit tumor growth *(6)*. Relative to tumor cells, most vascular endothelial cells are genetically stable and therefore unlikely to acquire mutations leading to resistance, in contrast to tumor cells where clonal resistance is common. In 1983, Dvorak and colleagues described a protein component of tumor supernatant, which they called vascular permeability factor (VPF), isolated from a guinea pig tumor cell line *(7)*. They suggested that VPF could be responsible for the hyperpermeability of tumor vasculature and possibly the development of ascites in cancer patients. In 1989, Napoleone Ferrara's group reported their identification, at the genetic and protein level, of a factor that selectively and potently stimulated the growth of vascular endothelial cells. They named this diffusible molecule vascular endothelial growth factor (VEGF) *(8)*. Subsequent studies demonstrated that VEGF and VPF were identical proteins and that the structure and sequence of VEGF was highly conserved across species *(9)*. Similar results were reported around the same time from Senger and colleagues *(10)*. In the 1990s, the first two major receptors for VEGF were discovered: VEGFR-1 (or FLT-1, FMS-like tyrosine kinase) and VEGFR-2 (or KDR, kinase insert domain-containing receptor, or FLK1, fetal liver kinase 1) *(11–14)*.

Alternative exon splicing of the VEGF gene results in multiple isoforms of VEGF: the principal isoforms $VEGF_{121}$, $VEGF_{165}$, $VEGF_{189}$, $VEGF_{206}$, and the less common $VEGF_{145}$, $VEGF_{183}$, $VEGF_{162}$, and $VEGF_{165b}$ *(15)*. These multiple isoforms each have variable bioavailability and different physiologic relevance. The transcription of the VEGF gene is regulated by many factors, one of the most important being hypoxia-inducible factor-1α (HIF-1α), which in turn is negatively regulated by the von Hippel-Lindau (VHL) tumor suppressor gene *(16)*. Mutations of VHL can lead to up-regulation of VEGF.

Among the dozens of known angiogenesis regulators, VEGF is currently the most potent and clinically relevant angiogenic factor. The prototype member of the VEGF family is VEGF-A, but additional related factors include VEGF-B, VEGF-C, VEGF-D, VEGF-E and the additional VEGFR ligand, placental growth factor (PlGF). These factors are bound by three VEGF receptors (VEGFR1, R2, and R3) and two coreceptors, neuropillin 1 and 2 (NP1 and NP2, respectively). VEGF gene expression is regulated by numerous factors including hypoxia (through HIF-1α), acidosis, tumor necrosis factor (TNF), interleukin-1 (IL-1) and other angiogenic factors such as platelet-derived growth factors A and B (PDGF-A and PDGF-B), and numerous tumor suppressor genes and oncogenes (e.g., ErbB2, bcl-2, p53, and VHL) *(17–19)*. Understanding this complex regulation will be needed to optimally develop combination anti-angiogenic regimen.

2. BIOLOGY AND MECHANISMS OF ACTION

In normal animal tissues, physiologic angiogenesis occurs primarily in reproductive physiology (ovarian and uterine changes in the menstrual cycle), embryonic and placental growth and development, and wound healing. In 1995, Fong et al. and

Shalaby et al. first described the central role of VEGF, VEGFR-1, and VEGFR-2 in embryonic vascular development *(20,21)* followed in 1996 by Carmeliet et al. and Ferrara et al. *(22,23)*.

Pathologic angiogenesis occurs most notably in tumorigenesis, as well as in inflammatory conditions (e.g., rheumatoid arthritis), infection, and pathologic intraocular neovascularization. Nearly all tumor types studied to date have demonstrated increased VEGF expression although the amount of overexpression may vary significantly *(24)*. Higher levels of VEGF expression have been shown in tumor types with greater vascularity, rapid growth, invasion, metastases, and worse overall clinical prognosis, lending support to the idea that VEGF expression correlates with more aggressive tumor progression and metastases. Intraocular neovascularization is another setting of abnormal vessel growth common in patients with diabetes mellitus (diabetic retinopathy), central retinal vein occlusion, retinopathy of prematurity, and age-related macular degeneration.

The exact mechanisms by which anti-angiogenic therapies exert their anti-tumor effects are still under investigation. VEGF inhibition may reduce the production of tumor proliferative and survival factors by endothelial and stromal cells. Similarly, the reduction of VEGF-mediated vascular permeability by VEGF inhibitors would be expected to reduce tumor exposure to serum, which contains many tumorigenic growth factors. As monotherapy, VEGF inhibitors have shown only modest therapeutic effects *(25–30)*. When used in combination with traditional cytotoxic therapies (chemotherapy and radiation), VEGF blockade increases efficacy of most these agents. These results of combination therapy may be explained by complimentary targeting of both the tumor and tumor-supporting stroma. Rakesh Jain has offered an explanation for this cooperative effect that focuses on the role of tumor vasculature in drug delivery *(31)*. Tumor vessels are typically chaotic and hyper-permeable, lacking the defined organization of a normal vascular network (arterioles, venules, capillaries, and normal connections between them). The vessels themselves have an irregular caliber, with areas of dilation and constriction. The endothelial cells forming these vessels are poorly organized and lack normal pericyte and basement membrane supports. Instead, vessel walls have gaps between endothelial cells, allowing fluid and large molecules to leak from the vessel lumen into the surrounding interstitium. High interstitial pressure resulting from capillary leakage may collapse thin-walled capillaries and venules and thereby decrease tumor blood flow, leading to hypoxia and acidosis.

The effects of hypoxia on the tumor microenvironment are multiple, but at a minimum include the up-regulation of many angiogenic and tumor growth and survival factors. High interstitial pressures may also reduce diffusion of drugs and oxygen, which may represent a mechanism of resistance to chemotherapy and radiation. VEGF blockade has been shown in preclinical models to "normalize" tumor vasculature, allowing better drug and oxygen delivery *(31)*. This improved drug delivery has been suggested as one explanation for the increased benefit seen when anti-VEGF therapy and chemotherapy are combined clinically.

3. CURRENT THERAPIES

3.1. Bevacizumab (Avastin®; Genentech)

Renal cell carcinoma (RCC) is among the most rational targets for anti-VEGF therapy and was among the first tumor types tested. Hereditary and sporadic RCCs are frequently characterized by loss of the VHL tumor suppressor gene. Loss of VHL results in loss of regulation of many hypoxia-responsive genes, leading to overexpression of VEGF and numerous other angiogenic factors *(32)*. In 2003, Yang and colleagues conducted a double-blind, randomized phase II trial of placebo versus monotherapy with bevacizumab at low (3 mg/kg) and high (10 mg/kg) doses every 14 days in patients with metastatic RCC *(25)*. One hundred sixteen patients were stratified to one of three treatment arms. Time to progression (TTP) was of borderline significance in the low-dose arm compared with placebo (hazard ratio = 1.26, $p = 0.053$) but significantly longer in the high-dose bevacizumab arm compared with placebo (hazard ratio = 2.55, $p < 0.001$). Despite the study closing accrual early because of a demonstrated benefit in progression-free survival (PFS), the primary study endpoint, the study showed a trend for improved overall survival (OS). This trend, however, did not reach statistical significance. A 10% response rate was noted in the high-dose bevacizumab group, suggesting that changes in the tumor microenvironment can even reverse the balance between tumor proliferation and apoptosis in some patients.

When used in combination with cytotoxic chemotherapy, some of the best results for bevacizumab were first reported in metastatic colorectal cancer (mCRC). In a randomized phase II study, 104 patients were randomized to one of three treatment arms: FU/LV alone, FU/LV plus low-dose bevacizumab (5 mg/kg every 2 weeks), or FU/LV plus high-dose bevacizumab (10 mg/kg every 2 weeks) *(33)*. Treatment on both bevacizumab-containing arms resulted in higher response rates, longer median times to disease progression, and longer median survival. Based on these findings, a phase III trial was conducted exploring the value of adding bevacizumab to the IFL chemotherapy regimen (irinotecan 125 mg/m^2, fluorouracil 500 mg/m^2, and leucovorin 20 mg/m^2) in patients with previously untreated Mcrc *(34)*. Eight hundred thirteen patients were randomized to receive either IFL plus bevacizumab (5 mg/kg) or IFL plus placebo. The bevacizumab arm demonstrated improvement in objective response rates (44.8 versus 34.8%, $p = 0.004$), median duration of response (10.4 versus 7.1 months, $p < 0.001$), median PFS (10.6 versus 6.2 months, $p < 0.001$) and median OS, the primary endpoint (20.3 versus 15.6 months, $p < 0.001$). These results represented a major advance in the treatment of mCRC, a tumor with a historically poor prognosis, and led to FDA approval of bevacizumab as part of the first-line treatment regimen for mCRC.

The value of bevacizumab in combination with chemotherapy has recently been demonstrated in second-line colorectal cancer (CRC), first-line non-small cell lung cancer (NSCLC), and first-line breast cancer. The Eastern Cooperative Oncology Group (ECOG) conducted a randomized phase III trial in patients with advanced CRC who had previously been treated with a fluoropyrimidine and irinotecan-based regimen. Patients were randomized to one of three treatment arms: FOLFOX4 alone (biweekly administration of oxiliplatin 85 mg/m^2, leucovorin 200 mg/m^2, and fluorouracil 400 mg/m^2 bolus followed by fluorouracil 600 mg/m^2 for 22 h), FOLFOX4 plus bevacizumab, or bevacizumab alone *(29)*. Compared to FOLFOX4 alone, the combination of FOLFOX4 plus bevacizumab demonstrated an improvement in median OS (12.5 versus 10.7 months, hazard ratio = 0.76, $p = 0.0018$) , median PFS (7.4 versus 5.5 months, hazard ratio = 0.64, $p < 0.0001$), and response rate (21.8 versus 9.2%, $p < 0.0001$). Bevacizumab monotherapy was associated with a biologically interesting but likely clinically unmeaningful response rate of 3%.

ECOG also conducted a randomized phase III trial to evaluate the combination of paclitaxel/carboplatin with or without bevacizumab in a total of 878 patients with advanced NSCLC *(35)*. Chemotherapy plus bevacizumab resulted in higher response rates (27.2 versus 10%, $p < 0.0001$), longer PFS (6.4 versus 4.5 months, $p < 0.0001$), and improved OS (12.5 versus 10.2 months, $p = 0.0075$) compared to chemotherapy alone *(36,37)*.

Two large randomized phase III trials have been conducted to evaluate the efficacy of bevacizumab in metastatic breast cancer, one in first line and one in refractory disease. The first study compared capecitabine alone versus capecitabine plus bevacizumab in a total of 462 patients who had undergone no more than two prior treatment regimens for metastatic disease *(38)*. Although objective response rates were significantly higher in the bevacizumab-treated arm (19.8 versus 9.1%, $p = 0.001$), neither PFS nor OS were significantly different between the two groups. The median PFS was 4.9 months versus 4.2 months, and the median OS was 15.1 versus 14.5 months for capecitabine/bevacizumab versus capecitabine/placebo, respectively.

The phase III trial of bevacizumab in first-line metastatic breast cancer (E2100) randomized patients to paclitaxel plus placebo versus paclitaxel plus bevacizumab with PFS as the primary endpoint *(38)*. An interim analysis of the data based on 355 patients (out of a total of 715 patients enrolled) showed a significantly improved PFS among patients treated with the combination as opposed to paclitaxel alone (11.0 versus 6.1 months, $p < 0.001$). Although the median survival had not yet been reached for the paclitaxel plus bevacizumab group, OS was improved with the addition of bevacizumab, as measured by a hazard ratio of 0.67 ($p = 0.01$).

3.1.1. SIDE EFFECTS OF BEVACIZUMAB

As a class, VEGF inhibitors have limited but notable toxicities compared with classical cytotoxic agents. The most common side effect is hypertension. Among patients with mCRC, hypertension of any severity occurs in approximately 22–32% of patients on bevacizumab *(33,34,39)*. Hypertension requiring the addition or adjustment of an anti-hypertensive regimen (grades 3–4) is more common in patients treated with bevacizumab, occurring in approximately 11–16% of patients. Hypertension appears readily manageable with standard anti-hypertensive agents. Chemotherapy-related side effects, such as neutropenia, nausea, vomiting and diarrhea, were similar in both the bevacizumab and placebo groups. Among all phase III studies of bevacizumab in colorectal, breast, and NSCLC, the only chemotherapy-related side effect that consistently increased in bevacizumab-treated patients was peripheral neuropathy, which has been attributed to a longer exposure to neurotoxic agents (oxiliplatin or paclitaxel) because of the longer TTP in the bevacizumab-treated patients.

In contrast to suggestions of increased risk in phase I/II studies, venous thromboembolic events (deep venous thrombosis and pulmonary embolus) were not increased with the addition of bevacizumab in any phase III studies of bevacizumab. In contrast to venous vascular events, however, arterial thromboembolic events were increased in the mCRC study and in other studies as well. These events included unstable angina, myocardial infarction, transient ischemic attack, cerebrovascular accidents, and retinal and mesenteric artery thromboses. In a pooled analysis of five Genentech-sponsored studies, the risk of these arterial events was increased by approximately twofold, from 1.7 to 3.8% of patients (or 3.1 versus 5.5 events per 100 person-years) *(40)*. Factors associated with increased risk were age greater than 65 years and history of atherosclerosis. Despite this increased risk of an arterial vascular event with bevacizumab, the survival benefit of bevacizumab has been shown to be independent of age in all randomized studies to date. This is likely due to the relatively low background rate of arterial events and the marked mortality risk from cancer in these patients. Although strategies to reduce the risk of arterial vascular events in this setting have not yet been studied prospectively, retrospective analyses suggest a potentially protective effect with the use of low-dose aspirin.

Congestive heart failure has been seen infrequently and predominantly limited to settings with prior anthracycline use or left chest wall irradiation. In the phase III study by Miller et al., which compared capecitabine alone versus capecitabine plus bevacizumab, a total of nine patients developed grade 3 or 4 congestive heart failure or cardiomyopathy. Seven of these patients were in the combination arm receiving both bevacizumab and chemotherapy. Two of the seven patients in the combination arm continued to receive bevacizumab. Cardiac symptoms were responsive to medical therapy in all but one in the total of nine patients. All patients who developed this toxicity in the bevacizumab arm had a history of prior anthracycline exposure, and three of the seven had previously received left chest wall irradiation. These patients' histories of prior cardiotoxic therapy made it difficult to draw conclusions about the possible contribution of bevacizumab to the observed cardiac toxicity (41,42). In a small phase II study of patients with sarcomas, none of whom had received prior anthracyclines, significantly decreased left ventricular ejection fraction (LVEF) by echo or multiple gated acquisition scan (MUGA) was noted in more than one third of patients (6 of 17 patients) (43). Clinically, symptomatic congestive heart failure (CHF) (grades 3 and 4 cardiac toxicity) was noted in two of the seven patients. The mechanism of this toxicity is difficult to explain given the lack of anthracycline exposure of all patients in this study. Proposed mechanisms include the impaired ability of cardiac tissue to repair itself following the toxic insult of doxorubicin exposure in the presence of simultaneous VEGF inhibition.

Bleeding risks related to bevacizumab are mostly mild and manageable, such as epistaxis and hemorrhoidal bleeding. Based on the first-line CRC study, the risk of bleeding with full dose anti-coagulation does not appear to increase with bevacizumab. Major bleeding events have been rare and were not increased with the majority of bevacizumab studies. The exception is the risk of severe or life-threatening hemoptysis in NSCLC. In the randomized phase II study of paclitaxel/carboplatin ± bevacizumab, several patients in the bevacizumab group had massive, and sometimes fatal, hemoptysis (35). These events occurred primarily in patients with centrally located lesions of squamous histology, who had tumor responses characterized by cavitation. For these reasons, the phase III study of paclitaxel/carboplatin in NSCLC excluded these patients. The rates of serious or life-threatening bleeding, particularly hemoptysis, in this study were 4.1 versus 1% in bevacizumab-treated patients.

Consistent with the known role of VEGF in wound healing, a small increase in wound-healing complications has been noted for patients on bevacizumab at the time of surgery (44). The risk in patients who have healed from surgery and who then start bevacizumab does not appear to be significantly increased (1.3 versus 0.5% in bevacizumab plus chemotherapy versus chemotherapy alone, respectively). However, for patients who underwent surgical procedures while taking bevacizumab, usually for complications of disease progression such as bowel obstruction, the rate of wound-healing complications was increased, from 3 to 13%. Similarly, bevacizumab is associated with a risk of bowel perforation of approximately 1–2%. Risks for this complication have not yet been identified. Potential risk factors include endoscopic biopsies and settings with potentially compromised bowel vasculature, such as peritoneal carcinomatosis and prior abdominal surgeries *(34,29)*.

Proteinuria is an uncommon complication of bevacizumab. Proteinuria was not increased in the CRC, breast, or NSCLC studies. However, proteinuria was more frequent in the renal cell study, perhaps related to the fact that most of these patients are uni-nephric and may have altered renal hemodynamics *(25)*.

3.2. Sorafenib, BAY 43-9006 (Nexavar®; Bayer)

Sorafenib is an orally administered inhibitor of several intracellular (CRAF, wild-type BRAF, and mutant BRAF) and receptor tyrosine kinases (Kit, Flt-3, VEGFR-2, VEGFR-3, and PDGFR-β). Sorafenib was FDA approved in December 2005 for the treatment of refractory RCC based on its improvement in PFS and OS, the primary endpoints of the study *(45)*. Sorafenib is also being investigated in multiple ongoing clinical trials on various tumor types including refractory/recurrent NSCLC, metastatic melanoma, recurrent/metastatic head and neck squamous cell carcinoma, prostate cancer, and other malignancies. In an initial phase II trial in several advanced malignancies including RCC, 202 patients with advanced RCC were treated with an induction phase of therapy with sorafenib (400 mg orally twice daily) for 12 weeks. Subsequently, 65 patients with stable disease by bidirectional tumor measurements were randomized to receive another 12 weeks of therapy with either sorafenib (400 mg orally twice daily) or placebo. Upon re-evaluation at 24 weeks, patients in the sorafenib arm showed improved median PFS (23 weeks versus 6 weeks, $p = 0.0001$), with an acceptable toxicity profile *(47)*.

This study was followed up with a randomized, phase III, double blind, placebo-controlled trial in patients with advanced RCC, which used a novel randomized discontinuation approach, a design that has been reported to have increased efficiency to detect benefit that is primarily related to disease stabilization *(30)*. Seven hundred sixty-nine patients were randomized to receive continuous oral sorafenib 400 mg twice daily or placebo with best supportive care (BSC). Assessment by independent review showed double the median PFS in the sorafenib arm (24 versus 12 weeks for placebo, $p < 0.000001$). Drug-related adverse events were tolerable and included rash (34%), diarrhea (33%), hand–foot skin reaction (27%), fatigue (26%), and hypertension (11%). Grade 3 and 4 adverse events were reported in 30% of sorafenib patients versus 22% of patients on placebo. The marked improvement in PFS prompted an interim analysis of survival. A strong trend for improved OS in the sorafenib group was noted with preliminary results demonstrating longer OS in the sorafenib-treated patients over BSC (hazard ratio = 0.72, based on 220 deaths) *(45)*. An analysis of mature survival data is expected soon. Results of phase III clinical trials of sorafenib in advanced hepatocellular carcinoma and metastatic melanoma are also expected soon.

3.3. Sunitinib Malate, SU11248 (Sutent®; Pfizer)

SU11248 is another orally administered, multitargeted small molecule tyrosine kinase inhibitor that is currently in phase III clinical development. SU11248 binds to and inactivates the VEGF tyrosine kinase receptor, as well as the platelet-derived growth factor receptor (PDGFR), c-kit, and FLT3 tyrosine kinases. Through its receptor tyrosine kinase inhibition, SU11248 is considered to have both anti-angiogenic and direct anti-tumor effects and is currently being studied in multiple malignancies including metastatic RCC, neuroendocrine tumors, and gastrointestinal stromal tumors (GIST).

Two open label, non-randomized phase II trials of SU11248 involving a total of 169 patients with metastatic RCC demonstrated a 40% response rate. Favorable TTP (8.7 months), duration of response (≥3 months), and median survival (16 months) *(27,28,47)* were also reported.

To date, the most promising target population for SU11248 appears to be metastatic GIST refractory to imatinib mesylate (Gleevec) *(48)*. This agent was approved for this indication in February 2006 based on the double-blind phase III trial led by Demetri et al. comparing SU11248 versus placebo in 312 patients with GIST who had failed treatment on Gleevec, with either disease progression or intolerable side effects. Using a 2:1 randomization, 207 patients were treated with SU11248 at 50 mg once daily for 4 weeks per 6-week cycle, with 2 weeks between each cycle, and 105 patients received placebo. The primary endpoint was TTP, and at the interim analysis, the investigators found a significant improvement in TTP among the patients on SU11248 compared with placebo (27.3 versus 6.4 weeks, respectively, $p < 0.0001$).

SU11248 is generally well tolerated. In phase I/II trials, grade 3 and 4 adverse events have included asymptomatic transient elevations in lipase ± amylase, hematologic toxicity (including uncomplicated neutropenia and thrombocytopenia), hypertension, diarrhea, nausea, asthenia, and skin rashes *(49)*. In the two multicenter phase II trials of SU11248 in a total of 169 patients with metastatic RCC, the most frequent adverse event was fatigue, which reached grade 3 severity in a total of 19% of patients. Other significant grade 3 toxicities included hypertension (10%), dyspnea (8%), and diarrhea (8%). Common grade 3–4 laboratory abnormalities included lymphopenia without infection (35%), elevated lipase (16%) and amylase (5%) without clinical pancreatitis, hypophosphatemia (10%), and hyperuricemia (10%) *(28,47)*. Two patients experienced grade 3 cardiac toxicity manifested by myocardial ischemia, and one patient experienced a fatal myocardial infarction while on treatment. Four patients had venous thromboembolic events, two with grade 4 pulmonary embolism and two with grade 3 deep venous thromboses. Rarely seizures have also been observed in patients with brain metastases undergoing treatment with SU11248, as well as radiologic findings consistent with reversible posterior leukoencephalopathy syndrome (RPLS), none of whom had a fatal outcome from their neurologic events. In the phase III trial of SU11248 in refractory GIST, most adverse events were mild to moderate (grade 1 or 2), with fatigue, diarrhea, nausea, hypertension, sore mouth, and skin discoloration being the most common non-hematologic events *(48)*. Grade 3 or 4 treatment-related adverse events were reported in 56 versus 51% of patients on SU11248 versus placebo, respectively. Grade 3 or 4 laboratory abnormalities seen more commonly in patients treated with SU11248 than placebo included myelosuppression, elevated liver function tests, elevated pancreatic enzymes, electrolyte disturbances of all types, and decreased left ventricular ejection fraction *(47)*. Acquired hypothyroidism was also observed in 4% of patients on SU11248 versus only 1% of patients on placebo. Thyroid dysfunction was not dose limiting, and all patients were treated effectively with thyroid hormone replacement. It has been theorized that the fatigue observed in many patients treated with SU11248 may be related at least in part to the acquired hypothyroidism seen in some of these patients *(50)*.

3.4. Vatalanib (PTK787/ZK222584; Novartis-Schering)

PTK787/ZK222584 (PTK) is an oral, multitargeted small molecule tyrosine kinase inhibitor that is also currently in phase III development. PTK targets all the known VEGF receptor tyrosine kinases, including VEGFR-1, VEGFR-2, VEGFR-3, as well as PDGFR and c-kit.

Following phase I/II trials demonstrating safety, the two major phase III trials in the development of PTK, CONFIRM-1 and CONFIRM-2, were launched. CONFIRM-1 sought to study the effects of adding PTK (1250 mg orally once daily) to the FOLFOX-4 regimen (oxiliplatin, 5-fluorouracil, and leucovorin) in previously untreated patients with mCRC *(51)*. The study failed to meet its primary endpoint of improved PFS as measured by an independent radiology review committee. Subgroup analysis noted a benefit for patients with high LDH values although the meaning of this finding is not yet understood. The study has yet to mature sufficient data for analysis of OS. A similar phase III study of FOLFOX4 ± PTK in second-line CRC, known as CONFIRM-2, also failed to meet its primary endpoint of improving PFS as assessed by an independent radiology review. As with CONFIRM-1, this study noted a benefit in the subgroup of patients with elevated LDH. Final survival data are expected soon *(51)*.

3.5. Others

Multiple additional anti-VEGF agents are now in clinical testing, many hitting additional potentially relevant kinases. These include ZD6474 (Zactima®; AstraZeneca), GW786034 (Pazopanib; GlaxoSmithKline), ZD2171 (AstraZeneca), AG3340 (Prinomastat, Pfizer), CHIR258 (Chiron), XL880 (Exelixis), and CEP7055 (Cephalon), among others. All of these agents have shown some hints of activity in early clinical testing. Phase II and III studies to better assess the clinical efficacy and toxicity of these agents are now in progress.

4. DISCUSSION

VEGF is now a well-validated target for anti-cancer therapy. However, several questions still need to be addressed by further preclinical and clinical testing. These include a more detailed elucidation of the mechanisms of action and resistance to anti-VEGF therapy. Remarkably, for agents that target the endothelial cell compartment, single-agent responses have been noted in several tumor types, including renal cell, GIST, breast, and ovarian cancers *(25,46,48)*. The mechanisms behind the favorable interaction with classical cytotoxic chemotherapy have also not yet been fully explained. To date, no clinical or molecular profile has identified those patients most or least likely to benefit for anti-VEGF therapy. Similarly, when patients progress on therapy, the mechanisms of resistance to anti-VEGF therapy clinically are not yet known. Although multiple angiogenesis factors other than VEGF have been implicated, which ones are most important is unknown. In addition, while targeting multiple factors may theoretically augment anti-tumor activity, this will likely increase the potential for vascular toxicities. Combination anti-angiogenesis therapies are now starting to be tested. The evaluation of anti-VEGF therapy in special populations, such as pediatric and geriatric cancer patients, is now just starting. Although anti-VEGF agents are among the most clinically useful anti-cancer drugs developed to date, the spectrum of activity and toxicity for the variety of VEGF agents suggest that the clinical value of each molecule must still be assessed on a case-by-case basis in well-designed clinical trials.

FINANCIAL DISCLOSURES

Dr. Saini is a fellow trainee whose salary is partially supported by GlaxoSmithKline. Dr. Hurwitz reports having received consulting fees from Genentech, Pfizer, and Imclone; lecture fees from Genentech and Pfizer; and research support from AstraZeneca, Cephalon, Bristol-Myers Squibb, Genentech, and GlaxoSmithKline.

REFERENCES

1. Ferrara N. VEGF and the quest for tumour angiogenesis factors. *Nat Rev Cancer* 2002; 2:795–803.
2. Lewis WH. The vascular pattern of tumors. *Johns Hopkins Hospital Bulletin* 1927; 41:156–162
3. Ide AG, et al. Vascularization of the Brown Pearce rabbit epithelioma transplant as seen in the transparent ear chamber. *Am J Roentgenol* 1939; 42:891–899.
4. Algire GH, et al. Vascular reactions of normal and malignant tissues *in vivo*. I. Vascular reactions of mice to wounds and to normal and neoplastic transplants. *J Natl Cancer Inst* 1945; 6:73–82.
5. Folkman J, et al. Isolation of a tumor factor responsible for angiogenesis. *J Exp Med* 1971; 133:275–288.
6. Folkman J. Tumor angiogenesis: therapeutic implications. *N Engl J Med* 1971;285: 1182–1186.
7. Senger DR, et al. Tumor cells secrete a vascular permeability factor that promotes accumulation of ascites fluid. *Science* 1983; 219: 983–985.
8. Ferrara N, et al. Pituitary follicular cells secrete a novel heparin-binding growth factor specific for vascular endothelial cells. *Biochem Biophys Res Commun* 1989;161: 851–858.
9. Senger DR, et al. A highly conserved vascular permeability factor secreted by a variety of human and rodent tumor cell lines. *Cancer Res* 1986; 46:5629–5632.
10. Senger DR, et al. Purification and NH2-terminal amino acid sequence of guinea pig tumor-secreted vascular permeability factor. *Cancer Res* 1990;50: 1774–1748.
11. de Vries C, et al. The FMS-like tyrosine kinase, a receptor for vascular endothelial growth factor. *Science* 1992; 255: 989–991.
12. Terman BI, et al. Identification of the KDR tyrosine kinase as a receptor for vascular endothelial growth factor. *Biochem Biophys Res Commun* 1992;187: 1579–1586.
13. Millauer B, et al. High affinity VEGF binding and developmental expression suggest Flk-1 as a major regulator of vasculogenesis and angiogenesis. *Cell* 1993; 72: 835–846.
14. Quinn TP, et al. Fetal liver kinase 1 is a receptor for vascular endothelial growth factor and is selectively expressed in vascular endothelium. *Proc Natl Acad Sci USA* 1993; 90:7533–7537.
15. Ferrara N, et al. The biology of VEGF and its receptors. *Nat Med* 2003; 9:669–676.
16. Pugh CW, et al. Regulation of angiogenesis by hypoxia: role of the HIF system. *Nat Med* 2003; 9:677–684.
17. Ferrara N. Vascular endothelial growth factor as a target for anticancer therapy. *Oncologist* 2004; 9(suppl):2–10.
18. Shweiki D, et al. Vascular endothelial growth factor induced by hypoxia may mediate hypoxia-initiated angiogenesis. *Nature* 1992; 359:843–845.
19. Josko J, et al. Transcription factors having impact on vascular endothelial growth factor (VEGF) gene expression in angiogenesis. *Med Sci Monitor* 2004;10: RA89–98.
20. Fong GH, et al. Role of the Flt-1 receptor tyrosine kinase in regulating the assembly of vascular endothelium. *Nature* 1995; 376:66–70.

21. Shalaby F, et al. Failure of blood-island formation and vasculogenesis in Flk-1 deficient mice. *Nature* 1995;376:62–66.

22. Carmeliet P, et al. Abnormal blood vessel development and lethality in embryos lacking a single VEGF allele. *Nature* 1996;380:435–439.

23. Ferrara N, et al. Heterozygous embryonic lethality induced by targeted inactivation of the VEGF gene. *Nature* 1996; 380:439–442.

24. Jubb A, et al. Expression of vascular endothelial growth factor, hypoxia inducible factor 1alpha, and carbonic anhydrase IX in human tumours. *J Clin Pathol* 2004; 57(5):504–512.

25. Yang JC, et al. A randomized trial of bevacizumab, an antivascular endothelial growth factor antibody, for metastatic renal cancer. *N Engl J Med* 2003; 349:427–434.

26. Cobleigh MA, et al. A phase I/II dose escalation trial of bevacizumab in previously treated metastatic breast cancer. *Semin Oncol* 2003; 30 (suppl 16): 117–124.

27. Motzer RJ, et al. Phase 2 trials of SU11248 show antitumor activity in second-line therapy for patients with metastatic renal cell carcinoma (RCC). *2005 ASCO Annual Meeting*, abstract #4508.

28. Motzer RJ, et al. Activity of SU11248, a multitargeted inhibitor of vascular endothelial growth factor receptor and platelet-derived growth factor receptor, in patients with metastatic renal cell carcinoma. *J Clin Orthod* 2006; 24:16–23.

29. Giantonio BJ, et al. High-dose bevacizumab improves survival when combined with FOLFOX4 in previously treated advanced colorectal cancer: results from the Eastern Cooperative Oncology Group (ECOG) study E3200. *J Clin Orthod* 2005; 23(suppl): 2s, abstract 2.

30. Escudier B, et al. Randomized phase III trial of the Raf kinase and VEGFR inhibitor sorafenib (BAY 43–9006) in patients with advanced renal cell carcinoma (RCC). *2005 ASCO Annual Meeting*, abstract #4510.

31. Jain RK. Antiangiogenic therapy for cancer: current and emerging concepts. *Oncology* 2005; 19(suppl): 7–16.

32. Semenza G. HIF-1 and mechanisms of hypoxia sensing. *Curr Opin Cell Biol* 2001;13:167–171.

33. Kabbinavar F, et al. Phase II, randomized trial comparing bevacizumab plus fluorouracil (FU)/leucovorin (LV) with FU/LV alone in patients with metastatic colorectal cancer. *J Clin Orthod* 2003; 21:60–65.

34. Hurwitz H, et al. Bevacizumab plus irinotecan, fluorouracil and leucovorin for metastatic colorectal cancer. *N Engl J Med* 2004; 350: 2335–2342.

35. Sandler AB, et al. Randomized phase II/III trial of paclitaxel (P) plus carboplatin (C) with or without bevacizumab (NSC #704865) in patients with advanced non-squamous non-small cell lung cancer (NSCLC): an Eastern Cooperative Group (ECOG) Trial – E4599. *J Clin Orthod* 2005; 23(suppl): abstract LBA4.

36. Laskin JJ, et al. First-line treatment for advanced NSCLC. Oncology, 19: 1671–76, 2005.

37. Tyagi P. Bevacizumab, when added to paclitaxel/carboplatin, prolongs survival in previously untrested patients with advanced NSCLC: preliminary results from the ECOG 4599 trial. Clin Lung Cancer, 6: 276–78, 2005.

38. Miller K, et al. Randomized phase III trial of capecitabine compared with bevacizumab plus capecitabine in patients with previously treated metastatic breast cancer. *J Clin Orthod* 2005; 23: 792–799.

39. Kabbinavar FF, et al. Addition of bevacizumab to bolus fluorouracil and leucovorin in first-line metastatic colorectal cancer: results of a randomized phase II trial. *J Clin Orthod* 2005; 23: 3697–3705.

40. Skillings JR, et al. Arterial thromboembolic events (ATEs) in a pooled analysis of 5 randomized, controlled trials (RCTs) of bevacizumab (BV) with chemotherapy. JCO, 23(suppl): abstract 3019, 2005.

41. Miller KD, et al. E2100: a randomized phase III trial of paclitaxel versus paclitaxel plus bevacizumab as first-line therapy for locally recurrent or metastatic breast cancer. Presented at the 41st Annual Meeting of the American Society of Clincial Oncology (ASCO), May 13–17, 2005, Orlando, FL.

42. Avastin (bevacizumab) package insert 2004.

43. D'Adamo DR, et al. Phase II study of doxorubicin and bevacizumab for patients with metastatic soft-tissue sarcomas. *J Clin Orthod* 2005; 23:7135–7142.

44. Scappaticci FA, et al. Surgical wound healing complications in metastatic colorectal cancer patients treated with bevacizumab. *J Surg Oncol* 2005; 91:173–180.

45. Nexavar (sorafenib) package insert 2005.

46. Ratain MJ, et al. Final findings from a phase II, placebo-controlled, randomized discontinuation trial (RDT) of sorafenib (BAY 43–9006) in patients with advanced renal cell carcinoma (RCC). *2005 ASCO Annual Meeting*, abstract #4544.
47. SU11248 (Sutent) package insert. Pfizer 2006.
48. Demetri et al. Improved survival and sustained clinical benefit with SU11248 (SU) in pts with GIST after failure of imatinib mesylate (IM) therapy in a phase III trial. *2006 ASCO Gastrointestinal Cancers Symposium*, abstract #8.
49. Faivre S, et al. Safety, pharmacokinetic, and antitumor activity of SU11248, a novel oral multitarget tyrosine kinase inhibitor, in patients with cancer. *J Clin Orthod* 2006; 24: 25–35.
50. Desai J, et al. Hypothyroidism may accompany SU11248 therapy in a subset of patients (pts) with metastatic (met) gastrointestinal stromal tumors (GIST) and is manageable with replacement therapy. *J Clin Orthod* 2005; 23(suppl): abstract 3040.
51. Hecht JR, et al. A randomized, double-blind, placebo-controlled, phase III study in patients with metastatic adenocarcinoma of the colon or rectum receiving first-line chemotherapy with oxaliplatin/5-fluorouracil/leucovorin and PTK787/ZK222584 or placebo (CONFIRM-1). *2005 ASCO Annual Meeting*, abstract #3.

25 Somatostatin Analogue Therapy

M. C. Champaneria, MD,
I. M. Modlin, MD, PhD, DSc, I. Latich, MD,
J. Bornschein, MD, I. Drozdov, BS,
and M. Kidd, PhD

SUMMARY

This chapter reviews the basis of somatostatin (SST) analogue therapy and the current analogues available and under investigation. Present analogues decrease the secretion of bioactive products from tumors but have a limited anti-proliferative effect. The advantage of analogues with a longer duration of action has increased the duration of therapeutic control from days up to 3–5 weeks. Ultimately the majority of patients experience loss of symptom control as drug tolerance develops or the tumor produces more secretory products. This loss of sensitivity may be associated with the emergence of cell clones lacking SST receptors (SSTRs) or down-regulation of SST receptors. Acromegalics appear to be less susceptible to this phenomenon in that tachyphylaxis is rare even after > 10 years of daily SST analogue injections.

The therapeutic effects of radiolabeled SST analogues are of interest since internalization of the isotope produces local cell death. A variety of isotopes ([111]Indium, γ emitter; [90]Yttrium, β emitter; [177]Lutetium, β and γ emitter) are under investigation. ß-emitting radionuclides exhibit a greater therapeutic potential since they emit sufficient energy to cause local tumor cell death without damaging surrounding tissue. In addition, systemic adverse effects are minimal although renal and bone marrow toxicity may occur. [177]Lu induces significant tumor regression and may be of particular benefit in the treatment of small tumors by minimizing radiation exposure of distant cells. A potential therapeutic strategy for individuals with micro-metastases or tumors of different sizes is treatment with a combination of different radionuclides with varying degrees of penetrance. Similarly, targeting co-expressed receptors (GRP, CCK, VIP) that have proliferative regulatory effects, in addition to SSTRs, may have therapeutic relevance. Since individual tumors exhibit heterogeneous expression of receptors, a combination of receptor-selective radiopeptides may further amplify therapeutic efficacy. The concept of "cocktail" isotope therapy designed to target different receptors with a variety of isotopes synchronously is a potentially attractive therapeutic prospect.

From: *Cancer Drug Discovery and Development: Molecular Targeting in Oncology*
Edited by: H. L. Kaufman, S. Wadler, and K. Antman © Humana Press, Totowa, NJ

Key Words: Acromegaly; carcinoid; carcinoid syndrome; growth hormone; Indium; inhibition; Lutetium; octreotide; neuroendocrine; proliferation; scintigraphy; somatostatin; somatostatin receptor; somatuline; tumor; Yttrium.

1. INTRODUCTION

1.1. Overview

This chapter reviews the basis of somatostatin (SST) analogue therapy and discusses the various analogues currently available and those under investigation (Fig. 1). In addition, the current status of somatostatin radio-peptide analogue therapy for neuroendocrine tumor disease and the utility of isotopic-labeled SST analogues in the imaging of tumors tissues are evaluated. Finally, a summary of current therapeutic information pertinent to both diagnosis and therapy with either unlabeled "cold" analogues or radiolabeled compounds is provided.

1.2. Discovery of SST and Its Receptors

Krulich et al. in 1968 provided the initial report of SST-like activity describing a factor in rat hypothalamic extracts that inhibited growth hormone (GH) secretion from anterior pituitary cultures (1,2). Based on this observation, they concluded that the secretion of GH was regulated by stimulatory growth hormone releasing hormone

---

Overview

Introduction
   Background
   Function
   Receptor distribution

Scintigraphy and autoradiography
   GEP-NETs
   Other neuroendocrine tumors
   CNS tumors
   Pituitary adenomas
   Adenocarcinomas
   Other tumors and paraneoplastic syndromes

Somatostatin Analogues
   Cold formulations
   Peptide receptor radionuclide therapy
Therapy

   GEP-NETs
   Other neuroendocrine tumors
   CNS tumors
   Pituitary adenomas
   Adenocarcinomas
   Other tumors and paraneoplastic syndromes

Coda

Fig. 1. Outline.

(GHRH) and inhibitory (SST) factors. Concurrent with this discovery, Hellman and Lernmark *(3)* independently observed similar activity in extracts of pigeon pancreatic islets that inhibited insulin secretion in vitro. In 1973, Brazeau et al. *(4)* described the molecular sequence of the hormone identified by the two groups as a 14-amino acid molecule (SST-14). Thereafter, a number of additional forms have been reported, including a 28-amino acid polypeptide (SST-28) with a NH_2-terminal extension of 14 amino acids corresponding to SST. SST-14 and SST-28 are the predominant physiologically active forms of SST *(5)*.

The first characterization of high-affinity, functional somatostatin receptors (SSTRs) was undertaken in rat pituitary cell cultures in 1978 *(6)*. Subsequent investigation suggested that SSTRs were likely coupled to multiple cellular effector pathways by pertussis toxin-sensitive and toxin-insensitive G proteins *(7)*. Clinical application of this information, however, was delayed by lack of delineation of the complexity of SSTR subtypes and sequences until the cloning and characterization of the *SSTR* gene family in 1995 *(2)*.

1.3. Function

SST is ubiquitous and has been identified in areas as diverse as the hypothalamus, cerebral cortex, brain stem, gastrointestinal (GI) tract, and the pancreas *(8)*. In the central nervous system (CNS), SST inhibits the release of dopamine, norepinephrine, thyrotropin-releasing hormone, thyroid-stimulating hormone (TSH), and corticotropin-releasing hormone (CRH), and feedback inhibits its own secretion *(9)*. Within the GI tract, SST inhibits gut endocrine and exocrine secretion including gastric acid, pepsin, gastrin, gut glucagon, secretin, cholecystokinin, bile, and colonic fluid *(10)*. In addition, it exhibits an inhibitory effect on gut motility including gastric emptying, gallbladder contraction, and small intestinal peristalsis and segmentation. At a cellular level, SST inhibits proliferation of both normal and tumor cells through hyperphosphorylation of the retinoblastoma gene product resulting in cell-cycle arrest in the G1 phase *(11)*. SST also downregulates angiogenesis, causes apoptosis (primarily mediated through SSTR3 but also SSTR2) *(2,12)*, and is considered to play a role in immunomodulation and neurotransmission.

1.4. Receptor Subtypes

Five principal SSTR subtypes have been identified and characterized. All are members of the Asp-Arg-Tyr (DRY) family of G-protein-coupled receptors whose ligands include neurotransmitters, neuropeptides, glycoprotein hormones, and olfactory molecules. Each subtype consists of seven transmembrane domains with an overall sequence homology of 45–61% *(2)*. Interestingly, the sequence of the seven α-helical transmembrane regions is most closely related to the opioid receptor family (~30% homology) *(13)*. All five receptor subtypes are functionally coupled to the inhibition of cyclic AMP and decreased calcium influx *(11)* although additional transduction pathways including protein phosphatases, cGMP-dependent protein kinases, phospholipase C, K^+ and Ca^{2+} channels, and a Na^+/H^+ exchanger have been described *(2)*. Agonist binding to SSTR subtypes 2, 3, and 5 can be reduced by GTP analogues and pertussis toxin treatment, indicating that these receptor subtypes are coupled to G proteins. SSTR1 and SSTR4, on the contrary, are not affected by GTP analogues or pertussis toxin treatment and do not effectively couple to adenylyl cyclase *(14,15)*. In addition, SSTR2 and SSTR5 have also been shown to mediate the antiproliferative effects of SST, with a 50% growth inhibition in animal tumor models and cultured cell lines *(16,17)*. SSTR2 couples to a tyrosine phosphatase resulting in reduced cell growth, but SSTR5 does not *(18)*. The antiproliferative effect of chronic SST analogue therapy and its site of action are as yet undefined in humans. Gel electrophysiological studies on brain neurons and in pituitary cell lines have demonstrated SSTR desensitization with exposure (3–4 h) to agonists *(2)*. This process is believed to result from receptor phosphorylation by the β-adrenergic receptor kinase (BARK) *(19)*.

1.5. Physiologic Distribution

1.5.1. CNS

All five SSTR subtypes are expressed in the CNS *(2)*. In situ hybridization studies have demonstrated expression of SSTR1–4 throughout the neocortex, hippocampus, and amygdala. This is consistent with the proposed role of SST in regulating complex integrative activities such as locomotion, learning, and memory. The expression of the same subtypes in the piriform cortex of the primary olfactory cortex in rat suggests that SSTRs may play a role in the processing of primary sensory information. In addition, high levels of SSTR2 and SSTR4 mRNA in the habenula (relay nucleus between the basal ganglia and mesolimbic structures and serotonin-containing cell bodies of the raphe nuclei) suggest that SST may play a role in maintaining communication between these brain regions *(20,21)*. All five SSTR mRNAs are expressed in the hypothalamus, suggesting that these receptors are involved in the regulation of autonomic and neuroendocrine function *(2,22,23)*. High levels of SSTR2 mRNA are present in the arcuate nucleus of the hypothalamus that is known to contain GH-releasing factor (GHRF) neurons that are typically feedback-regulated by SST *(24)*. Pan-receptor expression also occurs in the pituitary gland, where the most frequently expressed SSTR subtype is type 2, followed by types 1 and 3. The granular cell layer in the cerebellum, an area with inhibitory interneurons that modulate Purkinje neuron activity, is characterized by high levels of SSTR3 transcript, but the significance of this finding is unclear *(2,25)*.

1.5.2. PERIPHERY

SSTR mRNA has also been identified in several peripheral tissues, most notably tissue from the GI tract, pancreas, spleen, heart, lungs, skeletal muscle, placenta, kidneys, and adrenal glands *(26)*. In the GI tract, SSTR are found in high density in the gut mucosa and in the myenteric and submucosal plexi, where SST has been shown to inhibit cholinergic transmission *(16)*. SSTR is also expressed in the gut-associated lymphoid tissue (GALT), including palatine tonsils, ileal Peyer's patches, vermiform appendix, and colonic solitary lymphatic follicles, where SST may have an inhibitive role on immunoglobulin synthesis. The most frequently expressed receptor in the spleen is SSTR3, whereas the adrenal glands express high levels of SSTR2 and the heart expresses high levels of SSTR4 mRNA. Receptor types 1 and 4 are present in the lung, but kidneys (renal tubular cells and vasa recta) express SSTR1 and SSTR2 *(27)*. The SSTR subtype 1 is expressed by liver, whereas the human placenta as well as the fetal and adult lung display predominantly SSTR4 *(28,29)*. Skeletal muscle expresses SSTR5 *(2,11)*, whereas SSTR3 and SSTR5 have been identified in T lymphocytes *(30)*.

1.6. Distribution in Tumor Tissues

SSTRs are expressed in a variety of pathological conditions including neuroendocrine and non-neuroendocrine tumors as well as in tissues that are "inflamed" (Fig. 2). Neuroendocrine tumors [carcinoids, gastrinomas, pituitary tumors, pancreatic endocrine tumors (PETs), and paragangliomas] frequently overexpress SSTR2 and SSTR4, whereas intestinal adenocarcinomas predominantly overexpress one or both SSTR3 or SSTR4 *(31)*.

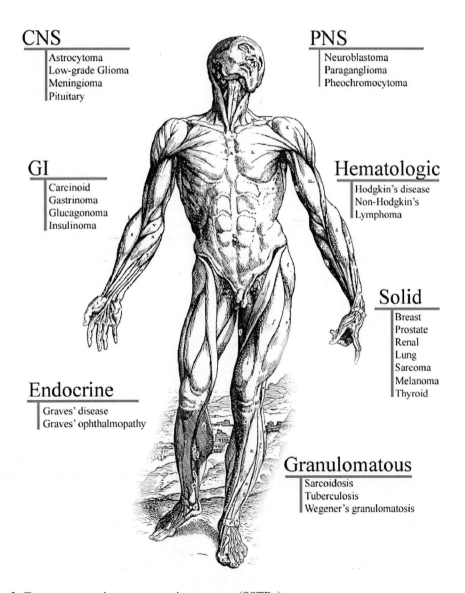

Fig. 2. Tumors expressing somatostatin receptors (SSTRs).

Elevated expression of SSTR1 occurs in prostate cancers and in many sarcomas, whereas the majority of neuroblastomas, medulloblastomas, breast cancers, meningiomas, paragangliomas, renal cell carcinomas, lymphomas, hepatocellular carcinomas (HCCs), and small-cell lung cancers (SCLCs) express SSTR2. Inactive pituitary adenomas frequently express SSTR3. Interestingly, SSTR4 is not often expressed in these human cancers. The simultaneous expression of multiple SSTRs subtypes is evident in GH-producing pituitary adenomas (especially SSTR2 and SSTR5), pheochromocytomas, hormone-producing gastroenteropancreatic (GEP) tumors, and gastric cancers *(32)*.

Other tumors with relatively high SSTR expression include medullary thyroid carcinomas (MTCs), adenocarcinomas of the breast and ovary, as well as CNS gliomas *(33)*. SSTRs may also be found on inflammatory and immune cells and in diseases such as sarcoidosis *(16)*.

SSTR positivity has been identified in primary human colonic adenocarcinomas although tumors with undifferentiated phenotypes do not express them frequently *(16)*. Malignant lymphomas in general, as well as GI lymphomas specifically, have been shown to express SSTR in 90% of cases investigated in vitro and in most in vivo studies *(34)*. Receptor-positive disease includes low-, intermediate-, and high-grade Hodgkin's disease (HD) and non-Hodgkin's lymphomas (NHLs).

Irrespective of tumor type, primary tumors and their metastases demonstrate fairly comparable receptor profiles, with between 90 and 100% expression of type 2 and 4 receptors and 60 and 70% expression of type 1 and 5 (Table 1). Type 3 receptors are expressed in approximately 40% of primary tumors and 50% of metastatic ones. The clinical implication of this observation is that both tumor primaries and metastases should be equally sensitive to SST analogue therapy *(16)*. By extension, small needle biopsies of liver metastases in patients with GEP tumors may be helpful in determining the SSTR status of the primary lesion.

Table 1
In Vitro Expression of Somatostatin Receptor (SSTR) Subtypes in Endocrine Pancreatic Tumors

Tumors	SSTR1	SSTR2	SSTR3	SSTR4	SSTR5
All tumors	68	86	46	93	57
Insulinoma	33	100	33	100	67
Gastrinoma	33	50	17	83	50
Glucagonoma	67	100	67	67	67
VIPoma	100	100	100	100	100
Non-functioning (well differentiated)	80	100	40	100	60
Non-functioning (poorly differentiated)	100	80	80	100	40
Well differentiated	61	87	39	91	61
Poorly differentiated	100	80	80	100	40
Primary lesions	70	80	40	100	60
Metastases	67	89	50	89	56
Previously treated	71	86	50	86	64
Untreated tumors	64	86	43	100	50

Reproduced with permission (11).

SSTRs are differentially distributed in well-differentiated tumors versus poorly differentiated tumors and as such may exert prognostic significance. Although 100% of poorly differentiated PETs express both type 1 and type 3 receptors, only 61% of well-differentiated malignancies express SSTR1 and 39% SSTR3. Receptor expression of subtypes 2 and 4 are, however, more comparable, with between 80 and 90% of tumors expressing type 2 and 90 and 100% expressing type 4 receptors, irrespective of the degree of differentiation.

In a series of colonic adenocarcinomas, a high density of vascular SSTRs is evident in vessels in immediate proximity to tumors (35). The receptor density decreases exponentially with increasing distance of the vessels from the tumor foci, suggesting a locally mediated phenomenon related to the tumor itself. A recent study that included 215 primary tumors and 25 metastases of various origin has suggested that the expression of SSTRs in peritumoral veins is a general phenomenon with all MTCs, colonic, and gastric cancers expressing SSTRs in peritumoral veins (36). The majority of parathyroid adenomas, renal cell cancers, melanomas, sarcomas, breast cancers, and prostate cancers had SSTRs in peritumoral veins, whereas GEP tumors or ovarian cancers rarely did. In addition, SSTRs were identified in veins surrounding lymph node, bone, and lung metastases of various tumor types. Of note, this study failed to identify receptors in arteries although other reports indicate that angiogenic vessels as well as peritumoral vessels predominantly express SSTR2 (37).

There is strong evidence that even non-tumoral lesions may express SSTRs. Active granulomas in sarcoidosis express SSTRs on the epithelioid cells (38), whereas inactive or successfully treated fibrosing granulomas devoid of epithelioid cells lack these receptors. Inflamed joints in active rheumatoid arthritis express SSTRs, preferentially located in the proliferating synovial vessels (39). SSTRs have also been demonstrated in the intestinal intramural veins of patients with inflammatory bowel disease (Crohn's disease and ulcerative colitis) (40). In addition, SSTR scintigraphy (SRS) occasionally identifies other non-neoplastic conditions and inflammatory processes such as gallbladder and thyroid abnormalities, accessory spleens, renal parapelvic cysts, and recent cerebro vascular accidents (CVAs) and activity at the site of a recent surgical incision (41). The expression of SSTRs is therefore not specific for tumoral pathologies but appears to be implicated in inflammation (infection) and healing.

2. SCINTIGRAPHY AND AUTORADIOGRAPHY

2.1. Carcinoid

Carcinoid tumors (a heterogenous miscellany of neuroendocrine neoplasms) are the most common type of neuroendocrine malignancy and are most frequently found in the GI tract but can also occur at numerous other sites including the lung and appendix (42). Systemic manifestations of carcinoid disease include flushing, sweating, diarrhea, paroxysms, or bronchospasm and are due to the secretion of bioactive substances by either the primary lesion or metastases. In addition, cardiac failure may occur as a result of tricuspid or pulmonary valvular fibrosis. In vitro studies indicate that approximately 90% of carcinoid tumors express SSTRs and that octreotide scintigraphy detects 90–100% of these lesions (Table 2) (8). However, SSTR expression is undetectable in certain lesions (either primary or metastatic) and reflects individual tumor biology.

Table 2
Somatostain Receptor (SSTR) Scintigraphy (SRS) Positivity in Neoplasia

Tumor	Scintigraphy positive (%)
Carcinoid	100
Insulinoma	60
Gastrinoma	60–90
Somatostatinoma	100
Glucagonoma	70
VIPoma	86
Non-functioning gastroenteropancreatic neuroendocrine tumors	84
Medullary thyroid	21–70
Small cell lung cancer	60–100
Pheochromocytoma	73–90
Neuroblastoma	65–90
Paraganglioma	92–95
Astrocytoma	0–84[a]
Meningioma	70–100
Pituitary adenoma	100
Non-Hodgkin's lymphoma	35–62
Melanoma	84
Merkel cell	80
Sarcoid	97
Gastric	67
Hepatocellular carcinoma	41
Breast	21–75
Renal cell	72
Prostate	14–86[b]
Ovarian	24–65[c]
Nasopharyngeal	75

[a] Receptor expression varies based on histologic grade and receptor type.
[b] SSTR1 (86%),
[c] SSTR2 (14%).

2.2. Pancreatic Islet Cell Tumors

Islet cell tumors arise from cells with some common characteristics [amine precursor uptake and decarboxylation (APUD)] and approximately 80% are hormonally active. Between 90 and 100% of islet cell tumors express SSTRs while pancreatic adenocarcinomas rarely if ever express these receptors (Table 1) (8,43). In general, these tumors behave in a malignant fashion with the exception of some lesions, particularly insulinomas. Metastatic disease is frequently present at the time of diagnosis.

2.2.1. INSULINOMA

This is the most common functioning islet cell tumor. Insulinomas arising from the β cells of the pancreas are frequently solitary and small (<2 cm) lesions of which 85–90% are benign (44). Ectopic insulinomas are mostly associated with the MEN-1 syndrome. Patients present with hypoglycemic symptoms or the sequelae thereof. In vitro studies demonstrate approximately 72% SSTR expression, whereas in vivo scintigraphy identifies these lesions in approximately 60% of cases *(8)*. About 50% of insulinomas express SSTR1, 70% express SSTR2 and SSTR5, and 15–20% express SSTR3 and SSTR4. However, as many as one-third of insulinomas do not express SSTR2 and SSTR5 *(45)*. Furthermore, non-malignant insulinomas rarely express SSTRs, thus the sensitivity of OctreoScan may be as low as 50% in this disease *(44)*. Nevertheless, in malignant insulinomas, SRS is recommended for detection and staging if feasible *(46)*.

2.2.2. GASTRINOMA

Gastrinomas are the second most common functional islet cell tumor and are principally located in the duodenal wall and pancreas. The majority are malignant and approximately 20% are associated with MEN-1 syndrome. Hypergastrinemia results in parietal cell hyperplasia and gastric acid hypersecretion with intractable ulcer disease (Zollinger–Ellison syndrome) *(47)*. SRS identifies these lesions in 80–90% of cases *(8)*. SRS is the principal imaging modality because of its high sensitivity (60–90%) in detecting the primary lesions as well as metastases *(48)*. Of note is the observation that OctreoScan alters patient management in 47% of cases *(47)*. About 50% of gastrinomas express SSTR2 and SSTR5, 33% express SSTR1, 17% express SSTR3, and 83% express SSTR4 *(11)*.

2.2.3. SOMATOSTATINOMA

Somatostatinomas are very rare (~100 cases described) neuroendocrine tumors of D-cell origin with excessive secretion of SST. About 70% are located in the pancreas, of which two-thirds arise in the head of the gland; the remaining 30% develop in the duodenum, ampulla, and small bowel *(49,50)*. Tumors are detectable by SRS suggesting the presence of functional SSTRs.

2.2.4. GLUCAGONOMA

Glucagonomas are rare endocrine pancreatic tumors associated with necrolytic migratory erythema, cheilitis, diabetes mellitus, anemia, weight loss, venous thrombosis, and neuropsychiatric symptoms *(51)*. As a consequence of this protean and non-specific symptomatology, diagnosis is often delayed. SRS is sensitive (70%) for tumor detection and facilitates management strategy by defining disease spread. Sixty-seven percent of glucagonomas express SSTR subtypes 1, 3, 4, and 5, whereas 100% of these tumors express SSTR subtype 2 *(11)*.

2.2.5. VIPOMAS

Pancreatic VIPomas are usually solitary, >3 cm in diameter, and 75% occur in the tail of the pancreas. Approximately 60–70% of VIPomas are metastatic at diagnosis and 5% are part of the MEN-1 syndrome *(52)*. SRS may be utilized for the detection of primary and distant metastases. OctreoScan sensitivity is about 86%, but it is much lower for lesions smaller than 1 cm *(53)*. The majority of VIPomas express all five SSTR subtypes *(11)*.

2.2.6. NON-FUNCTIONING ISLET CELL TUMORS

These lesions are relatively slow growing and are usually quite large (>6 cm) by the time of presentation. Approximately 5% of patients with von Hippel–Lindau disease have concurrent non-functioning PETs, and SRS identifies 84% of these lesions (Table 1) *(54)*.

2.3. Other Neuroendocrine Tumors

2.3.1. MTC

MTC arises from the parafollicular C cells that secrete calcitonin. The sensitivity of SRS in detecting MTC lesions is variable (50–70%) and may be related to loss of SSTRs as the tumor becomes less differentiated (Table 2) *(55,56)*. Carcinoembryonic antigen (CEA) and calcitonin levels are used to monitor disease, and pronounced elevations of the former are associated with tumors that are more aggressive. A higher ratio of calcitonin to CEA levels may also be associated with a greater likelihood for a positive SRS examination *(57)*. The addition of SST analogue imaging to the detection of MTC has, however, not been demonstrated to significantly increase detection of metastatic lesions *(58–60)*. In addition, scintigraphy is more frequently positive with high serum tumor markers and large tumors and therefore seems less suitable for showing microscopic disease *(56,61)*. Although papillary thyroid cancer (PTC), follicular thyroid cancer (FTC), and anaplastic thyroid cancer as well as Hürthle cell carcinomas do not belong to the group of traditional neuroendocrine tumors, the majority with these cancers show uptake of radiolabeled octreotide on SRS *(62,63)*.

Between 21 and 38% of MTCs express SSTRs in vitro *(8)*. MTCs from 19 patients were analyzed by receptor autoradiography with iodinated radioligands of octreotide and SST-28 *(64)*. Four of 19 cases were receptor positive with the octreotide radioligand, and an additional four tumors were imaged with the SST-28 radiotracer. However, no clinical correlation with tumor SSTR expression was evident for outcome or survival. High expression of SSTR subtypes 1, 3, 4, and 5 is present in both PTC and follicular thyroid adenoma *(65)*.

2.3.2. SCLCs

SSTRs are found in 50–75% of SCLCs *(66)*, and SRS detects between 63 and 100% of primary tumors, but the degree of uptake is variable and independent of lesion size *(8,66)*. Compared with conventional imaging, SRS visualized the primary tumor in 96% of examinations in a multicenter series of 100 patients *(66)*. Regional and distant metastases were detected in 60 and 45%, respectively. Some reports document the unexpected identification of brain metastases, whereas others failed to derive any additional information with SRS *(67–69)*. As SST uptake by the primary tumor was noted to be affected by chemotherapy, it has been suggested that SRS may be used to follow the course of SCLC *(66)*. Inclusion of SRS in the staging protocol of patients with SCLC may also lead to upstaging in some of the patients with limited disease, as illustrated by a group of 14 patients who appeared to have limited disease on conventional imaging, but five were noted to have brain metastases following SRS *(67)*. As SSTRs are absent (in vitro or in vivo) on most non-SCLCs, uptake by these lesions may be related to receptor-positive immune cells surrounding the tumor *(8)*.

Close to 60% of SCLCs demonstrate SSTR expression in vitro *(8)*. Biopsy specimens obtained from eight patients with lung cancer were tested for SSTRs by autoradiography *(70)*. SSTRs were detected in two of three patients with SCLC but in none of five patients with non-SCLC. In the former group, SSTR binding was evident only in tumor foci and not in surrounding stroma or normal lung parenchyma.

2.3.3. PHEOCHROMOCYTOMAS

SRS detected tumors in approximately 90% of patients with pheochromocytoma *(8)*. A limitation of SRS use for localization of these tumor types in the adrenal gland is the relatively high radioligand accumulation in the adjacent kidneys, which hampers visualization of small lesions *(57)*. An in vitro study demonstrated SSTR positivity in 73% of cases *(8)*. SSTR quantification by autoradiography in a sample of 33 pheochromocytomas and 5 normal adrenals demonstrated that all five SSTR subtype mRNAs were detectable although mRNA expression was highest for SSTR2 and SSTR4 *(71)*. The level of SSTR5 mRNA was, however, higher in normal adrenals (21%) than in pheochromocytomas (6%). In a separate study, 51 adrenal pheochromocytomas were evaluated for SSTR content with in vitro autoradiography on tissue sections from surgically removed tumors, using iodinated ^{125}I-labeled octreotide. Seventy-three percent of pheochromocytomas were SSTR positive *(72)*. There was no correlation between the receptor status and tumor size, benign versus malignant tumors or urinary metanephrine excretion.

2.3.4. NEUROBLASTOMAS

The detection of neuroblastomas by SRS is approximately 90%. It has been reported that patients with SSTR-negative tumors generally have a worse prognosis *(73)*. In vitro studies demonstrate 65% receptor expression *(8)*. Autoradiography with ^{125}I-labeled octreotide or SST-28 in 30 children with neuroblastoma identified 23 tumors that expressed SSTRs. Interestingly, receptor expression statistically correlated with survival. SSTRs were expressed more frequently in tumors with low disease staging than in those with no evidence of N-myc amplification *(74)*. Receptor analysis with radiolabeled ligands concluded that SST-14 mainly binds to SSTR2 in human neuroblastoma tumors *(75)*.

2.3.5. PARAGANGLIOMAS

Paragangliomas, often referred to as chemodectomas, are rare tumors of neural crest origin that most commonly arise from parasympathetic ganglia, most commonly in the head and neck *(76)*. Their origin is the chemoreceptor tissue of the carotid body, glomus jugular, or aortic body. As in other neuroendocrine tumors, paragangliomas overexpress SSTRs and, as such, may be visualized by SRS *(72,77–80)*. Using octreotide scintigraphy, 95% were correctly identified with detection of lesions as small as 1 cm *(81)*. Previously, unidentified sites were localized in 30–36% by SRS. SRS was also used to detect local recurrence or residual tumor following surgery that may occur in 15–30% *(82)*. In vitro autoradiography on surgically removed tissue sections from 14 paragangliomas demonstrated 93% to be SSTR positive *(72)*.

2.4. CNS Tumors

A prerequisite for the localization of CNS lesions with SRS is a locally open blood–brain barrier, otherwise octreotide, a polar water soluble compound, will not accumulate in the tumor *(57)*. About 70% of astrocytomas are detectable, with low-grade, well-differentiated astrocytomas (I and II) being best visualized, whereas more de-differentiated lesions (grades III and IV) were identified less frequently. This reflects the observation that a large percentage of low-grade tumors express SSTRs

(80%) compared with rare expression in high-grade lesions *(57,83,84)*. Immunohisto-chemical and western blot analysis of 50 astrocytomas (8 diffuse, 10 anaplastic, and 32 glioblastoma multiforme) demonstrated minimal expression of SSTR1 and SSTR2 in diffuse and anaplastic tumors *(85)*. In glioblastoma multiforme, however, SSTR1 and SSTR2 are present in 66 and 44% of cases, respectively. Loss of differentiation was thus significantly associated with increased expression of these receptor subtypes, and the presence of the five SSTR subtypes did not significantly influence survival time in 14 patients with glioblastoma multiforme. An inverse relationship between the presence of SST and epidermal growth factor receptors has been observed. In grade III astrocytomas, both receptors are present *(72)*.

One hundred percent of meningiomas (extra-axial lesions) were detected, and the intensity of the scintigraphic signal correlated well with the tumor SSTR density *(84)*. Fifty-nine of 63 meningiomas showed moderate to intense tracer uptake *(86)*. Immunore-active staining on 40 randomly selected meningiomas demonstrated the presence of SSTR2 in 70%. In contrast, all other SSTRs were only detected sporadically *(87)*.

SSTR1 was expressed in 83% and SSTR2 in 76% of central primitive neuroec-todermal tumors (cPNETs) including medulloblastomas *(88)*. This finding suggests that SST may be involved in the regulation of proliferation and differentiation in these developmental tumors. This observation was corroborated by a study of 52 brain tumors in which SSTRs were identified in most of the differentiated glial-derived tumors such as astrocytomas and oligodendrogliomas but not in the poorly differen-tiated glioblastomas *(89)*. Tumors originating from neuroblasts (ganglioneuroblastoma and medulloblastoma) are characterized by a high density of SSTRs, whereas neuri-nomas and neurofibromas as well as the ependymomas, one teratoma, and one plexus papilloma lacked SSTR.

The expression of the five SSTR subtypes was evaluated in tissue from glial tumors (glioblastomas or oligodendrogliomas), medulloblastomas, and normal human cortex *(90)*. Transcript levels of all receptors were high in the cortex and in oligo-dendroglioma tissue. Medulloblastoma tumoral cells expressed SSTR2 and SSTR3 transcripts at high levels in comparison with gliomas, where SSTR expression was restricted to endothelial cells on proliferating vessels. These cells displayed both SSTR2 and SSTR3 subtypes, whereas the parenchyma and reactive microglia only expressed SSTR2.

2.5. Pituitary Adenomas

Virtually all GH-producing pituitary adenomas express SSTRs (particularly SSTR2) in vitro, and the majority are SRS positive *(8,91,92)*. TSH-secreting pituitary tumors can also be visualized with nearly 100% sensitivity. Non-functioning pituitary tumors also express SSTRs, and 75% can be visualized by SRS *(8)*. However, metastases from SSTR-positive neoplasms, parasellar meningiomas, lymphomas, or granulomatous diseases of the pituitary may also be positive thereby limiting the diagnostic specificity of SRS in pituitary tumors *(92)*.

2.6. Adenocarcinomas

2.6.1. GI

Plasma membranes from specimens of tumor and normal mucosa from 51 patients undergoing surgical resection for malignancy (28 gastric and 23 colorectal) were assessed using a competitive displacement assay with ^{125}I-labeled SST-14 *(93)*. Low-affinity, high-capacity SST binding to the plasma membranes was observed in 79% of the gastric cancers and 74% of the colorectal cancers. A similar affinity and binding capacity was demonstrable in normal mucosal samples *(93)*. Similar studies with ^{125}I-labeled octreotide or [125]I-[LTT]-SS-28 in 27 surgically resected gastric tumor samples detected SSTRs in 67% of gastric adenocarcinomas, 9 of which were identified equally well with either ligand, suggesting that these tumors express SSTR2 *(94)*. In comparison, colon carcinomas expressed SSTRs only in a minority of cases (8%) and at low density.

2.6.2. Hepatocellular

In vitro receptor autoradiography with radiolabeled octreotide and [125]I-[LTT]-SS-28 in tissue sections from 59 cases of HCC demonstrated SSTR expression in 41% of HCC *(95)*. The SSTRs showed high affinity for SST and octreotide, but their density was low compared with that found in liver metastases of neuroendocrine tumors.

2.6.3. Breast

As many as 50% of breast carcinomas demonstrate SSTR positivity in vitro *(8)*. In a study of 52 patients with stages I and II breast cancers (lesions <2 cm and between 2 and 5 cm, respectively), SRS localized 75% (39/52) of tumors *(96)*, and images of the axillae in another study showed non-palpable cancer-containing lymph nodes in 4 of 13 patients with subsequent histologically proven metastases *(8)*.

SSTRs were measured with [125]I-[Tyr3]-octreotide in 342 breast tumor samples *(97)*. In a group of 158 "small" tumor samples (mean size 14 mm^2), 21% were SSTR positive. In a group of 72 "large" tumor samples (mean size 180 mm^2), 46% were SSTR positive. In this second group, >50% of the tumors had a heterogeneous distribution of SSTRs, but the receptors were invariably located on tumor tissue and were not seen on adjacent normal lobules and ducts. Furthermore, a retrospective study reported increased 5-year disease-free survival in patients with SSTR-positive tumors (82%) versus SSTR-negative ones (46%) *(98)*. The observation of faint, bilateral, and diffuse physiologic breast uptake in women with non-cancerous breasts in about 15% of patients 24 h after injection of [^{111}In-DTPA^0]octreotide is of interest although the underlying cause is unknown *(57)*. This distribution pattern is, however, different from the more localized accumulation at the site of breast cancer *(96)*.

2.6.4. Renal Cell

The SSTR status of 39 surgically removed human renal cell carcinomas was evaluated using ^{123}I-labeled octreotide *(99)*. Although 72% were SSTR positive, there was no correlation between the SSTR profile and the histopathological type or grade of the tumor or the tumor node metastasis stage of the disease. However, numerous cases considered to be of poor prognosis were SSTR positive.

2.6.5. Prostate

Benign as well as malignant human prostatic tissues have been evaluated to assess SSTR expression *(100)*. In vitro receptor autoradiography with ^{125}I-labeled octreotide and [125]I-[LTT]-SS-28 detected SSTRs in smooth muscles of the stroma from normal and hyperplastic prostates, whereas the glands themselves did not express the receptors.

Muscular nodules were strongly receptor positive as well. The receptors showed high affinity and high specificity for SST-14, SST-28, and octreotide, suggesting the presence of the SSTR2. Primary prostate cancers were negative for SSTRs when using ^{125}I-labeled octreotide; however, receptors were detected using ^{125}I-[LTT]-SS-28. These receptors were located on tumor cells, and in situ hybridization studies revealed preferential expression of SSTR1. Primary human prostate cancers, therefore, express a different SSTR subtype than benign prostate tissue. The expression of SSTR transcripts was investigated in 22 specimens of prostate cancers by RT-PCR *(101)*. Transcript for SSTR1 was detected in 86% of samples, whereas SSTR2 mRNA was detected in 14% of samples and SSTR5 mRNA in 64% of samples.

2.6.6. OVARY

Ligand competition assays with the SST analogue RC-160 (vapreotide) and RT-PCR were used to investigate the SSTR profile of 17 surgical specimens of human epithelial ovarian cancer and 2 human ovarian cancer lines *(102)*. Transcript for SSTR1 and SSTR2 was detected in 65% of the ovarian cancer specimens, whereas the incidence of SSTR3 and SSTR5 was 41 and 24%, respectively. Specific receptors for RC-160 were also found in xenografts of human ovarian cancer lines. The presence of these subtypes may be useful in assessing subtype-specific radiolabeled ligands complexed with SST analogues.

2.6.7. NASOPHARYNGEAL

Nasopharynx biopsies were obtained from 12 nasopharyngeal carcinoma (NPC) patients and 5 controls *(103)*. SSTR autoradiography was performed using radiolabeled octreotide and SST-28 ligands. Seventy-five percent of NPC samples demonstrated moderate to high expression of SSTRs, predominantly SSTR2. The five non-neoplastic samples, consisting mostly of granulomatous tissue, did not express measurable amounts of SSTRs.

2.7. Other Tumors and Paraneoplastic Syndromes

2.7.1. LYMPHOMAS

In a certain subpopulation of patients with HD and NHL, one or more lesions may be SSTR positive *(104)*. The mean regional tumor uptake of [^{111}In-DTPA]octreotide in lymphomas, however, can be as much as 10 times lower as compared with that in GEPs *(105,106)*.

The overall sensitivity of SRS for detecting HD is between 70 and 100% *(107)*. In 126 consecutive untreated patients with histologically proven HD, the results of SRS were compared with conventional examinations and SRS was positive in all *(108)*. The sensitivity varied from 98% for supradiaphragmatic lesions to 67% for infradiaphragmatic lesions, with a critical size of approximately 2 cm required for node detection *(107)*. Subdiaphragmatic lesion detection, however, can be limited by the large amount of background activity (liver/kidney/spleen). Above the diaphragm, SRS was superior to computerized tomography (CT) and ultrasonography for the detection of HD. In stages I and II supradiaphragmatic disease, SRS detected advanced disease in 18% of patients, resulting in upstaging of tumors to stage III or IV, thus directly influencing patient management. However, other series have shown less promising results, with detection rates of only 58% at confirmed extramedullary tumor sites *(105)*. Another limitation of SRS in lymphoma detection is that differentiation from adenopathy secondary to granulomatous or inflammatory disease is not possible.

The overall detection sensitivity of SRS in NHL was less than that for HD (35–62%). As for the HD studies, supradiaphragmatic abnormalities were better detected *(109)*. High-grade lesions were more readily imaged compared with low-grade ones (44 versus 29%), whereas detection of bone marrow involvement was poor. In a separate study, in vitro autoradiography on 30 surgical specimens demonstrated 87% SSTR positivity *(8)*. In these B-cell lymphomas, SSTR positivity was 10/11 in the low-grade group, 8/8 intermediate grade, and 7/10 high grade. Although SRS was positive in a large proportion of low-grade NHL, in most patients, only part of the lesion could be visualized, limiting the role of SRS in NHL *(57,109)*.

2.7.2. Melanomas

Analysis of SSTR subtypes in 17 patients revealed SSTR1 expression in 96% of tumors, SSTR2 in 83%, SSTR3 in 61%, SSTR4 in 57%, and SSTR5 in 9% *(110)*. A separate study demonstrated 16/19 positive octreotide scintigrams in melanomas *(111)*. However, the exact impact of SRS on staging and management remains to be determined.

2.7.3. Thymic Tumors

High uptake of indium-labeled octreotide has been noted in tumors of the thymus *(112)*. In contrast to neuroendocrine tumors, thymic tumors express high levels of SSTR3 in vitro. This may be relevant to the future use of receptor-specific ligands for these lesions *(113)*.

2.7.4. Mesenchymal Tumors

In vitro receptor autoradiography on cryostat sections was performed on 64 primary or metastatic human mesenchymal tumors using ^{125}I-labeled octreotide and SST-28 *(114)*. SSTRs were identified in bone and vascular/perivascular tumors in 3/3 osteosarcomas, 1/1 giant cell tumor, 2/2 angiosarcomas, 4/4 hemangiopericytomas, 2/2 synovial sarcomas, 2/5 histiocytomas, and in several muscle cell tumors (1/2 leiomyomas, 2/4 leiomyosarcomas, and 3/5 rhabdomyosarcomas) but were absent in 4 liposarcomas, 3 mesotheliomas, 3 chondrosarcomas, 10 Ewing sarcomas, 11 schwannomas, and 5 Wilms' tumors. The receptors were located on neoplastic cells and had high affinity and high specificity for SST-14 and SST-28 as well as for octreotide, indicating the expression of the SSTR2.

2.7.5. Merkel Cell Tumors

Trabecular carcinomas of the skin or Merkel cell tumors are aggressive neoplasms that tend to occur in sun-exposed skin, with the majority displaying neuroendocrine characteristics. Overall detection of tumor sites was 80% by SRS although small lesions (<0.5 cm) were not detected *(8)*. In an SRS study of five patients, all lesions previously imaged by CT, ultrasound (US), or both were positive, and in two cases, additional metastatic tumor sites were recognized *(115)*.

2.7.6. Cushing's Syndrome

SRS successfully identified the primary ectopic CRH-secreting tumors or their metastases in 8/10 patients, but in two CRH-secreting bronchial carcinoids, the tumors could not be visualized. SRS did not detect any lesions in eight patients with CRH-secreting pituitary tumors *(116)*. Whereas isolated case studies have reported the utility of SRS in localizing CRH-secreting bronchial carcinoids, others have concluded that although this modality may be helpful in selected instances, it offers no particular advantage over conventional imaging *(117–119)*.

2.7.7. Granulomatous Disease

In a cross-sectional assessment of 46 individuals with sarcoidosis, known mediastinal, hilar, and interstitial lesions were recognized in 36/37 *(120)*. Such pathology was also found in seven other patients who had normal chest X-rays. There was no correlation between the degree of radioactive accumulation in the thorax and specific patterns of pathological uptake with disease severity or clinical course. The degree of uptake of radioactivity in the parotid glands, however, was correlated with significantly higher serum angiotensin-converting enzyme levels *(118)*.

2.7.8. Graves' Disease

In Graves' hyperthyroidism, accumulation of radiolabeled octreotide in the thyroid gland was markedly increased and correlated with serum levels of free thyroxine- and thyrotropin-binding inhibiting immunoglobulins *(57)*. SRS demonstrated high orbital uptake of radioactivity in clinically active disease, and low uptake when ophthalmopathy was inactive *(121,122)*. Single photon emission computed tomography (SPECT) was found to be necessary for accurate interpretation of orbital scintigraphy. There was also a correlation between orbital SRS uptake and the Clinical Activity Score and Total Eye Score *(121,123)*. SRS in Graves' disease may be of clinical utility for identifying patients with ophthalmopathy who might benefit from treatment with octreotide *(122,123)*.

3. SST ANALOGUES

Currently, analogues of SST rather than the naturally occurring SST-14 and SST-28 peptides are used to study SSTR biochemistry, as SST-14 and SST-28 lack tyrosine residues that can be iodinated. Some of the first synthetic SST analogues to be synthesized were tyrosine-substituted [^{125}I-Tyr^1]SST, [^{125}I-Tyr^{11}]SST, and [^{125}I-N-Tyr^0]SST *(124)*. Subsequent functional studies demonstrated the necessity of the Trp-Lys dipeptide sequence for high-affinity receptor binding *(24)*, which has consequently led to the synthesis of stable SST analogues such as SMS 201-955 (octreotide) *(125)*, RC-160 (vapreotide) *(126)*, MK 678 (seglitide) *(127)*, BIM 23014 (lanreotide), and SOM 230 *(128)* (Table 3).

3.1. Cold

3.1.1. Short-Acting Formulations

Owing to the short half-life (1–3 min) of native SST, a principal focus of research between the years 1982 and 2001 has been synthesis of analogues with longer physiologic activity and broad receptor coverage. Octreotide, lanreotide, and vapreotide

Table 3
Human Somatostatin Receptor (SSTR)-Binding Affinities

Ligand	SSTR1	SSTR2	SSTR3	SSTR4	SSTR5
Somatostatin-14	1.1	1.3	1.6	0.5	0.9
Somatostatin-28	2.2	4.1	6.1	1.1	0.07
Octreotide	>1000	2.1	4.4	>1000	5.6
Vapreotide	>1000	5.4	30.9	45.0	0.7
Lanreotide	>1000	1.8	43.0	66.0	0.6
SOM 230 (128)	9.3	1.0	1.5	>100	0.16
^{111}InDTPA $-_d$-Phe^1-octreotide (OctreoScan)	>1000	1.5	30	>1000	1
[^{90}Y] DOTA-Tyr^3-octreotate (DOTATOC)	>10,000	1.6	>1000	523	187
[^{90Y}] DOTA-lanreotide (DOTALAN)	215	4.3	5.1	3.8	10

All values are IC_{50} SEM in nanometer. IC_{50} is the concentration of ligand causing half-maximal inhibition of somatostatin-14 binding to engineered cells expressing individual SSTRs. Lower values indicate stronger complex-receptor affinities. Reproduced with permission (11,154).

are all cyclic octapeptides resistant to peptidases with prolonged half-lives (1.5–2 h) (11,112). Currently, available SST analogues are characterized by exhibiting a very high affinity for SSTR2 and SSTR5, moderate affinity for SSTR3, and low affinity for SSTR1 and SSTR4 (Table 3).

The first commercially available SST analogue, octreotide, was synthesized in 1982 and formulated for subcutaneous (sc) administration (129). It is an eight amino acid structure, and its clinical utility is facilitated by a stereoisomer substitution at its first and fourth amino acids, conferring a half-life approximately 80 times greater than native SST (125). The compound has the highest affinity for SSTR2 and SSTR5, with the affinity for SSTR2 approximately 10 times greater than for SSTR5 (130). The sc formulation, administered by two to four injections daily, is the shortest acting of the available SST analogues (129). A medium-acting SST analogue, lanreotide SR, whose structure and binding profile are similar to octreotide with the exception of three amino acid substitutions: D-Phe is replaced by D-βnal, Phe by Tyr, and Thr by Val. This compound is longer acting than the sc octreotide because the drug is encapsulated in microspheres that provide release over 10–14 days after intramuscular (IM) administration (130). Lanreotide SR is provided in a 30-mg dose only, and the pharmacological effect is manipulated by changing the dosing interval between 7 and 14 days (131).

3.1.2. LONG-ACTING FORMULATIONS

Octreotide is also available in a long-acting formulation, octreotide long acting repeatable (LAR) (132). As with lanreotide SR, the active compound is encapsulated in microspheres of a biodegradable polymer. After an IM injection, drug levels begin to rise over 7–14 days and plateau for 20–30 days (133). The development of depot formulation of octreotide administered up to 30–60 mg once every 4 weeks has, to a large extent, eliminated the need for daily injections. However, symptom breakthrough is an issue and has been described in the time period before a steady state is achieved or in the last week of the cycle. This may necessitate "rescue" with an additional 50 or 100 μg (up to 1 mg) doses of a short-acting analogue such as sc octreotide or by increasing the dose and/or frequency of the depot injection.

A new slow-release depot preparation of lanreotide, Lanreotide Autogel, which is administered subcutaneously at doses of up to 120 mg once a month, has recently been introduced *(134)*. Like octreotide LAR, Lanreotide Autogel also has a monthly administration schedule. The active agent is the same as lanreotide SR and as such exhibits a higher affinity for SSTR2 than SSTR5 *(134)*. Of note is the alteration in formulation whereby the drug naturally congeals into a slow-release aqueous gel that can be given sc once monthly. The pharmacological effect can be manipulated by varying the dose from 60, 90, or 120 mg with a fixed monthly administration schedule *(130)*.

The most recently synthesized SST analogue, SOM 230, is a compound with high affinity for SSTR1, SSTR2, SSTR3, and SSTR5. In binding experiments, SOM 230 has a higher affinity to SSTR1, SSTR3, and SSTR5 and a slightly lower affinity to SSTR2 compared with octreotide *(135)*. In addition, this semi-universal ligand has a 30- to 40-fold higher affinity for SSTR1 and SSTR5 than octreotide or lanreotide *(128)* and has a >7-fold longer plasma half-life than octreotide (11 versus 1.5 h) *(135)*. In several animal species, SOM 230 was a more potent inhibitor of GH and insulin-like growth factor 1 (IGF-1) than octreotide *(135)*. In rats, SOM 230 caused a stronger inhibition of adrenocorticotropic hormone (ACTH) and corticosterone secretion than octreotide *(136)*. In contrast, octreotide was more potent than SOM 230 in the inhibition of ghrelin secretion in rats *(137)*.

In patient studies, SOM 230 inhibited IGF-1 more strongly than octreotide, whereas the latter stimulated IGFBP-1 suggesting differential effects of SST analogues on metabolic pathways in acromegalics *(138)*. These suggest modulatory effects by SOM 230 on glucose homeostasis. In proliferative studies, SOM 230 has been shown to inhibit proliferation of endothelial cells human vascular endothelial cells (HUVECs), which are unaffected by octreotide *(139)*. Based on these data, it is proposed that in tissues (or tumors), where several SSTRs are expressed, SOM 230 may be more effective than octreotide.

3.2. Peptide Receptor Radionuclide Therapy

The recent introduction of systemic receptor-targeted or metabolically directed radiotherapy using a variety of isotopes for the treatment of inoperable or metastatic GEP tumors has engendered considerable interest and even early optimism *(140)*. The technique involves the coupling of a radioisotope to an SST analogue such that the conjugate may then bind to tumor cells that express specific surface receptors and thereafter undergo endocytosis. The principal is to provide a focal and effective dose of radiation that can be administered to tumor or peritumoral cells only, leaving the majority of surrounding non-neoplastic tissue intact. In addition, by using individual isotopes with different emission wavelengths, the extent of local irradiation can be "tailored" to the size range of the lesions. Although renal exposure is of some concern, kidney irradiation can be substantially decreased (20–50%) by a pre-therapy IV infusion of positively charged amino acids (L-lysine and L-arginine) and intra-therapy IV fluid loading to "flush" the system *(141,142)*. The use of lanreotide as the parent molecule for isotopic therapy causes 25% less renal exposure than noted with octreotide analogues *(143)*.

3.2.1. PRINCIPLES

The 82 stable elements can be altered to produce approximately 275 isotopes that are considered as unstable atoms *(112)*. The latter represents an artificial combination of neutrons and protons (unstable atoms) called radioisotopes. With such entropy, stability of the radioisotope nucleus is usually achieved through emission of an α or a β particle. These particle emissions may be accompanied by emission of either electromagnetic radiation or γ rays and represent the process of radioactive decay.

3.2.1.1. β Electrons. The particle penetrance of β decay is between that of α and γ decay. β Electrons (from isotopes such as ^{131}I and ^{90}Y) can penetrate about 1 cm of tissue before being halted by local electrostatic forces *(112)*.

3.2.1.2. Auger Electrons. The term "Auger effect" describes the process in which an ionized atom emits a second electron rather than a photon *(112)*. Such ejected Auger electrons typically generate only a few thousand electron volts in energy as this process occurs principally in elements of low atomic number. The effective radius of travel of these electrons is typically less than 100 nm. As such, Auger emitters (such as ^{111}In) are generally most effective when internalized by a cell *(144)*.

3.2.1.3. γ Particles. A very highly charged γ ray is produced when a parent isotope falls into a lower energy state; γ radiation is the most penetrating type of radiation. Such photons thus exhibit extensive penetration and can cause damage by ionizing all molecules in their paths. [131]I and ^{111}In exhibit γ decay. The usage of high-energy particles is limited by the risk of myelosuppression as a consequence of exposure and accumulation of such circulating agents in the marrow *(145)*.

3.2.2. PRIMARY RADIOISOTOPES

Although ^{111}In has been the most frequently employed diagnostic and therapeutic isotope, ^{90}Y and ^{177}Lu, with their distinct radioemission profiles have more recently been examined in detail.

3.2.2.1. [111]In. This isotope, which has a half-life of 2.83 days and γ decay, has been extensively used for octreotide imaging bound to numerous SST analogues, including octreotide and its substitute molecule, octreotate. In addition to γ decay, ^{111}In also emits Auger and conversion electrons and is therefore only effective for therapy if internalized within a cell *(144)*.

3.2.2.2. [90]Y. Given its prominent β-emitting profile and half-life of 2.67 days, ^{90}Y has been primarily utilized as a therapeutic agent for SSTR-positive tumors. The carrier molecule most commonly used for yttrium has been the SST ligand DOTA-d-Phe(*1*)-Tyr(*3*)-octreotide (DOTATOC) as it has significant advantages in terms of stability and easy labeling *(146)*.

3.2.2.3. [177]Lu. Attempts to amplify the therapeutic index for radiopharmaceuticals has led to interest in lutetium, an emitter of β and γ particles *(147)*. Recent studies indicate that in the majority of SSTR-positive tumors, the uptake of lutetium-labeled SST analogues is threefold to fourfold higher than ^{111}In-octreotide *(148)*. Given the higher absorbance with equivalent dosages of isotopes, it has been suggested that lutetium should be the therapeutic agent of choice. A further advantage is the relatively long half-life (6.64 days) of reactor-produced lutetium that makes it useful for in situ receptor-mediated brachytherapy *(149)*. As ^{177}Lu has a lower tissue penetration range compared with ^{90}Y, it may also be advantageous in the treatment of "smaller" neuroendocrine tumors by minimizing the therapeutic-dose exposure of cells distant from the bound and internalized radiolabeled SSTRs *(112)*. These theoretical proposals are supported by reported cure rates in rat models that are strongly correlated to initial tumor size *(148)*. Thus, single- or dual-dose treatments resulted in cure rates of 75–100% for pancreatic tumors =1 cm in diameter, whereas only 40–50% cure rates were evident with larger tumors.

3.2.3. CHELATORS

Although SST analogues are able to selectively target SSTR-positive tissues, they are not able to effectively deliver the isotopes employed for the delivery of therapeutic radiation *(112)*. Complex chelating molecules, bound to one of the terminal peptides of the SST analogue, are utilized to sequester and deliver such energy-emitting isotopes. Initially, the octapeptide analogue octreotide was used as an iodinated (^{125}I- or ^{123}I-[Tyr^3]octreotide) compound *(150)*. The addition of linking chelators (DTPA and DOTA) to octreotide improved the biodistribution profile and shifted the excretion pathway from GI to predominantly renal *(32)*.

For scintigraphy, octreotide is frequently labeled with ^{111}In-DTPA. ^{111}In-DTPA-$_D$-Phe-octreotide (OctreoScan) at a dose of approximately 200 MBq is the most widely used tracer for SST scintigraphy as it emits γ rays that are optimal for scintigraphy (and Auger electrons that may be used for radiotherapy). Following internalization through the SSTR cascade, the agent is degraded in the lysosomes to a final radiolabeled metabolite, ^{111}In-DTPA-$_D$-Phe, but this latter product cannot pass through the lysosomal or other cell membranes and remains within the cell before being finally translocated into the nucleus *(57)*. One of the limitations of ^{111}In-DTPA-$_D$-Phe-octreotide for therapeutic purposes is its moderate binding affinity for SSTR2 and almost negligible affinity for SSTR subtypes 1, 3, 4, and 5 *(112)*. In addition, DTPA is not a suitable chelator for β-emitters such as ^{90}Y and ^{177}Lu. The optimal agent for these radiometals is the macrocyclic chelator DOTA that forms stable metal complexes, has high affinity for SSTR2, has moderately high affinity for SSTR5, has intermediate affinity for SSTR3, has high hydrophilicity, and demonstrates stable and facile labeling with ^{111}In and ^{90}Y *(142)*. The synthesis of octreotate, which is an octreotide derivative lacking an alcohol moiety, has resulted in a much improved SSTR2 affinity and biodistribution profile *(151)*. For radiotherapy, the most frequently used analogue has been ^{90}Y-DOTA-Tyr^3-octreotide (^{90}Y-DOTATOC), in addition to ^{90}Y-DOTA-lanreotide and ^{177}Lu-DOTA-octreotate *(86,144,152–164)*.

4. THERAPY

4.1. Carcinoid

4.1.1. COLD ANALOGUES

The SST analogue octreotide (Sandostatin®) was the first biotherapeutic agent utilized in the management of carcinoid tumors (Fig. 3) *(165)*. The most recent meta-analysis

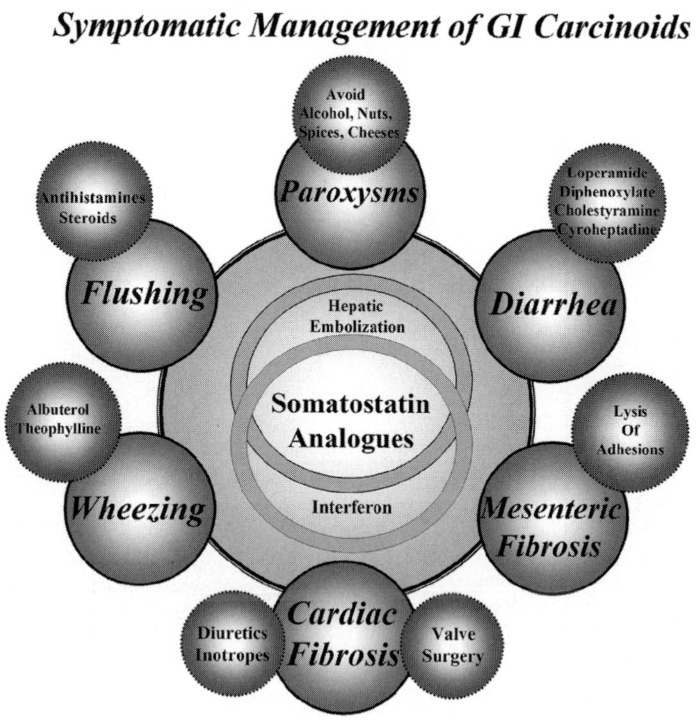

Symptomatic Management of GI Carcinoids

Fig. 3. Symptom management. Management of gastrointestinal (GI) carcinoids symptoms. Carcinoid patients often exhibit a constellation of symptoms, which may vary in intensity of both duration and expression. Some symptoms may be abrogated by avoiding inciting or provocative agents, and most can be ameliorated by pharmacological agents that specifically address a specific symptom. This usually results in polypharmacotherapy, which is difficult for patients to manage and has variable efficacy. Overall, somatostatin analogues have the broadest and most effective coverage for all symptoms particularly when utilized in a long-acting depot formulation. No effective pharmacological therapy is available for fibrosis, and this can currently be managed only by surgical intervention.

pooled data from 14 trials (20 years) and included close to 400 patients (Table 4) *(166–179)*. The calculated median biochemical response rate was approximately 37% (range 0–77%), and only four trials demonstrated decreased urinary 5-HIAA levels in >50% of patients. Objective tumor responses were evident in only three trials (range 3–9%), with a cumulative tumor response rate of 0%. In contrast, octreotide therapy appears somewhat effective at slowing the progression of carcinoid disease, with a median of 28 and 55% of patients manifesting biochemical or tumor stability in these studies, respectively. In general, the beneficial effects of octreotide are limited to symptom relief, with approximately 71% of patients experiencing resolution of diarrhea or flushing. There is minimal rigorous clinical data to support an inhibitory effect on tumor growth. To date, in all reported studies, about 30 patients may be regarded as having experienced partial tumor regression with SST analogue therapy *(180)*.

Lanreotide (Somatuline®), a long-acting SST analogue administered every 10–14 days, has a comparable efficacy to octreotide although its formulation is reported as easier and more comfortable to use *(181)*. The therapeutic effects in carcinoid have been studied in 11 groups totaling about 300 patients between 1994 and 2005, with

Table 4
Effects of Octreotide in Gastrointestinal (GI) Carcinoids

Investigator	Year	Number of patients	Biochemical response (%)	Tumor response (%)	No disease progression (%)		Symptomatic response (%)	
					Biochemical	Tumor	Diarrhea	Flushing
Kvols (166)	1986	25	72	0	28	62	88	92
Kvols (167)	1987	19	63	0	–	–	–	–
Vinik (168)	1989	14	75	0	25	50	75	100
Oberg (169)	1991	23	27	9	36	–	50	–
Saltz (170)	1993	20		0	–	50	71	–
Janson (171)	1992	24	45	0	17	62	–	–
Janson (172)	1993	55	37	2	49	–	69	70
Arnold (200)	1996	64	33	0	–	55	64	75
Di Bartolomeo (173)	1996	31	77	3	23	–	40	50
Nilsson (174)	1998	10	33		77	–	–	–
O'Toole (175)	2000	28	50		–	–	79	48
Garland (176)	2003	27	25	0	25	48	81	–
Kolby (178)	2003	35	0	0	46	–	–	–
Welin (179)	2004	12	17	0	75	75	–	–
Median (range)		24 (10–64)	37 (0–77)	0 (0–9)	28 (17–77)	55 (48–75)	71 (40–88)	71 (48–100)

–, not reported/no data available. Pooled data of 14 trials spanning the between 1986 and 2004 reflect a median biochemical response rate of 37% with only four trials showing decreased urinary 5-HIAA levels in more than half of the patients studied. Objective tumor responses were shown in only three trials (individual rates ranging between 3 and 9%), with the cumulative tumor response rate at 0%. Octreotide therapy has a better effect on slowing the progression of carcinoid disease, with 28 and 55% of patients manifesting biochemical or tumor stability. The beneficial effects of octreotide are limited to symptom relief, with 71% of patients experiencing resolution of diarrhea or flushing.

little overall improvement in responses over the short-acting octreotide, although the decreased need for injection was found to be advantageous (Table 5) *(175,182–192)*. The median biochemical response rate of the entire patient population was 42%, with only three trials reporting objective improvements in tumor size (5–9%). Compared with octreotide, lanreotide had somewhat better effects on slowing disease progression, with 46 and 81% of patients achieving biochemical and tumor size stability, respectively. The effects on symptom relief were comparable with those of octreotide, with reported decreases in diarrhea and flushing in between 75 and 80% of patients, respectively.

The efficacy and safety of the 28-day aqueous prolonged-release (PR) formulation of lanreotide (Lanreotide Autogel) was evaluated in 75 patients in a 6-month dose-titration study. Biochemical responses were reported in 35% of patients and resolution of diarrhea and flushing in 75 and 80% of patients, respectively, which are comparable with the reported effects of other lanreotide preparations. The median decrease in levels of urinary 5-HIAA and serum chromogranin A was 24 and 38%, respectively. The response was higher in patients who had not previously been treated with SST analogues (46 versus 34%) *(192)*.

An interim analysis of a phase II trial of the new SST analogue SOM 230 with pan-receptor selectivity in 21 patients with metastatic carcinoid tumors whose symptoms (diarrhea and flushing) were refractory/resistant to octreotide LAR demonstrated symptom relief in 33% *(193)*. A recent report comparing sc immediate-release octreotide with octreotide LAR reported an increased median survival from the time of metastatic carcinoid disease diagnosis (143 versus 229 months in favor of the LAR form) *(194)*. This represents a 66% lower risk of death among patients treated with the long-acting formulation. In addition, most recent data in carcinoid patients from a study with ultra-high-dose octreotide (Onco-LAR®) at 160 mg (IM) every 2 weeks for 2 months followed by the same dose once monthly suggested some advantage *(11)*. The preliminary results in 12 patients demonstrated tumor size stabilization in 9 and biochemical responses and/or stability in 11. No significant tumor reduction was noted. At 6 months, the median plasma concentrations of octreotide were 25–100 times higher than those obtained using octreotide LAR at regular doses. The protocol also demonstrated significant inhibition of angiogenesis through the downregulation of growth factors such as vascular endothelial growth factor (VEGF) and fibroblast growth factor (FGF) *(11)*. Overall, the highest response rates were reported using octreotide in doses greater than 30 mg/day or lanreotide in doses greater than 5 mg/day (and up to 15 mg) *(195)*. Such studies are, however, not adequately powered to allow for robust analysis and rigorous assessment of efficacy.

4.1.2. PEPTIDE RECEPTOR RADIONUCLIDE THERAPY

4.1.2.1. [111]In. ^{111}In-octreotide treatment was initially utilized on a compassionate-use basis for metastatic glucagonoma in an individual with no further treatment options and resulted in both symptom response and a decrease in tumor volume *(196)*. More recently, this isotope has become the most commonly utilized radiopeptide in the treatment of carcinoid disease. A recent meta-analysis indicates that the overall tumor response rates are between 13 and 20% (Table 6) *(144,155–159)*. Although the benefits of therapy are mostly limited to disease stabilization, the virtual lack of adverse events

Table 5
Effects of Lanreotide on Gastrointestinal (GI) Carcinoids

Investigator	Year	Number of patients	Biochemical response (%)	Tumor response (%)	No disease progression (%)		Symptomatic response (%)	
					Biochemical	Tumor	Diarrhea	Flushing
Canobbio (189)	1994	8	62	0	38	90	100	87
Scherubl (182)	1994	12	–	0	–	58	42	86
Eriksson (183)	1996	19	54	0	–	90	–	–
Ruszniewski (184)	1996	33	42	0	46	–	38	53
Wymenga (185)	1999	48	27	8	52	81	38	
Faiss (190)	1999	19	–	9	–	52	–	
Tomasetti (186)	2000	10	–	0	–	90	90	80
Ducreux (191)[a]	2000	38	40	5	24	54	–	40
O'Toole (175)	2000	28	50	–	–	–	89	41
Ricci (187)	2000	12	42	8	–	–	36	100
Rohaizak (188)	2002	10	0	0	83	–	90	
Ruszniewski (192)	2005	55	30	–	–	–	75	81
Median (range)		19 (8–55)	42 (0–62)	0 (0–9)	46 (46–83)	81 (58–90)	75 (36–100)	80 (38–100)

–, not reported/no data available. The median biochemical response rate of the entire patient population treated with lanreotide is 42% (versus 37% with octreotide), with only two trials reporting objective improvements in tumor size. Compared with octreotide, lanreotide has better effects on tumor stability, with 52 and 85% of patients maintaining biochemical and tumor size status quo (versus 28 and 55% of patients treated with octreotide). The effects of lanreotide on symptom relief, particularly diarrhea, are less pronounced compared with octreotide. Although 80% of patients treated with lanreotide reported decreased flushing, only 42% reported resolution or decrease of diarrhea. In the population treated with octreotide, the median rate of symptom relief of both was 71%.
[a]Overlapping patient population with ref. 184. Response defined as >30% decrease in biochemical markers.

Table 6
Effects of Peptide Receptor Radionuclide Therapy (PRRT) in Gastrointestinal (GI) Carcinoids

Investigator	Year	Number of patients	Agent	Tumor response (%)	No disease progression (%)	
					Biochemical	Tumor
Otte (152)	1999	9	^{90}Y	0	–	100
Waldherr (153)	2001	12	^{90}Y	8	–	92
Virgolini (154)	2002	34	^{90}Y	0	–	62
McCarthy (155)	1998	10	^{111}In	20	–	40
De Jong (156)	1999	10	^{111}In	17	–	50
Valkema (157)	2002	9	^{111}In	0	–	89
Anthony (144)	2002	17	^{111}In	13	–	81
Buscombe (158)	2003	10	^{111}In	20	–	5
Modlin (159)	2005	29	^{111}In	0	54	79
Hoefnagel (318)[a]	1994	52	^{131}MIBG	15	–	69
Taal (160)	1996	30	^{131}MIBG	0	52	74
Taal (160)	1996	20	^{131}MIBG	0	40	65
Mukherjee (161)	2001	18	^{131}MIBG	11	62	72
Safford (86)	2004	98	^{131}MIBG	15	–	–
Kwekkeboom (162)	2003	12	^{177}Lu	3	–	79
Teunissen (163)	2004	26	^{177}Lu	48	–	88
Kwekkeboom (164)[b]	2005	65	^{177}Lu	46	–	82
Median (range)				17 (9–98)	3 (0–48) 53 (40–62)	79 (5–100)

Four radiolabeled somatostatin analogues have been studied between 1994 and 2005 for their potential therapeutic effects on neuroendocrine tumors (^{90}Y, ^{111}In, ^{131}MIBG, and ^{177}Lu). Two of the three yttrium trials showed complete absence of objective tumor regression. Similarly, two of the four trials with ^{131}MIBG showed no effect on tumor size. Four of the six trials of ^{111}In showed response rates between 13 and 20%. The most recent ^{177}Lu trial (163) showed better tumor responses (48%), but the median response rate of the entire patient population remains at 2%. The effects of radionuclide therapy are better at maintaining the status quo, with 53 and 79% of patients achieving biochemical or tumor size stability, respectively.

[a]Collective review of five centers.

[b]Partial patient population overlap with refs 162,163.

and the minimal side effects of the therapy are major advantages. In an overall pool of 85 carcinoid patients, 65% exhibited disease stabilization and no increase in tumor size.

4.1.2.2. ^{90}Y. Following the introduction of indium-labeled SST analogues, the β-emitting ^{90}Y ($t_{1/2}$ = 2.67 days) was evaluated as a therapeutic agent for SSTR-positive tumors (112). DOTATOC has been reported to facilitate administration based on its enhanced stability and enhanced ^{90}Y-labeling facility. Overall, ^{90}Y-octreotide appears to provide greater disease stabilization compared with ^{111}In-labeled analogues. A median of 92% of patients in a pool of 55 evaluated subjects maintained stable tumor size and in 8% objective tumor regression was evident (152–154). The median disease-free survival and overall survival were 30 and 60 months, but these data were limited to one trial.

4.1.2.3. ^{177}Lu. The most promising advance in the field of peptide receptor radionuclide therapy (PRRT) has been the introduction of ^{177}Lu-octreotate that emits both β and γ radiation. This radionuclide has a half-life of 6.64 days and manifests a higher affinity for SSTR2 than other compounds. In 76 patients with GEP tumors, biochemical and tumor responses were reported in 30 and 35% of patients, respectively. Biochemical markers stabilized in 40% of patients, and 80–90% of patients experienced tumor stabilization *(162,163,197)*. The largest current lutetium-labeled SST analogue treatment series (n = 65 GI carcinoids) reported that remission rates were positively correlated with a high pre-therapy OctreoScan uptake and a limited hepatic tumor load *(164)*. Median disease-free survival (>36 months) was significantly shorter in patients with extensive liver spread (26 months). More recently, it has been identified that the DOTATATE form [(^{177}Lu-DOTA(0),Tyr3]octreotate) exhibited a longer tumor residence time than ^{177}Lu-DOTATOC that suggests that the more efficacious peptide for PRRT will be octreotate *(198)*.

4.2. Pancreatic Islet Cell Tumors

PETs are a heterogenous group of lesions including insulinomas, gastrinoma, VIPoma, and glucagonomas. Most are slow growing, present with symptoms of hormone secretion and have metastasized by the time of diagnosis. The majority (>80%) express SSTR, and SST analogue therapy is usually effective in the amelioration of symptomatology *(199)*. Thus, the various manifestations of excessive or paroxysmal hormone secretion can be substantially decreased, and quality of life (QOL) improved *(129)*. A review of 13 series that included more than 100 patients with islet cell tumors treated with "cold" SST analogues (octreotide or lanreotide) indicated median symptom improvement in approximately 70% of patients (range 45–90%) and median hormonal response of approximately 45% (range 33–77%) *(170,173,175,182,184,185,187,200–205)*. Tumor regression was far less evident, with partial responses reported in 8% of patients. Disease stabilization was achieved in a median of 50% (range 36–90%) of patients, and median time to progression was between 5 and 18 months. However, a complete interpretation of these data is limited by the fact that some series failed to differentiate between patients with GI carcinoid tumors and islet cell neoplasms.

The beneficial effects of SST analogues may be mediated both by a direct inhibitory effect on the tumor (decrease in hormone production) or by indirect effects on target organs such as the stomach or small intestine resulting in reabsorption of intestinal fluid, decrease in hydrochloric acid secretion, or diminished intestinal contractility *(206)*. As islet cell tumors usually are less aggressive than other metastatic tumors, early treatment with cold SST analogues has been proposed for individuals with significant symptoms from tumor hormonal excess but without evidence of tumor progression. The European Institute of Oncology Group has reported a >20% objective response rate in patients treated with varying levels of ^{90}Y-DOTATOC, including 12 patients with non-secretory tumors, 4 gastrinomas, 3 insulinomas, 2 glucagonomas, 1 somatostatinoma, and 1 VIPoma *(142)*.

4.2.1. INSULINOMAS

Insulinomas are the lesions least amenable to treatment with SST analogues as not all insulinomas express SSTR subtypes *(207)*. Thus, octapeptide SST analogues (like octreotide and lanreotide) are of limited use, and in addition, there is a risk of intensified hypoglycemia with SST analogue therapy *(208,209)*.

4.2.2. GASTRINOMAS

Although surgery was the initial therapeutic strategy proposed, it is evident that most gastrinomas are metastatic and target organ therapy with the proton pump inhibitor class of drugs is most effective *(210)*. Long-acting SST analogues such as octreotide LAR, lanreotide SR, or the Autogel preparation alone or in combination with α-interferon (given 3 times weekly) are recommended as initial anti-tumor treatments *(200,211)*. Although these agents cause a decrease in tumor size in only a small subset (10–20%), they have a tumoristic effect in 40–70% of patients. These agents are effective in a proportion of patients for considerable periods and are particularly effective in slower-growing lesions *(212,213)*.

4.2.3. SOMATOSTATINOMAS

There is no specific treatment for somatostatinomas apart from resection if feasible. Administration of SST analogues inhibit secretion from the tumor and may ameliorate symptomatology *(140)*. In a group of three patients, treatment with 0.5 mg/day of sc octreotide achieved resolution of diabetes and diarrhea and progressively decreased plasma levels of SST by 40–80% after 1 year *(214)*.

4.2.4. GLUCAGONOMAS

Numerous reports document the beneficial effects of SST analogues in decreasing peptide levels and symptoms in glucagonomas *(215–217)*. Although no change was noted in tumor size, good long-term relief for up to 33 months was achieved in a majority of patients treated with 150–500 µg sc octreotide three times daily *(215)*. Clinical remission of 1 year or more was noted in 7 patients (50%), despite signs of biochemical or radiologic progression. Preoperative use of SST analogues has also been advocated *(218–220)*. Octreotide is especially effective in the control of the necrolytic migratory erythema associated with glucagonoma. In some instances, marked and rapid improvement (within 48 h) and complete resolution of skin lesions may occur within 1 week *(221)*. Cessation of therapy often results in recurrence of the skin lesions *(222)*. Insulin requirements may also decrease substantially during SST analog therapy.

4.2.5. VIPOMAS

SST analogues are the principal agents effective in long-term management of these tumors as they effectively control the incapacitating hypokalemia and diarrhea *(223,224)*.

4.3. CNS Tumors

Treatment of CNS lesions with intrathecal injections of ^{90}Y-DTPA was initially utilized as early as 1976 *(225)*. However, early studies prior to the development of SST analogues failed to demonstrate any evidence of significant remission. Despite the availability of octreotide and metal-chelating complexes such as DTPA, the efficacy of receptor-mediated radiotherapy is to an extent limited by the presence of an intact blood-brain barrier. Thus, the accessibility of isotopic ligands to target receptors

(SSTR2) expressed by medulloblastomas, cPNETs, neurocytomas, gangliocytomas, olfactory neuroblastomas, and paragangliomas is a significant rate-limiting factor in determining therapeutic efficacy *(84,226)*.

4.3.1. Cold

An early study using up to 1.5 mg/24 h of octreotide for up to 16 weeks was reported three patients with unresectable meningiomas *(227)*. Therapeutic intervention failed to produce objective changes in tumor growth but did lead to alleviation of headaches (two patients) and a transient improvement in ocular movements (one patient).

4.3.2. PRRT

In an attempt to subvert the problem of limited blood-brain barrier penetrance by peptide ^{90}Y- or ^{177}Lu-labeled octreotide, brachytherapy has been utilized with some degree of success in the peripheral as well as CNS lesions *(226)*. Three patients with CNS tumors were treated with escalating dosages of ^{90}Y-DOTATOC *(228)*. In one meningioma, disease was stabilized for 13 months, whereas a second meningioma and an oligodendroma did not respond to therapy. Most medulloblastomas express high levels of SSTR2, for which DOTATOC has affinity in the low nanomolar range. The consolidating intrathecal brachytherapy using four cycles of 562.5 MBq ^{90}Y-DOTATOC resulted in complete remission in an 8-year-old male with a recurrent medulloblastoma of the cauda equina during a 3-year period *(229)*.

The European Institute of Oncology Group treated an overall number of 256 patients with varying levels of ^{90}Y-DOTATOC *(142)*. The group included 10 patients with meningiomas, 3 with astrocytomas, and 1 with ependymoma and oligodendroglioma. The overall objective response rate was >20% but it is difficult to extrapolate the data to the subgroup of CNS tumors as the results were presented as the effect of treatment in the entire group of SSTR-positive tumors.

Of the pool of 43 patients with glioblastomas treated with ^{90}Y-DOTA-lanreotide in the MAURITIUS trial, 12% achieved minor responses, 33% experienced disease stabilization, and 55% progressed *(154)*. In the same trial, 33% of patients with astrocytomas experienced minor responses, 50% had stable disease, and 17% progressed on therapy. None of the three patients with meningiomas had objective responses, two had stable disease, and one progressed.

A study of seven low-grade and four anaplastic gliomas loco-regionally injected with ^{90}Y-DOTATOC reported disease stabilization in six and noted inverse correlation between malignancy grade and SSTR expression *(230)*. A more recent study using the same vector halted tumor progression for 13–45 months in five patients with WHO grade II and III gliomas, allowing reduction or cessation of steroid medication *(231)*.

4.4. Pituitary Adenomas

4.4.1. Acromegaly

GH-producing pituitary adenomas are characterized by very high SSTR2 expression, and OctreoScan positivity is correlated with the ability of octreotide to suppress GH release *(8)*. Numerous studies support the efficacy of non-radioactive octreotide in the treatment of acromegaly that is not amenable to surgical therapy *(232)*. Radiolabeled octreotide is considered to be of limited value given the significant efficacy of established surgical or medical therapy.

Octreotide is of particular utility in reduction of acromegaly associated headache, which may be mediated by opioid receptors *(233)*. Octreotide is effective in the control of other symptoms of acromegaly, including joint pain, excessive perspiration, cardiomyopathy, and sleep apnea, which are reduced or eliminated in the majority of patients *(129)*. Octreotide also directly inhibits GH-stimulated production of IGF-1 and decreases serum prolactin concentrations as well as galactorrhea or (secondary) amenorrhea in patients who have pituitary adenomas secreting both GH and prolactin *(234)*.

4.4.1.1. Cold Analogues. The most recent meta-analysis reviewed 36 studies that included 921 patients with acromegaly treated with SST analogues *(130)*. Regardless of prior treatment history or the type of analogue utilized, approximately 42% of patients demonstrated tumor shrinkage. Individuals receiving SST analogue as the primary pharmacologic intervention exhibited even higher responses, with 52% showing tumor reduction versus 21% with adjuvant therapy. The average tumor reduction was around 50%, and fewer than 3% demonstrated tumor size progression on therapy for treatment periods of up to 3 years. Results vary slightly based on type of SST analogue used. The lower tumor reduction rates in the adjuvant groups were attributed to treatment-related changes in tumor anatomy (scarring and fibrosis) that rendered subsequent SST analogue therapy less effective. Biochemical response is another factor that has been used as a marker of whether SST analogue therapy will induce tumor shrinkage *(130)*. Some studies suggest that positive biochemical response to therapy (reductions in GH and IGF-1) predict tumor shrinkage *(235–239)*. However, a number of other studies failed to identify a correlation between hormone suppression and tumor shrinkage *(240–244)*.

The effects of Lanreotide Autogel in 25 patients with acromegaly reported significant improvement of the acromegalic symptom score and a small but significant reduction in the residual pituitary tumor volume after switching from octreotide LAR *(245)*. After 24 weeks of treatment, mean serum GH and IGF-1 concentrations remained statistically unchanged when compared with baseline values on octreotide LAR. Local side effects were observed less frequently, and no technical problems were encountered with the injections as opposed to those noted with octreotide LAR. Two additional studies compared the effects of octreotide LAR and Lanreotide Autogel in smaller numbers of acromegalics. In a group of 10 patients with well-controlled disease, switching from LAR 20 mg to Lanreotide Autogel 60–120 mg/month resulted in similar effects on GH levels but a slightly greater reduction of IGF-1 after 28 weeks *(246)*. A recent 12-month study of seven patients with good control on octreotide found no difference in GH and IGF-1 suppression between the two analogues *(247)*.

4.4.1.2. PRRT. Two patients with relapsed pituitary adenomas were treated with escalating dosages of ^{90}Y-DOTATOC and achieved disease stabilization for 4–19 months *(228)*. The European Institute of Oncology Group reported four pituitary adenomas treated with varying levels of ^{90}Y-DOTATOC *(142)*. The overall objective responses rate was >20%, but it is difficult to extrapolate the data to the subgroup of pituitary tumors as the results were presented as the effect of treatment in the whole group of SSTR-positive tumors ($n = 256$).

4.4.2. TSH-Secreting Pituitary Adenomas

Most TSH-secreting pituitary adenomas express SSTRs and are hence amenable to analogue therapy *(129)*. In 33 patients treated with sc octreotide (100–300 μg/day), 91% responded to short-term treatment with a decrease in thyrotropin secretion and in 73% serum thyroxine was normalized *(236)*.

4.4.3. Non-Secretory Pituitary Adenomas

Most non-secreting pituitary tumors express SSTRs *(248)*. Although octreotide therapy (300–1500 μg/day) infrequently results in tumor size reduction, 30–40% exhibit improvement in visual-field defects within days after treatment *(249)*. This response may reflect a direct effect of octreotide on the retina or optic nerve *(129)*.

4.5. Other Neuroendocrine Tumors

4.5.1. Thyroid Carcinomas

Patients with thyroid carcinoma who do not respond to radioiodine therapy or do not show uptake on radioiodine scintigraphy have few treatment options available *(250)*.

4.5.1.1. Cold Analogues. A recent series of 11 patients with advanced MTC treated with "cold" SST analogues (octreotide 30 mg or lanreotide 30 mg) reported disease stabilization in all for 12 months *(251)*. Data with respect to the treatment of differentiated thyroid cancer are for the most part limited to case reports. One of the first series examined the efficacy of long-term octreotide administration (4 mg daily for up to 12 months) in six patients with recurrent MTC, including one Hurthle cell, one medullary, and four papillary or mixed papillary/follicular cancers *(252)*. Treatment failed to significantly decrease tumor markers (e.g., thyroglobulin, calcitonin, and CEA), and all six patients exhibited disease progression. Two patients with widely metastatic PTC and ^{111}In-pentetreotide scans that demonstrated metastatic lesions with SSTR expression had tumor volume reductions and decreases in the standard uptake values of fluorodeoxyglucose (FDG) when treated with 3 or 4 months of Sandostatin LAR Depot therapy *(253)*.

4.5.1.2. PRRT.

[111]***In:*** Initial usage of ^{111}In-octreotide in thyroid malignancy involved treatment of two patients diagnosed with differentiated thyroid carcinoma (DTC) and reported disease stabilization in one *(254)*. A review of the use of PRRT in thyroid cancer from Rotterdam evaluated 25 patients with SSTR-positive tumors treated with cumulative doses of at least 20 GBq and up to 160 GBq ^{111}In-DTPA-octreotide follow-up for 1–28 months *(255)*. Three of the five patients with MTC, three of the four PTC patients, and one patient with FTC exhibited progression. The remaining group achieved stable disease as their best response.

In a single institution review evaluating ^{111}In-octreotide therapy in five patients with DTC, including four with PTC and one with FTC, four failed to respond and one, after initial stabilization, ultimately progressed *(157)*. A more recent series included 11 (9 evaluable) patients with thyroid carcinoma non-responsive to iodine treated with high-fixed doses of approximately 7.4 GBq ^{111}In-octreotide *(65)*. After a 6-month follow-up, four patients showed tumor stabilization, whereas the rest progressed. Six of nine evaluable patients demonstrated biochemical stabilization (plasma thyroglobulin reduction).

^{90}Y: Twenty patients with therapy-resistant and progressive thyroid cancer [12 with MTC, 7 with DTC including 4 with PTC and 3 with FTC, and 1 with anaplastic carcinoma (AC)] were treated with ^{90}Y-DOTATOC (1.7 GBq–7.4 GBq/m^2, administered in one to four injections every 6 weeks) (256). There were no objective responses to therapy. Disease stabilization was achieved in 35%, and progressive disease occurred in the remainder. The overall response rate (objective response + stable disease) was 42% in MTC, 29% in DTC, and 0% in AC. The median time to progression was 8 months, with a median follow-up of 15 months.

The European Institute of Oncology Group retrospectively evaluated the therapeutic efficacy of ^{90}Y-DOTATOC in 21 patients with SSTR-positive metastatic MTC (257). Patients received cumulative doses of 7.54–19.24 GBq in 2–8 cycles. Complete radiological response was observed in 10%, disease stabilization in 57%, and the remaining 33% did not respond. The duration of the objective responses ranged between 2 and 39 months. With regard to tumor markers, a complete response was observed in one, partial responses in five, and stabilization occurred in three.

Eight patients with MTC received two cycles of ^{90}Y-DOTATOC, with activity increased by 0.37 GBq/group, starting at 2.96 and terminating at 5.55 GBq (228). Of the seven evaluable patients, one developed a complete response that lasted through the follow-up period of 19 months, another achieved partial response, and in the other three, the disease stabilized. The duration of response to therapy was between 9 and 20 months.

The first report of ^{90}Y-DOTATOC treatment for Hürthle cell thyroid carcinoma (HCTC) (total dose, 1.7–9.6 GBq) reported disease stabilization for a period of nine months in one patient who received the highest dose, whereas the other two progressed despite treatment (258). The study concluded that the protocol was not ideal because of suboptimal radiation doses.

In the MAURITIUS trial, 25 patients with radioiodine-negative thyroid carcinoma were treated with 0.9–7.0 GBq SST analogue ^{90}Y-DOTA-lanreotide (154). Three had >25% reduction of tumor size, 11 disease stabilization, and another 11 progressed. More recently, all five patients with extensive DTC treated with 5.6–7.4 GBq ^{90}Y-DOTATOC exhibited disease stabilization for at least 5 months (247).

^{177}Lu: Five patients with recurrent or metastatic DTC (three with HCTC, one with papillary thyroid carcinoma, and one with follicular thyroid carcinoma) were treated with 22.4–30.1 GBq ^{177}Lu-DOTATATE (250). DOTATATE, in comparison with DOTATOC, improves binding to SSTR-positive tumors (259) and has a ninefold higher affinity for the SSTR2, whereas its affinity for SSTR3 and SSTR5 is lower (250). One patient with HCTC had stable disease as a maximum response, one patient had minor remission (tumor shrinkage between 25 and 50%), and one patient had partial remission (shrinkage ~50%). The responses in PTC and FTC were stable disease and progressive disease, respectively. Time to progression among responders was between 18 and 43 months. These results are of interest for HCTC as radioiodine therapy is ineffective in this non-iodine-avid tumor group.

Overall, early experience with receptor-mediated radionuclide therapy for non-medullary thyroid cancer is disappointing, with little evidence of adequate tumor control albeit in underpowered studies. At present therefore, it appears that traditional surgical or iodine-based radio-ablative procedures have yet to be supplanted *(112)*.

4.5.2. SCLC

SCLC and bronchial carcinoid tumors express SSTRs, and some studies have suggested that therapy with radiolabeled SST analogues may improve outcome when compared with treatment with traditional chemotherapy *(260)*. A variety of SSTR-positive and SSTR-negative tumors have been reported to benefit from SST analogue therapy *(112)*. Two lung carcinoids were treated with escalating dosages of ^{90}Y-DOTATOC, with one experiencing a partial response (3 months), whereas in the other, disease stabilization occurred for 13 months *(228)*. One patient with a non-SCLC failed to respond to therapy. Ten patients with lung cancer (non-SCLC and SCLC) were treated with four cycles of 23–114 mCi ^{90}Y-DOTA-lanreotide *(154)*. One experienced minor response, five disease stabilization, and the remaining four exhibited disease progression.

4.5.3. PHEOCHROMOCYTOMA

Pheochromocytomas are best treated surgically to obviate the associated dangerous cardiovascular catecholamine-mediated paroxysmal events. In the majority (~90%), tumor resection is successful unless metastasis or tumor spillage occurs with subsequent persistence in symptomatology and disease progression *(261)*.

4.5.3.1. Cold Analogues. Sporadic reports describe symptomatic or hormonal improvements following repeated sc injections of immediate-release octreotide *(262,263)*. One description of an unresectable multiple paraganglioma in the head and neck region successfully treated with Sandostatin LAR reported a 16-month almost complete abolition of paroxysmal events and stabilization of tumor growth *(264)*. Three studies evaluated the acute effects of intravenous or sc octreotide in a small number of patients with pheochromocytoma with conflicting results *(265)*. In six patients with chromaffin cell tumors, a 2-h 50 μg infusion of octreotide significantly decreased plasma norepinephrine and epinephrine levels and halved norepinephrine baseline levels with a prompt return of hormone concentrations to pre-infusion values thereafter *(266)*. The effects of octreotide infusion on blood pressure were, however, inconsistent, and in a separate series of 10 patients treated with three 100 μg sc injections of octreotide, no consistent reductions in mean 24-h ambulatory blood pressure or plasma or urinary catecholamine levels were evident *(267)*. In another study of 10 patients, plasma catecholamine responses to a single dose of 200 μg intravenous octreotide was monitored up to 5 h after injection *(268)*. A pronounced inhibition was not noted nor was there a relationship between SSTR density and catecholamine inhibition. In addition, in a further two patients treated with octreotide over several months, there was no correlation between octreotide scintigraphy and catecholamine suppression after octreotide infusion. Two patients on octreotide treatment showed symptomatic improvement but not mass reduction. A 3-month study of 20 mg slow-release octreotide Sandostatin LAR in 10 patients with malignant or recurrent pheochromocytomas did not result in significantly altered symptoms, blood pressure, blood glucose concentrations, plasma catecholamine, and chromogranin A concentrations or metanephrine excretion *(265)*. A positive OctreoScan was not an independent predictor of successful therapeutic outcome.

4.5.3.2. PRRT. The use of 22.65 GBq ^{111}In-DTPA-octreotide in an individual with an unresected primary, widespread lymph node, and bone metastases was not effective, and at 8 months, progressive disease was evident *(157)*. A second patient with an unresected primary tumor, regional nodal metastases, and extensive liver involvement received cumulative dose of 51.67 GBq and at the end of follow-up at 55 months continued to have stable disease.

4.6. Adenocarcinomas

4.6.1. BREAST

In vitro investigations of primary breast cancer and their LN metastases indicate that only 21% of tumors exhibit high-receptor density expression of SSTRs. Nevertheless, SSTR-expressing neuroendocrine tumors of the breast have been treated with SST analogues *(97)*. Phase I trials with three cycles of 120 mCi DOTATOC have demonstrated significant tumor shrinkage *(269)*. Other ligands, such as DOTA-lanreotide, may have superior binding profiles to breast tumors and prove efficacious in such therapy.

4.6.2. HCC

The antitumor effects of SST analogues in HCC are controversial with assessments ranging from substantial efficacy to no effect *(103,270–277)*. In the first major study on patients treated with SST analogues, SSTRs were measured in liver tissue homogenates from patients with acute and chronic hepatitis, cirrhosis, and HCC *(103)*. Various levels of SSTRs were identified in liver tissue of all patients including those with HCC. Fifty-eight patients with advanced HCC were randomized to receive either sc octreotide 250 μg twice daily or no treatment. Patients treated with octreotide had an increased median survival (13 versus 4 months, p = 0.002, log rank test) and an increased cumulative survival rate at 6 and 12 months (75 versus 37% and 56 versus 13%, respectively). Octreotide administration significantly reduced α fetoprotein levels at 6 months. Treated patients had a lower hazard (0.383) in the multivariate analysis.

Another study, comparing 32 patients treated with long-acting SST analogs with 27 untreated patients, noted an improved overall survival (15 versus 8 months) in the treated group *(270)*. Tumor stability or even regression was seen in 40% of the patients receiving SST analogues. A recent prospective study including 41 patients with HCC showed a median survival of 571 days, with one patient showing partial response to treatment, tumor stabilization in 63% of the patients, and tumor progression in 34% *(271)*. In contrast, two trials, using a sequenced regimen of octreotide (initially 2 weeks with 500–900 μg short-acting octreotide followed by 4–6 monthly administration of Sandostatin LAR), showed very limited beneficial results *(272,273)*. In one study, the median time to progression was 3.6 months with a median survival of 5.1 months and only 29% of the patients showing stable disease for a median period of 8.0 months *(272)*. The second study reported no changes in survival, tumor regression, reduction of α-fetoprotein (AFP), or QOL compared with the non-treated control group *(273)*. In a phase III trial of 272 patients with HCC unsuitable for surgery, percutaneous ablation or chemoembolization subjects received a monthly IM injection of 30 mg Sandostatin LAR *(274)*. Median overall survival was 6.5 months in the octreotide group versus 7.3 months in the placebo group. A separate study examined the effect of escalating doses (up to 30 mg every 4 weeks) of Sandostatin LAR on survival and QOL in 15 patients with advanced unresectable SSTR-positive HCC and compared the results with tamoxifen efficacy in a similar group *(275)*. One-third of the SST analogue group attained a moderate QOL and died 12–19 weeks after initiation of therapy, whereas the remainder survived 14–92 weeks with good QOL. No AFP reduction or decrease in tumor mass was noted. Although this study concluded that Sandostatin LAR improved the survival and QOL compared with tamoxifen in patients with unresectable HCC, appropriately powered more rigorous studies are required to support this conclusion.

A multicenter evaluation of 63 patients who have been treated with long- (20 patients) or short-acting SST analogues in 13 German centers showed partial remission of the tumor in two patients, no changes in 35%, and tumor progression in 41% of the patients at 3 months *(276)*. With a median survival of 9 months, the authors concluded that treatment with octreotide formulas has no marked effect on patient survival. In one study from Austria, analyzing the effect of long-acting lanreotide on hepatocellular malignancies in vitro and in vivo, patients showed poor results after treatment with 30 mg long-acting agent (partial response in 5%, tumor stability in 38%, and tumor progression in 57%) *(277)*. However, a dose-dependent induction of apoptosis in HepG2 cell lines was reported, suggesting that the administered dose (30 mg) might have been insufficient.

4.7. Miscellaneous Neoplasia and Disease

4.7.1. LYMPHOMAS

The role of SST analogues in management of immunoproliferative disease is critically related more to the process of initial imaging and staging. A recent review reported a sensitivity of SRS of 95% in case of HD, and about 80% in case of NHL *(278)*, while a study including 126 patients with HD, revealed an advantage of SRS versus conventional staging utilizing CT and/or ultrasound only for supradiaphragmatical lesions (sensitivity 98%), whereas SRS was inferior in imaging infradiaphragmatical tumor site (sensitivity 67%) *(108)*. Imaging using [^{111}In-DTPA-D-Phe(1)]-octreotide has emerged as excellent method for staging and therapy monitoring of extragastric MALT-lymphoma, which expresses SSTR subtypes that can be more easily targeted than gastric MALT-lymphomas *(277)*. A report from Rotterdam, the Netherlands, demonstrated a very low uptake of [^{125}I-Tyr(3)]-octreotide in six orbitals, two HDs, and two NHLs, confirming the limited possibilities for SST analogues in radionuclide treatment *(279)*.

There are only few reports about therapeutic use of radio-labeled SST analogues. One patient with HD was treated with escalating dosages of ^{90}Y-DOTATOC but did not demonstrate an objective response to therapy at 8 months *(228)*. In the MAURITIUS trial in which ^{90}Y-DOTA-lanreotide was administered at 50–100 mCi, two of four patients experienced disease stabilization, whereas the rest progressed *(154)*.

4.7.2. Melanomas

Although octreotide scintigraphy identifies only approximately 63% of melanomas, it has been suggested that these may be treatable with SST analogues radiolabeled with high-energy isotopes *(112)*. Although ^{90}Y-labeled lanreotide compounds demonstrate high-binding affinities for melanoma cell lines, further clinical trials are needed to assess efficacy *(280)*. In one study, one melanoma patient treated with escalating dosages of ^{90}Y-DOTATOC succumbed to disease at 3 months *(228)*. Biweekly infusions (480 μg/kg) of TT-232 (D-Phe-Cys-Tyr-D-Trp-Lys-Cys-Thr-NH$_2$) a novel SST analogue shown to bind to SSTR1 and SSTR4 have been used to treat metastatic malignant melanoma *(281)*. One patient had partial remission, and three patients achieved disease stabilization. No significant side effects except for transient fever were noted.

4.7.3. Merkel Cell Tumor

Merkel cell cancer (trabecular cancer or neuroendocrine skin cancer) is a rare neoplasm that presents in hair follicles or as a firm, painless skin lump *(112)*. In addition to other markers, these tumors express SSTRs, and both primary and metastases can be visualized using OctreoScan *(282,283)*. A case report documents the utility of ^{90}Y-DOTATOC in treating metastatic disease, with two complete remissions reported after four cycles *(282)*.

4.7.4. Chemotherapy-Induced Diarrhea

Diarrhea is a common side effect of chemotherapy and can be the rate-limiting event in the management of certain patients *(284)*. Chemotherapeutic regimens that contain fluoropyrimidines and irinotecan (CPT-11) tend to increase the risk of chemotherapy-induced diarrhea (CID) *(284)*. Octreotide has been shown to be effective in controlling CID refractory to conventional therapy and has been recommended in guidelines for the treatment of CID *(284,285)*. In one study, 37 colorectal cancer patients with grades 1–4 diarrhea caused by chemotherapy with 5-FU-containing regimens received oral loperamide at the initial dose of 4 mg followed by 4 mg every 8 h (total dose 16 mg/24 h). Twenty-five patients (69%) were diarrhea-free and were considered treatment responders *(286)*. Eight-four percent of the patients with grade 1 or 2 diarrhea achieved a response, but only 52% of those with grades 3–4 diarrhea.

The use of the microencapsulated, long-acting formulation (LAR) of octreotide for once-monthly IM dosing has been reported. An analysis of 11 case studies demonstrated complete resolution of diarrhea within 1–4 weeks from injection time in all patients with octreotide LAR 30 mg *(287)*. With a subsequent prophylactic injection once every 28 days, CID was limited to National Cancer Institute (NCI) grade 1. This resulted in increased patient QOL and allowed better patient compliance with therapy. Long-acting SST analogues therefore have the potential to be useful in the secondary prevention of diarrhea in patients undergoing chemotherapy.

4.7.5. Inflammatory, Infectious, and Granulomatous Diseases

SSTRs have been identified in vivo in the inflammatory lesions of patients with sarcoidosis, tuberculosis, Wegener's granulomatosis, DeQuervain's thyroiditis, aspergillosis, and giant-cell arteritis *(8,38,121,122,288)*. SSTR expression is evident on inflammatory cells as well as in areas of angiogenesis. Preliminary studies indicate SSTR2 as the predominant receptor subtype, and some reports of successful treatment with SST analogues suggest that receptor-specific radiolabeled analogues may be worthy of consideration in the management of such disease processes *(112)*.

4.7.6. Thyroid Ophthalmopathy

Thyroid-associated ophthalmopathy (TAO) is an autoimmune disorder characterized by orbital tissue swelling, proptosis, diplopia, and occasionally visual loss and is usually associated with Graves' hyperthyroidism *(289)*. When severe, it may be sight-threatening and require treatments such as systemic steroids, orbital irradiation, or orbital surgical decompression, but such measures have substantial morbidity *(290)*. Functional SSTRs are expressed on activated lymphocytes and orbital fibroblasts and are thought to be involved in the pathogenesis of the disease *(291)*.

A number of small, uncontrolled studies have suggested that SST analogues might be of benefit in TAO *(122,292–296)*. Five patients with severe symptoms of Graves' ophthalmopathy were treated with biweekly doses of 40 mg IM lanreotide *(297)*. After 3 months, four patients demonstrated significant improvement in ophthalmopathy in both eyes and one in one eye. However, a recent placebo-controlled series of 50 euthyroid patients with active thyroid eye disease who received either 30 mg LAR or placebo once a month for 4 months, followed by 30 mg LAR for another 16 weeks, and further 24 weeks of follow-up without treatment failed to demonstrate any significant therapeutic effects when compared with the placebo group *(298)*. The authors attributed the discrepancy between these results and the previous encouraging reports to the natural course of the illness that the previous studies had failed to control for.

4.7.7. Oncogenic Osteomalacia

Mesenchymal tumors are the most common cause of this hypophosphatemic syndrome characterized by bone-mineralization defects, reduced serum calcitriol, and phosphatemia caused by the paraneoplastic production of phosphatonins *(299,300)*. Mesenchymal tumors have recently been shown to express SSTR2 that may make these tumors responsive to octreotide therapy *(300–302)*. SRS has been employed to facilitate surgical treatment of the condition for widespread, diffuse, or otherwise unresectable disease *(112,300)*, and high-energy isotopes bound to SST analogues may provide equally effective therapy. Neuroendocrine tumors producing parathyroid hormone-related peptide are usually undifferentiated and can express SSTR2 *(181)*. These rare tumors present with various degrees of hypocalcemia. In selected cases, an octreotide trial may improve the clinical and biochemical profile *(303,304)*.

4.7.8. Cushing's Syndrome

In patients with the ectopic ACTH secretion and Cushing's syndrome, SST analogue therapy can result in a reduction of ACTH levels in some cases *(305,306)*. However, the variable responses as well as incomplete normalization of ectopic ACTH overproduction usually requires early bilateral adrenalectomy in these generally severely ill patients *(181)*.

4.7.9. Thymoma

Thymomas are rare, indolent neoplasms associated with myasthenia gravis, hypogammaglobulinemia, and pure red-cell aplasia. Seven patients with recurrent thymoma and positive SRS were treated with "cold" octreotide (Sandostatin LAR 20 mg twice a month) and showed stable disease on follow-up *(307)*. In the MAURITIUS trial with ^{90}Y-DOTA-lanreotide administered between 17 and 107 mCi, two of five patients with thymomas experienced disease stabilization, whereas the remaining three progressed despite therapy *(154)*.

5. CODA

The most striking effect of SST analogues is the control of hormone hypersecretion associated with tumors *(308)*. Available data on growth suppression indicate a limited anti-proliferative effect. The introduction of long-acting SST analogues (Somatuline Autogel® and Sandostatin LAR®) has dramatically increased the duration of therapeutic control from days to weeks and may even extend therapeutic control to months with the introduction of novel formulations. Thus, advances in drug delivery systems in conjunction with the development of more stable formulations and slow-release depot formulations have further facilitated symptom management and QOL.

However, tumor reduction or stabilization is often transient and limited to a minority of patients. Eventually, all patients escape from SST analogue therapy, and after weeks to months of treatment with octreotide, symptoms worsen and tumor hormone secretion increases in virtually all patients *(309)*. The only exceptions are patients with acromegaly who do not seem to experience tachyphylaxis even after more than 10 years of daily octreotide injections. The mechanism underlying the escape phenomenon is not yet known. This effect can initially be reversed by increasing the dose of octreotide, but eventually, the drug becomes ineffective in all patients. This loss of sensitivity is probably associated with the growth of clones of tumor cells that lack SSTRs rather than with a transient down-regulation of these receptors *(310,311)*.

Although promising, the therapeutic effects of radiolabeled SST analogues remain somewhat disappointing. In general, they may be considered to, at best, maintain the status quo by stabilizing disease progression. Nevertheless, they remain the most intellectually exciting therapeutic development in the treatment of neuroendocrine tumor disease. Overall, β-emitting radionuclides (e.g., ^{131}I) have a greater therapeutic potential because the particles they emit have sufficient energy to cause tumor cell damage without penetrating far into surrounding tissue *(254,269,312)*. Radionuclides emitting primarily γ radiation (e.g., ^{111}In) or Auger electrons exhibit anti-proliferative effects only if cellular DNA is within the particle range. Such compounds may therefore be most effective when given in combination with β-emitters or when used to eradicate micrometastases *(313)*. As an emitter of both β and γ particles, ^{177}Lu has been shown to induce significant tumor regression and may be of particular benefit in the treatment of small tumors by minimizing radiation exposure of cells distant from the bound SSTRs *(112)*.

Moreover, the co-expression of multiple receptors (GRP, CCK, and VIP) in addition to SST may have therapeutic relevance as many of the involved peptides are known to have growth-modulating properties *(32,314)*. Concomitant application of multiple radioligands may be extremely advantageous to improving the efficacy of peptide targeting by increasing the accumulation of radioactivity in the tumors. For instance, the use of a mixture of SSTR2, GLP-1, and GRP radioligands could offer optimal targeting of gastrinomas *(315)*. As some of the receptors are heterogeneously expressed, combining the corresponding receptor-selective radiopeptides may further improve the targeting efficacy by destroying more than one receptor-expressing tumor area.

Loss of efficacy during peptide radiotherapy may be due to tumor de-differentiation with a resulting loss of some peptide receptors. A cocktail of different peptides may cover such a loss by maximizing exposure to those remaining receptors. The issue of receptor heterogeneity, which may be responsible for some tumor resistance, could in theory be overcome by using cocktails of radiolabeled ligands in combination with one another or with other bio- or chemotherapeutics. One potentially promising approach in patients with micro-metastases or tumors of different sizes is treatment with a combination of radionuclides shown to be optimal for treating larger tumors (^{90}Y) and radionuclides shown to be optimal for smaller tumors (^{177}Lu) *(313,316)*.

Finally, an advantage of using a cocktail of radioligands is that they can each be labeled with different isotopes, for example, β-emitters with different ranges, which could optimize radiotherapy for large and small tumor lesions *(317)*. Prior to the use of multiple radiopeptide ligands in vivo, it would be beneficial to determine the individual peptide receptor profile of the tumor under consideration by in vitro receptor determination from biopsy samples. As the side effects of this type of therapy are minimal and the duration of response is more than 2 years, a combination "cocktail" isotope therapy is a potentially attractive therapeutic path in the future.

REFERENCES

1. Krulich L, Dhariwal A, McCann S. Stimulatory and inhibitory effects of purified hypothalamic extracts on growth hormone release from rat pituitary in vitro. *Endocrinology* 1968;83:783–90.
2. Reisine T, Bell GI. Molecular biology of somatostatin receptors. *Endocr Rev* 1995;16:427–42.
3. Hellman B, Lernmark A. Inhibition of the in vitro secretion of insulin by an extract of pancreatic α-1 cells. Endocrinology 1969;84:1484–8.
4. Brazeau P, Vale W, Burgus R, Ling N, Butcher M, Rivier J, Guillemin R. Hypothalamic polypeptide that inhibits the secretion of immunoreactive pituitary growth hormone. Science 1973;179:77–9.
5. Pradayrol L, Jornvall H, Mutt V, Ribet A. N-terminally extended somatostatin: the primary structure of somatostatin-28. FEBS Lett 1980;109:55–8.
6. Schonbrunn A, Tashjian H Jr. Characterization of functional receptors for somatostatin in rat pituitary cells in culture. J Biol Chem 1978;253:6473–83.
7. Raynor K, Reisine T. Somatostatin receptors. Crit Rev Neurobiol 1992;6:273–89.
8. Krenning EP, Kwekkeboom DJ, Bakker WH, Breeman WA, Kooij PP, Oei HY, van Hagen M, Postema PT, de Jong M, Reubi JC, et al. Somatostatin receptor scintigraphy with [111In-DTPA-D-Phe1]- and [123I-Tyr3]-octreotide: the Rotterdam experience with more than 1000 patients. Eur J Nucl Med 1993;20:716–31.
9. Patel YC. Somatostatin and its receptor family. Front Neuroendocrinol 1999;20:157–98.
10. Reichlin S. Somatostatin. N Engl J Med 1983;309:1495–501.
11. Oberg K. Future aspects of somatostatin-receptor-mediated therapy. Neuroendocrinology 2004;80 Suppl 1:57–61.
12. Patel YC, Greenwood MT, Panetta R, Demchyshyn L, Niznik H, Srikant CB. The somatostatin receptor family. Life Sci 1995;57:1249–65.
13. Yasuda K, Raynor K, Kong H, Breder CD, Takeda J, Reisine T, Bell GI. Cloning and functional comparison of kappa and delta opioid receptors from mouse brain. Proc Natl Acad Sci USA 1993;90:6736–40.

14. Raynor K, Murphy WA, Coy DH, Taylor JE, Moreau JP, Yasuda K, Bell GI, Reisine T. Cloned somatostatin receptors: identification of subtype-selective peptides and demonstration of high affinity binding of linear peptides. Mol Pharmacol 1993;43:838–44.

15. Raynor K, O'Carroll AM, Kong H, Yasuda K, Mahan LC, Bell GI, Reisine T. Characterization of cloned somatostatin receptors SSTR4 and SSTR5. Mol Pharmacol 1993;44:385–92.

16. Reubi JC, Laissue J, Waser B, Horisberger U, Schaer JC. Expression of somatostatin receptors in normal, inflamed, and neoplastic human gastrointestinal tissues. Ann N Y Acad Sci 1994;733: 122–37.

17. Buscail L, Esteve JP, Saint-Laurent N, Bertrand V, Reisine T, O'Carroll AM, Bell GI, Schally AV, Vaysse N, Susini C. Inhibition of cell proliferation by the somatostatin analogue RC-160 is mediated by somatostatin receptor subtypes SSTR2 and SSTR5 through different mechanisms. Proc Natl Acad Sci USA 1995;92:1580–4.

18. Buscail L, Delesque N, Esteve JP, Saint-Laurent N, Prats H, Clerc P, Robberecht P, Bell GI, Liebow C, Schally AV, et al. Stimulation of tyrosine phosphatase and inhibition of cell proliferation by somatostatin analogues: mediation by human somatostatin receptor subtypes SSTR1 and SSTR2. Proc Natl Acad Sci USA 1994;91:2315–9.

19. Mayor F Jr, Benovic JL, Caron MG, Lefkowitz RJ. Somatostatin induces translocation of the β-adrenergic receptor kinase and desensitizes somatostatin receptors in S49 lymphoma cells. J Biol Chem 1987;262:6468–71.

20. Kong H, DePaoli AM, Breder CD, Yasuda K, Bell GI, Reisine T. Differential expression of messenger RNAs for somatostatin receptor subtypes SSTR1, SSTR2 and SSTR3 in adult rat brain: analysis by RNA blotting and in situ hybridization histochemistry. Neuroscience 1994;59:175–84.

21. Perez J, Rigo M, Kaupmann K, Bruns C, Yasuda K, Bell GI, Lubbert H, Hoyer D. Localization of somatostatin (SRIF) SSTR-1, SSTR-2 and SSTR-3 receptor mRNA in rat brain by in situ hybridization. Naunyn Schmiedebergs Arch Pharmacol 1994;349:145–60.

22. Breder CD, Yamada Y, Yasuda K, Seino S, Saper CB, Bell GI. Differential expression of somatostatin receptor subtypes in brain. J Neurosci 1992;12:3920–34.

23. Beaudet A, Greenspun D, Raelson J, Tannenbaum GS. Patterns of expression of SSTR1 and SSTR2 somatostatin receptor subtypes in the hypothalamus of the adult rat: relationship to neuroendocrine function. Neuroscience 1995;65:551–61.

24. Epelbaum J. Somatostatin in the central nervous system: physiology and pathological modifications. Prog Neurobiol 1986;27:63–100.

25. Meyerhof W, Wulfsen I, Schonrock C, Fehr S, Richter D. Molecular cloning of a somatostatin-28 receptor and comparison of its expression pattern with that of a somatostatin-14 receptor in rat brain. Proc Natl Acad Sci USA 1992;89:10267–71.

26. Reubi J, Waser B, Schaer J, Laissue J. Somatostatin receptor sst1-sst5 expression in normal and neoplastic human tissues using receptor autoradiography with subtype-selective ligands. Eur J Nucl Med 2001;28:836–46.

27. Reubi JC, Horisberger U, Studer UE, Waser B, Laissue JA. Human kidney as target for somatostatin: high affinity receptors in tubules and vasa recta. J Clin Endocrinol Metab 1993;77:1323–8.

28. Caron P, Buscail L, Beckers A, Esteve JP, Igout A, Hennen G, Susini C. Expression of somatostatin receptor SST4 in human placenta and absence of octreotide effect on human placental growth hormone concentration during pregnancy. J Clin Endocrinol Metab 1997;82:3771–6.

29. Rohrer L, Raulf F, Bruns C, Buettner R, Hofstaedter F, Schule R. Cloning and characterization of a fourth human somatostatin receptor. Proc Natl Acad Sci USA 1993;90:4196–200.

30. Hofland L, van Hagen P, Lamberts S. Functional role of somatostatin receptors in neuroendocrine and immune cells. Ann Med 1999;31:23–7.

31. Pilichowska M, Kimura N, Schindler M, Kobari M. Somatostatin type 2A receptor immunoreactivity in human pancreatic adenocarcinomas. Endocr Pathol 2001;12:147–55.

32. Reubi JC. Peptide receptors as molecular targets for cancer diagnosis and therapy. Endocr Rev 2003;24:389–427.

33. Reubi JC, Maurer R, Klijn JG, Stefanko SZ, Foekens JA, Blaauw G, Blankenstein MA, Lamberts SW. High incidence of somatostatin receptors in human meningiomas: biochemical characterization. J Clin Endocrinol Metab 1986;63:433–8.

34. Reubi JC, Waser B, van Hagen M, Lamberts SW, Krenning EP, Gebbers JO, Laissue JA. In vitro and in vivo detection of somatostatin receptors in human malignant lymphomas. Int J Cancer 1992;50:895–900.

35. Reubi JC, Mazzucchelli L, Hennig I, Laissue JA. Local up-regulation of neuropeptide receptors in host blood vessels around human colorectal cancers. Gastroenterology 1996;110:1719–26.

36. Denzler B, Reubi JC. Expression of somatostatin receptors in peritumoral veins of human tumors. Cancer 1999;85:188–98.

37. Watson JC, Balster DA, Gebhardt BM, O'Dorisio TM, O'Dorisio MS, Espenan GD, Drouant GJ, Woltering EA. Growing vascular endothelial cells express somatostatin subtype 2 receptors. Br J Cancer 2001;85:266–72.

38. Vanhagen PM, Krenning EP, Reubi JC, Kwekkeboom DJ, Bakker WH, Mulder AH, Laissue I, Hoogstede HC, Lamberts SW. Somatostatin analogue scintigraphy in granulomatous diseases. Eur J Nucl Med 1994;21:497–502.

39. Vanhagen PM, Markusse HM, Lamberts SW, Kwekkeboom DJ, Reubi JC, Krenning EP. Somatostatin receptor imaging. The presence of somatostatin receptors in rheumatoid arthritis. Arthritis Rheum 1994;37:1521–7.

40. Reubi JC, Mazzucchelli L, Laissue JA. Intestinal vessels express a high density of somatostatin receptors in human inflammatory bowel disease. Gastroenterology 1994;106:951–9.

41. Gibril F, Reynolds JC, Chen CC, Yu F, Goebel SU, Serrano J, Doppman JL, Jensen RT. Specificity of somatostatin receptor scintigraphy: a prospective study and effects of false-positive localizations on management in patients with gastrinomas. J Nucl Med 1999;40:539–53.

42. Vinik AI, McLeod MK, Fig Shapiro B, Lloyd RV, Cho K. Clinical features, diagnosis, and localization of carcinoid tumors and their management. Gastroenterol Clin North Am 1989;18: 865–96.

43. Jamar F, Fiasse R, Leners N, Pauwels S. Somatostatin receptor imaging with indium-111-pentetreotide in gastroenteropancreatic neuroendocrine tumors: safety, efficacy and impact on patient management. J Nucl Med 1995;36:542–9.

44. Virgolini I, Traub-Weidinger T, Decristoforo C. Nuclear medicine in the detection and management of pancreatic islet-cell tumours. Best Pract Res Clin Endocrinol Metab 2005;19:213–27.

45. Bertherat J, Tenenbaum F, Perlemoine K, Videau C, Alberini JL, Richard B, Dousset B, Bertagna X, Epelbaum J. Somatostatin receptors 2 and 5 are the major somatostatin receptors in insulinomas: an in vivo and in vitro study. J Clin Endocrinol Metab 2003;88:5353–60.

46. Proye C, Malvaux P, Pattou F. Non-invasive imaging of insulinomas and gastrinomas with endoscopic ultrasonography and somatostatin receptor scintigraphy. Surgery 1998;124:1134–44.

47. Termanini B, Gibril F, Reynolds JC, Doppman JL, Chen CC, Stewart CA, Sutliff VE, Jensen RT. Value of somatostatin receptor scintigraphy: a prospective study in gastrinoma of its effect on clinical management. Gastroenterology 1997;112:335–47.

48. Gibril F, Doppman JL, Reynolds JC, Chen CC, Sutliff VE, Yu F, Serrano J, Venzon DJ, Jensen RT. Bone metastases in patients with gastrinomas: a prospective study of bone scanning, somatostatin receptor scanning, and magnetic resonance image in their detection, frequency, location, and effect of their detection on management. J Clin Oncol 1998;16:1040–53.

49. Konomi K, Chijiiwa K, Katsuta T, Yamaguchi K. Pancreatic somatostatinoma: a case report and review of the literature. J Surg Oncol 1990;43:259–65.

50. Harris GJ, Tio F, Cruz AB Jr. Somatostatinoma: a case report and review of the literature. J Surg Oncol 1987;36:8–16.

51. Zhang M, Xu X, Shen Y, Hu ZH, Wu LM, Zheng SS. Clinical experience in diagnosis and treatment of glucagonoma syndrome. Hepatobiliary Pancreat Dis Int 2004;3:473–5.

52. Smith SL, Branton SA, Avino AJ, Martin JK, Klingler PJ, Thompson GB, Grant CS, van Heerden JA. Vasoactive intestinal polypeptide secreting islet cell tumors: a 15-year experience and review of the literature. Surgery 1998;124:1050–5.

53. Thomason JW, Martin RS, Fincher ME. Somatostatin receptor scintigraphy: the definitive technique for characterizing vasoactive intestinal peptide-secreting tumors. Clin Nucl Med 2000;25:661–4.

54. Marcos HB, Libutti SK, Alexander HR, Lubensky IA, Bartlett DL, Walther MM, Linehan WM, Glenn GM, Choyke PL. Neuroendocrine tumors of the pancreas in von Hippel-Lindau disease: spectrum of appearances at CT and MR imaging with histopathologic comparison. Radiology 2002;225:751–8.

55. Kwekkeboom DJ, Reubi JC, Lamberts SW, Bruining HA, Mulder AH, Oei HY, Krenning EP. In vivo somatostatin receptor imaging in medullary thyroid carcinoma. J Clin Endocrinol Metab 1993;76:1413–7.

56. Ahlman H, Tisell LE, Wangberg B, Fjalling M, Forssell-Aronsson E, Kolby L, Nilsson O. The relevance of somatostatin receptors in thyroid neoplasia. Yale J Biol Med 1997;70:523–33.

57. Kwekkeboom D, Krenning EP, de Jong M. Peptide receptor imaging and therapy. J Nucl Med 2000;41:1704–13.

58. Kwekkeboom DJ, Lamberts SW, Habbema JD, Krenning EP. Cost-effectiveness analysis of somatostatin receptor scintigraphy. J Nucl Med 1996;37:886–92.

59. Baudin E, Lumbroso J, Schlumberger M, Leclere J, Giammarile F, Gardet P, Roche A, Travagli JP, Parmentier C. Comparison of octreotide scintigraphy and conventional imaging in medullary thyroid carcinoma. J Nucl Med 1996;37:912–6.

60. Kurtaran A, Scheuba C, Kaserer K, Schima W, Czerny C, Angelberger P, Niederle B, Virgolini I. Indium-111-DTPA-D-Phe-1-octreotide and technetium-99m-(V)-dimercaptosuccinic acid scanning in the preoperative staging of medullary thyroid carcinoma. J Nucl Med 1998;39:1907–9.

61. Adams S, Baum RP, Hertel A, Schumm-Draeger PM, Usadel KH, Hor G. Comparison of metabolic and receptor imaging in recurrent medullary thyroid carcinoma with histopathological findings. Eur J Nucl Med 1998;25:1277–83.

62. Postema PT, De Herder WW, Reubi JC, Oei HY, Kwekkeboom DJ, Bruining HJ, Bonjer J, van Toor H, Hennemann G, Krenning EP. Somatostatin receptor scintigraphy in non-medullary thyroid cancer. Digestion 1996;57 Suppl 1:36–7.

63. Gulec SA, Serafini AN, Sridhar KS, Peker KR, Gupta A, Goodwin WJ, Sfakianakis GN, Moffat FL. Somatostatin receptor expression in Hurthle cell cancer of the thyroid. J Nucl Med 1998;39: 243–5.

64. Reubi JC, Chayvialle JA, Franc B, Cohen R, Calmettes C, Modigliani E. Somatostatin receptors and somatostatin content in medullary thyroid carcinomas. Lab Invest 1991;64:567–73.

65. Stokkel MP, Verkooijen RB, Bouwsma H, Smit JW. Six month follow-up after 111In-DTPA-octreotide therapy in patients with progressive radioiodine non-responsive thyroid cancer: a pilot study. Nucl Med Commun 2004;25:683–90.

66. Reisinger I, Bohuslavitzki KH, Brenner W, Braune S, Dittrich I, Geide A, Kettner B, Otto HJ, Schmidt S, Munz DL. Somatostatin receptor scintigraphy in small-cell lung cancer: results of a multicenter study. J Nucl Med 1998;39:224–7.

67. Bombardieri E, Chiti A, Crippa F, Seregni E, Cataldo I, Maffioli L, Soresi E. 111In-DTPA-D-Phe-1-octreotide scintigraphy of small cell lung cancer. Q J Nucl Med 1995;39:104–7.

68. Kwekkeboom DJ, Kho GS, Lamberts SW, Reubi JC, Laissue JA, Krenning EP. The value of octreotide scintigraphy in patients with lung cancer. Eur J Nucl Med 1994;21:1106–13.

69. Kirsch CM, von Pawel J, Grau I, Tatsch K. Indium-111 pentetreotide in the diagnostic work-up of patients with bronchogenic carcinoma. Eur J Nucl Med 1994;21:1318–25.

70. Sagman U, Mullen JB, Kovacs K, Kerbel R, Ginsberg R, Reubi JC. Identification of somatostatin receptors in human small cell lung carcinoma. Cancer 1990;66:2129–33.

71. Epelbaum J, Bertherat J, Prevost G, Kordon C, Meyerhof W, Wulfsen I, Richter D, Plouin PF. Molecular and pharmacological characterization of somatostatin receptor subtypes in adrenal, extraadrenal, and malignant pheochromocytomas. J Clin Endocrinol Metab 1995;80:1837–44.

72. Reubi JC, Waser B, Khosla S, Kvols L, Goellner JR, Krenning E, Lamberts S. In vitro and in vivo detection of somatostatin receptors in pheochromocytomas and paragangliomas. J Clin Endocrinol Metab 1992;74:1082–9.

73. Manil L, Edeline V, Lumbroso J, Lequen H, Zucker JM. Indium-111-pentetreotide scintigraphy in children with neuroblast-derived tumors. J Nucl Med 1996;37:893–6.

74. Moertel CL, Reubi JC, Scheithauer BS, Schaid DJ, Kvols LK. Expression of somatostatin receptors in childhood neuroblastoma. Am J Clin Pathol 1994;102:752–6.

75. Prevost G, Veber N, Viollet C, Roubert V, Roubert P, Benard J, Eden P. Somatostatin-14 mainly binds the somatostatin receptor subtype 2 in human neuroblastoma tumors. Neuroendocrinology 1996;63:188–97.

76. Ellison DA, Parham DM. Tumors of the autonomic nervous system. Am J Clin Pathol 2001;115 Suppl:S46–55.

77. Lamberts SW, Krenning EP, Reubi JC. The role of somatostatin and its analogs in the diagnosis and treatment of tumors. Endocr Rev 1991;12:450–82.

78. Kwekkeboom DJ, van Urk H, Pauw BK, Lamberts SW, Kooij PP, Hoogma RP, Krenning EP. Octreotide scintigraphy for the detection of paragangliomas. J Nucl Med 1993;34:873–8.

79. Muros MA, Llamas-Elvira JM, Rodriguez A, Ramirez A, Gomez M, Arraez MA, Valencia E, Vilchez R. 111In-pentetreotide scintigraphy is superior to 123I-MIBG scintigraphy in the diagnosis and location of chemodectoma. Nucl Med Commun 1998;19:735–42.

80. Duet M, Sauvaget E, Guichard JP, Petelle B, Ledjedel S, Herman P, Tran Ba Huy P. [Scintigraphy by l'Octreoscan in the management of head and neck paragangliomas]. Ann Otolaryngol Chir Cervicofac 2002;119:315–21.

81. Myssiorek D, Palestro CJ. 111Indium pentetreotide scan detection of familial paragangliomas. Laryngoscope 1998;108:228–31.

82. Duet M, Sauvaget E, Petelle B, Rizzo N, Guichard JP, Wassef M, Le Cloirec J, Herman P, Tran Ba Huy P. Clinical impact of somatostatin receptor scintigraphy in the management of paragangliomas of the head and neck. J Nucl Med 2003;44:1767–74.

83. Lee JD, Kim DI, Lee JT, Chang JW, Park CY. Indium-111-pentetreotide imaging in intra-axial brain tumors: comparison with thallium-201 SPECT and MRI. J Nucl Med 1995;36:537–41.

84. Haldemann AR, Rosler H, Barth A, Waser B, Geiger L, Godoy N, Markwalder RV, Seiler RW, Sulzer M, Reubi JC. Somatostatin receptor scintigraphy in central nervous system tumors: role of blood-brain barrier permeability. J Nucl Med 1995;36:403–10.

85. Mawrin C, Schulz S, Pauli SU, Treuheit T, Diete S, Dietzmann K, Firsching R, Schulz S, Hollt V. Differential expression of sst1, sst2A, and sst3 somatostatin receptor proteins in low-grade and high-grade astrocytomas. J Neuropathol Exp Neurol 2004;63:13–9.

86. Safford SD, Coleman RE, Gockerman JP, Moore J, Feldman J, Onaitis MW, Tyler DS, Olson JA Jr. Iodine-131 metaiodobenzylguanidine treatment for metastatic carcinoid. Results in 98 patients. Cancer 2004;101:1987–93.

87. Schulz S, Pauli SU, Schulz S, Handel M, Dietzmann K, Firsching R, Hollt V. Immunohistochemical determination of five somatostatin receptors in meningioma reveals frequent overexpression of somatostatin receptor subtype sst2A. Clin Cancer Res 2000;6:1865–74.

88. Fruhwald MC, O'Dorisio MS, Pietsch T, Reubi JC. High expression of somatostatin receptor subtype 2 (sst2) in medulloblastoma: implications for diagnosis and therapy. Pediatr Res 1999;45:697–708.

89. Reubi JC, Lang W, Maurer R, Koper JW, Lamberts SW. Distribution and biochemical characterization of somatostatin receptors in tumors of the human central nervous system. Cancer Res 1987;47:5758–64.

90. Cervera P, Videau C, Viollet C, Petrucci C, Lacombe J, Winsky-Sommerer R, Csaba Z, Helboe L, Daumas-Duport C, Reubi JC, Epelbaum J. Comparison of somatostatin receptor expression in human gliomas and medulloblastomas. J Neuroendocrinol 2002;14:458–71.

91. Ur E, Mather SJ, Bomanji J, Ellison D, Britton KE, Grossman AB, Wass JA, Besser GM. Pituitary imaging using a labelled somatostatin analogue in acromegaly. Clin Endocrinol (Oxf) 1992;36:147–50.

92. Legovini P, De Menis E, Billeci D, Conti B, Zoli P, Conte N. 111Indium-pentetreotide pituitary scintigraphy and hormonal responses to octreotide in acromegalic patients. J Endocrinol Invest 1997;20:424–8.

93. Miller GV, Farmery SM, Woodhouse LF, Primrose JN. Somatostatin binding in normal and malignant human gastrointestinal mucosa. Br J Cancer 1992;66:391–5.

94. Reubi JC, Waser B, Schmassmann A, Laissue JA. Receptor autoradiographic evaluation of cholecystokinin, neurotensin, somatostatin and vasoactive intestinal peptide receptors in gastro-intestinal adenocarcinoma samples: where are they really located? Int J Cancer 1999;81:376–86.

95. Reubi JC, Zimmermann A, Jonas S, Waser B, Neuhaus P, Laderach U, Wiedenmann B. Regulatory peptide receptors in human hepatocellular carcinomas. Gut 1999;45:766–74.

96. van Eijck CH, Krenning EP, Bootsma A, Oei HY, van Pel R, Lindemans J, Jeekel J, Reubi JC, Lamberts SW. Somatostatin-receptor scintigraphy in primary breast cancer. Lancet 1994;343:640–3.

97. Reubi JC, Waser B, Foekens JA, Klijn JG, Lamberts SW, Laissue J. Somatostatin receptor incidence and distribution in breast cancer using receptor autoradiography: relationship to EGF receptors. Int J Cancer 1990;46:416–20.

98. Foekens JA, Portengen H, van Putten WL, Trapman AM, Reubi JC, Alexieva-Figusch J, Klijn JG. Prognostic value of receptors for insulin-like growth factor 1, somatostatin, and epidermal growth factor in human breast cancer. Cancer Res 1989;49:7002–9.

99. Reubi JC, Kvols L. Somatostatin receptors in human renal cell carcinomas. Cancer Res 1992;52:6074–8.

100. Reubi JC, Waser B, Schaer JC, Markwalder R. Somatostatin receptors in human prostate and prostate cancer. J Clin Endocrinol Metab 1995;80:2806–14.

101. Halmos G, Schally AV, Sun B, Davis R, Bostwick DG, Plonowski A. High expression of somatostatin receptors and messenger ribonucleic acid for its receptor subtypes in organ-confined and locally advanced human prostate cancers. J Clin Endocrinol Metab 2000;85:2564–71.

102. Halmos G, Sun B, Schally AV, Hebert F, Nagy A. Human ovarian cancers express somatostatin receptors. J Clin Endocrinol Metab 2000;85:3509–12.

103. Kouroumalis E, Skordilis P, Thermos K, Vasilaki A, Moschandrea J, Manousos ON. Treatment of hepatocellular carcinoma with octreotide: a randomised controlled study. Gut 1998;42:442–7.

104. Vanhagen PM, Krenning EP, Reubi JC, Mulder AH, Bakker WH, Oei HY, Lowenberg B, Lamberts SW. Somatostatin analogue scintigraphy of malignant lymphomas. Br J Haematol 1993;83:75–9.

105. Leners N, Jamar F, Fiasse R, Ferrant A, Pauwels S. Indium-111-pentetreotide uptake in endocrine tumors and lymphoma. J Nucl Med 1996;37:916–22.

106. Sarda L, Duet M, Zini JM, Berolatti B, Benelhadj S, Tobelem G, Mundler O. Indium-111 pentetreotide scintigraphy in malignant lymphomas. Eur J Nucl Med 1995;22:1105–9.

107. Lipp RW, Silly H, Ranner G, Dobnig H, Passath A, Leb G, Krejs GJ. Radiolabeled octreotide for the demonstration of somatostatin receptors in malignant lymphoma and lymphadenopathy. J Nucl Med 1995;36:13–8.

108. Lugtenburg PJ, Krenning EP, Valkema R, Oei HY, Lamberts SW, Eijkemans MJ, van Putten WL, Lowenberg B. Somatostatin receptor scintigraphy useful in stage I-II Hodgkin's disease: more extended disease identified. Br J Haematol 2001;112:936–44.

109. Lugtenburg PJ, Lowenberg B, Valkema R, Oei HY, Lamberts SW, Eijkemans MJ, van Putten WL, Krenning EP. Somatostatin receptor scintigraphy in the initial staging of low-grade non-Hodgkin's lymphomas. J Nucl Med 2001;42:222–9.

110. Lum SS, Fletcher WS, O'Dorisio MS, Nance RW, Pommier RF, Caprara M. Distribution and functional significance of somatostatin receptors in malignant melanoma. World J Surg 2001;25:407–12.

111. Hoefnagel CA, Rankin EM, Valdes Olmos RA, Israels SP, Pavel S, Janssen AG. Sensitivity versus specificity in melanoma imaging using iodine-123 iodobenzamide and indium-111 pentetreotide. Eur J Nucl Med 1994;21:587–8.

112. Modlin IM, Lye KD, Kidd M. Radio-labeled octreotide for treatment of endocrine and other tumors. In: Progress in Oncology, De Vita VT, Hellman S, Rosenberg SA, eds. Sudbury: Jones & Barlett, 2003:169–209.

113. Ferone D, van Hagen MP, Kwekkeboom DJ, van Koetsveld PM, Mooy DM, Lichtenauer-Kaligis E, Schonbrunn A, Colao A, Lamberts SW, Hofland LJ. Somatostatin receptor subtypes in human thymoma and inhibition of cell proliferation by octreotide in vitro. J Clin Endocrinol Metab 2000;85:1719–26.

114. Reubi JC, Waser B, Laissue JA, Gebbers JO. Somatostatin and vasoactive intestinal peptide receptors in human mesenchymal tumors: in vitro identification. Cancer Res 1996;56:1922–31.

115. Kwekkeboom DJ, Hoff AM, Lamberts SW, Oei HY, Krenning EP. Somatostatin analogue scintigraphy. A simple and sensitive method for the in vivo visualization of Merkel cell tumors and their metastases. Arch Dermatol 1992;128:818–21.

116. de Herder WW, Krenning EP, Malchoff CD, Hofland LJ, Reubi JC, Kwekkeboom DJ, Oei HY, Pols HA, Bruining HA, Nobels FR, et al. Somatostatin receptor scintigraphy: its value in tumor localization in patients with Cushing's syndrome caused by ectopic corticotropin or corticotropin-releasing hormone secretion. Am J Med 1994;96:305–12.

117. Phlipponneau M, Nocaudie M, Epelbaum J, De Keyzer Y, Lalau JD, Marchandise X, Bertagna X. Somatostatin analogs for the localization and preoperative treatment of an adrenocorticotropin-secreting bronchial carcinoid tumor. J Clin Endocrinol Metab 1994;78:20–4.

118. Weiss M, Yellin A, Husza'r M, Eisenstein Z, Bar-Ziv J, Krausz Y. Localization of adrenocorticotropic hormone-secreting bronchial carcinoid tumor by somatostatin-receptor scintigraphy. Ann Intern Med 1994;121:198–9.

119. Torpy DJ, Chen CC, Mullen N, Doppman JL, Carrasquillo JA, Chrousos GP, Nieman LK. Lack of utility of (111)In-pentetreotide scintigraphy in localizing ectopic ACTH producing tumors: follow-up of 18 patients. J Clin Endocrinol Metab 1999;84:1186–92.

120. Kwekkeboom DJ, Krenning EP, Kho GS, Breeman WA, Van Hagen PM. Somatostatin receptor imaging in patients with sarcoidosis. Eur J Nucl Med 1998;25:1284–92.

121. Postema PT, Krenning EP, Wijngaarde R, Kooy PP, Oei HY, van den Bosch WA, Reubi JC, Wiersinga WM, Hooijkaas H, van der Loos T, et al. [111In-DTPA-D-Phe1] octreotide scintigraphy in thyroidal and orbital Graves' disease: a parameter for disease activity? J Clin Endocrinol Metab 1994;79:1845–51.

122. Krassas GE, Dumas A, Pontikides N, Kaltsas T. Somatostatin receptor scintigraphy and octreotide treatment in patients with thyroid eye disease. Clin Endocrinol (Oxf) 1995;42:571–80.

123. Gerding MN, van der Zant FM, van Royen EA, Koornneef L, Krenning EP, Wiersinga WM, Prummel MF. Octreotide-scintigraphy is a disease-activity parameter in Graves' ophthalmopathy. Clin Endocrinol (Oxf) 1999;50:373–9.

124. Epelbaum J, Dournaud P, Fodor M, Viollet C. The neurobiology of somatostatin. Crit Rev Neurobiol 1994;8:25–44.

125. Bauer W, Briner U, Doepfner W, Haller R, Huguenin R, Marbach P, Petcher TJ, Pless J. SMS 201–995: a very potent and selective octapeptide analogue of somatostatin with prolonged action. Life Sci 1982;31:1133–40.

126. Czernik AJ, Petrack B. Somatostatin receptor binding in rat cerebral cortex. Characterization using a nonreducible somatostatin analog. J Biol Chem 1983;258:5525–30.

127. Veber DF, Saperstein R, Nutt RF, Freidinger RM, Brady SF, Curley P, Perlow DS, Paleveda WJ, Colton CD, Zacchei AG, et al. A super active cyclic hexapeptide analog of somatostatin. Life Sci 1984;34:1371–8.

128. Bruns C, Lewis I, Briner U, Meno-Tetang G, Weckbecker G. SOM230: a novel somatostatin peptidomimetic with broad somatotropin release inhibiting factor (SRIF) receptor binding and a unique antisecretory profile. Eur J Endocrinol 2002;146:707–16.

129. Lamberts SW, van der Lely AJ, de Herder WW, Hofland LJ. Octreotide. N Engl J Med 1996;334:246–54.

130. Bevan JS. Clinical review: The antitumoral effects of somatostatin analog therapy in acromegaly. J Clin Endocrinol Metab 2005;90:1856–63.

131. Freda PU. Somatostatin analogs in acromegaly. J Clin Endocrinol Metab 2002;87:3013–8.

132. Anthony LB. Long-acting formulations of somatostatin analogues. Ital J Gastroenterol Hepatol 1999;31 Suppl 2:S216–8.

133. Lancranjan I, Bruns C, Grass P, Jaquet P, Jervell J, Kendall-Taylor P, Lamberts SW, Marbach P, Orskov H, Pagani G, Sheppard M, Simionescu L. Sandostatin LAR: a promising therapeutic tool in the management of acromegalic patients. Metabolism 1996;45:67–71.

134. Lightman S. Somatuline Autogel: an extended release lanreotide formulation. Hosp Med 2002;63:162–5.

135. Schmid H, Silva A. Short- and long-term effects of octreotide and SOM230 on GH, IGF-I, ACTH, corticosterone and ghrelin in rats. J Endocrinol Invest 2005;28 Suppl 11:28–35.

136. Silva A, Schoeffter P, Weckbecker G, Bruns C, Schmid H. Regulation of CRH-induced secretion of ACTH and corticosterone by SOM230 in rats. Eur J Endocrinol 2005;153:R7–10.

137. Silva A, Bethmann K, Raulf F, Schmid H. Regulation of ghrelin secretion by somatostatin analogs in rats. Eur J Endocrinol 2005;152:887–94.

138. van der Hoek J, van der Lelij A, Feelders R, de Herder W, Uitterlinden P, Poon K, Boerlin V, Lewis I, Krahnke T, Hofland L, Lamberts S. The somatostatin analogue SOM230, compared with octreotide, induces differential effects in several metabolic pathways in acromegalic patients. Clin Endocrinol (Oxf) 2005;63:176–84.

139. Adams R, Adams I, Lindow S, Atkin S. Inhibition of endothelial proliferation by the somatostatin analogue SOM230. Clin Endocrinol (Oxf) 2004;61:431–6.

140. Kaltsas G, Besser G, Grossman A. The diagnosis and medical management of advanced neuroendocrine tumors. Endocr Rev 2004;25:458–511.

141. Rolleman EJ, Valkema R, de Jong M, Kooij PP, Krenning EP. Safe and effective inhibition of renal uptake of radiolabelled octreotide by a combination of lysine and arginine. Eur J Nucl Med Mol Imaging 2003;30:9–15.

142. Chinol M, Bodei L, Cremonesi M, Paganelli G. Receptor-mediated radiotherapy with Y-DOTA-DPhe-Tyr-octreotide: the experience of the European Institute of Oncology Group. Semin Nucl Med 2002;32:141–7.

143. Lambert B, Cybulla M, Weiner SM, Van De Wiele C, Ham H, Dierckx RA, Otte A. Renal toxicity after radionuclide therapy. Radiat Res 2004;161:607–11.

144. Anthony LB, Woltering EA, Espenan GD, Cronin MD, Maloney TJ, McCarthy KE. Indium-111-pentetreotide prolongs survival in gastroenteropancreatic malignancies. Semin Nucl Med 2002;32:123–32.

145. Buchsbaum DJ, Rogers BE, Khazaeli MB, Mayo MS, Milenic DE, Kashmiri SV, Anderson CJ, Chappell LL, Brechbiel MW, Curiel DT. Targeting strategies for cancer radiotherapy. Clin Cancer Res 1999;5:3048s–55s.

146. de Jong M, Bakker WH, Krenning EP, Breeman WA, van der Pluijm ME, Bernard BF, Visser TJ, Jermann E, Behe M, Powell P, Macke HR. Yttrium-90 and indium-111 labelling, receptor binding and biodistribution of [DOTA0,d-Phe1,Tyr3]octreotide, a promising somatostatin analogue for radionuclide therapy. Eur J Nucl Med 1997;24:368–71.

147. Breeman WA, de Jong M, Kwekkeboom DJ, Valkema R, Bakker WH, Kooij PP, Visser TJ, Krenning EP. Somatostatin receptor-mediated imaging and therapy: basic science, current knowledge, limitations and future perspectives. Eur J Nucl Med 2001;28:1421–9.

148. de Jong M, Breeman WA, Bernard BF, Bakker WH, Schaar M, van Gameren A, Bugaj JE, Erion J, Schmidt M, Srinivasan A, Krenning EP. [177Lu-DOTA(0),Tyr3] octreotate for somatostatin receptor-targeted radionuclide therapy. Int J Cancer 2001;92:628–33.

149. Lewis J, Wang M, Laforest R, Wang F, Erion J, Bugaj J, Srinivasan A, Anderson C. Toxicity and dosimetry of (177)Lu-DOTA-Y3-octreotate in a rat model. Int J Cancer 2001;94:873–7.

150. Reubi JC. New specific radioligand for one subpopulation of brain somatostatin receptors. Life Sci 1985;36:1829–36.

151. Reubi JC, Schar JC, Waser B, Wenger S, Heppeler A, Schmitt JS, Macke HR. Affinity profiles for human somatostatin receptor subtypes SST1-SST5 of somatostatin radiotracers selected for scintigraphic and radiotherapeutic use. Eur J Nucl Med 2000;27:273–82.

152. Otte A, Herrmann R, Heppeler A, Behe M, Jermann E, Powell P, Maecke HR, Muller J. Yttrium-90 DOTATOC: first clinical results. Eur J Nucl Med 1999;26:1439–47.

153. Waldherr C, Pless M, Maecke HR, Haldemann A, Mueller-Brand J. The clinical value of [90Y-DOTA]-D-Phe1-Tyr3-octreotide (90Y-DOTATOC) in the treatment of neuroendocrine tumours: a clinical phase II study. Ann Oncol 2001;12:941–5.

154. Virgolini I, Britton K, Buscombe J, Moncayo R, Paganelli G, Riva P. In- and Y-DOTA-lanreotide: results and implications of the MAURITIUS trial. Semin Nucl Med 2002;32:148–55.

155. McCarthy KE, Woltering EA, Espenan GD, Cronin M, Maloney TJ, Anthony LB. In situ radiotherapy with 111In-pentetreotide: initial observations and future directions. Cancer J Sci Am 1998;4:94–102.

156. De Jong M, Breeman WA, Bernard HF, Kooij PP, Slooter GD, Van Eijck CH, Kwekkeboom DJ, Valkema R, Macke HR, Krenning EP. Therapy of neuroendocrine tumors with radiolabeled somatostatin-analogues. Q J Nucl Med 1999;43:356–66.

157. Valkema R, De Jong M, Bakker WH, Breeman WA, Kooij PP, Lugtenburg PJ, De Jong FH, Christiansen A, Kam BL, De Herder WW, Stridsberg M, Lindemans J, Ensing G, Krenning EP. Phase I study of peptide receptor radionuclide therapy with [In-DTPA]octreotide: the Rotterdam experience. Semin Nucl Med 2002;32:110–22.

158. Buscombe JR, Caplin ME, Hilson AJ. Long-term efficacy of high-activity 111in-pentetreotide therapy in patients with disseminated neuroendocrine tumors. J Nucl Med 2003;44:1–6.

159. Modlin IMJ, Lye KD. 111Indium-labeled somatostatin analogues: a novel advance in the management of unresectable neuroendocrine tumors (NETs). Cancer 2007 (in press).

160. Taal BG, Hoefnagel CA, Valdes Olmos RA, Boot H, Beijnen JH. Palliative effect of metaiodobenzylguanidine in metastatic carcinoid tumors. J Clin Oncol 1996;14:1829–38.

161. Mukherjee JJ, Kaltsas GA, Islam N, Plowman PN, Foley R, Hikmat J, Britton KE, Jenkins PJ, Chew SL, Monson JP, Besser GM, Grossman AB. Treatment of metastatic carcinoid tumours, phaeochromocytoma, paraganglioma and medullary carcinoma of the thyroid with (131)I-meta-iodobenzylguanidine [(131)I-mIBG]. Clin Endocrinol (Oxf) 2001;55:47–60.

162. Kwekkeboom D, Bakker W, Kam B, Teunissen JJ, Kooij PP, de Herder WW, Feelders RA, van Eijck CH, de Jong M, Srinivasan A, Erion JL, Krenning EP. Treatment of patients with gastro-entero-pancreatic (GEP) tumours with the novel radiolabelled somatostatin analogue [177Lu DOTA(0),Tyr3]octreotate. Eur J Nucl Med Mol Imaging 2003;30:417–22.

163. Teunissen JJ, Kwekkeboom DJ, Krenning EP. Quality of life in patients with gastroenteropancreatic tumors treated with [177Lu-DOTA0,Tyr3]octreotate. J Clin Oncol 2004;22:2724–9.

164. Kwekkeboom DJ, Teunissen JJ, Bakker WH, Kooij PP, de Herder WW, Feelders RA, van Eijck CH, Esser JP, Kam BL, Krenning EP. Radiolabeled somatostatin analog [177Lu-DOTA0,Tyr3]octreotate in patients with endocrine gastroenteropancreatic tumors. J Clin Oncol 2005;23:2754–62.

165. Oberg K. Chemotherapy and biotherapy in neuroendocrine tumors. Curr Opin Oncol 1993;5:110–20.

166. Kvols LK, Moertel CG, O'Connell MJ, Schutt AJ, Rubin J, Hahn RG. Treatment of the malignant carcinoid syndrome. Evaluation of a long-acting somatostatin analogue. N Engl J Med 1986;315:663–6.

167. Kvols L, Moertel C, Schutt A, Rubin J. Treatment of the malignant carcinoid syndrome with a long-acting octreotide analogue (SMS 201–955): preliminary evidence that more is not better (abstract). Proc Am Soc Clin Oncol 1987:A27–95.

168. Vinik A, Moattari AR. Use of somatostatin analog in management of carcinoid syndrome. Dig Dis Sci 1989;34:14S–27S.

169. Oberg K, Norheim I, Theodorsson E. Treatment of malignant midgut carcinoid tumours with a long-acting somatostatin analogue octreotide. Acta Oncol 1991;30:503–7.

170. Saltz L, Trochanowski B, Buckley M, Heffernan B, Niedzwiecki D, Tao Y, Kelsen D. Octreotide as an antineoplastic agent in the treatment of functional and nonfunctional neuroendocrine tumors. Cancer 1993;72:244–8.

171. Janson EM, Ahlstrom H, Andersson T, Oberg KE. Octreotide and interferon alfa: a new combination for the treatment of malignant carcinoid tumours. Eur J Cancer 1992;28A:1647–50.

172. Janson ET, Oberg K. Long-term management of the carcinoid syndrome. Treatment with octreotide alone and in combination with α-interferon. Acta Oncol 1993;32:225–9.

173. di Bartolomeo M, Bajetta E, Buzzoni R, Mariani L, Carnaghi C, Somma L, Zilembo N, di Leo A. Clinical efficacy of octreotide in the treatment of metastatic neuroendocrine tumors. A study by the Italian Trials in Medical Oncology Group. Cancer 1996;77:402–8.

174. Nilsson O, Kolby L, Wangberg B, Wigander A, Billig H, William-Olsson L, Fjalling M, Forssell-Aronsson E, Ahlman H. Comparative studies on the expression of somatostatin receptor subtypes, outcome of octreotide scintigraphy and response to octreotide treatment in patients with carcinoid tumours. Br J Cancer 1998;77:632–7.

175. O'Toole D, Ducreux M, Bommelaer G, Wemeau JL, Bouche O, Catus F, Blumberg J, Ruszniewski P. Treatment of carcinoid syndrome: a prospective crossover evaluation of lanreotide versus octreotide in terms of efficacy, patient acceptability, and tolerance. Cancer 2000;88:770–6.

176. Garland J, Buscombe JR, Bouvier C, Bouloux P, Chapman MH, Chow AC, Reynolds N, Caplin ME. Sandostatin LAR (long-acting octreotide acetate) for malignant carcinoid syndrome: a 3-year experience. Aliment Pharmacol Ther 2003;17:437–44.

177. Gulanikar AC, Kotylak G, Bitter-Suermann H. Does immunosuppression alter the growth of metastatic liver carcinoid after orthotopic liver transplantation? Transplant Proc 1991;23:2197–8.

178. Kolby L, Persson G, Franzen S, Ahren B. Randomized clinical trial of the effect of interferon α on survival in patients with disseminated midgut carcinoid tumours. Br J Surg 2003;90:687–93.

179. Welin SV, Janson ET, Sundin A, Stridsberg M, Lavenius E, Granberg D, Skogseid B, Oberg KE, Eriksson BK. High-dose treatment with a long-acting somatostatin analogue in patients with advanced midgut carcinoid tumours. Eur J Endocrinol 2004;151:107–12.

180. Plockinger U, Rindi G, Arnold R, Eriksson B, Krenning EP, de Herder WW, Goede A, Caplin M, Wiedenmann B, Oberg K, Reubi JC, Nilsson O, Delle Fave G, Ruszniewski P. Guidelines for the diagnosis and treatment of neuroendocrine gastrointestinal tumours. Neuroendocrinology 2005;80:394–424.

181. Oberg K, Kvols L, Caplin M, Delle Fave G, de Herder W, Rindi G, Ruszniewski P, Woltering EA, Wiedenmann B.. Consensus report on the use of somatostatin analogs for the management of neuroendocrine tumors of the gastroenteropancreatic system. Ann Oncol 2004;15:966–73.

182. Scherubl H, Wiedenmann B, Riecken EO, Thomas F, Bohme E, Rath U. Treatment of the carcinoid syndrome with a depot formulation of the somatostatin analogue lanreotide. Eur J Cancer 1994;30A:1590–1.

183. Eriksson B, Janson ET, Bax ND, Mignon M, Morant R, Opolon P, Rougier P, Oberg KE. The use of new somatostatin analogues, lanreotide and octastatin, in neuroendocrine gastro-intestinal tumours. Digestion 1996;57 Suppl 1:77–80.

184. Ruszniewski P, Ducreux M, Chayvialle JA, Blumberg J, Cloarec D, Michel H, Raymond JM, Dupas JL, Gouerou H, Jian R, Genestin E, Bernades P, Rougier P. Treatment of the carcinoid

syndrome with the longacting somatostatin analogue lanreotide: a prospective study in 39 patients. Gut 1996;39:279–83.

185. Wymenga AN, Eriksson B, Salmela PI, Jacobsen MB, Van Cutsem EJ, Fiasse RH, Valimaki MJ, Renstrup J, de Vries EG, Oberg KE. Efficacy and safety of prolonged-release lanreotide in patients with gastrointestinal neuroendocrine tumors and hormone-related symptoms. J Clin Oncol 1999;17:1111.

186. Tomassetti P, Migliori M, Corinaldesi R, Gullo L. Treatment of gastroenteropancreatic neuroendocrine tumours with octreotide LAR. Aliment Pharmacol Ther 2000;14:557–60.

187. Ricci S, Antonuzzo A, Galli L, Orlandini C, Ferdeghini M, Boni G, Roncella M, Mosca F, Conte PF. Long-acting depot lanreotide in the treatment of patients with advanced neuroendocrine tumors. Am J Clin Oncol 2000;23:412–5.

188. Rohaizak M, Farndon JR. Use of octreotide and lanreotide in the treatment of symptomatic non-resectable carcinoid tumours. Aust N Z J Surg 2002;72:635–8.

189. Canobbio L, Cannata D, Miglietta L, Bocardo F. Use of long-acting somatostatin analogue, lanreotide, in neuroendocrine tumors. Oncol Rep Ort 1994:1.

190. Faiss S, Rath U, Mansmann U. Ultra-high dose lanreotide treatment in patients with metastatic neuroendocrine gastroenteropancreatic tumors. Digestion 1999;60:469–76.

191. Ducreux M, Ruszniewski M, Chayvialle J. The antitumoral effect of the long-acting somatostatin analog lanreotide in neuroendocrine tumors. Am J Gastroenterol 2000;95:3276–81.

192. Ruszniewski P, Ish-Shalom S, Wymenga M, O'Toole D, Arnold R, Tomassetti P, Bax N, Caplin M, Eriksson B, Glaser B, Ducreux M, Lombard-Bohas C, de Herder WW, Delle Fave G, Reed N, Seitz JF, Van Cutsem E, Grossman A, Rougier P, Schmidt W, Wiedenmann B. Rapid and sustained relief from the symptoms of carcinoid syndrome: results from an open 6-month study of the 28-day prolonged-release formulation of lanreotide. Neuroendocrinology 2004;80:244–51.

193. Kvols L, Oberg K, de Herder W. Early data on the efficacy and safety of the novel multi-ligand somatostatin analog, SOM230, in patients with metastatic carcinoid tumors refractory or resistant to octreotide LAR. In ASCO Annual Meeting, Orlando, FL, 2005.

194. Anthony LB, Kang Y, Shyr Y. Malignant carcinoid syndrome: survival in the octreotide LAR era. In ASCO Annual Meeting, Orlando, FL, 2005.

195. Faiss S, Rath U, Mansmann U, Caird D, Clemens N, Riecken EO, Wiedenmann B. Ultra-high-dose lanreotide treatment in patients with metastatic neuroendocrine gastroenteropancreatic tumors. Digestion 1999;60:469–76.

196. Krenning EP, Kooij PP, Bakker WH, Breeman WA, Postema PT, Kwekkeboom DJ, Oei HY, de Jong M, Visser TJ, Reijs AE, et al. Radiotherapy with a radiolabeled somatostatin analogue, [111In-DTPA-D-Phe1]-octreotide. A case history. Ann N Y Acad Sci 1994;733:496–506.

197. Kwekkeboom DJ, Mueller-Brand J, Paganelli G, Anthony LB, Pauwels S, Kvols LK, O'Dorisio T M, Valkema R, Bodei L, Chinol M, Maecke HR, Krenning EP. Overview of results of peptide receptor radionuclide therapy with 3 radiolabeled somatostatin analogs. J Nucl Med 2005;46 Suppl 1:62S–6S.

198. Esser JP, Krenning EP, Teunissen JJ, Kooij PP, van Gameren AL, Bakker WH, Kwekkeboom DJ. Comparison of [(177)Lu-DOTA (0),Tyr (3)]octreotate and [(177)Lu-DOTA (0),Tyr (3)]octreotide: which peptide is preferable for PRRT? Eur J Nucl Med Mol Imaging 2006;33:1346–51.

199. Reubi JC, Kvols L, Krenning E, Lamberts SW. Distribution of somatostatin receptors in normal and tumor tissue. Metabolism 1990;39:78–81.

200. Arnold R, Trautmann ME, Creutzfeldt W, Benning R, Benning M, Neuhaus C, Jurgensen R, Stein K, Schafer H, Bruns C, Dennler HJ. Somatostatin analogue octreotide and inhibition of tumour growth in metastatic endocrine gastroenteropancreatic tumours. Gut 1996;38:430–8.

201. Eriksson B, Renstrup J, Imam H, Oberg K. High-dose treatment with lanreotide of patients with advanced neuroendocrine gastrointestinal tumors: clinical and biological effects. Ann Oncol 1997;8:1041–4.

202. Tomassetti P, Migliori M, Gullo L. Slow-release lanreotide treatment in endocrine gastrointestinal tumors. Am J Gastroenterol 1998;93:1468–71.

203. Faiss S, Rath U, Mansmann U, Caird D, Clemens N, Riecken EO, Wiedenmann B. Ultra-high dose lanreotide treatment in patients with metastatic neuroendocrine gastroenteropancreatic tumors. Digestion 1999;60:469–76.

204. Ducreux M, Ruszniewski P, Chayvialle JA, Blumberg J, Cloarec D, Michel H, Raymond JM, Dupas JL, Gouerou H, Jian R, Genestin E, Hammel P, Rougier P. The antitumoral effect of the

long-acting somatostatin analog lanreotide in neuroendocrine tumors. Am J Gastroenterol 2000;95: 3276–81.

205. Aparicio T, Ducreux M, Baudin E, Sabourin JC, De Baere T, Mitry E, Schlumberger M, Rougier P. Antitumour activity of somatostatin analogues in progressive metastatic neuroendocrine tumours. Eur J Cancer 2001;37:1014–9.

206. Kvols LK, Buck M, Moertel CG, Schutt AJ, Rubin J, O'Connell MJ, Hahn RG. Treatment of metastatic islet cell carcinoma with a somatostatin analogue (SMS 201–995). Ann Intern Med 1987;107:162–8.

207. Maton PN. Use of octreotide acetate for control of symptoms in patients with islet cell tumors. World J Surg 1993;17:504–10.

208. Stehouwer CD, Lems WF, Fischer HR, Hackeng WH, Naafs MA. Aggravation of hypoglycemia in insulinoma patients by the long-acting somatostatin analogue octreotide (Sandostatin). Acta Endocrinol (Copenh) 1989;121:34–40.

209. de Herder WW. Insulinoma. Neuroendocrinology 2004;80 Suppl 1:20–2.

210. Modlin I, Sachs G. Acid Related Diseases-Biology and Treatment. Schnetzor Verlag, Konstanz 2004.

211. Gaztambide S, Vazquez JA. Short- and long-term effect of a long-acting somatostatin analogue, lanreotide (SR-L) on metastatic gastrinoma. J Endocrinol Invest 1999;22:144–6.

212. Arnold R, Wied M, Behr TH. Somatostatin analogues in the treatment of endocrine tumors of the gastrointestinal tract. Expert Opin Pharmacother 2002;3:643–56.

213. Shojamanesh H, Gibril F, Louie A, Ojeaburu JV, Bashir S, Abou-Saif A, Jensen RT. Prospective study of the antitumor efficacy of long-term octreotide treatment in patients with progressive metastatic gastrinoma. Cancer 2002;94:331–43.

214. Angeletti S, Corleto VD, Schillaci O, Marignani M, Annibale B, Moretti A, Silecchia G, Scopinaro F, Basso N, Bordi C, Delle Fave G. Use of the somatostatin analogue octreotide to localise and manage somatostatin-producing tumours. Gut 1998;42:792–4.

215. Wermers RA, Fatourechi V, Wynne AG, Kvols LK, Lloyd RV. The glucagonoma syndrome. Clinical and pathologic features in 21 patients. Medicine (Baltimore) 1996;75:53–63.

216. Boden G, Ryan IG, Eisenschmid BL, Shelmet JJ, Owen OE. Treatment of inoperable glucagonoma with the long-acting somatostatin analogue SMS 201–995. N Engl J Med 1986;314:1686–9.

217. Rosenbaum A, Flourie B, Chagnon S, Blery M, Modigliani R. Octreotide (SMS 201–995) in the treatment of metastatic glucagonoma: report of one case and review of the literature. Digestion 1989;42:116–20.

218. Fraker DL, Norton JA. The role of surgery in the management of islet cell tumors. Gastroenterol Clin North Am 1989;18:805–30.

219. Mozell E, Stenzel P, Woltering EA, Rosch J, O'Dorisio TM. Functional endocrine tumors of the pancreas: clinical presentation, diagnosis, and treatment. Curr Probl Surg 1990;27:301–86.

220. Vinik AI, Moattari AR. Treatment of endocrine tumors of the pancreas. Endocrinol Metab Clin North Am 1989;18:483–518.

221. Altimari AF, Bhoopalam N, O'Dorsio T, Lange CL, Sandberg L, Prinz RA. Use of a somatostatin analog (SMS 201–995) in the glucagonoma syndrome. Surgery 1986;100:989–96.

222. Bewley AP, Ross JS, Bunker CB, Staughton RC. Successful treatment of a patient with octreotide-resistant necrolytic migratory erythema. Br J Dermatol 1996;134:1101–4.

223. Park SK, O'Dorisio MS, O'Dorisio TM. Vasoactive intestinal polypeptide-secreting tumours: biology and therapy. Baillieres Clin Gastroenterol 1996;10:673–96.

224. Arnold R, Frank M. Gastrointestinal endocrine tumours: medical management. Baillieres Clin Gastroenterol 1996;10:737–59.

225. Smith S, Thomas RM, Steere HA, Beatty HE, Dawson KB, Peckham MJ. Therapeutic irradiation of the central nervous system using intrathecal 90Y–DTA. Br J Radiol 1976;49:141–7.

226. Cavalla P, Schiffer D. Neuroendocrine tumors in the brain. Ann Oncol 2001;12 Suppl 2:S131–4.

227. Garcia-Luna PP, Relimpio F, Pumar A, Pereira JL, Leal-Cerro A, Trujillo F, Cortes A, Astorga R. Clinical use of octreotide in unresectable meningiomas. A report of three cases. J Neurosurg Sci 1993;37:237–41.

228. Bodei L, Cremonesi M, Zoboli S, Grana C, Bartolomei M, Rocca P, Caracciolo M, Macke HR, Chinol M, Paganelli G. Receptor-mediated radionuclide therapy with 90Y-DOTATOC in association with amino acid infusion: a phase I study. Eur J Nucl Med Mol Imaging 2003;30:207–16.

229. Beutler D, Avoledo P, Reubi JC, Macke HR, Muller-Brand J, Merlo A, Kuhne T. Three-year recurrence-free survival in a patient with recurrent medulloblastoma after resection, high-dose chemotherapy, and intrathecal Yttrium-90-labeled DOTA0-D-Phe1-Tyr3-octreotide radiopeptide brachytherapy. Cancer 2005;103:869–73.

230. Merlo A, Hausmann O, Wasner M, Steiner P, Otte A, Jermann E, Freitag P, Reubi JC, Muller-Brand J, Gratzl O, Macke HR. Locoregional regulatory peptide receptor targeting with the diffusible somatostatin analogue 90Y-labeled DOTA0-D-Phe1-Tyr3-octreotide (DOTATOC): a pilot study in human gliomas. Clin Cancer Res 1999;5:1025–33.

231. Schumacher T, Hofer S, Eichhorn K, Wasner M, Zimmerer S, Freitag P, Probst A, Gratzl O, Reubi JC, Maecke R, Mueller-Brand J, Merlo A. Local injection of the 90Y-labelled peptidic vector DOTATOC to control gliomas of WHO grades II and III: an extended pilot study. Eur J Nucl Med Mol Imaging 2002;29:486–93.

232. Schally AV, Nagy A. Cancer chemotherapy based on targeting of cytotoxic peptide conjugates to their receptors on tumors. Eur J Endocrinol 1999;141:1–14.

233. Lamberts SW. The role of somatostatin in the regulation of anterior pituitary hormone secretion and the use of its analogs in the treatment of human pituitary tumors. Endocr Rev 1988;9:417–36.

234. Lamberts SW, Zweens M, Klijn JG, van Vroonhoven CC, Stefanko SZ, Del Pozo E. The sensitivity of growth hormone and prolactin secretion to the somatostatin analogue SMS 201–995 in patients with prolactinomas and acromegaly. Clin Endocrinol (Oxf) 1986;25:201–12.

235. Abe T, Ludecke DK. Effects of preoperative octreotide treatment on different subtypes of 90 GH-secreting pituitary adenomas and outcome in one surgical centre. Eur J Endocrinol 2001;145: 137–45.

236. Arosio M, Macchelli S, Rossi CM, Casati G, Biella O, Faglia G. Effects of treatment with octreotide in acromegalic patients–a multicenter Italian study. Italian Multicenter Octreotide Study Group. Eur J Endocrinol 1995;133:430–9.

237. Ezzat S, Snyder PJ, Young WF, Boyajy LD, Newman C, Klibanski A, Molitch ME, Boyd AE, Sheeler L, Cook DM, et al. Octreotide treatment of acromegaly. A randomized, multicenter study. Ann Intern Med 1992;117:711–8.

238. Lucas T, Astorga R, Catala M. Preoperative lanreotide treatment for GH-secreting pituitary adenomas: effect on tumour volume and predictive factors of significant tumour shrinkage. Clin Endocrinol (Oxf) 2003;58:471–81.

239. Verhelst JA, Pedroncelli AM, Abs R, Montini M, Vandeweghe MV, Albani G, Maiter D, Pagani MD, Legros JJ, Gianola D, Bex M, Poppe K, Mockel J, Pagani G. Slow-release lanreotide in the treatment of acromegaly: a study in 66 patients. Eur J Endocrinol 2000;143:577–84.

240. Bevan JS, Atkin SL, Atkinson AB, Bouloux PM, Hanna F, Harris PE, James RA, McConnell M, Roberts GA, Scanlon MF, Stewart PM, Teasdale E, Turner HE, Wass JA, Wardlaw JM. Primary medical therapy for acromegaly: an open, prospective, multicenter study of the effects of subcutaneous and intramuscular slow-release octreotide on growth hormone, insulin-like growth factor-I, and tumor size. J Clin Endocrinol Metab 2002;87:4554–63.

241. Lundin P, Eden Engstrom B, Karlsson FA, Burman P. Long-term octreotide therapy in growth hormone-secreting pituitary adenomas: evaluation with serial MR. AJNR Am J Neuroradiol 1997;18:765–72.

242. Plockinger U, Reichel M, Fett U, Saeger W, Quabbe HJ. Preoperative octreotide treatment of growth hormone-secreting and clinically nonfunctioning pituitary macroadenomas: effect on tumor volume and lack of correlation with immunohistochemistry and somatostatin receptor scintigraphy. J Clin Endocrinol Metab 1994;79:1416–23.

243. Amato G, Mazziotti G, Rotondi M, Iorio S, Doga M, Sorvillo F, Manganella G, Di Salle F, Giustina A, Carella C. Long-term effects of lanreotide SR and octreotide LAR on tumour shrinkage and GH hypersecretion in patients with previously untreated acromegaly. Clin Endocrinol (Oxf) 2002;56:65–71.

244. Colao A, Ferone D, Marzullo P, Cappabianca P, Cirillo S, Boerlin V, Lancranjan I, Lombardi G. Long-term effects of depot long-acting somatostatin analog octreotide on hormone levels and tumor mass in acromegaly. J Clin Endocrinol Metab 2001;86:2779–86.

245. Alexopoulou O, Abrams P, Verhelst J, Poppe K, Velkeniers B, Abs R, Maiter D. Efficacy and tolerability of lanreotide autogel therapy in acromegalic patients previously treated with octreotide LAR. Eur J Endocrinol 2004;151:317–24.

246. Ashwell SG, Bevan JS, Edwards OM, Harris MM, Holmes C, Middleton MA, James RA. The efficacy and safety of lanreotide Autogel in patients with acromegaly previously treated with octreotide LAR. Eur J Endocrinol 2004;150:473–80.

247. van Thiel SW, Romijn JA, Biermasz NR, Ballieux BE, Frolich M, Smit JW, Corssmit EP, Roelfsema F, Pereira AM. Octreotide long-acting repeatable and lanreotide Autogel are equally effective in controlling growth hormone secretion in acromegalic patients. Eur J Endocrinol 2004;150: 489–95.

248. de Bruin TW, Kwekkeboom DJ, Van't Verlaat JW, Reubi JC, Krenning EP, Lamberts SW, Croughs RJ. Clinically nonfunctioning pituitary adenoma and octreotide response to long term high dose treatment, and studies in vitro. J Clin Endocrinol Metab 1992;75:1310–7.

249. Warnet A, Timsit J, Chanson P, Guillausseau PJ, Zamfirescu F, Harris AG, Derome P, Cophignon J, Lubetzki J. The effect of somatostatin analogue on chiasmal dysfunction from pituitary macroadenomas. J Neurosurg 1989;71:687–90.

250. Teunissen JJ, Kwekkeboom DJ, Kooij PP, Bakker WH, Krenning EP. Peptide receptor radionuclide therapy for non-radioiodine-avid differentiated thyroid carcinoma. J Nucl Med 2005;46 Suppl 1:107S–14S.

251. Koussis H, Scola A, Tonello S, Karahontzitis P, Pagetta C, Toniato A, Pelizzo M, Casara D, Archonti A, Monfardini S. Characteristics and treatment of 35 patients affected by advanced medullary thyroid carcinoma. In ASCO Annual Meeting, Orlando, FL, 2005.

252. Zlock DW, Greenspan FS, Clark OH, Higgins CB. Octreotide therapy in advanced thyroid cancer. Thyroid 1994;4:427–31.

253. Robbins RJ, Hill RH, Wang W, Macapinlac HH, Larson SM. Inhibition of metabolic activity in papillary thyroid carcinoma by a somatostatin analogue. Thyroid 2000;10:177–83.

254. Krenning EP, de Jong M, Kooij PP, Breeman WA, Bakker WH, de Herder WW, van Eijck CH, Kwekkeboom DJ, Jamar F, Pauwels S, Valkema R. Radiolabelled somatostatin analogue(s) for peptide receptor scintigraphy and radionuclide therapy. Ann Oncol 1999;10 Suppl 2:S23–9.

255. Haslinghuis LM, Krenning EP, De Herder WW, Reijs AE, Kwekkeboom DJ. Somatostatin receptor scintigraphy in the follow-up of patients with differentiated thyroid cancer. J Endocrinol Invest 2001;24:415–22.

256. Waldherr C, Schumacher T, Pless M, Crazzolara A, Maecke HR, Nitzsche EU, Haldemann A, Mueller-Brand J. Radiopeptide transmitted internal irradiation of non-iodophil thyroid cancer and conventionally untreatable medullary thyroid cancer using. Nucl Med Commun 2001;22:673–8.

257. Grana C, Bodei L, Bartolomei M, Rocca P, Zampino G, Peccatori FA, Paganelli G. Advanced medullary thyroid carcinomas: the role of radiometabolic therapy with 90y-dotatoc. Proc Am Soc Clin Oncol 2003;22:247.

258. Gorges R, Kahaly G, Muller-Brand J, Macke H, Roser HW, Bockisch A. Radionuclide-labeled somatostatin analogues for diagnostic and therapeutic purposes in nonmedullary thyroid cancer. Thyroid 2001;11:647–59.

259. de Jong M, Breeman WA, Bakker WH, Kooij PP, Bernard BF, Hofland LJ, Visser TJ, Srinivasan A, Schmidt MA, Erion JL, Bugaj JE, Macke HR, Krenning EP. Comparison of (111)In-labeled somatostatin analogues for tumor scintigraphy and radionuclide therapy. Cancer Res 1998;58:437–41.

260. O'Byrne KJ, Schally AV, Thomas A, Carney DN, Steward WP. Somatostatin, its receptors and analogs, in lung cancer. Chemotherapy 2001;47 Suppl 2:78–108.

261. Bravo EL. Evolving concepts in the pathophysiology, diagnosis, and treatment of pheochromocytoma. Endocr Rev 1994;15:356–68.

262. Tenenbaum F, Schlumberger M, Lumbroso J, Parmentier C. Beneficial effects of octreotide in a patient with a metastatic paraganglioma. Eur J Cancer 1996;32A:737.

263. Koriyama N, Kakei M, Yaekura K, Okui H, Yamashita T, Nishimura H, Matsushita S, Tei C. Control of catecholamine release and blood pressure with octreotide in a patient with pheochromocytoma: a case report with in vitro studies. Horm Res 2000;53:46–50.

264. Tonyukuk V, Emral R, Temizkan S, Sertcelik A, Erden I, Corapcioglu D. Case report: patient with multiple paragangliomas treated with long acting somatostatin analogue. Endocr J 2003;50:507–13.

265. Lamarre-Cliche M, Gimenez-Roqueplo AP, Billaud E, Baudin E, Luton JP, Plouin PF. Effects of slow-release octreotide on urinary metanephrine excretion and plasma chromogranin A and catecholamine levels in patients with malignant or recurrent phaeochromocytoma. Clin Endocrinol (Oxf) 2002;57:629–34.

266. Invitti C, De Martin I, Bolla GB, Pecori Giraldi F, Maestri E, Leonetti G, Cavagnini F. Effect of octreotide on catecholamine plasma levels in patients with chromaffin cell tumors. Horm Res 1993;40:156–60.

267. Plouin PF, Bertherat J, Chatellier G, Billaud E, Azizi M, Grouzmann E, Epelbaum J. Short-term effects of octreotide on blood pressure and plasma catecholamines and neuropeptide Y levels in patients with phaeochromocytoma: a placebo-controlled trial. Clin Endocrinol (Oxf) 1995;42:289–94.

268. Kopf D, Bockisch A, Steinert H, Hahn K, Beyer J, Neumann HP, Hensen J, Lehnert H. Octreotide scintigraphy and catecholamine response to an octreotide challenge in malignant phaeochromo-cytoma. Clin Endocrinol (Oxf) 1997;46:39–44.

269. Smith MC, Liu J, Chen T, Schran H, Yeh CM, Jamar F, Valkema R, Bakker W, Kvols L, Krenning E, Pauwels S. OctreoTher: ongoing early clinical development of a somatostatin-receptor-targeted radionuclide antineoplastic therapy. Digestion 2000;62 Suppl 1:69–72.

270. Samonakis DN, Moschandreas J, Arnaoutis T, Skordilis P, Leontidis C, Vafiades I, Kouroumalis E. Treatment of hepatocellular carcinoma with long acting somatostatin analogues. Oncol Rep 2002;9:903–7.

271. Plentz RR, Tillmann HL, Kubicka S, Bleck JS, Gebel M, Manns MP, Rudolph KL. Hepatocellular carcinoma and octreotide: treatment results in prospectively assigned patients with advanced tumor and cirrhosis stage. J Gastroenterol Hepatol 2005;20:1422–8.

272. Slijkhuis WA, Stadheim L, Hassoun ZM, Nzeako UC, Kremers WK, Talwalkar JA, Gores GJ. Octreotide therapy for advanced hepatocellular carcinoma. J Clin Gastroenterol 2005;39:333–8.

273. Yuen MF, Poon RT, Lai CL, Fan ST, Lo CM, Wong KW, Wong WM, Wong BC. A randomized placebo-controlled study of long-acting octreotide for the treatment of advanced hepatocellular carcinoma. Hepatology 2002;36:687–91.

274. Barbare JC, Bouché O, Bonnetain F, Lombard-Bohas C, Faroux R, Dahan L, Raoul L, Cattan S, Lemoine A, Blanc JF. Treatment of advanced hepatocellular carcinoma with long-acting octreotide: preliminary results of a randomized placebo-controlled trial (FFCD-ANGH 2001–01 CHOC) (abstract). In ASCO Annual Meeting, Orlando, FL, 2005.

275. Dimitroulopoulos D, Xinopoulos D, Tsamakidis K, Andriotis E, Zisimopoulos A, Markidou S, Papadokostopoulou A, Paraskevas E. Octreotide acetate long-acting formulation in treating patients with advanced hepatocellular carcinoma (HCC): comparison with tamoxifen. In ASCO Annual Meeting, Orlando, FL, 2002.

276. Rabe C, Pilz T, Allgaier HP, Halm U, Strasser C, Wettstein M, Sauerbruch T, Caselmann WH. [Clinical outcome of a cohort of 63 patients with hepatocellular carcinoma treated with octreotide.]. Z Gastroenterol 2002;40:395–400.

277. Raderer M, Hejna MH, Muller C, Kornek GV, Kurtaran A, Virgolini I, Fiebieger W, Hamilton G, Scheithauer W. Treatment of hepatocellular cancer with the long acting somatostatin analog lanreotide in vitro and in vivo. Int J Oncol 2000;16:1197–201.

278. Ferone D, Semino C, Boschetti M, Cascini GL, Minuto F, Lastoria S. Initial staging of lymphoma with octreotide and other receptor imaging agents. Semin Nucl Med 2005;35:176–85.

279. Dalm VA, Hofland LJ, Mooy CM, Waaijers MA, van Koetsveld PM, Langerak AW, Staal FT, van der Lely AJ, Lamberts SW, van Hagen MP. Somatostatin receptors in malignant lymphomas: targets for radiotherapy? J Nucl Med 2004;45:8–16.

280. Smith-Jones PM, Bischof C, Leimer M, Gludovacz D, Angelberger P, Pangerl T, Peck-Radosavljevic M, Hamilton G, Kaserer K, Kofler A, Schlangbauer-Wadl H, Traub T, Virgolini I. DOTA-lanreotide: a novel somatostatin analog for tumor diagnosis and therapy. Endocrinology 1999;140:5136–48.

281. Gyergyay G, Ödény M, Sármay G, Kralovanszky J, Papp E, Gergye M, Vincze B, Kéri G, Bodrogi I. Antitumor activity and pharmacology of TT-232 (a novel somatostatin structural derivative) in malignant melanoma patients. J Clin Oncol, 2004 ASCO Annual Meeting Proceedings (Post-Meeting Edition) 2004;22:3151.

282. Meier G, Waldherr C, Herrmann R, Maecke H, Mueller-Brand J, Pless M. Successful targeted radiotherapy with 90Y-DOTATOC in a patient with Merkel cell carcinoma. A case report. Oncology 2004;66:160–3.

283. Guitera-Rovel P, Lumbroso J, Gautier-Gougis MS, Spatz A, Mercier S, Margulis A, Mamelle G, Kolb F, Lartigau E, Avril MF. Indium-111 octreotide scintigraphy of Merkel cell carcinomas and their metastases. Ann Oncol 2001;12:807–11.

284. Wadler S, Benson III A, Engelking C, Catalano R, Field M, Kornblau S, Mitchell E, Rubin J, Trotta P, Vokes E. Recommended guidelines for the treatment of chemotherapy-induced diarrhea. J Clin Oncol 1998;16:3169–78.

285. Zidan J, Haim N, Beny A, Stein M, Gez E, Kuten A. Octreotide in the treatment of severe chemotherapy-induced diarrhea. Ann Surg 2001;12:227–9.

286. Cascinu S, Bichisao E, Amadori D, Silingardi V, Giordani P, Sansoni E, Luppi G, Catalano V, Agostinelli R, Catalano G. High-dose loperamide in the treatment of 5-fluorouracil-induced diarrhea in colorectal cancer patients. Support Care Cancer 2000;8:65–7.

287. Rosenoff S. Resolution of refractory chemotherapy-induced diarrhea (CID) with octreotide long-acting formulation in cancer patients: 11 case studies. Support Care Cancer 2004;12:561–70.

288. ten Bokum AM, Hofland LJ, de Jong G, Bouma J, Melief MJ, Kwekkeboom DJ, Schonbrunn A, Mooy CM, Laman JD, Lamberts SW, van Hagen PM. Immunohistochemical localization of somatostatin receptor sst2A in sarcoid granulomas. Eur J Clin Invest 1999;29:630–6.

289. Bahn RS, Heufelder AE. Pathogenesis of Graves' ophthalmopathy. N Engl J Med 1993;329: 1468–75.

290. Bartalena L, Pinchera A, Marcocci C. Management of Graves' ophthalmopathy: reality and perspectives. Endocr Rev 2000;21:168–99.

291. Pasquali D, Notaro A, Bonavolonta G, Vassallo P, Bellastella A, Sinisi AA. Somatostatin receptor genes are expressed in lymphocytes from retroorbital tissues in Graves' disease. J Clin Endocrinol Metab 2002;87:5125–9.

292. Chang TC, Kao SC, Huang KM. Octreotide and Graves' ophthalmopathy and pretibial myxoedema. BMJ 1992;304:158.

293. Ozata M, Bolu E, Sengul A, Tasar M, Beyhan Z, Corakci A, Gundogan MA. Effects of octreotide treatment on Graves' ophthalmopathy and circulating sICAM-1 levels. Thyroid 1996;6:283–8.

294. Kung AW, Michon J, Tai KS, Chan FL. The effect of somatostatin versus corticosteroid in the treatment of Graves' ophthalmopathy. Thyroid 1996;6:381–4.

295. Uysal AR, Corapcioglu D, Tonyukuk VC, Gullu S, Sav H, Kamel N, Erdogan G. Effect of octreotide treatment on Graves' ophthalmopathy. Endocr J 1999;46:573–7.

296. Pichler R, Maschek W, Holzinger C. [Therapy with somatostatin analogs in patient with orbitopathy and positive Octreoscan]. Acta Med Austriaca 2001;28:99–101.

297. Krassas GE, Kaltsas T, Dumas A, Pontikides N, Tolis G. Lanreotide in the treatment of patients with thyroid eye disease. Eur J Endocrinol 1997;136:416–22.

298. Dickinson AJ, Vaidya B, Miller M, Coulthard A, Perros P, Baister E, Andrews CD, Hesse L, Heverhagen JT, Heufelder AE, Kendall-Taylor P. Double-blind, placebo-controlled trial of octreotide long-acting repeatable (LAR) in thyroid-associated ophthalmopathy. J Clin Endocrinol Metab 2004;89:5910–5.

299. De Herder WW, Lamberts SW. Octapeptide somatostatin-analogue therapy of Cushing's syndrome. Postgrad Med J 1999;75:65–6.

300. Rhee Y, Lee JD, Shin KH, Lee HC, Huh KB, Lim SK. Oncogenic osteomalacia associated with mesenchymal tumour detected by indium-111 octreotide scintigraphy. Clin Endocrinol (Oxf) 2001;54:551–4.

301. Seufert J, Ebert K, Muller J, Eulert J, Hendrich C, Werner E, Schuuze N, Schulz G, Kenn W, Richtmann H, Palitzsch KD, Jakob F. Octreotide therapy for tumor-induced osteomalacia. N Engl J Med 2001;345:1883–8.

302. Schiavi SC, Moe OW. Phosphatonins: a new class of phosphate-regulating proteins. Curr Opin Nephrol Hypertens 2002;11:423–30.

303. Wynick D, Ratcliffe WA, Heath DA, Ball S, Barnard M, Bloom SR. Treatment of a malignant pancreatic endocrine tumour secreting parathyroid hormone related protein. BMJ 1990;300: 1314–5.

304. Mantzoros CS, Suva LJ, Moses AC, Spark R. Intractable hypercalcaemia due to parathyroid hormone-related peptide secretion by a carcinoid tumour. Clin Endocrinol (Oxf) 1997;46:373–5.

305. Arnold R, Simon B, Wied M. Treatment of neuroendocrine GEP tumours with somatostatin analogues: a review. Digestion 2000;62 Suppl 1:84–91.

306. Lamberts SW, de Herder WW, Krenning EP, Reubi JC. A role of (labeled) somatostatin analogs in the differential diagnosis and treatment of Cushing's syndrome. J Clin Endocrinol Metab 1994;78:17–9.

307. Rosati MS, Longo F, Messina CG, Vitolo D, Venuta F, Anile M, Scopinaro F, Di Santo GP, Coloni GF. Targeting the therapy: octreotide in thymoma relapse (abstract). In ASCO Annual Meeting, Orlando, FL, 2005.

308. Froidevaux S, Eberle AN. Somatostatin analogs and radiopeptides in cancer therapy. Biopolymers 2002;66:161–83.

309. Wynick D, Anderson JV, Williams SJ, Bloom SR. Resistance of metastatic pancreatic endocrine tumours after long-term treatment with the somatostatin analogue octreotide (SMS 201–995). Clin Endocrinol (Oxf) 1989;30:385–8.

310. Lamberts SW, Pieters GF, Metselaar HJ, Ong GL, Tan HS, Reubi JC. Development of resistance to a long-acting somatostatin analogue during treatment of two patients with metastatic endocrine pancreatic tumours. Acta Endocrinol (Copenh) 1988;119:561–6.

311. Guillemin R. Peptides in the brain: the new endocrinology of the neuron. Science 1978;202:390–402.

312. Wiseman GA, Kvols LK. Therapy of neuroendocrine tumors with radiolabeled MIBG and somatostatin analogues. Semin Nucl Med 1995;25:272–8.

313. De Jong M, Valkema R, Jamar F, Kvols LK, Kwekkeboom DJ, Breeman WA, Bakker WH, Smith C, Pauwels S, Krenning EP. Somatostatin receptor-targeted radionuclide therapy of tumors: preclinical and clinical findings. Semin Nucl Med 2002;32:133–40.

314. Reubi JC, Waser B. Concomitant expression of several peptide receptors in neuroendocrine tumours: molecular basis for in vivo multireceptor tumour targeting. Eur J Nucl Med Mol Imaging 2003;30:781–93.

315. Reubi JC. Somatostatin and other peptide receptors as tools for tumor diagnosis and treatment. Neuroendocrinology 2004;80 Suppl 1:51–6.

316. Oberg K, Eriksson B. Nuclear medicine in the detection, staging and treatment of gastrointestinal carcinoid tumours. Best Pract Res Clin Endocrinol Metab 2005;19:265–76.

317. de Jong M, Breeman WA, Bernard BF, Bakker WH, Visser TJ, Kooij PP, van Gameren A, Krenning EP. Tumor response after [(90)Y-DOTA(0),Tyr(3)]octreotide radionuclide therapy in a transplantable rat tumor model is dependent on tumor size. J Nucl Med 2001;42:1841–6.

318. Hoefnagel C. Metaiodobenzylguanidine and somatostatin in oncology: role in the management of neural crest tumours. Eur J Nucl Med 1994;21:561–81.

V Challenges in Molecular Targeting in Oncology

26

Patient Selection for Rational Development of Novel Anticancer Agents

Grace K. Dy, MD,
and Alex A. Adjei, MD, PhD

SUMMARY

To understand the challenges a brief overview of the issues relevant to drug target and biomarker evaluation will highlight the key points involved in the process of patient selection for the rational development of novel anticancer agents.

Key Words: Targeted therapy; biomarkers; patient selection; gene expression; gene amplification.

1. INTRODUCTION

Recent advances in molecular biology have led to the development of a myriad of anticancer agents that specifically target aberrant pathways and other proteins that are relatively specific for tumor cells. These targets can be broadly classified as tumor stroma (blood and lymphatic vessels and other connective tissues), cell-cycle regulation, cell signaling, and cell death elements. A variety of approaches have been tested, all aimed at target inhibition to date. The most commonly used, and currently validated, are pharmacologic interventions using small molecule inhibitors and monoclonal antibodies. As anti-cancer therapeutics with distinct targeting capabilities against malignant cells become available for clinical evaluation, several critical issues in drug development need to be addressed. An ideal drug target should be present in cancer cells but not in normal cells, and its continued normal functioning must be essential to the survival of the cancer cells. Agents have to demonstrate inhibition of the intended target, and finally, patients need to be selected according to the presence or absence of specific tumor-related molecular signatures to enhance clinical benefit.

1.1. What are Valid Drug Targets for Cancer Therapy?

As attempts at inhibiting a number of potentially attractive targets in the clinic have yielded mixed results, a number of factors have emerged that can help identify targets

From: *Cancer Drug Discovery and Development: Molecular Targeting in Oncology*
Edited by: H. L. Kaufman, S. Wadler, and K. Antman © Humana Press, Totowa, NJ

that are more likely to be relevant for cancer therapy. Conceptually, one can set a list of rigorous criteria that if met may lead to successful cancer therapy. Apart from the concept of "druggability" *(1)*, the minimum requirement for a valid target in cancer therapy is its causal, rather than merely correlative, role in cancer development. A target should be validated as important to the maintenance of the malignant phenotype in vivo. In addition, this target and its homologues should not assume a similar critical role in normal cellular homeostasis. This is exemplified by the toxicities as well as issues pertaining to potential loss of anti-tumor activity that plagued the clinical development of matrix metalloproteinase inhibitors *(2)*.

The presence of a putative target in tumor tissue should correlate with clinical benefit when an inhibiting drug is employed, that is, blocking the target should impair tumor growth. It has now become clear, over the course of preclinical and clinical testing, that the relevance of any selected target is context-dependent as illustrated below.

1.1.1. MUTATED TARGETS

A mutated target may be the result of a "gain-of-function" process leading to an activated oncogene and/or deletions or inactivating mutations of tumor suppressor genes upon which the maintenance of the malignant phenotype is likely dependent. Such a target may be approached directly, such as inhibitors of the kinase domains of c-kit *(3)* and epidermal growth factor receptor (EGFR) *(4,5)*, or indirectly (e.g., mutated target is not readily "druggable") by inactivating key downstream signaling nodes, such as vascular endothelial growth factor (VEGF)-directed therapies or mTOR inhibition in tumors with inactivating mutations in the von Hippel–Lindau (*VHL*) gene. The *VHL* gene product encodes the ubiquitin ligase that targets hypoxia-inducible factor (HIF) for proteasomal degradation. VHL mutants overproduce HIF and the HIF-dependent transcriptional targets such as VEGF and platelet-derived growth factor (PDGF) *(6)*.

Not all mutations, however, confer growth or survival advantage to the tumor cell but instead represent one of the mechanisms of drug resistance. For example, mutations in the abl kinase domain that impede the binding of imatinib to the ATP pocket have been identified in patients with imatinib-resistant CML or Phi+ ALL *(7–10)*. Although these generally arise after therapy, cases of resistance in untreated patients have been described *(10)*. With the exception of the T315I mutation, newer generation dual src/abl kinase inhibitors are generally able to overcome most of the resistance to imatinib induced by this mechanism.

1.1.2. TARGET AMPLIFICATION

Gene amplification leads to multiple copies of a target, such as members of the EGFR family. Tumors with amplified oncogenic targets show exquisite sensitivity to target inhibition. Examples of such amplified targets are HER-2/neu in breast cancer *(11)* and EGFR in lung cancer *(12,13)*. Once again, effects are complex, and opposing effects may be seen as well. Bcr/abl gene amplification results in abundant active kinase that can overcome its inhibition by imatinib *(8–10)*. Although this also arises most prominently as a mechanism of resistance to prior exposure to imatinib, these can be seen even prior to imatinib therapy *(10)*.

1.1.3. TARGET EXPRESSION OR OVEREXPRESSION

Although patient selection based on target expression has been successfully applied to therapies such as rituximab (anti-CD20-expressing B lymphocytes) and hormonal ablation (anti-estrogen approaches in estrogen receptor-positive breast cancer and anti-androgen therapies in prostate cancer), an important lesson that has been learned over the last decade is the fact that mere protein overexpression does not imply a key role of such proteins in maintaining the malignant phenotype. In these circumstances, inhibiting an overexpressed protein target may not necessarily lead to a therapeutic benefit. Emerging evidence from various studies show that the paradigm of target overexpression in tumor samples as criteria for patient selection may not be relevant in all cases. Increased EGFR expression is common in colorectal cancers, but neither EGFR expression levels nor phosphorylation states correlate with response to EGFR-directed therapies *(14)*. Another example is the lack of clinical efficacy of imatinib in small cell lung cancer (SCLC) despite the high protein expression of c-kit *(15)*.

1.2. Biomarkers in Cancer Drug Development

The terms "surrogate markers" and "biomarkers" are often used interchangeably but do not have the same meaning. A recently organized workshop by the National Institutes of Health in Bethesda, MD, defined a biomarker (also referred to as a biological marker) as "a characteristic that is objectively measured and evaluated as an indicator of biologic processes, pathogenetic processes or pharmacologic responses to therapeutic intervention" *(16)*. The term surrogate marker has no generally recognizable meaning, is confusing, and should probably not be used in drug development, as in most instances, biomarkers are wrongly referred to as surrogate markers. On the contrary, the idea of a "surrogate endpoint" is widely accepted. A surrogate endpoint can be defined as a biomarker response intended to substitute for a clinical endpoint, where a clinical endpoint is a characteristic or variable that reflects how a patient feels or functions, the extent of tumor shrinkage, or how long a patient survives *(16)*. In oncology, it is generally accepted that the clinical endpoint of interest to investigators and patients is survival. Thus, objective response measured radiologically is a surrogate endpoint, where the real endpoint of interest is survival.

As the number of molecular targets for cancer therapy as well as agents inhibiting these targets rapidly increase, correlative biomarkers have become important for two main reasons. In phase I clinical trials, these assays are important for proof of target inhibition at achievable drug concentrations. When utilized in this setting, biomarkers are used to demonstrate pharmacodynamic effects. In phase II and III clinical trials, biomarker assays could potentially provide an early marker of drug efficacy if a strong correlation between assay results and clinical outcome can be established. Alternately, biomarkers can be used to select patients with a molecular profile characteristic of those who are likely to respond to a particular regimen. The challenges faced in designing these three types of studies are different and are discussed below.

1.2.1. THE USE OF BIOMARKERS TO DOCUMENT PHARMACODYNAMIC EFFECTS

In "first-in-human" phase I clinical trials, there is interest in demonstrating inhibition of a drug target in situ in patients. This information can complement the traditional pharmacokinetic information of levels of circulating drug. It is important to demonstrate that steady-state drug levels are consistent with drug concentrations needed to inhibit the drug target in vitro, but it is much more compelling if one can show that the target is inhibited in patient tissues. When evaluating the pharmacodynamic endpoints of anticancer agents, the use of surrogate patient tissues *(17)* as an indirect means to measure drug effects within tumor cells rather than obtaining tumor tissue per se is reasonable. There are a number of important variables in early phase I studies such as the varied stage of advanced disease of enrolled patients, tumor types, and small numbers of patients, which dictate that any information gathered will be preliminary, descriptive, and hypothesis-generating, and thus although evaluating biomarkers in tumor tissue is the gold standard for correlative studies, issues pertaining to patient risk and limited sampling arising from invasive procedures, and heterogeneity of tumor tissue obtained support the use of surrogate tissues. The utilization of tumor tissues is best suited to the phase I B level where doses have been defined and a reasonable number of patients can be treated with only one or two potentially therapeutic doses to allow for a more reasonable statistical analysis of results. The usefulness of evaluating biomarkers in surrogate tissues in phase II trials, however, is difficult to discern, as after proof-of-concept phase I studies demonstrating that the drug target is inhibited in surrogate tissue provides no additional information.

1.2.2. BIOMARKERS FOR RESPONSE PREDICTION AND PROGNOSTICATION

The expression of estrogen and progesterone receptors on breast cancer cells is a predictive marker for response to tamoxifen *(18)* and aromatase inhibitors *(19)*. The expression of HER-2/neu protein in breast cancer cells predicts for response to trastuzumab *(11)*. The presence of mutations in the kit gene predicts for response or resistance of (gastrointestinal stromal tumor GIST) tumors to imatinib *(3)*. These data clearly indicate that biomarkers can be used to identify patients likely to respond to a particular therapeutic agent. However, it is evident that the expression of ER/PR or HER-2/neu protein on breast cancer cells is necessary but not sufficient for these tumors to respond to therapy with the above agents. Response rates among patients whose tumors express these proteins are approximately 20–50% rather than close to 100%. On the contrary, patients with tumors that are devoid of these proteins almost never respond to agents that target them. These biomarkers therefore allow for the selection of a patient population for maximum response. Ideally, the endpoint of studies designed to evaluate predictive markers should be objective response rates, as the hypothesis being tested is usually the ability of the marker to predict for a response to therapy. Theoretically, a marker that is only predictive of response could select patient populations that would yield a high response rates in clinical trials without an effect on survival.

When used as prognostic markers, biomarkers stratify patients into groups with variations in survival. These biomarkers may not address the responsiveness of patients to a particular therapy. Rather, they may identify a subset that may not need therapy. Studies that evaluate a prognostic marker should ideally have survival as the endpoint.

1.2.3. IDEAL CHARACTERISTICS OF A BIOMARKER

There are a number of characteristics that would make a biomarker ideal for use in conjunction with clinical trials. In considering these properties, it is assumed that the biomarker has been validated in terms of assay-to-assay variation and reproducibility.

First, as is the case of any measures of biologic activity, the ideal biomarker should be sensitive and specific. Second, the biomarker assay should be simple and adaptable to clinical use, with a quick turnaround time between sample collection and availability of results. In this regard, even though immunohistochemical methods have been often criticized as subjective and unreliable, they remain a simple, widely available method that can be utilized in virtually all hospitals worldwide. This explains, in part, why the measurement of estrogen and progesterone receptor status in breast cancer evolved from quantitative measures of protein content to an immunohistochemistry method *(20)*. Third, the biomarker should be present in easily accessible tissues to make it simple for patients to consent to sample collection. The commonly accessible surrogate tissues that have been utilized in current clinical trials are peripheral blood mononuclear cells (PBMCs), skin biopsies, and buccal mucosa cells *(17)*. As discussed earlier, biomarker data obtained from surrogate tissues are limited in value beyond phase I trials, unless it can be shown that these correlate well with certain pre-specified clinical outcomes, thus obviating the need for invasive tumor biopsies for tissue samples. This is certainly the case for serum-based assays for tumor markers such as PSA, CA-125, and CEA.

2. SELECTION OF TARGET PATIENT POPULATIONS

The concept of targeted rationally based therapy is not new and has been exploited in hormone-dependent malignancies. A major challenge in administering new target-specific drugs is the ability to predict the outcome of therapy, which encompasses tumor response, clinical toxicity, and resistance. It cannot be overemphasized that the determinants of response to therapy are not only tumor-dependent but also defined by the inherent characteristics and limitations of the individual agents. For example, tumors that express truncated receptor variants or are located within the central nervous system may preclude the use of monoclonal antibodies because of mechanistic or physiologic/pharmacokinetic barriers (e.g., blood-brain barrier). A drug's high degree of target specificity may be a double-edged sword, as it is a distinct limitation when there are multiple relevant isoforms of the target, such as with antisense oligonucleotide strategies against only one of the family of receptors for VEGF.

The concept that "target population enrichment" in cancer clinical trials should be regularly employed is now generally accepted as a mantra. In theory, this is achieved by selection of subjects based on the presence or absence of one or more biologic markers, thought to increase the probability of demonstrating drug efficacy, to reduce the number of subjects required for statistical analyses, and thereby widening the benefit-risk ratio. An example is that of trastuzumab, a humanized IgG1 monoclonal antibody against the HER2/neu receptor overexpressed in approximately 25% of invasive breast cancer *(21)*. The pivotal studies that led to its regulatory approval by the Food and Drug Administration showed striking improvement in survival and objective responses limited to HER2-overexpressing cancers *(11,22)*. It is conceivable that such activity would be missed in an unselected group of patients. This poses the all-too-familiar dilemma in drug development, particularly of novel agents against cancers that are genetically heterogeneous, that treatment effects maybe diluted in an unselected population if only a small subgroup of patients is likely to respond.

The cogency of the argument above notwithstanding, experience thus far reminds us of the frequent lack of congruence between the knowledge of a drug's purported mechanism of action during the preliminary stages of clinical evaluation and its relation to the pertinent aspect of tumor biology. For example, farnesyltransferase inhibitors were designed to inhibit the ras protein, yet they exhibit activity against ras-independent tumors and ironically are ineffective in common tumors with mutant ras. Similarly, sorafenib was ostensibly developed as a C-raf kinase inhibitor, but its anti-tumor activity in clinical testing seems more consistent with its activity as an inhibitor of the VEGF receptor tyrosine kinase. The presence of activating mutations that involve the ATP-binding pocket of receptor tyrosine kinases may confer oncogenic dependence and, by inference, increased sensitivity to kinase inhibition, as demonstrated by the impressive objective responses of tumors with EGFR kinase domain mutations to EGFR tyrosine kinase inhibitors (4,5,13). Nevertheless, clinically meaningful responses in terms of objective tumor response and/or prolonged disease stabilization were also seen in tumor samples that do not harbor mutations. Whether these responses are attributable to gene amplification or some other mechanisms is currently unknown.

A pragmatic strategy that may enable investigators to evaluate biomarkers for patient selection in subsequent trials is drug administration for a short-period in the pre-operative setting (23), as in breast, colorectal, lung, and renal cancers. Alternatively, a trial design-dependent method of selecting patients most likely to respond to therapy is the randomized discontinuation trial design, most recently utilized in the evaluation of sorafenib (24,25). With this design, all patients receive drug treatment initially for a specified time frame. Patients who meet criteria for objective tumor response, tradi-tionally based upon radiologic tumor shrinkage, will continue therapy until disease progression or toxicity, whereas those with disease progression or unacceptable toxic-ities will discontinue therapy. On the contrary, all patients with radiologically stable disease after the initial therapy period are then randomized to continuing or discon-tinuing therapy in a double-blind placebo-controlled manner. This method preferentially "enriches" the randomized phase of the trial by excluding patients who failed therapy early while providing increased statistical power to discriminate prolonged stable disease because of the beneficial effects of drug therapy from an inherently indolent pattern of tumor growth, a clinically relevant difference. Moreover, the number of patients exposed to placebo is decreased. The total sample size required, however, could be larger than that required in a standard design. In addition, tumor response/toxicity maybe underestimated because of a carryover treatment effect through the random-ization stage from the first treatment stage.

3. CONCLUSION

An ever-present challenge in the clinical evaluation of novel target-based agents in cancer therapy arises from the fact that clinical outcomes often do not meet the expectations deduced from preclinical testing. A variety of factors can account for this such as the lack of proper in vivo models and the insufficient knowledge of a drug's spectrum of biologic targets. Attempts to select patients most likely to respond to drug therapy in early clinical testing, although well intentioned, are generally doomed to fail as the selection criteria are often based on non-validated biomarkers. Moreover, innate (in contrast to "escape" mechanisms induced by drug exposure) mechanisms of drug resistance are seldom understood prior to clinical testing although this represents a key aspect in the investigations regarding determinants of drug sensitivity. Establishing a system biology approach to characterize individual tumors according to their respective pathway-driven taxonomies *(26)* as well as incorporating resistance profiles will be a Herculean but a necessary endeavor, not only to optimize the "success rate" of the molecular profile chosen for patient selection for clinical trials but, more importantly, to realize the ultimate goals of tailored therapy.

REFERENCES

1. Hopkins AL, Groom CR. The druggable genome. *Nat Rev Drug Discov* 2002;1(9):727–30.
2. Overall CM, Kleifeld O. Validating matrix metalloproteinases as drug targets and anti-targets for cancer therapy. *Nat Rev Cancer* 2006;6:227–39.
3. Heinrich MC, Corless CL, Demetri GD, et al. Kinase mutations and imatinib response in patients with metastatic gastrointestinal stromal tumor. *J Clin Oncol* 2003;21(23):4342–9.
4. Lynch TJ, Bell DW, Sordella R, et al. Activating mutations in the epidermal growth factor receptor underlying responsiveness of non-small-cell lung cancer to gefitinib. *N Engl J Med* 2004;350(21):2129–39.
5. Paez JG, Janne PA, Lee JC, et al. EGFR mutations in lung cancer: correlation with clinical response to gefitinib therapy. *Science* 2004;304(5676):1497–500.
6. George DJ, Kaelin WG Jr. The von Hippel-Lindau protein, vascular endothelial growth factor, and kidney cancer. *N Engl J Med* 2003;349(5):419–21.
7. Miething C, Feihl S, Mugler C, et al. The Bcr-Abl mutations T315I and Y253H do not confer a growth advantage in the absence of imatinib. *Leukemia* 2006;20(4):650–7.
8. Gorre ME, Mohammed M, Ellwood K, et al. Clinical resistance to STI-571 cancer therapy caused by BCR-ABL gene mutation or amplification. *Science* 2001;293(5531):876–80.
9. Hochhaus A, Kreil S, Corbin AS, et al. Molecular and chromosomal mechanisms of resistance to imatinib (STI571) therapy. *Leukemia* 2002;16(11):2190–6.
10. Shah NP, Nicoll JM, Nagar B, et al. Multiple BCR-ABL kinase domain mutations confer polyclonal resistance to the tyrosine kinase inhibitor imatinib (STI571) in chronic phase and blast crisis chronic myeloid leukemia. *Cancer Cell* 2002;2(2):117–25.
11. Vogel CL, Cobleigh MA, Tripathy D, et al. Efficacy and safety of trastuzumab as a single agent in first-line treatment of HER2-overexpressing metastatic breast cancer. *J Clin Oncol* 2002; 20:719–26.
12. Cappuzzo F, Hirsch F, Rossi E, et al. Epidermal growth factor receptor gene and protein and gefitinib sensitivity in non-small cell lung cancer. *J Natl Cancer Inst* 2005;97:643–55.
13. Tsao MS, Sakurada A, Cutz JC, et al. Erlotinib in lung cancer – molecular and clinical predictors of outcome. *N Engl J Med* 2005;353(2):133–44.
14. Chung KY, Shia J, Kemeny NE, et al. Cetuximab shows activity in colorectal cancer patients with tumors that do not express the epidermal growth factor receptor by immunohistochemistry. *J Clin Oncol* 2005;23(9):1803–10.
15. Dy GK, Miller AA, Mandrekar S, et al. A phase II trial of imatinib (ST1571) in patients with c-kit expressing relapsed small cell lung cancer: a CALGB and NCCTG Study. *Ann Oncology* 2005;16(11):1811–6.
16. De Gruttola VG, Clax P, DeMets DL, et al. Considerations in the evaluation of surrogate endpoints in clinical trials: summary of a National Institutes of Health workshop. *Control Clin Trials* 2001; 22:485–502.
17. Adjei AA. Immunohistochemical assays of farnesyltransferase inhibition in patient samples. In: *Novel Anticancer Drug Protocols.* Buolamwinin J, Adjei AA (eds). Humana Press, Totowa, NJ, 2003, pp. 141–5.
18. Osborne CK. Tamoxifen in the treatment of breast cancer. *N Engl J Med* 1998;339:1609–18.
19. Smith IE, Dowsett M. Aromatase inhibitors in breast cancer. *N Engl J Med* 2003;348:2431–42.
20. Pertschuk LP, Kim YD, Axiotis CA, et al. Estrogen receptor immunocytochemistry: the promise and the perils. *J Cell Biochem Suppl* 1994;19:134–7.
21. Slamon DJ, Godolphin W, Jones LA, et al. Studies of the HER-2/neu proto-oncogene in human breast and ovarian cancer. *Science* 1989;244:707–12.
22. Slamon DJ, Leyland-Lones B, Shak S, et al. Use of chemotherapy plus a monoclonal antibody against HER2 for metastatic breast cancer that overexpressed HER2. *N Engl J Med* 2001;344:783–92.

23. Arteaga CL, Baselga J. Tyrosine kinase inhibitors: why does the current process of clinical development not apply to them? *Cancer Cell* 2004;5(6):525–31.
24. Ratain MJ, Eisen T, Stadler WM, et al. Final findings from a phase II, placebo-controlled, randomized discontinuation trial (RDT) of sorafenib (BAY 43–9006) in patients with advanced renal cell carcinoma (RCC). *J Clin Oncol* 2005;23(16s):4544.
25. Rosner GL, Stadler W, Ratain MJ. Randomized discontinuation design: application to cytostatic antineoplastic agents. *J Clin Oncol* 2002;20:4478–84.
26. Fishman MC, Porter JA. Pharmaceuticals: a new grammar for drug discovery. *Nature* 2005;437(7058):491–3.

27 Clinical Trial Design with Targeted Agents

Sarita Dubey, MD,
and Joan H. Schiller, MD

SUMMARY

Targeted agents differ from cytotoxic chemotherapy in their mechanism of action and clinical outcomes. Similarly, drug development for targeted agents may also differ. Targeted agents are generally better tolerated than chemotherapeutic agents and may have activity at doses lower than maximum tolerated dose (MTD). Optimum drug dosing needs to be based not only on toxicities but also on pharmacokinetic and pharmacodynamic parameters. As cytostatic-targeted agents may induce disease stabilization, time to progression and survival may be a more appropriate outcome of interest than disease regression. Translational research becomes an indispensable part of trials with targeted agents to identify optimum dosing and to identify predictive biomarkers. This includes advances in technology, better understanding of tumor biology, and validation of research techniques. Clinical trials with novel randomization designs have been developed in recent years to optimize identification of drug activity in a targeted population.

Key Words: Clinical trials; targeted agents.

1. INTRODUCTION

Targeted therapy implies treatment with a drug that interferes with a specific molecular target involved in tumor growth, whereas cytotoxic chemotherapy traditionally refers to drugs that indiscriminately affect all dividing cells. Thus, with targeted therapy, the hope is to convert cancer from a lethal disease to a chronic disease. As knowledge about targets and targeted therapies has grown, clinical trial methodology will need to be modified to accommodate the distinctive characteristics of targeted agents. In this review, we will discuss some of the challenges involved in the development of these drugs.

2. CLINICAL TRIAL DESIGNS

Traditional drug development in humans involves phase I, II, and III clinical trials. Phase I clinical trials evaluate the safety of a drug when studied for the first time in

From: *Cancer Drug Discovery and Development: Molecular Targeting in Oncology*
Edited by: H. L. Kaufman, S. Wadler, and K. Antman © Humana Press, Totowa, NJ

human subjects. The objective of these trials is to identify the dose that is maximally tolerated, as cytotoxic drugs are often administered at the maximally tolerated dose so as to achieve maximum benefit. Phase II trials look for preliminary activity of a drug in a specific malignancy and often use reduction in tumor size, or response rate, as the primary endpoint. A phase III clinical trial compares the efficacy of a drug or combination of drugs with other standard therapies, which may be placebo in situations where no standard or effective therapy exists. Traditional outcome measurements in phase III clinical trials include overall survival, time to progression, and quality of life assessments.

The classical trial designs that have been appropriate for cytotoxic drugs may not be as applicable for some of the new targeted therapies. For example, the optimal biological dose may not be the maximally tolerated dose. Response rates may not be an appropriate endpoint for cytostatic drugs and may not correlate with clinical benefit. Time to progression is dependent on frequency of patient assessment. And survival, the penultimate measure of benefit, is now clouded by cross-over trial designs and the advent of second, third, and even fourth line therapies.

Innovative trial designs will be necessary to allow us to identify active drugs to take forward. These may include "pick the winner" phase II/III trials, innovative early stopping rules, or randomized phase II studies *(1–4)*.

Another example of a unique trial design is that of a randomized discontinuation design. The purpose of this type of trial is to control for patients with disease that is growing so rapidly that the patient never has time to receive enough therapeutic doses of the drug before progressing *(5)* (Fig. 1). In this trial design, all eligible patients receive 2–3 months of the study drug. Those who respond continue on drug, those who progress discontinue drug, and those with stable disease are randomized to either continue the investigational drug or a placebo arm. By randomizing patients who do not have rapidly progressing disease, one will have enriched the study population for patients most likely to have derived a benefit. This design was recently adopted

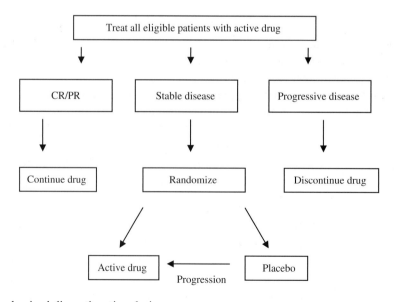

Fig. 1. Randomized discontinuation design.

by a trial evaluating BAY 43-9006, (sorafenib) a raf kinase pathway/tyrosine kinase inhibitor *(6)*. All eligible patients with renal cell carcinoma received 3 months of BAY 43-9006. Those with stable disease were randomized to either active drug or placebo. Those who received the study drug had a significantly longer progression free survival than those receiving placebo (23 versus 6 weeks, $p = 0.0001$) despite the fact the response rate was less than 5% in the investigational arm.

3. TO TARGET OR NOT TO TARGET

What are some of the challenges in developing a targeted therapy? First of all, the target must be not only identified but validated as important in the carcinogenesis pathway. This is not as straightforward as it might appear. For example, although the epidermal growth factor receptor (EGFR) pathway has been shown to be critical in the pathogenesis of many tumors, it is not clear which of the multiple molecules involved in the pathway should be targeted (The ligand? The receptor? Downstream molecules such as Akt or MAP kinase?). This is illustrated by studies that have shown that the level of expression of a presumed target does not always correlate with response. For example, in lung cancer, the degree of growth inhibition with EGFR inhibitor gefitinib did not always correlate with the level of EGFR expression *(7,8)*. In colorectal cancer, patients who were EGFR positive as well EGFR negative responded to the EGFR inhibitor cetuximab *(9)*.

Second, to successfully target a tumor with a targeted agent, the target must be present. However, all tumors, even tumors of a particular type, may not have the target in question. For example, the EGFR is expressed in only 62% of non-small cell lung cancers (non-SCLCs) *(10)*. Trastuzumab, a HER2 antibody, is effective only in HER2-overexpressing breast cancers, which accounts for about 20% of all breast cancers *(11)*. Hence, a specific targeted treatment may not be appropriate for all patients with a particular diagnosis and may be better suited for patients with a unique tumor signature.

So should patients without target overexpression be excluded from a targeted agent? One way of ascertaining the impact of the marker on treatment may be by conducting randomized marker design trials. Sargent et al. *(12)* have elegantly described four different randomization designs in marker-based trials. All four designs involve randomization into marker (+) and marker (–) groups. In addition, patients are randomized to two different treatments. In the more complex designs, patients with known marker status are compared with those with unknown marker status, who are also randomized to two different treatments. A significant difference in treatment outcomes with one treatment over the other or in the marker (+) groups versus the other groups validates marker-based treatment methodology.

It is also obvious that for a targeted agent to be effective in inhibiting tumor growth, the tumor should be dependent on the target for growth. In this aspect, imatinib may be used as a paradigm of targeted therapies. The progress of imatinib through clinical trials has been described in a review by Druker and David *(13)*. *BCR-ABL* gene has been shown to induce chronic myeloid leukemia (CML). A unique translocation between the long arms of chromosomes 9 and 22 t(9:22)(q34;q11) translocates c-ABL tyrosine kinase from chromosome 9 to chromosome 22. Imatinib inhibits ABL, KIT, and PDGFR tyrosine kinases. Through all stages of development—preclinical, phase I and II—imatinib demonstrated significant activity against CML. In a phase III direct comparison against interferon-α and Ara-C, which was then the standard treatment for CML, imatinib induced higher responses *(14)*. This study led to FDA approval in an astonishing 2.5 months.

Where can this process go wrong? Let us review the role of imatinib in SCLC. Between 30 and 70% of SCLCs express c-Kit, one of the targets of imatinib. However, in a phase II study of imatinib in SCLC, imatinib failed to display activity even in c-Kit-positive patients, indicating that c-Kit may not be the only oncogenic event in SCLC cells *(15)*. These two studies with contrasting results illustrate that a target needs to be a keystone in tumor development to increase the success of targeted therapy.

In addition to validating the target, validating the method to evaluate the target is also necessary. To ensure scientifically vigorous results, translational research techniques require validity and reproducibility. For example, expression of EGFR protein may be evaluated by means of ELISA or IHC. With IHC alone, there are no standard scoring criteria for the level of protein expression. Results are also prone to observer bias. Diversity in techniques and reagents, such as the antibody, can also lead to diversity in results. This disparity can be avoided by the use of a central laboratory in the case of multi-institutional trials and by standardizing evaluation criteria.

Last, if a targeted agent requires the pretreatment evidence of target in the tumor, the target should be easily detectable and measurable in the clinic. This relates to the availability of technology and the related costs. In addition, tissue acquisition in the case of cancers such as lung or colon involves interventions that may be physically challenging. The absence of these capabilities creates a hiatus between scientifically based clinical trials and clinical applicability.

4. DOSING DILEMMAS

Historically, malignancies have been treated with chemotherapy at the highest dose tolerated in human subjects. In phase I trials of cytotoxic chemotherapy, the maximum tolerated dose (MTD) (also known as the recommended phase II dose) is the dose below the dose at which dose-limiting toxicity (DLT) occurs.

As many other biological targeted agents are better tolerated than cytotoxic chemotherapy, they may hypothetically be active at doses below DLT. Doses for further study are then based on pharmacokinetic or pharmacodynamic parameters, so that the optimal biologic dose (OBD), and not the MTD, is chosen for future development.

For example, in phase I trials with cetuximab, the drug was so well tolerated that the MTD was not achieved *(16)*. Instead, pharmacokinetic parameters were used to identify the optimum dose. Saturation of clearance was seen at doses >200 mg/m^2/week. As preclinical studies demonstrated that saturation of serum drug levels correlated with maximum tumor regression, >200 mg/m^2/week was designated as the recommended phase II dose.

Although theoretically the optimum biological dose may be lower than the MTD, this may not always be the case. A case in point is the development of erlotinib and gefitinib *(17–21)*. Both are oral EGFR tyrosine kinase inhibitors, but erlotinib was found to have a survival benefit compared with best supportive care in patients with recurrent non-SCLC, whereas gefitinib did not *(22,23)*. Although other differences could account for this observation, one possibility remains the relative differences in doses that were used for each agent. In the case of gefitinib, a dose of 250 mg/day was chosen for development, even though the MTD was above 700 mg/day, whereas in the case of erlotinib, the MTD of 150 mg/day was used *(24)*. The different strategies in the development of these two related drugs illustrates the need for better understanding of the relationship between MTD and OBD.

Pharmacodynamic methods to identify the OBD require demonstration of drug activity in the target, which again requires tissue and is associated with the inherent problems with tissue acquisition in solid tumors. An alternative may be the demonstration of activity in a more easily accessible surrogate tissue. In the case of EGFR-expressing tumors, skin, which is EGFR positive, may be a reasonable surrogate to identify drug effect on target. In early phase I studies of EGFR murine monoclonal antibody RG83852, with doses of 200 mg/m^2 or higher, a high degree of receptor saturation >50% was observed in post-therapy tumor *(25)*. These data supported the use of >200 mg/m^2 in subsequent studies. Skin biopsies showed drug saturation of EGFR, with subsequent effects on EGFR tyrosine kinase phosphorylation. Similarly, with the EGFR tyrosine kinase inhibitor erlotinib, pre- and post-treatment skin biopsies showed significant decrease in phosphor-EGFR and increase in p27 with treatment *(26)*. These studies suggest that pharmacodynamic effects can be studied in surrogate tissues and may be an alternate to tumor studies. However, paired tumor and skin samples during early stages of drug development are needed to establish a correlation between target response in tumor and surrogate tissue.

Clinical pharmacodynamic changes may also predict drug activity in tumor. In the case of erlotinib, skin rash is associated with increased response to the drug *(27)*. In one trial, the median survival of patients without rash, grade 1, and grade 2 or 3 rash were 1.5, 8.5, and 19.6 months, respectively ($p = 0.0001$). Based on this concept, recent studies have been designed to treat patients starting with the standard dose of 150 mg daily but titrating the dose to achieve a grade 2 rash. These trials are ongoing, and if the results are positive, they may be proof of concept that skin is a valid surrogate marker for drug activity.

Alternatives to biopsies are functional imaging techniques such as positron emission tomography (PET) and dynamic contrast-enhanced magnetic resonance imaging (MRI). For example, in the case of anti-angiogenic agents, tumor regression has been associated with regression of tumor vasculature as seen on MRI *(28)*.

5. CONCLUSION

The advent of targeted agents has seen the development of novel clinical trial designs and sophisticated translational research. Although for many targeted agents, conventional clinical trial methodology is still valid for drug evaluation, and in the case of many others, pharmacodynamic and functional endpoints are replacing these methods. The low-toxicity profile of many of these agents challenges the identification of optimum dose based on methods employed in phase I trials with cytotoxic drugs. OBD identification may need to be based on demonstration of target activity in the tumor or on pharmacokinetic parameters. For many cytostatic agents that induce stable disease, the response rate may not be the ideal outcome of interest. Proteomics, genomics, and functional imaging techniques may become the new parameters to assess activity. Ongoing issues include validation of research techniques and the availability of such techniques in the clinic. Newer clinical trial designs such as the randomized discontinuation design may better elicit the benefit of cytostatic drugs in slow-growing tumors. A pending decision is the one to segregate patients into target expressing and non-expressing groups. New clinical trial paradigms will help to address this issue.

REFERENCES

1. Wieand S, Schroeder G, O'Fallon JR. Stopping when the experimental regimen does not appear to help. *Stat Med* 1994;13(13–14):1453–8.
2. Baum M, Houghton J, Abrams K. Early stopping rules–clinical perspectives and ethical considerations. *Stat Med* 1994;13(13–14):1459–69; discussion 71–2.
3. Inoue LY, Thall PF, Berry DA. Seamlessly expanding a randomized phase II trial to phase III. *Biometrics* 2002;58(4):823–31.
4. Steinberg SM, Venzon DJ. Early selection in a randomized phase II clinical trial. *Stat Med* 2002;21(12):1711–26.
5. Rosner GL, Stadler W, Ratain MJ. Randomized discontinuation design: application to cytostatic antineoplastic agents. *J Clin Oncol* 2002;20(22):4478–84.
6. Ratain M, Eisen T, Stadler WM, et al. Final findings from a phase II, placebo-controlled, randomized discontinuation trial (RDT) of sorafenib (BAY 43–9006) in patients with advanced renal cell carcinoma (RCC). *Proc Am Soc Clin Oncol* 2005;23 [abstract no. 3385–4544].
7. Pao W, Zakowski M, Cordon-Cardo C, Ben-Porat L, Kris MG, VA Miller. Molecular characteristics of non-small cell lung cancer (NSCLC) patients sensitive to gefitinib. *Proc Am Soc Clin Oncol* 2004;22:623s [abstract no. 6235–7025].
8. Cappuzzo F, Hirsch FR, Rossi E, et al. Epidermal growth factor receptor gene and protein and gefitinib sensitivity in non-small-cell lung cancer. *J Natl Cancer Inst* 2005;97(9):643–55.
9. Chung KY, Shia J, Kemeny NE, et al. Cetuximab shows activity in colorectal cancer patients with tumors that do not express the epidermal growth factor receptor by immunohistochemistry. *J Clin Oncol* 2005;23(9):1803–10.
10. Hirsch FR, Varella-Garcia M, Bunn PA Jr, et al. Epidermal growth factor receptor in non-small-cell lung carcinomas: correlation between gene copy number and protein expression and impact on prognosis. *J Clin Oncol* 2003;21(20):3798–807.
11. Vogel CL, Cobleigh MA, Tripathy D, et al. Efficacy and safety of trastuzumab as a single agent in first-line treatment of HER2-overexpressing metastatic breast cancer. *J Clin Oncol* 2002;20(3):719–26.
12. Sargent DJ, Conley BA, Allegra C, Collette L. Clinical trial designs for predictive marker validation in cancer treatment trials. *J Clin Oncol* 2005;23(9):2020–7.
13. Druker BJ. David A. Karnofsky Award lecture. Imatinib as a paradigm of targeted therapies. *J Clin Oncol* 2003;21(23 Suppl):239s–45s.
14. O'Brien SG, Guilhot F, Larson RA, et al. Imatinib compared with interferon and low-dose cytarabine for newly diagnosed chronic-phase chronic myeloid leukemia. *N Engl J Med* 2003;348(11):994–1004.
15. Johnson BE, Fischer T, Fischer B, et al. Phase II study of imatinib in patients with small cell lung cancer. *Clin Cancer Res* 2003;9(16 Pt 1):5880–7.
16. Baselga J, Pfister D, Cooper MR, et al. Phase I studies of anti-epidermal growth factor receptor chimeric antibody C225 alone and in combination with cisplatin. *J Clin Oncol* 2000;18(4):904–14.
17. Ranson M, Hammond LA, Ferry D, et al. ZD1839, a selective oral epidermal growth factor receptor-tyrosine kinase inhibitor, is well tolerated and active in patients with solid, malignant tumors: results of a phase I trial. *J Clin Oncol* 2002;20(9):2240–50.
18. Herbst RS, Maddox AM, Rothenberg ML, et al. Selective oral epidermal growth factor receptor tyrosine kinase inhibitor ZD1839 is generally well-tolerated and has activity in non-small-cell lung cancer and other solid tumors: results of a phase I trial. *J Clin Oncol* 2002;20(18):3815–25.
19. Baselga J, Rischin D, Ranson M, et al. Phase I safety, pharmacokinetic, and pharmacodynamic trial of ZD1839, a selective oral epidermal growth factor receptor tyrosine kinase inhibitor, in patients with five selected solid tumor types. *J Clin Oncol* 2002;20(21):4292–302.

20. Nakagawa K, Tamura T, Negoro S, et al. Phase I pharmacokinetic trial of the selective oral epidermal growth factor receptor tyrosine kinase inhibitor gefitinib ('Iressa', ZD1839) in Japanese patients with solid malignant tumors. *Ann Oncol* 2003;14(6):922–30.

21. Hidalgo M, Siu L, Nemunaitis J, et al. Phase I and pharmacologic study of OSI-774, an epidermal growth factor receptor tyrosine kinase inhibitor, in patients with advanced solid malignancies. *J Clin Oncol* 2001;19(13):3267–79.

22. Shepherd FA, Rodrigues Pereira J, Ciuleanu T, et al. Erlotinib in previously treated non-small cell lung cancer. *N Engl J Med* 2005;352(2):123–32

23. Thatcher N, Chang A, Parikh P, et al. plus best supportive care in previously treated patients with refractory advanced, non-small cell lung cancer results from a randomised, placebo-controlled, multicenter study (Iressa Survival Evaluation in Lung Cancer). Lancet 2005;366(9496):1527–37.

24. Kris MG, Natale RB, Herbst RS, et al. Efficacy of gefitinib, an inhibitor of the epidermal growth factor receptor tyrosine kinase, in symptomatic patients with non-small cell lung cancer: a randomized trial. *JAMA* 2003;290(16):2149–58.

25. Perez-Soler R, Donato NJ, Shin DM, et al. Tumor epidermal growth factor receptor studies in patients with non-small-cell lung cancer or head and neck cancer treated with monoclonal antibody RG 83852. *J Clin Oncol* 1994;12(4):730–9.

26. Malik SN, Siu LL, Rowinsky EK, et al. Pharmacodynamic evaluation of the epidermal growth factor receptor inhibitor OSI-774 in human epidermis of cancer patients. *Clin Cancer Res* 2003;9(7):2478–86.

27. Perez-Soler R, Chachoua A, Huberman M, Karp D, Rigas J. Final results from a phase II study of erlotinib monotherapy in patients with advanced non-small cell lung cancer following failure of platinum based chemotherapy. *Lung Cancer* 2003;41(S2):S246 [abstract no. p-611].

28. Pham CD, Roberts TP, van Bruggen N, et al. Magnetic resonance imaging detects suppression of tumor vascular permeability after administration of antibody to vascular endothelial growth factor. *Cancer Invest* 1998;16(4):225–30.

28

How to Define Treatment Success or Failure if Tumors Do Not Shrink

Consequences for Trial Design

J. J. E. M. Kitzen, MD,
M. J. A. de Jonge, MD, PhD,
and J. Verweij, MD, PhD

SUMMARY

The development of targeted anti cancer drugs in recent years puts a challenge on classical trial designs. In this chapter we point out that defining a biological relevant dose might become more important than a MTD and establishing the mechanism of action becomes pivotal early in drug development. Furthermore we present examples of trails that show the importance of defining the patient population likely to benefit from targeted drugs.

Furthermore, the use of surrogate tissues to evaluate the biological activity of the agents under study is described.

Non-invasive techniques like PET, CT or MRI imaging are addressed as well as their present lack of validity in predicting patient outcome.

Alternative endpoint in phase II trail designs in relation to the development of targeted anticancer drugs are described.

Key Words: Surrogate marker; surrogate tissue; Endothelial Growth Factor Receptor (EGFR); Farnesyl transferase inhibitor (FTI); Vascular Endothelial Growth Factor (VEGF); Patient selection; Imaging; 'Randomised discontinuation design'; 'Growth modulation index'; 'Progression rate'; 'Progression free rate'.

1. INTRODUCTION

Systemic treatment of advanced cancer traditionally involves cytotoxic drugs directed to a specific targets in the tumor cells. However, over the last few years, many processes involved in tumorigenesis and carcinogenesis have been unraveled. This knowledge enabled the development of a new generation of anti-cancer drugs that exert their effect by targeting extracellular, transmembrane, or intracellular processes. These new drugs include among others receptor tyrosine kinase inhibitors, angiogenesis

From: *Cancer Drug Discovery and Development: Molecular Targeting in Oncology*
Edited by: H. L. Kaufman, S. Wadler, and K. Antman © Humana Press, Totowa, NJ

inhibitors, farnesyltransferase inhibitors (FTIs), and matrix metalloproteinase inhibitors (MMPIs). In contrast to cytotoxic drugs, most of these new drugs have a cytostatic effect and therefore show tumor growth inhibition instead of regression and require prolonged administration of the drug to establish optimal effects. Chronic adminis-tration of a drug will require different formulations of drugs (oral versus intravenous) and a more tolerable toxicity profile. Furthermore, these newer agents act only in a subpopulation of patients whose tumor growth is driven by their target or a target-dependent process. To optimize the applicability of these drugs, it will be essential to define the patient population that is potentially susceptible to the new drug.

For years, the conventional cytotoxic drugs have been tested in phase I, II, and III trials.

However, can these traditional trial designs be applied to the new generation of anti-cancer drugs or do we need to adjust the trial design for these new agents to study the mechanism of action early in the development, to select a biological relevant dose instead of a maximum tolerated dose (MTD) and to identify the patient subpopulation that might benefit from treatment?

2. CURRENT CLINICAL TRIAL DESIGN—PHASE I, II, AND III

For cytotoxic drugs phase I studies focus on toxicity and are designed to determine the dose-limiting toxicity (DLT), the MTD, and the recommended dose for further phase II testing. Phase II trials can be viewed as a screening evaluation in which small groups of patients are used to assess the percentage of tumor regression and to screen for anti-tumor efficacy for the drug under development. Finally randomized phase III trials, the definitive evaluation, assess the anti-tumor efficacy using endpoints such as overall survival, time to progression (TTP), and disease-free survival. Also, quality of life assessments, symptom relief, functional status, sense of well being, and hospitalization rates are being used as secondary endpoints.

These traditional trial designs have been applied successfully for many decades. However, they are based on the assumptions that (i) anti-cancer drugs are non-specific and therefore toxic, (ii) there is a dose dependency for toxicity and anti-tumor activity resulting in dose–response curve, and (iii) benefit for the patient lies solely in observed tumor regression.

2.1. Phase I Trial Designs

New anti-cancer agents target specific processes outside or inside the tumor cell and importantly in models are mostly growth inhibitory. In other words, they do not shrink tumors. On the basis of their targeted specific mechanism of action, they are expected to have a more favourable toxicity profile. Therefore, limiting toxicity may not be seen and a DLT or MTD may not be defined. In phase I clinical trials of target-based drugs, determination of MTD might therefore not be appropriate or even possible. Furthermore, some growth-inhibiting drugs are non-toxic at doses that correspond to concentrations with desired biologic effects. Consequently, a biologic endpoint specific for the agent under investigation in addition to toxicity should be included as an endpoint in phase I studies.

As these drugs need prolonged administration to exert their effect, a dose that permits this prolonged administration should be defined. The recommended dose with optimal biological activity (OBD) can be far below the MTD, and the dose–response curve concerning tumor regression and toxicity is not as steep as for cytotoxic agents. In the absence of DLT, it is important to define other pharmacodynamic (PD) or pharmacokinetic (PK) endpoints. Aside from determining the optimal biologically active dose, a better selection of patients likely to benefit and defining proof of concept as soon as possible during the development of a drug should be incorporated in the phase I trial design.

2.1.1. DETERMINING DOSE

The use of endpoints such as PD and PK is often based on pre-clinical studies. The methods used should be validated and reproducible. An example of the use of PK endpoints is the pharmacologically guided dose escalation (PGDE) in which equal toxicity with equal drug exposure is assumed when comparing animal to human doses. With this knowledge in theory, more rapid dose escalations can be accomplished, which saves time and more importantly results in fewer patients receiving suboptimal doses. However, in practice, this turned out to be less favorable than expected, and the poor correlation between anti-tumor activity in models and anti-tumor activity in man precludes optimal use of PGDE from an activity point of view. For this reason, this approach is now largely abandoned.

When considering PD endpoints, the ultimate goal is to define the optimal biological dose (OBD). The definition of the OBD should be based on the demonstration in vivo of the desired biochemical effect on the target molecule. This can be done for example by measuring target enzyme inhibition in repeated tumor biopsies, which can be a problem because of difficult accessibility of tumor tissue. Aside from the measurement of target inhibition, measurement of downstream effects is also important in understanding the underlying mechanism of action. Instead of tumor biopsies other, more easily accessible tissues are used as surrogate tissue to measure biological effects of the drug under investigation, in the assumption that the drug effects in such surrogate tissues are similar to the effects in tumor tissue, which may be an overly optimistic assumption. Examples are skin, white blood cells, plasma, and buccal mucosa.

2.1.2. ASSESSING EARLY ACTIVITY

Figure 1 shows the epidermal growth factor receptor (EGFR) signaling pathways, and inhibition of EGFR represents a paradigm for a few principles related to determination of early drug activity. Skin biopsies have been suggested as a surrogate tissue in studies where EGFR inhibition was concerned as EGFR has an established role in renewal of the dermis. In studies on ZD1839 (gefitinib) and OSI-774 (erlotinib) for example, both EGF receptor tyrosine kinase inhibitors, multiple skin biopsies were performed.

Albanell et al. *(1)* studied 104 pre- and/or on-ZD1839 (gefitinib) therapy skin biopsies in 65 patients receiving doses of gefitinib ranging from 150–1000 mg/day. Total EGFR expression, being the most pronounced in the basal layer of the dermis and in the outer root sheet of the hair follicles, was not modified when comparing the pre- and on-therapy samples. Activated EGFR [phosphorylated EGFR (pEGFR)] however, only being present in cells that expressed EGFR in the first place, was significantly decreased in the on-therapy specimens (44% +/− 3.4 versus 2.1% +/− 0.8). Also, downstream effects such as the phosphorylation of mitogen-activated protein

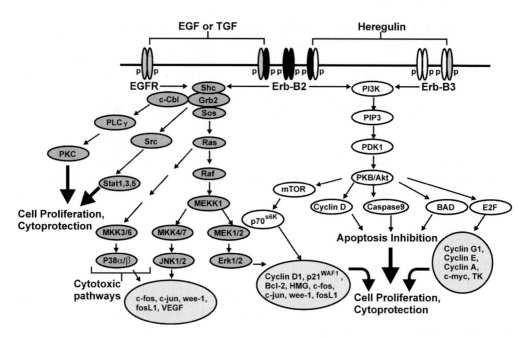

Fig. 1. EGFR signaling pathways.

kinase (MAPK) and the expression of Ki-67 and p27 were assessed. The activation of MAPK was significantly decreased in on-therapy specimens. Also, Ki-67, a marker for cell proliferation, showed a significant decrease. Measurement of p27 proved to be a hallmark of EGFR pathway inhibition and growth arrest in pre-clinical models. On-therapy samples showed a significant increase in p27 staining. Also markers of maturation and differentiation, Keratin I and STAT 3 respectively, showed a significant increase during therapy. Importantly and unfortunately, there was no relation between the dose of gefitinib and the on-therapy scores of activated EGFR, activated MAPK, p27, Keratin I, STAT3, or apoptosis. However, there was a relationship between increasing dose and Ki-67 decrease. The steady state plasma concentration, obtained at day 28, only showed a significant relationship with increasing levels of activated MAPK. There was no correlation between gefitinib-induced EGFR inhibition in the skin and the tumor.

The effect of OSI-774 (erlotinib) on human epidermis was also studied *(2)*. Skin biopsies were performed before and after therapy in 28 patients. Measurements included EGFR, pEGFR, extracytoplasmatic-regulated kinase (Erk, a marker of downstream signaling), and p27. Levels of pEGFR decreased significantly on treatment (22.91 +/− 10.45 to 16.75 +/− 10.03; $p = 0.001$). The level of total EGFR expression increased insignificantly (35.58 +/− 10.53 to 37.08 +/− 11.09; $p = 0.186$). Levels of Erk showed a non-significant decrease (2.29 +/− 0.71 versus 1.96 +/− 0.79; $p = 0.14$). The number of p27 positive nuclei increased from 185 +/− 101 to 253 +/− 111 ($p = 0.002$). This means that 56% of patients had up-regulation of p27 on therapy. A dose–response relationship existed between erlotinib and up-regulation of p27, in contrast to EGFR, pEGFR, and Erk. Although this study indicates that human skin could serve as a possible surrogate for tumor tissue and that p27 could serve as a possible biomarker of activity on EGFR-TK-blocking drugs, the relationship between PD effects in skin and tumor tissue has not been established.

This relationship has been addressed in patients with metastatic breast cancer in which the effect of erlotinib was studied by obtaining tumor, skin, and buccal mucosa biopsies before and after 1 month of treatment in 15 assessable patients *(3)*. Measurements of the stratum corneum height were performed, as were levels of Ki67, EGFR, pEGFR, phosphorylated MAPK (pMAPK), and phosphorylated AKT (pAKT). Of the 15 paired tumor samples, only one was EGFR positive. In this case, inhibition of EGFR, MAPK, and AKT phosphorylation was observed in the tumor samples. However, no inhibition of Ki67 was seen, which resulted in lack of clinical benefit, probably explained by the molecular heterogeneity in EGFR expression. In EGFR negative tumor biopsies, pEGFR showed a significant increase ($p = 0.001$). This paradoxical response might be explained by lack of sensitivity of the EGFR antibody or better sensitivity of the pEGFR antibody. pMAPK, pAKT, or Ki67 levels were not significantly increased however ($p = 0.19$, $p = 0.45$, and $p = 0.47$, respectively). There was no association between pEGFR in skin and tumor biopsies or between buccal mucosa and tumor biopsies in the patients with EGFR negative tumors.

An acne-like or maculopapular skin rash is a characteristic side effect of EGFR inhibition. The presence and severity of the skin rash has been shown to correlate with survival and response both for erlotinib and cetuximab *(4,5)*. These studies indicate that biological effects can be qualified and partially quantified. However, they can either not be correlated to drug-dose and concentration, or the biological effect in the surrogate tissue does not properly correlate to the effect in tumor tissue. Thus, in these studies, the PD endpoints could not be used to determine the optimal dose.

Whilst it cannot be used to identify OBD, skin can be used as a surrogate tissue to at least study presence of EGFR inhibition in patients. There are various possible explanations for the observed limitations. A difference in blood flow between tumor and skin can result in different concentrations of the drug and hence different inhibition of the EGFR in both tissues. In addition, the downstream effects can be different even after complete EGFR inhibition in both tumor and skin as shown by Vivanco and Sawyers *(6)*. They showed that activation of the phosphatidylinositol 3´-kinase-akt pathway by tumor-related mutations could make tumor cells independent of EGFR inhibition. Tumor tissue is known for its genetic instability and for mutations that result in activation of escape pathways, such as these in the case of EGFR inhibition. In conclusion, this implies that the EGFR inhibition observed in the skin mainly has a negative predictive value: when no EGFR inhibition is observed in the skin, there will definitively be no inhibition in the tumor.

The assumption that only patients with EGFR positive tumors might benefit from therapy with these new agents will lead to patient selection. There is, however, no apparent relationship between the efficacy of EGFR-blocking drugs and the level of EGFR expression in the tumor as shown for cetuximab and gefitinib *(5,7)*. This is in contrast to Her2 expression, a target for trastuzimab (herceptin) in breast cancer. Table 1 provides an overview of EGFR inhibitors currently in clinical development.

The possibility of patient selection to optimize clinical outcome of treatment with EGFR thyrosine kinase inhibitors (TKIs) was demonstrated by Lynch et al. *(8)* and Paez et al. *(9)*. Predictors of response to anti-EGFR therapy found in several trials include female gender, adenocarcinoma histology, and non-smoking. Furthermore,

Table 1
Overview of Epidermal Growth Factor Receptor (EGFR) Inhibitors Currently
in Clinical Development

Class	Inhibitor	Tumor type	Phase
Antibodies	C225	head and neck squamous cell carcinoma (HNSCC), colorectal carcinoma (CRC), NSCLC	III
	ABX-EGF	Metastatic renal carcinoma	II
	EMD 72000	EGFR positive tumors including CRC, head and neck, and cervix carcinoma	I
	h-R3	Head and neck	I
	IMC-C225	Solid tumors	I–III
Small molecule inhibitors	GW572016	Solid tumors	II
	CI 1033	Advanced solid tumors	I
	PKI-166	EGFR positive solid tumors	I
	OSI-774	HNSCC, ovarian carcinoma, NSCLC	II
	ZD1839	NSCLC	III
	EKB-569	Advanced solid tumors	I
	PD168393		Pre-clinical
	AG-1478		Pre-clinical
	CGP-59326A		Pre-clinical
Ligand-conjugated toxin	TP-38	Malignant brain tumors	I

NSCLC, non-small cell lung cancer; HNSCC, head and neck squamous cell carcinoma; CRC, colorectal carcinoma.

higher response rates were found in Japanese than in non-Japanese *(10)*. These clinical and pathological features are however far from perfect discriminators of patients prone to benefit from treatment with EGFR TKIs.

Lynch et al. *(8)* studied 275 patients who were treated with gefitinib monotherapy after being chemotherapy refractory. Twenty-five patients had a clinically significant response, defined as a partial response using the RECIST criteria. In 9 of 25 patients with a response to gefitinib, tumor tissue was available for evaluation. In 8 of 9 evaluable patients, heterozygous mutations were observed in the tyrosine kinase domain of EGFR. The ninth patient may have had a yet unknown mutation or a mutation in the heterodimerization partner of EGFR.

Seven non-responding patients showed no mutations ($p < 0.01$). These results suggests that these mutions account for the majority of responses to gefitinib reported in clinical studies.

These data were confirmed by Paez et al. *(9)*. Tumors with mutant receptors were much more sensitive to gefitinib than those with wild-type EGFRs. Furthermore, the mutations were strongly correlated with the clinical and pathological features found to predict for response to gefitinib. Mutations were more common in adenocarcinomas (21%) than in other non-small cell lung cancers (NSCLCs) (2%), more frequent in patients from Japan (26%) than in patients from the USA (2%), and more frequent in women (20%) than in men (9%).

These data show that the overexpression of EGFR is not essential but rather the presence of mutations in the tyrosine kinase domain of EGFR may be essential to tumor growth and the probability of a response to EGFR inhibitions. This knowledge can be used in prospectively identifying the subpopulation that may have clinical benefit from treatment with gefitinib. However, even patients with wild-type EGFR may respond to therapy with EGFR-blocking agents, albeit less likely. With the use of current dose recommendations *(1)*, the plasma concentration of gefitinib exceeded the autophosphorylation-suppressing dose but was below the dose required to suppress the wild-type EGFR. However, mutant EGFR transduces signals that are qualitatively distinct from those mediated by wild-type EGFR, implying that even if wild-type EGFR could be blocked completely, clinical benefit might not occur as the tumor is not dependent on signals mediated by the wild-type EGFR pathway.

2.1.3. SELECTING PATIENTS

As another alternative for tumor tissue, leukocytes can be used as surrogate tissue to measure biologic effects. Assessment of farnesyl transferase inhibition can, for example, be done in white blood cells but also in buccal mucosa. The farnesylation of retrovirus associated DNA sequences (RAS) oncoproteins, found to be essential in tumor progression, is blocked by FTIs. In animal models, the FTIs indeed showed anti-tumor activity, but the activity seen in humans turned out to be unrelated to RAS expression, suggesting other mechanisms to be of value *(11–13)*. A phase I trial in relapsed or refractory acute myeloid leuke mia (AML) reported a 32% response rate with tipifarnib (zarnestra) single agent therapy *(14)*. None of the patients however had ras mutations. A phase II trial of relapsed AML confirmed these results *(15)*.

A phase II trial on tipifarnib treatment in breast cancer also showed anti-tumor activity irrespective of ras mutational status or HER2 positivity *(16)*. These trials show that, in addition to oncogenic Ras, other farnesylated proteins may be important in the oncogenic transformation of cells. Indeed, more than 90 mammalian proteins are farnesylated, including the GTPase RhoB, phosphatases PRL-1, 2, and 3, and centromeric proteins CENP-E and CENP-F.

The development of FTIs indicates that sometimes our understanding of the target is insufficient. As a consequence, the targeted drug may be less targeted than predicted or even work through a completely unknown mechanism. Thus, the possibility that another target or downstream pathway is involved has its implication on screening for biological efficacy. More and other downstream effects should be assessed than assumed on the basis of the presumed mode of action.

Assessment of the cellular and subcellular effects of FTIs in the clinical setting can be achieved in various ways. Adjei et al. *(17)* compared several assays in four different human tumor cell lines (A549, HCT116, BxPC-3, and MCF-7) during exposure to SCH66336 or FTI-277. They showed by immunoblotting that the prenylated protein human Dnaj2 (HDJ)-2 demonstrated a shift in mobility upon treatment in all four cell lines. This shift was also present in non-cycling cells suggesting the possible utility of this assay in tumor tissue samples. Furthermore, processing of prelamin A was extensively inhibited in all cell lines examined. Accumulation of prelamin A could be assessed by immunoblotting or immunohistochemistry, rendering this assay particularly useful for clinical studies. In fact, in a phase I study on the FTI, SCH66336 inhibition of the farnesylation of prelamin A was detected in oral buccal mucosa cells *(18)*.

Moasser and Rosen *(19)* studied a panel of 10 breast cancer cell lines after FTI treatment to determine molecular characteristics that could serve as surrogate markers for sensitivity to FTIs. The fact that FTI sensitivity does not correlate with the relative expression of Ras and its isoforms or the inhibition of Ras processing was re-confirmed in this study. Farnesyl protein transferase (FPTase) was proposed as a possible marker because two cell lines sensitive to FTIs showed low FPTase activity. However, no linear relationship between growth inhibition and FPTase activity has been established, and larger analyses are necessary to establish its role as a potential surrogate marker for FTI sensitivity.

In conclusion, surrogate markers predictive for biologic efficacy and optimal patient selection when using FTIs have yet to be validated. Mononuclear peripheral blood cells or oral buccal mucosa cells potentially could serve this purpose. Yet, the mode of action of FTIs may well turn out to be completely different and more complex than initially assumed, which stresses the necessity to perform a wide efficacy screening including more targets or downstream pathways initially believed to be of interest. The development of FTIs further provides an example of the danger of too restricted patient selection in the earliest clinical trials, if our understanding of the target turns out to be insufficient. The activity of FTIs in AML and breast cancer would have been missed if study patient selection would have been based on presence of RAS expression.

2.1.4. APPROACHING THE VASCULATURE

Another crucial process for tumor growth and the formation of metastases is angiogenesis (Fig. 2). As the formation of new blood vessels is mandatory for the tumor to grow, blocking this process became a major challenge. Blocking the process of angiogenesis can be achieved in many ways *(20–23)*. Monoclonal antibodies can be used to bind to circulating vascular endothelial growth factor (VEGF) and thereby block its effect. Also, VEGF receptor tyrosine kinase inhibitors can be used. Furthermore, endothelial cell proliferation can be inhibited by mimicking the activity of thrombospondin (TSP)-1, endostatin, and angiostatin. These proteins are naturally occurring anti-angiogenic proteins. Another interaction between endothelial cells and the extracellular matrix can be blocked by inhibiting integrin $\alpha_v\beta_3$ and $\alpha_v\beta_5$. Furthermore, degradation of the extracellular matrix by MMPs is essential for angiogenesis and can be modulated by MMPIs. For angiogenesis to occur, an increase in the level of pro-angiogenic factors relative to a decrease of anti-angiogenic factors is needed. Pro-angiogenic factors include VEGF, platelet-derived growth factor (PDGF), basic fibroblast growth factor (bFGF), interleukin-8, insulin-like growth factor (IGF), transforming growth factors α and β, and others.

Because anti-angiogenic therapy was recognized as a possible way to target cancer, numerous anti-angiogenic compounds have been developed. Classical study endpoints used in cancer trials were short of discriminative power to monitor the effects inflicted by these drugs; so, again surrogate markers were needed. The ideal marker for anti-angiogenic therapies should (i) directly reflect the tumor vasculatures' angiogenic status, and indirect acting factors produced by the tumor itself should be excluded as they could be affected by other factors. Furthermore, the marker should be measurable in blood. The marker should (ii) decrease and eventually disappear after the tumor has been completely removed. It should also (iii) decrease during tumor regression

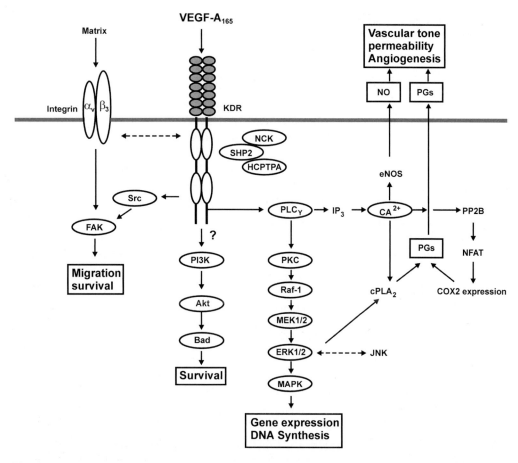

Fig. 2. Signaling pathways important in angiogenesis.

inflicted by classical anti-tumor therapy such as chemo- or radiotherapy but also during effective anti-angiogenic therapy.

The number of micro vessels per high power field, the micro vessel density (MVD), has been proposed as a surrogate marker for recurrence, metastasis, survival, and therapeutic efficacy. However, data are absolutely inconclusive and even contradictory. Furthermore, MVD measurement requires repeated tumor tissue biopsies, which creates practical limitations. Therefore, other techniques have been developed to try to monitor anti-angiogenic activity less invasively and without the need for repeated tumor biopsies.

Unfortunately, studies on tumor necrosis factor (TNF)-α in isolated peripheral blood cells *(24,25)*, or serum and plasma levels of circulating growth factors such as VEGF *(26)*, hepatocyte growth factor (HGF), FGF-2 *(20,26–28)*, vascular cell adhesion molecule (VCAM)-1 *(26)*, vWf, and E-selectin produced the same contradictory results as seen in MVD studies, and therefore, currently these are yet of little value as surrogate markers of drug activity and efficacy *(29,30)*.

Circulating vascular molecules released by proliferating endothelial cells such as growth factor receptors or cell adhesion molecules are, on the contrary, possible candidate surrogate markers *(31,32)*. However, initial findings on soluble VEGF-R1, MMP, thrombomodulin, or angiogenin amongst others, need to be confirmed, further explored, and validated before a final conclusion on their potential can be drawn.

Endothelial cell progenitors (ECPs) have been found to actively contribute to angiogenesis. They were mobilized from the bone marrow by a VEGF-R1-mediated process, and administration of VEGF-R1 monoclonal antibodies resulted in inhibited ECP mobilization and thus reduced tumor growth *(33)*. Also, circulating endothelial cells (CECs) are proposed as surrogate markers of anti-angiogenesis efficacy. It has been shown that higher levels of CEC are seen during rapid tumor growth and that tumor volume correlates with the level of CEC. The use of both ECP and CEC as valuable surrogate markers to monitor anti-angiogenic drug efficacy remains to be validated.

In conclusion, surrogate markers of anti-angiogenesis are urgently needed to early suggest activity of anti-angiogenic treatment. Better knowledge of the underlying processes and new techniques such as proteomics and gene expression profiling of tumor vasculature can provide us with other potential surrogate markers.

2.1.5. IMAGING

In trying to find ways to evaluate drug-induced effects in a non-invasive manner, other techniques including positron emission tomography (PET) scanning and dynamic contrast enhanced MRI (DEMRI) have been evaluated. [18]FDG-PET scanning assesses glucose metabolism and the proliferative activity of tumors. It was shown to correlate with symptom control and CT response in patients with gastrointestinal stromal tumor (GIST) treated with imatinib mesylate (gleevec) *(34)*. When PET responses (measured at day 8) were compared with CT responses, only for patients with stable disease or minimal response on PET, the subsequent CT response was less predictable. The PET response preceded the CT response by a median of 7 weeks (range 4–48 weeks). PET scans performed 24–48 h after the start of imatinib always predicted the PET response on day 8. More importantly, however, survival analysis revealed a significantly better 1-year progression-free survival in PET responders compared with non-responders: 92 versus 12%, respectively ($p = 0.00107$). Other PET-scanning techniques have been much less successful. ^{15}O-labeled water (H_2 ^{15}O)-PET- and ^{15}O-labeled carbon monoxide ($C^{15}O$)-PET-scanning techniques have been used to assess the effect of several drugs such as razoxane *(35)*, combretastatin A4 phosphate (CA4P) *(36)*, and human recombinant endostatin *(37)* amongst others. Although these trials mostly show reversible effects of the drug under investigation on the tumor blood volume and perfusion, none of the presented trials so far showed a relation between the observed effects and clinical tumor response.

DEMRI provides insights into tumor perfusion, capillary permeability, and leakage space. It measures the perfusion/blood volume of the target lesion, which can be correlated with tumor grade and MVD. A correlation between vascular permeability and tissue VEGF expression in breast tumors was first reported by Knopp et al. *(38)*. Also, in rectal tumors, a correlation between VEGF levels and K-trans (permeability-surface area product per unit volume of tissue) levels has been reported. Furthermore, DEMRI can detect suppression of vascular permeability after the administration of anti-VEGF antibodies. Fibrosis and stromal cellularity, tissue oxygenation, and tumor grade are other entities that can be assessed by DEMRI. In a phase I trial of the anti-vascular drug CA4P, DEMRI was used to measure changes in tumor perfusion *(39)*. In

8 of 10 patients at least a mild decrease in K-trans value was seen. Three patients even showed a marked loss of vascularity after CA4P administration. A more consistent pattern of alteration was seen in tumor leakage space (Ve) after CA4P. Furthermore, a significant negative correlation ($r = 549$, $p = 0.05$) existed between change in K-trans and day 5 CA4 AUC, but no consistent trend was seen between reduction in tumor Ve and PK indices of CA4P and CA4. Although not uniform, these results show a trend in which DEMRI can be a practical non-invasive way to assess drug-induced changes in tumor tissue, once further refinement of the technique can be employed. However, how all the observed changes relate to a possible gain in time to tumor progression or in overall survival is currently unknown.

Other techniques such as contrast-enhanced dynamic CT (CED-CT) can be used to assess tissue blood flow, blood volume, mean transit time, and permeability. With this technique, changes in blood flow and volume as small as 14 and 20%, respectively, could be measured *(40)*. Also, Doppler ultrasound techniques can be used to evaluate tumor perfusion, as long as the tumor is within the reach of the Doppler signal emitter/detector. A further limitation of this technique is the inter-observer variability. In conclusion, the described techniques are not yet validated to predict patient benefit and/or outcome. More and larger phase III evaluations are needed to serve this purpose.

In this section, numerous surrogate markers have been proposed. From the presented studies it can be concluded that (i) the ideal surrogate marker should be sensitive and specific, (ii) the surrogate marker should be present in accessible tissues other than tumor tissue, (iii) the use of a simple assay that can be adaptable to clinical use is mandatory, (iv) results in surrogate tissue should correlate with results in tumor tissue, and (v) a correlation between the surrogate marker and response rate, disease-free and overall survival, or TTP should exist. It is clear that to achieve this ideal situation for surrogate markers, more pre-clinical as well as clinical research has to be performed.

2.2. Phase II Trial Designs

Phase II trials screen drugs for their potential anti-tumor activity with the purpose of selecting the most promising agents for phase III trials. Because of this, most phase II study designs are aimed at minimizing the chance that an active agent is erroneously rejected. The possibility of a false-negative result should therefore be as low as possible. Also of importance is the need to limit the exposure of excessive numbers of patients to ineffective treatment.

Is phase II really necessary for cytostatic (growth inhibitory) agents that may not or only in limited numbers cause objective tumor responses? Skipping phase II studies would mean that a lot of new drugs would directly enter phase III studies that are time consuming, need a lot of patients, and cost a lot of money. Furthermore, phase II trials offer the chance to modify the recommended dose of an agent as results are based on a larger and often less pre-treated patient population than originally tested in phase I trials.

Examples of phase II studies on targeted agents are for instance available for gefitinib (IDEAL 1&2) and erlotinib. Phase II studies on gefitinib single agent therapy in patients with pre-treated advanced NSCLC demonstrated objective tumor response rates of 11.8–18.4% and maybe even more importantly, symptom improvement rates of 40.3–43.1% as shown in Fig. 3A and B (*10,41–43*). Almost similar data are available for erlotinib. These data suggest these agents may have some activity; thus, the phase

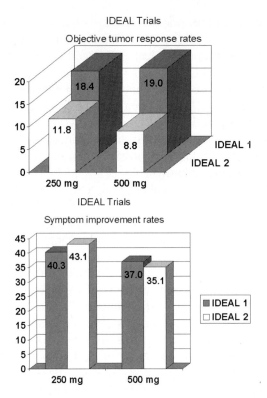

Fig. 3. (A) Objective tumor response rates in IDEAL 1&2 with gefitinib 250 or 500 mg/day; (B) Symptom improvement rates in IDEAL 1&2 with Gefitinib 250 or 500 mg/day.

II studies provided what they were intended for. And with subsequent knowledge of the relevance of EGFR mutations, the data of these phase II studies become even more important. These studies were not followed by randomized phase II studies on the combination of cytotoxic chemotherapy with EGFR-TKIs.

Despite this lack of information, phase III studies were performed for both gefitinib and erlotinib, combining each of them with either carboplatin/paclitaxel or cisplatin/gemcitabine *(44,45)*. The results were totally negative unfortunately, stressing the importance of intermediate randomized phase II studies before embarking on large phase III trails. The various potential reasons for the failure of the phase III trials are beyond the scope of this chapter.

The randomized phase II design prevents comparison of current results with historical data concerning the standard treatment and offers protection against possible selection bias, even though the design of such studies can never be powered to detect significant differences. In a randomized design, approximately twice the number of patients are needed, which can be considered a loss if at a certain point the development of a drug or combination is ended. Proceeding with a phase III trial without a proper phase II design can however, as shown, produce misleading results after including even more patients and spending more money. The patient numbers included in a randomized phase II design should therefore be considered an investment. In general, phase II trials are a very useful tool to assess activity of a drug or a drug combination as long as the proper trial design is used.

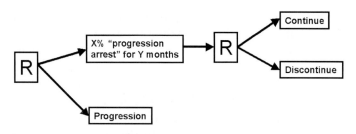

Fig. 4. The randomized discontinuation design.

2.3. The Randomized Discontinuation Design

An alternative strategy to perform a phase II clinical trial is the so-called randomized discontinuation design *(46)*. In this design, all patients are treated with the study agent for a pre-defined period of time (Fig. 4). If patients during or after this period of time do not show disease progression, they can be randomized to either continue treatment or to receive no drug or a placebo. TTP or progression-free rate (PFR) can be used as an endpoint. As all patients receive the proposed active medication, patient accrual can be relatively easy. Furthermore, the activity assessment includes a comparable control group to separate the drug's effect from the natural course of disease. Another advantage of this design is that generally fewer patients have to be randomized. However, the total number of patients needed to establish clinically relevant differences of activity frequently approaches that of a classic randomized phase III trial design.

The randomized discontinuation design also enables assessments of secondary endpoints such as surrogate marker inhibition and PKs and PDs in the entire study population. The major disadvantage of this design is that heterogeneity of tumor growth can reduce power substantially, causing the need for accrual of more patients. Furthermore, this design does not include available standard therapy as in the randomized phase II design. Also, generalizing the results to a larger population of patients can de difficult.

In conclusion, a positive result is indicated when a higher number of patients will have stable disease than expected during the first stage of the design, or fewer patients randomized to continue therapy will have disease progression after randomization, when compared with placebo. For this to be significant, often large numbers of patients are needed. Furthermore, because no control arm with standard therapy is included in the design, a possible positive result in the randomized discontinuation design means the initiation of a phase III trial in which vast numbers of patients will be needed once again. This randomized discontinuation design is therefore mainly useful as a screening tool.

2.4. Alternative Endpoints in Phase II Trial Design

Historically, the percentage of tumor regression (response rate) has been used as the most important endpoint in phase II trials. The advantage of this approach is that response is relatively easy to measure. The disadvantages however are, first, that response rate does not take into account the duration of response. For example, the use of DTIC in metastatic melanoma has an average response rate of approximately 17% with a duration of only 10 weeks. Thus, this agent should not be considered active in melanoma. Second, response rate does not take into account the percentage of patients that achieve stable disease on treatment. In several tumor types, it was noted that the prognosis of patients achieving an objective tumor response and patients with disease stabilization was comparable. However, several drugs have been discarded because of a low response rate, although treatment resulted in a high SD rate for a prolonged period of time and thus was likely beneficial to the patients. Because with cytostatic drugs tumor regression will be observed to an even more infrequent extent than with cytotoxic drugs, other alternative endpoints will be needed to assess potential anti-tumor activity in phase II studies.

2.4.1. Growth Modulation Index

An alternative endpoint to consider might be the use of a "growth modulation index." The hypothesis is that if a new agent has any anti-tumor effect, it will change the natural course of the disease. The index is defined as the ratio of the TTP in a current treatment (TTP2, the newly discovered agent) and the TTP of the previous treatment period (TTP1, conventional treatment). It is postulated that a 33% increase of TTP is proof of activity. Therefore, a ratio greater than 1.33 is considered to be indicative of the new drug having anti-tumor activity. Although 33% is an arbitrary value, the concept can be helpful in deciding whether a drug can enter phase III testing.

There are, however, some possible limitations to this approach. Often, there is a lack of previous data in the patient involved concerning TTP. Furthermore, according to natural history of the disease, TTP2 cannot be expected to equal TTP1 without treatment. It is more likely that TTP2

2.4.2. Progression Rate

To define the activity of a new agent in phase II studies with a minimal amount of patients, Gehan, Flemming, and Simon *48–50* developed many multistage designs in which two or more sequential stages of accrual are being used. Should, in the first stage, response rates be minimal and therefore activity be insufficient, the study will be terminated. In this format, the number of responding patients is the trigger to continue. The numbers of non-responding patients who must be included for a drug to be rejected depends on the target response rate hypothesized to be of interest. However, these designs do not take into account the amount of early progressors. Drugs that produce high response rates will also reduce the rate of early disease progression. It follows that drugs that show a high proportion of early disease progression can hardly be of importance for further testing. Low rates of early disease progression, on the contrary, may be a sign of an active drug being investigated, even if this goes along with stable disease in the other patients in the study. After van Oosterom *(51)* first described the possible value of assessing progression arrest, Zee et al. and Dent et al. *(52,53)* described a multinomial stopping rule for phase II trials that incorporated both response rate and early progression to determine whether the study should be continued or stopped after the first stage of accrual. This two-stage stopping rule is based on various hypotheses of response and progression rates thought to be relevant to specific tumor types. In breast cancer, for example, the drug under investigation will be considered active if at least 30% response rate is being achieved. The drug will be considered inactive once response rates of 10% or less are being found. The progression hypotheses would approve a drug that causes less than 30% early progression and would discard the drug once more than 50% of patients would experience early progression.

In glioma, a more chemo resistant tumor, drugs under investigation should be considered active with a response ratio of 10% or higher and be discarded when response rate (RR) <5%. The progression rates in glioma would accept the drug if early. progression rates <40% and would discard it if PD> 60%. If in a limited number of patients the drug under investigation seems active according to the hypothesized RR and PD, patient accrual in stage two of the trial can be started. Once again, the same hypotheses for RR and PD are being employed. If after the first stage the drug seems inactive, the trial can be stopped and no more patients have to be exposed to an inactive treatment. When applied to 23 phase II trials performed by the European Organisation for Research and Treatment of Cancer (EORTC), striking results were found when comparing the multinomial stopping rule to the Gehan design (53). Seven of 23 studies would have been continued using the Gehan design but would have been stopped according to the multinomial design. Eventually, these seven studies tested inactive drugs, which means that using the Gehan design, too many patients were exposed to this ineffective treatment whilst the multinomial design would have stopped further phase II testing.

The multinomial rule also holds the opportunity to draw an early conclusion on activity if such a decision would be desirable. In 8 of the 23 EORTC trials, the Gehan design and the multinomial design agreed that response activity was sufficient to continue to the second stage. All these trials tested active compounds, and although, according to the Gehan design and the multinomial design were in agreement that response activity was sufficient to continue to the second stage.

In conclusion, trial designs using only response rates are not sufficient to be used for efficacy testing of targeted therapy. Progression rate should be included in deciding whether the drug under investigation deserves further testing in phase III trials. Dent et al. and Zee et al. (52,53) provide an elegant way to test targeted therapies without a vast number of patients being exposed to ineffective treatment.

2.4.2.1. Progression-Free Rate. As another alternative endpoint in phase II clinical testing of targeted therapy, the PFR has been proposed (54). Before using this endpoint, a decision has to be made about what is considered an appropriate target progression-free survival concerning the tumor type under investigation, both for suggesting activity and inactivity. For this purpose one has to rely on historical databases, which has its obvious limitations. Also, the time point at which the PFR will be assessed is of importance. In the case of slowly growing tumors, determining the PFR at 6 months for example can be misleading, because in this timeframe the natural course of the disease might not even produce a measurable progression. Also, stable disease cannot be considered as evidence of treatment activity if no documented disease progression was present before the start of treatment. Thus, only truly progressive patients should be entered in a phase II trial where PFR is an important endpoint.

The PFR has been determined retrospectively in phase II trials concerning soft-tissue sarcomas by Van Glabbeke et al. (54). Reference values for the PFR were established using prior clinical trials performed by the EORTC. PFRs were determined in patients pre-treated with chemotherapy, as well as in non-pre-treated patients. It follows that no reference value could be established for patients treated with first-line ineffective treatment or treatment combinations. In pre-treated patients, the 3- and 6-month PFRs were 39 and 14%, respectively, for patients treated with an active drug. In patients treated with an inactive drug the 3- and 6-month PFRs were 21 and 8%, respectively.

This retrospective evaluation provides PFRs (with a standard error of approximately 5%) that can be used in the statistical evaluation of future phase II trials. Using PFR as an alternative endpoint in phase II clinical testing means that a proper statistical trial design should be used. The Gehan design, for example, tests the compatibility of the observed success rate with the rate of an active agent. Van Glabbeke et al. show that a small proportion of patients remain disease free at 3 or 6 months even when treated with an inactive agent *(55)*. This means that the Gehan design is not appropriate when activity is characterized by absence of progression. In this case, a design such as the "multinomial stopping rule" by Zee et al. and Dent et al. can better be applied.

3. CONCLUSION

The development of targeted anti-cancer drugs in recent years puts a challenge on classical trial designs. A biological relevant dose might become more important than MTD, and establishing the mechanism of action becomes pivotal early in drug development. Furthermore, results from clinical trials show the importance of defining the patient population likely to benefit from these new drugs. However, patient selection should not be too restricted in early clinical trials, as was shown for FTIs, because this will possibly limit the chance of finding an active drug when the mechanism of action is not fully unraveled.

Because repeated tumor biopsies to measure target inhibition can be a problem, the use of surrogate tissue-like skin, white blood cells, plasma, or buccal mucosa, should be explored. A relationship, however, between PD effects in surrogate and tumor tissue has not yet been established and therefore cannot be used as a validated endpoint in studies.

Using non-invasive techniques such as PET, CT, or MRI imaging is another challenging way to try to predict patient outcome. Although [18]FDG-PET used in patients with GIST tumors correlated with symptom control and CT response, other imaging techniques used in different tumor types are not yet validated to predict patient benefit and/or outcome.

To define and validate surrogate markers of efficacy of targeted drugs, more trials are needed in which PD and PK endpoints are studied in conjunction with repeated tumor biopsies.

REFERENCES

1. Albanell J, Rojo F, Averbuch S, et al. Pharmacodynamic studies of the epidermal growth factor receptor inhibitor ZD1839 in skin from cancer patients histopathologic and molecular consequences of receptor inhibition. *J Clin Oncol* 2002;20:110–124.
2. Malik S, Siu L, Rowinsky E. Pharmacodynamic evaluation of the epidermal growth factor receptor inhibitor OSI-774 in human epidermis of cancer patients. *Clin Cancer Res* 2003;9:2478–2486.
3. Tan A, Yang X, Hewitt S, et al. Evaluation of biologic end points and pharmacokinetics in patients with metastatic breast cancer after treatment with erlotinib, an epidermal growth factor receptor tyrosine kinase inhibitor. *J Clin Oncol* 2004;22:3080–3090.
4. Saltz LB, Meropol NJ, Loehrer PJ Sr, et al. Phase II trial of cetuximab in patients with refractory colorectal cancer that express the epidermal growth factor receptor. *J Clin Oncol* 2004; 22:1201–8.
5. Cunningham D, Humblet Y, Siena S, et al. Cetuximab monotherapy and cetuximab plus irinotecan in irinotecan-refractory metastatic colorectal cancer. *N Engl J Med* 2004; 351:337–45.
6. Vivanco I, Sawyers CL. The phosphatidylinositol 3-kinase-akt pathway in human cancer. *Nat Rev Cancer* 2002;2:489–501.

7. Ciardiello F, Tortora G. Epidermal growth receptor (EGFR) as a target in cancer therapy: understanding the role of receptor expression and other molecular determinants that could influence the response to anti-EGFR drugs. *Eur J Cancer* 2003;39:1348–1354.

8. Lynch T, Bell D, Sordella R, et al. Activating mutations in the epidermal growth factor receptor underlying responsiveness of non-small-cell lung cancer to gefitinib. *N Engl J Med* 2004;350: 2129–2139.

9. Paez J, Jänne P, Lee J, et al. EGFR mutations in lung cancer: correlation with clinical response to gefitinib therapy. *Science* 2004;304:1497–1500.

10. Douillard J, Giaccone G, Horai T, et al. Improvement in disease-related symptoms and quality of life in patients with advanced non-small-cell lung cancer treated with ZD1839. *Proc Am Soc Clin Oncol* 2002 (Abstract 1195).

11. Brunner T, Hahn S, Gupta A, et al. Farnesyltransferase Inhibitors: An overview of the results of preclinical and clinical investigations. *Cancer Res* 2003;63:5656–5668.

12. Eskens F, Stoter G, Verweij J. Farnesyltransferase inhibitors: current developments and future perspectives. *Cancer Treat Rev* 2000;26:319–332.

13. Rowinsky E, Windle J, Von Hoff D. Ras protein farnesyltransferase: a strategic target for anticancer therapeutic development. *J Clin Oncol* 1999;17:3631–3652.

14. Karp J, Lancet J, Kaufmann S. Clinical and biologic activity of the farnesyltransferase inhibitor R115777 in adults with refractory and relapsed acute leukemias: a phase I clinical-laboratory correlative trail. *Blood* 2001;97:3361–3369.

15. Harousseau J, Stone R, Thomas X, et al. Interim results from a phase II study of R115777 (Zarnestra) in patients with relapsed and refractory acute myelogenous leukaemia. *Proc Am Soc Clin Oncol* 2002 (Abstract 1056).

16. Johnston S, Kickish S, Houston S, et al. Efficacy and tolerability of two dosing regimens of R115777 (Zarnestra), a farnesylprotein transferase inhibitor, in patients with advanced breast cancer. *Proc Am Soc Clin Oncol* 2002 (Abstract 138).

17. Adjei A, Davis J, Erlichman C, et al. Comparison of potential markers of Farnesyltransferase inhibition. *Clin Cancer Res* 2000;6:2318–2325.

18. Awada A, Eskens F, Piccart M, et al. A clinical pharmacodynamic and pharmacokinetic phase I study of SCH66336, an oral inhibitor of the enzyme farnesyltransferase, given once daily in patients with solid tumors. *Clin Cancer Res* 1999;5(Suppl):3733s.

19. Moasser M, Rosen N. The use of molecular markers in farnesyltransferase inhibitor (FTI) therapy of breast cancer. *Breast cancer Res Treat* 2002;73:135–144.

20. Eskens F. Angiogenesis inhibitors in clinical development; where are we now and where are we going? *Br J Cancer* 2004;90:1–7.

21. Longo R, Sarmiento R, Fanelli M, et al. Anti-angiogenic therapy: rationale, challenges and clinical studies. *Angiogenesis* 2002;5:237–256.

22. McCarty M, Liu W, Fan F, et al. Promises and pitfalls of anti-angiogenic therapy in clinical trials. *Trends Mol Med* 2003; 9:53–58.

23. Rüegg C, Meuwly JY, Driscoll R, et al. The quest for surrogate markers of angiogenesis: a paradigm for translational research in tumor angiogenesis and anti-angiogenesis trials. *Curr Mol Med* 2003;3:673–691.

24. Levitt N, Eskens F, O'Byrne K, et al. Phase I and pharmacological study of the oral matrix metalloproteinase inhibitor MMI270, in patients with advanced solid cancer. *Clin Cancer Res* 2001;7: 1912–1922.

25. Gearing A, Becket P, Christodoulou M, et al. Processing of tumor necrosis factor-alpha precursor by metalloproteinases. *Nature* 1994;370:555–557.

26. Herbst R, Hess K, Tran H, et al. Phase I study of recombinant human endostatin in patients with advanced solid tumors. *J Clin Oncol* 2002;20:3792–3803.

27. Davis D, McConkey D, Abbruzzese J, et al. Surrogate markers in antiangiogenesis clinical trials. *Br J Cancer* 2003;89:8–14.

28. Fine H, Figg W, Jaeckle K, et al. Phase II trails of the antiangiogenic agent thalidomide in patients with recurrent high-grade gliomas. *J Clin Oncol* 2000;18:708–715.

29. Berglund A, Molin D, Larsson A, et al. Tumor markers as early predictors of response to chemotherapy in advanced colorectal carcinoma. *Ann Oncol* 2002;13:1430–1437.

30. Kerbel R. A cancer therapy resistant to resistance. *Nature* 1997;390:335–336.

31. Brown P, Bloxidge R, Stuart N, et al. Association between expression of activated 72-kilodalton gelatinase and tumor spread in non-small-cell lung carcinoma. *J Natl Cancer Inst* 1993;85:574–578.
32. Vihinen p, Kahari V. Matrix metalloproteinases in cancer: prognostic markers and therapeutic targets. *Int J Cancer* 2002;99:157–166.
33. Rafii S, Lyden D, Benezra R, et al. Vascular and haematopoietic stem cells: novel targets for anti-angiogenesis therapy? *Nat Rev Cancer* 2002;2:826–835.
34. Stroobants S, Goeminne J, Seegers M, et al. [18]FDG-Positron emission tomography for the early prediction of response in advanced soft tissue sarcoma treated with imatinib mesylate (Glivec). *Eur J Cancer* 2003;39:2012–2020.
35. Anderson H, Yap J, Wells P, et al. Measurement of renal tumor and normal tissue perfusion using positron emission tomography in a phase II clinical trial of razoxane. *Br J Cancer* 2003;89:262–267.
36. Anderson H, Yap J, Miller M, et al. Assessment of pharmacodynamic vascular response in a phase I trial of Combretastatin A4 phosphate. *J Clin Oncol* 2003;21:2823–2830.
37. Herbst R, Mullani N, Davis D, et al. Development of biologic markers of response and assessment of antiangiogenic activity in a clinical trial of human recombinant endostatin. *J Clin Oncol* 2002;20:3804–3814.
38. Knopp M, Weiss E, Sinn H, et al. Pathophysiologic basis of contrast enhancement in breast tumors. *J Magn Reson Imaging* 1999;10:260–266.
39. Stevenson J, Rosen M, Sun W, et al. Phase I trial of the antivascular agent Combretastatin A4 phosphate on a 5-day schedule to patients with cancer: magnetic resonance imaging evidence for altered tumor blood flow. *J Clin Oncol* 2003;21:4428–4438.
40. Purdie T, Henderson E, Lee T, et al. Functional CT imaging of angiogenesis in rabbit VX2 soft tissue tumor. *Phys Med Biol* 2001;46:3161–3175.
41. Natale RB, Skarin A, Maddox AM, et al. Improvement in symptoms and quality of life for advanced non-small-cell lung cancer patients receiving ZD1839 in IDEAL2. *Proc Am Soc Clin Oncol* 2002 (Abstract 1167).
42. Fukuoka M, Yano S, Giaccone G, et al. Multi-institutional randomized phase II trial of gefitinib for previously treated patients with advanced non-small-cell lung cancer. *J Clin Oncol* 2003;21:2237–2246.
43. Kris M, Natale RB, Herbst RS, et al. A phase II trial of ZD1839 ('Iressa') in advanced non-small cell lung cancer (NSCLC) patients who had failed platinum- and docetaxel-based regimens (IDEAL 2). *Proc Am Soc Clin Oncol* 21: 2002 (Abstract 1166)
44. Giaccone G, Herbst R, Manegold C, et al. Gefitinib in combination with gemcitabine and cisplatin in advanced non-small-cell lung cancer: a phase III trial - Intact 1. *J Clin Oncol* 2004;22:777–784.
45. Herbst R, Giaccone G, Schiller J, et al. Gefitinib in combination with paclitaxel and carboplatin in advanced non-small-cell lung cancer: a phase III trial – Intact 2. *J Clin Oncol* 2004;22:785–794.
46. Rosner G, Stadler W, Ratain M, et al. Randomized discontinuation design: application to cytostatic antineoplastic agents. *J Clin Oncol* 2002;20:4478–4484.
47. Mick R, Crowley J, Caroll R, et al. Phase II clinical trial design for noncytotoxic anticancer agents for which time to disease progression is the primary endpoint. *Control Clin Trials* 2000;21:343–359.
48. Gehan E. The determination of the number of patients required in a preliminary and a follow-up trial of a new chemotherapeutic agent. *J Chronic Dis* 1961;13:346–353.
49. Fleming T. One-sample multiple testing procedure for phase II clinical trials. *Biometrics* 1982;38:143–151.
50. Simon R. Optimal two-stage design for phase II clinical trials. *Control Clin Trials* 1989;10:1–10.
51. Van Oosterom AT. Progression arrest. In: *Clinical Management of Soft Tissue Sarcomas*. Martinus Nijhoff Publishers, den Haag (The Hague) 1986;131–138.
52. Zee B, Melnychuk D, Dancey J, et al. Multinomial phase II cancer trials incorporating response and early progression. *J Biopharm Stat* 1999;9:351–363.
53. Dent S, Zee B, Dancey J, et al. Application of a new multinomial phase II stopping rule using response and early progression. *J Clin Oncol* 2001;19:785–791.
54. Korn E, Arbuck S, Pluda J, et al. Clinical trial designs for cytostatic agents: are new approaches needed? *J Clin Oncol* 2001;19:265–272.
55. Van Glabbeke M, Verweij J, Judson I, et al. Progression-free rate as the principal end-point for phase II trials in soft-tissue sarcomas. *Eur J Cancer* 2002;38:543–549.

29   Molecular Imaging in Oncology

Lalitha K. Shankar, MD, PhD,
Anne Menkens, PhD,
and Daniel C. Sullivan, MD

SUMMARY

   The field of molecular imaging holds great promise in enhancing our understanding of the neoplastic processes and over the last few decades, has resulted in several useful tools for evaluating cancer biology in vivo and in vitro.

   This chapter discusses the broad principles governing the field of molecular imaging and provides a synopsis of the current activities pertaining to pre-clinical and clinical evaluation of molecular imaging agents and modalities applicable to oncology.

   Key Words: Molecular Imaging; Imaging Biomarkers: In vivo biomarkers; Cancer Trials; Neoplastic Processes; PET Imaging Agents; MR Contrast Agents; Optical technology.

In vivo medical imaging is performed by administering energy to the body and measuring, with spatial localization, the energy that is transmitted through, emitted from, or reflected back from various organs and tissues. The difference between the administered and the recorded energy provides information about some property of matter with which the energy interacted. The energy most commonly used is some form of electromagnetic energy, such as X-rays or light, but occasionally other forms are used, such as the mechanical energy used for ultra-sound scans.

   The information extracted from the energy detected in clinical-imaging technologies has generally been used to infer something about the underlying anatomy or structure. This has been, and continues to be, enormously important in oncology. For example, dramatic improvements have occurred in the past 25 years such that modern computed tomography (CT) and magnetic resonance imaging (MRI) scanners can now depict anatomic detail at sub-millimeter resolution. However, as oncology moves into the molecular era, the property of matter that oncologists will increasingly want to know about is the biochemical makeup of normal and abnormal tissues.

From: *Cancer Drug Discovery and Development: Molecular Targeting in Oncology*
Edited by: H. L. Kaufman, S. Wadler, and K. Antman © Humana Press, Totowa, NJ

There are a variety of imaging methods that can display information about a patient's biochemistry, and no single modality is superior to all others. Collectively, these methods are referred to as molecular imaging *(1)*. Molecular-imaging agents and methods have been developed for a variety of systems using different forms of energy. These include nuclear medicine methods—such as positron emission tomography (PET) and single photon emission-computed tomography (SPECT) MRI, ultrasound methods, CT, and optical technologies. Although the term "optical" implies the use of visible light, it is often more broadly applied to include near-infrared (NIR) methods as well. The term "photonics" is sometimes used to describe both visible and non-visible radiation. These technologies have different advantages and drawbacks. For example, CT and MRI are able to portray anatomical detail exceptionally well, whereas nuclear medicine and optical methods have very high sensitivity for detecting specific molecules but cannot portray anatomical detail with high spatial resolution. Increasingly, methods with complementary strengths are combined in clinical practice, such as the CT–PET systems that are now commercially available (Fig. 1).

In vivo molecular imaging can be thought of as a form of in vivo assay. Some in vivo imaging methodologies, such as magnetic resonance spectroscopy and optical spectroscopy, allow us to make direct inferences about underlying biochemistry simply from administering energy and analyzing the recorded energy. However, the extent of biochemical information that can be obtained from energy alone is currently limited. Therefore, we commonly give patients diagnostic drugs, referred to as contrast agents or molecular probes, that interact in a specific way with the patient's underlying biochemistry and thereby alter the recoded energy in a way that tells us more about the

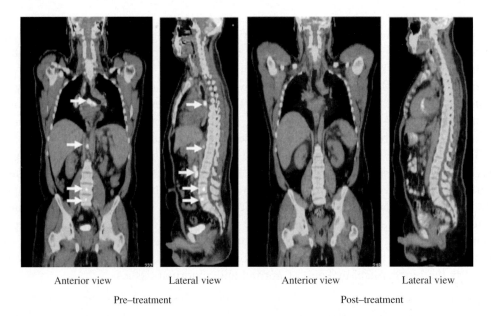

Anterior view   Lateral view   Anterior view   Lateral view

Pre–treatment   Post–treatment

Fig. 1. A 31 year-old man with recently diagnosed lymphoma. Pretreatment PET/CT scans (anterior and lateral views) demonstrated widespread foci of uptake throughout chest, abdomen and pelvis (white arrows). After treatment, PET/CT scans showed resolution of all the original foci and no new regions of suspicious uptake.
Courtesy: Dr. David Townsend, University of Tennessee.

patient's biochemistry than we could learn from administered energy alone. There is a wide variety of molecular mechanisms that can be used for developing imaging agents *(2)*. Some of these involve binding to cell-membrane structures, whereas others exploit transport mechanisms into the cell, and subsequent enzymatic or other biochemical reactions within the cytoplasm. Others may be localized to intracellular structures such as mitochondria or within the nucleus itself. No single molecular mechanism in the cell precludes all others for clinical utility.

In vivo imaging assays cannot currently provide the degree of genomic, proteomic, and other phenotypic information that can be obtained from various in vitro assays on biopsied tissue or body fluids. However, in vivo imaging has at least three potentially important advantages that complement information from in vitro tests. First, imaging provides spatially localized information over large volumes of tissue or the entire body, whereas in vitro tests are usually performed on a very small volume of tissue. The term "regional proteomics" is sometimes used, indicating that imaging may reflect the heterogeneity of cancer better than in vitro techniques. Second, in vivo imaging can give dynamic information by being obtained serially or continuously for periods of time. In vitro assays provide information from a single point in time. Third, in vivo imaging depicts information from a tumor in its usual milieu or microenvironment. In vitro assays, on the other hand, will reflect the changes in gene expression patterns that occur very quickly after tissue is removed by biopsy. Information from in vivo and in vitro studies is therefore complementary, and both are essential in modern oncology research and clinical care.

Investigators in drug development need in vivo assays ("imaging biomarkers") to tell whether a given patient has the appropriate molecular phenotype to benefit from a targeted therapy, to indicate whether the drug has hit its molecular target, to determine whether the drug has been given in the optimal biologic dose, and to ascertain whether the tumor is responding *(3)*. Similarly, clinicians increasingly will have a series of targeted therapies to choose from for any given tumor and will need in vivo assays to get an early determination as to whether their patient is responding to the chosen therapy. Early predictive assays will be important so that clinicians can change therapy quickly, thereby obviating unnecessary toxicity and expense and increasing the chances of matching the patient to an effective therapy. It is likely that functional imaging tests, such as 18-fluorodeoxyglucose (FDG)-PET, will be able to provide this information.

In addition to knowing whether a given biochemical event is occurring or not, researchers and clinicians need objective, quantitative information about the biochemical events and need to monitor them quantitatively over time, before and after interventions. That level of quantification is generally not yet available in clinical-imaging methods but is an area of active research and development.

As the fundamental basis of cancer is at the gene level, molecular-imaging methods that report directly on gene function would be particularly useful *(4)*. Coupling-imaging methods to reporter genes is often used in one of the two ways. One method is to insert a gene that codes for an enzyme. A labeled substrate for that enzyme could later be administered, and the signal from the trapped probe would identify those cells that express the reporter gene. The second method is to insert a gene that codes for a cell-surface receptor and later administer a labeled ligand specific to that receptor. Signal from the bound ligand identifies those cells that express the receptor. Unfortunately, in most diagnostic and therapeutic situations, it is not practical to insert reporter genes into a patient's native cell. However, reporter genes could be inserted into genetically engineered cells or viruses that will be administered to a patient, such as in stem-cell or vaccine therapies. Expression of the imaging reporter genes can be coupled to the expression of other genes of interest (such as therapeutic genes).

The concept of nanoparticles as generic platforms for imaging agents is also under intense investigation. A single particle, such as a dendrimer, liposome or other construct, acts as a general platform to which a variety of signaling moieties are attached, along with one or more targeting molecules that can be substituted as required. Such imaging agents have already been developed for nuclear medicine, ultra-sound, magnetic resonance, and optical applications *(5)*. A few have been tested in patients (such as iron-oxide nanoparticles for MR imaging), and a significant increase in human testing of nanoparticle-imaging agents is expected in the coming few years.

Photonic methodologies have also attracted much attention in recent years as they have the advantages of not involving ionizing radiation and of being cheaper than traditional clinical-imaging techniques. Photonic technologies that do not require the administration of any exogenous contrast materials and which measure photon reflection, transmission, refraction, or fluorescence from endogenous fluorophores have already been developed for use in humans. Other photonic methods use similar physical properties but require the administration of agents that either fluoresce or bioluminesce. Organic dyes that fluoresce have been used clinically for some time, but their value in the molecular imaging of humans is limited because of tissue absorption of the light emitted. There is also considerable interest in using quantum dots or other nano-constructed particles that fluoresce much more intensely than organic dyes *(6)*. Such particles can be engineered to fluoresce at a variety of wavelengths in response to stimulation by light of a single wavelength. A major drawback of bioluminescence is that it requires the combination of an enzyme (such as luciferase) and its substrate (luciferin), and most methods insert the enzyme's gene into target cells using genetic engineering *(7)*. As a result, this method is unlikely to be clinically useful, but it has become an enormously valuable technique for basic research.

Another potentially important characteristic of optical-imaging agents is the development of the so-called activatable agents, which exploit the phenomenon of fluorescence resonance energy transfer (FRET). Two fluorescent molecules can be held in a steric configuration such that they will not fluoresce when stimulated by the appropriate wavelength of light. If something disturbs that configuration, they will fluoresce. In

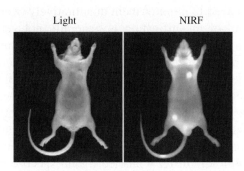

Fig. 2. Nude mouse, PC-3 tumor, 24 h after IV injection; 2.5 nmole broad cathepsin probe; Nature Biotech. 1999, 17, 375–378.
Courtesy: Dr. Ralph Weissleder, Massachusetts General Hospital.

biological situations, this is exploited by attaching the fluorophores to a substrate, such as a polymer, by peptide linkers, which are themselves the substrate for a particular enzyme, such as a protease. When the protease is present, for example, in cancer cells that overexpress it, the peptide linkers will be cleaved, and the fluorophores will move away from the substrate and fluoresce, thereby signaling that the protease of interest is present and active *(8)* (Fig. 2). Similar types of "activatable" agents that use magnetic resonance contrast materials have been reported *(9)*, but these are not yet clinically feasible. Activatable agents that use radioisotopes are not possible as the radioactive decay phenomenon cannot be controlled to respond to a particular molecular event as in photonic or magnetic resonance situations. The main drawback of photonic-imaging techniques is that visible light and NIR are highly absorbed by water and tissue. Therefore, the future role of photonic methods in patients, in which light may have to traverse many centimeters of tissue, is not yet clear. Nevertheless, photonic techniques could be valuable for the assessment of mucosal surfaces where the majority of human cancers originate. Engineering simulations have also suggested that it may be feasible to use photonic methods in deep tissue although such applications are still far from clinical use *(8)*.

At present, the most feasible modality for molecular imaging in humans is nuclear medicine. This is because the sensitivity of nuclear medicine detectors is such that clinically useful images can be obtained after administration of nanomolar amounts of radiolabeled tracers in humans. Millimolar amounts of CT or MR molecular contrast agents would have to be given, and the burden of proving the safety of such bulk quantities of imaging agents for diagnostic purposes has been an impediment to developing such agents. The primary limitation of optical methodologies in humans is the tissue absorption of light. Table 1 lists a wide variety of imaging agents that have been developed, but few have achieved clinical utility. Representative agents that are clinically useful or appear promising are described below.

1. FLUORODEOXYGLUCOSE

PET with FDG has been extremely helpful in characterization of lesions with indeterminate morphology on anatomic studies, as well as in staging and restaging of several malignancies. It is a widely used diagnostic modality in the diagnosis and management of cancer. Many tumors have an accelerated glucose metabolism and subsequently take up larger amounts of glucose and FDG (a glucose analog) compared with normal cells *(10,11)*. In addition, GLUT1—a transmembrane transporter that facilitates hexose uptake—is highly expressed in a number of cancers *(12)*. Another contributing factor to increased glucose uptake in tumor cells is the increased expression of hexokinases (predominantly HK-1 and HK-2), which catalyze the first phosphorylation step in glycolysis *(13)*.

Most malignancies demonstrate increased FDG accumulation in the tumor mass. Some studies suggest a correlation with tumor grade and prognosis. Clinical evaluations of FDG-PET scans are routinely assessed qualitatively (i.e., subjectively). Alternatively, a more objective measure—the standardized uptake value (SUV) or standardized uptake ratio (SUR)—can be used. The SUV is a semi-quantitative measure of FDG, which is derived by normalizing the FDG accumulation in a lesion (corrected for attenuation by tissue) to the injected dose and to some measure of the patient's body size, such as weight or surface area. SUVs are most often used in research studies for a more

Table 1
Imaging Probes Used to Visualize Molecular Targets and Processes in Cancer

Molecular target or process	Imaging probes
Small molecule probes	
Proliferation	2-[$_{11}$C]-thymidine, $_{18}$F-FLT, $_{18}$F-FMAU, $_{18}$F-FAU, and $_{124}$I-IUdR
Apoptosis	$_{99m}$Tc-annexin V and $_{18}$F-annexin V
Hypoxia	$_{18}$F-MISO, $_{18}$F-EF5, FETNIM, $_{18}$F-FETA, $_{64}$Cu-ATSM, $_{124}$I-IAZG, and $_{18}$F-FAZA
Pharmacokinetics	$_{18}$F-5-FU, $_{11}$C-DACA, $_{11}$C-BCNU, $_{11}$Ctemozolomide, and $_{13}$N-cisplatin
Multidrug resistance	$_{99m}$Tc-sestamibi, $_{11}$C-verapamil, $_{11}$Cdaunorubicin, $_{11}$C-colchicine, and $_{99m}$Tcmethoxyisobutylisonitrile
Breast cancer (estrogen receptor)	FES
Prostate cancer (androgen receptor)	FDHT
Peptide probes	
Somatostatin receptor	$_{90}$Y-DOTA-Tyr3-octreotide, $_{111}$In-DTPA- DPhe(*1*)-octreotide, and $_{90}$Y-DOTAlanreotide/vapreotide
Vasoactive intestinal peptide receptor	$_{123}$I-Vasoactive intestinal peptide and $_{99m}$Tc-TP3654
Gastrin-releasing peptide receptor	$_{99m}$Tc-Bombesin
Cholecysokinin receptor	$_{111}$In-DTPA-minigastrin
Angiogenesis	$_{18}$F-RGD peptide targeted to alphaVbeta3 integrin
Cathepsin proteases	Prosense (VM102)
Antibody probes	
Angiogenesis	Paramagnetic nanoparticles, using antibodies to integrin alphaVbeta3, the integrin alphaVbeta3 ligand, and VCAM-1, E-selectin
CEA	Arcitumomab (CEAscan) and Satumomab
Prostate Specific Membrane Antigen	Capromab pendetide (Prostascint)
Prostate Specific Membrane Antigen	Capromab pendetide (Prostascint)
CD20	$_{131}$I-labeled tositumomab (Bexxar) and $_{90}$Y-labeled ibritumomab tiuxetan (Zevalin)
CD22	Bectumomab

Abbreviations: ATSM, diacetyl-bis(*N*-4-methylthiosemicarbazone); BCNU, 1,3-bis(chloroethyl)-1-nitrosourea; CEA, carcinoembryonic antigen; DACA, *N*-[2-(dimethylamino)ethyl]acridine-4-carboxamide; DOTA, 1,4,7,10-tetraazacyclododecane-1,4,7,10-tetraacetic acid; DTPA, diethylenetriamine pentaacetic acid; EF5, 2-(2-nitro-1H imidazol-1-yl)-*N*-(2,2,3,3,3-pentafluoropropyl) acetamide; FAZA, fluoroazomycinarabinofuranoside; FDHT, 16-fluoro-5 dihydrotestosterone; FETA, fluoretanidazole; FETNIM, fluoroerythronitroimidazole; FES, 16-fluoroestradiol-17®); FLT, 3´-deoxy-3´-flurothymidine; FMAU, 1-(2´-deoxy-2´-fluoro-beta-D-arabinofuranosyl) thymine; 5-FU, 5-fluorouracil; IAZG, iodo-azomycin-galactoside; IUdR, iododeoxyuridine; VCAM-1, vascular cell dhesion molecule 1.

objective evaluation of FDG uptake. Patz and colleagues proposed using a cutoff value of 2.5 in the semi-quantitative SUR for lung nodules observing, in their study, that a SUR < 2.5 had a 100% specificity for benign lesions >1.2 cm *(14)*. A subsequent study by Lowe et al. suggested a similar diagnostic performance of FDG-PET in evaluation of lesions as small as 7 mm *(15)*. The increasing prominence of hybrid PET–CT scanner systems is expected to further improve the diagnostic performance of FDG-PET scans, both in terms of sensitivity and specificity.

The prognostic value of FDG-PET at diagnosis has been evaluated in several studies, based on the SUV. In these studies, the univariate analyses performed to determine a cutoff point for a SUV indicative of active tumor has ranged widely, from 5 to 20 *(16–19)*. The wide range of SUV values seen in these studies could be due to the variation in the acquisition protocols. For instance, there was considerable variability in the period between tracer administration and scanning as well as in blood glucose levels. There was variation in reconstruction and interpretive criteria, such as the lack of correction for partial volume effects in certain studies *(20)*. Despite this variability, SUV of a lesion on a FDG-PET scan was noted to provide independent prognostic information in several studies *(16,17,19)*.

Locoregional failure is a significant problem in many solid tumors for patients who are non-surgical candidates and are receiving radiation therapy either alone or with chemotherapy *(21,22)*. Several small studies have been performed to evaluate the impact of FDG-PET compared to simulation with CT scanning alone on radiation therapy target volume definitions *(23–27)*. These studies revealed that the use of FDG-PET findings altered the staging of the disease process and significantly changed the shape of radiation portals in 36–62% of the patients. Smaller treatment volumes treated with higher doses of radiation, sparing normal tissue, are possible with conformal radiation therapy *(28)*. Better or more accurate delineation of the target should mean better tumor coverage, leading presumably to better tumor control, but long-term outcome studies have not yet been performed.

As more options for second-line therapies become available, there is a growing need to find better ways to evaluate a patient's response to the first-line treatments—which often are toxic, expensive and not always beneficial *(29)*. Several studies have evaluated the role of FDG-PET in assessing response to chemotherapy or radiation therapy and have shown promising results *(30–33)*. FDG-PET in these studies showed an improved accuracy in predicting response compared with CT and did so at earlier time points. An EORTC group proposed criteria for FDG-PET assessment of response to treatment— analogous to the response evaluation criteria in solid tumors (RECIST) criteria *(34)*. A 15–25% decrease in SUV after one cycle of chemotherapy and greater than 25% after more than one treatment cycle were estimated to be signs of partial metabolic response. Further multicenter trials employing standardized scanning techniques for FDG-PET and with comparison to RECIST are needed to refine these evaluation criteria for assessing response.

2. FLUOROTHYMIDINE

F-18 flurothymidine (FLT) is a PET tracer used for non-invasive measurement of tumor proliferation *(35)* (Fig. 3). Buck et al. evaluated the proliferation rate of 30 solitary pulmonary nodules using FLT and found that FLT uptake was specific

FLT PET and CT demonstrating abnormal uptake in tumor and mediastinal node

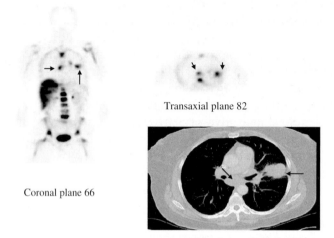

Transaxial plane 82

Coronal plane 66

Fig. 3. Coronal and transaxial image of FLT-PET with corresponding transaxial CT scan in a patient with non small cell lung cancer in the left lung, with mediastinal lymph node involvement. Serial FLT PET scans can be obtained for evaluation of the tumor's response to therapy.
Courtesy: Dr. Anthony Shields, Karmanos Cancer Institute.

for malignant lesions *(36)*. Halter et al. went on to compare FDG and FLT imaging in 28 patients with suspicious central focal lesions *(37)*. For staging of the primary tumor, FLT-PET had a sensitivity in this study of 86% and specificity of 100% compared with FDG-PET, which had a sensitivity of 95% and a specificity of 73%. For nodal staging, FLT-PET had a sensitivity of 57% and a specificity of 100% and FDG-PET had a sensitivity of 86% and a specificity of 100%. These preliminary studies suggest that there could be a role for FLT-PET in the evaluation of solitary pulmonary nodules, for example, in regions with a high prevalence of pulmonary fungal infection.

To investigate the potential of FLT-PET scans to provide tumor-specific information for staging or restaging purposes, Cobbens et al. performed both FDG-PET and FLT-PET on 16 patients with known or suspected recurrent non-small cell lung cancer (NSCLC). Using FDG-PET as a reference standard, The sensitivity on a lesion-by-lesion basis was 80% for eight patients who were being evaluated for tumor recurrence and was 27% in the eight patients who were treatment naïve *(38)*. This and other studies have shown no significant advantage in staging with FLT-PET as opposed to FDG-PET. Muzi et al. have compared compartmental modeling of FLT kinetics with simpler methods of estimating FLT flux, such as SUV. In their study, the flux estimates (KFLT) derived from the compartmental analysis of FLT-PET image data correlated well with in vitro measures of proliferation, such as Ki-67 immunohistochemistry. There was a significant underestimation of the KFLT when the simpler models were employed *(39)*.

Early comparative studies have been performed evaluating both FLT-PET and FDG-PET in the evaluation of response to therapy—in esophageal cancer, gliomas and

| MRI
T1+Gd | MRI
T2 | FDG
(75–90m) | FMISO
(127–147m) |

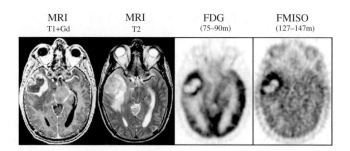

Fig. 4. Thirty-seven year-old man with progressive headaches. MRI scans performed after administration of gadolinium contrast agent (Tl+Gd and T2) showed a large right temporal ring-enhancing irregular lesion. FDG-PET revealed greater uptake in the lesion than in contralateral cortex. FMISO showed striking uptake with a maximum tumor/blood ratio of 2.8 and hypoxic volume of 36 cc indicating significant tissue hypoxia in the tumor. Glioblastoma multiforme was the pathological diagnosis.
Courtesy of Dr. Mark Muzi, University of Washington.

NSCLC *(40)*—but there are not yet enough data to determine whether either tracer is likely to be better than the other (Fig. 3).

3. HYPOXIA-IMAGING AGENTS

Hypoxia is an important predictor of the biological behavior of malignant tumors. Regional tumor hypoxia can increase radioresistance of tumors. Non-invasive evaluation of hypoxia levels in NSCLC has been performed with F-18 fluoromisonidazole-PET (FMISO-PET) and with copper-60-diacetyl-bis (N4-methylthiosemicarbazone) (Cu-64 ATSM-PET) *(41,42)*. Dedashti et al. assessed whether the pretherapy tumor uptake of Cu-60 ATSM in the tumor predicted response to treatment in 19 patients with NSCLC *(41)*. The follow-up period was 2–46 months. An arbitrarily selected tumor to muscle ratio threshold of 3.0 was found to identify those likely to respond to treatment. Studies evaluating hypoxia imaging with FMISO have shown that it is possible to use this non-invasive-imaging technique to evaluate oxygenation status in tumors (Fig. 4). This information could potentially be used in planning radiation therapy *(41,43)*. To date, only single-institution clinical trials of the hypoxia agents FMISO and Cu-ATSM have been done. Multicenter trials of hypoxia agents need to be done to better assess their clinical utility. Lack of facile access to the tracers has been an impediment to large-scale evaluation of these agents.

4. RECEPTOR IMAGING

Molecular imaging in its potential to image receptors and receptor-mediated actions can play a significant role in advancing the understanding of cancer biology and consequently transforming cancer care. A deepened understanding of the significance of tumor receptors is changing the clinical paradigm for assessing tumor burden as well as prognosis in the individual cancer patient. Several of the new molecular targeted therapies in development require sufficient expression or selective mutations of specific receptors, in order to obtain a significant therapeutic effect. In this section, synopses of current activities in representative receptor imaging classes are provided.

4.1. Imaging of HER2/NEU

Targeted molecular imaging can be used to directly detect changes in expression levels of key growth factor receptors. This approach is being applied to the detection of the epidermal growth factor receptor (EGFR/erbB) and a related receptor, the human epidermal growth factor receptor 2 (HER2/neu), among many other examples.

The transmembrane tyrosine kinase (tk) receptor HER-2/neu is overexpressed in 25–30% of breast cancers *(44)* and is often associated with poor prognosis. Efforts to identify a therapeutic agent specific for this target led to the discovery and development of a recombinant, humanized monoclonal antibody that recognizes the extracellular domain of HER-2/neu *(45)*. This agent, known as trastuzumab (trade name Herceptin®; Genentech, San Francisco, CA) was approved by the FDA as a therapeutic agent and is in clinical use.

There has been considerable interest in the development of trastuzumab as an imaging agent for the detection and staging of breast cancers expressing HER-2/neu, and a number of strategies have been employed *(46–55)*. One promising strategy is the use of truncated versions of the intact antibody (Fab fragments). The smaller Fab fragments allow for more rapid clearance and higher signal-to-noise ratio yet retain specificity for the HER2/neu target. For example, Olafsen et al. synthesized and tested a derivative of trastuzumab (^{64}Cu-DOTA hu4D5v8 scFv-Fc DM) in a small animal model *(56)*. Results from micro-PET imaging using this agent demonstrate characteristics favorable to development as a clinical-imaging agent.

In addition to the diagnostic or prognostic potential of an imaging agent based on trastuzumab, there is considerable interest in its use to monitor response to therapy. A study by Smith-Jones et al. describes the use of ^{68}Ga-DCHF, an F(ab')2 fragment of Herceptin to monitor response to the experimental therapeutic 17-allylaminogeldanamycin (17-AAG) in a small animal model *(52)*. Micro-PET images of mice bearing HER2-expressing xenografts were acquired prior to and after treatment with 17-AAG and clearly demonstrate that this agent can be used to quantitatively monitor response to therapy in animal models. This promising agent is currently being developed for use in the clinic.

4.2. Imaging of EGFR/ERBB

Another important therapeutic target is the EGFR. EGFR is a member of the ErbB family of receptor tks and, like HER2/neu, is often overexpressed in a variety of tumors *(57)*. Three main strategies have been pursued in the development of imaging agents targeted at EGFR. In one approach, antibodies that specifically recognize the EGFR have been labeled with ^{99m}Tc *(58,60)* or ^{111}In *(61,62)* and tested in limited clinical trials.

A second approach has been to directly radiolabel the EGF peptide, which is the native ligand for EGFR. For example, Cornelissen et al. describe the labeling and biodistribution imaging of ^{99m}Tc-HYNIC-hEGF, an analog of the human EGF peptide, in a small animal cancer model *(63)*. Using ^{99m}Tc-HYNIC-hEGF as an imaging agent, planar scintigraphic images were acquired prior to and 6–8 h following treatment with an experimental farnesyltransferase inhibitor (FTI). Although these results are preliminary, they demonstrate that ^{99m}Tc-HYNIC-hEGF is a selective imaging agent for the determination of EGF receptor status with SPECT. In addition, 99mTc-HYNIC-hEGF is also a possible tool for predicting early response to EGFR inhibitor therapy.

A third approach is to directly radiolabel small molecule EGFR inhibitors. One class of these small molecules is the 6-acrylamido-4-anilino-quinazolines *(64,65)*. These molecules bind irreversibly to the active site on the EGFR and inhibit downstream cell-signaling events. Synthetic procedures have been developed for radiolabeling derivatives of quinazoline with [11]C *(66,67)* and [18]F *(68,69)*, with varying levels of success in vitro. Recently, two quinazoline derivatives have been synthesized and radiolabeled with [124]I *(70)*. These two derivatives retain their inhibitory effect in vitro and are being pursued for development as clinical PET-imaging agents.

The majority of breast cancer patients (∼two-third) present with hormone receptor bearing tumors. The standard of care is to measure the levels of estrogen receptor (ER) and progesterone receptor (PR) positivity at initial diagnosis and triage patients to appropriate therapy based on the levels of ER and PR expressions *(71,72)*.

The level of ER expression in the primary tumor does not necessarily correlate with the level of ER expression in metastatic lesions—which could impact the therapeutic response to hormonal therapy *(73)*.

Studies by Mankoff and colleagues at the University of Washington evaluating quantitative PET with fluoroestradiol (FES) have shown a promising role for FES-PET in predicting response to hormonal therapy (Fig. 5) and can potentially help guide treatment selection *(73)*.

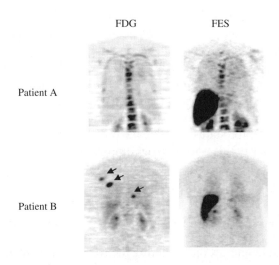

Fig. 5. Coronal PET images of FDG uptake (left) and ^{18}F-fluoroestradiol (FES) uptake (right) are shown for two patients with recurrent and metastatic disease from estrogen receptor positive (ER+) breast cancers. The top patient (Patient A) showed diffuse spinal metastases that are metabolically active by FDG PET with matched uptake of FES, indicating preserved ER expression. The bottom patient (Patient B) showed scattered lung and bone metastases by FDG PET (arrows) but no corresponding uptake by FES, suggesting a loss of ER expression. Both patients were treated with hormonal therapy subsequent to PET imaging. Patient A had an objective response while Patient B had disease progression. Normal liver and kidney uptake are also seen in all images. This example illustrates how uptake on FES PET can predict response of breast cancer to endocrine therapy. Courtesy of Dr. David Mankoff, University of Washington.

4.3. Imaging of Androgen Receptors (AR)

The majority of malignant prostate epithelial cells are androgen sensitive. Consequently, treatment paradigms in recurrent or metastatic prostate cancer include either orchiectomy or administration of drugs to suppress androgen suppression (74,75). Relapse or progression to hormone refractory prostate cancer occurs in the majority of cases and is attributed to a variety of factors including continual endogenous androgen production by the adrenal glands as well as the changing androgen receptor status within the tumor (76).

Dr. Larson and colleagues scanned patients with both 16-beta-18F-fluoro-5-alpha-dihydrotestosterone (FDHT-PET) and FDG in a pilot study. All patients had histologically documented progressive prostate cancer despite castrate levels of testosterone. There was a concordance of positive findings on both FDHT-PET and FDG-PET in approximately 80% of the lesions. The molecular signatures of the lesions exhibiting discordance—that is, demonstrating FDG uptake and absence of FDHT uptake—may potentially be linked to disease progression and remains a subject worthy of further study (77). A subsequent study on 20 patients by Dedashti et al. (76) at Washington University has shown that there appears to be a correlation between the PSA levels and positive FDHT-PET scans.

4.4. Somatostatin Receptor Imaging

The expression of somatostatin receptors in both neuroectodermal and non-neuroectodermal tumors has been related to biological behavior of these tumors. SPECT using radiolabeled somatostatin analogs such as In-111-octreotide is used clinically in the diagnosis, staging, and follow up of neuroectodermal tumors. The majority of the neuroectodermal tumors are visualized on In-111-octreotide SPECT and not on FDG-PET and demonstrate a slow growth and favorable prognosis. In contrast, the undifferentiated neuroectodermal tumors tend not to be visualized on In-111-octreotide SPECT but are seen on FDG-PET and tend to have a poor prognosis (78).

In addition to using In-111-octreotide SPECT for the non-invasive in vivo evaluation of the somatostatin receptor status, studies have demonstrated a role for radio-guided surgery with In-111 Octreotide—to improve intraoperative staging (79). Furthermore, the role of quantitative In-111-octreotide scintigraphy as a prognostic marker as well as in predicting response to therapy is being evaluated in prospective trials.

5. MR AGENTS: ULTRA-SMALL SUPERPARAMAGNETIC IRON OXIDE

There is considerable interest in evaluating the role of ultra-small superparamagnetic iron oxide (USPIO) complexes in staging solid tumors. These USPIO particles are taken up by functioning histicytes in normal lymph nodes but not in tumor-filled areas of lymph nodes. Thus, nodes or subsections of nodes that do not take up the contrast on MR studies have a high likelihood of representing tumor in the nodes (80). Because of the high spatial resolution of MR, it is possible to see millimeter deposits of tumor in some nodes. The specificity of findings on MR could thus potentially be enhanced by using one of these functional contrast agents, such as ferumoxtran which is a biodegradable USPIO particle coated with low-molecular weight dextran. There have been encouraging results in the staging of prostate cancer as well as several pelvic malignancies. There are ongoing studies evaluating this agent's role in staging the axilla in patients with breast cancer.

6. CONCLUSIONS

Molecular imaging is a relatively new field of clinical science and has not yet penetrated widely into routine clinical practice. The number of investigators involved in molecular imaging at academic medical centers throughout the world is small, but increasing. In recent years, there have developed two professional organizations for molecular imaging, and their annual meetings attract about 1000 participants. Many of the established radiology and imaging professional societies are embracing the concept of molecular imaging and devoting increasing resources to promote research in that field and to integrate molecular imaging into standard clinical practice. Thus, oncologists should continue to see an increase in new molecular-imaging devices, agents, and methodologies over the next several years. There are significant barriers to the commercialization of molecular-imaging agents and methods, because the potential markets for most of the agents and methods are small. Nevertheless, many companies involved in the diagnostic arena are trying to develop business models to make these new techniques available.

There are significant scientific challenges to making imaging agents that have the high degree of sensitivity and specificity that future targeted therapies are likely to demand. However, the concept of developing methods that will non-invasively display biochemical information in the living human has captured the imagination of innovative scientists in academia and industry. Their discoveries and developments are expected to have a major impact on cancer detection, individualized treatment, and drug development in the next two decades.

REFERENCES

1. Weissleder R. Molecular imaging in cancer. *Science* 2006;312(5777):1168–1171.
2. Danthi SN, Pandit SD, Li K. A primer on molecular biology for imagers: VII. Molecular imaging probes. *Acad Radiol* 2004;11(Suppl):77–84.
3. Kelloff GJ, Sigman CC. New science-based endpoints to accelerate oncology drug development. *Eur J Cancer* 2005;41: 491–501.
4. Min JJ, Gambhir SS. Gene therapy progress and prospects: noninvasive imaging of gene therapy in living subjects. *Gene Ther* 2004;11(2):115–125.
5. Sullivan DC, Ferrari M. Nanotechnology and tumor imaging: seizing an opportunity. Review. *Mol Imaging* 2004;3(4):364–369.
6. Michalet X, Pinaud FF, Bentolila LA, et al. Quantum dots for live cells, in vivo imaging, and diagnostics review. *Science* 2005;28:538–544.
7. Contag C, Bachmann MH. Advances in *in vivo* bioluminescence imaging of gene expression. *Annu Rev Biomed Eng* 2002;4:235–260.
8. Ntziachristos V, Bremer C, Weissleder R. Fluorescence imaging with near-infrared light: new technological advances that enable *in vivo* molecular imaging. *Eur Radiol* 2003;13:195–208.
9. Louie AY, Huber MM, Ahrens ET, Rothbacher U, Moats R, Jacobs RE, Fraser SE, Meade TJ. *In vivo* visualization of gene expression using magnetic resonance imaging. *Nat Biotechnol* 2000;18:321–325.
10. Warburg O. On the origin of cancer cells. *Science* 1956;123:309–314.
11. Flier J, Mueckler M, Usher P, et al. Elevated levels of glucose transport and transport messenger RNA are induced by ras or src oncogenes. *Science* 1987;235:1492–1495.
12. Smith TA. Facilitative glucose transporter expression in human cancer tissue. *Br J Biomed Sci* 1999;56:285–292.

13. Smith TA. Mammalian hexokinases and their abnormal expression in cancer. *Br J Biomed Sci* 2000;57:170–178.
14. Patz EF, Lowe VJ, Hoffman JM, et al. Focal pulmonary abnormalities: evaluation with F-18 fluorodeoxyglucose PET scanning. *Radiology* 1993;188:487–490.
15. Lowe VJ, Fletcher JW, Gobar L, et al. Prospective investigation of positron emission tomography in lung nodules. *J Clin Oncol* 1998;16:1075–1084.
16. Ahuja V, Coleman RE, Herndon J, et al. The prognostic significance of fluorodeoxyglucose positron emission tomography imaging for patients with nonsmall cell lung carcinoma. *Cancer* 1998;83:918–924.
17. Vansteenkiste JF, Stoorbants SG, De Leyn PR, et al. Prognostic importance of the standardized uptake value on FDG-PET scan in non-small cell lung cancer: an analysis of 125 cases. *J Clin Oncol* 1999;17:3201–3206.
18. Dhital K, Saunders CA, Seed PT, et al. [(18F)] Fluorodeoxyglucose positron emission tomography and its prognostic value in lung cancer. *Eur J Cardiothorac Surg* 2000;18:425–428.
19. Higashi K, Ueda Y, Arisaka Y, et al. 18F-FDG uptake as a biologic prognostic factor for recurrence in patients with surgically resected non-small cell lung cancer. *J Nucl Med* 2002;43:39–45.
20. Vansteenkiste J, Fischer BM, Dooms C, et al. Positron-emission tomography in prognostic and therapeutic assessment of lung cancer: systematic review. *Lancet Oncol* 2004;5:531–540.
21. Sibley GS, Jamieson TA, Marks LB, et al. Radiotherapy alone for medically inoperable stage I non-small cell lung cancer: The Duke experience. *Int J Radiat Oncol Biol Phys* 1998;40:149–154.
22. Furuse K, Fukoka M, Kawahara M, et al. Phase III study of concurrent versus sequential thoracic radiotherapy in combination with mitomycin, vindesine, and cisplatin in unresectable stage III non-small cell lung cancer. *J Clin Oncol* 1999;17:2692–2699.
23. Kiffer J, Berlangieri S, Scott A, et al. The contribution of 18F-fluro-deoxyglucose positron emission tomography imaging to radiotherapy planning in lung cancer. *Lung Cancer* 1998;19:161–177.
24. Nestle U, Walter K, Schmidt S, et al. 18F-deoxyglucose positron emission tomography (FDG-PET) for the planning of radiotherapy in lung cancer. *Int J Radiat Oncol Biol Phys* 1999;23:593–597.
25. Munley MT, Marks LB, Scarfone C, et al. Multimodality nuclear medicine imaging in three-dimensional radiation treatment planning for lung cancer: challenges and prospects. *Lung Cancer* 1999;23:105–114.
26. Vanuystel L, Vansteenkiste J, Stoorbants S, et al. The impact of 18F-fluoro-2-deoxy-D-glucose positron emission tomography (FDG-PET) lymph node staging on the radiation treatment volumes in patients with non-small cell lung cancer. *Radiother Oncol* 2000;55: 317–324.
27. Bradley J, Thorstad WL, Mutic S, et al. Impact of FDG-PET on radiation therapy volume delineation in non-small cell lung cancer. *Int J Radiat Oncol Biol Phys* 2004;59:78–86.
28. Paulino AC, Thorstad WL, Fox T. Role of fusion in radiotherapy treatment planning. *Semin Nucl Med* 2003;33:238–243.
29. Kelloff GJ, Hoffman JH, Johnson B, et al. Progress and Promise of FDG-PET imaging for cancer patient management and oncologic drug development. *Clin Cancer Res* 2005;11:2785–2808.
30. MacManus MP, Hicks RJ, Matthews JP, et al. Positron emission tomography is superior to computed tomography for scanning for response-assessment after radical radiotherapy or chemoradiotherapy in patients with non-small cell lung cancer. *J Clin Oncol* 2003;21:1285–1292.
31. Weber WA, Petersen V, Schmidt B, et al. Positron emission tomography in non-small-cell lung cancer: prediction of response to chemotherapy by quantitative assessment of glucose use. *J Clin Oncol* 2003;21:2651–2657.
32. Ichiya Y, Kuwabara Y, Sasaki M, et al. A clinical evaluation of FDG-PET to assess the response in radiation therapy for bronchogenic carcinoma. *Ann Nucl Med* 1996;10:193–200.
33. Choi NC, Fischman AJ, Niemerko, et al. Dose-response relationship between probability of pathologic tumor control and glucose metabolic rate measured with FDG-PET after preoperative chemoradiotherapy in locally advanced non-small cell lung cancer. *Int J Radiat Oncol Biol Phys* 2002;54:1024–1035.
34. Young H, Baum R, Cremerius U, et al., Measurement of clinical and subclinical tumour response using [18F]-fluorodeoxyglucose and positron emission tomography: review and 1999 EORTC recommendations. *Eur J Cancer* 1999;35:1773–1782.
35. Shields AF, Grierson JR, Dohmen BM, et al. Imaging proliferation in vivo with [F-18] FLT and positron emission tomography. *Nat Med* 1998;4:1334–1336.

36. Buck AK, Schirrmeister H, Hetzel M, et al. 3-Deoxy-3-[18F] fluorothymidine-positron emission tomography for noninvasive assessment of proliferation in pulmonary nodules. *Cancer Res* 2002;62:3331–3334.
37. Halter G, Buck AK, Schirrmeister H, et al. 3-Deoxy-3-[18F] fluorothymidine-positron emission tomography: alternative or diagnostic adjunct to 2-[18F] fluoro-deoxy-D-glucose positron emission tomography in the workup of suspicious central focal lesions? *J Thorac Cardiovasc Surg* 2004;127:1093–1099.
38. Cobben DCP, Elsinga PH, Hoekstra HJ, et al. Is F-18–3′-fluoro-3′-deoxy-L-thymidine useful for the staging and restaging of non-small cell lung cancer? *J Nucl Med* 2004;45:1677–1682.
39. Muzi M, Vesselle H, Grierson J, et al. The kinetic analysis of FLT (3'-deoxy-3'-fluorothymidine) PET studies: validation studies in patients with lung cancer. *J Nucl Med* 2005 Feb; 46(2):274–82.
40. Buck AK, Halter G, Schirrmeister H, et al. Imaging proliferation in lung tumors with PET: 18F-FLT versus 18F-FDG. *J Nucl Med* 2003;44(9):1426–1431.
41. Koh W, Bergman KS, Rasey JS, et al. Evaluation of oxygenation status during fractionated radiotherapy in human non-small cell lung cancers using [F-18] fluoromisonidazole positron emission tomography. *Int J Radiat Oncol Biol Phys* 1995;33:391–398.
42. Dedashti F, Mintun MA, Lewis JS, et al. In vivo assessment of tumor hypoxia in lung cancer with Cu60-ATSM. *Eur J Nucl Med Mol Imaging* 2003;30:844–850.
43. Rasey JS, Koh WJ, Evans ML, et al. Quantifying regional hypoxia in human tumors with positron emission tomography of [18F] fluoromisonidazole: a pretherapy study of 37 patients. *Int J Radiat Oncol Biol Phys* 1996;36:417–428.
44. Slamon DJ, Clark GM, Wong SG, Levin WJ, Ullrich A, McGuire WL. Human breast cancer: correlation of relapse and survival with amplification of the HER-2/neu oncogene. *Science* 1987;235: 177–182.
45. Emens LA. Trastuzumab: targeted therapy for the management of HER-2/neu-overexpressing metastatic breast cancer. *Am J Ther* 2005;12:243–253.
46. Blend MJ, Stastny JJ, Swanson SM, Brechbiel MW. Labeling anti-HER2/neu monoclonal antibodies with 111In and 90Y using a bifunctional DTPA chelating agent. *Cancer Biother Radiopharm* 2003;18:355–363.
47. Tang Y, Scollard D, Chen P, Wang J, Holloway C, Reilly RM. Imaging of HER2/neu expression in BT-474 human breast cancer xenografts in athymic mice using [(99m)Tc]-HYNIC-trastuzumab (Herceptin) Fab fragments. *Nucl Med Commun* 2005;26:427–432.
48. Tang Y, Wang J, Scollard DA, Mondal H, Holloway C, Kahn HJ, Reilly RM. Imaging of HER2/neu-positive BT-474 human breast cancer xenografts in athymic mice using (111)In-trastuzumab (Herceptin) Fab fragments. *Nucl Med Biol* 2005;32:51–58.
49. Copland JA, Eghtedari M, Popov VL, Kotov N, Mamedova N, Motamedi M, Oraevsky AA. Bioconjugated gold nanoparticles as a molecular based contrast agent: implications for imaging of deep tumors using optoacoustic tomography. *Mol Imaging Biol* 2004;6:341–349.
50. Gonzalez Trotter DE, Manjeshwar RM, Doss M, Shaller C, Robinson MK, Tandon R, Adams GP, Adler LP. Quantitation of small-animal (124)I activity distributions using a clinical PET/CT scanner. *J Nucl Med* 2004;45:1237–1244.
51. Olafsen T, Tan GJ, Cheung CW, Yazaki PJ, Park JM, Shively JE, Williams LE, Raubitschek AA, Press MF, Wu AM. Characterization of engineered anti-p185HER-2 (scFv-CH3)2 antibody fragments (minibodies) for tumor targeting. *Protein Eng Des Sel* 2004;17, 315–323.
52. Smith-Jones PM, Solit DB, Akhurst T, Afroze F, Rosen N, Larson SM. Imaging the pharmacodynamics of HER2 degradation in response to Hsp90 inhibitors. *Nat Biotechnol* 2004;22:701–706.
53. Palm S, Enmon RM, Jr., Matei C, Kolbert KS, Xu S, Zanzonico PB, Finn RL, Koutcher JA, Larson SM, Sgouros G. Pharmacokinetics and biodistribution of (86)Y-Trastuzumab for (90)Y dosimetry in an ovarian carcinoma model: correlative MicroPET and MRI. *J Nucl Med* 2003;44:1148–1155.
54. Garmestani K, Milenic DE, Plascjak PS, Brechbiel MW. A new and convenient method for purification of 86Y using a Sr(II) selective resin and comparison of biodistribution of 86Y and 111In labeled Herceptin. *Nucl Med Biol* 2002;29:599–606.
55. Kobayashi H, Shirakawa K, Kawamoto S, Saga T, Sato N, Hiraga A, Watanabe I, Heike Y, Togashi K, Konishi J, Brechbiel MW, Wakasugi H. Rapid accumulation and internalization of radiolabeled herceptin in an inflammatory breast cancer xenograft with vasculogenic mimicry predicted by the contrast-enhanced dynamic MRI with the macromolecular contrast agent G6-(1B4M-Gd)(256). *Cancer Res* 2002;62:860–866.

56. Olafsen T, Kenanova VE, Sundaresan G, Anderson AL, Crow D, Yazaki PJ, Li L, Press MF, Gambhir SS, Williams LE, Wong JY, Raubitschek AA, Shively JE, Wu AM. Optimizing radiolabeled engineered anti-p185HER2 antibody fragments for in vivo imaging. Cancer Res 2005;65:5907–5916.

57. Hynes NE and Lane HA. ERBB receptors and cancer: the complexity of targeted inhibitors. *Nat Rev Cancer* 2005;5:341–354.

58. Ramos-Suzarte M, Rodriguez N, Oliva JP, Iznaga-Escobar N, Perera A, Morales A, Gonzalez N, Cordero M, Torres L, Pimentel G, Borron M, Gonzalez J, Torres O, Rodriguez T, Perez R. 99mTc-labeled antihuman epidermal growth factor receptor antibody in patients with tumors of epithelial origin: Part III. Clinical trials safety and diagnostic efficacy. *J Nucl Med* 1999;40:768–775.

59. Vallis KA, Reilly RM, Chen P, Oza A, Hendler A, Cameron R, Hershkop M, Iznaga-Escobar N, Ramos-Suzarte M, Keane P. A phase I study of 99mTc-hR3 (DiaCIM), a humanized immunoconjugate directed towards the epidermal growth factor receptor. *Nucl Med Commun* 2002;23:1155–1164.

60. Iznaga-Escobar N, Ramos-Suzarte M, Morales-Morales A, Torres-Arocha L, Rodriguez-Mesa N, Perez-Rodriguez R. (99m)Tc-labeled murine ion C5 monoclonal antibody in colorectal carcinoma patients: pharmacokinetics, biodistribution, absorbed radiation doses to normal organs and tissues and tumor localization. *Methods Find Exp Clin Pharmacol* 2004;26:687–696.

61. Divgi CR, McDermott K, Johnson DK, Schnobrich KE, Finn RD, Cohen AM, Larson SM. Detection of hepatic metastases from colorectal carcinoma using indium-111 (111In) labeled monoclonal antibody (mAb): MSKCC experience with mAb 111In-C110. *Int J Rad Appl Instrum B* 1991;18: 705–710.

62. Divgi CR, Welt S, Kris M, Real FX, Yeh SD, Gralla R, Merchant B, Schweighart S, Unger M, Larson SM. Phase I and imaging trial of indium 111-labeled anti-epidermal growth factor receptor monoclonal antibody 225 in patients with squamous cell lung carcinoma. J Natl Cancer Inst 1991;83:97–104.

63. Cornelissen B, Kersemans V, Burvenich I, Oltenfreiter R, Vanderheyden JL, Boerman O, Vandewiele C, Slegers G. Synthesis, biodistribution and effects of farnesyltransferase inhibitor therapy on tumour uptake in mice of 99mTc labelled epidermal growth factor. *Nucl Med Commun* 2005;26: 147–153.

64. Fry DW, Bridges AJ, Denny WA, Doherty A, Greis KD, Hicks JL, Hook KE, Keller PR, Leopold WR, Loo JA, McNamara DJ, Nelson JM, Sherwood V, Smaill JB, Trumpp-Kallmeyer S, Dobrusin EM. Specific, irreversible inactivation of the epidermal growth factor receptor and erbB2, by a new class of tyrosine kinase inhibitor. *Proc Natl Acad Sci USA* 1998;95:12022–12027.

65. Richter M, Zhang H. Receptor-targeted cancer therapy. *DNA Cell Biol* 2005;24:271–282.

66. Mishani E, Abourbeh G, Rozen Y, Jacobson O, Laky D, Ben David I, Levitzki A, Shaul M. Novel carbon-11 labeled 4-dimethylamino-but-2-enoic acid [4-(phenylamino)-quinazoline-6-yl]-amides: potential PET bioprobes for molecular imaging of EGFR-positive tumors. *Nucl Med Biol* 2004;31:469–476.

67. Ortu G, Ben David I, Rozen Y, Freedman NM, Chisin R, Levitzki A, Mishani E. Labeled EGFr-TK irreversible inhibitor (ML03): in vitro and in vivo properties, potential as PET biomarker for cancer and feasibility as anticancer drug. *Int J Cancer* 2002;101:360–370.

68. Bonasera TA, Ortu G, Rozen Y, Krais R, Freedman NM, Chisin R, Gazit A, Levitzki A, Mishani E. Potential (18)F-labeled biomarkers for epidermal growth factor receptor tyrosine kinase. *Nucl Med Biol* 2001;28:359–374.

69. Vasdev N, Dorff PN, Gibbs AR, Nandanan E, Reid LM, O'Neill JP, VanBrocklin HF. Synthesis of 6-acrylamido-4-(2-[F-18]fluoroanilino)quinazoline: a prospective irreversible EGFR binding probe. *J Labelled Comp Radiopharm* 2005;48:109–115.

70. Shaul M, Abourbeh G, Jacobson O, Rozen Y, Laky D, Levitzki A, Mishani E. Novel iodine-124 labeled EGFR inhibitors as potential PET agents for molecular imaging in cancer. *Bioorg Med Chem* 2004;12:3421–3429.

71. Early Breast Cancer Trialists' Collaborative Group. Tamoxifen for early breast cancer: an overview of randomized trials. Lancet 1998;351:1451–1467.

72. Carlson RW, Anderson BO, Cox C, et al. *Breast Cancer Clinical Practice Guidelines in Oncology*, 1st edn. National Comprehensive Cancer Network, 2004.

73. Linden HM, Stekhova SA, Gralow JR, et al. Quantitative fluoroestradiol positron emission tomography imaging predicts response to endocrine treatment in breast cancer. *J Clin Oncol* 2006;24: 2793–2799.

74. Buchanan G, Irvine RA, Coetzee GA, Tilley WD. Contribution of the androgen receptor to prostate cancer predisposition and progression. *Cancer Metastasis Rev* 2001;20:207–223.

75. Loblaw DA, Mendelson DS, Talcott JA, et al. American Society of Clinical Oncology recommendations for the initial hormonal management of androgen-sensitive metastatic, recurrent, or progressive prostate cancer. 2004;22:2927–2941.

76. Dedashti F, Picus J, Dence CS, Siegel BA, Katzellenbogen JA, Welch MJ. Positron tomographic assessment of androgen receptors in prostatic carcinoma 2004;32:344–350.

77. Larson SM, Morris M, Gunther I, et al. Tumor localization of 16 beta-18F-fluoro-5alpha-dihydrotestosterone versus F-18 FDG in patients with progressive metastatic prostate cancer. 2004;45:366–373.

78. Rambaldi PF, Cuccurullo V, Briganti V, Mansi L. The present and future role of In-111 pentetreotide in the PET era. *QJ Nucl Med Mol Imaging* 2005;49:225–235.

79. Pastore V, Di Lieto E, Mansi L, Rambaldi PF, Santinin M, Mancusi R. Intraoperative detection of lung cancer by Octreotide labeled to Indium-111. *Semin Surg Oncol* 1998;15:220–222.

80. Torabi M, Aquino SL, Harisinghani MG. Current concepts in lymph node imaging. *J Nucl Med* 2004;45(9):1509–1518.

30

Combinations of Molecular-Targeted Therapies

Opportunities and Challenges

Helen X. Chen, MD,
and Janet E. Dancey, MD

SUMMARY

With our awareness of the complexity of cancer cell biology has come the appreciation that strategies aiming at simultaneous blockade of multiple molecular pathways will be critical to further therapeutic success. Given the availability of numerous targeted agents for clinical testing, there are now considerable interest and optimism for combination regimens with multiple targeted agents. Challenges to the task of evaluating such novel combinations are, however, unprecedented, in view of the almost limitless possibilities of combination regimens and the still inadequately understood complexity and heterogeneity of cancer biology.

In this chapter, we will discuss the critical elements of a development strategy for targeted agent combinations. Scientific, medical, and intellectual property issues that pose barriers to the rational preclinical and clinical evaluation of targeted combinations will be described. Possible means of overcoming these barriers will be discussed.

Key Words: targeted therapy; combination regimens; CTEP.

1. IMPORTANCE OF COMBINING MULTIPLE-TARGETED AGENTS

The use of drug combinations to circumvent tumor resistance is a well-established cancer therapeutics principle. It is based on the assumption that the probability of a given cancer cell being resistant to a combination of non-cross resistant drugs varies as the product of the probabilities of resistance to each of the individual agents. With few exceptions, traditional curative treatments for cancer are due to combining agents of known activity, with different mechanisms of resistance and minimally overlapping spectra of toxicity.

The combination strategy is equally and probably more important for molecular-targeted agents. Clinical data from studies evaluating single-agent activity of these drugs have yielded results that are modest except in rare circumstances where the

From: *Cancer Drug Discovery and Development: Molecular Targeting in Oncology*
Edited by: H. L. Kaufman, S. Wadler, and K. Antman © Humana Press, Totowa, NJ

tumor pathogenesis is dominated by a single molecular abnormality *(1–4)*. Demonstration of clinical benefit by targeted agents often required combining them with standard chemotherapy or radiation therapy regimens. Even so, the clinical benefit conferred by the addition of targeted agents, while clinically relevant, is still limited, particularly for patients with advanced disease *(1–4)*. The limited benefits provided by these agents are not surprising given that the complexity of the tumor biology dictates that the molecular pathways responsible for tumor growth and metastases may be redundant and adaptable, and variable between individual patients or between tumor cell subclones within the same patient. It is therefore unlikely that treatment focusing on a single target would offer durable tumor control in most patients. A logical approach would be to combine multiple targeted agents so as to overcome molecular heterogeneity of the tumors and resistance mechanisms. The belief that such combination regimen may offer greater antitumor activity than predicted by their single-agent activity or their activity in combination with standard cancer drugs has led to consideration and implementation of clinical tests for regimens with two or more targeted agents.

Because the number of drug combinations is limitless, it is essential to develop rational strategies for establishing priorities and for effective non-clinical and clinical proof-of-principle evaluations.

2. STRATEGIES FOR THE DEVELOPMENT OF MULTIPLE-TARGETED AGENT COMBINATION

2.1. What Have We Learned from Past Experience with Targeted Agent Combined with Conventional Cytotoxic Therapies?

As we attempt to optimize the development strategies for novel combination regimens, it is reasonable to review the experience from trials in the past of targeted agents combined with standard treatment as there are important insights to be gained that may inform the planning of trials of novel-targeted agents.

Although there are a number of trials evaluating combinations of targeted agents, success to date has been limited to a few specific agents combined with standard chemotherapy or radiation [e.g., trastuzumab in breast cancer *(4)*, bevacizumab in colorectal and non-small cell lung cancer *(5,6)*, and cetuximab in head and neck cancer *(7)*]. There have also been a number of notable failures: erlotinib and gefitinib with chemotherapy in lung cancer *(8–10)*, bevacizumab and chemotherapy in refractory breast cancer, and oblimersen with chemotherapy in melanoma *(12)*.

It is notable that many targeted agents had actually demonstrated inhibition of intended targets in patients. The reason for the failure of such targeted agents to demonstrated added clinical benefit when combined with standard chemotherapy may include any of three factors: (i) Antagonism between agents: There may be antagonism between the targeted and chemotherapy agents due to pharmacokinetic alteration or interference with agents' individual mechanisms of action. The latter, for example, might occur if the first agent induced G1 arrest in tumor cells, whereas the additional agent is active only in S or M phases of the cell cycle. (ii) Biological complexity and heterogeneity of tumors: For a given tumor, the intended target may be absent or biologically irrelevant or there are redundant pathways that provide alternative signaling to maintain critical functioning within cancer cells. Furthermore, tumor subclones or microenvironment heterogeneity within a patient may prevent the demonstration of significant clinical benefit due to the rapid emergence of these refractory clones. (iii) There may also be overlapping sensitivities of tumor cells to components within a combination regimen such that there is no net clinical gain when the agents are administered together.

Unfortunately, although listing possible causes of failure is straightforward, understanding the actual reasons for success/failure of particular targeted agents in specific trial settings is lacking. Owing to drug development time lines that commonly outpace scientific understanding of the agents and their targets, the assessment of these combinations of agents has largely been based on traditional development strategies. At the time clinical development is initiated, there has been a relative lack of mechanistic understanding of both sensitivity and resistance to the individual agents and the combination, thus limiting the ability to design clinical trials to optimally assess the experimental regimens in the most appropriate patient population. Given the availability of hundreds of novel agents that are able to perturb specific tumor signaling pathways and the need to be used in rational combinations to realize their therapeutic potential, it is imperative to move beyond empiric approaches to clinical studies for the evaluation of these targeted agents and their combinations.

3. SCIENTIFIC CONSIDERATIONS IN CLINICAL DEVELOPMENT OF TARGET AGENT COMBINATIONS

The questions to consider in the evaluation and prioritization of potential combination regimens should include (i) What molecules or signaling pathways should be targeted by the combination regimen; (ii) which agents should be selected for the intended targets; (iii) what is the optimal dose and sequence for the administration of multiple agents; and (iv) how patients should be selected such that additive or synergistic interactions between the agents are likely to be demonstrated. Answers to these questions are not straightforward and would require adequate understanding of the cancer biology, the clinical and biological effect of the agents, and carefully designed non-clinical studies. A general strategy for prioritizing clinical evaluation of targeted combinations has been proposed (Table 1).

3.1. What Targets Should be Considered in a Combination Regimen?

The choice of target/targeted agent for development can be made based on the knowledge of the presence and relevance of the target to the pathogenesis and growth of a given tumor. The target of the second agent in the combination may be selected with the goal to enhance the activity of the first agent by (i) more effectively inhibiting the initial target, (ii) inhibiting additional target(s) and/or pathways, and (iii) overcoming a compensatory cellular process that leads to resistance to the single agent. Such combined molecular targeting might induce apoptosis when single agents result in growth inhibition, prevent a secondary pathway from maintaining critical cellular function thus overcoming a mechanism of resistance, or inhibiting a process such as angiogenesis that is not impacted by the mechanism of action of the first agent, enhancing the overall antitumor effect.

Table 1
Prioritization of Combinations

Non-clinical
– Non-clinical supportive data demonstrating proof-of-principle for mechanism and potential to evaluate principle in clinical trial
– Synergy demonstrated in multiple human tumor models or within a biologically identifiable subset of tumors of relevance to human cancers
– Activity of the drug combination against primary human tumor cells at pharmacologically achievable or relevant concentrations
– In vitro selectivity against tumor cells—that is, the synergistic combination is nontoxic/minimally toxic to normal tissues
– Synergism in the presence of human plasma
– Clear, compelling, or at least reasonable demonstration of mechanism of action of the agents, with the drug combination based on a mechanistic approach that can be validated

Clinical
– Clinical data on pharmacology and toxicity of the individual agents/combination
– Clinical data confirming proof-of-principle for favorable interaction between the agent alone and/or in combination
– Clinical activity for the agent(s) in specific disease
– The means to select patients with tumor phenotype predicted to have a favorable response to the combination
– Agents are available to be used in combination

Adapted from ref.*38.*

3.2. Which Agents and Regimens Should be Selected?

For the selection of specific agents for the intended targets, priority should be given to those agents that have demonstrated ability to modulate the target and/or to induce a therapeutic effect, acceptable pharmacology, and safety. For the combination regimen, ideal credentials would include presence of clinical activity with the individual agents in the same indication, consistent evidence of synergy in multiple non-clinical studies, and an understanding of the molecular context required for synergy that could be utilized in clinical trails for patient selection.

Note that the list of credentials is "ideal," and the absence of one or more of these factors does not preclude the development of a particular agent/combination. For example, agents need not have single-agent clinical activity to be considered for development in combination if non-clinical studies consistently show striking synergy in models and the mechanisms of action/interaction are relevant to human. The transition from preclinical results to clinical trial for such a combination would require that appropriate concentrations of the agents are achievable, the schedule of administration is feasible, and the patient population for testing the combination reflects the molecular context of the preclinical models that demonstrated activity.

3.3. Choice Between a Regimen of One Drug with Many Targets or a Combination Regimen with Multiple Drugs with Individual Targets?

In theory, combined inhibition of multiple molecular targets could also be achieved by use of one or fewer agents that has a broad spectrum of target inhibition. A number of such agents, predominantly small molecule tyrosine kinase inhibitors, are available for clinical use. There are advantages and disadvantages to the development and evaluation of these agents. Although multiple targeted agents may overcome molecular heterogeneity within or between cancer patients and therefore have a better chance of successful clinical development (especially if tumor markers for patient selection are uncertain), they may not be ideal for rational combination strategies based on specific molecular profiles of tumors. As these agents are not "designed" to interact with a particular array of targets, the potency against specific targets as well as the tolerability to individual target inhibition may vary such that the fixed concentration may not be optimal for all relevant targets. They may also preclude "validation" of individual targets, as the antitumor activity could be due to its effect on any or all of the proposed targets. In contrast, combinations of specific targeted agents would allow greater flexibility for tailoring regimens to specific patients and molecular profiles of their cancers. Doses/schedules of the agents within a combination may also be tailored for desired concentrations/exposures to optimize interactions between agents. The toxicity of the combination may also be more predictable because of the limited off-target effects.

4. ISSUES OF INTELLECTUAL PROPERTY IN STUDIES OF COMBINATION INVESTIGATIONAL AGENTS

Besides scientific issues, additional challenges for evaluation of targeted agent combinations are regulatory, intellectual property, and data sharing concerns, as many of the targeted agents are investigational and under development by different pharmaceutical companies. Although the pharmaceutical industry have recognized the limitation of single-targeted agents and are interested in the concept of targeted agent combinations, negotiation and agreement on such collaboration are not straightforward. To this end, the Cancer Therapy Evaluation Program (CTEP) of NCI has been able to provide a common platform for implementation of such novel studies. With access to a variety of investigational agents under collaborative clinical development with pharmaceutical partners, CTEP is uniquely positioned to facilitate collaboration between companies and has established common intellectual property language for combination studies that have been accepted by many pharmaceutical companies. In the template language of agreement for combination of investigational agents, it stipulates that each collaborator shall receive non-exclusive royalty-free licenses to the combination IP for all purposes including that of commercial use.

5. CURRENT EXPERIENCE WITH THE EVALUATION OF TARGETED AGENT COMBINATIONS

Combination strategies for targeted agents that have been implemented to date may be divided into three broad categories (Table 2) (i) combinations to maximize the inhibition of a specific target [e.g., antibody and small molecule kinase inhibitor to the same target such as epidermal growth factor receptor (EGFR)], (ii) combinations to maximize the inhibition of molecular signaling by blocking both the ligand and the receptor [e.g., vascular endothelial growth factor (VEGF) and vascular endothelial growth factor receptor 2 (VEGFR-2)], (iii) combinations to maximize the inhibition of a pathway by targeting the signaling and the downstream molecules (e.g., HER-2/EGFR and mTOR), and (iv) combinations to expand the inhibition of multiple cellular

Table 2
Clinical Trials for Combinations of Novel Target Agents

Targets	Clinical trials	Cancer
VEGR + EGFR	Bevacizumab + cetuximab	Colon, pancreas, lung, and head and neck
	Bevacizumab + erlotinib	Breast, head and neck, kidney, lung, pancreas, ovary, and HCC
VEGF + PDGFR/c-kit	Bevacizumab + imatinib	Melanoma, GIST, and RCC
VEGF + mTOR	Bevacizumab + temsirolimus	Kidney and melanoma
VEGF + VEGFR/raf	Bevacizumab + sorafenib	Kidney, ovarian, and melanoma
HER2 + CDK	Trastuzumab + flaopiridol	Breast
EGFR + VEGF/raf	Cetuximab + sorafenib	Colon
EGFR + EGFR TKI	Cetuximab + erlotinib	Colon and lung
	Cetuximab + gefitinib	
EGFR + mTOR	Erlotinib + temsirolimus	Lung and glioma
	Gefitinib + everolimus	
	Gefitinib + sirolimus	
EGFR + FTI	Erlotinib + tipifarnib	Phase 1
HER2 + mTOR	Trastuzumab + temsirolimus	Breast
HER2 + EGFR	Trastuzumab + gefitinib	Breast
	Trastuzumab + erlotinib	
HER-2 + VEGF	Trastuzumab + bevacizumab	Breast
HDAC + VEGF	SAHA + bevacizumab	Kidney
HDAC + CDK	SAHA + flavopiridol	Phase 1
Proteosome + Hsp90	Bortezomib + 17AAG	Phase 1

Source: www.clinicaltrials.gov.
Abbreviations: EGFR, epidermal growth factor receptor; VEGF/VEGFR, vascular endothelial cell growth factor/receptor; PDGFR, platelet derived growth factor receptor; HER2, human epidermal growth factor receptor-2; mTOR, mammalian target of rapamycin; CDK, cyclin dependent kinase; HDAC, histone deacetylase; Hsp90, heat shock protein 90; 17AAG, 17-allylamino 17-demethoxy geldanamycin.

mechanisms (e.g., EGFR and VEGFR). These strategies have been proposed based on our limited understanding of cancer targets/pathways, availability of the agents, and limited preclinical experiments that suggest at least additivity of the combination.

Preliminary results from clinical trials assessing these strategies have been reported. Although data are limited, they could provide insights to further development of combination regimens. Combining agents to inhibit a single target may augment target inhibition and toxicity associated with such inhibition. For example, combining VEGF-neutralizing monoclonal antibody bevacizumab and VEGFR inhibitor sorafenib has been reported to enhance toxicities of hypertension and proteinuria associated with VEGF inhibition at relatively low doses of each agent (13). Alternatively, agents that do not have overlapping toxicities because of non-cross reacting targets seem to be able to be combined at full doses (e.g., EGFR + VEGFR) (14). These data suggest that agents are more likely to be combined at doses and schedules established for the single agents if toxicity of the individual agents (due to inhibition of purported targets) is non-overlapping. Enhanced toxicity is not surprising if the established doses for individual agents are maximally tolerable based on toxicities directly related to target inhibition, and the toxicities are overlaps between the agents. Such overlapping toxicity profile may lead to unacceptable adverse events requiring dose reduction of individual agents to ensure tolerability. It remains to be seen whether a combination of agents at reduced dose will still result in enhanced activity as compared with single agents at full doses.

There is as yet limited data to assess the benefits that may be derived from combination-targeted agents. Initial assessments of antitumor activity have yielded mixed results. For example, the first report of a phase 2 clinical trial assessing the combination of EGFR inhibitor gefitinib with trastuzumab in patients with metastatic breast cancer did not identify a favorable interaction between the agents (15). Similarly, although initial results from uncontrolled early clinical trials suggested promising response data for the combination of bevacizumab with EGFR small molecule kinase inhibitor erlotinib in renal cell cancer (16–17), preliminary analysis of the first randomized phase 2 trial failed to demonstrate improvement in objective response rate or progression-free survival compared with bevacizumab alone (18). On the other hand, preliminary results for the combination of bevacizumab and the EGFR targeting antibody cetuximab in colorectal cancer (CRC) were promising (14), with the response rate and the progression-free survival endpoints exceeding the historical data from studies of cetuximab alone; the definitive trial of the combination of these two targeted agents is now ongoing.

It seems clear that success or failure of the clinical trials of combinations that have been evaluated to date has not been clearly predicted by published preclinical data.

The reasons for the lack of correlation between laboratory and clinical studies are uncertain. The actual differences between the mechanism(s) of action in cell lines compared with human cancer are unknown. Furthermore, preclinical studies are often conducted in limited number of tumor models given limited considerations to how they may be applied clinically.

6. NON-CLINICAL EVALUATIONS OF COMBINATIONS

Non-clinical studies with in vitro and in vivo tumor models have played an essential role in the development of cancer therapy in the past, and it will continue to provide critical guidelines for the development of combination studies of targeted agents. However, limitation of the preclinical models should be recognized, and meticulous attention to the experimental design and appropriate interpretation of the outcome data are critical to application of the results to clinical evaluation.

6.1. Factors that limit the Applicability of Non-Clinical Studies

Cancer journals are filled with preclinical studies describing "synergistic" combinations of agents that authors then propose for testing in the clinic. However, as have already been observed, there is a poor concordance between preclinical and clinical results. Such a poor concordance is not surprising, as non-clinical combination studies have generally involved limited in vitro cell line and in vivo human tumor xenograft models in experiments designed with limited consideration of the predictive value of the model for translation to cancer patients. Such evaluations can only provide insight into mechanism of action and interaction between the agents within the parameters of the experiment. However, interpretation of data for subsequent design of clinical studies must consider a number of variables of the preclinical experiments. Major variables include the origin of the tumor (i.e., cell line versus patient biopsy), target/receptor status of the tumor, the site of tumor implantation (e.g., subcutaneous, intraperitoneal, and orthotopic), the size of tumor at the onset of agent treatment, growth rate and growth characteristics, doses of the agent required, and the experimental endpoints *(21)*.

In addition to the above considerations, experiments assessing combinations must be designed to determine optimal concentration/exposures and sequencing. Preclinical studies of combinations require the generation of accurate dose–response curves for the agents tested, both alone and in combination (in fixed-ratio concentrations), when depend on the proper dynamic range of the assay and accurate assessment of the endpoint of interest is critical. Effective combinations should achieve a two to three log improvement in cytotoxicity or growth inhibition *(22)*. Formal quantitative analysis of experimental data may be done through a number of methods *(23–25)*. Drug interactions can also be studied in vivo, and elegant demonstrations of synergy in mouse xenografts have been reported *(26–27)*. However, the variables that may influence the outcomes of synergy studies in xenograft studies are multiple, requiring large numbers of animals to achieve statistically valid results, and relatively few laboratories have the resources to perform such evaluations *(28)*. Thus, resource constraints have led to a strategy of defining synergy in robust tissue culture models and then confirming a beneficial interaction between the two drugs in more limited xenograft studies.

Even when the evaluations of combinations have been meticulously performed, there remains a lack of strong correlation to clinical outcomes because of the limitations of in vitro and in vivo tumor models. Cancer cell lines show considerable alterations in biological properties and chemosensitivity pattern as compared with the original tumors *(29–31)*. They are also highly variable among one another, even within the same class of tumor histology. Limiting experimentation to a few model cell lines cannot reflect the diversity of human organisms. Although we are in the era of "molecularly targeted" therapy, in vitro and in vivo cell lines have often been used without characterization at the molecular level for the particular target or pathway being addressed. Such a lack of understanding of the molecular context of tumor cell lines limits the value of these models for predicting the clinical activity of combinations or proper patient selection.

In absence of knowledge about potential markers to predict efficacy, demonstration of consistent activity or synergism across various models would be essential for a given regimen. In two studies that assessed the correlation between in vitro or in vivo testing in NCI 60 cell lines and clinical activity of single cytotoxic agents *(32–33)*, it was suggested that tests in human cell-line models are predictive of clinical activity of cytotoxic agents but only if the activity was seen across a number of models. Although there has not been a systematic evaluation of the predictive value of preclinical combination studies reported to date, it seems likely that the benefit of combinations is likely to correlate to consistent observation of synergy when the combination is tested in a number of models that reflect the molecular heterogeneity of human cancers.

There are thus a number of limitations immediately apparent to the rational application of synergy data from a combination of agents evaluated in the laboratory to a clinical trial. First, the link between observed synergy and the underlying biochemical, molecular, and physiological mechanisms is very difficult to discover.

Thus, determining the mechanisms of synergy in laboratory models and correlating to clinical situations is daunting. Second, the in vitro synergistic activity depends on specific drug ratios, and the in vivo activity depends on maintaining the synergistic ratios. In general, however, clinical testing is still largely the empirical application of a combination of agents administered at maximum tolerable doses to patients. Third, combinations may improve outcome by circumventing tumor heterogeneity within a patient or between patients; however, laboratory experiments are generally not designed to assess for improved outcomes across models (reflecting greater tumor cell heterogeneity). Last, there is no direct laboratory endpoint for measuring magnitude of enhanced cytotoxicity or inhibition of cell proliferation that is known to directly correlate to patient benefit.

6.2. Value and Directions of Non-clinical Studies with Targeted Agents and Their Combination

Despite their limitations, non-clinical models are still the most valuable tools that may provide critical guidance for the clinical development of a molecule. First, laboratory studies may uncover mechanisms of resistance to a single agent and thus identify the additional targets for the combination regimen. For example, preclinical studies based on parental EGFR-sensitive cell lines and EGFR-resistant subclones demonstrated that activation of additional pathways such as those stimulated by VEGF (34–35) and insulin like growth factor receptor 1 (IGF-IR) (36) may be responsible for acquired resistance to EGFR inhibitors. Specifically inhibiting these signaling pathways in combination with EGFR inhibition may provide greater efficacy. Second, laboratory studies can also provide information on optimal sequencing. For example, recent laboratory studies have suggested that EGFR inhibitor gefitinib may be more effective in combination with standard cytotoxic agents given as a high-dose "pulse" prior to cytotoxic agent compared to continuous, concurrent administration (37). This novel schedule is being tested in a clinical trial. Last, in parallel with efficacy determinations, the xenograft model is useful in assessing the pharmacokinetics and pharmacodynamics and toxicity of the agents in combination. Thus, non-clinical studies can provide valuable information for design of appropriate clinical trials to test combinations.

However, if laboratory studies are to inform clinical trial designs, it is important that experimental conditions reflect clinical constraints and clinical trials are designed to mirror the parameters of laboratory studies as much as possible. A number of principles for this type of testing have been proposed that may improve the overall quality and interpretability of such non-clinical experiments. For in vitro assessment of synergy (i) the assay system should have a wide dynamic range, ideally three to four logs of cell kill, (ii) the cell line panel should employ multiple cell lines, including drug-resistant lines, (iii) major mechanisms of resistance should be identified and used to structure the cell line panel, (iv) exposure to drugs should be at clinically achievable levels, (v) the combination should be tested under hypoxic conditions because hypoxia may antagonize drug action (38), and (vi) possible certain caveats related to culture conditions should be considered such as effects of growth factor supplementation or protein binding differences (39) between culture conditions and the patients plasma/sera. Similarly, in vivo experiments should use the most appropriate tumor models with known target/receptor status of the tumor and use the doses/schedules of the agents that result in achievable concentrations/exposures in patients *(21)*. In addition, as most agents and combinations are tested in patients with advanced/metastatic disease resistant to standard agents, use of such models to demonstrate activity of an investigational combination would be ideal. Similarly, if a combination is to be compared with a standard treatment regimen in a clinical trial, non-clinical experiments might be improved if an active control representative of the standard drug treatment were included.

7. FUTURE DIRECTIONS

An efficient screening process to identify the most promising combinations among a myriad of possible agents/targets is needed. The most exciting prospect for a significant advance in cancer therapeutics development is the possibility of identifying and evaluating combinations that induce "synthetic lethality" due to specific exploitable genetic abnormalities in cancer cells *(40)*. This paradigm has not been exploited in the past because of the lack of robust methods for identifying synthetic lethal genes. With the availability of chemical and genetic tools for perturbing gene function in somatic cells, there is every reason to be optimistic that the systematic evaluation of combinations of targeted agents will lead to the identification of the specific genotypic or phenotypic milieu leading to synthetic lethality. Subsequently, the results of such experiments will result in the evaluation of combinations of agents within an appropriately selected population of cancer patients.

Another approach to rapid evaluation and identification of promising combinations of agents involves the use of systematic high-throughput screening of combinations of small molecules to reveal favorable interactions between compounds, presumably because of interactions between the pathways on which they act. A practical application of this methodology is the creation of combination drugs through systematic screening of compounds in disease-relevant phenotypic assays. A synergistic combination discovered empirically may subsequently be evaluated to understand the mechanisms of action and interaction of the agents *(41)*.

8. CONCLUSIONS

Combination of targeted agents is believed to be an important therapeutic strategy that capitalizes on the emergence of multiple molecularly targeted agents and holds the promise for further therapeutic success by overcoming resistant/redundant tumor growth and survival signals. Currently, dozens of clinical studies have been implemented to evaluate the safety and efficacy of regimens combining two or more targeted agents. It remains to be determined whether such novel combinations represent a new paradigm that is superior to traditional combinations between conventional cytotoxic agents or between conventional and novel agents.

The number of possible combination regimens is virtually limitless, which presents both opportunities of potential therapeutic interventions and challenges for their clinical evaluations. Strategies to rationally evaluate the activity of combination regimens should take into consideration both the knowledge of the single agents—including mechanisms of action, clinical pharmacology, toxicity profile, and antitumor activity—and non-clinical evaluation of the combinations. To maximize translatability of preclinical study results, the proposed regimen should be tested in a variety of tumor models at clinical achievable doses/exposures, and if possible, the target status and molecular context of the tumor models that correspond to treatment outcomes should be defined.

Our current ability to move beyond empiric selection and evaluation of molecular-targeted agents and their combinations is limited due to inadequate knowledge of the tumor biology, inadequate understanding of the mechanisms of action/resistance for individual agents and their combinations, as well as inadequate tools to measure and compare treatment effects in laboratory models that can be predictably used in clinical development. Fortunately, such limitations are increasingly recognized by both the clinical and laboratory scientists and more research efforts are now directed toward systematic approach to studies of targeted agents and combinations. These efforts should improve our strategy in the clinical development of these novel combinations.

REFERENCES

1. Heinrich MC, Corless CL, Demetri GD, Blanke CD, von Mehren M, Joensuu H, et al. Kinase mutations and imatinib response in patients with metastatic gastrointestinal stromal tumor. *J Clin Oncol* 2003;21(23):4342–9.
2. Druker BJ, Talpaz M, Resta DJ, Peng B, Buchdunger E, Ford JM, et al. Efficacy and safety of a specific inhibitor of the BCR-ABL tyrosine kinase in chronic myeloid leukemia. *N Engl J Med* 2001;344(14):1031–7.
3. Mass RD, Press MF, Anderson S, Cobleigh MA, Vogel CL, Dybdal N, et al. Evaluation of clinical outcomes according to HER2 detection by fluorescence in situ hybridization in women with metastatic breast cancer treated with trastuzumab. *Clin Breast Cancer* 2005;6(3):240–6.
4. Slamon DJ, Leyland-Jones B, Shak S, Fuchs H, Paton V, Bajamonde A, et al. Use of chemotherapy plus a monoclonal antibody against HER2 for metastatic breast cancer that overexpresses HER2. *N Engl J Med* 2001;344(11):783–92.
5. Hurwitz H, Fehrenbacher L, Novotny W, Cartwright T, Hainsworth J, Heim W, et al. Bevacizumab plus irinotecan, fluorouracil, and leucovorin for metastatic colorectal cancer. *N Engl J Med* 2004;350(23):2335–42.
6. Sandler AB, Gray R, Brahmer J, Dowlati A, Schiller JH, Perry MC, et al. Randomized phase II/III trial of paclitaxel (P) plus carboplatin (C) with or without bevacizumab (NSC # 704865) in patients with advanced non-squamous non-small cell lung cancer (NSCLC): an Eastern Cooperative Oncology Group (ECOG) Trial - E4599. *J Clin Oncol* 2005;24:Abstract 4.
7. Bonner JA, Harari PM, Giralt J, Azarnia N, Shin DM, Cohen RB, et al. Radiotherapy plus cetuximab for squamous-cell carcinoma of the head and neck. *N Engl J Med* 2006;354(6):567–78.
8. Herbst RS, Giaccone G, Schiller JH, Natale RB, Miller V, Manegold C, et al. Gefitinib in combination with paclitaxel and carboplatin in advanced non-small-cell lung cancer: a phase III trial–INTACT 2. *J Clin Oncol* 2004;22(5):785–94.
9. Giaccone G, Herbst RS, Manegold C, Scagliotti G, Rosell R, Miller V, et al. Gefitinib in combination with gemcitabine and cisplatin in advanced non-small-cell lung cancer: a phase III trial–INTACT 1. *J Clin Oncol* 2004;22(5):777–84.
10. Herbst RS, Prager D, Hermann R, Fehrenbacher L, Johnson BE, Sandler A, et al. TRIBUTE: a phase III trial of erlotinib hydrochloride (OSI-774) combined with carboplatin and paclitaxel chemotherapy in advanced non-small-cell lung cancer. *J Clin Oncol* 2005;23(25):5892–9.
11. Burtness B, Goldwasser MA, Flood W, Mattar B, Forastiere AA. Phase III randomized trial of cisplatin plus placebo compared with cisplatin plus cetuximab in metastatic/recurrent head and neck cancer: an Eastern Cooperative Oncology Group study. *J Clin Oncol* 2005;23(34):8646–54.
12. Kirkwood JM, Bedikian A. Y., Millward MJ, Conry RM, Gore ME, Pehamberger HE, et al. Long-term survival results of a randomized multinational phase 3 trial of dacarbazine (DTIC) with or without Bcl-2 antisense (oblimersen sodium) in patients (pts) with advanced malignant melanoma (MM). *J Clin Oncol* 2005;24:Abstract 7506.
13. Posadas EM, Kwitkowski V, Liel M, Kotz H, Minasian L, Sarosy G, et al. Clinical synergism from combinatorial VEGF signal transduction inhibition in patients with advanced solid tumors – early

results from a phase I study of sorafenib (BAY 43–9006) and bevacizumab. *Eur J Cancer* 2005;3 (suppl 2):Abstract 1450.

14. Saltz LB, Lenz H, Hochster H, Wadler S, Hoff P, Kemeny N, et al. Randomized phase II trial of cetuximab/bevacizumab/irinotecan (CBI) versus cetuximab/bevacizumab (CB) in irinotecan-refractory colorectal cancer. *Proc Am Soc Clin Oncol* 2005;24:Abstract 3508.

15. Moulder SL, Arteaga CL. A Phase I/II Trial of trastuzumab and gefitinib in patients with metastatic breast cancer that overexpresses HER2/neu (ErbB-2). *Clin Breast Cancer* 2003;4(2):142–5.

16. Hainsworth JD, Sosman JA, Spigel DR, Edwards DL, Baughman C, Greco A. Treatment of metastatic renal cell carcinoma with a combination of bevacizumab and erlotinib. *J Clin Oncol* 2005;23(31):7889–96.

17. Herbst RS, Johnson DH, Mininberg E, Carbone DP, Henderson T, Kim ES, et al. Phase I/II trial evaluating the anti-vascular endothelial growth factor monoclonal antibody bevacizumab in combination with the HER-1/epidermal growth factor receptor tyrosine kinase inhibitor erlotinib for patients with recurrent non-small-cell lung cancer. *J Clin Oncol* 2005;23(11):2544–55.

18. Genentech. *Genentech Announces Preliminary Results from Randomized Phase II Trial of Avastin and Tarceva in Kidney Cancer*; Press Release October 18, 2005. Available at http://www.gene.com/ gene/news/press-releases/display.do?method=detail&id=8967&categoryid=4.

19. Normanno N, Campiglio M, De LA, Somenzi G, Maiello M, Ciardiello F, et al. Cooperative inhibitory effect of ZD1839 (Iressa) in combination with trastuzumab (Herceptin) on human breast cancer cell growth. *Ann Oncol* 2002;13(1):65–72.

20. Warburton C, Dragowska WH, Gelmon K, Chia S, Yan H, Masin D, et al. Treatment of HER-2/neu overexpressing breast cancer xenograft models with trastuzumab (Herceptin) and gefitinib (ZD1839): drug combination effects on tumor growth, HER-2/neu and epidermal growth factor receptor expression, and viable hypoxic cell fraction. *Clin Cancer Res* 2004;10(7):2512–24.

21. Kelland LR. Of mice and men: values and liabilities of the athymic nude mouse model in anticancer drug development. *Eur J Cancer* 2004;40(6):827–36.

22. Reynolds CP, Maurer BJ. Evaluating response to antineoplastic drug combinations in tissue culture models. *Methods Mol Med* 2005;110:173–83.

23. Tallarida RJ. Drug synergism: its detection and applications. *J Pharmacol Exp Ther* 2001;298(3): 865–72.

24. Teicher BA. Assays for in vitro and in vivo synergy. *Methods Mol Med* 2003;85:297–321.

25. Chou TC. Drug combinations: from laboratory to practice. *J Lab Clin Med* 1998;132(1):6–8.

26. Houghton PJ, Stewart CF, Cheshire PJ, Richmond LB, Kirstein MN, Poquette CA, et al. Antitumor activity of temozolomide combined with irinotecan is partly independent of O6-methylguanine-DNA methyltransferase and mismatch repair phenotypes in xenograft models. *Clin Cancer Res* 2000;6(10):4110–8.

27. Meco D, Colombo T, Ubezio P, Zucchetti M, Zaffaroni M, Riccardi A, et al. Effective combination of ET-743 and doxorubicin in sarcoma: preclinical studies. *Cancer Chemother Pharmacol* 2003;52(2):131–8.

28. Tan M, Fang HB, Tian GL, Houghton PJ. Experimental design and sample size determination for testing synergism in drug combination studies based on uniform measures. *Stat Med* 2003;22(13):2091–100.

29. Engelholm SA, Vindelov LL, Spang-Thomsen M, Brunner N, Tommerup N, Nielsen MH, et al. Genetic instability of cell lines derived from a single human small cell carcinoma of the lung. *Eur J Cancer Clin Oncol* 1985;21(7):815–24.

30. Ferguson PJ, Cheng YC. Phenotypic instability of drug sensitivity in a human colon carcinoma cell line. *Cancer Res* 1989;49(5):1148–53.

31. Smith A, van Haaften-Day C, Russell P. Sequential cytogenetic studies in an ovarian cancer cell line. *Cancer Genet Cytogenet* 1989;38(1):13–24.

32. Johnson JI, Decker S, Zaharevitz D, Rubinstein LV, Venditti JM, Schepartz S, et al. Relationships between drug activity in NCI preclinical in vitro and in vivo models and early clinical trials. *Br J Cancer* 2001;84(10):1424–31.

33. Voskoglou-Nomikos T, Pater JL, Seymour L. Clinical predictive value of the in vitro cell line, human xenograft, and mouse allograft preclinical cancer models. *Clin Cancer Res* 2003;9(11):4227–39.

34. Ciardiello F, Bianco R, Damiano V, Fontanini G, Caputo R, Pomatico G, et al. Antiangiogenic and antitumor activity of anti-epidermal growth factor receptor C225 monoclonal antibody in combination

with vascular endothelial growth factor antisense oligonucleotide in human GEO colon cancer cells. *Clin Cancer Res* 2000;6(9):3739–47.

35. Ciardiello F, Caputo R, Bianco R, Damiano V, Fontanini G, Cuccato S, et al. Inhibition of growth factor production and angiogenesis in human cancer cells by ZD1839 (Iressa), a selective epidermal growth factor receptor tyrosine kinase inhibitor. *Clin Cancer Res* 2001;7(5):1459–65.

36. Jones HE, Goddard L, Gee JM, Hiscox S, Rubini M, Barrow D, et al. Insulin-like growth factor-I receptor signalling and acquired resistance to gefitinib (ZD1839; Iressa) in human breast and prostate cancer cells. *Endocr Relat Cancer* 2004;11(4):793–814.

37. Solit DB, She Y, Lobo J, Kris MG, Scher HI, Rosen N, et al. Pulsatile administration of the epidermal growth factor receptor inhibitor gefitinib is significantly more effective than continuous dosing for sensitizing tumors to paclitaxel. *Clin Cancer Res* 2005;11(5):1983–9.

38. Gitler MS, Monks A, Sausville EA. Preclinical models for defining efficacy of drug combinations: mapping the road to the clinic. *Mol Cancer Ther* 2003;2(9):929–32.

39. Shinn C, Larsen D, Suarez JR. Flavopiridol sensitivity of chronic lymphocytic leukemia (CLL) cells in vitro varies based upon species specific drug protein binding. *Blood* 2000;96:294b.

40. Kaelin WG, Jr. The concept of synthetic lethality in the context of anticancer therapy. *Nat Rev Cancer* 2005;5(9):689–98.

41. Borisy AA, Elliott PJ, Hurst NW, Lee MS, Lehar J, Price ER, et al. Systematic discovery of multicomponent therapeutics. *Proc Natl Acad Sci USA* 2003;100(13):7977–82.

31

Preclinical Development of Molecularly Targeted Agents in Oncology

Joseph E. Tomaszewski, PhD,
and James H. Doroshow, MD

SUMMARY

The discovery and development of molecularly targeted agents to treat cancer requires a new paradigm for moving novel drugs more rapidly through the development pipeline into the clinical setting. We must clearly understand the pharmacology (kinetics and dynamics) of each agent and the effect of the drug on its purported target(s). To optimize the development process, validated pharmacokinetic (PK) and pharmacodynamic (PD) assays must be available and used in the preclinical setting before advancing a new drug to clinical trials for first in-human studies. We present a new paradigm for moving through the discovery-developmental continuum that supports performing human pharmacologic studies earlier in drug development to shorten the time from target discovery to clinical evaluation.

Key Words: Antitumor agents; drug development; molecular targets; preclinical; pharmacokinetics; pharmacodynamics; toxicology.

1. INTRODUCTION

The focus of discovery for new therapies to treat cancer has changed markedly over the past 5 years. This is due, in part, to the dramatic increase in our knowledge of the signal transduction pathways that can be modulated by small molecular species and antibodies *(1)* and to the success of some of the first generation of molecularly targeted agents, such as imatinib *(2,3)* and gefitinib *(4)*. Developing new agents targeting specific molecular pathways requires a significant modification in the methodological approach used in the past for traditional cytotoxic agents *(5)*.

The usual practice for cytotoxic agents is to determine the "maximum-tolerated dose" (MTD) in the preclinical setting using a rodent and a non-rodent animal model, which in turn establishes the starting dose (SD) in humans that is suitable for filing

From: *Cancer Drug Discovery and Development: Molecular Targeting in Oncology*
Edited by: H. L. Kaufman, S. Wadler, and K. Antman © Humana Press, Totowa, NJ

an Investigational New Drug Application (IND) with the US Food and Drug Administration (FDA). Cancer patients have been dosed with cytotoxic agents in phase I clinical trials up to an MTD, because the MTD was assumed to be the biologically effective dose (BED). Dosing animals to what would clearly be unacceptable degrees of toxicity in humans somehow assured greater confidence that the lower doses selected as the starting point for the first in-human studies will not be harmful to humans. Furthermore, the range of toxic effects observed at these admittedly supra-physiological dose levels was assumed to provide a basis for more informed patient consent despite the lack of relevance to the doses that initially would be explored in humans. However, some of the thinking behind this strategy is shifting (see discussion in the Section 5).

2. DEVELOPMENT OF CYTOTOXIC AGENTS

The historical experience of the National Cancer Institute (NCI) with the development of cytotoxic anticancer agents confirms that assumptions about the BED being equivalent to the MTD have merit. Results from toxicology studies conducted by the NCI across different species (mice, rats, and dogs) have enabled the estimation of a safe SD in 98% of the clinical studies conducted since 1983 with NCI-developed drugs *(6)*. The SD in those studies was typically one-tenth of the rodent MTD or one sixth of the non-rodent MTD, whichever species was more sensitive to the toxicity of the agent. SDs could also be safely estimated by considering the PK aspects of essentially all drugs examined, with the exception of the most potent cytotoxins, such as bizelesin and Dolastatin 10, for which appropriately sensitive pharmacologic assays were not available at the inception of the trials. In these latter examples, however, the use of animal toxicity data, in conjunction with in vitro granulocyte–macrophage colony-forming unit (CFU-GM) bone marrow assay data from the mouse, dog, and human marrow, correctly predicted human sensitivity in relation to animal models and safe SDs *(7)*.

The published methodological details of the toxicology studies used by the NCI over 1980s and 1990s provide numerous examples *(7,8)*. The principles delineated in these publications are still applicable today despite much more comprehensive recent study designs. The FDA requests preclinical toxicology studies conducted in two species, a rodent and a non-rodent, for small well-defined molecules *(9,10)* but do not emphasize the need for pharmacologic data. The FDA indicates that PK studies on the agent in question are useful but not required for entry into the clinic. However, current clinical research practice involves the development of detailed preclinical PK information prior to the first in-human trials.

Past anticancer drug-development procedures were characterized by a lack of surrogate marker use for either efficacy or toxicity. Plasma drug levels are typically measured in multiple species in the preclinical setting; however, dose escalation in phase I clinical trials is generally performed without reference to this pharmacological information despite pioneering recommendations by Collins and coworkers 20 years ago *(11,12)*. Instead of simply using the Fibonacci series for dose escalation, Collins et al. recommended using animal pharmacology data in conjunction with human data to modify the dose-escalation scheme to reach the MTD as rapidly as possible. Pharmacokinetic (PK) testing may predict the likelihood of being at or near the desired peak concentration or the area under the concentration time curve (AUC) associated with an efficacy endpoint. PK samples, generally obtained during phase I clinical trials, are not routinely analyzed in real time but after completion of dosing. The current practice evolved because the classic cytotoxic agents initially developed under this paradigm did not have an obvious plateau in efficacy as a function of dose prior to the occurrence of toxicity or because no clear delineation of efficacious versus toxic concentrations or AUCs had been defined in preclinical models or simply because of the inconvenience and cost of analyzing each sample as soon as it was obtained from the patient.

3. DEVELOPMENT OF MOLECULARLY TARGETED AGENTS

In contrast to this approach of empirical drug development, modern cancer therapeutics originates from a very different set of assumptions. Beneficial therapeutic effects from cytotoxic agents are understood to arise from the induction of a variety of death cascades in the tumoral compartment. Molecularly targeted agents act on one or more of the myriad of defined signal transduction or differentiation pathways. In appropriate preclinical efficacy models, the relationships between the effects of the drug on the tumor and those on the surrogate organ compartments (e.g., peripheral blood mononuclear cells, skin or hair samples, or buccal keratinocytes), including relationships between the induction of apoptosis, differentiation or altered signal transduction, and pharmacological parameters (e.g., peak and threshold concentrations and AUC), can be defined. Therefore, the inception of the first in-human clinical trials is easy to imagine based on dosing strategies that seek to emulate well-tolerated pharmacologic approaches in animal models rather than dosing based on arbitrary fractions of toxic doses in animals. Human clinical trials could be stopped when a predefined concentration or AUC is reached, and that decision could be supported or modified depending on the effects of the drug on surrogate or tumor compartments. Consequently, the endpoint is a BED rather than the traditional MTD (Fig. 1).

The critical point to consider in this developmental strategy is whether spending the time and resources on pharmacological studies, both kinetics and dynamics, rather than toxicity studies, is more important to provide a more scientific rationale for moving into the clinic. Pharmacology studies are intrinsically more useful, as preclinical pharmacology evaluations contribute to the development and characterization of assays that could be used in human clinical trials. Likewise, evidence of efficacy in surrogate or tumor compartments is now routinely sought as part of a molecule's qualification for further development. Thus, estimation of SDs from pharmacologic information, followed by modification of the escalation strategy once initial human pharmacology becomes available, should obviate the need for extensive toxicology studies at doses likely to be irrelevant to human use.

Should detailed safety testing protocols be completely eliminated from the drug developmental cycle? The answer is "No!" Before a drug is advanced to late phase II or phase III trials, higher dose ranges and longer toxicology studies are necessary to provide a more complete picture of a drug's toxicity at useful concentration ranges, particularly in chronic settings. This is especially important for agents that are intended for prolonged daily administration. These longer toxicology studies would be reserved for agents that have completed phase I clinical trials, in which drug levels have been measured and are consistent with tumor stasis or regression, and for target modulation in animal efficacy studies.

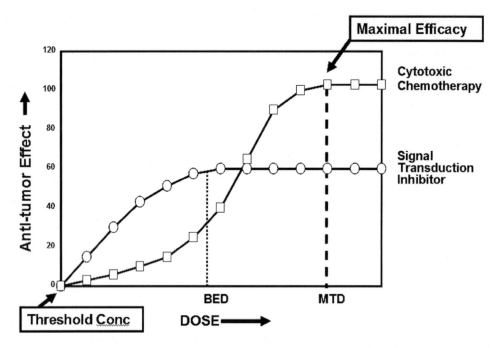

Fig. 1. Hypothetical dose–response curves: maximum-tolerated dose (MTD) versus biologically effective dose (BED). A typical dose–response curve is used to depict the effects of cytotoxic chemotherapy, in which the BED is generally MTD. For signal transduction inhibitors or molecularly targeted new molecular entities, the BED is typically a non-toxic dose between a threshold concentration at which some measure of efficacy is first observed and the maximally effective concentration above which no further increase in efficacy can be observed. Reprinted from The Lancet Oncology *(43)*.

4. PROPOSED EFFICACY, PK/PHARMACODYNAMIC, SAFETY STUDIES FOR ENTRY OF SMALL, WELL-DEFINED ONCOLOGY AGENTS INTO PHASE I CLINICAL TRIALS

4.1. Efficacy Studies

Lack of efficacy in the clinic has emerged as the leading cause of attrition for novel therapeutics *(13)*. As a result, considerable debate *(14,15)* surrounds the utility of various animal tumor models, including subcutaneous human tumor xenografts *(16,17)*, orthotopic xenografts *(18)*, genetically engineered mouse models *(19,20)*, and spontaneous tumor models *(20)*. Each model has several advantages and disadvantages, which have been reviewed extensively elsewhere.

Efficacy studies conducted in vitro to support this new therapeutic paradigm should include the determination of the concentration that is 50% effective (EC_{50}) in biochemical/enzyme molecular target assays, as well as traditional tumor cell assays. The specific molecular target assays to be used will depend on the agent in question, and they are generally very useful in establishing the potency of a series of analogs for selection of lead candidates. Owing to the importance of developing a time course for activity, determining how rapidly the target is inhibited and the duration of this inhibition in molecular target assays is necessary. Once activity against the target is established in an appropriate biochemical assay, assessing antitumor activity in an in vitro human tumor cell line, such as the NCI 60 cell-line panel, is critical *(21,22)*. This should be followed by the determination of a time versus concentration tumor-inhibition profile, estimating the type of exposure (peak, AUC, or threshold) that results in maximal tumor-growth inhibition and that correlates with the degree of target inhibition as previously determined (e.g., inhibition of kinase activity).

Simultaneously or prior to this time course study, the NCI typically performs an in vivo hollow fiber (HF) assay *(23)* to determine whether the compound can survive in an intact animal model and inhibit tumor growth. As this model *(24)* uses both the intraperitoneal (i.p.) and subcutaneous (s.c.) compartments in the same animal, same site and distant site activity can be determined. The latter indicates successful absorption and distribution of the compound to a remote site and provides some preliminary information on the PKs of the compound. Once activity is determined in the HF model, assess whether activity exists in a s.c. or orthotopic xenograft or in spontaneous or genetically engineered models. During the course of efficacy testing in these models, determine schedule dependence (time versus concentration profile) in the animal tumor model(s) and the minimally effective dose (MED), the BED, and the MTD.

Although these studies are ongoing, identify assay(s) evaluating pharmacodynamic (PD) effects on a tumor and/or surrogate marker to develop the linkage between tumor and surrogate marker response. Many techniques can be used, for example, immunohistochemistry (IHC), enzyme-linked immunosorbent assay (ELISA), real time–polymerase chain reaction (RT–PCR), genomics, proteomics, and metabonomics. Consider the use of imaging as a biomarker, if possible. (See the essential components of these efficacy studies in Table 1 and Figs. 2 and 3.)

4.2. PK Studies

Although distant site activity in animal tumor models provides some indication of compound kinetics, develop and validate a sensitive assay to determine the drug concentration in biological matrices, such as cell-culture media, plasma, urine, tissues, and most importantly, tumor. Evaluate this assay in normal and tumored animals. Metabolism is a critical feature of a lead compound. Thus, evaluate drug metabolism

Table 1
Proposed Efficacy Studies

- Determine IC/EC_{50} in biochemical molecular target and cellular assays (is an IC/EC_{50} sufficient?)
- Determine time versus concentration profile in vitro
- Determine time versus concentration profile in animal tumor model(s)
- Determine MED, BED, and MTD; define schedule
- Develop biomarker assay(s)
- Incorporate imaging where possible
- Incorporate "Omics" studies

Abbreviations: BED, biologically effective dose; EC_{50}, concentration that is 50% effective; IC_{50}, concentration that inhibits tumor cell growth by 50%; MED, minimally effective dose; and MTD, maximum tolerated dose.

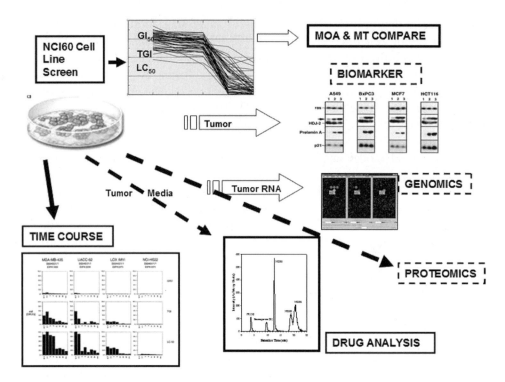

Fig. 2. New model for in vitro efficacy studies. This is a representation of the type of pharmacological testing and data associated with in vitro efficacy studies. The images shown in the solid boxes (e.g., time course) refer to studies that are typically performed at this point in time. The examples shown in the dotted boxes (e.g., proteomics) are considered important studies that should be routinely performed at this stage. COMPARE, compare analysis, as typically performed on the NCI60 tumor cell line data; GI_{50}, drug concentration which inhibits tumor growth by 50%; LC_{50}, concentration which kills 50% of the tumor cells; MOA, mechanism of action; MT, molecular target; NCI60, tumor cell-line screen from the Developmental Therapeutics Program, NCI (see: http://dtp.nci.nih.gov/branches/btb/ivclsp.html); and TGI, drug concentration which inhibits tumor growth completely.

in vitro in various species, including humans, using different liver preparations, such as S-9, microsomes, hepatocytes, or liver slices.

As protein binding can also affect whether a compound is available to interact with the target and the concentration at which binding of free drug target occurs, evaluate protein binding in various species, including human. However, because binding to human proteins can be considerably different than those of various animal species *(25)*, carefully evaluate binding to a variety of human proteins, such as albumin and α1-acid glycoprotein and not protein binding in general. Determine the stability of the drug in blood and plasma in vitro in various species, as well as human. These in vitro studies help to define the type of in vivo studies that should be designed and completed and whether animal studies are warranted.

Once a method has been established and all the other in vitro studies have been completed, the first in vivo PK studies to perform include animal tumor models and not normal animals. Knowing the extent of drug exposure necessary to induce either tumor stasis or regression is essential.

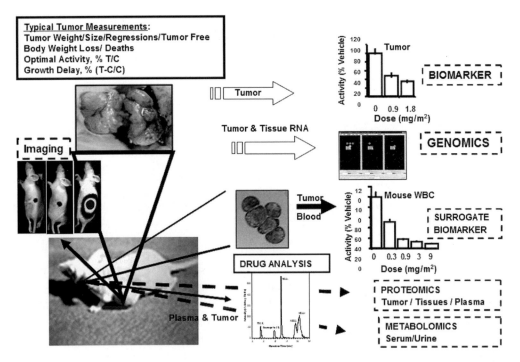

Fig. 3. New model for in vivo efficacy studies. This is a representation of the type of pharmacological testing and data associated with in vivo efficacy studies. The examples depicted in the solid boxes (e.g., typical tumor analysis) refer to studies that are typically performed at this point in time. The examples depicted in the dotted boxes (e.g., proteomics) are considered important studies that should be routinely performed at this stage. C, tumor weight of control animals; T, tumor weight of treated animals; and WBC, white blood cells.

Although knowing the circulating drug concentration (assuming that this is a useful surrogate for the tumor concentration) is important, knowing the extent of drug exposure in the tumor and how this relates to the in vitro concentrations that are associated with antitumor activity is more important. Thus, perform PK studies in animal tumor models at the MED, BED, and MTD to establish the effective concentration [plasma concentrations associated with efficacy (C_{Peff}), peak, steady state or threshold] and the effective exposure (AUC_{Eff}). Similar studies can be performed in non-tumored mice, but performing these initial PK studies at MED, BED, and MTD, in both the rodent and non-rodent toxicology models is more important, because mice are typically not used for toxicity evaluations. These studies will define the safety of the drug exposure associated with efficacy and whether a therapeutic index exists. This paradigm may be even more important for molecularly targeted drugs, which may be administered over a lifetime like antihypertensives or lipid-lowering drugs, than for cytotoxins. The essential components of these studies are shown in Table 2 and Fig. 3.

4.3. PD Studies

Similar to PK testing, if the intent is to perform PK/PD evaluations in phase I clinical trials, perform preclinical PD evaluations in the same manner as discussed above for PK studies. Therefore, develop and validate PD/biomarker assay(s) as early as possible

Table 2
Proposed Pharmacokinetic (PK) Studies

- Develop and validate assay for drug in biological matrices (plasma, tumor, urine, and tissues)
- Evaluate metabolism in vitro in various species including man
- Evaluate protein binding in various species including man
- Evaluate blood/plasma stability in vitro in various species including man
- Perform PK studies in animal tumor model at MED, BED, and MTD; determine $C_{P/T\ Eff}$ (max, steady state, or threshold) and AUC_{Eff}
- Perform PK studies at MED, BED, and MTD in rodent toxicology model
- Perform PK Studies at MED, BED, and MTD in non-rodent toxicology model

Abbreviations: AUC_{Eff}, area under the concentration x time curve associated with efficacy; BED, biologically effective dose; $C_{P/T\ Eff}$, plasma/tumor concentrations associated with efficacy; MED, minimally effective dose; and MTD, maximum tolerated dose.

in the discovery/developmental program. These assays must be suitable for use in biological matrices, such as tumor and surrogate compartments [e.g., white blood cells (WBCs), saliva, skin, or buccal compartments]. In general, each agent may require the development of a specific marker assay. At the very least, biomarkers should be developed for a critical pathway related to the proposed mechanism of action of the agent and for the predicted molecular endpoint if the agent is effective.

Establish the time course for marker inhibition and recovery, which will have as great an impact on the proposed clinical schedule as the kinetics of the drug. Thus, PD studies must be performed at relevant concentrations in appropriate in vitro studies; the results of these studies should provide approximate time points for sampling in animal studies. After these in vitro studies, treat animal efficacy model(s) at the MED, BED, and MTD in the same manner as studies evaluating PK testing and determine the impact of the drug on the biomarkers in both tumor and surrogate compartments if possible. Ideally, conduct both PK and PD determinations in the same studies. Perform these PD studies at the MED, BED, and MTD, in both rodent and non-rodent toxicology models, provided that the target effect is evaluable in normal animals. The essential components of these studies are shown in Table 3 and Fig. 4.

Table 3
Proposed Pharmacodynamic (PD) Studies

- Develop and validate PD/biomarker assay(s) in tumor and surrogate tissues
- Perform PD studies at relevant concentrations in vitro and in animal efficacy model(s) at MED, BED, and MTD
- Determine time course for inhibition of marker and recovery
- Incorporate "Omics" studies
- Perform PD studies at MED, BED, and MTD in rodent toxicology model and correlate effect to dose, C_p, or AUC
- Perform PD studies at MED, BED, and MTD in non-rodent toxicology model and correlate effect to dose, C_p, or AUC

Abbreviations: AUC, area under the concentration time curve; BED, biologically effective dose; CP, peak plasma concentration; MED, minimally effective dose; and MTD, maximum tolerated dose.

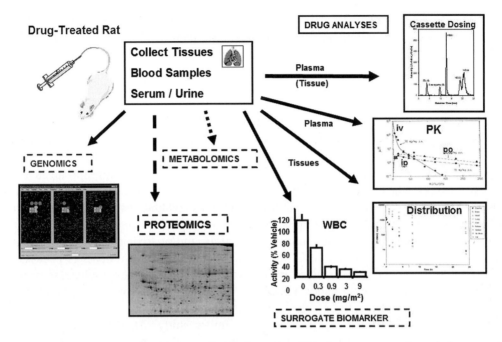

Fig. 4. New model for pharmacology (PK/PD) studies. This is a representation of the type of pharmacological testing and data associated with PK and PD studies. The images shown in the solid boxes (e.g., drug analyses) refer to studies that are typically performed at this point in time. The images shown in the dotted boxes (e.g., proteomics) are considered important studies that should be routinely performed at this stage.

Based on prior clinical studies, developing predictive therapeutic correlations with single biomarkers is difficult. Therefore, investigate multiple approaches, multiple targets, and multiple assays in both preclinical and clinical studies. Types of markers to consider include the following:

1. PD (dosing) marker: To detect the presence of a drug and its interaction with a specific target (e.g., receptor occupancy, and enzyme product levels).
2. Response or mechanistic marker: To measure the molecular signal transduction response of a patient to the presence of a drug [e.g., response of 20S proteosome activity in WBCs to treatment with bortezomib *(26)*].
3. Efficacy or clinical response marker: A surrogate endpoint in response to a drug (e.g., decreased oral mucositis incidence/severity).
4. Disease progression marker: To measure the stage of disease progression [often the reciprocal of the efficacy marker (e.g., increased urine protein-to-creatinine ratios in nephropathy)] *(27)*.
5. Predictive marker for patient stratification: To allow the prospective selection of potential responders from non-responders in a seemingly uniform patient population [e.g., trastuzumab and the Herceptest® *(28)*].
6. Toxicity marker: To measure the toxic side effects of a drug (e.g., liver function tests such as alanine aminotransferase (ALT) and aspartate aminotransferase (AST)).

In the past, assays measuring the PD effects have not been developed as frequently as PK assays. If the response of the assay to the drug in patient tumor or surrogate samples is uninformative, one needs to have confidence that a lack of effect is not due to an insensitive or poorly controlled assay. Thus, for molecularly targeted agents, developing rigorous, validated assays is essential, with a reasonable signal-to-noise ratio, an acceptable coefficient of variation, and an appropriate limit of quantitation, just as one develops and validates a PK assay.

4.4. Toxicity/Safety Studies

As previously discussed in the PK and PD sections, assess the effects of each agent in various species, including humans, whenever possible (Table 4). Because toxicity in humans cannot be evaluated at the preclinical stage, carry out appropriate in vitro toxicity assays when available to predict human sensitivity in relation to the various animal species used. Similar to the traditional toxicity evaluations of cytotoxic agents, two species are generally used, unless the use of one species can be justified as being the most appropriate. Thus, determining the safety and/or toxicity of the efficacious doses and/or drug concentrations/exposures in rodents and non-rodents is necessary. These studies are usually conducted in rats and dogs for most new molecular entities (NMEs), with the following major objectives: to define the toxicokinetics (TK), MTD, dose-limiting toxicities (DLT), schedule-dependency of toxicity, reversibility of side effects, and a safe clinical SD for the agent under study (5). When rats and dogs are deemed inappropriate for valid scientific reasons, other species can be used [e.g., mice, rabbits, miniature swine, hamsters, guinea pigs, or non-human primates (NHPs); cynomolgus; rhesus monkeys; and marmosets). Hence, better methods to better predict human toxicity from novel therapeutic agents are urgently needed if the process of developing targeted chemotherapeutic agents is to be streamlined (Fig. 5).

Owing to their small size, cost, and ease of use and manipulation, mice are favored for many preclinical studies, especially efficacy. However, they are not typically the rodent species of choice for toxicology studies (Fig. 5). The small size of the mouse precludes serial blood sampling needed to determine the kinetics of an agent and its effects on various hematology and clinical chemistry parameters and biomarkers. Furthermore, mice in general tend to be worse predictors of human toxicity (29,30). The use of mice rather than rats also requires a larger number of animals due to

Table 4
Proposed Toxicity/Safety Studies (Most Appropriate Species)

- Determine toxicity/sensitivity in various species including man in appropriate in vitro toxicity assays if available
- Determine safety/toxicity of efficacious doses and/or drug concentrations after (single dose[a]) in rodents
- Determine safety/toxicity of efficacious doses and/or drug concentrations after (single dose[a]) in non-rodents
- Determine impact of efficacious drug concentrations on PD/biomarker/"Omics" after (single dose[a]) in rodents
- Determine impact of efficacious drug concentrations on PD/biomarker/"Omics" after (single dose[a]) in non-rodents

[a]The actual number of doses will depend on the intended schedule.
Abbreviation: PD, pharmacodynamic.

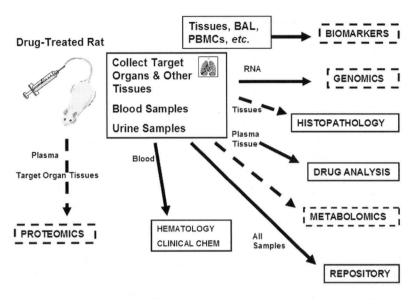

Fig. 5. New model for toxicology studies. This is a representation of the type of pharmacological testing and data associated with toxicology studies. The examples depicted in the solid boxes (e.g., histopathology) refer to studies that are typically performed at this point in time. The examples depicted in the dotted boxes (e.g., proteomics) are considered important studies that should be routinely performed at this stage. BAL, bronchial alveolar lavage and PBMCs, peripheral blood mononuclear cells.

the small blood volumes, which is contrary to the current thrust within the research community to reduce, refine, and replace the use of animals.

The beagle dog is typically the non-rodent animal of choice, rather than other species (e.g., non human primates), due to the long history of success using this species for pharmaceutical evaluations. The beagle dog is extremely useful for very intensive procedures (serial bleeding for clinical pathology, PK, toxico-kinetics, and biomarkers and long-term continuous intravenous infusion). NHPs are difficult and expensive to obtain and house. NHPs probably should be reserved for situations in which the beagle is simply inappropriate *(31,32)*. The use of rats and dogs has generally been successful, especially in defining a safe SD for phase I clinical trials *(6)*. However, the oncology community has expressed a desire to streamline this process and make it easier to move into phase I trials. From a toxicity/safety perspective, learning to design and conduct preclinical studies more intelligently is key to maximizing the amount of useful samples and information that are obtained from each animal study. Additionally, we must develop more in vitro assays predictive of human sensitivity.

In contrast to older, cytotoxic antineoplastics, new molecularly targeted agents may have more limited, or at least less non-specific, toxicity profiles, and they are more appropriately studied at or near their BED rather than the MTD *(33)*. Furthermore, studies in rodents, dogs, and other species that lead to an IND do not typically permit in depth evaluations of mechanisms of toxicity prior to the initiation of human clinical trials. Evaluating numerous analogs of a class of compounds in multiple species prior to the selection of the best developmental candidate is generally impractical, as animal studies are expensive and time consuming. Finally, predicting drug-related side effects in humans from traditional toxicology studies is imprecise at best *(6,30)*. We urgently need better methods to better predict human toxicity from novel therapeutic agents if we are to streamline the process of developing targeted chemo-therapeutic agents.

Recently developed and validated in vitro bone marrow assays using rodent, canine, human CFU-GM, and other stem cells have demonstrated utility in predicting toxicities in animals and humans *(34–38)*. Molecular endpoints to evaluate toxicity and high throughput-toxicity screening has facilitated exploration of toxicity at an earlier stage in the drug development process *(39)*. Thus, developing assays (molecular, cell based, or organ culture) to predict other DLT, such as cardiotoxicity, gastrointestinal toxicity, hepatotoxicity, nephrotoxicity, neuron-toxicity, and pulmonary toxicity, would be a very useful adjunct to the FDA-required in vivo toxicology studies. Such assays could assist in the evaluation and prediction of human sensitivity, provide cost-efficient evaluation of numerous analogs prior to selection of the developmental candidate, and reduce the use of animals, thereby reducing the cost of therapeutic development while improving the predictability of these studies.

Essential safety/toxicity studies for NMEs would include

1. Conducting single-dose safety/toxicity studies in rodents to determine whether efficacious drug concentrations [peak plasma concentration (C_p), AUC, or threshold] can be safely attained and maintained for the appropriate time period.

   a. Assessing the impact of these concentrations on selected biomarkers (metabolites, proteins, or nucleic acids), safety, and toxicity.
   b. Evaluating reversibility of toxicity or whether it is delayed.
   c. Determining histopathology within the BED range.

2. Conducting combined single-dose range-finding/PK studies in non-rodents rather than separate studies.

   a. Determining if efficacious drug concentrations (peak C_p, AUC or threshold) can be safely attained and maintained for the appropriate time period.
   b. Assessing the impact of these concentrations on selected biomarkers (metabolites, proteins, or nucleic acids), safety, and toxicity.
   c. Evaluating reversibility of toxicity or whether toxicity is delayed. Biomarker studies in normal animals will only be applicable to situations in which the marker is not tumor specific.

5. EXPLORATORY INVESTIGATIONAL NEW DRUG STUDIES

In January 2006, the US FDA published a new guidance entitled "Exploratory IND Studies" *(40)*, which clarifies the preclinical and clinical requirements for clinical trials that: (i) are conducted early in phase I, (ii) involve very limited human exposure, and (iii) have no therapeutic or diagnostic intent (e.g., screening studies of closely related drugs or therapeutic biological products, microdose studies).

As described in the FDA *Critical Path Report (41)*, this approach should reduce the time and resources that are expended on potential drugs that are unlikely to succeed. New tools are needed earlier in the drug development process to distinguish promising candidates from those that are not. These exploratory IND studies will be conducted prior to the traditional dose-escalation, safety, and tolerance studies that ordinarily initiate a clinical drug development program in phase I. The duration of dosing in an exploratory IND study probably will be limited (e.g., no longer than 7 days).

In this guidance, the FDA envisions a number of scenarios in which preclinical safety requirements can be reduced and sponsors can be helped in the developmental process; for example:

1. Determining whether a mechanism of action defined in experimental systems can also be observed in humans (e.g., a binding property or inhibition of an enzyme).
2. Providing important information on PK.
3. Selecting the most promising lead product from a group of candidates designed to interact with a particular therapeutic target in humans, based on PK or PD properties.

The three safety programs described are applicable to

1. PK or imaging studies.
2. Studies of pharmacologically relevant doses.
3. Studies of mechanisms of action related to efficacy.

What is the advantage of embarking on such exploratory studies? This paradigm allows the investigator to obtain the most sorely needed data in a development program (e.g., human pharmacology) without producing large amounts of drug product and conducting long-term toxicity studies. As most new molecularly targeted agents for oncology are cytostatic, the duration of exposure to such compounds could be for a lifetime.

For continuous administration of an oncology agent in phase I, the toxicologist is faced with conducting a variety of range-finding studies in two species to permit dose selection for the definitive IND-enabling, 28-day studies. Because the intent of phase I trials is to define safety, the toxicology program to support this effort generally takes at least 9–12 months and $1–2 million to complete. Producing sufficient drug for toxicology and clinical studies could consume the same amount of resources. Knowing that a PD effect on the molecular target of interest can be measured after a single dose in animal studies would greatly simplify the development program for both the production of drug product (much less) and the required toxicity studies (much shorter). However, knowing that the molecular target can be favorably modulated after a single dose is not enough for this strategy to be successfully evaluated in a clinical setting. The effect on the target should also be directly linked to an impact on tumor growth (stasis or regression), which should be defined in both in vitro and in vivo studies, as described earlier. Why is this important? To reduce the number of clinical failures because of the lack of efficacy, investigators need to be reasonably certain that tumor growth is dependent on the target of interest and that its inhibition will lead to a positive outcome. Thus, proper application of these principles in an exploratory IND setting can reduce the time to get an agent to the clinic, the resources required, and perhaps even the number of drug failures.

6. CONCLUSIONS

Why is the drug development paradigm discussed here different from those used in the past and why has PK/PD testing failed in the past? The answers to these questions are relatively clear. PK analyses were rarely performed in real time and thus, rarely influenced dose escalation. Most samples collected from patients on clinical trials are frozen and analyzed en masse at the end of the clinical trial. Because the results are calculated after the fact using the same standards that produce a single uniform data set, they provide no assistance to the clinician. Pharmacokinetic testing is not used to guide dose escalation, as proposed by Collins and coworkers in the 1980s *(11,12)*. Thus, dose escalation is an empirical process based on the Fibonnaci sequence or some other paradigm. Patient data are not used to determine the next dose escalation, with the exception of toxicity modifications. PD testing is rarely evaluated in real time, and when used typically involves correlations, such as C_P or AUC versus toxicity (e.g., WBCs or neutropenia) and C_P or AUC versus tumor response. This is due, in part, to uncertainty about defining BED versus MTD, as preclinical studies are not designed and performed with these endpoints in mind.

So, why should the paradigm proposed in this chapter work any better than those used in the past? The studies described here are designed not only to maximize therapeutic effects while minimizing toxicity but also to maximize knowledge of the pharmacological effects necessary to produce an antitumor effect. The investigations proposed evaluate all aspects of pharmacology (PK and PD) related to anti-tumor activity, producing potentially safer and more effective regimens. As pharmacological effects are determined sooner in clinical development, early trials should be less costly and more importantly, require fewer patients. We can succeed in discovery by "failing early and failing often" *(42)*.

REFERENCES

1. Irish JM, Kotecha N, Nolan GP. Mapping normal and cancer cell signaling networks: towards single-cell proteomics. *Nat Rev Cancer* 2006;6:146–155.
2. Druker BJ, Talpaz M, Resta DJ, et al. Efficacy and safety of a specific inhibitor of BCR-ABL tyrosine kinase in chronic myeloid leukemia. *N Engl J Med* 2001;344:1031–1037.
3. Druker BJ, Sawyers CL, Kantarjian H, et al. Activity of a specific inhibitor of the BCR-ABL tyrosine kinase in the blast crisis of chronic myeloid leukemia and acute lymphoblastic leukemia with the Philadelphia Chromosome. *N Engl J Med* 2001;344:1038–1042.
4. Fukuoka M, Yano S, Giaccone G, et al. Multi-institutional randomized phase II trial of gefitinib for previously treated patients with advanced non-small-cell lung cancer (The IDEAL 1 Trial). *J Clin Oncol* 2003;21:2237–2246.
5. Tomaszewski JE. Multi-species toxicology approaches for oncology drugs: the US perspective. *Eur J Cancer* 2004;40:907–913.
6. Smith AC, Tomaszewski JE. *Correlation of Human Toxicity and MTDs with Preclinical Animal Data.* Presented at the Preclinical Pharmacology and Toxicology Workshop at the EORTC/NCI/AACR Meeting, Frankfurt, Germany, 2002.
7. Tomaszewski JE, Smith AC, Covey JM, Donohue SJ, Rhie JK, Schweikart KM. Relevance of preclinical pharmacology and toxicology to phase I trial extrapolation techniques. Relevance of animal toxicology. In: Baguley BC, ed. *Anti-Cancer Drug Design.* San Diego, CA: Academic Press, 2001:301–328.
8. Tomaszewski JE, Smith AC. Safety testing of antitumor agents. In: Williams PD, Hottendorf GH, eds. *Comprehensive Toxicology, Toxicity Testing and Evaluation,* vol 2. Oxford, England: Elsevier Science Ltd., 1997:299–309.
9. ICH Guidance to Industry: M3, Nonclinical safety studies for the conduct of human clinical trials for pharmaceuticals. July 1997.
10. DeGeorge JJ, Ahn C, Andrews PA, et al. Regulatory considerations for preclinical development of anticancer drugs. *Cancer Chemother Pharmacol* 1998;41:173–185.
11. Collins JM, Zaharko DS, Dedrick RL, Chabner BA. Potential roles for preclinical pharmacology in phase I clinical trials. *Cancer Treat Rep* 1986;70:73–80.
12. Collins JM, Leyland-Jones B, Grieshaber CK. Role of preclinical pharmacology in phase I clinical trials: considerations of schedule-dependence. In: Muggia FM, ed. *Concepts, Clinical Developments,*

and Therapeutic Advances in Cancer Chemotherapy. Boston, MA: Martinus Nijhoff Publishers, 1987:129–140.

13. Kennedy T. Managing the drug discovery/development interface. *Drug Discov Today* 1997;2: 436–444.

14. Johnson JI, Decker S, Zaharevitz D, et al. Relationships between drug activity in NCI preclinical in vitro and in vivo models and early clinical trials. *Br J Cancer* 2001;84:1424–1431.

15. Frijhoff AF, Conti CJ, Senderowicz AM. Advances in molecular carcinogenesis: current and future use of mouse models to screen and validate molecularly targeted anticancer drugs. *Mol Carcinog* 2004;39:183–194.

16. Kelland LR. "Of mice and men": values and liabilities of the athymic nude mouse model in anticancer drug development. *Eur J Cancer* 2004;40:827–836.

17. Fiebig HH, Maiera A, Burger AM. Clonogenic assay with established human tumor xenografts: correlation of in vitro to in vivo activity as a basis for anticancer drug discovery. *Eur J Cancer* 2004;40:802–820.

18. Bibby MC. Orthotopic models of cancer for preclinical drug evaluation: advantages and disadvantages. *Eur J Cancer* 2004;40:852–857.

19. Green JE, Hudson T. The promise of genetically engineered mice for cancer prevention studies. *Nat Rev Cancer* 2005;5:184–198.

20. Hansen K, Khanna C. Spontaneous and genetically engineered animal models: use in preclinical cancer drug development. *Eur J Cancer* 2004;40:858–880.

21. Alley MC, Scudiero DA, Monks PA, et al. Feasibility of drug screening with panels of human tumor cell lines using a microculture tetrazolium assay. *Cancer Res* 1988;48:589–601.

22. Boyd MR, Paull KD. Some practical considerations and applications of the National Cancer Institute in vitro anticancer drug discovery screen. *Drug Dev Res* 1995;34:91–109.

23. Hollingshead MG, Alley MC, Camalier RF, et al. *In vivo* cultivation of tumor cells in hollow fibers. *Life Sci* 1995;57:131–141.

24. Decker S, Hollingshead M, Bonomib CA, Carterb JP, Sausville EA. The hollow fibre model in cancer drug screening: the NCI experience. *Eur J Cancer* 2004; 40:821–826.

25. Sausville EA, Lush RD, Headlee D, et al. Clinical pharmacology of UCN-01: initial observations and comparison to preclinical models. *Cancer Chemother Pharmacol* 1998;42(Supp):S54–S59.

26. Adams J, Palombella VJ, Sausville EA, et al. Proteasome inhibitors: a novel class of potent and effective antitumor agents. *Cancer Res* 1999;59:2615–2622.

27. Tune BM, Kirpekar R, Sibley RK, Reznik VM, Griswold WR, Mendoza SA. Intravenous methylprednisolone and oral alkylating agent therapy of prednisone-resistant pediatric focal segmental glomerulosclerosis: a long-term follow up. *Clin Nephrol* 1995;43:84–88.

28. Gouvea AP, Milanezi F, Olson SJ, Leitao D, Schmitt FC, Gobbi H. Selecting antibodies to detect HER2 overexpression by immunohistochemistry in invasive mammary carcinomas. Appl Immunohistochem. *Mol Morphol* 2006;14:103–108.

29. Grieshaber CK, Marsoni S. Relation of preclinical toxicology to findings in early clinical trials. *Cancer Treat Rep* 1986;70:65–72.

30. Olson H, Betton G, Robinson D, et al. Concordance of the toxicity of pharmaceuticals in humans and in animals. *Regul Toxicol Pharmacol* 2000; 32:56–67.

31. Parkinson C, Grasso P. The use of the dog in toxicity tests on pharmaceutical compounds. *Hum Exp Toxicol* 1993;12:99–109.

32. Broadhead CL, Benton G, Combes R, et al. Prospects for reducing and refining the use of dogs in the regulatory toxicity testing of pharmaceuticals. *Hum Exp Toxicol* 2000;19:440–447.

33. Johnston SRD. Farnesyl transferase inhibitors: a novel targeted therapy for cancer. *Lancet Oncol* 2001;2:18–26.

34. Schweikart KM, Parchment R, Osborn B, et al. *In vitro/in vivo* bone marrow toxicity: comparison of clinical and preclinical data for selected anticancer drugs. [Abstract No. 2185]. *Proc Am Assoc Cancer Res* 1995;36:367.

35. Tomaszewski JE, Schweikart KM, Parchment RE. Correlation of clinical and preclinical bone marrow toxicity data for selected anticancer drugs. *First International Symposium on Hematotoxicology in New Drug Development*, Lugano, Switzerland, 1997.

36. Tomaszewski JE, Schweikart KM, Parchment RE. Correlation of clinical and preclinical bone marrow toxicity data for selected anticancer drugs. *In Vitro Human Tissue Models in Risk Assessment Workshop of the Society of Toxicology*, Ellicot Ciity, MD, 1999.

37. Parchment R, Tomaszewski J, Schweikart K, LoRusso P. Success in predicting human maximum tolerated dose from *in vitro* data and mouse MTD. [Abstract No. 1046]. *Proc Am Assoc Cancer Res* 2002;43:208–209.
38. Pessina A, Albella B, Bayo M, et al. Application of the CFU-GM assay to predict acute drug-induced neutropenia: an international blind trial to validate a prediction model for the maximum tolerated dose (MTD) of myelosuppressive xenobiotics. *Toxicol Sci* 2003;75:355–367.
39. MacGregor JT, Collins JM, Sugiyama Y, et al. In vitro human tissues models in risk assessment: report of a consensus-building workshop. *Toxicol Sci* 2001; 59:17–36.
40. US FDA Guidance for Industry, Investigators, and Reviewers. *Exploratory IND Studies*. January 2006.
41. US FDA Critical Path Report, *Innovation or Stagnation, Challenge and Opportunity on the Critical Path to New Medical Products*. March 2004.
42. Cribbs C. Succeeding in discovery by failing early and often. *Molecular Connection* 1999;18:4–5.
43. Johnston SRD. Farnesyl transferase inhibitors: a novel targeted therapy for cancer. *Lancet Oncol* 2001;2:18–26. Reprinted from *Lancet Oncol* with permission from Elsevier.

Index

Printed in the United States of America.